T0388098

Food Process Engineering

Safety Assurance and Complements

Food Process Engineering

Safety Assurance and Complements

by

F. Xavier Malcata

CRC Press is an imprint of the
Taylor & Francis Group, an **informa** business

CRC Press
Taylor & Francis Group
6000 Broken Sound Parkway NW, Suite 300
Boca Raton, FL 33487-2742

Printed on acid-free paper

International Standard Book Number-13: 978-0-367-35105-2 (Hardback)

Library of Congress Cataloging-in-Publication Data

Names: Malcata, F. Xavier, author.
Title: Food process engineering : safety assurance and complements / by F. Xavier Malcata.
Description: Boca Raton : CRC Press, [2020] | Includes bibliographical references. | Summary: "Food Process Engineering: Safety Assurance and Complements pursues a logical sequence of coverage of industrial processing of food and raw materials, where safety and complementary issues are germane. A brief overview of useful mathematical tools from algebra, calculus, and statistics is conveyed for convenience."
-- Provided by publisher.
Identifiers: LCCN 2019023635 (print) | LCCN 2019023636 (ebook) | ISBN 9780367351052 (hardback) | ISBN 9780429329760 (ebook)
Subjects: LCSH: Food--Preservation. | Food--Safety measures. | Cooking.| Food service--Sanitation.
Classification: LCC TP371 .M3495 2020 (print) | LCC TP371 (ebook) | DDC 664/.028--dc23
LC record available at https://lccn.loc.gov/2019023635
LC ebook record available at https://lccn.loc.gov/2019023636

Visit the Taylor & Francis Web site at
http://www.taylorandfrancis.com

and the CRC Press Web site at
http://www.crcpress.com

Out of clutter, find simplicity; from discord, find harmony; in the middle of difficulty, lies opportunity.

Albert Einstein (1879–1955)

To my family: Angela, Filipa and Diogo.

For their everlasting understanding, unselfish support and endless love.

Contents

Preface

Food science and technology is the discipline of knowledge in which physicochemical and biological sciences, wrapped by engineering sciences are used to study the nature of foods, the causes of their deterioration, the principles underlying their processing, and the possibilities for their improvement – so as to adequately respond to the general goal of obtaining safe, nutritious, appealing, convenient, and useful foods. An essential strategy for successful product and process development and optimization is integration – throughout the whole food chain, according to a "from farm to fork" approach; this encompasses selection, preservation, transformation, packaging, and distribution of foods as a whole.

For pedagogical reasons, the aforementioned engineering sciences should, to advantage, resort to a more eloquent and specific heading – *Food Process Engineering*, so as to clearly differentiate them from e.g. agricultural and zootechnical engineering (concerned with production of foods by plants and animals) or management engineering (concerned with logistics and marketing). The proposed trilogy entails a wide and relevant area of knowledge and application – which permeates everyday life, and effectively responds to a basic set of physiological needs of the human race; focus (i.e. *food*), approach to that focus (i.e. *process*), and tools for that approach (i.e. *engineering*) are thus its underlying key elements. This book directly addresses this trilogy, and places a particular emphasis on comprehensive and phenomenological understanding of industrial operations; it accordingly conveys a guided study based on extensive mathematical modeling of the physicochemical changes undergone by liquid/solid food items – during transformation from their original form as provided by nature, into elaborated forms that will eventually become available to consumers.

In order to permit undergraduate and graduate food science and technology students acquire and improve skills that respond directly to their intended professional profile, several core competencies are aimed at – and even required. These may be broken down into five major groups, following the standard guidelines set forth by the Institute of Food Technologists (USA): (i) food chemistry and analysis; (ii) food safety and microbiology; (iii) food processing and engineering; (iv) applied food science; and (v) success skills. This book responds specifically to item (iii); the principles of process engineering (e.g. conservation of momentum, enthalpy, and mass), coupled with thermodynamics of limitation, and kinetics of transport and transformation accordingly constitute its core, with an obvious focus on food – complemented, to the extent found necessary, by supporting information under items (i) and (ii). After taking a course that uses this book as major text and/or reference text, the student will be expected to: understand the most important transport and transformation operations in mechanical, thermal, and chemical processing of food, chiefly from a conceptual point of view; be able to apply momentum, enthalpy, and mass balances (as appropriate) to a given food process; and understand the various unit operations required to produce a given food, via classical or novel processing techniques (or a combination thereof) – including the effects of processing parameters on final product specifications and quality. To better convey the germane engineering concepts, major types of commercial equipment are briefly reviewed, and meaningful situations are proposed as illustration problems – in attempts to increase students' awareness of current topics of relevance, including applicability; these constitute competencies entailed by item (iv) above. The foregoing tools will aid in demonstrating how to define a problem or situation in the food field, e.g. via hypothesizing possible solutions and thoughtful recommendations, upon identification of potential causes and with the ultimate goal of managing or designing/tuning processes thereto. Therefore, the competencies normally postulated under item (v) are also (indirectly) addressed.

Evolution of food process engineering into a separate and unique engineering discipline has been taking place worldwide, over the last decades; it has accordingly become a major core component of undergraduate and graduate curricula in food science and technology. With the advent of increasingly more powerful and available, and less and less expensive automatic computation devices, application of models with higher degrees of complexity has become an unavoidable trend. However, students often feel frustrated with the size, and especially the form of the underlying equations – hardly suitable for simple hand computation; even worse, they lose track of the fundamentals behind those equations, as more time is devoted to solve than to understand them. This book effectively addresses the said problem, by resorting to analytical solutions as often as possible, and by keeping the mathematical forms of the equations as general and simple as possible – yet always constructing then from the mechanisms followed by the germane phenomena; and by insisting on simple, concentrated-parameter models laid out on a single dimension, along with enough derivation steps to support a self-paced approach. Therefore, starting from the general phenomenological forms of the equations have set the boundaries for complexity right from the beginning. Moreover, applying exact (or as approximate as feasible) methods of solution, while formatting them to the food product or unit operation/process under scrutiny, will turn into a motivating game – in which applicability overrides cumbersomeness, and innovativeness balances standard routine.

Resorting to fundamental phenomena in an educated and logical fashion is indeed in order – in attempts to rationalize the major driving forces for occurrence of phenomena, and simultaneously quantitate the rate at, and the extent to which they take place. This book follows such a philosophy: it always recalls first principles – presented in the simplest form possible, and avoiding complex equations that disturb understanding and distract from the major physicochemical routes. This intent departs from as simple and intelligible as possible functionalities for the said equations – which typically correspond to zero-th or first-order behaviors. More complex, nonlinear patterns (often demanding correction factors) are criteriously introduced whenever relevant – which tend to keep the original functionality essentially intact, while permitting a better approximation to (industrial) practice. In every case, however, a greater importance is placed upon setting up the underlying general equations and attempting to reach analytical (closed-form) solutions thereof; a graphical rationale is emphasized, as well as logical reasoning whenever deemed appropriate. Closure is always provided by a representative (nontrivial) problem, included for illustrative purposes – carefully designed to reinforce or complement the previous concept, and solved in a stepwise fashion to facilitate self-learning.

From a critical review of competing literature in the field, two sorts of books can essentially be found that are relevant for food process engineering: some pertain to recent advances and ongoing efforts in novel (or classical) processing techniques, with refined reasoning and integrated discussion – but are hardly suitable for undergraduate students, who feel overturned by excessive information and detail, and by the numerical manipulation required. Other books emphasize instead the technological component, with ample description of typical devices and facilities, together with provision of charts and correlations – thus materializing excessively descriptive approaches, oftentimes at the expense of empirical views that lack sufficient generality, while turning away from exact methods of solution (regardless of their usefulness, accuracy, or functionality). Although most such books are obviously helpful and have accordingly made their way over the years into (basic or advanced) training guides for professionals-to-be in the field, they leave space for an alternative, more fundamental approach – which builds on the principles themselves, and applies them with a clear focus on engineering rather than technology. Without discouraging correlations *tout court*, this book proposes mathematical models as simple as possible, and systematically minimizes the number of parameters left, via lumping to generate dimensionless and normalized versions thereof – coupled with explanation of their physical meaning, typically via order of magnitude arguments and graphical illustrations of their behavior. This strategy frees the approach from the limitations of inconsistent and different systems of units, and permits schematic representation of dominant (often asymptotic) patterns – which possess much wider limits of applicability, while minimizing the relevant information to be retained (by resorting to visual memory or the like).

On the other hand, several textbooks available propose a number of solved, and an even larger number of unsolved problems – oftentimes brought together as a mere compilation of putative applications with more or less trivial solutions, namely, calculating the missing variable when a set of equations and the remainder data are given on purpose. Conversely, the strategy followed in this book is to work out carefully selected examples, topically presented as concept illustrations – ranging, in a balanced fashion, from theoretical (and thus generic) manipulations to industrial (and thus focused) cases. A long teaching history has unfolded that students typically attempt to excel in routine-solving problems (which may support high grading in preformatted exams), but invest less and less in understanding the basic physicochemical (including biological) phenomena behind the said problems; the latter are, nevertheless, the crucial features when facing nonconventional situations in the real professional life.

This book is mainly directed at use in academia – especially in teaching of, and learning by undergraduate or graduate students enrolled in food engineering, food science, or food technology courses. In order to gain full advantage from the book, a few background courses are recommended (yet not mandatory): a course in general chemistry, a course in general biology/microbiology, and a course in basic nutrition are thus suggested, to consolidate the general process applications described. Moreover, a course in elementary calculus and a course in general physics are in order, to more easily follow the mathematical formulations and simulations proposed.

F. Xavier Malcata

Author

F. Xavier Malcata was born in Malange (Angola) in 1963. He earned a B.Sc. degree in *Chemical Engineering* (5-year program) from the College of Engineering, University of Oporto (Portugal) in 1986; a Ph.D. degree in Chemical Engineering/Food Science from the University of Wisconsin – Madison (UW), in 1991; and a Habilitation/Professorship degree in Food Science and Engineering from the Portuguese Catholic University (UCP) in 2004.

During his outstanding career in teaching and research, Prof. Malcata has held academic appointments as: Teaching Assistant at UCP (1985–1987) and UW (1988); Lecturer at UW (1989); Assistant Professor at UCP (1991–1998); Associate Professor at UCP (1998–2004); and Full Professor at UCP (2004–2010), Instituto Superior da Maia (2010–2012), and University of Oporto (2012–). His teaching responsibilities have encompassed (among others) thermodynamics, transfer phenomena, fluid mechanics, system dynamics and control, applied biochemistry, enzyme engineering, bioprocess engineering, bioreactor technology, bioresource engineering, food technology, and industrial projects of B.Sc. and M.Sc. degrees in Food Engineering and in Bioengineering.

Prof. Malcata has held professional appointments as: Dean of the College of Biotechnology of UCP (1998–2008) – with major responsibilities for the B.Sc. and M.Sc. degrees in Food Engineering; President of the Portuguese Society of Biotechnology (2003–2008); Coordinator of the Northern Chapter of Chemical Engineering, and Funding Member of the Specialization in Food Engineering of the Portuguese Engineering Accreditation Board (2004–2009); Official Delegate of the Portuguese Government to the VI and VII Framework R&D Programs of the European Union, in the key areas of Food Quality and Safety (2004–2006) and Food, Agriculture, and Biotechnology (2006–2013); and Chief Executive Officer of the nonprofit University/Industry Extension Associations AESBUC (1998–2008) and INTERVIR+ (2006–2008), and the nonprofit Entrepreneurial Biotechnological Support Associations CiDEB (2005–2008) and INOVAR&CRESCER (2006–2008).

He has been the recipient of several international distinctions and awards, namely: election for membership in *Phi Tau Sigma* (honor society of Food Science) in 1990, *Sigma Xi* (honor society of scientific and engineering research) in 1990, *Tau Beta Pi* (honor society of engineering) in 1991, and New York Academy of Sciences in 1992; the *Cristiano P. Spratley Award* in 1985 and the *Centennial Award* in 1986, by the University of Oporto; the *Ralph H. Potts Memorial Award* in 1991, the *Young Scientist Research Award* in 2001, and election as Fellow in 2014, by the American Oil Chemists' Society; the *Foundation Scholar Award – Dairy Foods Division* in 1998, the *Danisco International Dairy Science Award* in 2007, the *Distinguished Service Award* in 2012, election as Fellow in 2013, and the *International Dairy Foods Association Teaching Award in Dairy Manufacturing* in 2020, by the American Dairy Science Association; the *Canadian/International Constituency Investigator Award in Physical Sciences and Engineering* in 2002 and 2004, by Sigma Xi; the *Excellence Promotion Award* in 2005, by the Portuguese Foundation for Science and Technology; the *Edgar Cardoso Innovation Award* in 2007, by the County of Gaia (Portugal); the *Scientist of the Year Award* in 2007, and election as Fellow in 2012, by the European Federation of Food Science and Technology; the *Samuel C. Prescott Award* in 2008, election as Fellow in 2011, and the *William V. Cruess Award* in 2014, by the Institute of Food Technologists; the *International Leadership Award* in 2008 and the *Elmer Marth Educator Award* in 2011, by the International Association of Food Protection; and decoration as *Chevalier dans l'Ordre des Palmes Académiques* in 1999, by the Government of France.

Among various complementary areas of research, Prof. Malcata has placed a major emphasis on: microbiological and biochemical characterization, and technological improvement of traditional foods; development of nutraceutical ingredients and functional foods; characterization of grape pomace, and improvement of spirits obtained therefrom; and rational application of unit operations to specific agri-food processes. To date, he has published more than 450 papers in peer-reviewed science journals, which have witnessed 400,000+ downloads and received more than 15,000 documented citations by the scientific community – supporting an *h*-index of 59. He has supervised 31 Ph.D. dissertations that were successfully concluded, and written 12 monographic books and edited 4 multiauthored books; furthermore, he has authored more than 70 chapters in edited books and 35 papers in trade journals, besides approximately 50 technical publications. He has been a member of 74 peer-review committees of research projects and fellowships, and has acted as supervisor of 90 individual fellowships. He has collaborated in 45 research and development projects – of which he served as principal investigator in 36. He has participated in 50 organizing and scientific committees of professional meetings. He has delivered 150 invited lectures worldwide, besides more than 500 volunteer presentations in professional meetings, congresses, and workshops. He has served on the editorial boards of 5 major journals in the food science and applied biotechnology areas. He has also reviewed more than 600 manuscripts for international journals and encyclopedias.

Prof. Malcata has been a longstanding professional member of the Institute of Food Technologists, as well as the American Institute of Chemical Engineers, the American Chemical Society, the American Association for the Advancement of Science, the American Oil Chemists' Society, the American Dairy Science Association, and the International Association of Food Protection.

1

Safety Assurance

Owing to their biological nature, foods deteriorate over time; a number of chemical reactions occur indeed in a spontaneous manner – and a great many of them are accelerated by adventitious or contaminating enzymes and microorganisms. Even worse is the case of food-borne microorganisms that pose a threat to consumer health upon ingestion – known as pathogens; hazards result from either excessively high viable numbers thereof, or previous release of toxins thereby. Viable numbers of food-borne pathogens and spoilage microorganisms must accordingly be kept below specific thresholds, so as to assure an acceptable risk in terms of public health (safety assurance), while guaranteeing that the expected shelf-life is not compromised (quality assurance); this constitutes a nuclear goal of food processing. Note that spoilage may be perceived as failure to abide to one (or more) specific quality parameters, usually of a sensory nature; hence, this criterion may be subjective – due to different perceptions of the same food by distinct consumers arising from their physiological condition or even ethno-cultural background.

The EU and most Western countries require most prepacked foods to carry a date of minimum durability. Some information on the expected shelf-life is normally conveyed by the food label via such expressions as "best before (date)" or "sell by (date)," suitable for relatively stable foods and concerned with optimal sensory performance; or "use by (date)," suitable for microbiologically perishable foods, expected to raise a health risk if consumed beyond the said date (on the hypothesis that recommended storage conditions were followed). The latter foods are usually meant for storage at refrigeration temperatures to enforce safety (rather than quality); they typically encompass foods labile to growth of pathogens (including formation of toxins thereby) up to poisoning levels, or else foods intended for consumption without (sufficiently severe) thermal treatment that may destroy putative pathogens (i.e. either consumed as fresh or upon mild-temperature reheating).

A wide number of foods fall under the safety-related deterioration class: cooked ready-to-eat products (e.g. cooked meats, eggs, fermented sausages, fish, hams, and poultry, as well as sandwiches and combos containing these type of ingredients); dairy products (e.g. cheeses, fromage frais, milk, and whipped cream); (cooked) cereals, coleslaw (and other mayonnaise-containing foods), fresh vegetables (and salads prepared therewith), pulses, smoked or cured fish; uncooked meat products; uncooked or partly cooked pastry dough (e.g. pizzas, sausage rolls), and fresh pasta containing fish, meat, poultry, or vegetables; and foods packaged under vacuum or modified atmospheres. Another group of foods undergo quality-related deterioration – as is the case of bread (and cakes) and butter (and margarines). Some prepacked foods are even not legally required to carry a date mark – for instance, chocolate, coffee, and honey (stored at room temperature) and frozen foods (stored in a frozen state); further to foods sold loose or packed for direct sale.

The ultimate goal of food preservation and safety is to minimize microbial contamination and growth, besides physicochemical decay throughout a preset period (known as shelf-life); hence, one resorts to one (or more) hurdles against microorganism viability – of either the intrinsic or extrinsic type. Minimal processing approaches include ultra-high pressure, pulsed electric or magnetic fields, pulsed light, and power ultrasound; as well as irradiation – where such issues as nature of radioactive decay and penetration depth are germane. In the specific case of hyperbaric treatments, it has indeed been found that high pressures play a role only upon weak bonds – as is the case of electrostatic and hydrophobic bonds; these are the dominant bonds causing molecules to unfold or aggregate – which may lead to gelation in the case of food proteins, or inactivation in the case of food enzymes. Hence, this mode of processing has been (increasingly) sought to modulate the occurrence of enzymatic reactions and bring about structural changes in proteins to improve food functional properties.

Classical approaches have, however, resorted to thermal treatments – either at high temperature, as happens with blanching, pasteurization, and sterilization (sorted by increasing lethality in the final product); or at low temperature, as is the case of chilling and freezing (sorted by increasing stability of the final product). On the hypothesis of linear death kinetics of target unwanted microorganisms, heat processing is characterized by such microbial descriptors as decimal reduction time – as a measure of lethality at a given temperature; and thermal resistance factor – as a measure of sensitivity to temperature. The latter supports the rationale for thermal processing – in terms of minimum conditions of temperature/time to assure microbial safety, along with maximum conditions thereof to assure sensory/nutritional quality.

Besides resorting to the concept of thermal death time as a tool in industrial practice, three cases should be considered in detail when addressing thermal processing: direct and indirect heating of (uniform temperature) food, and direct heating of (uniform temperature) liquid food. Liquid foods normally resort to heat exchanger equipment for thermal processing – with specific design equations and performance features. In the case of solid foods, direct heating occurs throughout cooking – and a temperature gradient will in general buildup within the food matrix; in the case of packaged liquid and semi-liquid foods, such a gradient may often be neglected due to internal natural convection, so batch evolution of (uniform) temperature throughout time – along with continuous supply of heat via an (essentially isothermal) fluid is to be considered. Indirect heating – where air filling a chamber is externally heated, while the (usually packaged) foods inside the said chamber are in turn heated by that air; and direct heating – where water steam is directly bubbled inside a liquid food, also deserve special attention.

Chilling may resort to cold water or air as utility; freezing requires much colder utilities, as is the case of single-use cryogenic fluids, or enhanced (often nested) vapor compression systems operating in a cycle. The reverse of freezing – normally necessary before ingestion of a frozen food, is normally much more detrimental than freezing itself, and adds to decay throughout (long-term) storage; distinct behaviors are, however, expected in fast or slow freezing/thawing.

Manipulation of intrinsic conditions, with food preservation as the bottom line, entails such relevant factors as water activity (i.e. water available for cell growth) and *pH*; while the most useful extrinsic factors are temperature and pressure (i.e. conditions appropriate for cell metabolism). The goal in either case is to attain levels for such factors well apart from those regularly found in natural foods, thus taking advantage of the susceptibility of unwanted, adventitious (or contaminating) microorganisms thereto. Low water activity – a thermodynamic parameter easy to ascertain in practice, constrains microbial metabolism and growth due to the lower humidity content, coupled to the bound form taken by water molecules – via surface adsorption, capillary condensation, or osmotic pressure. A number of compounds originally present in food or deliberately added thereto exhibit *pH* buffering capacity – i.e. they resist *pH* changes arising from (minor) addition of H^+ or OH^- ions to the food. The issue of an essentially constant (optimal) *pH* accordingly requires two major features exhibited by the buffer: useful *pH* range, typically centered at the intended *pH*; and capacity to resist changes of the intended *pH*, usually termed buffering stability (or power). Thermodynamic and kinetic temperature effects upon food performance have traditionally been described by van't Hoff and Arrhenius' laws, respectively; whereas thermodynamic pressure effects obey Joule and Thomson's law.

When physical and chemical reactions are at stake, Le Châtelier's principle provides a simple rule as to the direction of (deliberate) change in environmental parameters, in order to repress the said reactions – from thermodynamic or kinetic points of view. One illustrative application of such a principle pertains to hyperbaric processing: reactions undergoing a volume decrease are accelerated or occur to a larger extent – while those producing a volume increase are inhibited.

1.1 Preservation

1.1.1 Introduction

The desire to make food last longer stems back before Biblical times – yet refrigerators are no more than 150 yr old. Over history, food safety has indeed provided a bridge between civilizations – and refrigeration, in particular, may be seen as the watershed for division into two major epochs, the pre- and the postrefrigeration eras.

HISTORICAL FRAMEWORK

People of all times and cultures have realized that bad food can make one sick; even Confucius, in 500 BC, warned against eating sour rice – a notable practical advice from an eminently spiritual leader. Ancient Egyptians developed the silo – a storage tank designed to hold grain previously harvested from the agricultural fields; storing grain kept it cool and dry, so it could last into the nonharvest months (and even longer). The Bible speaks of the Hebrews receiving manna from heaven every morning – which was apparently similar to a flat wafer (or cracker), and tasted like honey- or oil-based cake; it was quick to gather every day, nourishing, and easy to carry while traveling in the wilderness – yet it spoiled between 1 and 2 days, due to worm breeding. The Romans were famous for their focus not only on food freshness, but also on salting foods in attempts to preserve them; this process dries out the food, so microbial contamination becomes unfeasible. They also invented *garum*, a concentrated fish pickle sauce, while ancient Greeks resorted to concentrated honey (another form of dehydration) to preserve quince. However, it was not until the 1600s that scientists were able to isolate germs, and claimed them to be the source of illness; Francisco Redi, an Italian physician, demonstrated in 1668 that maggots (or fly larvae) were not created by decaying meat itself, but rather by flies that laid their eggs on and in the meat – thus pioneering the belief that living things cannot arise spontaneously from nonliving matter. In 1835, James Paget and Richard Owen described the pig parasite *Trichinela spiralis* – the cause of trichinosis. Louis Pasteur, with his work on fermentation and pasteurization in the 1860s, and August Gartner, who diagnosed a food-borne illness bacterium, *Bacillus enteritidis*, in 1888, demonstrated beyond doubt that there are (micro)organisms in the air, soil, animals, and water – which, although invisible to the naked eye, can be vectors of sickness.

Combination of common knowledge about making food safer by keeping it colder, with a growing understanding of the behavior of germs and bacteria led inventors to look for ways to keep foods cold easily and anywhere. The change of society from agricultural to industrial that came along with the Industrial Revolution, and the demographic increase of urban agglomerations urged the need to keep food fresh throughout shipping and storage. From the trivial use of ice in early iceboxes, a long (but faster and faster) technological path has been tracked – which eventually led to development of cryogenic fluids, as well as compact and handy mechanical vapor compression systems for refrigeration.

Despite the improvements in food safety at large, scarce standards and regulations existed on processing and selling food by the end of the 19th century; meat-processing plants were for long an example of filthy food processes. Upton Sinclair went undercover into one such plant in Chicago to write his novel *The Jungle* – which vividly portrayed the reality of health violations and unsanitary practices therein; such conditions were accordingly brought to the open public for the first time. Despite his goal of preaching socialism, he triggered a full investigation of the meat-packing industry, under the auspices of U.S. President Theodore Roosevelt – the basis of the *Pure Food and Drug Act* and the *Meat Inspection Act* of 1906; they set the foundations for the thrust of safety regulations in the modern food industry. In 1938, approval of the *Federal Food, Drug and Cosmetic Act* supported authority, by the recently founded Food and Drug Administration, upon food safety issues. Sparked by an *Escherichia coli* outbreak in 1997, food safety shifted drastically from the existing sight–smell–touch food inspection into a science-based approach known as *HACCP*; this was soon recognized by the Centers for Disease Control and Prevention as able to dramatically reduce food-borne illness. Despite stricter and stricter control of food processes, mass-distributed items with spotty (or low-level) contamination are nowadays consumed far away from their sources; this has led to a new, insidious kind of epidemic – characterized by low attack rates, but huge numbers of dispersed victims. Some of the most fearsome threats faced today – e.g. *Campylobacter*, *Listeria*, and *E. coli* O157:H7, were just overlooked (or yet to be discovered) as recently as three decades ago; but food still serves as an unwanted vehicle for countless mysterious, lurking microorganisms and viruses.

Food may undergo spoiling, or even become unsafe for consumption due to physical, chemical, and biological reasons (e.g. infestation by pests, contamination by microorganisms); the most common examples are summarized in Table 1.1. Physical damage due to poor handling is relatively frequent in crisp products (e.g. extruded, baked, fried, or frozen) – and bruising of fresh fruits and vegetables similarly entails a form of physical spoilage; in the latter case, microbial growth and enzymatic browning are facilitated, as well as loss of moisture that causes wilting. Another example is destabilization due to breakdown of suspensions – i.e. agglomeration of droplets into larger clusters; for instance, mayonnaise may undergo this mode of spoilage upon freezing, heating, or (extreme) vibration.

A major form of physical spoilage is due to moisture migration – a temperature-dependent phenomenon driven by differences in water activity, a_w, and involving also glass transition. In operational terms, glass transition encompasses changes from a glassy (and thus brittle) state to a rubbery (and thus softer) state; this is observed in staling of bakery products, attributable in part to moisture migration from the crumb (characterized by higher a_W) to the crust (characterized by lower a_W). On the other hand, moisture migration in the crust drives change from a hard and crisp state to a tough and rubbery state; this phenomenon also justifies hard, dry, baked, or fried items becoming soft – or else confectionery becoming sticky, or food powders undergoing caking upon absorption of moisture from the atmosphere. Crystallization also causes physical spoilage; this may occur in the form of growth of ice crystals, which disrupt cellular tissues in food owing to volume expansion of water upon freezing. Another form of (re)crystallization occurs when amylose and amylopectin chains in cooked, gelatinized starch realign themselves via hydrogen bonding; it may occur upon cooling or sitting at room temperature for a long time, as this allows said molecules to retrograde from the viscous solution to a more crystalline structure. Retrogradation can also expel water from the polymer network (syneresis) – and is directly related to staling (or aging) of bread.

A number of chemical reactions contribute to spoilage of food, as is apparent from inspection of Table 1.1; this includes transformation of original fats, carbohydrates, proteins, and even micronutrients – with concomitant changes in color, flavor, or texture perceived as unacceptable by the final consumer. The most important examples are (zero-th order) autooxidation – also known as oxidative rancidity of fats; and (first-order) lipid oxidation in fish, dairy products, and meat – which leads to off-flavors and colors. The former is a chain reaction – catalyzed by Fe and Cu, moisture and ultraviolet light, and slowed down by free radical scavengers; it yields free radicals that, in turn, decompose to generate volatile hydrocarbons, alcohols, and aldehydes implicated in the rancid flavor. Another form of rancidity proceeds via (first-order) hydrolysis, which releases free fatty acids possessing pungent flavors. Oxidation of (proteins) myoglobin and oxymyoglobin may also take place in meat, with simultaneous decay of red to the brown color characteristic of metmyoglobin. Reducing sugars are, in turn, quite reactive with free amino acids or amino acid residues of proteins – according to (zero-th order) Maillard's reactions, often known as nonenzymatic browning as a whole. A large number of different compounds may accordingly form – namely volatile pyrazines, pyridines, furans, and thiazoles associated with off-odors, low-molecular weight compounds responsible for off-tastes, and antioxidants and melanoidin pigments that affect food appearance. Despite being causes of spoilage in beer, dairy produts, and fruit juices, the aforementioned reactions are sought for the development of typical flavors in bakery and fried foods, as well as in roasted cocoa or coffee, and even in soy sauce. Adventitious enzymes in food, and contaminating microorganisms of food that excrete extracellular enzymes or release endocellular enzymes upon lysis also bring about a number of unwanted chemical reactions responsible for spoilage.

High temperature is a major driver of food spoilage – namely via acceleration of oxidation of lipids and pigments; browning reactions, vitamin losses, increase of catalytic efficiency of enzymes, and enhancement of rates of microbial growth and metabolism. Further events favored by high temperatures include melting of fat that eventually leaks off as oil, and respiration of fresh fruits and vegetables – which speeds up not only ripening, but also senescence at postharvest stage. Conversely, some fruits and vegetables are labile to chilling injury that may lead to off-flavors, texture changes, and discoloration; freezer burn, or severe moisture loss, may be observed as a consequence of fluctuating freezing temperatures, and grittiness may develop in ice cream when frozen water recrystallizes. Overall, maintenance of

TABLE 1.1

Major mechanisms of spoilage of foods – with specification of target components and/or outcomes thereof.

Food Type		Spoilage Type									
		Physical				Chemical			Microbiological		
		Breakage	**Caking**	**Crystallization**	**Migration**	**Browning**	**Oxidation**	**Hydrolysis**	**Bacteria**	**Yeasts**	**Molds**
Bakery products	Bread	–	–	Starch (retrogradation)	Moisture (staling)	–	–	–	–	–	Growth
	Biscuits	Matrix	–	–	Moisture (softening)	–	Fat	–	–	–	–
	Cakes	–	–	Starch (retrogradation)	Moisture (staling, drying)	–	–	–	–	–	Growth
Beer		–	–	–	–	–	Flavors	–	Growth		–
Cereals		Matrix	–	Starch (retrogradation)	Moisture (softening)	–	–	–	–	–	–
Chocolate		–	–	Fat (bloom)	–	–	Fat	–	–	–	–
Coffee		–	–	–	Volatiles (loss)	–	Flavors	–	–	–	–
Confectionery		–	–	Sugar	Moisture (stickiness)	–	–	–	–	–	–
Cooked meats		–	–	–	Moisture (drying)	–	Fat	–	Growth	–	–
Dairy products		–	–	Lactose	–	–	Fat			–	–
Dried products		–	Matrix	–	Moisture (pickup)	Reduced sugars, amino acids	Fat	–	–	–	–
Fresh	Fish, meat, seafood	–	–	–	Moisture (drying)	–		–	Growth	–	–
	Fruits	Bruising	–	–		Reduced sugars, amino acids	–	Carbohydrates (softening)	–	Growth	
	Vegetables	–	–	–	Moisture (wilting)	–	–		–		
Fried/extruded products		Matrix	–	–	Moisture (softening)	–	Fat	–	–	–	–
Frozen products		–	–	Water	Moisture (freezer burning)	–		–	–	–	–
Teas		–	–	–	Volatiles (loss)	–	Flavors	–	–	–	–
Wines		–	–	–	–	–		–	Growth		–

the recommended temperatures throughout storage and transportation of foods is crucial to avoid unexpected spoilage – knowing that the effect of temperature comes as rather unselective, i.e. a great many physicochemical events will take place in the same or opposite direction, thus making the outcome rather difficult to predict.

Water is by far the dominant component of foods, so its equilibrium pattern is relevant – because, upon comparison with free, pure water, it gives a clue as to the degree of nonideality of the food as a mixture. Furthermore, the performance of microorganisms and enzymes in food – either adventitious or added via spurious contamination or deliberate inoculation, responds to free (or solvent-like) water, instead of bound water. The fraction of free water is termed water activity and is denoted by a_W – yet a correlation with moisture content, applicable to all foods, does not exist. For illustrative purposes, note the moisture content of 70%(w/w) for a_W=0.985 in fresh meat, 40%(w/w) for a_W=0.96 in bread, 35%(w/w) for a_W=0.86 in marmalade, 27%(w/w) for a_W=0.60 in raisins, 14.5%(w/w) for a_W=0.72 in wheat flour, 10%(w/w) for a_W=0.45 in macaroni, 5%(w/w) for a_W=0.40 in cocoa powder, 5.0%(w/w) for a_W=0.20 in biscuits, 3.5%(w/w) for a_W=0.11 in dried milk, or 1.5%(w/w) for a_W=0.08 in potato crisps.

The rate of deterioration of any food depends in fact on water available therein – to either play the role of solvent or directly engage is some physicochemical transformation (e.g. gelation, hydrolysis). This is confirmed by the stability of most foods failing to correlate with the overall moisture content thereof – for instance, peanut oil deteriorates when its moisture content exceeds 0.6%(w/w), whereas potato starch is stable up to 20%(w/w) moisture. Bound water – e.g. to hydroxyl groups of polysaccharides, carbonyl groups, and amino acid residues of proteins, and involved in hydrogen bonding at large, is a part of the overall moisture inventory of a food, but hardly participates in the aforementioned spoilage mechanisms. Typical values of a_W for selected foods are provided in Fig. 1.1. Water-rich foods are characterized by $0.9 < a_W < 1.0$ – and often are relatively dilute solutions of sugars/salts; the package is currently used to prevent moisture loss therefrom. Intermediate moisture foods are characterized by $0.55 < a_W < 0.90$, and typically encompass more concentrated solutions of sugar and salt (up to saturation thereof); in this case, marginal protection is required (or no package at all). Dried foods correspond to $0.0 < a_W < 0.55$, and are accounted for, namely, by dried matrices and powders; the package is accordingly needed, but to prevent moisture uptake in this case.

The concept of water activity is of relevance for the food industry; for instance, the rate of drying depends on the actual vapor pressure of water, measured in turn by a_W. Moreover, the enthalpy of vaporization – equal to 2.4×10^6 J.kg^{-1} in the case of free water, gradually increases as a_W decreases – and will reach almost twice that value when a_W attains 0.1; therefore, a larger and larger heat supply per unit mass of food will be necessary as drying proceeds. Knowledge of water activity patterns also allows prediction of storage stability and moisture evolution, shelf-life, optimum frozen storage temperature, and moisture barrier properties required for packaging materials.

In view of the ultimate focus on food preservation, the ranges of a_W for occurrence of chemical and metabolic reactions are crucial; evidence in this regard is conveyed by Fig. 1.2. Almost all microbial metabolism is inhibited below a_W=0.6: bacteria are the most sensitive, with a_W=0.9 as a lower threshold for significant activity (including production of toxins), yeasts with an a_W=0.85 (except for osmophilic yeasts), and molds with a_W=0.8.

Germane a_W requirements for growth of bacteria in food include > 0.97 for *Clostridium botulinum* E and *Pseudomonas fluorescens*, >0.95 for *Escherichia coli*, *Clostridium perfringens*, *Salmonella* spp., and *Vibrio cholerae*, >0.94 for *Cl. botulinum* A, B and *Vibrio parahaemolyticus*, >0.93 for *Bacillus cereus*, >0.92 for *Listeria monocytogenes*, >0.91 for *Bacillus subtilis*, >0.90 for anaerobic *Staphylococcus aureus*, and >0.87 for aerobic *Staph. aureus*. In the case of yeasts, a_W should abide to >0.90 for *Saccharomyces cerevisiae*, >0.88 for *Candida* spp., and >0.83 for *Debaryomyces hansenii*. Finally, a_W suitable for growth of molds encompasses > 0.94 for *Stachybotrysatra* spp., > 0.93 for *Rhizopus nigricans*, > 0.90 for *Trichothecium roseum*, > 0.85 for *Aspergillus clavatus*, > 0.84 for *Byssochlamys nivea*, >0.83 for *Penicillium expansum*, *Penicillium islandicum*, and *Penicillium viridicatum*, >0.82 for *Aspergillus fumigatus* and *Aspergillus parasiticus*, >0.81 for *Penicilium cyclopium* and *Penicillium patulum*, > 0.79 for

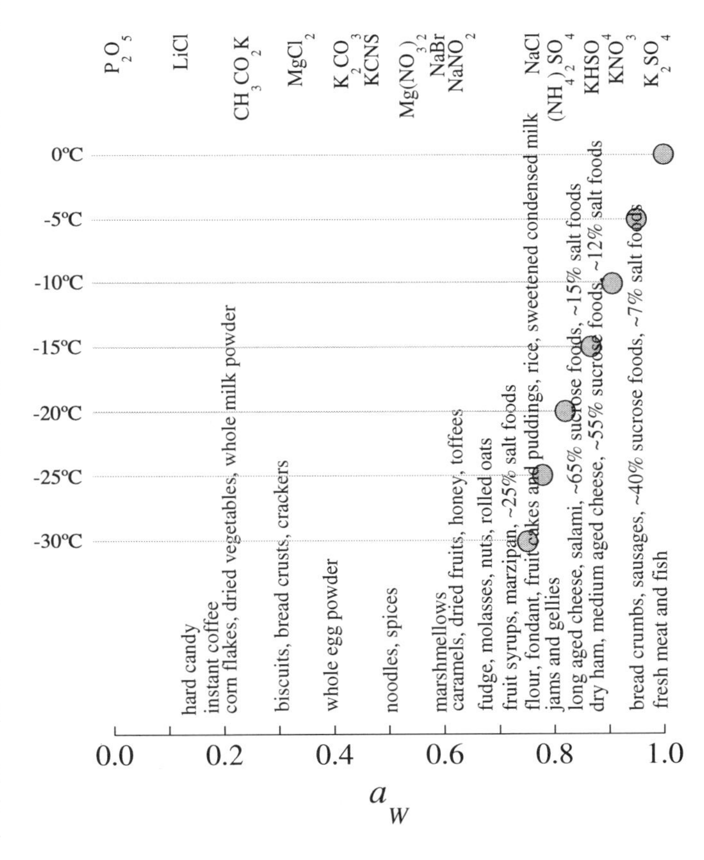

FIGURE 1.1 Water activity (a_W) in selected foods (bottom) and saturated aqueous solutions of selected salts (top) – with indication of variation of water activity of ice with temperature (●).

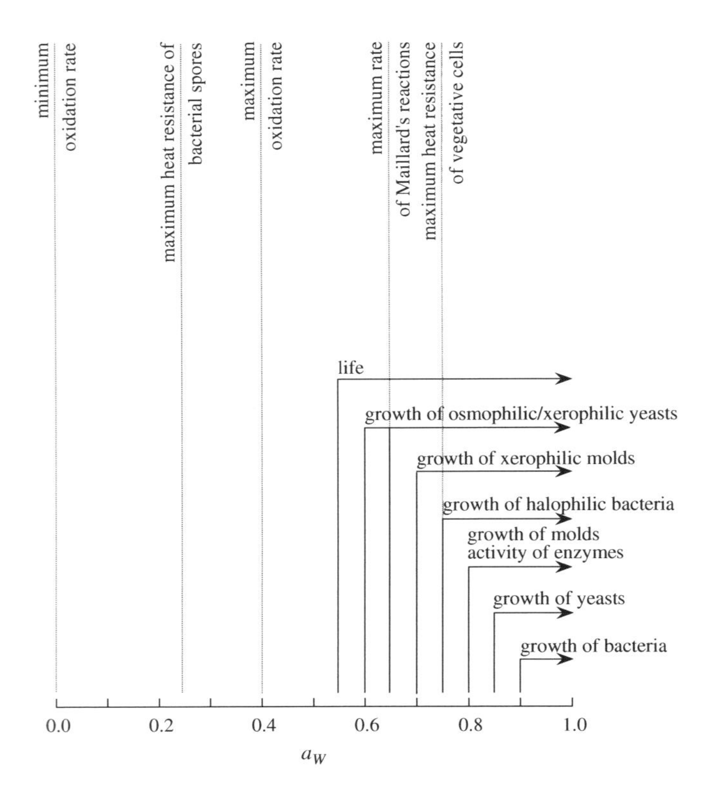

FIGURE 1.2 Ranges (⌐→) of water activity for occurence of physicochemical transformations – with indication of maxima/minima (|).

Penicillium martensii, >0.78 for *Aspergillus flavus*, and >0.77 for *Aspergillus niger*, *Aspergillus ochraceous*, *Aspergillus restrictus*, and *Aspergillus candidus*.

However, the degree of inhibition brought about by a given a_W depends also on prevailing temperature and *pH*, as well as presence of O_2 and CO_2, or any chemical preservatives – knowing that a suboptimal value of any such factor considered *per se* enhances the detrimental effect of a reduced a_W. For instance, when a_W exceeds 0.95, one must resort to chilling for preservation in the case of fresh meat (for days) or milk (also for days) – while fresh vegetables remain stable provided that respiration thereof is allowed (for weeks); and (packaged) cooked meat (for weeks) or bread (for days) do not even require special storage conditions. One should resort to salt and low *pH* for preservation in the case of dry sausages (for months), or just low *pH* in the case of pickles (also for months); and to thermal processing in the case of fruit cakes (for weeks), chilling in the case of yogurt (for weeks), or none in the case of dry milk (for months). Food operations that increase the fraction of immobile water lead to decreases in a_W; this is the case of evaporation and drying, and freeze-drying and -concentration – as well as addition of humectants in intermediate-moisture foods.

Although growth of microorganisms is influenced chiefly by a_W, the nature of food solutes also plays a role; grains and sausages are preserved via addition of a salty brine, while jams and jellies resort to addition of a sugar brine – both causing considerable reduction in a_W. Reduction of a_W holds a bactericidal action *per se*, yet thermal destruction of bacterial spores becomes more difficult in such low-a_W environments. The maximum resistance of vegetative cells to heat occurs within $0.75 < a_W < 0.85$, and that of spores occurs at 0.2–0.3; the protection brought about by fats and oils may thus be rationalized on the basis of a decrease in water activity effected by the hydrophobic phase. As a heuristic rule, foods characterized by $0.85 < a_W < 0.9$ – accounting for a major fraction of foods available in the market, can be stabilized via vacuum packaging, provided that some sort of fungicide is added (e.g. sorbic acid, pimaricin, calcium propionate, esters of *p*-hydroxybenzoic acid, polyalcohols); coupled with moderate thermal treatment to destroy osmophilic yeasts.

Enzymatic activity is essentially discontinued when a_W lies below the monolayer value – due to poor substrate mobility, and associated inability to diffuse to the enzyme active site; this means, in practice, that most enzymes are inactive at $a_W < 0.8$ – even though a few esterases remain active at as low a_W as 0.2–0.3, and certain enzymes in legumes can withstand even lower activities. Enzymes and proteins at large are normally less susceptible to thermal treatment if in a dehydrated state. Maillard's reactions – which account for cooked taste, brown color, or loss of available Lys, exhibit maximum rates at $0.6 < a_W < 0.7$, with some variability among foods; a lower a_W reduces occurrence of this phenomenon, again due to reduced mobility of reactants. On the other end, water being formed as product justifies why higher a_W slows down browning as per product inhibition – and significant dilution of reactants may account for the extra slowing down at even higher a_W. Lipid oxidation occurs at low a_W due to activation of free radicals; however, antioxidants and chelating agents – able to sequester trace metals acting as potential catalysts for this reaction, can solubilize at higher a_W, thus gaining access to such free radicals and promoting inactivation thereof. Conversely, a low a_W favors stability of ascorbic acid. Formation of a hydration shell and insoluble hydroxides may account for the decreasing rates of metal-mediated catalysis observed at high a_W – while hydrolysis of protopectin, transformation of chlorophyll to pheophytin, and spontaneous (autocatalytic) hydrolysis of triglycerides are strongly influenced by water activity. Finally, a_W affects the degree of retention of aroma compounds; extreme concentration of solutes substantially reduces their diffusivity, yet the actual binding of some compounds requires a minimum amount of water be present.

Food sorption isotherms describe the thermodynamic relationship between water activity and moisture concentration prevailing in food at a given temperature (and pressure); their shape reflects the way water binds within the system. Illustrative examples are presented in Fig. 1.3. Despite the (expected) variability between foods – due to unique matrix composition and structure, a general trend appears consisting of a (concave) fast rise toward a horizontal asymptote, followed by a (convex) sigmoidal pattern toward a vertical asymptote. In some cases (e.g. yam, pineapple, mango, walnut kernels), the said horizontal asymptote essentially coincides with the horizontal axis, while the vertical asymptote is placed at a_W between ca. 0.6 and 0.8 in most foods (although it may appear earlier, e.g. yam, or later, e.g. walnut kernels). An increase in temperature reduces a_W to a slight degree – although the opposite may be observed in fat- or sugar-rich foods.

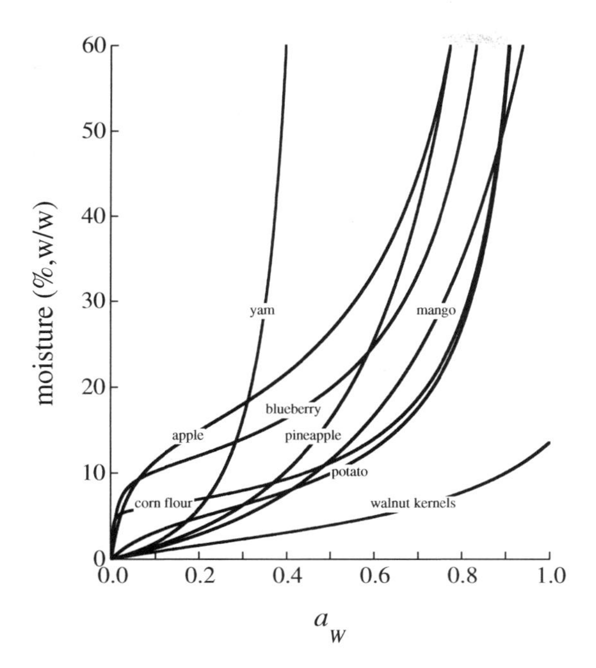

FIGURE 1.3 Sorption isotherms of selected foods at room temperature, as mass fraction of moisture versus water activity, a_W.

Three approaches have been in regular use to experimentally generate sorption isotherms of foods: (i) gravimetric methods, (ii) manometric methods, and (iii) hygrometric methods. In the first case, a small quantity of product sample is placed in contact with a small quantity of gas (usually air); the weight of the sample is monitored with a balance until it plateaus. In manometric methods, a small quantity of material is placed under an amount of gas of known relative humidity – and the partial pressure of water vapor is measured after a sufficiently long time (so as to allow equilibrium be reached); the overhead gas may be obtained by saturating it at a given pressure and temperature, and then reheating or expanding it to the chosen value of relative humidity. In hygrometric methods, a small quantity of material is placed in a confined atmosphere in the presence of a saturated solution containing excess (solid) salt – see Fig. 1.1; it suffices to measure the concentration of water in the product at equilibrium with the vapors of these solutions to ascertain the dependence of a_W on the moisture concentration. More sophisticated techniques of measurement include impedance spectroscopy, as well as light refraction/attenuation (e.g. infrared spectroscopy); the specific form in which water molecules are held within the food matrix can be ascertained via magnetic resonance, differential scanning calorimetry, or X-ray diffraction.

Sorption isotherms obtained by gradually increasing humidity or by diminishing humidity are not normally superimposable; this realization is illustrated in Fig. 1.4. The hysteresis loop, or difference between desorption and adsorption, is particularly notorious in some foods (e.g. rice) – and relevant in attempts to determine the degree of protection required against moisture uptake.

Hysteresis is apparently related to nature and state of the components of a food – and reflects their potential for structural and conformational rearrangements, which alter accessibility of energetically favorable polar sites; several theories have been put forward to rationalize such an observation – namely the ink bottle theory, the molecular shrinkage theory, the capillary condensation theory, and the swelling fatigue theory. The rate of displacement of water vapor from a food to its surroundings, or *vice versa*, depends indeed on both the moisture content (and remainder composition) of the food and the moisture content (and corresponding temperature) of the outer atmosphere; for any given temperature, the moisture content of a food evolves until it reaches chemical equilibrium with water vapor in the surrounding air.

Another relevant factor toward food preservation is pH – defined as the negative of the decimal logarithm of activity (usually replaced by mole concentration) of hydrogen ions H^+ in solution, associated with water molecules in the form H_3O^+. Many compounds in food tend to release hydrogen ions, namely (weak) acids containing –COOH groups; conversely, amine-containing foods (as well as ammonia) abstract hydrogen ions, thus acting as (weak) bases – while metal hydroxides release OH^- upon dissolution that combine, in turn, with free H^+ to form H_2O. Typical pH values for several foods are conveyed in Fig. 1.5. Pure water exhibits a pH of 7.0 at room temperature – and is said to be neutral because the concentration of H^+ balances that of OH^-, i.e. 10^{-7} mol.L^{-1}. However, when exposed to air for a sufficiently long time, its pH tends to fall due to dissolution of atmospheric CO_2 that reacts with H_2O to form H_2CO_3; the latter dissociates, in turn, as HCO_3^- and H^+, with possibility for further dissociation of the former to CO_3^{2-} and H^+ (so pH may decrease down to 5.7). When pH lies below 7.0, the food is said to be acidic; whereas $pH > 7$ unfolds a (less frequent) alkaline food.

In foods from plant and animal sources, pH is intrinsically controlled by their native constituents with buffering properties – namely proteins and amino acids, carboxylic acids (including fatty acids), and phosphate salts. However,

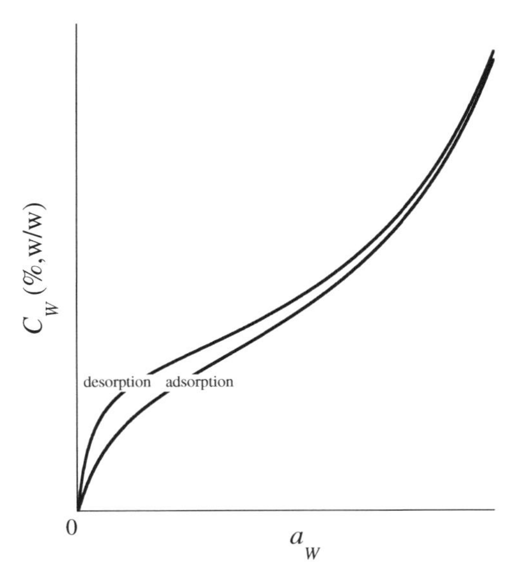

FIGURE 1.4 Adsorption and desorption isotherms of typical food product.

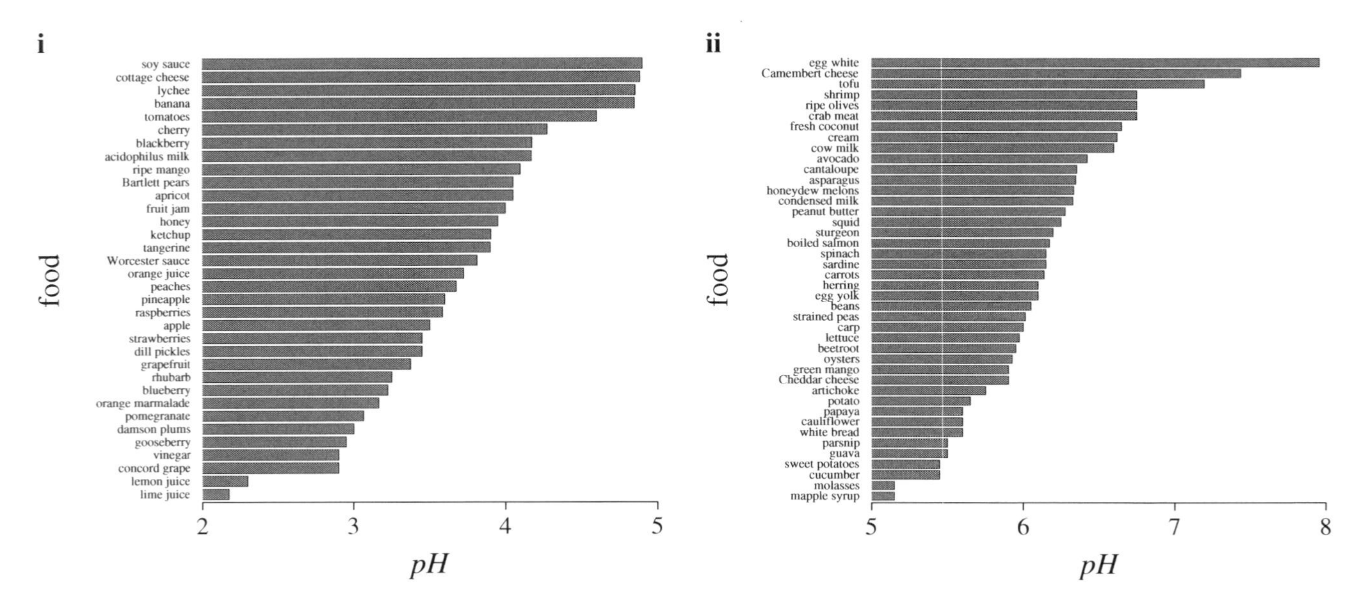

FIGURE 1.5 Average *pH* values of selected foods, within (i) 2–5 and (ii) 5–8 ranges.

extraneous buffering compounds, either extracted from other biological sources or obtained via chemical synthesis, may be added to foods to set *pH* at some intended value (and maintain it within a narrow range). Such additives are targeted at improving nutritional value, color, flavor, or texture, or else aid merely in the preservation of food; this is the case of additives labeled between E330 and E343, see Table 1.8 of *Food Proc. Eng.: basics & mechanical operations*, which can also act as *pH* buffers. A few examples are shown in Table 1.2. A buffer is frequently an aqueous solution containing a mixture of a weak acid and its conjugate base supplied in salt form (e.g. citric acid and sodium citrate) – although two salts with protolysis capacity connecting their forms and pK_as at least 2 units apart from each other (e.g. Na_2HPO_4 and NaH_2PO_4), or an amphiprotic compound and a strong acid or base (e.g. Gly with either HCl or NaOH) will also do. The germane pK_a decreases normally with temperature – as a direct consequence of the endothermic nature of proton dissociation, in agreement with Le Châtelier's principle.

Sodium, calcium, and potassium citrate have found a wide range of applications as *pH* buffers (further to their antioxidant roles); they can all reduce the rate of occurrence of reactions implicated in unwanted discoloration of fruits. Sodium citrate (E331) is used as a buffer chiefly in jams and jellies; potassium citrate (E332) is preferred in cakes and biscuits, cheese, and jam; and calcium citrate (E333) is an important acidity regulator present in carbonated drinks. Monopotassium phosphate (E340) is mainly utilized for its antioxidant features, yet it also exhibits buffering capability; it is thus used in pudding products, e.g. custard and milk powder, as well as jelly mixes – and may even be added to cooked meat. Monopotassium phosphate is an important ingredient in sports drinks because it provides potassium required as electrolyte; however, its buffering action is also relevant. Finally, potassium tartrate (E336) is produced from grapes during winemaking; it is accordingly used as a buffer – not only in winemaking, but also in breadmaking, besides formulation of fruit pie mixes. Further to its buffering action, potassium tartrate helps consistent bread rising.

The optimum growth *pH* is the most favorable for growth – even though a minimum and a maximum of *pH* tolerance exist for each microorganism, beyond which no significant growth occurs; most bacteria of interest grow well at *pH* between 6.0 and 8.0, most yeasts between 4.5 and 6, and most molds between 3.5 and 4. Illustrative examples pertaining to microorganisms with implication in food safety are listed in Table 1.3.

The great majority of microorganisms prefer a neutral *pH* for growth, but can still grow in more acidic media; this is the case of bacteria (e.g. *Listeria monocytogenes*, *Clostridium botulinum*, *Salmonella* spp.), hence termed neutrophiles – as they grow optimally at *pH* 7±2. Most neutrophiles stop growing at *pH* 5.0, even though some can withstand as low as 4.5 – which has accordingly been elected as upper boundary for high-acid foods (i.e. *pH* < 4.5), usually resistant to dangerous contaminations. Conversely, acidophiles (e.g. *Staphylococcus typhi*) grow optimally around *pH* 3.0. These microorganisms display a number of adaptations for survival in strongly acidic

TABLE 1.2

Typical buffers in food systems, with qualitative composition, protolysis constant(s), and useful *pH* range.

Compound(s)	pK_a	Buffer *pH* range
Gly/HCl	2.4	2.2–3.6
Na_2HPO_4/citrate	7.2 5.2	2.2–8.0
citric acid/sodium citrate	5.2	3.0–6.2
acetic acid/sodium acetate	4.8	3.6–5.6
Na_2HPO_4/NaH_2PO_4	7.2 2.1	5.8–8.0
Gly/NaOH	9.8	8.6–10.6

TABLE 1.3

Ranges of *pH* for growth of selected food-borne pathogens.

	pH range for growth		
Microorganism	**Minimum**	**Optimum**	**Maximum**
Staphylococcus aureus	4	6–7	10
(enterohemorrhagic) *Escherichia coli*	4.4	6–7	9
Bacillus cereus	4.9	6–7	8.8
Shigella spp.	5	6–8	9.3
Campylobacter spp.	4.9	6.5–7.5	9
Listeria monocytogenes	4.4	7	9.4
Clostridium botulinum	4.6	~7	8.5
Salmonella spp.	4.2	7–7.5	9.5
Yersinia enterocolitica	4.2	7.2	9.6
Clostridium perfringens	5.5–5.8	7.2	8.9
Vibrio vulnificus	5	7.8	10.2
Vibrio parahaemolyticus	4.8	7.8–8.6	11

TABLE 1.4

Ranges of temperature for growth of selected food-borne pathogens.

	Temperature range for growth (°C)		
Microorganism	**Minimum**	**Optimum**	**Maximum**
Yersinia enterocolitica	–1	28–30	42
Aeromonas hydrophila	–4	28–35	42–45
Bacillus cereus	5	28–40	55
Clostridium botulinum E, F	4	29	45
Plesiomonas shigelloides	8	30	45
Aspergillus flavus	10	33	43
Salmonella spp.	5	35–37	45–47
(enterotoxigenic) *Escherichia coli*	7	35–40	46
Listeria monocytogenes	–0.5	37	45
Cl. botulinum B	3	37	50
Cl. botulinum A	4	37	50
Vibrio parahaemolyticus	5	37	43
Brucella spp.	6	37	42
Shigella spp.	7	37	45–47
Vibrio vulnificus	8	37	43
Vibrio cholerae	10	37	43
Staphylococcus aureus	10	40–45	46
Campylobacter spp.	32	42–45	45
Clostridium perfringens	12	46	50

environments – such as proteins with increased negative surface charge, crucial to stabilize their active conformation; coupled with changes in the phospholipid profile of their membrane, as required for fluidity at low *pH*. Alkaliphiles appear at the other end of the spectrum – with *pH* optima between 8.0 and 10.5 (e.g. *Vibrio cholerae*). These microorganisms have adapted to their harsh environment through evolutionary modification of lipid and protein structure, complemented by compensatory mechanisms to maintain the proton motive force; they may even resort to an Na^+ gradient, rather than the proton motive force itself, as source of energy for transport reactions and motility. In addition, many of their enzymes have a higher isoelectric point than their neutrophile counterparts, due to richnness in basic amino acid residues.

Special mention in this regard is deserved by *Helicobacter pylori* – a slim, corkscrew-shaped bacterium, recently implicated in most peptic ulcers; its isolation and characterization were indeed awarded with the Nobel Prize of Medicine in 2005. Despite its ability to survive the low *pH* prevailing in the stomach, it is a neutrophile; in fact, *H. pylori* creates a microenvironment of neutral *pH* via production of large amounts of enzyme urease – which breaks down urea to NH_4^+ and CO_2, with the former ion being responsible for the local raise in *pH*.

Microbial growth can occur over a temperature range from -8°C under deep water up to 100°C at atmospheric pressure; the critical requirement is that water is present in the liquid state. However, the microorganisms of relevance in food processing/consumption reduce to only those that can grow around the normal body temperature – and thus will raise a health hazard upon ingestion; or those that can grow under regular processing/cooking operations, and leave a trace of metabolites that spoil or even poison the final food. Data pertaining to selected food pathogens are tabulated in Table 1.4.

Within the permissive temperature range for growth of each genus, each microorganism exhibits a minimum, an optimum, and a maximum – which may be species- and even strain-dependent such so-called cardinal temperatures are also affected by environmental factors, viz. nutrient profile, *pH*, a_w, and redox potential. Mesophiles (or middle-loving) are adapted to moderate temperatures – with optimal growth temperatures ranging from room temperature (ca. 20°C) to ca. 45°C; inspection of Table 1.4 indicates that most pathogens carried by food are mesophiles, in view of the body temperature being 37°C. They can usually withstand freezing and freeze-drying, provided that some biocompatible agent (e.g. glycerol) is added as protectant. Psychotrophs, or psychotolerant microorganisms prefer cooler environments, from a high temperature of 25°C down to refrigeration temperatures of the order of 4°C; ubiquitous in temperate climates, they account for spoilage of refrigerated foods. Organisms that grow above 50°C up to a maximum of 80°C are termed thermophiles (or heat-loving); they cannot multiply at room temperature – and are widely distributed in hot springs and geothermal soils, as well as garden compost piles.

Very low temperatures affect the cells in many ways; their membranes lose fluidity and become damaged by ice crystal formation – while chemical reactions and diffusional transport slow down considerably, and proteins gain too rigid a conformation that is unsuitable for catalysis and prompts denaturation. At the opposite end of the temperature scale, heat promotes denaturation of proteins and nucleic acids, and the increased fluidity impairs metabolic processes within membranes. It has been found that proteins in psychrophiles are richer in hydrophobic residues, display unusual flexibility, and possess a larger number of secondary stabilizing bonds when compared to regular proteins from mesophiles; antifreeze proteins and solutes that decrease the freezing point of the cytoplasm are also common, and their membranes are richer in unsaturated lipids for enhanced fluidity. Macromolecules in thermophiles exhibit some unique structural differences from their mesophile counterparts; the lower ratio of saturated to polyunsaturated lipids constrains membrane fluidity, additional secondary ionic and covalent bonds are present, and key amino acid residues are replaced that stabilize folding of proteinaceous backbones, whereas their DNA sequences possess a higher proportion of guanine–cytosine nitrogenous bases – held together by three hydrogen bonds, unlike adenine–thymine that are connected in the double helix by only two hydrogen bonds.

Since many biochemical processes involve redox reactions – based on transfer of an electron from, or to a molecule or ion, the accompanying oxidation state may change. An accepted electron reduces the oxidation state of its acceptor (thus termed oxidizing agent), whereas a donated electron increases the oxidation state of its donor (thus termed reducing agent); oxidation and reduction are thus simultaneous phenomena, because electrons cannot exist free on their own. The capacity of a compound to donate electron(s) in an aqueous medium is measured by its redox potential, E_h (usually expressed in mV).

Redox reactions are nuclear in life because they form the metabolic basis for release of energy (as in fermentation and respiration) and storage of energy (as in photosynthesis); the ability of microorganisms to carry out these types of reactions depends on the prevailing value of E_h. It appears that E_h affects the structural composition of a few sensitive components on the surface of the cell, which are enrolled in transport and energy-yielding mechanisms, e.g. enzymes and phospholipids; this is especially relevant when the apoenzyme portion of the former is rich in sulfur-containing amino acid residues – and they require Fe, Zn, Mg, or Cu as cofactors, since their contribution to the catalytic process hinges critically on oxidation state. Furthermore, redox potential affects intracellular *pH* – which in turn modulates proton-motive force. Strictly aerobic microorganisms can grow only at $E_h > 0$, whereas strict anaerobes require $E_h < 0$; the current value of E_h actually gives a clue to the nature of metabolism expected to take place during food fermentation, and thus of the physiological state of the microbial culture. In fact, monitoring and control of dissolved O_2 resort to measurement of E_h, although redox-sensitive pigments may also be used for this purpose. The actual redox potential exhibited by a food is the sum of E_h values for all components therein – and it affects solubility of specific nutrients, e.g. mineral ions; antioxidants possess, in particular, the ability to alter E_h within a food. Both animal and plant pigments are sensitive to E_h, so redox changes may affect the color of foods; this is typically the case of redox interactions between Fe^{3+} and polyphenols containing *o*-dihydroxyl groups.

As emphasized above, injury to consumers upon ingestion of contaminated foods may arise from physical vectors (e.g. foreign bodies), chemical vectors (e.g. chemical contaminants), or microbiological vectors (e.g. pathogenic microorganisms); however, preservation techniques have been chiefly focused on reduction of viable numbers of pathogen and spoilage microorganisms in food to sufficiently low levels – resorting to either mild technologies or classical thermal treatments, where modulation of water activity often plays a role. One example of the former is high-pressure (or hyperbaric) processing – while the latter encompasses pasteurization and sterilization.

HISTORICAL FRAMEWORK

The first use of hyperbaric processing dates back to 1899 in the United States, and was aimed at preserving milk, fruit juice, and meat; application of 660 MPa for 10 min was, in fact, found effective in destroying microorganisms. Research conducted during the early 20th century unfolded alterations in the structure of egg white when subjected to high pressure – yet several operational constraints, arising from inadequate equipment and packaging materials available at the time, hampered investigation in this area for several decades. Studies were resumed in the 1980s, namely in Japan, and industrial exploitation started in 1990 in the form of high-pressure processed jams – including apple, kiwi, strawberry, and raspberry; their shelf-life could be extended up to 2 months, under chilled storage. More recently, bulk orange and grapefruit juices became commercially available, as well as premium fruit jellies and yogurts, purées, sauces, and salad dressings; the associated success derived chiefly from their intrinsic low *pH* – which does not allow spoilage by pressure-resistant spores, while active vegetative forms are destroyed by the high-pressure treatment. Pressure-treated chicken strips, fruit smoothies, guacamole, hummus, and oysters are now easily found in the US market, while orange juice is available in the UK, and sliced cooked ham in France and Spain; commercial prospects are quite favorable for extension of hyperbaric processing in the coming future to cheese, foie gras, liquid egg, poultry, red meats, and sea foods.

Absence of physical or chemical contaminants is only achievable through careful handling practices, including control at critical points of the process; although chemical contaminants may be extremely poisonous – e.g. microbial toxins, the typical contamination of foods of the chemical type tends to lead to chronic manifestations (e.g. pesticide accumulation in adipose tissue or heavy metal accumulation in kidneys). Conversely, microbial contamination tends to produce acute effects – especially due to the multiplication capacity of pathogens under the favorable environmental conditions prevailing in the human body. Fully representative data are unfortunately not available on microbial-mediated food intoxications, because most events and situations are not even reported as such to the official authorities; when reliable data exist on reported cases – as happens in the United States with a population of the order of 300 million people, the distribution of events per source is as depicted in Fig. 1.6. The most common reports pertain to *Staphylococcus aureus*, of the order of 10 million per year; this corresponds to about

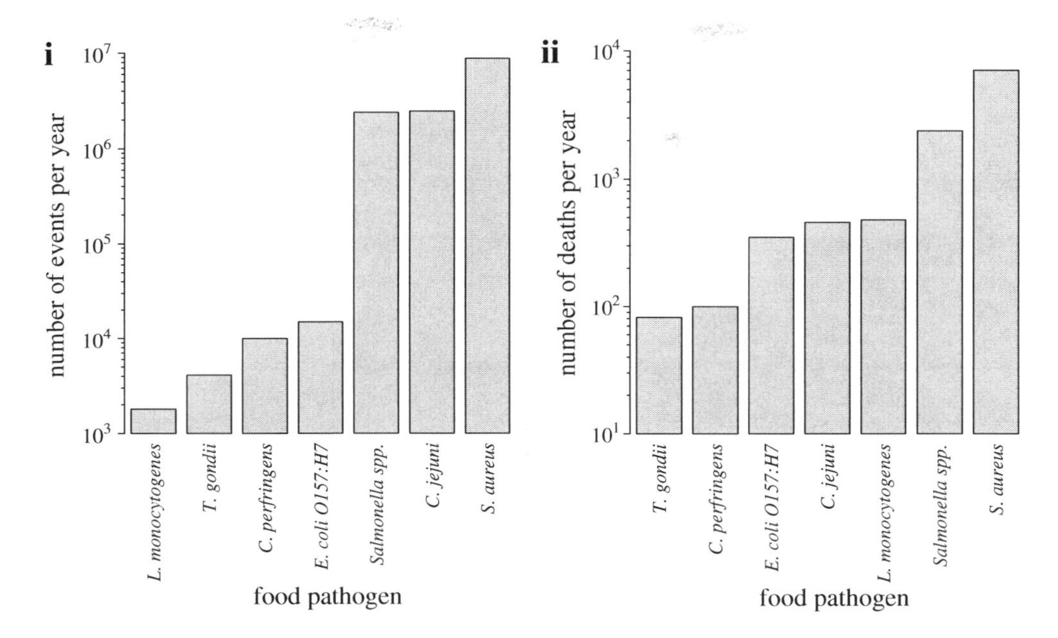

FIGURE 1.6 Public health panorama of foodborne intoxications in the United States by the end of the 20th century, in terms of (i) morbidity and (ii) mortality.

one intoxication event per every 30 people – a figure conveying an underestimate, since most cases are not formally recorded (as mentioned above). Fortunately, most cases cause only morbidity, with temporary personal health and well-being implications besides work absenteeism. However, about 1 out of every 1000 events caused by this agent degenerates to death – usually within the more fragile subpopulations, viz. the younger and the elder, or otherwise immunocompromised individuals (e.g. patients taking immunosuppressant drugs after organ transplants, infected with HIV, or undergoing cancer treatments).

The demand by consumers for high-quality foods – exhibiting fresh or natural features yet bearing an extended shelf-life, has led to the development of mild technologies to preserve them; the oldest, and still the most frequent is chilling – yet several nonconventional technologies have meanwhile reached commercial expression, or show at least a market potential. Although it is conceptually simpler to apply a single preservation technique, there is no reason why more than one should not be applied, simultaneously or sequentially; this realization has led to the multiple hurdle concept, introduced by Leistner in 1995 – and also known as combined preservation. Design of such systems requires a fundamental understanding of the own effect and the mutual interactions of temperature, pressure, a_w, pH, E_h, chemical preservatives, and the like – in attempts to effectively control numbers of viable spoilage and pathogenic microorganisms, and eventually guarantee microbiological safety of the final product. The central idea is that microorganisms – as all alive beings, maintain a stable internal environment (or homeostasis); hurdles can disturb one or more homeostasis mechanisms, thus hampering their regular metabolic activity and ultimately leading to death thereof. If multiple hurdles are utilized, the intensity of each one can be comparatively lower – so less detrimental effects upon the final quality of food are expected, especially if they act synergistically. When a number of techniques are used in combination with mild heating, less product damage results indeed in terms of color and flavor retention, thus enhancing consumer appeal for the final (processed) product. For instance, high-pressure processing has been effective to eliminate *Clostridium sporogenes* in chicken breast when combined with irradiation, the decimal reduction time of *Yersinia enterocolitica* reduces from 1.4 min at 59°C during classical pasteurization, to a mere 0.3 min when simultaneously treated with (gauge) pressure at 0.3 MPa, and ultrasound at 150 dB and 20 kHz at that temperature; and sterilization with regard to spore-forming *Bacillus coagulans* is possible within 30 min via hyperbaric treatment at 400 MPa and 70°C, following previous addition of nisin.

The variety of hurdles available to preserve foods is quite wide; in terms of physical hurdles, one may resort to high temperatures (blanching, pasteurization, heat sterilization, evaporation, extrusion, frying), low temperatures (chilling, freezing), pressure (ultra-high pressure and ultrasonication), electromagnetic energy (microwave or radiofrequency, and pulsed magnetic or electric fields), radiation (ionizing and ultraviolet, and photodynamic inactivation), and packaging (including aseptic, as well as active and modified atmosphere packaging). Chemical hurdles include low pH, low E_h, low a_w, addition (or generation *in situ*) of carbon dioxide, ethanol, lactic acid, lactoperoxidase, Maillard's reaction products, organic acids, oxygen, ozone, phenols, phosphates, salt, smoke compounds, sodium nitrite/nitrate, sodium or potassium sulfite, sulfur dioxide, and spices and herbs, as well as surface-treatment agents. Microbial hurdles encompass antibiotics, bacteriocins, competitive microflora, and protective cultures. Nevertheless, utilization of one over the other, or combinations thereof are to be decided on a case-by-case basis.

1.1.2 Product and process overview

1.1.2.1 Food-borne agents of disease

The main types of food-borne pathogens are bacteria, viruses, and parasites, as well as a few molds; although they may cause spoilage, yeasts are seldom implicated with food poisoning. The microbial groups with potential risk are *Aeromonas* spp., *Bacilus cereus*, *Brucella* spp., *Campylobacter* spp., *Arcobacter* spp., *Clostridium botulinum*, *Clostridium perfringens*, (enteropathogenic) *Escherichia coli*, *Listeria monocytogenes*, *Mycobacterium paratuberculosis*, *Plesiomonas* spp., *Salmonella* spp., *Shigella* spp., *Staphylococcus aureus*, *Vibrio* spp., *Yersinia enterocolitica*, toxigenic molds, *Giardia duodenalis*, *Cryptosporidium parvum*, *Cyclospora cayetanensis*, and *Toxoplasma gondii*. Harmful viruses include Norwalk- and Sapporo-like caliciviruses; enteric adenovirus 40 and 41; rotaviruses A, B, and C; astrovirus; hepatitis A and E viruses; and enteroviruses. Such parasites as *Taenis saginata* and *T. solium*, and *Trichinela spiralis* are also of concern from a public health perspective. Their typical morphologies are illustrated in Fig. 1.7.

Aeromonas hydrophilia HG1 and A (see Fig. 1.7i), *A. caviae* HG4 and A, and *A. veronii* HG8 have been implicated with gastroenteritis, particularly in immunocompromised individuals and cancer patients, as well as in children under 5 years old. Typical symptoms include vomiting, fever, and diarrhea within 12–36 hr upon ingestion – although the former may also cause inflammation of gallbladder, septicemia, and meningitis, with a death toll that may reach 60%; this pathogen invades cells of the human intestine, where it produces toxins. The usual reservoir is fresh or brackish water; and the major food vehicles are (obviously) water, as well as fresh vegetables, raw milk and cheese, poultry, lamb, fish, and shellfish – which were exposed to contaminated water for watering or drinking purposes, as living medium, or just for processing. Their minimum growth conditions are 1–5°C and *pH* of 6.5, and they are characterized by $D_{60} = 3–7$ min – with D_x denoting time required to reduce viable counts by tenfold at x°C.

Arcobacter butzleri and *A. cryaerophilus* (previously known as aerotolerant *Campylobacter*) have been associated with enteritis in humans (besides causing enteritis and abortions in animals), with abdominal pain and diarrhea appearing within 24–72 hr upon contamination. Being a normal constituent of the intestinal microflora of warm-blooded animals, the major source of contamination thereby is ingestion of recontaminated, or else raw or inadequately cooked poultry.

Two modes of food poisoning have been claimed for *Bacillus cereus* (see Fig. 1.7ii), both observed with a minimum of 10^5 viable cell numbers: nausea and vomiting within 1–5 hr, similar to symptoms caused by *Staphylococcus aureus*, due to an emetic toxin; or diarrhea and abdominal pain within 8–16 hr (without vomiting), similar to symptoms caused by

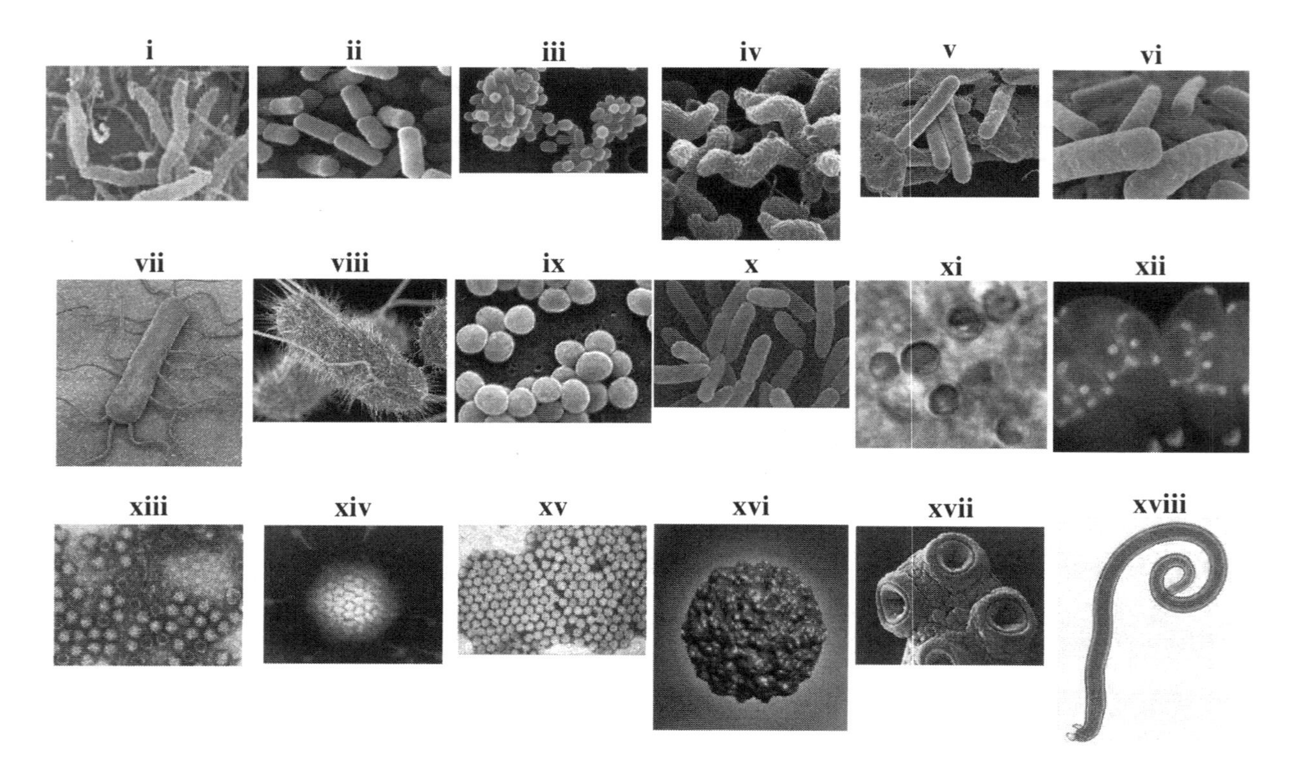

FIGURE 1.7 Examples of (i) *Aeromonas hydrophilia*, (ii) *Bacillus cereus*, (iii) *Brucella abortis*, (iv) *Campylobacter jejunii*, (v) *Clostridium botulinum*, (vi) *Clostridium perfringens*, (vii) *Listeria monocytogenes*, (viii) *Salmonella* spp., (ix) *Staphylococcus aureus*, (x) *Yersinia enterocolitica*, (xi) *Cryptosporidium parvum*, (xii) *Toxoplasma gondii*, (xiii) norovirus, (xiv) enteric adenovirus, (xv) hepatitis A virus, (xvi) enterovirus, (xvii) *Taenia saginata*, and (xviii) *Trichinella spiralis*.

Clostridium perfringens, owing to diarrheagenic toxins. In both cases, spores survive heating and eventually germinate during cooling – thus attaining sufficiently high numbers to produce toxins once in the intestine. Being relatively mild and short lived these symptoms seldom lead to formal reporting to public health authorities. Four psychrotrophic strains and 35 mesophilic strains are known; related species include *Bacillus subtilis* (causing acute vomiting within 2–3 hr, followed by diarrhea), *B. lichenformis* (causing diarrhea within 8 hr, sometimes with vomiting), and *B. thuringiensis* (producing toxins – and thus raising concerns as per its use as insecticide in cabbages). The usual reservoirs are soil, surfaces of cereals, vegetables, and meat; the emetic type is chiefly transmitted by ingestion of reheated rice and cereal-containing products and spices, whereas the diarrheal type is found mainly in meat, milk, vegetable products, and sauces. Their minimum growth conditions are *pH* of 4.3 and a_W of 0.92, and they are characterized by $D_{100}=18$ s–27 min and $D_{121}=$ 1.2–3.6 s, complemented with a temperature sensitivity given by $z=10°C$ – where z denotes temperature increase required for a tenfold increase in activity.

The strains of *Brucella* genus responsible for brucellosis in humans are *B. abortis* from cows (see Fig. 1.7iii), *B. melitensis* from sheep and goats, and *B. suis* from pigs. Their natural reservoirs are farm animals; farm workers may thus contract the disease directly from contaminated animals, although it can also be transmitted by consumption of (contaminated) raw milk or unpasteurized dairy products. Acute symptoms include fatigue, weakness, muscle and joint pain, and weight loss within 2 months of infection; chronic health problems include inflammation of joints, genitourinary, cardiovascular and neurological conditions, and insomnia and depression. *Brucella* spp. are characterized by $D_{63}=2.5$ min and $z=4.1°C$.

Campylobacter jejuni (see Fig. 1.7iv) – a (curved) Gram$^-$, microaerophilic, moderate thermophilic rod belonging to the Campylobacteriaceae family, is the most important pathogen within the *Campylobacter* genus; it is responsible for more than 90% of campylobacteriosis infections. Although cells cannot survive for long periods in foods due to their sensitivity to heating, drying, freezing, acidity, oxygen, and salt, they are quite virulent – with 10^2 being sufficient to trigger infection. Widespread outbreaks are rare; typical symptoms are vomiting within 1–14 hr, diarrhea (from mild to severe, including bloody) within 4–16 hr, and headache and abdominal pain within 24–72 hr following ingestion. A higher incidence has been recorded in young children – whereas immunocompromised individuals may undergo a severe and prolonged illness that can degenerate to septicemia; campylobacteriosis has been associated with Guillán-Barré syndrome – an autoimmune (potentially fatal) disease that causes limb weakness and paralysis, and it can also cause chronic arthritis, meningitis, abortion, and neonatal sepsis. The usual reservoir is the intestinal microflora of warm-blooded birds and mammals; it can be transmitted by ingestion of recontaminated foods, or raw or inadequately cooked products from poultry, as well as meat, dairy products, water, and shellfish to a lesser extent. The minimum growth conditions are 25°C, *pH* of 4.9 and a_W of 0.98; and they are characterized by $D_{60}=6$ s, complemented with $z=5°C$.

Clostridium botulinum (see Fig. 1.7v) is a strictly anaerobic bacterium, able to sporulate only under anaerobic conditions; despite its 4 phenotypes, only group I (proteolytic) and group II (nonproteolytic) cause significant poisoning – with the former producing A, B, and F toxins, and the latter producing B toxin (see Fig. 1.8i), besides E and F toxins. These toxins are among the most poisonous natural toxins known because they block acetylcholine release across nerve synapses, thus causing muscular paralysis. The major symptoms are nausea and vomiting, readily followed by blurred vison, speech impediment, and increasing difficulty in swallowing, general muscular weakness, lack of coordination, and respiratory failure within 12–36 hr (even though they can be delayed by up to 10 days); the death toll may attain 70% – yet an antitoxin has been developed, which requires quick administration for effectiveness. It is ubiquitous in soils and marine and freshwater sediments – and conveyed via canned vegetables, fish and meat, and smoked fish; hence, special care is routinely exercised by the canning industry. Most outbreaks have in fact been due to home vegetable canning, improperly cured or undercooked fish or meat, and inclusion of fresh herbs and spices in cooking oils. Their minimum growth conditions are 10 and 3°C, *pH* 4.6 and 5.0, and a_W of 0.94 and 0.97, for group I and group II, respectively; and they are characterized by $D_{100}=25$ min and $D_{121}=12$ s, or $D_{80}=1$ min, $D_{100}=6$–12 s and $D_{121}=0.1$ s, for group I and group II, respectively – complemented with $z=10°C$.

Clostridium perfringens (see Fig. 1.7vi), a Gram$^+$, endospore-forming, anaerobic mesophilic rod belonging to the Clostridiaceae family, synthesizes some of four main types of enterotoxins – α-toxin (see Fig. 1.8ii), and β-, ε-, and ι-toxins. Contamination often yields mild poisoning (except in rare cases of type C clostridial necrotizing enteritis), which typically lasts no longer than 24 hr – thus justifying its being sometimes called 24-hr flu; hence, it is believed to be severely under-reported. With 43–45°C for optimum growth temperature, it can double in viable numbers every 8–10 min – thus making it one of the fastest growing food pathogens; typical symptoms of contamination are minor nausea, fever, vomiting, stomach crumps, flatulence, and acute diarrhea within 8–18 h. It is ubiquitous in soil and decaying vegetation, marine sediments, insects, and vertebrates – as well as in raw, dried, and cooked foods; the main sources are meat, poultry, fish, dairy products, and dried foods (e.g. pasta, soups, spices) in raw form, or undercooked, or cooked but subjected to temperature abuse, or even subjected to inadequate refrigeration upon cooking; improper cooking may also stimulate spore germination throughout cooling. Although vegetative cells are killed by refrigeration and freezing, surviving spores may germinate and rapidly grow during reheating or thawing; such a growth is inhibited by 6–8%(w/w) salt and up to 400 $mg.kg^{-1}$ nitrite. Its minimum growth conditions are 12°C, *pH* 5.5, and a_W of 0.95; $D_{95}=1$–3 min applies to heat-sensitive spores, and $D_{95}=18$–64 min to heat-resistant spores – and thermal sensitivity is described by $z=10.3°C$ and 16.8°C, respectively.

Escherichia coli – a Gram$^-$, mesophilic motile rod belonging to the Enterobacteriaceae family, is the most common nonpathogenic species in the human intestinal microflora; however, some

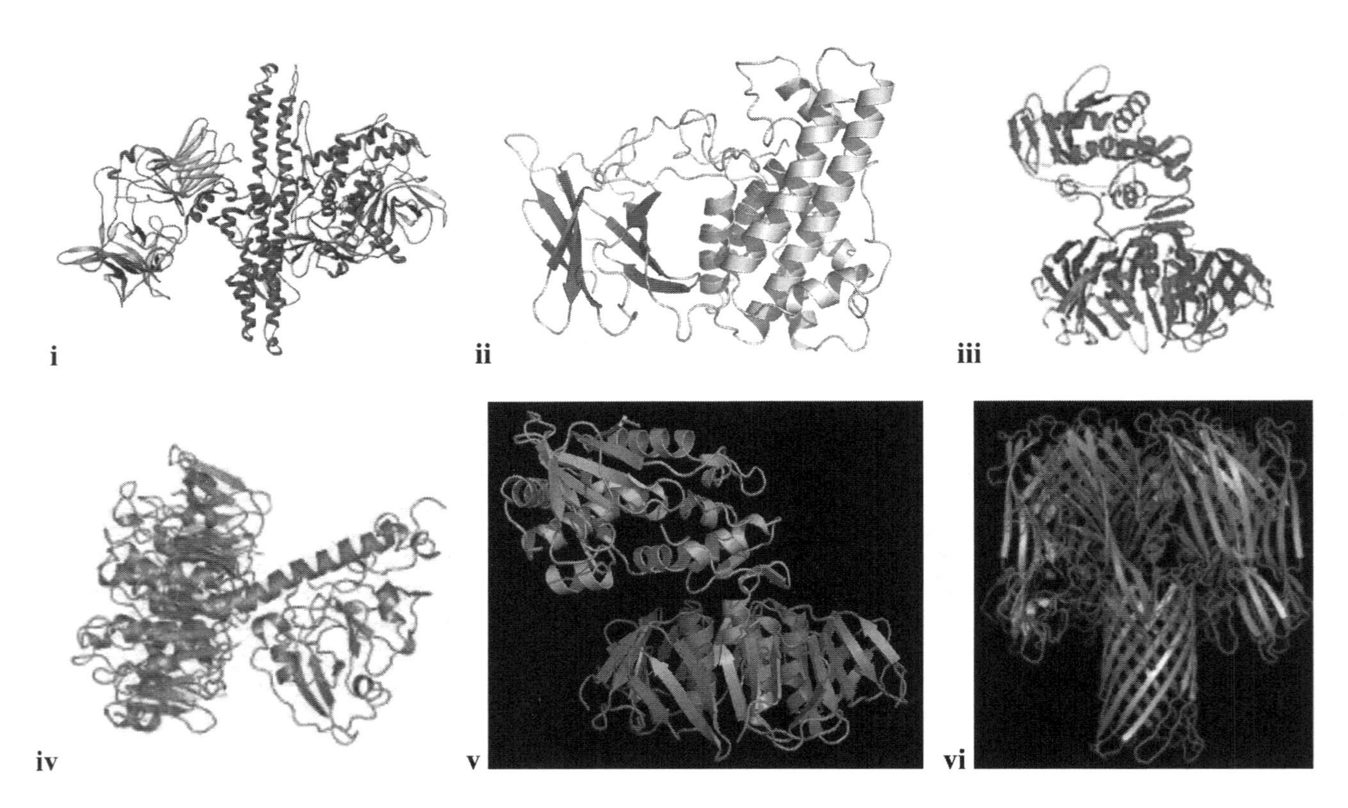

FIGURE 1.8 Tertiary structures of (i) B toxin from *Clostidium botulinum*, (ii) α-toxin from *Clostridium perfringens*, (iii) Shiga toxin from *Escherichia coli* O157:H7, (iv) enterotoxin from *Plesiomonas shigelloides*, (v) Shiga toxin from *Shigella dysenteriae*, and (vi) α-toxin from *Staphylococcus aureus*.

strains have developed capacity to cause disease and illness – and a few of them are characterized by significant morbidity and mortality, particularly when ability to produce Vero- or Shiga-cytotoxin (see Fig. 1.8iii) is at stake. Six types of pathogenicity have been identified so far – based on differences in virulence genes: (i) enteropathogenic *E. coli*, with onset within 9–72 hr and duration from 6 hr up to 3 days – with invasion of mucosal cells that causes fever, vomiting, abdominal cramps, and severe diarrhea; (ii) enterotoxigenic *E. coli*, with onset within 8–44 hr and duration from 3 up to 19 days – with adhesion to small intestinal mucosa, and production of toxins that cause nausea, cramps, and diarrhea; (iii) enteroinvasive *E. coli*, with onset within 8–72 hr, and duration ranging from days to weeks – with invasion of epithelial cells in the colon that causes vomiting, fever, headache, abdominal cramps, and diarrhea; (iv) enterohemorrhagic *E. coli*, with onset within 3–9 days and duration from 2 to 9 days, and with as low an infective dose as 50 cells – with attachment to mucosal tissues, and production of very potent toxins that cause vomiting, severe abdominal pain, and bloody diarrhea (but not fever), hemolytic uremic syndrome (i.e. hemolytic anemia, or destruction of red blood cells, followed by thrombocytopenia, or destruction of platelets) that may degenerate to acute renal failure, septicemia, seizures, blood clots in the brain, coma, and eventual death in young children and the elderly within 7 hr–4 days; (v) enteroaggregative *E. coli*, with onset within 7–48 hr and duration from 2 up to several weeks – with clump-binding to cells of the small intestine, and production of toxins that cause persistent diarrhea (but not fever or vomiting); and (vi) diffusely adherent *E. coli*, with variable onset and duration – with epidemiology and clinical profiles not yet available, but known to cause diarrhea in older children. Major attention has been paid in industrialized countries to *E. coli* serotype O157:H7 – one of the Shiga-like Verotoxin-producing types, belonging to the enterohemorrhagic group; remember that O denotes the somatic antigen number (associated with the body of the bacterium), whereas H denotes the flagellar antigen number (associated with the flagellum used by the said bacterium for propulsion through liquid media). This toxin cleaves a specific adenine from the 28S rRNA domain of the 60S subunit of the ribosome – and so disrupts protein synthesis. The usual reservoir of the pathogenic strains of *E. coli* is the intestinal tract of warm-blooded animals; they can be transmitted by ingestion of contaminated meat, poultry, fish, vegetables (e.g. lettuce, spinach), soft cheeses, water, and alfalfa shoots. The minimum growth conditions are 6.5°C for *E. coli* O157:H7 and 7–8°C for the remaining pathogenic serotypes (e.g. *E. coli* O26, O103, O111, O118, O121:H19, and O145, *pH* 4.0 and 4.4, respectively, and a_w of 0.95 and 0.93, respectively; and they are characterized by $D_{63}=30$ s and $D_{64}=22$–28 s, respectively.

Listeria monocytogenes (see Fig. 1.7vii), a Gram$^+$, nonsporulating, mesophilic motile rod belonging to the Listeriaceae family, is the most important pathogen under the *Listeria* genus – and appears as serotypes 4b and 1/2a; it is the food-borne pathogen with the highest associated risk in the United States. Listeriosis is relatively rare in healthy adults – but, when it occurs, causes mild flu-like symptoms (i.e. fever and muscle aches) or vomiting/diarrhea within 24–48 hr, should large numbers of cells be ingested. The situation is, however, quite different in the case of the elderly or immunocompromised individuals – since they may develop meningitis and encephalitis (with headache, stiff neck, confusion, loss of balance, and convulsions) or even septicemia, thus leading to quite high mortality rates. Following contamination during pregnancy, subsequent infection of the uterus, bloodstream, or central nervous system may lead to spontaneous miscarriage and stillbirth – or even infection of the fetus, with severe consequences following birth (responsible for 25–50% death rate of newborns). *Listeria* is ubiquitous in soil, water, most foods, material surfaces, and even healthy humans and animals; cross-infection in maternity hospitals and food-borne infection are the major causes of transmission – the latter chiefly via ingestion of contaminated milk and soft cheeses, raw or cooked meats and pâtés, poultry products, smoked or marinated fish, seafoods, sandwiches and raw vegetables, salads, and coleslaw, as well as chilled, ready-to-eat foods. Its minimum growth conditions are –0.5°C, *pH* 4.5, and a_W of 0.92; and it is characterized by D_{60}=4.3–5.5 min, complemented with z=2.0–7.5°C.

Mycobacterium avium subsp. *paratuberculosis* constitutes the cause of Johne's disease in cattle, as well as its human version, Crohn's disease – an incurable, highly debilitating chronic inflammation of the gastrointestinal tract. Its classical reservoir is milk; hence, milk and its products are the most likely vectors of transmission – knowing that it may survive regular pasteurization. The minimum growth conditions are 25°C and *pH* 5.5; and it is characterized by D_{71}= 12 s, together with z=8.6°C.

Within 24–48 hr following contamination, *Plesiomonas shigelloides* causes nausea and abdominal pain, coupled with one of the three types of diarrheal symptoms – secretory, shigella-like, or cholera-like (the rarest); however, it may degenerate to meningitis, with a mortality rate of the order of 80%. Its mechanism of action is invasion of the intestinal tissue, followed by production of enterotoxins (see Fig. 1.8iv), protease, elastase, and hemolysin therein; people at the highest risk are young children, the elderly, and cancer patients. Its natural reservoir is freshwater, so illness usually results from drinking the said water, or else eating raw shellfish previously grown (or at least maintained) in such water; it is fully eliminated after thermal treatment at 60°C for at least 30 min, with 8°C and *pH* 4.0 for minimum growth conditions.

Salmonella spp. (see Fig. 1.7viii) – Gram⁻, mesophilic motile rods of the Enterobacteriaceae family, are among the most frequent causes of food-borne diseases worldwide. A single outbreak may easily affect quite a large number of people simultaneously, namely participants in banquets during summertime where mayonnaise and chocolate mousses (produced with contaminated raw eggs) are part of the menu. Typical symptoms range from mild/severe gastroenteritis, with nausea, vomiting, high fever, and abdominal pain arising within 12–48 hr and lasting for 2–7 days; through severe typhoid or paratyphoid fever, with nausea, high fever, and abdominal pain arising within 7–28 days and lasting for 14 days; to septicaemia, with high fever, and abdominal and thoracic pains, associated with overall high rates of morbidity, and even mortality. The underlying mechanism is infection of intestinal tissue, followed by release of an enterotoxin that directly causes inflammation and diarrhea; in more complex cases, invasion of blood vessels and lymphatic system may occur – with reactive arthritis, pancreatitis, osteomyelitis, and meningitis as likely outcomes. The most dangerous species is *S. enterica*, with more than 2400 serotypes known, e.g. serotypes *typhimurium* and *enteritidis*; however, it has been evolving rapidly – and meanwhile developed multiple antibiotic resistance traits. The typical reservoirs are poultry, cattle, and pigs – with poultry, milk, cooked meats, salami, and cheeses serving as main vehicles, second only to eggs; feces from a previously contaminated individual are also an important vector, especially in the case of food-handling staff. Viable cell numbers above 10^4 usually suffice for the manifestation of illness – although such a threshold may decrease to 10–10^2 in fatty foods; this is why the requirement of absence in 25 g-samples has been incorporated in national legislation and international safety standards. Their minimum growth conditions are 5.2°C, *pH* 3.8, and a_W of 0.94; and they are characterized by D_{71}=0.3–10 s in milk, D_{71}=24 s in ground beef, D_{71}=36 s–9.5 min in liquid egg, and D_{71}=4.5–6.6 hr in chocolate.

Shigella dysenteriae, *S. flexneri*, *S. boydii*, and *S. sonnei* are known to infect people chiefly via personal contact; however, ingestion of contaminated water or else salads washed with that water, milk and soft cheese, poultry, or even cooked rice in regions characterized by poor hygienic standards (e.g. crops with fecal contamination) also constitute causes of infection. The major symptoms are fever, severe abdominal pain, and diarrhea within 12–50 hr upon contamination, and may last for 3–4 days; it is normally not life-threatening, except in the case of immunocompromised people. Their mechanism of action encompasses multiplication in the colon, followed by invasion of the epithelial tissue – along with development of ulcerative lesions. The first species also releases a (heat-sensitive) cytotoxin known as Shiga toxin (see Fig. 1.8v), able to kill colon cells – besides an enterotoxin and a neurotoxin; it thus accounts for such symptoms as convulsions and delirium. Their minimum growth conditions are 6.1°C, *pH* 4.8, and a_W of 0.96; and they are characterized by D_{60}= 1 min and D_{80}= 10 s.

Staphylococcus aureus (see Fig. 1.7ix) is a normal part of the microflora established on the skin and in the nasal cavities of humans; nausea and vomiting, and sometimes diarrhea arise within 2–24 hr upon contamination, but seldom last for more than 24 hr and are hardly fatal – so it is also believed to be quite under-reported to health authorities. This species is able to synthesize up to 11 enterotoxins if grown on food, and their threshold for intoxication following ingestion is 95–185 ng; hence, contamination is possible by pasteurized foods (e.g. dried milk or salami) – even though the most common vehicles are recontaminated heat-processed foods, cheese, meats, and sandwiches. Fortunately, the yield of toxin on biomass is rather low, and the former is more susceptible than its producing cells to unfavorable *pH* and a_W. On the other hand, said enterotoxins (e.g. α-toxin, see Fig. 1.8vi) do not directly affect intestinal cells, but instead challenge nerve endings that, in turn, stimulate the vomiting center in the brain. Its minimum growth conditions are 6.5°C, *pH* 4.0, and a_W of 0.83 for vegetative cells, and 10°C, *pH* 4.0, and a_W of 0.85 for toxin production; the former is further characterized by D_{77}=0.1–1 s and z=8–12°C. Although refrigeration is the best way to control its growth, *S. aureus* is immune to the salt and nitrite contents normally prevailing in cured meats; and it can also grow on foods packaged under vacuum or modified atmosphere.

Ten species of the *Vibrio* genus exist that are able to cause gastrointestinal illness, yet the most relevant for food safety purposes are *V. parahaemolyticus*, *V. vulnificus*, and *V. cholerae*; the second can lead to septicemia, whereas the last one is the direct cause of Asiatic cholera. Their natural reservoirs are inshore marine waters, so the chief vectors for contamination are water itself, as well as improperly cooked or recontaminated fish and shellfish. *Vibrio cholerae* has two serotypes O1 and O139 – both of which can produce hemolysins and cytotoxins; nausea, fever, abdominal cramps, and abundant diarrhea (cholera) accordingly arise within 6 hr–3 days upon contamination, thus leading to a rapid loss of body fluids and mineral salts that can cause dehydration – the most likely reason for eventual death, if not treated in time. *Vibrio parahaemolyticus* is an invasive non-toxigenic pathogen responsible for diarrhea that lasts for 2–3 days. *Vibrio vulnificus* – strongly associated with consumption of raw oysters in Summertime, secretes hemolysins, proteinases, collagenases, and lipases, so it can destroy body tissues; it normally causes mild gastroenteritis, but skin lesions and septicemia (with a death toll of the order of 50%) may arise in individuals suffering from such medical conditions as hepatitis, cirrhosis, or gastric disease. Their minimum growth conditions are 5.0°C, pH 4.8, and a_W of 0.94; and they are characterized by $D_{50}=2.4$–9.9 min.

Yersinia enterocolitica (see Fig. 1.7x) is an infectious foodborne pathogen carried by pigs, so consumption of pork meat or intestines (chitterlings), besides sausages and tofu, are the major vehicles of contamination. Typical symptoms include vomiting, fever, joint pain, severe abdominal pain, and diarrhea within 24–36 hr following contamination – with young infants and teenagers being particularly susceptible thereto; such complications as autoimmune thyroid disease, liver abscesses, pneumonia, conjunctivitis, pharyngitis, and even septicemia may arise in the case of the elderly or immunocompromised individuals. Its minimum growth conditions are –1.3°C, pH 4.2, and a_W of 0.945; it is further characterized by $D_{60}=1$ min, as well as sensitivity to carbon dioxide – as often used in modified atmosphere packaging.

Toxigenic molds synthesize a number of toxins when they grow on cereals (e.g. maize) or legumes (e.g. peanuts), as well as on nuts, spices, pulses, and oilseeds in general; although mold growth may occur when the crop is subjected to drought stress (and is thus more fragile, from an immunity point of view), it is more likely when the harvested crop has been inadequately dried and subsequently stored under humid conditions. Unlike bacterial toxins – characterized by acute poisoning, mold toxins produce chronic toxicity that may eventually degenerate to cancer, liver damage, or immunosuppression; the most dangerous are aflatoxins, fumonisins, ochratoxin A, and patulin, as illustrated in Fig. 1.9. Aflatoxins are produced by a few strains of *Aspergillus* spp., namely *A. parasiticus*, *A. flavus*, *A. nomius*, and *A. ochraceoroseus*; the most toxic is aflatoxin B_1 (see Fig. 1.9i) – which, besides being carcinogenic, may also be acutely toxic (and thus fatal). It indeed generates a reactive epoxide intermediate in the liver (when contacting cytochrome P450) – which, in turn, forms a covalent bond to the N7 atom of guanine in codon 249 of the p53 gene, a tumor-suppressor gene. Upon ingestion, it is metabolized to aflatoxin M_1 (see Fig. 1.9ii), and eventually secreted in mother's milk during breastfeeding – thus posing an extra problem to public health, due to the risk of ingestion by newborns. Since the optimum temperature for production of these toxins is ca. 30°C, contamination preferentially occurs in tropical regions.

Fumonisins are synthesized by *Fusarium moniliforme* (and other species) growing on maize, namely when damp conditions prevail during formation of the cob or in insect-damaged grains; they have also been found in asparagus, beer, and rice. Their specific cellular target is dihydrosphingosine *N*-acyl transferase, thus inhibiting sphingolipid biosynthesis – and causing cell deregulation and death; fumonisin B_1 (see Fig. 1.9iii) has specifically been linked to esophageal cancer in humans.

Ochratoxin A (see Fig. 1.9iv) is produced by *Penicillium verrucosum*, mainly on barley in temperate climates; as well as by *Aspergillus ochraceus* (and other species) chiefly on cocoa, coffee, grapes, and spices in tropical and subtropical regions. It remains in the carcasses of animals fed with contaminated crops and bred for meat, and survives coffee roasting and wine manufacture if the original coffee beans or grapes were contaminated. Ochratoxin inhibits carboxypeptidase *A*, renal phosphoenolpyruvate carbokinase, phenylalanine-tRNA synthetase, and phenylalanine hydroxylase; hence, it is a potent nephrotoxin that causes kidney damage (including tumor development), as well as liver necrosis.

Patulin (see Fig. 1.9v) is synthesized by *Penicillium expansum* (and other species), and some strains of *Aspergillus* spp. and *Byssochlamys* spp. – and has been implicated in soft rot of fruits.

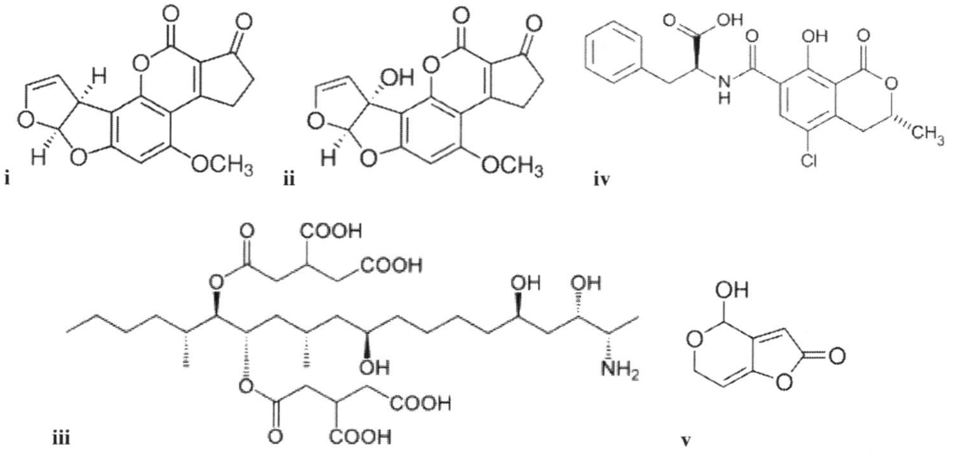

FIGURE 1.9 Structural formulae of (i) aflatoxin B_1 and (ii) aflatoxin M_1, (iii) fumonisin B_1, (iv) ochratoxin A, and (v) patulin.

Being electrophilic, it covalently binds to sulfhydryl groups on proteins – thus inhibiting many enzymes, namely in the intestine, liver, and brain; symptoms following its intake include agitation, tremors, convulsions, and dyspnea – and, eventually, premature death. Apples represent a particularly critical case, since mold contamination may occur inside and pass the toxin to apple juice, since the fruit is processed as a whole; in fact, contamination invisible from the outside will not lead to rejection of the fruit in the first place. Due to the acidic nature of this juice, the toxin will withstand pasteurization – thus representing a higher risk to young children, as they drink a higher proportion of juice compared to their body weight; however, it is destroyed by fermentation during cidermaking, or treatment with sulfur dioxide.

Other mycotoxins worth mentioning for their pathogenicity are trichothecenes – with immunosuppression capacity, and produced by several *Fusarium* spp. on cereals; T-2 toxin, characterized by acute effects and synthesized by *F. sporotrichioides*; vomitoxin (or deoxynivalenol), produced by *F. graminearum*; citrinin, responsible for kidney damage, and synthesized by some *Penicillium* spp.; and sterigmatocystin, present on the surface of cheeses stored at low temperature for long periods, and produced specifically by *A. versicolor.*

Such protozoa as *Giardia duodenalis*, *Cryptosporidium parvum* (see Fig. 1.7xi), *Cyclospora cayetanensis*, and *Toxoplasma gondii* (see Fig. 1.7xii) are pathogenic intestinal parasites – often conveyed by contaminated water or food; *T. gondii*, in particular, is a unicellular, spore-forming intracellular parasite possessing an organelle apicoplast – bearing an apical complex structure, crucial (and especially suited) for cell penetration. Examples of infection routes include crops contaminated by infected animals, birds, or insects, or sprayed with manure slurry; raw fruits and salad vegetables handled by already infected processors, under poor conditions of hygiene; and contaminated water used to manufacture ice or wash salad vegetables and fruits (or other foods to be eaten raw). The aforementioned protozoa produce cysts (or oocysts) that are excreted in feces, and can survive for long periods in damp and dark environments outside their host. While oocysts from *G. duodenalis* and *C. parvum* may reinfect people at once, oocysts from *C. cayetanensis* require a maturation period before they become infective.

Acute giardiasis is characterized by flatulence, abdominal extension, cramps, and diarrhea – whereas its chronic version entails malabsorption of nutrients, which induces weight loss and general malaise; as little as 25 cysts may suffice as infectious dose.

Cryptosporidiosis causes acute, self-limiting gastroenteritis; symptoms are observed within 3–14 days upon infection, and last for up to 2 weeks – and include nausea, vomiting, flatulence, severe abdominal pain, fever, and even weight loss. Such symptoms may persist and develop into serious infections of the gastrointestinal and respiratory tracts, gall bladder, and pancreas in the case of immunocompromised individuals or if immunosuppressive drugs are taken.

The symptoms of cyclosporiasis include nausea, vomiting, fatigue, abdominal pain, and flu-like fever; the incubation takes between 2 and 11 days, and lasts for more than 6 weeks in healthy people (or even longer in immunocompromised individuals).

Cats (or felids in general) are the definitive reservoir of oocysts of *T. gondii*, a member of the Sarcocystidae family – and only host for the sexual component of the lifecycle of the said parasite; however, they may remain infective in water or soil for up to 1 year. Cattle, sheep, goats, pigs, and even birds may, in turn, serve as intermediate hosts – thus carrying the asexual component of that lifecycle. This is one of the world's most common parasites – and ranks second as a cause of death, and third as a cause of hospitalization attributed to food-borne illnesses in the United States. The typical symptoms resemble a mild flu, with headache, fever, muscle aches, and sore lymph nodes. However, there is a 30% chance of a mother infected with toxoplasmosis passing the disease, through her placenta, to the unborn child during pregnancy; this is prone to cause stillbirth or perinatal death. Surviving babies will likely experience damages to sight, hearing, or the central nervous system (e.g. seizures), and may suffer from hydrocephalus, or mental retardation later in life, enlarged liver and spleen, jaundice, and even blindness. Raw and undercooked meat are the usual vectors of infection.

Viruses are much smaller than bacteria – exhibiting typical diameters within 22–110 nm; for lacking metabolic machinery of their own, they must resort to using that of true living beings upon infection of their cells. However, they can remain dormant for undefined periods, in and on a multiplicity of matrices. Although food contamination may result from viruses already present in processing water and food handlers' skin (or after previous contact with farm animals or insects), a major source is fecal material; contaminated products usually exhibit a normal appearance and flavor, so *a priori* rejection of problematic food items is quite difficult to implement (if not impossible at all). There are three major classes of food-borne viruses: (i) gastroenteritis viruses; (ii) hepatitis viruses; and (iii) viruses that replicate in the intestine, but migrate to other organs where illness eventually develops.

The most common cause of viral gastroenteritis is, by far, the Norwalk-like calicivirus (or norovirus, see Fig. 1.7xiii), a member of the Caliciviridae family; it is often known as small-round-structured virus – for its positive-sense, single-stranded RNA, nonenveloped morphology. It is highly contagious from person to person, but is also carried by contaminated water or food (thus serving as a likely cause for extended outbreaks); its incubation time lies within 24–48 hr, whereas subsequent illness lasts for days. Its infection is characterized by mild severity; it multiplies in the small intestine, thus causing nausea, vomiting, diarrhea, abdominal pain, headache, and low-grade fever as major symptoms; institutionalized or hospitalized people are the most susceptible groups to contamination. Less common virus types include enteric adenovirus types 40 and 41 (see Fig. 1.7xiv), with 7–14 days as incubation time and illness from days to weeks, mild severity, diarrhea as major symptom, and children under 5 years old as the highest risk group; rotaviruses groups A, B, and C, with 24–48 hr of incubation time and illness lasting for days – with vomiting, fever, and diarrhea as major symptoms, and children under 5 years old again as the highest risk group (with quite high a death toll in developing countries); and astrovirus, with 24–48 hr of incubation time, and illness lasting for a few days – with mild severity, diarrhea as major symptom, and children under 10 years old as the highest risk group.

In the second group, hepatitis A virus (see Fig. 1.7xv) appears most frequently; it is transmitted by water contaminated by feces, and shellfish grown in the said water or infected by contaminated food handlers. Typical symptoms include nausea, vomiting, and fever within 2–6 weeks of contamination, followed by hepatitis and liver damage; incubation may take up to 50 days, while illness may last for weeks. Although all age groups share essentially the same risk if endemic, its severity increases with age of first infection. A similar description applies to hepatitis E virus – except that incubation may go up to 70 days following contamination, and that infection is relatively mild (but in pregnant women).

Enteroviruses (see Fig. 1.7xvi) belong to the third group – but are a less-common cause of illness, despite its severity; in fact, they may produce not only diarrhea, but also meningitis, encephalitis, and ultimately paralysis. Its incubation period ranges from 1 to 2 weeks, and may last from days to a lifetime; children under 15 years old have the highest risk of infection.

No general antiviral treatments exist at present; in fact, most food- and water-borne viruses are relatively resistant to heat, acidity ($pH > 3$), and disinfection chemicals – and may survive for extended periods on equipment surfaces and in the air. However, heat and chlorine-based disinfectants have proven successful in the particular case of norovirus.

Other parasites in food include flatworms (or trematodes), most often found in watercress and salad plants, as well as raw or undercooked freshwater fish and shellfish. They are responsible for a number of acute infections, and cause fatigue, fever, abdominal pain, diarrhea, and jaundice; chronic outcomes of the said infections range from liver, spleen, and pancreas damage to dwarfism and retardation of sexual development.

Two species of *Taenia*, namely *T. saginata* (see Fig. 1.7xvii) from cattle and *T. solium* from pigs, may be carried by undercooked food, or water and food previously contaminated by animal feces – and cause infection in the human intestine. Intestinal infection by the eggs of *T. solium* (or cysticercosis) may also occur; after they hatch in the intestine, the larvae migrate to such other organs as eye, heart, and central nervous system. Manifestation of physical symptoms may take between days and years – but they are normally quite severe, and include seizures, psychiatric disturbance, and eventually death.

Raw or undercooked meat may also contain *Trichinella spiralis* (see Fig. 1.7xviii); infection arises when the larvae grow into adults in the human intestine – thus causing nausea, vomiting, fever, and diarrhea. The adult worm produces, in turn, larvae that migrate through the blood and lymph toward the muscles, where they set and cause rheumatic conditions; retinal hemorrhage and photophobia often arise afterward, combined with intense sweating and prostration. Cardiac and neurological complications may occur within 3–6 weeks – and may lead to death as per heart failure.

1.1.2.2 Nonconventional hurdle strategies

To control growth/reproduction (and toxin production, if applicable) of the aforementioned biological agents in food, one should apply some physicochemical hurdle. This may take the form of thermal processing at high temperature – as in blanching, pasteurization, and sterilization implemented *per se*; or as a result of cooking via boiling, baking, roasting/grilling, ohmic heating, dielectric heating, and extrusion, or even as a consequence of dehydration via frying, evaporation, drying, and freeze-drying. Processing at low temperatures constitutes an alternative possibility – such as in chilling and freezing, or else via irradiation, (aseptic and active) packaging, or modified atmospheres (to be covered in dedicated sections), besides cold smoking (already covered in section 3.4 of *Food Proc. Eng.: thermal & chemical operations*). Nonconventional approaches, via pulsed magnetic and electric fields, photodynamic inactivation, ultrasonication, ultra-high pressure, and ultraviolet-based processes are often classified under the common heading of minimal processing technologies, since they do not entail significant heating – yet they are able to inactivate microorganisms and often also enzymes, while retaining most of the original sensory and nutritional features.

Adequate food safety and shelf-life, or instead deliberate alteration of sensory characteristics of a given food cannot often be attained via a single unit operation; hence, a combination of more than one method is in order – and the concept of hurdle has indeed expanded and strengthened the idea of multiple preservation factors to inhibit microbial growth. One example pertains to fermented sausages (as is the case of salami): added salt and sodium nitrite inhibit many contaminating bacteria, while permitting multiplication of aerobic bacteria that cause a reduction in E_h, due to O_2 taken up by respiration; growth of lactic acid bacteria is meanwhile promoted, which release lactic acid that lowers pH; finally, ripening decreases water content, and concomitantly decreases a_W – all of which contribute to extend shelf-life at room temperature. Another example is smoked foods; the combination of heat, reduced a_W (as a consequence of drying), and deposition of smoke-containing antimicrobial compounds on the surface account already for a combination of three hurdles. Some smoked foods are further dipped or soaked in brine, or else rubbed with salt prior to smoking – which also contribute to preservation. Finally, smoked foods may be chilled or packed in modified atmospheres *a posteriori*, to further expand their shelf-life. The success of combined preservation hinges critically upon the quantitative and qualitative profiles of the (anticipated) initial microbial load; hurdles should be strong enough to reduce viable numbers of microorganisms to the intended threshold – but inadequately cleaned raw materials, or extra richness thereof in specific nutrients may compromise their effectiveness.

1.1.2.2.1 Ultra-high pressure

In high-pressure processing – also known as high hydrostatic pressure or ultra-high pressure processing, foods are exposed to pressures of 100–1000 MPa for several seconds to minutes; enzymes and microorganisms are inactivated, with outcomes similar to those of pasteurization – while retaining sensory features and nutritional content in full, and sometimes even improving functional properties. Fruit-based products (e.g. jams, purées, sauces, yogurts) have been accordingly processed in Japan at 400 MPa, for 10–30 min at 20°C, by taking advantage of improved gelation, faster sugar penetration, and limited activity of pectin methylesterase; grapefruit juice at 200 MPa, for 10–15 min at 5°C, for reduced bitterness; sugared fruits for ice cream and sorbet at 50–200 MPa, for faster sugar penetration and water removal; raw pork ham at 250 MPa, for 3 hr at 20°C, to promote faster maturation and tenderization by endogenous proteases, and improve water retention; fish sausages, terrines, and puddings at 400 MPa, for gelation and improved gel texture; and rice cake and hypoallergenic precooked rice at 400–600 MPa, for 10 min at 45°C or 70°C, toward fresh taste/flavor retention, enhanced porosity, and salt-mediated extraction of allergenic proteins. All such effects take place beyond the reduction in viable numbers of pathogens and spoilage microorganisms. The product portfolio is not yet so wide in Europe, though: fruit juices and sliced (processed) ham have been processed at 400 MPa for a few minutes, and squeezed orange juice at 500 MPa, both at room temperature. Avocado paste (guacamole and salsa) has been processed in the United States at 700 MPa, for 10–15 min at 20°C, with concomitant polyphenol oxidase inactivation; main meals (macaroni cheese, salmon fettuccine, ravioli, and beef stroganoff) and ready-to-eat meats (pastrami and Cajun-beef) have been exposed to 600 MPa, for 3 min at 20°C, with a resulting shelf-life extended to up to 100 days at 4°C; and oysters have been subjected to 300–400 MPa, for 10 min at room temperature, retaining raw taste/flavor besides native shape and size. Further to better microbiological quality, oyster shells become easier to open (thus avoiding the labor-intensive shucking process), and exhibit a higher yield due to water absorption following exposure to hyperbaric conditions. Hyperbarically processed vegetables acquire a crisp texture, whereas cocoa butter gains the stable crystal form preferred in chocolate tempering.

This form of unconventional processing is gradually gaining popularity – not only due to its minimal effects upon food quality, but also for its potential in high-pressure freezing or thawing, reduced processing times, marginal energy consumption, and production of virtually no effluents at all. The dominant shortcoming arises from the high capital cost – coupled with unsuitability to handle dry foods (since water is a *sine qua non* for microbial destruction) and air-rich foods (e.g. entrapped air in strawberries leads to food crushing).

The major effect of hyperbaric processing is felt by macromolecules, due to their three-dimensional architecture – as is notably the case of starch, which experiences changes in structure and reactivity; however, lipids undergo faster oxidation and hydrolysis, especially in meat and other fatty foods. Starch is indeed opened and partially degraded – thus producing increased sweetness and becoming more susceptible to amylase action; potato and sweet potato matrices become softer, more pliable, and more transparent. In the case of proteins, the higher levels of structure, namely secondary and tertiary, due to folding of the otherwise linear peptide backbone, are as well affected by hyperbaric processing – because the molar volume of the native molecule is distinct from that of its unfolded counterpart. Pressures above 300–400 MPa bring about the unfolding of proteins, so subsequent aggregation and refolding during decompression may alter food texture; gel formation has accordingly been observed in soy, gluten, and egg albumin. Unlike heat-treated gels, pressure-induced gels are smooth, glossy, soft and elastic, and retain their natural color and flavor; β-lactoglobulin from milk whey, in particular, refolds to a more hydrophobic configuration, with increased surface activity – which is useful in foam stabilization. Meat is tenderized at lower pressures, ca. 100 MPa at 35°C during pre-, or 150 MPa at 60°C during post-rigor mortis – due to effects upon myofibril proteins, but not collagen of connective tissues; its red color may be affected by changes in myoglobin – but not the green color due to chlorophyll in vegetables. The type of effect of ultra-high pressure upon proteins depends, however, on their hydrophobicity and pressure operating range; hydrophobic interaction tends to increase volume up to 100 MPa, but the opposite holds above that threshold. For instance, polyphenoloxidase in peas, mushroom, and potato, and pectin methylesterase (responsible for cloud destabilization in juices, gelation of fruit concentrates, and loss of consistency of solid fruits) in strawberries can withstand 900–1200 MPa; while lipoxygenase (responsible for off-flavors and undesirable color changes in legumes) is inactivated at 400–600 MPa at room temperature, and polyphenol oxidase from apricot, strawberry, and grape is labile to 100, 400, and 600 MPa, respectively. High pressure has been found to easily disrupt the weak interactions that stabilize structure of microbial cells – although a negligible outcome is experienced by covalent bonds, and no effect at all is undergone by the small molecules normally associated with odor and taste. Reactions involving formation of hydrogen bonds are favored by high pressures because they cause a volume decrease; remember that water is the dominant component in most foods, and that its molecules normally engage in hydrogen bonding due to their electronegative oxygen atom covalently linked to their electropositive (and small) hydrogen atoms.

Although far less compressible than regular vapor or gas, liquid water undergoes volume reduction as pressure is increased; such other physicochemical properties of food as density, viscosity, thermal conductivity, ionic dissociation, *pH*, and freezing point are also affected by large pressure increases. Ice acquires different polymorphic forms, depending on the prevailing pressure; higher pressures distort the angles formed by the hydrogen bonds, and produce more compact molecular structures. For instance, regular ice (or polymorph ice I) forms at a lower and lower freezing point as pressure is increased – with a minimum of –22°C at 207 MPa. If pressure is further increased, ice I becomes another polymorph, ice III, which in turn gives rise to ice II, and then to ice V – each one possessing a higher melting point; the melting point accordingly increases to 20°C at 880 MPa, and further to 30°C at 1036 MPa. This sequence of phenomena may be taken advantage of for fast thawing via pressurizing foods above their melting point, or instead for fast freezing via decompressing foods below their melting point.

It is believed that the major consequences of ultra-high pressure upon microorganism viability come from the change in conformation of proteins directly involved in replication, cellular integrity, and metabolism; these include irreversible denaturation of membrane-bound ATPases, DNA polymerases, and RNA transcriptases, as well as of membrane proteins responsible for amino acid uptake. In addition, there are physical effects of hyperbaric treatments, namely damage to cell walls (with consequent softening), reduction in cell volume, increase in membrane permeability (with consequent loss of cytoplasmic constituents, including leakage of such relevant cations as K^+, Mn^{2+}, and Ca^2, and release of intracellular enzymes in active form), and collapse of intracellular vacuoles. Bacteria possessing more rigid cell membranes are more labile to hyperbaric inactivation, while those containing compounds that enhance fluidity thereof (e.g. unsaturated fatty acids instead of saturated ones) are more resistant. The barosensitivity of microorganisms appears to be related to their capacity to repair cell leaks following decompression;

cells undergoing active growth (as during the exponential phase) are thus particularly sensitive to high pressures, unlike their stationary phase counterparts. Representative examples are provided in Fig. 1.10. Due to their spherical shape, cocci undergo fewer morphological changes than rod-shaped bacteria, and are thus more resistant to high pressures; inactivation of bacteria normally exhibits a lag period, followed by first-order death kinetics – which eventually tails off, likely due to sporulation. The recovery of microorganisms from hyperbaric injury is enhanced at temperatures close to their optimum temperature, and somehow inhibited by the organic acids normally present in foods – and is only possible when repair proteins have not been denatured themselves.

More evolutionary developed life forms (e.g. parasites and protozoa) are more sensitive to high pressures; simpler microorganisms and viruses exhibit barosensitivities according to yeasts > Gram$^-$ bacteria > complex viruses > molds > Gram$^+$ bacteria > bacterial endospores. As a heuristic rule, 350 MPa applied for 30 min, or 400 MPa for 5 min causes a tenfold reduction in vegetative cells of bacteria, yeasts, and molds, provided that at least 40% of free water is present; in general, increases in pressure, processing time, or temperature promote inactivation. Note that pH undergoes reduction upon application of very high pressures – of the order of 0.5 units per 100 MPa increase in pressure; this aids in microbial inactivation, since microorganisms become intrinsically more labile, and fewer sub-lethally injured cells will be able to recover. Conversely, low a_W arising from high salt or sugar concentrations may convey extra protection to microorganisms, while inhibiting recovery of injured cells. Bacterial spores are particularly pressure-resistant, and some remain unaffected up to 1000 MPa; therefore, heat is required to eliminate them in low-acid foods – for instance, 90–110°C for 30 min at 500–700 MPa in the case of *Clostridium botulinum*. An alternative strategy is to induce their germination at mild temperatures (25–40°C) and relatively low pressures (60–100 MPa), and then increase temperature (to 50–70°C) and pressure (to 600 MPa) to kill germinated spores and vegetative cells together. Note that long exposure to pressures of the order of 500 MPa may induce pressure resistance, as happens with spores of *Bacillus subtilis*. Viruses exhibit a wide range of pressure resistances – from bacteriophages that are inactivated at 300–400 MPa, to Sindbis virus that remains relatively unaffected upon exposure to 300–700 MPa at –20°C.

The basic technology for hyperbaric equipment has arisen from the expertise developed by the electronic component industry to press metal and ceramic components, or manufacture quartz crystals; the main apparatuses are a pressure vessel with end closures (and yoke, threads, or pins to restrain them), a pressure generation system (usually a cascade of pumps), apparatuses for handling materials, and controls as deemed necessary. Hyperbaric equipment can be arranged to operate batchwise with packaged foods, or semicontinuous-wise to treat (unpackaged) liquid foods – as illustrated in Fig. 1.11. Industrial vessels operate at pressures of 400–700 MPa – and are either monobloc cylinders made of a high-tensile steel alloy, or cylinders prestressed by winding with several kilometers of wire to prevent expansion during pressurization; the latter configuration is usually selected, because it allows the vessel to leak and relieve pressure prior to catastrophic failure and eventual explosion. The isostatic compression fluid is usually water; pressure vessels are sealed by either a threaded steel closure with an interrupted thread to facilitate removal, or a retractable prestressed frame located over the vessel. Pressure is generated via direct or indirect compression; a piston moved by hydraulic pressure compresses the fluid in the former, thus requiring dynamic pressure seals between piston and internal cylinder surface. Indirect compression resorts to an intensifier, or high-pressure pump after a low-pressure pump – which simplifies materials handling, and is more affordable namely due to the requirement of only static pressure seals. Operating temperatures may range from –20°C to more than 100°C – with thermostatting assured by electric heating elements wrapped around the pressure vessel, or pumping of heating/cooling medium through an outer jacket.

Batch operation resembles the cycle followed in sterilization retorts – encompassing filling the process vessel with product (usually transported in metal baskets via conveyors), closing, producing the intended pressure, holding for some time,

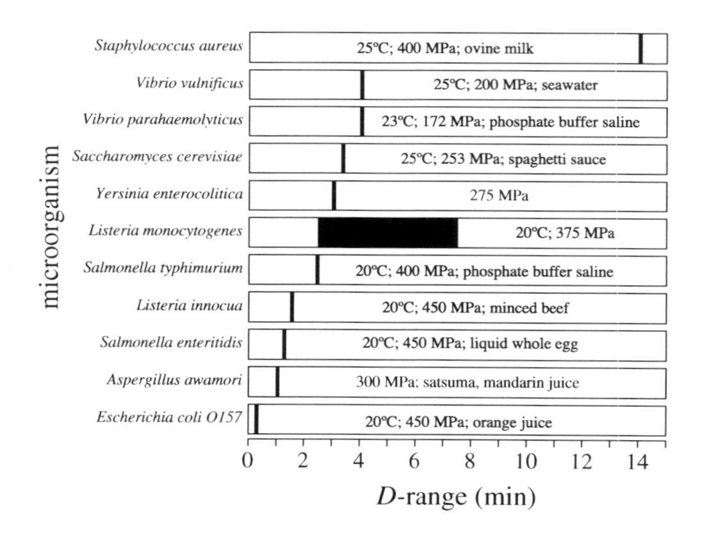

FIGURE 1.10 Resistance ranges (■) of selected pathogens or spoilage microorganisms upon exposure to high pressures, in selected foods and under selected processing conditions, expressed as decimal reduction times, *D*.

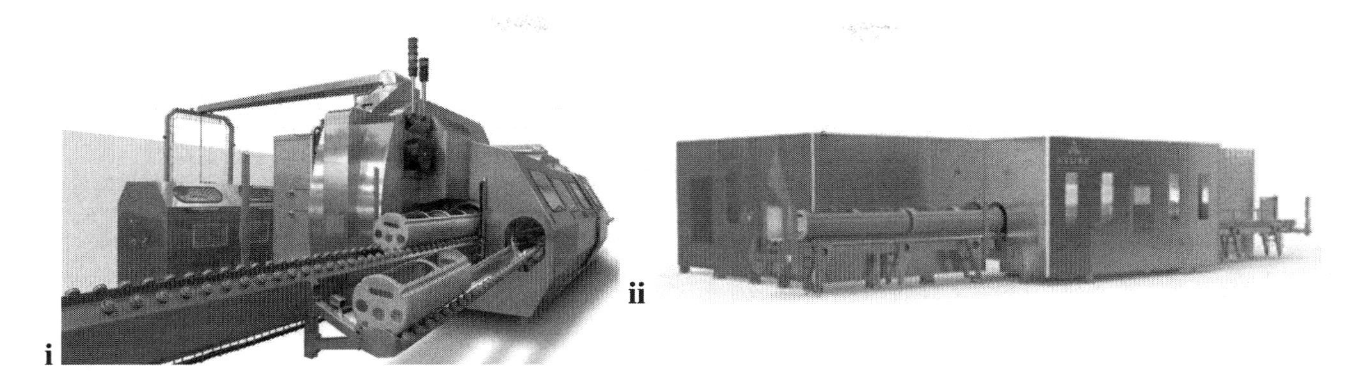

FIGURE 1.11 Examples of high-pressure equipment: (i) for batch operation (courtesy of Hiperbaric, Burgos, Spain) and (ii) for semicontinuous operation (courtesy of Avure Technologies, Middletown, OH).

decompressing the vessel, and finally removing the product. Typical vessel volumes do not go beyond 10 m^3, and cycles can be made as short as 2–3 min – thus lying well below thermal sterilization cycles that take up to 60 min; the pressure operating range is 200–500 MPa – with equipment cost increasing exponentially with target pressure, but operation cost increasing linearly with processing time. After loading the packaged food, the vessel is filled with water to displace any air, and pressure is then increased as much as needed after having the relief valve closed; decompression is made gradually via opening the said relief valve. Due to the volume decrease in foods during compression, and subsequent expansion after decompression, considerable distortion of the package and stress on the seal will occur; hence, the package (often glass containers, or ethylene–vinyl alcohol copolymer or polyvinyl alcohol films) must be designed so as to accommodate up to 15% changes in volume of its content, without compromising integrity of seal and barrier properties of package.

Semicontinuous systems for treating liquid foods consist of a low-pressure pump to fill the pressure vessel, and a freely moving piston to compress the food via water pumping into the vessel once the inlet valve is closed; after the holding time, decompression is in order – and the treated food is discharged, and aseptically filled into pre-sterilized containers. Compared to batch operation, materials handling is simpler (as only pumps, pipework, and valves are required), several vessels may be operated in parallel, a higher efficiency can be attained, and dead time is minimized; however, all components in contact with food must have an aseptic design, besides being suitable for cleaning-in-place and sterilizing-in-place systems – while only pumpable foods may be processed. Typical throughput rates range from 4 to 6 $m^3.hr^{-1}$, at pressures of 150–500 MPa.

Recent developments have unfolded a higher effectiveness toward death of spores, as well as vegetative bacteria and yeasts when pressure is applied in pulses – with critical dependence on pulse waveform and frequency, and on ratio of time under pressure to time without pressure. Very rapid (controlled) decompression is also particularly detrimental to (unwanted) cells and spores, due to cavitation and increased shearing associated with high-velocity turbulent flow. High pressure-assisted freezing and thawing, on the other hand, allow up to 70% increase in freezing or thawing rates as compared to conventional methods. In the former case, smaller ice crystals form that reduce damage to the cellular structure of food, with concomitant minimization of biochemical changes (and, in some cases, even improving texture) – while reduced loss of color and dripping rates are observed during fish and meat thawing.

1.1.2.2.2 Pulsed electric fields

High voltages (typically above 18 kV), when applied as microsecond-pulses, have been found in the 1960s to promote destruction of microorganisms – caused by electricity itself, unlike what happens with electromagnetic fields used for ohmic or dielectric heating. To date, a variety of liquid foods, e.g. fruit juices, soft drinks, alcoholic beverages, soups, liquid egg, and milk, have been successfully processed via this technology as a form of cold pasteurization; a similar approach has been utilized to disrupt plant and animal cells, with the goal of improving oil and fat recovery. The conductivity of most foods ranges within 0.1–0.5 $S.m^{-1}$, but food products with higher salt contents possess a higher ionic strength and thus a higher electrical conductivity; this reduces resistance, so more energy is required for a given electrical field strength – with microbial inactivation becoming more difficult. A lower food conductivity increases indeed the electric potential between food and microbial cytoplasm as driving force, and magnifies ion flow across the membrane that weakens its structure.

Application of periodic discharge of high-voltage electrical pulses follows storage of direct current in a bank of capacitors; when the liquid food is placed between two electrodes – and subjected to an electric field of 20–80 $kV.cm^{-1}$ as short pulses of 1–100 μs duration each, viable numbers of vegetative microorganisms undergo a significant and rapid decline. Examples of equipment able to perform pulsed electric field processing are depicted in Fig. 1.12. Although batch operation is possible, industrial equipment usually operates continuouswise. Their major components are a repetitive pulse generator, delivering a high voltage (ca. 40 kV of difference of potential, and 17 MW of power); capacitors to store the charge; inductors to modify the shape and width of the electrical pulse; discharge switches to release charge to the electrodes; a fluid handling system for product flow control; and a treatment chamber, where the pulsed electric field is applied to the food – which is then fed to an aseptic packaging line. Original difficulties in scaling up were meanwhile overcome by resorting to solid-state switching systems.

FIGURE 1.12 Examples of pulsed electric field equipment: (i) at pilot scale (courtesy of Wageningen Food & Biobased Research, Wageningen, The Netherlands) and (ii) at industrial scale (courtesy of Wernsing Feinkost, Addrup, Germany).

The most common designs encountered for the treatment chamber are coaxial and cofield arrangements; in the former, the product flows between inner and outer cylindrical electrodes – whereas cofield chambers possess two hollow cylindrical (and concentric) electrodes, separated by an insulator, so as to form a tube through which the liquid product flows. Although the system is normally intended to operate at room temperature, increases in temperature up to 30°C are relatively common; hence, temperature control resorts to cooling coils fitted to the processing equipment, or to passing food through a heat exchanger upon treatment. If operating in a single chamber, then food is recirculated the required number of times – whereas multichamber systems entail two (or more) chambers connected in series, with cooling system(s) in between. The entire apparatus is contained within a restricted-access area, with interlocked gates for human operator protection; all connections to the chamber – including product pipework and cooling units, must be isolated and earthed to prevent (dangerous) energy leakage.

Advantages of this form of processing include reduction of vegetative cells within relatively short treatment times, thus making it suitable for decontamination of heat-sensitive foods – while preserving original colors, flavors, and nutrient integrity. Examples of destruction of microorganisms using this method are provided in Fig. 1.13. The decimal reduction time is hereby expressed as the total duration of pulses (i.e. the number of pulses multiplied by the duration of each one); note the great variability observed among degrees of microorganism survival.

Enzymes are, in general, more resistant than microorganisms, but again a notorious variability is observed: for instance, alkaline phosphatase reduces its activity to 65% of the native form in milk when subjected, at 20–50°C, to a field intensity of 18–22 kV.cm^{-1} in the form of 70 pulses of 0.7–0.8 μs duration each; lipase, glucose oxidase, and α–amylase undergo reduction of their original activity to 70–85% when subjected to 13–87 kV.cm^{-1} in the form of 30 pulses of 2 μs duration each; and pectin methylesterase loses 90% of its native activity when exposed to 35 kV.cm^{-1} at 900 kHz for a total duration of 59 μs. Proper controls have confirmed that inactivation arises because of the pulsed electrical field, rather than the high temperature or extreme *pH* generated locally; in mechanistic terms, it is likely due to disruption of the secondary/tertiary structure of the proteinaceous backbone of enzymes that is critical for activity. Conversely, vitamins are almost not destroyed at all.

The major limitations of the foregoing form of nonthermal processing are its restriction to liquid foods, or foods containing small solid particles in suspension – since dielectric breakdown may occur at particle/liquid interfaces, owing to differences in the corresponding dielectric constants; to liquid foods that withstand high electric fields, and are thus characterized by low electrical conductivity – so salt (if required as per product specifications) is to be added only after processing; and to degassed liquid foods, because discharges inside the bubbles may occur if the electric field exceeds the dielectric strength of the said bubbles – which would cause volatilization of liquid that increases, in turn, their volume, until bridging the gap between electrodes (with consequent sparking).

Two mechanisms have been hypothesized to rationalize pulsed electric field effects upon microorganisms: electrical breakdown of cells and electroporation.

According to the former, the membrane of the microbial cell can be considered as a capacitor filled with a dielectric; the charge separation across the membrane creates a potential

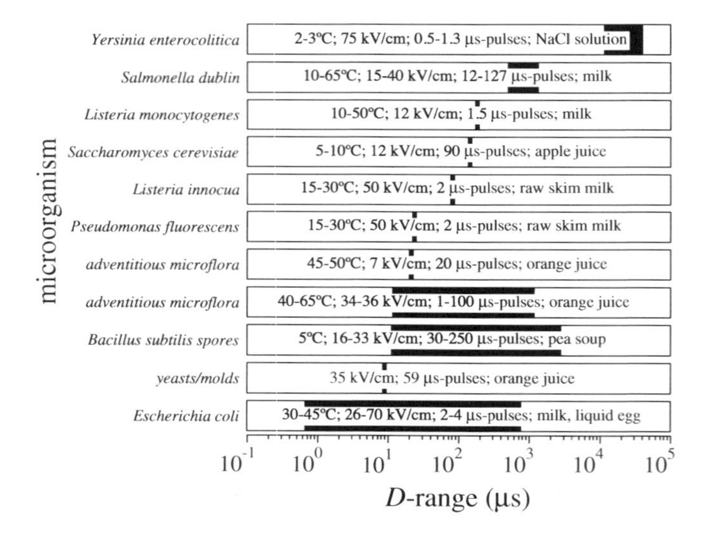

FIGURE 1.13 Resistance ranges (■) of selected pathogens or spoilage microorganisms upon exposure to pulsed electric fields, in selected foods and under selected processing conditions, expressed as decimal reduction times, *D*.

difference of the order of 10 mV (although proportional to actual cell radius). An increase in membrane potential – occurring as a consequence of an increase in field strength due to the electrical pulses, reduces cell membrane thickness; transmembrane pores filled with conductive solution then form if (and when) the breakdown voltage (ca. 1 V) is attained, which causes an immediate discharge and concomitant disruption of the membrane. Such a breakdown is reversible if pores are few, and each one is small relative to the overall membrane surface area; above the critical field strength and after longer exposure times, the extra size and number of pores bring about irreversible mechanical destruction of the membrane. The critical field strength for destruction of bacteria, with a diameter of ca. 1 μm and the aforementioned breakdown voltage across the cell membrane, is of the order of 10 $kV.cm^{-1}$, for pulses of 0.01–1 μs duration.

Conversely, electroporation – or formation of pores in the cell membrane, takes place when high electric field pulses temporarily destabilize the constitutive lipid bilayer of, or proteins inserted in the cell membrane; this increases permeability, due to membrane compression and poration. Cell swelling and rupture ensue, followed by leakage of cytoplasmic materials and eventual cell death.

Other confirmed effects of pulsed electrical fields encompass disruption of cellular organelles (chiefly ribosomes), formation of electrolysis products or highly reactive free radicals, and induced redox reactions within the cell structure that hamper metabolic pathways; complemented by a certain degree of heating, arising from dissipation of induced electrical energy.

Inactivation kinetics appears to be first order on the logarithm of overall treatment time, or number of equal-sized pulses; electric field intensity, pulse waveform and frequency, duration and temperature of treatment, and type, concentration, and physiological status of the adventitious microflora also play a role – further to such food properties as *pH*, electrical conductivity, ionic strength, and presence of antimicrobial compounds. The degree of inactivation of vegetative cells increases with electric field intensity, and number and duration of pulses, besides higher temperature, lower ionic strength, and lower *pH*; however, actively growing cells are more susceptible than stationary phase cells. As expected, bacterial spores and yeast ascospores are considerably more resistant than vegetative cells; the former may indeed withstand up to 30 $kV.cm^{-1}$. $Gram^-$ bacteria are more sensitive than their $Gram^+$ counterparts; and yeasts are more sensitive than bacteria, probably due to their larger size and cellular complexity.

It is more economical to use higher field strengths and shorter pulses for a given overall amount of electrical work supplied; such pulses may be mono- or bipolar, while the waveform is usually sinusoidal, square, rectangular, or exponentially decaying; oscillatory pulses are the least efficient toward microbial inactivation – probably due to their gradual evolution in time, whereas square wave pulses (characterized by periodic discontinuities) have proven the most effective. Bipolar pulses cause a greater damage to the cell membrane, due to the said reversal in orientation of electric field – which causes a similar change in direction of charged molecules; this extra stress in cell membranes enhances their breakdown rate – besides causing less deposition of solids on electrodes and less extensive electrolysis, so the final food product turns more appealing. Bipolar pulses exhibit a relaxation time between consecutive pulses; hence, a lower critical electric field strength may be required for electroporation, because alternating stresses imposed on cells bring about structure fatigue and inactivation.

Pulsed electric field processing has been successful in pasteurizing milk, and has led to reduced viscosity but enhanced β-carotene-mediated color in the case of liquid whole eggs (but not scrambled eggs); sponge cakes manufactured with the latter exhibited a lower volume, but a similar sensory profile. The shelf-life of fresh apple juice can be extended to beyond 55 days at 22–25°C, with negligible change in physicochemical properties, following treatment at 30–40 $kV.cm^{-1}$; vitamin C and color were better retained in reconstituted orange juice *vis a vis* with thermally treated one. Synergisms are often found between pulsed electric field treatment and such antimicrobials as nisin and benzoic (or sorbic) acid; or with hyperbaric treatment, e.g. in the case of *Bacillus subtilis* at 200 MPa for 10 min.

High-voltage arc discharge is a variant of pulsed electric field, which operates via application of rapid voltage discharges between electrodes, located below the surface of liquid foods – and causes negligible increase in temperature. Intense pressure waves and electrolysis are accordingly generated (causing a so-called electrohydraulic shock) when a current of the order of 1500 A, produced by a voltage difference of ca. 40 kV, is applied for 50–300 μs. Irreversible damage to cell membrane consequently results, and reactive radicals and other strongly oxidizing compounds are generated that inactivate microorganisms and enzymes present in food. For a given degree of lethality, this process uses less energy than thermal pasteurization, and is more energy-efficient than hyperbaric or conventional pulsed electric field processes; the major drawback is the range of unknown compounds formed by electrolysis, which may be seen as food contaminants – while shock waves may cause disintegration of food particles, with a negative impact upon texture. A recent improvement, although not yet available at commercial scale, consists in passing oxygenated products through an arc plasma chamber; the submerged arc discharge takes place within the gas bubbles and (partial) breakdown of gas causes ionization – which produces, in turn, reactive ozone and UV radiation, both known antimicrobial agents.

1.1.2.2.3 Pulsed magnetic fields

When an alternated current is passed through electromagnets, oscillating magnetic fields are generated, which may reach 5–100 T (for the sake of comparison, Earth's magnetic field ranges within 25–70 μT); their typical frequencies are 5–500 kHz, and pulse duration normally varies from micro- to millisecond level. This type of treatment inactivates most vegetative cells, but frequencies beyond 500 kHz are less effective and tend to heat up the food; both spores and enzymes are not apparently affected. Available examples of decimal reduction times for microorganisms are conveyed by Fig. 1.14. The extent of microbial inactivation depends on magnetic field intensity, number, and frequency of pulses, and specific properties of food; a high electrical resistivity (above 10–25 $\Omega.cm^{-1}$) is usually needed, as penetration in electrically conductive materials is poor.

The underlying mechanism is not fully understood to date – but it is believed to reside on the magnetic field itself, or else derived from the induced electric field (or both); effects are likely

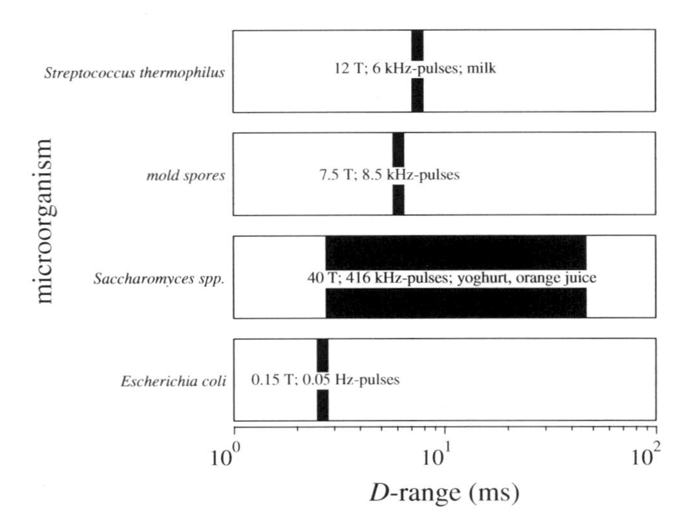

FIGURE 1.14 Resistance ranges (■) of selected pathogens or spoilage microorganisms upon exposure to oscillating magnetic fields, in selected foods and under selected processing conditions, expressed as decimal reduction times, *D*.

associated with translocation of free radicals, changes in cell membrane fluidity or even disruption thereof, covalent breakdown of DNA molecules, or weakened bonds between ions and proteins.

The aforementioned high field strengths are achieved using superconducting liquid helium-refrigerated coils, or else coils energized via discharge of a capacitor; oscillating magnetic fields are applied as pulses, with polarity reversed in each pulse. The intensity of each pulse varies periodically, according to frequency and type of wave in the magnet – and decreases with time by ca. 10% of the initial intensity. The process normally involves subjecting packaged food to 1–100 pulses, for a total exposure time of 25–100 ms at 0–50°C; temperature increases in food are not expected to go beyond 5°C. Nutritional and sensory properties remain essentially unaffected; however, concerns may be raised with regard to creating so strong magnetic fields in industrial environments – in view of their adverse health effects upon human operators.

1.1.2.2.4 Pulsed light

Pulsed white light was used in the 1980s for the first time, namely to inactivate vegetative cells and spores present on the surfaces of foods and packaging materials; it is currently employed in commercial processing of fresh grapes exported by Chile, besides disinfection of drinking water, prevention of mold growth on the surface of baked products, and reduction of microbial contamination in cheeses, fruit juices, meats, and seafood. This is a relatively inexpensive process, involving a low-energy input that brings about little or no changes to the nutritional and organoleptic quality of food – due to its marginal depth of penetration (except in the case of transparent water, where bulk sterilization becomes possible); it is also suitable for dry foods. However, its effective action is restricted to convex surfaces, and it has not been fully proven against spores; furthermore, some adverse chemical effects may result, and reliability of the equipment is still to be established.

Pulsed light resorts obviously to the same (visible) part of the electromagnetic spectrum as sunlight, from 170 nm pertaining to ultraviolet to 2600 nm associated with infrared; hence, it is not expected to bring about major chemical changes, in view of its (essentially) nonionizing nature. It proceeds through 1–100,000 μs pulses, with frequencies of the order of 1–20 Hz and intensities close to 20,000-fold that of sunlight at sea level; such high intensities at topical times are more effective toward destruction of microbial viability than their average value supplied uniformly throughout time. The relatively broad spectrum of light inactivates microorganisms, due to a combination of photochemical and -thermal effects; the former are associated especially with UVC, i.e. 200–280 nm. Such effects include chemical modification of proteins and nucleic acids due to absorption of radiant energy by their highly conjugated double bonds, namely cross-linking between pyrimidine nucleoside bases in the latter; these modifications disrupt cellular metabolism, as well as repair and reproduction mechanisms. Illustrative light fluencies (also referred to as energy fluxes or energy densities), for a given degree of reduction in viable numbers of microorganisms, are conveyed by Fig. 1.15 – where the extreme lability of *Vibrio cholera* is worth mentioning. The lethality of pulsed light to a given microorganism increases with increasing fluency; in bulk sterilization of water, oocysts of *Klebsiella* sp. and *Cryptosporidium* sp. – which do not respond to traditional chlorination or (continuous) UV treatment, can be reduced by 6–7 logarithmic cycles under a fluency of 1 J.cm^{-2}.

Practical examples include extension by 11 d of the shelf-life, at room temperature, of bread, cakes, pizza, and bagels, packaged in transparent films; and by 7 d of shrimp stored under refrigeration, following exposure to pulsed light. The levels of reduction of microbial viability do not, however, go much beyond 3 log cycles (unlike happens with water); the presence of fissures on the surface of foods that shield some microorganisms from light has been implicated with such an observation. An UV fluency of at least 0.4 J.cm^{-2} on all parts of the product may be taken as a heuristic rule for effective microbial inactivation. The typical death curve exhibits a sigmoidal shape, with an initial plateau where pulsed light causes only injury; followed by a rapid decline in number of survivors, as exposure to additional light

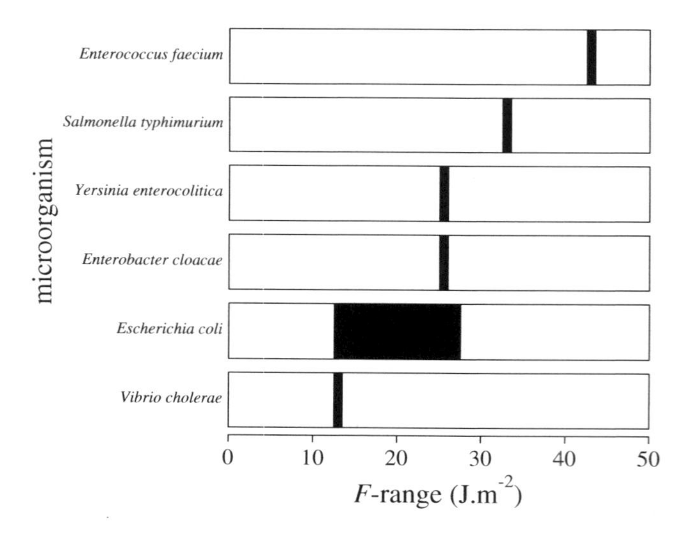

FIGURE 1.15 Resistance ranges (■) of selected pathogens or spoilage microorganisms upon exposure to pulsed light yielding a 4-log decimal reduction in viability, expressed as fluency, *F*.

becomes lethal to already injured microorganisms; and a final tailing off, accounted for by UV-resistant microorganisms – or by photo-reactivation, a repair mechanism present in some bacteria against UV-damage, triggered by exposure to light.

Electricity at normal mains voltage is used to charge capacitors – which periodically release the accumulated power as high-voltage, high-current pulses of electricity to lamps filled with low-medium pressure xenon or low-pressure mercury vapor that generate light flashes. Lasers may be used instead – due to a higher controllability, although their much narrower wavelength may be inconvenient (further to their higher price). Since ordinary glass is not transparent to UV wavelengths, special quartz is usually employed as outer lamp material; in either case, the electric arc generated ionizes the gas (following previous rapid vaporization, should mercury be employed) – and UV is produced when the excited electrons return to their fundamental states. Ammeters are regularly employed to measure lamp current at each flash, thus allowing estimation of overall light intensity – complemented with (silicon-based) photodiode detectors to ascertain fluency within the UV range specifically. Examples of equipment designed for this unit operation are given in Fig. 1.16. Foods are typically exposed to 1–20 pulses, each with energy density in the range 0.1–50 J.cm^{-2} on the food surface; hence, high throughput rates are feasible. The light spectrum is usually adjusted to < 300 nm for transparent foods and packaging materials; this is the case of water, which is flown through an annular clearance between a quartz sleeve that contains the lamp and the outside chamber wall. In the case of opaque foods, shorter wavelengths are filtered out to reduce color loss and lipid oxidation. Since most absorption occurs up to 1 mm from the surface, equipment for pasteurization of fruit juices resorts to a transparent helical tube that creates a continuously renewed surface; juice pumped under turbulent flow undergoes extra random motion as it passes through the coiled tube – via a secondary eddy flow effect, which standardizes velocity and thus residence time.

1.1.2.2.5 Power ultrasound

High power ultrasound processing (or ultrasonication) is a novel technology that holds great promise in the food industry; it basically resorts to a form of mechanical energy, transported by sound waves with frequencies not below 16 kHz (and thus inaudible to the human ear). In nature, bats and dolphins use low-intensity ultrasound to locate preys, whereas some marine animals use high-intensity pulses of ultrasound to stun their preys. A similar classification is possible encompassing application to food processing – with low-intensity ultrasound, i.e. up to 1 W.cm^{-2}, useful as a (nondestructive) analytical method for assessment of composition, structure, or flow rate; and high-intensity ultrasound, i.e. between 10 and 1000 W.cm^{-2}, at frequencies within the range 20–100 kHz, to bring about physical disruption of tissues, cutting or homogenization of foods, creation of emulsions, and cleaning of processing equipment.

During ultrasonication, pressure waves hit the surface of the food, thus generating a force thereon; if the force applies perpendicularly to the surface, then a compression wave results that moves throughout the food bulk – whereas a force applying tangentially to the said surface produces a shearing wave. Both wave types become attenuated as they propagate along the food material; their depth of penetration – and thus the efficacy of their effect against (chiefly spoilage) microorganisms, depends on wave intensity and frequency, as well as food composition itself (especially with regard to dissolved solids and air). Changes in pressure – when very strong and very fast, cause shear disruption and create microbubbles that rapidly collapse (a phenomenon known as cavitation); they also promote thinning of cell membranes, development of localized heating, production of free radicals, and disintegration of organelles. All these phenomena are lethal to microorganisms; shearing and compression, in particular, denaturate such elaborately shaped proteins as enzymes, thus compromising their catalytic activity *in vivo*. However, short bursts of ultrasound may have the opposite effect, believed to result from breaking down of supramolecular structures – since substrates may become more accessible to enzyme active sites. Myofibrillar proteins are also released upon long exposure to ultrasound, thus resulting in tenderization of meat, and increase in water-holding capacity thereof; little effect is, however, observed upon smaller molecules, responsible namely for color or flavor, and upon vitamins – except if extensive production of free radicals occurs.

Large cells are, in general, more susceptible to ultrasonication than small ones – in view of Kolmogorov's minimum eddy length scale, see Eq. (3.301) of *Food Proc. Eng.: basics & mechanical operations*; rod-shaped bacteria are more sensitive than cocci – due to unbalanced effects of vibration as per their nonsymmetrical shape; and Gram$^+$ bacteria are more sensitive than their Gram$^-$ counterparts – probably because of distinct membrane compositions. As expected, (bacterial) spores are the most difficult to disrupt by ultrasonication. Examples of ultrasonication equipment are provided in Fig. 1.17. A typical apparatus consists

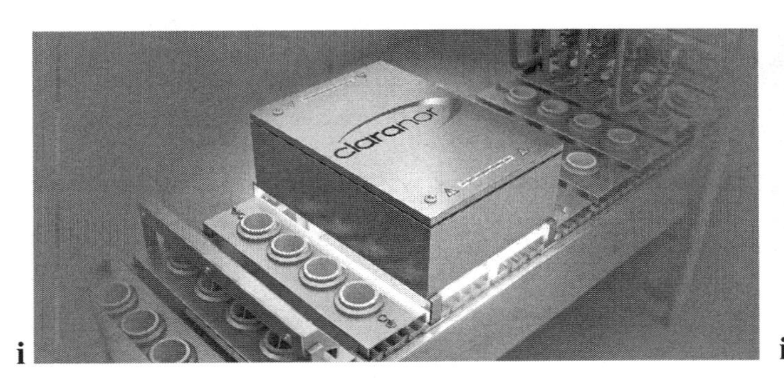

FIGURE 1.16 Examples of pulsed-light equipment: (i) for sterilization of liquid products in small glass packages (courtesy of Claranor, Avignon, France) and (ii) for sanitation of shredded products and mushroom powder (courtesy of SteriBeam Systems, Kehl-am-Rhein, Germany).

of an ultrasound generator – usually a piezoelectric transducer that creates ultrasound from an alternate electrical current; the ultrasound is then transmitted to the food via a horn submerged in the liquid. The ultrasound transducer/horn may be placed inside heating equipment for thermosonication, or inserted in a pressurized vessel to effect manothermosonication.

Power ultrasound is not yet a widespread industrial technique owing to scale-up difficulties – coupled with realization that the duration and intensity of ultrasonication required for sterilization would first produce adverse effects upon structure, and thus textural features of the food – likely to compromise sensory acceptability by prospective consumers. However, a combination of ultrasonication with mild thermal processing and mild hyperbaric treatment, or with low a_w and pH, have proven successful toward microbial destruction or enzyme deactivation; in fact, physical damage to cellular structures and disruption of proteins essential at the cell level reduce resistance of microorganisms to high temperature and other unfavorable environmental conditions. For instance, the D-value of *Yersinia enterocolitica* reduces eightfold when hyperbaric conditions are combined with application of ultrasound – but reduces exponentially up to 11-fold when the soundwave amplitude is changed from 20 to 150 μm; further combination with (mild) heat treatment allows an extra reduction of up to 30-fold. Evidence also exists on the response of polyphenoloxidase, lipoxygenase, lipase, protease, and pectin methylesterase; these enzymes undergo increases in inactivation rates up to 400-fold when simultaneously submitted to heating at 72°C, compression at 200 MPa, and ultrasonication at 20 kHz with 117 μm-pulses.

A number of other applications for ultrasonication are worth mentioning: enhanced rates of crystallization of sugars and fats (due to faster homogeneous nucleation); easier degassing of fermentation liquors (due to formation of tiny bubbles that promptly detach toward the overhead atmosphere); breaking of foams and emulsions (due to vibrational stress at the interface); accelerated extraction of solutes (due to microeddying that favors diffusional dispersion); accelerated filtration (due to better compaction of pellet particles upon energetic vibration); and better cleaning of equipment (due to disruption of dead zones, and subsequent detachment of fouling films from surfaces and fissures). Ultrasound may also assist in acoustic drying; for instance, gelatine, yeast, and orange powders are obtained twice as fast than via classical (plain) drying. This is thought to result from microscopic channels created within the solid food matrix by the oscillating compression waves – which facilitate water migration toward the surface; complemented by a changing pressure gradient at the air–liquid interface – which increases the rate of mass transfer of water.

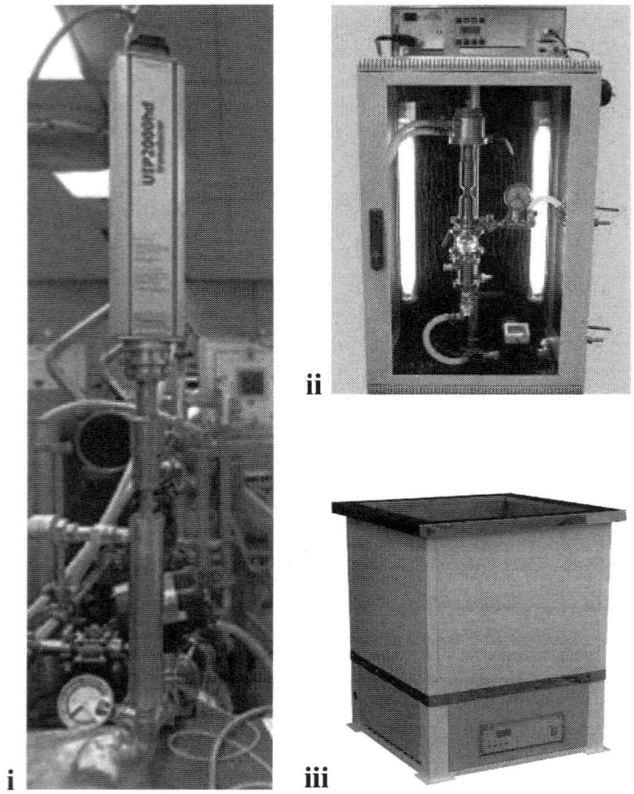

FIGURE 1.17 Examples of ultrasonic equipment: (i) for flow-through sterilization (courtesy of Hielscher Ultrasonics, Teltow, Germany), (ii) continuous liquid processing (courtesy of Industrial Sonomechanics, New York, NY), and (iii) batch sterilization (courtesy of Romer, Skarżysko-Kamienna, Poland).

1.1.3 Mathematical simulation

1.1.3.1 Le Châtelier's principle

Thermodynamics is a well-established discipline, with strong and consistent theories beneath – and useful for mathematical prediction of the maximum (or minimum) extent ever attainable by a given physicochemical phenomenon, irrespective of the kinetic characteristics of the path leading thereto. Therefore, the issue arises as to how a system, initially at (chemical) equilibrium, responds to a change in one (or more) environmental conditions. Empirical observations in this regard have eventually led to formulation of Le Châtelier's principle – a qualitative statement; it claims that when a system at equilibrium experiences a change in temperature, pressure, or concentration of any of its components, its composition will shift toward a new equilibrium in a direction that counteracts the imposed change. A mathematical proof of this useful principle is presented below.

Consider a system encompassing conversion between reactants and products (i.e. chemical reaction) or conversion between components in one phase to components in another phase (i.e. phase change) – with a total of N components. If the said chemical or physical transformation takes place spontaneously, at constant temperature and pressure (as usually happens), then the associated molar Gibbs' energy change, Δg – defined as

$$\Delta g \equiv \sum_{i=1}^{N} \nu_i \mu_i, \tag{1.1}$$

with ν_i denoting algebraic stoichiometric coefficient and μ_i denoting chemical potential, should decrease as per the second law of thermodynamics. This process will take place, at some finite rate, until the minimum is attained – corresponding to chemical equilibrium, described by

$$(\Delta g)_{P,T,eq} = 0 \tag{1.2}$$

and, consequently,

$$d(\Delta g) = 0; \tag{1.3}$$

in view of Eq. (1.1), one may rewrite Eq. (1.3) as

$$\boxed{d\sum_{i=1}^{N} \nu_i \mu_i = 0.} \tag{1.4}$$

Due to linearity of the differential operator, Eq. (1.4) may be redone to

$$\sum_{i=1}^{N} d(\nu_i \mu_i) = 0, \tag{1.5}$$

where constancy of ν_i ($i = 1, 2, \ldots, N$) during the course of a given physicochemical transformation supports, in turn,

$$\sum_{i=1}^{N} \nu_i d\mu_i = 0; \tag{1.6}$$

upon multiplication and division of μ_i by (absolute) temperature, T, Eq. (1.6) turns to

$$\sum_{i=1}^{N} \nu_i d\left(T\frac{\mu_i}{T}\right) = 0, \tag{1.7}$$

where the rule of differentiation of a product may be invoked to write

$$\sum_{i=1}^{N} \nu_i \left(\frac{\mu_i}{T} dT + T d\left(\frac{\mu_i}{T}\right)\right) = 0. \tag{1.8}$$

After splitting the summation, Eq. (1.8) becomes

$$\sum_{i=1}^{N} \nu_i \frac{\mu_i}{T} dT + \sum_{i=1}^{N} \nu_i T d\left(\frac{\mu_i}{T}\right) = 0, \tag{1.9}$$

where a further division of both sides by T yields

$$\frac{1}{T^2}\left(\sum_{i=1}^{N} \nu_i \mu_i\right) dT + \sum_{i=1}^{N} \nu_i d\left(\frac{\mu_i}{T}\right) = 0 \tag{1.10}$$

– with dT/T^2 meanwhile taken off the first summation. Chemical potential, defined as

$$\mu_i \equiv \left(\frac{\partial G}{\partial n_i}\right)_{P,T,x_{j\neq i}} \tag{1.11}$$

in parallel to Eq. (3.36) of *Food Proc. Eng.: basics & mechanical operations* – with G denoting Gibbs' energy, and n_i and x_i denoting number of moles and mole fraction, respectively, of i-th component, is in general a function of P, T, and $x_1, x_2, \ldots, x_N$ as per Gibbs' phase rule, see Eq. (3.125) also of *Food Proc. Eng.: basics & mechanical operations*; division by T does not change the underlying $N+2$ degrees of freedom, so its differential may be expressed as

$$d\left(\frac{\mu_i}{T}\right) = \left(\frac{\partial}{\partial P}\left(\frac{\mu_i}{T}\right)\right)_{T,x} dP + \left(\frac{\partial}{\partial T}\left(\frac{\mu_i}{T}\right)\right)_{P,x} dT + \sum_{j=1}^{N}\left(\frac{\partial}{\partial x_j}\left(\frac{\mu_i}{T}\right)\right)_{P,T,x_{k\neq j}} dx_j; \quad i = 1, 2, \ldots, N \tag{1.12}$$

– where T being held constant during differentiation of the first and third terms of the right-hand side permits simplification of Eq. (1.12) to

$$d\left(\frac{\mu_i}{T}\right) = \frac{1}{T}\frac{\partial \mu_i}{\partial P} dP + \frac{\partial}{\partial T}\left(\frac{\mu_i}{T}\right) dT + \frac{1}{T}\sum_{j=1}^{N} \frac{\partial \mu_i}{\partial x_j} dx_j. \tag{1.13}$$

Partial differentiation of μ_i/T with regard to T may now proceed as

$$\frac{\partial}{\partial T}\left(\frac{\mu_i}{T}\right) = \frac{T\frac{\partial \mu_i}{\partial T} - \mu_i}{T^2} = -\frac{\mu_i - T\frac{\partial \mu_i}{\partial T}}{T^2}; \tag{1.14}$$

Eq. (3.21) of *Food Proc. Eng.: thermal & chemical operations* allows, in turn, one to write

$$\frac{\partial \mu_i}{\partial T} = -s_i \tag{1.15}$$

with s_i denoting specific entropy of i-th component, where insertion in Eq. (1.14) leads to

$$\frac{\partial}{\partial T}\left(\frac{\mu_i}{T}\right) = -\frac{\mu_i - T(-s_i)}{T^2} = -\frac{\mu_i + Ts_i}{T^2}; \tag{1.16}$$

furthermore, combination with Eq. (3.13) of *Food Proc. Eng.: thermal & chemical operations* unfolds

$$\frac{\partial}{\partial T}\left(\frac{\mu_i}{T}\right) = -\frac{h_i}{T^2}, \tag{1.17}$$

with h_i denoting specific enthalpy of i-th component. Equation (3.17) of *Food Proc. Eng.: thermal & chemical operations* and Eq. (1.17) permit transformation of Eq. (1.13) to

$$d\left(\frac{\mu_i}{T}\right) = \frac{v_i}{T} dP - \frac{h_i}{T^2} dT + \frac{1}{T}\sum_{j=1}^{N} \frac{\partial \mu_i}{\partial x_j} dx_j, \tag{1.18}$$

where combination with Eqs. (3.155) and (3.156) of *Food Proc. Eng.: basics & mechanical operations* yields

$$d\left(\frac{\mu_i}{T}\right) = \frac{v_i}{T} dP - \frac{h_i}{T^2} dT + \frac{1}{T}\sum_{j=1}^{N} \frac{\partial}{\partial x_j}\left(\mu_i^\theta + \mathcal{R}T \ln \gamma_i x_i\right) dx_j \tag{1.19}$$

– with γ_i denoting activity coefficient of i-th component and superscript θ referring to pure compound at the same temperature; due to the linearity of $\partial/\partial x_j$ as operator and the constancy of μ_i^θ as pertaining to a pure compound, Eq. (1.19) may be rewritten as

$$d\left(\frac{\mu_i}{T}\right) = \frac{v_i}{T} dP - \frac{h_i}{T^2} dT + \frac{\mathcal{R}T}{T}\sum_{j=1}^{N} \frac{\partial}{\partial x_j}\left(\ln x_i \gamma_i\right) dx_j, \tag{1.20}$$

because $\partial/\partial x_j$ is applied under constant temperature and $\mathcal{R}T$ is independent of the summation counting variable. Cancellation of T between numerator and denominator in Eq. (1.20) finally generates

$$d\left(\frac{\mu_i}{T}\right)=\frac{v_i}{T}dP-\frac{h_i}{T^2}dT+\mathcal{R}\sum_{j=1}^{N}\frac{\partial}{\partial x_j}\left(\ln x_i\gamma_i\right)dx_j. \quad (1.21)$$

The operational features of a logarithmic function allow transformation of Eq. (1.21) to

$$d\left(\frac{\mu_i}{T}\right)=\frac{v_i}{T}dP-\frac{h_i}{T^2}dT+\mathcal{R}\sum_{j=1}^{N}\frac{\partial}{\partial x_j}\left(\ln x_i+\ln\gamma_i\right)dx_j \quad (1.22)$$

that obviously becomes

$$d\left(\frac{\mu_i}{T}\right)=\frac{v_i}{T}dP-\frac{h_i}{T^2}dT+\mathcal{R}\sum_{j=1}^{N}\frac{\partial\ln x_i}{\partial x_j}dx_j+\mathcal{R}\sum_{j=1}^{N}\frac{\partial\ln\gamma_i}{\partial x_j}dx_j \quad (1.23)$$

in view of the rules of differentiation of a sum of functions. Note that $\partial \ln x_i/\partial x_j=0$ when $i\neq j$, so the first summation in Eq. (1.23) reduces to

$$d\left(\frac{\mu_i}{T}\right)=\frac{v_i}{T}dP-\frac{h_i}{T^2}dT+\mathcal{R}\frac{\partial\ln x_i}{\partial x_i}dx_i+\mathcal{R}\sum_{j=1}^{N}\frac{\partial\ln\gamma_i}{\partial x_j}dx_j; \quad (1.24)$$

the rule of differentiation of a logarithm allows, in turn, conversion of Eq. (1.24) to

$$d\left(\frac{\mu_i}{T}\right)=\frac{v_i}{T}dP-\frac{h_i}{T^2}dT+\frac{\mathcal{R}}{x_i}dx_i+\mathcal{R}\sum_{j=1}^{N}\frac{\partial\ln\gamma_i}{\partial x_j}dx_j. \quad (1.25)$$

If an ideal solution is considered, then $\gamma_i=1$ for $i=1, 2, \ldots, N$; this is normally valid with concentrated and not too dissimilar substrates – and permits simplification of Eq. (1.25) to

$$d\left(\frac{\mu_i}{T}\right)=\frac{v_i}{T}dP-\frac{h_i}{T^2}dT+\frac{\mathcal{R}}{x_i}dx_i; \quad (1.26)$$

insertion of Eq. (1.26) converts Eq. (1.10) to

$$\frac{1}{T^2}\left(\sum_{i=1}^{N}\nu_i\mu_i\right)dT+\sum_{i=1}^{N}\nu_i\left(\frac{v_i}{T}dP-\frac{h_i}{T^2}dT+\frac{\mathcal{R}}{x_i}dx_i\right)=0 \quad (1.27)$$

– where splitting of the second summation, followed by algebraic regrouping of terms containing dT yield

$$\frac{1}{T^2}\left(\sum_{i=1}^{N}\nu_i\mu_i-\sum_{i=1}^{N}\nu_i h_i\right)dT+\frac{1}{T}\sum_{i=1}^{N}\nu_i v_i dP+\mathcal{R}\sum_{i=1}^{N}\frac{\nu_i}{x_i}dx_i=0. \quad (1.28)$$

After dividing both sides by $\mathcal{R}$ and isolating the last term in the left-hand side, Eq. (1.28) becomes

$$\sum_{i=1}^{N}\frac{\nu_i}{x_i}dx_i=\frac{1}{\mathcal{R}T^2}\left(\sum_{i=1}^{N}\nu_i h_i-\sum_{i=1}^{N}\nu_i\mu_i\right)dT-\frac{1}{\mathcal{R}T}\left(\sum_{i=1}^{N}\nu_i v_i\right)dP, \quad (1.29)$$

while combination with Eqs. (1.1) and (1.2) gives rise to

$$\sum_{i=1}^{N}\frac{\nu_i}{x_i}dx_i\approx\frac{1}{\mathcal{R}T^2}\left(\sum_{i=1}^{N}\nu_i h_i\right)dT-\frac{1}{\mathcal{R}T}\left(\sum_{i=1}^{N}\nu_i v_i\right)dP \quad (1.30)$$

– because only differential changes around chemical equilibrium are of interest here; upon recalling

$$\Delta h\equiv\sum_{i=1}^{N}\nu_i h_i \quad (1.31)$$

and

$$\Delta v\equiv\sum_{i=1}^{N}\nu_i v_i \quad (1.32)$$

using Eq. (1.1) as template, one may rewrite Eq. (1.30) as

$$\sum_{i=1}^{N}\frac{\nu_i}{x_i}dx_i=\frac{\Delta h}{\mathcal{R}T^2}dT-\frac{\Delta v}{\mathcal{R}T}dP, \quad (1.33)$$

where Δh and Δv denote molar enthalpy and volume changes, respectively, accompanying the physicochemical process. If the j-th component of the summation in Eq. (1.33) is singled out, one gets

$$\sum_{\substack{i=1\\i\neq j}}^{N}\frac{\nu_i}{x_i}dx_i=\frac{\Delta h}{\mathcal{R}T^2}dT-\frac{\Delta v}{\mathcal{R}T}dP-\frac{\nu_j}{x_j}dx_j. \quad (1.34)$$

On the other hand, upon definition of an extent of (chemical) reaction or (physical) phase change, χ, via

$$\boxed{d\chi\equiv\frac{1}{\nu_i}\frac{dn_i}{V};\quad i=1,2,\ldots,N} \quad (1.35)$$

with V denoting volume, one may proceed with division of both sides by (constant) total mole inventory, n, to get

$$\frac{d\chi}{n}=\frac{1}{\nu_i V}\frac{dn_i}{n}\equiv\frac{dx_i}{\nu_i V} \quad (1.36)$$

also with the aid of Eq. (3.44) of *Food Proc. Eng.: basics & mechanical operations* – or else

$$\nu_i dx_i=\nu_i^2\frac{V}{n}d\chi=\nu_i^2\,v\,d\chi, \quad (1.37)$$

upon isolation of dx_i and multiplication of both sides by ν_i afterward – where v denotes (overall) molar volume. Insertion of Eq. (1.37) transforms Eq. (1.34) to

$$\sum_{\substack{i=1\\i\neq j}}^{N}v\frac{\nu_i^2}{x_i}d\chi=\frac{\Delta h}{\mathcal{R}T^2}dT-\frac{\Delta v}{\mathcal{R}T}dP-\frac{\nu_j}{x_j}dx_j, \quad (1.38)$$

where v and $d\chi$ may be factored out as

$$\nu\left(\sum_{\substack{i=1\\i\neq j}}^{N}\frac{\nu_i^2}{x_i}\right)d\chi=\frac{\Delta h}{\mathcal{R}T^2}dT-\frac{\Delta v}{\mathcal{R}T}dP-\frac{\nu_j}{x_j}dx_j; \tag{1.39}$$

isolation of $d\chi$ finally gives

$$d\chi=\frac{\Delta h}{\mathcal{R}T^2\nu\sum_{\substack{i=1\\i\neq j}}^{N}\frac{\nu_i^2}{x_i}}dT-\frac{\Delta v}{\mathcal{R}T\nu\sum_{\substack{i=1\\i\neq j}}^{N}\frac{\nu_i^2}{x_i}}dP-\frac{\nu_j}{x_j\nu\sum_{\substack{i=1\\i\neq j}}^{N}\frac{\nu_i^2}{x_i}}dx_j. \tag{1.40}$$

Since all terms ν_i^2/x_i are positive, their summation – and multiplication thereof by either $\mathcal{R}T$ or $\mathcal{R}T^2$, is positive as well, so one readily concludes that the sign of $d\chi$ coincides with that of $\Delta h dT$ if only temperature is changed, with that of $-\Delta v dP$ is only pressure is changed, and with that of $-\nu_j dx_j$ if only molar fraction of j-th component is changed. In other words,

$$\boxed{\frac{1}{\Delta h}\left(\frac{\partial\chi}{\partial T}\right)_{P,x}>0} \tag{1.41}$$

– i.e. an increase in temperature raises the extent of an endothermic reaction (i.e. $\Delta h>0$) and depresses the extent of an exothermic reaction (i.e. $\Delta h<0$), should the system be initially at chemical equilibrium;

$$\boxed{\frac{1}{\Delta v}\left(\frac{\partial\chi}{\partial P}\right)_{T,x}<0} \tag{1.42}$$

– i.e. an increase in pressure depresses the extent of an expansive reaction (i.e. $\Delta v>0$), but raises the extent of a contractive reaction (i.e. $\Delta v<0$), departing once more from chemical equilibrium; and

$$\boxed{\frac{1}{\nu_j}\left(\frac{\partial\chi}{\partial x_j}\right)_{P,T,x_{k\neq j}}<0} \tag{1.43}$$

– i.e. an increase in molar fraction of a reactant (i.e. $\nu_j<0$) raises the extent of reaction, whereas it acts otherwise in the case of a product (i.e. $\nu_j>0$). Equations (1.41)–(1.43) constitute, as a set, the mathematical statement of Le Châtelier's principle – that any change in *status quo* of a system (at equilibrium) prompts a spontaneous opposing response thereby: in addition, it proves its validity from first principles – while providing a quantitative measure of the tendency of response in the vicinity of an original equilibrium.

PROBLEM 1.1

Prove that Le Châtelier's principle, in the simplest form conveyed by Eq. (1.43), may be misleading in the case of a reacting mixture with nonideal behavior. Illustrate the reasoning for a binary mixture following Margules' equation of state.

Solution

If a mixture is nonideal, then the last term in Eq. (1.25) cannot be neglected; hence, the correct form thereof reads

$$d\left(\frac{\mu_i}{T}\right)=\frac{v_i}{T}dP-\frac{h_i}{T^2}dT+\frac{\mathcal{R}}{x_i}dx_i+\mathcal{R}\sum_{j=1}^{2}\frac{\partial\ln\gamma_i}{\partial x_j}dx_j, \tag{1.1.1}$$

valid for $i=1, 2$, under the specific case of $N=2$. Insertion of Eq. (1.1.1) transforms Eq. (1.10) to

$$\frac{1}{T^2}\left(\sum_{i=1}^{2}\nu_i\mu_i\right)dT+\sum_{i=1}^{2}\nu_i\left(\begin{array}{l}\frac{v_i}{T}dP-\frac{h_i}{T^2}dT+\frac{\mathcal{R}}{x_i}dx_i\\+\mathcal{R}\sum_{j=1}^{2}\frac{\partial\ln\gamma_i}{\partial x_j}dx_j\end{array}\right)=0; \tag{1.1.2}$$

the outer summation in the second term may then be split, and alike terms lumped afterward to produce

$$\frac{1}{T^2}\left(\sum_{i=1}^{2}\nu_i\mu_i-\sum_{i=1}^{2}\nu_i h_i\right)dT+\frac{1}{T}\sum_{i=1}^{2}\nu_i v_i dP$$
$$+\mathcal{R}\sum_{i=1}^{2}\frac{\nu_i}{x_i}dx_i+\mathcal{R}\sum_{i=1}^{2}\nu_i\sum_{j=1}^{2}\frac{\partial\ln\gamma_i}{\partial x_j}dx_j=0, \tag{1.1.3}$$

along with factoring out of $\mathcal{R}$. In view of the definition of molar Gibbs' energy change, Δg, as per Eq. (1.1), molar enthalpy change, Δh, as per Eq. (1.31), and volume change, Δv, as per Eq. (1.32), accompanying conversion of reactant to product via the chemical reaction, one may simplify notation in Eq. (1.1.3) to

$$\frac{\Delta g-\Delta h}{\mathcal{R}T^2}dT+\frac{\Delta v}{\mathcal{R}T}dP+\sum_{i=1}^{2}\frac{\nu_i}{x_i}dx_i+\sum_{i=1}^{2}\nu_i\sum_{j=1}^{2}\frac{\partial\ln\gamma_i}{\partial x_j}dx_j=0, \tag{1.1.4}$$

where both sides were also divided by $\mathcal{R}$ for convenience; in the vicinity of equilibrium, Eq. (1.2) holds, thus implying no relevant change in Gibbs' energy – so Eq. (1.1.4) simplifies to

$$\frac{\Delta v}{\mathcal{R}T}dP-\frac{\Delta h}{\mathcal{R}T^2}dT+\frac{\nu_1}{x_1}dx_1+\frac{\nu_2}{x_2}dx_2+\nu_1\frac{\partial\ln\gamma_1}{\partial x_1}dx_1$$
$$+\nu_1\frac{\partial\ln\gamma_1}{\partial x_2}dx_2+\nu_2\frac{\partial\ln\gamma_2}{\partial x_1}dx_1+\nu_2\frac{\partial\ln\gamma_2}{\partial x_2}dx_2=0, \tag{1.1.5}$$

where the terms contained in the summations were meanwhile made explicit. According to Margules' equation of state,

$$\ln\gamma_1=x_2^2\left(A+2(B-A)x_1\right) \tag{1.1.6}$$

and

$$\ln\gamma_2=x_1^2\left(B+2(A-B)x_2\right), \tag{1.1.7}$$

with A and B denoting (experimentally adjusted) parameters; Eq. (1.1.6) prompts

$$\frac{\partial \ln \gamma_1}{\partial x_1} = 2(B-A)x_2^2 \tag{1.1.8}$$

and

$$\frac{\partial \ln \gamma_1}{\partial x_2} = 2x_2\left(A+2(B-A)x_1\right), \tag{1.1.9}$$

while Eq. (1.1.7) similarly supports

$$\frac{\partial \ln \gamma_2}{\partial x_1} = 2x_1\left(B+2(A-B)x_2\right) \tag{1.1.10}$$

and

$$\frac{\partial \ln \gamma_2}{\partial x_2} = 2(A-B)x_1^2 \tag{1.1.11}$$

upon partial differentiation with regard to x_1 (keeping x_2 constant) and x_2 (keeping x_1 constant), respectively. Insertion in Eqs. (1.1.8)–(1.1.11) transforms Eq. (1.1.5) to

$$\begin{aligned}&\frac{\Delta v}{\mathcal{R}T}dP - \frac{\Delta h}{\mathcal{R}T^2}dT + \frac{v_1}{x_1}dx_1 + \frac{v_2}{x_2}dx_2 \\ &+v_1 2(B-A)x_2^2 dx_1 + v_1 2x_2\left(A+2(B-A)x_1\right)dx_2 \\ &+v_2 2x_1\left(B+2(A-B)x_2\right)dx_1 + v_2 2(A-B)x_1^2 dx_2 = 0\end{aligned} \tag{1.1.12}$$

– where dx_1 and dx_2 may be factored out (as appropriate) to obtain

$$\begin{aligned}&\frac{\Delta v}{\mathcal{R}T}dP - \frac{\Delta h}{\mathcal{R}T^2}dT + \begin{pmatrix}\frac{v_1}{x_1} + 2v_1(B-A)x_2^2 \\ +2v_2 x_1\left(B+2(A-B)x_2\right)\end{pmatrix}dx_1 \\ &+\begin{pmatrix}\frac{v_2}{x_2} + 2v_1 x_2\left(A+2(B-A)x_1\right) \\ +2v_2(A-B)x_1^2\end{pmatrix}dx_2 = 0.\end{aligned} \tag{1.1.13}$$

After retrieving Eq. (1.36) as

$$dx_i = w_i d\chi, \tag{1.1.14}$$

one may convert Eq. (1.1.13) to

$$\begin{aligned}&\frac{\Delta v}{\mathcal{R}T}dP - \frac{\Delta h}{\mathcal{R}T^2}dT \\ &+\left(\frac{v_1}{x_1} + 2v_1(B-A)x_2^2 + 2v_2 x_1\left(B+2(A-B)x_2\right)\right)w_1 d\chi \\ &+\left(\frac{v_2}{x_2} + 2v_1 x_2\left(A+2(B-A)x_1\right) + 2v_2(A-B)x_1^2\right)dx_2 = 0\end{aligned} \tag{1.1.15}$$

after having (arbitrarily) chosen $i=2$ for disturbance in composition; Eq. (1.1.15) reduces to

$$\begin{aligned}&w_1\left(\frac{v_1}{x_1} + 2v_1(B-A)x_2^2 + 2v_2 x_1\left(B+2(A-B)x_2\right)\right)d\chi \\ &= -\left(\frac{v_2}{x_2} + 2v_1 x_2\left(A+2(B-A)x_1\right) + 2v_2(A-B)x_1^2\right)dx_2\end{aligned} \tag{1.1.16}$$

under constant temperature and pressure, which is equivalent to writing

$$\frac{1}{v_2}\left(\frac{\partial \chi}{\partial x_2}\right)_{P,T} = -\frac{1}{w_1^2}\frac{\frac{1}{x_2} + 2\frac{v_1}{v_2}x_2\left(A+2(B-A)x_1\right) + 2(A-B)x_1^2}{\frac{1}{x_1} + 2(B-A)x_2^2 + 2\frac{v_2}{v_1}x_1\left(B+2(A-B)x_2\right)} \tag{1.1.17}$$

after isolation of $v_2^{-1}\partial\chi/\partial x_2$ and factoring out of ν_1. Therefore, the direction of evolution of a previous chemical equilibrium, upon a small increase in the mole fraction of component 2, depends on the relative sign of numerator and denominator in Eq. (1.1.17) – and is actually opposite to such a sign because $-1/w_1^2 < 0$; the right hand side is a function not only of the ratio of algebraic stoichiometric coefficients, ν_1/ν_2, and state equation parameters, A and B – but also of current composition, x_1 or x_2. As expected, when $A=B=0$, Eq. (1.1.17) degenerates to

$$\frac{1}{v_2}\left(\frac{\partial \chi}{\partial x_2}\right)_{P,T} = -\frac{1}{w_1^2}\frac{x_1}{x_2} < 0 \tag{1.1.18}$$

in agreement with Eq. (1.43) – corresponding to an ideal mixture as per $\ln \gamma_1 = \ln \gamma_2 = 0$, based on Eqs. (1.1.6) and (1.1.7), or $\gamma_1 = \gamma_2 = 1$ for that matter.

1.1.3.2 Joule and Thomson's coefficient

When a (high) hydrostatic pressure is applied to a packaged food submerged in a liquid, the new pressure distributes instantaneously and uniformly throughout the whole outer surface of the food; this is so because of the unique ability of a fluid to convey the same pressure throughout all its points – unlike happens with a solid, which instead conveys the same force throughout all its points. Such a realization constitutes an advantage over classical methods of thermal processing (e.g. conductive or radiative heating), and even new methods thereof (e.g. dielectric heating) – in that food can undergo uniform treatment, irrespective of its shape, thickness, or composition. Even though hyperbaric handling is considered a nonthermal process, an increase in temperature of the order of 2–3°C per every 100 MPa of increase in pressure is expected, depending on food composition – with a maximum of 9°C/100 MPa, recorded for soybean oil and similar fats.

The aforementioned variation in temperature may be estimated based on Joule and Thomson's coefficient, pertaining to adiabatic and reversible conditions, i.e. $(\partial T/\partial P)_S$; note that pressure is gradually applied onto a food – which essentially behaves as an insulator, by a compressor during hyperbaric treatment – so no heat is in principle exchanged, neither are irreversible (and thus dissipative) processes present. The said isentropic coefficient should not be mistaken for its isenthalpic form, $(\partial T/\partial P)_H$ – widely applicable in mechanical cooling, brought about by adiabatic expansion of a vapor through an insulated nozzle in standard refrigeration systems.

According to the general definition of a bivariate total differential, (small) variations of pressure P may indeed be expressed as

$$dP \equiv \left(\frac{\partial P}{\partial T}\right)_V dT + \left(\frac{\partial P}{\partial V}\right)_T dV \tag{1.44}$$

– since, for a single-component (or, equivalently, constant composition), single-phase system, Gibbs' phase rule enforces two degrees of freedom, say temperature T and volume V, in general agreement with Eq. (3.125) of *Food Proc. Eng.: basics & mechanical operations*; upon division of both sides by dT, under constant S, Eq. (1.44) turns to

$$\left(\frac{\partial P}{\partial T}\right)_S = \left(\frac{\partial P}{\partial T}\right)_V + \left(\frac{\partial P}{\partial V}\right)_T \left(\frac{\partial V}{\partial T}\right)_S. \tag{1.45}$$

On the other hand, implicit differentiation has it that

$$\left(\frac{\partial V}{\partial T}\right)_S = -\frac{\left(\frac{\partial S}{\partial T}\right)_V}{\left(\frac{\partial S}{\partial V}\right)_T}, \tag{1.46}$$

so insertion of Eq. (1.46) permits conversion of Eq. (1.45) to

$$\left(\frac{\partial P}{\partial T}\right)_S = \left(\frac{\partial P}{\partial T}\right)_V - \left(\frac{\partial P}{\partial V}\right)_T \frac{\left(\frac{\partial S}{\partial T}\right)_V}{\left(\frac{\partial S}{\partial V}\right)_T} = \left(\frac{\partial P}{\partial T}\right)_V - \left(\frac{\frac{\partial P}{\partial V}}{\frac{\partial S}{\partial V}}\right)_T \left(\frac{\partial S}{\partial T}\right)_V; \tag{1.47}$$

upon cancellation of dV between numerator and denominator, under constant T, Eq. (1.47) reduces to

$$\left(\frac{\partial P}{\partial T}\right)_S = \left(\frac{\partial P}{\partial T}\right)_V - \left(\frac{\partial P}{\partial S}\right)_T \left(\frac{\partial S}{\partial T}\right)_V. \tag{1.48}$$

It is instructive, at this stage, to recall Eq. (3.23) of *Food Proc. Eng.: basics & mechanical operations* and write

$$\left(\frac{\partial V}{\partial S}\right)_P = \left(\frac{\partial}{\partial S}\left(\frac{\partial H}{\partial P}\right)_S\right)_P \equiv \frac{\partial^2 H}{\partial S \partial P}, \tag{1.49}$$

following differentiation with regard to S; and likewise

$$\left(\frac{\partial T}{\partial P}\right)_S = \left(\frac{\partial}{\partial P}\left(\frac{\partial H}{\partial S}\right)_P\right)_S \equiv \frac{\partial^2 H}{\partial P \partial S}, \tag{1.50}$$

after taking the derivative of T with regard to P – in much the same way Eqs. (3.31) and (3.32) were obtained from Eq. (3.28), all from *Food Proc. Eng.: basics & mechanical operations.* Since the second derivatives of H are continuous functions, one may invoke Schwarz's theorem to state

$$\frac{\partial^2 H}{\partial P \partial S} = \frac{\partial^2 H}{\partial S \partial P} \tag{1.51}$$

– so one obtains

$$\left(\frac{\partial T}{\partial P}\right)_S = \left(\frac{\partial V}{\partial S}\right)_P \tag{1.52}$$

from Eqs. (1.49) and (1.50). On the other hand, the left-hand side of Eq. (1.48) may be combined with the various partial derivatives involving T, P, and S as either dependent, independent, or constant variable as

$$\left(\frac{\partial T}{\partial P}\right)_S \left(\frac{\partial P}{\partial S}\right)_T \left(\frac{\partial S}{\partial T}\right)_P = -1; \tag{1.53}$$

isolation of $(\partial T/\partial P)_S$ yields

$$\left(\frac{\partial T}{\partial P}\right)_S = -\frac{1}{\left(\frac{\partial P}{\partial S}\right)_T \left(\frac{\partial S}{\partial T}\right)_P}, \tag{1.54}$$

where the rule of differentiation of the inverse function may be used to produce

$$\left(\frac{\partial T}{\partial P}\right)_S = -\left(\frac{\partial S}{\partial P}\right)_T \left(\frac{\partial T}{\partial S}\right)_P. \tag{1.55}$$

Elimination of $(\partial T/\partial P)_S$ between Eqs. (1.52) and (1.55) gives rise to

$$\left(\frac{\partial V}{\partial S}\right)_P = -\left(\frac{\partial S}{\partial P}\right)_T \left(\frac{\partial T}{\partial S}\right)_P, \tag{1.56}$$

where condensation of partial derivatives, taken under constant P, may proceed as

$$\left(\frac{\partial S}{\partial P}\right)_T = -\left(\frac{\frac{\partial V}{\partial S}}{\frac{\partial T}{\partial S}}\right)_P \tag{1.57}$$

– or else

$$\boxed{\left(\frac{\partial S}{\partial P}\right)_T = -\left(\frac{\partial V}{\partial T}\right)_P,} \tag{1.58}$$

after dropping dS from both numerator and denominator; Eq. (1.58) is but one of Maxwell's relations. Equation (1.58) may instead appear as

$$\left(\frac{\partial P}{\partial S}\right)_T = -\left(\frac{\partial T}{\partial V}\right)_P, \tag{1.59}$$

after taking reciprocals of both sides – while Eq. (3.141) of *Food Proc. Eng.: thermal & chemical operations* allows one to write

$$\left(\frac{\partial S}{\partial T}\right)_V = \frac{C_V}{T}, \tag{1.60}$$

with C_V denoting isobaric heat capacity; combination of Eqs. (1.48), (1.59), and (1.60) gives then rise to

$$\left(\frac{\partial P}{\partial T}\right)_S = \left(\frac{\partial P}{\partial T}\right)_V + \frac{C_V}{T}\left(\frac{\partial T}{\partial V}\right)_P. \tag{1.61}$$

After invoking again the rule of differentiation of an implicit function, one may write $(\partial T/\partial V)_P$ as

$$\left(\frac{\partial T}{\partial V}\right)_P = -\frac{\left(\frac{\partial P}{\partial V}\right)_T}{\left(\frac{\partial P}{\partial T}\right)_V}, \tag{1.62}$$

so Eq. (1.61) can be redone to

$$\left(\frac{\partial P}{\partial T}\right)_S = \left(\frac{\partial P}{\partial T}\right)_V - \frac{C_V}{T}\frac{\left(\frac{\partial P}{\partial V}\right)_T}{\left(\frac{\partial P}{\partial T}\right)_V}. \tag{1.63}$$

Equation (3.161) of *Food Proc. Eng.: thermal & chemical operations* may now be retrieved as

$$C_V = C_P - T\left(\frac{\partial V}{\partial T}\right)_P\left(\frac{\partial P}{\partial T}\right)_V, \tag{1.64}$$

so C_V can be eliminated from Eq. (1.63) to obtain

$$\left(\frac{\partial P}{\partial T}\right)_S = \left(\frac{\partial P}{\partial T}\right)_V - \frac{1}{T}\left(C_P - T\left(\frac{\partial V}{\partial T}\right)_P\left(\frac{\partial P}{\partial T}\right)_V\right)\frac{\left(\frac{\partial P}{\partial V}\right)_T}{\left(\frac{\partial P}{\partial T}\right)_V}; \tag{1.65}$$

elimination of parenthesis permits simplification to

$$\left(\frac{\partial P}{\partial T}\right)_S = \left(\frac{\partial P}{\partial T}\right)_V - \frac{C_P}{T}\frac{\left(\frac{\partial P}{\partial V}\right)_T}{\left(\frac{\partial P}{\partial T}\right)_V} + \left(\frac{\partial V}{\partial T}\right)_P\left(\frac{\partial P}{\partial T}\right)_V\frac{\left(\frac{\partial P}{\partial V}\right)_T}{\left(\frac{\partial P}{\partial T}\right)_V}, \tag{1.66}$$

where cancellation of similar factors between numerator and denominator unfolds

$$\left(\frac{\partial P}{\partial T}\right)_S = \left(\frac{\partial P}{\partial T}\right)_V - \frac{C_P}{T}\frac{\left(\frac{\partial P}{\partial V}\right)_T}{\left(\frac{\partial P}{\partial T}\right)_V} + \left(\frac{\partial V}{\partial T}\right)_P\left(\frac{\partial P}{\partial V}\right)_T \tag{1.67}$$

– or else

$$\left(\frac{\partial P}{\partial T}\right)_S = \left(\frac{\partial P}{\partial T}\right)_V + \frac{C_P}{T}\left(\frac{\partial T}{\partial V}\right)_P + \left(\frac{\partial V}{\partial T}\right)_P\left(\frac{\partial P}{\partial V}\right)_T, \tag{1.68}$$

following insertion of Eq. (1.62). On the other hand, the combined derivatives of P, V, and T must satisfy

$$\left(\frac{\partial P}{\partial T}\right)_V\left(\frac{\partial T}{\partial V}\right)_P\left(\frac{\partial V}{\partial P}\right)_T = -1 \tag{1.69}$$

as happens with any other set of three (continuous and differentiable) functions – meaning that isolation of $(\partial P/\partial T)_V$ produces

$$\left(\frac{\partial P}{\partial T}\right)_V = -\frac{1}{\left(\frac{\partial T}{\partial V}\right)_P\left(\frac{\partial V}{\partial P}\right)_T}; \tag{1.70}$$

in view of the rule of differentiation of the inverse function, Eq. (1.70) is equivalent to

$$\left(\frac{\partial P}{\partial T}\right)_V = -\left(\frac{\partial V}{\partial T}\right)_P\left(\frac{\partial P}{\partial V}\right)_T. \tag{1.71}$$

Upon insertion of Eq. (1.71), one gets

$$\left(\frac{\partial P}{\partial T}\right)_S = -\left(\frac{\partial V}{\partial T}\right)_P\left(\frac{\partial P}{\partial V}\right)_T + \frac{C_P}{T}\left(\frac{\partial T}{\partial V}\right)_P + \left(\frac{\partial V}{\partial T}\right)_P\left(\frac{\partial P}{\partial V}\right)_T \tag{1.72}$$

from Eq. (1.68) – which becomes merely

$$\left(\frac{\partial P}{\partial T}\right)_S = \frac{C_P}{T}\left(\frac{\partial T}{\partial V}\right)_P \tag{1.73}$$

following cancellation of $(\partial V/\partial T)_P(\partial P/\partial V)_T$ with its negative; if reciprocals are taken of both sides, then Eq. (1.73) turns to

$$\frac{1}{\left(\frac{\partial P}{\partial T}\right)_S} = \frac{T}{C_P\left(\frac{\partial T}{\partial V}\right)_P} \tag{1.74}$$

– or, once more with the aid of the rule of differentiation of an inverse function,

$$\boxed{\left(\frac{\partial T}{\partial P}\right)_S = \frac{T}{C_p}\left(\frac{\partial V}{\partial T}\right)_P}. \tag{1.75}$$

Inspection of Eq. (1.75) – classically taken as definition of Joule and Thomson's (isentropic) coefficient, unfolds $(\partial T/\partial P)_S > 0$ because $C_P > 0$ by definition and $(\partial V/\partial T)_P > 0$ for all fluids; this positive change in T – as measure of thermal kinetic energy, upon a pressure increase accompanies supply of external work by the compressor.

All that is needed at this stage to proceed further is a relationship between volume and temperature (under a constant pressure) – as may be provided by a (standard) equation of state; in the case of liquids, however, one preferably resorts to the thermal expansion coefficient, α_P, defined as

$$\boxed{\alpha_P \equiv \frac{1}{V}\left(\frac{\partial V}{\partial T}\right)_P}, \tag{1.76}$$

since values thereof are more readily available. After having isolated $(\partial V/\partial T)_P$ in Eq. (1.76), one obtains

$$\left(\frac{\partial T}{\partial P}\right)_S = \frac{\alpha_P V}{C_P}T \tag{1.77}$$

from Eq. (1.75) – or, upon division of both numerator and denominator by mass m,

$$\left(\frac{\partial T}{\partial P}\right)_S = \alpha_P \frac{\frac{V}{m}}{\frac{C_P}{m}}T; \tag{1.78}$$

since ρ is defined as m/V and c_P is the intensive counterpart of C_P, Eq. (1.78) may be reformulated to

$$\left(\frac{\partial T}{\partial P}\right)_S = \frac{\alpha_P}{\rho c_P}T. \tag{1.79}$$

A reference state can meanwhile be defined as

$$T\big|_{P=P_0} = T_0, \tag{1.80}$$

corresponding to room conditions of pressure and temperature; assuming that α_P, c_P, and ρ do not change significantly over the working pressure range, one may integrate Eq. (1.79), through separation of variables, as

$$\left(\int_{T_0}^{T}\frac{d\tilde{T}}{T} = \frac{\alpha_P}{\rho c_P}\int_{P_0}^{P}d\tilde{P}\right)_S, \tag{1.81}$$

with Eq. (1.80) serving as a boundary condition. Equation (1.81) gives rise to

$$\ln\tilde{T}\Big|_{T_0}^{T} = \frac{\alpha_P}{\rho c_P}\tilde{P}\Big|_{P_0}^{P} \tag{1.82}$$

via the fundamental theorem of integral calculus, or else

$$\ln\frac{T}{T_0} = \frac{\alpha_P}{\rho c_P}(P-P_0); \tag{1.83}$$

upon taking exponentials of both sides and multiplying them by T_0 afterward, Eq. (1.83) gives rise to

$$\boxed{T = T_0\exp\left\{\frac{\alpha_P}{\rho c_P}(P-P_0)\right\}} \tag{1.84}$$

– as plotted in Fig. 1.18, for the specific case of water. Temperature increases almost linearly with pressure applied – at a rate of the order of 2.3 K per 100 MPa.

The simple graphical pattern in Fig. 1.18 is a consequence of the very small value of $\alpha_P/\rho c_P$, meaning that the argument of the exponential function in Eq. (1.84) is small – so this function may be safely expanded via Taylor's series around $P=P_0$, truncated after the linear term, viz.

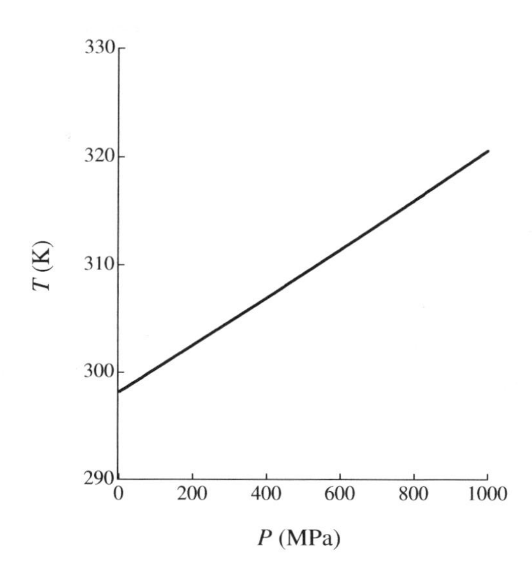

FIGURE 1.18 Variation of temperature, T, versus applied pressure, P, induced by hyperbaric treatment of water – characterized by thermal expansion coefficient $\alpha_P=3.03\times10^{-4}$ K^{-1}, mass density $\rho=9.97\times10^{2}$ kg.m^{-3}, and specific heat capacity $c_P=4.184\times10^{3}$ J.kg^{-1}.K^{-1}, originally at room temperature ($T_0=298$ K) and pressure ($P_0=1.01\times10^{5}$ Pa).

$$T \approx T_0\exp\left\{\frac{\alpha_P}{\rho c_P}(P-P_0)\right\}\Bigg|_{P=P_0} + T_0\exp\left\{\frac{\alpha_P}{\rho c_P}(P-P_0)\right\}\frac{\alpha_P}{\rho c_P}\Bigg|_{P=P_0}(P-P_0) \tag{1.85}$$

or, equivalently,

$$T = T_0 + \frac{\alpha_P T_0}{\rho c_P}(P-P_0); \tag{1.86}$$

straightforward algebraic rearrangement then leaves

$$T = T_0\left(1-\frac{\alpha_P P_0}{\rho c_P}\right) + \frac{\alpha_P T_0}{\rho c_P}P \tag{1.87}$$

as descriptor of the approximately straight line $T\equiv T\{P\}$ plotted in Fig. 1.18.

PROBLEM 1.2

Calculate Joule and Thomson's isenthalpic coefficient, compare with its isentropic counterpart, and discuss its physical meaning.

Solution

After recalling that the coefficient under scrutiny reads $(\partial T/\partial P)_H$ using Eq. (1.75) as template, one may resort to the general property of combined partial derivatives to write

$$\left(\frac{\partial T}{\partial P}\right)_H\left(\frac{\partial P}{\partial H}\right)_T\left(\frac{\partial H}{\partial T}\right)_P = -1 \tag{1.2.1}$$

– which readily yields

$$\left(\frac{\partial T}{\partial P}\right)_H = -\frac{1}{\left(\frac{\partial P}{\partial H}\right)_T \left(\frac{\partial H}{\partial T}\right)_P}, \tag{1.2.2}$$

upon isolation of $(\partial T/\partial P)_H$; the rule of differentiation of the inverse function then supports

$$\left(\frac{\partial T}{\partial P}\right)_H = -\left(\frac{\partial H}{\partial P}\right)_T \left(\frac{\partial T}{\partial H}\right)_P. \tag{1.2.3}$$

On the other hand, one may retrieve Eq. (3.21) and transform it with the aid of Eq. (3.19), both from *Food Proc. Eng.: basics & mechanical operations*, to get

$$dH = (TdS - PdV) + PdV + VdP, \tag{1.2.4}$$

or else

$$dH = TdS + VdP \tag{1.2.5}$$

after dropping symmetrical terms; division of both sides by dP, while keeping T constant, gives then rise to

$$\left(\frac{\partial H}{\partial P}\right)_T = T\left(\frac{\partial S}{\partial P}\right)_T + V. \tag{1.2.6}$$

Equation (1.58) may now be recalled to transform Eq. (1.2.6) to

$$\left(\frac{\partial H}{\partial P}\right)_T = V - T\left(\frac{\partial V}{\partial T}\right)_P, \tag{1.2.7}$$

thus allowing conversion of Eq. (1.2.3) to

$$\left(\frac{\partial T}{\partial P}\right)_H = -\left(V - T\left(\frac{\partial V}{\partial T}\right)_P\right)\left(\frac{\partial T}{\partial H}\right)_P. \tag{1.2.8}$$

After revisiting Eq. (7.121) of *Food Proc. Eng.: thermal & chemical operations* as

$$\frac{1}{C_P} = \left(\frac{\partial T}{\partial H}\right)_P, \tag{1.2.9}$$

one obtains

$$\left(\frac{\partial T}{\partial P}\right)_H = \frac{1}{C_P}\left(T\left(\frac{\partial V}{\partial T}\right)_P - V\right) \tag{1.2.10}$$

from Eq. (1.2.8) – or, equivalently,

$$\left(\frac{\partial T}{\partial P}\right)_H = \frac{T}{C_P}\left(\frac{\partial V}{\partial T}\right)_P - \frac{V}{C_P}, \tag{1.2.11}$$

upon elimination of parenthesis. In view of Eqs. (1.75) and (1.2.11), one realizes that

$$\left(\frac{\partial T}{\partial P}\right)_H = \left(\frac{\partial T}{\partial P}\right)_S - \frac{V}{C_P}, \tag{1.2.12}$$

i.e. Joule and Thomson's isenthalpic coefficient is lower than the corresponding isentropic coefficient – because $V>0$ and $C_P>0$, so $-V/C_P<0$; hence, temperature will increase at a slower pace under isenthalpic conditions for a given pressure increase – or (more frequently) cooling will be less pronounced upon free (throttling) expansion of a fluid, than if work were recovered from the system during the said expansion. Temperature is a measure of degraded energy, or energy associated with random motion of molecules; whereas pressure is a measure of upgraded energy, or energy associated with ordered motion of, and interaction among molecules. Even when heat as form of energy transfer is absent – as happens in both isenthalpic and isentropic processes, irreversible dissipation of mechanical energy in the former case will not allow a variation in system temperature as large as in the isentropic case (where no such dissipation exists).

1.1.3.3 Pressure and temperature effects

A few physicochemical processes are rather fast – either in absolute terms, or (more often) relative to other physicochemical phenomena taking place in series; chemical equilibrium will thus prevail therein at essentially all times. Examples include reactions consisting of minor structural modifications of a molecule – e.g. slight conformational changes of the polypeptide backbone of enzyme molecules, which affect catalytic activity but are usually fast and reversible (as a consequence of the associated low activation energies); or change in intensity of intermolecular interactions that drive phase changes (again associated to relatively low activation energies); or even transformations involving very small substrates – as happens with proton exchange in acid/base reactions (because protons diffuse quite fast). In equilibrium-controlled reactions, one may resort to Le Châtelier's principle to qualitatively describe the expected evolution – as per Eq. (1.41) for temperature, Eq. (1.42) for pressure, and Eq. (1.43) for mole fraction; an accurate quantitative description of the ultimate (or current) state will in either case require retrieval of the concept of (chemical) equilibrium constant.

Recall the fundamental definition of Gibbs' energy in differential form, i.e.

$$dG = V\,dP - S\,dT + \sum_{i=1}^{N} \mu_i\,dn_i \tag{1.88}$$

in parallel to Eq. (3.35) of *Food Proc. Eng.: basics & mechanical operations*; under constant temperature and pressure, Eq. (1.88) reduces to

$$\left(dG = \sum_{i=1}^{N} \mu_i\,dn_i\right)_{P,T}. \tag{1.89}$$

One may, for convenience, rewrite Eq. (1.89) as

$$dG = \sum_{i=1}^{N} \mu_i\,dn_i - \sum_{i=1}^{N} \mu_i^{\theta}\,dn_i + \sum_{i=1}^{N} \mu_i^{\theta}\,dn_i, \tag{1.90}$$

via subtraction and addition of $\sum_{i=1}^{N} \mu_i^\theta dn_i$, with superscript θ referring to pure compound at the same temperature – where the first two summations may be combined to give

$$dG = \sum_{i=1}^{N} \mu_i^\theta dn_i + \sum_{i=1}^{N} \left(\mu_i - \mu_i^\theta\right) dn_i; \tag{1.91}$$

dG may now be eliminated from Eqs. (1.89) and (1.91) as

$$\sum_{i=1}^{N} \mu_i dn_i = \sum_{i=1}^{N} \mu_i^\theta dn_i + \sum_{i=1}^{N} \left(\mu_i - \mu_i^\theta\right) dn_i. \tag{1.92}$$

After revisiting Eq. (1.35) as

$$dn_i = V\nu_i d\chi, \tag{1.93}$$

one may redo Eq. (1.92) to

$$V\left(\sum_{i=1}^{N} \nu_i \mu_i\right) d\chi = V\left(\sum_{i=1}^{N} \nu_i \mu_i^\theta\right) d\chi + V\left(\sum_{i=1}^{N} \nu_i \left(\mu_i - \mu_i^\theta\right)\right) d\chi; \tag{1.94}$$

once both sides have been divided by $Vd\chi$, Eq. (1.94) reduces to

$$\sum_{i=1}^{N} \nu_i \mu_i = \sum_{i=1}^{N} \nu_i \mu_i^\theta + \sum_{i=1}^{N} \nu_i \left(\mu_i - \mu_i^\theta\right), \tag{1.95}$$

or else

$$\Delta g = \Delta g^\theta + \sum_{i=1}^{N} \nu_i \left(\mu_i - \mu_i^\theta\right) \tag{1.96}$$

– with the aid of Eq. (1.1) applied in general, and to the reference state in particular. After revisiting Eq. (3.155) of *Food Proc. Eng.: basics & mechanical operations* as

$$\left(\ln a_i \equiv \frac{\mu_i - \mu_i^\theta}{\mathcal{R}T}\right)_T, \tag{1.97}$$

one may redo Eq. (1.96) to

$$\Delta g = \Delta g^\theta + \mathcal{R}T \sum_{i=1}^{N} \nu_i \ln a_i \tag{1.98}$$

– where $\mathcal{R}T$ was meanwhile factored out for being constant; Eq. (1.98) can be algebraically manipulated to read

$$\frac{\Delta g - \Delta g^\theta}{\mathcal{R}T} = \sum_{i=1}^{N} \nu_i \ln a_i. \tag{1.99}$$

The operational features of a logarithmic function support transformation of Eq. (1.99) to

$$\sum_{i=1}^{N} \ln a_i^{\nu_i} = \frac{\Delta g - \Delta g^\theta}{\mathcal{R}T}, \tag{1.100}$$

and further to

$$\ln \prod_{i=1}^{N} a_i^{\nu_i} = \frac{\Delta g - \Delta g^\theta}{\mathcal{R}T}; \tag{1.101}$$

at chemical equilibrium (denoted by subscript *eq*),

$$\boxed{\left.(\Delta g = 0)\right|_{eq}} \tag{1.102}$$

in agreement with Eq. (1.2). In this case, Eq. (1.101) becomes

$$\left.\left(\ln \prod_{i=1}^{N} a_i^{\nu_i} = -\frac{\Delta g^\theta}{\mathcal{R}T}\right)\right|_{eq}, \tag{1.103}$$

or instead

$$\boxed{K_{eq} = \prod_{i=1}^{N} a_{i,eq}^{\ \nu_i}} \tag{1.104}$$

after taking exponentials of both sides – and known as Guldberg and Waage's law of mass action; provided that

$$\boxed{K_{eq} \equiv \exp\left\{-\frac{\Delta g^\theta}{\mathcal{R}T}\right\},} \tag{1.105}$$

with $K_{eq} \equiv K_{eq}\{T\}$ denoting (dimensionless) equilibrium constant. On the other hand, Eq. (3.156) of *Food Proc. Eng.: basics & mechanical operations* may be recalled to transform Eq. (1.104) to

$$K_{eq} = \prod_{i=1}^{N} \left(\gamma_{i,eq} x_{i,eq}\right)^{\nu_i} = \prod_{i=1}^{N} \gamma_{i,eq}^{\ \nu_i} x_{i,eq}^{\ \nu_i}. \tag{1.106}$$

The molar fraction of the i-th component, as per Eq. (3.44) of *Food Proc. Eng.: basics & mechanical operations*, may be used to write

$$x_i \equiv \frac{n_i}{n} = \frac{\frac{n_i}{V}}{\frac{n}{V}} \equiv \frac{C_i}{C} \tag{1.107}$$

upon division of both numerator and denominator by V – where C denotes total molar concentration (usually constant, especially due to the contribution of water as solvent), i.e.

$$C \equiv \frac{n}{V}, \tag{1.108}$$

together with

$$\boxed{C_i \equiv \frac{n_i}{V}; \quad i = 1, 2, \ldots, N} \tag{1.109}$$

standing as molar concentration of i-th component; hence, one may reformulate Eq. (1.106) to

$$K_{eq} = \prod_{i=1}^{N} \gamma_{i,eq}^{\ \nu_i} \prod_{i=1}^{N} \left(\frac{C_{i,eq}}{C}\right)^{\nu_i} \tag{1.110}$$

– with the aid of Eq. (1.107), which becomes

$$K_{eq} \prod_{i=1}^{N} C^{\nu_i} = \prod_{i=1}^{N} \gamma_{i,eq}^{\ \nu_i} \prod_{i=1}^{N} C_{i,eq}^{\ \nu_i} \tag{1.111}$$

after multiplying both sides by $\prod_{i=1}^{N} C^{\nu_i}$. The definition of power allows notation be simplified to

$$K_{eq}C^{\sum_{i=1}^{N}\nu_i} = \prod_{i=1}^{N} \gamma_{i,eq}{}^{\nu_i} \prod_{i=1}^{N} C_{i,eq}{}^{\nu_i} \tag{1.112}$$

in Eq. (1.111), which is equivalent to

$$K_{eq}C^{\Delta\nu} = \prod_{i=1}^{N} \gamma_{i,eq}{}^{\nu_i} \prod_{i=1}^{N} C_{i,eq}{}^{\nu_i} \tag{1.113}$$

as long as

$$\Delta\nu \equiv \sum_{i=1}^{N} \nu_i; \tag{1.114}$$

a pseudo-chemical equilibrium constant, K_{eq}^{*} (with units of $mol^{\Delta\nu}.m^{-3\Delta\nu}$ in SI), has classically been defined as

$$K_{eq}^{*} \equiv K_{eq}C^{\Delta\nu}, \tag{1.115}$$

thus allowing Eq. (1.113) be rephrased as

$$\boxed{\prod_{i=1}^{N} \gamma_{i,eq}{}^{\nu_i} \prod_{i=1}^{N} C_{i,eq}{}^{\nu_i} = K_{eq}^{*}.} \tag{1.116}$$

A general functional form for γ_i on P, T, and x_i's is seldom available – yet oftentimes the effect of individual γ_i's is diluted in $\prod_{i=1}^{N} \gamma_{i,eq}{}^{\nu_i}$; hence, it is common practice to simplify Eq. (1.116) to

$$\prod_{i=1}^{N} C_{i,eq}{}^{\nu_i} \approx K_{eq}^{*} \tag{1.117}$$

that implicitly assumes $\prod_{i=1}^{N} \gamma_{i,eq}{}^{\nu_i} \approx 1$. In view of Eq. (1.115), one may write

$$\boxed{\ln K_{eq}^{*}\{P,T\} \equiv \Delta\nu \ln C - \frac{\Delta g^{\theta}\{P,T\}}{\mathcal{R}T}} \tag{1.118}$$

after taking logarithms of both sides, combined with Eq. (1.105).

The composition of the system at equilibrium, i.e. $C_{1,eq}$, $C_{2,eq}$, …, $C_{N,eq}$, is described by K_{eq}^{*} based on Eq. (1.116) – so all that is left is ascertaining how $\ln K_{eq}^{*}$ varies with pressure and temperature, i.e.

$$\left(\frac{\partial \ln K_{eq}^{*}}{\partial T}\right)_P = -\frac{1}{\mathcal{R}}\left(\frac{\partial}{\partial T}\left(\frac{\Delta g^{\theta}\{P,T\}}{T}\right)\right)_P \tag{1.119}$$

and

$$\left(\frac{\partial \ln K_{eq}^{*}}{\partial P}\right)_T = -\frac{1}{\mathcal{R}T}\left(\frac{\partial \Delta g^{\theta}\{P,T\}}{\partial P}\right)_T, \tag{1.120}$$

respectively, both based on partial differentiation of both sides of Eq. (1.118). After invoking Eq. (1.1) at the reference state, one may rewrite Eq. (1.119) as

$$\frac{\partial \ln K_{eq}^{*}}{\partial T} = -\frac{1}{\mathcal{R}}\frac{\partial}{\partial T}\left(\frac{\sum_{i=1}^{N} \nu_i \mu_i^{\theta}}{T}\right) = -\frac{1}{\mathcal{R}}\sum_{i=1}^{N} \nu_i \frac{\partial}{\partial T}\left(\frac{\mu_i^{\theta}}{T}\right), \tag{1.121}$$

as per the linearity of the differential operator; insertion of Eq. (1.17) gives rise to

$$\frac{\partial \ln K_{eq}^{*}}{\partial T} = \frac{1}{\mathcal{R}}\sum_{i=1}^{N} \nu_i \frac{h_i^{\theta}}{T^2} = \frac{1}{\mathcal{R}T^2}\sum_{i=1}^{N} \nu_i h_i^{\theta}, \tag{1.122}$$

along with realization that T^2 is not a function of the counting variable of the summation – or, in view of Eq. (1.31),

$$\boxed{\left(\frac{\partial \ln K_{eq}^{*}}{\partial T}\right)_P = \frac{\Delta h^{\theta}}{\mathcal{R}T^2}.} \tag{1.123}$$

Equation (1.123) is known as van't Hoff's law; in other words, K_{eq}^{*} increases with temperature if the reaction is endothermic (i.e. $\Delta h^{\theta} > 0$), and decreases otherwise – in plain agreement with Le Châtelier's principle, see Eq. (1.41). By the same token, one may retrieve Eqs. (1.1) and (1.120) as

$$\frac{\partial \ln K_{eq}^{*}}{\partial P} = -\frac{1}{\mathcal{R}T}\frac{\partial}{\partial P}\left(\sum_{i=1}^{N} \nu_i \mu_i^{\theta}\right) = -\frac{1}{\mathcal{R}T}\sum_{i=1}^{N} \nu_i \frac{\partial \mu_i^{\theta}}{\partial P}, \tag{1.124}$$

again at the expense of the derivative of a linear combination of functions coinciding with the linear combination of derivatives thereof; insertion of Eq. (3.17) of *Food Proc. Eng.: thermal & chemical operations* produces

$$\frac{\partial \ln K_{eq}^{*}}{\partial P} = -\frac{1}{\mathcal{R}T}\sum_{i=1}^{N} \nu_i v_i^{\theta}, \tag{1.125}$$

where Eq. (1.32) may be recalled to obtain

$$\boxed{\left(\frac{\partial \ln K_{eq}^{*}}{\partial P}\right)_T = -\frac{\Delta v^{\theta}}{\mathcal{R}T}.} \tag{1.126}$$

Therefore, a physicochemical transformation undergoing expansion, i.e. $\Delta v^{\theta} > 0$, decreases its characteristic equilibrium constant K_{eq}^{*} when pressure increases – because $\partial \ln K_{eq}^{*}/\partial P < 0$ in Eq. (1.126); conversely, K_{eq}^{*} increases with pressure when Δv^{θ} is negative, since $\partial \ln K_{eq}/\partial P > 0$ – again in general agreement with Le Châtelier's principle, see Eq. (1.42).

PROBLEM 1.3

Despite the (relative) simplicity of the thermodynamic approach developed previously, most reactions actually proceed at finite rates – so dependence of their kinetic constants upon pressure and temperature is germane. Unfortunately, the theoretical background in this case lags far behind in generality and mathematical form, and experimental validation is also far scarcer – with the activation state theory providing the most widely accepted rationale. Derive expressions for $\partial k/\partial T$ and $\partial k/\partial P$ based on Eyring's theory for the activated state, i.e.

$$k = \frac{k_b}{k_p} T \exp\left\{-\frac{\Delta g_{act}^{\theta}}{\mathcal{R}T}\right\} \quad (1.3.1)$$

– where k denotes intrinsic kinetic constant of an (elementary) chemical step, k_b and k_p denote Boltzmann's and Planck's constants, respectively, and Δg_{act}^{θ} denotes Gibbs' activation energy of the activated complex (under standard conditions).

Solution

As done previously with regard to molar Gibbs' energy change accompanying a whole physicochemical transformation under chemical equilibrium, one may compute molar Gibbs' energy change associated only with an elementary step, described by Δg_{act}^{θ}. One should accordingly resort to Eq. (1.3.1) and apply logarithms to both sides to get

$$\ln k\{P,T\} = \ln\frac{k_b}{k_p} + \ln T - \frac{\Delta g_{act}^{\theta}\{P,T\}}{\mathcal{R}T}; \quad (1.3.2)$$

Arrhenius' (empirical) approximation is often used, which basically neglects the linear effect of T upon k, and thus lumps k_b, k_p, and T in Eq. (1.3.1) into a single pre-exponential, constant parameter – say, k_0. The variation of k, as per Eq. (1.3.2), with temperature should thus be given by

$$\left(\frac{\partial \ln k}{\partial T}\right)_P = \left(\frac{\partial \ln T}{\partial T}\right)_P - \frac{1}{\mathcal{R}}\left(\frac{\partial}{\partial T}\left(\frac{\Delta g_{act}^{\theta}}{T}\right)\right)_P, \quad (1.3.3)$$

upon (partial) differentiation of both sides under constant pressure; a similar result may be obtained with regard to isothermal dependence on pressure, i.e.

$$\left(\frac{\partial \ln k}{\partial P}\right)_T = -\frac{1}{\mathcal{R}T}\left(\frac{\partial \Delta g_{act}^{\theta}}{\partial P}\right)_T. \quad (1.3.4)$$

Since Δg_{act}^{θ} is, in this case, given by

$$\Delta g_{act}^{\theta} = \mu_{\#}^{\theta}\{P,T\} + \sum_{i=1}^{R} \nu_i \mu_i^{\theta}\{P,T\} \quad (1.3.5)$$

– where subscript # denotes activated state and R denotes number of reactants leading thereto, one may redo Eq. (1.3.3) as

$$\frac{\partial \ln k}{\partial T} = \frac{1}{T} - \frac{1}{\mathcal{R}}\frac{\partial}{\partial T}\left(\frac{\mu_{\#}^{\theta} + \sum_{i=1}^{R} \nu_i \mu_i^{\theta}}{T}\right), \quad (1.3.6)$$

which degenerates to

$$\frac{\partial \ln k}{\partial T} = \frac{1}{T} - \frac{1}{\mathcal{R}}\frac{\partial}{\partial T}\left(\frac{\mu_{\#}^{\theta}}{T}\right) - \frac{1}{\mathcal{R}}\sum_{i=1}^{R} \nu_i \frac{\partial}{\partial T}\left(\frac{\mu_i^{\theta}}{T}\right). \quad (1.3.7)$$

Insertion of the analogue to Eq. (1.17) supports transformation of Eq. (1.3.7) to

$$\frac{\partial \ln k}{\partial T} = \frac{1}{T} + \frac{1}{\mathcal{R}}\frac{h_{\#}^{\theta}}{T^2} + \frac{1}{\mathcal{R}}\sum_{i=1}^{R} \nu_i \frac{h_i^{\theta}}{T^2} = \frac{1}{T} + \frac{h_{\#}^{\theta}}{\mathcal{R}T^2} + \frac{1}{\mathcal{R}T^2}\sum_{i=1}^{R} \nu_i h_i^{\theta} \quad (1.3.8)$$

– where the linearity of the differential operator was meanwhile taken advantage of, and constants with regard to the summation counting variable were factored out; after inserting Eq. (1.31), one finally obtains

$$\left(\frac{\partial \ln k}{\partial T}\right)_P = \frac{1}{T} + \frac{\Delta h_{\#}^{\theta}}{\mathcal{R}T^2}. \quad (1.3.9)$$

In the case of Eq. (1.3.4), one finds that

$$\frac{\partial \ln k}{\partial P} = -\frac{1}{\mathcal{R}T}\frac{\partial}{\partial P}\left(\mu_{\#}^{\theta} + \sum_{i=1}^{R} \nu_i \mu_i^{\theta}\right) \quad (1.3.10)$$

with the aid again of Eq. (1.3.5), which may be algebraically rearranged to read

$$\frac{\partial \ln k}{\partial P} = -\frac{1}{\mathcal{R}T}\frac{\partial \mu_{\#}^{\theta}}{\partial P} - \frac{1}{\mathcal{R}T}\sum_{i=1}^{R} \nu_i \frac{\partial \mu_i^{\theta}}{\partial P}; \quad (1.3.11)$$

in view of Eq. (3.17) of *Food Proc. Eng.: thermal & chemical operations*, one obtains

$$\frac{\partial \ln k}{\partial P} = -\frac{1}{\mathcal{R}T} v_{\#}^{\theta} - \frac{1}{\mathcal{R}T}\sum_{i=1}^{R} \nu_i v_i^{\theta}, \quad (1.3.12)$$

where Eq. (1.32) supports further transformation to

$$\left(\frac{\partial \ln k}{\partial P}\right)_T = -\frac{\Delta v_{\#}^{\theta}}{\mathcal{R}T}. \quad (1.3.13)$$

The conclusions drawn for kinetic performance are thus identical to those drawn with regard to thermodynamic performance; an endothermic transformation (characterized by $\Delta h_{\#}^{\theta} > 0$) occurs faster, due to $\partial \ln k/\partial T > 0$ in Eq. (1.3.9), should temperature be raised – while an exothermic transformation behaves otherwise. In the case of pressure, a physicochemical transformation leading to expansion of the activated state relative to reactants, i.e. $\Delta v_{\#}^{\theta} > 0$, becomes slower upon a pressure increase, because $\partial \ln k/\partial P < 0$ in Eq. (1.3.13) – while k increases with pressure when $\Delta v_{\#}^{\theta}$ is negative owing to $\partial \ln k/\partial P > 0$.

1.1.3.4 Water activity

1.1.3.4.1 Thermodynamic approach

Consider, for the sake of simplicity, that a food item is constituted solely by water W and an essentially nonvolatile solute S. According to Eq. (3.103) of *Food Proc. Eng.: basics & mechanical operations*, chemical equilibrium of water between liquid and gaseous phases is described by

$$\boxed{\left(\mu_{W,l}\Big|_P = \mu_{W,g}\Big|_P\right)_T}, \tag{1.127}$$

at the given temperature and pressure – where the same temperature in both phases guarantees simultaneous thermal equilibrium as per Eq. (3.105) of *Food Proc. Eng.: basics & mechanical operations*, and the same pressure in both phases ensures mechanical equilibrium as per Eq. (3.104) also of *Food Proc. Eng.: basics & mechanical operations*. Here subscript l refers to liquid phase, and subscript g refers to vapor phase – while superscript θ refers to pure compound, at the same pressure and temperature of the food (mixture). Recalling Eq. (3.155) of *Food Proc. Eng.: basics & mechanical operations*, one may redo Eq. (1.127) to

$$\mu_{W,l}^{\theta}\Big|_P + \mathcal{R}T\ln a_{W,l} = \mu_{W,g}^{\theta}\Big|_P + \mathcal{R}T\ln a_{W,g} \tag{1.128}$$

– explicit on the activity of water, a_W, in either liquid or gaseous form, and applicable at a given temperature T; insertion of Eq. (3.156) of *Food Proc. Eng.: basics & mechanical operations* in the right-hand side allows reformulation of Eq. (1.128) to

$$\mu_{W,l}^{\theta}\Big|_P + \mathcal{R}T\ln a_{W,l} = \mu_{W,g}^{\theta}\Big|_P + \mathcal{R}T\ln \gamma_{W,g}x_{W,g}, \tag{1.129}$$

where $\gamma_{W,g}$ and $x_{W,g}$ denote activity coefficient and mole fraction, respectively, of water vapor. Under the pressures and temperatures usually of interest in food processing, water vapor behaves ideally when mixed with air – so $\gamma_{W,g} \approx 1$, and Eq. (1.129) simplifies to

$$\mu_{W,l}^{\theta}\Big|_P + \mathcal{R}T\ln a_{W,l} = \mu_{W,g}^{\theta}\Big|_P + \mathcal{R}T\ln x_{W,g}; \tag{1.130}$$

the same statement does not hold for the liquid phase in general, however – due to the much higher complexity of liquid behavior, in terms of intermolecular interactions. The (isothermal) variation of chemical potential of a pure liquid compound with pressure abides to

$$\left(\frac{\partial \mu_{W,l}^{\theta}}{\partial P}\right)_T = v_{W,l}^{\theta} \tag{1.131}$$

in agreement with Eq. (3.17) of *Food Proc. Eng.: thermal & chemical operations* – yet $v_{W,l}^{\theta}$ is, in practice, sufficiently small to support

$$\boxed{\left(\frac{\partial \mu_{W,l}^{\theta}}{\partial P}\right)_T \approx 0}; \tag{1.132}$$

Eq. (1.132) obviously implies

$$\mu_{W,l}^{\theta}\Big|_P \approx \mu_{W,l}^{\theta}\Big|_{P_W^{\sigma}}, \tag{1.133}$$

where P_W^{σ} denotes vapor pressure of water at the given temperature. By the same token, one realizes that

$$\boxed{\left(\frac{\partial \mu_{W,g}^{\theta}}{\partial P}\right)_T = v_{W,g}^{\theta}} \tag{1.134}$$

is applicable to the gaseous phase – but $v_{W,g}^{\theta}$ can hardly be neglected in this case, as $v_{W,g}^{\theta} >> v_{W,l}^{\theta}$; integration of Eq. (1.134) should then proceed via separation of variables as

$$\int_{\mu_{W,g}^{\theta}|_P}^{\mu_{W,g}^{\theta}|_{P_W^{\sigma}}} d\mu_{W,g}^{\theta} = \int_P^{P_W^{\sigma}} v_{W,g}^{\theta}d\tilde{P}. \tag{1.135}$$

In view of Eq. (3.135) of *Food Proc. Eng.: basics & mechanical operations* – describing ideal state behavior of water as vapor, one has it that

$$\boxed{v_{W,g}^{\theta} = \frac{\mathcal{R}T}{P}} \tag{1.136}$$

with $v_{W,g}^{\theta}$ given by the ratio of volume available to the number of moles of water vapor; one may accordingly reformulate Eq. (1.135) to read

$$\int_{\mu_{W,g}^{\theta}|_P}^{\mu_{W,g}^{\theta}|_{P_W^{\sigma}}} d\mu_{W,g}^{\theta} = \mathcal{R}T\int_P^{P_W^{\sigma}} \frac{d\tilde{P}}{\tilde{P}}, \tag{1.137}$$

where $\mathcal{R}T$ was moved off the kernel for its constancy (since T has been hypothesized as constant). Application of the fundamental theorem of integral calculus to Eq. (1.137) unfolds

$$\mu_{W,g}^{\theta}\Big|_{\mu_{W,g}^{\theta}|_P}^{\mu_{W,g}^{\theta}|_{P_W^{\sigma}}} = \mathcal{R}T\ln\tilde{P}\Big|_P^{P_W^{\sigma}}, \tag{1.138}$$

or else

$$\mu_{W,g}^{\theta}\Big|_{P_W^{\sigma}} - \mu_{W,g}^{\theta}\Big|_P = \mathcal{R}T\ln\frac{P_W^{\sigma}}{P}; \tag{1.139}$$

upon isolation of $\mu_{W,g}^{\theta}\big|_P$, Eq. (1.139) becomes

$$\mu_{W,g}^{\theta}\Big|_P = \mu_{W,g}^{\theta}\Big|_{P_W^{\sigma}} + \mathcal{R}T\ln\frac{P}{P_W^{\sigma}}, \tag{1.140}$$

where the negative sign preceding the logarithm was removed while taking the reciprocal of its argument. Insertion of Eqs. (1.133) and (1.140) transforms Eq. (1.130) to

$$\mu_{W,l}^{\theta}\Big|_{P_W^{\sigma}} + \mathcal{R}T\ln a_{W,l} = \mu_{W,g}^{\theta}\Big|_{P_W^{\sigma}} + \mathcal{R}T\ln\frac{P}{P_W^{\sigma}} + \mathcal{R}T\ln x_{W,g}, \tag{1.141}$$

where the original (arbitrary) pressure, P, was related to the vapor pressure of water, P_W^{σ}. Equation (3.103) of *Food Proc. Eng.: basics & mechanical operations* enforces

$$\boxed{\mu_{W,l}^{\theta}\Big|_{P_W^{\sigma}} = \mu_{W,g}^{\theta}\Big|_{P_W^{\sigma}}} \tag{1.142}$$

pertaining to liquid/vapor equilibrium of water, as implicit in the definition of P_W^{σ} at the same temperature considered so far; Eq. (1.141) accordingly reduces to

$$\mathcal{R}T\ln a_{W,l} = \mathcal{R}T\ln\frac{P}{P_W^{\sigma}} + \mathcal{R}T\ln x_{W,g}, \tag{1.143}$$

where division of both sides by $\mathcal{R}T$ permits further simplification to

$$\ln a_{W,l} = \ln\frac{P}{P_W^{\sigma}} + \ln x_{W,g} = \ln\frac{P}{P_W^{\sigma}}x_{W,g} \tag{1.144}$$

– also with the aid of the operational features of a logarithm. After taking exponentials of both sides, Eq. (1.144) becomes

$$a_{W,l} = \frac{P}{P_W^\sigma} x_{W,g}; \tag{1.145}$$

since the partial pressure of a gaseous component satisfies

$$P_W = x_{W,g} P \tag{1.146}$$

in an ideal solution, in parallel to Eq. (3.141) of *Food Proc. Eng.: basics & mechanical operations*, one may rewrite Eq. (1.145) as

$$\boxed{a_{W,l} = \frac{P_W}{P_W^\sigma}} \tag{1.147}$$

– which serves as definition for water activity in food. The concept of water activity possesses a few intrinsic advantages; for instance, both P_W and P_W^σ depend quite strongly on temperature – yet their ratio is relatively insensitive thereto; this is so because such a ratio approaches the exponential of the difference of the original arguments, as per the exponential dependence of P_W^σ upon T.

If $a_{W,l}$ is to be theoretically estimated, then some model will be required; one of the simplest is van Laar's model, viz.

$$\ln \gamma_{S,l} = \frac{A x_{W,l}^2}{\left(\frac{A}{B} x_{S,l} + x_{W,l}\right)^2} \tag{1.148}$$

and

$$\ln \gamma_{W,l} = \frac{B x_{S,l}^2}{\left(x_{S,l} + \frac{B}{A} x_{W,l}\right)^2} \tag{1.149}$$

pertaining to the activity coefficients of solute and water, respectively, in the liquid phase – where A and B denote adjustable parameters;

$$x_{S,l} = 1 - x_{W,l} \tag{1.150}$$

appears as a consequence of the definition of mole fraction of the two components in a binary mixture – see Eq. (3.47) of *Food Proc. Eng.: basics & mechanical operations*, so Eq. (1.149) becomes

$$\ln \gamma_{W,l} = \frac{B(1 - x_{W,l})^2}{\left(1 - x_{W,l} + \frac{B}{A} x_{W,l}\right)^2} = B\left(\frac{1 - x_{W,l}}{1 + \left(\frac{B}{A} - 1\right) x_{W,l}}\right)^2. \tag{1.151}$$

Recalling Eq. (3.156) of *Food Proc. Eng.: basics & mechanical operations* applied to liquid water, i.e.

$$a_{W,l} = \gamma_{W,l} x_{W,l}, \tag{1.152}$$

one may insert Eq. (1.151) in Eq. (1.152) to generate

$$\boxed{a_{W,l} = \exp\left\{B\left(\frac{1 - x_{W,l}}{1 + \left(\frac{B}{A} - 1\right) x_{W,l}}\right)^2\right\} x_{W,l};} \tag{1.153}$$

a graphical illustration of Eq. (1.153), in the form $x_{W,l} \equiv x_{W,l}\{a_{W,l}\}$, is provided in Fig. 1.19. The difference between $x_{W,l}$ and $a_{W,l}$ is accounted for by $\gamma_{W,l}\{x_{W,l}\}$ as per Eq. (1.152); in particular, when the liquid mixture is infinitely concentrated in S (and thus $x_{W,l}$ approaches zero), Eq. (1.151) is driven by

$$\lim_{x_{W,l} \to 0} \ln \gamma_{W,l} = B\left(\frac{1 - 0}{1 + \left(\frac{B}{A} - 1\right) 0}\right)^2 = B\left(\frac{1}{1}\right)^2 = B, \tag{1.154}$$

and consequently

$$\gamma_{W,l}^\infty \equiv \lim_{x_{W,l} \to 0} \gamma_{W,l} = \mathrm{e}^B \tag{1.155}$$

after taking exponentials of both sides – where $\gamma_{W,l}^\infty$ is known as activity coefficient at infinite dilution. In view of Eq. (1.155), one may redo Eq. (1.152) to

$$\lim_{x_{W,l} \to 0} a_{W,l} = \lim_{x_{W,l} \to 0} \gamma_{W,l} x_{W,l}, \tag{1.156}$$

or else

$$\lim_{x_{W,l} \to 0} \frac{a_{W,l}}{x_{W,l}} = \mathrm{e}^B \tag{1.157}$$

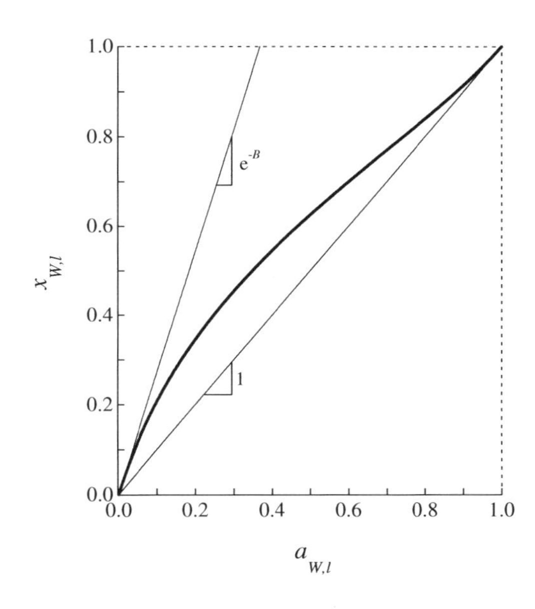

FIGURE 1.19 Variation of water mole fraction, $x_{W,l}$, versus water activity, $a_{W,l}$, in liquid system following van Laar's equation of state, characterized by parameters $A = -1.5$ and $B = -1$, with indication of low- and high-$a_{W,l}$ asymptotes, and corresponding slopes.

that justifies the slope of the low-$a_{W,l}$ asymptote in Fig. 1.19, i.e. $\lim_{a_{W,l}\to 0}\frac{x_{W,l}}{a_{W,l}}=\lim_{x_{W,l}\to 0}\frac{1}{a_{W,l}/x_{W,l}}=\frac{1}{e^{B}}=e^{-B}$ (since $x_{W,l}\to 0$ when $a_{W,l}\to 0$). Conversely, an infinitely concentrated liquid mixture in W (or infinitely diluted in S) supports $x_{W,l}\to 1$, so Eq. (1.151) becomes

$$\lim_{x_{W,l}\to 1}\ln\gamma_{W,l}=B\left(\frac{1-1}{1+\left(\frac{B}{A}-1\right)1}\right)^2=B\left(\frac{0}{\frac{B}{A}}\right)^2=0; \tag{1.158}$$

after taking exponentials of both sides, Eq. (1.158) turns to

$$\lim_{x_{W,l}\to 1}\gamma_{W,l}=e^{0}=1. \tag{1.159}$$

Equation (1.152) may thus be rewritten as

$$\lim_{x_{W,l}\to 1}a_{W,l}=\lim_{x_{W,l}\to 1}\gamma_{W,l}x_{W,l}, \tag{1.160}$$

which gives rise to

$$\lim_{x_{W,l}\to 1}\frac{a_{W,l}}{x_{W,l}}=1 \tag{1.161}$$

in view of Eq. (1.159); the unit slope of the large-a_W asymptote in Fig. 1.19 is described by Eq. (1.161), because $\lim_{a_{W,l}\to 1}\frac{x_{W,l}}{a_{W,l}}=\lim_{a_{W,l}\to 1}\frac{1}{a_{W,l}/x_{W,l}}=\frac{1}{1}=1$ (along again with $\lim_{x_{W,l}\to 1}a_{W,l}=1$). This asymptotic pattern has classically been known as Randall's rule.

In the case of multicomponent mixtures, constituted by, say, N solutes, one may resort to

$$\boxed{a_{W,l}\approx\prod_{i=1}^{N}a_{W,i,l}} \tag{1.162}$$

– as long as interactions between all components (except with water) are neglected; hence, the water activity of the food becomes merely the product of the water activities of every single solute, as if binary solutions were considered independently.

The above derivation departed from a uniphasic liquid solution, containing an (otherwise solid) solute – with water serving as solvent; deviation of $\gamma_{W,l}$ from unit is a consequence solely of interactions between the molecules of water with the molecules of the said solute – which also respond in magnitude to temperature (and pressure, to a lesser extent). However, a great many foods are biphasic, and constituted by a solid, porous matrix filled with an aqueous solution; hence, interactions between molecules of water and molecules of that solid material will also play a role – and will cause the activity coefficient of those aqueous solutions be distinct from unity. A number of factors contribute to this observation: water can adsorb on hydrophilic groups of the food matrix, thus sequentially forming adjacent layers that are less and less tightly bound; water in a wetted capillary is subjected to suction pressure; and osmotic pressure of solute diminishes vapor pressure of water. These three phenomena will be discussed below to some extent – via Brunauer, Emmett, and Teller's model for surface adsorption, Laplace's equation for capillary action, and van't Hoff's equation for osmotic pressure, respectively. The curves representing water concentration versus water activity (corresponding to an exchange of axes in Fig. 1.19) are often designated by sorption isotherms – because they are valid for a given temperature, and encompass not only (true) adsorption, but also any of the other two phenomena; however, attempts to apply a thermodynamic approach often resort only to the predominant phenomenon at stake. As will be seen, a sorption isotherm covering the whole range of water activities exhibits a shape compatible with the curve shown in Fig. 1.19, after converting mole fraction to concentration – and looks like the curves depicted in Fig. 1.3.

On the other hand, crystals may also appear in solid products, containing hydration water that exhibits a low vapor pressure; this is relevant in the case of lactose, or else crystalline sucrose – since it is much less hydrophilic than amorphous sucrose. In this case, the water activity of the food is controlled directly by $a_{W,l}$ of the said crystals – which is to be determined experimentally in each case.

PROBLEM 1.4

In freezing operations (including freeze-drying), the water activity still follows Eq. (1.147) – except that the numerator pertains to pure ice and the denominator to supercooled liquid water. Unfortunately, Antoine's equation is not accurate for either form within such a temperature range, so one should resort to Murphy and Koop's correlations, i.e.

$$\ln P_i^{\sigma}=9.5504-\frac{5723.3}{T}+3.5307\ln T-0.0072833T \tag{1.4.1}$$

for the vapor pressure of ice, and

$$\ln P_{sc}^{\sigma}=54.843-\frac{6763.2}{T}-4.210\ln T+0.000367T+\begin{pmatrix}52.878-\frac{1331.22}{T}\\-9.4452\ln T+0.014025T\end{pmatrix}\tanh 0.0415(T-218.8) \tag{1.4.2}$$

for the vapor pressure of supercooled liquid water. Plot a_W as a function of temperature, and comment on the result.

Solution

Insertion of Eqs. (1.4.1) as numerator and (1.4.2) as denominator in Eq. (1.147) produces the plots below, laid out as $a_W\equiv P_i^{\sigma}/P_{sc}^{\sigma}$ vs. T.

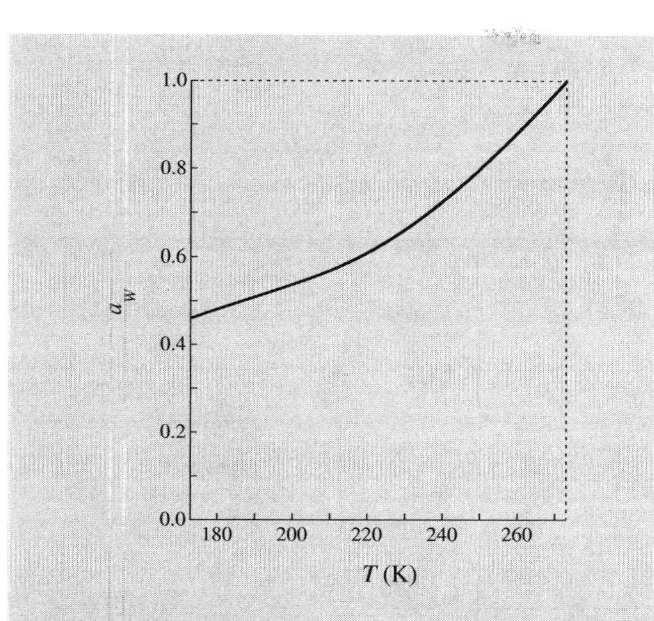

As expected, a_w becomes unity under true equilibrium between solid and liquid water that occurs at 1 atm and 0°C, or 273.15 K. Below that temperature, $a_w < 1$; a faster, linear decrease with decreasing temperature is observed in the neighborhood of the normal freezing point – consistent with the data points superimposed in Fig. 1.1, followed by a slower decrease at lower temperatures. The nature and concentration of solutes in aqueous solutions in equilibrium with ice serve only to fix the equilibrium temperature of the system; a_w accordingly depends just on the resulting temperature, irrespective of the constituents in the said aqueous mixture.

1.1.3.4.2 *Surface adsorption*

At low moisture content, Brunauer, Emmett, and Teller's model for adsorption of a vapor is quite useful – despite having been originally derived for adsorption of nonpolar gases on homogeneous surfaces, rather than polar water vapor adsorbing on heterogeneous food surfaces. This model assumes multilayer adsorption; the first layer is expected to undergo Langmuir's adsorption – with associated enthalpy change arising from interactions between adsorbate and adsorbent. The following layers are (putatively) taken as equivalent to each other – but characterized by lower enthalpy changes of adsorption, resulting from adsorbate/adsorbate interactions only. Therefore, the molecules of vapor distribute over various layers on top of each other; adsorption on the surface (or on a previous layer) is considered to be an elementary process, which occurs at a rate proportional to amount of vapor, and also proportional to amount of surface still available for intermolecular interactions – while desorption from the actual layer is also considered as an elementary process, occurring at a rate proportional to amount of adsorbed molecules.

If the fraction of total area, A, occupied by 0, 1, …, i, …, n layers of adsorbed water molecules is denoted as α_0, α_1, …, α_i, …, α_n, respectively, under an overhead pressure P of water vapor, then equilibrium conditions require

$$k_1 P \alpha_0 A = k_{-1} \alpha_1 A \tag{1.163}$$

– where the left-hand side represents rate of adsorption onto the bare surface, characterized by kinetic constant k_1, and the right-hand side represents rate of desorption from the first layer, characterized by kinetic constant k_{-1}. After cancellation of A between sides, and definition of a monolayer adsorption constant, $K_{1st,ads}$, as

$$\boxed{K_{1st,ads} \equiv \frac{k_1}{k_{-1}},} \tag{1.164}$$

one may rewrite Eq. (1.163) as

$$\alpha_1 = K_{1st,ads} P \alpha_0. \tag{1.165}$$

By the same token, the rate of adsorption onto the first layer, $k_2 P \alpha_1 A$, should be balanced by the rate of desorption from the second layer, $k_{-2}\alpha_2 A$, according to

$$k_2 P \alpha_1 A = k_{-2} \alpha_2 A \tag{1.166}$$

– where k_2 and k_{-2} denote kinetic constants describing adsorption and desorption, respectively; A may once again drop off both sides to give

$$\alpha_2 = K_{ith,ads} P \alpha_1, \tag{1.167}$$

provided that the adsorption constant characterizing layers other than the first one, $K_{ith,ads}$, abides in particular to

$$K_{ith,ads} \equiv \frac{k_2}{k_{-2}}. \tag{1.168}$$

Note that $K_{ith,ads}$ should be distinct from $K_{1st,ads}$ – because the former results from intramolecular interactions between adsorbate molecules, while the latter corresponds to interactions of adsorbate directly with the surface. Due to the functional form of Eqs. (1.163) and (1.166), one may write

$$k_i P \alpha_{i-1} A = k_{-i} \alpha_i A; \quad i = 1, 2, \ldots, n \tag{1.169}$$

in general, and consequently

$$\alpha_i = \frac{k_i}{k_{-i}} P \alpha_{i-1} \tag{1.170}$$

upon isolation of α_i; the hypothesized equivalence of interaction between adsorbate molecules in the first and second layers, in the second and third layers, and so on eventually support

$$\boxed{K_{ith,ads} \equiv \frac{k_i}{k_{-i}}; \quad i = 2, 3, \ldots, n} \tag{1.171}$$

– which implies

$$\alpha_i = K_{ith,ads} P \alpha_{i-1}; \quad i = 2, 3, \ldots, n \tag{1.172}$$

as per Eq. (1.170). The recursive relationship labeled as Eq. (1.172) may be applied from $i=2$ up to $i=n$, to eventually give

$$\alpha_i = \left(K_{ith,ads} P\right)^{i-1} \alpha_1, \tag{1.173}$$

whereas further combination with Eq. (1.165) leads to

$$\boxed{\alpha_i = K_{1st,ads} K_{ith,ads}{}^{i-1} P^i \alpha_0; \quad i = 1, 2, \ldots, n.} \tag{1.174}$$

The total volume of water vapor adsorbed, V_{ads}, should accordingly be given by

$$V_{ads} = \beta A \sum_{i=1}^{n} i\alpha_i, \tag{1.175}$$

where β denotes volume of an adsorbed molecule per unit surface; the fraction of overall area corresponding to the i-th layer has necessarily another $i-1$ molecules below – which justifies the (cumulative) summation. A particular case of Eq. (1.175) pertains to monolayer coverage of the surface, characterized by volume $V_{1st,ads}$, viz.

$$V_{1st,ads} = \beta A \sum_{i=1}^{n} \alpha_i \tag{1.176}$$

– because the total area occupied by adsorbed molecules in a single layer (although extending to all possible distances from the surface) is to be taken into account, with the same unit weight (i.e. $i=1$). The maximum possible value for $V_{1st,ads}$ should consider also the fraction of surface not occupied by adsorbed molecules, α_0, so the maximum value for $V_{1st,ads}$ actually reads

$$V_{1st,ads,\max} = \beta A \sum_{i=0}^{n} \alpha_i \tag{1.177}$$

– where, by definition of fractions of overall surface,

$$\sum_{i=0}^{n} \alpha_i = 1; \tag{1.178}$$

hence, Eq. (1.177) reduces to

$$\boxed{V_{1st,ads,\max} = \beta A.} \tag{1.179}$$

After ordered division of Eq. (1.175) by Eq. (1.177), one obtains

$$\frac{V_{ads}}{V_{1st,ads,\max}} = \frac{\sum_{i=1}^{n} i\alpha_i}{\sum_{i=0}^{n} \alpha_i} \tag{1.180}$$

since βA dropped off both numerator and denominator; insertion of Eq. (1.174) transforms Eq. (1.180) to

$$\frac{V_{ads}}{V_{1st,ads,\max}} = \frac{\sum_{i=1}^{n} i K_{1st,ads} K_{ith,ads}{}^{i-1} P^i \alpha_0}{\alpha_0 + \sum_{i=1}^{n} K_{1st,ads} K_{ith,ads}{}^{i-1} P^i \alpha_0}, \tag{1.181}$$

where α_0 may be factored out in both numerator and denominator, and $K_{1st,ads}P$ further taken off both summations to produce

$$\frac{V_{ads}}{V_{1st,ads,\max}} = \frac{\alpha_0 K_{1st,ads} P \sum_{i=1}^{n} i \left(K_{ith,ads} P\right)^{i-1}}{\alpha_0 \left(1 + K_{1st,ads} P \sum_{i=1}^{n} \left(K_{ith,ads} P\right)^{i-1}\right)}. \tag{1.182}$$

Upon cancelling α_0 between numerator and denominator, Eq. (1.182) will appear as

$$\frac{V_{ads}}{V_{1st,ads,\max}} = \frac{K_{1st,ads} P}{1 + K_{1st,ads} P \sum_{i=1}^{n} \left(K_{ith,ads} P\right)^{i-1}} \sum_{i=1}^{n} \frac{d\left(K_{ith,ads} P\right)^i}{d\left(K_{ith,ads} P\right)} \tag{1.183}$$

– in view of $d\xi^i/d\xi = i\xi^{i-1}$ for every real ξ and integer i above zero; linearity of the differential operator supports transformation to

$$\frac{V_{ads}}{V_{1st,ads,\max}} = \frac{K_{1st,ads} P}{1 + K_{1st,ads} P \sum_{i=1}^{n} \left(K_{ith,ads} P\right)^{i-1}} \frac{d\sum_{i=1}^{n} \left(K_{ith,ads} P\right)^i}{d\left(K_{ith,ads} P\right)}. \tag{1.184}$$

Equation (1.184) can be reformulated as

$$\frac{V_{ads}}{V_{1st,ads,\max}} = \frac{K_{1st,ads} P}{1 + K_{1st,ads} P \sum_{i=1}^{n} \left(K_{ith,ads} P\right)^{i-1}} \frac{d\sum_{i=0}^{n} \left(K_{ith,ads} P\right)^i}{d\left(K_{ith,ads} P\right)}, \tag{1.185}$$

because the derivative of the (unit) constant consubstantiated in $(K_{ith,ads}P)^0$ is nil. Both lower and upper limits of the summation in the denominator of Eq. (1.185) may, for convenience, be decreased by one unit, viz.

$$\frac{V_{ads}}{V_{1st,ads,\max}} = \frac{K_{1st,ads} P}{1 + K_{1st,ads} P \sum_{j=0}^{n-1} \left(K_{ith,ads} P\right)^{j}} \frac{d\sum_{i=0}^{n} \left(K_{ith,ads} P\right)^i}{d\left(K_{ith,ads} P\right)}; \tag{1.186}$$

inspection of Eq. (1.186) indicates that the two summations represent geometric progressions, with ratio $K_{ith,ads}P$ between consecutive terms and first term equal to unity – so one should resort to the expression for their sum and write

$$\frac{V_{ads}}{V_{1st,ads,\max}} = \frac{K_{1st,ads} P}{1 + K_{1st,ads} P \dfrac{1 - \left(K_{ith,ads} P\right)^n}{1 - K_{ith,ads} P}} \frac{d}{d\left(K_{ith,ads} P\right)} \left(\frac{1 - \left(K_{ith,ads} P\right)^{n+1}}{1 - K_{ith,ads} P}\right). \tag{1.187}$$

Since the number of adsorption layers may theoretically grow unbounded, Eq. (1.187) should be more appropriately formulated as

$$\frac{V_{ads}}{V_{1st,ads,\max}} = \lim_{n\to\infty} \frac{K_{1st,ads}P}{1+K_{1st,ads}P\dfrac{1-\left(K_{ith,ads}P\right)^n}{1-K_{ith,ads}P}} \frac{d}{d\left(K_{ith,ads}P\right)}\left(\frac{1-\left(K_{ith,ads}P\right)^{n+1}}{1-K_{ith,ads}P}\right); \tag{1.188}$$

Eq. (1.188) reduces to

$$\frac{V_{ads}}{V_{1st,ads,max}} = \frac{K_{1st,ads}P}{1+K_{1st,ads}P\dfrac{1}{1-K_{ith,ads}P}} \frac{d}{d\left(K_{ith,ads}P\right)}\left(\frac{1}{1-K_{ith,ads}P}\right), \tag{1.189}$$

provided that $K_{1st,ads}P < 1$ – which constitutes a realistic hypothesis. After multiplying both numerator and denominator by $1-K_{ith,ads}P$, and calculating the indicated derivative, Eq. (1.189) becomes

$$\frac{V_{ads}}{V_{1st,ads,\max}} = \frac{K_{1st,ads}P\left(1-K_{ith,ads}P\right)}{1-K_{ith,ads}P+K_{1st,ads}P} \frac{1}{\left(1-K_{ith,ads}P\right)^2}; \tag{1.190}$$

cancellation of $1-K_{ith,ads}P$ between numerator and denominator permits simplification to

$$\frac{V_{ads}}{V_{1st,ads,\max}} = \frac{K_{1st,ads}P}{\left(1-K_{ith,ads}P\right)\left(1-K_{ith,ads}P+K_{1st,ads}P\right)}. \tag{1.191}$$

Following multiplication of both sides of Eq. (1.191) by $V_{1st,ads,\max}$, and defining K_{bet} as

$$\boxed{K_{bet} \equiv \frac{K_{1st,ads}}{K_{ith,ads}},} \tag{1.192}$$

one may rephrase Eq. (1.191) as

$$V_{ads} = \frac{V_{1st,ads,\max}K_{bet}K_{ith,ads}P}{\left(1-K_{ith,ads}P\right)\left(1+\left(K_{bet}-1\right)K_{ith,ads}P\right)}. \tag{1.193}$$

At saturation pressure P^σ, an infinite amount of adsorbate should build up on the surface, i.e.

$$\lim_{P\to P^\sigma} V_{ads} = \infty, \tag{1.194}$$

so Eq. (1.193) has it that

$$\lim_{P\to P^\sigma} \frac{V_{1st,ads,\max}K_{bet}K_{ith,ads}P}{\left(1-K_{ith,ads}P\right)\left(1+\left(K_{bet}-1\right)K_{ith,ads}P\right)} = \frac{V_{1st,ads,\max}K_{bet}K_{ith,ads}P^\sigma}{\left(1-K_{ith,ads}P^\sigma\right)\left(1+\left(K_{bet}-1\right)K_{ith,ads}P^\sigma\right)} = \infty; \tag{1.195}$$

since P^σ is finite, Eq. (1.195) can hold only if either

$$1-K_{ith,ads}P^\sigma = 0 \tag{1.196}$$

or

$$1+\left(K_{bet}-1\right)K_{ith,ads}P^\sigma = 0 \tag{1.197}$$

is satisfied. Adsorption directly on the surface is typically stronger than adsorption over sequential layers of fluid – meaning that $K_{1st,ads} > K_{ith,ads}$, and thus $K_{bet} > 1$ as per Eq. (1.192); therefore, solution of Eq. (1.197) for $K_{ith,ads}$ would lead to a negative value thereof – meaning that Eq. (1.196) conveys the only possible solution, i.e.

$$K_{ith,ads} = \frac{1}{P^\sigma}, \tag{1.198}$$

which may, in turn, be inserted in Eq. (1.193) to yield

$$V_{ads} = \frac{V_{1st,ads,\max}K_{bet}\dfrac{P}{P^\sigma}}{\left(1-\dfrac{P}{P^\sigma}\right)\left(1+\left(K_{bet}-1\right)\dfrac{P}{P^\sigma}\right)}. \tag{1.199}$$

In view of the definition of water activity conveyed by Eq. (1.147), one can redo Eq. (1.199) to

$$V_{ads} = \frac{V_{1st,ads,\max}K_{bet}a_W}{\left(1-a_W\right)\left(1+\left(K_{bet}-1\right)a_W\right)}, \tag{1.200}$$

since water is the compound under scrutiny; multiplication of both sides by ρ_W/m_{tot} then gives rise to

$$\frac{\rho_W V_{ads}}{m_{tot}} = \frac{\dfrac{\rho_W V_{1st,ads,\max}}{m_{tot}}K_{bet}a_W}{\left(1-a_W\right)\left(1+\left(K_{bet}-1\right)a_W\right)}, \tag{1.201}$$

where m_{tot} denotes total mass (of water) and ρ_W denotes mass density of water. In view of the definition of mass concentration of water in adsorbed state, on a total mass basis, i.e.

$$C_W \equiv \frac{\rho_W V_{ads}}{m_{tot}}, \tag{1.202}$$

and thus

$$C_{W,ml} \equiv \frac{\rho_W V_{1st,ads,\max}}{m_{tot}} \tag{1.203}$$

pertaining to a single monolayer of adsorbed water, one may finally reformulate Eq. (1.201) to

$$\boxed{C_W = \frac{C_{W,ml}K_{bet}a_W}{\left(1-a_W\right)\left(1+\left(K_{bet}-1\right)a_W\right)}.} \tag{1.204}$$

Typical values of $C_{W,ml}$ are 11%(w/w) for gelatin or starch, 6%(w/w) for (amorphous) lactose, and 3%(w/w) for (spray) dried milk. Such a monolayer value indeed represents the moisture content at which food tends to be most stable – with lipid oxidation likely occurring below this threshold; and higher values promoting Maillard's browning, as well as enzyme-mediated reactions and microbial metabolism at large – for encompassing less and less tightly-bound water molecules. The variation of $C_W/C_{W,ml}$ with a_W, as per Eq. (1.204), is depicted in Fig. 1.20; splitting between monolayer and multilayer coverage obviously occurs at $C_W/C_{W,ml} = 1$. Note the existence of an inflection of the curve close

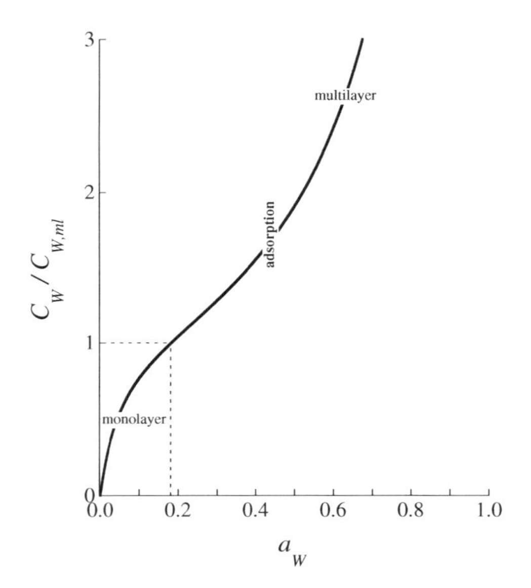

FIGURE 1.20 Variation of ratio of water concentration, C_W, to saturating water concentration in monolayer, $C_{W,ml}$, versus water activity, a_W, in typical solid-like food – accounted for by surface adsorption as dominating phenomenon.

to monolayer coverage – thus producing a sigmoidal shape that tends to a vertical asymptote at $P=P^\sigma$, as per Eqs. (1.199) and (1.202); the portion of the curve with concavity facing upward represents indeed water adsorbed as multilayers within the food. Brunauer, Emmett, and Teller's law is particularly accurate when P/P^σ lies in the range 0.05–0.35. When a_W lies within 0.6–0.7, a sharp increase in water content is observed despite the accompanying small increase in a_W; this unfolds capillary condensation – a phenomenon to be described shortly.

The germane constant K_{bet} may be ascribed an alternative fundamental meaning, in terms of energies of adsorption and desorption at stake. In fact, Eq. (1.192) conveys a ratio of two phase equilibrium constants, so one may write

$$K_{bet} = \frac{\exp\left\{-\dfrac{\Delta g^\theta_{1st,ads}}{\mathcal{R}T}\right\}}{\exp\left\{-\dfrac{\Delta g^\theta_{ith,ads}}{\mathcal{R}T}\right\}} \tag{1.205}$$

in general agreement with Eq. (1.105) – where $\Delta g^\theta_{1st,ads}$ and $\Delta g^\theta_{ith,ads}$ denote standard Gibbs' energy change accompanying adsorption of water in the first monolayer next to the solid surface, and of adsorption in the upper layers, respectively. In view of Eq. (3.40) of *Food Proc. Eng.: basics & mechanical operations* and Eq. (3.13) of *Food Proc. Eng.: thermal & chemical operations*, one may relate those two Gibbs' energy changes to the corresponding enthalpy and entropy changes via

$$\Delta g^\theta_{1st,ads} = \Delta h^\theta_{1st,ads} - T\Delta s^\theta_{1st,ads}, \tag{1.206}$$

and likewise

$$\Delta g^\theta_{ith,ads} = \Delta h^\theta_{ith,ads} - T\Delta s^\theta_{ith,ads} \tag{1.207}$$

– where $\Delta h^\theta_{1st,ads}$ denotes molar heat of adsorption of water directly onto the solid (outer) surface of the food, and $\Delta h^\theta_{ith,ads}$ denotes molar heat of adsorption of water onto pre-existing layers thereof; by the same token, $\Delta s^\theta_{1st,ads}$ denotes molar entropy of adsorption of water onto the food surface, and $\Delta s^\theta_{ith,ads}$ denotes molar entropy of adsorption thereof onto the layers of water molecules already adsorbed. Upon insertion of Eqs. (1.206) and (1.207), one obtains

$$K_{bet} = \frac{\exp\left\{\dfrac{\Delta s^\theta_{1st,ads}}{\mathcal{R}}\right\}\exp\left\{-\dfrac{\Delta h^\theta_{1st,ads}}{\mathcal{R}T}\right\}}{\exp\left\{\dfrac{\Delta s^\theta_{ith,ads}}{\mathcal{R}}\right\}\exp\left\{-\dfrac{\Delta h^\theta_{ith,ads}}{\mathcal{R}T}\right\}} \tag{1.208}$$

from Eq. (1.205) – which may be rewritten as

$$\boxed{K_{bet} = K_{bet,0}\exp\left\{\frac{\Delta h^\theta_{ith,ads} - \Delta h^\theta_{1th,ads}}{\mathcal{R}T}\right\}}, \tag{1.209}$$

after lumping the exponential functions appearing second in numerator and denominator; the entropic accommodation factor, $K_{bet,0}$, is likewise given by

$$K_{bet,0} \equiv \exp\left\{\frac{\Delta s^\theta_{1st,ads} - \Delta s^\theta_{ith,ads}}{\mathcal{R}}\right\}. \tag{1.210}$$

Inspection of Eq. (1.209) unfolds a useful relationship of K_{bet} to the difference in adsorption enthalpy changes between upper layers and first layer.

Each food is characterized by a sorption isotherm for a given temperature; its precise shape depends on its unique physical structure, chemical composition, and extent of water binding within the matrix – yet sorption isotherms share, in general, the form depicted in Fig. 1.20. As mentioned above, the first portion of the curve corresponds to monolayered water – which is strongly bound, and thus quite stable, unfreezable, and not susceptible to removal via regular drying. Water molecules are, in this case, strongly bound to specific sites (e.g. hydroxyl groups of polysaccharides, carbonyl and amino groups of proteins, and other moieties via hydrogen bonding), and not available for chemical reaction or to serve as plasticizer. The portion of the curve labeled as multilayer in Fig. 1.20 entails water adsorbed as multiple layers, each on top of the previous one; water is more loosely bound, so the associated enthalpy change of vaporization is only slightly higher than that of plain water.

Despite the wide applicability of Brunauer, Emmett, and Teller's model, several other models have been proposed – either possessing some theoretical basis, arising from the simplification of more elaborate models, or being just empirical correlations; the most important ones are tabulated in Table 1.5. Langmuir's model departing hypothesis is physical adsorption of water on a single (or unimolecular) layer, with identical and equivalent adsorption sites. Oswin's model is of an empirical nature, and stems from the series expansion of sigmoidal-shaped sorption isotherms; its validity breaks down at $a_W > 0.5$. Smith's model is also empirical, and appropriate to describe the final curved portion of the water sorption isotherm of high-molecular weight

TABLE 1.5

Selected models for water sorption isotherms, encompassing dependence of moisture content, C_W, on water activity, a_W.

Equation	Denomination	(lumped) Parameters description	number
$C_W = \frac{K C_{W,ml} a_W}{1 + K a_W}$	Langmuir's model	$C_{W,ml}$ K	2
$C_W = k \left(\frac{a_W}{1 - a_W} \right)^n$	Oswin's model	k n	
$C_W = k_1 + k_2 \ln\{1 - a_W\}$	Smith's model	k_1 k_2	
$C_W = \sqrt[n]{-\frac{\ln\{1 - a_W\}}{k}}$	Henderson's model	k n	
$\ln\left\{C_W + \sqrt{C_W{}^2 + C_{W,1/2}{}^2}\right\} = k_1 + k_2 a_W$	Iglesias and Chirife's model	k_1 k_2	
$C_W = C_{W,ml} \sqrt[n]{-\frac{k}{\mathcal{R}T \ln a_W}}$	Halsey's model	$C_{W,ml}$ k n	3
$C_W = k_1 a_W{}^{k_3} + k_2 a_W{}^{k4}$	Peleg's model	k_1 k_2 k_3 k_4	4

biopolymers; two fractions of water have been hypothesized – one, associated with k_1, characterized by a larger heat of condensation than expected via Langmuir's model, and a second fraction, characterized by k_2, that only forms after the first fraction has been sorbed. This model focuses on the second fraction, with postulation that it is proportional to the logarithm of the difference between a_W and pure water. Halsey's model conveys an expression for condensation of multilayers, which assumes that the potential energy is proportional to the reciprocal of the n-th power of its distance to the surface – and is particularly appropriate for starch-containing products. On the basis of Henderson's model, a plot of $\ln\{-\ln\{1-a_W\}\}$ vs. $\ln C_W$ should lead to a straight line. Iglesias and Chirife's empirical model takes advantage of the moisture content at a_W=0.5, i.e. $C_{W,1/2}$ – and was found suitable for sugar-rich foods. Although devoid of any theoretical background, Peleg's model permits quite good fits, probably due to its power functional form and large number of parameters (i.e. four); it should, however, abide to $k_3 < 1$ and $k_4 > 1$ as constraints.

According to Blahovec and Yanniotis' criterion, all sorption isotherms breakdown to one out of three categories according to shape – and objectively based on S_0, or sign of slope of moisture content versus water activity curve at a_W=0, and S_1, or sign of ratio of slope at a_W=1 to that at a_W=0. Type *I*, or Langmuir's type isotherms, are described by $S_0 > 0$ and $S_1 > 0$; type *II*, or miscellaneous type isotherms, are characterized by $S_0 > 0$ and $S_1 < 0$; and type *III*, or solution-like isotherms, satisfy $S_0 < 0$ and $S_1 > 0$.

PROBLEM 1.5

Brunauer, Emmett, and Teller's model has, by far, been the most widely used to model water adsorption within food systems – and indeed represents a milestone in interpreting sequential adsorption of water molecules as multiple layers. However, the assumptions of uniform adsorbent surface, coupled with the absence of lateral interactions between adsorbed molecules have been challenged – in view of the heterogeneous and compact nature of typical food microstructure. This has set the basis for its refinement via Guggenheim, Anderson, and de Boer's model, viz.

$$C_W = \frac{C_{W,ml} K_{bet} a_W}{\left(1 - K_{gab} a_W\right)\left(1 + \left(K_{bet} - 1\right) K_{gab} a_W\right)}, \quad (1.5.1)$$

arising from the postulate that the state of adsorbed molecules in the second layer is (again) identical to that in upper layers, but distinct from that of free liquid – with (equilibrium constant), K_{gab}, measuring the difference in chemical potential between them. Confirm that Brunauer, Emmett, and Teller's model is a particular case of Guggenheim, Anderson, and de Boer's model; and suggest a form for the temperature dependence of K_{gab}.

Solution

Inspection of Eq. (1.5.1) *vis-à-vis* with Eq. (1.204) unfolds a mere replacement of a_W by $K_{gab} a_W$ in denominator; therefore, setting K_{gab}=1 transforms Eq. (1.5.1) to Eq. (1.204).

If, by hypothesis, K_{gab} represents an equilibrium constant that measures differences in chemical potential between higher-layer adsorbed liquid, $\Delta g_{ith,ads}$, and free liquid, Δg_{free}, then one may write

$$K_{gab} = \frac{\exp\left\{-\dfrac{\Delta g_{ith,ads}}{\mathcal{R}T}\right\}}{\exp\left\{-\dfrac{\Delta g_{free}}{\mathcal{R}T}\right\}} = \exp\left\{\frac{\Delta g_{free} - \Delta g_{ith,ads}}{\mathcal{R}T}\right\}; \quad (1.5.2)$$

in view of the analogy between Eqs. (1.205) and (1.5.2), one may jump to analogs of Eqs. (1.209) and (1.210), i.e.

$$K_{gab} = K_{gab,0} \exp\left\{\frac{\Delta h^{\theta}_{free} - \Delta h^{\theta}_{ith,ads}}{\mathcal{R}T}\right\} \quad (1.5.3)$$

and

$$K_{gab,0} \equiv \exp\left\{\frac{\Delta s^{\theta}_{ith,ads} - \Delta s^{\theta}_{free}}{\mathcal{R}}\right\} \quad (1.5.4)$$

respectively. Experimental evidence has indicated that introduction of K_{gab} extends validity of Brunauer, Emmett, and Teller's model up to approximately a_W=0.9 – yet C_W gets underestimated within [0.95,1] as interval for water activity.

1.1.3.4.3 Capillary condensation

Adsorption of water, as described by Brunauer, Emmett, and Teller's equation, is a consequence of surface energy – a concept also underlying surface tension; therefore, it is somehow expected that adsorption will gradually give rise to capillary condensation – especially within the range 0.6–0.7 of a_W (see Fig. 1.20), where many foods exhibit a shallow increase in a_W along with a sudden increase in water concentration.

In attempts to simulate the two constitutive phases underlying capillary condensation, one should first realize that total Gibbs' energy change, dG, undergone by molar amount dn initially in vapor form outside a pore, at equilibrium pressure P^{σ}, should encompass change to a new equilibrium pressure $P < P^{\sigma}$ inside the pore, and condensation at that new pressure – at constant temperature T^{θ}. In mathematical terms, this corresponds to stating

$$dG = dG_{l\backslash g}\Big|_{P^{\sigma},T^{\theta}} + \left.\left(\int_{P^{\sigma}}^{P}\left(\frac{\partial g_g}{\partial \tilde{P}}\right)_T d\tilde{P}\right)\right|_{T^{\theta}} dn + dG_{g\backslash l}\Big|_{P,T^{\theta}}, \quad (1.211)$$

where the first term in the right-hand side represents elementary change of Gibbs's energy during vaporization at pressure P^{σ} and temperature T^{θ}, the third term represents elementary change in Gibbs' energy during condensation of vapor back to liquid at pressure P and temperature T^{θ}, and the second term represents elementary change of molar Gibbs' energy, g_g, incurred in (isothermally) bringing vapor from P^{σ} to P, at temperature T^{θ}. This process may be analyzed from the point of view of the solid system rather than the point of view of the fluids – in which case a change dA_l in surface contact between phases will occur at the solid/liquid interface, associated with a simultaneous change dA_l at the solid/gas interface; hence, the corresponding change in Gibbs' energy, dG, should look like

$$dG = \sigma_{s\backslash l}\Big|_{P,T^{\theta}} dA_l - \sigma_{s\backslash g}\Big|_{P,T^{\theta}} dA_l, \quad (1.212)$$

where the first term in the right-hand side represents elementary change in Gibbs' energy to expand the interface of condensed liquid in contact with the solid wall of the pore, and the second term represents elementary change in Gibbs' energy accompanying the concomitant contraction of the gas/solid interface – both at pressure P and temperature T^{θ}. Here $\sigma_{s\backslash l}$ denotes surface tension prevailing at solid/liquid interface, and $\sigma_{s\backslash g}$ likewise denotes surface tension prevailing at solid/gas interface. Since evaporation at P^{σ} and condensation at P in Eq. (1.211) are both phase equilibrium processes, one finds that

$$dG_{l\backslash g}\Big|_{P^{\sigma},T^{\theta}} = 0, \quad (1.213)$$

and similarly

$$dG_{g\backslash l}\Big|_{P,T^{\theta}} = 0 \quad (1.214)$$

due to the definition of equilibrium between phases as per Eq. (3.103) of *Food Proc. Eng.: basics & mechanical operations* – so Eq. (1.211) breaks down to merely

$$dG = \left.\left(\int_{P^{\sigma}}^{P}\left(\frac{\partial g_g}{\partial \tilde{P}}\right)_T d\tilde{P}\right)\right|_{T^{\theta}} dn; \quad (1.215)$$

insertion of Eq. (1.215) permits transformation of Eq. (1.212) to

$$\left.\left(\int_{P^{\sigma}}^{P}\left(\frac{\partial g_g}{\partial \tilde{P}}\right)_T d\tilde{P}\right)\right|_{T^{\theta}} dn = \sigma_{s\backslash l}\Big|_{P,T^{\theta}} dA_l - \sigma_{s\backslash g}\Big|_{P,T^{\theta}} dA_l, \quad (1.216)$$

where dA_l may, in turn, be factored out as

$$\left.\left(\int_{P^{\sigma}}^{P}\left(\frac{\partial g_g}{\partial \tilde{P}}\right)_T d\tilde{P}\right)\right|_{T^{\theta}} dn = \left(\sigma_{s\backslash l}\Big|_{P,T^{\theta}} - \sigma_{s\backslash g}\Big|_{P,T^{\theta}}\right) dA_l. \quad (1.217)$$

On the other hand, Eq. (3.40) of *Food Proc. Eng.: basics & mechanical operations* and Eq. (3.17) of *Food Proc. Eng.: thermal & chemical operations* apply here in the form

$$\left(\frac{\partial g_g}{\partial P}\right)_T = v_g, \quad (1.218)$$

where v_g denotes molar volume of vapor phase; remember that μ and g coincide as per Eq. (3.40) of *Food Proc. Eng.: basics & mechanical operations*, because a pure compound (i.e. water) is under scrutiny. If the vapor is assumed to approximately behave as an ideal gas, after Eq. (3.135) of *Food Proc. Eng.: basics & mechanical operations*, then one may replace Eq. (1.218) by

$$\left.\left(\frac{\partial g_g}{\partial P}\right)_T\right|_{T^\theta} = \frac{\mathcal{R}T^\theta}{P} \tag{1.219}$$

– so integration, under (constant) temperature T^θ, between pressures P^σ and P gives rise to

$$\left.\left(\int_{P^\sigma}^{P}\left(\frac{\partial g_g}{\partial \tilde{P}}\right)_T d\tilde{P}\right)\right|_{T^\theta} = \int_{P^\sigma}^{P}\frac{\mathcal{R}T^\theta}{\tilde{P}}\,d\tilde{P}; \tag{1.220}$$

once constants are taken off the kernel, Eq. (1.220) becomes

$$\left.\left(\int_{P^\sigma}^{P}\left(\frac{\partial g_g}{\partial \tilde{P}}\right)_T \partial\tilde{P}\right)\right|_{T^\theta} = \mathcal{R}T^\theta\int_{P^\sigma}^{P}\frac{d\tilde{P}}{\tilde{P}}. \tag{1.221}$$

The fundamental theorem of integral calculus may be invoked to transform the right-hand side of Eq. (1.221) to

$$\left.\left(\int_{P^\sigma}^{P}\left(\frac{\partial g_g}{\partial \tilde{P}}\right)_T d\tilde{P}\right)\right|_{T^\theta} = \mathcal{R}T^\theta \ln \tilde{P}\Big|_{P^\sigma}^{P}, \tag{1.222}$$

or else

$$\left.\left(\int_{P^\sigma}^{P}\left(\frac{\partial g_g}{\partial \tilde{P}}\right)_T d\tilde{P}\right)\right|_{T^\theta} = \mathcal{R}T^\theta \ln \frac{P}{P^\sigma}; \tag{1.223}$$

combination of Eqs. (1.217) and (1.223) then unfolds

$$\boxed{\mathcal{R}T^\theta \ln \frac{P}{P^\sigma}\,dn = \left(\sigma_{s\backslash l} - \sigma_{s\backslash g}\right)dA_l,} \tag{1.224}$$

with both $\sigma_{s\backslash l}$ and $\sigma_{s\backslash g}$ taken at P and T^θ. On the other hand, the situation of liquid intruding a capillary pore, at the expense of extruding the (atmospheric) gas previously therein, may be tackled via the work undergone by either phase – together with the simplifying assumption of negligible volumetric work on the gas and negligible gravitational work on the liquid. If the liquid rises and wets an extra elementary area $2\pi R dL$, of length dL, on the inner wall of a cylindrical pore of radius R, then the active surface work performed thereby, $\text{đ}W_l$, reads

$$\text{đ}W_l = 2\pi R\sigma_{s\backslash l}dL; \tag{1.225}$$

a simultaneous resistive surface work, $\text{đ}W_g$, is undergone by the gaseous phase, according to

$$\text{đ}W_g = -2\pi R\sigma_{s\backslash g}dL, \tag{1.226}$$

as it is expelled from the pore. The overall surface work performed, $\text{đ}W_{sf}$, may consequently be calculated via ordered addition of Eqs. (1.225) and (1.226), i.e.

$$\text{đ}W_{sf} \equiv \text{đ}W_l + \text{đ}W_g = 2\pi R\sigma_{s\backslash l}dL - 2\pi R\sigma_{s\backslash g}dL, \tag{1.227}$$

where $2\pi RdL$ will be factored out to produce

$$\boxed{\text{đ}W_{sf} = 2\pi R\left(\sigma_{s\backslash l} - \sigma_{s\backslash g}\right)dL.} \tag{1.228}$$

The aforementioned work may instead be ascertained as the difference between work, $\text{đ}W_i$, performed by (contact) bulk force at the inlet, i.e.

$$\text{đ}W_i = -\pi R^2 P_i dL, \tag{1.229}$$

and its outlet counterpart, $\text{đ}W_o$, viz.

$$\text{đ}W_o = -\pi R^2 P_o dL \tag{1.230}$$

– where πR^2 denotes cross-sectional area of (capillary) pore, and P_i and P_o denote pressure at its inlet and outlet, respectively; hence, the net bulk work, $\text{đ}W_{bk}$, will be equal to their difference, according to

$$\text{đ}W_{bk} \equiv \text{đ}W_i - \text{đ}W_o = -\pi R^2 P_i dL - \left(-\pi R^2 P_o dL\right) \tag{1.231}$$

obtained via ordered subtraction of Eq. (1.230) from Eq. (1.229). After factoring $\pi R^2 dL$ out, Eq. (1.231) becomes

$$\boxed{\text{đ}W_{bk} = -\pi R^2 \Delta P dL,} \tag{1.232}$$

where ΔP abides to

$$\Delta P \equiv P_i - P_o. \tag{1.233}$$

Since either of the two aforementioned approaches should hold on its own, one promptly realizes that

$$\text{đ}W_{sf} = \text{đ}W_{bk} \tag{1.234}$$

– where insertion of Eqs. (1.228) and (1.232) gives rise to

$$2\pi R\left(\sigma_{s\backslash l} - \sigma_{s\backslash g}\right)dL = -\pi R^2 \Delta P dL; \tag{1.235}$$

after having divided both sides by $-\pi R^2 dL$, Eq. (1.235) reduces to

$$\frac{2}{R}\left(\sigma_{s\backslash g} - \sigma_{s\backslash l}\right) = \Delta P. \tag{1.236}$$

One may now revisit Eq. (4.183) of *Food Proc. Eng.: basics & mechanical operations* as

$$\Delta P = \frac{2\sigma_{l\backslash g}\cos\theta}{R}, \tag{1.237}$$

where the fact that surface tension, $\sigma_{l\backslash g}$, refers specifically to the gas/liquid interface was highlighted; elimination of ΔP between Eqs. (1.236) and (1.237) unfolds

$$\frac{2}{R}\left(\sigma_{s\backslash g} - \sigma_{s\backslash l}\right) = \frac{2\sigma_{l\backslash g}\cos\theta}{R}, \tag{1.238}$$

while further dropping of $2/R$ between sides eventually leaves

$$\boxed{\cos\theta = \frac{\sigma_{s\backslash g} - \sigma_{s\backslash l}}{\sigma_{l\backslash g}}} \tag{1.239}$$

– known as Yang and Dupré's law. Equation (1.239) stresses that the surface tensions prevailing at the various interfaces at stake are not independent of each other, but are instead related via angle θ; this denotes contact angle of liquid with pore wall, easily accessible via experimentation. Since the meniscus angle of water in food pores is normally close to zero, cos θ approaches unity, so one may safely write

$$\frac{\sigma_{s\backslash g}-\sigma_{s\backslash l}}{\sigma_{l\backslash g}}\approx 1 \tag{1.240}$$

in lieu of Eq. (1.239); this supports

$$\sigma_{s\backslash g}-\sigma_{s\backslash l}=\sigma_{l\backslash g}, \tag{1.241}$$

after isolation of $\sigma_{l\backslash g}$. In view of Eq. (1.241), one may reformulate Eq. (1.224) to

$$\mathcal{R}T^{\theta}\ln\frac{P}{P^{\sigma}}dn=-\sigma_{l\backslash g}dA_l; \tag{1.242}$$

the volume condensed in the pores, dV_l, satisfies in turn

$$dV_l=v_l dn, \tag{1.243}$$

where v_l denotes molar volume of liquid phase – so insertion in Eq. (1.242) supports conversion to

$$\frac{\mathcal{R}T^{\theta}}{v_l}\ln\frac{P}{P^{\sigma}}dV_l=-\sigma_{l\backslash g}dA_l. \tag{1.244}$$

Upon isolation of the logarithmic function, Eq. (1.244) becomes

$$\ln\frac{P}{P^{\sigma}}=-\frac{\sigma_{l\backslash g}v_l}{\mathcal{R}T^{\theta}}\frac{dA_l}{dV_l} \tag{1.245}$$

– where the underlying cylindrical shape of the pore gives

$$\frac{dA_l}{dV_l}=\frac{d\left(4\pi R^2\right)}{d\left(\frac{4}{3}\pi R^3\right)}=\frac{4\pi 2RdR}{\frac{4}{3}\pi 3R^2dR}=\frac{2}{R}, \tag{1.246}$$

based on the definition of differential and the rules of differentiation, complemented with cancellation of common factors between numerator and denominator; insertion of Eq. (1.246) converts Eq. (1.245) to

$$\ln\frac{P}{P^{\sigma}}=-\frac{\sigma_{l\backslash g}v_l}{\mathcal{R}T^{\theta}}\frac{2}{R}, \tag{1.247}$$

which can be written as

$$a_W=\exp\left\{-\frac{2\sigma_{l\backslash g}v_l}{\mathcal{R}T^{\theta}R}\right\}, \tag{1.248}$$

after taking exponentials of both sides and recalling Eq. (1.147). By definition,

$$v_l\equiv\frac{V_l}{n_l} \tag{1.249}$$

where n_l denotes number of moles of liquid – while molecular weight, M_l, abides, by definition, to

$$M_l\equiv\frac{m_l}{n_l}, \tag{1.250}$$

with m_l denoting mass of liquid; ordered division of Eq. (1.250) by Eq. (1.249), followed by cancellation of n_l between numerator and denominator unfold

$$\frac{M_l}{v_l}=\frac{\frac{m_l}{n_l}}{\frac{V_l}{n_l}}=\frac{m_l}{V_l}\equiv\rho_l, \tag{1.251}$$

with ρ_l denoting mass density of liquid – which is equivalent to writing

$$v_l=\frac{M_l}{\rho_l}. \tag{1.252}$$

Combination with Eq. (1.252) transforms Eq. (1.248) to

$$\boxed{a_W=\exp\left\{-\frac{2M_l\sigma_{l\backslash g}}{\mathcal{R}T^{\theta}\rho_l R}\right\}} \tag{1.253}$$

– widely known as Kelvin's, or Thompson's law. In a sense, Eq. (1.253) describes reduction, at constant temperature T^{θ}, in vapor pressure from the saturation vapor pressure P^{σ} to the actual pressure P, inside the pore of inner radius R, due to a curved liquid/vapor interface (or meniscus). The overlapping potential of the wall overcomes the translational energy of the adsorbate molecules, so condensation (or vaporization) occurs at pressures lower than P^{σ}; as pressure is increased, condensation takes place first in the pores of smaller radius, and then progresses toward the larger pores.

Inspection of Eq. (1.253) finds no explicit dependence of water activity upon moisture concentration; in mathematical terms, this corresponds to

$$\boxed{\lim_{a_W\to\exp\left\{-\frac{2M_l\sigma_{l\backslash g}}{\mathcal{R}T^{\theta}\rho_l R}\right\}}C_W=\infty,} \tag{1.254}$$

as highlighted in Fig. 1.21. This asymptotic behavior justifies why a_W remains almost constant in a vicinity contained normally within [0.6,0.7], irrespective of moisture concentration; this corresponds indeed to a vertical asymptote centered at a_W, with

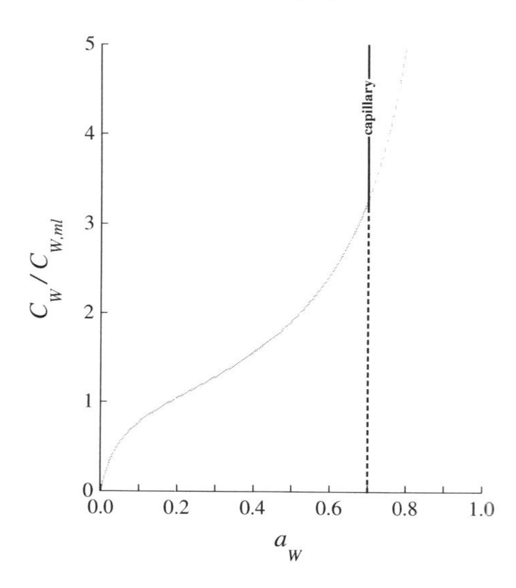

FIGURE 1.21 Variation of ratio of water concentration, C_W, to saturating water concentration in monolayer, $C_{W,ml}$, versus water activity, a_W, in typical solid-like food – accounted for by capillary condensation as dominating phenomenon (——), overlaid on result of the previous phenomenon (——).

horizontal intercept given by the exponential of the negative of $2M_l\sigma_{l\backslash g}/\mathcal{R}T^\theta\rho_l R$. As emphasized above, the germane form of water here encompasses water condensed within the capillary structure or in the empty cells of the food; it is mechanically trapped, and thus held only by weak forces – so it can be easily removed via drying with hot air, or freezing under low temperature (as implied by the steep slope of the $C_W/C_{W,ml}$ vs. a_W curve).

An obvious issue in modeling attempts arises when the pore size distribution of the food fails to be uniform, or is not known at all; furthermore, the pore structure (and thus the average pore size) may change with increased water content, and concomitant swelling of the food itself. In practice, capillary action is of great importance for porous foods at relatively significant water concentrations.

PROBLEM 1.6

Capillary condensation may itself be used to ascertain the pore size distribution in porous matrices, by resorting to nitrogen – specifically in the case of micropores; however, the said distribution is better assessed via liquid intrusion resorting to (nonwetting) mercury, due to dominance of meso- and macropores in most matrices relevant as foods. A typical assay involves measurement of the (cumulative) volume of mercury, V_f, ranging from 0 up to V_0, which has entered a porous matrix upon a gradual rise in pressure, departing from atmospheric pressure up to P_f, as illustrated below.

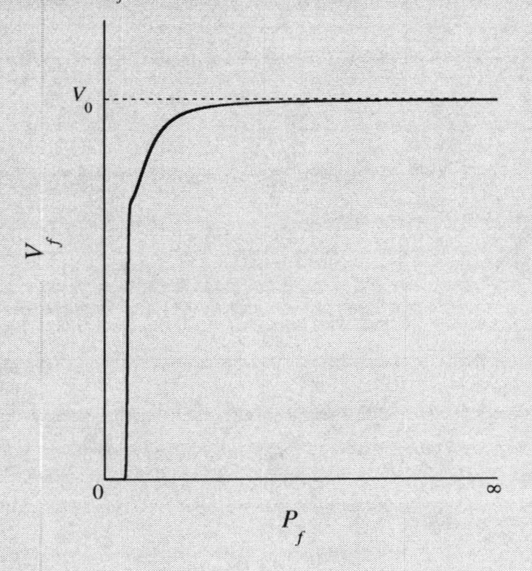

Mercury is the liquid of election, owing to its very low surface tension, 0.48 N.m[-1] (at 293 K), coupled with an average contact angle of 140°.

a) Assume Young and Laplace's law to be valid; plot pore radius, R_p, as a function of P_f, for the above V_f vs. P_f curve.
b) Using the plot generated in a), construct the curve V_f vs. R_p.
c) Based on the curve produced in b), obtain the (graphical) pore size distribution.
d) Departing from the pore size distribution obtained in c), estimate overall porosity and average pore radius – knowing that the total (outer) volume occupied by the matrix under scrutiny is V_{tot}.

Solution

a) The pressure differential exerted upon the fluid, P_f, to enter the matrix pores satisfies Eq. (4.47) of *Food Proc. Eng.: basics & mechanical operations*, i.e.

$$P_f = \sigma\left(\frac{1}{R_y} + \frac{1}{R_z}\right), \tag{1.6.1}$$

where R_y and R_z denote curvature radii perpendicular to each other, and σ denotes surface tension; assuming cylindrical pores, Eq. (4.179) of *Food Proc. Eng.: basics & mechanical operations* holds – so Eq. (1.6.1) simplifes to

$$P_f = \frac{2\sigma}{R_y}. \tag{1.6.2}$$

Equation (4.182) of *Food Proc. Eng.: basics & mechanical operations* may now be invoked to convert Eq. (1.6.2) to

$$P_f = \frac{2\sigma\cos\theta}{R_p}, \tag{1.6.3}$$

with θ denoting contact angle of fluid with pore wall; isolation of R_p finally gives

$$R_p = \frac{2\sigma\cos\theta}{P_f}. \tag{1.6.4}$$

The variation of R_p with P_f, as per Eq. (1.6.4), is sketched below.

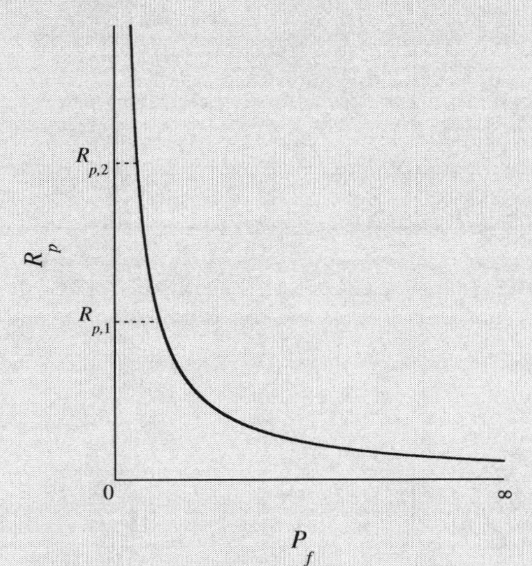

As expected, a hyperbolic behavior is found for R_p vs. P_f, in view of the functional form of Eq. (1.6.4) – with steepness determined by both σ and θ; two (nominal) pore radii, $R_{p,1}$ and $R_{p,2} > R_{p,1}$, were highlighted to serve as reference hereafter.

b) Upon elimination of P_f between the plot given and the plot produced in a), one obtains the curve $V_f \equiv V_f\{R_p\}$ depicted below.

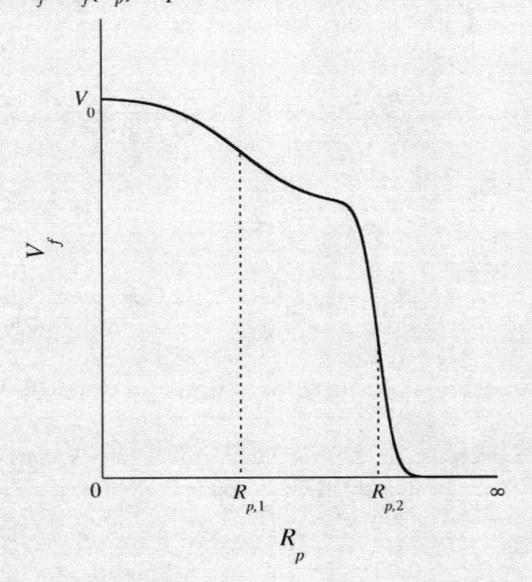

The original curve above has become geometrically more informative when V_f is taken as a (univariate) function of R_p – with inflection points appearing explicitly, namely at $R_{p,1}$ and $R_{p,2}$, between a region of concave shape and a region of convex shape as R_p increases.

c) Since V_f, at a given pressure P_f, denotes total amount of fluid that was able to penetrate pores with mouth radius above R_p, throughout the whole compression assay up to P_f, the pore size distribution, $\Phi\{R_p\}$, should abide to

$$\Phi\{R_p\} \equiv -\frac{dV_f}{dR_p}; \tag{1.6.5}$$

the minus sign is required because V_f increases at the expense of smaller and smaller radii R_p penetrated by working fluid, as pressure gets higher and higher. Graphical differentiation of the curve $V_f\{R_p\}$ obtained in b), in agreement with Eq. (1.6.5), leads to the curve plotted below.

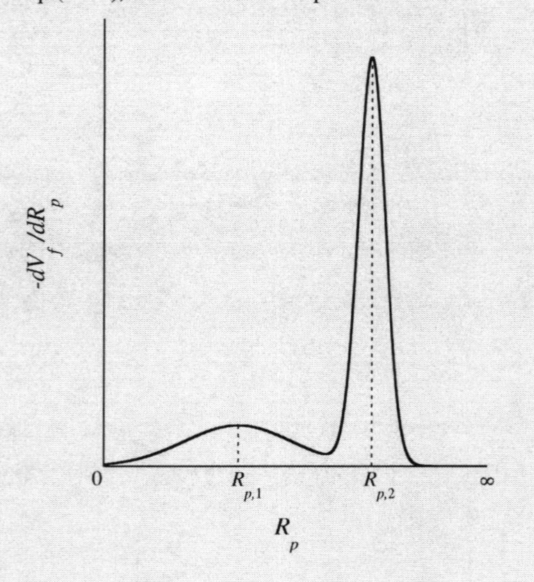

Inspection of this curve unfolds a bimodal distribution – with two distinct (nominal) pore sizes, $R_{p,1}$ and $R_{p,2}$, around which most pores clump.

d) Based on Eq. (1.6.5), one may write

$$\int_{\infty}^{0} \Phi\{R_p\}\,dR_p = -\int_{0}^{V_0} dV_f \tag{1.6.6}$$

following integration by separation of variables – where

$$V_f\big|_{R_p=\infty} = 0 \tag{1.6.7}$$

and

$$V_f\big|_{R_p=0} = V_0 \tag{1.6.8}$$

were used as (trivial) boundary conditions; V_0 obviously represents the total amount of voids, accounted for by the inner volume of the pores. Application of the fundamental theorem of integral calculus to the right-hand side transforms Eq. (1.6.6) to

$$V_0 = \int_0^{\infty} \Phi\{R_p\}\,dR_p, \tag{1.6.9}$$

where the minus sign was removed at the expense of exchanging upper and lower limits of the outstanding integral; in view of the definition of porosity, i.e.

$$\varepsilon \equiv \frac{V_0}{V_{tot}}, \tag{1.6.10}$$

one may insert Eq. (1.6.9) to get

$$\varepsilon = \frac{\int_0^{\infty} \Phi\{R_p\}\,dR_p}{V_{tot}} \tag{1.6.11}$$

as an estimate of (average) porosity of the matrix of interest. The average pore radius, $\overline{R}_p$, will in turn be given by

$$\overline{R}_p \equiv \frac{\int_0^{\infty} R_p \Phi\{R_p\}\,dR_p}{\int_0^{\infty} \Phi\{R_p\}\,dR_p} \tag{1.6.12}$$

– which may be rewritten as

$$\overline{R}_p = \frac{\int_0^{\infty} R_p \Phi\{R_p\}\,dR_p}{V_0}, \tag{1.6.13}$$

after taking Eq. (1.6.9) on board.

1.1.3.4.4 Osmotic pressure

In view of the angle θ made by water with the inner wall of a capillary pore, in a food matrix, being typically small, one finds that $\cos\theta \approx 1$ (as discussed previously); hence, Eq. (1.237) may be rephrased as

$$\Delta P = \frac{2\sigma_{l\backslash g}}{R} \tag{1.255}$$

for practical purposes – which allows reformulation of Eq. (1.253) to

$$a_W = \exp\left\{-\frac{M_l \Delta P}{\mathcal{R}T^\theta \rho_l}\right\}. \tag{1.256}$$

The effect of osmotic solutes can now be calculated from Eq. (1.256), via replacement of ΔP, arising from surface tension effects, by osmotic pressure of solute, Π_s – thus giving rise to

$$\boxed{a_W = \exp\left\{-\frac{M_l \Pi_S}{\mathcal{R}T^\theta \rho_l}\right\};} \tag{1.257}$$

in fact, the pressure differential pushing water into narrow pores due to surface tension is essentially equivalent to the pressure differential pushing (solvent) water toward a solution bearing a higher concentration of solute.

Equation (4.901) of *Food Proc. Eng.: basics & mechanical operations* may to advantage be revisited as

$$\frac{\Pi_s v_l}{\mathcal{R}T^\theta} = \ln\frac{1}{x_W}, \tag{1.258}$$

after multiplying both sides by $v_l/\mathcal{R}T^\theta$ and realizing that the molar fraction of water in the external phase is unity; after rewriting Eq. (1.258) as

$$\frac{\Pi_s M_l}{\mathcal{R}T^\theta \rho_l} = -\ln x_W \tag{1.259}$$

with the aid of Eq. (1.252), one may redo Eq. (1.257) as

$$a_W = \exp\{-(-\ln x_W)\} = \exp\{\ln x_W\}, \tag{1.260}$$

which breaks down to merely

$$a_W = x_W \tag{1.261}$$

since an exponential is the inverse function of a logarithm. The molar fraction of water in the food is the ratio of moisture concentration, C_W, to total molar concentration, C, so one obtains

$$a_W = \frac{C_W}{C} \tag{1.262}$$

from Eq. (1.261) – where isolation of C_W, complemented with division of both sides by $C_{W,ml}$, give promptly rise to

$$\boxed{\frac{C_W}{C_{W,ml}} = \frac{C}{C_{W,ml}} a_W;} \tag{1.263}$$

Eq. (1.263) is plotted in Fig. 1.22. The linear behavior of a_W with C_W in the vicinity of $a_W=1$ is expected, as per Randall's rule – which states that the activity coefficient of water tends to unity when $C_W \to C$; this means that a_W becomes coincident with x_W, as per Eq. (3.156) of *Food Proc. Eng.: basics & mechanical operations*. Therefore, the solid line in Fig. 1.22 drives the sorption isotherm at high water activities, and intercepts the origin of coordinates if extrapolated toward $a_W \to 0$ (see dashed portion);

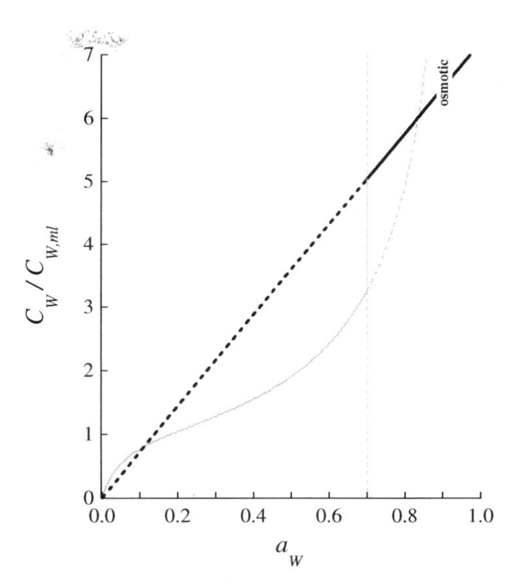

FIGURE 1.22 Variation of ratio of water concentration in food, C_W, to saturating water concentration in monolayer, $C_{W,ml}$, versus water activity, a_W, in typical solid-like food – accounted for by osmotic pressure as dominating phenomenon (——), overlaid on result of previous phenomena (——).

the vertical asymptote arising from capillary condensation accordingly drives $C_W/C_{W,ml}$ just until the said asymptote crosses the solid line in Fig. 1.22 – with the latter leading the behavior of $C_W/C_{W,ml}$ vs. a_W thereafter.

PROBLEM 1.7

Osmotic pressure is a result of the tendency for coordinated movement of solvent molecules in a single direction through a membrane impermeable to solute molecules, driven by diffusion down a concentration gradient of solvent – which will eventually cause bulk displacement of the system. Such a phenomenon will take place until chemical equilibrium is reached, characterized in general by Eq. (3.112) of *Food Proc. Eng.: basics & mechanical operations*; it will be stable if

$$\left(\frac{\partial^2 G}{\partial n_i^2}\right)_{P,T,n_{j\neq i}} > 0, \tag{1.7.1}$$

further to $\left(\partial G / \partial n_i\right)_{P,T,n_{j\neq i}} = 0$ are satisfied – corresponding to a minimum of G with regard to n_i, under constant pressure, P, and temperature, T. Find the alias of Eq. (1.7.1) applying to a stable chemical equilibrium of a system formed by N components, resorting only to partial derivatives of molar Gibbs' energy with regard to the various mole fractions.

Solution

After rewriting Eq. (1.7.1) as

$$\frac{\partial}{\partial n_i}\left(\frac{\partial G}{\partial n_i}\right)_{P,T,n_{j\neq i}} > 0 \tag{1.7.2}$$

as per the definition of second-order derivative, one may retrieve Eq. (3.36) of *Food Proc. Eng.: basics & mechanical operations* to get

$$\frac{\partial \mu_i}{\partial n_i} > 0 \tag{1.7.3}$$

– with μ_i denoting chemical potential of i-th component. On the other hand, the combination of Eqs. (3.36) and (3.40) of *Food Proc. Eng.: basics & mechanical operations* has it that

$$\mu_i = \frac{\partial}{\partial n_i}(ng), \tag{1.7.4}$$

where n denotes total number of moles as per Eq. (3.38) of *Food Proc. Eng.: basics & mechanical operations*, and g denotes molar Gibbs' energy; application of the classical rules of differentiation gives then rise to

$$\mu_i = g\frac{\partial n}{\partial n_i} + n\frac{\partial g}{\partial n_i}. \tag{1.7.5}$$

Differentiation of both sides of Eq. (3.38) of *Food Proc. Eng.: basics & mechanical operations* produces

$$\frac{\partial n}{\partial n_i} = 1, \tag{1.7.6}$$

so Eq. (1.7.5) simplifies to

$$\mu_i = g + n\frac{\partial g}{\partial n_i}; \tag{1.7.7}$$

in view of Eq. (1.7.7), one may redo Eq. (1.7.3) as

$$\frac{\partial}{\partial n_i}\left(g + n\frac{\partial g}{\partial n_i}\right) > 0, \tag{1.7.8}$$

or else

$$\frac{\partial g}{\partial n_i} + \frac{\partial}{\partial n_i}\left(n\frac{\partial g}{\partial n_i}\right) > 0 \tag{1.7.9}$$

after applying the rule of differentiation of a sum. The partial derivative of g with regard to n_i may be computed as

$$\left(\frac{\partial g}{\partial n_i}\right)_{n_{j\neq i}} = \sum_{\substack{j=1\\ j\neq i}}^{N}\left(\frac{\partial g}{\partial x_j}\right)_{x_{k\neq i,j}}\left(\frac{\partial x_j}{\partial n_i}\right)_{n_{k\neq i,j}} \tag{1.7.10}$$

via the chain (partial) differentiation rule; x_j may, in turn, appear as

$$x_j = \frac{n_j}{n_i + \sum_{\substack{k=1\\ k\neq i}}^{N} n_k}, \tag{1.7.11}$$

upon combining Eqs. (3.38) and Eq. (3.44) of *Food Proc. Eng.: basics & mechanical operations*, and making the i-th term explicit afterward in denominator – so one may proceed to differentiation as

$$\left(\frac{\partial x_j}{\partial n_i}\right)_{n_{k\neq i,j}} = -\frac{n_j}{\left(n_i + \sum_{\substack{k=1\\ k\neq i}}^{N} n_k\right)^2} \tag{1.7.12}$$

since $n_{i\neq j}$ appears only in the denominator of Eq. (1.7.11) – or, equivalently,

$$\left(\frac{\partial x_j}{\partial n_i}\right)_{n_{k\neq i,j}} = -\frac{n_j}{n^2} \tag{1.7.13}$$

with the aid again of Eq. (3.38) of *Food Proc. Eng.: basics & mechanical operations*. Insertion of Eq. (1.7.13) transforms Eq. (1.7.10) to

$$\left(\frac{\partial g}{\partial n_i}\right)_{n_{j\neq i}} = \sum_{\substack{j=1\\ j\neq i}}^{N}\left(\frac{\partial g}{\partial x_j}\right)_{x_{k\neq i,j}}\left(-\frac{n_j}{n^2}\right) = -\frac{1}{n}\sum_{\substack{j=1\\ j\neq i}}^{N}\frac{n_j}{n}\left(\frac{\partial g}{\partial x_j}\right)_{x_{k\neq i,j}}; \tag{1.7.14}$$

Eq. (3.44) of *Food Proc. Eng.: basics & mechanical operations* may be invoked once more to obtain

$$\frac{\partial g}{\partial n_i} = -\frac{1}{n}\sum_{\substack{j=1\\ j\neq i}}^{N} x_j\frac{\partial g}{\partial x_j} \tag{1.7.15}$$

using a simplified notation – or, in operator form,

$$\left(\frac{\partial}{\partial n_i}\right)g = \left(-\frac{1}{n}\sum_{\substack{j=1\\ j\neq i}}^{N} x_j\left(\frac{\partial}{\partial x_j}\right)\right)g. \tag{1.7.16}$$

Multiplication of both sides of Eq. (1.7.15) by n gives rise to

$$n\frac{\partial g}{\partial n_i} = -\sum_{\substack{j=1\\ j\neq i}}^{N} x_j\frac{\partial g}{\partial x_j}, \tag{1.7.17}$$

as all partial differential operators assume n to remain constant. The partial derivative of $n\partial g/\partial n_i$, with regard to n_i, may be seen as the outcome of applying the operator conveyed by Eq. (1.7.16) to Eq. (1.7.17), according to

$$\frac{\partial}{\partial n_i}\left(n\frac{\partial g}{\partial n_i}\right) = \left(-\frac{1}{n}\sum_{\substack{j=1\\ j\neq i}}^{N} x_j\left(\frac{\partial}{\partial x_j}\right)\right)\left(-\sum_{\substack{k=1\\ k\neq i}}^{N} x_k\frac{\partial g}{\partial x_k}\right); \tag{1.7.18}$$

cancellation of minus signs with each other, followed by application of the rule of differentiation of a sum yield

$$\frac{\partial}{\partial n_i}\left(n\frac{\partial g}{\partial n_i}\right) = \frac{1}{n}\sum_{\substack{j=1\\ j\neq i}}^{N} x_j\sum_{\substack{k=1\\ k\neq i}}^{N}\frac{\partial}{\partial x_j}\left(x_k\frac{\partial g}{\partial x_k}\right). \tag{1.7.19}$$

The rule of differentiation of a product then converts Eq. (1.7.19) to

$$\frac{\partial}{\partial n_i}\left(n\frac{\partial g}{\partial n_i}\right) = \frac{1}{n}\sum_{\substack{j=1\\ j\neq i}}^{N} x_j\sum_{\substack{k=1\\ k\neq i}}^{N}\left(\frac{\partial x_k}{\partial x_j}\frac{\partial g}{\partial x_k} + x_k\frac{\partial^2 g}{\partial x_j\partial x_k}\right), \tag{1.7.20}$$

where the summation of a sum coinciding with the corresponding sum of summations permits further transformation to

$$\frac{\partial}{\partial n_i}\left(n\frac{\partial g}{\partial n_i}\right)=\frac{1}{n}\sum_{\substack{j=1\\j\neq i}}^{N}x_j\left(\sum_{\substack{k=1\\k\neq i}}^{N}\frac{\partial x_k}{\partial x_j}\frac{\partial g}{\partial x_k}+\sum_{\substack{k=1\\k\neq i}}^{N}x_k\frac{\partial^2 g}{\partial x_j\partial x_k}\right);\qquad(1.7.21)$$

the first inner summation reduces to just its j-th term, since all other molar fractions are taken as constant when applying $\partial/\partial x_j$ – so Eq. (1.7.21) reduces to

$$\frac{\partial}{\partial n_i}\left(n\frac{\partial g}{\partial n_i}\right)=\frac{1}{n}\sum_{\substack{j=1\\j\neq i}}^{N}x_j\left(\frac{\partial g}{\partial x_j}+\sum_{\substack{k=1\\k\neq i}}^{N}x_k\frac{\partial^2 g}{\partial x_j\partial x_k}\right),\qquad(1.7.22)$$

which may be rewritten as

$$\frac{\partial}{\partial n_i}\left(n\frac{\partial g}{\partial n_i}\right)=\frac{1}{n}\sum_{\substack{j=1\\j\neq j}}^{N}x_j\frac{\partial g}{\partial x_j}+\frac{1}{n}\sum_{\substack{j=1\\j\neq i}}^{N}\sum_{\substack{k=1\\k\neq i}}^{N}x_jx_k\frac{\partial^2 g}{\partial x_j\partial x_k}\qquad(1.7.23)$$

upon elimination of parenthesis. Equations (1.7.15) and (1.7.22) may finally be inserted in Eq. (1.7.9) to get

$$-\frac{1}{n}\sum_{\substack{j=1\\j\neq i}}^{N}x_j\frac{\partial g}{\partial x_j}+\frac{1}{n}\sum_{\substack{j=1\\j\neq i}}^{N}x_j\frac{\partial g}{\partial x_j}+\frac{1}{n}\sum_{\substack{j=1\\j\neq i}}^{N}\sum_{\substack{k=1\\k\neq i}}^{N}x_jx_k\frac{\partial^2 g}{\partial x_j\partial x_k}>0,\qquad(1.7.24)$$

where cancellation of the first two terms for being the negative of each other permits simplification to

$$\frac{1}{n}\sum_{\substack{j=1\\j\neq i}}^{N}\sum_{\substack{k=1\\k\neq i}}^{N}x_jx_k\frac{\partial^2 g}{\partial x_j\partial x_k}>0;\qquad(1.7.25)$$

upon multiplication of both sides by $n>0$, Eq. (1.7.25) becomes

$$\sum_{\substack{j=1\\j\neq i}}^{N}\sum_{\substack{k=1\\k\neq i}}^{N}x_jx_k\frac{\partial^2 g}{\partial x_j\partial x_k}>0;\;\; i=1,2,\ldots,N.\qquad(1.7.26)$$

Since no restriction has been placed upon any mole fraction in Eq. (1.7.26), all N equations at stake will yield a positive result as long as every term is positive as well, i.e.

$$x_jx_k\frac{\partial^2 g}{\partial x_j\partial x_k}>0;\qquad(1.7.27)$$

this is equivalent to stating

$$\frac{\partial^2 g}{\partial x_j\partial x_k}>0;\;\; j=1,2,\ldots,N;k=1,2,\ldots,N\qquad(1.7.28)$$

because, by definition, a mole fraction is always positive. The N^2 inequalities labeled as Eq. (1.7.28) guarantee that the chemical equilibrium, described by $\partial G/\partial n_i=0$, is stable – i.e. associated with a minimum of G; and resort only to (second-order) partial derivatives of g with regard to the mole fractions of all combinations of components, as initially intended.

1.1.3.5 pH

1.1.3.5.1 Buffer range

A weak acid, HA, may undergo protolysis, and accordingly release H^+ and A^- – as described by its protolysis constant, K_a; the definition of equilibrium constant, as per Eq. (1.113) applying to an ideal solution (i.e. $\gamma_i=1$), conveys a relationship between mole concentrations of those species that looks like

$$K_a\equiv\frac{\frac{C_{H^+}C_{A^-}}{C_{HA}}}{C}.\qquad(1.264)$$

Since total mole concentration, C, is normally dominated by the concentration of solvent water (ca. 55 mol.L^{-1}) and remains essentially constant, one has classically lumped it with K_a to produce

$$\boxed{K_a^*=\frac{C_{H^+}C_{A^-}}{C_{HA}}}\qquad(1.265)$$

– with K_a^* given by

$$K_a^*\equiv K_aC,\qquad(1.266)$$

in close agreement with Eq. (1.115), and thus bearing units of mol.L^{-1}. An identical reasoning applies to (auto)protolysis of water, generating H^+ and OH^-, viz.

$$\boxed{K_w^*=C_{H^+}C_{OH^-};}\qquad(1.267)$$

remember that this spontaneous chemical process accounts for the non-nil (although low, ca. 5.5×10^{-6} S.m^{-1}) conductivity of water.

If HA represents a (uniprotic) strong acid – hereafter referred to via subscript sa, at initial concentration C_{bf}, it completely ionizes in water upon dissolution; molar concentrations $C_{H^+,sa}$ of H^+ and $C_{A^-,sa}$ of its conjugate anion, A^-, accordingly satisfy

$$C_{H^+,sa}=C_{A^-,sa}=C_{bf}.\qquad(1.268)$$

Since this protolysis occurs quantitatively, the only relevant chemical equilibrium is that associated with Eq. (1.267) – which allows one to write

$$K_w^*=\left(C_{H^+,w}+C_{H^+,sa}\right)C_{OH^-,w},\qquad(1.269)$$

knowing that the total hydrogen ion concentration reads

$$C_{H^+}=C_{H^+,w}+C_{H^+,sa};\qquad(1.270)$$

here $C_{H^+,w}$ and $C_{OH^-,w}$ denote molar concentration of H^+ and OH^-, respectively, arising from dissociation of water. The said autoprotolysis is characterized by 1:1 stoichiometry, i.e.

$$C_{OH^-,w} = C_{H^+,w}; \tag{1.271}$$

hence, combination of Eqs. (1.268), (1.269), and (1.271) generates

$$K_w^* = \left(C_{H^+,w} + C_{bf}\right)C_{H^+,w}. \tag{1.272}$$

Since K_w^* and C_{bf} are known and given, in agreement with prevailing environmental conditions (namely temperature that affects K_w^*, and buffer initial composition measured by C_{bf}), Eq. (1.272) is univariate on $C_{H^+,w}$; upon elimination of parenthesis, one obtains

$$C_{H^+,w}{}^2 + C_{bf}C_{H^+,w} - K_w^* = 0. \tag{1.273}$$

Inspection of Eq. (1.273) unfolds a quadratic equation (in its canonical form) – which may be solved as

$$C_{H^+,w} = \frac{-C_{bf} \pm \sqrt{C_{bf}{}^2 + 4K_w^*}}{2}; \tag{1.274}$$

$\sqrt{C_{bf}{}^2 + 4K_w^*} > \sqrt{C_{bf}{}^2} = C_{bf}$, so one promptly realizes that only the plus sign preceding the square root will produce a physically meaningful value for $C_{H^+,w}$ – i.e. Eq. (1.274) reduces in practice to

$$C_{H^+,w} = \frac{\sqrt{C_{bf}{}^2 + 4K_w^*} - C_{bf}}{2} \tag{1.275}$$

or, after taking the denominator inside the root sign,

$$C_{H^+,w} = \sqrt{K_w^* + \left(\frac{C_{bf}}{2}\right)^2} - \frac{1}{2}C_{bf}. \tag{1.276}$$

Combination with Eqs. (1.268) and (1.276) converts Eq. (1.270) to

$$C_{H^+} = \sqrt{K_w^* + \left(\frac{C_{bf}}{2}\right)^2} - \frac{1}{2}C_{bf} + C_{bf}, \tag{1.277}$$

or else

$$C_{H^+} = \frac{1}{2}C_{bf} + \sqrt{K_w^* + \left(\frac{C_{bf}}{2}\right)^2} \tag{1.278}$$

upon pooling terms alike. The classical definition of *pH*, i.e.

$$\boxed{pH \equiv -\log_{10} C_{H^+}}, \tag{1.279}$$

may now be invoked – preferably in the form

$$C_{H^+} = 10^{-pH} \tag{1.280}$$

obtained after taking exponentials of the negatives both sides; in view of Eq. (1.280), one may redo Eq. (1.278) to

$$10^{-pH} = \frac{1}{2}C_{bf} + \sqrt{K_w^* + \left(\frac{C_{bf}}{2}\right)^2} \tag{1.281}$$

– while logarithms can now be taken of both sides as

$$\boxed{pH\Big|_{sa} = -\log_{10}\left\{\frac{1}{2}C_{bf} + \sqrt{K_w^* + \left(\frac{C_{bf}}{2}\right)^2}\right\}}, \tag{1.282}$$

applicable to an aqueous solution of a strong acid. The variation of *pH* with C_{bf} is illustrated in Fig. 1.23. Inspection of this figure indicates that *pH* essentially coincides with the negative of the decimal logarithm of the strong acid concentration, C_{bf}, above ca. 10^{-7} mol.L^{-1}; below said threshold, the buffering effect of water overrides, and thus sets *pH* to ca. 7. This realization is expected because $K_w^* + (C_{bf}/2)^2 \approx (C_{bf}/2)^2$ in the former range, so its square root becomes essentially equal to $C_{bf}/2$ that adds to $C_{bf}/2$ as per Eq. (1.282) to give C_{bf}, i.e.

$$\left(pH\Big|_{sa} \approx -\log_{10} C_{bf}\right)_{\frac{C_{bf}}{2} >> \sqrt{K_w^*}}; \tag{1.283}$$

the opposite situation corresponds to $K_w^* >> (C_{bf}/2)^2$, so the square root in Eq. (1.282) reduces to $\sqrt{K_w^*}$ and thus $C_{bf}/2 + \sqrt{K_w^*} \approx \sqrt{K_w^*}$, according to

$$\left(pH\Big|_{sa} \approx \frac{1}{2}pK_w^*\right)_{\frac{C_{bf}}{2} << \sqrt{K_w^*}} \tag{1.284}$$

– or, in other words, the resulting *pH* coincides with $-\frac{1}{2}\log_{10} K_w^*$, which gives ca. 7 since $K_w^* \approx 10^{-14}$ mol^2.L^{-2}.

Consider now the general case of a buffer constituted by a uniprotic acid and a salt of the corresponding anion, with overall concentration C_{bf} abiding to

$$C_{bf} = C_{HA}\Big|_{t=0} + C_{AB}\Big|_{t=0} \tag{1.285}$$

– where $C_{HA}\big|_{t=0}$ and $C_{AB}\big|_{t=0}$ denote starting molar concentrations of (weak) acid and (strongly soluble) salt AB used to formulate

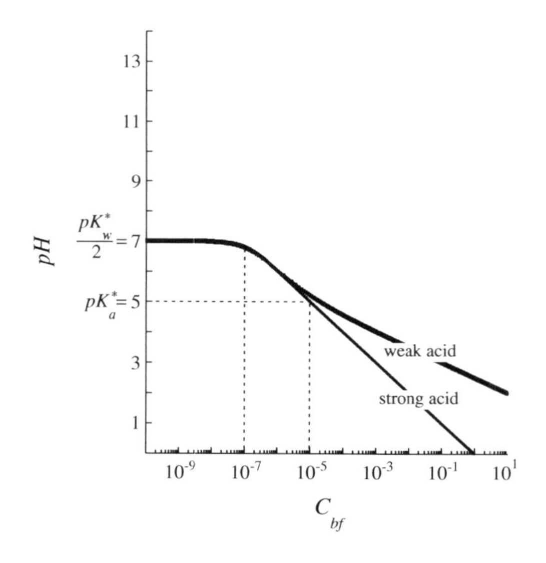

FIGURE 1.23 Variation of *pH* versus concentration of buffer, C_{bf}, for solution of strong acid (——) or weak acid (............, with $pK_a^*=5$) in water (with $pK_w^*=14$).

such a buffer; once chemical equilibrium is attained (normally a fast process), one obtains

$$C_{bf} = C_{HA} + C_{A^-,wa} + C_{A^-,st} \tag{1.286}$$

to be used *in lieu* of Eq. (1.285). Here $C_{A^-,wa}$ denotes molar concentration of A^- resulting from (partial) ionization of HA, and $C_{A^-,st}$ denotes molar concentration of extra A^- in solution resulting from (complete) ionization of AB; upon isolation of C_{HA}, Eq. (1.286) produces

$$C_{HA} = C_{bf} - C_{A^-,wa} - C_{A^-,st}. \tag{1.287}$$

On the other hand,

$$C_{A^-,wa} = C_{H^+,wa} \tag{1.288}$$

– in parallel to the first equality in Eq. (1.268), and again a consequence of the 1:1 stoichiometry of anion and cation released by acid dissociation; here $C_{H^+,wa}$ denotes concentration of H^+ resulting from ionization of HA. The other contribution to the total inventory of H^+ ions is $C_{H^+,w}$ – arising from auto-ionization of water, and already described by Eq. (1.271). The equilibrium constants K_a^* and K_w^* provide extra relationships between total concentrations of H^+ and A^- – according to

$$K_a^* = \frac{\left(C_{H^+,wa} + C_{H^+,w}\right)\left(C_{A^-,wa} + C_{A^-,st}\right)}{C_{HA}} \tag{1.289}$$

based on Eq. (1.265), coupled with realization that C_{H^+} receives contributions from ionization of weak acid ($C_{H^+,wa}$) and water ($C_{H^+,w}$), and that C_{A^-} receives contributions from ionization of weak acid ($C_{A^-,wa}$) and salt ($C_{A^-,st}$); by the same token, one finds

$$K_w^* = \left(C_{H^+,wa} + C_{H^+,w}\right)C_{OH^-} \tag{1.290}$$

stemming from Eq. (1.267). Upon combination with Eqs. (1.287) and (1.288), one obtains

$$K_a^* = \frac{\left(C_{H^+,wa} + C_{H^+,w}\right)\left(C_{H^+,wa} + C_{A^-,st}\right)}{C_{bf} - C_{H^+,wa} - C_{A^-,st}}, \tag{1.291}$$

from Eq. (1.289), and Eq. (1.290) likewise becomes

$$K_w^* = \left(C_{H^+,wa} + C_{H^+,w}\right)C_{H^+,w} \tag{1.292}$$

at the expense of Eq. (1.271). Elimination of parenthesis in Eq. (1.292) produces

$$C_{H^+,w}^{\ \ 2} + C_{H^+,wa}C_{H^+,w} - K_w^* = 0, \tag{1.293}$$

so a quadratic equation emerges again; its solving formula has it that

$$C_{H^+,w} = \frac{-C_{H^+wa} \pm \sqrt{C_{H^+,wa}^{\ \ 2} + 4K_w^*}}{2}, \tag{1.294}$$

where the function under the root sign being larger than $C_{H^+,wa}^{\ \ 2}$ implies that the root itself is larger than $C_{H^+,wa}$ – and thus rules out the negative sign preceding the root for physical consistency, i.e. Eq. (1.294) reduces to

$$C_{H^+,w} = \frac{\sqrt{C_{H^+,wa}^{\ \ 2} + 4K_w^*} - C_{H^+,wa}}{2}. \tag{1.295}$$

Since K_w^* is normally small compared to $C_{H^+,wa}^{\ \ 2}$, one may resort to the very definition of differential of function $\sqrt{\xi}$ to write

$$\begin{aligned} \left.\sqrt{\xi}\right|_{\xi=C_{H^+,wa}^{\ \ 2}+4K_w^*} &\approx \left.\sqrt{\xi}\right|_{\xi=C_{H^+,wa}^{\ \ 2}} \\ &+ \left.\frac{d\sqrt{\xi}}{d\xi}\right|_{\xi=C_{H^+,wa}^{\ \ 2}} \left(\left.\xi\right|_{\xi=C_{H^+,wa}^{\ \ 2}+4K_w^*} - \left.\xi\right|_{\xi=C_{H^+,wa}^{\ \ 2}}\right); \end{aligned} \tag{1.296}$$

selective replacement of ξ by the values indicated as subscript, followed by calculation of the indicated derivative yield

$$\begin{aligned} &\sqrt{C_{H^+,wa}^{\ \ 2} + 4K_w^*} - \sqrt{C_{H^+,wa}^{\ \ 2}} \\ &= \left.\frac{1}{2\sqrt{\xi}}\right|_{\xi=C_{H^+,wa}^{\ \ 2}} \left(C_{H^+,wa}^{\ \ 2} + 4K_w^* - C_{H^+,wa}^{\ \ 2}\right) \end{aligned} \tag{1.297}$$

Cancellation of symmetrical terms, and of square root with square power permit simplification of Eq. (1.297) to

$$\sqrt{C_{H^+,wa}^{\ \ 2} + 4K_w^*} - C_{H^+,wa} = \frac{1}{2C_{H^+,wa}}4K_w^*, \tag{1.298}$$

or else

$$\sqrt{C_{H^+,wa}^{\ \ 2} + 4K_w^*} - C_{H^+,wa} = \frac{2K_w^*}{C_{H^+,wa}}. \tag{1.299}$$

Insertion of Eq. (1.299) converts Eq. (1.295) to

$$C_{H^+,w} = \frac{\dfrac{2K_w^*}{C_{H^+,wa}}}{2}, \tag{1.300}$$

or merely

$$C_{H^+,w} = \frac{K_w^*}{C_{H^+,wa}} \tag{1.301}$$

once similar factors have been canceled between numerator and denominator; Eq. (1.301) permits elimination of C_{H^+w}, from Eq. (1.291), i.e.

$$K_a^* = \frac{\left(C_{H^+,wa} + \dfrac{K_w^*}{C_{H^+,wa}}\right)\left(C_{H^+,wa} + C_{A^-,st}\right)}{C_{bf} - C_{H^+,wa} - C_{A^-,st}}. \tag{1.302}$$

For the acids, and corresponding initial concentrations relevant in food practice, one realizes that $C_{H^+,wa} >> C_{H^+,w}$, so one may write

$$C_{H^+,wa} + C_{H^+,w} = C_{H^+,wa} + \frac{K_w^*}{C_{H^+,wa}} \approx C_{H^+,wa} \tag{1.303}$$

at the expense of Eq. (1.301); therefore, Eq. (1.302) simplifies to

$$K_a^* \approx \frac{C_{H^+,wa}\left(C_{H^+,wa}+C_{A^-,st}\right)}{C_{bf}-C_{H^+,wa}-C_{A^-,st}}. \quad (1.304)$$

Upon elimination of denominators, Eq. (1.304) becomes

$$C_{H^+,wa}\left(C_{H^+,wa}+C_{A^-,st}\right)=K_a^*\left(C_{bf}-C_{A^-,st}\right)-K_a^*C_{H^+,wa}, \quad (1.305)$$

while removal of parenthesis in the left-hand side followed by factoring out of $C_{H^+,wa}$ give rise to

$$C_{H^+,wa}{}^2+\left(K_a^*+C_{A^-,st}\right)C_{H^+,wa}-K_a^*\left(C_{bf}-C_{A^-,st}\right)=0; \quad (1.306)$$

the solving formula of a quadratic equation may again be invoked to write

$$C_{H^+,wa} = \frac{-\left(K_a^*+C_{A^-,st}\right)\pm\sqrt{\left(K_a^*+C_{A^-,st}\right)^2+4K_a^*\left(C_{bf}-C_{A^-,st}\right)}}{2}. \quad (1.307)$$

Since $\left(K_a^*+C_{A^-,st}\right)^2+4K_a^*\left(C_{bf}-C_{A^-,st}\right)$ lies above $\left(K_a^*+C_{A^-,st}\right)^2$, one realizes that $\sqrt{\left(K_a^*+C_{A^-,st}\right)^2+4K_a^*\left(C_{bf}-C_{A^-,st}\right)}$ is greater than $K_a^*+C_{A^-,st}$; hence, Eq. (1.307) should be replaced by

$$C_{H^+,wa}=\frac{\sqrt{\left(K_a^*+C_{A^-,st}\right)^2+4K_a^*\left(C_{bf}-C_{A^-,st}\right)}-\left(K_a^*+C_{A^-,st}\right)}{2} \quad (1.308)$$

for physical significance, which may be algebraically rearranged to read

$$C_{H^+,wa}=\sqrt{\left(\frac{K_a^*+C_{A^-,st}}{2}\right)^2+K_a^*\left(C_{bf}-C_{A^-,st}\right)}-\frac{K_a^*+C_{A^-,st}}{2}. \quad (1.309)$$

After retrieving the analog to Eq. (1.270) as

$$C_{H^+}=C_{H^+,w}+C_{H^+,wa} \quad (1.310)$$

applying to a weak acid, one can proceed to simplification as

$$C_{H^+,wa}\approx C_{H^+} \quad (1.311)$$

in view of Eq. (1.303); based on Eq. (1.280), one may redo Eq. (1.311) to

$$C_{H^+,wa}=10^{-pH}, \quad (1.312)$$

so Eq. (1.309) will be eventually reformulated to

$$10^{-pH}=\sqrt{\left(\frac{K_a^*+C_{A^-,st}}{2}\right)^2+K_a^*\left(C_{bf}-C_{A^-,st}\right)}-\frac{K_a^*+C_{A^-,st}}{2}. \quad (1.313)$$

Upon applying decimal logarithms to both sides, and taking their negatives afterward, Eq. (1.313) will appear as

$$\boxed{pH\big|_{wa}=-\log_{10}\left\{\sqrt{\left(\frac{K_a^*+C_{A^-,st}}{2}\right)^2+K_a^*\left(C_{bf}-C_{A^-,st}\right)}-\frac{K_a^*+C_{A^-,st}}{2}\right\}} \quad (1.314)$$

– which exhibits a functional form quite distinct from that of Eq. (1.282) pertaining to a strong acid.

If no conjugate base is initially added to the acid while formulating the buffer, then $C_{A^-,st}=0$ – so Eq. (1.314) will reduce to

$$\boxed{pH\big|_{C_{AB}|_{t=0}=0}=-\log_{10}\left\{\sqrt{\left(\frac{K_a^*}{2}\right)^2+K_a^*C_{bf}}-\frac{K_a^*}{2}\right\}} \quad (1.315)$$

as plotted in Fig. 1.23; pH does not change so abruptly with C_{bf}, at larger concentrations, as if a strong acid were at stake. In fact, expansion of the square root in Eq. (1.315) via Taylor's series, about $C_{bf}=0$, truncated after the quadratic term, gives rise to

$$\sqrt{\left(\frac{K_a^*}{2}\right)^2+K_a^*C_{bf}}\approx\sqrt{\left(\frac{K_a^*}{2}\right)^2+K_a^*C_{bf}}\,\Bigg|_{C_{bf}=0}+\frac{1}{2}\left(\left(\frac{K_a^*}{2}\right)^2+K_a^*C_{bf}\right)^{-\frac{1}{2}}K_a^*\Big|_{C_{bf}=0}C_{bf}+\frac{1}{2}K_a^*\left(-\frac{1}{2}\right)\left(\left(\frac{K_a^*}{2}\right)^2+K_a^*C_{bf}\right)^{-\frac{3}{2}}K_a^*\Big|_{C_{bf}=0}\frac{C_{bf}{}^2}{2}, \quad (1.316)$$

which becomes merely

$$\sqrt{\left(\frac{K_a^*}{2}\right)^2+K_a^*C_{bf}}=\frac{K_a^*}{2}+\frac{K_a^*}{2}\left(\frac{K_a^*}{2}\right)^{-1}C_{bf}-\frac{K_a^{*2}}{8}\left(\frac{K_a^*}{2}\right)^{-3}C_{bf}{}^2 \quad (1.317)$$

when C_{bf} is set equal to zero (as indicated); Eq. (1.317) degenerates to

$$\sqrt{\left(\frac{K_a^*}{2}\right)^2+K_a^*C_{bf}}=\frac{K_a^*}{2}+C_{bf}-\frac{C_{bf}{}^2}{K_a^*} \quad (1.318)$$

after lumping factors alike. Subtraction of $K_a^*/2$ from both sides transforms Eq. (1.318) to

$$\sqrt{\left(\frac{K_a^*}{2}\right)^2+K_a^*C_{bf}}-\frac{K_a^*}{2}=C_{bf}-\frac{C_{bf}{}^2}{K_a^*}<C_{bf}, \quad (1.319)$$

since C_{bf}^2/K_a^* is positive; decimal logarithms may now be taken of both sides to obtain

$$\log_{10}\left\{\sqrt{\left(\frac{K_a^*}{2}\right)^2 + K_a^* C_{bf}} - \frac{K_a^*}{2}\right\} < \log_{10} C_{bf}, \quad (1.320)$$

– where advantage was taken of the monotonically increasing nature of a logarithm function. If negatives of both sides are, in turn, taken, then Eq. (1.320) becomes

$$-\log_{10}\left\{\sqrt{\left(\frac{K_a^*}{2}\right)^2 + K_a^* C_{bf}} - \frac{K_a^*}{2}\right\} > -\log_{10} C_{bf}, \quad (1.321)$$

or else

$$\left(pH\Big|_{C_{AB}|_{t=0}=0} > pH\Big|_{sa}\right)\Bigg|_{C_{bf}|_{t=0} >> \sqrt{K_w^*}} \quad (1.322)$$

with the help of Eqs. (1.283) and (1.315); one accordingly concludes that the pH associated with a weak acid is larger than that associated with a strong acid, for the same overall concentration of acid (C_{bf}) – as easily grasped in Fig. 1.23. Note that Eq. (1.314) fails to be applied when $C_{bf} < 10^{-7}$ mol.L^{-1}, because the simplification conveyed by Eq. (1.303) is no longer valid – due to dominance of protolysis of water within that range. The correct equation to be used is then Eq. (1.302), instead of Eq. (1.304) – with the former retrieved as

$$K_a^* = \frac{\left(C_{H^+,wa}^{\ 2} + K_w^*\right)\left(C_{H^+,wa} + C_{A^-,st}\right)}{C_{H^+,wa}\left(C_{bf} - C_{H^+,wa} - C_{A^-,st}\right)}, \quad (1.323)$$

following multiplication of both numerator and denominator by $C_{H^+,wa}$; this ultimately leads to a cubic equation in $C_{H^+,wa}$, viz.

$$\begin{aligned} &K_a^*\left(C_{bf} - C_{A^-,st}\right)C_{H^+,wa} - K_a^* C_{H^+,wa}^{\ 2} \\ &= C_{H^+,wa}^{\ 3} + C_{A^-,st} C_{H^+,wa}^{\ 2} + K_w^* C_{H^+,wa} + K_w C_{A^-,st}, \end{aligned} \quad (1.324)$$

after elimination of denominators and application of the distributive property – which is equivalent to

$$\begin{aligned} &C_{H^+,wa}^{\ 3} + \left(K_a^* + C_{A^-,st}\right)C_{H^+,wa}^{\ 2} \\ &+ \left(K_w^* + K_a^*\left(C_{A^-,st} - C_{bf}\right)\right)C_{H^+,wa} + K_w^* C_{A^-,st} = 0, \end{aligned} \quad (1.325)$$

following straightforward algebraic rearrangement.

Inspection of Fig. 1.23 also suggests that the buffering effect of a weak acid is particularly strong when pH is close to its pK_a – note the concave curving of the pH vs. C_{bf} dependence in that region; this realization is due to the underlying equilibrium between H$^+$ and A$^-$, related by a finite equilibrium constant when a weak acid is at stake.

However, a regular buffer resorts to a weak acid and a salt of its conjugate base, in terms of initial formulation; the variation of pH in this case follows Eq. (1.314) within the region of practical interest – as graphically illustrated in Fig. 1.24, for a reasonable buffer concentration. Note that most buffering capacity – visible as lower rate of change of pH with $C_{A^-,st}$, sits again around pK_a; this corresponds to $C_{A^-,st}$ of the order of half C_{bf} – i.e. when $C_{bf} - C_{A^-,st} \approx C_{A^-,st}$ or, equivalently, $C_{HA} + C_{A^-,wa} \approx C_{A^-,st}$ as per Eq. (1.286). If $C_{HA} + C_{A^-,wa} << C_{A^-,st}$ or $C_{HA} + C_{A^-,wa} >> C_{A^-,st}$, then pH undergoes unwanted sudden and considerable changes – which are driven by vertical asymptotes in Fig. 1.24.

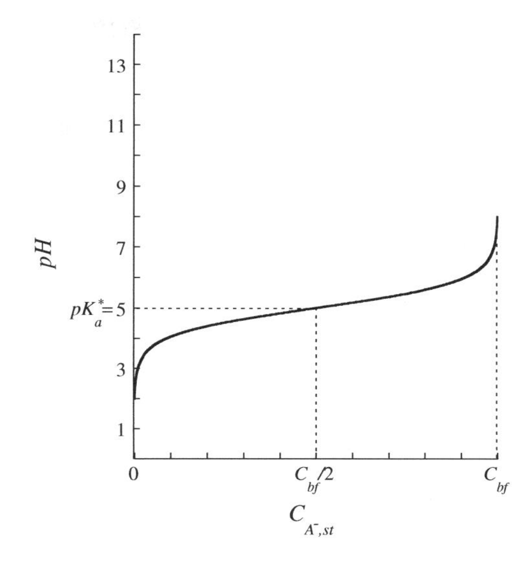

FIGURE 1.24 Variation of pH versus concentration of conjugate base supplied by salt, $C_{A^-,st}$, for solution of weak acid (with $pK_a^*=5$) in water (with $pK_w^*=14$), at an overall buffer concentration C_{bf}.

PROBLEM 1.8

As emphasized previously, chemical reactions do occur at finite rates due to the resistance offered by the energy barrier consubstantiated as activation energy; however, in some cases, they are sufficiently fast to be considered at a state of chemical equilibrium at all times – as is notably the case of protolysis reactions. When a solution is pH-buffered, the concentration of ions, arising from the (weak) acid and its conjugate salt, is relatively high; the major contribution for nonunit activity coefficients, γ_i, is then the electrical field generated by the solvated charges – and may be approximately described by

$$\ln\gamma_i = -\alpha z_i^2\sqrt{\vartheta};\quad i = 1, 2, \ldots, N, \quad (1.8.1)$$

classically referred to as Debye and Huckel's model. Here ϑ, known as ionic strength, is given by

$$\vartheta \equiv \frac{1}{2}\sum_{j=1}^{M} C_j z_j^2, \quad (1.8.2)$$

and represents an average square charge using mole concentration as weight factor; α denotes a constant, C_j denotes mole concentration of j-th ion species (with $M > N$ in general), and z_j denotes its molecular charge. Obtain $K_{eq}^* \equiv K_{eq}^*\{\vartheta\}$ in its simplest form.

Solution

After retrieving Eq. (1.116), one may insert Eq. (1.8.1) to get

$$K_{eq}^* = \prod_{i=1}^{N}\left(e^{-\alpha z_i^2\sqrt{\vartheta}}\right)^{\nu_i}\prod_{i=1}^{N}C_i^{\nu_i} \tag{1.8.3}$$

where both C_i and ϑ refer to equilibrium conditions; after combination of exponents in the composite powers, Eq. (1.8.3) becomes

$$K_{eq}^* = \prod_{i=1}^{N}e^{-\alpha\nu_i z_i^2\sqrt{\vartheta}}\prod_{i=1}^{N}C_i^{\nu_i}, \tag{1.8.4}$$

where the operational features of the exponential function allow further transformation to

$$K_{eq}^* = \exp\left\{-\alpha\sqrt{\vartheta}\sum_{i=1}^{N}\nu_i z_i^2\right\}\prod_{i=1}^{N}C_i^{\nu_i}. \tag{1.8.5}$$

Inspired by Eqs. (1.31) or (1.32), one may define Δz^2 as

$$\Delta z^2 \equiv \sum_{i=1}^{N}\nu_i z_i^2, \tag{1.8.6}$$

since it permits notation in Eq. (1.8.5) be simplified to

$$K_{eq}^* = e^{-\alpha\sqrt{\vartheta}\Delta z^2}\prod_{i=1}^{N}C_i^{\nu_i}; \tag{1.8.7}$$

Eq. (1.8.7) conveys $K_{eq}^* \equiv K_{eq}^*\{\vartheta\}$ as intended – and should be preferably used when buffer strength is high. Note, however, that $\vartheta \equiv \vartheta\{C_1, C_2, \ldots, C_N\}$ as per Eq. (1.8.2) – so one ultimately finds that $K_{eq}^* \equiv K_{eq}^*\{C_1, C_2, \ldots, C_N\}$, for a given temperature; when $\vartheta \to 0$, the exponential function in Eq. (1.8.7) approaches unity, and thus

$$\lim_{\vartheta\to 0} K_{eq}^* = \prod_{i=1}^{N}C_i^{\nu_i} \tag{1.8.8}$$

in parallel to Eq. (1.117) – i.e. the functional form commonly used to model protolysis equilibria.

1.1.3.5.2 *Buffer stability*

The usefulness of a buffer in a food hinges, in general, on how efficiently it can resist changes in *pH*; if a strong acid is deliberately or inadvertently added, the equilibrium is shifted in the HA direction in agreement with Le Châtelier's principle, see Eq. (1.43) – yet H^+ is supposed to increase less than expected based on the quantity added of the said acid. A more quantitative statement is, however, convenient in attempts to determine the amount of buffer necessary to avoid extensive *pH* changes – or else to estimate *pH* change in native foods upon some extraneous/unwanted increase or decrease in acid concentration.

Toward this goal, one may revisit Eq. (1.291) simply as

$$K_a^* = \frac{C_{H^+,wa}\left(C_{A^-,wa} + C_{A^-,st}\right)}{C_{HA}}, \tag{1.326}$$

at the expense of Eqs. (1.287), (1.288), and (1.303); isolation of C_{HA} readily ensues as

$$C_{HA} = \frac{C_{H^+,wa}\left(C_{A^-,wa} + C_{A^-,st}\right)}{K_a^*}. \tag{1.327}$$

Equations (1.286) and (1.327) thus support

$$C_{bf} = \frac{C_{H^+,wa}}{K_a^*}\left(C_{A^-,wa} + C_{A^-,st}\right) + C_{A^-,wa} + C_{A^-,st}$$

$$= \left(\frac{C_{H^+,wa}}{K_a^*} + 1\right)\left(C_{A^-,wa} + C_{A^-,st}\right) \tag{1.328}$$

along with appropriate factoring out, where isolation of $C_{A^-,wa} + C_{A^-,st}$ gives rise to

$$C_{A^-,wa} + C_{A^-,st} = \frac{C_{bf}}{1 + \dfrac{C_{H^+,wa}}{K_a^*}}; \tag{1.329}$$

upon multiplication and division of both numerator and denominator of the right-hand side by K_a^*, Eq. (1.329) turns to

$$C_{A^-,wa} + C_{A^-,st} = \frac{K_a^* C_{bf}}{K_a^* + C_{H^+,wa}}. \tag{1.330}$$

Consider now addition of a strong base, say DOH – which promptly dissociates in full as D^+ and OH^-, described by concentrations C_{D^+} and $C_{OH^-,sb}$, respectively; the associated buffering power, β, may then be defined via

$$\boxed{\beta \equiv \frac{dC_{D^+}}{dpH}} \tag{1.331}$$

– and describes how high the concentration $C_{DOH}\big|_{t=0} = C_{D^+}$ of an added strong base must be to bring about a unit change in solution *pH*. The underlying charges must add up to zero to assure solution neutrality, in agreement with

$$C_{A^-,wa} + C_{A^-,st} + C_{OH^-,sb} + C_{OH^-,w} = C_{H^+,wa} + C_{H^+,w} + C_{B^+} + C_{D^+} \tag{1.332}$$

encompassing ions H^+ and A^- from (weak) acid HA, B^+ and A^- from salt AB, D^+ and OH^- from (strong) base DOH, and H^+ and OH^- from H_2O; here $C_{OH^-,w}$ denotes concentration of hydroxyl ions released via protolysis of water. Based again on Eq. (1.303), one may simplify Eq. (1.332) to

$$C_{A^-,wa} + C_{A^-,st} + C_{OH^-,sb} + C_{OH^-,w} \approx C_{H^+,wa} + C_{B^+} + C_{D^+}, \tag{1.333}$$

thus allowing isolation of C_{D^+} as

$$C_{D^+} = C_{A^-,wa} + C_{A^-,st} + C_{OH^-,sb} + C_{OH^-,w} - C_{H^+,wa} - C_{B^+}. \tag{1.334}$$

Since $C_{OH^-,sb} + C_{OH^-,w}$ must satisfy Eq. (1.267) describing protolysis of water, one has it that

$$C_{OH^-} = C_{OH^-,sb} + C_{OH^-,w} = \frac{K_w}{C_{H^+,wa}} \tag{1.335}$$

since $C_{H^+} = C_{H^+,wa} + C_{H^+,w} \approx C_{H^+,wa}$; insertion of Eqs. (1.330) and (1.335) supports transformation of Eq. (1.334) to

$$C_{D^+} = \frac{K_a^* C_{bf}}{K_a^* + C_{H^+,wa}} + \frac{K_w^*}{C_{H^+,wa}} - C_{H^+,wa} - C_{B^+}, \tag{1.336}$$

where C_{D^+} becomes a univariate function of $C_{H^+,wa}$ because C_{bf} and $C_{AB}|_{t=0} = C_{B^+}$ are (given) constants. Upon differentiation of both sides with regard to $C_{H^+,wa}$, Eq. (1.336) generates

$$\frac{dC_{D^+}}{dC_{H^+,wa}} = -\frac{K_a^*}{\left(K_a^* + C_{H^+,wa}\right)^2} C_{bf} - \frac{K_w^*}{C_{H^+,wa}^{\ 2}} - 1; \tag{1.337}$$

one may also rephrase Eq. (1.279) as

$$pH = -\ln C_{H^+,wa} \log_{10} e \tag{1.338}$$

based on the mathematical rule supporting change of base in a logarithm, coupled with Eq. (1.303) – so differentiation of both sides with regard to $C_{H^+,wa}$ gives rise to

$$\frac{dpH}{dC_{H^+,wa}} = -\frac{\log_{10} e}{C_{H^+,wa}} \tag{1.339}$$

or, after taking reciprocals of both sides,

$$\frac{dC_{H^+,wa}}{dpH} = -\frac{C_{H^+wa}}{\log_{10} e}. \tag{1.340}$$

Reformulation of Eq. (1.331) is possible at the expense of the chain differentiation rule, viz.

$$\beta = \frac{dC_{D^+}}{dC_{H^+,wa}} \frac{dC_{H^+,wa}}{dpH}, \tag{1.341}$$

where insertion of Eqs. (1.337) and (1.340) produces

$$\beta = -\left(-\frac{K_a^*}{\left(K_a^* + C_{H^+,wa}\right)^2} C_{bf} - \frac{K_w^*}{C_{H^+,wa}^{\ 2}} - 1 \right) \frac{C_{H^+,wa}}{\log_{10} e}; \tag{1.342}$$

algebraic manipulation of Eq. (1.342) gives rise to

$$\beta = \frac{1}{\log_{10} e} \left(\frac{K_w^*}{C_{H^+,wa}} + C_{H^+,wa} + \frac{K_a^* C_{H^+,wa}}{\left(K_a^* + C_{H^+,wa}\right)^2} C_{bf} \right). \tag{1.343}$$

Since Eq. (1.279) may alternatively appear as

$$C_{H^+,wa} = 10^{-pH} \tag{1.344}$$

with the aid again of Eq. (1.303), one may redo Eq. (1.343) to

$$\boxed{\beta = \frac{1}{\log_{10} e} \left(K_w^* 10^{pH} + 10^{-pH} + \frac{K_a^* 10^{-pH}}{\left(K_a^* + 10^{-pH}\right)^2} C_{bf} \right);} \tag{1.345}$$

Eq. (1.345) gives $\beta \equiv \beta\{C_{bf}, pH\}$, as plotted in Fig. 1.25 for a fixed value of pK_a^* and selected values of C_{bf}. At low pH, the value of β tends to be overriden by 10^{-pH}, so an exponential decrease is apparent when pH increases – whereas $K_w^* 10^{pH}$ dominates at high pH, and accordingly unfolds an increasing trend with increasing pH, see Eq. (1.345); therefore, solutions are intrinsically resistant to pH changes at extreme pH values, regardless of the presence of (adventitious or added) buffer.

In the intermediate region in Fig. 1.25 – typically $[pK_a - 1, pK_a + 1]$, the last term in Eq. (1.345) becomes the most important; the effect of C_{bf} then rises (unlike previous situations, where it was essentially absent), and pH now appears as an exponent in both numerator and denominator – thus creating the possibility for existence of a maximum. The necessary condition for the said maximum reads

$$\frac{\partial \beta}{\partial pH} = 0, \tag{1.346}$$

or else

$$\frac{\partial \beta}{\partial 10^{-pH}} \frac{d10^{-pH}}{dpH} = 0 \tag{1.347}$$

– obtained via the chain differentiation rule, using 10^{-pH} as intermediate variable; given the non-nil nature of the second factor (actually equal to $10^{-pH}(-\ln 10) \neq 0$, since 10^{-pH} coincides with $e^{-pH \ln 10}$), Eq. (1.347) reduces to

$$\frac{\partial \beta}{\partial 10^{-pH}} = 0. \tag{1.348}$$

After recalling that the dominant term under the parenthesis in Eq. (1.345), within the intermediate range under scrutiny, is $K_a^* 10^{-pH} C_{bf} / (K_a^* + 10^{-pH})^2$, one obtains

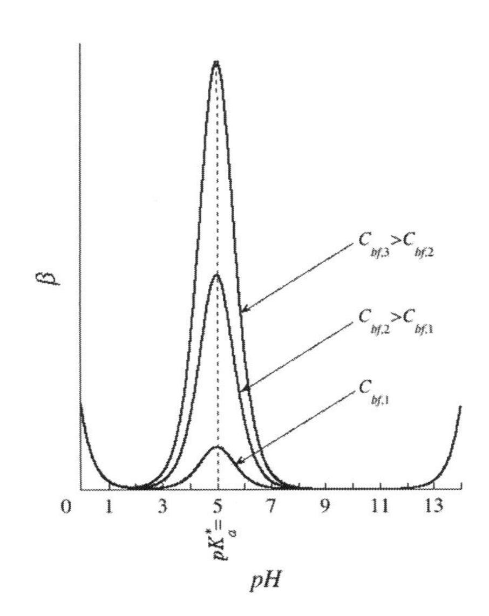

FIGURE 1.25 Variation of buffering capacity, β, versus pH, for solution of weak acid (with $pK_a^* = 5$) in water (with $pK_w^* = 14$), at various overall buffer concentrations, i.e. $C_{bf,1}$, $C_{bf,2}$ and $C_{bf,3}$.

$$\frac{\partial \beta}{\partial 10^{-pH}} \approx \frac{K_a^*}{\log_{10} e} \frac{d}{d10^{-pH}} \left(\frac{10^{-pH}}{\left(K_a^* + 10^{-pH}\right)^2} \right) C_{bf}, \tag{1.349}$$

it can be redone to

$$\frac{\partial \beta}{\partial 10^{-pH}} = \frac{K_a^*}{\log_{10} e} \frac{\left(K_a^* + 10^{-pH}\right)^2 - 10^{-pH} 2\left(K_a^* + 10^{-pH}\right)}{\left(K_a^* + 10^{-pH}\right)^4} C_{bf} \tag{1.350}$$

– where $K_a^* + 10^{-pH}$ may be dropped from both numerator and denominator as

$$\frac{\partial \beta}{\partial 10^{-pH}} = \frac{K_a^*}{\log_{10} e} \frac{K_a^* + 10^{-pH} - 2 \cdot 10^{-pH}}{\left(K_a^* + 10^{-pH}\right)^3} C_{bf} \tag{1.351}$$

or, equivalently,

$$\frac{\partial \beta}{\partial 10^{-pH}} = \frac{K_a^*}{\log_{10} e} \frac{K_a^* - 10^{-pH}}{\left(K_a^* + 10^{-pH}\right)^3} C_{bf} \tag{1.352}$$

after lumping terms alike. Combination of Eqs. (1.348) and (1.352) yields simply

$$K_a^* - 10^{-pH} = 0, \tag{1.353}$$

since $(K_a^* + 10^{-pH})^3$ remains finite and $K_a^*/\log_{10} e \neq 0$; Eq. (1.353) accepts

$$10^{-pH} = K_a^* \tag{1.354}$$

as single solution. Once decimal logarithms are taken of both sides, Eq. (1.354) becomes

$$-pH = \log_{10} K_a^*, \tag{1.355}$$

which may be rewritten as

$$\boxed{pH_{opt} = pK_a^*} \tag{1.356}$$

as long as pK_a^* is defined as

$$\boxed{pK_a^* \equiv -\log_{10} K_a^*} \tag{1.357}$$

– in much the same way Eq. (1.279) was put forward. When pH is low, one finds that

$$\lim_{pH \to 0} \frac{K_a^*}{\log_{10} e} \frac{10^{-pH}}{\left(K_a^* + 10^{-pH}\right)^2} C_{bf} = \frac{K_a^*}{\log_{10} e} \frac{10^{-pH}}{10^{-2pH}} C_{bf} = \frac{K_a^* C_{bf}}{\log_{10} e} 10^{pH} \tag{1.358}$$

because $K_a^* + 10^{-pH} \approx 10^{-pH}$ under such circumstances; hence, the function under scrutiny increases with pH, within the low pH range. Conversely, high values of pH support $K_a^* + 10^{-pH} \approx K_a^*$ that implies

$$\lim_{pH \to \infty} \frac{K_a^*}{\log_{10} e} \frac{10^{-pH}}{\left(K_a^* + 10^{-pH}\right)^2} C_{bf} = \frac{K_a^*}{\log_{10} e} \frac{10^{-pH}}{K_a^{*2}} C_{bf} = \frac{C_{bf}}{K_a^* \log_{10} e} 10^{-pH} \tag{1.359}$$

– meaning that the function at stake decreases with pH, within the high pH range. These two pieces of information, combined with Eq. (1.346) support the claim that Eq. (1.356) describes a maximum. In other words, relatively concentrated solutions, in terms of weak acid/base, are required to assure a good buffering capacity at a given pH – which should, in turn, match the pK_a of the said conjugated acid/base. Equation (1.345) is plotted in Fig. 1.26, for selected values of pK_a^*, and a fixed value of C_{bf}. Inspection of this figure indicates that the bell shape of the curve is kept, but undergoes a horizontal translation by a number of pH units equal to the variation in pK_a^*; moreover, the peak value of β remains the same for a given C_{bf}, although centered at the appropriate pK_a^*.

The latter realization may be mathematically proven after combining Eqs. (1.345) and (1.356) in the form

$$\beta\Big|_{pH_{opt}} \approx \frac{C_{bf}}{\log_{10} e} \frac{K_a^* 10^{-pK_a^*}}{\left(K_a^* + 10^{-pK_a^*}\right)^2}, \tag{1.360}$$

which may be rewritten as

$$\beta\Big|_{pH_{opt}} = \frac{C_{bf}}{\log_{10} e} \frac{10^{-pK_a^*} 10^{-pK_a^*}}{\left(10^{-pK_a^*} + 10^{-pK_a^*}\right)^2} \tag{1.361}$$

– at the expense of Eq. (1.357), upon taking decimal exponentials of its both sides; after identical factors are lumped, Eq. (1.361) becomes

$$\beta\Big|_{pH_{opt}} = \frac{C_{bf}}{\log_{10} e} \frac{10^{-2pK_a^*}}{\left(2 \cdot 10^{-pK_a^*}\right)^2} = \frac{C_{bf}}{\log_{10} e} \left(\frac{10^{-pK_a^*}}{2 \cdot 10^{-pK_a^*}} \right)^2, \tag{1.362}$$

or merely

$$\beta\Big|_{pH_{opt}} = \frac{C_{bf}}{\log_{10} e} \left(\frac{1}{2} \right)^2 = \frac{C_{bf}}{4 \log_{10} e} \tag{1.363}$$

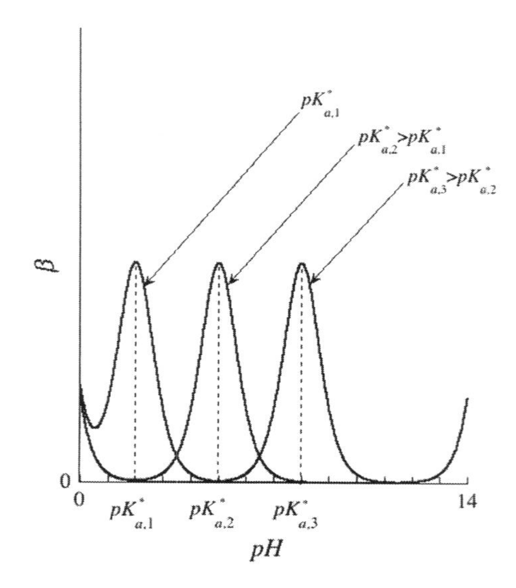

FIGURE 1.26 Variation of buffering capacity, β, versus pH, for solution of weak acid (with given overall buffer concentration C_{bf}) in water (with $pK_w^* = 14$), at various values of protolysis constants, i.e. $pK_{a,1}^*$, $pK_{a,2}^*$ and $pK_{a,3}^*$.

upon cancellation of $10^{-pK_a^*}$ between numerator and denominator. Note that $\beta\big|_{pH_{opt}}$ is indeed independent of pK_a^*, but proportional to $C_{b_f^-}$ – thus confirming the graphical pattern in Fig. 1.26; a similar reasoning may be set up for addition of a strong acid instead of a strong base.

PROBLEM 1.9

Consider a polytropic (weak) acid, H_mA, able to release up to m ions H^+, and originally dissolved in pure water at concentration $C_{H_mA}\big|_{t=0}$; together with salts $X_iH_{m-i}A$ ($i=1, 2, \ldots, m$) of a monovalent cation, X^+, at initial concentrations $C_{H_{m-i}A^{i-},st}\big|_{t=0}$ ($i=1, 2, \ldots, m$).

a) Obtain an (implicit) relationship of the form $f\{C_{A^{m-}},C_{H^+}\}=0$, where f denotes a bivariate function and $C_{A^{m-}}$ denotes mole concentration of the anion with the highest charge.
b) Obtain a second independent equation, again in the form of an (implicit) relationship, viz. $g\{C_{A^{m-}},C_{H^+}\}=0$, where g denotes also a bivariate function – so that numerical solution of $f\{C_{A^{m-}},C_{H^+}\}=0$ and $g\{C_{A^{m-}},C_{H^+}\}=0$ simultaneously will unfold the concentrations of all the germane chemical species.

Solution

a) The salt anions will distribute between $H_{m-1}A^-$, $H_{m-2}A^{2-}$, …, $HA^{(m-1)-}$ and eventually A^{m-}, so the resulting concentrations will satisfy equilibrium relationships of the form

$$K_{a,i}^* \equiv \frac{C_{H_{m-i}A^{i-}}C_{H^+}}{C_{H_{m-i+1}A^{(i-1)-}}};\quad i=1,2,\ldots,m. \tag{1.9.1}$$

With regard to moiety A, a mass balance looks like

$$C_{H_mA}\big|_{t=0} + \sum_{i=1}^{m} C_{X_iH_{m-i}A}\big|_{t=0} = \sum_{i=0}^{m} C_{H_{m-i}A^{i-}} \tag{1.9.2}$$

between initial conditions (represented in the left-hand side) and conditions at any time (listed in the right-hand side); this is equivalent to writing

$$C_{H_mA}\big|_{t=0} - C_{H_mA} + \sum_{i=1}^{m}\left(C_{X_iH_{m-i}A}\big|_{t=0} - C_{H_{m-i}A^{i-}}\right) = 0, \tag{1.9.3}$$

once the initial and current terms, pertaining to every species $H_{m-i}A^{i-}$ ($i=0,1,\ldots,m$), are duly paired with each other. A similar mass balance to H^+, between the initial and current states, takes the form

$$C_{H^+} - C_{H^+}\big|_{t=0} = \sum_{i=1}^{m} i\left(C_{H_{m-i}A^{i-}} - C_{X_iH_{m-i}A}\big|_{t=0}\right) + \left(C_{OH^-} - C_{OH^-}\big|_{t=0}\right), \tag{1.9.4}$$

since each species $H_{m-i}A^{i-}$ ($i=1,2,..,m$) – formed in excess of its initial concentration for being supplied in salt form, has meanwhile contributed i protons to the system (using H_mA as reference for both $C_{X_iH_{m-i}A}\big|_{t=0}$ and $C_{H_{m-i}A^{i-}}$); these are to be added to those formed via self-protolysis of H_2O to OH^- and H^+. Since one normally departs from pure water, Eq. (1.271) implies

$$C_{H^+}\big|_{t=0} = C_{OH^-}\big|_{t=0}, \tag{1.9.5}$$

in which case Eq. (1.9.4) reduces to

$$\sum_{i=1}^{m} i\left(C_{H_{m-i}A^{i-}} - C_{X_iH_{m-i}A}\big|_{t=0}\right) + C_{OH^-} - C_{H^+} = 0; \tag{1.9.6}$$

on the other hand, Eq. (1.267) may be solved for C_{OH^-} to get

$$C_{OH^-} = \frac{K_w^*}{C_{H^+}}, \tag{1.9.7}$$

so Eq. (1.9.6) will degenerate to

$$\sum_{i=1}^{m} i\left(C_{H_{m-i}A^{i-}} - C_{X_iH_{m-i}A}\big|_{t=0}\right) + \frac{K_w^*}{C_{H^+}} - C_{H^+} = 0. \tag{1.9.8}$$

After decreasing the counting variable of the summation by one unit, Eq. (1.9.8) becomes

$$\sum_{i=2}^{m+1} (i-1)\left(C_{H_{m-i+1}A^{(i-1)-}} - C_{X_{i-1}H_{m-i+1}A}\big|_{t=0}\right) + \frac{K_w^*}{C_{H^+}} - C_{H^+} = 0, \tag{1.9.9}$$

which is equivalent to

$$\sum_{i=2}^{m} (i-1)\left(C_{H_{m-i+1}A^{(i-1)-}} - C_{X_{i-1}H_{m-i+1}A}\big|_{t=0}\right) + m\left(C_{A^{m-}} - C_{X_mA}\big|_{t=0}\right) + \frac{K_w^*}{C_{H^+}} - C_{H^+} = 0 \tag{1.9.10}$$

after making explicit the last term of the summation. Isolation of $C_{H_{m-i+1}A^{(i-1)-}}$ in Eq. (1.9.1) gives rise to

$$C_{H_{m-i+1}A^{(i-1)-}} = C_{H_{m-i}A^{i-}}\frac{C_{H^+}}{K_{a,i}^*}, \tag{1.9.11}$$

and sequential application of Eq. (1.9.11) for i up to m yields

$$C_{H_{m-i+1}A^{(i-1)-}} = C_{A^{m-}}\frac{C_{H^+}^{\;m}}{\prod_{j=1}^{i} K_{a,j}^*};\quad i=1,2,\ldots,m; \tag{1.9.12}$$

insertion of Eq. (1.9.12) converts Eq. (1.9.10) to

$$\sum_{i=2}^{m}(i-1)\left(C_{A^{m-}}\frac{C_{H^+}{}^{m}}{\prod_{j=1}^{i}K_{a,j}^{*}}-C_{X_{i-1}H_{m-i+1}A}\Big|_{t=0}\right) + m\left(C_{A^{m-}}-C_{X_mA}\Big|_{t=0}\right)+\frac{K_w^*}{C_{H^+}}-C_{H^+}=0, \quad (1.9.13)$$

which entails the implicit relationship sought between $C_{A^{m-}}$ and C_{H^+} for given protolysis constants and initial concentrations.

b) By the same token, Eq. (1.9.3) may be rewritten as

$$C_{H_mA}\Big|_{t=0}-C_{H_mA}+\sum_{i=2}^{m}\left(C_{X_{i-1}H_{m-i+1}A}\Big|_{t=0}-C_{H_{m-i+1}A^{(i-1)-}}\right) + C_{X_mA}\Big|_{t=0}-C_{A^{m-}}=0, \quad (1.9.14)$$

following again decrement of the counting variable by one unit and explicitation of the uppermost term; insertion of Eq. (1.9.12) then transforms Eq. (1.9.14) to

$$C_{H_mA}\Big|_{t=0}+C_{X_mA}\Big|_{t=0}-C_{H_mA}-C_{A^{m-}} + \sum_{i=2}^{m}\left(C_{X_{i-1}H_{m-i+1}A}\Big|_{t=0}-C_{A^{m-}}\frac{C_{H^+}{}^{m}}{\prod_{j=1}^{i}K_{a,j}^{*}}\right)=0, \quad (1.9.15)$$

thus materializing the intended second (independent) implicit relationship between $C_{A^{m-}}$ and C_{H^+} – since $C_{H_mA}\big|_{t=0}$, $C_{X_mA}\big|_{t=0}$, $C_{X_{i-1}H_{m-i+1}A}\big|_{t=0}$ ($i = 2, 3, \ldots, m$), and $K_{a,j}^*$ ($j = 1, 2, \ldots, m$) are constants known in advance. After Eqs. (1.9.13) and (1.9.15) are simultaneously (and numerically) solved for variables $C_{A^{m-}}$ and C_{H^+}, the concentrations of all other charged species can be obtained via Eq. (1.9.7) in the case of OH^- and Eq. (1.9.12) in the case of $H_{m-i}A^{i-}$ ($i=0, 1, \ldots, m-1$).

1.1.3.6 Thermal resistance of enzymes and microorganisms

1.1.3.6.1 Decimal reduction time

Enzymes are the functional units of cell metabolism – and accordingly exist in both cell tissues eventually serving as food, and in microorganisms present in food (either adventitious, or following environmental contamination or deliberate addition). As proteins (and similarly to other susceptible biomolecules), enzymes undergo unimolecular thermal deactivation – with rate, r_d, described by

$$r_d = k_d C_E, \quad (1.364)$$

where k_d denotes deactivation constant and C_E denotes molar concentration of (intact) enzyme; this first-order rate expression reflects the elementary nature of the decay process. Since foods are normally taken as closed systems – as long as they do not exchange mass with their surroundings, one may use the mass balance to a batch reactor as template to write

$$\frac{dC_E}{dt} = -r_d, \quad (1.365)$$

with accumulation term (as left-hand side) coinciding with consumption term (as right-hand side). Equation (1.364) may then be brought on board to generate

$$\boxed{\frac{dC_E}{dt} = -k_d C_E;} \quad (1.366)$$

with a suitable initial condition reading

$$\boxed{C_E\big|_{t=0} = C_{E,0},} \quad (1.367)$$

provided that $C_{E,0}$ denotes the original (given) concentration of active enzyme under scrutiny. Integration of Eq. (1.366), via separation of variables, leads to

$$\int_{C_{E,0}}^{C_E}\frac{d\tilde{C}_E}{\tilde{C}_E} = -k_d\int_0^t d\tilde{t} \quad (1.368)$$

with the aid of Eq. (1.367), where the fundamental theorem of integral calculus allows transformation to

$$\ln\tilde{C}_E\Big|_{C_{E,0}}^{C_E} = -k_d\,\tilde{t}\,\Big|_0^t; \quad (1.369)$$

Eq. (1.369) degenerates to

$$\ln C_E - \ln C_{E,0} = \ln\frac{C_E}{C_{E,0}} = -k_d(t-0) = -k_d t \quad (1.370)$$

or, equivalently,

$$\boxed{\frac{C_E}{C_{E,0}} = e^{-k_d t}} \quad (1.371)$$

after taking exponentials of both sides – as plotted in Fig. 1.27. The concentration of active enzyme departs from its initial value, $C_{E,0}$ as per Eq. (1.367), and evolves down to zero – reached only in an asymptotic fashion, i.e.

$$\lim_{t\to\infty} C_E = 0; \quad (1.372)$$

this means that a nil value for C_E will not be attained, unless time grows unbounded.

The curve in Fig. 1.27 becomes a straight line when a logarithmic scale is used for vertical axis, as apparent in Fig. 1.28i – due to the exponential function in the right-hand side of Eq. (1.371), revisited as

$$C_E = C_{E,0}e^{-k_d t} \quad (1.373)$$

for convenience. The reduction of C_E to one-tenth (or by 90% of) its original value, i.e. $C_{E,0}/10$, may thus be calculated via

$$\frac{1}{10} = e^{-k_d D} \quad (1.374)$$

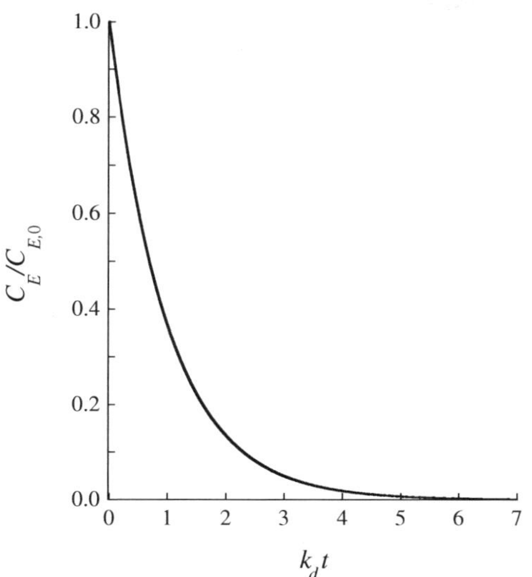

FIGURE 1.27 Evolution of normalized concentration of active enzyme, $C_E/C_{E,0}$, versus dimensionless time, $k_d t$, for first-order decay – plotted in bilinear scale.

based directly on Eq. (1.371); here D denotes the time required thereto, or decimal reduction time – defined hereafter as

$$\boxed{D \equiv t\big|_{C_E=\frac{C_{E,0}}{10}} - t\big|_{C_E=C_{E,0}}}. \tag{1.375}$$

After taking reciprocals of both sides, Eq. (1.374) becomes

$$e^{k_d D} = 10, \tag{1.376}$$

where logarithms may, in turn, be applied to both sides to generate

$$k_d D = \ln 10; \tag{1.377}$$

isolation of D from Eq. (1.377) finally yields

$$\boxed{D = \frac{\ln 10}{k_d}} \tag{1.378}$$

– so D (with units of time) is inversely proportional to the deactivation constant, k_d, with ln 10 (or 2.303) serving as proportionality constant.

If decimal logarithms were taken of both sides of Eq. (1.373), one would get

$$\log_{10} C_E = \log_{10} C_{E,0} + \log_{10} e^{-k_d t}, \tag{1.379}$$

where the logarithm of a power being equal to the product of the exponent by the logarithm of its base supports conversion to

$$\log_{10} C_E - \log_{10} C_{E,0} = -k_d t \log_{10} e; \tag{1.380}$$

after solving for t, Eq. (1.380) becomes

$$t = \frac{\log_{10} C_{E,0} - \log_{10} C_E}{k_d \log_{10} e}. \tag{1.381}$$

Inspection of Eq. (1.381) indicates that t can be obtained from the difference of decimal logarithms of $C_{E,0}$ and C_E, corrected by $1/k_d \log_{10} e$ as multiplicative factor; since

$$\frac{1}{\log_{10} e} = \ln 10, \tag{1.382}$$

one may rewrite Eq. (1.381) as

$$t = \frac{\ln 10}{k_d}\left(\log_{10} C_{E,0} - \log_{10} C_E\right) \tag{1.383}$$

– or, in view of Eq. (1.378),

$$\boxed{\frac{t}{D} = \log_{10} C_{E,0} - \log_{10} C_E} \tag{1.384}$$

where both sides were meanwhile divided by D. Equation (1.384) means that the difference between initial and current

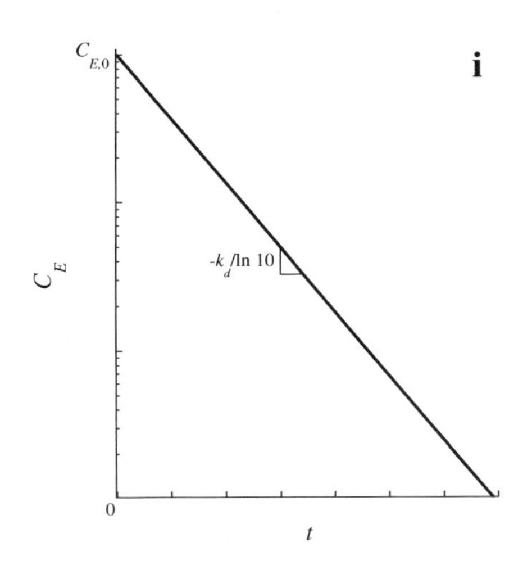

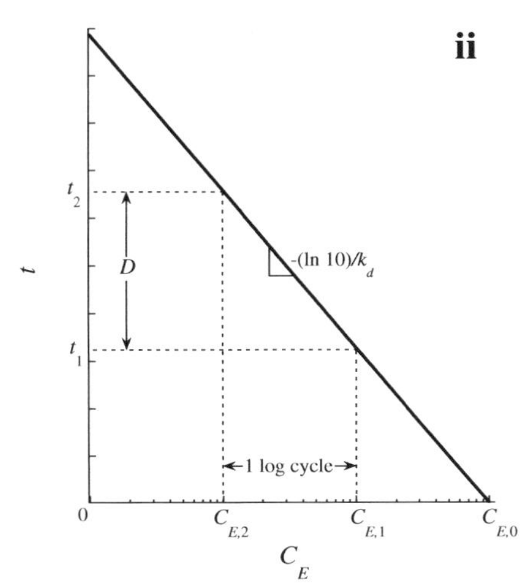

FIGURE 1.28 Evolution of concentration of active enzyme, C_E, versus time, t, from its initial value, $C_{E,0}$, for first-order decay characterized by kinetic constant k_d – (i) plotted in logarithmic/linear scale, or (ii) reversewise, with specific indication of two enzyme concentrations, $C_{E,1}$ and $C_{E,2}=C_{E,1}/10$, and corresponding times, t_1 and $t_2>t_1$, such that their difference equals the decimal reduction time, D.

(logarithmic) concentrations of active enzyme equals the ratio between time required to achieve the said difference, to decimal reduction time; consequently, $C_E=C_{E,0}/10$, or $\log_{10}C_{E,0}-\log_{10}C_E=\log_{10}C_{E,0}/C_E=\log_{10}10=1$, implies a unit value for the left-hand side of Eq. (1.384) – and one obtains $t=D$ (by definition). The reasoning underlying Eq. (1.384) is graphically illustrated in Fig. 1.28ii, with the straight line described by $-(\ln 10)/k_d$ as slope – where D represents distance along the time (linear) vertical axis, corresponding to one logarithmic cycle along the enzyme concentration (logarithmic) horizontal axis. If Eq. (1.381) is applied to two points, with coordinates $(t_1,C_{E,1})$ and $(t_2,C_{E,2})$, i.e.

$$t_1=\frac{\log_{10}C_{E,0}-\log_{10}C_{E,1}}{k_d\log_{10}\mathrm{e}} \tag{1.385}$$

and

$$t_2=\frac{\log_{10}C_{E,0}-\log_{10}C_{E,2}}{k_d\log_{10}\mathrm{e}}, \tag{1.386}$$

respectively, then ordered subtraction of Eq. (1.385) from Eq. (1.386) gives rise to

$$t_2-t_1=\frac{\log_{10}C_{E,0}-\log_{10}C_{E,2}}{k_d\log_{10}\mathrm{e}}-\frac{\log_{10}C_{E,0}-\log_{10}C_{E,1}}{k_d\log_{10}\mathrm{e}}; \tag{1.387}$$

after factoring out $1/k_d\log_{10}\mathrm{e}$, Eq. (1.387) becomes

$$t_2-t_1=\frac{\log_{10}C_{E,0}-\log_{10}C_{E,2}-\log_{10}C_{E,0}+\log_{10}C_{E,1}}{k_d\log_{10}\mathrm{e}} \tag{1.388}$$

that readily simplifies to

$$t_2-t_1=\frac{\log_{10}C_{E,1}-\log_{10}C_{E,2}}{k_d\log_{10}\mathrm{e}} \tag{1.389}$$

following cancellation of symmetrical terms – which mimics Eq. (1.383) in functional form, due to Eq. (1.382). If the logarithmic distance described by $\log_{10}C_{E,1}$ – $\log_{10}$ $C_{E,2}$ is again unity, then t_2-t_1 must equal D – so D represents decimal reduction time, irrespective of whether $C_{E,1}=C_{E,0}$ or $C_{E,1}<C_{E,0}$, as long as $C_{E,2}/C_{E,1}=0.1$ as highlighted also in Fig. 1.28iii. This constant time required for one logarithmic cycle reduction, everywhere along the process reaction curve, is a unique mathematical feature of any first-order process.

Enzymes are key catalysts of the metabolic pathways underlying survival and growth of microorganisms (and cells at large), so deactivation of critical enzymes in such pathways will disrupt their regular functioning – and eventually lead to death of those microorganisms. Therefore, it is expected that cell death kinetics abides to an analog of Eq. (1.366), viz.

$$\boxed{\frac{dC_X}{dt}=-k_dC_X;} \tag{1.390}$$

k_d now denotes first-order rate constant of cell death, whereas C_X denotes concentration of (viable) biomass. For convenience, C_X is expressed in units of mass per unit volume in both sides of Eq. (1.390), so the units of reciprocal time of k_d in Eq. (1.366) are shared by those of k_d in Eq. (1.390). By the same token, the initial condition reads

$$\boxed{C_X\big|_{t=0}=C_{X,0}} \tag{1.391}$$

in parallel to Eq. (1.367), where $C_{X,0}$ denotes initial biomass concentration; all other conclusions derived from Eqs. (1.366) and (1.367) are therefore valid when departing from Eqs. (1.390) and (1.391) – including the analog of Eq. (1.378), in particular. Values for decimal reduction times of selected food pathogens are listed in Fig. 1.29. Quite a wide variety of food pathogens are characterized by D values in the 0–10 min range – encompassing pasteurization/sterilization temperatures and cells in their vegetative form. When spores are considered, however, the ranges tend to be higher, typically up to 40 min; such higher values are expected, because spores materialize the resistance form taken by (pathogenic) microorganisms when exposed to unfavorable (e.g. excessively hot) environments.

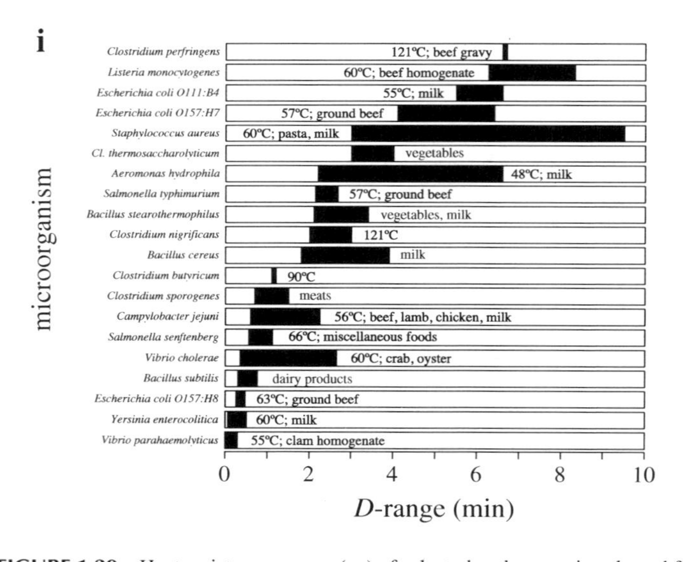

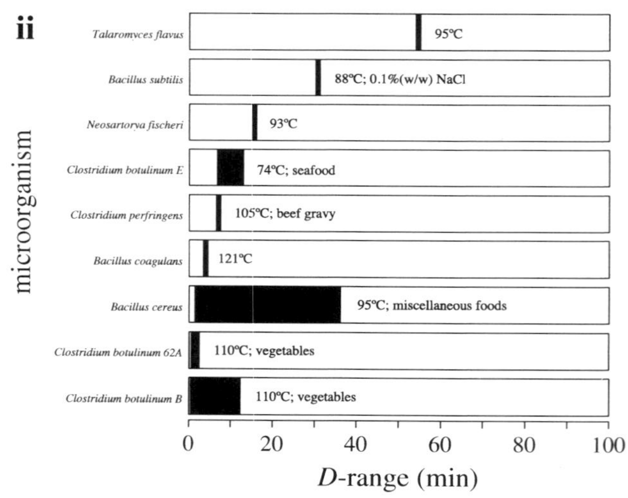

FIGURE 1.29 Heat resistance ranges (■) of selected pathogens, in selected foods and under selected processing conditions, expressed as decimal reduction times, D, for (i) vegetative cells and (ii) spores (i.e. bacterial spores or mold ascospores).

As a consequence of the underlying first-order death process, all cells will not be destroyed unless heating occurs for an infinite time, in agreement with Eq. (1.372); thermal processing is therefore aimed at reducing number of surviving microorganisms to a predetermined, yet finite level. Furthermore, the time required to attain that level depends on the initial microbial load – so a higher number of microorganisms requires a longer processing time, for a given final specification in terms of viability thereof (and associated safety for the consumer).

The underlying rationale for a first-order process of decay comes from the microheterogeneity of the system constituents; if they were uniform in features (and, particularly, in resistance to heat), then they should decay at a constant rate, irrespective of the amount of active entities left – as typical of a zero-th order process. In practice, however, one observes that the more labile entities decay first, and the more resistant ones will decay afterward – once they have been exposed longer to heat; this process can be iterated in a continuous fashion, thus giving rise to a linear, first-order ordinary differential equation. Deviations from the linear model entertained by Eq. (1.390) do sometimes arise, when more than one critical enzyme exists – possessing distinct thermal labilities; under such circumstances, kinetic control of the death process will shift between enzymes, so k_d will no longer be constant in time.

PROBLEM 1.10

The concept of thermal death time is particularly useful when death kinetics follows first order; it considers a constant probability of death of a microbial cell referred to the current population, irrespective of the time when it happens. To encompass the possibility that heat lability changes with processing time, Weibull proposed

$$-\frac{dC_X}{dt} = k_d f\{t\} C_X; \tag{1.10.1}$$

here $f\{t\}$ is a probability density function that may be formulated as

$$f\{t\} = p\left(\frac{t}{\delta}\right)^{p-1}, \tag{1.10.2}$$

with p being an indicator of cell population behavior and δ denoting a destruction ratio (which resembles D). Discuss this model *vis-à-vis* with the classical first-order model – and give physical significance to $p<1$, $p=1$ and $p>1$.

Solution

In view of the analogy between Eqs. (1.366) and (1.390) as differential mass balances, as well as Eqs. (1.367) and (1.391) for initial conditions, one may write

$$\log_{10}\frac{C_{X,0}}{C_X} = \frac{t}{D} \tag{1.10.3}$$

using Eq. (1.384) as a template – where D abides to Eq. (1.378); note that Eq. (1.10.3) is associated with first-order death kinetics. On the other hand, when Eq. (1.10.1) is integrated via separation of variables, one obtains

$$-\int_{C_{X,0}}^{C_X} \frac{d\tilde{C}_X}{\tilde{C}_X} = k_d\delta \int_0^t p\left(\frac{\tilde{t}}{\delta}\right)^{p-1} \frac{1}{\delta} d\tilde{t} \tag{1.10.4}$$

with the aid of Eq. (1.10.2) as probability descriptor – along with multiplication and division of the right-hand side by δ, coupled with Eq. (1.391) for initial condition. Equation (1.10.4) gives rise to

$$-\ln \tilde{C}_X \Big|_{C_{X,0}}^{C_X} = k_d\delta \left(\frac{\tilde{t}}{\delta}\right)^p \Bigg|_0^t \tag{1.10.5}$$

at the expense of the fundamental theorem of integral calculus – or else

$$-\ln\frac{C_X}{C_{X,0}} = k_d\delta\left(\left(\frac{t}{\delta}\right)^p - 0^p\right) \tag{1.10.6}$$

that breaks down to

$$\ln\frac{C_{X,0}}{C_X} = k_d\delta\left(\frac{t}{\delta}\right)^p, \tag{1.10.7}$$

after taking the reciprocal of the argument of the logarithmic function on account of the preceding minus sign. Recalling the rule of change of base of a logarithm, one may rewrite Eq. (1.10.7) as

$$\log_{10}\frac{C_{X,0}}{C_X}\ln 10 = k_d\delta\left(\frac{t}{\delta}\right)^p \tag{1.10.8}$$

or, after dividing both sides by ln 10,

$$\log_{10}\frac{C_{X,0}}{C_X} = \frac{k_d\delta}{\ln 10}\left(\frac{t}{\delta}\right)^p. \tag{1.10.9}$$

When $p=1$, Eq. (1.10.9) degenerates to

$$\left(\log_{10}\frac{C_{X,0}}{C_X} = \frac{k_d}{\ln 10}\delta\frac{t}{\delta}\right)\Bigg|_{p=1}, \tag{1.10.10}$$

so Eq. (1.378) can be retrieved to write

$$\left(\log_{10}\frac{C_{X,0}}{C_X} = \frac{1}{D}t\right)\Bigg|_{p=1}, \tag{1.10.11}$$

together with cancellation of δ between numerator and denominator; Eq. (1.10.11) coincides with Eq. (1.10.3), so first-order kinetics is indeed a particular case of Weibull's model. When $p>1$, Eq. (1.10.9), revisited as

$$\log_{10} C_X = \log_{10} C_{X,0} - \frac{\delta}{D}\left(\frac{t}{\delta}\right)^p, \tag{1.10.12}$$

is represented by a convex curve – meaning that surviving cells become more susceptible to destruction as processing time elapses, i.e. the cumulative degree of damage lessens their possibility of further survival; conversely, $p < 1$ produces a concave curve – so the rate of destruction of cells decreases, as the more resistant fraction survives longer than originally expected.

1.1.3.6.2 Thermal resistance factor

Besides decreasing exponentially with time elapsed, as per Eq. (1.371) for a given k_d, the fraction of enzyme deactivated after a given period of time is temperature-dependent – via change of k_d itself; classical description has resorted to Arrhenius' model, labeled as Eq. (5.23) of *Food Proc. Eng.: thermal & chemical operations*, or else

$$\boxed{\frac{k_d}{k_{d,0}} = \exp\left\{-\frac{E_{act}}{\mathcal{R}T}\right\}} \tag{1.392}$$

following division of both sides by $k_{d,0}$ – where $k_{d,0}$ denotes pre-exponential factor and E_{act} denotes activation energy for enzyme decay, T denotes absolute temperature, and $\mathcal{R}$ denotes ideal gas constant. A plot of $k_d/k_{d,0}$, as a function of $\mathcal{R}T/E_{act}$, is made available in Fig. 1.30. Note the fast increase in k_d with an increase in T – due to the exponential dependence on the negative of the reciprocal of the latter as per Eq. (1.392), which hyperbolically slows down at large T.

It is common practice to present the above relationship in a logarithmic/reciprocal plot, as done in Fig. 1.31. An estimate of the activation energy is graphically apparent in Fig. 1.31i, as the negative of the slope of the resulting straight line, following multiplication by $\mathcal{R}$.

For a given initial temperature T_1, the corresponding kinetic constant reads

$$k_{d,1} = k_{d,0}\exp\left\{-\frac{E_{act}}{\mathcal{R}T_1}\right\} \tag{1.393}$$

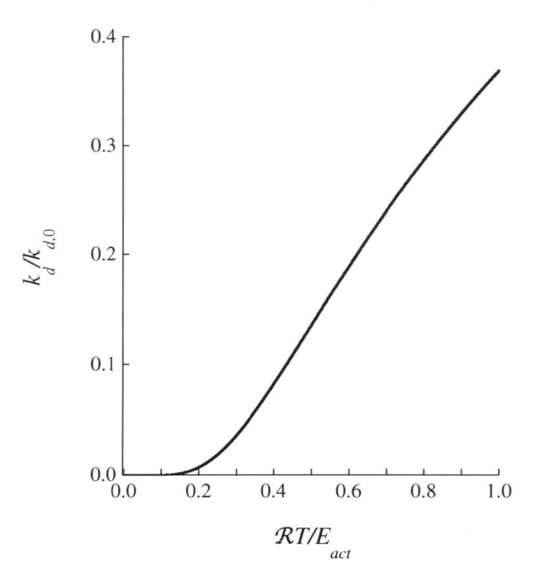

FIGURE 1.30 Variation of dimensionless kinetic constant of decay of active enzyme, $k_d/k_{d,0}$, versus dimensionless (absolute) temperature, $\mathcal{R}T/E_{act}$, for first-order reaction – plotted in bilinear scale.

based on Eq. (1.392); by the same token, the kinetic constant at another temperature T_2 reads

$$k_{d,2} = k_{d,0}\exp\left\{-\frac{E_{act}}{\mathcal{R}T_2}\right\}, \tag{1.394}$$

and likewise for a third temperature, T_3, i.e.

$$k_{d,3} = k_{d,0}\exp\left\{-\frac{E_{act}}{\mathcal{R}T_3}\right\}. \tag{1.395}$$

Ordered division of Eq. (1.394) by Eq. (1.393) produces

$$\frac{k_{d,2}}{k_{d,1}} = \frac{k_{d,0}\exp\left\{-\frac{E_{act}}{\mathcal{R}T_2}\right\}}{k_{d,0}\exp\left\{-\frac{E_{act}}{\mathcal{R}T_1}\right\}} \tag{1.396}$$

that readily simplifies to

$$\frac{k_{d,2}}{k_{d,1}} = \exp\left\{\frac{E_{act}}{\mathcal{R}T_1} - \frac{E_{act}}{\mathcal{R}T_2}\right\}, \tag{1.397}$$

after dropping off similar factors between numerator and denominator, and lumping exponential functions; application of logarithms to both sides, followed by factoring out of $E_{act}/\mathcal{R}$ then lead to

$$\ln\frac{k_{d,2}}{k_{d,1}} = \frac{E_{act}}{\mathcal{R}}\left(\frac{1}{T_1} - \frac{1}{T_2}\right). \tag{1.398}$$

Once the fundamental relationship between logarithms of different bases is recalled, one may redo Eq. (1.398) to

$$\frac{\log_{10}\frac{k_{d,2}}{k_{d,1}}}{\log_{10}\mathrm{e}} = \frac{E_{act}}{\mathcal{R}}\left(\frac{1}{T_1} - \frac{1}{T_2}\right) \tag{1.399}$$

or, equivalently,

$$\log_{10}\frac{k_{d,2}}{k_{d,1}} = \frac{E_{act}\log_{10}\mathrm{e}}{\mathcal{R}}\left(\frac{1}{T_1} - \frac{1}{T_2}\right) \tag{1.400}$$

upon multiplication of both sides by $\log_{10}$e. Denote by z the increase in absolute temperature, consistent with

$$T_2 = T_1 + z, \tag{1.401}$$

required to increase the kinetic constant by a factor of 10, i.e.

$$\boxed{z \equiv T\Big|_{k_d=10k_{d,1}} - T\Big|_{k_d=k_{d,1}};} \tag{1.402}$$

here z is termed thermal resistance factors. Together with

$$k_{d,2} = 10\,k_{d,1} \tag{1.403}$$

by hypothesis, one may apply Eq. (1.401) to transform Eq. (1.400) to

$$\log_{10}10 = \frac{E_{act}\log_{10}\mathrm{e}}{\mathcal{R}}\left(\frac{1}{T_1} - \frac{1}{T_1 + z}\right), \tag{1.404}$$

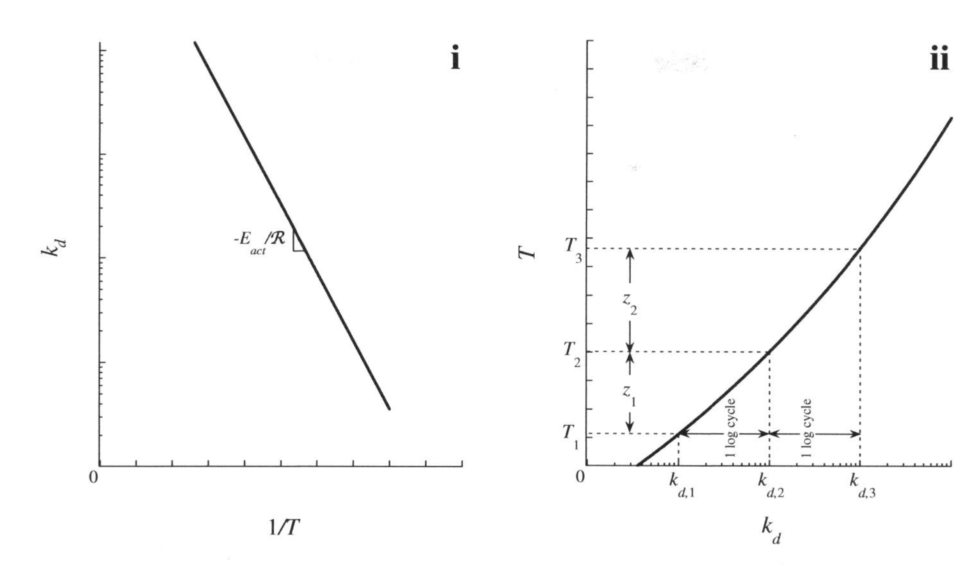

FIGURE 1.31 Variation of kinetic constant of first-order decay of active enzyme, k_d, versus (i) reciprocal of absolute temperature, $1/T$, plotted in logarithmic/linear scale, or (ii) plain absolute temperature, T, plotted in linear/logarithmic scale reversewise – with specific indication of (i) underlying activation energy, E_{act}, or (ii) three temperatures, T_1, $T_2>T_1$, and $T_3>T_2$, and corresponding kinetic constants, $k_{d,1}$, $k_{d,2}=10k_{d,1}$ and $k_{d,3}=10k_{d,2}$, such that their differences abide to $T_2-T_1=z_1$ and $T_3-T_2=z_2$, where z denotes thermal resistance factor.

which breaks down to merely

$$1=\frac{E_{act}\log_{10}e}{\mathcal{R}}\frac{T_1+z-T_1}{T_1\left(T_1+z\right)} \tag{1.405}$$

after elimination of parenthesis. Multiplication of both sides by $\mathcal{R}/E_{act}\log_{10}e$ and cancellation of symmetrical terms in numerator of the right-hand side produce

$$\frac{z}{T_1\left(T_1+z\right)}=\frac{\mathcal{R}}{E_{act}\log_{10}e} \tag{1.406}$$

– where multiplication of both sides by $T_1(T_1+z)$ and elimination of the outstanding parenthesis lead to

$$z=\frac{\mathcal{R}}{E_{act}\log_{10}e}T_1^2+\frac{\mathcal{R}}{E_{act}\log_{10}e}T_1z; \tag{1.407}$$

z may then be isolated as

$$z=\frac{\dfrac{\mathcal{R}}{E_{act}\log_{10}e}T_1^2}{1-\dfrac{\mathcal{R}}{E_{act}\log_{10}e}T_1} \tag{1.408}$$

or, after multiplying both numerator and denominator by $E_{act}\log_{10}e/\mathcal{R}$,

$$z=\frac{T_1^2}{\dfrac{E_{act}\log_{10}e}{\mathcal{R}}-T_1}. \tag{1.409}$$

Typical values found for E_{act} and T_1 pertaining to thermal processing operations of practical relevance for the food sector indicate that

$$\frac{E_{act}\log_{10}e}{\mathcal{R}}>>T_1, \tag{1.410}$$

so Eq. (1.409) can be well approximated by

$$\boxed{z\approx\frac{\mathcal{R}T_1^2}{E_{act}\log_{10}e}} \tag{1.411}$$

– where multiplication of both numerator and denominator by $\mathcal{R}$ was performed for convenience; one therefore concludes that z (with units of temperature) is inversely proportional to the underlying activation energy – with $\mathcal{R}T_1^2/\log_{10}e$ (or $19.14/T_1^2$, in SI units) serving as proportionality constant.

Despite the specific dependence of z on T_1^2 as per Eq. (1.411), one realizes that the range of operating absolute temperatures germane in food processing is relatively narrow – so T^2 does not change that much, and one may safely assume that it remains essentially constant; this point is corroborated by plotting $k_d\equiv k_d\{T\}$ in a logarithmic/linear scale, as done in Fig. 1.31ii. The dependence of $\log_{10}k_d$ is indeed almost linear on absolute temperature, throughout the temperature range normally of interest in practice.

Values of z for selected enzymes are conveyed by Fig. 1.32; since the mechanisms of decay of such important food ingredients as enzyme cofactors (analogues to/precursors of vitamins), amino acids (for nutrition purposes), and pigments (for sensory purposes) parallel those of enzymes, a similar rationale may be followed with these molecules. Note the typical range 10–100 for z factors associated with compounds of nutritional relevance in foods – namely enzymes and vitamins, in fresh fruits and vegetables.

Once Eq. (1.400) is revisited as

$$\log_{10}\frac{k_d}{k_{d,1}}=\frac{E_{act}\log_{10}e}{\mathcal{R}}\left(\frac{1}{T_1}-\frac{1}{T}\right) \tag{1.412}$$

pertaining to a given initial temperature T_1 and a generic temperature T, one may expand the logarithmic function and collapse the terms in parenthesis to get

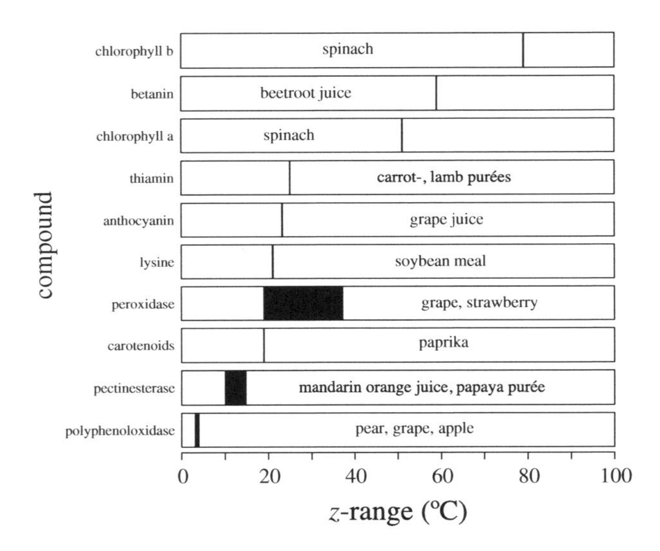

FIGURE 1.32 Heat resistance ranges (■) of selected components in selected foods, expressed as thermal resistance factors, z.

$$\log_{10} k_d - \log_{10} k_{d,1} = \frac{E_{act} \log_{10} \mathrm{e}}{\mathcal{R}} \frac{T - T_1}{TT_1}; \tag{1.413}$$

as stressed above,

$$TT_1 \approx T_1^2 \tag{1.414}$$

within the range of practical interest around the initial temperature (note that absolute, rather than relative temperature is at stake), so one will be able to transform Eq. (1.413) to

$$\log_{10} k_d - \log_{10} k_{d,1} = \frac{E_{act} \log_{10} \mathrm{e}}{\mathcal{R} T_1^2}\left(T - T_1\right). \tag{1.415}$$

If Eq. (1.415) is solved for $T-T_1$, then one obtains

$$T - T_1 = \frac{\mathcal{R} T_1^2}{E_{act} \log_{10} \mathrm{e}}\left(\log_{10} k_d - \log_{10} k_{d,1}\right) \tag{1.416}$$

that provides a linear approximation to the (slightly convex) curve in Fig. 1.31ii, after relocating the origin of coordinates to $(k_{d,1}, T_1)$; if $T = T_2$ under the hypothesis that $k_{d,2} = 10k_{d,1}$, then Eq. (1.416) becomes

$$\begin{aligned} T_2 - T_1 &= \frac{\mathcal{R} T_1^2}{E_{act} \log_{10} \mathrm{e}}\left(\log_{10} k_{d,2} - \log_{10} k_{d,1}\right) \\ &= \frac{\mathcal{R} T_1^2}{E_{act} \log_{10} \mathrm{e}}\left(\log_{10} 10 + \log_{10} k_{d,1} - \log_{10} k_{d,1}\right) \end{aligned} \tag{1.417}$$

with the aid of the operational features of a logarithmic function – which breaks down to

$$z_1 = \frac{\mathcal{R} T_1^2}{E_{act} \log_{10} \mathrm{e}} \tag{1.418}$$

due to the definition of z_1 as per Eq. (1.402), and compatible with Eq. (1.411). By the same token, one may use T_2 as starting temperature, and then consider $T = T_3$ such that $k_{d,3} = 10k_{d,2}$ – so Eq. (1.416) will take the form

$$\begin{aligned} T_3 - T_2 &= \frac{\mathcal{R} T_2^2}{E_{act} \log_{10} \mathrm{e}}\left(\log_{10} k_{d,3} - \log_{10} k_{d,2}\right) \\ &= \frac{\mathcal{R} T_2^2}{E_{act} \log_{10} \mathrm{e}}\left(\log_{10} 10 + \log_{10} k_{d,2} - \log_{10} k_{d,2}\right), \end{aligned} \tag{1.419}$$

or merely

$$z_2 = \frac{\mathcal{R} T_2^2}{E_{act} \log_{10} \mathrm{e}}, \tag{1.420}$$

again based on Eq. (1.402) and following cancellation of symmetrical terms. Both z_1 and z_2 are plotted in Fig. 1.31ii – and the almost coincidence of their magnitudes can be easily grasped, owing to $T_2^2 \approx T_1^2$ consistent with Eq. (1.414). Therefore, one concludes that z is essentially invariant within the temperature range of interest for food processing purposes, say $[T_1, T_2]$ – so the same (additive) increase of T is approximately required for a given (multiplicative) increase of k_d, irrespective of the actual temperature chosen as departure point (i.e. a linear behavior is hypothesized throughout the whole working range). On the other hand, one may insert Eq. (1.418) to transform Eq. (1.416) to

$$T - T_1 = z_1\left(\log_{10} k_d - \log_{10} k_{d,1}\right), \tag{1.421}$$

where essential constancy of the thermal resistance factor (i.e. $z_1 \approx z$) allows reformulation of Eq. (1.421)

$$\boxed{\frac{T - T_1}{z} = \log_{10} k_d - \log_{10} k_{d,1}.} \tag{1.422}$$

In other words, the difference in (logarithm of) kinetic constant for deactivation, between current and initial temperature, is equal to the ratio between the said temperature difference to that required to bring about a decimal increase in k_d (i.e. z); the functional resemblance of Eq. (1.422) to Eq. (1.384) is noteworthy.

As discussed before, enzymes being key players in life-supporting metabolic pathways implies that the death of microorganisms correlates directly to the heat lability of critical enzymes in one (or more) such pathways – so the z factor, originally rationalized for enzymes, may be extended to related cofactors (viz. vitamins), and ultimately to microorganisms. An analog to Eq. (1.392) therefore applies to microorganisms – as long as the pre-exponential factor and the activation energy are now associated with microbial death; Eqs. (1.411) and (1.422) remain valid as well. Values for z factors of relevant food pathogens are conveyed by Fig. 1.33. Note the outstanding resistance of spores, namely, from *Clostridium botulinum* – one of the most lethal food-borne pathogens, thus justifying thermal processing at very high temperatures as done in canning of fish- and meat-based matrices. Most food pathogens in vegetative cell form possess a z value within the range 0–10°C, see Fig. 1.33i – i.e. about one order of magnitude lower than happens with food nutrients (e.g. enzymes, vitamins, amino acids, colorants), see Fig. 1.32; hence, an increase in temperature of a food is much more detrimental to its resident (pathogen) microflora than to its intrinsic (chemical) constituency – which sets the basis for thermal processing of food *sensu latu*, as expanded below to some length.

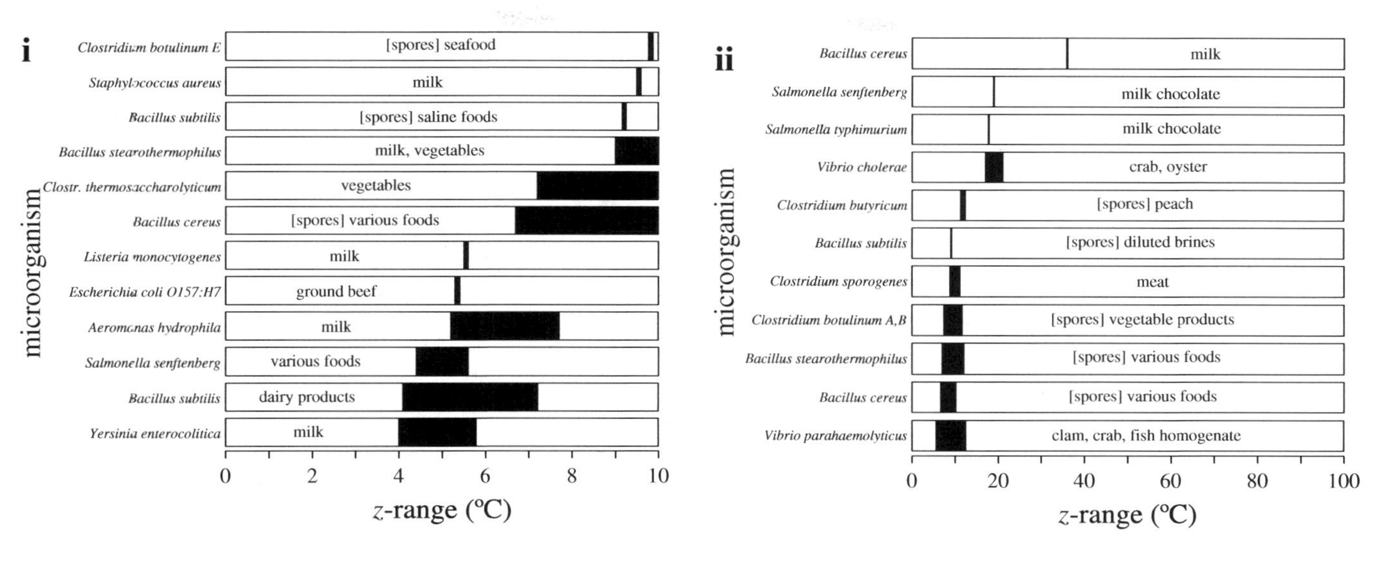

FIGURE 1.33 Heat resistance ranges (■) of selected pathogens in selected foods, expressed as thermal resistance factors, z, within (i) 0–10°C and (ii) 0–100°C.

An overall picture, in terms of families of compounds/microorganisms, to better support the above point is conveyed by Fig. 1.34. Inspection of this figure confirms indeed that z values of chemicals responsible for sensory characteristics (i.e. 25–50°C) are ca. fivefold those of microorganisms, in either vegetative or spore form (i.e. 4–12°C); proteins span an intermediate range (i.e. 15–37°C), and enzymes an even wider range (i.e. 3–50°C). As a rule of thumb, an increase of 10°C in processing temperature doubles, on average, the cooking effect – while the extent of microbial inactivation undergoes a tenfold increase. The reasons for these discrepancies stem from the basics of life – as a complex, highly coordinated set of enzyme-catalyzed chemical reactions; even small changes in physicochemical integrity of a single enzyme, or related compound (e.g. cofactor/vitamin, or even intermediate substrate) may disrupt a life-supporting biochemical pathway, due to interference with its intrinsic implementation and control mechanisms – so a higher sensitivity to temperature is anticipated for life forms (i.e. a lower z) than chemical entities (i.e. a higher z).

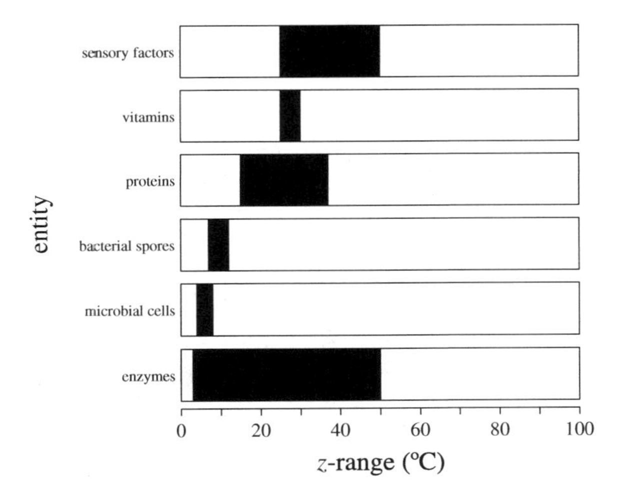

FIGURE 1.34 Heat resistance ranges (■) of chemical and microbial entities responsible for quality of foods, expressed as thermal resistance factors, z.

Classical literature in food engineering has sometimes resorted to the definition of (dimensionless) Q_{10} to describe the effect of temperature upon reaction rate – according to

$$\boxed{Q_{10} \equiv \frac{k_d\big|_{T=T_1+10}}{k_d\big|_{T=T_1}}} \tag{1.423}$$

In this case, a measure of the effect of an (additive) decimal increment in T_1 upon k_d is under scrutiny, rather than the effect of a (multiplicative) decimal increment in $k_{d,1}$ upon T, as in Eq. (1.422). After setting

$$T_2 = T_1 + 10, \tag{1.424}$$

one may retrieve Eq. (1.400) as

$$\log_{10} Q_{10} = \frac{E_{act} \log_{10} e}{\mathcal{R}} \left(\frac{1}{T_1} - \frac{1}{T_1 + 10} \right) \tag{1.425}$$

upon elimination of the formal ratio of k_ds between Eqs. (1.400) and (1.423). Removal of parenthesis converts Eq. (1.425) to

$$\log_{10} Q_{10} = \frac{E_{act} \log_{10} e}{\mathcal{R}} \frac{T_1 + 10 - T_1}{T_1 (T_1 + 10)}, \tag{1.426}$$

or else

$$\log_{10} Q_{10} = \frac{10 E_{act} \log_{10} e}{\mathcal{R} T_1^2} \tag{1.427}$$

– after canceling symmetrical terms in numerator, and simplifying the denominator to T_1^2 in agreement with the approximation conveyed by Eq. (1.414); therefore, the decimal logarithm of Q_{10} is proportional to the activation energy, with $10\log_{10}e/\mathcal{R}T_1^2$ (or $0.5223/T_1^2$ in SI units) serving as proportionality constant. Upon insertion of Eq. (1.411), one gets

$$\boxed{\log_{10} Q_{10} = \frac{10}{z}} \tag{1.428}$$

from Eq. (1.427) – thus providing a conversion rule between (archaic) Q_{10} and (preferred) z.

PROBLEM 1.11

The concept of decimal reduction time applies to a single (arbitrarily chosen) microbial species, characterized by D as alias of k_d, and z as alias of E_{act}; however, when two microbial species are relevant – described by $k_{d,1}$ and $k_{d,2} > k_{d,1}$, the actual decimal reduction time becomes a function of the initial fraction of biomass accounted for by either (and both) species, as well as $k_{d,1}$ and $k_{d,2}$ themselves. Derive an expression for D in the case of such a binary population, neglecting interactions between species; produce a plot, in dimensionless form, and discuss the patterns found.

Solution

The mass balance to biomass of the first species may be setup as

$$-\frac{dC_{X,1}}{dt} = k_{d,1} C_{X,1} \tag{1.11.1}$$

as per Eq. (1.390), coupled with

$$C_{X,1}\big|_{t=0} = C_{X,1,0} \tag{1.11.2}$$

as initial condition of the type of Eq. (1.391); by the same token, the evolution in biomass of the second species satisfies

$$-\frac{dC_{X,2}}{dt} = k_{d,2} C_{X,2}, \tag{1.11.3}$$

where

$$C_{X,2}\big|_{t=0} = C_{X,2,0} \tag{1.11.4}$$

serves as initial condition. Integration of Eq. (1.11.1), with the aid of Eq. (1.11.2), produces

$$-\int_{C_{X,1,0}}^{C_{X,1}} \frac{d\tilde{C}_{X,1}}{\tilde{C}_{X,1}} = k_{d,1} \int_0^t d\tilde{t}, \tag{1.11.5}$$

which degenerates to

$$\ln \tilde{C}_{X,1}\Big|_{C_{X,1,0}}^{C_{X,1}} = -k_{d,1}\,\tilde{t}\,\Big|_0^t \tag{1.11.6}$$

as per the fundamental theorem of integral calculus; Eq. (1.11.6) breaks down to

$$\ln \frac{C_{X,1}}{C_{X,1,0}} = -k_{d,1} t, \tag{1.11.7}$$

or else

$$C_{X,1} = C_{X,1,0} e^{-k_{d,1} t} \tag{1.11.8}$$

upon isolation of $C_{X,1}$. In view of the analogy between Eqs. (1.11.1) and (1.11.3), and between Eqs. (1.11.2) and (1.11.4), one may use Eq. (1.11.8) as a template to write

$$C_{X,2} = C_{X,2,0} e^{-k_{d,2} t} \tag{1.11.9}$$

pertaining to the second microbial species; ordered addition of Eqs. (1.11.8) and (1.11.9) produces

$$C_{X,1} + C_{X,2} = C_{X,1,0} e^{-k_{d,1} t} + C_{X,2,0} e^{-k_{d,2} t}, \tag{1.11.10}$$

where division of both sides by $C_{X,1,0} + C_{X,2,0}$ unfolds

$$\frac{C_{X,1} + C_{X,2}}{C_{X,1,0} + C_{X,2,0}} = \frac{C_{X,1,0}}{C_{X,1,0} + C_{X,2,0}} e^{-k_{d,1} t} + \frac{C_{X,2,0}}{C_{X,1,0} + C_{X,2,0}} e^{-k_{d,2} t}. \tag{1.11.11}$$

After defining α as

$$\alpha \equiv \frac{C_{X,1,0}}{C_{X,1,0} + C_{X,2,0}}, \tag{1.11.12}$$

which represents the initial fraction of species 1 in the overall biomass load – and realizing that

$$\frac{C_{X,2,0}}{C_{X,1,0} + C_{X,2,0}} = \frac{(C_{X,1,0} + C_{X,2,0}) - C_{X,1,0}}{C_{X,1,0} + C_{X,2,0}} = 1 - \frac{C_{X,1,0}}{C_{X,1,0} + C_{X,2,0}} = 1 - \alpha \tag{1.11.13}$$

can be obtained upon adding and subtracting $C_{X,1,0}$ in numerator, splitting the fraction thus obtained, and combining the outcome back with Eq. (1.11.12), one may rewrite Eq. (1.11.11) as

$$\frac{C_{X,1} + C_{X,2}}{C_{X,1,0} + C_{X,2,0}} = \alpha e^{-k_{d,1} t} + (1-\alpha) e^{-k_{d,2} t}; \tag{1.11.14}$$

the decimal reduction time, D, implies, in this case,

$$\frac{1}{10} = \alpha e^{-k_{d,1} D} + (1-\alpha) e^{-k_{d,2} D}, \tag{1.11.15}$$

stemming from Eqs. (1.375) and (1.11.14). One may reformulate Eq. (1.11.15) to

$$\frac{1}{10} = \alpha e^{-\frac{k_{d,1}}{k_{d,2}} k_{d,2} D} + (1-\alpha) e^{-k_{d,2} D} = \alpha \left(e^{-k_{d,2} D}\right)^{\frac{k_{d,1}}{k_{d,2}}} + (1-\alpha) e^{-k_{d,2} D} \tag{1.11.16}$$

after multiplying and dividing the argument of the first exponential by $k_{d,2}$, and recalling the operational rules involving the power of an exponential function; note that

$0 < k_{d,1}/k_{d,2} < 1$, by hypothesis. Upon elimination of parenthesis, Eq. (1.11.16) becomes

$$\frac{1}{10} = \alpha\left(e^{-k_{d,2}D}\right)^{\frac{k_{d,1}}{k_{d,2}}} + e^{-k_{d,2}D} - \alpha e^{-k_{d,2}D}, \quad (1.11.17)$$

where isolation of α proves possible as

$$\alpha = \frac{\frac{1}{10} - e^{-k_{d,2}D}}{\left(e^{-k_{d,2}D}\right)^{\frac{k_{d,1}}{k_{d,2}}} - e^{-k_{d,2}D}}; \quad (1.11.18)$$

Eq. (1.11.18) is of the form $\alpha \equiv \alpha\{k_{d,2}D, k_{d,1}/k_{d,2}\}$, i.e. the dimensionless value of D – looking like $k_{2,d}D$, is given as an implicit function of parameters α and $k_{d,1}/k_{d,2}$. A plot of $k_{d,2}D$ vs. α is provided below, for selected values of ratio $k_{d,1}/k_{d,2}$.

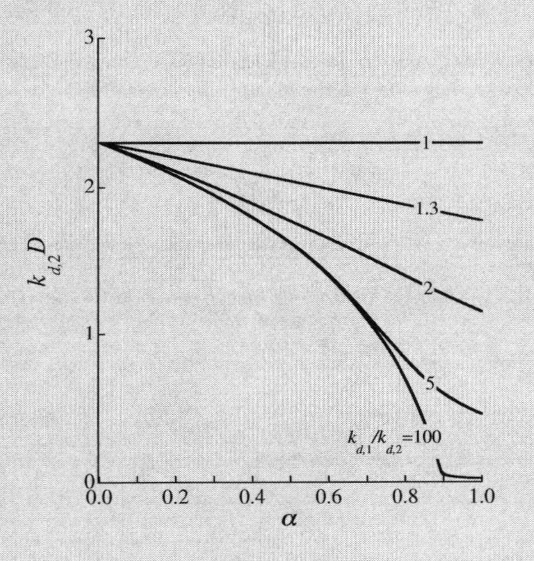

When α increases, the dimensionless decimal reduction time decreases as well – from its initial value of ln 10≈2.3, as conversion factor associated with

$$\lim_{\alpha \to 0} \alpha = \lim_{k_{d,2}D \to \ln 10} \left(\frac{1}{10} - e^{-k_{d,2}D}\right); \quad (1.11.19)$$

α=0 represents indeed an initial population of only species 2, i.e. $C_{X,1,0}$=0 due to Eq. (1.11.12) – so the associated decimal reduction time is directly given by (ln 10)/$k_{d,2}$. A similar situation, irrespective of α, will occur when $k_{d,1}$ approaches $k_{d,2}$, and thus $k_{d,1}/k_{d,2} \approx 1$ – which reduces Eq. (1.11.16) to

$$\left(\frac{1}{10} = \alpha e^{-k_{d,2}D} + (1-\alpha)e^{-k_{d,2}D} = e^{-k_{d,2}D}\right)\Bigg|_{\frac{k_{d,1}}{k_{d,2}} \to 1}; \quad (1.11.20)$$

this is likewise equivalent to $k_{d,2}t \approx \ln 10$ – describing now the horizontal line with ln 10 for vertical intercept. As expected, a larger difference between $k_{d,1}$ and $k_{d,2}$ magnifies the effect of composition of the initial mixed population, α.

1.1.3.6.3 Rationale of thermal processing

In order to assure microbial safety of a food, and consequently render it appropriate for ingestion by consumers at large, any putative microbial load of pathogens must be decreased to an acceptable threshold, say $C_{X,\max}$; although several species may be of concern, one should start by simulating how a given increase in time/temperature of processing affects each one of them (assumed to exhibit similar harmful effects), taking into account their intrinsic thermal resistance (measured by z) and initial microbial load (i.e. $C_{X,0}$). One should then select the one lying farther above the maximum acceptable value in the final product as indicator microorganism – and design equipment and operating conditions in a conservative fashion, i.e. for thermal processing using that microorganism as reference. Nevertheless, industrial practice has indicated that, in most foods, there is a specific pathogen more likely to appear at highest initial loads, to be more resistant to thermal decay, or to cause the worst health damage to the consumer upon ingestion – thus critically controlling food safety; or else there is a specific detrimental microorganism more likely to degrade the quality of the food, from manufacture to consumption, due to its initial load and characteristic activity, thus critically defining its shelf-life.

Under the above circumstances, the single indicator microorganism of the maximum acceptable level, referred to via subscript M, should abide to

$$\boxed{C_M \le C_{M,\max}} \quad (1.429)$$

in the final food product as a criterion of microbiological safety; here C_M denotes actual (viable) biomass of the said microorganism, and $C_{M,\max}$ denotes maximum acceptable level thereof. At the same time, any thermal process will also damage to some degree the chemical components of the food under scrutiny, thus affecting not only its nutritional features but also its organoleptic appeal. A similar preliminary analysis may be conducted on how a given increase in time/temperature of processing affects each such compound of relevance (again under the hypothesis of similar degrading effects); and eventually pinpoint the one lying closer to the minimum acceptable value in the final product as indicator. Once again, there is normally a specific food chemical constituent more likely to undergo unacceptable reduction, owing either to its low initial level, high thermal lability, or strong deleterious effect upon nutritional or sensory quality once damaged or even destroyed. The critical constituent should thus be chosen as indicator compound of minimum acceptable level, referred to via subscript Q – and

$$\boxed{C_Q \ge C_{Q,\min}} \quad (1.430)$$

should be established as criterion for nutritional/sensory quality, where C_Q denotes actual concentration of compound in the final food, and $C_{Q,\min}$ denotes its minimum acceptable threshold.

Recall that the (biomass) concentration of M should follow Eq. (1.371) in terms of time evolution, when some thermal processing takes place at a given (constant) temperature, T; hence, k_d may be taken off the kernel as done in Eq. (1.368), and validity of Eq. (1.371) will in principle be guaranteed. Insertion of Eq. (1.392) meanwhile converts Eq. (1.373) to

$$C_M = C_{M,0} \exp\left\{ -k_{d,M,0} \exp\left\{ -\frac{E_{act,M}}{\mathcal{R}T} \right\} t \right\} \tag{1.431}$$

– where $C_{M,0}$ denotes initial load of reference microorganism in the food, and $k_{d,M,0}$ and $E_{act,M}$ denote pre-exponential factor and activation energy, respectively, describing (the temperature-dependence of) its viability decay. Combination of Eqs. (1.429) and (1.431) leads to

$$C_{M,0} \exp\left\{ -k_{d,M,0} \exp\left\{ -\frac{E_{act,M}}{\mathcal{R}T} \right\} t \right\} \le C_{M,\max}, \tag{1.432}$$

where division of both sides by $C_{M,\max}$ and isolation of the term in $C_{M,0}$ afterward produce

$$\exp\left\{ k_{d,M,0} \exp\left\{ -\frac{E_{act,M}}{\mathcal{R}T} \right\} t \right\} \ge \frac{C_{M,0}}{C_{M,\max}}; \tag{1.433}$$

after taking logarithms of both sides, Eq. (1.433) becomes

$$k_{d,M,0} \exp\left\{ -\frac{E_{act,M}}{\mathcal{R}T} \right\} t \ge \ln \frac{C_{M,0}}{C_{M,\max}}, \tag{1.434}$$

which may be solved for t as

$$t \ge \frac{1}{k_{d,M,0}} \exp\left\{ \frac{E_{act,M}}{\mathcal{R}T} \right\} \ln \frac{C_{M,0}}{C_{M,\max}}. \tag{1.435}$$

Equation (1.435) is normally used in its logarithmic version, i.e.

$$\boxed{\ln t \ge \ln\left\{ \ln \frac{C_{M,0}}{C_{M,\max}} \right\} - \ln k_{d,M,0} + \frac{E_{act,M}}{\mathcal{R}} \frac{1}{T}}, \tag{1.436}$$

since it is expected that $C_{M,0} > C_{M,\max}$; Eq. (1.436) defines the portion of the ($\ln t$)0T plane satisfying Eq. (1.429), sometimes known as minimum time/temperature loci for thermal processing – as graphically depicted by the shaded area in Fig. 1.35i. The curve providing the lower boundary for thermal processing that guarantees satisfaction of the microbiological safety criterion entails a decrease of t with T – as expected, because a higher temperature is more time-effective in destroying microorganisms. The logarithmic time, $\ln t$, required to exactly satisfy Eq. (1.429) thus varies linearly with reciprocal temperature, $1/T$ – with a slope equal to $E_{act,M}/\mathcal{R}$, and vertical intercept given by $\ln\{\ln C_{M,0}/C_{M,\max}\} - \ln k_{d,M,0}$.

By the same token, one may use Eq. (1.371) as template for the evolution in concentration of compound Q; once again, thermal processing will be considered to take place along time at a constant temperature, so combination with Eq. (1.392) will convert the analog of Eq. (1.371) to

$$C_Q = C_{Q,0} \exp\left\{ -k_{d,Q,0} \exp\left\{ -\frac{E_{act,Q}}{\mathcal{R}T} \right\} t \right\}, \tag{1.437}$$

where $C_{Q,0}$ denotes concentration of indicator compound in the original food, while $k_{d,Q,0}$ and $E_{act,Q}$ denote pre-exponential factor and activation energy, respectively, describing molecular decay. Insertion of Eq. (1.437) transforms Eq. (1.430) to

$$C_{Q,0} \exp\left\{ -k_{d,Q,0} \exp\left\{ -\frac{E_{act,Q}}{\mathcal{R}T} \right\} t \right\} \ge C_{Q,\min}; \tag{1.438}$$

division of both sides by $C_{Q,\min}$, followed by multiplication thereof by $\exp\{k_{d,Q,0}\exp\{-E_{act,Q}/\mathcal{R}T\}\}$ lead to

$$\exp\left\{ k_{d,Q,0} \exp\left\{ -\frac{E_{act,Q}}{\mathcal{R}T} \right\} t \right\} \le \frac{C_{Q,0}}{C_{Q,\min}}, \tag{1.439}$$

where logarithms may be taken of both sides to produce

$$k_{d,Q,0} \exp\left\{ -\frac{E_{act,Q}}{\mathcal{R}T} \right\} t \le \ln \frac{C_{Q,0}}{C_{Q,\min}} \tag{1.440}$$

in view of the monotonically increasing nature of the said logarithmic function. The solution of Eq. (1.440) for t unfolds

$$t \le \frac{1}{k_{d,Q,0}} \exp\left\{ \frac{E_{act,Q}}{\mathcal{R}T} \right\} \ln \frac{C_{Q,0}}{C_{Q,\min}}, \tag{1.441}$$

where logarithms may again be taken of both sides to generate

$$\boxed{\ln t \le \ln\left\{ \ln \frac{C_{Q,0}}{C_{Q,\min}} \right\} - \ln k_{d,Q,0} + \frac{E_{act,Q}}{\mathcal{R}} \frac{1}{T}} \tag{1.442}$$

– with $C_{Q,0} > C_{Q,\min}$ justifying $\ln C_{Q,0}/C_{Q,\min} > 0$; therefore, one obtains a mathematical descriptor of the maximum time/temperature loci for thermal processing illustrated in Fig. 1.35ii, as the shaded portion of plane located below the curve defined by the right-hand side of Eq. (1.442). The upper boundary of the aforementioned portion of plane providing acceptable final nutritional/sensory quality of the food product is described once more by a monotonically decreasing $t \equiv t\{T\}$ curve; the (logarithmic) time required to exactly satisfy Eq. (1.442) also varies linearly with reciprocal temperature $1/T$ – with slope equal to $E_{act,Q}/\mathcal{R}$, and vertical intercept given by $\ln\{\ln C_Q/C_{Q,\min}\} - \ln k_{d,Q,0}$. Comparative inspection of Figs. 1.35i and ii indicates that the (upper) boundary curve in the latter typically departs from a lower time than the former, in the low temperature range – but reverses relative positioning in the high temperature range, due to a decrease at a slower pace; in regular food systems, $E_{act,M} > E_{act,Q}$ accounts for this pattern.

Feasible thermal operations should, therefore, abide to both microbiological safety and nutritional/sensory quality thresholds; this graphically corresponds to the intersection of the dashed planes in Figs. 1.35i and ii, as represented in Fig. 1.35iii. The region highlighted therein is the one to be selected for thermal processing – because it assures that the indicator microorganism does not exceed in viability its maximum allowable threshold upon processing, while simultaneously guaranteeing that the chemical compound directly responsible for nutritional/sensory quality does not fall below its minimum acceptable threshold. Once the minimum processing temperature, T_{crt}, to satisfy both criteria is found – as intercept

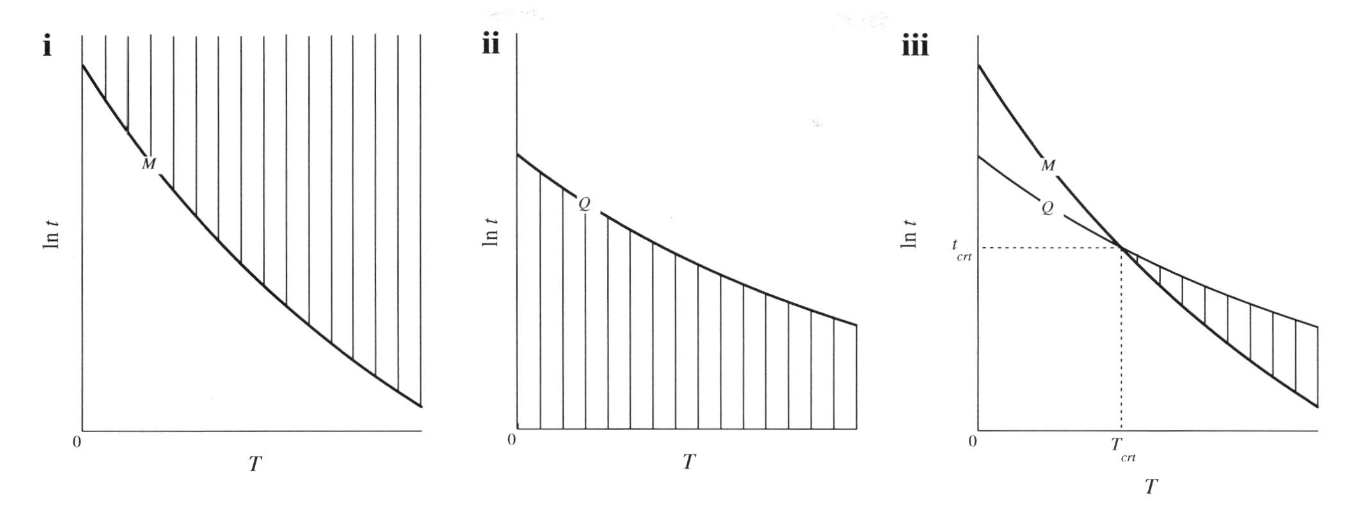

FIGURE 1.35 Processing time, t, required (shaded) to satisfy (i) criterion of microbiological safety, $C_M \leq C_{M,\max}$, (ii) criterion of nutritional/sensory quality, $C_Q \geq C_{Q,\min}$, and (iii) both criteria simultaneously, versus absolute temperature, T – with specific indication of intercept of M- and Q-curves, with coordinates (T_{crt}, t_{crt}).

of the foregoing two plane boundaries, one may proceed more or less freely to higher processing temperatures within the shaded area – but a range of times, rather than a single time (equal to t_{crt} when $T=T_{crt}$) will then become possible. Economic reasons associated with shorter processing times, coupled with the possibility of minimizing nutritional/sensory quality decay while guaranteeing microbiological safety, usually dictate the lowest possible processing time – so curve M is normally followed. Under these circumstances, T_{crt} provides the minimum processing temperature (as seen above), while t_{crt} provides the maximum processing time. The decision on how high a temperature to select will then depend on other types of constraints, such as cost of steam as utility or rate of heat transfer; the latter becomes the limiting factor when very high processing temperatures are sought, as in high temperature/short time processes (or HTST, for short).

It is common practice to replace $E_{act,M}/\mathcal{R}$ in the equality portion of Eq. (1.436) by Eq. (1.411), in agreement with

$$\ln t = \ln\left\{\ln\frac{C_{M,0}}{C_{M,\max}}\right\} - \ln k_{d,M,0} + \frac{T_1^2}{z_M \log_{10} e}\frac{1}{T}, \tag{1.443}$$

and proceed likewise with regard to Eq. (1.442), viz.

$$\ln t = \ln\left\{\ln\frac{C_{Q,0}}{C_{Q,\min}}\right\} - \ln k_{d,Q,0} + \frac{T_1^2}{z_Q \log_{10} e}\frac{1}{T}; \tag{1.444}$$

elimination of ln t between Eqs. (1.443) and (1.444) – possible only when the corresponding curves cross each other (i.e. at $t=t_{crt}$ and $T=T_{crt}$), generates

$$\ln\left\{\ln\frac{C_{M,0}}{C_{M,\max}}\right\} - \ln k_{d,M,0} + \frac{T_1^2}{z_M \log_{10} e}\frac{1}{T_{crt}} = \ln\left\{\ln\frac{C_{Q,0}}{C_{Q,\min}}\right\} - \ln k_{d,Q,0} + \frac{T_1^2}{z_Q \log_{10} e}\frac{1}{T_{crt}}, \tag{1.445}$$

while condensation of terms alike complemented by factoring out of $T_1^2/\log_{10}e$ give rise to

$$\frac{T_1^2}{\log_{10} e}\left(\frac{1}{z_M} - \frac{1}{z_Q}\right)\frac{1}{T_{crt}} = \ln\left\{\ln\frac{C_{Q,0}}{C_{Q,\min}}\right\} - \ln\left\{\ln\frac{C_{M,0}}{C_{M,\max}}\right\} + \ln k_{d,M,0} - \ln k_{d,Q,0}. \tag{1.446}$$

After recalling the operational features of a logarithm, Eq. (1.446) may be rewritten as

$$\frac{T_1^2}{\log_{10} e}\left(\frac{1}{z_M} - \frac{1}{z_Q}\right)\frac{1}{T_{crt}} = \ln\frac{\ln\dfrac{C_{Q,0}}{C_{Q,\min}}}{\ln\dfrac{C_{M,0}}{C_{M,\max}}} + \ln\frac{k_{d,M,0}}{k_{d,Q,0}}; \tag{1.447}$$

a further application of the said rules unfolds

$$\frac{T_1^2}{\log_{10} e}\frac{1}{T_{crt}} = \frac{\ln\dfrac{k_{d,M,0}\ln\dfrac{C_{Q,0}}{C_{Q,\min}}}{k_{d,Q,0}\ln\dfrac{C_{M,0}}{C_{M,\max}}}}{\dfrac{1}{z_M} - \dfrac{1}{z_Q}} \tag{1.448}$$

– or, in view of Eq. (1.382) and upon algebraic rearrangement of the outer logarithm,

$$T_1^2 \ln 10 \frac{1}{T_{crt}} = \frac{\ln\dfrac{\dfrac{\ln\dfrac{C_{Q,0}}{C_{Q,\min}}}{k_{d,Q,0}}}{\dfrac{\ln\dfrac{C_{M,0}}{C_{M,\max}}}{k_{d,M,0}}}}{\dfrac{1}{z_M} - \dfrac{1}{z_Q}}; \tag{1.449}$$

the critical temperature accordingly satisfies

$$T_{crt}=\frac{\left(\frac{1}{z_M}-\frac{1}{z_Q}\right)T_1^2\ln 10}{\ln\frac{\ln\frac{C_{Q,0}}{C_{Q,\min}}}{\frac{k_{d,Q,0}}{\frac{\ln\frac{C_{M,0}}{C_{M,\max}}}{k_{d,M,0}}}}},\tag{1.450}$$

or else

$$\boxed{\frac{T_{crt}}{T_1}=\frac{T_1\left(\frac{1}{z_M}-\frac{1}{z_Q}\right)\ln 10}{\ln\chi_{QM}}}.\tag{1.451}$$

Here χ_{QM} denotes the logarithmic reduction in chemical indicator relative to microbiological indicator, according to

$$\boxed{\chi_{QM}\equiv\frac{\frac{\ln\frac{C_{Q,0}}{C_{Q,\min}}}{k_{d,Q,0}}}{\frac{\ln\frac{C_{M,0}}{C_{M,\max}}}{k_{d,M,0}}}};\tag{1.452}$$

Eq. (1.451) is plotted in Fig. 1.36. Inspection of this figure unfolds a decrease in T_{crt} with χ_{QM}, meaning that lower processing temperatures are possible when the ratio of $(\ln C_{Q,0}/C_{Q,\min})/k_{d,Q,0}$ to $(\ln C_{M,0}/C_{M,\max})/k_{d,M,0}$ is larger; as well as an increase with $1/z_M-1/z_Q$, meaning that a target microorganism more sensitive to thermal processing (i.e. characterized by a lower z_M) or a target quality factor less sensitive

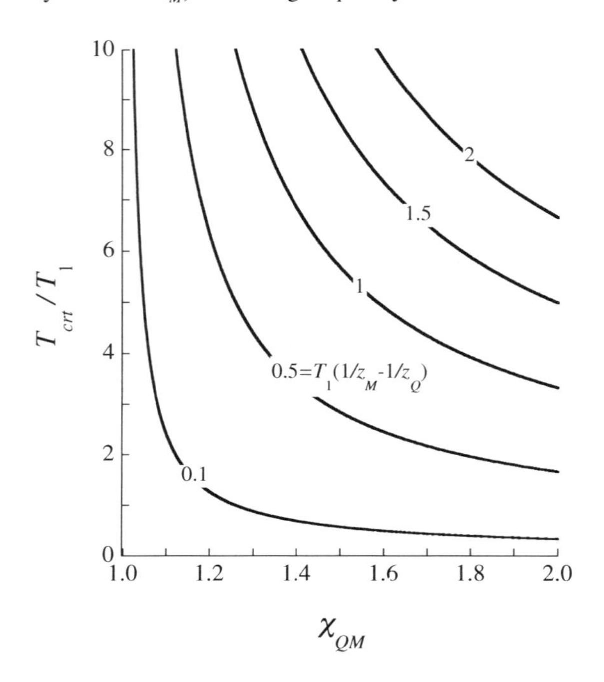

FIGURE 1.36 Variation of dimensionless minimum processing temperature, T_{crt}/T_1, versus ratio of logarithmic reduction of nutritional/sensory quality factor to microbiological quality factor, χ_{QM}, for selected values of germane parameter $T_1(1/z_M-1/z_Q)$ – with indicator microorganism characterized by z_M, and indicator compound characterized by z_Q as thermal resistance factors.

thereto (i.e. associated to a higher z_Q) favors higher-temperature processing.

Insertion of Eq. (1.448) permits, in turn, transformation of Eq. (1.443) to

$$\ln t_{crt}=\ln\left\{\ln\frac{C_{M,0}}{C_{M,\max}}\right\}-\ln k_{d,M,0}+\frac{\ln\frac{k_{d,M,0}\ln\frac{C_{Q,0}}{C_{Q,\min}}}{k_{d,Q,0}\ln\frac{C_{M,0}}{C_{M,\max}}}}{z_M\left(\frac{1}{z_M}-\frac{1}{z_Q}\right)}\tag{1.453}$$

in attempts to ascertain the corresponding t_{crt} – which may be algebraically rearranged to read

$$\ln t_{crt}=\ln\left\{\ln\frac{C_{M,0}}{C_{M,\max}}\right\}-\ln k_{d,M,0}+\frac{\ln\frac{\frac{\ln\frac{C_{Q,0}}{C_{Q,\min}}}{k_{d,Q,0}}}{\frac{\ln\frac{C_{M,0}}{C_{M,\max}}}{k_{d,M,0}}}}{1-\frac{z_M}{z_Q}};\tag{1.454}$$

after lumping the first two logarithmic functions and bringing $1-z_M/z_Q$ to exponent of the argument in the last logarithm, Eq. (1.454) will appear as

$$\ln t_{crt}=\ln\frac{\ln\frac{C_{M,0}}{C_{M,\max}}}{k_{d,M,0}}+\ln\left(\frac{\frac{\ln\frac{C_{Q,0}}{C_{Q,\min}}}{k_{d,Q,0}}}{\frac{\ln\frac{C_{M,0}}{C_{M,\max}}}{k_{d,M,0}}}\right)^{\frac{1}{1-\frac{z_M}{z_Q}}}\tag{1.455}$$

that is equivalent to

$$\ln t_{crt}=\ln\frac{\ln\frac{C_{M,0}}{C_{M,\max}}}{k_{d,M,0}}\sqrt[1-\frac{z_M}{z_Q}]{\frac{\frac{\ln\frac{C_{Q,0}}{C_{Q,\min}}}{k_{d,Q,0}}}{\frac{\ln\frac{C_{M,0}}{C_{M,\max}}}{k_{d,M,0}}}}\tag{1.456}$$

once the arguments of the outer logarithms in the right-hand side are lumped – or, after taking exponentials of both sides,

$$t_{crt}=\frac{\ln\frac{C_{M,0}}{C_{M,\max}}}{k_{d,M,0}}\sqrt[1-\frac{z_M}{z_Q}]{\frac{\frac{\ln\frac{C_{Q,0}}{C_{Q,\min}}}{k_{d,Q,0}}}{\frac{\ln\frac{C_{M,0}}{C_{M,\max}}}{k_{d,M,0}}}}.\tag{1.457}$$

Multiplication of both sides by $k_{d,M,0}/\ln C_{M,0}/C_{M,\max}$ finally transforms Eq. (1.457) to

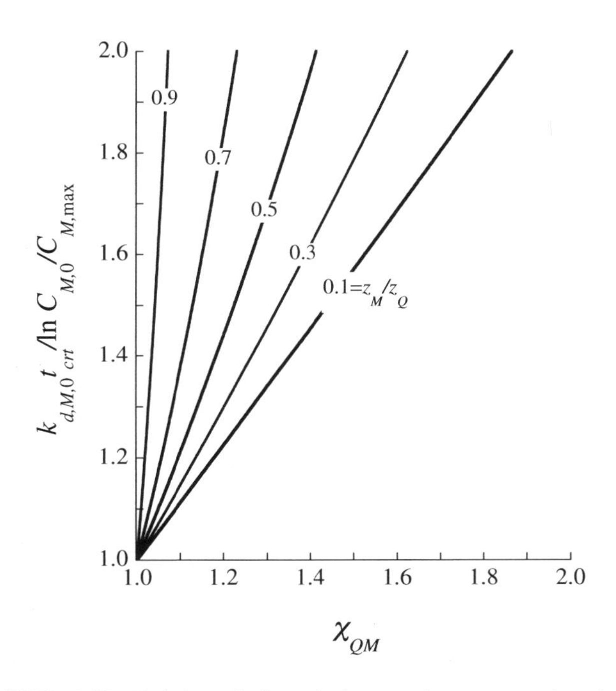

FIGURE 1.37 Variation of dimensionless maximum processing time, $k_{d,M,0}t_{crt}/\ln C_{M,0}/C_{M,\max}$, versus ratio of logarithmic reduction of nutritional/sensory quality factor to microbiological quality factor, χ_{QM}, for selected values of germane parameter z_M/z_Q – with indicator microorganism characterized by z_M, and indicator compound characterized by z_Q as thermal resistance factors.

$$\boxed{\frac{k_{d,M,0}t_{crt}}{\ln\dfrac{C_{M,0}}{C_{M,\max}}} = \sqrt[1-\frac{z_M}{z_Q}]{\chi_{QM}}}, \tag{1.458}$$

with the aid of Eq. (1.452); a plot of Eq. (1.458) is provided in Fig. 1.37. Note the increase in t_{crt} with χ_{QM}, opposite to the decrease of T_{crt} with χ_{QM} as per Fig. 1.36 – but in agreement with the general shape of the curves in Fig. 1.35; furthermore, a target microorganism more sensitive to heat (i.e. described by a lower z_M), or a reference quality factor less sensitive thereto (i.e. associated to a higher z_Q) produces a shallower variation of the t_{crt} vs. χ_{QM} curve (in view of the lower value of z_M/z_Q).

PROBLEM 1.12

The rationale of thermal processing, from the point of view of minimum safe microbial load, implies that one should operate above the curve described by

$$\ln t = \alpha + \frac{E_{act,M}}{\mathcal{R}T} \tag{1.12.1}$$

in parallel to Eq. (1.436), see Fig. 1.35i – where α denotes a constant, defined as

$$\alpha \equiv \ln\left\{\ln\frac{C_{M,0}}{C_{M,\max}}\right\} - \ln k_{d,M,0}; \tag{1.12.2}$$

on the other hand, the rationale for thermal processing, from the point of view of maximum acceptable quality decay, requires operation below the curve described by

$$\ln t = \beta + \frac{E_{act,Q}}{\mathcal{R}T} \tag{1.12.3}$$

as per Eq. (1.442), see Fig. 1.35ii – where β denotes another constant, abiding to

$$\beta \equiv \ln\left\{\ln\frac{C_{Q,0}}{C_{Q,\min}}\right\} - \ln k_{d,Q,0}. \tag{1.12.4}$$

However, two degrees of freedom are left, which span the dashed portion of the plane in Fig. 1.35iii; find the best processing temperature (and time), from an economic point of view – knowing that the processing costs, C_{pr}, per unit mass of food, abide to

$$C_{pr} = \varepsilon t + \xi\lambda_{vap}^{\theta} + \psi c_P\left(T - T^{\theta}\right), \tag{1.12.5}$$

where ε, ξ, and ψ denote (known) constants.

Solution

Inspection of Eq. (1.12.5) indicates that costs scale linearly on processing time, corresponding to the first term in the right-hand side; as well as on latent heat supplied to vaporize water under standard conditions (via latent heat of vaporization, λ_{vap}^{θ}), as per the second term; and on sensible heat supplied to increase temperature of steam from the reference temperature (T^{θ}) up to the desired temperature (T), as consubstantiated in the third term (via isobaric heat capacity, c_P). The necessary condition for a minimum in processing costs accordingly reads

$$\frac{dC_{pr}\{t,T\}}{dT} = 0, \tag{1.12.6}$$

using (absolute) temperature as independent variable – where application of the chain (partial) differentiation rule unfolds

$$\frac{\partial C_{pr}}{\partial T} + \frac{\partial C_{pr}}{\partial t}\frac{dt}{dT} = 0, \tag{1.12.7}$$

since $t \equiv t\{T\}$; upon insertion of Eq. (1.12.5), one obtains

$$\psi c_P + \varepsilon\frac{dt}{dT} = 0 \tag{1.12.8}$$

where multiplication and division of the second term by t gives rise to

$$\psi c_P + \varepsilon t\frac{d\ln t}{dT} = 0, \tag{1.12.9}$$

in view of the rule of differentiation of a logarithm. For any chosen temperature, the processing time should be as short as possible for economic reasons, and in agreement with Eq. (1.12.5); furthermore, one should somehow select a point along the curve described by Eq. (1.12.1) – since the minimum safety threshold will be ensured when sliding along the said curve, while sensory decay will be minimized for being the farthest possible from the curve described by Eq. (1.12.3). Resorting to Eq. (1.12.1) as a relationship between ln t and T, one may apply derivatives, with regard to T, to get

$$\frac{d \ln t}{dT} = -\frac{E_{act,M}}{\mathcal{R}T^2}; \tag{1.12.10}$$

while exponentials can be taken of both sides of Eq. (1.12.1) to obtain

$$t = e^{\alpha} \exp\left\{\frac{E_{act,M}}{\mathcal{R}T}\right\}. \tag{1.12.11}$$

Insertion of Eqs. (1.12.10) and (1.12.11) transforms Eq. (1.12.9) to

$$\psi c_P - \varepsilon e^{\alpha} \exp\left\{\frac{E_{act,M}}{\mathcal{R}T}\right\} \frac{E_{act,M}}{\mathcal{R}T^2} = 0, \tag{1.12.12}$$

or else

$$\psi c_P - \frac{\varepsilon e^{\alpha} \mathcal{R}}{E_{act,M}} \left(\frac{E_{act,M}}{\mathcal{R}T}\right)^2 \exp\left\{\frac{E_{act,M}}{\mathcal{R}T}\right\} = 0 \tag{1.12.13}$$

after multiplying and dividing the second term by $E_{act,M}/\mathcal{R}$; isolation of the (dimensionless) term in $E_{act,M}/\mathcal{R}T$ then yields

$$\frac{\psi c_P E_{act,M}}{\varepsilon \mathcal{R} e^{\alpha}} = \left(\frac{E_{act,M}}{\mathcal{R}T_{opt}}\right)^2 \exp\left\{\frac{E_{act,M}}{\mathcal{R}T_{opt}}\right\}, \tag{1.12.14}$$

where $\frac{E_{act,M}}{\mathcal{R}T_{opt}} \equiv \frac{E_{act,M}}{\mathcal{R}T_{opt}}\left\{\frac{\psi c_P E_{act,M}}{\varepsilon \mathcal{R} e^{\alpha}}\right\}$ is provided in implicit form. The variation of (dimensionless) optimum temperature, $\mathcal{R}T_{opt}/E_{act,M}$, versus $\psi c_P E_{act,M}/\varepsilon \mathcal{R} e^{\alpha}$ is depicted below.

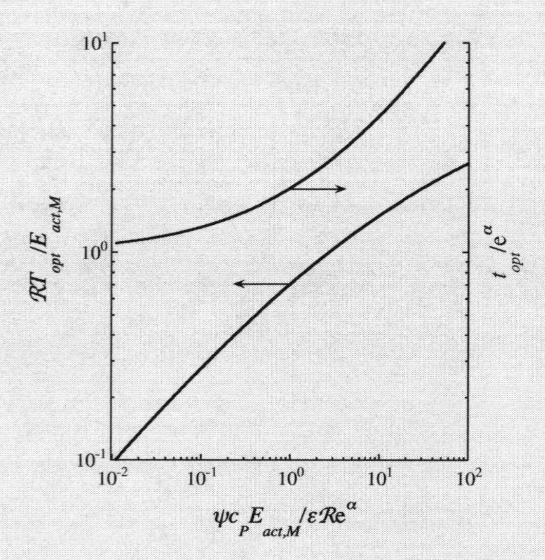

The dimensionless optimum temperature for thermal processing increases almost linearly with $\psi c_P E_{act,M}/\varepsilon \mathcal{R} e^{\alpha}$; it starts bending off only within the range of high values for the said parameter. The optimum processing time may finally be ascertained via

$$t_{opt} = e^{\alpha} \exp\left\{\frac{E_{act,M}}{\mathcal{R}T_{opt}}\right\}, \tag{1.12.15}$$

stemming from Eq. (1.12.11) – after setting $T = T_{opt}$, as obtained via Eq. (1.12.14); upon multiplication of both sides by $e^{-\alpha}$, one gets

$$e^{-\alpha} t_{opt} = \exp\left\{\frac{E_{act,M}}{\mathcal{R}T_{opt}}\right\}, \tag{1.12.16}$$

which is overlaid on the plot above, using again $\psi c_P E_{act,M}/\varepsilon \mathcal{R} e^{\alpha}$ as independent variable. The optimum time exhibits a distinct pattern – consisting of a slow increase within the low-value range of $\psi c_P E_{act,M}/\varepsilon \mathcal{R} e^{\alpha}$, which becomes faster as the latter variable increases.

1.2 Blanching, pasteurization, and sterilization

1.2.1 Introduction

Classical heat treatments resorting to water (in either liquid or vapor form) and intended at extending shelf-life and/or assuring safety of food encompass blanching, pasteurization, and sterilization; the former two use sub-boiling temperatures, whereas the latter resorts to superheated steam – with blanching and sterilization typically lasting for a shorter period than pasteurization.

Blanching (or scalding) is a mild thermal treatment, somewhat related to pasteurization in terms of time/temperature intensity; although designed primarily to inactivate enzymes that catalyze degradation reactions in vegetables and fruits prior to further processing (e.g. drying, canning, freezing, frying), microorganisms also lose viability to some extent. Remember that the time/temperature combination *per se* in drying or freezing may be insufficient to inactivate enzymes or destroy microorganism viability to a significant extent. If a food from plant origin has not undergone previous blanching, the adventitious enzymes in its tissues can indeed cause undesirable changes in sensory and nutritional features during storage afterward – and microorganisms in a state of latency may easily grow upon rehydration or thawing, respectively. On the other hand, the time taken to reach the sterilizing temperature in canned vegetables and fruits may be sufficient to permit such an enzyme activity to manifest – thus also justifying application of blanching prior to sterilization. For economic reasons, blanching is often combined with cleaning of raw plant material and/or peeling of fruits – thus saving energy, space, and equipment time.

Despite the aforementioned advantages of blanching, under-blanching may turn more detrimental to food than no blanching at all. When heat is applied, tissues are normally disrupted, and enzymes are thus released to the intracellular space; if the thermal treatment is not severe enough, the said enzymes will exhibit enhanced activity, along with higher availability to contact substrates – so food decay will actually be accelerated. Furthermore, inactivation of only some types of enzymes may induce unbalanced enzymatic activity later – which also contributes to faster deterioration of food quality.

Boiling water or steam are employed in blanching for a relatively short period of time, say 5–15 min (depending on the product); as emphasized above, this operation often takes place prior to thermal sterilization, dehydration, or freezing. Beneficial effects of blanching – other than enzyme deactivation and (limited) reduction of microbial viability, include cleansing; color improvement (or, at least, reduced rate of discoloration); removal of air from food tissues, which increases brightness; softening of tissues, which facilitates packaging in food containers; and increase in bioavailability of some food constituents (e.g. partially released β-carotene).

Thermal pasteurization is a relatively mild heat treatment, during which liquid, semi-liquid, or particulate liquid foods are heated at a specific temperature below 100°C, for a stated period of time – with many combinations of time and temperature being possible. More recently, the term has been extended to solid foods as well, e.g. almond pasteurization via oil- or dry-roasting, or else steam processing. It is aimed at severely reducing the viability of vegetative pathogens and spoilage microorganisms – i.e. to a level unlikely to raise a public health risk, under normal conditions of food distribution and storage; while conveying a reduced loss of nutritional and sensory features of the original food.

In low-acid foods (i.e. characterized by $pH > 4.5$, as is the case of plain milk), the major goal of pasteurization is to keep safety hazards associated to pathogenic microorganisms to a minimum – as well as prolonging shelf-life for periods ranging from a few days to a couple of weeks. In acidic foods (i.e. $pH \leq 4.5$, as is the case of fruit juices), pasteurization is aimed at extending shelf-life for several weeks, via destruction of spoilage microorganisms (namely yeasts and molds) and/or inactivation of enzymes; their low pH (either original, or deliberately lowered via addition of some acidulant) provides an intrinsic protection against growth of (pathogenic) bacteria.

In addition to destroying pathogenic and spoilage microorganisms, pasteurization attains almost complete inactivation of undesirable enzymes (e.g. lipase in milk). This being a moderate heat treatment generally causes acceptable changes in sensory properties of foods – yet it only permits a limited extension of shelf-life, as noted above.

HISTORICAL FRAMEWORK

The process of heating wine for preservation purposes has been known in China since 1117, and was documented in Japan in the diary *Tamonin-nikki* by several generations of monks, between 1478 and 1618. The Italian priest and scientist Lazzaro Spallanzani proved experimentally, in 1768, that heat kills small (microscopic) germs – which become unable to reappear, should the treated product be kept hermetically sealed. In 1795, a Parisian chef and confectioner, Nicholas Appert, began experimenting with ways to preserve foodstuffs – and was successful when dairy products, jams, jellies, juices, soups, syrups, and vegetables were placed in glass jars, sealed with cork and wax, and placed in boiling water for a period of time. Eventually, he won the 12,000-franc prize offered by Napoleon Bonaparte in 1809, for a new method to preserve food suitable for long expeditions of military and naval forces, via an improved version of his early invention. In 1810, Peter Durand patented a related method using a tin can as container – thus launching the modern process of canning, with *in situ* final sterilization. In 1812, Bryan Donkin and John Hall purchased both patents and started industrial manufacture of preserves; their company, Donkin, Hall & Gamble, soon resorted to a controlling oven to check sterility of the said preserves – which marked the beginning of quality control in this field. Canning was introduced in Australia and the United States in 1818, but failed to become a widespread practice for many decades – partly because hammer and chisel were required to open most cans; this constraint was eventually overcome by the can opener invented by Yates in 1855. The appertization process has remained essentially unchanged until now, as it permits freshness be retained and maximum safety be guaranteed – while providing consumers with reasonable taste and good nutritional value; for instance, vitamins keep more than 70% of their original potency in most canned foods.

Although effective in safety terms, Appert's method implied heating food at a temperature so high and for so long that it became detrimental to flavor – especially in the case of fruits and vegetables (and the like). A less aggressive method was meanwhile devised by Louis Pasteur, during a Summer vacation in Arbois in 1864, when he discovered that heating a young wine to 50–60°C for several minutes was sufficient to prevent acidity development, without hampering regular aging afterward or compromising good final quality; he also provided a scientific justification, based on death of acidifying adventitious microflora. The earlier appertization technique has thereafter been known as pasteurization in his honor. Although initially designed to prevent wine and beer souring, in 1870 pasteurization found an application in fresh milk and the dairy industry. In 1892, chemist Earnest Lederle proved that consumption of unpasteurized milk was a major cause for the large death toll due to tuberculosis; two decades later, he introduced mandatory pasteurization of milk in New York, in his role of Commissioner of Health.

Thermal sterilization involves heating the food to a sufficiently high temperature, i.e. above 100°C, and holding the product at this temperature for a prespecified period of time; the ultimate goal is to quantitatively kill microorganisms in both vegetative and spore forms, besides inactivating enzymes. The typical processes encompass retorting, where overatmospheric water steam is employed for solid or liquid foods; and ultra-high temperature aseptic processes, where steam is bubbled directly in liquid food. Sterilization renders food safe, and permits shelf-lives in excess of 6 months at room temperature. Oftentimes foods are precooked, in which case they will require marginal heating to assure safety and stability.

Although prolonged exposure to heat, namely in the case of in-container sterilization (e.g. canning, bottling), can substantially degrade the sensory and nutritional quality of food products, the higher the temperature, the shorter the processing time required for a given viable number reduction – see Fig. 1.35. In addition, aseptic packaging has been on the rise – where food is rapidly heated, under pressure, to temperatures above 130°C for a time of the order of second, and then filled into presterilized containers; the associated quality gains, however, only become significant if a higher profit margin is possible for the product – so as to allow recovery of the extra capital investment required by the said process. Nonconventional container geometries have also been tested to process thinner layers of product in flexible pouches or trays – or else combine sterilization with (previous) acidification of the food product.

For successful operation, the microbial load of raw materials should be as low as possible anyway – which demands hygienic handling and appropriate preparation procedures upstream, complemented with blanching (where appropriate). Thermal death times are then calculated based on an estimated maximum allowable/expected initial microbial load, and the corresponding reduction in viable count number required for consumer safety by the end of the intended shelf-life. Moreover, correct sterilization should guarantee that all containers receive the same amount of heat; failure to abide by these requirements may result in higher rates of incidence of spoilage, and ultimately compromise safety as a whole.

1.2.2 Product and process overview

1.2.2.1 Blanching

A few vegetables, namely green peppers and onions, do not require blanching to prevent enzyme activity throughout storage; however, most common vegetables undergo significant decay in quality if underblanched, or not blanched at all. Enzymes that cause loss of texture, production of off-tastes and -odors, or breakdown of nutrients in vegetables and fruits include (but are not limited to) chlorophyllase, lipoxygenase, pectic enzymes, polygalacturonase, polyphenoloxidase, and tyrosinase. On the other hand, blanching removes intercellular gases from plant tissues – which, together with removal of dust from the food surface, changes the wavelength of light reflected thereby, and thus improves the brightness of its color; these effects enhance consumer appeal. Blanching effectiveness is affected by size and shape, and thermal conductivity (a function of type, cultivar, and maturation degree) of food, as well as temperature and heat transfer coefficient of blanching vehicle. For instance, viable numbers of bacteria may be reduced by 10^2–10^3 in whole cucumbers treated at 80°C for as little as 15 s.

In addition to a number of physical and metabolic changes within cells of food tissues – dependent on type and maturity of the source plant(s), specific area (associated, in turn, to form and degree of cutting, slicing, or dicing), method, time and temperature of processing, and ratio of water to food, blanching causes cell death to some extent, especially on the surface of the food and layers in its vicinity. Typical outcomes of blanching are loss of cell turgor and nutrients, as the cytoplasmic membrane (but not the cell wall) becomes more permeable – while the associated higher permeability of subcellular organelles allows unconstrained interaction of their active constituents with the remainder of the cytoplasm. Therefore, cellular starch granules may undergo gelatinization, pectins may experience modification, proteins in both cytoplasm and nucleous may be denatured, and chloroplasts and chromoplasts may be distorted. Food pigments may suffer color changes as well; this is why sodium carbonate or calcium oxide are often added to blanching water, to protect chlorophyll – and thus retain color in green vegetables (even though the concomitant increase in *pH* favors loss of ascorbic acid). Previous dipping of foods (e.g. cut apples or potatoes) in dilute salt brine prior to blanching also prevents enzymatic browning afterward. Unlike water-soluble components (e.g. phenolic antioxidants and vitamins) – which may undergo leaching, thermal decay, or even oxidation, fat-soluble components are quite resistant to blanching; this includes carotenoids and sterols, further to dietary fiber and a few minerals.

Adequate enzyme inactivation requires rapid heating to a preset temperature, holding for some time at such a target temperature, and then cooling rapidly to near room temperature; therefore, one possible check for the success of blanching is whether a specified temperature has been attained at the thermal center of the food pieces for a preset time. To circumvent the need to employ thermocouples for this purpose, one has classically resorted to assay for peroxidase activity – which, not being responsible for relevant decay during storage (and thus not a target of blanching), is especially heat-stable. Therefore, it can be used as an indicator to ascertain blanching effectiveness, since inactivation of peroxidase implies inactivation of essentially all other enzymes of relevance; catalase may also be used for this purpose, despite its being more sensitive to heat. As an alternative, one may assay for intact vitamin C (or ascorbic acid) – as reference nutrient; note that typical losses of vitamins due to blanching are 10%–30% for ascorbic acid, besides 15%–20% for riboflavin and ca. 10% for niacin. Over-blanching may cause excessive softening and flavor loss in food – yet such a thermal treatment is still less aggressive than pasteurization.

Typical time and temperature ranges employed in blanching are 1–15 min and 70–100°C, respectively; the two-most widespread methods involve passing food through an atmosphere of saturated steam or a bath of hot water – as illustrated in Figs. 1.38i and ii, respectively. Both types of equipment are relatively simple to operate and inexpensive to purchase. However, steam blanchers are normally preferred to hot-water blanchers when the food product exhibits a high specific (outer) surface area (as leaching occurs to a lesser extent) – besides generating smaller volumes of effluent (and thus lower disposal costs thereof); exhibiting better energy efficiency; raising lower risk of contamination by thermophilic bacteria; allowing higher retention of color and flavor; and causing less extensive physical damage to texture (due to lower turbulence). Programmable logic controllers are often used to control space time and temperature – thus reducing operator-based decision-making and manual machine adjustments, further to minimizing energy consumption and effluent generation.

The simplest configuration of a steam blancher is a mesh conveyor that carries food through a steam atmosphere along a tunnel, typically 15 m long and 1–1.5 m wide – with space time

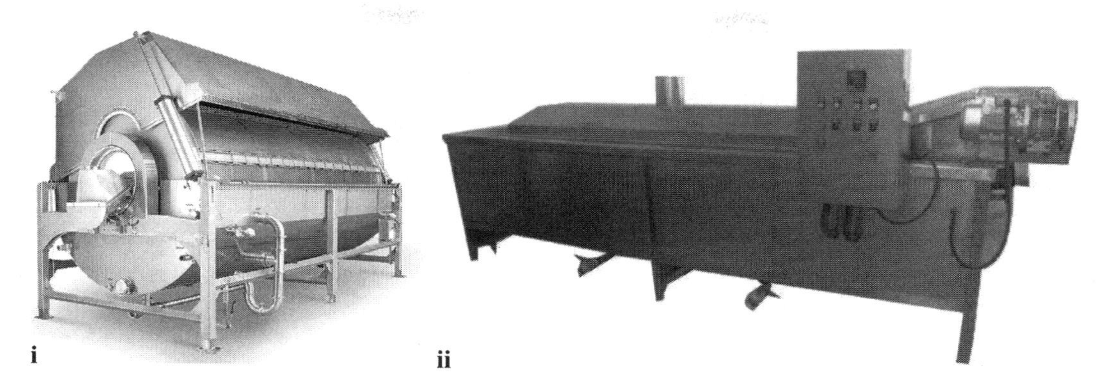

FIGURE 1.38 Examples of blanching equipment: apparatuses based on (i) continuous flow of steam (courtesy of Lyco Manufacturing, Columbus WI) or (ii) continuous submersion in hot water (courtesy of Gelgoog Machinery, Zhengzhou City, China).

determined by the ratio of tunnel length to conveyor speed. If water sprays are used at inlet and outlet to condense any escaping steam, an extra efficiency of energy consumption of up to 20% becomes possible. Food may instead enter and leave the blancher through rotary valves or hydrostatic seals, in attempts to bring steam losses to a minimum – thus allowing supplementary efficiency in energy consumption of ca. 25%; if, in addition, steam is reused by passing it through Venturi valves, then the said supplementary efficiency may reach 30%. Peroxidase is likely to undergo quantitative inactivation under these conditions; and degree of retention of ascorbic acid normally ranges within 75–85%. Drawbacks of steam blanchers include the limited degree of food cleaning (so extra washers are normally necessary), uneven blanching (if food is piled too high), larger size requiring more complex devices (and thus higher maintenance costs), and cleaning difficulties.

Blanching is also possible via a fluidized bed, by resorting to a mixture of air and water steam moving at velocities typically above 4.5 $m.s^{-1}$ – i.e. the minimum fluidization velocity, but below the terminal velocity of several particulate vegetables of interest, namely peas or diced carrots. Such a gaseous stream not only fluidizes the bed, but also heats the product without promoting water evaporation. Advantages of this configuration include faster and more uniform heating – thus causing smaller losses of vitamins and other heat-sensitive components, coupled to a substantial reduction in effluent volume; however, their commercial expression is still small.

Several designs are commercially available for hot-water blanchers – which basically hold the target food under water at 70–100°C for a prespecified time, before proceeding to the dewatering/cooling stage. The two best designs, in terms of food handling capacity per unit of floor area occupied, are the reel and the pipe blanchers. Food enters a slowly rotating cylindrical mesh drum that is partially submerged in water in the former configuration; the food is moved through the drum by internal flights – with residence time determined by drum length and rotation speed. Pipe blanchers consist of insulated metal pipe, fitted with feed and discharge ports – with hot water recirculated through the pipe and food metered in; in this case, the residence time is fixed by pipe length and water velocity. Advantages of hot-water blanching include lower capital cost and more uniform product heating, further to extra cleaning ability.

It should be stressed that the most important factor for commercial success of a blancher is food yield; oftentimes, greater losses of product or nutrients therein do occur in the final cooling stage – so flowing cold air or spraying cold water are frequently the methods of choice, rather than just immersing the blanched food in cold water. The former still brings about weight losses in the product due to water evaporation, which may outweigh any advantages gained in nutrient retention. Cooling with running water (sometimes termed fluming) induces also considerable leaching losses, since a few food components are readily washed away; conversely, overall yield may be improved due to absorption of water. Blanching at subatmospheric pressure reduces decay by oxidation, due to the lower partial pressure of oxygen – with consequent lower pigment degradation; however, blanching is faster when pressurized steam is utilized.

Improvements have been implemented since the 1980s toward fewer losses of soluble components and better energy use – in attempts to not only increase product yield, but also reduce environmental impact of processing; this includes recycling hot water, since it substantially reduces the volume of effluents generated – while barely affecting product quality or yield. Another improvement is individual quick blanching, or IQB technique; this entails two-stage blanching, with food heated in a single layer to a sufficiently high temperature to inactivate enzymes in the first stage – followed by adiabatic holding in the second stage, with a deep bed of food kept for a sufficiently long time to allow temperature at the thermal center of each piece reach the said target temperature. Such an approach overcomes poor heating uniformity of food laid out in multiple layers, while avoiding overheating of food at the edges; for instance, conventional blanching takes 3 min for carrots sliced as 1-cm pieces, but only 75 s via IQB (corresponding to 25 s for heating and 50 s for holding), with associated improvements in color and flavor retention of the final food – along with increases in productivity from ca. 0.5 up to 7 kg_{food}/kg_{steam}. A third improvement entails preconditioning, or exposure of food to warm air (ca. 65°C) prior to blanching; this permits evaporation of moisture from the food itself, thus improving steam absorption throughout subsequent blanching – with concomitant lower losses in both nutrients and overall weight. Adequate hygienic standards are a must, encompassing both product and equipment – to prevent buildup of unwanted bacteria in the cooling water, which unavoidably carries some nutrients in its way out; the extra costs of hygiene control may thus outweigh savings in energy and effluent generated.

The apparatuses for IQB using steam basically consist of a heating section – where a single layer of food is heated on a conveyor, and then held on a holding conveyor before cooling; food loading and unloading are carried out by bucket elevators located in closed-fitting tunnels, and the blancher chamber is fitted with rotary valves aimed at minimizing heat losses. The cooling section employs a fog spray to saturate the cold air with moisture – so as to reduce evaporative losses from the food, while retaining the effluent generated. A hot-water blancher–cooler, integrating the IQB principle, consists of three sections – a pre-heating stage, a blanching stage proper, and a cooling stage. The first uses water previously heated via a heat exchanger, where it is heated at the expense of the cooling undergone by water to be used in the third stage – so up to 70% of heat may be recovered; it is possible to process up to 20 kg_{food}/kg_{steam}, while material savings amount to ca. 0.1 m^3_{water}/ton_{food}. The food remains in a single conveyor belt throughout each stage, so it does not experience physical damage caused by turbulence as in conventional designs; a recirculated water/steam mixture is used to blanch the food, whereas final cooling is provided by cold air.

A novel strategy for blanching entails microwave heating; it can bring about faster and more uniform heating – thus reducing start-up and shut-down periods, as well as processing time and nutrient losses. This approach also saves energy; it allows for more accurate process control; and it reduces nutrient leaching, by requiring less extraneous water. However, equipment cost is far higher than that of conventional blanchers; and quantity and shape of food play an important role. Successful applications so far include blanching mushrooms to inactivate polyphenoloxidase (and thus delay browning) – as well as bell peppers, carrots, peanuts, spinach, and sweetcorn kernels (but not green turnip leaves), with the possibility of reducing both temperature and time required; special care is to be exercised with mushroom piece size, so as to avoid vaporization of constitutive water and concomitant texture damage. Although microwave-assisted blanching alone provides a fresh vegetable flavor, combining it with initial hot-water or steam blanching conveys an economic advantage – because (low cost) hot water or steam is used first to externally raise temperature, while (more expensive) microwave power does the harder task of internally heating the matrix.

Another possibility is ohmic heating – again utilized with success with mushrooms, besides pea purée; however, being normally restricted to fluid-like foods, it proves inadequate for processing of fruits or (leafy) vegetables. Finally, high-pressure blanching has been attempted, and can further reduce nutrient loss – yet insufficient enzyme inactivation seems to occur.

1.2.2.2 Pasteurization

The effects of heat upon enzymes and microorganisms have been discussed above – namely in terms of decimal reduction times for a given temperature, and *z*-factors for variable temperature; however, a major determinant of the efficacy of heat treatment is the (intrinsic or final) *pH* of the processed food. To design a pasteurization process, the most heat-resistant enzyme, or pathogen or spoilage microorganism that is likely to be initially present, has to be identified in advance; in the case of (low-acid) liquid whole egg, *Salmonella senftenberg* is the reference microorganism – whereas *Brucella abortis*, *Mycobacterium tuberculosis*, and *Coxiella burnetii* account for the target microorganisms in (low-acid) milk. Enterotoxicosis (via *Staphylococcus aureus*), several infections (via *Streptococus* spp.), listeriosis (via *Listeria monocytogenes*), paratyphoid fever (via *Salmonella* spp.), shigellosis (via *Shigella* spp.), and toxoplasmosis (via *Toxoplasma gondii*) – besides brucellosis (via *B. abortis*), tuberculosis (via *M. tuberculosis*), and flu-like symptom *Q* fever (via *C. burnetii*) are indeed diseases potentially transmitted to Man upon ingestion of milk. However, assurance that the latter three harmful microorganisms have been eliminated automatically guarantees that harmful microorganisms responsible for the remaining diseases have been eliminated as well – for their being more heat-labile. Some spoilage microorganisms that are more heat-resistant than *B. abortis*, *M. tuberculosis*, and *C. burnetii* may nevertheless remain in milk – so pasteurized milk is to be stored under refrigeration anyway, to attain a reasonable shelf-life.

Bulk liquid foods are normally pasteurized by passing them continuously through a heat exchanger, and then pursuing to aseptic packaging – or else via batchwise heating; if previously divided in discrete portions and then filled into appropriate containers (as is the case of beers and fruit juices), pasteurization will take place via hot water to reduce the risk of fracture of the glass container – or air/steam mixtures, in the case of metal or plastic containers, as the risk of thermal shock is marginal. Examples of pasteurizers are provided in Fig. 1.39. Temperature differences between hot water and food container are often kept below 20°C for heating and 10°C for cooling; containers are usually cooled to just 40°C to help evaporate surface water – so as to minimize chance for external corrosion of the container or cap, as well as accelerate setting of label adhesives (when present).

Hot-water pasteurizers may be operated batch- or continuous-wise; the simplest version of the former is a water bath, where crates of packaged foods are dumped – to be heated to, and then held at the preset temperature for a given time period, followed by cooling via (room temperature or cold) water. Its continuous counterpart is a (long and narrow) trough fitted with a conveyor (see Fig. 1.39iv), which carries containers along heating and cooling stages. More sophisticated designs entail a tunnel formally divided into preheating, heating, and cooling zones – with atomized, high-velocity water sprayed to heat or cool the food containers as they pass through the corresponding zone, thus permitting incremental changes thereof in temperature. Air/steam tunnels permit faster heating, and consequently demand smaller space times and space requirements; the temperature in the heating zone is gradually increased by reducing the fraction of air in the heating mixture. Final cooling is brought about via cold (or room temperature) water – either via spraying onto, or immersion of the containers.

Bulk pasteurization of (some) liquid foods may be carried out in open jacketed pans, especially if small amounts are to be processed at a time (see Fig. 1.39i); larger-scale operation involving liquids – as is the case of beer, fruit juices, liquid egg, milk, and wine, usually resorts to continuous heat exchangers, of tube-and-shell (see Fig. 1.39ii) or plate types. Energy and water savings are possible if the water cooled by the inlet food in the preheating zone is recirculated to the cooling zone – where it heats up at the expense of the hot outlet food (see Fig. 1.39iii). Advantages

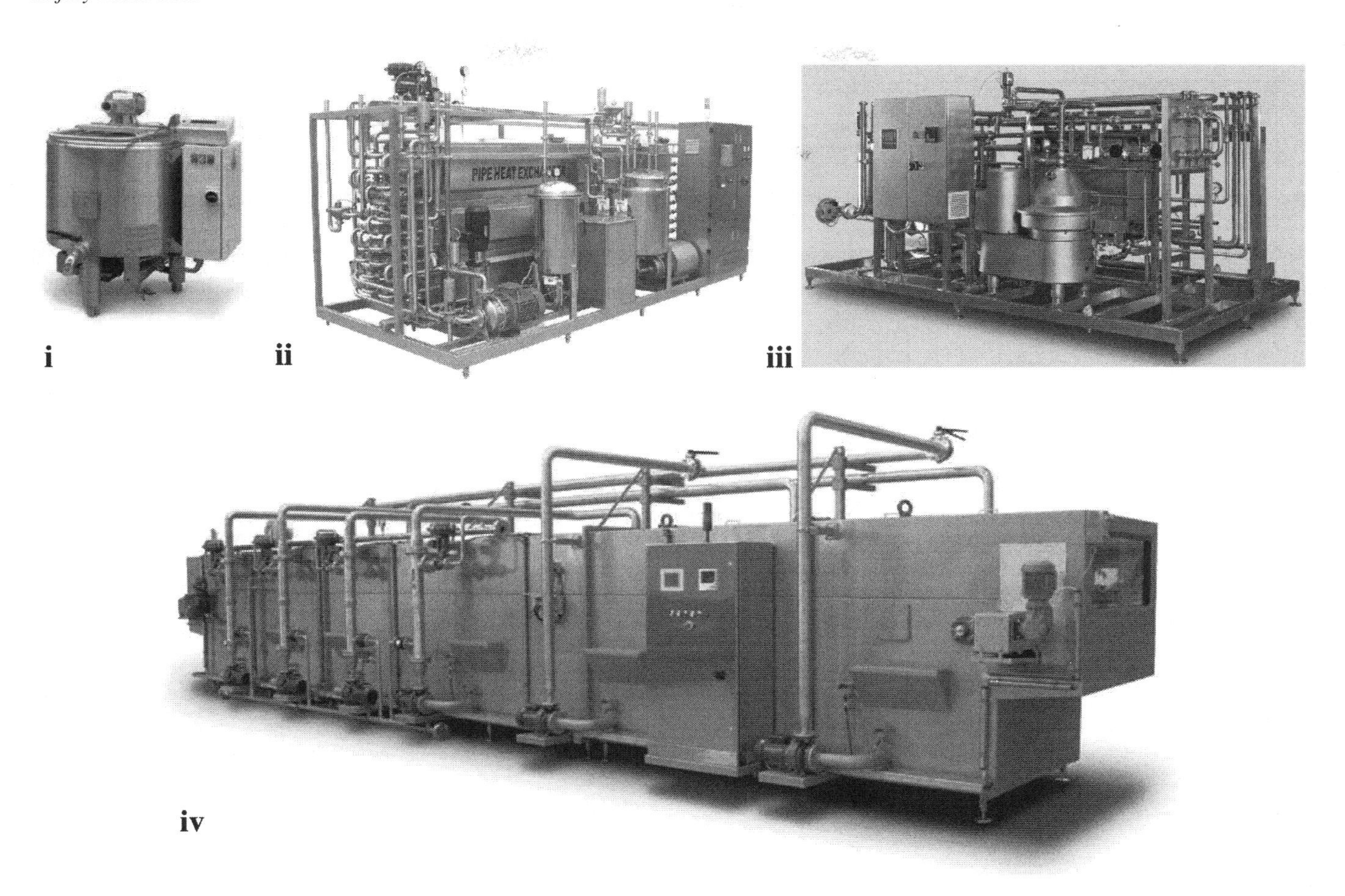

FIGURE 1.39 Examples of pasteurizing equipment: (i) batch liquid pasteurizer (courtesy of Swanwick Catering Equipment, Glos, UK), (ii) tubular pasteurizer (courtesy of Shangai Changlong Industrial Equipment, Shangai, China), (iii) compact milk pasteurizer (courtesy of GEA, Dusseldorf, Germany), and (iv) tunnel pasteurizer for cans (courtesy of Co.Mac, Sotto, Italy).

of continuous equipment include more uniform heat treatment, lower space requirement, greater control over pasteurization conditions (with associated lower labor costs), and better efficiency of energy use. Shell-and-tube and concentric tube heat exchangers are used especially to pasteurize viscous, non-Newtonian fluids – as is the case of baby foods, dairy products, mayonnaise, and tomato ketchup; (compact) plate heat exchangers are general-purpose apparatuses for continuous pasteurization. When in regular operation, the liquid food is pumped from a balance tank to a regeneration section, frequently using a positive displacement pump; it is preheated there by food that has already been pasteurized. It then passes to a heating section, where it reaches the intended pasteurizing temperature, and finally through a holding tube – where it remains for the time required to complete pasteurization, as per the specification of the said temperature. If, by any chance, the pasteurization temperature is not attained, a temperature sensor normally activates an inlet valve – which diverts flow and returns the food to the balance tank for repasteurization. The processed product is cooled in the regeneration section, while simultaneously preheating incoming food (as seen above); finally, the pasteurized liquid is further cooled by cold water, or chilled water (if necessary) in a cooling section. Following pasteurization, food is filled at once into cartons or bottles – duly sealed to prevent recontamination. Typical pretreatments for better performance encompass deaeration of fruit juices and wines, via application of a partial vacuum, to prevent oxidative changes; and homogenization of milk, via forced passage through a tiny nozzle, to minimize phase separation during heat treatment and storage afterward.

Various combinations of processing temperature and time are normally feasible in pasteurization processes – as illustrated in Fig. 1.40. The slightly concave-bended alignment in the log/lin plot is consistent with the patterns apparent in Fig. 1.35iii. Pasteurization of fruit juice has enzyme inactivation as the primary aim, and destruction of spoilage microorganisms as subsidiary purpose; while pasteurization of beer has destruction of spoilage microorganisms (e.g. wild yeasts, *Lactobacillus* spp.) and residual added yeasts (e.g. *Saccharomyces* spp.) as the primary aim. The main goal of milk pasteurization is destruction of pathogens (viz. *B. abortis*, *M. tuberculosis*, *C. burnetii*) as discussed previously, while destruction of spoilage microorganisms comes as subsidiary purpose; a similar situation arises with such dairy products as ice cream, ice milk, and eggnog, as well as cream or chocolate milk – which require higher pasteurization temperature, due to their higher sugar content and viscosity. The chief purpose of pasteurization of liquid egg is destruction of pathogens (*S. senftenberg*), with elimination of spoilage microorganisms coming as secondary target.

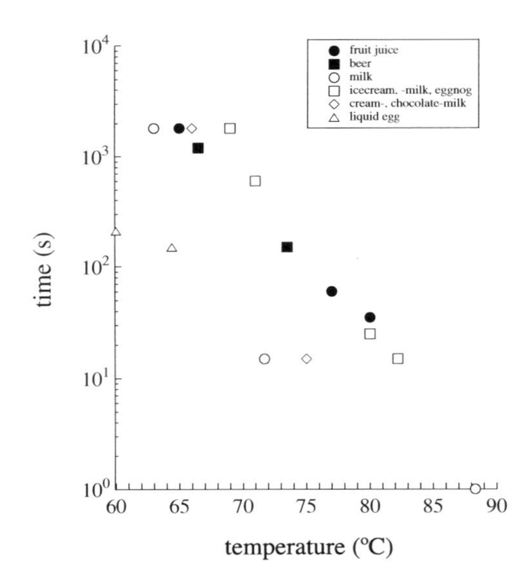

FIGURE 1.40 Typical pasteurization conditions for selected acidic foods (●, ■) and low-acid foods (○, □, ◇, △), expressed as processing temperature and time.

A holding time of 30 min at 60°C has classically been elected by cheese manufacturers for batch pasteurization of raw milk – known as Holder's process, or low-temperature, long-time process (LTLT for short); higher temperatures have in fact a detrimental effect upon behavior of curds, to be formed during subsequent cheesemaking – which may compromise the final cheese texture. Postclotting pasteurization is out of the question, because of the (otherwise avoidable) multiplication of unwanted microorganisms; this would magnify the public health problem associated with cheese consumption. In addition, it would probably be incompatible with use of commercial starter cultures for specific sensory development – not only owing to ecological reasons of dominance, but also because uniform distribution throughout the cheese matrix would become impossible. Conversely, such long processing times are not suitable for drinking milk – due to the significant changes in flavor and loss of vitamins that would be observed; 15 s at 71.8°C is instead preferred, often labeled high-temperature, short-time process (or HTST, for short) – yet higher temperatures and associated shorter times permit even better results, in general agreement with Fig. 1.35iii. During the aforementioned 15 s-heating period, reduction in viable numbers not below 6.8 logarithmic cycles is possible for *Salmonella* serotype *typhimurium* at 61.5°C, *Yersinia enterocolitica* at 62.5°C, (enteropathogenic) *Escherichia coli* at 65°C, *Listeria monocytogenes* at 65.5°C, *Staphylococcus aureus* at 66.5°C, and *Cronobacter* (previously *Enterobacter) sakazakii* at 67.5°C; this is consistent with previous information on easier elimination of pathogens other than the three ones used as reference. Inspection of Fig. 1.40 also emphasizes that intensity of thermal treatment, for a given product, depends on other intrinsic factors, namely *pH*; this is why fruit juices require 65°C for just 30 min, while low acid vegetables require temperatures in the vicinity of 100°C for the same time.

Sous-vide thermal processing, within 65–95°C, has meanwhile been on the rise – especially in the food service industry; besides being quite convenient, it has proven a successful form of pasteurization for meat, ready-to-eat meals, fish stews, and salmon fillets, provided that the treated product is stored not above 3°C and for no more than a couple of days.

Complete destruction of reference microorganisms for public health in low-acid foods is normally envisaged by regular pasteurization – and has classically been set at 12*D*-level; this means that, for an initial viable number of 10^2 in a given container, pasteurization should guarantee a probability lower than 10^{-10} of finding a viable cell in that container afterward – or, equivalently, the probability of survival would not exceed 1 viable cell in 10^{10} containers alike. Since assays for viable microorganisms are usually expensive and time-consuming – and thus incompatible with in-line information on raw materials, indirect tests have been developed and duly validated. This is notably the case of alkaline phosphatase, a native enzyme in milk – which coincidentally possesses a *D*-value similar to those of *B. abortis*, *M. tuberculosis*, and *C. burnetii* as target microorganisms; hence, a quick and fiable test results, suitable for routine analysis. A similar test, based on native α-amylase, has been in use to check pasteurization of liquid egg. Although lack of alkaline phosphatase activity indicates successful pasteurization of milk, the said enzyme may be reactivated in cream or cheese – whereas microorganisms used in cheesemaking may also synthesize, and eventually release that enzyme. Therefore, special care is needed to avoid false negatives – namely via testing immediately after pasteurization, and calibrating with milk source; in fact, the amount and activity vary widely between dairying species – and such enzymes are adsorbed on milkfat globules, thus being affected by milk skimming.

Since relevant pathogens are unable to grow in acidic foods, the target of pasteurization is in this case heat-resistant enzymes and acid-tolerant spoilage microorganisms (e.g. lactic acid bacteria, besides yeasts and molds). Therefore, the elimination of yeasts and Gram$^+$, non-spore forming bacteria is targeted in fruit products – and pectin esterase may serve as indicator enzyme, due to its higher thermal resistance.

Once the correct combination of time and temperature has been found – usually along the lower boundary of the shaded region in Fig. 1.35iii, one must guarantee that all portions of the food have been exposed to at least that temperature, and for at least that time. The controlling factors to establish the correct residence time for a liquid food in a continuous pasteurizer are velocity of the fastest moving portion or particle of food, and length of the holding tube. To work on the safe side, the heating and cooling stages are not taken into account when estimating lethality; hence, isothermal conditions may be assumed in the holding section – with the critical line coinciding with the axis of the tube, where velocity is twice the nominal (average) velocity as per Eqs. (2.50) and (2.60) of *Food Proc. Eng.: basics & mechanical operations* pertaining to laminar regime, or ca. 1.2-fold in the case of fully developed turbulent flow as per Eq. (2.3.9) in the same book. In other words, the design should assure the indicated residence time for the fluid along the axis of the tube (at the reference temperature) – so the remaining fluid will unavoidably be overprocessed.

Pigments in foods, from both animal and plant sources, are essentially unaffected by pasteurization; in the case of fruit juices, the chief reason for color decay is polyphenol oxidase-mediated browning – which may be prevented by previous deaeration to remove dissolved oxygen. Pasteurization has been claimed to accelerate juice browning due to formation of 5-hydroxymethylfurfural throughout storage; fortification with ascorbic acid and fructose is useful to retain color at large – and protect carotenoids, in particular. Although pasteurized milk is typically whiter than raw milk, this effect is not due to pasteurization – but instead to the homogenization that usually precedes it.

Loss of volatile aromas during juice pasteurization obviously reduces quality of the final product – even though the cooked flavor meanwhile developed may be disguised thereby; deaeration appears to play a major role in the said loss. Although volatile recovery, followed by reincorporation, is technically feasible, this is not employed on a routine basis for economic reasons. In the case of raw milk, loss of volatiles may even be a favorable trait – as it allows removal of hay-like keynotes, thus leaving a blander product behind.

The most significant losses in nutritional quality incurred in by pasteurization pertain to thermally labile vitamins; for instance, milk pasteurization causes up to 25% loss of ascorbic acid, up to 10% loss of folate, cyanocobalamin, and riboflavin, and up to 7% loss of thiamine. In the case of fruit juices, deaeration prior to pasteurization is useful to keep losses of ascorbic acid and carotene to a minimum.

In any case, classical techniques of pasteurization using heat supplied by conduction suffer from a number of limitations – namely regarding degradation to some extent of sensory and nutritional features. Such novel techniques as pulsed electric field and electric arcing, ultrasound, hyperbaric treatment, and membrane processing have been under active scrutiny (as referred to above). In the specific case of orange juice, 60 μs-pulsed electric fields at 35 $kV.cm^{-1}$ yield 7 logarithmic cycle-reduction in viable numbers of aerobic bacteria, yeasts, and molds; and 1 logarithmic cycle in pectin methylesterase activity. High-pressure pasteurization has proven feasible in the case of jams, juices, milk, and wines – with rates of microbial destruction much above those found for enzyme inactivation or color/flavor decay; in the particular case of carbonated juices, treatment at 35 MPa can effect a 10^5-fold reduction in viable numbers of several pathogens. For liquid matrices requiring previous homogenization – as is the case of milk (for being constituted by a suspension of milkfat globules of a wide size range), homogenization at ultrahigh pressures within 200–300 MPa at 30–40°C produces results analogous to classical thermal pasteurization for 15 s at 90°C, namely 3–4 logarithmic reduction in numbers of viable lactococci, and essentially complete destruction of lactobacilli, enterococci, and coliforms. Despite their potential, none of these approaches has reached commercial scale to date.

1.2.2.3 Sterilization

The time required to sterilize a food depends on the thermal resistance of the microorganisms and enzymes (expected to be) present in the said food – and their numbers and activity, as well as *pH*, physical state, and thermal conductivity of food, size and shape of container, and heating conditions.

In low-acid foods, the heat-resistant, spore-forming microorganism *Clostridium botulinum* is the most dangerous pathogen likely to be present; under the anaerobic conditions prevailing inside a sealed container, it can grow and produce a powerful exotoxin – botulin, which is sufficiently potent to cause 65% mortality in humans following ingestion. This microorganism is ubiquitous in soil, and thus likely to be found in small numbers on any raw material that has previously contacted it (via adventitious contamination), or be transferred by equipment or operators to other foods (via cross-contamination). Because of the extreme hazard arising from ingestion of botulin, destruction of this microorganism is a minimum requirement for heat sterilization – although, for precaution, foods are normally exposed to more intense thermal processing, since other more heat-resistant (yet less aggressive) pathogens or spoilage bacteria, in either vegetative or spore form, may also be present. Consequently, canning food processors have resorted to 12*D*-processes using *C. botulinum* as reference pathogen; in other words, their target is to reduce the viable numbers of that microorganism by a factor of 10^{12} – which normally suffices to guarantee a negligible probability of incidence, for realistic levels of initial contamination. The severity of this treatment is required by the high *pH* of canned foods – typically above 4.5, and often in the neighborhood of neutrality in the case of meat and fish matrices; coupled with the expected long shelf-life – usually beyond 2 years, at room temperature. Such a pathogen cannot grow in acidic foods, but other microorganisms – namely spoilage yeasts and molds, as well as heat-resistant enzymes are of concern with regard to shelf-life.

The shelf-life of a sterilized food depends, at least in part, on the ability of its container to completely isolate it from the surrounding environment; metals cans, glass jars or bottles, flexible pouches, and rigid trays have all successfully been used for this purpose. Before further processing of filled containers, air is to be removed by exhaustion; this unit operation prevents expansion of air later during heating – which would buildup useless strain upon the container seals. Exhaustion can be brought about by hot filling of the container with food (which also preheats the food, and thus saves processing time afterward); cold filling of food, followed by heating of both container and contents to 80–95°C with lid partially sealed (or clinched); partial vacuum, using a vacuum pump; or steam flow closure, with a blast of steam carrying air away from the overhead space in the container, immediately prior to sealing. The latter method allows replacement of air by steam – so a partial vacuum develops upon cooling of the sealed container as steam condenses; a typical click will thus be heard when the container is opened at the time of consumption, as the metal lid regains its original shape once atmospheric air is allowed into the container. Oxygen is removed with air as a constituent thereof – and this prevents not only oxidation of the food itself, but also corrosion of the containing metal cans; filled and sealed containers will eventually be loaded to a retort. Remember that blanching also serves the purpose of removing most entrapped air from vegetable foods prior to filling.

Sterilization of canned foods is normally carried out in a retort using (water) steam as heating fluid, and later tap water as cooling fluid – and either utility is flown through its outer casing; this heats up the air inside the retort – which actively circulates, and transfers enthalpy, in turn, to the cans (and their contents) placed inside. An example of this type of equipment is put forward in Figs. 1.41i and ii. Note the sealed cover to be screwed in place after loading and prior to steaming (see Fig. 1.41i), and the clearance between the load and the inner surface of the retort to allow for free circulation of hot fluid (see Fig. 1.41ii). Sterilizing retorts may be operated batch- or continuouswise, and the former may

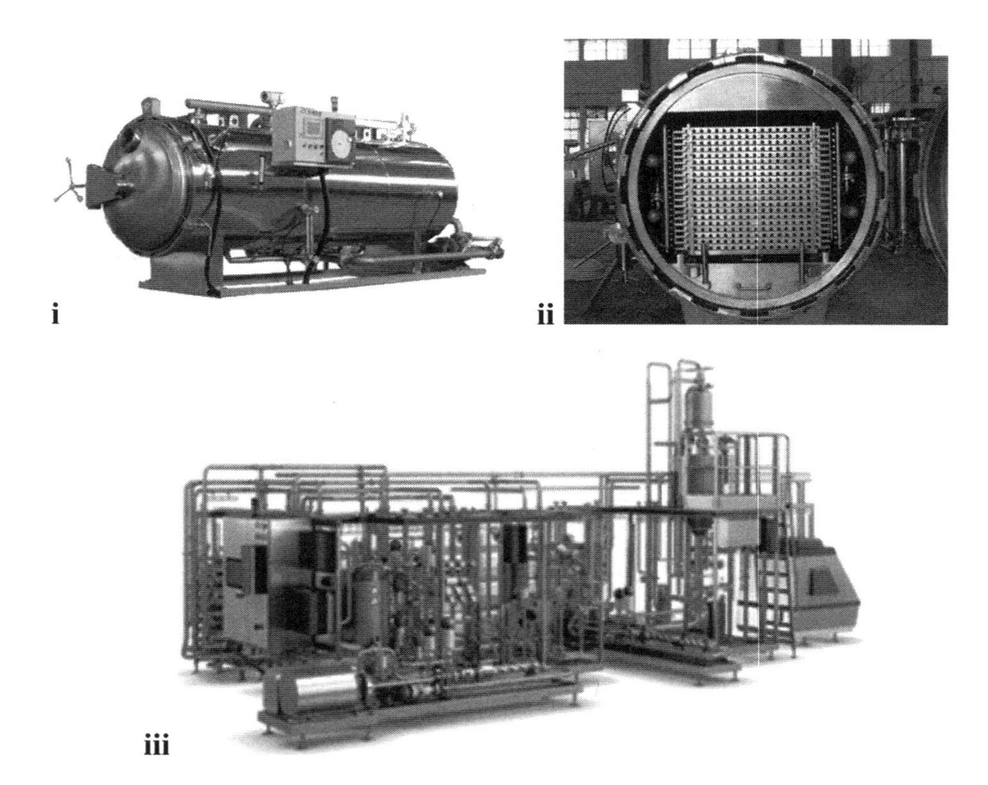

FIGURE 1.41 Examples of sterilizing equipment: steam retort, (i) as overall view including supporting devices (courtesy of Alibaba Group, Hangzhou, China) and (ii) entrance side view with trays for (canned) foods (courtesy of Wenzhou Longqiang Machinery, Zhejiang, China); and (iii) integrated unit for UHT processing of milk (courtesy of TetraPak, Lund, Sweden).

take a horizontal or vertical layout; batch apparatuses are easier to load and unload, and may be equipped for end-over-end agitation – which nevertheless takes up more floor space. Batteries of retorts can be operated simultaneously, each at its own stage, so the overall process may be viewed as semicontinuous; in fact, an almost continuous supply of food by the preparation section upstream, and an almost continuous supply of processed containers to the packing section are to take place. This hybrid configuration resorts to the intrinsic flexibility of individual batchwise processing, and couples it to the ability of computer controls to synchronize collective operation; however, transient heating at startup and cooling at shutdown do not allow the said batteries to attain the thermal and operational efficiencies characteristic of true continuous retorts, operated under steady-state conditions.

Continuous retorts are characterized by more gradual thermal transitions than their batch counterparts – and are less prone to developing strain caused by pressure buildup inside the containers; in addition, no heating or cooling cycles are necessary, which adds to process efficiency. Conversely, their capital cost is much higher; and a larger in-process stock will be lost if some form of breakdown occurs. Three major designs are available: cooker/cooler, for lower throughput rates and more adequate if regular change in container size of processing conditions occurs; rotary, able to handle up to 300 cans per minute; and hydrostatic, suitable for up to 1000 cans per minute. The (static) cooker/cooler design carries cans on a roller or chain conveyor through three sections along a tunnel – intended for sequential preheating, sterilizing, and cooling; they are duly interconnected by pressure locks. Rotary sterilizers consist of a slowly rotating cage of cylindrical shape – with cans loaded horizontally into the annular space between cage and pressure vessel, and guided by a static spiral track; rotation promotes forced convection, and also causes the headspace bubble to move through the can so as to mix its contents – with an accompanying decrease in resistance to heat transfer. Hydrostatic sterilizers possess two columns of water on either side of a steam chamber – sufficiently high to create hydrostatic pressure that balances steam pressure, with water sealing the chamber; cans are loaded horizontally, end to end, on carriers held between two chains that pass through the sequential sections.

Retorts are fitted with equipment for monitoring and control – and manage such variables as temperature and pressure of steam, temperature of cooling water, time and temperature of processing, and pressure of compressed air. Computer-based control of actuators on steam, water, and compressed air valves automatically corrects for spurious deviations from preset values. Artificial intelligence has been used to continuously compare accumulated to target lethality (as required for commercial sterility); process deviations are detected in real time, the potential risk to public health is estimated, and the required control effected also in real time – thus minimizing overall damage.

When food is sterilized in containers, the outer layers of food shield the inner layers from externally supplied heat; this adds to the package itself in creating a higher resistance to heat transfer. Aseptic processing overcomes these limitations, as food is heated through thin layers in the first place, and lethality is attained before filling the presterilized containers (under a sterile atmosphere). To attain higher quality final products, from nutritional and sensory standpoints, temperatures within the range 130–150°C are to be employed – which calls for direct injection of steam, in equipment of the type shown in Fig. 1.41iii; this approach reduces processing times to the order of second, or fraction thereof. Such features have overridden in-container sterilization of such liquid foods as cream, fruit juices and concentrates, icecream mix, liquid egg, milk, salad dressings, and yogurt. Aseptic processing is also more and more frequently employed with baby foods, cottage cheese, rice desserts, soups, and tomato products – all of which contain particulate solids. Liquid foods containing larger solid pieces are more difficult to process by UHT, because enzyme inactivation at the thermal center of the said pieces brings about overcooking on their surface; in addition, energetic stirring is required to improve the rate of heat transfer, ensure uniform temperature, and avoid settling in the holding tube. Multiphasic fluid foods are difficult to handle in heat exchanger-type of equipment – chiefly due to phase separation, and disparate thermal capacities and conductivities prone to cause nonuniform thermal treatment.

The quality of UHT foods compares well with that of chilled or frozen foods that undergo no thermal processing at all – with the unique advantage of a much longer shelf-life, without need for refrigeration. Compared with conventional can sterilization, UHT processing retains sensory characteristics and nutritional value much better, saves more energy, is easier to automate, permits poststerilization addition of heat-sensitive components, and does not hold a maximum package size; in fact, the time for aseptic processing is essentially independent of package size, unlike classical can sterilization. However, survival of heat-resistant enzymes is rare in conventional sterilization, further to low-acid particulate liquids being effectively handled, and operation proving simple – unlike happens with aseptic processing.

The criteria for UHT processing are essentially the same as for in-container sterilization, in terms of commercial sterility; once again, the come-up time and cooling periods are normally so short that they often constitute a safety margin – with an estimation of lethality based on the holding period alone. Both the time required for heat to reach the thermal center and the time taken by fluid therein to move through the holding section of the equipment are to be taken into account; turbulence is advantageous – not only because it facilitates heat transfer, but also because it reduces the spread of residence times around their average (or space time). Furthermore, the size of the solid particles suspended in the liquid food undergoing treatment is critical; should the heating process be designed for a given lethality of particles with a reference size, larger particles will be underprocessed while smaller particles will be overprocessed.

In a typical UHT process, preheated food is pumped via a positive displacement (metering) pump through a vacuum chamber to remove dissolved air, and then passed to a heat exchanger; the former step assures a constant volume, and thus a preset space time in the holding section (otherwise air might expand upon heating) – while keeping oxidative changes to a minimum. Heating takes place via thin layers, with pressure created by the said pump and maintained by a back-pressure device, e.g. piston, diaphragm valve, or pressurized tank; the said device is placed after the cooler to avoid boiling. The target temperature lies typically above 132°C; therefore, large surface areas are to be used compared to food volume, turbulence is to be promoted to decrease boundary layer thickness, and heating surfaces (when present) are to be constantly cleaned – all intended to maximize rate of heat transfer. Heating itself is either accomplished via direct steam injection or steam infusion (with throughput rates up to 9 ton.hr^{-1}) – or else via plate, concentric tube, shell-and-tube, or scraped-surface heat exchangers; dielectric, ohmic, or induction heating have also been utilized, but (still) to a much lesser extent than steam supply.

Steam injection (also known as upérisation) and steam infusion are each used to effect intimate contact of food product with potable (or culinary) steam. The former is introduced at ca. 9.65×10^5 Pa in the preheated food via an injector, thus forming fine bubbles; the temperature of the liquid food rises rapidly to 140–150°C. The preheated liquid is then passed to the holding tube – inclined upward at a shallow angle, so as to remain full at all times and thus avoid formation of air pockets. Typical time/temperature conditions necessary to destroy *C. botulinum* are 1.8 s at 141°C; in the case of dairy products, the minimum treatments are 1 s at 135°C for milk, 2 s at 140°C for cream and milk-based products, and 2 s at 148.9°C for icecream mixes. Flash-cooling of the sterilized product to (a little) above room temperature takes place fast in a vacuum chamber – thus allowing removal of volatiles, as well as an equivalent amount of otherwise condensed steam. This process is feasible only with low-viscosity liquids – besides requiring expensive potable steam, and allowing no more than ca. 50% recovery of energy. In steam infusion, the food is sprayed as a free-falling film into potable steam at 4.5×10^5 Pa, thus allowing temperature to quickly reach 140–145°C; heat released during flash-cooling to 65–70°C is reutilized to preheat the feed material. Such a mode of operation does not allow the liquid to ever contact hot surfaces – so burning-on is reduced, and even high-viscosity liquids may be manipulated; there is also a lower risk of localized overheating; and heating and cooling are faster than via steam injection, thus permitting higher retention of sensory and nutritional features. However, the spray nozzles are susceptible to blockage, and some fluids undergo phase separation. In alternative, a second heat exchanger may be used for cooling – without making up for steam-mediated dilution or getting rid of cooked odors, but allowing up to 90% energy regeneration.

Plate heat exchangers and tube-and-shell heat exchangers possess a number of limitations with regard to sterilization. The latter is less liable to fouling and more easily cleaned than the former, it can operate at higher pressures and entails better developed turbulent conditions, and it does not suffer from the shortcoming of nonuniform expansion of plate stack (which may distort and damage plates or seals). Plate heat exchangers are instead flexible to changes in throughput rate, more energy-efficient, and more economical in floor space occupied and water consumed. Improved designs include double- or triple-tube heat exchangers – with a corrugated tube concentrically (and axially) positioned in a larger diameter outer tube; as well as scraped-surface heat exchangers. The former type is quite suitable for liquid products with high pulp content or containing particulates, or else viscous fluids and high pressures – since the clearance between tubes may be manufactured at will; if three tubes are utilized, then heat transfer is improved, because it takes place through both annular surfaces. Scraped-surface heat exchangers are also appropriate for viscous and particulate (with piece diameter not exceeding 1 cm) fluid foods, including fruit sauces and fruit bases for yogurt and pies; however, they are expensive in terms of investment and operation, and heat recovery is not possible.

The constraints associated with straight tube heat exchangers may be essentially overcome by molding the said tube into a helix or coil, with a carefully defined coil-to-tube diameter ratio; this configuration promotes secondary flow of liquid in the tube that, in turn, promotes local turbulence at relatively low flow rates – thus allowing two- to fourfold improvement in heat transfer rate, relative to conventional straight designs. Hence, heat-sensitive and fouling-prone food products can be efficiently manipulated, as well as cheese sauces, fruit purées, and salad dressings – due to the gentle mixing action that ensures uniform distribution of particles; cleaning-in-place is also possible, without loss of sterility – so processing may resume immediately afterward.

In the case of fluids containing large particles, separate thermal treatment of liquid and solid particulates is a solution – as is the case of Jupiter™ double-cone heat exchanger, which combines indirect heating by a rotating jacketed double cone with direct heating by steam or superheated liquor; solid pieces are tumbled through steam, while liquor is sterilized separately and added to the solids to minimize mechanical damage during the aforementioned tumbling. Particulate food is heated by direct steam injection in a pressurized horizontal cylindrical vessel, undergoing slow rotation, as part of the Twintherm™ system; once sterilized, the particles are cooled with liquid presterilized via conventional UHT heat exchangers, and then subjected to aseptic filling. The FSTP™ system resorts to a cylindrical vessel containing slowly rotating blades mounted on a shaft – which forms cages to hold particles, as they are rotated around the cylinder from inlet to outlet; liquid, after presterilization in a conventional heat exchanger, moves freely through the cages where it assures rapid heat transfer. Nonconventional systems include Muli-therm™ and Achiles™– resorting to a combination of hot liquid and dielectric heating, or direct ohmic heating of the liquid food.

Following temporary storage in a sterile surge tank under pressurized nitrogen, sterilized liquid food is to be aseptically packaged (without need for specific pumping through the filler) in a chamber under UV light and filter-sterilized, pressurized air. Since containers do not need to stand the high temperature (and pressure) prevailing during food sterilization – as sterilization via UV or ionizing radiation, or even via hydrogen peroxide suffices, a very many packaging materials are suitable, namely pouches, cups, sachets, and bulk packs, further to laminated microwaveable cartons. Besides being more economical than cans or jars/bottles, such packages are also less expensive to transport and store.

When steam is used as heating source, latent heat is transferred to the canned food – once saturated steam condenses on the outside of the container; if air is trapped inside the retort, it forms an insulating film around the cans that prevents steam condensation – and may cause underprocessing of food, besides generating a lower temperature than that obtainable with saturated steam. Therefore, as much air as possible should be removed from the retort; when operating batchwise, steam itself may be used to do so, in a process known as venting. Following sterilization, the containers are cooled via sprays of cold water; steam readily condensates, yet food cools down at a slower pace. The pressure imbalance between the inside and the outside of the container is overcome by pumping compressed air into the retort, thus keeping strain on the container seams as low as possible; this process is discontinued when the food attains ca. 100°C, since all water vapor inside the can should have already condensed by then. The crates of containers are taken off when temperature reaches ca. 40°C; the remaining moisture on the surface of the containers evaporates promptly at this temperature – thus avoiding surface corrosion thereafter, while permitting label adhesives to set properly.

Sterilization of glass containers and flexible pouches usually resorts to hot water as heat source. The former must be thicker than metal cans, so as to convey adequate mechanical strength – which worsens the thermal resistance already offered by the container glass material relative to the can metal; hence, a higher risk of thermal shock arises, associated to a slower heat penetration – which precludes utilization of steam as heating medium. In order to attain the reference sterilization temperature, i.e. 121°C, the boiling point of water must be accordingly elevated via application of overpressure with air – an extra 10^5 Pa is indeed necessary, to be more precise.

Rigid polymer trays of flexible pouches are thinner than glass containers, besides exhibiting smaller overall cross-section; the associated higher specific surface area available for heat transfer, coupled with the shorter path to be tracked thereby permit energy thereby permit energy savings – besides causing marginal overheating on the outside of the container. However, their constitutive polymeric materials may undergo physicochemical changes throughout thermal processing; for instance, the container volume may change due to stretching or shrinking – and the heat seals may soften, or the headspace gas pressure may increase sufficiently to weaken the (heat) seals, and thus increase the probability of failure. For this set of reasons, overpressure is to be applied before the headspace gas inflates the container. Liquid or semiliquid foods may be processed horizontally to ensure a constant thickness of food across the pouch; vertical packs promote better circulation of hot water in the retort, yet their tendency for bulging at the bottom has to be counteracted by resorting to special frames – with consequent change in heating pattern, and thus in degree of lethality achieved.

Another possibility for sterilization is to use direct flame heating – as happens with cans containing mushrooms, sweetcorn, green beans, or beef cubes, in the absence of brine or syrup, and aimed at high-quality final products. The flame temperature may go above 1700°C, thus permitting very fast heat transfer via both conduction and radiation; however, the said cans have to undergo energetic spinning to promote uniform heating, and thus avoid hot spots. Since the internal pressure may reach 2.8×10^5 Pa at 130°C, only small cans, able to withstand the said stress, can be utilized.

Although extensive destruction of vegetative cells starts once temperature has reached ca. 70°C (as typical of pasteurization), spores are much more resistant – so correct monitoring of temperature inside a retort is a must; this usually resorts to copper/constantan thermocouples, platinum resistance thermometers, or nonmetallic thermistors as thermal probes. More sophisticated (and accurate) approaches include such noninvasive techniques as magnetic resonance imaging, microwave radiometry, or optic fiber thermometry. Validation of sterilization models in canning factories involves inoculating an enzyme or a nonpathogenic test microorganism into the food, bearing D- and z-values similar to those of the reference microorganism; or encapsulating it in gel beads, with thermal and physical properties similar to those of the food under scrutiny. Upon processing, the test containers are recovered – and residual enzyme activity is assayed for, or viable cells are enumerated.

Since sterility conditions have to be guaranteed at the thermal center (or critical point) of any given food, overprocessing of its outer portions will usually occur when heat transfer takes place via conventional conduction from an externally imposed higher temperature; this unfavorably affects nutritional and sensory features, especially of the fluid near the vessel walls and particles suspended therein. Therefore, reduction in thickness of the layer directly exposed to heat, or provision of some form of stirring are useful measures to alleviate the said shortcoming. The former requires smaller and tailor-shaped cans – whereas the latter calls for end-over-end tumbling of cans containing particulate liquid foods; improvement of up to 30% in can-to-fluid heat transfer coefficient, or up to 50% in fluid-to-particle heat transfer coefficient can be attained by just doubling the rotational frequency. Heuristic rules indicate that the thermal center of a food in a cylindrical can lies at ca. 1/5 of its height above the base, if heating is delivered by a hot fluid – or coincides with its geometric center, when heating occurs by conduction through immobile (solid) materials; this is valid for height-to-diameter ratios ranging from 0.3 to 0.95. Note that a change in heat transfer patterns may take place during the process – as happens when starch suspended in a given liquid food gelatinizes, thus behaving like a solid thereafter.

A higher temperature of the heating fluid would permit a much faster death rate of unwanted microorganisms than decay in nutrients and sensory characteristics, and thus a reduction in processing time – as per the general strategy outlined in Fig. 1.35. However, this approach cannot be carried out *ad libitum*, because the heating steam would be at too high a pressure – which would, in turn, be impractical for demanding substantially stronger, and thus more expensive equipment, besides conformity to stricter safety precautions.

Besides information on resistance of target microorganisms/enzymes in food, the pattern of heat penetration should be accessible in attempts to correctly estimate sterilization time; product-, process-, and package-related factors are indeed to be taken into account. Consistency is a key concept concerning the former; natural convection develops inside plain liquid or particulate liquid foods, which aid in heat transfer relative to plain solid (or very viscous liquid) foods. This phenomenon is strongest in broths, juices, and milk, with soups and tomato juice coming second; as well as thick purées, rice and spaghetti, meat pastes and corned beef, and high sugar products and low-moisture puddings – sorted by decreasing order of importance of natural convection. In any case, foods typically possess poor thermal conductivities – which, by themselves, pose a severe constraint upon heat transfer during sterilization. Process-related factors encompass: retort temperature – knowing that a higher temperature difference as driving force leads to faster heat transfer *tout court*; type of heat transfer medium – with steam mixed with air being the most common, and thickness of boundary layer being dependent on the linear velocity of the carrying stream; and agitation of containers – where internal spontaneous currents, due to natural convection, may be complemented by forced convection arising from end-over-end or (to a lesser extent) axial agitation of cans. Finally, package-related factors include: container size – with heat reaching the center faster in smaller containers; container shape – with convection promoted in taller containers, and less resistance offered in thin containers, leading to faster heat transfer in both cases; container material – with faster heat transfer in metallic than glass or plastic containers; and headspace gas – which creates a resistance to heat transfer above that created by filling liquid, unless container agitation is provided that serves as a mixing aid.

Some types of sterilization problems – namely *C. botulinum*-mediated spoilage and interaction of food acids with can metal, may cause can swelling due to accumulation of gas; carbon dioxide is released by the former due to catabolism of glucose (or other fermentable carbohydrates) via the Embden, Meyerhof, and Parnas' pathway, see Fig. 5.58i in *Food Proc. Eng.: thermal & chemical operations*, together with ammonia, due to Strickland's degradation of proteinaceous amino acids – whereas the latter involves release of molecular hydrogen. The aforementioned phenomenon occurs gradually as storage time elapses; increasingly severe can swelling has been termed flipper, springer, soft swell, and hard swell in canning jargon. Routine quality assurance measures accordingly include detection of swollen or bloated cans, which are to be immediately rejected; the same decision should be taken at domestic consumption, should any of those abnormalities be found. Note that overfilling, denting, poor can closure after cooling, and storage at high temperature or low pressure (e.g. at high altitudes) may also account for can bloating.

The combinations of time and temperature always play a role upon native components in food; for instance, sterilization contributes to convert red oxymyoglobin and purplish myoglobin to brown metmyoglobin and red-brown myohemichromogen, respectively, in meat. When packaged in cans, meat may also undergo pyrolysis, as well as deamination and decarboxylation of amino acids – further to Maillard's browning, caramelization of carbohydrates to furfural and hydroxymethylfurfural, and oxidation and decarboxylation of lipids. When sodium nitrate and nitrite are added to meats in attempts to reduce the risk of growth of *C. botulinum*, the resulting red-pink coloration is accounted for by nitric oxide myoglobin and metmyoglobin nitrite. Coagulation and loss of water-holding capacity cause textural decay – namely shrinkage and stiffening of muscle tissues. Hydrolysis of collagen, complemented by solubilization of the resulting gelatin, and melting and dispersion of fats cause softening; addition of (water-binding) polyphosphates can reverse this trend. Hydrolysis of lipids and carbohydrates also takes place, yet the overall nutritional value remains essentially unchanged; conversely, chemical losses of amino acids may add to 20%, as well as water-soluble and oxygen-labile vitamins B and C. Note that canned fish and meat are normally richer in sodium, due to addition of salt for seasoning purposes.

In milk and milk products, color changes upon sterilization are normally due to caramelization, Maillard's reactions, and changes in reflectivity of its constitutive casein micelles; and development of cooked flavors results from production of hydrogen sulfide from denatured whey proteins; whereas lactones and methyl ketones are generated from lipids. The said changes are less severe when UHT processing takes place *in lieu* of in-container sterilization.

In vegetables and fruits, chlorophyll is converted to pheophytin, carotenoids are isomerized from 5,6-epoxides to poorly colored 5,8-epoxides, and anthocyanins decay to brown pigments; if in canned form, anthocyanins may react with the iron or tin of the container and seal to form a purple pigment – whereas colorless leucoanthocyanins in pear and quinces form pink anthocyanin complexes. Degradation, recombination, and volatilization of aldehydes, ketones, sugars, lactones, and amino and organic acids also take place, with concomitant development of off-flavors. Softening arises from hydrolysis of pectic compounds, gelatinization of starches, and partial solubilization of hemicelluloses, together with loss of cell turgor. This justifies addition of calcium salts to blanching water, or to brine or syrup to form insoluble calcium pectate at large – both meant to increase firmness; or else calcium hydroxide in the case of cherries, calcium chloride in the case of tomatoes, or calcium lactate in the case of apples, to match their different proportions of demethylated pectin. In all these cases, the canned product becomes obviously richer in calcium, which entails an advantage. Nutrient losses are strongly dependent on a number of factors: cultivar and maturity at harvest; type of processing water used (especially in terms of calcium content); and presence of residual oxygen in the container. Since the accompanying brine or syrup is normally consumed together with the food, leaching of water-soluble vitamins is inconsequent for nutritional purposes. Isomerization of naturally occurring *trans*-β-carotene may occur to its (less biologically active) *cis*-counterpart; this is beneficial in the case of conversion of *trans*- to *cis*-lycopene in tomatoes, since the latter exhibits a greater biological activity – and a similar advantage results from the increased level of α-tocopherol. Soy-meat products also gain nutritional value upon sterilization, owing to a less stable trypsin inhibitor.

Meat pigments tend to change color when subjected to UHT treatment – yet little caramelization or Maillard's reactions are observed; carotenes and betanin remain almost unaffected, and chlorophyll and anthocyanins are even better retained in plant-based foods. However, a few heat-resistant enzymes secreted in milk by psychrotrophic microorganisms do not lose all their activity, and may accordingly contribute to decay upon prolonged storage. For instance, modifications in κ-casein facilitate precipitation/coagulation thereof in the presence of Ca^{2+}. This is followed by breakdown of α- and β-caseins by adventitious proteases; as well as polymerization thereof via Maillard's reactions, and formation of κ-casein-β-lactoglobulin complexes – thus causing changes in viscosity of sterilized milk and milk products, further to age-gelation throughout storage. On the other hand, toughening of meat may occur in UHT preserves, because collagen does not undergo hydrolysis to a significant extent. Texture of solid fruit and vegetable pieces is normally softer than prior to UHT treatment, owing to solubilization of pectic materials and loss of cell turgor – but certainly less than their canned counterparts. Thiamin and pyridoxine in meat and vegetable products are gradually lost as time elapses, but the remaining vitamins remain essentially unaffected; negligible vitamin losses are observed in UHT processed milk, except ascorbic acid – although to a lesser degree than in bottled milk; and denaturation of whey proteins is also less extensive.

1.2.3 Mathematical simulation

1.2.3.1 Thermal death time

Remember that cell death kinetics is normally described by Eq. (1.390) – so the decimal reduction time abides to an analog to Eq. (1.375) that unfolds a logarithmic decrease in cell viability with time; this mathematical pattern implies that a completely sterile product cannot be ever obtained, irrespective of how long thermal processing is carried out (as stressed previously). Nevertheless, the probability of survival of a single microorganism can be predicted based on its inherent heat resistance, on the one hand, and temperature and duration of thermal processing, on the other – thus leading to the concept of commercial sterility.

In practice, if a decrease in microbial viability or enzyme activity by n (decimal) logarithmic cycles suffices for commercial sterility, then one should apply an nD-process, where D denotes decimal reduction time of the reference microorganism/enzyme (as usual). Take *C. botulinum* as reference microorganism, with $n=12$ as outlined before; this means that for 10^3 spores initially present in a food container, commercial sterility will be characterized by 10^{-12}-fold that level, i.e. 1 surviving spore in every 10^9 containers processed. In view of the direct relationship between cell behavior and controlling-enzyme behavior, one may rewrite Eq. (1.384) as

$$\frac{t}{D} = \log_{10} C_{X,0} - \log_{10} C_X, \tag{1.459}$$

or else

$$\frac{t}{D} = \log_{10} \frac{C_{X,0}}{C_X} \tag{1.460}$$

valid for cells (rather than enzymes); if $C_{X_z,cst}$ denotes required final viability level for commercial sterility and $C_{X_z,0}$ denotes initial load of viable (reference) microorganism, then Eq. (1.460) may be rewritten as

$$\boxed{F_y^z \equiv D_y \log_{10} \frac{C_{X_z,0}}{C_{X_z,cst}},} \tag{1.461}$$

with F_y^z denoting thermal death time at temperature y of microorganism characterized by decimal reduction time D_y at that temperature and thermal resistance factor z. Equation (1.461) – obtainable directly from Eq. (1.460) upon setting $t = F_y^z$, $D = D_y$, $C_{X,0} = C_{X_z,0}$, and $C_X = C_{X_z,cst}$, and solving for F_y^z afterward, may be used to estimate the number of containers that can be processed before one is found contaminated; the process time needed to attain a preselected viable number of cells of target contaminant; or the level of contamination to be expected from a process with a known thermal death time.

The reference temperature for sterilization has classically been 121°C – in which case Eq. (1.461) becomes

$$F_{121}^z = D_{121} \log_{10} \frac{C_{X_z,0}}{C_{X_z,cst}}, \tag{1.462}$$

for a reference microorganism characterized by D_{121} as decimal reduction time and z as thermal resistance factor; if, in particular, $z=10$°C and commercial sterility is to be assured by a $12D$-process, then Eq. (1.462) supports

$$F_{121}^{10} = D_{121} \log_{10} 10^{12}, \tag{1.463}$$

or merely

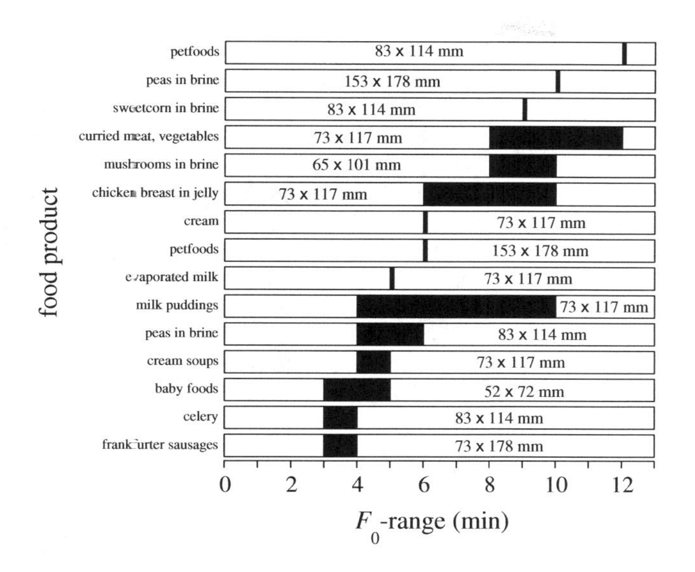

FIGURE 1.42 Safety ranges (■) of selected foods, packaged in selected diameter × height cans, expressed as reference thermal death times, F_0.

$$\boxed{F_0 \equiv F_{121}^{10} = 12D_{121}} \tag{1.464}$$

– where F_0 denotes reference thermal death time. Typical values for F_0, pertaining to selected foods and regular can sizes are conveyed by Fig. 1.42. A wide variety of (reference) thermal death value ranges is found – depending not only on food composition, but also on can characteristics; the same food is normally characterized by a higher F_0 if packaged in a larger can (e.g. peas in brine), since heat would take longer to reach its geometric center by conduction.

For the same level of commercial sterility – and departing from tabulated values of F_{121}^z for an entity characterized by a generic z factor, one would readily estimate thermal death time at any other temperature y; one may accordingly proceed to ordered division of Eq. (1.461) by Eq. (1.462), viz.

$$\frac{F_y^z}{F_{121}^z} = \frac{D_y \log_{10} \dfrac{C_{X_z,0}}{C_{X_z,cst}}}{D_{121} \log_{10} \dfrac{C_{X_z,0}}{C_{X_z,cst}}}, \tag{1.465}$$

which simplifies to

$$\frac{F_y^z}{F_{121}^z} = \frac{D_y}{D_{121}} \tag{1.466}$$

upon cancellation of common factors between numerator and denominator – or, equivalently,

$$F_y^z = \frac{D_y}{D_{121}} F_{121}^z. \tag{1.467}$$

As expected, the ratio of thermal death times, at two given temperatures, coincides with the ratio of decimal reduction times at the same temperatures, as emphasized by Eq. (1.466). After having reformulated Eq. (1.422) to

$$\frac{121 - T_y}{z} = \log_{10} \frac{k_d\big|_{121^\circ\mathrm{C}}}{k_d\big|_{T_y}}, \tag{1.468}$$

one may retrieve Eq. (1.378) to write

$$\frac{121 - T_y}{z} = \log_{10} \frac{\dfrac{\ln 10}{D_{121}}}{\dfrac{\ln 10}{D_y}} = \log_{10} \frac{D_y}{D_{121}}, \tag{1.469}$$

along with straightforward algebraic manipulation; upon isolation of D_y/D_{121}, Eq. (1.469) unfolds

$$\frac{D_y}{D_{121}} = 10^{\frac{121-T_y}{z}}. \tag{1.470}$$

Combination of Eqs. (1.467) and (1.470) gives finally rise to

$$F_y^z = 10^{\frac{121-T_y}{z}} F_{121}^z, \tag{1.471}$$

which may instead be coined as

$$\boxed{\frac{F_y^z}{F_{121}^z} = 10^{-\frac{T_y-121}{z}}} \tag{1.472}$$

– with graphical interpretation provided in Fig. 1.43. As expected, a temperature below 121°C demands a death time longer than F_{121}^z – and the opposite applies to $T_y > 121$°C; this justifies the monotonically decreasing trend in Fig. 1.43, linear in the log/lin scale elected on account of the exponential form apparent in Eq. (1.472).

When Eq. (1.471) is rewritten as

$$\frac{1}{F_y^z} = 10^{\frac{T_y-135}{z}} \frac{1}{F_{135}^z} \tag{1.473}$$

– after taking reciprocals of both sides and choosing 135°C as alternative reference sterilization temperature, one can further simplify it to

$$F_{135}^z = 10^{\frac{T_y-135}{z}} F_y^z; \tag{1.474}$$

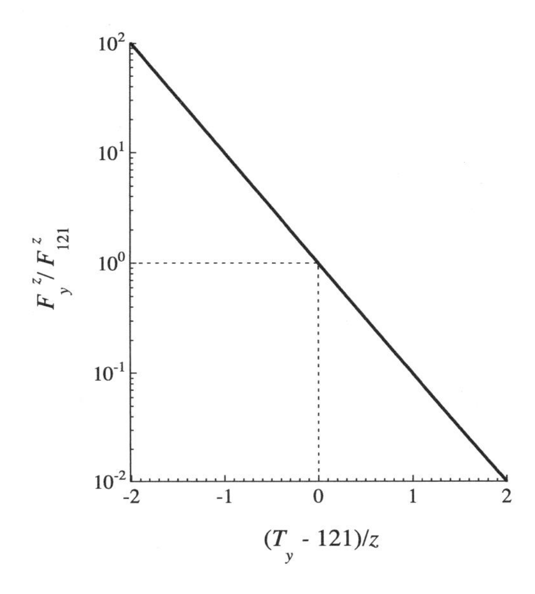

FIGURE 1.43 Variation of dimensionless thermal death time, F_y^z/F_{121}^z, versus dimensionless temperature, $(T_y - 121)/z$.

this relationship is more convenient for UHT processes. As highlighted before, thermal death occurs almost exclusively during the holding time, t_{hold}, in such case – so one may recoin Eq. (1.474) as

$$F_{135}^{z} = 10^{\frac{T_y - 135}{z}} t_{hold}. \tag{1.475}$$

Two dimensionless parameters have classically been employed in dairy UHT processing – B^* that measures microbial lethal effect for a z factor of 10.5°C, and C^* that measures chemical damage effect for a z factor of 31.4°C; when the former is unity, a 9D-reduction in spore count takes 10.1 s at 135°C, with an equivalent 30.5 s for the nutritional/sensory indicator. Under these circumstances, Eq. (1.475) has it that

$$\boxed{B^* = 10^{\frac{T_y - 135}{10.5}} \frac{t_{hold}}{10.1}} \tag{1.476}$$

with B^* formally defined as

$$B^* \equiv \frac{F_{135}^{10.5}}{10.1}; \tag{1.477}$$

as well as

$$\boxed{C^* = 10^{\frac{T_y - 135}{31.4}} \frac{t_{hold}}{30.5},} \tag{1.478}$$

with C^* abiding to

$$C^* \equiv \frac{F_{135}^{31.4}}{30.5} \tag{1.479}$$

as formal definition.

Equation (1.461) permits calculation of thermal death time for commercial sterility when temperature is held essentially constant – as is the case of the holding section in UHT treatment. However, in classical in-container sterilization using a retort, temperature varies considerably along the heating and cooling processes – which may even become the major contributors to the sterilization effect proper; hence, actual lethality should take those two periods into account. In other words, integration of the biomass analog to Eq. (1.366), with the aid of the corresponding analog to Eq. (1.367) should take place as

$$\int_{C_{X,0}}^{C_X} \frac{d\tilde{C}_X}{\tilde{C}_X} = -\int_0^t k_d\left\{T\{\tilde{t}\}\right\} d\tilde{t} \tag{1.480}$$

– where $k_d \equiv k_d\{T\{t\}\}$ cannot be taken off the kernel here for not being constant throughout the integration interval, unlike happened in Eq. (1.368); application of the fundamental theorem of integral calculus to the left-hand side produces

$$-\ln \tilde{C}_X \Big|_{C_{X,0}}^{C_X} = \int_0^t \frac{\ln 10}{D\left\{T\{\tilde{t}\}\right\}} d\tilde{t}, \tag{1.481}$$

with Eq. (1.378) meanwhile inserted in the outstanding integral – which further simplifies to

$$\int_0^t \frac{1}{D\left\{T\{\tilde{t}\}\right\}} d\tilde{t} = \frac{\ln \frac{C_{X,0}}{C_X}}{\ln 10}. \tag{1.482}$$

In view of the operational rules of logarithms, Eq. (1.482) may be redone to

$$\int_0^t \frac{d\tilde{t}}{D\left\{T\{\tilde{t}\}\right\}} = \log_{10} \frac{C_{X,0}}{C_X} \tag{1.483}$$

– which would obviously retrieve an alias to Eq. (1.384) if D remained constant, i.e.

$$\frac{\int_0^t d\tilde{t}}{D} = \frac{t}{D} = \log_{10} \frac{C_{X,0}}{C_X}; \tag{1.484}$$

for commercial sterility characterized by $C_{X_{z,cst}}$ and using a generic z factor, Eq. (1.483) will become

$$\boxed{\int_0^{F^z} \frac{d\tilde{t}}{D\left\{T\{\tilde{t}\}\right\}} = \log_{10} \frac{C_{X_z,0}}{C_{X_z,cst}},} \tag{1.485}$$

where the variable temperature in the left-hand side, i.e. $T \equiv T\{\tilde{t}\}$, leads to $D \equiv D\{\tilde{t}\}$, up to a cumulative thermal death time, F^z, obtained upon integration – to be used *in lieu* of Eq. (1.461). The final relationship attained – involving F^z associated with a given $C_{X_z,0}/C_{X_z,cst}$, depends in this case on the temperature vs. time path followed.

PROBLEM 1.13

Due to the nonuniform velocity profiles within liquids when subjected to processing, attempts have been made to simulate their hydrodynamic pattern departing from a combination of ideal units; one of the most successful models is the cascade of identical stirred tanks – due to its mathematical simplicity that resorts to just two parameters (space time, τ, and number of units, N), and its remarkable ability to simulate convective flow associated with dispersion. This model is useful for estimation of the extent of thermal death in thermal processing of liquid foods – especially when first-order death kinetics prevails, as well as essentially isothermal conditions.

a) Determine the fraction of remaining viable microorganisms in one such apparatus, assuming first-order death kinetics characterized by kinetic constant k_d.
b) Find the asymptotic behavior of the expression produced in a), when N grows unbounded.
c) Based on the results obtained in a) and b), comment on the efficiency of thermal treatment of a liquid food subjected to hydrodynamic dispersion.
d) Obtain the residence time distribution of the said cascade of stirred tanks.

e) What will the estimate be for the fraction of viable microorganisms if the total segregation model is combined with the result conveyed by d)?

Solution

a) If the volumetric flow rate of either inlet or outlet stream to or from the i-th stirred tank of volume V_i is denoted by Q, then a steady-state mass balance to biomass reads

$$Q\,C_{X,i-1} - V_i k_d C_{X,i} = Q\,C_{X,i}, \tag{1.13.1}$$

where $C_{X,i-1}$ and $C_{X,i}$ denote (viable) biomass concentration at inlet and outlet streams, respectively, and $k_d C_{X,i}$ denotes rate of biomass death; after dividing both sides by Q, Eq. (1.13.1) reduces to

$$C_{X,i-1} - k_d \tau_i C_{X,i} = C_{X,i}, \tag{1.13.2}$$

as long as space time of the i-th unit, τ_i, satisfies

$$\tau_i \equiv \frac{V_i}{Q}. \tag{1.13.3}$$

Solution of Eq. (1.13.2) for $C_{X,i}$ unfolds

$$C_{X,i} = \frac{C_{X,i-1}}{1+k_d\tau_i};\quad i=1,2,\ldots,N, \tag{1.13.4}$$

or else

$$\frac{C_{X,i}}{C_{X,i-1}} = \frac{1}{1+k_d\tau_i} \tag{1.13.5}$$

after dividing both sides by $C_{X,i-1}$; since the ratio of $C_{X,N}$ to $C_{X,0}$ may be written as a telescopic (sequential) product, viz.

$$\frac{C_{X,N}}{C_{X,0}} = \frac{C_{X,N}}{C_{X,N-1}}\frac{C_{X,N-1}}{C_{X,N-2}}\cdots\frac{C_{X,2}}{C_{X,1}}\frac{C_{X,1}}{C_{X,0}}, \tag{1.13.6}$$

one may insert Eq. (1.13.5) for each value of i between 1 and N to get

$$\frac{C_{X,N}}{C_{X,0}} = \frac{1}{1+k_d\tau_N}\frac{1}{1+k_d\tau_{N-1}}\cdots\frac{1}{1+k_d\tau_2}\frac{1}{1+k_d\tau_1} \tag{1.13.7}$$

– where a condensed notation is possible as

$$\frac{C_{X,N}}{C_{X,0}} = \prod_{i=1}^{N}\frac{1}{1+k_d\tau_i}. \tag{1.13.8}$$

The N units in the cascade are identical by hypothesis, thus enforcing

$$\tau_i = \frac{\tau}{N}, \tag{1.13.9}$$

where τ denotes space time of the overall system; insertion of Eq. (1.13.9) transforms Eq. (1.13.8) to

$$\frac{C_{X,N}}{C_{X,0}} = \prod_{i=1}^{N}\frac{1}{1+k_d\dfrac{\tau}{N}}, \tag{1.13.10}$$

which is equivalent to

$$\frac{C_{X,N}}{C_{X,0}} = \frac{1}{\left(1+\dfrac{k_d\tau}{N}\right)^N} \tag{1.13.11}$$

in view of the (mathematical) definition of power – thus leaving an expression for the degree of lethality brought about by the apparatus.

b) When N is very large, Eq. (1.13.11) will abide to

$$\lim_{N\to\infty}\frac{C_{X,N}}{C_{X,0}} = \lim_{N\to\infty}\frac{1}{\left(\left(1+\dfrac{k_d\tau}{N}\right)^{\frac{N}{k_d\tau}}\right)^{k_d\tau}} = \frac{1}{\left(\lim\limits_{\frac{N}{k_d\tau}\to\infty}\left(1+\dfrac{1}{\dfrac{N}{k_d\tau}}\right)^{\frac{N}{k_d\tau}}\right)^{k_d\tau}} \tag{1.13.12}$$

– where advantage was gained from the redundant composition of the $k_d\tau$-th power with the $k_d\tau$-th root, and the classical theorems on limits including realization that $N/k_d\tau \to \infty$ when $N\to\infty$; after recalling the definition of Neper's number, Eq. (1.13.12) becomes

$$\lim_{N\to\infty}\frac{C_{X,N}}{C_{X,0}} = \frac{1}{(\mathrm{e})^{k_d\tau}} = \mathrm{e}^{-k_d\tau} \tag{1.13.13}$$

that drives the lethality function when N grows unbounded.

c) Equation (1.13.13) guarantees full lethality, as long as $k_d\tau$ is sufficiently large – because the exponential function tends eventually to zero; a measure of the degree of approach to the maximum level of lethality degree, η_{let}, at any given k_d and τ can be ascertained using the said long-N asymptotic value as normalizing factor, according to

$$\eta_{let} \equiv \frac{\dfrac{C_{X,N}}{C_{X,0}}}{\lim\limits_{N\to\infty}\dfrac{C_{X,N}}{C_{X,0}}}. \tag{1.13.14}$$

Insertion of Eqs. (1.13.11) and (1.13.13) consequently transforms Eq. (1.13.14) to

$$\eta_{let} = \frac{\dfrac{1}{\left(1+k_d\dfrac{\tau}{N}\right)^N}}{e^{-k_d\tau}} = \frac{e^{k_d\tau}}{\left(1+\dfrac{k_d\tau}{N}\right)^N}; \tag{1.13.15}$$

a plot of η_{let} vs. $k_d\tau$ is provided below, for various values of N.

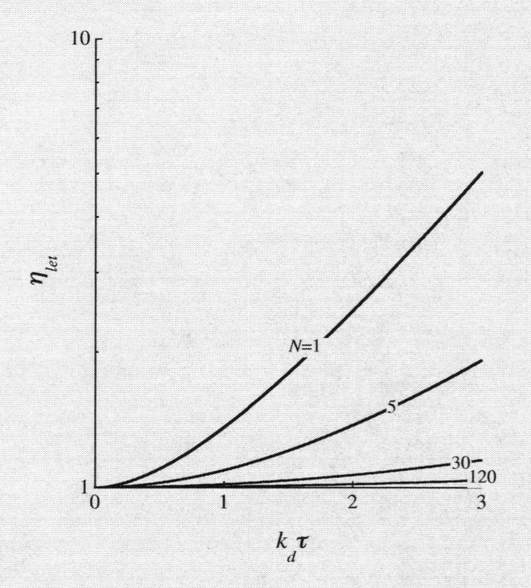

Therefore, one concludes that a higher dispersion, measured by a smaller N, worsens the degree of lethality relative to what would be expected in absence thereof – even though the same space time is kept; for $k_d\tau=3$, for instance, the final fraction of viable microorganisms in a plain stirred tank is fivefold the minimum value attained when all fluid elements spend the same space time τ, and reduces to ca. twofold when $N=5$. All in all, the underprocessing of fluid elements remaining in the vessel for a shorter residence time cannot be balanced by overprocessing of the corresponding fluid elements remaining in the vessel for a longer residence time due to the underlying nonzeroth-order process – even if the residence times of those elements are balanced (otherwise the given average residence time, or space time τ, would not be met).

d) It is possible to derive the residence time distribution of the apparatus under scrutiny directly from its theoretical model – since one such model is available *a priori*. If tracer were injected at the entrance of the i-th unit, then a balance would accordingly read

$$QC_{T,i-1} = QC_{T,i} + V_i\frac{dC_{T,i}}{dt}, \tag{1.13.16}$$

analogous to Eq. (1.13.1), but with accumulation term *in lieu* of reaction term (as per the definition of tracer, in the first place); here $C_{T,i-1}$ and $C_{T,i}$ denote tracer concentration at inlet and outlet streams, respectively. After dividing both sides by Q and taking Eq. (1.13.3) on board, Eq. (1.13.16) gives

$$C_{T,i-1} = C_{T,i} + \tau_i\frac{dC_{T,i}}{dt}, \tag{1.13.17}$$

while

$$C_{T,i}\Big|_{t=0} = 0 \tag{1.13.18}$$

may serve as initial condition. In Laplace's domain, Eq. (1.13.17) looks like

$$\overline{C}_{T,i-1} = \overline{C}_{T,i} + \tau_i\left(s\overline{C}_{T,i} - C_{T,i}\Big|_{t=0}\right), \tag{1.13.19}$$

yet combination with Eq. (1.13.18) permits simplification to

$$\overline{C}_{T,i-1} = \overline{C}_{T,i} + \tau_i s\overline{C}_{T,i}; \tag{1.13.20}$$

isolation of $\overline{C}_{T,i}$ yields

$$\overline{C}_{T,i} = \frac{\overline{C}_{T,i-1}}{1+\tau_i s}, \tag{1.13.21}$$

or else

$$\frac{\overline{C}_{T,i}}{\overline{C}_{T,i-1}} = \frac{1}{1+\tau_i s} \tag{1.13.22}$$

after dividing both sides by $\overline{C}_{T,i-1}$. The left-hand side of Eq. (1.13.22) has the form of a transfer function – meaning that a telescopic product mimicking Eq. (1.13.6) may be set as

$$\frac{\overline{C}_{T,N}}{\overline{C}_{T,0}} = \frac{\overline{C}_{T,N}}{\overline{C}_{T,N-1}}\frac{\overline{C}_{T,N-1}}{\overline{C}_{T,N-2}}\cdots\frac{\overline{C}_{T,2}}{\overline{C}_{T,1}}\frac{\overline{C}_{T,1}}{\overline{C}_{T,0}}, \tag{1.13.23}$$

thus allowing $\overline{C}_{T,N}$ be directly related to $\overline{C}_{T,0}$; insertion of Eq. (1.13.22) supports transformation of Eq. (1.13.23) to

$$\frac{\overline{C}_{T,N}}{\overline{C}_{T,0}} = \frac{1}{1+\tau_N s}\frac{1}{1+\tau_{N-1}s}\cdots\frac{1}{1+\tau_2 s}\frac{1}{1+\tau_1 s}, \tag{1.13.24}$$

while Eq. (1.13.9) permits further simplification to

$$\frac{\overline{C}_{T,N}}{\overline{C}_{T,0}} = \prod_{i=1}^{N}\frac{1}{1+\dfrac{\tau}{N}s} = \frac{1}{\left(1+\dfrac{\tau s}{N}\right)^N}. \tag{1.13.25}$$

If tracer were instantaneously injected at the inlet stream to the first unit, then $C_{T,0}$ would abide to

$$C_{T,0}\{t\} = C_T^0\delta\{t\} \tag{1.13.26}$$

or, equivalently,

$$\overline{C}_{T,0} = C_T^0 \tag{1.13.27}$$

in Laplace's domain – where C_T^0 denotes a constant; insertion of Eq. (1.13.27) simplifies Eq. (1.13.25) to

$$\frac{\overline{C}_{T,N}}{C_T^0}=\frac{1}{\left(1+\frac{\tau s}{N}\right)^N}, \tag{1.13.28}$$

so $\overline{C}_{T,N}$ can be readily isolated as

$$\overline{C}_{T,N}=\frac{C_T^0}{\left(1+\frac{\tau s}{N}\right)^N}. \tag{1.13.29}$$

The evolution of tracer at the outlet of the cascade of stirred tanks will then be ascertained upon application of inverse Laplace's transforms to Eq. (1.13.29), i.e.

$$\begin{aligned} C_{T,N}\{t\}&\equiv\mathcal{L}^{-1}\left(\overline{C}_{T,N}\right)=\mathcal{L}^{-1}\left(\frac{C_T^0}{\left(1+\frac{\tau s}{N}\right)^N}\right)\\ &=\mathcal{L}^{-1}\left(\frac{\left(\frac{N}{\tau}\right)^N C_T^0}{(N-1)!}\frac{(N-1)!}{\left(s+\frac{N}{\tau}\right)^N}\right), \end{aligned} \tag{1.13.30}$$

where both numerator and denominator of the argument thereof were meanwhile multiplied by $(N/\tau)^N/(N-1)!$; in view of the linearity of Laplace's transform, Eq. (1.13.30) may be redone as

$$C_{T,N}\{t\}=\frac{\left(\frac{N}{\tau}\right)^N C_T^0}{(N-1)!}\mathcal{L}^{-1}\left(\frac{(N-1)!}{\left(s+\frac{N}{\tau}\right)^N}\right). \tag{1.13.31}$$

The translation theorem in Laplace's domain may now be invoked to transform Eq. (1.13.31) to

$$C_{T,N}\{t\}=\frac{\left(\frac{N}{\tau}\right)^N C_T^0}{(N-1)!}\exp\left\{-\frac{N}{\tau}t\right\}\mathcal{L}^{-1}\left(\frac{(N-1)!}{s^N}\right), \tag{1.13.32}$$

where the outstanding inverse Laplace's transform will finally look like

$$C_{T,N}\{t\}=\frac{\left(\frac{N}{\tau}\right)^N C_T^0}{(N-1)!}\exp\left\{-\frac{N}{\tau}t\right\}t^{N-1}. \tag{1.13.33}$$

The residence time distribution satisfies Eq. (5.201) of *Food Proc. Eng.: thermal & chemical operations*, so Eq. (1.13.33) may be retrieved to write

$$E\{t\}=\frac{\frac{\left(\frac{N}{\tau}\right)^N C_T^0}{(N-1)!}t^{N-1}\exp\left\{-\frac{N}{\tau}t\right\}}{\displaystyle\int_0^\infty\frac{\left(\frac{N}{\tau}\right)^N C_T^0}{(N-1)!}t^{N-1}\exp\left\{-\frac{N}{\tau}t\right\}dt}, \tag{1.13.34}$$

or else

$$E\{t\}=\frac{t^{N-1}\exp\left\{-\frac{N}{\tau}t\right\}}{\displaystyle\int_0^\infty t^{N-1}\exp\left\{-\frac{N}{\tau}t\right\}dt} \tag{1.13.35}$$

upon taking constant factors off the kernel, and dropping $(N/\tau)^N C_T^0/(N-1)!$ from both numerator and denominator afterward; integration is simplified after defining an auxiliary variable ξ as

$$\xi\equiv\frac{N}{\tau}t, \tag{1.13.36}$$

because it supports

$$\begin{aligned} \int_0^\infty t^{N-1}\exp\left\{-\frac{N}{\tau}t\right\}dt&=\int_0^\infty\left(\frac{\tau}{N}\right)^{N-1}\xi^{N-1}e^{-\xi}\frac{\tau}{N}d\xi\\ &=\left(\frac{\tau}{N}\right)^N\int_0^\infty\xi^{N-1}e^{-\xi}d\xi \end{aligned} \tag{1.13.37}$$

– since the lower and upper limits of integration remain the same, i.e. $\xi\rightarrow 0$ when $t\rightarrow 0$, and $\xi\rightarrow\infty$ when $t\rightarrow\infty$, besides $dt=\frac{\tau}{N}d\xi$. After retrieving the general result

$$\lim_{\xi\rightarrow\infty}\frac{\xi^{N-1}}{e^{\xi}}=0 \tag{1.13.38}$$

for every (integer) $N>1$, one may proceed to integration by parts of the outstanding integral in Eq. (1.13.37) as

$$\begin{aligned} \int_0^\infty\xi^{N-1}e^{-\xi}d\xi&=\xi^{N-1}\frac{e^{-\xi}}{(-1)}\bigg|_0^\infty-\int_0^\infty(N-1)\xi^{N-2}\frac{e^{-\xi}}{(-1)}d\xi\\ &=\xi^{N-1}e^{-\xi}\Big|_0-\xi^{N-1}e^{-\xi}\Big|_\infty+(N-1)\int_0^\infty\xi^{N-2}e^{-\xi}d\xi; \end{aligned} \tag{1.13.39}$$

in view of Eq. (1.13.38), one may indeed reduce Eq. (1.13.39) to

$$\int_0^\infty \xi^{N-1} e^{-\xi} d\xi = 0^{N-1} e^{-0} - \lim_{\xi \to \infty} \frac{\xi^{N-1}}{e^{\xi}} + (N-1)\int_0^\infty \xi^{N-2} e^{-\xi} d\xi$$

$$= 0 \cdot 1 - 0 + (N-1)\int_0^\infty \xi^{N-2} e^{-\xi} d\xi, \tag{1.13.40}$$

and thus

$$\int_0^\infty \xi^{N-1} e^{-\xi} d\xi = (N-1)\int_0^\infty \xi^{N-2} e^{-\xi} d\xi. \tag{1.13.41}$$

Equation (1.13.41) consubstantiates a recursive relationship, and may be applied to every single (integer) exponent of ξ, say $N-2$, $N-3$, ..., 2, 1 – to eventually yield

$$\int_0^\infty \xi^{N-1} e^{-\xi} d\xi = (N-1)(N-2)\cdots 1 \int_0^\infty e^{-\xi} d\xi; \tag{1.13.42}$$

after recalling the definition of factorial and the fundamental theorem of integral calculus, Eq. (1.13.42) becomes

$$\int_0^\infty \xi^{N-1} e^{-\xi} d\xi = (N-1)! \left.\frac{e^{-\xi}}{(-1)}\right|_0^\infty$$

$$= (N-1)!\left(e^0 - e^{-\infty}\right) = (N-1)!(1-0) \tag{1.13.43}$$

that breaks down to

$$\int_0^\infty \xi^{N-1} e^{-\xi} d\xi = (N-1)!. \tag{1.13.44}$$

Combination of Eqs. (1.13.37) and (19.13.44) gives then rise to

$$\int_0^\infty t^{N-1} \exp\left\{-\frac{N}{\tau}t\right\} dt = \left(\frac{\tau}{N}\right)^N (N-1)!, \tag{1.13.45}$$

which may in turn be inserted in Eq. (1.13.35) to generate

$$E\{t\} = \frac{t^{N-1} \exp\left\{-\frac{N}{\tau}t\right\}}{\left(\frac{\tau}{N}\right)^N (N-1)!}; \tag{1.13.46}$$

after lumping factors alike, Eq. (1.13.46) finally yields

$$E\{t\} = \frac{N}{\tau} \frac{\left(\frac{N}{\tau}t\right)^{N-1}}{(N-1)!} \exp\left\{-\frac{N}{\tau}t\right\} \tag{1.13.47}$$

to serve as residence time distribution of the vessel. A graphical interpretation of Eq. (1.13.47) is provided below – after having multiplied both sides by τ, so as to render it dimensionless.

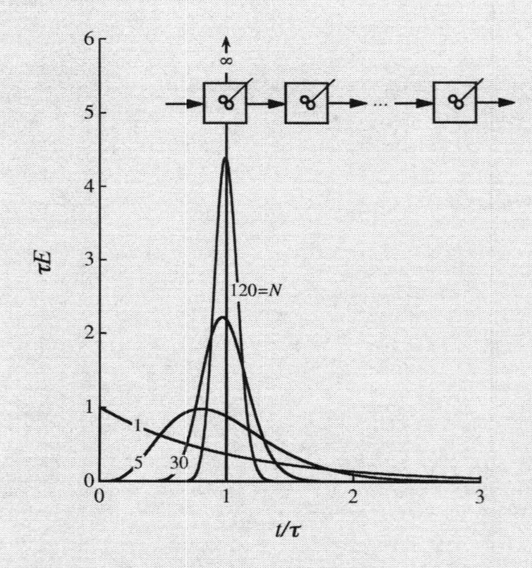

As N increases, dispersion becomes smaller and smaller; when $N \to \infty$, all liquid elements will share essentially the same residence time, i.e.

$$\lim_{N \to \infty} E\{t\} = \delta\{t - \tau\}, \tag{1.13.48}$$

characteristic of a plug flow vessel with τ for space time.

e) The total segregation model requires knowledge of the extent of viability loss in a plug flow reactor, following the same first-order death kinetics under scrutiny; the associated mass balance reads

$$QC_X\big|_V - k_d C_X\big|_V dV = QC_X\big|_{V+dV}, \tag{1.13.49}$$

where V denotes longitudinal volumetric coordinate. After dividing both sides by Q – and realizing that

$$dt \equiv \frac{dV}{Q} \tag{1.13.50}$$

using Eq. (1.13.3) as template, one is led to

$$C_X\big|_t - k_d C_X dt = C_X\big|_{t+dt}; \tag{1.13.51}$$

once terms alike are lumped, and both sides are divided by dt, Eq. (1.13.51) turns to

$$-k_d C_X = \frac{C_X\big|_{t+dt} - C_X\big|_t}{dt}, \tag{1.13.52}$$

or else

$$\frac{dC_X}{dt} = -k_d C_X \tag{1.13.53}$$

in view of the definition of derivative. Upon taking

$$C_X\big|_{t=0} = C_{X,0} \quad (1.13.54)$$

for initial condition, integration of Eq. (1.13.53) will ensue via separation of variables as

$$\int_{C_{X,0}}^{C_X} \frac{d\tilde{C}_X}{\tilde{C}_X} = -k_d \int_0^t d\tilde{t}; \quad (1.13.55)$$

the fundamental theorem of integral calculus has it that

$$\ln \tilde{C}_X\Big|_{C_{X,0}}^{C_X} = -k_d\,\tilde{t}\Big|_0^t, \quad (1.13.56)$$

or else

$$\ln \frac{C_X}{C_{X,0}} = -k_d t \quad (1.13.57)$$

– where exponentials may be taken of both sides to get

$$C_X = C_{X,0}\,\mathrm{e}^{-k_d t}, \quad (1.13.58)$$

followed by multiplication thereof by $C_{X,0}$. One may finally retrieve Eq. (5.235) from *Food Proc. Eng.: thermal & chemical operations* as descriptor of death extent as per the total segregation model, i.e.

$$C_{X,ts} = \int_0^\infty C_X\{t\}E\{t\}dt, \quad (1.13.59)$$

and combine it with Eqs. (1.13.47) and (1.13.58) to generate

$$C_{X,ts} = \int_0^\infty C_{X,0}\,\mathrm{e}^{-k_d t}\,\frac{N}{\tau}\,\frac{\left(\frac{N}{\tau}t\right)^{N-1}}{(N-1)!}\exp\left\{-\frac{N}{\tau}t\right\}dt; \quad (1.13.60)$$

after taking $1/(N-1)!$ and $C_{X,0}$ off the kernel, multiplying and dividing the argument of the first exponential function by N/τ, and taking N/τ under the differential sign, Eq. (1.13.60) becomes

$$C_{X,ts} = \frac{C_{X,0}}{(N-1)!}\int_0^\infty \exp\left\{-\frac{k_d\tau}{N}\left(\frac{N}{\tau}t\right)\right\}\left(\frac{N}{\tau}t\right)^{N-1}\exp\left\{-\frac{N}{\tau}t\right\}d\left(\frac{N}{\tau}t\right) = \frac{C_{X,0}}{(N-1)!}\int_0^\infty \xi^{N-1}\mathrm{e}^{-(1+\Xi)\xi}d\xi, \quad (1.13.61)$$

with the aid of Eq. (1.13.36) and

$$\Xi \equiv \frac{k_d\tau}{N}, \quad (1.13.62)$$

besides lumping of exponential functions. The integral in Eq. (1.13.61) may be more easily manipulated after defining an auxiliary variable ζ as

$$\zeta \equiv (1+\Xi)\xi, \quad (1.13.63)$$

since one promptly gets

$$\int_0^\infty \xi^{N-1}\mathrm{e}^{-(1+\Xi)\xi}d\xi = \int_0^\infty \left(\frac{\zeta}{1+\Xi}\right)^{N-1}\mathrm{e}^{-\zeta}d\left(\frac{\zeta}{1+\Xi}\right); \quad (1.13.64)$$

$\left(1/(1+\Xi)\right)^{N-1}$ and $1/(1+\Xi)$ may now be taken off kernel and off differential, viz.

$$\int_0^\infty \xi^{N-1}\mathrm{e}^{-(1+\Xi)\xi}d\xi = \left(\frac{1}{1+\Xi}\right)^N \int_0^\infty \zeta^{N-1}\mathrm{e}^{-\zeta}d\zeta. \quad (1.13.65)$$

Since the integral left in the right-hand side of Eq. (1.13.65) coincides with that labeled as Eq. (1.3.44), one may promptly redo Eq. (1.13.65) to

$$\int_0^\infty \xi^{N-1}\mathrm{e}^{-(1+\Xi)\xi}d\xi = \frac{(N-1)!}{(1+\Xi)^N} \quad (1.13.66)$$

– so Eq. (1.13.61) becomes

$$C_{X,ts} = \frac{C_{X,0}}{(N-1)!}\frac{(N-1)!}{(1+\Xi)^N} = \frac{C_{X,0}}{(1+\Xi)^N}, \quad (1.13.67)$$

along with the cancellation of $(N-1)!$ between numerator and denominator. The definition of Ξ conveyed by Eq. (1.13.62) permits Eq. (1.13.67) be rewritten as

$$C_{X,ts} = \frac{C_{X,0}}{\left(1+\frac{k_d\tau}{N}\right)^N}, \quad (1.13.68)$$

where division of both sides by $C_{X,0}$ unfolds

$$\frac{C_{X,ts}}{C_{X,0}} = \frac{1}{\left(1+\frac{k_d\tau}{N}\right)^N}. \quad (1.13.69)$$

Equation (1.13.69) coincides with Eq. (1.13.11), so the same result was attained via two independent routes; one should, however, be warned that such a coincidence is restricted to first-order processes – as is the (hypothesized) death kinetics. In this specific case, the total segregation model leads to the same result as the maximum mixedness model, see Eq. (5.251) of *Food Proc. Eng.: thermal & chemical operations*, as discussed with regard to pseudo-first-order enzyme-mediated reactions; remember that modeling of (ideal) well-stirred vessels via their basic design equation leads, in turn, to the same result as the maximum mixedness model, using the corresponding residence time distribution.

1.2.3.2 Direct heating of uniform-temperature solid

The thermal conductivity of a solid (food) is typically much higher than that of the heating/cooling liquid or vapor utilized, so one may neglect gradients of temperature, T_f, within the food in the direction of heat flow – associated with transfer of heat by conduction from the heating/cooling utility. Presence of turbulence and good stirring may, in turn, be hypothesized for the heating/cooling fluid in the direction of flow, so gradients of temperature, $T_{h\backslash c}$, will also be neglected – except at the very neighborhood of the entrance. The associated overall problem can accordingly be sketched as done in Fig. 1.44. The enthalpic fluid is utilized during the heating phase, with temperature $T_h > T_f$ (where T_f varies in time, but not in space) – or instead during the cooling phase, with temperature $T_c < T_f$. Furthermore, $T_{h\backslash c} \equiv T_{h\backslash c,out}$ may either coincide with $T_{h\backslash c,in}$, if saturated steam is employed and only latent heat of condensation is at stake – or else $T_{h\backslash c} \equiv T_{h\backslash c,out} < T_{h\backslash c,in}$ if hot water is used for heating, or $T_{h\backslash c} \equiv T_{h\backslash c,out} > T_{h\backslash c,in}$ if cold water is used for cooling.

If only latent heat is at stake with regard to the heating fluid utility, then $T_{h\backslash c}$ remains essentially constant from entrance to exit; conversely, only sensible heat is exchanged by the food, so $T_f \equiv T_f\{t\}$. Under these circumstances, an enthalpic balance to food reads

$$\boxed{U_{h\backslash c}A_f\left(T_{h\backslash c} - T_f\right) = V_f \rho_f c_{P,f}\frac{dT_f}{dt},} \tag{1.486}$$

where $U_{h\backslash c}$ denotes overall heat transfer coefficient for heating (or cooling) process (encompassing conductances of both food material and heating fluid), A_f and V_f denote outer area and volume of food, respectively, ρ_f and $c_{P,f}$ denote mass density and (isobaric) specific heat capacity of food, respectively, and t denotes time;

$$\boxed{T_f\Big|_{t=0} = T_{f,0}} \tag{1.487}$$

may, in turn, play the role of initial condition – provided that $T_{f,0}$ denotes initial (usually room) temperature of food. Since the differential of an algebraic sum of a function with a constant coincides with the differential of the former, Eq. (1.486) may be rewritten as

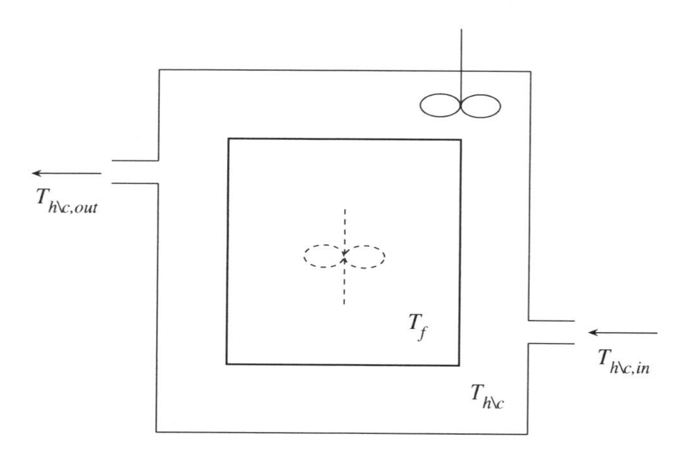

FIGURE 1.44 Schematic representation of blancher/pasteurizer/sterilizer, operated with direct heating/cooling by enthalpic fluid utility at temperature $T_{h\backslash c}$, putatively dropping/jumping from $T_{h\backslash c,in}$ at entrance to $T_{h\backslash c,out} \equiv T_{h\backslash c}$ thereafter – with temperature T_f of food to be processed.

$$U_{h\backslash c}A_f\left(T_{h\backslash c} - T_f\right) = -V_f \rho_f c_{P,f}\frac{d}{dt}\left(T_{h\backslash c} - T_f\right), \tag{1.488}$$

so integration may proceed, via separation of variables, according to

$$\int_{T_{h\backslash c}-T_{f,0}}^{T_{h\backslash c}-T_f} \frac{d\left(T_{h\backslash c} - \tilde{T}_f\right)}{T_{h\backslash c} - \tilde{T}_f} = -\frac{U_{h\backslash c}A_f}{\rho_f c_{P,f} V_f}\int_0^t d\tilde{t}; \tag{1.489}$$

Eq. (1.487) was meanwhile taken advantage of in setting the (new) limits of integration for variable $T_{h\backslash c} - \tilde{T}_f$ – while both sides were divided by $-V_f\rho_f c_{P,f}$ (for convenience), under the hypothesis that the said product, as well as product $U_{h\backslash c}A_f$ remain constant in time and with temperature. Application of the fundamental theorem of integral calculus transforms Eq. (1.489) to

$$\ln\left\{T_{h\backslash c} - \tilde{T}_f\right\}\Big|_{T_{h\backslash c}-T_{f,0}}^{T_{h\backslash c}-T_f} = -\frac{U_{h\backslash c}A_f}{\rho_f c_{P,f} V_f}\tilde{t}\Big|_0^t, \tag{1.490}$$

which breaks down to

$$\ln\frac{T_{h\backslash c} - T_f}{T_{h\backslash c} - T_{f,0}} = -\frac{U_{h\backslash c}A_f}{\rho_f c_{P,f} V_f}t; \tag{1.491}$$

after taking exponentials of both sides, Eq. (1.491) becomes

$$\frac{T_{h\backslash c} - T_f}{T_{h\backslash c} - T_{f,0}} = \exp\left\{-\frac{U_{h\backslash c}A_f}{\rho_f c_{P,f} V_f}t\right\}, \tag{1.492}$$

where the left-hand side represents dimensionless change in food temperature, and the argument of the exponential in the right-hand side represents dimensionless time. (The situation of cooling fluid utility undergoing exchange of only sensible heat will be explored later, when discussing chilling operations.)

In the case of heating, Eq. (1.492) should to advantage be coined as

$$\boxed{\frac{T_h - T_f}{T_h - T_{f,0}} = \exp\left\{-\frac{U_h A_f}{\rho_f c_{P,f} V_f}t\right\},} \tag{1.493}$$

as depicted in the left portion of Fig. 1.45. A typical exponential decay of the normalized temperature difference is found during heating; a (lower) horizontal asymptote would be reached after a sufficiently long time, described by

$$\lim_{t\to\infty}\frac{T_h - T_f}{T_h - T_{f,0}} = \mathrm{e}^{-\infty} = 0 \tag{1.494}$$

based on Eq. (1.493), and coincident with the horizontal axis – or, equivalently,

$$\lim_{t\to\infty} T_f = T_h. \tag{1.495}$$

In practice, however, heating occurs for a finite period, so one finds that

$$\frac{T_h - T_f}{T_h - T_{f,0}}\bigg|_{t=\Delta t_h} = \frac{T_h - T_{f,\infty}}{T_h - T_{f,0}} > 0, \tag{1.496}$$

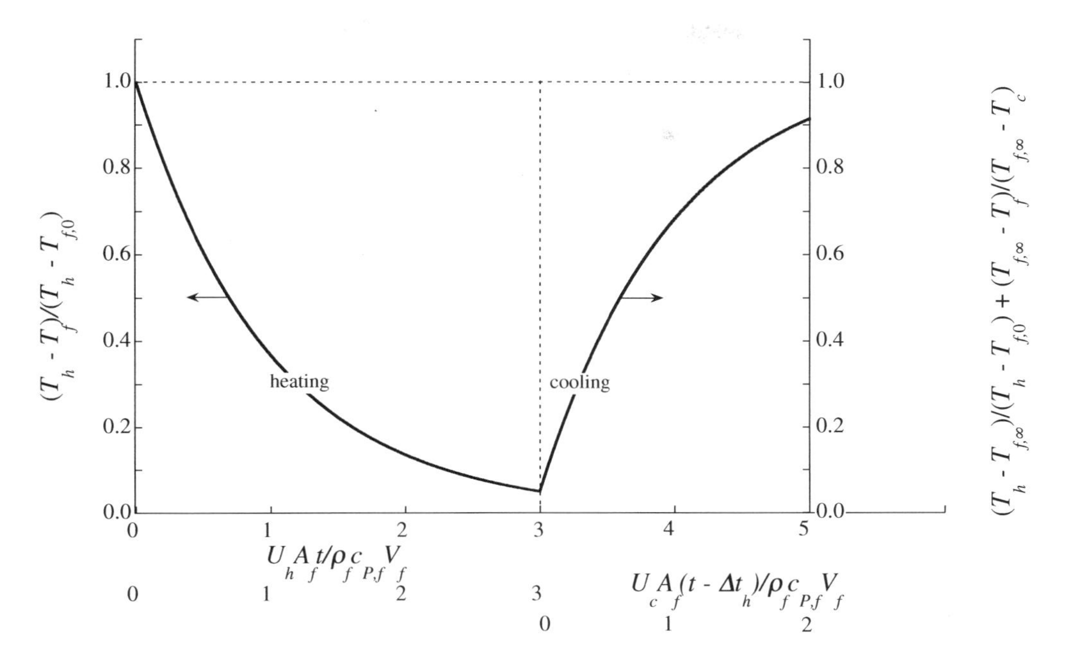

FIGURE 1.45 Evolution of normalized temperature, $(T_h-T_f)/(T_h-T_{f,0})$ for heating, or $(T_h-T_{f,\infty})/(T_h-T_{f,0})+(T_{f,\infty}-T_f)/(T_{f,\infty}-T_c)$ for cooling, versus dimensionless time, $U_hA_ft/(\rho_f c_{P,f}V_f)$ for heating or $U_cA_f(t-\Delta t_h)/\rho_f c_{P,f}V_f$ for cooling, in blanching/pasteurization/sterilization, under transfer of only latent heat by heating/cooling fluid utility.

where $T_{f,\infty}$ denotes maximum temperature reached by food during the heating phase, and Δt_h denotes duration of the said phase.

In the case of cooling, Eq. (1.492) should look like

$$\frac{T_c - T_f}{T_c - T_{f,0}} = \frac{T_f - T_c}{T_{f,0} - T_c} = \exp\left\{-\frac{U_c A_f}{\rho_f c_{P,f} V_f} t\right\} \tag{1.497}$$

if Eq. (1.487) remained valid at $t=\Delta t_h$ – where negatives were taken of both numerator and denominator so as to generate positive quantities, and $U_c \neq U_h$ in general due to distinct patterns of natural convection. However, cooling always takes place after heating (for Δt_h), so one should redo Eq. (1.497) as

$$\frac{T_f - T_c}{T_{f,\infty} - T_c} = \exp\left\{-\frac{U_c A_f}{\rho_f c_{P,f} V_f}\left(t - \Delta t_h\right)\right\}, \tag{1.498}$$

corresponding to use of

$$\boxed{T_f\big|_{t=\Delta t_h} = T_{f,\infty}} \tag{1.499}$$

as initial condition – consistent with Eq. (1.496), where $T_{f,\infty}$ is to be employed as departing condition *in lieu* of $T_{f,0}$ as in Eq. (1.487). To facilitate graphical interpretation of temperature evolution – and knowing that an exponential function of negative argument only takes (positive) values between 0 and 1, one may take negatives of both sides of Eq. (1.498) and then add unity and $(T_h-T_{f,\infty})/(T_h-T_{f,0})$, according to

$$\frac{T_h - T_{f,\infty}}{T_h - T_{f,0}} + 1 - \frac{T_f - T_c}{T_{f,\infty} - T_c} = \frac{T_h - T_{f,\infty}}{T_h - T_{f,0}} + 1 - \exp\left\{-\frac{U_c A_f}{\rho_f c_{P,f} V_f}\left(t - \Delta t_h\right)\right\}; \tag{1.500}$$

after merging the last two terms in the left-hand side, one gets

$$\frac{T_h - T_{f,\infty}}{T_h - T_{f,0}} + \frac{T_{f,\infty} - T_c - T_f + T_c}{T_{f,\infty} - T_c} = \frac{T_h - T_{f,\infty}}{T_h - T_{f,0}} + 1 - \exp\left\{-\frac{U_c A_f}{\rho_f c_{P,f} V_f}\left(t - \Delta t_h\right)\right\} \tag{1.501}$$

which reduces to

$$\boxed{\frac{T_h - T_{f,\infty}}{T_h - T_{f,0}} + \frac{T_{f,\infty} - T_f}{T_{f,\infty} - T_c} = \frac{T_h - T_{f,\infty}}{T_h - T_{f,0}} + 1 - \exp\left\{-\frac{U_c A_f}{\rho_f c_{P,f} V_f}\left(t - \Delta t_h\right)\right\}} \tag{1.502}$$

upon cancellation of symmetrical terms – as sketched in the right portion of Fig. 1.45. The evolution pattern has essentially been reversed relative to heating, along with a cooling period typically shorter than the previous heating period for a warm product left in the end, due to the hypothesized $U_c \neq U_h$, different normalizations result for t, i.e. via $\rho_f c_{P,f} V_f/U_h A_f$ throughout heating and $\rho_f c_{P,f} V_f/U_c A_f$ throughout cooling. The dimensionless temperature will eventually be driven by

$$\lim_{t\to\infty}\left(\frac{T_h - T_{f,\infty}}{T_h - T_{f,0}} + \frac{T_{f,\infty} - T_f}{T_{f,\infty} - T_c}\right) = \frac{T_h - T_{f,\infty}}{T_h - T_{f,0}} + 1 - e^{-\infty} = 1 + \frac{T_h - T_{f,\infty}}{T_h - T_{f,0}} \tag{1.503}$$

at long times, in the form of an (upper) horizontal asymptote; Eq. (1.503) may be rewritten as

$$\lim_{t\to\infty}\frac{T_{f,\infty} - T_f}{T_{f,\infty} - T_c} = 1 \tag{1.504}$$

after dropping (constant) $(T_h-T_{f,\infty})/(T_h-T_{f,0})$ from both sides, which readily implies

$$\lim_{t\to\infty} T_f = T_c \tag{1.505}$$

– meaning that the temperature of the food, T_f, will not reach the temperature of the cooling medium, T_c, unless a (theoretically) infinite time is allowed.

If the heating (or cooling) medium – characterized by mass density $\rho_{h\backslash c}$ and specific heat capacity $c_{P,h\backslash c}$, and flown at (average) velocity $v_{h\backslash c}$ as plug flow through fluid cross-section $A_{c,h\backslash c}$, releases (or takes up) sensible heat, then it will undergo variation of its temperature $T_{h\backslash c}$ along linear coordinate x (with $0 \leq x \leq L$) according to

$$U_{h\backslash c}\left(T_{h\backslash c}-T_f\right)P_f dx=-v_{h\backslash c}A_{c,h\backslash c}\rho_{h\backslash c}c_{P,h\backslash c}dT_{h\backslash c}, \quad (1.506)$$

which serves as (differential) enthalpy balance to the heating/cooling utility; here P_f denotes outer perimeter of food. Division of both sides of Eq. (1.506) by dx supports transformation thereof to

$$\boxed{U_{h\backslash c}P_f\left(T_{h\backslash c}-T_f\right)=-v_{h\backslash c}A_{c,h\backslash c}\rho_{h\backslash c}c_{P,h\backslash c}\frac{dT_{h\backslash c}}{dx},} \quad (1.507)$$

along with

$$\boxed{T_{h\backslash c}\Big|_{x=0}=T_{h\backslash c,in}} \quad (1.508)$$

for initial condition – to be complemented by Eq. (1.486) as enthalpy balance to the food. Variation of T_f is, in practice, (much) slower than that experienced by the heating/cooling fluid – so it is reasonable to assume that T_f remains essentially constant during a single pass of utility over the food; this consubstantiates a pseudo steady state. Equation (1.507) yields

$$\frac{dT_{h\backslash c}}{T_{h\backslash c}-T_f}=-\frac{U_{h\backslash c}P_f}{v_{h\backslash c}A_{c,h\backslash c}\rho_{h\backslash c}c_{P,h\backslash c}}dx \quad (1.509)$$

upon separation of variables, where integration ensues as

$$\int_{T_{h\backslash c,in}}^{T_{h\backslash c}}\frac{d\tilde{T}_{h\backslash c}}{\tilde{T}_{h\backslash c}-T_f}=-\frac{U_{h\backslash c}P_f}{v_{h\backslash c}A_{c,h\backslash c}\rho_{h\backslash c}c_{P,h\backslash c}}\int_0^x d\tilde{x} \quad (1.510)$$

with the aid of Eq. (1.508); since T_f is to be treated as constant for the sake of integration (as per the above pseudo steady state), Eq. (1.510) degenerates to

$$\ln\left\{\tilde{T}_{h\backslash c}-T_f\right\}\Big|_{T_{h\backslash c,in}}^{T_{h\backslash c}}=-\frac{U_{h\backslash c}P_f}{v_{h\backslash c}A_{c,h\backslash c}\rho_{h\backslash c}c_{P,h\backslash c}}\tilde{x}\Big|_0^x \quad (1.511)$$

via the fundamental theorem of integral calculus, or else

$$\ln\frac{T_{h\backslash c}-T_f}{T_{h\backslash c,in}-T_f}=-\frac{U_{h\backslash c}P_f}{v_{h\backslash c}A_{c,h\backslash c}\rho_{h\backslash c}c_{P,h\backslash c}}x. \quad (1.512)$$

If the temperature of heating/cooling fluid at the exit, i.e. at $x=L_f$, abides to

$$T_{h\backslash c}\Big|_{x=L_f}=T_{h\backslash c,out}, \quad (1.513)$$

then one obtains

$$\ln\frac{T_{h\backslash c,out}-T_f}{T_{h\backslash c,in}-T_f}=-\frac{U_{h\backslash c}P_fL_f}{v_{h\backslash c}A_{c,h\backslash c}\rho_{h\backslash c}c_{P,h\backslash c}} \quad (1.514)$$

based on Eq. (1.512); after taking exponentials of both sides, Eq. (1.514) becomes

$$\frac{T_{h\backslash c,out}-T_f}{T_{h\backslash c,in}-T_f}=\exp\left\{-\frac{U_{h\backslash c}P_fL_f}{v_{h\backslash c}A_{c,h\backslash c}\rho_{h\backslash c}c_{P,h\backslash c}}\right\}, \quad (1.515)$$

where isolation of $T_{h\backslash c,out}$ – coupled with realization that

$$P_f\ L_f=A_f, \quad (1.516)$$

give rise to

$$T_{h\backslash c,out}=T_f+\left(T_{h\backslash c,in}-T_f\right)\exp\left\{-\frac{U_{h\backslash c}A_f}{v_{h\backslash c}A_{c,h\backslash c}\rho_{h\backslash c}c_{P,h\backslash c}}\right\}. \quad (1.517)$$

The right-hand side of Eq. (1.517) may be algebraically rearranged to

$$\boxed{\begin{aligned}T_{h\backslash c,out}&=T_{h\backslash c,in}\exp\left\{-\frac{U_{h\backslash c}A_f}{Q_{h\backslash c}\rho_{h\backslash c}c_{P,h\backslash c}}\right\}\\&\quad+\left(1-\exp\left\{-\frac{U_{h\backslash c}A_f}{Q_{h\backslash c}\rho_{h\backslash c}c_{P,h\backslash c}}\right\}\right)T_f,\end{aligned}} \quad (1.518)$$

where

$$v_{h\backslash c}A_{c,h\backslash c}=Q_{h\backslash c} \quad (1.519)$$

was taken on board – with $Q_{h\backslash c}$ denoting volumetric flow rate of heating/cooling fluid; hence, a (linear) relationship of the type $T_{h\backslash c,out}\equiv T_{h\backslash c,out}\left\{T_f\right\}$ is obtained. In other words, Eq. (1.518) is of the form

$$T_{h\backslash c,out}=\Omega T_{h\backslash c,in}+(1-\Omega)T_f, \quad (1.520)$$

provided that

$$\Omega\equiv\exp\left\{-\frac{U_{h\backslash c}A_f}{Q_{h\backslash c}\rho_{h\backslash c}c_{P,h\backslash c}}\right\} \quad (1.521)$$

is defined as auxiliary parameter; $-\ln\,\Omega$ represents the ratio of heat transported by conduction as per Fourier's law under an (arbitrary) driving force ΔT, i.e. $U_{h\backslash c}A_f\Delta T$, to the sensible heat exchanged with the heating/cooling fluid associated with the same temperature difference, i.e. $Q_{h\backslash c}\rho_{h\backslash c}c_{P,h\backslash c}\Delta T$.

PROBLEM 1.14

Departing from Eq. (1.514), obtain an expression for the enthalpic power in transit – and compare it with performance of a regular (plug flow) heat exchanger.

Solution

The enthalpic power in transit between phases, Θ, satisfies

$$\Theta=-Q_{h\backslash c}\rho_{h\backslash c}c_{P,h\backslash c}\left(T_{h\backslash c,out}-T_{h\backslash c,in}\right), \quad (1.14.1)$$

which may be redone to

$$-Q_{h\backslash c}\rho_{h\backslash c}c_{P,h\backslash c}=\frac{\Theta}{T_{h\backslash c,out}-T_f-\left(T_{h\backslash c,in}-T_f\right)} \quad (1.14.2)$$

after isolating $Q_{h\backslash c}\rho_{h\backslash c}c_{P,h\backslash c}$, and adding and subtracting T_f to the denominator. On the other hand, algebraic rearrangement of Eq. (1.514) unfolds

$$-Q_{h\backslash c}\rho_{h\backslash c}c_{P,h\backslash c}=\frac{U_{h\backslash c}A_f}{\ln\dfrac{T_{h\backslash c,out}-T_f}{T_{h\backslash c,in}-T_f}}, \quad (1.14.3)$$

with the aid of Eqs. (1.516) and (1.519); elimination of $-Q_{h\backslash c}\rho_{h\backslash c}c_{P,h\backslash c}$ between Eqs. (1.14.2) and (1.14.3) produces

$$\frac{\Theta}{T_{h\backslash c,out}-T_f-\left(T_{h\backslash c,in}-T_f\right)}=\frac{U_{h\backslash c}A_f}{\ln\dfrac{T_{h\backslash c,out}-T_f}{T_{h\backslash c,in}-T_f}}, \quad (1.14.4)$$

where isolation of Θ gives rise to

$$\Theta=U_{h\backslash c}A_f\frac{\left(T_{h\backslash c,out}-T_f\right)-\left(T_{h\backslash c,in}-T_f\right)}{\ln\dfrac{T_{h\backslash c,out}-T_f}{T_{h\backslash c,in}-T_f}}. \quad (1.14.5)$$

Equation (1.14.5) is consistent with Eq. (1.312) in *Food Proc. Eng.: thermal & chemical operations*, or Eq. (1.366) therein for that matter – since the relative direction of flow will be redundant if the other phase undergoes essentially no change in temperature; the last factor in the right-hand side of Eq. (1.14.5) represents indeed the logarithmic mean of $T_{h\backslash c}-T_f$, between entrance and exit of heating/cooling fluid.

The enthalpy balance to food, in the case currently under scrutiny, should preferably be expressed as

$$Q_{h\backslash c}\rho_{h\backslash c}c_{P,h\backslash c}\left(T_{h\backslash c,in}-T_{h\backslash c,out}\right)=V_f\rho_f c_{P,f}\frac{dT_f}{dt}, \quad (1.522)$$

in lieu of Eq. (1.486) – since all sensible heat exchanged with food must come at the expense of heat lost/gained by the enthalpic fluid. Combination of Eqs. (1.520) and (1.522) gives rise to

$$Q_{h\backslash c}\rho_{h\backslash c}c_{P,h\backslash c}\left(T_{h\backslash c,in}-\Omega T_{h\backslash c,in}-(1-\Omega)T_f\right)=V_f\rho_f c_{P,f}\frac{dT_f}{dt}, \quad (1.523)$$

where $T_{h\backslash c,in}$ may be factored out to get

$$Q_{h\backslash c}\rho_{h\backslash c}c_{P,h\backslash c}\left(T_{h\backslash c,in}(1-\Omega)-(1-\Omega)T_f\right)=V_f\rho_f c_{P,f}\frac{dT_f}{dt}; \quad (1.524)$$

a further factoring out of $1-\Omega$ generates

$$Q_{h\backslash c}\rho_{h\backslash c}c_{P,h\backslash c}(1-\Omega)\left(T_{h\backslash c,in}-T_f\right)=V_f\rho_f c_{P,f}\frac{dT_f}{dt}. \quad (1.525)$$

Inspection of Eq. (1.525) *vis-à-vis* with Eq. (1.486) unfolds the same derivative-containing side; but the other side is also analogous, as long as $U_{h\backslash c}A_f$ is replaced by $Q_{h\backslash c}\rho_{h\backslash c}c_{P,h\backslash c}(1-\Omega)$ and $T_{h\backslash c}$ is replaced by $T_{h\backslash c,in}$; hence, one may jump immediately to the corresponding analog to Eq. (1.492), i.e.

$$\frac{T_{h\backslash c,in}-T_f}{T_{h\backslash c,in}-T_{f,0}}=\exp\left\{-(1-\Omega)\frac{\rho_{h\backslash c}c_{P,h\backslash c}Q_{h\backslash c}}{\rho_f c_{P,f}V_f}t\right\}. \quad (1.526)$$

The heating stage accordingly looks like

$$\boxed{\frac{T_{h,in}-T_f}{T_{h,in}-T_{f,0}}=\exp\left\{-\left(1-\exp\left\{-\frac{U_hA_f}{Q_h\rho_h c_{P,h}}\right\}\right)\frac{\rho_h c_{P,h}Q_h}{\rho_f c_{P,f}V_f}t\right\}}, \quad (1.527)$$

which stems from Eq (1.526) after recalling Eq. (1.521); once again, T_f increases monotonically with t, up to a maximum value $T_{f,\infty}$ attained by $t=\Delta t_h$ – as illustrated in Fig. 1.46, with a more complex definition of dimensionless time than that underlying Eq. (1.493). As expected, the left portion of Fig. 1.46 coincides in shape with the left portion of Fig. 1.45 – since T_h has simply been replaced by $T_{h,in}$, and the dimensionless independent variable $U_hA_ft/\rho_fc_{P,f}V_f$ has been replaced by $(1-\exp\{-U_hA_f/Q_h\rho_hc_{P,h}\})\rho_hc_{P,h}Q_ht/\rho_fc_{P,f}V_f$.

When $U_hA_f/Q_h\rho_hc_{P,h}$ is small, the nested exponential function in Eq. (1.527) may be expanded via Taylor's series about zero, truncated after the linear term, to get

$$\lim_{\frac{U_hA_f}{Q_h\rho_hc_{P,h}}\to 0}\frac{T_{h,in}-T_f}{T_{h,in}-T_{f,0}}=\exp\left\{-\left(1-\left(1-\frac{U_hA_f}{Q_h\rho_h c_{P,h}}\right)\right)\frac{\rho_h c_{P,h}Q_h}{\rho_f c_{P,f}V_f}t\right\}, \quad (1.528)$$

where cancellation of symmetrical terms allows simplification to

$$\lim_{\frac{U_hA_f}{Q_h\rho_hc_{P,h}}\to 0}\frac{T_{h,in}-T_f}{T_{h,in}-T_{f,0}}=\exp\left\{-\frac{U_hA_f}{Q_h\rho_h c_{P,h}}\frac{\rho_h c_{P,h}Q_h}{\rho_f c_{P,f}V_f}t\right\}; \quad (1.529)$$

dropping of common factors, between numerator and denominator, permits further simplification to

$$\lim_{\frac{U_hA_f}{Q_h\rho_hc_{P,h}}\to 0}\frac{T_{h,in}-T_f}{T_{h,in}-T_{f,0}}=\exp\left\{-\frac{U_hA_f}{\rho_f c_{P,f}V_f}t\right\} \quad (1.530)$$

that retrieves Eq. (1.493). This result is expected because $Q_h\to\infty$ or $c_{P,h}\to\infty$ implies an essentially constant temperature for the heating fluid, i.e. $T_{h,in}=T_{h,out}=T_h$ – formally equivalent to the transfer of latent heat only; besides knowing that $U_hA_f/Q_h\rho_hc_{P,h}\to 0$ when either $Q_h\to\infty$ or $c_{P,h}\to\infty$.

If t grows unbounded, then Eq. (1.527) is driven by

$$\lim_{t\to\infty}\frac{T_{h,in}-T_f}{T_{h,in}-T_{f,0}}=e^{-\infty}=0, \quad (1.531)$$

so the horizontal axis plays again the role of horizontal asymptote; Eq. (1.531) is equivalent to

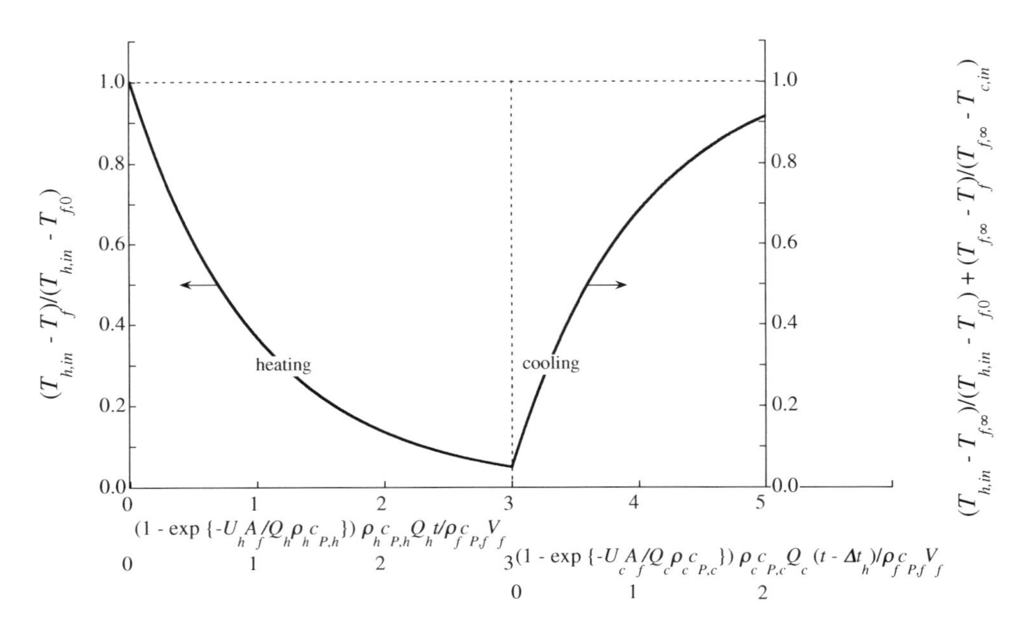

FIGURE 1.46 Evolution of normalized temperature, $(T_{h,in}-T_f)/(T_{h,in}-T_{f,0})$ for heating, or $(T_{h,in}-T_{f,\infty})/(T_{h,in}-T_{f,0})+(T_{f,\infty}-T_f)/(T_{f,\infty}-T_{c,in})$ for cooling, versus dimensionless time, $(1-\exp\{-U_hA_f/Q_h\rho_hc_{P,h}\})\rho_hc_{P,h}Q_ht/\rho_fc_{P,f}V_f$ for heating or $(1-\exp\{-U_cA_f/Q_c\rho_cc_{P,c}\})\rho_cc_{P,c}Q_c(t-\Delta t_h)/\rho_fc_{P,f}V_f$ for cooling, in blanching/pasteurization/sterilization, under transfer of only sensible heat by heating/cooling fluid utility.

$$\lim_{t\to\infty} T_f = T_{h,in}, \tag{1.532}$$

meaning that the food will eventually reach the temperature of the inlet heating fluid. If heating takes place for only a (finite) period Δt_h, then Eq. (1.527) approaches

$$\left.\frac{T_{h,in}-T_f}{T_{h,in}-T_{f,0}}\right|_{t=\Delta t_h} = \frac{T_{h,in}-T_{f,\infty}}{T_{h,in}-T_{f,0}} > 0, \tag{1.533}$$

with $T_{f,\infty}$ denoting again maximum temperature reached by food during the heating phase – which obeys

$$\frac{T_{h,in}-T_{f,\infty}}{T_{h,in}-T_{f,0}} = \exp\left\{-\left(1-\exp\left\{-\frac{U_hA_f}{Q_h\rho_hc_{P,h}}\right\}\right)\frac{\rho_hc_{P,h}Q_h}{\rho_fc_{P,f}V_f}\Delta t_h\right\} \tag{1.534}$$

as per Eq. (1.527), after setting $t=\Delta t_h$.

Should cooling of food be under scrutiny departing from temperature $T_{f,0}$, Eq. (1.526) would then take the form

$$\frac{T_{c,in}-T_f}{T_{c,in}-T_{f,0}} = \frac{T_f-T_{c,in}}{T_{f,0}-T_{c,in}} = \exp\{-\Theta(1-\Omega)t\}, \tag{1.535}$$

where negatives were again taken of both numerator and denominator so as to generate positive quantities – besides defining Θ as auxiliary constant, viz.

$$\Theta \equiv \frac{\rho_cc_{P,c}Q_c}{\rho_fc_{P,f}V_f}, \tag{1.536}$$

and using the alias of Eq. (1.521) applicable specifically to cooling (i.e. Ω replaced by $\exp\{-U_cA_f/Q_c\rho_cc_{P,c}\}$). Cooling coming (obviously) after heating for a period Δt_h enforces use of Eq. (1.533) to write

$$\frac{T_f-T_{c,in}}{T_{f,\infty}-T_{c,in}} = \exp\{-\Theta(1-\Omega)(t-\Delta t_h)\} \tag{1.537}$$

based on Eq. (1.535), where $T_{f,\infty}$ satisfies Eq. (1.534); note that $T_{f,0}$, as the initial condition of heating given by Eq. (1.487), was replaced by $T_{f,\infty}$, as the initial condition of cooling (or the final condition of heating, for that matter). Graphical interpretation of temperature evolution throughout the cooling period suggests replacement of $(T_f-T_{c,in})/(T_{f,\infty}-T_{c,in})$ by $(T_{h,in}-T_{f,\infty})/(T_{h,in}-T_{f,0})+1-(T_f-T_{c,in})/(T_{f,\infty}-T_{c,in})$ in Eq. (1.537), as done before when going from Eq. (1.498) to Eq. (1.500) – which promptly generates

$$\frac{T_{h,in}-T_{f,\infty}}{T_{h,in}-T_{f,0}}+1-\frac{T_f-T_{c,in}}{T_{f,\infty}-T_{c,in}} = \frac{T_{h,in}-T_{f,\infty}}{T_{h,in}-T_{f,0}}+1-\exp\{-\Theta(1-\Omega)(t-\Delta t_h)\}; \tag{1.538}$$

the second and third terms in the left-hand side may then be lumped to get

$$\frac{T_{h,in}-T_{f,\infty}}{T_{h,in}-T_{f,0}}+\frac{T_{f,\infty}-T_{c,in}-T_f+T_{c,in}}{T_{f,\infty}-T_{c,in}} = \frac{T_{h,in}-T_{f,\infty}}{T_{h,in}-T_{f,0}}+1-\exp\{-\Theta(1-\Omega)(t-\Delta t_h)\}, \tag{1.539}$$

which simplifies to

$$\frac{T_{h,in}-T_{f,\infty}}{T_{h,in}-T_{f,0}}+\frac{T_{f,\infty}-T_f}{T_{f,\infty}-T_{c,in}} = \frac{T_{h,in}-T_{f,\infty}}{T_{h,in}-T_{f,0}}+1-\exp\{-\Theta(1-\Omega)(t-\Delta t_h)\} \tag{1.540}$$

upon cancellation of symmetrical terms – or, after recalling Eq. (1.521) associated specifically with cooling and Eq. (1.536),

$$\frac{T_{h,in}-T_{f,\infty}}{T_{h,in}-T_{f,0}}+\frac{T_{f,\infty}-T_f}{T_{f,\infty}-T_{c,in}}=\frac{T_{h,in}-T_{f,\infty}}{T_{h,in}-T_{f,0}} +1-\exp\left\{-\left(1-\exp\left\{-\frac{U_cA_f}{Q_c\rho_c c_{P,c}}\right\}\right)\frac{\rho_c c_{P,c}Q_c}{\rho_f c_{P,f}V_f}\left(t-\Delta t_h\right)\right\}. \tag{1.541}$$

The corresponding evolution pattern is illustrated as the right portion of Fig. 1.46; note the reversal of the heating pattern, including both monotony and orientation of concavity. Figure 1.46 as a whole accordingly echoes Fig. 1.45 – except for the distinct definition of dimensionless (independent and dependent) variables. The dimensionless temperature throughout cooling will tend to

$$\lim_{t\to\infty}\left(\frac{T_{h,in}-T_{f,\infty}}{T_{h,in}-T_{f,0}}+\frac{T_{f,\infty}-T_f}{T_{f,\infty}-T_{c,in}}\right)=\frac{T_{h,in}-T_{f,\infty}}{T_{h,in}-T_{f,0}}+1-e^{-\infty} =1+\frac{T_{h,in}-T_{f,\infty}}{T_{h,in}-T_{f,0}} \tag{1.542}$$

when time grows unbounded – which consubstantiates a horizontal asymptote; Eq. (1.542) is equivalent to

$$\lim_{t\to\infty}\frac{T_{f,\infty}-T_f}{T_{f,\infty}-T_{c,in}}=1 \tag{1.543}$$

after taking $(T_{h,in}-T_{f,\infty})/(T_{h,in}-T_{f,0})$ off both sides. Equation (1.543) also implies

$$\lim_{t\to\infty}T_f=T_{c,in}, \tag{1.544}$$

so the temperature of the food will reach the inlet temperature of the cooling fluid after a sufficiently long time.

1.2.3.3 Indirect heating of uniform-temperature solid

Consider a retort, where some (packaged) food undergoes pasteurization or sterilization. The mixture of air and steam in the retort may be assumed well-mixed, since its circulation (forced, or arising from natural convection) prevents relevant temperature gradients from building up – so a uniform temperature T_2, evolving in time, t, will be hypothesized. The contents of the jar/can containing the food may be subjected to some form of rotational mixing as well, or else be static but undergo natural convection through the filling liquid inside – on the hypothesis that the container is relatively thin. In any case, the thermal conductivities of the (solid and liquid) constituents of the food product are expected to be larger than those of the gaseous stream on their outside – so again an essentially uniform temperature T_1 (although varying in time) will hold. A graphical interpretation of the underlying problem is provided in Fig. 1.47. The associated model is equivalent to a batch stirred tank – corresponding to each can/jar and its food content; placed inside another batch stirred tank – accounted for by the inner fluid in the retort. The heating/cooling utility is, in turn, continuously flown through the casing of the latter, and will be assumed to exchange enthalpy only in the form of latent heat (i.e. $T_{h\backslash c,in}=T_{h\backslash c,out}\equiv T_{h\backslash c}$) – since the high target temperatures normally require use of steam.

An enthalpy balance consistent with Fig. 1.47 reads

$$U_1A_1\left(T_2-T_1\right)=V_1\rho_1c_{P,1}\frac{dT_1}{dt}, \tag{1.545}$$

encompassing the (packaged) food item – where U_1 denotes overall heat transfer coefficient between retort fluid and food, A_1 and V_1 denote outer area and total volume of food, respectively, and ρ_1 and $c_{P,1}$ denote mass density and specific heat capacity of food; the left-hand side of Eq. (1.545) accordingly represents the rate of enthalpy input by conduction to the food, whereas the right-hand side represents the rate of accumulation of enthalpy therein. By the same token, an enthalpy balance to the retort contents takes the form

$$U_2A_2\left(T_{h\backslash c}-T_2\right)=U_1A_1\left(T_2-T_1\right)+V_2\rho_2c_{P,2}\frac{dT_2}{dt}, \tag{1.546}$$

where U_2 denotes overall heat transfer coefficient between heating/cooling utility and retort fluid, A_2 and V_2 denote outer area and total volume of retort, respectively, and ρ_2 and $c_{P,2}$ denote mass density and specific heat capacity, respectively, of retort fluid – with $T_{h\backslash c}$ denoting temperature of heating/cooling utility. Hence, the left-hand side of Eq. (1.546) represents the rate of enthalpy input by conduction to the fluid in the retort – whereas the first term in the right-hand side represents the rate of enthalpy output by conduction to the food, and the second term represents the rate of accumulation of enthalpy in the said retort fluid.

If steady-state conditions prevailed (i.e. $dT_1/dt=dT_2/dt=0$), then Eq. (1.545) would reduce to

$$U_1A_1\left(T_{2,0}-T_{1,0}\right)=0, \tag{1.547}$$

and Eq. (1.546) would likewise break down to

$$U_2A_2\left(T_{h\backslash c,0}-T_{2,0}\right)=U_1A_1\left(T_{2,0}-T_{1,0}\right); \tag{1.548}$$

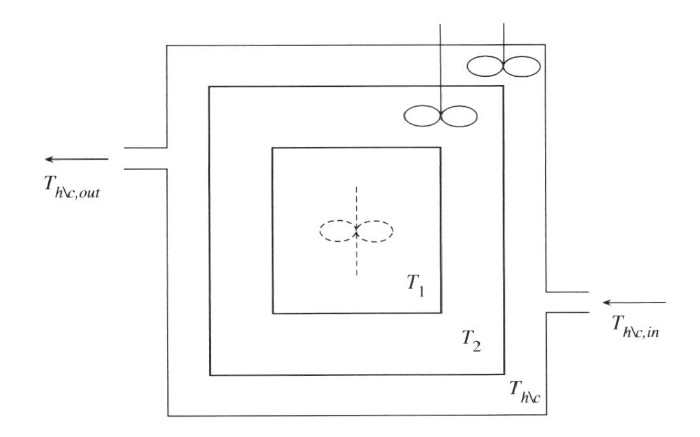

FIGURE 1.47 Schematic representation of pasteurizer/sterilizer, operated with indirect heating/cooling by enthalpic fluid utility at temperature $T_{h\backslash c}$, putatively dropping/jumping from $T_{h\backslash c,in}$ at entrance to $T_{h\backslash c}\equiv T_{h\backslash c,out}$ thereafter – with temperature T_2 prevailing inside the retort, and temperature T_1 of food to be processed.

subscript 0 refers to steady-state conditions, assumed to eventually hold prior to startup. Ordered addition of Eqs. (1.547) and (1.548) yields

$$U_2A_2\left(T_{h\backslash c,0}-T_{2,0}\right)=0, \quad (1.549)$$

so one readily concludes that

$$T_{2,0}=T_{h\backslash c,0}; \quad (1.550)$$

insertion of Eq. (1.550) transforms Eq. (1.547) to

$$U_1A_1\left(T_{h\backslash c,0}-T_{1,0}\right)=0, \quad (1.551)$$

which implies, in turn,

$$T_{1,0}=T_{h\backslash c,0}. \quad (1.552)$$

Inspection of Eqs. (1.550) and (1.552) indicates that at (the original) steady state, both retort fluid and food share the temperature of the heating/cooling fluid, i.e. $T_{h\backslash c,0}$ (or room temperature, as usual).

On the other hand, one may write

$$\boxed{T_1\big|_{t=0}=T_{h\backslash c,0}} \quad (1.553)$$

as per Eq. (1.552), and similarly

$$\boxed{T_2\big|_{t=0}=T_{h\backslash c,0}} \quad (1.554)$$

based on Eq. (1.550) – and then use these results as initial conditions to integrate Eqs. (1.545) and (1.546), respectively. After abstracting Eq. (1.547) from Eq. (1.545), one gets

$$U_1A_1\left(T_2-T_1\right)-U_1A_1\left(T_{2,0}-T_{1,0}\right)=V_1\rho_1c_{P,1}\frac{dT_1}{dt}, \quad (1.555)$$

where U_1A_1 may be factored out in the left-hand side and constant $-T_{1,0}$ inserted under the derivative sign to produce

$$U_1A_1\left(T_2-T_{2,0}-\left(T_1-T_{1,0}\right)\right)=V_1\rho_1c_{P,1}\frac{d\left(T_1-T_{1,0}\right)}{dt}. \quad (1.556)$$

For convenience of algebraic manipulation hereafter, deviation variables will be defined as

$$\hat{T}_1\equiv T_1-T_{1,0} \quad (1.557)$$

and

$$\hat{T}_2\equiv T_2-T_{2,0}, \quad (1.558)$$

thus allowing Eq. (1.556) be reformulated to

$$U_1A_1\left(\hat{T}_2-\hat{T}_1\right)=V_1\rho_1c_{P,1}\frac{d\hat{T}_1}{dt} \quad (1.559)$$

in a simpler notation; upon division of both sides by U_1A_1, one obtains

$$\boxed{\zeta_1\frac{d\hat{T}_1}{dt}=\hat{T}_2-\hat{T}_1} \quad (1.560)$$

from Eq. (1.559) – provided that parameter ζ_1 is defined as

$$\zeta_1\equiv\frac{V_1\rho_1c_{P,1}}{U_1A_1}. \quad (1.561)$$

After having rewritten Eq. (1.561) as

$$\zeta_1=\frac{\dfrac{V_1}{A_1}}{\dfrac{U_1}{\rho_1c_{P,1}}}, \quad (1.562)$$

one may view ζ_1 – or thermal inertia of food (i.e. $V_1\rho_1c_{P,1}$) normalized by thermal conductance (i.e. U_1A_1), as the ratio of a length (encompassing two geometric characteristics of the system), V_1/A_1, to a pseudo thermal diffusivity (encompassing a set of physicochemical characteristics of the system), $U_1/\rho_1c_{P,1}$.

A similar subtraction of Eq. (1.548) from Eq. (1.546) unfolds

$$\begin{aligned}&U_2A_2\left(T_{h\backslash c}-T_2\right)-U_2A_2\left(T_{h\backslash c,0}-T_{2,0}\right)\\&=U_1A_1\left(T_2-T_1\right)-U_1A_1\left(T_{2,0}-T_{1,0}\right)+V_2\rho_2c_{P,2}\frac{dT_2}{dt},\end{aligned} \quad (1.563)$$

where U_1A_1 and U_2A_2 may be factored out to generate

$$\begin{aligned}&U_2A_2\left(T_{h\backslash c}-T_{h\backslash c,0}-\left(T_2-T_{2,0}\right)\right)\\&=U_1A_1\left(T_2-T_{2,0}-\left(T_1-T_{1,0}\right)\right)+V_2\rho_2c_{P,2}\frac{d\left(T_2-T_{2,0}\right)}{dt}\end{aligned} \quad (1.564)$$

while taking advantage of $d(T_2-T_{2,0})$ coinciding with dT_2 due to the constancy of $T_{2,0}$; Eq. (1.564) translates to

$$U_2A_2\left(\hat{T}_{h\backslash c}-\hat{T}_2\right)=U_1A_1\left(\hat{T}_2-\hat{T}_1\right)+V_2\rho_2c_{P,2}\frac{d\hat{T}_2}{dt} \quad (1.565)$$

with the aid of Eqs. (1.557) and (1.558) – coupled with

$$\hat{T}_{h\backslash c}\equiv T_{h\backslash c}-T_{h\backslash c,0}, \quad (1.566)$$

consistent with the said equations. Division of both sides of Eq. (1.565) by U_2A_2 leads to

$$\boxed{\zeta_2\frac{d\hat{T}_2}{dt}=\hat{T}_{h\backslash c}-\hat{T}_2-\xi_{12}\left(\hat{T}_2-\hat{T}_1\right)} \quad (1.567)$$

as long as parameter ζ_2 is defined as

$$\zeta_2\equiv\frac{V_2\rho_2c_{P,2}}{U_2A_2} \quad (1.568)$$

in parallel to Eq. (1.561), and parameter ξ_{12} satisfies

$$\xi_{12}\equiv\frac{U_1A_1}{U_2A_2}. \quad (1.569)$$

After rearranging Eq. (1.569) to

$$\xi_{12} = \frac{A_1}{A_2}\frac{U_1}{U_2}, \tag{1.570}$$

one may view (dimensionless) ξ_{12} as the product of a geometric parameter (encompassing two geometric characteristics of the system), A_1/A_2, by a physicochemical parameter (encompassing thermal kinetic characteristics of the system), U_1/U_2. The initial conditions relevant for Eqs. (1.560) and (1.567) may be obtained from Eqs. (1.553) and (1.554) as

$$\boxed{\hat{T}_1\Big|_{t=0} = 0} \tag{1.571}$$

and

$$\boxed{\hat{T}_2\Big|_{t=0} = 0,} \tag{1.572}$$

respectively – with the aid of Eqs. (1.550), (1.552), (1.557), and (1.558).

Solution of the set of differential equations labeled as Eqs. (1.560) and (1.567) may to advantage be pursued in Laplace's domain, according to

$$\zeta_1\left(s\overline{T}_1 - \hat{T}_1\Big|_{t=0}\right) = \overline{T}_2 - \overline{T}_1 \tag{1.573}$$

and

$$\zeta_2\left(s\overline{T}_2 - \hat{T}_2\Big|_{t=0}\right) = \overline{T}_{h\backslash c} - \overline{T}_2 - \xi_{12}\left(\overline{T}_2 - \overline{T}_1\right), \tag{1.574}$$

respectively; after recalling Eq. (1.571), it is possible to simplify Eq. (1.573) to

$$\zeta_1 s\overline{T}_1 = \overline{T}_2 - \overline{T}_1, \tag{1.575}$$

and Eq. (1.574) may likewise be reformulated to

$$\zeta_2 s\overline{T}_2 = \overline{T}_{h\backslash c} - \overline{T}_2 - \xi_{12}\overline{T}_2 + \xi_{12}\overline{T}_1 \tag{1.576}$$

at the expense of Eq. (1.572), complemented by elimination of parenthesis. Condensation of terms alike supports transformation of Eqs. (1.575) and (1.576) to

$$\boxed{\left(1+\zeta_1 s\right)\overline{T}_1 - \overline{T}_2 = 0} \tag{1.577}$$

and

$$\boxed{\xi_{12}\overline{T}_1 - \left(1+\xi_{12}+\zeta_2 s\right)\overline{T}_2 = -\overline{T}_{h\backslash c},} \tag{1.578}$$

respectively; solution of the system of linear algebraic equations in $\overline{T}_1$ and $\overline{T}_2$, with s serving as independent variable – labeled as Eqs. (1.577) and (1.578), will take the form

$$\overline{T}_1 = \frac{\begin{vmatrix} 0 & -1 \\ -\overline{T}_{h\backslash c} & -\left(1+\xi_{12}+\zeta_2 s\right) \end{vmatrix}}{\begin{vmatrix} 1+\zeta_1 s & -1 \\ \xi_{12} & -\left(1+\xi_{12}+\zeta_2 s\right) \end{vmatrix}} \tag{1.579}$$

and

$$\overline{T}_2 = \frac{\begin{vmatrix} 1+\zeta_1 s & 0 \\ \xi_{12} & -\overline{T}_{h\backslash c} \end{vmatrix}}{\begin{vmatrix} 1+\zeta_1 s & -1 \\ \xi_{12} & -\left(1+\xi_{12}+\zeta_2 s\right) \end{vmatrix}}, \tag{1.580}$$

at the expense of Cramer's rule. Recalling the definition of a second-order determinant, one may redo Eq. (1.579) to

$$\overline{T}_1 = \frac{-\overline{T}_{h\backslash c}}{-\left(1+\zeta_1 s\right)\left(1+\xi_{12}+\zeta_2 s\right)+\xi_{12}} \tag{1.581}$$

or, after taking negatives of both numerator and denominator,

$$\boxed{\overline{T}_1 = \frac{1}{\left(1+\zeta_1 s\right)\left(1+\xi_{12}+\zeta_2 s\right)-\xi_{12}}\overline{T}_{h\backslash c};} \tag{1.582}$$

by the same token, Eq. (1.580) becomes

$$\overline{T}_2 = \frac{-\overline{T}_{h\backslash c}\left(1+\zeta_1 s\right)}{-\left(1+\zeta_1 s\right)\left(1+\xi_{12}+\zeta_2 s\right)+\xi_{12}} \tag{1.583}$$

that is equivalent to

$$\boxed{\overline{T}_2 = \frac{1+\zeta_1 s}{\left(1+\zeta_1 s\right)\left(1+\xi_{12}+\zeta_2 s\right)-\xi_{12}}\overline{T}_{h\backslash c},} \tag{1.584}$$

upon similar algebraic rearrangement.

The first (and more relevant) stage of pasteurization/sterilization is heating – with a pattern often described by

$$\boxed{T_h = T_{wst}\mathrm{H}\{t\}} \tag{1.585}$$

and, consequently,

$$\hat{T}_h = \hat{T}_{wst} \equiv T_{wst} - T_{h,0} \tag{1.586}$$

for $t>0$, in agreement with Eq. (1.566); here T_{wst} denotes (constant) temperature of water steam used as enthalpic utility to heat the food system. The counterpart of Eq. (1.586), in Laplace's domain, looks like

$$\overline{T}_h = \frac{\hat{T}_{wst}}{s}, \tag{1.587}$$

so Eq. (1.582) becomes

$$\overline{T}_1 = \frac{\hat{T}_{wst}}{s\left(\left(1+\zeta_1 s\right)\left(1+\xi_{12}+\zeta_2 s\right)-\xi_{12}\right)} \tag{1.588}$$

or, after eliminating inner parentheses and lumping similar terms,

$$\overline{T}_1 = \frac{\hat{T}_{wst}}{s\left(1+\left(\zeta_2+\zeta_1\left(1+\xi_{12}\right)\right)s+\zeta_1\zeta_2 s^2\right)}. \tag{1.589}$$

The right-hand side of Eq. (1.589) can be decomposed as a sum of three partial fractions, viz.

$$\frac{\hat{T}_{wst}}{\zeta_1\zeta_2 s(s-s_1)(s-s_2)}=\frac{b_1}{s}+\frac{b_2}{s-s_1}+\frac{b_3}{s-s_2} \quad (1.590)$$

with the aid of the rule of factorization of a polynomial, coupled with Heaviside's expansion of a (regular) rational fraction as partial fractions – where b_1, b_2, and b_3 denote constants to be calculated, and s_1 and s_2 denote the two real, distinct solutions of

$$1+\left(\zeta_2+\zeta_1\left(1+\xi_{12}\right)\right)s+\zeta_1\zeta_2 s^2=0. \quad (1.591)$$

Elimination of denominators supports transformation of Eq. (1.590) to

$$b_1(s-s_1)(s-s_2)+b_2 s(s-s_2)+b_3 s(s-s_1)=\frac{\hat{T}_{wst}}{\zeta_1\zeta_2}, \quad (1.592)$$

where setting $s=0$ permits simplification to

$$b_1(-s_1)(-s_2)=\frac{\hat{T}_{wst}}{\zeta_1\zeta_2}; \quad (1.593)$$

b_1 may thus be obtained as

$$b_1=\frac{\hat{T}_{wst}}{\zeta_1\zeta_2 s_1 s_2}. \quad (1.594)$$

Since the product of the two roots of a quadratic equation equals the ratio of its zero-th order to its second-order coefficient, one may reformulate Eq. (1.594) to

$$b_1=\hat{T}_{wst}, \quad (1.595)$$

based on the functional form of Eq. (1.591). If s were instead made equal to s_1, then Eq. (1.592) would have led to

$$b_2 s_1(s_1-s_2)=\frac{\hat{T}_{wst}}{\zeta_1\zeta_1} \quad (1.596)$$

– or, equivalently,

$$b_2=\frac{\hat{T}_{wst}}{\zeta_1\zeta_1 s_1(s_1-s_2)}; \quad (1.597)$$

finally, one may set $s=s_2$ to get

$$b_3 s_2(s_2-s_1)=\frac{\hat{T}_{wst}}{\zeta_1\zeta_2} \quad (1.598)$$

from Eq. (1.592), where isolation of b_3 is possible as

$$b_3=\frac{\hat{T}_{wst}}{\zeta_2\zeta_2 s_2(s_2-s_1)}. \quad (1.599)$$

Once in possession of constants b_1, b_2, and b_3 as per Eqs. (1.594), (1.597), and (1.599), one can rewrite Eqs.(1.589) and (1.590) as

$$\frac{\hat{T}_{wst}}{s\left(1+\left(\zeta_2+\zeta_1\left(1+\xi_{12}\right)\right)s+\zeta_1\zeta_2 s^2\right)}=\frac{\frac{\hat{T}_{wst}}{\zeta_1\zeta_2 s_1 s_2}}{s}+\frac{\frac{\hat{T}_{wst}}{\zeta_1\zeta_2 s_1(s_1-s_2)}}{s-s_1}+\frac{\frac{\hat{T}_{wst}}{\zeta_1\zeta_2 s_2(s_2-s_1)}}{s-s_2}; \quad (1.600)$$

hence, Eq. (1.589) will read

$$\bar{T}_1=\frac{\hat{T}_{wst}}{\zeta_1\zeta_2}\left(\frac{\frac{1}{s_1 s_2}}{s}+\frac{\frac{1}{s_1(s_1-s_2)}}{s-s_1}+\frac{\frac{1}{s_2(s_2-s_1)}}{s-s_2}\right) \quad (1.601)$$

after $\hat{T}_{wst}/\zeta_1\zeta_2$ has been factored out. Reversal to the time domain may now proceed at the expense of the theorem of translation in Laplace's domain and the linearity of Laplace's transform, i.e.

$$\hat{T}_1=\frac{\hat{T}_{wst}}{\zeta_1\zeta_2}\left(\frac{1}{s_1 s_2}\mathcal{L}^{-1}\left(\frac{1}{s}\right)+\frac{e^{s_1 t}}{s_1(s_1-s_2)}\mathcal{L}^{-1}\left(\frac{1}{s}\right)+\frac{e^{s_2 t}}{s_2(s_2-s_1)}\mathcal{L}^{-1}\left(\frac{1}{s}\right)\right) \quad (1.602)$$

or, equivalently,

$$T_1\{t\}-T_{1,0}=\frac{T_{wst}-T_{1,0}}{\zeta_1\zeta_2}\left(\frac{1}{s_1 s_2}+\frac{e^{s_1 t}}{s_1(s_1-s_2)}+\frac{e^{s_2 t}}{s_2(s_2-s_1)}\right)\mathrm{H}\{t\} \quad (1.603)$$

upon explicitation of the stated inverse Laplace's transforms, and retrieval of Eqs. (1.552), (1.557), and (1.586); division of both sides by T_{wst}–$T_{1,0}$ finally gives

$$\boxed{\frac{T_1-T_{1,0}}{T_{wst}-T_{1,0}}=\frac{1}{\zeta_1\zeta_2}\left(\frac{1}{s_1 s_2}+\frac{e^{s_1 t}}{s_1(s_1-s_2)}+\frac{e^{s_2 t}}{s_2(s_2-s_1)}\right)\mathrm{H}\{t\}, \quad 0\le t\le \Delta t_h} \quad (1.604)$$

where Δt_h denotes again the full duration of the heating period. Equation (1.604) is graphically plotted in Fig. 1.48, as $T_1 \equiv T_1\{t\}$ – where it describes the rising portion of the curve, on the left side. There is a second-order behavior for $T_1\{t\}$, from the departing value, $T_{1,0}$, toward the ultimate value, T_{wst}, approached in an asymptotic mode – and going through an inflection point somewhere in between. In fact, for a sufficiently long time, one may recall the final value theorem to write

$$\lim_{t\to\infty}\hat{T}_1=\lim_{s\to 0}s\bar{T}_1, \quad (1.605)$$

and accordingly obtain

$$\lim_{t\to\infty}\hat{T}_1=\frac{\hat{T}_{wst}}{\zeta_1\zeta_2}\lim_{s\to 0}\left(\frac{1}{s_1 s_2}+\frac{1}{s_1(s_1-s_2)}\frac{s}{s-s_1}+\frac{1}{s_2(s_2-s_1)}\frac{s}{s-s_2}\right) \quad (1.606)$$

from Eq. (1.601); meanwhile, Eq. (1.606) degenerates to

$$\lim_{t\to\infty}\hat{T}_1=\frac{\hat{T}_{wst}}{\zeta_1\zeta_2 s_1 s_2} \quad (1.607)$$

due to $s=0$, so the $\hat{T}_1$ vs. t curve will eventually plateau at $\hat{T}_{wst}/\zeta_1\zeta_2 s_1 s_2$. Once again, one may resort to the general properties of a quadratic equation in the canonical form labeled as Eq. (1.591), viz.

$$s_1 s_2 = \frac{1}{\zeta_1 \zeta_2}; \tag{1.608}$$

insertion of Eq. (1.608) allows simplification of Eq. (1.607) to

$$\lim_{t \to \infty} \hat{T}_1 = \hat{T}_{wst} \tag{1.609}$$

that becomes merely

$$\lim_{t \to \infty} T_1 = T_{wst} \tag{1.610}$$

in view of Eqs. (1.552), (1.557), and (1.586) – and after taking $T_{h,0}$ off both sides. Equation (1.610) confirms that the food will eventually reach the temperature of the enthalpic utility employed, as suspected from inspection of Fig. 1.48.

Remember that the solving formula for a quadratic equation supports

$$s_1, s_2 = \frac{-\left(\zeta_2 + \zeta_1\left(1+\xi_{12}\right)\right) \pm \sqrt{\left(\zeta_2 + \zeta_1\left(1+\xi_{12}\right)\right)^2 - 4\zeta_1\zeta_2}}{2\zeta_1\zeta_2}, \tag{1.611}$$

following application to Eq. (1.591); after algebraically rearranging the square under the root sign as

$$\left(\zeta_2 + \zeta_1\left(1+\xi_{12}\right)\right)^2 - 4\zeta_1\zeta_2 = \left(\left(\zeta_1 + \zeta_2\right) + \zeta_1\xi_{12}\right)^2 - 4\zeta_1\zeta_2, \tag{1.612}$$

one may invoke Newton's binomial to write

$$\left(\zeta_2 + \zeta_1\left(1+\xi_{12}\right)\right)^2 - 4\zeta_1\zeta_2 = \left(\zeta_1 + \zeta_2\right)^2 + 2\left(\zeta_1 + \zeta_2\right)\zeta_1\xi_{12} + \zeta_1^2\xi_{12}^2 - 4\zeta_1\zeta_2. \tag{1.613}$$

A second application of Newton's binomial allows transformation of Eq. (1.613) to

$$\left(\zeta_2 + \zeta_1\left(1+\xi_{12}\right)\right)^2 - 4\zeta_1\zeta_2 = \zeta_1^2 + 2\zeta_1\zeta_2 + \zeta_2^2 + 2\left(\zeta_1 + \zeta_2\right)\zeta_1\xi_{12} + \zeta_1^2\xi_{12}^2 - 4\zeta_1\zeta_2, \tag{1.614}$$

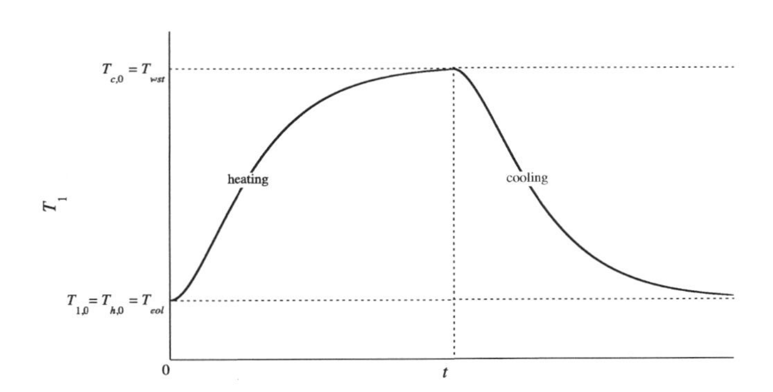

FIGURE 1.48 Evolution of temperature of food, T_1, versus time, t, in pasteurization/sterilization, for heating up from temperature $T_{1,0} = T_{h,0}$ via water steam at temperature T_{wst}, followed by cooling down from temperature $T_{wst} = T_{c,0}$ via excess water/air at temperature $T_{col} = T_{h,0}$, under negligible transfer of sensible heat by heating/cooling fluid utility.

where condensation of terms alike generates

$$\left(\zeta_2 + \zeta_1\left(1+\xi_{12}\right)\right)^2 - 4\zeta_1\zeta_2 = \left(\zeta_1^2 - 2\zeta_1\zeta_2 + \zeta_2^2\right) + 2\left(\zeta_1 + \zeta_2\right)\zeta_1\xi_{12} + \zeta_1^2\xi_{12}^2; \tag{1.615}$$

since the first parenthesis in the right-hand side represents expansion of the square of a difference as per Newton's rule, one may redo Eq. (1.615) to

$$\left(\zeta_2 + \zeta_1\left(1+\xi_{12}\right)\right)^2 - 4\zeta_1\zeta_2 = \left(\zeta_1 - \zeta_2\right)^2 + 2\left(\zeta_1 + \zeta_2\right)\zeta_1\xi_{12} + \zeta_1^2\xi_{12}^2, \tag{1.616}$$

where addition and subtraction of $2(\zeta_1 - \zeta_2)\zeta_1\xi_{12}$ lead to

$$\left(\zeta_2 + \zeta_1\left(1+\xi_{12}\right)\right)^2 - 4\zeta_1\zeta_2 = \left(\left(\zeta_1 - \zeta_2\right)^2 + 2\left(\zeta_1 - \zeta_2\right)\zeta_1\xi_{12} + \zeta_1^2\xi_{12}^2\right) + 2\left(\zeta_1 + \zeta_2\right)\zeta_1\xi_{12} - 2\left(\zeta_1 - \zeta_2\right)\zeta_1\xi_{12}. \tag{1.617}$$

The first three terms in the right-hand side of Eq. (1.617) constitute, as a whole, the expansion of a square, viz.

$$\left(\zeta_2 + \zeta_1\left(1+\xi_{12}\right)\right)^2 - 4\zeta_1\zeta_2 = \left(\left(\zeta_1 - \zeta_2\right) + \zeta_1\xi_{12}\right)^2 + 2\left(\zeta_1 + \zeta_2\right)\zeta_1\xi_{12} - 2\left(\zeta_1 - \zeta_2\right)\zeta_1\xi_{12}, \tag{1.618}$$

whereas elimination of the first inner parenthesis, followed by factoring out of ζ_1 or $2\zeta_1\xi_{12}$ (as appropriate) yield

$$\left(\zeta_2 + \zeta_1\left(1+\xi_{12}\right)\right)^2 - 4\zeta_1\zeta_2 = \left(\zeta_1\left(1+\xi_{12}\right) - \zeta_2\right)^2 + 2\left(\zeta_1 + \zeta_2 - \zeta_1 + \zeta_2\right)\zeta_1\xi_{12}; \tag{1.619}$$

upon cancellation of symmetrical terms, Eq. (1.619) reduces to

$$\left(\zeta_2 + \zeta_1\left(1+\xi_{12}\right)\right)^2 - 4\zeta_1\zeta_2 = \left(\zeta_1\left(1+\xi_{12}\right) - \zeta_2\right)^2 + 4\zeta_1\zeta_2\xi_{12}. \tag{1.620}$$

With regard to the right-hand side of Eq. (1.620), one realizes that

$$\left(\zeta_1\left(1+\xi_{12}\right) - \zeta_2\right)^2 + 4\zeta_1\zeta_2\xi_{12} > 0, \tag{1.621}$$

even if $\zeta_1(1+\xi_{12})-\zeta_2<0$ for its being squared, and because $4\zeta_1\zeta_2\xi_{12}>0$ as per Eqs. (1.561), (1.568) and (1.569); therefore,

$$\left(\zeta_2 + \zeta_1\left(1+\xi_{12}\right)\right)^2 - 4\zeta_1\zeta_2 > 0 \tag{1.622}$$

based on Eq. (1.620), and the square root in Eq. (1.611) will necessarily hold only real values. In addition,

$$\left(\zeta_2+\zeta_1\left(1+\xi_{12}\right)\right)^2-4\zeta_1\zeta_2<\left(\zeta_2+\zeta_1\left(1+\xi_{12}\right)\right)^2, \tag{1.623}$$

since $\zeta_1>0$ and $\zeta_2>0$ as just seen – which, together with $\xi_{12}>0$, imply

$$\begin{aligned}&\sqrt{\left(\zeta_2+\zeta_1\left(1+\xi_{12}\right)\right)^2-4\zeta_1\zeta_2}<\sqrt{\left(\zeta_2+\zeta_1\left(1+\xi_{12}\right)\right)^2}\\&=\left|\zeta_2+\zeta_1\left(1+\xi_{12}\right)\right|=\zeta_2+\zeta_1\left(1+\xi_{12}\right).\end{aligned} \tag{1.624}$$

The result conveyed by Eq. (1.624) means that s_1 – corresponding to the minus sign preceding the square root in Eq. (1.611), and s_2 – corresponding to the plus sign, abide to

$$s_1<s_2<0 \tag{1.625}$$

because $\zeta_1\zeta_2>0$; hence, the arguments of the two exponential functions in Eq. (1.604) are negative, and both accordingly decay to zero as time elapses. Furthermore, Eq. (1.625) implies that

$$\frac{1}{s_1\left(s_1-s_2\right)}>0 \tag{1.626}$$

and

$$\frac{1}{s_2\left(s_2-s_1\right)}<0, \tag{1.627}$$

so the first pre-exponential constant in Eq. (1.604) is positive and the other is negative; this justifies the sigmoidal behavior of the left portion of the T_1 vs. t curve, apparent in Fig. 1.48.

In attempts to ascertain $T_2\{t\}$, one may redo the reasoning above – but now departing from Eqs. (1.584) and (1.587), according to

$$\overline{T}_2=\frac{\hat{T}_{wst}\left(1+\zeta_1 s\right)}{s\left(\left(1+\zeta_1 s\right)\left(1+\xi_{12}+\zeta_2 s\right)-\xi_{12}\right)} \tag{1.628}$$

with $\overline{T}_{h\backslash c}\equiv\overline{T}_h$ once more – or, upon elimination of parenthesis in denominator followed by condensation of terms alike,

$$\overline{T}_2=\frac{\hat{T}_{wst}\left(1+\zeta_1 s\right)}{s\left(1+\left(\zeta_2+\zeta_1\left(1+\xi_{12}\right)\right)s+\zeta_1\zeta_2 s^2\right)}. \tag{1.629}$$

Decomposition of the right-hand side of Eq. (1.629) as partial fractions unfolds

$$\frac{\hat{T}_{wst}\left(1+\zeta_1 s\right)}{\zeta_1\zeta_2 s\left(s-s_1\right)\left(s-s_2\right)}=\frac{b_4}{s}+\frac{b_5}{s-s_1}+\frac{b_6}{s-s_2} \tag{1.630}$$

– where b_4, b_5, and b_6 denote constants, yet to be calculated; elimination of denominators allows transformation of Eq. (1.630) to

$$b_4\left(s-s_1\right)\left(s-s_2\right)+b_5 s\left(s-s_2\right)+b_6 s\left(s-s_1\right)=\frac{\hat{T}_{wst}\left(1+\zeta_1 s\right)}{\zeta_1\zeta_2}, \tag{1.631}$$

while setting $s=0$ permits simplification

$$b_4\left(-s_1\right)\left(-s_2\right)=\frac{\hat{T}_{wst}}{\zeta_1\zeta_2} \tag{1.632}$$

or, upon isolation of b_4,

$$b_4=\frac{\hat{T}_{wst}}{\zeta_1\zeta_1 s_1 s_2}. \tag{1.633}$$

By the same token, $s=s_1$ simplifies Eq. (1.631) to

$$b_5 s_1\left(s_1-s_2\right)=\frac{\hat{T}_{wst}\left(1+\zeta_1 s_1\right)}{\zeta_1\zeta_2}, \tag{1.634}$$

or else

$$b_5=\frac{\hat{T}_{wst}\left(1+\zeta_1 s_1\right)}{\zeta_1\zeta_2 s_1\left(s_1-s_2\right)} \tag{1.635}$$

after solving for b_5; one may finally set $s=s_2$ – in which case Eq. (1.631) turns to

$$b_6 s_2\left(s_2-s_1\right)=\frac{\hat{T}_{wst}\left(1+\zeta_1 s_2\right)}{\zeta_1\zeta_2}, \tag{1.636}$$

so b_6 ends up as

$$b_6=\frac{\hat{T}_{wst}\left(1+\zeta_1 s_2\right)}{\zeta_1\zeta_2 s_2\left(s_2-s_1\right)}. \tag{1.637}$$

Equations (1.633), (1.635), and (1.637) permit reformulation of Eqs. (1.629) and (1.630) to

$$\begin{aligned}\frac{\hat{T}_{wst}\left(1+\zeta_1 s\right)}{\zeta_1\zeta_2 s\left(s-s_1\right)\left(s-s_2\right)}&=\frac{\dfrac{\hat{T}_{wst}}{\zeta_1\zeta_2 s_1 s_2}}{s}+\frac{\dfrac{\hat{T}_{wst}\left(1+\zeta_1 s_1\right)}{\zeta_1\zeta_2 s_1\left(s_1-s_2\right)}}{s-s_1}\\&+\frac{\dfrac{\hat{T}_{wst}\left(1+\zeta_1 s_2\right)}{\zeta_1\zeta_2 s_2\left(s_2-s_1\right)}}{s-s_2},\end{aligned} \tag{1.638}$$

where insertion in Eq. (1.629), along with factoring out of $\hat{T}_{wst}/\zeta_1\zeta_2$ yield

$$\overline{T}_2=\frac{\hat{T}_{wst}}{\zeta_1\zeta_2}\left(\frac{\dfrac{1}{s_1 s_2}}{s}+\frac{\dfrac{1+\zeta_1 s_1}{s_1\left(s_1-s_2\right)}}{s-s_1}+\frac{\dfrac{1+\zeta_1 s_2}{s_2\left(s_2-s_1\right)}}{s-s_2}\right); \tag{1.639}$$

reversal of Laplace's transform, with the aid of the theorem of translation in Laplace's domain generate

$$\frac{\hat{T}_2\{t\}}{\hat{T}_{wst}}=\frac{1}{\zeta_1\zeta_2}\left(\begin{aligned}&\frac{1}{s_1 s_2}\mathcal{L}^{-1}\left(\frac{1}{s}\right)+\frac{1+\zeta_1 s_1}{s_1\left(s_1-s_2\right)}e^{s_1 t}\mathcal{L}^{-1}\left(\frac{1}{s}\right)\\&+\frac{1+\zeta_1 s_2}{s_2\left(s_2-s_1\right)}e^{s_2 t}\mathcal{L}^{-1}\left(\frac{1}{s}\right)\end{aligned}\right), \tag{1.640}$$

along with division of both sides by $\hat{T}_{wst}$. Finalization of the reversal procedure in Eq. (1.640) produces

$$\boxed{\begin{aligned}&\frac{T_2 - T_{2,0}}{T_{wst} - T_{2,0}}\\&= \frac{1}{\zeta_1\zeta_2}\left(\frac{1}{s_1 s_2} + \frac{1+\zeta_1 s_1}{s_1(s_1 - s_2)}e^{s_1 t} + \frac{1+\zeta_1 s_2}{s_2(s_2 - s_1)}e^{s_2 t}\right)\mathrm{H}\{t\},\\&0 \le t \le \Delta t_h\end{aligned}} \tag{1.641}$$

with the help of Eqs. (1.550), (1.558), and (1.586); the curve T_2 vs. t is plotted as a thick solid line in Fig. 1.49, and accounts for its ascending portion – where the curve T_1 vs. t has been overlaid as a thin solid line to facilitate comparison. The dominant pattern observed is similar to that of a first-order behavior; the sigmoidicity observed in the T_1 vs. t curve is no longer present.

To ascertain the sign of $1+\zeta_1 s_1$, one may resort to

$$1+\zeta_1 s_1 = 1 - \frac{\sqrt{\left(\zeta_1(1+\xi_{12}) - \zeta_2\right)^2 + 4\zeta_1\zeta_2\xi_{12}} + \zeta_2 + \zeta_1(1+\xi_{12})}{2\zeta_2} \tag{1.642}$$

based on Eqs. (1.611) and (1.620) – pertaining to calculation of the roots of Eq. (1.591), and also of the poles of the right-hand side of Eq. (1.629); one may likewise write

$$1+\zeta_1 s_2 = 1 + \frac{\sqrt{\left(\zeta_1(1+\xi_{12}) - \zeta_2\right)^2 + 4\zeta_1\zeta_2\xi_{12}} - \zeta_2 - \zeta_1(1+\xi_{12})}{2\zeta_2}, \tag{1.643}$$

bearing now in mind the plus sign preceding the square root in Eq. (1.611), besides Eq. (1.620). Elimination of denominators leads to

$$1+\zeta_1 s_1 = \frac{\left(\begin{array}{c}2\zeta_2 - \zeta_2 - \zeta_1(1+\xi_{12})\\ -\sqrt{\left(\zeta_1(1+\xi_{12}) - \zeta_2\right)^2 + 4\zeta_1\zeta_2\xi_{12}}\end{array}\right)}{2\zeta_2} \tag{1.644}$$

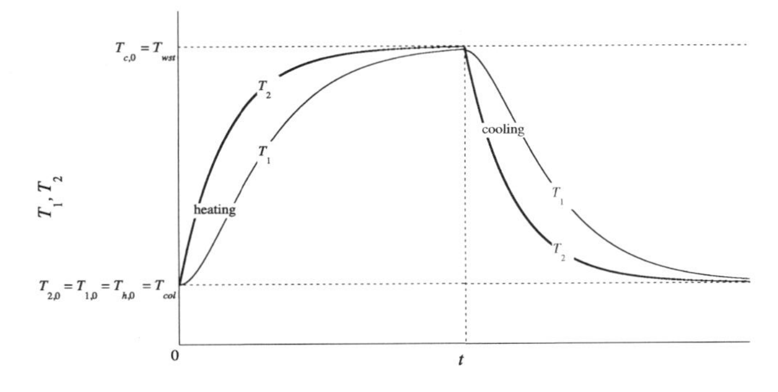

FIGURE 1.49 Evolution of temperature of retort fluid, T_2 (——), superimposed on temperature of food, T_1 (——), versus time, t, in pasteurization/sterilization, for heating up from temperature $T_{1,0}=T_{2,0}=T_{h,0}$, via water steam at temperature T_{wst}, followed by cooling down from temperature $T_{wst}=T_{c,0}$, via excess water/air at temperature $T_{col}=T_{h,0}$, under negligible transfer of sensible heat by heating/cooling fluid utility.

stemming from Eq. (1.642); and likewise

$$1+\zeta_1 s_2 = \frac{\left(\begin{array}{c}2\zeta_2 - \zeta_2 - \zeta_1(1+\xi_{12})\\ +\sqrt{\left(\zeta_1(1+\xi_{12}) - \zeta_2\right)^2 + 4\zeta_1\zeta_2\xi_{12}}\end{array}\right)}{2\zeta_2} \tag{1.645}$$

arising from Eq. (1.643). Upon lumping terms alike and taking negatives of both sides afterward, one gets

$$-(1+\zeta_1 s_1) = \frac{\sqrt{\left(\zeta_1(1+\xi_{12}) - \zeta_2\right)^2 + 4\zeta_1\zeta_2\xi_{12}} + \zeta_1(1+\xi_{12}) - \zeta_2}{2\zeta_2} \tag{1.646}$$

from Eq. (1.644), whereas Eq. (1.645) generates

$$1+\zeta_1 s_2 = \frac{\sqrt{\left(\zeta_1(1+\xi_{12}) - \zeta_2\right)^2 + 4\zeta_1\zeta_2\xi_{12}} - \left(\zeta_1(1+\xi_{12}) - \zeta_2\right)}{2\zeta_2} \tag{1.647}$$

after condensing similar terms; one easily finds

$$\begin{aligned}&\sqrt{\left(\zeta_1(1+\xi_{12}) - \zeta_2\right)^2 + 4\zeta_1\zeta_2\xi_{12}} > \sqrt{\left(\zeta_1(1+\xi_{12}) - \zeta_2\right)^2}\\&= \left|\zeta_1(1+\xi_{12}) - \zeta_2\right|\end{aligned} \tag{1.648}$$

because $\zeta_1 > 0$, $\zeta_2 > 0$, and $\xi_{12} > 0$, so insertion in Eq. (1.646) has it that

$$-(1+\zeta_1 s_1) > \frac{\left|\zeta_1(1+\xi_{12}) - \zeta_2\right| + \zeta_1(1+\xi_{12}) - \zeta_2}{2\zeta_2}. \tag{1.649}$$

The distinct definition of absolute value depending on the sign of its argument allows reformulation of Eq. (1.649) to

$$\begin{aligned}-(1+\zeta_1 s_1) &> \frac{\zeta_1(1+\xi_{12}) - \zeta_2 + \left(\zeta_1(1+\xi_{12}) - \zeta_2\right)}{2\zeta_2}\\&= \frac{\zeta_1(1+\xi_{12}) - \zeta_2}{\zeta_2} \ge 0 \wedge \zeta_1(1+\xi_{12}) - \zeta_2 \ge 0\end{aligned} \tag{1.650}$$

or, alternatively,

$$\begin{aligned}&-(1+\zeta_1 s_1) > \frac{-\left(\zeta_1(1+\xi_{12}) - \zeta_2\right) + \left(\zeta_1(1+\xi_{12}) - \zeta_2\right)}{2\zeta_2} = 0\\&\wedge \zeta_1(1+\xi_{12}) - \zeta_2 < 0;\end{aligned} \tag{1.651}$$

in both cases, one concludes that

$$-(1+\zeta_1 s_1) > 0, \tag{1.652}$$

and consequently

$$\frac{1+\zeta_1 s_1}{s_1(s_1 - s_2)} < 0 \tag{1.653}$$

following combination with Eq. (1.626). By the same token, insertion of Eq. (1.648) permits transformation of Eq. (1.647) to

$$1+\zeta_1 s_2 > \frac{\left|\zeta_1\left(1+\xi_{12}\right)-\zeta_2\right|-\left(\zeta_1\left(1+\xi_{12}\right)-\zeta_2\right)}{2\zeta_2}; \qquad (1.654)$$

once again, the definition of absolute value leads to

$$1+\zeta_1 s_2 > \frac{\left(\zeta_1\left(1+\xi_{12}\right)-\zeta_2\right)-\left(\zeta_1\left(1+\xi_{12}\right)-\zeta_2\right)}{2\zeta_2}=0 \qquad (1.655)$$
$$\wedge \zeta_1\left(1+\xi_{12}\right)-\zeta_2 \geq 0,$$

and

$$1+\zeta_1 s_2 > \frac{-\left(\zeta_1\left(1+\xi_{12}\right)-\zeta_2\right)-\left(\zeta_1\left(1+\xi_{12}\right)-\zeta_2\right)}{2\zeta_2}$$
$$=-\frac{\zeta_1\left(1+\xi_{12}\right)-\zeta_2}{\zeta_2}>0 \wedge \zeta_1\left(1+\xi_{12}\right)-\zeta_2<0 \qquad (1.656)$$

otherwise. Therefore,

$$1+\zeta_1 s_2>0 \qquad (1.657)$$

based on either Eqs. (1.655) or (1.656), and irrespective of the actual values of ζ_1, ζ_2, and ξ_{12}; upon combination with Eq. (1.627), one gets

$$\frac{1+\zeta_1 s_2}{s_2\left(s_2-s_1\right)}<0 \qquad (1.658)$$

from Eq. (1.657). Since the pre-exponential coefficients in Eq. (1.641) share the same (negative) sign as per Eqs. (1.653) and (1.658), the term containing $e^{s_1 t}$ adds up to that containing $e^{s_2 t}$, and the resulting sum is subtracted from $1/s_1 s_2$; hence, the $T_2\{t\}$ curve experiences a less damped, first-order behavior that accelerates evolution toward the final value of the ascending part of the curve in Fig. 1.49, when compared to the dynamics of $T\{t\}$ – while removing sigmoidicity. This less sluggish pattern arises because not all heat received from the heating utility is used up, as a portion is meanwhile transferred to the food – unlike the latter that uses the whole heat received to heatup, but in a delayed fashion.

For a sufficiently long time, the final value theorem – consubstantiated in

$$\lim_{t\to\infty}\hat{T}_2=\lim_{s\to 0} s\overline{T}_2 \qquad (1.659)$$

in parallel to Eq. (1.605), allows one to write

$$\lim_{t\to\infty}\hat{T}_2=\frac{\hat{T}_{wst}}{\zeta_1\zeta_2}\lim_{s\to 0}\left(\frac{1}{s_1 s_2}+\frac{1+\zeta_1 s_1}{s_1\left(s_1-s_2\right)}\frac{s}{s-s_1}+\frac{1+\zeta_1 s_2}{s_2\left(s_2-s_1\right)}\frac{s}{s-s_2}\right) \qquad (1.660)$$

departing from Eq. (1.639); direct application of the classical theorems on limits supports transformation to

$$\lim_{t\to\infty}\hat{T}_2=\frac{\hat{T}_{wst}}{\zeta_1\zeta_2 s_1 s_2} \qquad (1.661)$$

or, in view Eq. (1.608), to

$$\lim_{t\to\infty}\hat{T}_2=\hat{T}_{wst} \qquad (1.662)$$

– which, in turn, becomes

$$\lim_{t\to\infty}T_2=T_{wst} \qquad (1.663)$$

due to Eqs. (1.550), (1.558), and (1.586). Inspection of Eq. (1.663) *vis-à-vis* with Eq. (1.610) unfolds the same long-t horizontal asymptote for temperatures T_2 and T_1, respectively – as expected, since thermal equilibrium of three media would be at stake; this point is corroborated by inspection of the common plateau of the two rising curves in Fig. 1.49.

Equations (1.610) and (1.663) describe indeed a new steady state, and may thus serve as the initial condition for the cooling period prevailing after Δt_h, i.e.

$$\boxed{T_1\big|_{\hat{t}=0}=T_{wst},} \qquad (1.664)$$

and similarly

$$\boxed{T_2\big|_{\hat{t}=0}=T_{wst}} \qquad (1.665)$$

– as long as a deviation time, $\hat{t}$, is defined as

$$\hat{t}\equiv t-\Delta t_h; \qquad (1.666)$$

Eqs. (1.545) and (1.546) remain both valid, provided that t is swapped for $\hat{t}$, and a sufficiently high volumetric flow rate of cooling fluid is employed to keep its temperature change (between inlet and outlet) to a negligible value – so one may resort to Eqs. (1.582) and (1.584) again. The new thermal processing pattern is described by

$$\boxed{T_c=T_{col}\mathrm{H}\{\hat{t}\},} \qquad (1.667)$$

where T_{col} denotes temperature of cooling utility (usually water or air at room temperature); the corresponding deviation variable reads

$$\hat{T}_c=\hat{T}_{col}\equiv T_{col}-T_{wst}=T_{col}-T_{c,0} \qquad (1.668)$$

in lieu of Eq. (1.586), so application of Laplace's transforms unfolds

$$\overline{T}_c=\frac{\hat{T}_{col}}{s}. \qquad (1.669)$$

Combination of Eqs. (1.582) and (1.669) gives rise to

$$\overline{T}_1=\frac{\hat{T}_{col}}{s\left(\left(1+\zeta_1 s\right)\left(1+\xi_{12}+\zeta_2 s\right)-\xi_{12}\right)}; \qquad (1.670)$$

and Eqs. (1.584) and (1.669) likewise support

$$\overline{T}_2=\frac{\hat{T}_{col}\left(1+\zeta_1 s\right)}{s\left(\left(1+\zeta_1 s\right)\left(1+\xi_{12}+\zeta_2 s\right)-\xi_{12}\right)}. \qquad (1.671)$$

In view of the analogy of Eqs. (1.670) and (1.671) to Eqs. (1.588) and (1.628), respectively, one may jump directly to the corresponding analogue of Eq. (1.604), viz.

$$\boxed{\begin{aligned}&\frac{T_{wst}-T_1}{T_{wst}-T_{col}}\\&=\frac{1}{\zeta_1\zeta_2}\left(\frac{1}{s_1 s_2}+\frac{e^{s_1\left(t-\Delta t_h\right)}}{s_1\left(s_1-s_2\right)}+\frac{e^{s_2\left(t-\Delta t_h\right)}}{s_2\left(s_2-s_1\right)}\right)\mathrm{H}\left\{t-\Delta t_h\right\}\\&\qquad t\geq\Delta t_h,\end{aligned}} \qquad (1.672)$$

and likewise of Eq. (1.641), viz.

$$\frac{T_{wst}-T_2}{T_{wst}-T_{col}}=\frac{1}{\zeta_1\zeta_2}\left(\frac{1}{s_1 s_2}+\frac{(1+\zeta_1 s_1)e^{s_1(t-\Delta t_h)}}{s_1(s_1-s_2)}+\frac{(1+\zeta_1 s_2)e^{s_2(t-\Delta t_h)}}{s_2(s_2-s_1)}\right)H\{t-\Delta t_h\}$$

$$t\geq \Delta t_h \tag{1.673}$$

– where the negatives of the numerators were taken, so as to work with only positive quantities; remember that

$$T_{wst}=T_{c,0} \tag{1.674}$$

in this case, see Eq. (1.668). The curves pertaining to T_1 vs. t and T_2 vs. t are plotted in Figs. 1.48 and 1.49, respectively, as their decreasing portions placed on the right; the patterns of such curves are reversed relative to the corresponding heating curves – as expected. Holding times at high and low temperature become apparent upon inspection of Fig. 1.49 – as the vicinity of the upper and lower plateaus, respectively. If T_{wst} is sufficiently high and T_2 comes sufficiently close thereto, then the contribution toward lethality comes chiefly from the said holding time – as heating and cooling transients would then be relatively fast events, besides entailing much lower average temperatures.

PROBLEM 1.15

If resistance to heat transfer within a solid food cannot be neglected, or the natural convection pattern of cover liquid inside a packaged solid food is insufficient to maintain uniformity of temperature, then the underlying enthalpic balance should resort to a partial differential equation in both time and space coordinates. While all points in the food still share the same temperature at startup of heating, the same would not apply, however, at startup of cooling – thus making mathematical modeling of the whole thermal process much harder to develop. Flambert and Deltour's model has been proposed, in this regard, to describe temperature evolution in cylindrical cans during the cooling period – and looks like

$$\frac{T-T_c}{T_h-T_c}=\sum_{j=1}^{\infty}\sum_{i=1}^{\infty}\left(\begin{array}{c}\dfrac{1-\dfrac{T_h-T_0}{T_h-T_c}\exp\left\{-\left(\left(\dfrac{Z}{R}\right)^2 B_j^2+(2i-1)^2\left(\dfrac{\pi}{2}\right)^2\right)\dfrac{kt}{\rho c_P Z^2}\right\}}{\dfrac{(2i-1)\pi}{8}B_j J_1\{B_j\}}\\ \times\exp\left\{-\left(\left(\dfrac{Z}{R}\right)^2 B_j^2+(2i-1)^2\left(\dfrac{\pi}{2}\right)^2\right)\dfrac{k(t-t_h)}{\rho c_P Z^2}\right\}\\ \times J_0\left\{B_j\dfrac{r}{R}\right\}\sin(2i-1)\dfrac{\pi}{2}\dfrac{Z-z}{Z}\end{array}\right). \tag{1.15.1}$$

Here $T\equiv T\{z,r,t\}$ denotes temperature within food at time t, and axial coordinate z and radial coordinate r (both counted from the geometric center); T_h and T_c denote (constant) temperature of heating and cooling medium, respectively; T_0 denotes initial (room) temperature of food; Z denotes half height and R denotes radius of can; k, ρ, and c_P denote thermal conductivity, mass density, and specific heat capacity, respectively, of food; t_h denotes heating time (with $t>t_h$ obviously); and B_j denotes the j-th root of

$$J_0\{B_j\}=0, \tag{1.15.2}$$

with J_0 and J_1 denoting Bessel's functions of the first kind, and zero-th or first order, respectively.

a) Knowing that $B_1=2.4048$ and $J_1\{B_1\}=0.5191$, plot dimensionless temperature $(T-T_c)/(T_h-T_c)$ at the geometric center as a function of dimensionless time $kt/\rho c_P Z^2$, using Eq. (1.15.1) truncated after the linear terms – and assuming $(T_h-T_0)/(T_h-T_c)=1$, $Z/R=1$, and $kt_h/\rho c_P Z^2=1$.
b) Repeat a), but upon truncation of Eq. (1.15.1) after the quadratic term – with $B_2=5.5201$ and $J_1\{B_2\}=-0.3404$.

Solution

a) For the given data, Eq. (1.15.1) reduces to

$$\left.\frac{T-T_c}{T_h-T_c}\right|_{i=j=1}=\frac{1-\exp\left\{-\left(B_1^2+\left(\dfrac{\pi}{2}\right)^2\right)\dfrac{kt}{\rho c_P Z^2}\right\}}{\dfrac{\pi}{8}B_1 J_1\{B_1\}}\times\exp\left\{-\left(B_1^2+\left(\frac{\pi}{2}\right)^2\right)\left(\frac{kt}{\rho c_P Z^2}-1\right)\right\}J_0\{0\}\sin\frac{\pi}{2} \tag{1.15.3}$$

after considering only $i=1$ and $j=1$ – and since the geometric center is characterized by $z=0$ and $r=0$; Eq. (1.15.3) simplifies further to

$$\left.\frac{T-T_c}{T_h-T_c}\right|_{i=j=1}=\frac{1-\exp\left\{-\left(2.4048^2+\left(\dfrac{\pi}{2}\right)^2\right)\dfrac{kt}{\rho c_P Z^2}\right\}}{\dfrac{\pi}{8}2.4048\cdot 0.5191}\times\exp\left\{-\left(2.4048^2+\left(\frac{\pi}{2}\right)^2\right)\left(\frac{kt}{\rho c_P Z^2}-1\right)\right\} \tag{1.15.4}$$

in view of $J_0\{0\}=1$ and $\sin\pi/2=1$, or merely

$$\left.\frac{T-T_c}{T_h-T_c}\right|_{i=j=1} = \frac{1-\exp\left\{-8.250\dfrac{kt}{\rho c_P Z^2}\right\}}{0.4902}\exp\left\{-8.250\left(\frac{kt}{\rho c_P Z^2}-1\right)\right\} \tag{1.15.5}$$

– as plotted below.

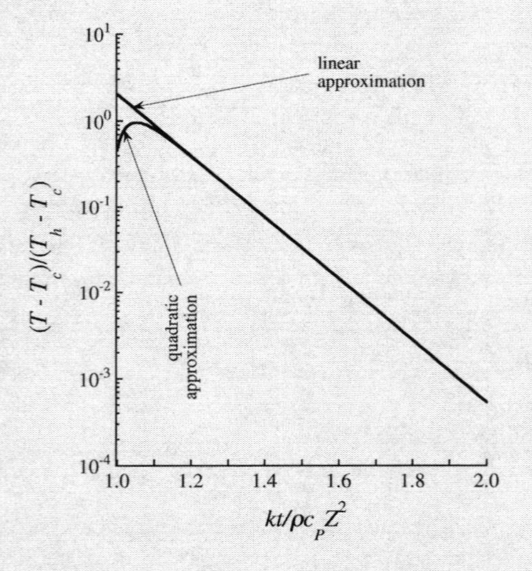

Note the almost linear decay in temperature with time, explained mainly by the exponential dependence appearing second in Eq. (1.15.5), when the linear approximation is under scrutiny.

b) If two terms of the summation are to be considered, then Eq. (1.15.1) becomes

$$\left.\frac{T-T_c}{T_h-T_c}\right|_{i,j=1,2} = \begin{pmatrix} \dfrac{1-\exp\left\{-\left(B_1^2+\left(\dfrac{\pi}{2}\right)^2\right)\dfrac{kt}{\rho c_P Z^2}\right\}}{\dfrac{\pi}{8}B_1 J_1\{B_1\}} \\ \times\exp\left\{-\left(B_1^2+\left(\dfrac{\pi}{2}\right)^2\right)\left(\dfrac{kt}{\rho c_P Z^2}-1\right)\right\}J_0\{0\}\sin\dfrac{\pi}{2} \\ +\dfrac{1-\exp\left\{-\left(B_1^2+9\left(\dfrac{\pi}{2}\right)^2\right)\dfrac{kt}{\rho c_P Z^2}\right\}}{\dfrac{3}{8}\pi B_1 J_1\{B_1\}} \\ \times\exp\left\{-\left(B_1^2+9\left(\dfrac{\pi}{2}\right)^2\right)\left(\dfrac{kt}{\rho c_P Z^2}-1\right)\right\}J_0\{0\}\sin\dfrac{3\pi}{2} \end{pmatrix} + \begin{pmatrix} \dfrac{1-\exp\left\{-\left(B_2^2+\left(\dfrac{\pi}{2}\right)^2\right)\dfrac{kt}{\rho c_P Z^2}\right\}}{\dfrac{\pi}{8}B_2 J_1\{B_2\}} \\ \times\exp\left\{-\left(B_2^2+\left(\dfrac{\pi}{2}\right)^2\right)\left(\dfrac{kt}{\rho c_P Z^2}-1\right)\right\}J_0\{0\}\sin\dfrac{\pi}{2} \\ +\dfrac{1-\exp\left\{-\left(B_2^2+9\left(\dfrac{\pi}{2}\right)^2\right)\dfrac{kt}{\rho c_P Z^2}\right\}}{\dfrac{3}{8}\pi B_2 J_1\{B_2\}} \\ \times\exp\left\{-\left(B_2^2+9\left(\dfrac{\pi}{2}\right)^2\right)\left(\dfrac{kt}{\rho c_P Z^2}-1\right)\right\}J_0\{0\}\sin\dfrac{3\pi}{2} \end{pmatrix}, \tag{1.15.6}$$

which degenerates to

$$\left.\frac{T-T_c}{T_h-T_c}\right|_{i,j=1,2} = \begin{pmatrix} \dfrac{1-\exp\left\{-\left(2.4048^2+\left(\dfrac{\pi}{2}\right)^2\right)\dfrac{kt}{\rho c_P Z^2}\right\}}{\dfrac{\pi}{8}2.4048\cdot 0.5191} \\ \times\exp\left\{-\left(2.4048^2+\left(\dfrac{\pi}{2}\right)^2\right)\left(\dfrac{kt}{\rho c_P Z^2}-1\right)\right\} \\ -\dfrac{1-\exp\left\{-\left(2.4048^2+9\left(\dfrac{\pi}{2}\right)^2\right)\dfrac{kt}{\rho c_P Z^2}\right\}}{\dfrac{3}{8}\pi 2.4048\cdot 0.5191} \\ \times\exp\left\{-\left(2.4048^2+9\left(\dfrac{\pi}{2}\right)^2\right)\left(\dfrac{kt}{\rho c_P Z^2}-1\right)\right\} \end{pmatrix} + \begin{pmatrix} \dfrac{1-\exp\left\{-\left(5.5201^2+\left(\dfrac{\pi}{2}\right)^2\right)\dfrac{kt}{\rho c_P Z^2}\right\}}{\dfrac{\pi}{8}5.5201(-0.3404)} \\ \times\exp\left\{-\left(5.5201^2+\left(\dfrac{\pi}{2}\right)^2\right)\left(\dfrac{kt}{\rho c_P Z^2}-1\right)\right\} \\ -\dfrac{1-\exp\left\{-\left(5.5201^2+9\left(\dfrac{\pi}{2}\right)^2\right)\dfrac{kt}{\rho c_P Z^2}\right\}}{\dfrac{3}{8}\pi 5.5201(-0.3404)} \\ \times\exp\left\{-\left(5.5201^2+9\left(\dfrac{\pi}{2}\right)^2\right)\left(\dfrac{kt}{\rho c_P Z^2}-1\right)\right\} \end{pmatrix} \tag{1.15.7}$$

since $\sin 3\pi/2 = -1$; calculation of the constants as indicated finally yields

$$\left.\frac{T - T_c}{T_h - T_c}\right|_{i,j=1,2} = \frac{1 - \exp\left\{-8.250\dfrac{kt}{\rho c_P Z^2}\right\}}{0.4092} \exp\left\{-8.250\left(\frac{kt}{\rho c_P Z^2} - 1\right)\right\} - \frac{1 - \exp\left\{-27.99\dfrac{kt}{\rho c_P Z^2}\right\}}{1.471} \exp\left\{-27.99\left(\frac{kt}{\rho c_P Z^2} - 1\right)\right\} - \frac{1 - \exp\left\{-32.94\dfrac{kt}{\rho c_P Z^2}\right\}}{0.7379} \exp\left\{-32.94\left(\frac{kt}{\rho c_P Z^2} - 1\right)\right\} + \frac{1 - \exp\left\{-52.68\dfrac{kt}{\rho c_P Z^2}\right\}}{2.214} \exp\left\{-52.68\left(\frac{kt}{\rho c_P Z^2} - 1\right)\right\}, \tag{1.15.8}$$

with graphical representation overlaid above. One realizes that the linear approximation suffices in most of the time range of interest – since the linear and quadratic forms of Flambert and Deltour's model essentially coincide. Relevant differences appear only at early cooling – with temperature still going through a maximum; this is so because presence of cooling fluid on the food outer surface was not yet felt in full at the geometric center of the can, as some residual heat transfer in remains in progress owing to thermal inertia.

1.2.3.4 Direct heating of uniform-temperature liquid

The benefits of using a very high temperature, during a rather short period, in thermal processing have been discussed previously – and lie chiefly on the lower rate of decay of nutritional and sensory components, for a prespecified degree of reduction in viability of spoilage/pathogen microorganisms. This calls for particularly fast heat transfer – hardly compatible with conventional heat exchange through a solid (metallic) barrier; the corresponding short times are possible only under continuous operation. The best approach is via direct contact with fine bubbles of (food-grade) steam above atmospheric pressure – usually injected inside the bulk of a liquid food; they bring about a quite rapid temperature increase, owing to the very high turbulence and specific interfacial area, prior to the (preset) holding time under adiabatic conditions. A complementary step encompasses flash vaporization of the excess water resulting from the said steam condensation.

The (multiple) nozzles employed in UHT processes are typically very narrow in inner diameter, and linear velocities of steam are very high so that only tiny bubbles form; in addition, bulk liquid is flown up, so buoyancy effects are favorable – while drag by the liquid essentially carries said bubbles at the average liquid velocity (irrespective of their smaller and smaller size). The distance tracked by the bubbles carried by the target liquid food is normally small, so change in hydrostatic pressure is negligible during their path – and such a short heating zone prior to the holding tube makes hydrodynamic pressure drop (due to viscous dissipation) also of little relevance; furthermore, the plug flow hypothesis, in the absence of radial gradients of any sort, is in principle reasonable. Under these circumstances, a steady-state enthalpic balance to the system formed by steam bubbles and liquid food may be set as

$$Q_f \rho_f c_{P,f}\left(T_f - T^\theta\right)\Big|_x + UA_b\left(T_s - T_f\right)NA_f dx = Q_f \rho_f c_{P,f}\left(T_f - T^\theta\right)\Big|_{x+dx}; \tag{1.675}$$

here Q_f and A_f denote volumetric flow rate and cross-section, respectively, of liquid food, ρ_f and $c_{P,f}$ denote mass density and specific (isobaric) heat capacity, respectively, of liquid food, U denotes overall heat transfer coefficient between phases, A_b denotes outer area of a single bubble, T_s and T_f denote temperature of water steam and liquid food, respectively, T^θ denotes reference (arbitrary) temperature, N denotes number concentration of steam bubbles, and x denotes axial coordinate. Upon lumping of terms alike and division of both sides by dx, Eq. (1.675) becomes

$$UA_b\left(T_s - T_f\right)NA_f = Q_f \rho_f c_{P,f} \frac{\left(T_f - T^\theta\right)\Big|_{x+dx} - \left(T_f - T^\theta\right)\Big|_x}{dx}; \tag{1.676}$$

cancellation of symmetrical terms and division of both sides by $Q_f \rho_f c_{P,f}$, coupled to the definition of derivative allow further simplification to

$$\boxed{\frac{dT_f}{dx} = \frac{UNA_f A_b}{Q_f \rho_f c_{P,f}}\left(T_s - T_f\right).} \tag{1.677}$$

Since heat is supplied only at the expense of water steam condensation – with prompt mixing of water condensate with the aqueous liquid food, a mass balance to each bubble is in order, according to

$$UA_b\left(T_s - T_f\right)dt = -\lambda dm_b, \tag{1.678}$$

where m_b denotes mass of a steam bubble and λ denotes latent heat of vaporization; division of both sides by $-\lambda dt$ gives then rise to

$$\boxed{\frac{dm_b}{dt} = -\frac{UA_b}{\lambda}\left(T_s - T_f\right).} \tag{1.679}$$

The ideal gas equation, as per Eq. (3.135) in *Food Proc. Eng.: basics & mechanical operations*, may be revisited here as

$$\boxed{PV_b = \frac{m_b}{M_s}\mathcal{R}T_s} \tag{1.680}$$

as approximant to the behavior of the steam phase – where M_s denotes molecular weight of water (or steam, for that matter), and P and V_b denote bubble pressure and volume, respectively. After isolating m_b, Eq. (1.680) yields

$$m_b = \frac{M_s P}{\mathcal{R}T_s}V_b, \tag{1.681}$$

and thus

$$dm_b = \frac{M_s P}{\mathcal{R} T_s} dV_b \tag{1.682}$$

following application of differentials to both sides – since T_s (for the underlying phase change) and P (for the hydrostatic and hydrodynamic arguments referred to above) remain essentially constant; insertion of Eq. (1.682) transforms Eq. (1.679) to

$$\frac{M_s P}{\mathcal{R} T_s} \frac{dV_b}{dt} = -\frac{UA_b}{\lambda}\left(T_s - T_f\right), \tag{1.683}$$

which may be recoined as

$$\frac{dV_b}{dt} = -\frac{\mathcal{R} U A_b T_s}{\lambda M_s P}\left(T_s - T_f\right). \tag{1.684}$$

On the other hand, geometrical considerations pertaining to a sphere have it that

$$\boxed{V_b = \frac{4}{3}\pi R_b^3,} \tag{1.685}$$

with R_b denoting bubble radius – coupled with

$$\boxed{A_b = 4\pi R_b^2;} \tag{1.686}$$

upon differentiation, Eq. (1.685) becomes

$$dV_b = \frac{4}{3}\pi 3 R_b^2 dR_b = 4\pi R_b^2 dR_b, \tag{1.687}$$

so insertion of Eqs. (1.686) and (1.687) converts Eq. (1.684) to

$$4\pi R_b^2 \frac{dR_b}{dt} = -\frac{\mathcal{R} U 4\pi R_b^2 T_s}{\lambda M_s P}\left(T_s - T_f\right) \tag{1.688}$$

– where cancellation of $4\pi R_b^2$ between sides leaves merely

$$\boxed{\frac{dR_b}{dt} = -\frac{\mathcal{R} U T_s}{\lambda M_s P}\left(T_s - T_f\right).} \tag{1.689}$$

By the same token, Eq. (1.677) takes the form

$$\boxed{\frac{dT_f}{dx} = 4\pi \frac{UNA_f}{Q_f \rho_f c_{P,f}} R_b^2\left(T_s - T_f\right)} \tag{1.690}$$

following combination with Eq. (1.686); the set of Eqs. (1.689) and (1.690) is now to be simultaneously solved for its two dependent variables, T_f and R_b, with x and t, respectively, serving as independent variables. Equation (1.689) is complemented by

$$\boxed{R_b\big|_{t=0} = R_{b,0}} \tag{1.691}$$

as initial condition, whereas

$$\boxed{T_f\big|_{x=0} = T_{f,0}} \tag{1.692}$$

plays the role of boundary condition for Eq. (1.690).

Solution of the aforementioned equations is facilitated after ordered division of Eq. (1.689) by Eq. (1.690), i.e.

$$\frac{\dfrac{dR_b}{dt}}{\dfrac{dT_f}{dx}} = -\frac{\dfrac{\mathcal{R} U T_s}{\lambda M_s P}}{\dfrac{4\pi UNA_f}{Q_f \rho_f c_{P,f}}} \frac{1}{R_b^2}, \tag{1.693}$$

which eliminates T_f as explicit variable due to cancellation of $T_s - T_f$ between numerator and denominator; Eq. (1.693) reduces, in turn, to

$$R_b^2 \frac{dR_b}{dT_f} \frac{dx}{dt} = -\frac{\mathcal{R} Q_f \rho_f c_{P,f} T_s}{4\pi\lambda M_s NPA_f} \tag{1.694}$$

upon straightforward algebraic rearrangement. An extra intrinsic relationship will be required between spatial position of liquid food and time elapsed with steam bubbles therein, viz.

$$\boxed{\frac{dx}{dt} = \frac{Q_f}{A_f},} \tag{1.695}$$

because bubbles are carried by liquid at its own flow velocity (as emphasized above). Insertion of Eq. (1.695) converts Eq. (1.694) to

$$R_b^2 \frac{dR_b}{dT_f} \frac{Q_f}{A_f} = -\frac{\mathcal{R} \rho_f c_{P,f} T_s}{4\pi\lambda M_s NP} \frac{Q_f}{A_f}, \tag{1.696}$$

where Q_f and A_f may be dropped from both sides to give

$$\boxed{R_b^2 \frac{dR_b}{dT_f} = -\frac{\mathcal{R} \rho_f c_{P,f} T_s}{4\pi\lambda M_s NP};} \tag{1.697}$$

Eq. (1.697) will hereafter replace Eq. (1.690), and entails an expression for dR_b/dT_f independent of T_f. Integration of Eq. (1.697) is thus feasible via separation of variables, viz.

$$\int_{R_{b,0}}^{R_b} \tilde{R}_b^2 \, d\tilde{R}_b = -\frac{\mathcal{R} \rho_f c_{P,f} T_s}{4\pi\lambda M_s NP} \int_{T_{f,0}}^{T_f} d\tilde{T}_f, \tag{1.698}$$

on the hypothesis that food properties (i.e. ρ_f and c_{Pf}) and steam properties (i.e. T_s and λ), besides process conditions (i.e. N and P) remain essentially constant within the temperature range of interest – and where Eqs. (1.691) and (1.692) were meanwhile taken advantage of. The fundamental theorem of integral calculus may now be invoked to rewrite Eq. (1.698) as

$$\left.\frac{\tilde{R}_b^3}{3}\right|_{R_{b,0}}^{R_b} = -\frac{\mathcal{R} \rho_f c_{P,f} T_s}{4\pi\lambda M_s NP} \left.\tilde{T}_f\right|_{T_{f,0}}^{T_f} \tag{1.699}$$

that degenerates to

$$\frac{1}{3}\left(R_b^3 - R_{b,0}^3\right) = -\frac{\mathcal{R} \rho_f c_{P,f} T_s}{4\pi\lambda M_s NP}\left(T_f - T_{f,0}\right), \tag{1.700}$$

or else

$$\boxed{T_f = T_{f,0} + \frac{4\pi\lambda M_s NP}{3\mathcal{R} \rho_f c_{P,f} T_s}\left(R_{b,0}^3 - R_b^3\right)} \tag{1.701}$$

following isolation of T_f; one thus obtains $T_f \equiv T_f\{R_b\}$. Note that the liquid food will attain its maximum temperature, $T_{f,\max}$, when the right-hand side of Eq. (1.701) is maximum, i.e. when R_b is nil – in agreement with

$$\boxed{T_{f,\max} = T_{f,0} + \frac{4\pi}{3\mathcal{R}}\frac{\lambda M_s NPR_{b,0}^3}{\rho_f c_{P,f} T_s},} \quad (1.702)$$

so there is no practical advantage in extending the heating section beyond bubble collapse; calculation of the time it takes to reach $T_{f,\max}$ then requires integration of Eq. (1.689) to get $R_b \equiv R_b\{t\}$.

To proceed along the aforementioned lines, one should first insert Eq. (1.701) to transform Eq. (1.689) to

$$-\frac{dR_b}{dt} = \frac{\mathcal{R}UT_s}{\lambda M_s P}\left(T_s - T_{f,0} - \frac{4\pi\lambda M_s NP}{3\mathcal{R}\rho_f c_{P,f} T_s}\left(R_{b,0}^3 - R_b^3\right)\right), \quad (1.703)$$

where removal of inner parenthesis and factoring in of $\mathcal{R}UT_s/\lambda M_s P$ produce

$$-\frac{dR_b}{dt} = \frac{\mathcal{R}UT_s}{\lambda M_s P}\left(T_s - T_{f,0}\right) - \frac{4\pi UN}{3\rho_f c_{P,f}}R_{b,0}^3 + \frac{4\pi UN}{3\rho_f c_{P,f}}R_b^3; \quad (1.704)$$

the notation in Eq. (1.704) may be simplified to

$$-\frac{dR_b}{dt} = \xi_1 + \xi_2 R_b^3, \quad (1.705)$$

after defining

$$\xi_1 \equiv \frac{\mathcal{R}UT_s}{\lambda M_s P}\left(T_s - T_{f,0}\right) - \frac{4\pi UN}{3\rho_f c_{P,f}}R_{b,0}^3 \quad (1.706)$$

and

$$\xi_2 \equiv \frac{4\pi UN}{3\rho_f c_{P,f}} \geq 0 \quad (1.707)$$

as auxiliary parameters. Note that

$$T_s \geq T_{f,\max} \quad (1.708)$$

– stemming from the definition of $T_{f,\max}$ and the whole purpose of steam-mediated heating, so

$$T_s - T_{f,0} \geq T_{f,\max} - T_{f,0} \quad (1.709)$$

upon subtraction of $T_{f,0}$ from both sides; in view of Eq. (1.702), this is equivalent to state

$$T_s - T_{f,0} \geq \frac{4\pi}{3\mathcal{R}}\frac{\lambda M_s NP}{\rho_f c_{P,f} T_s}R_{b,0}^3. \quad (1.710)$$

Equation (1.710) may be rewritten as

$$\frac{\mathcal{R}UT_s}{\lambda M_s P}\left(T_s - T_{f,0}\right) \geq \frac{4\pi UN}{3\rho_f c_{P,f}}R_{b,0}^3, \quad (1.711)$$

following multiplication of both sides by $\mathcal{R}UT_s/\lambda M_s P$ – thus implying

$$\xi_1 \geq 0 \quad (1.712)$$

as per Eq. (1.706); this is indeed required for physical consistency, i.e. to get $dR_b/dt \leq 0$ at all times, see Eq. (1.705), as required by a shrinking steam bubble that loses its vapor content as condensate – in view also of Eq. (1.707). Integration of Eq. (1.705) may proceed, via separation of variables, as

$$-\int_{R_{b,0}}^{R_b}\frac{d\tilde{R}_b}{\xi_1 + \xi_2\tilde{R}_b^3} = \int_0^t d\tilde{t} \quad (1.713)$$

with the aid of Eq. (1.691), which is equivalent to writing

$$t = \tilde{t}\Big|_0^t = \int_{R_b}^{R_{b,0}}\frac{d\tilde{R}_b}{\xi_1 + \xi_2\tilde{R}_b^3} \quad (1.714)$$

upon exchange of integration limits at the expense of the preceding minus sign; calculation of the outstanding integral requires expansion of its kernel as partial fractions – so one should first calculate the (or one) real root of the denominator, upon setting

$$\xi_1 + \xi_2 R_b^3 = 0. \quad (1.715)$$

Equation (1.715) is equivalent to

$$R_b = \sqrt[3]{-\frac{\xi_1}{\xi_2}} = -\sqrt[3]{\frac{\xi_1}{\xi_2}} \quad (1.716)$$

upon isolation of R_b, or else

$$R_b = -\Xi \quad (1.717)$$

as long as Ξ is defined by

$$\Xi \equiv \sqrt[3]{\frac{\xi_1}{\xi_2}}. \quad (1.718)$$

The above reasoning implies that $R_b-(-\Xi)$ may be factored out from the polynomial in denominator of Eq. (1.714), i.e. $\xi_1 + \xi_2 R_b^3 = \xi_2 R_b^3 + 0R_b^2 + 0R_b + \xi_1$; the (exact) quotient can be ascertained via Ruffini's rule as

$$\xi_1 + \xi_2 R_b^3 = \xi_2\left(R_b^3 + 0R_b^2 + 0R_b + \frac{\xi_1}{\xi_2}\right)$$
$$= \xi_2\left(R_b - \left(-\left(\frac{\xi_1}{\xi_2}\right)^{\frac{1}{3}}\right)\right)\left(R_b^2 - \left(\frac{\xi_1}{\xi_2}\right)^{\frac{1}{3}}R_b + \left(\frac{\xi_1}{\xi_2}\right)^{\frac{2}{3}}\right), \quad (1.719)$$

with the aid of Eq. (1.716) and after factoring ξ_2 out – which simplifies to

$$\xi_2\left(R_b^3 + \Xi^3\right) = \xi_2\left(R_b + \Xi\right)\left(R_b^2 - \Xi R_b + \Xi^2\right) \quad (1.720)$$

when Eq. (1.718) is taken into account. The last trinomial may, in turn, be factored out once its two roots have been calculated – based on

$$R_b^2 - \Xi R_b + \Xi^2 = 0, \quad (1.721)$$

where the canonical solving formula for a quadratic equation can be applied to give

$$R_b = \frac{\Xi \pm \sqrt{(-\Xi)^2 - 4\Xi^2}}{2} = \frac{\Xi \pm \sqrt{\Xi^2 - 4\Xi^2}}{2}; \tag{1.722}$$

Eq. (1.722) simplifies to

$$R_b = \frac{\Xi \pm \sqrt{-3\Xi^2}}{2} = \frac{\Xi \pm \iota\sqrt{3\Xi^2}}{2}, \tag{1.723}$$

with $\iota \equiv \sqrt{-1}$ denoting imaginary unit. Equation (1.723) may be further redone as

$$R_b = \frac{\Xi \pm \iota\sqrt{3}\Xi}{2}, \tag{1.724}$$

or else

$$R_b = \frac{1 \pm \iota\sqrt{3}}{2}\Xi \tag{1.725}$$

after factoring Ξ out. In view of the complex nature of the above two roots, expansion of the kernel of Eq. (1.714) should resort to a partial fraction of type I and a partial fraction of type III, according to

$$\frac{1}{\xi_2 R_b^3 + \xi_1} = \frac{1}{\xi_2\left(R_b^3 + \Xi^3\right)} = \frac{1}{\xi_2}\left(\frac{\kappa_1}{R_b + \Xi} + \frac{\kappa_2 + \kappa_3 R_b}{R_b^2 - \Xi R_b + \Xi^2}\right) \tag{1.726}$$

with the aid of Eq. (1.720) – where κ_1, κ_2 and κ_3 are constants to be determined; Eq. (1.726) may be reformulated to

$$\frac{1}{R_b^3 + \Xi^3} = \frac{\kappa_1}{R_b + \Xi} + \frac{\kappa_3}{2}\frac{2R_b - \Xi + \Xi + \frac{2\kappa_2}{\kappa_3}}{R_b^2 - \Xi R_b + \Xi^2}, \tag{1.727}$$

following cancellation of ξ_2 between sides, factoring out of $\kappa_3/2$ in the second term of the right-hand side, and addition and subtraction of Ξ to its numerator afterward. Equation (1.727) can be algebraically rearranged to read

$$\frac{1}{R_b^3 + \Xi^3} = \frac{\kappa_1}{R_b + \Xi} + \frac{\kappa_3}{2}\frac{2R_b - \Xi}{R_b^2 - \Xi R_b + \Xi^2} + \frac{\frac{\kappa_3}{2}\Xi + \kappa_2}{R_b^2 - \Xi R_b + \Xi^2} \tag{1.728}$$

after having splitted the last fraction – where the denominator of the last term may, in turn, be rewritten as

$$\frac{1}{R_b^3 + \Xi^3} = \frac{\kappa_1}{R_b + \Xi} + \frac{\kappa_3}{2}\frac{2R_b - \Xi}{R_b^2 - \Xi R_b + \Xi^2} + \frac{\kappa_2 + \frac{\kappa_3}{2}\Xi}{\left(R_b^2 - \Xi R_b + \frac{1}{4}\Xi^2\right) - \frac{1}{4}\Xi^2 + \Xi^2} \tag{1.729}$$

upon addition and subtraction of $\Xi^2/4$; in view of Newton's binomial formula, Eq. (1.729) simplifies to

$$\frac{1}{R_b^3 + \Xi^3} = \frac{\kappa_1}{R_b + \Xi} + \frac{\kappa_3}{2}\frac{2R_b - \Xi}{R_b^2 - \Xi R_b + \Xi^2} + \frac{\kappa_2 + \frac{\kappa_3}{2}\Xi}{\left(R_b - \frac{1}{2}\Xi\right)^2 + \frac{3}{4}\Xi^2}, \tag{1.730}$$

whereas division of both numerator and denominator of the final term by $\frac{3}{4}\Xi^2 = \left(\sqrt{\frac{3}{4}\Xi^2}\right)^2 = \frac{\sqrt{3}}{2}\Xi$ unfolds

$$\frac{1}{R_b^3 + \Xi^3} = \frac{\kappa_1}{R_b + \Xi} + \frac{\kappa_3}{2}\frac{2R_b - \Xi}{R_b^2 - \Xi R_b + \Xi^2} + \frac{\dfrac{\kappa_2 + \frac{\kappa_3}{2}\Xi}{\frac{3}{4}\Xi^2}}{1 + \left(\dfrac{R_b - \frac{1}{2}\Xi}{\frac{\sqrt{3}}{2}\Xi}\right)^2}. \tag{1.731}$$

The integral of the first term in the right-hand side of Eq. (1.731) looks like

$$\int \frac{\kappa_1}{R_b + \Xi}\, dR_b = \kappa_1 \ln\left\{R_b + \Xi\right\} \tag{1.732}$$

(integration constants will be skipped hereafter for simplicity, and since the final result will have the form of a definite integral); while the second term gives rise to

$$\int \frac{\kappa_3}{2}\frac{2R_b - \Xi}{R_b^2 - \Xi R_b + \Xi^2}\, dR_b = \frac{\kappa_3}{2}\ln\left\{R_b^2 - \Xi R_b + \Xi^2\right\}, \tag{1.733}$$

and the last term in Eq. (1.731) integrates to

$$\int \frac{\dfrac{\kappa_2 + \frac{\kappa_3}{2}\Xi}{\frac{3}{4}\Xi^2}}{1 + \left(\dfrac{R_b - \frac{1}{2}\Xi}{\frac{\sqrt{3}}{2}\Xi}\right)^2} dR_b = \frac{\kappa_2 + \frac{\kappa_3}{2}\Xi}{\frac{\sqrt{3}}{2}\Xi} \int \frac{d\left(\dfrac{R_b - \frac{1}{2}\Xi}{\frac{\sqrt{3}}{2}\Xi}\right)}{1 + \left(\dfrac{R_b - \frac{1}{2}\Xi}{\frac{\sqrt{3}}{2}\Xi}\right)^2}, \tag{1.734}$$

at the expense of the operational rules of differentiation and after splitting of $3\Xi^2/4$ as $\sqrt{3}\Xi/2$ multiplied by itself – which eventually gives

$$\int \frac{\frac{\kappa_2+\frac{\kappa_3}{2}\Xi}{\frac{3}{4}\Xi^2}}{1+\left(\frac{R_b-\frac{1}{2}\Xi}{\frac{\sqrt{3}}{2}\Xi}\right)^2}dR_b = \frac{\kappa_2+\frac{\kappa_3}{2}\Xi}{\frac{\sqrt{3}}{2}\Xi}\tan^{-1}\frac{R_b-\frac{1}{2}\Xi}{\frac{\sqrt{3}}{2}\Xi} = \frac{2\kappa_2+\kappa_3\Xi}{\sqrt{3}\Xi}\tan^{-1}\frac{2R_b-\Xi}{\sqrt{3}\Xi}, \tag{1.735}$$

along with multiplication of both numerator and denominator by 2. After revisiting Eq. (1.714) as

$$t = \frac{1}{\xi_2}\int_{R_b}^{R_{b,0}} \frac{d\tilde{R}_b}{\tilde{R}_b^{\,3}+\frac{\xi_1}{\xi_2}} = \frac{1}{\xi_2}\int_{R_b}^{R_{b,0}} \frac{d\tilde{R}_b}{\tilde{R}_b^{\,3}+\Xi^3} \tag{1.736}$$

upon factoring ξ_2 out and combining with Eq. (1.718), one may insert Eq. (1.731) to obtain

$$t = \frac{1}{\xi_2}\int_{R_b}^{R_{b,0}} \frac{\kappa_1}{\tilde{R}_b+\Xi}d\tilde{R}_b + \frac{1}{\xi_2}\int_{R_b}^{R_{b,0}} \frac{\kappa_3}{2}\frac{2\tilde{R}_b-\Xi}{\tilde{R}_b^{\,2}-\Xi\tilde{R}_b+\Xi^2}d\tilde{R}_b + \frac{1}{\xi_2}\int_{R_b}^{R_{b,0}} \frac{\frac{\kappa_2+\frac{\kappa_3}{2}\Xi}{\frac{3}{4}\Xi^2}}{1+\left(\frac{\tilde{R}_b-\frac{1}{2}\Xi}{\frac{\sqrt{3}}{2}\Xi}\right)^2}d\tilde{R}_b; \tag{1.737}$$

in view of Eqs. (1.732), (1.733), and (1.735), one attains

$$t = \frac{1}{\xi_2}\kappa_1 \ln\left\{\tilde{R}_b+\Xi\right\}\Big|_{R_b}^{R_{b,0}} + \frac{1}{\xi_2}\frac{\kappa_3}{2}\ln\left\{\tilde{R}_b^{\,2}-\Xi\tilde{R}_b+\Xi^2\right\}\Big|_{R_b}^{R_{b,0}} + \frac{1}{\xi_2}\frac{2\kappa_2+\kappa_3\Xi}{\sqrt{3}\Xi}\tan^{-1}\frac{2\tilde{R}_b-\Xi}{\sqrt{3}\Xi}\Bigg|_{R_b}^{R_{b,0}} \tag{1.738}$$

based on Eq. (1.737), where replacement by upper and lower limits of integration unfolds

$$t = \frac{1}{\xi_2}\kappa_1 \ln\frac{R_{b,0}+\Xi}{R_b+\Xi} + \frac{1}{\xi_2}\frac{\kappa_3}{2}\ln\frac{R_{b,0}^{\,2}-\Xi R_{b,0}+\Xi^2}{R_b^{\,2}-\Xi R_b+\Xi^2} + \frac{1}{\xi_2}\frac{2\kappa_2+\kappa_3\Xi}{\sqrt{3}\Xi}\left(\tan^{-1}\frac{2R_{b,0}-\Xi}{\sqrt{3}\Xi} - \tan^{-1}\frac{2R_b-\Xi}{\sqrt{3}\Xi}\right). \tag{1.739}$$

To calculate the pending constants in Eq. (1.739), one should revisit Eq. (1.726) as

$$\frac{1}{R_b^{\,3}+\Xi^3} = \frac{\kappa_1}{R_b+\Xi} + \frac{\kappa_2+\kappa_3 R_b}{R_b^{\,2}-\Xi R_b+\Xi^2}, \tag{1.740}$$

upon multiplication of both sides by ξ_2 and use of Eq. (1.718) – where elimination of denominators leaves

$$1 = \kappa_1\left(R_b^{\,2}-\Xi R_b+\Xi^2\right)+\left(\kappa_2+\kappa_3 R_b\right)\left(R_b+\Xi\right), \tag{1.741}$$

with Eq. (1.720) utilized to advantage. Equation (1.741) must be valid irrespective of the specific value attributed to R_b – so it should, in particular, be valid when $R_b=-\Xi$, as suggested by Eq. (1.717); under such circumstances, one gets

$$1 = \kappa_1\left((-\Xi)^2-\Xi(-\Xi)+\Xi^2\right)+\left(\kappa_2+\kappa_3(-\Xi)\right)(-\Xi+\Xi) \tag{1.742}$$

that readily simplifies to

$$1 = \kappa_1\left(\Xi^2+\Xi^2+\Xi^2\right) \tag{1.743}$$

after symmetrical terms have been cancelled out – where isolation of κ_1 yields

$$\kappa_1 = \frac{1}{3\Xi^2}. \tag{1.744}$$

By the same token, one may recall Eq. (1.725), and accordingly set R_b equal to, say, $\left(1+\iota\sqrt{3}\right)\Xi/2$ – thus converting Eq. (1.741) to

$$1 = \kappa_1\left(\left(\frac{1+\iota\sqrt{3}}{2}\Xi\right)^2 - \Xi\frac{1+\iota\sqrt{3}}{2}\Xi+\Xi^2\right) + \left(\kappa_2+\kappa_3\frac{1+\iota\sqrt{3}}{2}\Xi\right)\left(\frac{1+\iota\sqrt{3}}{2}\Xi+\Xi\right); \tag{1.745}$$

expansion of the square in the first term, lumping of similar factors, elimination of the last parentheses, and realization that $\iota^2=-1$ then support transformation of Eq. (1.745) to

$$1 = \kappa_1\left(\frac{1+2\iota\sqrt{3}-3}{4}\Xi^2 - \frac{1+\iota\sqrt{3}}{2}\Xi^2+\Xi^2\right) + \kappa_2\frac{1+\iota\sqrt{3}}{2}\Xi+\kappa_2\Xi+\kappa_3\left(\frac{1+\iota\sqrt{3}}{2}\right)^2\Xi^2+\kappa_3\frac{1+\iota\sqrt{3}}{2}\Xi^2 \tag{1.746}$$

– where condensation of terms alike, coupled with factoring out of Ξ^2, $\kappa_2\Xi$, and $k_3\Xi^2$ (as appropriate) further give

$$1 = \kappa_1\left(\frac{-2+\iota 2\sqrt{3}}{4} - \frac{1+\iota\sqrt{3}}{2}+1\right)\Xi^2 + \kappa_2\left(\frac{1+\iota\sqrt{3}}{2}+1\right)\Xi+\kappa_3\left(\frac{(1+\iota\sqrt{3})^2}{4}+\frac{1+\iota\sqrt{3}}{2}\right)\Xi^2. \tag{1.747}$$

After an extra application of Newton's binomial formula and elimination of parentheses, Eq. (1.747) becomes

$$1 = \kappa_1\frac{-2+2\iota\sqrt{3}-2-\iota 2\sqrt{3}+4}{4}\Xi^2+\kappa_2\frac{1+\iota\sqrt{3}+2}{2}\Xi + \kappa_3\frac{1+\iota 2\sqrt{3}-3+2+\iota 2\sqrt{3}}{4}\Xi^2; \tag{1.748}$$

lumping of terms alike produces

$$1=\kappa_2\frac{3+\iota\sqrt{3}}{2}\Xi+\kappa_3\iota\sqrt{3}\Xi^2, \tag{1.749}$$

where (as expected) the κ_1-containing term vanished. Equation (1.749) may be algebraically reorganized as

$$\frac{3}{2}\Xi\kappa_2-1+\iota\left(\frac{\sqrt{3}}{2}\Xi\kappa_2+\sqrt{3}\Xi^2\kappa_3\right)=0 \tag{1.750}$$

– which obviously requires

$$\frac{3}{2}\Xi\kappa_2-1=0, \tag{1.751}$$

coupled with

$$\frac{\sqrt{3}}{2}\Xi\kappa_2+\sqrt{3}\Xi^2\kappa_3=0 \tag{1.752}$$

for universal validity, no matter the value taken by Ξ. Equation (1.751) promptly produces

$$\kappa_2=\frac{2}{3\Xi}, \tag{1.753}$$

which may be inserted in Eq. (1.752) to generate

$$\frac{\sqrt{3}}{2}\Xi\frac{2}{3\Xi}+\sqrt{3}\Xi^2\kappa_3=0; \tag{1.754}$$

isolation of κ_3 finally gives

$$\kappa_3=-\frac{1}{3\Xi^2}. \tag{1.755}$$

In view of Eqs. (1.744), (1.753), and (1.755), one may revisit Eq. (1.739) as

$$\xi_2 t=\frac{1}{3\Xi^2}\ln\frac{R_{b,0}+\Xi}{R_b+\Xi}+\frac{-\frac{1}{3\Xi^2}}{2}\ln\frac{R_{b,0}^2-\Xi R_{b,0}+\Xi^2}{R_b^2-\Xi R_b+\Xi^2}+\frac{2\frac{2}{3\Xi}+\left(-\frac{1}{3\Xi^2}\right)\Xi}{\sqrt{3}\Xi}\left(\tan^{-1}\frac{2R_{b,0}-\Xi}{\sqrt{3}\Xi}-\tan^{-1}\frac{2R_b-\Xi}{\sqrt{3}\Xi}\right), \tag{1.756}$$

where both sides were previously multiplied by ξ_2 for convenience; Eq. (1.756) breaks down to

$$\xi_2 t=\frac{1}{3\Xi^2}\ln\frac{R_{b,0}+\Xi}{R_b+\Xi}-\frac{1}{6\Xi^2}\ln\frac{R_{b,0}^2-\Xi R_{b,0}+\Xi^2}{R_b^2-\Xi R_b+\Xi^2}+\frac{1}{\sqrt{3}\Xi^2}\left(\tan^{-1}\frac{2R_{b,0}-\Xi}{\sqrt{3}\Xi}-\tan^{-1}\frac{2R_b-\Xi}{\sqrt{3}\Xi}\right) \tag{1.757}$$

upon lumping terms alike, and dropping similar factors from numerator and denominator. A final multiplication of both sides by Ξ^2, complemented by division of both numerator and denominator of the argument of the first and third terms by $R_{b,0}$, and of the second term by $R_{b,0}^2$ transform Eq. (1.757) to

$$\boxed{\Gamma t=\frac{1}{3}\ln\frac{1+\vartheta}{\frac{R_b}{R_{b,0}}+\vartheta}-\frac{1}{6}\ln\frac{1-\vartheta+\vartheta^2}{\left(\frac{R_b}{R_{b,0}}\right)^2-\vartheta\frac{R_b}{R_{b,0}}+\vartheta^2}+\frac{1}{\sqrt{3}}\left(\tan^{-1}\frac{2-\vartheta}{\sqrt{3}\vartheta}-\tan^{-1}\frac{2\frac{R_b}{R_{b,0}}-\vartheta}{\sqrt{3}\vartheta}\right)} \tag{1.758}$$

– which conveys $R_b\equiv R_b\{t\}$ implicitly, and eventually $T_b\equiv T_b\{t\}$ upon insertion in Eq. (1.701); conversion to spatial location x along the plug-flow unit readily ensues, after bringing Eq. (1.695) on board, via mere multiplication of t by Q_f/A_f upon integration from $x|_{t=0}=0$. Auxiliary parameter ϑ is hereby defined as

$$\vartheta\equiv\frac{\Xi}{R_{b,0}}=\frac{\left(\frac{\xi_1}{\xi_2}\right)^{\frac{1}{3}}}{R_{b,0}}=\sqrt[3]{\frac{\xi_1}{\xi_2 R_{b,0}^3}}, \tag{1.759}$$

where Eq. (1.718) was taken into account – or else

$$\vartheta\equiv\sqrt[3]{\frac{\frac{\mathcal{R}UT_s}{\lambda M_s P}\left(T_s-T_{f,0}\right)-\frac{4\pi UN}{3\rho_f c_{P,f}}R_{b,0}^3}{\frac{4\pi UN}{3\rho_f c_{P,f}}R_{b,0}^3}} \tag{1.760}$$

in view of Eqs. (1.706) and (1.707), which further produces

$$\boxed{\vartheta\equiv\sqrt[3]{\frac{3\mathcal{R}\rho_f c_{P,f}T_s\left(T_s-T_{f,0}\right)}{4\pi N\lambda M_s P R_{b,0}^3}-1}} \tag{1.761}$$

following multiplication of numerator and denominator by $3\rho_f c_{P,f}/4\pi UNR_{b,0}^3$; while Γ is defined as

$$\Gamma\equiv\xi_2\Xi^2=\xi_2\left(\frac{\xi_1}{\xi_2}\right)^{\frac{2}{3}}=\xi_1^{\frac{2}{3}}\xi_2^{\frac{1}{3}} \tag{1.762}$$

as per Eq. (1.718) – thus giving rise to

$$\boxed{\Gamma=\left(\frac{\mathcal{R}UT_s}{\lambda M_s P}\left(T_s-T_{f,0}\right)-\frac{4\pi UN}{3\rho_f c_{P,f}}R_{b,0}^3\right)^{\frac{2}{3}}\left(\frac{4\pi UN}{3\rho_f c_{P,f}}\right)^{\frac{1}{3}}}, \tag{1.763}$$

again at the expense of Eqs. (1.706) and (1.707). The functional dependence of R_b on t is conveyed by Eq. (1.758), in implicit form; a reverse plot, of $R_b/R_{b,0}$ versus Γt, is provided as Fig. 1.50 – and encompasses selected values of parameter ϑ. When this parameter takes small vaues, say, below unity, there is initially a fast decrease in steam bubble radius, followed by a slower and slower decrease – with curvature decreasing with larger ϑ; the trend substantially differs when ϑ is large, say, above unity, with an almost linear decrease in bubble radius with time – along with a higher slope with increasing ϑ.

The time taken for the bubble to vanish, or collapse time t_{coll} – when $T_{f,\max}$ as per Eq. (1.702) is reached, may be obtained after setting $R_b=0$ in Eq. (1.758), viz.

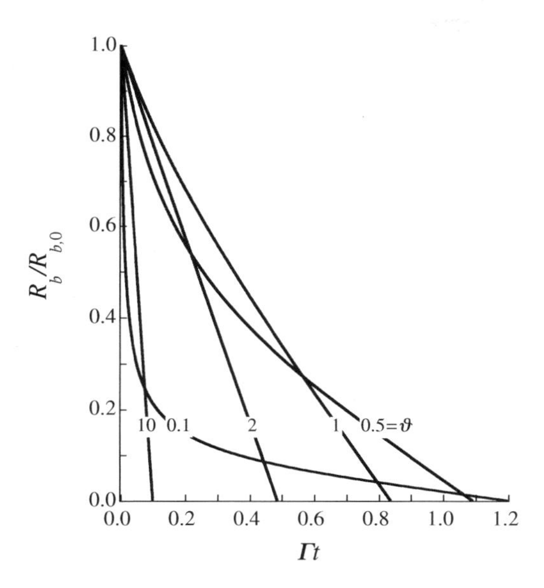

FIGURE 1.50 Evolution of normalized radius of steam bubbles, $R_b/R_{b,0}$, versus dimensionless time, Γt, for selected values of dimensionless parameter ϑ, in the heating stage of sterilization via direct heating.

$$\Gamma t_{coll} = \frac{1}{3}\ln\frac{1+\vartheta}{\vartheta} - \frac{1}{6}\ln\frac{1-\vartheta+\vartheta^2}{\vartheta^2} + \frac{1}{\sqrt{3}}\left(\tan^{-1}\frac{2-\vartheta}{\sqrt{3}\vartheta} - \tan^{-1}\frac{-\vartheta}{\sqrt{3}\vartheta}\right), \tag{1.764}$$

which degenerates to

$$\Gamma t_{coll} = \frac{1}{3}\ln\frac{1+\vartheta}{\vartheta} - \frac{1}{6}\ln\frac{1-\vartheta+\vartheta^2}{\vartheta^2} + \frac{\sqrt{3}}{3}\left(\tan^{-1}\left\{\frac{2\sqrt{3}}{3\vartheta} - \frac{\sqrt{3}}{3}\right\} + \tan^{-1}\frac{\sqrt{3}}{3}\right) \tag{1.765}$$

after multiplying numerator and denominator by $\sqrt{3}$ and splitting fractions – besides recalling that the trigonometric tangent (and thus its reverse) is an odd function; 1/3, and then 1/2 may be factored out, and the operational rules of a logarithm applied to get

$$\Gamma t_{coll} = \frac{1}{3}\left(\begin{array}{l}\frac{1}{2}\left(2\ln\frac{1+\vartheta}{\vartheta} - \ln\frac{1-\vartheta+\vartheta^2}{\vartheta^2}\right) + \\ \sqrt{3}\left(\tan^{-1}\left\{\frac{2\sqrt{3}}{3\vartheta} - \frac{\sqrt{3}}{3}\right\} + \tan^{-1}\frac{\sqrt{3}}{3}\right)\end{array}\right)$$
$$= \frac{1}{3}\left(\frac{1}{2}\ln\frac{\left(\frac{1+\vartheta}{\vartheta}\right)^2}{\frac{1-\vartheta+\vartheta^2}{\vartheta^2}} + \sqrt{3}\left(\tan^{-1}\left\{\frac{2\sqrt{3}}{3\vartheta} - \frac{\sqrt{3}}{3}\right\} + \tan^{-1}\frac{\sqrt{3}}{3}\right)\right), \tag{1.766}$$

where further algebraic rearrangement unfolds

$$\boxed{\Gamma t_{coll} = \frac{1}{3}\left(\ln\sqrt{\frac{(1+\vartheta)^2}{1-\vartheta+\vartheta^2}} + \sqrt{3}\left(\tan^{-1}\left\{\frac{2\sqrt{3}}{3\vartheta} - \frac{\sqrt{3}}{3}\right\} + \tan^{-1}\frac{\sqrt{3}}{3}\right)\right)} \tag{1.767}$$

as graphically depicted in Fig. 1.51 – with time ordinate proportional to (longitudinal) coordinate along flow, as per Eq. (1.695). The curve in this figure represents the loci of the horizontal intercepts in Fig. 1.50, with a monotonically lower t_{coll} at higher ϑ; in addition, low- and high-ϑ horizontal asymptotes are apparent. The former is described by

$$\lim_{\vartheta\to 0}\Gamma t_{coll} = \frac{1}{3}\left(\ln\sqrt{\frac{1}{1}} + \sqrt{3}\left(\tan^{-1}\left\{\frac{2\sqrt{3}}{0} - \frac{\sqrt{3}}{3}\right\} + \tan^{-1}\frac{\sqrt{3}}{3}\right)\right), \tag{1.768}$$

stemming directly from Eq. (1.767) – which simplifies to

$$\lim_{\vartheta\to 0}\Gamma t_{coll} = \frac{1}{3}\left(\ln 1 + \sqrt{3}\left(\tan^{-1}\infty + \tan^{-1}\frac{\sqrt{3}}{3}\right)\right). \tag{1.769}$$

Equation (1.769) reduces further to

$$\lim_{\vartheta\to 0}\Gamma t_{coll} = \frac{\sqrt{3}}{3}\left(\frac{\pi}{2} + \frac{\pi}{6}\right) = \frac{\sqrt{3}}{3}\frac{3\pi+\pi}{6} = \frac{\sqrt{3}}{3}\frac{4\pi}{6}, \tag{1.770}$$

since ln 1 is nil and $\tan^{-1}\infty$ equals $\pi/2$; and eventually to

$$\lim_{\vartheta\to 0}\Gamma t_{coll} = \frac{2\sqrt{3}\pi}{9} \approx 1.21, \tag{1.771}$$

thus justifying the vertical intercept in Fig. 1.51 and the amplitude chosen for the horizontal scale in Fig. 1.50. By the same token, one may search for the asymptotic behavior of Eq. (1.767) when ϑ grows unbounded, viz.

$$\lim_{\vartheta\to\infty}\Gamma t_{coll} = \frac{1}{3}\left(\lim_{\vartheta\to\infty}\ln\sqrt{\frac{\vartheta^2}{\vartheta^2}} + \sqrt{3}\left(\tan^{-1}\left\{\frac{2\sqrt{3}}{\infty} - \frac{\sqrt{3}}{3}\right\} + \tan^{-1}\frac{\sqrt{3}}{3}\right)\right) \tag{1.772}$$

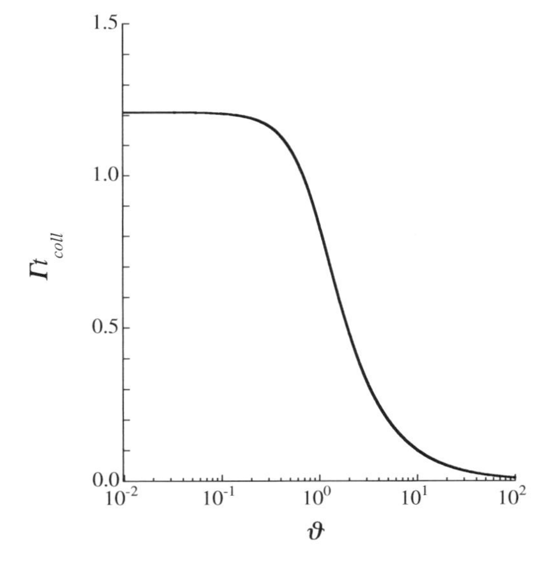

FIGURE 1.51 Variation of dimensionless bubble collapse time, Γt, versus dimensionless parameter ϑ, in the heating stage of sterilization via direct heating.

because $\vartheta^2 >> 1 > 1 - \vartheta$, which further reduces to

$$\lim_{\vartheta \to \infty} \Gamma t_{coll} = \frac{1}{3}\left(\ln 1 + \sqrt{3}\left(\tan^{-1}\left\{ 0 - \frac{\sqrt{3}}{3} \right\} + \tan^{-1}\frac{\sqrt{3}}{3} \right) \right); \quad (1.773)$$

Eq. (1.773) is equivalent to

$$\lim_{\vartheta \to \infty} \Gamma t_{coll} = \frac{\sqrt{3}}{3}\left(-\tan^{-1}\frac{\sqrt{3}}{3} + \tan^{-1}\frac{\sqrt{3}}{3} \right), \quad (1.774)$$

due to the odd nature of inverse tangent and the nil value of logarithm of unity – which becomes merely

$$\lim_{\vartheta \to \infty} \Gamma t_{coll} = 0, \quad (1.775)$$

i.e. the horizontal axis in Fig. 1.51 drives Γt_{coll} at large ϑ.

After the short transient of initial heating, the hot liquid food at temperature $T_{f,h}$ is kept under adiabatic conditions, at elevated pressure, in a holding tube – for the (holding) time necessary to assure the intended lethality. A sudden pressure release is applied afterward, also adiabatically, so water is forced to partially vaporize, with concomitant (spontaneous and chaotic) formation of bubbles of steam in its bulk as typical of boiling – and the associated latent heat of vaporization is obtained at the expense of cooling the liquid food left behind. In fact, the exact amount of high-pressure steam, Δm_s, condensed to liquid water and cooled down to increase the liquid food temperature from $T_{f,0}$ up to $T_{f,h}$ satisfies

$$\lambda \Delta m_s + \Delta m_s c_{P,l}\left(T_s - T_{f,h}\right) = m_{f,0} c_{P,f}\left(T_{f,h} - T_{f,0}\right), \quad (1.776)$$

where λ denotes latent heat of condensation of steam, $c_{P,l}$ denotes specific heat capacity of the resulting condensate, and $m_{f,0}$ denotes mass of liquid food; its specific heat capacity, $c_{P,f}$ is assumed to remain constant over the temperature and pressure changes at stake, while λ pertains to condensation at a single temperature, T_s. The first term in the left-hand side of Eq. (1.776) represents latent heat of condensation of steam, and the second term represents sensible heat released by the corresponding condensate until it reaches the temperature of the liquid food – while the right-hand side represents sensible heat gained by the original liquid food itself. In practice, Eq. (1.776) may be well approximated by

$$\lambda \Delta m_s \approx m_{f,0} c_{P,f}\left(T_{f,h} - T_{f,0}\right); \quad (1.777)$$

because the final holding temperature is normally close to the steam temperature in the first place. The purpose of the cooling stage, with target (final) temperature $T_{f,c}$, is to essentially reverse both the gain in enthalpy and the gain in mass by the liquid food – so as to retrieve the departing physicochemical conditions as close as possible. Therefore, Eq. (1.777) will serve as template to write

$$\lambda \Delta m_s' = m_{f,0} c_{P,f}\left(T_{f,h} - T_{f,c}\right) + \Delta m_s' c_{P,l}\left(T_{f,h} - T_{f,c}\right), \quad (1.778)$$

on the hypothesis that λ (now regarded as heat of vaporization) does not significantly depend on temperature – at least within the range of interest, since vaporization now takes place at an increasingly lower temperature; here m_s' denoting mass of vaporized water. Since $\Delta m_s' << m_{f,0}$ in general, Eq. (1.778) simplifies to

$$\lambda \Delta m_s' \approx m_{f,0} c_{P,f}\left(T_{f,h} - T_{f,c}\right); \quad (1.779)$$

upon comparative inspection of Eqs. (1.777) and (1.779) – and because $T_{f,c} \approx T_{f,0}$ is normally used as a goal, one readily concludes that

$$\Delta m_s' \approx \Delta m_s \quad (1.780)$$

when the extra enthalpy added to the food owing to steam condensation has been withdrawn – which implies removal of essentially the same amount of water added as steam condensate in the first place.

The underlying mass balance during the flash period – considering a well-stirred container downstream, takes the form

$$\boxed{-\frac{dm_f}{dt} = k_s A_{fsh}\left(P_s^{\sigma}\{T_f\} - P_0\right),} \quad (1.781)$$

where the left-hand side represents (negative) rate of accumulation of food mass and the right-hand side represents mass flow rate of vaporization of water from the said food as per Darcy's law; here m_f denotes mass of liquid food, k_s denotes a corrected permeability of steam through the vessel (assumed to be essentially independent of temperature, and given by the ratio of permeability to viscosity), A_{fsh} denotes surface area of liquid food undergoing flash, P_s^{σ} denotes saturation pressure of steam at fluid temperature T_f, and P_0 denotes imposed (outer) overall pressure. The corresponding initial condition reads

$$\boxed{m_f\big|_{t=0} = m_{f,h} = m_{f,0} + \Delta m_s,} \quad (1.782)$$

where $m_{f,h}$ denotes mass of liquid food after holding; remember that steam condensate, Δm_s, was added to the initial mass of liquid food, $m_{f,0}$, to yield the actual amount of liquid subjected to holding, and subsequently to cooling. The associated enthalpic balance looks like

$$\boxed{\lambda \frac{dm_f}{dt} = m_f c_{P,f}\frac{dT_f}{dt},} \quad (1.783)$$

as a result of plain conversion of sensible heat to latent heat – coupled with assumption of a pseudo steady state, described by $T_f dm_f/dt << m_f dT_f/dt$ so that $d(m_f c_{P,f} T_f) \approx m_f c_{P,f} dT_f/dt$; the missing initial condition may thus be coined as

$$\boxed{T_f\big|_{t=0} = T_{f,h},} \quad (1.784)$$

where $T_{f,h}$ denotes temperature prevailing at holding stage. Equation (1.783) may reappear as

$$\frac{dm_f}{m_f} = \frac{c_{P,f}}{\lambda} dT_f \quad (1.785)$$

upon multiplication of both sides by $dt/m_f\lambda$; integration ensues as

$$\int_{m_{f,h}}^{m_f} \frac{d\tilde{m}_f}{\tilde{m}_f} = \frac{c_{P,f}}{\lambda}\int_{T_{f,h}}^{T_f} d\tilde{T}_f, \quad (1.786)$$

using Eqs. (1.782) and (1.784) as initial conditions. The fundamental theorem of integral calculus permits transformation of Eq. (1.786) to

$$\ln \tilde{m}_f\Big|_{m_{f,h}}^{m_f} = \frac{c_{P,f}}{\lambda}\tilde{T}_f\Big|_{T_{f,h}}^{T_f}, \quad (1.787)$$

which is equivalent to

$$\ln \frac{m_f}{m_{f,h}} = \frac{c_{P,f}}{\lambda}\left(T_f - T_{f,h}\right); \tag{1.788}$$

after taking exponentials of both sides, Eq. (1.788) becomes

$$\boxed{\frac{m_f}{m_{f,h}} = \exp\left\{-\frac{c_{P,f}}{\lambda}\left(T_{f,h} - T_f\right)\right\}} \tag{1.789}$$

that entertains a direct relationship between m_f and T_f. Application of differentials to both sides of Eq. (1.789) leads, in turn, to

$$dm_f = m_{f,h} \exp\left\{\frac{c_{P,f}}{\lambda}\left(T_f - T_{f,h}\right)\right\}\frac{c_{P,f}}{\lambda} dT_f, \tag{1.790}$$

or else

$$-\frac{dm_f}{dt} = -\frac{m_{f,h}c_{P,f}}{\lambda}\exp\left\{-\frac{c_{P,f}}{\lambda}\left(T_{f,h} - T_f\right)\right\}\frac{dT_f}{dt} \tag{1.791}$$

after dividing both sides by $-dt$; elimination of $-dm_f/dt$ between Eqs. (1.781) and (1.791) gives then rise to

$$k_s A_{fsh}\left(P_s^\sigma\{T_f\} - P_0\right) = -\frac{m_{f,h}c_{P,f}}{\lambda}\exp\left\{-\frac{c_{P,f}}{\lambda}\left(T_{f,h} - T_f\right)\right\}\frac{dT_f}{dt}, \tag{1.792}$$

where isolation of the differential term unfolds

$$\boxed{\frac{dT_f}{dt} = -\frac{\lambda k_s A_{fsh}\left(P_s^\sigma\{T_f\} - P_0\right)}{m_{f,h}c_{P,f}\exp\left\{\frac{c_{P,f}}{\lambda}\left(T_f - T_{f,h}\right)\right\}}} \tag{1.793}$$

– thus leading to a differential equation that will produce $T_f \equiv T_f\{t\}$ explicitly upon solution. All that is needed now is a relationship between P_s^σ and T_f – and one may accordingly resort to Antoine's equation, viz.

$$\boxed{\log_{10} P_s^\sigma = A_s - \frac{B_s}{C_s + T_f},} \tag{1.794}$$

which may instead appear as

$$\frac{\ln P_s^\sigma}{\ln 10} = A_s - \frac{B_s}{C_s + T_f} \tag{1.795}$$

given the operational property underlying change in the base of a logarithm; upon multiplying both sides by ln 10 and taking exponentials thereafter, Eq. (1.795) becomes

$$P_s^\sigma = \exp\left\{\left(A_s - \frac{B_s}{C_s + T_f}\right)\ln 10\right\} \tag{1.796}$$

– and this rationale may as well be applied to pressure P_0 to get

$$P_0 = \exp\left\{\left(A_s - \frac{B_s}{C_s + T_0^\sigma}\right)\ln 10\right\}, \tag{1.797}$$

where $T_0^\sigma < T_{f,0}$ denotes temperature at which the vapor pressure of water equals P_0. Note that P_0 is often set equal to the vapor pressure of water at room temperature, at the expense of some form of vacuum pump; whereas the target temperature of the final product lies somewhat above the said room temperature – to guarantee sufficient driving force for low pressure-mediated boiling, and associated convection. Expansion of P_s^σ via Taylor's series about $T_f = T_0^\sigma$, truncated after the linear term, takes the form

$$P_s^\sigma \approx \exp\left\{\left(A_s - \frac{B_s}{C_s + T_f}\right)\ln 10\right\}\Bigg|_{T_f = T_0^\sigma} + \exp\left\{\left(A_s - \frac{B_s}{C_s + T_f}\right)\ln 10\right\}\frac{B_s \ln 10}{\left(C_s + T_f\right)^2}\Bigg|_{T_f = T_0^\sigma}\left(T_f - T_0^\sigma\right), \tag{1.798}$$

based on Eq. (1.796); insertion of Eq. (1.797), complemented with algebraic simplification give rise to

$$P_s^\sigma = P_0 + \frac{B_s P_0 \ln 10}{\left(C_s + T_0^\sigma\right)^2}\left(T_f - T_0^\sigma\right), \tag{1.799}$$

which is typically valid within the range $[P_0, P_s^\sigma]$ of interest in industrial practice. By the same token, one may expand the outstanding exponential function in Eq. (1.793) about $T_f = T_{f,h}$, again truncated after the linear term, to get

$$\exp\left\{\frac{c_{P,f}}{\lambda}\left(T_f - T_{f,h}\right)\right\} \approx \exp\left\{\frac{c_{P,f}}{\lambda}\left(T_f - T_{f,h}\right)\right\}\Bigg|_{T_f = T_{f,h}} + \exp\left\{\frac{c_{P,f}}{\lambda}\left(T_f - T_{f,h}\right)\right\}\frac{c_{P,f}}{\lambda}\Bigg|_{T_f = T_{f,h}}\left(T_f - T_{f,h}\right), \tag{1.800}$$

which readily becomes

$$\exp\left\{\frac{c_{P,f}}{\lambda}\left(T_f - T_{f,h}\right)\right\} = 1 + \frac{c_{P,f}}{\lambda}\left(T_f - T_{f,h}\right). \tag{1.801}$$

Insertion of Eqs. (1.799) and (1.801) transforms Eq. (1.793) to

$$\frac{dT_f}{dt} = -\frac{\lambda k_s A_{fsh}\left(P_0 + \frac{B_s P_0 \ln 10}{\left(C_s + T_0^\sigma\right)^2}\left(T_f - T_0^\sigma\right) - P_0\right)}{m_{f,h}c_{P,f}\left(1 + \frac{c_{P,f}}{\lambda}\left(T_f - T_{f,h}\right)\right)}, \tag{1.802}$$

where cancellation of symmetrical terms permits simplification to

$$-\frac{dT_f}{dt} = -\frac{\lambda k_s A_{fsh} B_s P_0 \ln 10\left(T_f - T_0^\sigma\right)}{m_{f,h}c_{P,f}\left(C_s + T_0^\sigma\right)^2\left(1 + \frac{c_{P,f}}{\lambda}\left(T_f - T_{f,h}\right)\right)}; \tag{1.803}$$

integration becomes now possible via separation of variables coupled with straightforward algebraic reorganization, i.e.

$$-\int_{T_{f,h}}^{T_f} \frac{\frac{c_{P,f}}{\lambda}\tilde{T}_f + \left(1 - \frac{c_{P,f}}{\lambda}T_{f,h}\right)}{\tilde{T}_f - T_0^\sigma} d\tilde{T}_f = \frac{\lambda k_s A_{fsh} B_s P_0 \ln 10}{m_{f,h}c_{P,f}\left(C_s + T_0^\sigma\right)^2}\int_0^t d\tilde{t}, \tag{1.804}$$

written also with the help of Eq. (1.784). Since the kernel in the left-hand side of Eq. (1.804) is an irregular rational fraction (i.e. with degree of numerator equal to degree of denominator), polynomial division should first be applied to get

$$\frac{\frac{c_{P,f}}{\lambda}T_f+\left(1-\frac{c_{P,f}}{\lambda}T_{f,h}\right)}{T_f-T_0^\sigma}=\frac{c_{P,f}}{\lambda}+\frac{1-\frac{c_{P,f}}{\lambda}T_{f,h}+\frac{c_{P,f}}{\lambda}T_0^\sigma}{T_f-T_0^\sigma}; \tag{1.805}$$

Eq. (1.804) accordingly becomes

$$-\int_{T_{f,h}}^{T_f}\frac{c_{P,f}}{\lambda}d\tilde{T}_f-\int_{T_{f,h}}^{T_f}\frac{1-\frac{c_{P,f}}{\lambda}\left(T_{f,h}-T_0^\sigma\right)}{\tilde{T}_f-T_0^\sigma}d\tilde{T}_f=\frac{\lambda k_s A_{fsh}B_s P_0\ln 10}{m_{f,h}c_{P,f}\left(C_s+T_0^\sigma\right)^2}\int_0^t d\tilde{t}, \tag{1.806}$$

after splitting the integral in the left-hand side – where constant terms may then be taken off the kernels, and limits of integration reversed on account of taking negatives of their kernels to obtain

$$\frac{c_{P,f}}{\lambda}\int_{T_f}^{T_{f,h}}d\tilde{T}_f+\left(1-\frac{c_{P,f}}{\lambda}\left(T_{f,h}-T_0^\sigma\right)\right)\int_{T_f}^{T_{f,h}}\frac{d\tilde{T}_f}{\tilde{T}_f-T_0^\sigma}=\frac{\lambda k_s A_{fsh}B_s P_0\ln 10}{m_{f,h}c_{P,f}\left(C_s+T_0^\sigma\right)^2}\int_0^t d\tilde{t}. \tag{1.807}$$

The fundamental theorem of integral calculus may finally be invoked to write

$$\frac{c_{P,f}}{\lambda}\tilde{T}_f\Big|_{T_f}^{T_{f,h}}+\left(1-\frac{c_{P,f}}{\lambda}\left(T_{f,h}-T_0^\sigma\right)\right)\ln\left\{\tilde{T}_f-T_0^\sigma\right\}\Big|_{T_f}^{T_{f,h}}=\frac{\lambda k_s A_{fsh}B_s P_0\ln 10}{m_{f,h}c_{P,f}\left(C_s+T_0^\sigma\right)^2}\tilde{t}\Big|_0^t \tag{1.808}$$

stemming from Eq. (1.807), which is equivalent to

$$\frac{\lambda k_s A_{fsh}B_s P_0\ln 10}{m_{f,h}c_{P,f}\left(C_s+T_0^\sigma\right)^2}t=\frac{c_{P,f}}{\lambda}\left(T_{f,h}-T_f\right)+\left(1-\frac{c_{P,f}}{\lambda}\left(T_{f,h}-T_0^\sigma\right)\right)\ln\frac{T_{f,h}-T_0^\sigma}{T_f-T_0^\sigma}; \tag{1.809}$$

division of both sides by $1-c_{P,f}(T_{f,h}-T_0^\sigma)/\lambda$ generates

$$\frac{\lambda k_s A_{fsh}B_s P_0\ln 10}{m_{f,h}c_{P,f}\left(C_s+T_0^\sigma\right)^2\left(1-\frac{c_{P,f}}{\lambda}\left(T_{f,h}-T_0^\sigma\right)\right)}t=\frac{\frac{c_{P,f}}{\lambda}\left(T_{f,h}-T_f\right)}{1-\frac{c_{P,f}}{\lambda}\left(T_{f,h}-T_0^\sigma\right)}+\ln\frac{T_{f,h}-T_0^\sigma}{T_f-T_0^\sigma}, \tag{1.810}$$

where $T_{f,h}$ is to be added to and subtracted from the denominator of the argument of the logarithm function to give

$$\frac{\lambda k_s A_{fsh}B_s P_0\ln 10}{m_{f,h}c_{P,f}\left(C_s+T_0^\sigma\right)^2\left(1-\frac{c_{P,f}}{\lambda}\left(T_{f,h}-T_0^\sigma\right)\right)}t=\frac{\frac{c_{P,f}}{\lambda}\left(T_{f,h}-T_f\right)}{1-\frac{c_{P,f}}{\lambda}\left(T_{f,h}-T_0^\sigma\right)}+\ln\frac{T_{f,h}-T_0^\sigma}{\left(T_{f,h}-T_0^\sigma\right)-\left(T_{f,h}-T_f\right)}. \tag{1.811}$$

A final division of both numerator and denominator of the argument of the logarithm by $T_{f,h}-T_0^\sigma$ supports transformation of Eq. (1.811) to

$$\boxed{\Phi t=\frac{1}{\Psi-1}\frac{T_{f,h}-T_f}{T_{f,h}-T_0^\sigma}-\ln\left\{1-\frac{T_{f,h}-T_f}{T_{f,h}-T_0^\sigma}\right\},} \tag{1.812}$$

together with a further division of both numerator and denominator of the first term in the right-hand side by $c_{P,f}(T_{f,h}-T_0^\sigma)/\lambda$, and imposition of a minus sign prior to the logarithm on account of taking the reciprocal of its argument; here parameters Ψ and Φ are defined as

$$\boxed{\Psi\equiv\frac{\lambda}{c_{P,f}\left(T_{f,h}-T_0^\sigma\right)}} \tag{1.813}$$

and

$$\boxed{\Phi\equiv\frac{\lambda k_s A_{fsh}B_s P_0\ln 10}{m_{f,h}c_{P,f}\left(C_s+T_0^\sigma\right)^2\left(1-\frac{c_{P,f}}{\lambda}\left(T_{f,h}-T_0^\sigma\right)\right)},} \tag{1.814}$$

respectively. The variation of $(T_{f,h}-T_f)/(T_{f,h}-T_0^\sigma)$ versus Φt is plotted in Fig. 1.52, for selected values of parameter Ψ. Cooling of the liquid food accordingly supports increase in the corresponding

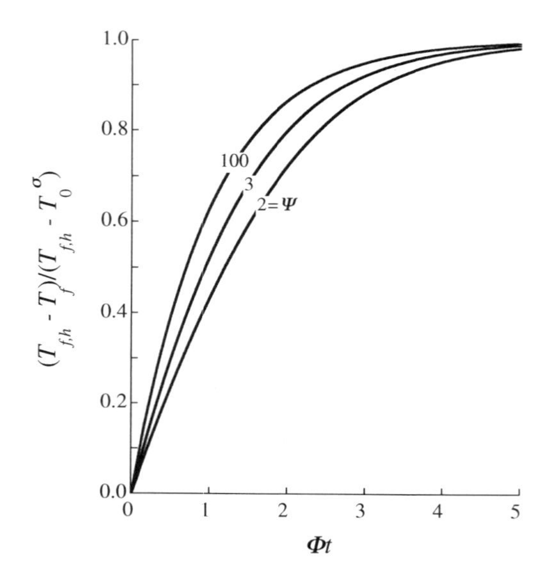

FIGURE 1.52 Evolution of normalized temperature in liquid food, $(T_{f,h}-T_f)/(T_{f,h}-T_0^\sigma)$, versus dimensionless time, Φt, for selected values of dimensionless parameter Ψ, in the cooling stage of sterilization via direct heating.

dimensionless version of temperature to its inlet value; the earlier portion of the said variation is relatively fast due to a large driving force for mass transfer – which gradually slows down with the decrease in P_s^σ as temperature falls, in agreement with Eq. (1.796), thus pushing the water vapor pressure toward the overall reduced pressure externally imposed. As expected, the change in temperature of the liquid food is faster at larger Ψ – corresponding to a higher heat of vaporization (and consequently a larger absorption of latent heat through vaporization of previously condensated steam) or a lower specific heat capacity (i.e. easier conversion of latent heat to temperature drop of liquid food left behind), note that the discriminating effect of Ψ is relevant only within its low-value range (say, around unity). Finally, it should be stressed that plain Fourier's law, with $T_f - T_0^\sigma$ for driving force of heat transfer multiplied by UA_{fsh}/λ, cannot be used *in lieu* of Eq. (1.781) – because the cooling process is adiabatic from the point of view of the liquid food; however, the system is open in that water vapor is eliminated once formed – thus allowing constant P_0 appear in Eq. (1.781).

PROBLEM 1.16

In food plants, holding time turns out to be the only practically relevant parameter when implementing an UHT process – meaning that heating up and cooling down periods are sufficiently short, and associated with much lower median temperatures.

a) Estimate heating time, assuming that $U=1.39\times10^3$ W.m^{-2}.K^{-1}, $T_s=131$°C, $M_s=1.80\times10^{-2}$ kg.mol^{-1}, $\lambda=2.17\times10^6$ J.kg^{-1}, $P=2.82\times10^5$ Pa, $T_{f,0}=25$°C, $N=2\times10^6$ m^{-3}, $\rho_f=10^3$ kg.m^{-3}, and $c_{P,f}=4.18\times10^3$ J.kg^{-1}.K^{-1}; as well as $\sigma=3.76\times10^{-2}$ N.m^{-1}, $\rho_g=1.55$ kg.m^{-3}, and $R_n=5.9\times10^{-4}$ m.
b) For pure water, within the range of interest, one may take $A_s=10.19$ (log$_{10}$)Pa, $B_s=1731$ (log$_{10}$) Pa.K, and $C_s=-$ 39.72 *K* as per Antoine's equation; besides $k_s=2.25\times10^{-4}$ kg.Pa^{-1}.m^{-2}.s^{-1}. For a reduced pressure $P_0=2.32\times10^3$ Pa and a final temperature $T_f=35$°C, estimate the cooling time – with $A_{fsh}/m_{f,h}\approx 1$ m^2.kg^{-1}.

Solution

a) The bubble radius at release from the nozzle may be estimated via Eq. (3.293) from *Food Proc. Eng.: basics & mechanical operations* as

$$R_{b,0}=\sqrt[3]{\frac{3\sigma R_n}{2(\rho_f-\rho_g)}}$$

$$=\sqrt[3]{\frac{3\cdot3.76\times10^{-2}\cdot5.9\times10^{-4}}{2(10^3-1.55)}}=3.24\times10^{-3}\text{ m;}\quad(1.16.1)$$

one may now resort to Eq. (1.761), also with $\mathcal{R}=8.314$ J.mol^{-1} K^{-1}, to obtain

$$\vartheta=\sqrt[3]{\frac{3\cdot8.314\cdot10^3\cdot4.18\times10^3\cdot404(404-298)}{4\pi\cdot2\times10^6\cdot2.17\times10^6\cdot1.80\times10^{-2}\cdot2.82\times10^5\cdot\left(3.24\times10^{-3}\right)^3}-1}$$

$$=7.79,\quad(1.16.2)$$

since $T_s=131+273=404$ K and $T_{f,0}=25+273=298$ K, coupled with Eq. (1.16.1) for $R_{b,0}$. One may likewise calculate Γ via Eq. (1.763), viz.

$$\Gamma=\left(\frac{8.314\cdot1.39\times10^3\cdot404(404-298)}{2.17\times10^6\cdot1.80\times10^{-2}\cdot2.82\times10^5}-\frac{4\pi\cdot1.39\times10^3\cdot2\times10^6\left(3.24\times10^{-3}\right)^3}{3\cdot10^3\cdot4.18\times10^3}\right)^{\frac{2}{3}}\quad(1.16.3)$$

$$\times\left(\frac{4\pi\cdot1.39\times10^3\cdot2\times10^6}{3\cdot10^3\cdot4.18\times10^3}\right)^{\frac{1}{3}}=1.78\text{ s}^{-1};$$

inspection of the curve in Fig. 1.51, for the value of ϑ conveyed by Eq. (1.16.2), indicates that $\Gamma t_{coll}=0.12$, so

$$t_{coll}=\frac{\Gamma t_{coll}}{\Gamma}=\frac{0.12}{1.78}=0.067\text{ s}\quad(1.16.4)$$

with the aid of Eq. (1.16.3) – meaning that heating up is indeed almost instantaneous, and occurs at a median temperature of $(T_s+T_{f,0})/2=(131+25)/2=78$°C. Therefore, most lethality is expected to occcur during the holding stage, usually ranging within 2–20 s; this constitutes a major operational difference relative to classical retorting.

b) Based on Eq. (1.797), one finds

$$T_0^\sigma=\frac{B_s}{A_s-\log_{10}P_0}-C_s$$

$$=\frac{1731}{10.19-\log_{10}2.32\times10^3}-(-39.72)\quad(1.16.5)$$

$$=293\text{K},$$

as temperature at which the vapor pressure of water equals the reduced pressure imposed. If $\lambda=2.17\times10^6$J.kg^{-1} and $c_{P,f}=4.18\times10^3$ J.kg^{-1}.K^{-1} as in a), as well as $T_{f,h}=T_s=131+273=404$ K and $T_f=35+273=308$ K, one promptly obtains

$$\frac{T_{f,h}-T_f}{T_{f,h}-T_0^\sigma}=\frac{404-308}{404-293}=0.86,\quad(1.16.6)$$

also at the expense of Eq. (1.16.5) – together with

$$\Psi=\frac{2.17\times10^6}{4.18\times10^3(404-293)}=4.7\quad(1.16.7)$$

stemming from Eq. (1.813). Inspection of Fig. 1.52, with the aid of Eqs. (1.16.6) and (1.16.7) unfold $\Phi t=2.2$, upon interpolation between adjacent curves; Eq. (1.814) has it that

$$\Phi = \frac{2.17\times10^{6}\cdot2.25\times10^{-4}\cdot1\cdot1731\cdot2.32\times10^{3}\ln10}{4.18\times10^{3}\left(-39.72+293\right)^{2}\left(1-\frac{4.18\times10^{3}}{2.17\times10^{6}}(404-293)\right)}$$

$$=21\,\mathrm{s}^{-1} \tag{1.16.8}$$

for the given data – thus supporting

$$t=\frac{\Phi t}{\Phi}=\frac{2.2}{21}=0.10\ \mathrm{s}. \tag{1.16.9}$$

The resulting cooling is also quite brief, only slightly longer than its heating counterpart, see Eq. (1.16.4), for a median temperature of $(T_f+T_s)/2=(35+131)/2=83$°C – thus confirming that most lethality of UHT treatments comes indeed from the holding time, between 1 and 2 orders of magnitude longer. In view of its constant temperature, the period of holding under adiabatic conditions also permits *per se* a better control of heat applied.

1.3 Chilling

1.3.1 Introduction

Foods at large require cooling, and even refrigeration upon harvest (in the case of plant crops) or slaughter (in the case of animal meats) – to reduce both metabolic activity of the original tissues and growth of adventitious/contaminant microorganisms; this helps minimize decay of sensory and nutritional properties, while allowing extension of shelf-life and ultimately contributing to food safety. Chilling is the unit operation in which temperature of a food is reduced to between 0 and 10°C; it is often combined with such other unit operations as fermentation or pasteurization, or one of several techniques of minimal processing.

HISTORICAL FRAMEWORK

Seasonal harvesting of snow and ice was an ancient practice, estimated to have begun prior to 1000 BC in China – and became a routine practice by Jews for a long time, as a means to cool beverages; ancient Greeks and Romans dug large snow pits insulated with grass, chaff, or branches of trees, for cooling drinks and cold-storing them afterward. Egyptians, and later also Indians cooled water by pouring it in shallow earthen jars, and placing them on the roofs of their houses at night; slaves kept the outside of such jars moistened, so the resulting evaporation would cool the water inside. Persians stored ice in a pit called Yahkhchal – and may have been the first people to resort to cold storage of food for preservation purposes. Under the hot and dry Australian weather, many farmers resorted to a coolgardie safe, or room with hessian curtains hanging from the ceiling and permanently soaked in water; upon evaporation, water would cool the said curtains, and thus the air circulating in the room – where perishables such as fruits, butter, or cured meats were kept.

The history of artificial refrigeration began, however, with William Cullen in 1755, who used a pump to create a partial vacuum over an insulated container of diethyl ether – which then boiled up, and concomitantly absorbed heat from the surrounding air. In 1758, Benjamin Franklin and John Hadley confirmed that evaporation of highly volatile liquids could be used to drive down the temperature of an object close to (and even below) the freezing temperature of water. Michael Faraday was able to liquefy ammonia and other gases in 1820, using high pressures and low temperatures; and in 1834, Jacob Perkins built the first working vapor compression-based, closed-cycle refrigeration system in the world. Nevertheless, the first practical version of such a system was patented only in 1856, by James Harrison; and was soon introduced in breweries and meat-packing houses in Australia. The first gas absorption refrigeration system – using gaseous ammonia dissolved in water, was developed by Ferdinand Carré in 1859 in France; while Carl von Linde patented in 1876 an improved method for liquefying gases in Germany. In 1911, General Electric released a household refrigeration machine powered by gas, and the Monitor Top™ in 1927 – the first refrigerator run on electricity. In 1930, Frigidaire synthesized Freon™, the first synthetic refrigerant – which made domestic refrigerators safer, smaller, lighter, and cheaper ever since.

Low temperatures reduce the rate of microbial growth, and extend its preceding lag time – as a consequence of a complex series of events encompassing changes to cell structure, coupled to decceleration of rate-controlling steps of respiration and metabolism at large. Chilling prevents indeed growth of many mesophilic and all thermophilic microorganisms – since their minimum growth temperatures usually lie within 5–10°C and 30–40°C, respectively. However, psychrotrophs (with maximum growth temperature of 35–40°C) and psychrophiles (with maximum growth temperature of ca. 20°C) can grow, since their minimum growth temperatures range from 0 to 5°C, i.e. well within the chilling range; remember that the former may appear in virtually all foods, whereas the latter are restricted to deep-sea fish. The spoilage microorganisms most commonly found in chilled foods are such Gram$^-$ bacteria as *Acinetobacter* spp., *Aeromonas* spp., *Flavobacterium* spp., and *Pseudomonas* spp. – which produce pigmentation, slime, and off-flavors; and contaminating yeasts – e.g. *Candida* spp., *Debaryomyces* spp., *Kluyveromyces* spp., and *Saccharomyces* spp. Molds – e.g. *Aspergillus* spp., *Cladosporium* spp., *Geotrichum* spp., *Penicillium* spp., and *Rhizopus* spp., tolerate chill temperatures, but are eventually out-competed for their slower growth than bacteria. The sources of such microorganisms are typically contaminated water, and inadequately cleaned equipment or working surfaces. When stored below their minimum growth temperatures, microorganisms in food may gradually die; however, surviving cells will resume growth once the food temperature has been increased to ambient (or even higher, not lethal) temperature.

Chilled foods are normally classified, according to their storage temperature range, as low, medium, and high – as depicted

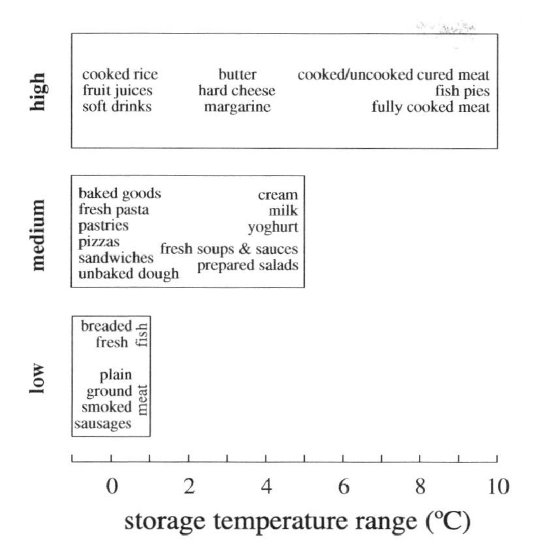

FIGURE 1.53 Classification of chilled foods per range of storage temperatures (▭), with selected examples.

in Fig. 1.53. It should be emphasized that previous implementation of some form of thermal treatment or *pH*-lowering fermentation (or buffer addition) allows storage of food at higher chilling temperatures; low-acid chilled foods (e.g. fresh and precooked meats, pizzas, and unbaked dough) are of particular concern, so they must be prepared, packaged, and stored under strict conditions of hygiene and temperature control – due to the possibility of contamination thereof by pathogens, which would develop (even though slowly) throughout long storage times.

Although chilling to and storage at appropriate sub-ambient temperatures cause little or no reduction in eating quality and nutritional wholesomeness of food, incorrect chilling temperatures, excessive storage times, or presence of mechanical damages to the original crop may promote changes with negative impacts. Such changes include enzymatic browning, lipolysis, wilting, color and flavor deterioration, vitamin decay, and weight loss (namely due to transpiration); besides retrogradation of starch, which specifically causes staling of baked products.

Since the rate of biochemical reactions in fresh foods, as catalyzed by adventitious enzymes, increases with temperature, chilling reduces the said rate – thus retarding respiration and senescence therein. However, a number of other factors constrain shelf-life of chilled foods, associated with type of food and source cultivar, portion of crop (faster-growing parts accordingly exhibit higher metabolic rates), physiological condition at harvest (including degree of maturity, as well as presence of mechanical damage or microbial contamination), temperature at harvest, and relative humidity and atmosphere composition prevailing throughout the postharvest period. If chilling is combined with a modified atmosphere – usually poorer in O_2 and/or richer in CO_2 than the regular atmosphere, then a much greater preservation effect results; this is due to inhibition of insect growth by asphyxiation, coupled with reduction of respiration (and transpiration) rate of fresh produce itself.

HISTORICAL FRAMEWORK

Much of the pioneering work on modified atmosphere packaging was done in Australia and New Zealand in the 1930s – when beef and lamb carcasses were shipped to the United Kingdom under carbon dioxide, to help maintain their freshness. By the same time, it was observed that ships transporting fruits in their enclosed holding rooms experienced an increased shelf-life; in fact, the intrinsic respiratory metabolism of the said fruits reduced the oxygen inventory, and concomitantly increased the CO_2 level in the storage areas – which, in turn, gradually slowed down tissue respiration. During the 1940s and 1950s, apples and pears were stored in warehouses under modified atmospheres, right upon harvest – and could be kept for as much as 6 months, i.e. for twice as long as if plain chilling were employed. Modified atmosphere packages were first used in the 1970s, for some retail packs of bacon and fish in Mexico; microperforation and antifogging layers were meanwhile developed to improve product visibility. However, the major commercial boost was brought about by Marks & Spencer retail chain in the United Kingdom, which introduced a wide range of meat products packaged under modified atmosphere in 1981.

Carbon dioxide also inhibits microbial activity through dissolution in water – via formation of carbonic acid, a mild acid that lowers *pH* on the food surface, and accordingly reduces chances for contamination by bacterial pathogens; further to its direct unfavorable effect upon aerobic activity in food cells. These two factors may, however, bring about physiological disorders in food tissues, for permitting preferential growth of unwanted anaerobic microorganisms – which justify the need for close monitoring and control of CO_2 concentration, when employing modified atmospheres.

Regarding respiration rate, foods have classically been divided between very low, low, moderate, high, very high, and extremely high – as outlined in Fig. 1.54. Note the particularly high rates of respiration of asparagus, broccoli, peas, spinach, and sweetcorn – typically leafy vegetables (or the like); conversely, such dry fruits as nuts or dates exhibit rather low rates of respiration. Typical storage times range from 1 to 3 d in the case of foods possessing extremely high respiration rates, 1 to 2 wk for very high, 2 to 3 wk for high, 5 to 20 wk for moderate, 25 to 50 wk for low, and beyond 50 wk for very low respiration rates.

As emphasized before, marginal changes occur to sensory and nutritional features of food as a consequence of chilling; chilled foods are indeed rated by consumers using such qualifiers as healthy, natural, or fresh – usually synonyms of high-quality, and thus strongly appealing foods. Following wider dissemination of technically more sophisticated cold distribution networks (including chill stores, refrigerated transport, retail chilled display cabinets, and ownership of larger capacity domestic refrigerators) from the 1980s on, substantial product development and concomitant expansion of the chilled food market resulted; desserts, fresh pasta, ready- and organic meals, Oriental (namely, Chinese, Malaysian, Singaporean, and Thai) dishes, prepared and healthy salads, pizza, and sandwiches are but a few cases of continued success.

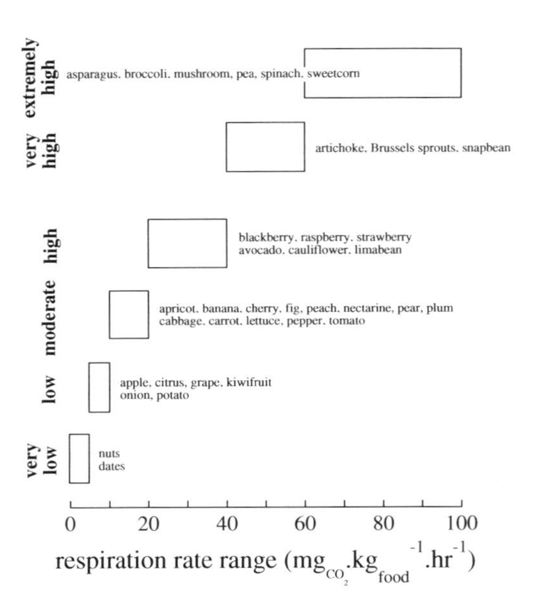

FIGURE 1.54 Classification of chilled foods per range of respiration rates (▭) throughout storage at 5°C, with selected examples of fruits and vegetables.

1.3.2 Product and process overview

Cooling and refrigeration of raw materials slows down spoilage brought about by adventitious enzymes, and native or contaminating microorganisms (as outlined above); as a consequence, shelf-life can be considerably extended. However, rapid cooling of animal carcasses immediately upon slaughter is to be avoided – otherwise, the state known as *rigor mortis* cannot be attained, and cold shortening will eventually prevail.

There are basically two methods of chilling foods – mechanical vapor compression and cryogenics; the principles and mechanisms of generation of cold by either approach will be explained later to further length.

The former encompasses a compressor, a condenser, an expansion valve, and an evaporator as key processing devices – working in a cyclic sequence, and effecting (reversible) physical changes on a given refrigerant fluid. As will be detailed in due course, such changes involve increase in enthalpy of refrigerant vapor, due to an increase in pressure effected by the compressor; decrease in enthalpy of refrigerant vapor, due to a loss (chiefly) of latent heat of condensation at the condenser; maintenance of enthalpy during flash expansion of refrigerant liquid at the (adiabatic) expansion valve; and finally increase in enthalpy, due to a gain as latent heat of vaporization at the evaporator.

A cryogen is a total-loss refrigerant, which cools a food by absorbing sensible heat therefrom, and transforming it to its own latent heat of vaporization (or sublimation); the most common system is solid carbon dioxide – with sublimation point of −78.5°C and latent heat of sublimation of 571.3 $kJ.kg^{-1}$, besides a specific heat capacity of vapor equal to 850 $J.kg^{-1}.K^{-1}$. Therefore, production of solid CO_2, starting from a room temperature of 25°C, entails removal of $0.850(25-(-78.5)) = 88.0$ $kJ.kg^{-1}$ as sensible heat (of cooling), and another 571 $kJ.kg^{-1}$ as latent heat (of sublimation). This means that a mere $88.0/(88.0+571) = 13\%$ of its total cooling power to refrigerate food below room temperature is accounted for by sensible heat – so CO_2 does not require gas-handling equipment to extract most of the said power, while control of chilling temperatures is simpler. Besides obvious risks of cold burns, frostbite, and hypothermia upon exposure to intense cold, the major drawback of using CO_2 is asphyxiation – witha maximum operating threshold of 0.5%(v/v), for operator safety; excess gas is to be removed (if and when necessary) from the processing premises by a powerful exhaust system.

To keep such living commodities as fresh fruits and vegetables under chilled storage, it is necessary to permanently remove sensible heat (also known as field heat) – because thermal insulation is never perfect, and loading/unloading operations admit air at room temperature into the storage room; however, the heat continuously generated by respiratory activity is also to be withdrawn – as illustrated in Fig. 1.55, for glucose as the starting (and also most efficient) energy-yielding nutrient. The said respiration occurs in all food cells, because they are not expected to perform photosynthesis any more (for lack of light and liquid water); since little cell development takes place, most energy made available by respiration will actually be released as heat.

The rate at which the above set of chemical reactions takes place varies with the type of commodity – including cultivar and maturation stage at harvest; furthermore, not all tissues undergo respiration at the same rate, and these rates are far from being identically affected by temperature. The overall behavior of selected fruits and vegetables, with regard to respiration heat, is sketched in Fig. 1.56. Inspection of this figure indicates that fruits are storage commodities exhibiting significant rates of release of heat of respiration – unlike dry plant products, such as seeds and nuts. The curves deviate from linearity because a different enzymatic process becomes rate-controlling at distinct temperatures – characterized by a higher activation energy (and thus a larger slope) as temperature is decreased.

Young, actively growing tissues (e.g. asparagus, broccoli, spinach) exhibit high rates of respiration – as do immature seeds (e.g. green peas, sweet corn); fast developing fruits (e.g. peaches, plums, or black-, blue-, and raspberries) have much higher respiration rates than fruits that develop slowly (e.g. apples, citrus fruits, grapes, pears). In general, most vegetables (other than root crops) have a high initial respiration rate, for the first 1–2 d following harvest; within a few days, however, the respiration rate drops to the so-called steady-state rate.

Once a food product has been chilled, its low temperature is to be kept throughout (refrigerated) storage afterward; sustained cooling is normally conveyed by cold air, produced by mechanical refrigeration units and circulated by air fans. Foods are stored on pallets or racks, in the case of fruits and vegetables; or hung from hooks, in the case of meat carcasses – with sufficient clearance in stacking to enable uniform air circulation. Transport of chilled foods into and out of cold stores may be done manually, or via human-operated forklift trucks, or automated robotic devices. Control of air composition, in the case of modified/controlled atmosphere storage, resorts to sensors that measure thermal conductivity – which changes with fraction of CO_2 (characterized by 0.015 $W.m^{-1}.K^{-1}$ if pure), N_2 (characterized by

$$C_6H_{12}O_6 + 6O_2 \rightarrow 6CO_2 + 6H_2O + 2.667\times10^6\,J$$

FIGURE 1.55 Chemical equation describing aerobic respiration, with indication of heat released per mole of glucose, $C_6H_{12}O_6$.

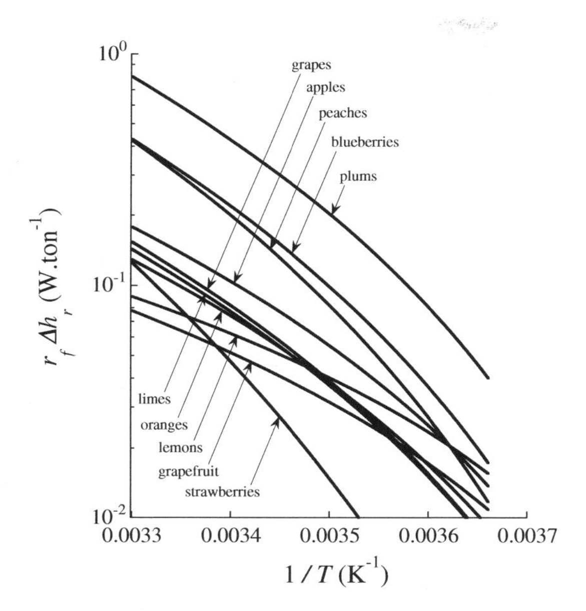

FIGURE 1.56 Variation of specific rate of production of heat associated with respiration, $r_f \Delta h_r$, versus reciprocal absolute temperature, $1/T$, for selected fruits under regular atmosphere and within the usual range of chilling temperatures.

0.024 W.m^{-1}.K^{-1} if pure), and O_2 (characterized by 0.025 W.m^{-1}.K^{-1} if pure); control may also take advantage of infrared absorption. Probes to measure temperature, and devices to monitor and control air moisture can thus be installed *in situ*. Special care is to be exercised in the case of respiring foods – due to production of heat of respiration that must be removed (as outlined above), besides concomitant changes in composition of overhead atmosphere in terms of oxygen, carbon dioxide, and water vapor. If dead spots develop – prone to localized increases in humidity, growth of molds may occur; they not only degrade food from its surface in, but also multiply rapidly upon sporulation – with spores promptly carried away by circulating air. Another critical food in this regard is cheese – for which important flavor development, associated with ripening, occurs throughout storage, even though tissue respiration is not present.

Optimal storage temperatures and typical (maximum) storage periods of selected foods are provided in Fig. 1.57. The optimal chilling temperature lies in the vicinity of 0°C for most fruits and vegetables; the long period of storage of apples is noteworthy, followed by lemons – while blackberries and mushrooms are unique for lying on the other limit of the scale. Optimal humidity in the circulating air normally ranges within 90–100% saturation to avoid drying of food – except in the case of grapes (85%), garlic (65–70%), peppers (60–70%), and squashes (50–70%).

Retail chill storage and display cabinets also resort to chilled air – which circulates internally via forced convection, or natural convection in the case of horizontal apparatuses. The two most common designs are serve-over – as delicatessen cabinets that exhibit food on a chilled base; or vertical, multideck display cabinets – with open-fronted or side glass doors. Besides inexistence of perfect thermal insulators, display lights warm up, and frequent open/close operations by (trained) staff and (relaxed) customers produce losses of cold air – with concomitant exchange with air at (higher) room temperature; therefore, provision for chilling in permanence becomes critical. A central plant that distributes refrigerant to all cabinets proves more economical than a variety of scattered chilling apparatuses – while the heat generated in the condenser may be used for in-store heating; plastic curtains and night blinds on the front of (or above) cabinets to trap cold air have commonly been employed as energy-saving devices.

Temperature monitoring along the whole cold chain is of the utmost relevance, in attempts to guarantee food product quality and safety until consumption; data-loggers are important as well – not only to access the temperature history of storage and thus trace problems of food products downstream, but also to sound alarms if preset limits are exceeded. Such devices are connected to temperature probes that normally measure air temperature, or product temperature to a lesser extent (obviously more accurate, but also likely to cause damage to package/food); these include thermocouples, semiconductor-based thermistors, and Pt-resistance thermometers. The most frequently used thermocouples are of type K (i.e. based on a junction of nickel–chromium to nickel–aluminum), or type T (i.e. based on a junction of copper to copper–nickel); these sensors are rather inexpensive, respond rapidly, and cover linearly a wide temperature range (typically from −180 to 1600°C). Thermistors are, in general, more accurate than thermocouples, yet their useful working range is much narrower (from −40 to 140°C); while Pt-resistances cover from −270 to 850°C, but respond more slowly – while their cost is far higher.

Undesirable changes to some fruits and vegetables are observed when storage temperature lies below a given threshold, specific for each individual crop; the set of unwanted (bio)chemical events that take place below the said threshold are, as a whole, termed chilling injury – detectable as physiological changes caused by imbalanced metabolic activity, with overproduction of metabolites that are toxic to the plant tissues. The lowest safe temperatures of selected plant foods are depicted in Fig. 1.58; lemons, sweet potatoes, and bananas are particularly sensitive to low chilling temperatures throughout storage. Easily perceptible symptoms of chilling injury include surface scald (e.g. aubergines, grapefruit); grey discoloration of flesh (e.g. avocados);

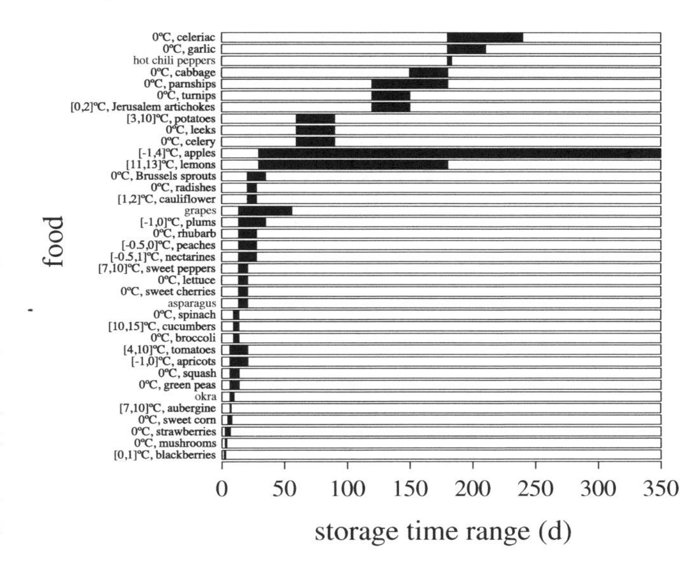

FIGURE 1.57 Ranges of maximum periods of chilled storage (■), for selected foods of plant origin – with indication of recommended temperature interval.

Alternaria rot (e.g. aubergines, ripe tomatoes, sweet peppers); dull, grey/brown skin color (e.g. bananas, honeydew melons, pineapples): pitting and russeting (e.g. cucumbers, green beans, lemons, limes, oranges, papaya, sweet peppers, watermelon); water-soaked spots and softening (e.g. cucumbers, green tomatoes, okra); watery breakdown (e.g. grapefruit, oranges); membrane stain (e.g. lemons, oranges); red blotch (lemons); failure to ripen (e.g. honeydew melon, papaya); discoloration (e.g. okra, ripe tomatoes); poor and off-flavors (e.g. papaya, pineapple); internal discoloration (e.g. potatoes, sweet potato); sweetening (e.g. potatoes); and general decay (e.g. cucumbers, green tomatoes, honeydew melon, papaya, pumpkins, sweet potato). Many of the above symptoms result from loss of cell membrane integrity, and concomitant leakage of solutes – as a consequence of changes in membrane lipid structure, regulatory enzyme activity, and conformation of structural proteins. On the other hand, fresh-cut fruits spoil visually before significant nutrient decay occurs, along with limited losses of total phenolics; while poor appearance is accelerated by excessive trimming.

In animal tissues, aerobic respiration declines rapidly once supply of (oxygenated) blood is discontinued at slaughter. However, muscles contain glycogen, creatinine phosphate, and sugar phosphates – which will still be used to synthesize ATP via glycolysis. Anaerobic respiration, of glycogen to lactic acid, causes *pH* in meat fall from ca. 7 in living animals to 5.4–5.6 after slaughter; when supply of ATP stops, the muscle tissue acquires firmness and becomes inextensible – a status usually termed *rigor mortis*. This condition is attained by 1–30 hr postmortem, depending on animal species and breed, physiological condition at slaughter, and prevailing room temperature; if glycogen is not rate-limiting, a lower ambient temperature extends the time required to reach the said minimum *pH*. A reduced *pH* causes protein denaturation and drip losses; while both lactic acid and inosine monophosphate (a breakdown product of ATP) contribute to final flavor.

In view of the above considerations, carcass cooling is recommended during anaerobic respiration, as it can slow down the associated biochemical processes – thus leading to a final product with the required color and texture. However, rapid chilling to temperatures below 12°C, before anaerobic glycolysis has come to a halt, brings about permanent contraction of muscles – a phenomenon known as cold shortening, and responsible for toughening of meat (besides several other unwanted changes). If, on the other hand, the animals are exhausted/starving at slaughter, their glycogen inventory is reduced – so generation of lactic acid will be hampered, and the final *pH* will be higher; for instance, pork with *pH* within 6.0–6.2 yields a dark, firm, and dry meat – which is, however, more labile to bacterial spoilage due to a *pH* closer to neutrality. Conversely, a too rapid fall in *pH* or a too slow decrease in temperature within the first postmortem hours produce pale, soft, and exudative meat; the said paleness results from light reflection by the white powdered precipitate formed by denaturation of (otherwise soluble) sarcoplasmic proteins. In addition, changes to membrane-bound myofibrillar proteins damage the cell membranes themselves – so leakage of intracellular contents takes place, which accounts for drip losses and causes cell softening; microbial growth and phospholipid oxidation are also promoted, thus compromising shelf-life. Although the reduced *pH* of muscle tissues offers some protection against bacterial contamination, this does not hold with such nonmuscular organs as liver and kidneys; in this case, rapid chilling is a must to prevent microbial growth.

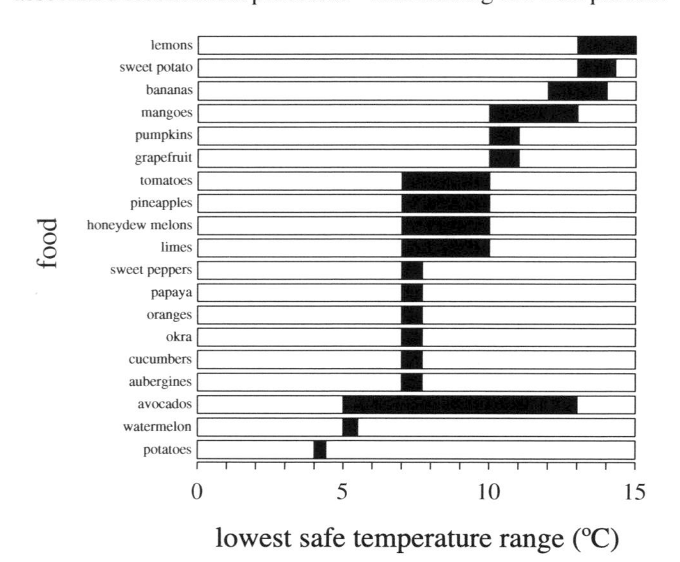

FIGURE 1.58 Ranges of lowest safe temperatures of chilled storage (■), for selected foods of plant origin.

Adverse changes to color, flavor, texture, and nutritional value in chill-stored meat and meat products are often a consequence of lipid oxidation; this is why preslaughter dietary supplementation with vitamin E – a strong (natural) antioxidant, prevents such shortcomings to a considerable extent. Proteases adventitious in meat are important to decrease muscle stiffness upon *rigor mortis* – a process classically known as conditioning; large carcasses are indeed hung at chill temperatures for 2–3 wk to become tender (although this process is obviously accelerated if carcasses are kept at room temperature). Conversely, gut proteases in fish and crustaceans weaken the gut wall upon death, thus permitting leakage of its contents into the surrounding tissues (a phenomenon referred to as belly burst); this calls for gutting within hours upon fish capture, while all seafood is to be promptly chilled to prevent such a form of deterioration (as gutting is harder to perform).

The chilled foods mentioned so far encompass raw or uncooked ingredients (e.g. salads, cheese, meats, fish, ready-to-eat foods) – which form class 1 foods, and require cooking (i.e. thermal processing yielding a minimum reduction of target pathogens) prior to consumption, or instead are prompt to eat after warming up to ambient temperature, or reheating for a short period; and products obtained as a mixture of cooked and low-risk raw ingredients, also termed class 2 foods. Both these classes may pose microbiological risks due to their original composition and possibility of contamination prior to chilling; hence, decay throughout chilling (storage) may come from a variety of biochemical sources, conveyed by adventitious enzymes and contaminating microflora. Handling of such foods must consequently occur under strict conditions of hygiene, and specific chilling temperatures and upper time thresholds have to be enforced – since they are critical for consumer safety upon ingestion, further to keeping sensory decay to a minimum.

However, a great many (processed) foods have been already cooked and packaged afterward (e.g. desserts, pizzas, sandwiches) – class 3 foods; or have been cooked after packaging – class 4 foods. Such products already underwent thermal treatment, so their physicochemical decay during chilling (storage) hardly derives from adventitious enzymes or microorganisms. One major example is individual foods or complete meals manufactured via cook–chill processes – as is the case of *sous-vide* products, vacuum-packed prior to pasteurization and subjected to chilled storage afterward for typical periods of 2–3 weeks. The said products have been originally developed for institutional catering, in attempts to replace warm-holding approaches – where food is kept hot (via lying on heated plates, or under infrared lamps) for long periods prior to consumption; in fact, the former undergo less extensive losses of nutritional value and eating quality, besides being less expensive overall. After cooking, foods intended for cook–chill are portioned within 30 min, and chilled to 3°C within no more than 90 min – with storage afterward between 0 and 3°C. In cook–pasteurize–chill products, the hot food is filled into a flexible container, a partial vacuum is applied to remove oxygen, and the pack is then heat-sealed; pasteurization ensues at 80°C for 10 min (at least), followed by immediate cooling to 3°C.

Class 4 foods, as well as class 1 and class 2 foods, demand a dedicated hygienic area – planned for easy cleaning, in attempts to avoid establishment of such pathogenic bacteria as *Listeria* spp. In the case of ready-to-eat foods, an extra high-care area is necessary – physically separated from the others, and designed for isolation of cooked foods during preparation, assembly of meals, and packaging. Hygienic requirements include (but are not limited to): positive pressure ventilation, to avoid environmental contamination; admittance of fully processed foods only through hatches or air-locks; entry and exit of staff only if wearing protective clothing (e.g. boots, coats, hairnets) and through changing rooms; no-touch washing facilities; use of easily cleaned building materials for walls, floors, and all surfaces meant for (or expected to be in) contact with food; and periodic stops for cleaning and disinfection, no less frequent than once every 2 hr.

Hardening is one the most significant effects of chilling upon the sensory profile of processed foods – and entails solidification of fats and oils; whereas lipid oxidation is a leading cause of quality loss of cook-chilled products – namely, through development of an oxidized flavor in cooked meats known as warmed-over flavor. Other changes due to chilling and implicated in quality decay include migration of oils from mayonnaise to cabbage in chilled coleslaw; evaporation of moisture from unpackaged chilled meats and cheeses; and moisture migration from sandwich fillings to the outer bread piece(s), or pie fillings or pizza toppings into the pastry and crust. Another example is syneresis in sauces and gravies, brought about by changes in starch thickeners used in their formulation; in the case of amylose-rich starches, amylose leaches out into solution to form aggregates via hydrogen bonding that end up expelling water. Therefore, chilled products should resort to modified starches, containing blocking molecules able to prevent amylose aggregation; or use amylose-poor starches in the first place.

From a microbiological point of view, the main safety concerns with chilling pertain to a number of pathogens that can grow slowly during extended periods of refrigerated storage below 5°C, or as a result of temperature abuse; this is the case of *Listeria monocytogenes* (with minimum growth temperature of −0.5°C), *Clostridium botulinum* (3–4°C), and *Aeromonas hydrophila* (−4°C). *Yersinia enterocolitica* (−1°C), *Salmonella* spp., *Bacillus cereus*, and *Vibrio parahaemolyticus* (with a minimum growth temperature of 5°C – see Table 1.4), enteropathogenic *Escherichia coli* (7°C), and *Campylobacter* spp. (32°C) will not significantly grow under regular chilling temperatures; however, they can do so should temperature abuse occur to a sufficiently high temperature, and for a sufficiently long period. Minimizing the levels of pathogens in incoming ingredients, and ensuring that storage procedures are appropriately carried out are thus *sine qua non* for safe chilled foods. One food raising an unusually high risk is sliced or minced meat; in fact, blade- or screw-mediated contamination (by e.g. *Clostridium perfringens*, *Escherichia coli*, *Staphylococcus aureus*) is likely to occur through the outer surface of the meat pieces, if the comminuting equipment has not been properly disinfected – especially because the specific surface area has been increased by orders of magnitude relative to the departing feedstock.

Remember that the normal composition of air is 78%(v/v) N_2 and 21%(v/v) O_2 – with CO_2 present at ca. 0.035%(v/v), and water vapor and other minor gases balancing up to 100%(v/v). If the proportion of oxygen is decreased and/or the proportion of carbon dioxide is increased in the surrounding atmosphere, within preset limits, then the rate of occurrence of respiration as per Fig. 1.56 will be reduced. In addition, growth of aerobic bacteria and molds is constrained, biochemical activity accompanying ripening and senescence (or aging) becomes slower, insect infestation is avoided, and oxidative changes (and even moisture loss) may be kept to a minimum. All these effects combined together permit preservation of the original quality of fresh produce for a longer time, with consequent extension of its shelf-life. When the rate of respiration remains high even at the lowest safe temperature of chilled storage, chill injuries will eventually result; hence, composition of the atmosphere has to be modified, so as to reduce the said rate of respiration – a common practice in large-scale storage of fresh apples, pears, or cabbage. Remember that respiration basically entails carbohydrate oxidation – and glucose was accordingly used in Fig. 1.56 as reference, because it releases the maximum amount of heat of respiration. However, other monosaccharides (as well as lipids and proteins) can also be taken up for energy if glucose becomes limiting; in any case, the average molar ratio of CO_2 formed per O_2 consumed remains of the order of unity for most fresh products.

When the level of O_2 is reduced from 21 to 3%(v/v), irrespective of accompanying changes in the level of CO_2, the rate of (aerobic) respiration will be reduced to about one-third of the corresponding rate under regular air. Although this rule may be linearly extrapolated for even lower levels of O_2, other biochemical phenomena would then come into play – namely anaerobic respiration (as discussed above), which generates compounds responsible for off-flavors; in addition, ethylene may build up in the atmosphere, which accelerates ripening of the whole batch – thus putatively leading to physiological defects. Hence, minimum thresholds for O_2 exist for the various foods – ranging from 0.8%(v/v) for spinach up to 2.3%(v/v) for asparagus, Brussels sprouts, and cabbage in the case of leafy vegetables. Typical compositions of modified atmospheres used in chilled storage of a few foods are conveyed by Fig. 1.59. Most modified atmospheres suitable for food storage cluster around 3%(w/v) oxygen and 7%(v/v) carbon dioxide. This heuristic rule is clearly broken by mold-ripened cheeses – most stable in total absence of oxygen and carbon dioxide, thus guaranteeing that both (aerobic) molds and (anaerobic) spoilage bacteria cannot grow; one specific cultivar of apples – most stable at a quite high oxygen fraction, to avoid anaerobic respiration that would eventually lead to formation of alcohols associated with off-flavors; and cured meat and hard cheese, and fresh pasta – most stable at very high fractions of carbon dioxide. Absence of O_2, coupled with CO_2 fractions up to 20%(v/v) have also been employed for grain and cocoa storage – where both insect destruction and mold inhibition are sought, thus avoiding the need for fumigation (with such toxic gases as phosphine or methyl bromide).

The rate of respiration of fresh fruits may undergo relevant changes, even when stored at constant chilling temperatures coupled with constant composition of overhead atmosphere in O_2 and CO_2; some fruits indeed experience climacteric ripening – consubstantiated by a short, yet sudden increase in respiration rate, induced by (plant hormone) ethylene. As a consequence, the rate of production of CO_2 suddenly jumps to the point of maximum ripeness. This observation may be taken advantage of when handling climacteric fruits – which include apple, apricot, melon, pear, peach, plum, tomato, and watermelon, coming from temperate climates; as well as avocado, banana, breadfruit, cherimoya, fig, guava, jackfruit, kiwifruit, mango, nectarine, papaya, passion fruit, persimmon, soursop, and sapote, originating in subtropical climates. In fact, they can be picked at full size and maturity (but before they are ripe), and then allowed to undergo controlled ripening only after chilled storage for the required period; enhancement in flavor quality at consumption is thus feasible, including higher sugar content and juice yield. Conversely, nonclimacteric fruits produce little (or no) ethylene, and consequently no major increases in CO_2 released are expected. This is

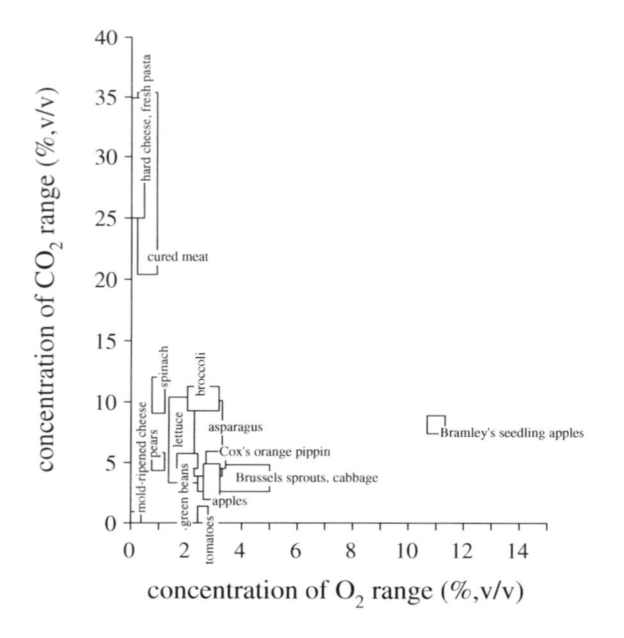

FIGURE 1.59 Ranges of composition of recommended modified atmospheres (▭) of chilled storage for selected foods – expressed as concentrations of carbon dioxide and oxygen.

the case of blueberry, cherry, cucumber, grape, olive, and strawberry from temperate climates; as well as cashew apple, grapefruit, java plum, lemon, lime, litchi, orange, pepper, pineapple, and tamarillo from subtropical climates. Consequently, their original quality at harvest will be maintained throughout storage. Vegetables respire similarly to nonclimacteric fruits.

The most relevant drawbacks associated to modified atmosphere storage are of an economic nature. Only short-season crops, with sufficient demand off-season to reach premium prices and large sales, and without alternative suppliers with complementary schedules (as is notably the case of apples, and pears and cabbage to a lesser extent) justify the extra investment incurred in; and processing costs of supply and maintenance of modified atmospheres, which may reach twice those of plain cold storage. Furthermore, tolerance to the modified atmosphere depends on cultivar, as well as previous growth conditions and maturity at harvest. Unfortunately, the same storage facilities and equipment cannot be routinely used throughout the whole year, nor can they handle mixed crops – due not only to distinct modified atmosphere requirements, but also to the risk of unwanted odor transfer. As an alternative, storage may be carried out under a partial vacuum, since it reduces overhead concentration of O_2 by the same factor, while eliminating ethylene once it is released; however, the associated costs are still prohibitive.

In modified atmosphere storage, the cold store is made airtight via metal cladding and carefully sealed doorways; the respiratory activity of the fresh produce is then allowed to change the overhead atmosphere, as oxygen is taken up and CO_2 concomitantly released. To speed-up the process of establishing the required gas composition, in either modified or controlled atmosphere storage, CO_2 may be added from pressurized cylinders to increase its concentration – or scrubbers, relying on bags of activated carbon or hydrated calcium hydroxide (or lime), under sprays of sodium hydroxide, may be used to decrease it; controlled air vents may instead be employed to admit O_2 as deemed necessary. This type of storage is appropriate for crops that ripen after harvest, or else deteriorate quickly; a high relative humidity helps retain crispness thereof, while reducing weight losses.

Most wholesale buyers frequently demand fresh produce be cooled prior to transportation (also under chilled conditions) to distribution warehouses or retailers. Chilling equipment per se is designed to reduce temperature of a given food product, at a preset rate, to the final target temperature; whereas cold storage apparatuses (or transportation vehicles) are designed to hold the chilled food at the said target temperature. The former requires a larger power, since sensible heat is to be removed, in addition to a larger respiration heat load (due to an average higher temperature); temperature should also be lowered as fast as possible from (at most) 50 down to 10°C, as this is the interval most prone to (unwanted) microbial growth. Either type of equipment may be operated batch- or continuouswise; air and water (or brine and metal surface, to a lesser degree) may be used to chill solid foods. Liquid foods are chilled by passing them through a suitable heat exchanger; this extends to such semiliquid foods as butter and margarine, via contact with refrigerated metal surfaces – specifically in scraped-surface heat exchangers, due to the need of removing solidified product.

Air cooling in air-blast chillers encompasses forcing refrigerated air (at ca. −10°C) at high speed (of ca. 4 $m.s^{-1}$) over the food products, to reduce their temperature without freezing (see Fig. 1.60iv); the low temperature is required by a driving force large enough for (fast) heat transfer, while the high velocity is meant at reducing thickness of the thermal boundary layer. The two main designs are static tunnels – where trolleys or pallets of food are placed for a convenient period; or continuous tunnels, where the said trolleys or pallets are moved therethrough, at a speed consistent with the required space time – or where food as individual pieces is placed on a conveyor belt, for the same purpose. In either case, a cycle of loading, chilling, and automatic defrosting (to remove ice from evaporators and containers) is routinely followed – which may be automatically operated, with data received from air- or product-inserted probes. Alarms for personnel trapped inside, or temperature rise/mains failure (with corresponding data-logging) are a requirement for safety and traceability.

Eutectic plate systems are an alternative form of *in situ* cooling, particularly appropriate for refrigerated vehicles aimed at local distribution. Salt solutions, e.g. potassium, sodium, or ammonium chloride, are frozen to their eutectic temperature (usually between –3 and –21°C), thus giving rise to a single, stable phase; air from the car chamber is circulated across the plates that absorb heat therefrom – with plate regeneration to be accomplished by external refreezing, at car stops specifically infrastructured to do so.

As an alternative, one may resort to hydrocooling – via spraying with, or submerging in water (or brine) previously chilled to ca. 1.5°C in a parallel refrigeration facility; heat withdrawal will speed up to 15-fold, relative to air-mediated cooling. This is a common practice in removing field heat from fruits and vegetables, prechilling of meat and poultry (prior to freezing), on-board chilling of fish (using refrigerated seawater), and cooling cheese (by direct immersion in cool brine). The latter possesses the extra advantages of avoiding food dehydration and permitting revival of slightly wilted produce – even though wetting may promote microbial growth. Hydrocooling is especially suitable for foods characterized by low specific surface area (e.g. apples, peaches, whole sweetcorn). Fresh produce is susceptible to postharvest infection when stressed (by too little or too much water), or somehow damaged by bruises or abrasion; to prevent spreading of infection, the (recirculated) cooling water is pretreated with 2–3 mg.L^{-1} chlorine (ineffective, however, against surface bacteria).

A growing concern over use of chlorine (Cl_2) has been recorded, due to its reactivity with organic materials that can yield trihalomethanes and halogenated acetic acids (among other noxious compounds). Such chemical residues have been associated with a number of environmental and health problems, including ecosystem disruption, fish death, cancer development, and several physiological diseases. Ozone (O_3) has been claimed as alternative to chlorine – in view of its 1.5-fold oxidation potential, complemented by ready decomposition to (harmless) water and molecular oxygen; further advantages encompass much wider bactericidal spectrum, as well as removal of pesticide traces and a few microbial-mediated off-flavors. Soluble ferrous and manganese salts responsible for water discoloration are also oxidized by O_3 to insoluble salts, which can subsequently be removed via filtration; and H_2S is likewise oxidized to (elemental) sulfur, which forms a solid phase susceptible of mechanical removal as well. Ozone is further able to oxidize bacterial membranes, thus weakening them – and eventually causing cell rupture

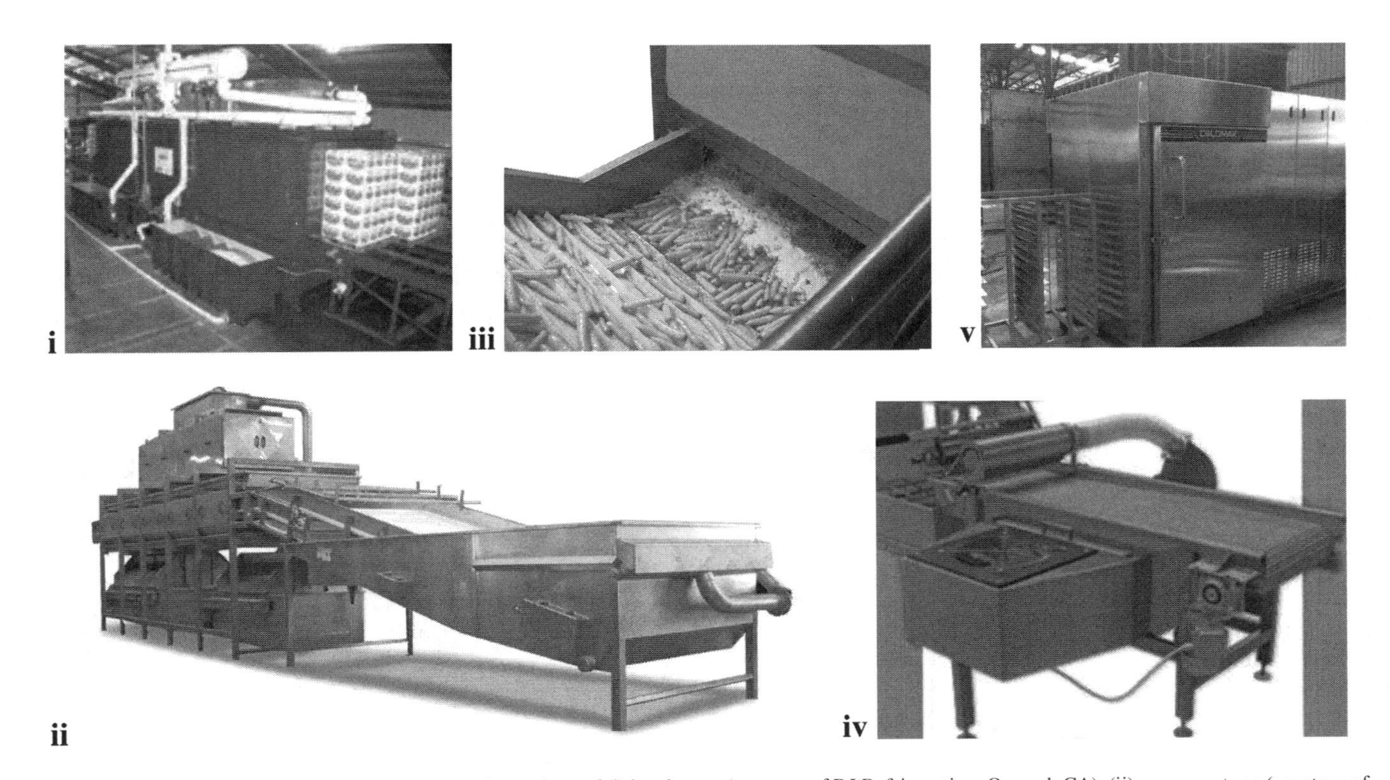

FIGURE 1.60 Examples of chilling equipment: hydrocoolers, of (i) batch type (courtesy of RJ Refrigeration, Oxnard, CA), (ii) conveyor type (courtesy of Meyer Industries, San Antonio, TX), and (iii) immersion type (courtesy of Wyma Engineering, Hornby, New Zealand); (iv) air cooler (courtesy of NEAEN, Malacky, Slovakia); and (v) vacuum cooler (courtesy of Coldmax, Dongguan, China).

and concomitant death. Ozone is normally obtained via corona discharge in air or oxygen, or via UV light; water is collected, filtered, and treated (or retreated) with ozone before being recycled to the cooling equipment – and can be discharged at will, with negligible environmental impact.

Produce is normally packed into wire-bound wooden crates, mesh bags, or perforated bins – characterized by a large portion of open space; to maximize rate of heat removal under hydrocooling, container design and stacking layout on pallets should enable water to flow through, rather than around the packages. There are four basic types of hydrocoolers – which differ in typical cooling rates and processing efficiencies: batch, conveyor, immersion, and truck. In batch hydrocoolers (see Fig. 1.60i), palletized bins of produce are loaded into an enclosure with a forklift truck – and chilled water is sprayed over them, collected at drain, filtered, recooled, and recycled; these apparatuses are relatively inexpensive, and thus suitable for small-scale operation. A more efficient design resorts to a high-capacity fan employed to suck a fine mist of chilled water through the bin stack – thus giving rise to more uniform cooling. Conveyor hydrocoolers (see Fig. 1.60ii) – with length up to 15 m and width up to 2.5 m, move containers of produce under a shower of chilled water, and are normally employed for large throughput rates; regular water volumetric flow rates are of the order of 45 $m^3.hr^{-1}.m^{-2}$, with an average rate of water recirculation of 1800 $m^3.hr^{-1}$. In submerged hydrocoolers (see Fig. 1.60iii), crates of produce are moved by a submerged conveyor through a large, shallow tank of recirculated chilled water; this design exhibits the largest rate of cooling, because of the greater contact of chilled water with the outer surface of food. In truck hydrocooling, packaged produce is loaded into a trailer and moved to a cooling system – where perforated pipes are inserted above the load; a shower of chilled water is then generated, at up to 250 $m^3.hr^{-1}$, and the water flowing out of the trailer is collected, recooled, and recycled. After cooling, the pipes are removed and a layer of crushed ice is placed above the crates – calculated to last throughout the entire transportation time.

A final option is vacuum, or flash cooling (see Fig. 1.60v) – appropriate for such large specific area, leafy vegetables as cabbage, lettuce and spinach, but also for mushrooms; the surface of the produce is previously moistened with plain water, and then subjected to vacuum (down to 500 Pa) in a chamber to induce evaporative cooling – so the latent heat of vaporization of the said water is obtained at the expense of sensible heat of the food material itself.

In the case of cryogenic chilling, solid CO_2 is supplied in the form of dry-ice pellets, or injected into an air stream in powder (or snow) form – as illustrated in Fig. 1.61; pellets or snow are deposited onto, or mixed with food in combo bins, trays, cartons, or conveyors. Excess of solid CO_2 is usually provided for a continued cooling effect during transportation and early storage, or prior to further processing. Its major advantages reside on simplicity and low cost – since it replaces expensive on-site cold stores, when products are meant to be dispatched in insulated containers within short periods (thus saving space and reducing labor costs). The snow form has been replacing its pellet counterpart due to easier handling and storage, along with higher operational safety; in addition, a more constant low temperature is possible, because a uniform layer can be generated via a snow horn as the food pieces are loaded – rather than inhomogeneous packing of pellets with food. By adjusting the amount of dry snow added, storage at the intended chilling temperature will be feasible for up to 24 hr – and distinct loads can be carried by the same standard vehicle.

Cryogenic cooling has also found application in sausage manufacture – to remove heat produced during mixing; in cryogenic grinding – to remove dust levels (and thus reduce explosion hazards) and increase yield of mills, or else to minimize loss of odoriferous compounds in spice milling; and in manufacture of multilayered chilled foods – where a first layer of food will promptly harden upon filling, thus allowing addition of the second layer at a faster pace than if conventional setting time had to elapse.

Liquid nitrogen (to be discussed in further depth when employed for freezing purposes) may also be used in chilling; generated onsite or supplied in pressurized containers, it is useful toward rapid chilling of 100–200 kg-batches of foods inside an insulated steel cabinet – containing centrifugal fans and a liquid nitrogen injector. Immediate vaporization of N_2 follows its injection, and the fan movement distributes the cold gas around the cabinet for a uniform cooling effect. Thermocouples installed inside the cabinet aid in controlling amount of liquid N_2 injected, so as to cope with different amounts or heat capacities of the food batches. In its continuous version, food (e.g. diced meat or vegetables) is passed on a variable speed conveyor to an inclined (insulated) cylindrical barrel; this barrel rotates slowly, and internal flights accordingly lift the food and tumble it – while liquid nitrogen is continuously injected (with sticking of food pieces prevented by the said tumbling action). This type of equipment is suitable to improve texture and binding capacity of mechanically reformed meat products – via solubilization of proteins that increases their water-holding features; forming and coating operations downstream will accordingly become easier. An alternative design resorts to a screw conveyor – where such foods as minced beef, sauce mixes, mashed potato, or diced vegetables are rapidly chilled as they are conveyed through a cold stream of nitrogen; the food material then undergoes firming, prior to partitioning or forming operations.

1.3.3 Mathematical simulation

1.3.3.1 Chilling of fresh produce

The thermal conductivity of a solid is typically higher than that of a liquid – since more atoms/molecules are available per unit

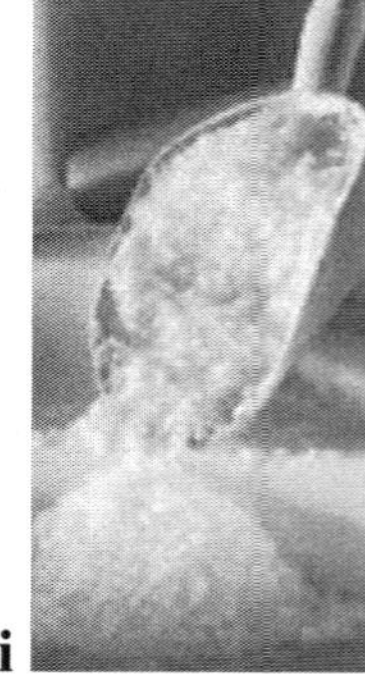

FIGURE 1.61 Examples of solid carbon dioxide used as cryogenic utility, in the form of (i) pellets and (ii) snow.

volume that transmit kinetic energy (namely in rotational and vibrational modes) from one to another. Hence, simulation of chilling of fruits and vegetables right upon harvest, when cold water is used as enthalpic utility (as often occurs), will hereafter resort to a simple model of two well-stirred entities contacting each other through their common surface – see Fig. 1.62. For modeling purposes, absence of significant temperature gradients within the food is indeed equivalent to a well-stirred system, characterized by uniform temperature T_f – expected, nevertheless, to evolve in time; while stirring within the complex chaotic pattern of (cold) water around the food pieces will be approached via a well-mixed pattern proper – described by temperature $T_{c,in}$ at entrance (constant in time) and $T_{c,out} \equiv T_c > T_{c,in}$ at exit (varying in time).

The enthalpic balance to the cooling water takes the form

$$\boxed{Q_c \rho_c c_{P,c}\left(T_{c,in} - T^\theta\right) + UA_f\left(T_f - T_c\right) = Q_c \rho_c c_{P,c}\left(T_c - T^\theta\right)} \quad (1.815)$$

where Q_c, ρ_c, and $c_{P,c}$ denote volumetric flow rate, mass density, and specific heat capacity, respectively, of cooling water, U denotes overall heat transfer coefficient, A_f denotes outer surface area of food, $T_{c,in}$ and T_c denote inlet and inside/outlet temperature of cooling water, respectively, T_f denotes temperature of food, and T^θ denotes an (arbitrarily chosen) reference temperature. The first term in either side of Eq. (1.815) represents inlet or outlet, respectively, rate of enthalpy transported by convection, and the other term in the left-hand side represents inlet rate of enthalpy transported by conduction. A pseudo steady state has been implicitly hypothesized for cooling water – in that $V_c dT_c/dt << Q_c(T_c - T_{c,in})$ with V_c denoting volume of cold water in apparatus; this is normally encountered in practice. Therefore, T_c does not appreciably change during a single pass of cold water, although T_c will eventually change in response to the gradual (cumulative) change of T_f in time. The corresponding enthalpy balance to the food reads, in turn,

$$\boxed{V_f \rho_f r_f \left\{T_f\right\} \Delta h_r = UA_f\left(T_f - T_c\right) + V_f \rho_f c_{P,f} \frac{dT_f}{dt}}, \quad (1.816)$$

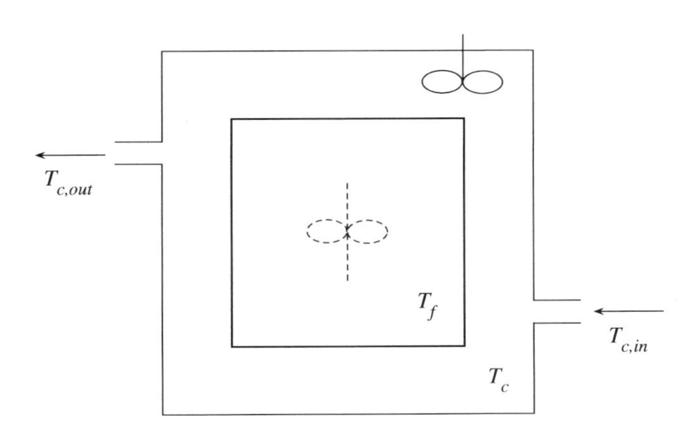

FIGURE 1.62 Schematic representation of chiller, operated with direct cooling by enthalpic fluid utility at temperature T_c – putatively jumping from $T_{c,in}$ at entrance to $T_c \equiv T_{c,out}$ thereafter, and temperature T_f of food to be processed.

where the left-hand side represents rate of production of enthalpy by respiration in the food matrix, the first term in the right-hand side represents outlet rate of enthalpy by conduction, and the last term represents rate of accumulation of enthalpy; V_f, ρ_f, and $c_{P,f}$ denote volume, mass density, and specific heat capacity, respectively, of food – while r_f denotes respiration rate of food, and Δh_r denotes specific enthalpy change accompanying respiration. Equation (1.816) is to be complemented by

$$\boxed{T_f\big|_{t=0} = T_{f,0}} \quad (1.817)$$

as initial condition – where $T_{f,0}$ denotes initial (usually room) temperature of food.

Equation (1.815) may be algebraically reorganized as

$$UA_f\left(T_f - T_c\right) = Q_c \rho_c c_{P,c}\left(T_c - T_{c,in}\right), \quad (1.818)$$

where factoring out of $Q_c \rho_c c_{P,c}$, followed by cancellation of symmetrical terms took place; elimination of parentheses then unfolds

$$UA_f T_f - UA_f T_c = Q_c \rho_c c_{P,c} T_c - Q_c \rho_c c_{P,c} T_{c,in}, \quad (1.819)$$

while algebraic regrouping gives rise to

$$\left(Q_c \rho_c c_{P,c} + UA_f\right) T_c = UA_f T_f + Q_c \rho_c c_{P,c} T_{c,in}. \quad (1.820)$$

Isolation of T_c from Eq. (1.820) prompts

$$T_c = \frac{Q_c \rho_c c_{P,c} T_{c,in}}{UA_f + Q_c \rho_c c_{P,c}} + \frac{UA_f}{UA_f + Q_c \rho_c c_{P,c}} T_f, \quad (1.821)$$

whereas division of both numerator and denominator of either term by UA_f yields

$$\boxed{T_c = \frac{\Omega}{1+\Omega} T_{c,in} + \frac{1}{1+\Omega} T_f} \quad (1.822)$$

– provided that characteristic (dimensionless) parameter Ω is defined as

$$\boxed{\Omega \equiv \frac{Q_c \rho_c c_{P,c}}{UA_f};} \quad (1.823)$$

a linear relationship between T_c and T_f accordingly arises – which may to advantage replace Eq. (1.815) hereafter. Once in possession of Eq. (1.822), one may redo Eq. (1.816) to

$$\rho_f V_f r_f \Delta h_r = UA_f\left(T_f - \frac{\Omega}{1+\Omega} T_{c,in} - \frac{1}{1+\Omega} T_f\right) + \rho_f c_{P,f} V_f \frac{dT_f}{dt}, \quad (1.824)$$

where terms explicit in T_f can be pooled together as

$$\rho_f V_f r_f \Delta h_r = UA_f\left(1 - \frac{1}{1+\Omega}\right) T_f - UA_f \frac{\Omega}{1+Q} T_{c,in} + \rho_f c_{P,f} V_f \frac{dT_f}{dt}; \quad (1.825)$$

upon removal of outstanding parenthesis, Eq. (1.825) becomes

$$\rho_f V_f r_f \Delta h_r = UA_f \frac{1+\Omega-1}{1+\Omega} T_f - UA_f \frac{\Omega}{1+\Omega} T_{c.in} + \rho_f c_{P,f} V_f \frac{dT_f}{dt}, \tag{1.826}$$

or else

$$\rho_f V_f r_f \Delta h_r = UA_f \frac{\Omega}{1+\Omega} T_f - UA_f \frac{\Omega}{1+\Omega} T_{c.in} + \rho_f c_{P,f} V_f \frac{dT_f}{dt} \tag{1.827}$$

after dropping symmetrical terms – where $UA_f\Omega/(1+\Omega)$ may be factored out to produce

$$\rho_f V_f r_f \Delta h_r = UA_f \frac{\Omega}{1+\Omega}\left(T_f - T_{c.in}\right) + \rho_f c_{P,f} V_f \frac{dT_f}{dt}. \tag{1.828}$$

Remember that the rate of respiration of a food is a function of its current temperature; Becker's (empirical) model may be retrieved in this particular, i.e.

$$r_f \Delta h_r = \phi_f \left(\frac{9}{5} T_f - 460\right)^{\gamma_f} \tag{1.829}$$

valid in SI units (with T_f specifically expressed in K) – where ϕ_f and γ_f denote parameters characteristic of each food item; Eq. (1.829) fits well the data conveyed by Fig. 1.56, for instance. Equation (1.829) may be linearized via expansion in Taylor's series about $T_f = T_{f,0}$, according to

$$r_f \Delta h_r \approx \phi_f \left(\frac{9}{5} T_f - 460\right)^{\gamma_f}\Bigg|_{T_{f,0}} + \phi_f \gamma_f \left(\frac{9}{5} T_f - 460\right)^{\gamma_f - 1} \frac{9}{5}\Bigg|_{T_{f,0}} \left(T_f - T_{f,0}\right), \tag{1.830}$$

which is equivalent to

$$r_f \Delta h_r = r_{f,0} \Delta h_r + \beta_f r_{f,0} \Delta h_r \left(T_f - T_{f,0}\right); \tag{1.831}$$

this implies definition of $r_{f,0}$ via

$$r_{f,0} \Delta h_r \equiv \phi_f \left(\frac{9}{5} T_{f,0} - 460\right)^{\gamma_f} > 0, \tag{1.832}$$

coupled with β_f via

$$\beta_f \equiv \frac{\frac{9}{5}\gamma_f}{\frac{9}{5} T_{f,0} - 460} > 0 \tag{1.833}$$

since $T_{f,0} > 5{\cdot}460/9 = 256$ K in practice – consistent with

$$\frac{9}{5}\phi_f \gamma_f \left(\frac{9}{5} T_{f,0} - 460\right)^{\gamma_f - 1} = \frac{\frac{9}{5}\gamma_f}{\frac{9}{5} T_{f,0} - 460}\phi_f \left(\frac{9}{5} T_{f,0} - 460\right)^{\gamma_f} \tag{1.834}$$

as algebraic reformulation of the last term in Eq. (1.830). Insertion of Eq. (1.831) transforms Eq. (1.828) to

$$\rho_f V_f r_{f,0} \Delta h_r \left(1 + \beta_f \left(T_f - T_{f,0}\right)\right) = UA_f \frac{\Omega}{1+\Omega}\left(T_f - T_{c.in}\right) + \rho_f c_{P,f} V_f \frac{dT_f}{dt}, \tag{1.835}$$

after having $r_{f,0}\Delta h_r$ factored out, while division of both sides by (constant) $UA_f(T_{f,0} - T_{c,in})$, complemented with multiplication and division of β_f also by $T_{f,0} - T_{c,in}$ give rise to

$$\frac{\rho_f V_f r_{f,0} \Delta h_r}{UA_f\left(T_{f,0} - T_{c.in}\right)}\left(1 + \beta_f\left(T_{f,0} - T_{c.in}\right)\frac{T_f - T_{f,0}}{T_{f,0} - T_{c.in}}\right) = \frac{\Omega}{1+\Omega}\frac{T_f - T_{c,in}}{T_{f,0} - T_{c,in}} + \frac{d\left(\frac{T_f - T_{c,in}}{T_{f,0} - T_{c,in}}\right)}{d\left(\frac{UA_f}{\rho_f c_{P,f} V_f} t\right)}, \tag{1.836}$$

at the expense of the rule of differentiation of a product by, and of a sum to a constant; Eq. (1.836) may be rewritten as

$$\frac{\rho_f V_f r_{f,0} \Delta h_r}{UA_f\left(T_{f,0} - T_{c.in}\right)}\left(1 + \beta_f\left(T_{f,0} - T_{c.in}\right)\frac{\left(T_f - T_{c.in}\right) - \left(T_{f,0} - T_{c.in}\right)}{T_{f,0} - T_{c.in}}\right) = \frac{\Omega}{1+\Omega}\frac{T_f - T_{c,in}}{T_{f,0} - T_{c,in}} + \frac{d\left(\frac{T_f - T_{c,in}}{T_{f,0} - T_{c,in}}\right)}{d\left(\frac{UA_f}{\rho_f c_{P,f} V_f} t\right)}, \tag{1.837}$$

upon addition and subtraction of $T_{c,in}$ in numerator. Equation (1.837) may instead appear as

$$\frac{d\left(\frac{T_f - T_{c,in}}{T_{f,0} - T_{c,in}}\right)}{d\left(\frac{UA_f}{\rho_f c_{P,f} V_f} t\right)} = \varsigma\left(1 - \Lambda + \Lambda\frac{T_f - T_{c,in}}{T_{f,0} - T_{c,in}}\right) - \frac{\Omega}{1+\Omega}\frac{T_f - T_{c,in}}{T_{f,0} - T_{c,in}}, \tag{1.838}$$

after defining a second characteristic parameter ς as

$$\boxed{\varsigma \equiv \frac{\rho_f V_f r_{f,0} \Delta h_r}{UA_f\left(T_{f,0} - T_{c,in}\right)} > 0} \tag{1.839}$$

and a third characteristic parameter as

$$\boxed{\Lambda \equiv \beta_f\left(T_{f,0} - T_{c,in}\right) > 0,} \tag{1.840}$$

once the ratio $((T_f - T_{c,in}) - (T_{f,0} - T_{c,in}))/(T_{f,0} - T_{c,in})$ is splitted as $(T_f - T_{c,in})/(T_{f,0} - T_{c,in}) - 1$; both such parameters are positive due to the exothermic nature of respiration (i.e. $\Delta h_r > 0$), combined with the cooling nature of the process (i.e. $T_{c,in} < T_{f,0}$). Factoring out of $(T_f - T_{c,in})/(T_{f,0} - T_{c,in})$ in Eq. (1.838) finally yields

$$\boxed{\frac{d\left(\dfrac{T_f - T_{c,in}}{T_{f,0} - T_{c,in}}\right)}{d\left(\dfrac{UA_f}{\rho_f c_{P,f} V_f}t\right)} = \varsigma(1-\Lambda) + \left(\varsigma\Lambda - \frac{\Omega}{1+\Omega}\right)\frac{T_f - T_{c,in}}{T_{f,0} - T_{c,in}};} \quad (1.841)$$

Eq. (1.841) is a linear, ordinary differential equation of (dimensionless) dependent variable $(T_f - T_{c,in})/(T_{f,0} - T_{c,in})$ on (dimensionless) independent variable $UA_f t/\rho_f c_{P,f} V_f$ – and will be used from now on *in lieu* of Eq. (1.816). By the same token, Eq. (1.817) should instead appear as

$$\boxed{\left.\frac{T_f - T_{c,in}}{T_{f,0} - T_{c,in}}\right|_{\frac{UA_f t}{\rho_f c_{P,f} V_f}=0} = 1,} \quad (1.842)$$

following subtraction of $T_{c,in}$ from both sides and division thereof by $T_{f,0} - T_{c,in}$ afterward, coupled with multiplication of both sides of the subscript by $UA_f/\rho_f c_{P,f} V_f$.

For simplicity of algebraic handling hereafter, Eq. (1.841) will be rewritten as

$$\frac{dT^*}{dt^*} = \varsigma(1-\Lambda) - \left(\frac{\Omega}{1+\Omega} - \varsigma\Lambda\right)T^* \quad (1.843)$$

– as long as auxiliary variables T^* and t^* abide to

$$T^* \equiv \frac{T_f - T_{c,in}}{T_{f,0} - T_{c,in}} \quad (1.844)$$

and

$$t^* \equiv \frac{UA_f}{\rho_f c_{P,f} V_f}t; \quad (1.845)$$

Eq. (1.842) will likewise appear as

$$T^*\big|_{t^*=0} = 1. \quad (1.846)$$

To integrate Eq. (1.843), one may first solve the corresponding homogeneous equation, i.e.

$$\frac{dT^*}{dt^*} = -\left(\frac{\Omega}{1+\Omega} - \varsigma\Lambda\right)T^*, \quad (1.847)$$

where separation of variables promptly leads to

$$\int \frac{dT^*}{T^*} = -\left(\frac{\Omega}{1+\Omega} - \varsigma\Lambda\right)\int dt^* \quad (1.848)$$

on the (reasonable) hypothesis that ρ_c, ρ_f, $c_{P,c}$, $c_{P,f}$, U, A_f, Q_c, and Δh_r remain constant; Eq. (1.848) degenerates to

$$\ln T^* = k_0 - \left(\frac{\Omega}{1+\Omega} - \varsigma\Lambda\right)t^*, \quad (1.849)$$

with k_0 denoting an arbitrary constant. After taking exponentials of both sides, Eq. (1.849) turns to

$$T^* = k_1 \exp\left\{-\left(\frac{\Omega}{1+\Omega} - \varsigma\Lambda\right)t^*\right\}, \quad (1.850)$$

with constant k_1 obviously given by

$$k_1 \equiv e^{k_0}. \quad (1.851)$$

The full solution of Eq. (1.843) should now be constructed by assuming that $k_1 \equiv k_1\{t^*\}$, instead of being a constant as hypothetically considered before; under these circumstances, one may differentiate both sides of Eq. (1.850) to get

$$\frac{dT^*}{dt^*} = \frac{dk_1}{dt^*}\exp\left\{-\left(\frac{\Omega}{1+\Omega} - \varsigma\Lambda\right)t^*\right\} - k_1\left(\frac{\Omega}{1+\Omega} - \varsigma\Lambda\right)\exp\left\{-\left(\frac{\Omega}{1+\Omega} - \varsigma\Lambda\right)t^*\right\}. \quad (1.852)$$

Elimination of dT^*/dt^* between Eqs. (1.843) and (1.852) unfolds

$$\varsigma(1-\Lambda) - \left(\frac{\Omega}{1+\Omega} - \varsigma\Lambda\right)T^* = \frac{dk_1}{dt^*}\exp\left\{-\left(\frac{\Omega}{1+\Omega} - \varsigma\Lambda\right)t^*\right\} - k_1\left(\frac{\Omega}{1+\Omega} - \varsigma\Lambda\right)\exp\left\{-\left(\frac{\Omega}{1+\Omega} - \varsigma\Lambda\right)t^*\right\}, \quad (1.853)$$

whereas insertion of Eq. (1.850) permits further transformation to

$$\varsigma(1-\Lambda) - \left(\frac{\Omega}{1+\Omega} - \varsigma\Lambda\right)k_1\exp\left\{-\left(\frac{\Omega}{1+\Omega} - \varsigma\Lambda\right)t^*\right\} = \frac{dk_1}{dt^*}\exp\left\{-\left(\frac{\Omega}{1+\Omega} - \varsigma\Lambda\right)t^*\right\} - k_1\left(\frac{\Omega}{1+\Omega} - \varsigma\Lambda\right)\exp\left\{-\left(\frac{\Omega}{1+\Omega} - \varsigma\Lambda\right)t^*\right\}; \quad (1.854)$$

after dropping common terms between sides, Eq. (1.854) simplifies to

$$\varsigma(1-\Lambda) = \frac{dk_1}{dt^*}\exp\left\{-\left(\frac{\Omega}{1+\Omega} - \varsigma\Lambda\right)t^*\right\}. \quad (1.855)$$

Integration of Eq. (1.855) may proceed via separation of variables as

$$\int dk_1 = \varsigma(1-\Lambda)\int \exp\left\{\left(\frac{\Omega}{1+\Omega} - \varsigma\Lambda\right)t^*\right\}dt^*, \quad (1.856)$$

which leads to

$$k_1 = k_2 + \varsigma(1-\Lambda)\frac{\exp\left\{\left(\dfrac{\Omega}{1+\Omega} - \varsigma\Lambda\right)t^*\right\}}{\dfrac{\Omega}{1+\Omega} - \varsigma\Lambda}; \quad (1.857)$$

Eq. (1.857) is equivalent to

$$k_1\left\{t^*\right\} = k_2 + \frac{\varsigma(1-\Lambda)(1+\Omega)}{\Omega-\varsigma\Lambda(1+\Omega)}\exp\left\{\left(\frac{\Omega}{1+\Omega}-\varsigma\Lambda\right)t^*\right\}, \quad (1.858)$$

upon multiplication of both numerator and denominator of the pre-exponential factor by $1+\Omega$. Therefore, Eq. (1.850) should be reformulated to

$$T^* = \left(k_2 + \frac{\varsigma(1-\Lambda)(1+\Omega)}{\Omega-\varsigma\Lambda(1+\Omega)}\exp\left\{\left(\frac{\Omega}{1+\Omega}-\varsigma\Lambda\right)t^*\right\}\right) \times \exp\left\{-\left(\frac{\Omega}{1+\Omega}-\varsigma\Lambda\right)t^*\right\} \quad (1.859)$$

with the aid of Eq. (1.858), where $\exp\{-(\Omega/(1+\Omega)-\varsigma\Lambda)t^*\}$ may be factored in to give

$$T^* = \frac{\varsigma(1-\Lambda)(1+\Omega)}{\Omega-\varsigma\Lambda(1+\Omega)} + k_2\exp\left\{-\left(\frac{\Omega}{1+\Omega}-\varsigma\Lambda\right)t^*\right\}. \quad (1.860)$$

Equation (1.860) will not satisfy Eq. (1.846) unless

$$1 = \frac{\varsigma(1-\Lambda)(1+\Omega)}{\Omega-\varsigma\Lambda(1+\Omega)} + k_2 e^0 = \frac{\varsigma(1-\Lambda)(1+\Omega)}{\Omega-\varsigma\Lambda(1+\Omega)} + k_2, \quad (1.861)$$

which readily yields

$$k_2 = 1 - \frac{\varsigma(1-\Lambda)(1+\Omega)}{\Omega-\varsigma\Lambda(1+\Omega)}; \quad (1.862)$$

insertion of Eq. (1.862) finally transforms Eq. (1.860) to

$$T^* = \frac{\varsigma(1-\Lambda)(1+\Omega)}{\Omega-\varsigma\Lambda(1+\Omega)} + \left(1 - \frac{\varsigma(1-\Lambda)(1+\Omega)}{\Omega-\varsigma\Lambda(1+\Omega)}\right)\exp\left\{-\left(\frac{\Omega}{1+\Omega}-\varsigma\Lambda\right)t^*\right\}. \quad (1.863)$$

After realizing that

$$1 - \frac{\varsigma(1-\Lambda)(1+\Omega)}{\Omega-\varsigma\Lambda(1+\Omega)} = \frac{\Omega-\varsigma\Lambda-\varsigma\Lambda\Omega-\varsigma+\varsigma\Lambda-\varsigma\Omega+\varsigma\Lambda\Omega}{\Omega-\varsigma\Lambda(1+\Omega)} = \frac{\Omega-\varsigma-\varsigma\Omega}{\Omega-\varsigma\Lambda(1+\Omega)} = \frac{\Omega-\varsigma(1+\Omega)}{\Omega-\varsigma\Lambda(1+\Omega)} \quad (1.864)$$

upon conversion to a single fraction, followed by dropping symmetrical terms and factoring out ς in numerator, one may redo Eq. (1.863) to

$$\boxed{\frac{T_f-T_{c,in}}{T_{f,0}-T_{c,in}} = \frac{\varsigma(1-\Lambda)(1+\Omega)}{\Omega-\varsigma\Lambda(1+\Omega)} + \frac{\Omega-\varsigma(1+\Omega)}{\Omega-\varsigma\Lambda(1+\Omega)}\exp\left\{-\left(\frac{\Omega}{1+\Omega}-\varsigma\Lambda\right)\frac{UA_f}{\rho_f c_{P,f} V_f}t\right\}} \quad (1.865)$$

– once the original notation is retrieved via Eqs. (1.844) and (1.845); a graphical interpretation of Eq. (1.865) is made available in Fig. 1.63, for selected values of (dimensionless) parameters ς, Ω, and Λ. A typical exponential decay is apparent in all curves – consistent with the functional form of Eq. (1.865); cooling is faster at higher Ω – due to the higher mass flow rate ($\rho_c Q_c$) and specific heat capacity ($c_{P,c}$) of cooling fluid as per Eq. (1.823), and also when ς decreases – owing to a less unfavorable contribution of rate of heat production by respiration ($r_{f,0}\Delta h_r$) and a larger driving force for heat transfer ($T_{f,0}-T_{c,in}$), as per Eq. (1.839). Note that physical constraints enforce $\Omega>\varsigma$ and $\varsigma<1$, otherwise $T_f>T_{f,0}$ – and heating, rather than chilling would take place, due to an excessively high rate of production of heat by respiration compared to rate of removal of heat via conduction and convection.

The short-time thermal behavior of the food is driven by

$$\lim_{\frac{UA_f}{\rho_f c_{P,f} V_f}t\to 0}\frac{T_f-T_{c,in}}{T_{f,0}-T_{c,in}} = \frac{\varsigma(1-\Lambda)(1+\Omega)}{\Omega-\varsigma\Lambda(1+\Omega)} + \frac{\Omega-\varsigma(1+\Omega)}{\Omega-\varsigma\Lambda(1+\Omega)}e^0, \quad (1.866)$$

which reduces to

$$\lim_{\frac{UA_f}{\rho_f c_{P,f} V_f}t\to 0}\frac{T_f-T_{c,in}}{T_{f,0}-T_{c,in}} = \frac{\varsigma(1-\Lambda)(1+\Omega)}{\Omega-\varsigma\Lambda(1+\Omega)} + \frac{\Omega-\varsigma(1+\Omega)}{\Omega-\varsigma\Lambda(1+\Omega)} = \frac{\varsigma(1+\Omega)-\varsigma\Lambda(1+\Omega)+\Omega-\varsigma(1+\Omega)}{\Omega-\varsigma\Lambda(1+\Omega)} = \frac{\Omega-\varsigma\Lambda(1+\Omega)}{\Omega-\varsigma\Lambda(1+\Omega)} \quad (1.867)$$

upon lumping fractions, and cancelling out symmetrical terms afterward; Eq. (1.867) becomes merely

$$\boxed{\lim_{\frac{UA_f}{\rho_f c_{P,f} V_f}t\to 0}\frac{T_f-T_{c,in}}{T_{f,0}-T_{c,in}} = 1,} \quad (1.868)$$

as enforced by Eq. (1.846) in the first place. Conversely, a horizontal asymptote exists that drives the behavior of $(T_f-T_{c,in})/(T_{f,0}-T_{c,in})$ when time grows unbounded, viz.

$$\lim_{\frac{UA_f}{\rho_f c_{P,f} V_f}t\to\infty}\frac{T_f-T_{c,in}}{T_{f,0}-T_{c,in}} = \frac{\varsigma(1-\Lambda)(1+\Omega)}{\Omega-\varsigma\Lambda(1+\Omega)} + \frac{\Omega-\varsigma(1+\Omega)}{\Omega-\varsigma\Lambda(1+\Omega)}e^{-\infty}, \quad (1.869)$$

which reduces to

$$\boxed{\lim_{\frac{UA_f}{\rho_f c_{P,f} V_f}t\to\infty}\frac{T_f-T_{c,in}}{T_{f,0}-T_{c,in}} = \frac{\varsigma(1-\Lambda)(1+\Omega)}{\Omega-\varsigma\Lambda(1+\Omega)} > 0} \quad (1.870)$$

since $e^{-\infty}$ is nil. Inspection of Eq. (1.870) makes it clear that T_f will not attain $T_{c,in}$ – even after a sufficiently long time has elapsed, due to existence of active respiration that serves as source of heat; if $\Delta h_r \to 0$, then $\varsigma \to 0$ as per Eq. (1.839) – in which case Eq. (1.870) would degenerate to

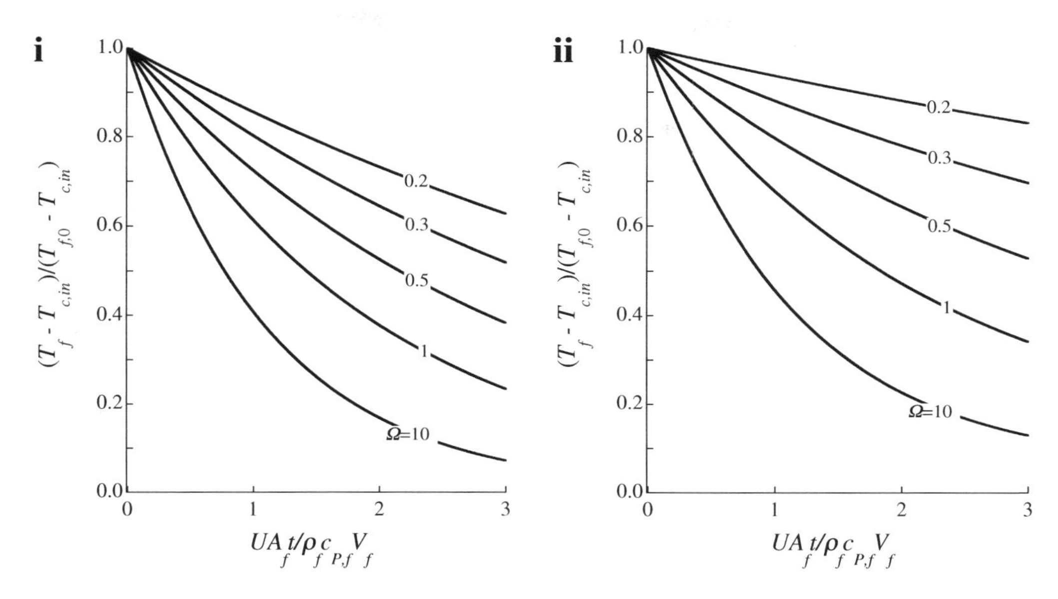

FIGURE 1.63 Evolution of normalized temperature, $(T_f-T_{c,in})/(T_{f,0}-T_{c,in})$, versus dimensionless time, $UA_f t/\rho_f c_{P,f} V_f$, for selected values of dimensionless parameters Ω, $\Lambda=0.5$, and (i) $\varsigma=0.01$ and (ii) $\varsigma=0.1$, in cold water-mediated chilling.

$$\lim_{\substack{\frac{UA_f}{\rho_f c_{P,f} V_f} t \to \infty \\ \varsigma \to 0}} \frac{T_f - T_{c,in}}{T_{f,0} - T_{c,in}} = \lim_{\varsigma \to 0} \frac{\varsigma(1-\Lambda)(1+\Omega)}{\Omega - \varsigma\Lambda(1+\Omega)} = \frac{0(1-\Lambda)(1+\Omega)}{\Omega - 0\Lambda(1+\Omega)} = \frac{0}{\Omega} = 0 \tag{1.871}$$

– as expected for a new steady state characterized by thermal equilibrium of food with cooling utility, along with absence of a heat source.

PROBLEM 1.17

In the case of cold air-mediated chilling of fresh produce, the difference in thermal conductivities of solid food and refrigerant is even higher than when cold water is used as refrigeration vector; therefore, temperature gradients may *a fortiori* be neglected on the food side, yet evaporation of water will have to be taken into account as heat sink, further to respiration as heat source.

a) Based on the hypothesis that temperature of food subjected to chilling, T_f, and temperature of cold air at chamber outlet, T_c, are (univariate) functions of time, t – in addition to constant temperature of incoming cold air, $T_{c,in}$, write down the mass balance to water vapor, and the enthalpic balances to both food and cold air.
b) Find a suitable strategy to solve the set of algebraic/differential equations produced in a).
c) Redo the equations in b) in dimensionless form – and comment qualitatively on the results expected from the first-order ordinary differential equation, *vis-à-vis* with chilling via cold water.

Solution

a) A schematic representation of the cold air-operated chiller proposed is provided below – which accounts for evaporation of water, with consequent change in moisture content of the cold air utility.

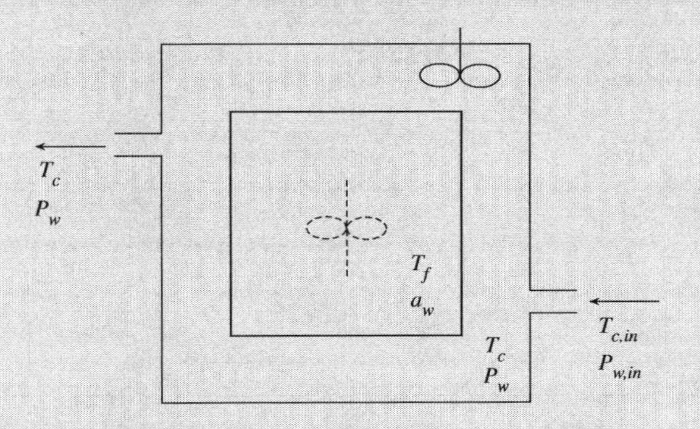

The water activity in food, a_w, is considered not to vary appreciably over time – neither due to change in temperature nor to change in moisture content. Cold air enters at (constant) temperature $T_{c,in}$ and (constant) partial pressure of water vapor $P_{w,in}$, and leaves at (variable) temperature $T_c \equiv T_c\{t\} > T_{c,in}$ and (variable) partial pressure of water steam $P_w \equiv P_w\{t\} > P_{w,in}$ – both coincident with those prevailing in the cooling chamber, seen as a whole; the food exhibits no temperature gradients by hypothesis, but is described by (variable) temperature T_f. While absence of temperature gradients within the solid food matrix is theoretically equivalent to a well-stirred device, the chilling chamber itself may be simulated as a well-stirred tank – due to the complex random pattern of cold air around the food pieces and within the container. The mass balance to water vapor carried by the cold air stream looks like

$$W_c w_{w,in} + KM_w A_f \left(P_{w,f}^{\sigma}\left\{a_w, T_f\right\} - P_w\right) = W_c w_w, \qquad (1.17.1)$$

where W_c denotes (constant) mass flow rate of air, $w_{w,in}$ and w_w denote mass concentration of water vapor (referred to mass of air) at inlet and outlet, respectively, M_w denotes molecular weight of water, and A_f denotes (exposed) outer surface of food; K denotes mass transfer coefficient of water vapor in air (with SI units mol.s^{-1}.m^{-2}.Pa^{-1}), and $P_{w,f}^{\sigma}$ denotes pressure of water vapor in equilibrium with food. After revisiting Eq. (1.147) as

$$P_{w,f}^{\sigma} = a_w P_w^{\sigma}, \qquad (1.17.2)$$

one may redo Eq. (1.17.1) as

$$KM_w A_f \left(a_w P_w^{\sigma}\left\{T_f\right\} - P_w\right) = W_c\left(w_w - w_{w,in}\right) \qquad (1.17.3)$$

together with collapse of terms alike – where P_w^{σ} denotes vapor pressure of water at equilibrium with pure liquid water, and thus dependent solely on temperature T_f; by hypothesis, a_w does not change appreciably with decreasing temperature T_f or the (minor) amount of water removed by evaporation. An extra relationship is accordingly required between P_w and w_w – and the ideal gas equation may serve such a purpose; one may indeed recall

$$P_w = x_w P \qquad (1.17.4)$$

using Eq. (3.141) of *Food Proc. Eng.: basics & mechanical operations* as template, with x_w denoting mole fraction of water in air and P denoting overall pressure of the latter – based, in turn, on Eq (3.132) applied to water vapor and Eq. (3.135) applied to air, both also from *Food Proc. Eng.: basics & mechanical operations*. Since

$$x_w \equiv \frac{n_w}{n_w + n_a} \qquad (1.17.5)$$

by definition of molar fraction of a binary mixture, see Eq. (3.43) of *Food Proc. Eng.: basics & mechanical operations,* with n_a and n_w denoting number of moles of air and water, respectively, one may proceed to

$$x_w \approx \frac{n_w}{n_a} = \frac{\dfrac{m_w}{M_w}}{\dfrac{m_a}{M_a}} = \frac{M_a}{M_w}\frac{m_w}{m_a} \approx \frac{M_a}{M_w}\frac{m_w}{m_w + m_a} \equiv \frac{M_a}{M_w} w_w \qquad (1.17.6)$$

at the expense of $n_w << n_a$ and the definition of mass fraction – where m_a and m_w denote mass of air and water, respectively, and M_a denotes (average) molecular weight of air. Insertion of Eq. (1.17.6) converts Eq. (1.17.4) to

$$P_w = \frac{M_a}{M_w} w_w P, \qquad (1.17.7)$$

or else

$$w_w = \frac{M_w}{M_a P} P_w \qquad (1.17.8)$$

upon isolation of w_w; if $P_{w,in}$ denotes partial pressure of water vapor in inlet stream, then one may as well write

$$w_{w,in} = \frac{M_w}{M_a P} P_{w,in} \qquad (1.17.9)$$

pertaining to inlet conditions, after taking P as approximately constant between entrance and exit of chiller. Combination of Eqs. (1.17.3), (1.17.8), and (1.17.9) generates

$$KM_w A_f \left(a_w P_w^{\sigma}\left\{T_f\right\} - P_w\right) = W_c\left(\frac{M_w}{M_a P} P_w - \frac{M_w}{M_a P} P_{w,in}\right), \qquad (1.17.10)$$

where both sides may be multiplied by $M_a P/M_w$ to give

$$KPM_a A_f \left(a_w P_w^{\sigma}\left\{T_f\right\} - P_w\right) = W_c\left(P_w - P_{w,in}\right) \qquad (1.17.11)$$

as final form of the mass balance to water vapor. The enthalpic balance to the cooling air is analogous to that conveyed by Eq. (1.815), but after accounting for water vaporization, viz.

$$W_c c_{P,a}\left(T_{c,in} - T^{\theta}\right) + UA_f\left(T_f - T_c\right) + KM_w A_f\left(a_w P_w^{\sigma} - P_w\right)\lambda_w = W_c c_{P,a}\left(T_c - T^{\theta}\right) \qquad (1.17.12)$$

– hereafter referred to a mass flow rate basis for convenience (i.e. with $Q_c\rho_c$ replaced by W_c); T^{θ} was used to denote reference temperature for calculation of enthalpy, U to denote (constant) overall heat transfer coefficient, $c_{P,a}$ and $c_{P,w}$ to denote (constant) isobaric specific heat capacity of air and water vapor, respectively, and λ_w to denote heat of vaporization of water at temperature T_f – with $c_{P,w,l}(T_f - T^{\theta})$, $c_{P,w,g}(T_c - T_f) << \lambda_w$, where $c_{P,w,l}$ and $c_{P,w,g}$ denote specific heat capacity of water liquid and vapor, respectively. A pseudo-steady state approach was implicitly assumed in Eq. (1.17.12) – based on realization that the change in food temperature during a single pass of cold air is rather small compared to the jump in temperature of the said air between inlet and outlet. The first term in Eq. (1.17.12) represents rate of enthalpy in via convection of air, the second term represents rate of enthalpy in via conduction, the third term represents rate of enthalpy in via convection of water vapor (previously formed from liquid water, and entering the air stream at temperature T_f), and the right-hand side represents rate of enthalpy out via convection of air. Due to the small mass flow rate of water vapor following evaporation, i.e. $KM_w A_f(a_w P_w^{\sigma} - P_w)$, compared to the mass flow rate of air, W_c, the contribution of enthalpy of vaporized water may be discarded – so Eq. (1.17.12) simplifies, in practice, to

$$UA_f\left(T_f - T_c\right) = W_c c_{P,a}\left(T_c - T_{c,in}\right), \qquad (1.17.13)$$

where $W_c c_{P,a}$ meanwhile factored out and T^θ canceled with its negative. Finally, the enthalpy balance to food reads

$$V_f \rho_f r_f \Delta h_r = UA_f\left(T_f - T_c\right) + KM_w A_f\left(a_w P_w^\sigma - P_w\right)\lambda_w + V_f \rho_f c_{P,f}\frac{dT_f}{dt} \tag{1.17.14}$$

using Eq. (1.816) as template, but taking vaporization of (constituent) food water on board; the enthalpy of vaporization of water was included, and will be retained hereafter as sink term – owing to its relative magnitude, and to the cumulative, transient nature of the balance at stake. The first term of Eq. (1.17.14) represents rate of heat produced by respiration, the second term represents rate of enthalpy loss by conduction to the cooling air, the third term represents rate of enthalpy consumption by vaporization of water, and the fourth (and last) term represents rate of accumulation of enthalpy associated with sensible heat. The amount of water evaporated is considered as sufficiently small to marginally affect V_f (or $r_f\Delta h_r$, ρ_f, and A_f, for that matter).

b) Upon elimination of parentheses, Eq. (1.7.13) will read

$$UA_f T_f - UA_f T_c = W_c c_{P,a} T_c - W_c c_{P,a} T_{c,in}, \tag{1.17.15}$$

where condensation of terms in T_c leads to

$$\left(UA_f + W_c c_{P,a}\right)T_c = UA_f T_f + W_c c_{P,a} T_{c,in}; \tag{1.17.16}$$

Eq. (1.17.16) may be solved for T_c as

$$T_c = \frac{W_c c_{P,a} T_{c,in}}{UA_f + W_c c_{P,a}} + \frac{UA_f}{UA_f + W_c c_{P,a}} T_f, \tag{1.17.17}$$

thus providing $T_c \equiv T_c\left\{T_f\right\}$ in linear form. By the same token, elimination of parentheses transforms Eq. (1.17.11) to

$$KPM_a A_f a_w P_w^\sigma\left\{T_f\right\} - KPM_a A_f P_w = W_c P_w - W_c P_{w,in}, \tag{1.17.18}$$

where P_w can be factored out as

$$\left(KPM_a A_f + W_c\right)P_w = KPM_a A_f a_w P_w^\sigma\left\{T_f\right\} + W_c P_{w,in}; \tag{1.17.19}$$

isolation of P_w gives

$$P_w = \frac{W_c P_{w,in}}{KPM_a A_f + W_c} + \frac{KPM_a A_f a_w}{KPM_a A_f + W_c} P_w^\sigma\left\{T_f\right\}, \tag{1.17.20}$$

so a relationship of the type $P_w \equiv P_w\left\{T_f\right\}$ has arisen. Equation (1.17.3) may now be inserted in Eq. (1.17.14) to obtain

$$\rho_f V_f r_f \Delta h_r = UA_f\left(T_f - T_c\right) + W_c\left(w_w - w_{w,in}\right)\lambda_w + \rho_f V_f c_{P,f}\frac{dT_f}{dt}, \tag{1.17.21}$$

while further combination with Eqs. (1.17.8) and (1.17.9) unfolds

$$\rho_f V_f r_f \Delta h_r = UA_f\left(T_f - T_c\right) + \lambda_w W_c\left(\frac{M_w}{M_a P}P_w - \frac{M_w}{M_a P}P_{w,in}\right) + \rho_f V_f c_{P,f}\frac{dT_f}{dt} \tag{1.17.22}$$

– or, upon factoring $M_w/M_a P$ out,

$$\rho_f V_f r_f \Delta h_r = UA_f\left(T_f - T_c\right) + \frac{\lambda_w M_w W_c}{M_a P}\left(P_w - P_{w,in}\right) + \rho_f V_f c_{P,f}\frac{dT_f}{dt}; \tag{1.17.23}$$

Eqs. (1.17.17) and (1.17.20) may then be brought on board to produce

$$\rho_f V_f r_f \Delta h_r = UA_f\left(T_f - \frac{W_c c_{P,a} T_{c,in}}{UA_f + W_c c_{P,a}} - \frac{UA_f}{UA_f + W_c c_{P,a}}T_f\right) + \frac{\lambda_w M_w W_c}{M_a P}\left(\frac{W_c P_{w,in}}{KPM_a A_f + W_c} + \frac{KPM_a A_f a_w}{KPM_a A_f + W_c}P_w^\sigma\left\{T_f\right\} - P_{w,in}\right) + \rho_f V_f c_{P,f}\frac{dT_f}{dt}, \tag{1.17.24}$$

where straightforward algebraic rearrangement unfolds

$$\rho_f V_f c_{P,f}\frac{dT_f}{dt} = \rho_f V_f\left(r_f \Delta h_r\right)\left\{T_f\right\} + UA_f\left(\frac{W_c c_{P,a}}{UA_f + W_c c_{P,a}}T_{c,in} - \left(1 - \frac{UA_f}{UA_f + W_c c_{P,a}}\right)T_f\right) + \frac{\lambda_w M_w W_c}{M_a P}\left(\left(1 - \frac{W_c}{KPM_a A_f + W_c}\right)P_{w,in} - \frac{KPM_a A_f}{KPM_a A_f + W_c}a_w P_w^\sigma\left\{T_f\right\}\right) \tag{1.17.25}$$

upon isolation of the derivative term in the left-hand side, complemented by regrouping of T_f- and $P_{w,in}$-dependent terms. Integration of Eq. (1.17.25), with the aid of

$$T_f\big|_{t=0} = T_{f,0} \tag{1.17.26}$$

for initial condition consistent with Eq. (1.817), will eventually lead to $T_f \equiv T_f\{t\}$, since $r_f\Delta h_r$ and P_w^σ are (known) functions solely of T_f – as emphasized in Eq. (1.17.25), and previously quantified via Eqs. (1.829) and (1.794), respectively; upon insertion of $T_f\{t\}$ in Eq. (1.17.20), one will obtain $P_w \equiv P_w\left\{T_f\right\} \equiv P_w\{t\}$, while insertion of the same result in Eq. (1.17.17) will permit calculation of $T_c \equiv T_c\left\{T_f\right\} \equiv T_c\{t\}$.

c) One may, for convenience, start by defining an auxiliary parameter ω as

$$\omega \equiv \frac{W_c c_{P,a}}{UA_f} \tag{1.17.27}$$

using Eq. (1.823) as template – which supports

$$\frac{W_c c_{P,a}}{UA_f + W_c c_{P,a}} = \frac{\dfrac{W_c c_{P,a}}{UA_f}}{1+\dfrac{W_c c_{P,a}}{UA_f}} = \frac{\omega}{1+\omega}, \tag{1.17.28}$$

upon division of both numerator and denominator of the left-hand side by UA_f; and similarly

$$\frac{UA_f}{UA_f + W_c c_{P,a}} = \frac{1}{1+\dfrac{W_c c_{P,a}}{UA_f}} = \frac{1}{1+\omega}, \tag{1.17.29}$$

following division of both numerator and denominator of the left-hand side by UA_f, and combination again with Eq. (1.17.27). One may further proceed to

$$1-\frac{UA_f}{UA_f + W_c c_{P,a}} = \frac{UA_f + W_c c_{P,a} - UA_f}{UA_f + W_c c_{P,a}} = \frac{W_c c_{P,a}}{UA_f + W_c c_{P,a}} = \frac{\dfrac{W_c c_{P,a}}{UA_f}}{1+\dfrac{W_c c_{P,a}}{UA_f}} = \frac{\omega}{1+\omega}, \tag{1.17.30}$$

after lumping fractions in the left-hand side, dividing both numerator and denominator by UA_f, and combining with Eq. (1.17.27) in the end. By the same token, one can define another auxiliary parameter as

$$\varpi \equiv \frac{W_c}{KPM_a A_f}, \tag{1.17.31}$$

and consequently obtain

$$\frac{KPM_a A_f}{KPM_a A_f + W_c} = \frac{1}{1+\dfrac{W_c}{KPM_a A_f}} = \frac{1}{1+\varpi} \tag{1.17.32}$$

following division of both numerator and denominator of the left-hand side by $KPM_a A_f$, and insertion of Eq. (1.17.31) afterward; complemented by

$$\frac{W_c}{KPM_a A_f + W_c} = \frac{\dfrac{W_c}{KPM_a A_f}}{1+\dfrac{W_c}{KPM_a A_f}} = \frac{\varpi}{1+\varpi}, \tag{1.17.33}$$

upon division of both numerator and denominator of the left-hand side by $KPM_a A_f$, and then combination with Eq. (1.17.31). One similarly gets

$$1-\frac{W_c}{KPM_a A_f + W_c} = \frac{KPM_a A_f + W_c - W_c}{KPM_a A_f + W_c} = \frac{KPM_a A_f}{KPM_a A_f + W_c} = \frac{1}{1+\dfrac{W_c}{KPM_a A_f}} = \frac{1}{1+\varpi} \tag{1.17.34}$$

from Eq. (1.17.33), after lumping fractions, cancelling symmetrical terms, dividing both numerator and denominator by $KPM_a A_f$, and finally recalling Eq. (1.17.31). In view of Eqs. (1.17.28), (1.17.30), (1.17.32), and (1.17.34), one may rewrite Eq. (1.17.25) as

$$\rho_f V_f c_{P,f}\frac{dT_f}{dt} = \rho_f V_f r_f \Delta h_r + UA_f\left(\frac{\omega}{1+\omega}T_{c,in} - \frac{\omega}{1+\omega}T_f\right) + \frac{\lambda_w M_w W_c}{M_a P}\left(\frac{1}{1+\varpi}P_{w,in} - \frac{1}{1+\varpi}a_w P_w^\sigma\right); \tag{1.17.35}$$

another division of both sides by $UA_f T_{f,0}$ gives rise to

$$\frac{\rho_f V_f c_{P,f}}{UA_f T_{f,0}}\frac{dT_f}{dt} = \frac{\rho_f V_f}{UA_f T_{f,0}} r_f \Delta h_r + \frac{\omega}{1+\omega}\left(\frac{T_{c,in}}{T_{f,0}} - \frac{T_f}{T_{f,0}}\right) + \frac{\lambda_w M_w W_c}{M_a UA_f T_{f,0}}\frac{1}{1+\varpi}\left(\frac{P_{w,in}}{P} - a_w\frac{P_w^\sigma}{P}\right), \tag{1.17.36}$$

where $\omega/(1+\omega)$ or $1/(1+\varpi)$ were meanwhile factored out, and $T_{f,0}$ and P factored in. Equation (1.17.36) can be reformulated as

$$\frac{dT_f^*}{dt^*} = \left(r_f \Delta h_r\right)^* - \frac{\omega}{1+\omega}\left(T_f^* - T_{c,in}^*\right) - \frac{\psi}{1+\varpi}\left(a_w P_w^{\sigma^*} - P_{w,in}^*\right) \tag{1.17.37}$$

– provided that

$$T_f^* \equiv \frac{T_f}{T_{f,0}} \tag{1.17.38}$$

stands for normalized food temperature, and likewise

$$T_{c,in}^* \equiv \frac{T_{c,in}}{T_{f,0}} \tag{1.17.39}$$

stands for inlet air temperature,

$$P_w^{\sigma^*} \equiv \frac{P_w^\sigma}{P} \tag{1.17.40}$$

stands for normalized saturation pressure of water,

$$P_{w,in}^* \equiv \frac{P_{w,in}}{P} \tag{1.17.41}$$

stands for inlet normalized partial pressure of water vapor,

$$t^* \equiv \frac{UA_f}{\rho_f V_f c_{P,f}} t \quad (1.17.42)$$

stands for dimensionless time,

$$\left(r_f \Delta h_r\right)^* \equiv \frac{\rho_f V_f}{UA_f T_{f,0}} r_f \Delta h_r \quad (1.17.43)$$

stands for dimensionless rate of release of heat by respiration, and

$$\psi \equiv \frac{\lambda_w M_w W_c}{M_a UA_f T_{f,0}} \quad (1.17.44)$$

entails a characteristic dimensionless parameter. Following an identical rationale, one will redo Eq. (1.17.20) to

$$P_w = \frac{\varpi}{1+\varpi} P_{w,in} + \frac{1}{1+\varpi} a_w P_w^\sigma \quad (1.17.45)$$

with the aid of Eqs. (1.17.32) and (1.17.33); while Eq. (1.17.17) will read

$$T_c = \frac{\omega}{1+\omega} T_{c,in} + \frac{1}{1+\omega} T_f \quad (1.17.46)$$

after taking Eqs. (1.17.28) and (1.17.29) on board. Division of both sides by P converts Eq. (1.17.45) to

$$P_w^* = \frac{\varpi}{1+\varpi} P_{w,in}^* + \frac{1}{1+\varpi} a_w P_w^{\sigma^*} \quad (1.17.47)$$

at the expense of Eqs. (1.17.40) and (1.17.41), besides their analog encompassing P_w^*, i.e.

$$P_w^* \equiv \frac{P_w}{P} \quad (1.17.48)$$

– or else

$$P_w^* = \frac{\varpi P_{w,in}^* + a_w P_w^{\sigma^*}}{1+\varpi}, \quad (1.17.49)$$

after factoring $1/(1+\varpi)$ out in Eq. (1.17.47); while Eq. (1.17.46) becomes

$$T_c^* = \frac{\omega}{1+\omega} T_{c,in}^* + \frac{1}{1+\omega} T_f^* \quad (1.17.50)$$

upon division of both sides by $T_{f,0}$, and combination with Eqs. (1.17.38) and (1.17.39), complemented by

$$T_c^* \equiv \frac{T_c}{T_{f,0}} \quad (1.17.51)$$

consistent therewith – with Eq. (1.17.50) being equivalent to

$$T_c^* = \frac{\omega T_{c,in}^* + T_f^*}{1+\omega}, \quad (1.17.52)$$

obtained upon lumping the two fractions in Eq. (1.17.50). Inspection of Eq. (1.17.37) unfolds a chilling effect brought about by cold air as per the second term in its right-hand side – analogous to the chilling effect associated with cold water discussed previously, after replacing Ω by ω, and $(T_f - T_{c,in})/(T_{f,0} - T_{c,in})$ by $T_f^* - T_{c,in}^*$ in Eq. (1.838), and realizing that $\varsigma(1 - \Lambda + \Lambda(T_f - T_{c,in})/(T_{f,0} - T_{c,in}))$ coincides with $(r_f \Delta h_r)^*$ in view of Eq. (1.830); complemented by the chilling effect associated with vaporization of water, as per the third term in its right-hand side. If chemical equilibrium existed between water (mixed) in food and in (pure) vapor forms, i.e. $a_w P_w^{\sigma^*} = P_{w,in}^*$, then Eq. (1.17.37) would degenerate to Eq. (1.838); as expected, Eq. (1.17.46) mimics Eq. (1.822), after Ω is again replaced by ω. Conversely, when resistance to mass transfer of water vapor in air is small, i.e. K is large and thus ϖ is small as per Eq. (1.17.31), then $\psi\left(a_w P_w^{\sigma^*} - P_{w,in}^*\right)/(1+\varpi) \approx \psi\left(a_w P_w^{\sigma^*} - P_{w,in}^*\right)$ in Eq. (1.17.37). If the overall coefficient of heat transfer in air, U is large, or the heat of vaporization of water, λ_w, is small, then ψ is small as per Eq. (1.17.44); hence, the last term in the right-hand side of Eq. (1.17.37) can be discarded, as vaporization-mediated chilling will be negligible compared to cool air-mediated chilling. Conversely, a large ψ will magnify the effect of a reduced pressure upon $a_w P_w^{\sigma^*} - P_{w,in}^*$ in Eq. (1.17.36), owing to Eqs. (1.17.40) and (1.17.41); hence, full drying of inlet air may actually suffice to attain the intended chilling effect, without the need to cool air in advance – should the target food possess enough mobile water, in the first place.

1.4 Freezing/storage/thawing

1.4.1 Introduction

Freezing is a unit operation intended to preserve foods for a long time, without causing significant changes in their sensory qualities and nutritional value – by resorting to refrigeration below the freezing point of water (in theory), or $\leq -18°C$ (in practice); this causes a fraction of the food constitutive water to undergo a phase change to the solid state, in the form of ice crystals. The immobilization of water as ice, and the concomitant increase of solute concentration in (the still) unfrozen water lowers water activity; hence, preservation is achieved at the expense of a low temperature complemented by a low a_w, both of which hamper occurrence of (bio)chemical, enzymatic, and microbial reactions – and thus help keep decay to a minimum. In this regard, freezing is generally regarded as superior to canning and drying of fruits and vegetables, if long-term storage is intended – despite its higher cost, and need of available refrigeration network(s).

HISTORICAL FRAMEWORK

The Chinese were apparently the first people to harness the power of freezing food beyond wintertime, by resorting to ice cellars by as early as 1000 BC. Greeks and Romans also stored compressed snow in insulated cellars – while Egyptians and Indians discovered that rapid evaporation through the porous walls of clay vessels produces ice crystals in the water stored therein.

Modern (quick) freezing of food is due to American taxidermist Clarence Birdseye – who observed, while conducting fieldwork in the Arctic, how the Inuit natives preserved freshly caught fish and meat in barrels of seawater that quickly froze due to the frigid climate. Those iced items were thawed months later and cooked (when needed) – and still tasted fresh, unlike happened with frozen foods at large in the Western world by that time. In 1923, following investment in an electric fan, buckets of brine, and cakes of ice, Birdseye invented (and later improved) a system of packing fresh food into waxed cardboard boxes, coupled with flash-freezing at high pressure. By 1927, his company, General Seafoods, successfully applied such a new technology to preserve beef, poultry, fruits, and vegetables. The Goldman-Sachs Trading Co. and the Postum Co. (later General Food Co.) purchased the associated patents from him in 1929; and the first quick-frozen vegetables, fruits, seafoods, and meat were sold to the public in 1930 for the first time, in Springfied, MA – under the trade name Birds Eye Frosted Foods. The products were a hit; further expansion focused on transportation of frozen products in refrigerated boxcars to more distant locations. Development of blanching in the 1940s conveyed a further impetus for the frozen vegetable industry, due to the more appealing final products (namely mixed vegetables and corn) in terms of appearance and texture.

The booming ownership of domestic freezers in the 1970s–1990s for storage, and microwave ovens for thawing/cooking led to a notable expansion in sales of frozen foods; currently, total frozen sales worldwide add up to more than 75 B€. The intrinsic image of higher quality and freshness has supported most marketing efforts; frozen meats, fruits, and vegetables have meanwhile outsold their canned and dried counterparts. On the other hand, increase in the percentage of working women and decrease in family size – with concomitant reduction in time spent preparing food, have led to improvements in kitchen appliances and widened the portfolio of ready-to-eat foods, normally in frozen form. However, distribution of frozen foods is relatively expensive (accounting for ca. 10% of their final cost), because a rather low temperature is to be maintained throughout the whole food chain – thus leading to preferential selection of commodities as the most perishable ones, and which bear a premium price. This realization has prompted a shift toward chilled foods in recent decades, especially in more developed countries.

In the case of foods particularly rich in water, freezing of this compound plays a nuclear role in their final thermophysical features, especially below ca. −10°C; density falls by 9% (because more distant molecules permit stronger hydrogen bonding, as per a more favorable relative orientation), thermal conductivity increases (because of the solid state matrix) by fourfold compared to pure liquid water, and specific heat capacity decreases. Although virtually all foods may be subjected to freezing, the temperature required to do so depends on their initial content of water – as illustrated in Fig. 1.64. The major groups of foods available in frozen form are baked foods (e.g. bread, cakes, and fruit and meat pies); fish fillets (e.g. cod, plaice, hake) and seafood (e.g. shrimp, crab meat); fruits (e.g. blackcurrants, oranges, raspberries, strawberries), either as whole or in purée or concentrate forms; meats, in the form of carcasses, boxed joints, or cubes, and derived products (e.g. beef burgers, reformed steaks, sausages); vegetables (e.g. green beans, peas, potatoes, spinach, sprouts, sweetcorn); and prepared foods or full meals (e.g. cook-freeze dishes, fish cakes and fingers, desserts, ice cream, pizzas, prepared dishes with accompanying sauces, ready meals).

As is common practice in chilling, freezing processes resort normally to vapor compression, or else to cryogenic fluids; the former requires refrigerants with lower boiling point – since temperature must remain (well) below the freezing temperature of the food, otherwise the associated driving force will not permit heat removal (at least at reasonable rates). Cryogenic freezing is a one-way operation, which takes advantage of the single use of a utility available at very low temperature – preferably characterized by a large latent heat of vaporization, so as to minimize the amount of cryogenic fluid needed for a given heat load; it occurs much faster than when vapor compression is utilized, due to both large heat transfer coefficient at boiling and large temperature difference as driving force thereto. Estimation of freezing time for foods is relevant, not only to ensure their final quality, but also to anticipate throughput rate in a freezing plant; note that latent heat is released at the (moving) freezing front, to be transferred thereafter through the frozen layer to the surface of the food, and then through the boundary film to the freezing utility fluid. Freezing time increases with higher food density and size – as a higher amount of heat is to be removed; but decreases with larger thermal conductivity of food and surface heat transfer coefficient (besides conductivity of package material, if present), outer

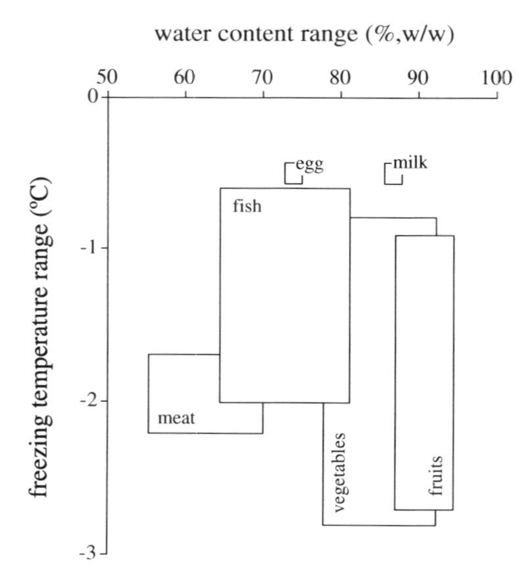

FIGURE 1.64 Ranges of physicochemical features (▭) of selected families of foods – expressed as freezing temperatures and water contents.

surface area of food, and temperature difference between food and freezing medium – in general agreement with Fourier's law. Once frozen, the issue arises as to the power necessary to keep the food under frozen state throughout storage.

The reverse operation of freezing is thawing; when air or water is utilized thereto, ice melts from the surface toward the bulk of the food matrix, thus forming a layer of liquid water with increasing thickness. Such a liquid possesses a lower thermal conductivity than ice (as liquids always do, compared to their solid counterparts), so the said surface layer reduces the rate at which heat is conducted into the food; as the outer layer of liquid water grows thicker, its insulating effect is magnified. Note that this phenomenon is the opposite of what is observed during freezing – where heat transfer accelerates as ice is formed, due to the higher thermal conductivity of ice relative to liquid water. One immediate consequence is that thawing takes longer to occur than freezing, between the same initial and final conditions (although in reverse order) – as a typical example of hysteresis.

1.4.2 Product and process overview

1.4.2.1 Freezing

As emphasized previously when discussing crystallization (see Fig. 4.12 of *Food Proc. Eng.: basics & mechanical operations*) and freeze concentration (see Fig. 3.45 of *Food Proc. Eng.: thermal & chemical operations*), pure water eventually separates out as ice (crystals) if temperature of the mother aqueous solution is decreased sufficiently – provided that its departing solute concentration lies below the concentration of the eutectic point; this corresponds to the ice + liquid solution region in Fig. 1.65, pertaining to a solution of sucrose in water. As ice forms, freezing temperature decreases – as the resulting solution becomes more and more concentrated in its solute, see the gradually decreasing curve on the left of Fig. 1.65. This process continues along the equilibrium freezing line (also known as liquidus), until the eutectic mass fraction, w_{eut}, is reached – characterized by eutectic temperature, T_{eut}, with liquid and solid phases sharing the same composition, w_{eut}. The solubility of solute line – see fast increasing curve on the right of Fig. 1.65 – crosses the freezing line also at the eutectic point; hence, solute has been freeze-concentrated to its saturation concentration at this (invariant) point.

Where freeze-concentration is concerned, crystallizing as much ice as possible (to be eventually rejected) is the goal, so as to obtain a concentrated liquor; the eutectic point is usually not approached and seldom crossed – so the thermodynamic diagram labeled as Fig. 3.45 in *Food Proc. Eng.: thermal & chemical operations* suffices. The situation is, however, quite distinct when freezing of a food is envisaged – since no portion of the whole food is to be removed. In fact, the food evolves as a mixture of ice crystals, unfrozen solute, frozen eutectic mixture(s), and the still unfrozen remainder of the food matrix, as the temperature lowers throughout the whole process of freezing; more information is accordingly required regarding phase behavior – as provided in Fig. 1.65, known as supplemented state diagram (for including kinetic information, in addition to thermodynamic information). As heat is further withdrawn, it is unlikely that solute will co-crystallize at T_{eut} (–14°C for sucrose) – due to the very high viscosity of the accompanying solution at the said low temperature; this causes freeze-concentration to proceed beyond T_{eut} in a nonequilibrium state of supersaturation – since attainment of equilibrium would take excessively long a time to be of practical interest in food processing. Hence, the freezing curve extrapolates toward higher mass fractions of solute, thus including a new region containing not only ice and liquid solution, but also (solid) solute; this extrapolation takes place until no more water exists in solution that can freeze within a practical timeframe. A characteristic transition of the remaining solution to a rubbery state then occurs, which defines a three-state region containing ice, rubber, and solute. This transition is represented by the small vertical dashed line in Fig. 1.65, because no change in system concentration occurs; and ends when the glass transition curve is crossed (see gradually increasing curve), at a glass transition temperature, T_{gls}, characteristic of each solute (–39°C in the case of sucrose in water). Examples of T_{gls} are made available in Fig. 1.66, pertaining to a selection of single compounds dissolved in water and actual foods; note that pure solute and pure solvent also exhibit glass transition temperatures – see $T_{gls,s}$ and $T_{gls,w}$ for sucrose and water, respectively, in Fig. 1.65, at the intercept of the glass transition curve of the corresponding mixture with the vertical lines defined by unit and nil mass fraction, respectively, of sucrose. If further removal of heat takes place, the unfrozen liquid (with quite poor mobility) undergoes a change in physical state from viscoelastic (or rubbery) liquid to brittle, amorphous solid glass; at lower mass fractions, ice, glass, and solute will instead come to simultaneous existence.

A glass is defined as a nonequilibrium, metastable, amorphous, and disordered solid – which, upon heating, undergoes smooth transition to a rubbery material, or a liquid of outstandingly high

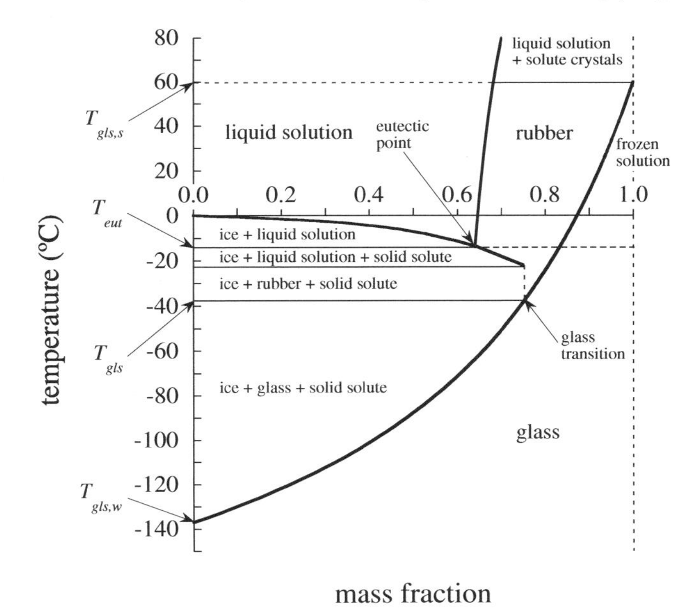

FIGURE 1.65 Supplemented state diagram of sucrose–water mixture, encompassing freezing curve (left, slowly decreasing), solubility curve (right, quickly increasing), and glass transition curve (left and right, slowly increasing), as temperature versus sucrose mass fraction – with indication of temperature of eutectic point, T_{eut}, temperature of glass transition of mixture, T_{gls}, temperature of glass transition of pure water, $T_{gls,w}$, and temperature of glass transition of pure solute, $T_{gls,s}$.

viscosity (somewhere between 10^{10} and 10^{14} Pa.s, dependent on temperature). As discussed above, the glass transition curve extends from the glass transition temperature of pure water, $T_{gls,w}$ (–136°C), up to the glass transition temperature of pure solute, $T_{gls,s}$ (60°C in the case of sucrose) – see Fig. 1.65. The evolution in between abides to

$$T_{gls} = \frac{w_s T_{gls,s} + k_s\left(1 - w_s\right)T_{gls,w}}{w_s + k_s\left(1 - w_s\right)}, \tag{1.872}$$

known as Gordon and Taylor's model – where T_{gls} appears as the arithmetic average of $T_{gls,s}$ and $T_{gls,w}$, weighed by w_s and $k_s(1-w_s)$, respectively; here parameter k_s is characteristic of each solute (e.g. $k_s \approx 3$ in the case of sucrose), and w_s denotes mass fraction of solute. The supersaturated solute takes on solid properties because of its extremely reduced molecular motion – responsible for a tremendous reduction in translational (but not rotational) mobility; this outstanding slowness of molecular reorganization below T_{gls} is sought by food engineers to avoid molecular migration – via creation of a pseudo-shell surrounding other food constituents. However, warming up from a glassy state to above T_{gls} produces a substantial increase in rate of diffusion – not only due to transition from amorphous solid to viscous liquid, but also because of increased dilution, since melting of small ice crystals occurs almost simultaneously.

The glass transition line goes above the horizontal line bearing T_{eut} as ordinate in Fig. 1.65, so the observed (kinetically controlled) behavior between T_{eut} and $T_{gls,s}$ actually overrides the theoretical (thermodynamically controlled) behavior – should the original mass fraction be greater than w_{eut}. Therefore, two regions result – one enclosed by the solubility line on the left and the glass transition line on the right, and constituted by rubber; and another defined by the latter on the left and the vertical line described by $w_s = 1$ on the right, and constituted by frozen solution. A biphasic region made up of liquid solution and solute crystals only happens above $T_{gls,s}$.

When storage temperature lies below the glass transition temperature (as normally happens with meat and vegetables), foods experience a particularly high stability – because formation of the said glass preserves texture. The same does not hold with regard to many fruits, however, owing to their particularly low T_{gls} – so texture losses due to frozen storage itself add to textural losses due to formation of ice crystals; this problem has been regularly overcome in the case of such processed foods as icecream or surimi via addition of fructose, maltodextrin, or sucrose – which raise the glass transition temperature to above the recommended frozen storage temperature.

As emphasized before, a glass is not in an equilibrium state – so the actual value of T_{gls} depends somewhat on the temperature history of the food; T_{gls} typically lies below its melting temperature, often by ca. 100°C (or more). This unique state is not restricted to the freezing range; for instance, biopolymer-containing foods (e.g. starch, with a melting point of ca. 225°C) may readily form a glass upon baking or extrusion ($T_{gls,s} \approx 122$°C for starch) besides freezing and freeze-drying – as happens with high-boiled sweets, dried pasta, skimmed milk, hard biscuits, and some breakfast cereals. The lower the water content of a food, the higher its glass transition temperature; the main reason is water acting as plasticizer. Another advantage of the glassy state, from a food processing point of view, is that its inherent brittleness and crispiness are appealing to consumers at large; however, such textural features may suddenly disappear upon slight heating, or increase of a mere 1–2%(w/w) in water content. On the other hand, the extremely high viscosity of a glassy material essentially discontinues all molecular motion (as outlined above), and thus halts food change, namely, with regard to texture. However, in a glass of mixed composition, molecules smaller than those of the component responsible for such a state may still diffuse (even though very slowly); therefore, full chemical stability cannot be guaranteed – despite most physical changes (e.g. crystallization) being prevented below T_{gls}.

The above rationale was focused on the freezing behavior of a simple, binary solution (of sucrose in water); however, most foods subjected to freezing are complex aqueous solutions of multiple components, or else colloids (e.g. suspended protein micelles, dispersed oil droplets); therefore, more stages occur during freezing – as illustrated in Fig. 1.67. Evolution between A and B of the slow-freezing curve has classically been termed

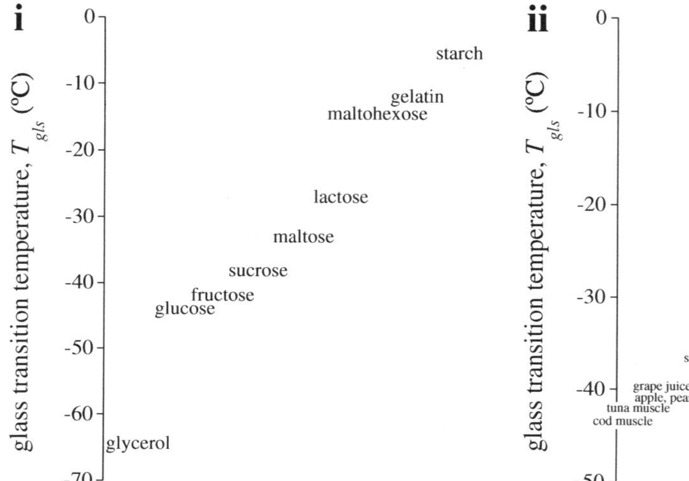

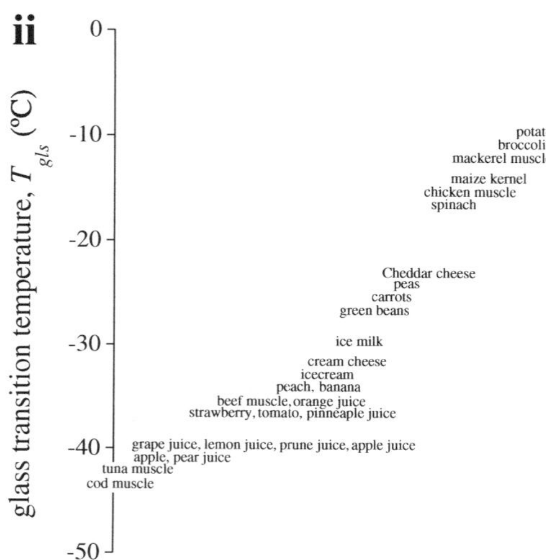

FIGURE 1.66 Glass transition temperature, pertaining to freezing of selected (i) pure compounds dissolved in liquid water and (ii) complex foods.

prefreezing, between B and D as freezing proper (see essential coincidence with freezing curve in Fig. 1.65), and from D onwards as reduction to storage temperature. The first step of the overall process is removal of sensible heat from the fresh food – which often has already undergone some form of chilling, until it reaches the freezing point, T_f; the associated heating load corresponds roughly to 4.18 kJ.kg^{-1}.K^{-1}, i.e. the isobaric thermal heat capacity of pure liquid water. Latent heat is removed next, consubstantiated in freezing of (pure) water to ice – since the mass concentration of its solutes lies usually below the corresponding eutectic point (see Fig. 1.65); the heat of crystallization at stake is ca. 335 kJ.kg^{-1}. Sensible and latent heat of fusion of water represent the major fraction of heat to be removed during freezing – even though the latent heat of other food components is to be added; however, their being much lower than that of water, and their mass fraction being much below that of water justify the minor contributions thereof.

By operational definition, the freezing point, T_f (referred to above), is the temperature at which the first (necessarily minute) crystal of water forms, under a state of chemical equilibrium with the surrounding liquid water. Since all foods contain a number of solutes (e.g. salts, carbohydrates, globular proteins), a freezing point depression is always observed – similar to Duhring's rule applying to boiling point elevation. Water is found both intra- and extracellularly in animal and plant tissues; their extracellular fluids exhibit a lower concentration of solutes – so the first ice crystals appear here, as per a higher freezing temperature.

Physicochemical processes encompassing discontinuity in properties, as is the case of phase change and chemical reaction, are characterized by an energy barrier that must first be overcome – which, in the former case, is accounted for by the energy required to form nuclei. Nucleation may occur in a homogeneous fashion, as a result of random formation of clusters of water molecules – containing a minimum number of entities, and in the correct relative orientation; or (more likely) take place in a heterogeneous fashion, at the expense of already suspended particles or microorganisms. Higher rates of heat withdrawal cause faster freezing – yet it is energetically more favorable for liquid water molecules to migrate to already existing nuclei and crystallize as ice there, than to form new nuclei. The reason for this observation comes from surface tension (σ) arguments, because a growing crystal implies a gradually larger pocket in water – which pushes liquid water molecules away, while enforcing expansion of the water/ice interface. According to Young and Laplace's law, see Eq. (4.47) of *Food Proc. Eng.: basics & mechanical operations* with $R_y = R_z = R$, a semihemispherical surface of curvature radius R implies a pressure difference between inside and outside given by $2\sigma/R$; hence, a change dR in the said radius due to ice crystal growth causes a change $2\sigma dR/R^2$ in the associated pressure difference. This effect is accordingly much larger when R is smaller, and theoretically infinite when $R \rightarrow 0$ – and thus unfavorable to homogeneous nucleation. Hence, there is a natural tendency for spontaneous formation of a small number of large ice crystals, for requiring weaker pressure differentials for growth that consume Gibbs' energy – should enough time be allowed for freezing, because reversible conditions are attained only in the vicinity of thermodynamic equilibrium. However, the reverse, i.e. formation of a very many small crystals, occurs when

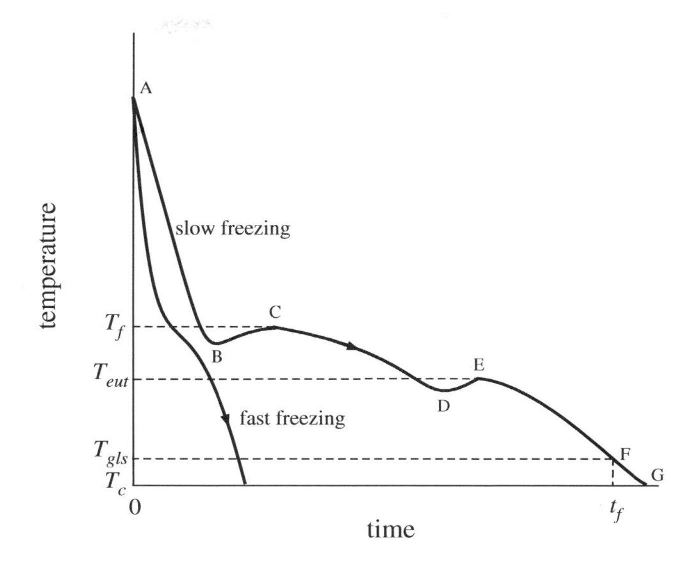

FIGURE 1.67 Typical evolution of temperature of food throughout slow and fast freezing, with indication of starting state (A); as well as first cooling period ([AB]), first maximum subcooled state (B), freezing of water at and below freezing point at temperature T_f ([CD]), second maximum subcooled state (D), freezing of water and solutes at and below eutectic point at temperature T_{eut} ([EF]), formation of glassy state at temperature T_{gls} (F) after freezing time t_f, and final cooling period below T_{gls} until reaching temperature of (cryogenic) fluid T_c – in slow freezing; and monotonic, concave decrease in temperature between A and T_f, followed by monotonic, convex decrease in temperature below T_f – in fast freezing.

freezing is sufficiently fast, since a kinetically controlled process would result (see distinct curves in Fig. 1.67); fast freezing not only makes less time available for growth of existing crystals, but also promotes spontaneous formation of nuclei. Nucleation phenomena are also largely affected by type and content of solutes in food, as well as prefreezing treatments putatively received.

When freezing is sufficiently slow, the upper curve plotted in Fig. 1.67 results; the thermal center of the food is a good candidate for the said type of process. The food first undergoes an exponential decrease in temperature, typical of a first-order process that follows Fourier's law – due to a narrower and narrower temperature difference between food and external heat sink; this is well approximated by a linear trend, see segment [AB] in Fig. 1.67, when such a heat sink is sufficiently cold. The said temperature decrease should in principle stop at the freezing point of food, T_f – yet a lower temperature is attained, labeled as point B in Fig. 1.67. This supercooling (which may be as high as 10°C below freezing temperature) occurs because nucleation is a *sine qua non* for crystal birth prior to growth; and since it is not an instantaneous phenomenon, especially when heterogeneous nuclei are scarce – since solid foods hardly undergo internal motion.

When ice crystals start forming, thermodynamic equilibrium is rapidly restored – so temperature rises fast to attain T_f, see point C in Fig. 1.67. Temperature does not, however, remain constant – in agreement with the upper curve bounding the ice+ liquid solution region in Fig. 1.65, where a gradual depression in freezing point of the still unfrozen liquor is observed as solute becomes more and more concentrated – due to removal of (otherwise liquid) solvent as ice. This process continues until the eutectic point, characterized by temperature T_{eut}, is reached – see Fig. 1.65.

Once again, some supercooling normally occurs (and for the same reasons) until point D in Fig. 1.67, followed by a rapid recovery toward T_{eut} as per point E. Crystallization of water and solute(s) continues, with temperature undergoing further gradual decrease along the upper curve bounding the ice+liquid solution+solute region region in Fig. 1.65 – until crossing the glass transition curve, at temperature T_{gls}, after an overall time period t_f (so-called freezing time) has elapsed; this corresponds to an inflection point, see F in Fig. 1.67.

The rate of heat transfer controls the rate of ice crystal growth for most of the freezing period; the rate of mass transfer, of water molecules moving to the surface of growing crystals and solutes moving away therefrom, does not constrain the rate of growth until near the end of the freezing period – as solute becomes particularly concentrated, thus causing great increases in viscosity and surface tension that hamper molecular transport. After t_f, the unfrozen portion of the food behaves no more as a viscoelastic liquid, and concomitantly acquires a brittle, amorphous glass structure – so no further phase change takes place; plain cooling will eventually bring its temperature down to the temperature of the cryogenic utility employed, T_c, in the vicinity of point G in Fig. 1.67.

The above rationale resorted to a single solute, say sucrose, as reference – and accompanied its evolution, until reaching saturation and crystallizing out at the corresponding eutectic temperature; other examples of this temperature are –5°C for glucose, –21°C for sodium chloride, and –55°C for calcium chloride. However, a typical aqueous food is composed of several solutes, so the final (or lowest) eutectic temperature – rather the various intermediate eutectic temperatures, is normally used for simplicity; for instance, the final eutectic temperature is ca. –55°C for ice cream, ranges from –60 to –50°C for meat, and is ca. –70°C for bread. Ice crystal formation is maximized below this temperature, yet several commercial foods do not reach such low temperatures; hence, a fraction of the original liquid water usually remains unfrozen, e.g. 12% in lamb, 9% in fish, and 7% in egg albumin at –20°C. The enthalpy to be removed, Δh_f (in kJ.kg^{-1}), during the overall freezing process of fruits and vegetables may resort to

$$\Delta h_f = \left(1 - x_s\right)\Delta h_j + 1.21 w_s \Delta T \tag{1.873}$$

as initial guess, known as Riedel's law – and obtained by putting together latent heat of freezing of water and sensible heat of solids (weighed by their relative amounts); here w_s denotes mass fraction of dry matter, Δh_j denotes freezing enthalpy change of juice fraction, and ΔT (in °C) denotes difference between initial and final temperature.

To reduce the adverse effects of freezing, cryoprotectants are frequently used as additives. As soon as ice forms, water activity and osmotic pressure do indeed depend on temperature only; for instance, a_W=0.8 at –23°C, irrespective of food composition, see Fig. 1.1. By adding a nonionic solute, the extent of freeze-concentration, measured by x_w, is reduced at a given temperature, because γ_w changes due to the different ionic strength – under a fixed $a_w = \gamma_w x_w$, see Eq. (1.152); since a colligative property is at stake, the smaller the molar weight of the said solute, the stronger effect a given weight will have – for containing a larger number of solute molecules. Besides being utilized to reduce (or even suppress) ice crystal growth, such additives inhibit recrystallization at temperatures below the freezing point of water, and even form stronger gels at freezing temperatures and hamper further changes in ionic strength. In addition to stabilizing protein conformation, cryoprotectants decrease damage to cell membranes that would otherwise permit release of cellular components – as will eventually become apparent during thawing. The most common examples of cryoprotectants are amino acids, carbohydrates, methylamines, polyols, and sugars; less common, yet more effective cryoprotectants are unique glycoproteins found in several bacteria, fungi, plants, and invertebrates, as well as in such fish as Antarctic cod and winter flounder. The latter have evolved as mechanisms of protection against freeze injury, namely by acting as scaffold for ice formation or inhibiting formation of ice nuclei by foreign particles. These proteins of a kind will eventually be manufactured to large scale as recombinant proteins by vector microorganisms – but a few constraints have still to be overcome, viz. cytotoxic action above a certain level, dependent on food composition and storage conditions.

Freezers have classically been categorized as mechanical refrigerators or cryogenic freezers; the former apparatuses sequentially evaporate and compress a refrigerant fluid along a continuous cycle, and then resort to the resulting cooled air, cooled liquid, or even cooled surface to remove heat from food – whereas solid (or liquid) CO_2 or (preferably) liquid N_2 are employed in the latter case, via direct contact with food. Mechanical freezers operate at ca. –40°C, while cryogenic freezers operate between –70 and –50°C; the higher capital investment required by the former, especially associated with compressor and heat exchanger, is balanced by the higher operating costs of the second – because refrigerant is usually not recirculated, and is thus lost to the atmosphere after a single use.

Depending on rate of progression of the ice front, an alternative classification of freezers is possible, as depicted in Fig. 1.68. Slow (also known as sharp) freezers include still-air freezers (see Fig. 1.69i) and cold stores; quick freezers encompass air-blast (see Fig. 1.69ii), cooled liquid freezers (see Fig. 1.69iv), and plate freezers (see Fig. 1.69v); rapid freezers are accounted for by fluidized bed freezers (see Fig. 1.69iii); and ultrarapid freezers

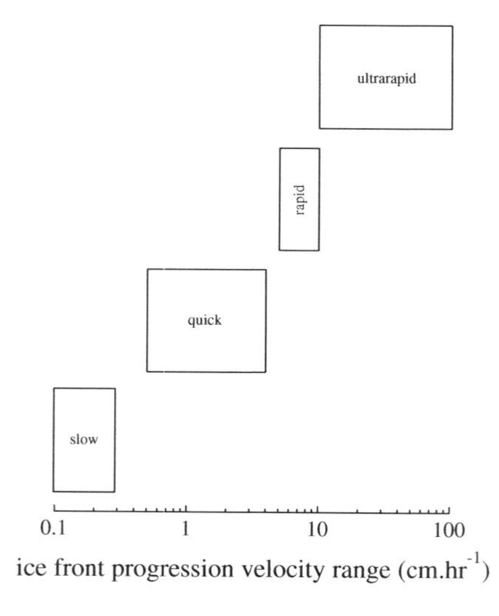

FIGURE 1.68 Classification of freezing equipment per range of progression rates (▭) of ice front.

require cryogenic fluids (see Fig. 1.69vi). Freezers are typically built in stainless steel for corrosion-proof features, and well insulated on the outer walls with e.g. polystyrene or polyurethane (due to their poor thermal conductivities) to minimize leakage of heat in. Most large-scale equipment includes programmable logic devices – which, besides monitoring process parameters and equipment status, assure automatic control and keep track of (and correct) putative faults. In the case of mechanical refrigeration, the corresponding compressors have often solid-state controls, so that one compressor will be able to meet the heat load of several freezers; in the case of multiple compressors, a computerized integrated control is normally preferred.

Cooled air freezers normally come in either chest or air-blast designs; the former freeze food using stationary air at between −30 and −20°C, undergoing (at most) some natural convection – yet some degree of forced circulation of air is sometimes provided by low-power fans, aimed at keeping temperature uniform. Due to their long freezing times – which may reach 72 hr, they are used chiefly for storage of previously frozen products. Ranges of heat transfer coefficients and regular freezing times are provided in Fig. 1.70; as expected, freezing time is almost inversely proportional (beware of the log/log scales utilized) to the underlying heat transfer coefficient. These type of freezers are, nevertheless, appropriate for freezing of carcass meat and hardening of ice cream – for which slow rates of freezing suffice.

A major drawback of air-cooled freezers is ice formation on floors, walls, and evaporator coils, resulting from air moisture condensing and freezing afterward; atmospheric air at 10°C and 60% relative humidity contains ca. 4.5 $g_{water}.kg_{air}^{-1}$ – meaning that 100 $m^3_{air}.hr^{-1}$ entering the cold store, through loading doors and the like, will deposit 13 $kg_{ice}.d^{-1}$. The accumulated ice reduces rate of heat transfer, due to the extra resistance associated with the layer of ice; hence, the compressors have to be more powerful or operate for longer periods. To minimize this problem, defrosting cycles of evaporation coils are to be carried out on a regular basis – but the energy required thereby adds to the energy used for refrigerated storage proper. In addition, slippery surfaces may arise during defrosting that constitute a hazard to circulation of human operators. To reduce the frequency of defrosting periods, desiccant dehumidifiers have to be employed – which reduce moisture content of air prior to admission.

In attempts to increase the rates of heat transfer, forced (re)circulation of cold air (between −50 and −30°C) over the food is in order, at linear velocities ranging within 1.5–6 $m.s^{-1}$ – which are high enough to reduce thickness of the boundary layer of air contacting the food, and thus allow increase of the heat transfer coefficients by up to one order of magnitude (see Fig. 1.70). This procedure is known in practice as air-blast. It may be brought about in batch equipment, e.g. trays in rooms or cabinets – with extensive manpower required for handling, cleaning, and transportation of the said trays; or continuouswise, at throughput rates of up to 20 $ton.hr^{-1}$, via trolleys stacked with trays of food or (solid or meshed) conveyor belts carrying food – moving along an insulated tunnel, with air supplied along the direction of movement, or perpendicular thereto for more efficient heat transfer. This is indeed one of the oldest, but still most common freezer designs; choice thereof lies chiefly on its versatility of operation and stability of temperature.

More efficient freezing can be attained by passing food through a sequence of air-blast apparatuses, with air flown co- or countercurrentwise, at velocities of 3–10 $m.s^{-1}$. The first in the sequence is aimed at quickly generating a thin layer of frozen product on its surface – meant to avoid further weight loss, as well as individual distortion and sticking to other pieces of food. Subsequent freezers, characterized by higher space times, bring about thickening of the frozen layer – with putative clumps broken between

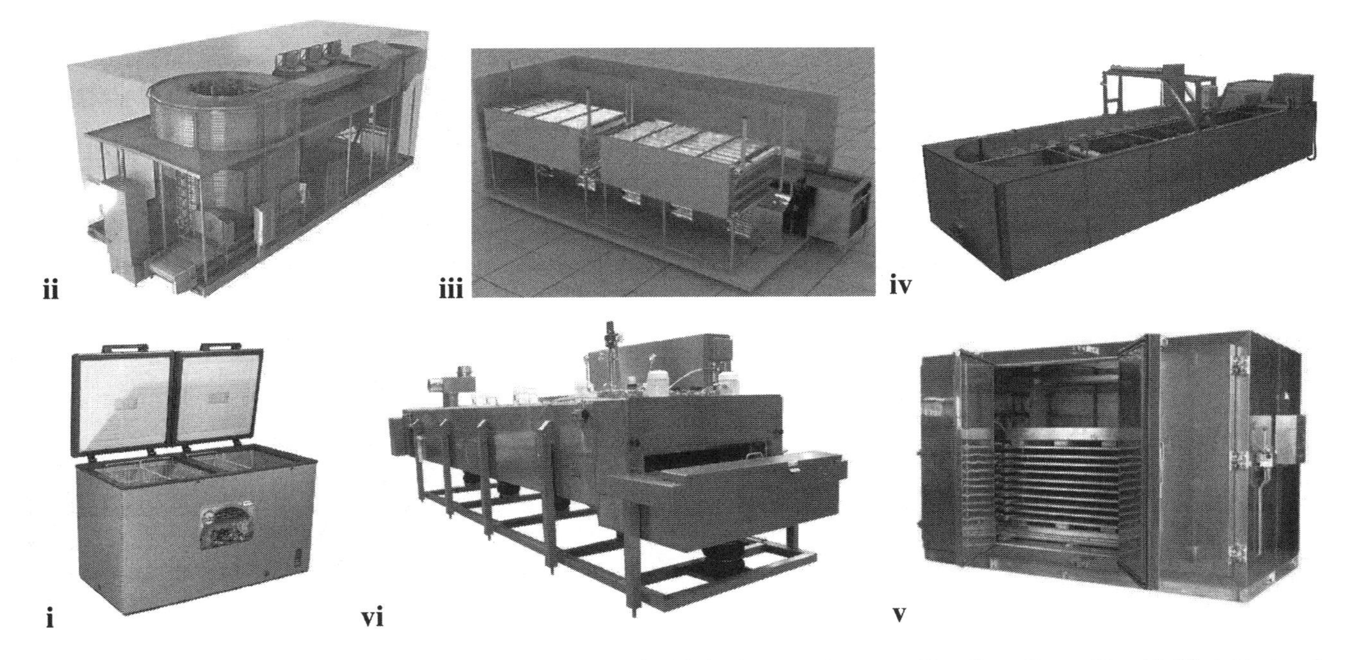

FIGURE 1.69 Examples of freezing equipment: cooled air freezer, as (i) still-air chest type (courtesy of ScanFrost, Enugu, Nigeria), (ii) air-blast configuration (courtesy of SanYod M&D Technologies, Singapore) and (iii) fluidized bed (courtesy of Chenguan Machinery & Freezing Technologies, Shaoxing, China); (iv) cooled liquid freezer (courtesy of Palinox, Sabadell, Spain); (v) cooled surface freezer (courtesy of AS Klenz BioTechnology, Chennai, India); and (vi) cryogenic freezer (courtesy of CES Freezing Technology, Kortrijk, Belgium).

stages; a tighter control of progression of the ice front is also feasible. The need of supporting rails may be circumvented by self-stacking, up to 30-fold – with each tier (50–75 cm wide) of the belt resting on the vertical (lateral) sides of the tier beneath. The belt may be bent laterally around a rotating drum, in attempts to maximize belt surface area per unit area of floor space occupied; this spiral design, suitable for both packaged and unpackaged food items, possesses the further advantage of eliminating product damage at transfer points – especially in the case of products that require gentle handling.

In view of the relatively high rates of heat transfer, air-blast freezing is quite flexible in terms of type, size, and shape of foods that can be handled (as stressed above), besides decreasing the size of the required equipment; routine cases include unpackaged vegetables and fish fingers. However, the need for defrosting of refrigeration coils remains – and may even be aggravated in the case of unpackaged foods, due to enhanced dehydration; this phenomenon not only leads to losses in weight of the food and potential freezer burn thereof, but also accelerates oxidative decay. Countercurrent flow of cold air reduces dehydration – since the most humid food supplied at the entrance contacts cold air richest in moisture at its exit.

Further increase in heat transfer rate is possible by pumping cold air at high velocity (ca. 35 $m.s^{-1}$) – as obtained with specially designed nozzles, placed perpendicularly to the food surface. Such an impingement freezing is convenient when the surrounding film accounts for the higher fraction of thermal resistance compared to the bulk of food – as happens in the case of meat patties, which are normally thin and exhibit a high specific surface area.

Uniform freezing may, in alternative, be effected in the case of (unpackaged) particulate food via a fluidized bed – of thickness within the range 2–12 cm; typical air velocities are 2–6 $m.s^{-1}$, while temperatures range between −35 and −25°C. Fluidization will occur provided that air is fed above the fluidization velocity of the food particles – but below their terminal velocity, otherwise pneumatic transport results. The bed is supplied on a perforated

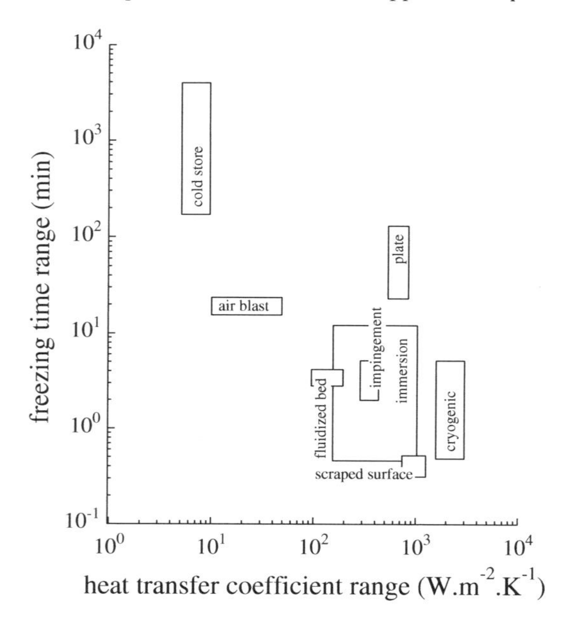

FIGURE 1.70 Ranges of performance features (▭) of freezing equipment for selected configurations – expressed as freezing times and heat transfer coefficients.

tray or conveyor, through which cold air is pumped from beneath – and the products are discharged over a weir, with height adjustable for space time control. Since food particles are well mixed, uniform freezing takes place – as it occurs the same way from their whole outer surface; dehydration is also less extensive, thus reducing the frequency of defrosting. Furthermore, particles remain as individual entities – a deed relevant for particulate foods with a tendency to stick to each other. As discussed above for air-blast, two-stage freezing is also convenient in this case – with the first resorting to a shallow bed, designed to generate an ice glaze on the surface of the food particles that reduces stickiness to other particles and adds to mechanical robustness; followed by a second stage as a thick bed, say 10–15 cm, meant to complete the freezing process. Fluidized bed freezing is appropriate for shrimps, pasta, meat cubes, and fruit and vegetable pieces (e.g. diced carrots, French fries, green beans, sliced fruits, and sliced potatoes); as well as vegetables and fruits that come in small sizes – as is the case of cooked beans, Brussels sprouts, cooked rice, peppers, mushrooms, peas, small onions, strawberries and other berries, and sweetcorn kernels. In the case of larger pieces of food (e.g. fish fillets), a similar approach is possible – but without reaching fluidization proper, owing to the excessively high velocity required for fluidization.

In the case of immersion freezers, packaged food, previously placed in perforated containers, is passed through a bath of refrigerant or brine, supplied at from −55 to 0°C; displacement is via a rotating drum or a submerged mesh conveyor, while wrapping films or laminated card/polyethylene cans may serve as packaging materials. Unlike cryogenic freezing, the said refrigerant remains liquid, and is thus not lost as gas throughout the freezing operation; aqueous solutions containing soluble monosaccharides (e.g. glucose, fructose) or disaccharides (e.g. sucrose), alcohols (ethanol, glycerol, propylene glycol), or salts (e.g. sodium, potassium, or calcium chloride) are eligible as refrigerants – in view of their safety, coupled with moderate contribution to odor, taste, and color. This mode of freezing was pioneered by the fishing industry, in view of its ease of implementation in high sea and low cost – chiefly because no high velocities of refrigerant are needed, and its large specific heat capacity and thermal conductivity call for low degrees of recirculation; in fact, the rate of freezing often outruns its air-blast counterpart – so food texture is better retained and dehydration is less extensive. Current uses encompass freezing of ice cream and orange juice, as well as prefreezing of poultry resorting to Freon™.

Applicability of immersion freezing to unpackaged foods has been hampered by the possibility of chemical/microbiological contamination and uncontrolled rates of mass transfer – so preliminary application of a flexible membrane onto the food is in order. Even when clean/sterile refrigerants/brines are used, entrainment thereof by the food is still to be minimized – to reduce rate of fluid makeup and degree of food contamination; hence, foods possessing smooth, nonporous surfaces (e.g. peas), fat-rich foods, or foods after preliminary surface treatment by freezing qualify for immersion freezing with hydrophilic refrigerants/brines. Of particular concern is uptake of sodium chloride, in the case of brines containing this salt; such a phenomenon may, however, be substantially reduced if sucrose is also present in the brine – probably due to formation of a protective layer on the food surface. Conversely, fruits frozen in syrup solutions give rise to desserts possessing the characteristic sensory attributes of the original fruits.

Hydrofluidization is a recent improvement of immersion freezing; it consists of pumping the refrigerating liquid through orifices or nozzles in the vessel, so as to create jets that induce strong turbulence therein – characterized by high coefficients of heat transfer. Another approach is use of pumpable ice slurries as refrigerating medium, where very small ice particles absorb heat rapidly as they thaw on the product surface, thus entailing high heat transfer coefficients (of the order of 1 $kW.m^{-2}.K^{-1}$); the freezing time accordingly approaches that typical of cryogenic freezing.

Plate freezers consist of a vertical or horizontal stack of several (up to 25) hollow stainless steel or aluminum plates with 2.5–5 cm for thickness, through which refrigerant is pumped at temperatures lying between –5 and –30°C; food is placed between such plates – thus guaranteeing high rates of heat transfer altogether. Therefore, only flat, relatively thin (not above 8 cm) foods are eligible for this type of freezer – where they are placed in frames, as single layers; packaged foods experience an extra resistance to heat transfer due to their packaging material, while metal packages may instead contribute to corrode the cold plates – due to formation of pseudo-batteries of non-negligible electromotive force, when distinct electrochemical potentials are exposed to each other. The plates are to be pushed together – not only to improve contact between food surfaces and plates that facilitates heat transfer, but also to reduce pack bulging owing to water expansion during freezing. Common applications encompass bags of foods, blocks of beef patties, cartons of vegetables and seafood, fish sticks, hamburgers, and molded products (e.g. fruit pulps, ready meals). Besides compactness of design, the said freezers are prone to marginal dehydration – as available space between foods and plates is kept to a minimum, and allow throughput rates of up to 3 $ton.hr^{-1}$ at relatively low operating costs; however, the underlying capital investment is relatively high.

In the case of foods that are sticky by nature (e.g. pasta, pulped fruits), likely to dehydrate (e.g. chicken breasts, shrimps), or delicate in terms of shape (e.g. cakes), a single refrigerated plate is employed – but often for prefreezing purposes only. The frozen crust formed sets the shape of the food – and, in particular, prevents sticking to the conveyor belt, or imprinting of marks thereof on the food surface; therefore, further freezing may proceed in a conventional freezer. The major difficulty comes from adhesion of the frozen product to the supporting metallic belt – caused by the surface force established between ice and metal, which becomes stronger as temperature is decreased. Eventually, the said force overcomes mechanical strength of ice itself, so removal of the frozen product will imply its breakage; this hampers food appearance, besides being costly to effect. Such a problem can be avoided if temperature is brought below –80°C, as happens in cryogenic freezers – owing to a remarkable reduction in intensity of the said force.

A variant in design is scraped-surface freezers, used for liquid and semiliquid foods; although conceptually similar to scraped-surface heat exchangers, they operate with liquid ammonia or refrigerants alike. This design is quite suitable for icecream manufacture, known to aim at smoothness and creaminess of the final product; in fact, most water freezes in a matter of seconds, thus forming tiny ice crystals – too small in size to be detectable in the mouth upon ingestion. The rotor accordingly scrapes frozen material from the barrel wall, where it builds up to 1 mm-thick layers; while air is incorporated, via rotor action solely or direct injection on purpose. After the said prefreezing stage, hardening may be produced in a regular cold store.

A quite different approach to freezing consists of providing intimate contact of the food with a very cold solid/liquid produced elsewhere (rather than generated *in situ* in the food process); common examples include beefburgers, cakes, fruits, and vegetables (including green beans, peas, and sweetcorn) and seafood. Solid (or liquid) carbon dioxide may be used, as happens in chilling – yet the preferred cryogenic fluid is liquid nitrogen; it exhibits a boiling point of –195.4°C (i.e. much lower than that of CO_2) and a latent heat of vaporization of 198.3 $kJ.kg^{-1}$, besides a specific heat capacity of vapor equal to 1.04 $kJ.kg^{-1}.K^{-1}$. Therefore, to produce liquid N_2 departing from room temperature, $1.04\times(25-(-195.4))=229$ $kJ.kg^{-1}$ is to be extracted as sensible heat (of cooling), and another 198.3 $kJ.kg^{-1}$ as latent heat (of condensation). In other words, $229/(229+198.3)=54\%$ of the overall cooling power of liquid N_2 is accounted for by sensible heat, if used to refrigerate foods departing from room temperature – thus calling for gas-handling equipment to extract a major fraction thereof. Handling procedures for safety are essentially the same described previously for CO_2 – although the risk of asphyxiation is much lower. Dichlorodifluoromethane (or Freon™) was once extensively utilized, namely for sticky and fragile foods with a tendency to form clumps (e.g. meat paste, shrimps, tomato slices); however, its contribution to depleting the UV-protective ozone layer in the upper atmosphere has led to its phasing out, as per the Montréal Protocol.

Besides shorter freezing times and generation of ice crystals responsible for less extensive textural decay (including reduced drip losses *a posteriori*), cryogenic freezing is quite flexible to handle different products without major process changes; lower investment on equipment and maintenance then comes along. Conversely, processing costs are six- to eightfold those of vapor compression freezing – on account chiefly of the single use of cryogenic fluids. The eventual loss to the atmosphere enforces use of fluids previously obtained therefrom (or, at least, present therein); this is the case of (liquid) nitrogen, or carbon dioxide to a lesser extent. Both are suitable for being inert – unlike oxygen that holds a high explosion risk, further to accelerating oxidative degradation of food.

Liquid nitrogen is industrially obtained as a by-product of manufacture of liquid O_2 – sought for clinical purposes. Liquid N_2 and CO_2 may be applied likewise to food toward freezing, yet the latter requires pressures above ca. 5 atm and cannot go below –56°C – so its solid counterpart is normally more useful.

When working in overpressure, liquid CO_2 is injected along the freezer length, produces roughly equal mass fractions of solid and vapor forms (a unique feature exhibited by this compound) accompanying pressure drop to below 5 atm, and simultaneous temperature drop to below –78°C caused by (adiabatic) vaporization – and the associated expansion creates forced convection inside the chamber; solid particles then sublime rapidly upon contact with food, thus accounting for ca. 85% of the total heat withdrawn.

Liquid N_2 is normally sprayed as a very fine mist of droplets, via nozzles located at a short distance (ca. 15 cm) from the food surface; such a distance is the best compromise between uniform (and total) coverage of food surface, and low loss as vapor along the path tracked by the food. Separation as a liquid/vapor mixture ensues, with ca. 50% of heat being withdrawn through vaporization upon contact with food; enhancement of convective currents is attained by recirculating the (still cold) vapor to the freezing chamber. Spraying along the chamber is the preferred mode of operation, as it circumvents the need for circulating fans; older designs resorted to spraying in the vicinity of the exit, coupled with fans blowing the vapor backward in a countercurrent mode. Typical degrees of consumption of this cryogenic fluid lie in the vicinity of 1.2 $kg_{nitrogen}.kg_{food}^{-1}$; oftentimes, it is used after prefreezing with CO_2.

Immersion of foods directly in the cryogenic liquid is also possible, and actually accounts for the fastest freezing process (up to 1 $ton.hr^{-1}$ in a 1 m-long bath); after dropping the foods (e.g. chicken portions, diced meat, shrimps), a conveyor lifts them right after generation of the frozen crust on their surface. Both prefrozen food and vapor nitrogen formed are then passed to a tunnel to complete freezing; although sticking is prevented due to violent boiling, overfreezing is difficult to avoid – and extreme stresses incurred in by the strong thermal shock may degenerate to cracking or splitting of the food matrix.

Freezing equipment for use with liquid N_2 follows essentially the same designs as that resorting to vapor compression – and throughput rates up to 1.5 $ton.hr^{-1}$ are possible; however, smaller units are required as space times are shorter, and there is no need for heat exchange coils or defrosting cycles. Dehydration (and flavor loss) is reduced by one order of magnitude relative to air cooling systems – because sprayed N_2 promptly forms a protective crust on the surface of the food; oxidative decay is also reduced, due to absence of oxygen inside the chamber. When premium value, and often delicate foods (e.g. diced fruit, diced poultry, meatballs, pizza toppings, scallops, sliced mushrooms, strawberries) are to be cryogenically frozen, they are transported through the freezer via flighted conveyors – together with gentle tumbling for maximum exposure to cold fluid; hence, instant formation of an ice crust is promoted, and agglomeration between pieces is prevented. Owing to the very low temperatures utilized when compared to regular storage temperatures (lying from −30 to −18°C), a setting period is allowed for equilibration of food temperature prior to leaving the freezer. Automatic control is normally provided, which manipulates rate of injection of cryogenic fluid (as well as its droplet size) so as to respond to rate of incoming food and corresponding temperature; as well as pressure, so as to minimize outlet of cryogen and inlet of warm room air. Strict safety devices are required to avoid dangerous buildup of gaseous nitrogen in the surroundings of the freezer – namely, composition probes and extraction fans.

Hybrids of cryogenic freezing and air-blast have met with success in the recent past; this is the case of high-velocity impingement jets applied above and below such foods as chicken, fish fillets, and hamburger patties – which accompany movement across the freezer tunnel, distribute the air/vapor more evenly, and reduce thickness of the boundary layer on the product surface. As a result, heat transfer coefficients may experience increases up to 25%, while dehydration losses remain below 0.1% – and floorspace required for a given throughput rate may be reduced to one-third. Further developments include an apparatus where flavored liquid products are dripped as droplets into the freezing chamber filled with cryogen – thus producing solid beads of flavored icecream.

The low-temperature phase diagram of water unfolds a polymorphic behavior, with several distinct types – the most important of which are represented in Fig. 1.71. Under atmospheric pressure, ice I is formed; however, as pressure increases, nucleation rate increases and freezing point decreases. Hence, smaller and more uniform ice crystals are generated – with the freezing point falling to −22°C at 200 MPa. High pressure-assisted freezing has been conceived to take advantage of this unique behavior, and consists of cooling foods to −21°C at ca. 200 MPa – slightly above the freezing point of liquid water; release of pressure will then lead to prompt formation of ice I (see Fig. 1.71) – but crystals formed are smaller than if regular freezing had taken place at atmospheric pressure (i.e. 0.1 MPa). Another strategy, known as high-pressure shift freezing, is to effect cooling to just below the freezing point, at pressures in the range 200–300 MPa – so ice III forms (see Fig. 1.71); pressure is then suddenly released, and conversion to ice I occurs – but crystals are smaller than via high pressure-assisted freezing, and between one and two orders of magnitude smaller in size than via air-blast freezing. So small crystals (of the order of 50 μm), homogeneously distributed throughout the food matrix, minimize mechanical damage to cells – so drip losses will be essentially absent throughout thawing afterward; even smaller crystals may be obtained if a hydrocolloid is added in advance, due to the associated reduction in water mobility as required for ice crystal growth. Although enzymatic and microbial activities are constrained, and less extensive biochemical damage permits better retention of nutrients, color, texture, and flavor in the final product – aggregation of myofibrilar proteins may still take place that increases toughness of meat.

The nuclei required by ice crystals to grow during freezing may be generated by microbubbles formed by cavitation inside

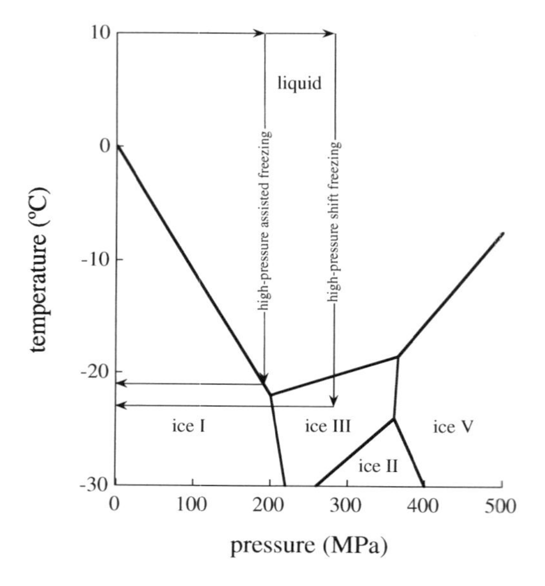

FIGURE 1.71 State diagram of water, covering extended ranges of pressure and temperature.

a liquid – as induced by power ultrasound; higher applied power and longer exposure times increase the rate of freezing, but an upper limit is imposed by the rate of *in situ* generation of heat. The resulting ice crystals are quite small, and form preferentially in the interstitial spaces – thus causing less cell disruption; it is believed that the alternating acoustic stress associated with ultrasound contributes to fracture the ice crystals, thus distorting the size distribution downwards.

One intrinsic difficulty in performing freezing is water migration off the cells, and mass transfer afterward that supports ice crystal growth; this may be circumvented if formation of ice crystals is decoupled from decrease of temperature within the freezing range. Following this concept, magnetic freezing exposes food to continuous oscillating electromagnetic waves as provided by a magnetic resonance device – to prevent ice crystallization, as food is (super)cooled below freezing point; once the desired temperature is reached, the magnetic field is switched off – and food undergoes almost instantaneous (and uniform) flash-freezing. Only very tiny ice crystals can form under these conditions; water migration does not have time to take place, so crystals cannot grow and cells will not become dehydrated. This system has been successfully employed aboard fishing vessels, namely, to freeze tuna in Japan, and cod milt and roe in Alaska; as well as in French cuisine, to preserve gourmet dough, duck meat (magret), foie gras, and truffles. All these products reach premium prices in the market, thus justifying use of such a (still expensive) technology with regard to investment; note, however, that energy costs may actually be lower than following conventional approaches.

Remember that freeze-drying resorts to dehydration after freezing to preserve foods; if such operations are employed reverse ways, then dehydrofreezing results – with proven advantages in retaining food color and flavor, reducing drip losses upon thawing, and permitting cost savings. In fact, preliminary (osmotic) dehydration reduces the inventory of water to freeze – which not only decreases the energy required afterward, but also the weight of food to be transported and stored. Incorporation of specific solutes in the dipping solution also raises the glass transition temperature (as explained before) – so the food more easily acquires (and keeps) a glassy state, which decelerates biochemical decay throughout storage at common refrigeration temperatures; if the dipping solution is formulated with antioxidants, then browning can be reduced as well.

1.4.2.2 Storage

After freezing by one of the methods described above, foods enter the cold chain: from intramural storage in the factory premises; through transport to and storage by wholesalers, followed by transport to, storage, and display by retailers; to transport to and storage by final consumers. Foods are to be stored at (or below) −18°C, while distribution temperatures should not (legally) exceed −15°C, and retail display should not occur above −12°C. Therefore, special care is to be exercised so as to minimize fluctuation (especially increase) in temperature relative to specifications. This normally resorts to accurate thermostatted control, besides refrigeration facilities in stores and transport trucks, automatic doors and airtight curtains for loading to and unloading from stores, correct stock rotation rate, and as short as possible transportation time between stores.

Determination of the maximum storage time depends critically on the definition of acceptable quality/safety of a food product by the moment of final consumption; criteria may range from as vague as "product unsuitable or unacceptable for consumption" to as strict as "preservation of the original, intrinsic characteristics of foods". In the latter case, decision is to be based on a taste panel – and should resort to statistical analysis, to determine the storage time after which differences can be detected, at some low level of significance (say, 5% or 1%). Examples of maximum storage times for selected foods are provided in Fig. 1.72 – pertaining to the standard frozen storage temperature, i.e. −18°C. As a statistically-based decision supported by a somewhat subjective evaluation, storage time below the maximum recommended cannot guarantee quality/safety in all cases; neither can it enforce unsuitability for consumption when such a period is exceeded (at least for not too long).

The maximum period of storage (and corresponding temperature) is normally mentioned in the product label; accurate control should use some form of automatic record of violation of upper (critical) temperature, or its time/temperature history (to be discussed shortly). Extension of the said period depends on the application of specific hurdles; which and to what extent are to be determined as a compromise between price reached by the final food product (determined by market) and cost incurred thereby (determined by industrial processing).

The major detrimental effect of frozen storage is dehydration of cells, especially over long storage periods – with a negative effect upon final food texture; the degree of damage typically increases with crystal size. During slow freezing, sufficient time is allowed for migration of extracellular water toward incipient crystals, and for loss of water by cells via diffusion through their

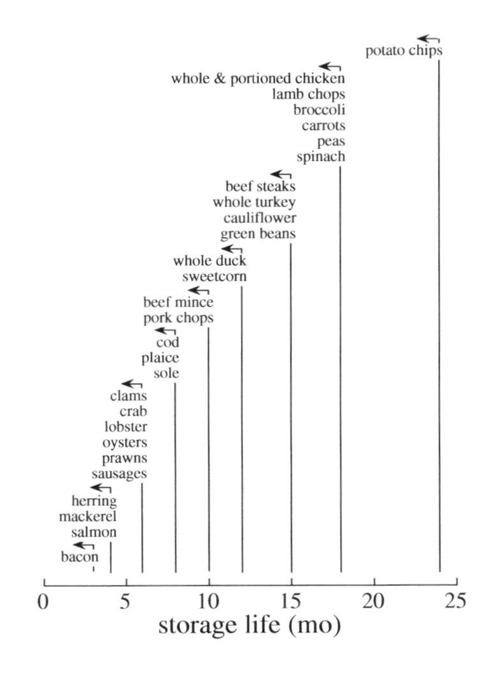

FIGURE 1.72 Ranges of maximum frozen storage times at −18°C (↰), for selected foods.

membrane; the latter process is driven by the increasing osmotic pressure difference between intra- and extracellular fluid, since solutes become gradually more concentrated in the latter. As the process of cell dehydration progresses, irreversible shrinking takes place; overall food shrinkage may occur to such a high degree as 5%(v/v). The growing ice crystals in the outside of the cells exert, in turn, an increasing pressure upon their (flexible) cell membranes – so cells become more and more distorted; this process is often irreversible, i.e. the tissue matrix will not be able to recover its original shape upon thawing. Note that ice crystals cannot normally spear through the cell membrane, because the most likely portions of the crystal to do so would lie adjacent to the cell membrane itself, thus being somewhat shielded thereby. The lipophilic composition of the cell membrane and stereochemical hindrance would indeed prevent regular addition of water molecules on that specific fraction of the outer surface of the ice crystal – and would certainly not promote sharp aggregation thereon. Cell shrinking combined with cell distortion – both a result of dehydration as seen above, may be so severe as to cause cell bursting. When this happens, the cytoplasm is released – but such an event will not become noticeable until thawing has occurred, during which it accounts for drip losses.

Animal and plant tissues in food do not respond similarly to storage under freezing conditions; warm blooded animal meats possess a more flexible fibrous structure that separates during freezing – thus making room for ice crystal growth, and reducing the severity of textural damage. Conversely, fish meat becomes tougher as freezing-induced drying evolves, due to protein structural rearrangement and aggregation; its water-holding capacity is also affected, thus causing drip losses once the tissue is thawed. A disparate pattern is found in fruits and vegetables, owing to the existence of a relatively rigid external cell wall – further to the relatively flexible cell membrane; this justifies enhanced irreversibility of their textural decay due to dehydration – including loss of turgor that arises from a compromised cell wall network.

Other changes undergone by food as time elapses under freezing conditions include destabilization of emulsions (due to progressive clumping of fat globules), precipitation of globular proteins from solution (due to unfolding), and retrogradation and staling of starch; this is why frozen storage of sauces/gravies, milk, and low-amylopectin bread, respectively, seriously compromises the final quality of the product once thawed. Although low temperatures slow down chemical reactions appreciably, the increase in solute concentration caused by water crystallization acts in the opposite way; the associated rise of ionic strength may, in turn, lead to salting out of proteins – often claimed to produce irreversible changes in muscle tissue. A volume expansion also results from freezing, since ice by itself is 9% less dense than liquid water; however, the actual degree of expansion seldom reaches that threshold, because no food contains 100%(w/w) water (except plain water) and every other component contracts upon freezing – whereas not all water will freeze due to freezing point depression, brought about by presence of nonvolatile solutes. In the case of plant tissues, (intact) intercellular spaces are normally occupied by tiny bubbles of air that can absorb part of the ice-mediated volume expansion – thus justifying why coarsely ground strawberries increase ca. 8% in volume upon freezing, but a mere 3% when intact; while formation of a surface crust hampers volume expansion of the remaining food material – thus leading to internal stress buildup, which makes frozen fruits particularly susceptible to cracking and shattering after long storage times.

When stored for a sufficiently long time, the original ice crystals may undergo recrystallization – yielding alteration of overall shape, size, or orientation. This includes changes in surface shape or internal structure of individual crystals – which produce a lower specific surface area (or isomass recrystallization); joining of adjacent crystals – which yields a lower number of, and larger crystals (or accretive recrystallization); and growth of larger crystals at the expense of smaller crystals, or Ostwald's ripening – which also generates fewer but larger crystals (or migratory recrystallization). Migratory recrystallization is probably the most important phenomenon, and is enhanced by temperature fluctuations – as those likely to occur during loading/unloading of cold stores. Slight warming causes ice crystals to partially melt – with sharper surfaces becoming gradually smoother, because flatter surfaces are thermodynamically more stable than sharper ones; over time, smaller crystals tend to become even smaller, and the smallest ones may eventually disappear. However, water vapor pressure increases with ice crystal melting, and extra moisture is driven to regions of lower vapor pressure – thus causing local dehydration of the warmer portions of the food. When temperature falls again, water vapor hardly forms new nuclei, but joins onto existing ice crystals upon solidification – thereby increasing their size; the cyclic process of alternated warming/cooling leads to preferential growth of larger crystals – so any advantage gained from fast freezing in the first place, in terms of small average crystal size, ends up being lost.

In unpackaged foods, moisture may leave their surface toward the surrounding atmosphere – and eventually create areas of visible damage, traditionally known as freezer burn; this phenomenon may also take place in cartons of food, where leaving moisture forms ice on the inside of the pack. Areas affected by freezer burn are readily detectable owing to their lighter color – caused by a change in wavelength of incident light, when it strikes the microcavities once occupied by ice microcrystals and is reflected back. The internal voids thus created become sites for more likely occurrence of oxidation or crosslinking reactions in biopolymers, with negative outcomes upon flavor or texture. The said changes may be kept to a minimum by maintaining storage temperature as low and constant as possible – as well as employing close-fit, moisture-proof packaging materials. Freezer burn, and dehydration at large are enhanced by large specific surface area and porosity – as exhibited by most foods quick-frozen individually. The advantages exhibited by such items – e.g. faster freezing, possibility of single use without thawing the whole pack, and better portion control, may be easily offset by their being particularly sensitive to temperature abuse, in view of the delicate matrix of ice crystals originally formed.

Another source of decay is partial thawing of frozen products sometime during storage – with their surface temperature approaching that of surrounding air, namely, throughout defrosting cycles in horizontal, open display cabinets at retail level; this constitutes a form of (inadvertent) temperature abuse prone to shorten shelf-life.

Remember that the ultimate purpose of freezing is to extend the shelf-life of foods – by slowing down microbial growth and (intra- and extracellular) enzymatic activity, as well as chemical and biochemical decay of components responsible for relevant sensory features. However, freezing *per se* does not inactivate enzymes, and will hardly compromise viability of most microorganisms; in addition, lipid oxidation takes place at measurable rates even at −18°C, thus constraining storage due to development of rancid flavors – see the rather short (maximum) storage time of (fatty fish) herring, mackerel, and samon, versus (non-fatty fish) cod, plaice, and sole in Fig. 1.72. On the other hand, ice crystal formation, and consequent increase of solute concentration in the remaining unfrozen material promote chemical kinetics (as outlined above); and the associated reduction in a_W may selectively accelerate some chemical reactions as well, further to the accompanying changes in *pH* (favorable to loss of color of native anthocyanins) and E_h (favorable to loss of vitamin C by oxidation). This is particularly critical in the case of enzymes that can act individually, rather than as part of specific biochemical pathways (as happens in viable cells) – either extracellular by nature, or intracellular but released upon freezing-induced lysis. Examples encompass decay of (green) chlorophyll to (brown) pheophytin when chloroplasts and chromoplasts become compromised, polyphenoloxidase-mediated browning, and lipoxyenase-catalyzed decay of carotenes yielding off-flavors; blanching prior to freezing of fruits and vegetables accordingly finds full justification. Further examples include protein and lipid breakdown, and associated development of off-flavors in meats effected by endogenous proteases and lipases; as well as enzymatic release of formaldehyde from trimethylamine oxide, which causes off-odors and precipitation of myofibrillar proteins in fish.

In view of the above, the effect of freezing storage upon control of microorganism activity at large is the result of several contributions – namely temperature shock, higher concentration of solutes, dehydration, and ice crystal formation and expansion. Fast freezing is less detrimental to bacterium survival, probably because the ice crystals generated are smaller on average; storage at lower temperature is also less detrimental, since lower enzymatic activities are expected. Bacterial spores, especially of *Bacillus* spp. and *Clostridium botulinum*, are essentially unaffected by low-temperature storage; mold spores, and Gram$^+$ bacteria (e.g. *Staphylococcus* spp. and *Enterococcus* spp.) in vegetative forms come next in resistance – whereas vegetative cells of Gram$^-$ bacteria (e.g. coliforms and *Salmonella* spp.), molds, and yeasts are the most susceptible. A number of pretreatments reduce the probability of microbial-mediated decay during freezing and frozen storage afterward – as is the case of blanching (for being a thermal treatment that also restricts microorganism viability), acidification (for reducing *pH* to levels unfavorable to unwanted microorganisms), or treatment with sulfur dioxide (for being toxic to microorganisms at large).

Nowadays, vegetables constitute one of the largest and most important food groups; and freezing is the simplest way to preserve them, from the point of view of food processors – while being regarded as the most natural way of preservation by consumers. Freezing never improves the quality of the original vegetables, so special attention is to be paid to cultivar, maturity at harvest, growth and harvest practices, chilled storage, transport, and factory reception in the first place. Among the characteristics to be sought in selecting crop cultivar, suitability for mechanical harvesting, uniform maturity, outstanding flavor, even color, resistance to diseases, and high yield are to be outlined. A number of factors prevailing throughout crop growth may play a significant role upon vegetable quality – e.g. site selection, watering, nutrition, and use of agrochemicals for pest and disease control. When approaching maturity, physiological changes start taking place rapidly in most vegetables – so accurate monitoring close to maturity is crucial to find the optimal time for harvest; for instance, green peas and sweet corn retain their prime quality over just a rather short period. Another concern is the degree of bruising throughout mechanical harvesting. Since vegetables close to peak flavor are those selected for freezing and subsequent frozen storage, postharvest delays should be kept to a minimum – and chilling should be done at once, so as to permit vegetables to withstand transport (oftentimes from remote locations to the food plant). With the notorious exception of herbs and green peppers, essentially all vegetables are to be blanched (and promptly cooled thereafter, to avoid compromising nutritional value) prior to freezing. Besides inactivating most enzymes present to higher and higher numbers as maturation progresses – so as to prevent toughening and off-flavor and -color development afterwards during prolonged frozen storage, blanching releases air bubbles entrapped between cells, helps remove surface dirt, and brings about wilting that facilitates packaging.

Another leading group of frozen foods is fruits – for which the final quality demanded after freezing and frozen storage is mainly driven by the intended use. If the fruit is to be eaten upon thawing without further processing, then textural characteristics are more important than if it will be used as raw material by the food industry; hence, conventional methods of freezing that compromise turgidity are not suitable for the former fate. Fruits possess a poorer fibrous structure than most vegetables, and are harvested in a fully ripe state when meant for freezing (unlike happens with vegetables); hence, fruits are intrinsically softer, and thus more susceptible to decay during frozen storage. Consequently, such factors as genetic makeup, growing area climate, fertilization type, and degree of maturity at harvest – further to ability to withstand rough handling, resistance to virus- and mold-mediated diseases, uniformity at ripening and yield, are critical in attempts to obtain high-quality fruits suitable for preservation by freezing. For instance, hand-picking is preferred to mechanical harvesting, since bruising is less likely and maturity sorting is possible; while touch-pressure testing and color inspection are useful to ascertain the right degree of ripening for harvest. Since fruits normally undergo intermediate chilled storage prior to frozen storage, a modified atmosphere rich in CO_2 and poor in O_2 is recommended – as it extends the shelf-life of respiring fruits. Once again, the heuristic rule that poor-quality fruits at harvest never originate high-quality fruits upon freezing is to be consistently borne in mind. On the other hand, washing and cutting normally result in significant losses if applied after thawing – so fruits should be prepared prior to freezing, in terms of peeling, slicing, and cutting; the resulting smaller sizes aid also in heat removal, yet they are associated to larger specific areas that promote dehydration – so a compromise is to be found. Since any heat treatment affects flavor and color negatively, it should be avoided when the food item relies specifically on such attributes; this is why fruits are not typically blanched prior to freezing – unlike done with vegetables in general. Conversely, the intrinsic sweetness of foods allows further pretreatment with sugar solutions; they exclude oxygen from direct contact with the fruit – thus retarding oxidative decay that causes browning, and simultaneously remove water via osmosis – thus depressing freezing point in cell tissues, and concomitantly preventing excessive structural decay.

1.4.2.3 Thawing

Thawing as unit operation is an obvious consequence of freezing and frozen storage having taken place at temperatures well below those normally prevailing throughout preparation and consumption of food; since it requires supply of heat, the bottom line is to avoid overheating – or, at least, assure a temperature as uniform as possible throughout the whole food. This operation is quite less demanding than the original freezing – and is usually carried out at room temperature or moderately above, say 20–40°C; in addition, it frequently takes place at short scale and in a quite distributed manner (i.e. at home and food servicing locations) – so its relevance upon quality of food at final consumption is often overlooked, and clearly off the reach of process engineers working in industrial facilities.

At a commercial scale, foods are thawed under partial vacuum using condensing steam, or warm water, or even moist air within a temperature range of 50–80°C; hence, excessive dehydration is prevented. The typical evolution in temperature of a frozen food throughout thawing is illustrated in Fig. 1.73. At early thawing, absence of a significant layer of liquid water on the food surface supports a fast rise in temperature associated to the high thermal conductivity of the solid (frozen) food, see curve [AB]; a relatively long period ensues, with essentially constant temperature around that of melting ice, T_f. After most ice has turned to liquid water, at point C, an increase in temperature becomes again noticeable, but for a longer period owing to the lower thermal conductivity of the thawed matrix – corresponding to a second heating stage, see curve [CD], until essentially attaining the temperature of the heating fluid, T_h. In practice, thawing time, t_t, is taken as duration of the period represented by [AC] in Fig. 1.73.

Decay arises in many forms during thawing of a previously frozen food; their severity increases with slower thawing and longer maintenance in the proximity of the warming up temperature, as well as with occurrence of temperature abuse as part of the storage history. As emphasized before, damages undergone by the cellular structure during freezing and frozen storage usually do not become perceptible until thawing takes place; the typical outcome is excessive dripping, which may lead to losses up to 10% of the original weight. The said drip losses unfavorably affect nutritional wholesomeness of food, especially in terms of vitamins lost – which may reach ca. 30% vitamin C in the case of fruits, or ca. 30% pyridoxine and ca. 15% niacin in the case of beef. Dripping fluids consequently constitute a rich broth for growth of contaminating microorganisms – either contributed by the environment during thawing or the result of prefreezing contamination, chiefly by psychrotrophic pathogenic or spoilage microorganisms. This problem may be alleviated by cooking foods right after thawing, or even before thawing is complete (e.g. French fries, hamburgers); or else by consuming the food within a short period upon thawing (e.g. cakes, cream).

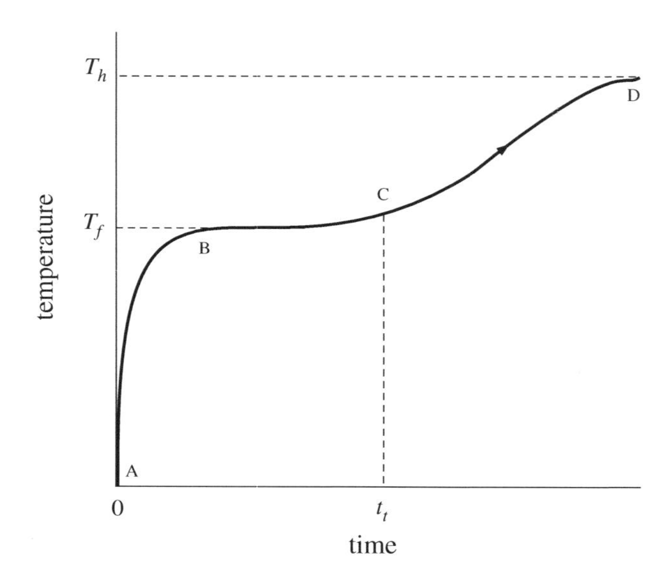

FIGURE 1.73 Typical evolution of temperature of food throughout thawing, with indication of starting state (A); as well as first heating period ([AB]), melting of ice around freezing point at temperature T_f ([BC]) until thawing time t_t, and final heating period ([CD]) above T_f, until reaching temperature of (heating) fluid T_h.

Thawing can be accelerated by applying dielectric or ohmic heating, or power ultrasound instead; in this case, heat is generated inside-out and uniformly throughout the food matrix, so no temperature gradients are necessary to drive conduction of heat outside-in as per Fourier's law. However, the former needs special scrutiny, because the loss factor of ice is much lower than that of liquid water – which may lead to development of hot spots (as discussed previously), and thus uneven thawing. Similar to high-pressure freezing, thawing under such extreme conditions has advantages over conventional techniques – namely a two- to fivefold reduction in t_t, and partial destruction or viability reduction of pathogens, along with better retention of sensory features; however, the advantages of high-pressure thawing hardly outweigh the extra capital costs incurred in. Magnetic resonance devices are also useful to control thawing – since temperatures can be kept at the same value throughout the whole frozen food, thus preventing the traditional time lag between surface and bulk thawing.

1.4.3 Mathematical simulation

1.4.3.1 Freezing pattern

Remember that the freezing temperature, T_f, of an aqueous solution (as is the case of food at large) decreases as the number concentration, e.g. X_s (or molality, in mol$_s$.kg$_{water}^{-1}$), of its solute(s) S increases – for being a colligative property. This pattern becomes apparent upon inspection of Fig. 3.45 of *Food Proc. Eng.: thermal & chemical operations*, where mass fraction is used as independent variable – with T_f decreasing from the freezing point of pure water, $T_{f,0}$, down to the eutectic temperature, T_{eut}; such a curve is not strictly linear – and the straight line conveyed by Eq. (3.941) in *Food Proc. Eng.: thermal & chemical operations*, using mole fraction as independent variable, will not hold unless a dilute solution is at stake. However, even when molality itself is elected as independent variable, a concave trend is perceived – as illustrated via the bold solid line in Fig. 1.74. One may accordingly resort to Taylor's expansion of T_f about $X_{s,0}$, truncated after the quadratic term, according to

$$\boxed{T_f = T_{f,0} - \alpha_w X_s - \beta_w X_s^2} \tag{1.874}$$

– provided that parameters α_w and β_w are defined as

$$\alpha_w \equiv -\left.\frac{dT_f}{dX_s}\right|_{X_s=0} = -\left.\frac{dT_f}{dX_s}\right|_{T_f=T_{f,0}} > 0 \tag{1.875}$$

and

$$\beta_w \equiv -\left.\frac{1}{2}\frac{d^2T_f}{dX_s^2}\right|_{X_s=0} = -\left.\frac{1}{2}\frac{d^2T_f}{dX_s^2}\right|_{T_f=T_{f,0}} > 0, \tag{1.876}$$

respectively; note that α_w and β_w are characteristic of water as solvent, but independent of solute – whereas parameters a and b in Eq. (3.955) of *Food Proc. Eng.: thermal & chemical operations* depend on both solvent (usually water) and solute. The values found experimentally for α_w and β_w read

$$\alpha_w \equiv K_{mtg,w} = 1.853 \text{ K.kg.mol}^{-1} \tag{1.877}$$

and

$$\beta_w = 0.09268 \text{ K.kg}^2\text{.mol}^{-2}, \tag{1.878}$$

with $K_{mtg,w}$ denoting molal cryoscopic constant of water; while molal concentration abides obviously to

$$\boxed{X_s \equiv \frac{n_s}{m_w},} \tag{1.879}$$

where n_s denotes number of moles of solute and m_w denotes mass of (solvent) water.

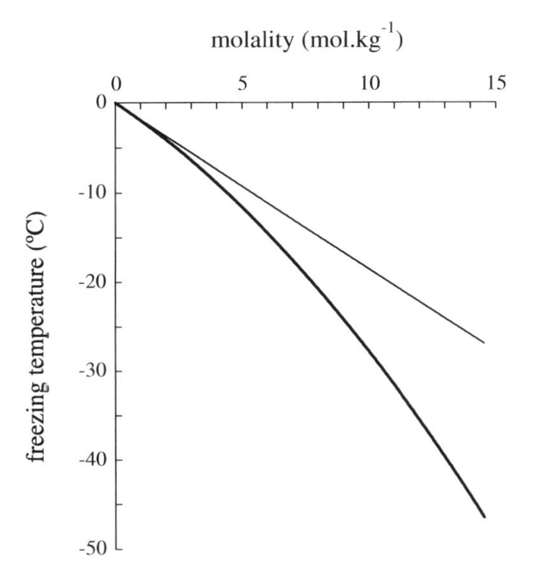

FIGURE 1.74 Variation of freezing temperature of aqueous solution of sucrose versus molality thereof, using Duhring's rule (——) or second-order Taylor's approach (——).

For simplicity of mathematical modeling of freezing processes, truncation of the aforementioned Taylor's expansion after the linear term – or Duhring's rule, has been in common use; inspection of Fig. 1.74 unfolds a good accuracy thereof, namely up to a concentration of solute of ca. 3 mol.kg^{-1}.

A correct account of the phenomena involved in freezing requires consideration of the whole supplemented state diagram of the food at stake – materialized in the specific trajectory throughout the ice+liquid solution region, extrapolated within the rubbery state region with appearance of solid solute, until the glass transition line is crossed toward a glassy state – see Fig. 1.65.

PROBLEM 1.18

A food company plans to manufacture a flavorful candy, starting with sucrose and a volatile flavor. The product should exhibit a fine granular (or powdered) form, be readily soluble in cold water, and possess good stability – especially with regard to aroma loss. What manufacturing method would be suitable – knowing that sucrose melts at 192°C, and its glass transition temperature in solution is –37°C?

Solution

The first conceptual approach to the problem would be to grind crystalline sucrose, and then add the aroma compounds; however, most such compounds are probably hydrophobic (thus justifying their volatility) – so they will not at all, or only slightly adsorb onto a (hydrophilic) sucrose matrix, despite its large specific surface area. Even if a sufficient amount of aroma compounds were adsorbed, they would soon be lost by evaporation. A more realistic route is thus to incorporate the said volatile compounds into a sucrose glass; in fact, formation of co-crystals of sucrose with volatile compounds is highly unlikely, for possessing disparate chemical nature. Toward that goal, one may melt sucrose at 192°C; unfortunately, sugar will undergo caramelization and browning at such a high temperature – besides readily losing every volatile compound whatsoever, assuming that they would not meanwhile breakdown. One should instead resort to dissolving sucrose in water, and then adding the aroma compounds; sucrose concentration should be as high as possible, yet water should be enough to still solvate the said compounds. To obtain the final product as a powder, some form of dehydration will be in order; one possibility could be spray-drying – since the outside of the drop would soon become glassy, thus greatly slowing down diffusion out of the volatile compounds – while water diffusivity would remain considerably high. However, this form of drying uses the difference between vapor pressure of water in the drop, $P^\sigma_{w,d}$, to the partial pressure of water in the overhead atmosphere, $P_{w,0}$, as driving force (i.e. $\Delta P_w = P^\sigma_{w,d} - P_{w,0}$). If dry air were employed, with $P_{w,0}=0$, and vapor pressure of pure water, $P^\sigma_w > P^\sigma_{w,d}$, were considered (thus implying $a_w = 1$), then one would get $0.1 \text{ atm} = P^\sigma_w = P^\sigma_w - 0 > \Delta P_w$ for an operating temperature of ca. 45°C – below which the rate of drying would be excessively low (owing to excessively small a value for P_w^σ) to be practical; yet 45°C>>–37°C=T_{gls} of sucrose in solution, so the final product would be unstable. Still another – and actually the only feasible option, is freeze-concentrate the sucrose solution; according to the pattern shown in Fig. 1.65, one may easily obtain a mixture of ice and sucrose/aroma compounds glass, provided that cooling goes below T_{gls}; all is left then is to remove the ice crystals via sublimation, always carried at a temperature below –37°C. The resulting porous, glassy product should finally be ground, as needed to generate the intended fine powder.

Although molality is the most appropriate measure of solute concentration – in view of the independence of freezing point depression on chemical nature of the solute, many food components are macromolecules (e.g. polysaccharides, proteins), with ill-defined molecular weights due to wide ranges of the underlying polymerization degree; since analytical determination of number concentration turns out difficult (if not impossible in these cases), one has classically resorted to mass fraction of solute in liquid phase, w_l. When $T_f \equiv T_f\{w_l\}$ is at stake, a different curve results for each solute, with distinctive values for parameters a and b as descriptors in Eq. (3.955) of *Food Proc. Eng.: thermal & chemical operations*; an example is provided in Fig. 1.75i for the specific case of sucrose, and in Fig. 1.75ii for common food mono- and disaccharides, alcohol, and proteins. In the particular case of sucrose as solute, a=7.7143 and b=20.161. It should be emphasized that Fig 1.75i is but a magnification of the freezing curve of sucrose in Fig. 1.65 – covering the (thermodynamically stable) region of ice and liquid solution in equilibrium, and the initial portion of the (thermodynamically metastable) region of coexisting ice, liquid solution, and solid solute.

Equivalent curves for other solutes are observed in practice, see Fig. 1.75ii; these curves complement those provided in Fig. 3.44 of *Food Proc. Eng.: thermal & chemical operations*, pertaining to (complex) juices. In particular, the curves in the former figure reflect (in an inverse manner) the molecular weight of the solute under scrutiny; this is why ethanol lies on the left, α-lactalbumin and β-lactoglobulin (as soluble proteins of milk whey) appear on the (far) right, and fructose and glucose, or maltose, lactose, and sucrose exhibit coincident lines (for being isomers). The distinct lines representing alcohol and mono- and disaccharides somehow justify the spread observed in Fig. 3.44 of *Food Proc. Eng.: thermal & chemical operations*; remember that ethanol is the major component of wine, whereas the various (alcohol-free) fruit juices appear closer due to their major content in (some of) the above sugars.

In terms of freezing behavior, two extreme patterns are normally considered – slow and fast freezing; they were highlighted previously in the (experimental) curves plotted in Fig. 1.67. The latter resorts to the concept of freezing front that progresses as heat is withdrawn fast and in a sustained manner; this agrees with the quick (and tendentially linear) decrease in temperature labeled as [CD] in Fig. 1.67, as a consequence of an essentially linear temperature profile within the already frozen portion of food – accounting for a kinetically-controlled process. Slow freezing entails instead freezing that takes place throughout the whole bulk of the food – considered as a continuum, and in a gradual fashion as time elapses; since enough time is allowed for chemical equilibrium be reached due to the small rate of heat transfer, Eq. (1.874) appears as a good model for the downward-oriented concave curve between consecutive eutectic points in Fig. 1.67, knowing that molality increases with time. Two alternative approaches are, in turn, possible in the case of slow freezing. The concentrated-parameter approach postulates no temperature gradient within the food – and resorts to the full quadratic form, labeled as Eq. (1.874), to describe the underlying transient of heat transfer; this permits an analytical solution be attained. According to the distributed-parameter approach, temperature is changing in both time and space – as a consequence of the low temperatures imposed on the food surface, and the heat of freezing released throughout the bulk of the food; one normally sticks to only the linear version of Eq. (1.874), which still needs extensive numerical work to produce true profiles of temperature and ice fraction.

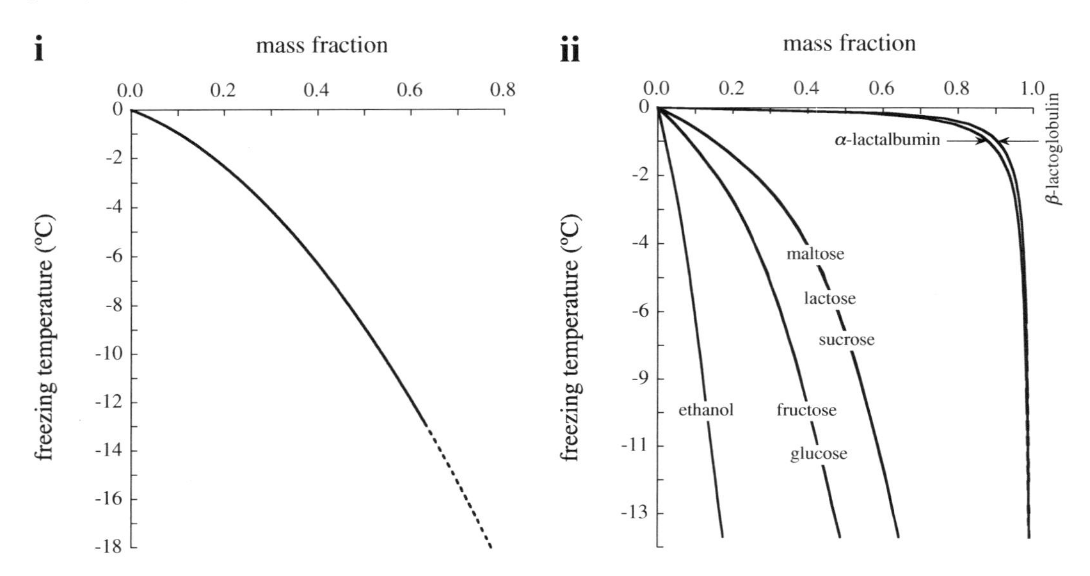

FIGURE 1.75 Freezing temperature of water in (i) aqueous solution of sucrose, spanning thermodynamically stable (——) and metastable (- - - -) regions, and in (ii) aqueous solutions of selected solutes, versus solute mass fraction.

1.4.3.2 Fast freezing and thawing

Consider the classical situation of tray (or chamber) air-blast freezing – which may be conceived as a batch operation, from the point of view of the food item. Freezing takes place from the top of the food – along which the cold air stream is circulated; although technically possible, bottom freezing, via refrigerated (metallic) trays is inconvenient because of sticking – an outcome of formation of ice crystals at the food/tray contact surface.

In a typical case, food is not originally at the Variation of freezing temperature of water, so cooling/refrigeration is to take place in advance – thus bringing about a thermal transient. Consider, in addition, that: the food matrix holds constant and identical densities and specific heat capacities throughout the unfrozen and frozen portions, but thermal conductitivies distinct from each other; no freezing point depression occurs, so a single freezing point is at stake – along with a constant heat of freezing; phase change takes place at a sharply localized interface, which steadily advances deeper and deeper in the food as heat is withdrawn – thus requiring very low refrigerant temperatures, and relatively large contact surface area and external heat transfer coefficient, besides keeping a shape similar to that of the food; and a (pseudo) steady state holds regarding heat transfer through the frozen layer – with sensible heat effects neglected when compared to latent heat in transit, so a linear temperature profile will build up in its entirety. The associated system is schematically represented in Fig. 1.76. Transport of enthalpy out takes place through the frozen portion of the food; note that a frozen matrix offers better thermal conductivity than its unfrozen counterpart. As ice forms, the ice front (at $x=x_f$) progresses toward the base of the food (i.e. $x=0$), and the surface temperature, T_s, decreases toward T_∞.

As mentioned above, the first freezing period actually encompasses fast cooling/chilling of the surface – formally equivalent to baking (except of evolution toward lower and lower, rather than higher and higher temperatures), until the externally imposed temperature on the top surface reaches the freezing point; the enthalpy balance thereafter looks like

$$\boxed{\frac{k_f}{\rho_f c_{P,f}}\frac{\partial^2 T}{\partial x^2}=\frac{\partial T}{\partial t}} \tag{1.880}$$

that mimics Eq. (2.525) of *Food Proc. Eng.: thermal & chemical operations*, where k_f denotes thermal conductivity of original (ice-free) food, and ρ_f and c_{Pf} denote its mass density and specific heat capacity, respectively. Equation (1.880) may be integrated with the aid of

$$\boxed{T\big|_{x=L}=T_f,} \tag{1.881}$$

accordingly serving as boundary condition valid at the top of the food – where T_f denotes (the unique) temperature at which freezing occurs; complemented by

$$\boxed{\left.\frac{\partial T}{\partial x}\right|_{x=0}=0} \tag{1.882}$$

that reflects the insulated nature of the tray holding the food (namely coated with some anti-adhesion, poorly conductive agent, e.g. Teflon™), as per Fourier's law. The initial condition may be coined as

$$\boxed{T\big|_{t=0}=T_{in},} \tag{1.883}$$

for the temperature prevailing throughout the whole food – where T_{in} denotes the starting (uniform) temperature thereof. One may hereby resort to Eqs. (3.1158) and (3.1159) in *Food Proc. Eng.: thermal & chemical operations* as definition of dimensionless space coordinate, x^*, and time, t^*, respectively, while defining a dimensionless temperature, T^*, via

$$T^*\equiv\frac{T-T_f}{T_{in}-T_f}; \tag{1.884}$$

Eq. (1.880) will accordingly be reformulated to

$$\frac{\partial^2 T^*}{\partial x^{*2}}=\frac{\partial T^*}{\partial t^*} \tag{1.885}$$

– whereas Eq. (1.881) becomes

$$T^*\big|_{x^*=1}=0, \tag{1.886}$$

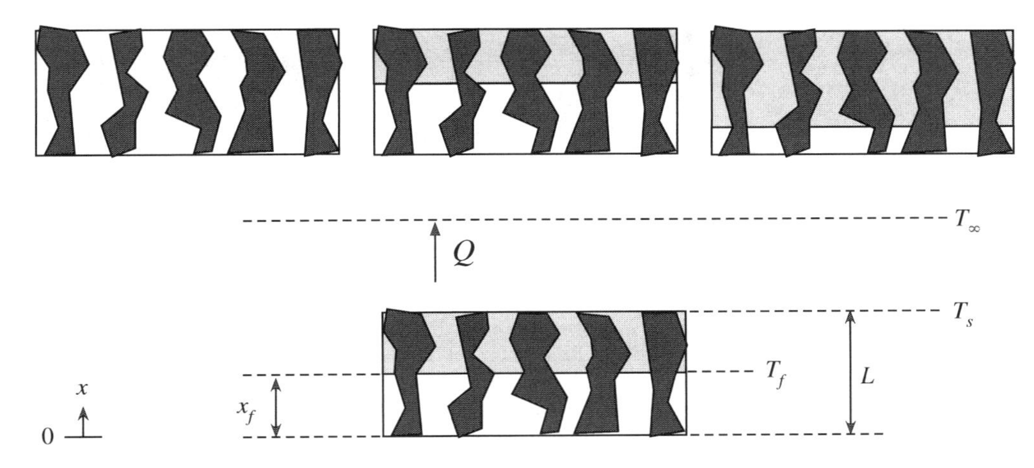

FIGURE 1.76 Evolution in time (top, left to right) of frozen portion (▭), within structure of parallelepipedal food formed by porous solid matrix (■) embedded with liquid (aqueous) fluid (▭), as fast freezing progresses – with indication (bottom) of food thickness, L, ice layer thickness, $L-x_f$, freezing temperature, T_f, surface temperature, T_s, refrigerating fluid bulk temperature, T_∞, and outlet heat flow, Q, in the upward x-direction via top cooling.

Eq. (1.882) turns to

$$\left.\frac{\partial T^*}{\partial x^*}\right|_{x^*=0} = 0, \tag{1.887}$$

and Eq. (1.883) takes the form

$$\left.T^*\right|_{t^*=0} = 1. \tag{1.888}$$

Since Eqs. (1.885)–(1.888) mimic Eqs. (2.551), (2.553), (2.605), and (2.552), respectively, in *Food Proc. Eng.: thermal & chemical operations*, one realizes that

$$\frac{T - T_f}{T_{in} - T_f} = \frac{4}{\pi}\sum_{i=0}^{\infty}\frac{\sin\dfrac{(2i+1)\pi\left(1-\dfrac{x}{L}\right)}{2}}{2i+1}e^{-\frac{(2i+1)^2\pi^2}{4}\frac{k_f t}{\rho_f c_{P,f} L^2}} \tag{1.889}$$

should hold as solution – for being an analog to Eq. (2.627) of *Food Proc. Eng.: thermal & chemical operations*; the original notation was meanwhile retrieved as per Eqs. (3.1158) and (3.1159) of *Food Proc. Eng.: thermal & chemical operations*, complemented by Eq. (1.884). Since $(T-T_f)/(T_{in}-T_f)$ may be expressed as a linear function of $(T_{in}-T)/(T_{in}-T_f)$, according to

$$\frac{T - T_f}{T_{in} - T_f} = \frac{(T_{in} - T_f) - (T_{in} - T)}{T_{in} - T_f} = 1 - \frac{T_{in} - T}{T_{in} - T_f} \tag{1.890}$$

via adding and subtracting T_{in} to the numerator, one may reformulate Eq. (1.889) to

$$\boxed{\frac{T_{in} - T}{T_{in} - T_f} = 1 - \frac{4}{\pi}\sum_{i=0}^{\infty}\frac{\sin\dfrac{(2i+1)\pi\left(1-\dfrac{x}{L}\right)}{2}}{2i+1}e^{-\frac{(2i+1)^2\pi^2}{4}\frac{k_f t}{\rho_f c_{P,f} L^2}};} \tag{1.891}$$

a graphical interpretation is provided in Fig. 1.77. The initial uniform temperature profile – coincident with the horizontal axis, takes off gradually; and eventually reaches a new uniform temperature – equal to that externally imposed, labeled as $k_f t/\rho_f c_{P,f} L^2 = \infty$.

The rate of heat removal from the food during this chilling period, q_{chl}, may be estimated via

$$q_{chl} = -k_f A_f \left.\frac{\partial T}{\partial x}\right|_{x=L^-}, \tag{1.892}$$

following plain application of Fourier's law – with A_f denoting the outer area of food available for heat transfer; differentiation of T with regard to x may then be coined as

$$\frac{\partial T}{\partial t} = -\frac{(T_{in} - T_f)\,\partial\left(\dfrac{T_{in} - T}{T_{in} - T_f}\right)}{L\,\partial\left(\dfrac{x}{L}\right)} = -\frac{T_{in} - T_f}{L}\frac{\partial\left(\dfrac{T_{in} - T}{T_{in} - T_f}\right)}{\partial\left(\dfrac{x}{L}\right)}, \tag{1.893}$$

with the aid of the linearity of a differential operator. Insertion of Eq. (1.893) transforms Eq. (1.892) to

$$q_{chl} = k_f A_f \frac{T_{in} - T_f}{L}\left.\frac{\partial\left(\dfrac{T_{in} - T}{T_{in} - T_f}\right)}{\partial\left(\dfrac{x}{L}\right)}\right|_{\frac{x}{L}=1^-}, \tag{1.894}$$

where isolation of the derivative in the right-hand side leaves

$$\frac{q_{chl}}{k_f A_f \dfrac{T_{in} - T_f}{L}} = \left.\frac{\partial\left(\dfrac{T_{in} - T}{T_{in} - T_f}\right)}{\partial\left(\dfrac{x}{L}\right)}\right|_{\frac{x}{L}=1^-}. \tag{1.895}$$

On the other hand, Eq. (1.891) supports

$$\frac{\partial\left(\dfrac{T_{in} - T}{T_{in} - T_f}\right)}{\partial\left(\dfrac{x}{L}\right)} = -\frac{4}{\pi}\sum_{i=0}^{\infty}\frac{-\dfrac{(2i+1)\pi}{2}\cos\dfrac{(2i+1)\pi\left(1-\dfrac{x}{L}\right)}{2}}{2i+1}e^{-\frac{(2i+1)^2\pi^2}{4}\frac{k_f t}{\rho_f c_{P,f} L^2}}, \tag{1.896}$$

where x/L may be set equal to unity to leave merely

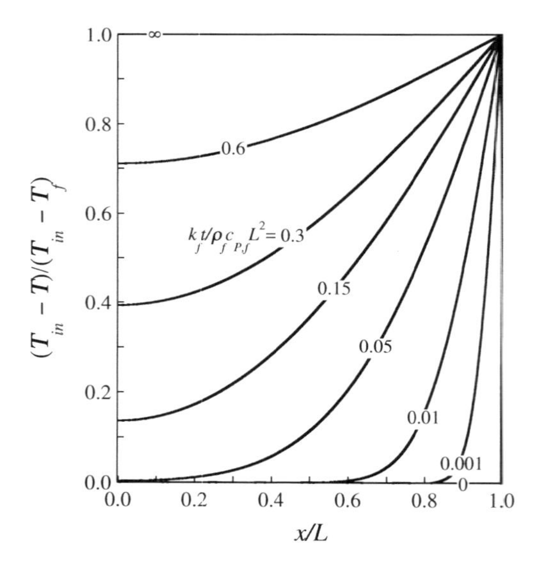

FIGURE 1.77 Variation of normalized temperature, $(T_{in}-T)/(T_{in}-T_f)$, versus normalized position, x/L, for selected dimensionless times, $k_f t/\rho_f c_{P,f} L^2$, throughout the first stage of fast freezing.

$$\left.\frac{\partial\left(\dfrac{T_{in}-T}{T_{in}-T_f}\right)}{\partial\left(\dfrac{x}{L}\right)}\right|_{\frac{x}{L}=1^-}=\frac{4}{\pi}\sum_{i=0}^{\infty}\frac{\pi}{2}\cos 0\,\mathrm{e}^{-\frac{(2i+1)^2\pi^2}{4}\frac{k_f t}{\rho_f c_{P,f} L^2}} \tag{1.897}$$

$$=2\sum_{i=0}^{\infty}\mathrm{e}^{-\frac{(2i+1)^2\pi^2}{4}\frac{k_f t}{\rho_f c_{P,f} L^2}}$$

along with cancellation of $2i+1$ and $\pi/2$ between numerator and denominator; insertion of Eq. (1.897) in Eq. (1.895) finally yields

$$\boxed{\frac{q_{chl}}{k_f A_f \dfrac{T_{in}-T_f}{L}}=2\sum_{i=0}^{\infty}\mathrm{e}^{-\frac{(2i+1)^2\pi^2}{4}\frac{k_f t}{\rho_f c_{P,f} L^2}}.} \tag{1.898}$$

The cumulative heat removed, Q_{chl}, within an initial chilling period of duration t – defined as

$$Q_{chl}\equiv\int_0^t q_{chl}\,d\tilde{t}, \tag{1.899}$$

will then satisfy

$$\frac{Q_{chl}}{k_f A_f \dfrac{T_{in}-T_f}{L}}=2\int_0^t\sum_{i=0}^{\infty}\mathrm{e}^{-\frac{(2i+1)^2\pi^2}{4}\frac{k_f \tilde{t}}{\rho_f c_{P,f} L^2}}d\tilde{t}$$

$$=2\sum_{i=0}^{\infty}\int_0^t\mathrm{e}^{-\frac{(2i+1)^2\pi^2}{4}\frac{k_f \tilde{t}}{\rho_f c_{P,f} L^2}}d\tilde{t} \tag{1.900}$$

after division of both sides by $k_f A_f(T_{in}-T_f)/L$ and insertion of Eq. (1.898) – where advantage was meanwhile taken of the linearity of the integral and summation operators; the fundamental theorem of integral calculus may now be invoked to write

$$\frac{Q_{chl}}{k_f A_f \dfrac{T_{in}-T_f}{L}}=2\sum_{i=0}^{\infty}\frac{1}{\dfrac{(2i+1)^2\pi^2}{4}\dfrac{k_f}{\rho_f c_{P,f} L^2}}\left.\mathrm{e}^{-\frac{(2i+1)^2\pi^2}{4}\frac{k_f \tilde{t}}{\rho_f c_{P,f} L^2}}\right|_t^0, \tag{1.901}$$

where the limits of integration were already swapped on account of the outstanding minus sign. Equation (1.901) may instead appear as

$$\frac{Q_{chl}}{\dfrac{\rho_f c_{P,f} L^2}{k_f}k_f A_f\dfrac{T_{in}-T_f}{L}}=\frac{8}{\pi^2}\sum_{i=0}^{\infty}\frac{1}{(2i+1)^2}\left(1-\mathrm{e}^{-\frac{(2i+1)^2\pi^2}{4}\frac{k_f t}{\rho_f c_{P,f} L^2}}\right) \tag{1.902}$$

upon multiplying both sides by $k_f/\rho_f c_{P,f} L^2$, and moving constant factors to the left-hand side; or simply

$$\boxed{\frac{Q_{chl}}{\rho_f c_{P,f} L A_f\left(T_{in}-T_f\right)}=\frac{8}{\pi^2}\sum_{i=0}^{\infty}\frac{1}{(2i+1)^2}\left(1-\mathrm{e}^{-\frac{(2i+1)^2\pi^2}{4}\frac{k_f t}{\rho_f c_{P,f} L^2}}\right)} \tag{1.903}$$

after collapsing factors alike in the left-hand side. The variation of dimensionless heat taken out during chilling, $Q_{chl}/\rho_f c_{P,f} L A_f(T_{in}-T_f)$, with dimensionless chilling time, $k_f t/\rho_f c_{P,f} L^2$, is plotted in Fig. 1.78. An exponential evolution toward a horizontal asymptote with finite (normal) vertical intercept can be readily grasped as in a typical cumulative process, which approximates a first-order response (associated with the first exponential function as overriding term) – slightly distorted by the next terms, as per its rounder shape.

If chilling time grew unbounded, then the total amount of heat withdrawn would look like

$$\frac{Q_{chl,tot}}{\rho_f c_{P,f} L A_f(T_{in}-T_f)}\equiv\lim_{\frac{k_f t}{\rho_f c_{P,f} L^2}\to\infty}\frac{Q_{chl}}{\rho_f c_{P,f} L A_f\left(T_{in}-T_f\right)}$$

$$=\frac{8}{\pi^2}\sum_{i=0}^{\infty}\frac{1}{(2i+1)^2}\left(1-\mathrm{e}^{-\infty}\right)=\frac{8}{\pi^2}\sum_{i=0}^{\infty}\frac{1}{(2i+1)^2}, \tag{1.904}$$

stemming from Eq. (1.903); after recalling the (exact) mathematical result

$$\sum_{i=0}^{\infty}\frac{1}{(2i+1)^2}\equiv\lim_{n\to\infty}\sum_{i=0}^{n}\frac{1}{\left(2i+1\right)^2}=\frac{\pi^2}{8}, \tag{1.905}$$

one may retrieve Eq. (1.904) as

$$\frac{Q_{chl,tot}}{\rho_f c_{P,f} L A_f(T_{in}-T_f)}=\frac{8}{\pi^2}\frac{\pi^2}{8}=1 \tag{1.906}$$

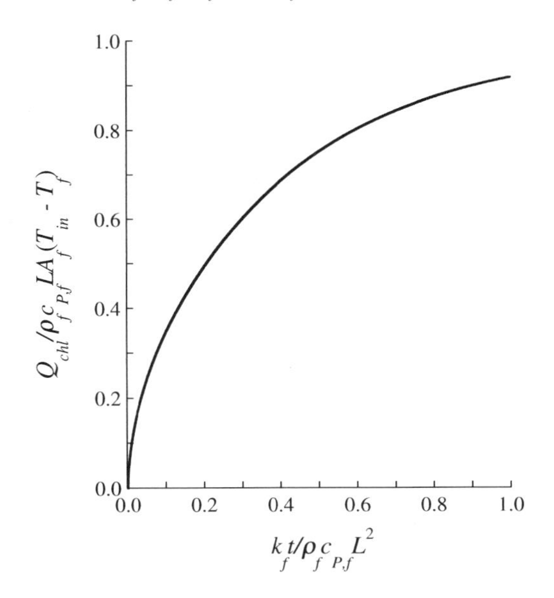

FIGURE 1.78 Evolution of normalized heat withdrawn, $Q_{chl}/\rho_f c_{P,f} L A_f(T_{in}-T_f)$, versus dimensionless time, $k_f t/\rho_f c_{P,f} L^2$, throughout the first stage of fast freezing.

– thus justifying why the variable used for the vertical axis of Fig. 1.78 was termed normalized. Upon isolation of $Q_{chl,tot}$, Eq. (1.906) transforms to

$$Q_{chl,tot} = \rho_f c_{P,f} L A_f \left(T_{in} - T_f\right); \tag{1.907}$$

this expected result corresponds to the sensible heat needed to bring a whole (parallelepipedal) body, with specific heat capacity $c_{P,f}$, volume LA_f, and mass density ρ_f, from temperature T_{in} down to temperature T_f.

Once food has been brought to (an essentially) uniform temperature – equal to (the reference) freezing temperature, T_f, one may establish an enthalpy balance to the frozen layer of food, extending from the location of the freezing front, x_f, up to the slab thickness, L; this is meant to estimate the corresponding temperature profile. Remember the aforementioned postulates – while assuming that temperature remains essentially constant in the (still) unfrozen layer of food, and equal to its initial value, T_f (so no heat is expected to flow through the said unfrozen layer); under these conditions, Eq. (3.784) of *Food Proc. Eng.: thermal & chemical operations* will apply to the frozen layer, and eventually lead to

$$\boxed{k_l \frac{d^2T}{dx^2} = 0} \tag{1.908}$$

– where $k_l \neq k_f$ denotes thermal conductivity of the frozen portion of food (even though it will turn redundant in terms of shape of the said profile, for dropping off on account of the nil right-hand side). Equation (3.793) of *Food Proc. Eng.: thermal & chemical operations* remains valid, whereas Eq. (3.794) in the same volume should be replaced by

$$T\Big|_{x=x_f} = T_f; \tag{1.909}$$

one may thus jump to

$$\boxed{T = \frac{T_f L - T_s x_f}{L - x_f} + \frac{T_s - T_f}{L - x_f} x} \tag{1.910}$$

as an alias of Eq. (3.801) of *Food Proc. Eng.: thermal & chemical operations* – following replacement of x^σ by x_f and T^σ by T_f, and valid for $x_f < x < L$. Equation (7.802), again of *Food Proc. Eng.: thermal & chemical operations,* still holds as descriptor of heat transfer rate, q_{frz}, to the outer cold air stream, after reshaping as

$$q_{frz} = h_a A_f \left(T_s - T_\infty\right) \tag{1.911}$$

so as to match the direction of heat flow; the bulk temperature and heat transfer coefficient of air are hereby denoted by $T_\infty < T_f$ and h_a, respectively, for temperature T_s of the food surface (still to be found, but no longer constant in this case). Since no accumulation of thermal energy, in the form of sensible heat, is considered (as per the original pseudo steady state assumption), q_{frz} must also satisfy

$$q_{frz} = \frac{k_l}{L - x_f} A_f \left(T_f - T_s\right); \tag{1.912}$$

Eqs. (1.911) and (1.912) are identical in form to Eqs. (3.802) and (3.805), respectively, both from *Food Proc. Eng.: thermal & chemical operations* – so a similar reasoning as that leading from the latter two equations to Eq. (3.812) in the said volume may be followed here, to ultimately get

$$\boxed{T_s = T_f - \frac{L - x_f}{\frac{k_l}{h_a} + L - x_f}\left(T_f - T_\infty\right),} \tag{1.913}$$

where the original T^σ, x^σ, T_0, and k_d were swapped for T_f, x_f, T_∞, and k_l, respectively. By the same token, one may resort to

$$\boxed{\frac{T_f - T}{T_f - T_\infty} = \frac{\frac{x}{L} - \frac{x_f}{L}}{\frac{k_l}{h_a L} + 1 - \frac{x_f}{L}}; \quad \frac{x_f}{L} \leq \frac{x}{L} \leq 1} \tag{1.914}$$

as an analog to Eq. (7.819) of *Food Proc. Eng.: thermal & chemical operations* – which describes the temperature profile throughout the iced portion of the food; as emphasized above, the remainder temperature profile should read

$$\frac{T_f - T}{T_f - T_\infty} = 0; \quad 0 \leq \frac{x}{L} \leq \frac{x_f}{L}. \tag{1.915}$$

A graphical account of Eqs. (1.914) and (1.915) is provided in Fig. 1.79; temperature drops linearly from the ice front, T_f, to the surface of the food, T_s – in a manner similar to that portrayed in Fig. 3.48 of *Food Proc. Eng.: thermal & chemical operations.* In view of Eq. (1.914), one may ascertain the temperature gradient via

$$\frac{dT}{dx} = -\left(T_f - T_\infty\right)\frac{d}{dx}\left(\frac{T_f - T}{T_f - T_\infty}\right) = -\frac{\frac{T_f - T_\infty}{L}}{\frac{k_l}{h_a L} + 1 - \frac{x_f}{L}}; \tag{1.916}$$

the temperature profile is steeper at early freezing times (i.e. larger x_f/L), as expected. In fact, the (constant) temperature difference available for heat transfer, $T_f - T_\infty$, supports a decreasing driving force in time, once divided by the growing ice thickness, $L - x_f$. The reason why the straight lines in Fig. 1.79 do not reach unity at $x/L = 1$ is a consequence of air itself offering resistance to heat transfer; a higher resistance implies a lower h_a and thus a higher $k_l/h_a L$, so $(T_f - T)/(T_f - T_\infty)\big|_{x/L=1} = (T_f - T_s)/(T_f - T_\infty)$ becomes lower when going from Fig. 1.79i to Fig. 1.79ii, with $T_\infty < T_s < T_f$.

A mass balance to ice formed within the food, m_i, looks like

$$\boxed{\lambda_f \frac{dm_i}{dt} = -k_l A_f \frac{dT}{dx}\bigg|_{x=x_f^+},} \tag{1.917}$$

with λ_f denoting heat of freezing – where m_i is, in turn, given by

$$\boxed{m_i = \rho_f A_f \left(L - x_f\right);} \tag{1.918}$$

after differentiating with regard to time, Eq. (1.918) gives rise to

$$\frac{dm_i}{dt} = -\rho_f A_f \frac{dx_f}{dt} \tag{1.919}$$

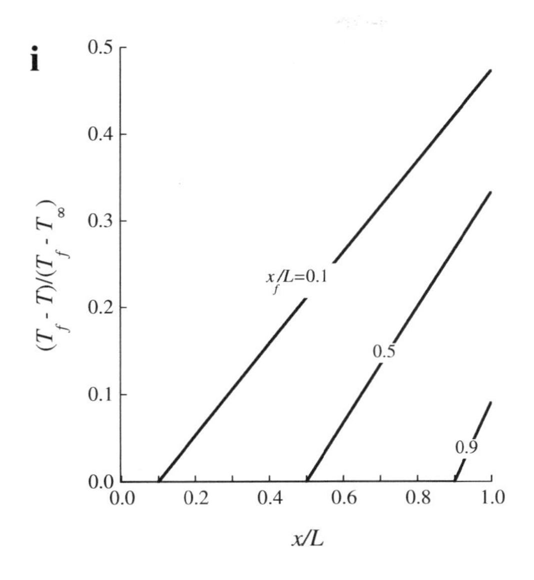

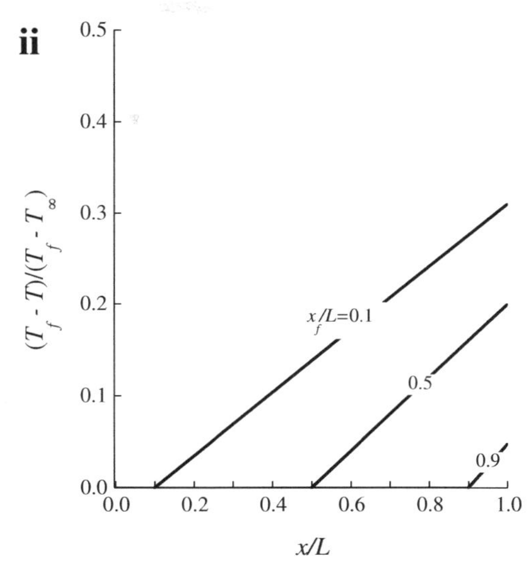

FIGURE 1.79 Variation of normalized temperature, $(T_f-T)/(T_f-T_\infty)$, versus normalized position, x/L, for selected normalized locations of the ice front, x_f/L, and selected values of dimensionless parameter, (i) $k_l/h_aL=1$ and (ii) $k_l/h_aL=2$, throughout the second stage of fast freezing.

– so Eq. (1.917) may be rewritten as

$$\lambda_f \rho_f A_f \frac{dx_f}{dt} = k_l A_f \left.\frac{dT}{dx}\right|_{x=x_f^+} \tag{1.920}$$

once Eq. (1.919) is inserted, and negatives are taken of both sides afterward. Combination with Eq. (1.916) transforms Eq. (1.920) to

$$\rho_f \lambda_f A_f \frac{dx_f}{dt} = -k_l A_f \frac{\dfrac{T_f - T_\infty}{L}}{\dfrac{k_l}{h_a L} + 1 - \dfrac{x_f}{L}}, \tag{1.921}$$

where division of both sides by $\lambda_f \rho_f A_f$ produces, in turn,

$$\frac{dx_f}{dt} = -\frac{k_l}{\rho_f \lambda_f} \frac{\dfrac{T_f - T_\infty}{L}}{\dfrac{k_l}{h_a L} + 1 - \dfrac{x_f}{L}}; \tag{1.922}$$

Eq. (1.922) is subjected to

$$\boxed{\left.x_f\right|_{t=0} = L} \tag{1.923}$$

as condition satisfied at the beginning of freezing proper. Integration of Eq. (1.922) can proceed via separation of variables, with the aid of Eq. (1.923), viz.

$$\int_L^{x_f} \left(\frac{k_l}{h_a L} + 1 - \frac{\tilde{x}_f}{L} \right) d\tilde{x}_f = -\frac{k_l}{\rho_f \lambda_f} \frac{T_f - T_\infty}{L} \int_0^t d\tilde{t}, \tag{1.924}$$

since all physical parameters were hypothesized to remain constant in time and throughout the frozen phase; the fundamental theorem of integral calculus may now be invoked to write

$$\left. \left(\left(1 + \frac{k_l}{h_a L} \right) \tilde{x}_f - \frac{1}{L} \frac{\tilde{x}_f^2}{2} \right) \right|_L^{x_f} = -\frac{k_l (T_f - T_\infty)}{\rho_f \lambda_f L} \left. \tilde{t} \right|_0^t, \tag{1.925}$$

which readily simplifies to

$$-\left(1 + \frac{k_l}{h_a L}\right)(x_f - L) + \frac{1}{2L}\left(x_f^2 - L^2\right) = \frac{k_l (T_f - T_\infty)}{\rho_f \lambda_f L} t. \tag{1.926}$$

Upon dividing both sides by L, and factoring in $-1/L$ at the second term of the left-hand side, Eq. (1.926) transforms to

$$\left(1 + \frac{k_l}{h_a L}\right)\left(1 - \frac{x_f}{L}\right) - \frac{1}{2}\left(1 - \left(\frac{x_f}{L}\right)^2\right) = \frac{k_l (T_f - T_\infty)}{\rho_f \lambda_f L^2} t; \tag{1.927}$$

a further factoring out of $1 - x_f/L$ supports conversion to

$$\frac{k_l (T_f - T_\infty)}{\rho_f \lambda_f L^2} t = \left(1 - \frac{x_f}{L}\right)\left(1 + \frac{k_l}{h_a L} - \frac{1}{2}\left(1 + \frac{x_f}{L}\right)\right), \tag{1.928}$$

along with splitting of $1 - (x_f/L)^2$ as the product of $1 - x_f/L$ by $1 + x_f/L$. Following elimination of inner parentheses, Eq. (1.928) becomes

$$\frac{k_l (T_f - T_\infty)}{\rho_f \lambda_f L^2} t = \left(1 - \frac{x_f}{L}\right)\left(1 + \frac{k_l}{h_a L} - \frac{1}{2} - \frac{1}{2}\frac{x_f}{L}\right), \tag{1.929}$$

where condensation of terms alike unfolds

$$\frac{k_l (T_f - T_\infty)}{\rho_f \lambda_f L^2} t = \left(1 - \frac{x_f}{L}\right)\left(\frac{1}{2} + \frac{k_l}{h_a L} - \frac{1}{2}\frac{x_f}{L}\right); \tag{1.930}$$

½ may finally be factored out to give

$$\boxed{\frac{k_l (T_f - T_\infty)}{\rho_f \lambda_f L^2} t = \left(1 - \frac{x_f}{L}\right)\left(\frac{k_l}{h_a L} + \frac{1}{2}\left(1 - \frac{x_f}{L}\right)\right).} \tag{1.931}$$

Equation (1.931) is widely known as Plank's equation, or else Ede's equation if applied specifically to foods; a graphical illustration is conveyed by Fig. 1.80, for chosen values of parameter k_l/h_aL – which may be viewed as a Biot's number for heat transfer between a solid phase (characterized by pseudo heat transfer coefficient k_l/L) and an adjacent fluid phase (characterized by heat transfer coefficient h_a). As anticipated, Fig. 1.80 resembles

Fig. 3.51 in *Food Proc. Eng.: thermal & chemical operations* – with a quadratic dependence of freezing time on thickness of the (already) frozen layer, $L-x_f$. The time it takes for the ice front to reach the bottom surface of the food, i.e.

$$t_{frz} \equiv t\big|_{x_f=0}, \tag{1.932}$$

is finite – and can be ascertained, in dimensioness form, by setting $x_f = 0$, which produces

$$\frac{k_l\left(T_f-T_\infty\right)}{\rho_f\lambda_f L^2}t_{frz} = (1-0)\left(\frac{k_l}{h_a L}+\frac{1}{2}(1-0)\right) = \frac{1}{2}+\frac{k_l}{h_a L} \tag{1.933}$$

based on Eq. (1.931); the vertical intercepts of the curves in Fig. 1.80 are accordingly described by Eq. (1.933).

Equation (1.931) can be generalized to other (regular and simple) food shapes, after rewriting it as

$$\frac{k_l\left(T_f-T_\infty\right)}{\rho_f\lambda_f L^2}t = \frac{1}{\gamma}\left(1-\frac{x_f}{L}\right)\left(\frac{k_l}{h_a L}+\frac{1}{2}\left(1-\frac{x_f}{L}\right)\right), \tag{1.934}$$

where γ denotes a (dimensionless) correction factor accounting for the specific geometry of the food; $\gamma=1$ for an infinite slab – thus turning Eq. (1.934) identical to Eq. (1.931), $\gamma=2$ for an infinite cylinder, and $\gamma=3$ for a sphere. There is a number of limitations to Plank's method, however, that constrain its accuracy to no better than 10% error – related primarily to the assignment of quantitative values; density values for foods are indeed difficult to estimate and not strictly contant across their constitutive material, and thermal conductivities of the frozen product are not available for most foods. Furthermore, sensible heat has been neglected along the derivation – a relevant deed, since the freezing temperature actually drops as ice forms, in agreement with Fig. 1.65. Nevertheless, Eq. (1.931) still constitutes the most popular method to predict freezing time, owing to its intrinsic simplicity and mechanistic basis; more accurate methods have

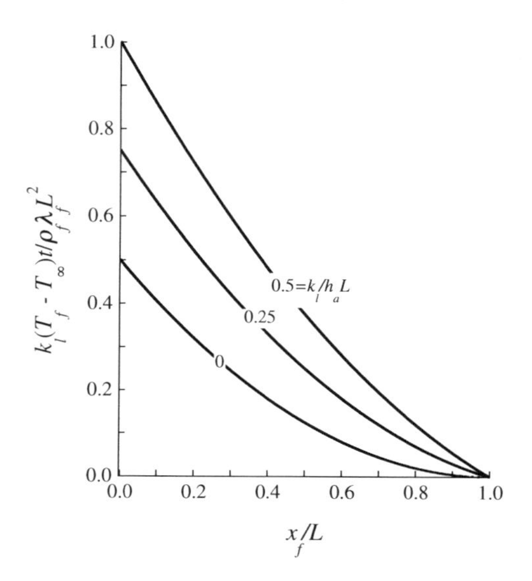

FIGURE 1.80 Variation of dimensionless time, $k_l(T_f-T_\infty)t/\rho_f\lambda_f L^2$, required to bring the ice front to dimensionless position x_f/L, versus x_f/L, for selected values of dimensionless parameter k_l/h_aL, throughout the second stage of fast freezing.

indeed been built as improvements using the said equation as departure point.

The instantaneous heat transfer rate, q_{frz}, in this stage of freezing may be accessed via

$$q_{frz} = \lambda_f\frac{dm_i}{dt}, \tag{1.935}$$

since sensible heat is not accounted for – which turns to

$$q_{frz} = -\rho_f\lambda_f A_f\frac{dx_f}{dt}, \tag{1.936}$$

after inserting Eq. (1.919); combination with Eq. (1.921) gives, in turn, rise to

$$q_{frz} = k_l A_f\frac{\dfrac{T_f-T_\infty}{L}}{\dfrac{k_l}{h_a L}+1-\dfrac{x_f}{L}} \tag{1.937}$$

that may be rephrased as

$$\boxed{\frac{q_{frz}}{k_l A_f\dfrac{T_f-T_\infty}{L}} = \frac{1}{\dfrac{k_l}{h_a L}+1-\dfrac{x_f}{L}}}. \tag{1.938}$$

The cumulative heat removed, Q_{frz}, accordingly abides to

$$Q_{frz} \equiv \int_0^t q_{frz}\left\{x_f\right\}d\tilde{t} \tag{1.939}$$

in parallel to Eq. (1.899), which can be reformulated to

$$Q_{frz} = \int_L^{x_f} q_{frz}\frac{dt}{d\tilde{x}_f}d\tilde{x}_f = \int_L^{x_f}\frac{q_{frz}}{\dfrac{d\tilde{x}_f}{dt}}d\tilde{x}_f \tag{1.940}$$

via multiplication and division of the kernel by dx_f, coupled with Eq. (1.923) and the rule of differentiation of an inverse function. Insertion of Eqs. (1.922) and (1.937) transforms Eq. (1.940) to

$$Q_{frz} = -\int_L^{x_f}\frac{k_l A_f\dfrac{\dfrac{T_f-T_\infty}{L}}{\dfrac{k_l}{h_a L}+1-\dfrac{x_f}{L}}}{\dfrac{k_l}{\rho_f\lambda_f}\dfrac{\dfrac{T_f-T_\infty}{L}}{\dfrac{k_l}{h_a L}+1-\dfrac{x_f}{L}}}d\tilde{x}_f, \tag{1.941}$$

where factors alike will be dropped from numerator and denominator, and integration limits exchanged at the expense of the minus sign preceding the integral to obtain merely

$$Q_{frz} = \rho_f\lambda_f A_f\int_{x_f}^{L}d\tilde{x}_f. \tag{1.942}$$

Trivial integration of Eq. (1.942) then unfolds

$$Q_{frz} = \rho_f \lambda_f A_f \tilde{x}_f \Big|_{x_f}^{L} = \rho_f \lambda_f A_f \left(L - x_f\right) = \lambda_f m_i, \tag{1.943}$$

in view of the underlying relationship between mass of ice formed and (latent) heat removed, see Eq. (1.918); upon division of all sides by $\rho_f\lambda_f LA_f$, Eq. (1.943) may be rephrased to

$$\boxed{\frac{Q_{frz}}{\rho_f \lambda_f LA_f} = 1 - \frac{x_f}{L}} \tag{1.944}$$

– as plotted in Fig. 1.81, after insertion of $x_f \equiv x_f\{t\}$ as obtained from solution of Eq. (1.931). As expected from the linear relationship between Q_{frz} and x_f – labeled as Eq. (1.944), Fig. 1.81 may be obtained by appropriate vertical flipping of Eq. 1.80 about the straight horizontal line of vertical intercept 0.5, followed by flipping of the outcome about the bisector line.

The maximum value of $Q_{frz}/\rho_f\lambda_f LA_f$ may be (trivially) calculated via

$$\frac{Q_{frz,tot}}{\rho_f \lambda_f LA_f} \equiv \lim_{\frac{k_l(T_f - T_\infty)t}{\rho_f \lambda_f L^2} \to \frac{k_l(T_f - T_\infty)t_{frz}}{\rho_f \lambda_f L^2}} \frac{Q_{frz}}{\rho_f \lambda_f LA_f} \tag{1.945}$$

$$= \lim_{\frac{k_l(T_f - T_\infty)t}{\rho_f \lambda_f L^2} \to \frac{1}{2} + \frac{k_f}{h_a L}} \frac{Q_{frz}}{\rho_f \lambda_f LA_f} = \lim_{\frac{x_f}{L} \to 0} \frac{Q_{frz}}{\rho_f \lambda_f LA_f} = 1$$

as per Eqs. (1.932), (1.933), and (1.944); it describes the maximum ordinates reached by the curves in Fig. 1.81, corresponding to $Q_{frz,tot}$ given by $\rho_f\lambda_f LA_f$ in agreement with Eq. (1.943), i.e. the latent heat released by the whole food piece upon freezing. On the other hand, the cumulative heat withdrawn may be readily converted to mass of ice formed using latent heat of freezing as conversion factor, i.e.

$$m_i = \frac{Q_{frz}}{\lambda_f} \tag{1.946}$$

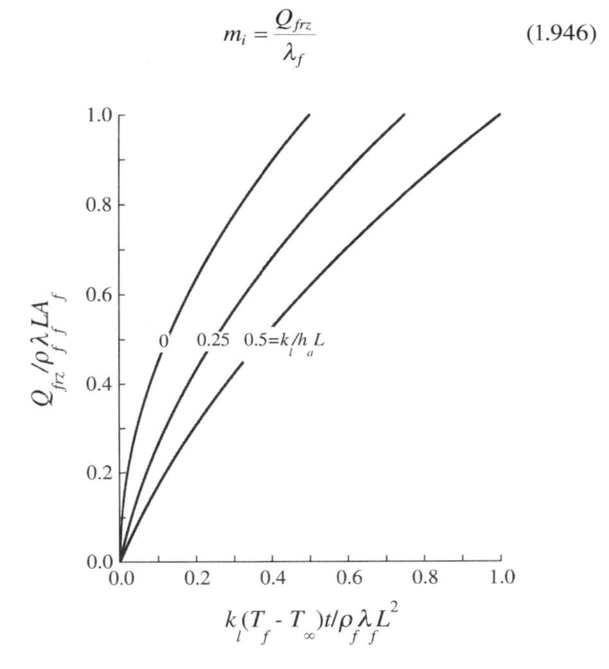

FIGURE 1.81 Evolution of normalized heat withdrawn, $Q_{frz}/\rho_f\lambda_f LA_f$, versus dimensionless time, $k_l(T_f - T_\infty)t/\rho_f\lambda_f L^2$, throughout the second stage of fast freezing.

departing from Eq. (1.943), where insertion of Eq. (1.944) produces

$$m_i = \rho_f LA_f \left(1 - \frac{x_f}{L}\right); \tag{1.947}$$

Eq. (1.918) will then be retrieved (as expected), after factoring L in.

PROBLEM 1.19

When the nominal velocity of a liquid pumped upward through a bed of solid particles is increased, fluidization will eventually occur – with the mixture of particles and liquid exhibiting properties typical of a homogeneous fluid. When cold air is pumped through a bed of small particulate foods, e.g. peas aimed at freezing – at sufficiently high a velocity, the system will eventually fluidize as well. This mode of operation is particularly useful for freezing – not only due to the very large heat transfer coefficients at stake, but also because the individual pieces are prevented from clumping with each other as caused by frozen moisture. Meanwhile, the gradually increasing toughness of the frozen pieces makes them less and less susceptible to mechanical damage thereafter.

Unlike happens with fluidization by a liquid – where the bed of particles expands uniformly so that pressure drop remains the same as at the minimum velocity of fluidization, and the overall system behaves like a liquid; gas-mediated fluidization is intrinsically heterogeneous. This so-called aggregative fluidization is observed when the difference in densities of the discrete and continuous phases is quite large – so the gas passes through the bed mainly in the form of (large) bubbles that burst at the surface, while the suspended bed still behaves as a liquid. Kunii and Levenspiel's model for the aforementioned phenomenon assumes that all bubbles formed are of one size; the solids in the suspended bed flow smoothly downward in plug flow; the fluidized bed remains at minimum fluidization conditions; the gas occupies the same void fraction of the suspension as it did in the entire bed at the minimum fluidization point; the bubble comprises a wake at its bottom, and is surrounded by a thin cloud of gas before contacting the suspension of solid particles proper; and the concentration of solids in the wake mimics that in the bulk suspension, so the gaseous fraction in the wake coincides with that prevailing in the fluidized bed. In attempts to simulate the germane phenomena, one may recall Davidson's formula for bubble rising (terminal) velocity in a quiescent liquid, $v_{b,t}$, viz.

$$v_{b,t} = \sqrt{gR_b} \tag{1.19.1}$$

– where R_b denotes radius of bubble and g denotes acceleration of gravity. One may also retrieve Mori and Wen's correlation for bubble (average) diameter in fluidized beds, viz.

$$\frac{R_{b,M} - R_b}{R_{b,M} - R_{b,m}} = \exp\left\{-0.3\frac{h}{D_t}\right\}, \tag{1.19.2}$$

where R_b denotes bubble radius at height h above the distributor plate in a bed of diameter D_t, and $R_{b,m}$ denotes bubble radius just above the distributor plate – whereas $R_{b,M}$ denotes maximum bubble radius attained if all bubbles in a horizontal plane would coalesce to form a single bubble (as they will, should the bed be high enough). These two formulae may then be complemented by

$$R_{b,M} = 0.819\left(A_n\left(v_0 - v_{m,f}\right)\right)^{0.4} \quad (1.19.3)$$

as correlation to estimate $R_{b,M}$, and

$$R_{b,m} = 0.436\left(\frac{A_n\left(v_0 - v_{m,f}\right)}{n_n}\right)^{0.4} \quad (1.19.4)$$

as correlation to estimate $R_{b,m}$ – with A_n denoting area of (perforated) distributor and n_n denoting number of openings thereof, and v_0 and v_{mf} denoting (average) velocity of fluid and minimum fluidization velocity, respectively.

a) Produce a correlation for the rise velocity of a bubble during heterogeneous fluidization, as a function of v_0, $v_{m,f}$, g, A_n, h, D_t, and n_n.
b) Estimate the fraction of overall bed volume occupied by gas in bubble form during fluidization.

Solution

a) In view of many bubbles present together, the actual rise velocity of a bubble, v_b, will not coincide with $v_{b,t}$ as given by Eq. (1.19.1); the more bubbles present, the less drag on each individual bubble – while bubbles will move faster for larger overall amounts of gas passing through the bed, at velocity v_0. Other factors playing a role are viscosity of gas, as well as size and density of solid particles that make up the bed; however, both these terms already determine the minimum fluidization velocity (so they do not need to be taken specifically into account on their own) – with a lower rise velocity resulting from a higher minimum fluidization velocity, $v_{m,f}$. Therefore, one may write

$$v_b = v_{b,t} + v_0 - v_{m,f}, \quad (1.19.5)$$

where elimination of $v_{b,t}$ becomes possible as

$$v_b = v_0 - v_{m,f} + \sqrt{gR_b} \quad (1.19.6)$$

at the expense of Eq. (1.19.1). On the other hand, R_b can be obtained from Eq. (1.19.2) as

$$R_b = R_{b,M} - \left(R_{b,M} - R_{b,m}\right)\exp\left\{-0.3\frac{h}{D_t}\right\}$$
$$= R_{b,M}\left(1 - \exp\left\{-0.3\frac{h}{D_t}\right\}\right) + R_{b,m}\exp\left\{-0.3\frac{h}{D_t}\right\}, \quad (1.19.7)$$

while insertion of Eqs. (1.19.3) and (1.19.4) unfolds

$$R_b = 0.819\left(A_n\left(v_0 - v_{m,f}\right)\right)^{0.4}\left(1 - \exp\left\{-0.3\frac{h}{D_t}\right\}\right)$$
$$+ 0.436\left(\frac{A_n\left(v_0 - v_{m,f}\right)}{n_n}\right)^{0.4}\exp\left\{-0.3\frac{h}{D_t}\right\}; \quad (1.19.8)$$

hence, the final expression for v_b will be obtained after plugging Eq. (1.19.8) into Eq. (1.19.6).

b) Denoting as ε_b the fraction of total bed volume occupied by bubbles (excepting wake), and as ε_w the fraction of bubble volume occupied by wake, one concludes that the bed fraction in the suspension (including clouds), η_s, is given by

$$\eta_s = 1 - \varepsilon_b - \varepsilon_w\varepsilon_b \quad (1.19.9)$$

– since the bed fraction in the wakes themselves reads $\varepsilon_w\varepsilon_b$; although a function of particle size, ε_w usually ranges within 0.25–1.0, with typical values close to 0.4. A mass balance to the solids then takes the form

$$A\rho_s\eta_s v_s = A\rho_s\left(1 - \varepsilon_b - \eta_s\right)v_b, \quad (1.19.10)$$

where A denotes cross-sectional area of bed, ρ_s denotes mass density of solid particles, and v_s denotes velocity of solids; the left-hand side represents mass flow rate of solids seen from the point of view of the suspension, while the right-hand side represents mass flow rate of solids seen from the point of view of the bubble wake, see Eq. (1.19.9). Insertion of Eq. (1.19.9), cancellation of symmetrical terms, and dropping of $A\rho_s$ from both sides support indeed transformation of Eq. (1.19.10) to

$$\left(1 - \varepsilon_b - \varepsilon_w\varepsilon_b\right)v_s = \varepsilon_w\varepsilon_b v_b, \quad (1.19.11)$$

where isolation of v_s unfolds, in turn,

$$v_s = \frac{\varepsilon_w\varepsilon_b}{1 - \varepsilon_b - \varepsilon_w\varepsilon_b}v_b. \quad (1.19.12)$$

A mass balance to gas is now in order, according to

$$Av_0 = A\varepsilon_b v_b + A\varepsilon_{m,f}\varepsilon_w\varepsilon_b v_b + A\varepsilon_{m,f}\eta_s v_e \quad (1.19.13)$$

– where the left-hand side represents total gas flow rate, the first term in the right-hand side represents gas flow via bubbles, the second term represents gas flow via wakes, and the third term represents gas flow via suspension; here $\varepsilon_{m,f}$ denotes porosity of bed at minimum fluidization velocity, whereas v_e denotes velocity of overall suspension. On the other hand, the velocity of rise of gas as part of the suspension is given by

$$v_e = \frac{v_{m,f}}{\varepsilon_{m,f}} - v_s, \quad (1.19.14)$$

since solids are flowing downward – and recalling that $v_{m,f}$ refers to gas velocity relative to moving solids; correction by $\varepsilon_{m,f}$ in denominator is justified because $v_{m,f}$ is, by definition, a surface velocity (i.e. based on an empty-tube cross-section). After dropping A from both sides and factoring out $\varepsilon_b v_b$, Eq. (1.19.13) reduces to

$$v_0 = \left(1+\varepsilon_{m,f}\varepsilon_w\right)\varepsilon_b v_b + \varepsilon_{m,f}\eta_s v_e; \tag{1.19.15}$$

insertion of Eq. (1.19.14) gives then rise to

$$v_0 = \left(1+\varepsilon_{m,f}\varepsilon_w\right)\varepsilon_b v_b + \varepsilon_{m,f}\eta_s\left(\frac{v_{m,f}}{\varepsilon_{m,f}} - v_s\right), \tag{1.19.16}$$

or else

$$v_0 = \left(1+\varepsilon_{m,f}\varepsilon_w\right)\varepsilon_b v_b + \eta_s\left(v_{m,f} - \varepsilon_{m,f}v_s\right) \tag{1.19.17}$$

after factoring $\varepsilon_{m,f}$ in – while Eqs. (1.19.9) and (1.19.12) may be brought on board as

$$v_0 = \left(1+\varepsilon_{m,f}\varepsilon_w\right)\varepsilon_b v_b + \left(1-\varepsilon_b-\varepsilon_w\varepsilon_b\right)\left(v_{m,f} - \varepsilon_{m,f}\frac{\varepsilon_w\varepsilon_b}{1-\varepsilon_b-\varepsilon_w\varepsilon_b}v_b\right). \tag{1.19.18}$$

Once $1-\varepsilon_b-\varepsilon_w\varepsilon_b$ is factored in, and the first parenthesis is eliminated, Eq. (1.19.18) becomes

$$v_0 = \varepsilon_b v_b + \varepsilon_{m,f}\varepsilon_w\varepsilon_b v_b + \left(1-\varepsilon_b-\varepsilon_w\varepsilon_b\right)v_{m,f} - \varepsilon_{m,f}\varepsilon_w\varepsilon_b v_b, \tag{1.19.19}$$

so cancellation of symmetrical terms permits simplification to

$$v_0 = \varepsilon_b v_b + \left(1-\varepsilon_b-\varepsilon_w\varepsilon_b\right)v_{m,f}; \tag{1.19.20}$$

upon elimination of the outstanding parenthesis, Eq. (1.19.20) gives

$$v_0 = \varepsilon_b v_b + v_{m,f} - \varepsilon_b v_{m,f} - \varepsilon_w\varepsilon_b v_{m,f}, \tag{1.19.21}$$

where ε_b can be factored out as

$$v_0 - v_{m,f} = \varepsilon_b\left(v_b - v_{m,f} - \varepsilon_w v_{m,f}\right). \tag{1.19.22}$$

Solution of Eq. (1.19.22) for ε_b finally produces

$$\varepsilon_b = \frac{v_0 - v_{m,f}}{v_b - \left(1+\varepsilon_w\right)v_{m,f}}, \tag{1.19.23}$$

where $v_{m,f}$ was meanwhile factored out in denominator. As expected, the fraction of overall bed of particulate food occupied by gas bubbles (ε_b) increases with overall velocity of gas (v_0) and decreases with bubble velocity (v_b) – with a dependence on minimum fluidization velocity ($v_{m,f}$) as such, and also hanging on fraction of wake in bubble volume (ε_w). Although the true porosity of the fluidized bed remains essentially equal to the porosity of the bed under minimum fluidization conditions ($\varepsilon_{m,f}$), the dispersed gas phase causes an overall expansion of bed volume compatible with ε_b as per Eq. (1.19.23). Finally, note that aggregative fluidization exhibits considerable bypassing and slugging – i.e. intermittent and unstable flow of large gas bubbles through the bed; both such behavior patterns are strongly affected by design of the distributor that holds up the packed bed and provides gas flow thereinto.

An alternative formulation of the second stage in fast freezing is possible by resorting to Biot's number for heat transfer during freezing, defined as $Bi_f \equiv h_a L / k_l$ – in parallel to Eq. (5.521) of *Food Proc. Eng.: thermal & chemical operations* pertaining to mass transfer (and as mentioned previously); if an auxiliary space variable, ξ, is defined as

$$\boxed{\xi \equiv 1 - \frac{x_f}{L},} \tag{1.948}$$

and an auxiliary time variable, ω, abides to

$$\boxed{\omega \equiv \frac{k_l\left(T_f - T_\infty\right)}{\rho_f\lambda_l L^2}t,} \tag{1.949}$$

then Eq. (1.931) can be rewritten as

$$\omega = \xi\left(\frac{1}{Bi_f} + \frac{1}{2}\xi\right). \tag{1.950}$$

Equation (1.950) can take the form of a canonical quadratic equation, viz.

$$\xi^2 + \frac{2}{Bi_f}\xi - 2\omega = 0, \tag{1.951}$$

after eliminating parenthesis and multiplying both sides by 2; once the solving formula for a quadratic equation is applied, Eq. (1.951) becomes

$$\xi = \frac{-\frac{2}{Bi_f} \pm \sqrt{\left(\frac{2}{Bi_f}\right)^2 + 8\omega}}{2} = -\frac{1}{Bi_f} \pm \sqrt{\left(\frac{1}{Bi_f}\right)^2 + 2\omega}, \tag{1.952}$$

where 2 in denominator was meanwhile melted into the numerator. The inequality $1/Bi_f^{\,2} + 2\omega > 1/Bi_f^{\,2}$ also holds for the corresponding square roots, so a positive value of ξ will result only if the positive sign preceding the said root is considered in Eq. (1.952), viz.

$$\boxed{\xi = \sqrt{\frac{1}{Bi_f^{\,2}} + 2\omega} - \frac{1}{Bi_f};} \tag{1.953}$$

in view of the definitions conveyed by Eqs. (1.948) and (1.949), x_f (under alias ξ) may thus be expressed as a function of t (under alias ω). On the other hand, Eq. (1.914) will be reformulated as

$$\frac{T_f - T}{T_f - T_\infty} = \frac{\left(1 - \frac{x_f}{L}\right) - 1 + \frac{x}{L}}{\frac{1}{Bi_f} + 1 - \frac{x_f}{L}} \tag{1.954}$$

upon addition and subtraction of unity to the numerator, and retrieval of Bi_f; insertion of Eqs. (1.948) and (1.949) then supports transformation to

$$\boxed{\frac{T_f - T}{T_f - T_\infty} = \frac{\xi - 1 + \frac{x}{L}}{\frac{1}{Bi_f} + \xi},} \tag{1.955}$$

entailing $T \equiv T\{x,t\}$, or dimensionless dependent variable $(T_f - T)/(T_f - T_\infty)$ as a function of dimensionless location variable x/L, and dimensionless time variable $\xi \equiv \xi\{\omega\}$, see Eqs. (1.949) and (1.953). The evolution in time of the temperature profiles under scrutiny is presented in Fig. 1.82, for two orders of magnitude of germane Biot's number. As expected, a linear profile of temperature prevails along the slab-shaped food at any time – more explicitly given by

$$\frac{T_f - T}{T_f - T_\infty} = \frac{\xi - 1}{\frac{1}{Bi_f} + \xi} + \frac{1}{\frac{1}{Bi_f} + \xi} \frac{x}{L}, \tag{1.956}$$

stemming from Eq. (1.955) upon algebraic rearrangement – since ξ is constant when ω (besides Bi_f) is fixed, see Eq. (1.953); this is consistent with the trends shown in Fig. 1.79. Note that the model developed above is valid only until the ice front reaches the base of the food – so lines leading to $(T_f - T)/(T_f - T_\infty) > 0$ at $x/L = 0$ are meaningless (and were accordingly omitted), see Eqs. (1.948) and (1.956). The horizontal intercept of the straight lines described by Eq. (1.955) reads

$$\frac{\xi - 1 + \frac{x}{L}}{\frac{1}{Bi_f} + \xi} = 0, \tag{1.957}$$

which unfolds

$$\left.\frac{x}{L}\right|_{\frac{T_f - T}{T_f - T_\infty} = 0} = 1 - \xi; \tag{1.958}$$

the said straight lines in Fig. 1.82 cross the vertical line defined by $x/L = 1$ at

$$\left.\frac{T_f - T}{T_f - T_\infty}\right|_{\frac{x}{L} = 1} = \frac{\xi - 1 + 1}{\frac{1}{Bi_f} + \xi} = \frac{\xi}{\frac{1}{Bi_f} + \xi}, \tag{1.959}$$

based on Eq. (1.955). One may rewrite Eq. (1.955) with the aid of Eq. (1.959) as

$$\left.\frac{T_f - T}{T_f - T_\infty}\right|_{\frac{x}{L} = 1} = \left.\frac{T_f - T}{T_f - T_\infty}\right|_{T = T_s} \equiv \frac{T_f - T_s}{T_f - T_\infty} = \frac{\xi}{\frac{1}{Bi_f} + \xi}, \tag{1.960}$$

since the temperature prevailing on the surface of the food is denoted by T_s – see Fig. 1.76. Upon extrapolation toward x/L above unity, the straight lines in Fig. 1.82 meet at a common (nodal) point – with abscissa, say, x_n dependent only on Bi_f; the said intercept should occur at the lowest temperature ever reachable, i.e. T_∞. In fact, when $T = T_\infty$, Eq. (1.955) becomes

$$\frac{T_f - T_\infty}{T_f - T_\infty} = 1 = \frac{\xi - 1 + \frac{x_n}{L}}{\frac{1}{Bi_f} + \xi}, \tag{1.961}$$

which is equivalent to

$$\xi - 1 + \frac{x_n}{L} = \frac{1}{Bi_f} + \xi \tag{1.962}$$

after getting rid of denominators; elimination of ξ between sides simplifies Eq. (1.962) to

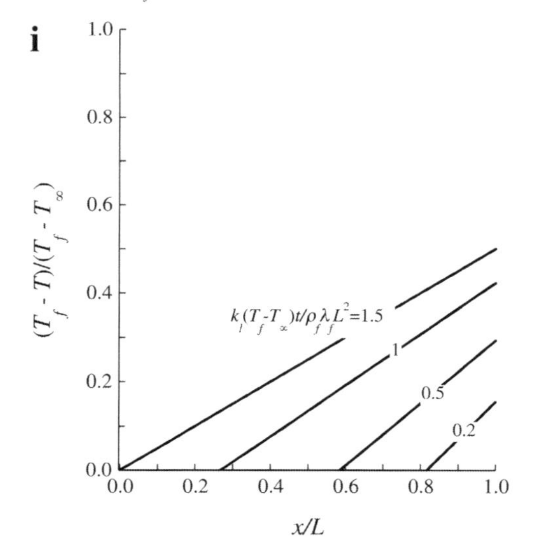

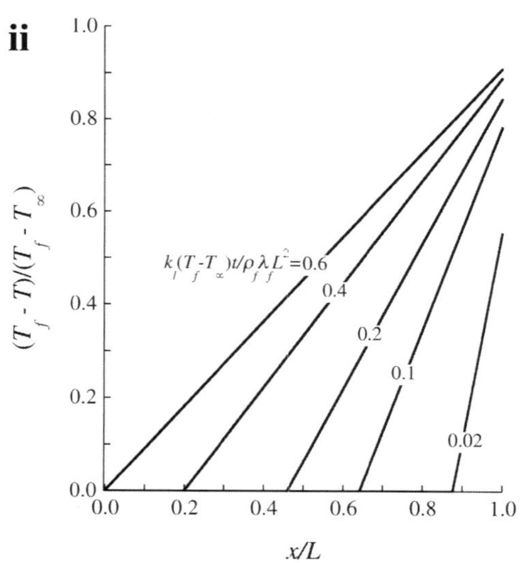

FIGURE 1.82 Variation of normalized temperature, $(T_f - T)/(T_f - T_\infty)$, versus normalized position, x/L, for selected dimensionless times, $k_l(T_f - T_\infty)t/\rho_f\lambda_f L^2$, and selected values of Biot's number, (i) $Bi_f = 1$ or (ii) $Bi_f = 10$.

$$\frac{x_n}{L} - 1 = \frac{1}{Bi_f}, \tag{1.963}$$

or else

$$\frac{x_n}{L} = 1 + \frac{1}{Bi_f} \tag{1.964}$$

that confirms the above claim. When h_a (i.e. conductance of the cold air phase) increases relative to k_f/L (as equivalent conductance in the frozen food phase), such a hypothetical x_n/L where lines merge approaches unity – owing to a higher and higher Bi_f, which causes $1/Bi_f$ to eventually vanish in Eq. (1.964).

The information conveyed by Eq. (1.955) may be grasped in a complementary manner by plotting the evolution of temperature with time at selected points along the food – as done in Fig. 1.83, again covering a range of Bi_f with practical interest. The curves appear truncated because the second freezing stage entails a finite duration – as it lasts just until the freezing front reaches the base of the food; the normalized temperatures, $(T_f - T)/(T_f - T_\infty)$, prevailing deeper in the food are obviously lower, at any given time – corresponding to lower values of $T_f - T$, along with $0 < T_f - T < T_f - T_\infty$. Should T, rather than its dimensionless counterpart be plotted, then the curves equivalent to those in Fig. 1.83 would correspond to a relatively rapid decrease in temperature departing from T_f – followed by a sluggish evolution toward some value T_s, comprised between T_∞ and T_f.

The exact shape of the evolution curves in Fig. 1.83 – holding a rectangular hyperbola-type of pattern, arises from

$$\frac{T_f - T}{T_f - T_\infty} = \frac{\sqrt{\frac{1}{Bi_f^2} + 2\omega} - \left(\frac{1}{Bi_f} + 1 - \frac{x}{L}\right)}{\frac{1}{Bi_f} + \sqrt{\frac{1}{Bi_f^2} + 2\omega} - \frac{1}{Bi_f}}, \tag{1.965}$$

following combination of Eqs. (1.953) and (1.955); elimination of symmetrical terms in denominator, and splitting of the outstanding fraction afterward lead to

$$\boxed{\frac{T_f - T}{T_f - T_\infty} = 1 - \frac{\frac{1}{Bi_f} + 1 - \frac{x}{L}}{\sqrt{\frac{1}{Bi_f^2} + 2\omega}}} \tag{1.966}$$

– which justifies their lying below unity (as Bi_f and $1 - x/L$ are both positive). Unity will be approached after a sufficiently long time had elapsed – due to $\omega \to \infty$ and per Eq. (1.949), and thus $\sqrt{1/Bi_f^2 + 2\omega} \to \infty$ in denominator; however, x_f would become nil before – see Fig. 1.80, thus precluding validity of the postulates describing the second freezing stage.

The curves in Fig. 1.83 depart from an abscissa x/L satisfying

$$0 = 1 - \frac{\frac{1}{Bi_f} + 1 - \frac{x}{L}}{\sqrt{\frac{1}{Bi_f^2} + 2\omega}}, \tag{1.967}$$

corresponding to $T = T_f$ in Eq. (1.966); after elimination of denominators, Eq. (1.967) becomes

$$\frac{1}{Bi_f} + 1 - \frac{x}{L} = \sqrt{\frac{1}{Bi_f^2} + 2\omega}, \tag{1.968}$$

which transforms to

$$\left(\left(\frac{1}{Bi_f} + 1\right) - \frac{x}{L}\right)^2 = \frac{1}{Bi_f^2} + 2\omega \tag{1.969}$$

upon squaring both sides. Newton's formula for the square of a binomial supports conversion of Eq. (1.969) to

$$\left(\frac{1}{Bi_f} + 1\right)^2 - 2\left(\frac{1}{Bi_f} + 1\right)\frac{x}{L} + \left(\frac{x}{L}\right)^2 = \frac{1}{Bi_f^2} + 2\omega, \tag{1.970}$$

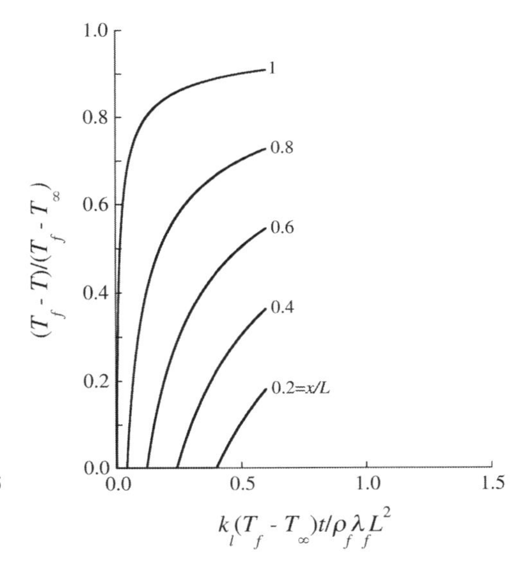

FIGURE 1.83 Variation of normalized temperature, $(T_f - T)/(T_f - T_\infty)$, versus dimensionless time, $k_l(T_f - T_\infty)t/\rho_f\lambda_f L^2$, for selected normalized positions, x/L, and selected values of Biot's number, (i) $Bi_f = 1$ or (ii) $Bi_f = 10$.

and a second application of this rule to the first term unfolds

$$\left(\frac{1}{Bi_f}\right)^2+\frac{2}{Bi_f}+1-\frac{2}{Bi_f}\frac{x}{L}-2\frac{x}{L}+\left(\frac{x}{L}\right)^2=\frac{1}{Bi_f^{\,2}}+2\omega \tag{1.971}$$

along with elimination of the second parenthesis; after dropping identical terms from both sides and then isolating ω, Eq. (1.971) becomes

$$\omega=\frac{1}{Bi_f}-\frac{1}{Bi_f}\frac{x}{L}+\left(\frac{1}{2}-\frac{x}{L}+\frac{1}{2}\left(\frac{x}{L}\right)^2\right) \tag{1.972}$$

– where Newton's formula may now be applied backward, and $1/Bi_f$ factored out to obtain

$$\omega=\frac{1}{Bi_f}\left(1-\frac{x}{L}\right)+\frac{1}{2}\left(1-\frac{x}{L}\right)^2. \tag{1.973}$$

A final factoring out of $1-x/L$ in Eq. (1.973) gives rise to

$$\omega\Big|_{\frac{T_f-T}{T_f-T_\infty}=0}=\left(1-\frac{x}{L}\right)\left(\frac{1}{Bi_f}+\frac{1}{2}\left(1-\frac{x}{L}\right)\right); \tag{1.974}$$

Eq. (1.974) resembles Eq. (1.931), in view of Eq. (1.949) and the definition of Bi_f – since $T=T_f$ implies $x=x_f$.

Although the model labeled as Eq. (1.966) can be mathematically extrapolated to $\omega\to\infty$, it would break in terms of validity after an upper maximum – as already emphasized; more specifically, this model holds as long as

$$\left(\frac{T_f-T}{T_f-T_\infty}\le 0\right)\Bigg|_{\frac{x}{L}=0}, \tag{1.975}$$

in agreement with Fig. 1.82 – which implies

$$\frac{\xi-1+0}{\dfrac{1}{Bi_f}+\xi}\le 0 \tag{1.976}$$

as per Eq. (1.955), or else

$$\xi\le 1 \tag{1.977}$$

after removal of denominators and isolation of ξ. Recalling Eq. (1.953), one can redo Eq. (1.977) to

$$\sqrt{\frac{1}{Bi_f^{\,2}}+2\omega}-\frac{1}{Bi_f}\le 1 \tag{1.978}$$

that is equivalent to

$$\sqrt{\frac{1}{Bi_f^{\,2}}+2\omega}\le 1+\frac{1}{Bi_f}; \tag{1.979}$$

after taking squares of both sides, one gets

$$\frac{1}{Bi_f^{\,2}}+2\omega\le\left(1+\frac{1}{Bi_f}\right)^2=1+\frac{2}{Bi_f}+\left(\frac{1}{Bi_f}\right)^2 \tag{1.980}$$

in view of Newton's binomial theorem – which breaks down to

$$2\omega\le 1+\frac{2}{Bi_f}, \tag{1.981}$$

once common terms are dropped off both sides. Isolation of ω in Eq. (1.981) finally gives

$$\omega\le\frac{1}{2}+\frac{1}{Bi_f}; \tag{1.982}$$

this justifies the curves in Fig. 18.83i being discontinued at $\omega=1.5$, and in Fig. 1.83ii at $\omega=0.6$ – corresponding to $Bi_f=1$ and $Bi_f=10$, respectively.

During the last freezing period, the temperature inside the food is no longer controlled by the (constant) freezing temperature of the freezing front – which has meanwhile reached the bottom (insulated) surface of the food; but instead by the temperature of the cold air stream flowing over the food. For simplicity, the temperature prevailing on the top surface of the food by the end of the freezing stage proper may be considered to be close to T_∞ (especially if h_a is large), while its bottom counterpart is equal to T_f; hence, the initial condition of this stage looks like

$$\boxed{T\big|_{t=0}=T_f-\left(T_f-T_\infty\right)\frac{x}{L},} \tag{1.983}$$

in lieu of Eq. (1.883). The boundary conditions read

$$\boxed{\frac{\partial T}{\partial x}\bigg|_{x=0}=0,} \tag{1.984}$$

still in parallel to Eq. (1.882); coupled with

$$\boxed{T\big|_{x=L}=T_\infty,} \tag{1.985}$$

instead of Eq. (1.881). The enthalpy balance may be setup as

$$\boxed{\frac{k_l}{\rho_f c_{P,f}}\frac{\partial^2 T}{\partial x^2}=\frac{\partial T}{\partial t},} \tag{1.986}$$

which mimics Eq. (1.880) after replacing k_f by k_l – since the matrix is now frozen in full; variable x^* may still be used as per Eq. (3.1313) in *Food Proc. Eng.: thermal & chemical operations*, while one should resort to

$$t^*\equiv\frac{k_l}{\rho_f c_{P,f}L^2}t \tag{1.987}$$

as replacement for Eq. (3.1159) also from *Food Proc. Eng.: thermal & chemical operations* – as well as

$$T^*\equiv\frac{T_f-T}{T_f-T_\infty}, \tag{1.988}$$

meant hereafter to play the role of Eq. (1.884). Under these circumstances, Eq. (1.986) may be rewritten as

$$\frac{\partial^2 T^*}{\partial x^{*2}}=\frac{\partial T^*}{\partial t^*} \tag{1.989}$$

or, equivalently,

$$\frac{\partial^2\left(1-T^*\right)}{\partial\left(1-x^*\right)^2}=\frac{\partial\left(1-T^*\right)}{\partial t^*} \tag{1.990}$$

because $d^2(1-T^*)=d(d(1-T^*))=d(-dT^*)=-d(dT^*)=-d^2T^*$ and $d(1-T^*)=-dT^*$, complemented by $d(1-x^*)^2=(d(1-x^*))^2=(-dx^*)^2=(dx^*)^2=dx^{*2}$; while Eq. (1.985) should take the form

$$T^*\Big|_{x^*=1}=1, \tag{1.991}$$

or else

$$\left(1-T^*\right)\Big|_{1-x^*=0}=0 \tag{1.992}$$

– whereas Eq. (1.984) is to look like

$$\left.\frac{\partial T^*}{\partial x^*}\right|_{x^*=0}=0 \tag{1.993}$$

that is the same as writing

$$\left.\frac{\partial\left(1-T^*\right)}{\partial\left(1-x^*\right)}\right|_{1-x^*=1}=0. \tag{1.994}$$

By the same token, Eq. (1.983) may be recoined as

$$T^*\Big|_{t^*=0}=x^* \tag{1.995}$$

or, equivalently,

$$\left(1-T^*\right)\Big|_{t^*=0}=1-x^*. \tag{1.996}$$

Equation (1.990) coincides in functional form with Eq. (2.551) of *Food Proc. Eng.: thermal & chemical operations*, whereas Eqs. (1.992), (1.994), and (1.996) mimic Eqs. (3.1162), (3.1163), and (3.1242), respectively, also from *Food Proc. Eng.: thermal & chemical operations* – provided that x^* is swapped for $1-x^*$ and T^* for $1-T^*$. Hence, one may resort to

$$1-T^*=\frac{8}{\pi^2}\sum_{i=0}^{\infty}\frac{(-1)^i}{(2i+1)^2}\sin\frac{(2i+1)\pi\left(1-x^*\right)}{2}e^{-\frac{(2i+1)^2\pi^2}{4}t^*} \tag{1.997}$$

as solution in parallel to Eq. (3.1253) of *Food Proc. Eng.: thermal & chemical operations*, where x^* was accordingly replaced by $1-x^*$, and T^* likewise by $1-T^*$. After solving for T^*, Eq. (1.997) becomes

$$T^*=1-\frac{8}{\pi^2}\sum_{i=0}^{\infty}\frac{(-1)^i}{(2i+1)^2}\sin\frac{(2i+1)\pi\left(1-x^*\right)}{2}e^{-\frac{(2i+1)^2\pi^2}{4}t^*}, \tag{1.998}$$

where the original notation can be retrieved as

$$\boxed{\frac{T_f-T}{T_f-T_\infty}=1-\frac{8}{\pi^2}\sum_{i=0}^{\infty}\frac{(-1)^i}{(2i+1)^2}\sin\frac{(2i+1)\pi\left(1-\frac{x}{L}\right)}{2}e^{-\frac{(2i+1)^2\pi^2}{4}\frac{k_lt}{\rho_fc_{P,f}L^2}}} \tag{1.999}$$

– with the aid of Eqs. (1.987) and (1.988), coupled with Eq. (3.1313) of *Food Proc. Eng.: thermal & chemical operations*; a graphical interpretation of Eq. (1.999) is conveyed by Fig. 1.84. The starting linearity of the temperature profile described by Eq. (1.983) is progressively distorted toward a horizontal straight line of unit vertical intercept, as time elapses; in fact,

$$\begin{aligned}\lim_{t\to\infty}\frac{T_f-T}{T_f-T_\infty}&=1-\frac{8}{\pi^2}\sum_{i=0}^{\infty}\frac{(-1)^i}{(2i+1)^2}\sin\frac{(2i+1)\pi\left(1-\frac{x}{L}\right)}{2}e^{-\infty}\\&=1-\frac{8}{\pi^2}\sum_{i=0}^{\infty}\frac{(-1)^i}{(2i+1)^2}\sin\frac{(2i+1)\pi\left(1-\frac{x}{L}\right)}{2}0\\&=1-\frac{8}{\pi^2}\sum_{i=0}^{\infty}0=1-0=1\end{aligned} \tag{1.1000}$$

stands as asymptotic behavior of Eq. (1.999) – corresponding to $T=T_\infty$ all over the food matrix.

The rate of heat withdrawal during this subfreezing period, q_{sfz}, throughout the top surface of the food abides to

$$q_{sfz}=-k_lA_f\left.\frac{\partial T}{\partial x}\right|_{x=L}, \tag{1.1001}$$

as per Fourier's law; since $\partial T/\partial x$ can be rewritten as

$$\frac{\partial T}{\partial x}=\frac{dT}{dT^*}\frac{\partial T^*}{\partial x^*}\frac{dx^*}{dx}=\frac{\frac{dx^*}{dx}}{\frac{dT^*}{dT}}\frac{\partial T^*}{\partial x^*} \tag{1.1002}$$

with the aid of the chain differentiation rule and the rule of differentiation of an inverse function, one may retrieve Eq. (3.1313)

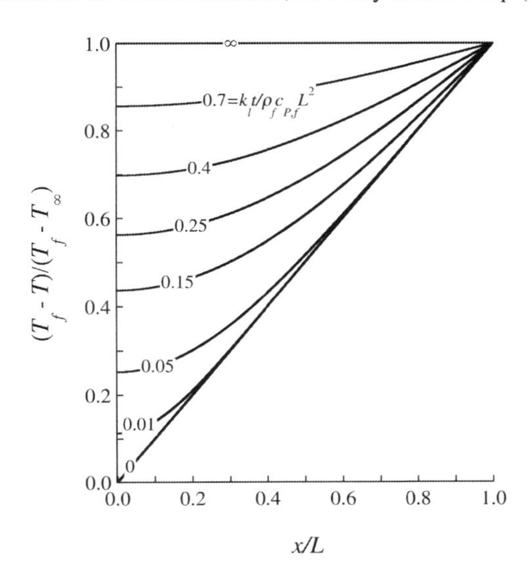

FIGURE 1.84 Variation of normalized temperature, $(T_f-T)/(T_f-T_\infty)$, versus normalized position, x/L, for selected dimensionless times, $k_lt/\rho_fc_{P,f}L^2$, throughout the third stage of fast freezing.

of *Food Proc. Eng.: thermal & chemical operations* and Eq. (1.988) to get

$$\frac{\partial T}{\partial x}=\frac{\frac{1}{L}}{\frac{1}{-T_f-T_\infty}}\frac{\partial T^*}{\partial x^*}=-\frac{T_f-T_\infty}{L}\frac{\partial T^*}{\partial x^*}. \tag{1.1003}$$

Insertion of Eq. (1.1003) converts Eq. (1.1001) to

$$q_{sfz}=k_lA_f\frac{T_f-T_\infty}{L}\left.\frac{\partial T^*}{\partial x^*}\right|_{x^*=1^-}, \tag{1.1004}$$

which is equivalent to writing

$$\frac{q_{sfz}}{k_lA_f\dfrac{T_f-T_\infty}{L}}=\left.\frac{\partial\left(\dfrac{T_f-T}{T_f-T_\infty}\right)}{\partial\left(\dfrac{x}{L}\right)}\right|_{\frac{x}{L}=1^-} \tag{1.1005}$$

upon division of both sides by $k_lA_f(T_f-T_\infty)/L$, and with the aid again of Eq. (1.988) coupled with Eq. (3.1313) of *Food Proc. Eng.: thermal & chemical operations*; the outstanding derivative in Eq. (1.1005) may, in turn, be obtained from Eq. (1.999) as

$$\frac{\partial\left(\dfrac{T_f-T}{T_f-T_\infty}\right)}{\partial\left(\dfrac{x}{L}\right)}=-\frac{8}{\pi^2}\sum_{i=0}^{\infty}\frac{(-1)^i\dfrac{(2i+1)\pi}{2}(-1)}{(2i+1)^2}\cos\frac{(2i+1)\pi\left(1-\dfrac{x}{L}\right)}{2}e^{-\frac{(2i+1)^2\pi^2}{4}\frac{k_lt}{\rho_fc_{P,f}L^2}}, \tag{1.1006}$$

which reduces to

$$\frac{\partial\left(\dfrac{T_f-T}{T_f-T_\infty}\right)}{\partial\left(\dfrac{x}{L}\right)}=\frac{4}{\pi}\sum_{i=0}^{\infty}\frac{(-1)^i}{2i+1}\cos\frac{(2i+1)\pi\left(1-\dfrac{x}{L}\right)}{2}e^{-\frac{(2i+1)^2\pi^2}{4}\frac{k_lt}{\rho_fc_{P,f}L^2}} \tag{1.1007}$$

after condensing factors alike. Once x/L is set equal to unity, Eq. (1.1007) reduces to

$$\left.\frac{\partial\left(\dfrac{T_f-T}{T_f-T_\infty}\right)}{\partial\left(\dfrac{x}{L}\right)}\right|_{\frac{x}{L}=1^-}=\frac{4}{\pi}\sum_{i=0}^{\infty}\frac{(-1)^i}{2i+1}\cos 0\,e^{-\frac{(2i+1)^2\pi^2}{4}\frac{k_lt}{\rho_fc_{P,f}L^2}} \tag{1.1008}$$

that may be inserted in Eq. (1.1005) to yield

$$\boxed{\frac{q_{sfz}}{k_lA_f\dfrac{T_f-T_\infty}{L}}=\frac{4}{\pi}\sum_{i=0}^{\infty}\frac{(-1)^i}{2i+1}e^{-\frac{(2i+1)^2\pi^2}{4}\frac{k_lt}{\rho_fc_{P,f}L^2}},} \tag{1.1009}$$

along with the realization that cos 0 equals unity. The amount of heat removed up to time t, during the last stage of freezing, abides to

$$Q_{sfz}\equiv\int_0^t q_{sfz}\,d\tilde{t}, \tag{1.1010}$$

which gives rise to

$$\frac{Q_{sfz}}{k_lA_f\dfrac{T_f-T_\infty}{L}}=\int_0^{t_{sfz}}\frac{4}{\pi}\sum_{i=0}^{\infty}\frac{(-1)^i}{2i+1}e^{-\frac{(2i+1)^2\pi^2}{4}\frac{k_l\tilde{t}}{\rho_fc_{P,f}L^2}}d\tilde{t} \tag{1.1011}$$

upon division of both sides by $k_lA_f(T_f-T_\infty)/L$ and insertion of Eq. (1.1009) afterward; exchange of the integral and summation operators, due to their linearity, unfolds

$$\frac{Q_{sfz}}{k_lA_f\dfrac{T_f-T_\infty}{L}}=\frac{4}{\pi}\sum_{i=0}^{\infty}\frac{(-1)^i}{2i+1}\int_0^t e^{-\frac{(2i+1)^2\pi^2}{4}\frac{k_l\tilde{t}}{\rho_fc_{P,f}L^2}}d\tilde{t}. \tag{1.1012}$$

Application of the fundamental theorem of integral calculus supports transformation of Eq. (1.1012) to

$$\frac{Q_{sfz}}{k_lA_f\dfrac{T_f-T_\infty}{L}}=\frac{4}{\pi}\sum_{i=0}^{\infty}\left.\frac{\dfrac{(-1)^i}{2i+1}}{-\dfrac{(2i+1)^2\pi^2}{4}\dfrac{k_l}{\rho_fc_{P,f}L^2}}e^{-\frac{(2i+1)^2\pi^2}{4}\frac{k_l\tilde{t}}{\rho_fc_{P,f}L^2}}\right|_0^t, \tag{1.1013}$$

which breaks down to

$$\frac{Q_{sfz}}{k_lA_f\dfrac{T_f-T_\infty}{L}}=\frac{16}{\pi^3}\frac{\rho_fc_{P,f}L^2}{k_l}\sum_{i=0}^{\infty}\left.\frac{(-1)^i}{(2i+1)^3}e^{-\frac{(2i+1)^2\pi^2}{4}\frac{k_l\tilde{t}}{\rho_fc_{P,f}L^2}}\right|_t^0 \tag{1.1014}$$

after pooling together similar factors – while swapping integration limits at the expense of the outstanding minus sign; upon multiplication of both sides $k_l/\rho_fc_{P,f}L^2$, Eq. (1.1014) degenerates to

$$\boxed{\frac{Q_{sfz}}{\rho_fc_{P,f}LA_f\left(T_f-T_\infty\right)}=\frac{16}{\pi^3}\sum_{i=0}^{\infty}\frac{(-1)^i}{(2i+1)^3}\left(1-e^{-\frac{(2i+1)^2\pi^2}{4}\frac{k_lt}{\rho_fc_{P,f}L^2}}\right),} \tag{1.1015}$$

on account of $e^0 = 1$. The variation of cumulative dimensionless heat taken out during subfreezing, $Q_{sfz}/\rho_f c_{P,f} LA_f(T_f - T_\infty)$, with dimensionless subfreezing time, $k_l t/\rho_f c_{P,f} L^2$, is depicted in Fig. 1.85. As observed during the chilling period, the cumulative nature of the cooling effect below freezing temperature leads to a monotonically increasing trend, which grows slower and slower as time elapses – until a horizontal asymptote, of vertical intercept equal to ½, is reached; note that the early rise of Q_{sfz} is not as fast as Q_{chl} in Fig. 1.78, because the departing profile spans $[T_\infty, T_f]$ – instead of uniformly coinciding with T_{in}.

When time grows unbounded, Eq. (1.1015) will be driven by

$$\frac{Q_{sfz,tot}}{\rho_f c_{P,f} LA_f\left(T_f - T_\infty\right)} \equiv \lim_{\frac{k_l t}{\rho_f c_{P,f} L^2} \to \infty} \frac{Q_{sfz}}{\rho_f c_{P,f} LA_f\left(T_f - T_\infty\right)}$$

$$= \frac{16}{\pi^3} \sum_{i=0}^{\infty} \frac{(-1)^i}{(2i+1)^3}\left(1 - e^{-\infty}\right) = \frac{16}{\pi^3} \sum_{i=0}^{\infty} \frac{(-1)^i}{(2i+1)^3}, \tag{1.1016}$$

because $e^{-\infty} = 0$; in view of the exact mathematical equality

$$\sum_{i=0}^{\infty} \frac{(-1)^i}{(2i+3)^3} \equiv \lim_{n \to \infty} \sum_{i=0}^{n} \frac{(-1)^i}{(2i+1)^3} = \frac{\pi^3}{32}, \tag{1.1017}$$

one may redo Eq. (1.1016) to

$$\frac{Q_{sfz,tot}}{\rho_f c_{P,f} LA_f\left(T_f - T_\infty\right)} = \frac{16}{\pi^3} \frac{\pi^3}{32} = \frac{1}{2} \tag{1.1018}$$

that accounts for the aforementioned horizontal asymptote. After having multiplied both sides by $\rho_f c_{P,f} LA_f(T_f - T_\infty)$, Eq. (1.1018) turns to

$$Q_{sfz,tot} = \frac{1}{2} \rho_f c_{P,f} LA_f\left(T_f - T_\infty\right) \tag{1.1019}$$

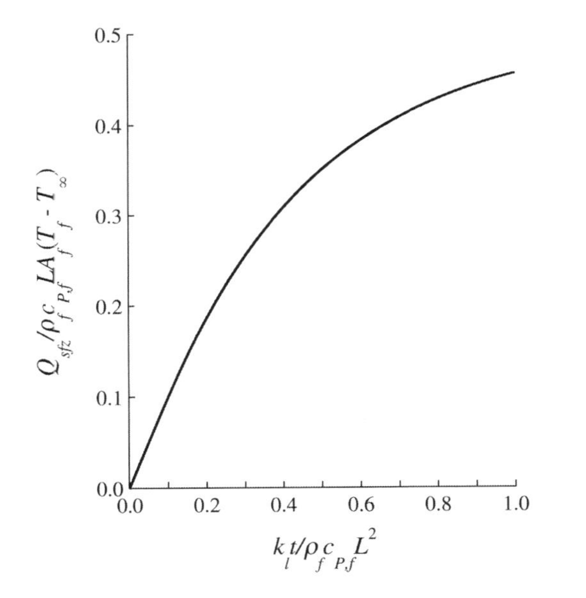

FIGURE 1.85 Evolution of normalized heat withdrawn, $Q_{sfz}/\rho_f c_{P,f} LA_f(T_f - T_\infty)$, versus dimensionless time, $k_l t/\rho_f c_{P,f} L^2$, throughout the third stage of fast freezing.

that resembles Eq. (1.907) in functional form; an initial linear profile, with T_f and T_∞ as extreme temperatures, implies that the total amount of heat to be withdrawn during subfreezing is just one-half the total volume, LA_f, multiplied by the mass density ρ_f to get mass of food, further multiplied by specific heat capacity $c_{P,f}$ to get heat capacity, and finally multiplied by temperature change $T_f - T_\infty$ to get heat transferred – in much the same way the area of a triangle (as planar projection of a parallelepiped) is one-half the area of the corresponding rectangle.

When thawing is concerned, essentially the reverse of freezing occurs – as schematized in Fig. 1.86. In this case, the whole food starts at subfreezing temperature, undergoes a transient until reaching T_f, and then the ice front experiences regression from the top of the food due to exposure to warm air (or water), at temperature $T_\infty > T_f$.

The sequence of steps in thawing is accordingly reversed relative to freezing – yet a similar reasoning applies, so the same expressions and plots may in principle be utilized. During heating, Eq. (1.986) is satisfied as enthalpic balance during the first stage of thawing – but subjected to Eqs. (1.881) and (1.882) as boundary conditions, coupled with Eq. (1.883) as initial condition, with $T_{in} < T_f$. The temperature profile is as in Fig. 1.77, provided that the vertical axis represents $(T - T_{in})/(T_f - T_{in})$, and the curves are labeled with the corresponding values of $k_l t/\rho_f c_{P,f} L^2$; while Eq. (1.898) will represent q_{htg}, as long as $T_f - T_{in}$ is used in the normalizing factor and again $k_l t/\rho_f c_{P,f} L^2$ is used as dimensionless time; with the same applying to Eq. (1.903), which would stand for cumulative heat of thawing, Q_{htg}.

The thawing period proper will be described by Eq. (1.914), in terms of temperature profile prevailing in the liquid water portion – as long as $(T - T_f)/(T_\infty - T_f)$ is used as left-hand side, and $k_f/h_a L$ in the right-hand side; the frozen portion still abides to Eq. (1.915), after similarly taking negatives of both numerator and denominator in its left-hand side. The position of the ice front, x_f, evolves in time again as depicted in Eq. (1.931), using $k_f(T_\infty - T_f)$ in the left-hand side and in the caption to the vertical axis in Fig. 1.80 – with $k_f/h_a L$ serving as label to the curves. The heat transfer rate required by melting, q_{mtg}, can be estimated via Eq. (1.938), using again $k_f(T_\infty - T_f)$ in the left-hand side instead, and $k_f/h_a L$ in the right-hand side; while Eq. (1.944) remains valid for cumulative heat of thawing, Q_{mtg} – with Fig. 1.81 to be relabeled vertically as $Q_{mtg}/\rho_f \lambda_f LA_f$ and horizontally as $k_f(T_\infty - T_f)t/\rho_f \lambda_f L^2$, and curves relabeled once more as $k_f/h_a L$.

For the period of overheating, one should resort to Eq. (1.880) as enthalpic balance – complemented by Eqs. (1.984) and (1.985) as boundary conditions, and Eq. (1.983) as initial condition. The corresponding transient temperature profile will be described by Eq. (1.999) – with the left-hand side replaced by $(T - T_f)/(T_\infty - T_f)$, and with a similar alias as caption to the vertical axis in Fig. 1.84; while the curves in the latter are to be relabeled with the appropriate values of $k_f t/\rho_f c_{P,f} L^2$. The associated rate of heat transfer, q_{ovh}, again satisfies Eq. (1.1009), with $T_\infty - T_f$ in the left-hand side and $k_f t/\rho_f c_{P,f} L^2$ similarly in the argument of the exponential function in the right-hand side; the same changes apply with regard to Q_{ovh} in Eq. (1.1015), with $Q_{ovh}/\rho_f c_{P,f} LA_f(T_\infty - T_f)$ serving also as label to the vertical axis, and $k_f t/\rho_f c_{P,f} L^2$ likewise as label to its horizontal counterpart in Fig. 1.85.

The above considerations assume that liquid water does not drip out of the food to a significant extent; this represents indeed

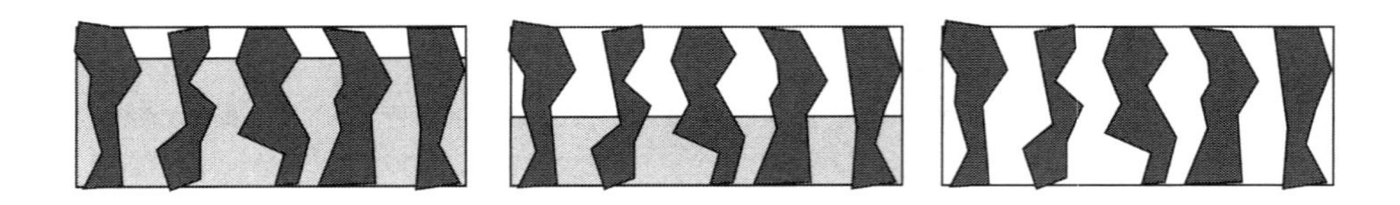

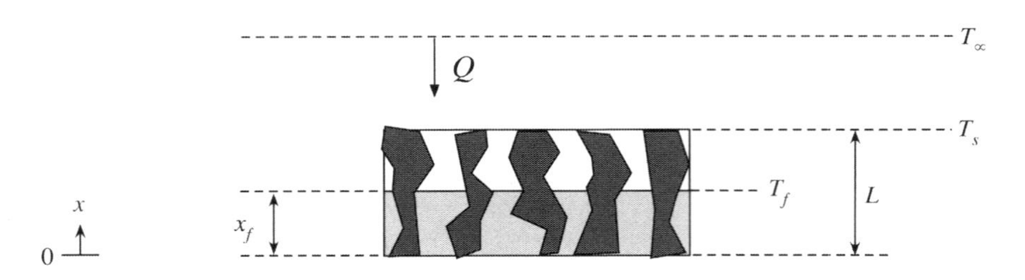

FIGURE 1.86 Evolution in time (top, left to right) of frozen portion (▭), within structure of parallelepipedal food formed by porous solid matrix (▬) embedded with liquid (aqueous) fluid (▭), as fast thawing progresses – with indication (bottom) of food thickness, L, ice layer thickness, x_f, freezing temperature, T_f, surface temperature, T_s, heating fluid bulk temperature, T_∞, and inlet heat flow, Q, in the downward x-direction via top heating.

an ideal situation, since freezing may have disrupted the cellular network of the food – thus (partially) hampering its ability to retain the initial amount of water, and concomitantly resume the original food microstructure. Finally, it should be emphasized that freezing is a positively cooperative process – because $k_i > k_f$, so heat flows faster out of the food as freezing progresses. The opposite consequently occurs during thawing – which inherently takes longer to happen, as heat flows in through a liquid water-type food matrix characterized by a lower thermal conductivity than its iced food counterpart. Hence, a negatively cooperative process arises – hidden mathematically in the disparate factor (i.e. k_f instead of k_i) used to normalize variables in the previous analysis. This means that a more sluggish evolution of temperature in time takes place during heating, thawing proper, and overheating, as per the corresponding enthalpy balances discussed above – when compared to the corresponding chilling, freezing proper, and subcooling, respectively.

1.4.3.3 Slow freezing and thawing

Consider a homogeneous food, in the bulk of which temperature gradients do not essentially build up; if not that low a refrigerant temperature is employed, or a large external resistance to heat transfer exists, then slow freezing will be prone to occur. Under these circumstances, the said phase change will take place more or less uniformly throughout the food – subjected to depression in freezing point, as molality of solute(s), X_s, increases, owing to essential prevalence of chemical equilibrium between solid and liquid phases.

After solving for amount of solute in solution, n_s, Eq. (1.879) turns to

$$n_s = m_w X_s, \tag{1.1020}$$

where m_w denotes mass of water remaining in liquid form; differentiation of both sides then unfolds

$$dn_S = X_S dm_w + m_w dX_s. \tag{1.1021}$$

The total mass of solvent, $m_{w,0}$, can in turn be equated as

$$m_{w,0} = m_w + m_i, \tag{1.1022}$$

where m_i denotes mass of ice (i.e. pure water in solid form); Eq. (1.1022) meanwhile reads

$$dm_w + dm_i = 0 \tag{1.1023}$$

in differential form, since $m_{w,0}$ is constant. Only water originally serving as solvent freezes out, so n_s is also constant – and dn_s is consequently nil; therefore, Eq. (1.1021) degenerates to

$$\boxed{X_s dm_w + m_w dX_s = 0.} \tag{1.1024}$$

After solving Eq. (1.1023) for dm_w, viz.

$$dm_w = -dm_i, \tag{1.1025}$$

and Eq. (1.1022) for m_w, viz.

$$m_w = m_{w,0} - m_i, \tag{1.1026}$$

one will be able to rewrite Eq. (1.1024) as

$$-X_s dm_i + \left(m_{w,0} - m_i\right) dX_s = 0. \tag{1.1027}$$

Integration of Eq. (1.1027) is now in order, via separation of variables, i.e.

$$\int_{X_{s,0}}^{X_s} \frac{d\tilde{X}_s}{\tilde{X}_s} = \int_0^{m_i} \frac{d\tilde{m}_i}{m_{w,0} - \tilde{m}_i}, \tag{1.1028}$$

since

$$\boxed{X_s\big|_{m_i=0} = X_{s,0}} \tag{1.1029}$$

may be used as boundary condition – with $X_{s,0}$ denoting solute molality in original (unfrozen) food; Eq. (1.1028) is equivalent to

$$\ln \tilde{X}_s \Big|_{X_{s,0}}^{X_s} = -\ln\left\{m_{w,0} - \tilde{m}_i\right\}\Big|_0^{m_i} = \ln\left\{m_{w,0} - \tilde{m}_i\right\}\Big|_{m_i}^0, \tag{1.1030}$$

in view of the fundamental theorem of integral calculus complemented by exchange of integration limits – or else

$$\ln\frac{X_S}{X_{S,0}}=\ln\frac{m_{w,0}}{m_{w,0}-m_i}. \tag{1.1031}$$

Equation (1.1031) may instead appear as

$$\boxed{\frac{X_s}{X_{s,0}}=\frac{1}{1-\dfrac{m_i}{m_{w,0}}}} \tag{1.1032}$$

once exponentials are taken of both sides, and numerator and denominator of the right-hand side are divided by $m_{w,0}$ afterward. Equation (1.874) may now be revisited as

$$\frac{T_f-T_{f,0}}{\beta_w}=-\frac{\alpha_w}{\beta_w}X_s-X_s^2, \tag{1.1033}$$

upon addition of the negative of $T_{f,0}$ to both sides, followed by division thereof by β_w; a quadratic equation accordingly results, which reads

$$X_s^2+\frac{\alpha_w}{\beta_w}X_s-\frac{T_{f,0}-T_f}{\beta_w}=0 \tag{1.1034}$$

in canonical form. The solving formula for a quadratic equation has it that

$$X_s=\frac{-\dfrac{\alpha_w}{\beta_w}\pm\sqrt{\left(\dfrac{\alpha_w}{\beta_w}\right)^2-4\left(-\dfrac{T_{f,0}-T_f}{\beta_w}\right)}}{2} \tag{1.1035}$$

when applied to Eq. (1.1034), where straightforward algebraic rearrangement unfolds

$$X_s=\frac{\pm\sqrt{\left(\dfrac{\alpha_w}{\beta_w}\right)^2+4\dfrac{T_{f,0}-T_f}{\beta_w}}-\dfrac{\alpha_w}{\beta_w}}{2}. \tag{1.1036}$$

Since the argument of the square root is greater than $(\alpha_w/\beta_w)^2$ because $T_{f,0}>T_f$ as per Eqs. (1.874)–(1.876), then the corresponding square root is larger than $\alpha_w/\beta_w>0$ – so only the plus sign preceding the square root is to be taken in Eq. (1.1036) for physical realizability, i.e.

$$X_s=\frac{\sqrt{\left(\dfrac{\alpha_w}{\beta_w}\right)^2+4\dfrac{T_{f,0}-T_f}{\beta_w}}-\dfrac{\alpha_w}{\beta_w}}{2}; \tag{1.1037}$$

insertion of Eq. (1.1037) converts Eq. (1.1032) to

$$\frac{\sqrt{\left(\dfrac{\alpha_w}{\beta_w}\right)^2+4\dfrac{T_{f,0}-T_f}{\beta_w}}-\dfrac{\alpha_w}{\beta_w}}{2X_{s,0}}=\frac{1}{1-\dfrac{m_i}{m_{w,0}}}, \tag{1.1038}$$

or else

$$\sqrt{\left(\frac{\alpha_w}{\beta_w}\right)^2+4\frac{T_{f,0}-T_f}{\beta_w}}=\frac{\alpha_w}{\beta_w}+\frac{2X_{s,0}}{1-\dfrac{m_i}{m_{w,0}}} \tag{1.1039}$$

upon isolation of the square root. After squares are taken of both sides, Eq. (1.1039) becomes

$$\left(\frac{\alpha_w}{\beta_w}\right)^2+4\frac{T_{f,0}-T_f}{\beta_w}=\left(\frac{\alpha_w}{\beta_w}+\frac{2X_{s,0}}{1-\dfrac{m_i}{m_{w,0}}}\right)^2, \tag{1.1040}$$

with isolation of $(T_{f,0}-T_f)/\beta_w$ unfolding

$$\frac{T_{f,0}-T_f}{\beta_w}=\frac{\left(\dfrac{\alpha_w}{\beta_w}+\dfrac{2X_{s,0}}{1-\dfrac{m_i}{m_{w,0}}}\right)^2-\left(\dfrac{\alpha_w}{\beta_w}\right)^2}{4}; \tag{1.1041}$$

a final division of both sides by $X_{s,0}^2$ yields

$$\boxed{\frac{T_{f,0}-T_f}{\beta_w X_{s,0}^2}=\frac{\left(\dfrac{\alpha_w}{\beta_w X_{s,0}}+\dfrac{2}{1-\dfrac{m_i}{m_{w,0}}}\right)^2-\left(\dfrac{\alpha_w}{\beta_w X_{s,0}}\right)^2}{4}} \tag{1.1042}$$

– as graphically depicted in Fig. 1.87i, for selected values of germane (dimensionless) parameter $\alpha_w/\beta_w X_{s,0}$. The freezing point depression grows faster and faster as ice formation progresses, in agreement with Eq. (1.874) – especially in the vicinity of $m_i=m_{w,0}$, due to the hyperbolic jump in solute molality as solvent gets depleted, see Eq. (1.1032). The increase of the logarithm of $T_{f,0}-T_f$ with m_i becomes less notorious when parameter $\alpha_w/\beta_w X_{s,0}$ increases, yet the spread among curves reduces as $m_i/m_{w,0}$ approaches unity.

If $\beta_w X_{s,0}<<\alpha_w$ – as happens in a number of situations of practical interest (see Fig. 1.74), then Eq. (1.874) reduces to

$$T_f-T_{f,0}=-\alpha_w X_s; \tag{1.1043}$$

isolation of X_s promptly gives

$$X_s=\frac{T_{f,0}-T_f}{\alpha_w}. \tag{1.1044}$$

Combination of Eqs. (1.1032) and (1.1044) generates

$$\boxed{\frac{T_{f,0}-T_f}{\alpha_w X_{s,0}}=\frac{1}{1-\dfrac{m_i}{m_{w,0}}}} \tag{1.1045}$$

that is sketched in Fig. 1.87ii; the general trend observed coincides with the one already apparent in Fig. 1.87i – but no discriminating parameter appears in this case.

Although the linear form for $T_f\equiv T_f\{X_s\}$ implies $\beta_w\to 0$ in Eq. (1.874), and thus $(\alpha_w/\beta_w X_{s,0})\to\infty$, the curve in Fig. 1.87ii represents the asymptotic behavior of the curves in Fig. 1.87i when β_w approaches zero – provided that appropriate conversion of vertical coordinates is performed, i.e. $\alpha_w X_{s,0}$ is used instead of $\beta_w X_{s,0}^2$ as normalizing factor. In fact,

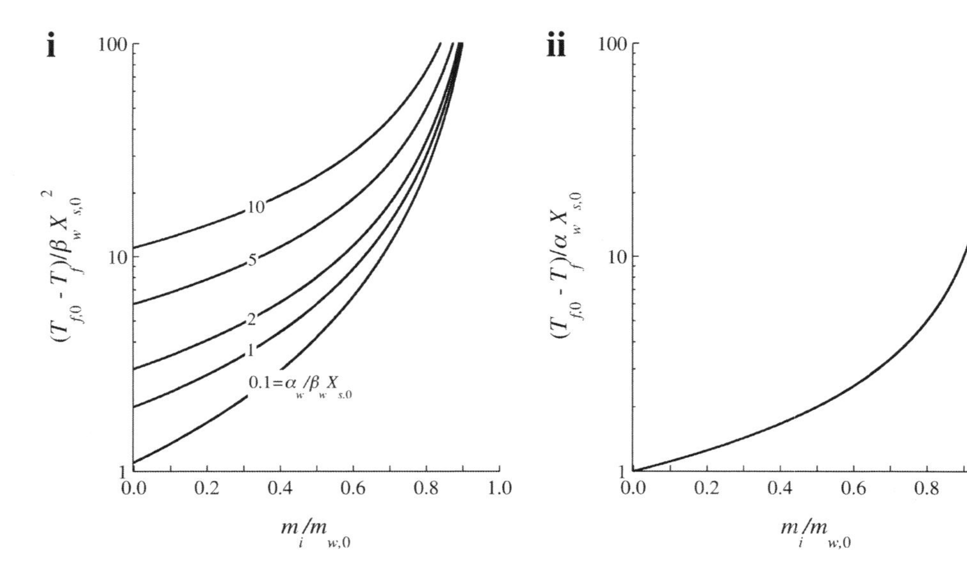

FIGURE 1.87 Variation of dimensionless freezing temperature, (i) $(T_{f,0}-T_f)/\beta_w X_{s,0}^2$ or (ii) $(T_{f,0}-T_f)/\alpha_w X_{s,0}$, versus normalized mass of ice formed, $m_i/m_{w,0}$ – assuming (i) quadratic freezing point depression, for selected values of dimensionless parameter $\alpha_w/\beta_w X_{s,0}$, and (ii) linear freezing point depression with regard to solute molality, throughout slow freezing.

$$\lim_{\beta_w \to 0} \frac{T_{f,0}-T_f}{\beta_w X_{s,0}^2} = \lim_{\beta_w \to 0} \frac{\left(\frac{\alpha_w}{\beta_w X_{s,0}} + \frac{2}{1-\frac{m_i}{m_{w,0}}} \right)^2 - \left(\frac{\alpha_w}{\beta_w X_{s,0}} \right)^2}{4} \quad (1.1046)$$

stems from Eq. (1.1042), where Newton's binomial theorem supports transformation to

$$\lim_{\beta_w \to 0} \frac{T_{f,0}-T_f}{\beta_w X_{s,0}^2} = \lim_{\beta_w \to 0} \frac{\left(\frac{\alpha_w}{\beta_w X_{s,0}} \right)^2 + 2\frac{\alpha_w}{\beta_w X_{s,0}} \frac{2}{1-\frac{m_i}{m_{w,0}}} + \left(\frac{2}{1-\frac{m_i}{m_{w,0}}} \right)^2 - \left(\frac{\alpha_w}{\beta_w X_{s,0}} \right)^2}{4}; \quad (1.1047)$$

upon cancellation of symmetrical terms in numerator, Eq. (1.1047) reduces to

$$\lim_{\beta_w \to 0} \frac{T_{f,0}-T_f}{\beta_w X_{s,0}^2} = \lim_{\beta_w \to 0} \frac{\frac{\alpha_w}{\beta_w X_{s,0}} \frac{4}{1-\frac{m_i}{m_{w,0}}} + \frac{4}{\left(1-\frac{m_i}{m_{w,0}}\right)^2}}{4}$$
$$= \frac{1}{1-\frac{m_i}{m_{w,0}}} \lim_{\beta_w \to 0} \left(\frac{\alpha_w}{\beta_w X_{s,0}} + \frac{1}{1-\frac{m_i}{m_{w,0}}} \right) \quad (1.1048)$$

once the reciprocal of $1-m_i/m_{w,0}$ is factored out, and 4 is dropped from both numerator and denominator. Since $\beta_w \to 0$ implies $(\alpha_w/\beta_w X_{s,0}) \to \infty$ (as noted before), then $\alpha_w/\beta_w X_{s,0}$ will override $1/(1-m_i/m_{w,0})$; hence, Eq. (1.1048) will be driven by

$$\lim_{\beta_w \to 0} \frac{T_{f,0}-T_f}{\beta_w X_{s,0}^2} = \frac{1}{1-\frac{m_i}{m_{w,0}}} \lim_{\beta_w \to 0} \frac{\alpha_w}{\beta_w X_{s,0}}, \quad (1.1049)$$

which may be rewritten as

$$\left. \left(\frac{T_{f,0}-T_f}{\beta_w X_{s,0}^2} = \frac{1}{1-\frac{m_i}{m_{w,0}}} \frac{\alpha_w}{\beta_w X_{s,0}} \right) \right|_{\beta_w \to 0}. \quad (1.1050)$$

Cancellation of $\beta_w X_{s,0}$ between sides supports simplification of Eq. (1.1050) to

$$\left. \left(\frac{T_{f,0}-T_f}{X_{s,0}} = \frac{1}{1-\frac{m_i}{m_{w,0}}} \alpha_w \right) \right|_{\beta_w \to 0}, \quad (1.1051)$$

to be rephrased as

$$\left. \left(\frac{T_{f,0}-T_f}{\alpha_w X_{s,0}} = \frac{1}{1-\frac{m_i}{m_{w,0}}} \right) \right|_{\beta_w \to 0} \quad (1.1052)$$

upon division of both sides by α_w; Eq. (1.1052), or Eq. (1.1045) for that matter accordingly set the behavior of Eq. (1.1042) at negligible β_w, as claimed previously.

The decrease in temperature of a food, T_f, during freezing is brought about by heat withdrawn through its outer surface of area, A_f, via a cooling fluid characterized by h_a for heat transfer coefficient and maintained at temperature $T_\infty < T_f$; Fourier's law has it that

$$\boxed{h_a A_f \left(T_f - T_\infty\right) = \lambda_f \frac{dm_i}{dt},} \tag{1.1053}$$

which mimics Eq. (1.917) when a uniform temperature prevails in the food matrix (as postulated). Equation (1.1053) holds as long as latent heat of fusion of ice, λ_f, accounts for the whole heat transferred – and is subjected to

$$\boxed{m_i\big|_{t=0} = 0} \tag{1.1054}$$

as initial condition. In this regard, a simplification has been introduced relative to Eq. (3.971) in *Food Proc. Eng.: thermal & chemical operations*, pertaining to freeze-concentration and encompassing formation of ice as well – since terms describing sensible heat have been discarded here, on account of $d(m_i c_{P,i}(T_f - T^\theta)) << \lambda_f dm_i$ pertaining to the ice phase, and likewise $d(m_f c_{P,l}(T_f - T^\theta)) << \lambda_f dm_i$ encompassing the food phase left behind; this approach simplifies solution of the enthalpic balance if duly combined with the mass balance, with the final outcome being sufficiently accurate in practice for modeling purposes. Consider, in this regard, that Eq. (1.1045) is valid; upon isolation of T_f as

$$T_f = T_{f,0} - \frac{\alpha_w X_{s,0}}{1 - \dfrac{m_i}{m_{w,0}}}, \tag{1.1055}$$

one may use it to convert Eq. (1.1053) to

$$\frac{dm_i}{dt} = \frac{h_a A_f}{\lambda_f}\left(T_{f,0} - \frac{\alpha_w X_{s,0}}{1 - \dfrac{m_i}{m_{w,0}}} - T_\infty\right), \tag{1.1056}$$

complemented by division of both sides by λ_f. Further division of both sides by $m_{w,0}$, followed by factoring out of $T_{f,0}$ in the right-hand side produce

$$\frac{d\left(\dfrac{m_i}{m_{w,0}}\right)}{dt} = \frac{h_a A_f T_{f,0}}{\lambda_f m_{w,0}}\left(1 - \frac{T_\infty}{T_{f,0}} - \frac{\dfrac{\alpha_w X_{s,0}}{T_{f,0}}}{1 - \dfrac{m_i}{m_{w,0}}}\right), \tag{1.1057}$$

where the operational rules of differentiation and the distributive property of multiplication, coupled with division of both sides by $h_a A_f T_{f,0}/\lambda_f m_{w,0}$ allow further transformation to

$$\boxed{\frac{d\left(1 - \dfrac{m_i}{m_{w,0}}\right)}{d\left(\dfrac{h_a A_f T_{f,0}}{\lambda_f m_{w,0}} t\right)} = \frac{\dfrac{\alpha_w X_{s,0}}{T_{f,0}}}{1 - \dfrac{m_i}{m_{w,0}}} - \left(1 - \frac{T_\infty}{T_{f,0}}\right).} \tag{1.1058}$$

A redefinition of lumped variables and constants is convenient at this stage to facilitate manipulation hereafter, namely,

$$\chi \equiv 1 - \frac{m_i}{m_{w,0}} \tag{1.1059}$$

and

$$\zeta \equiv \frac{h_a A_f T_{f,0}}{\lambda_f m_{w,0}} t, \tag{1.1060}$$

for dependent and independent variables, respectively; as well as

$$\Omega \equiv \frac{\alpha_w X_{s,0}}{T_{f,0}} \tag{1.1061}$$

and

$$\Theta \equiv 1 - \frac{T_\infty}{T_{f,0}}, \tag{1.1062}$$

for germane physicochemical parameters. Equation (1.1058) will accordingly look like

$$\frac{d\chi}{d\zeta} = \frac{\Omega}{\chi} - \Theta \tag{1.1063}$$

following insertion of Eqs. (1.1059)–(1.1062), while Eq. (1.1054) takes the form

$$\boxed{\chi\big|_{\zeta=0} = 1.} \tag{1.1064}$$

Integration of Eq. (1.1063) may now proceed, via separation of variables, as

$$\int \frac{\chi}{\Omega - \Theta\chi} d\chi = \int d\zeta, \tag{1.1065}$$

where multiplication of both numerator and denominator of the kernel in the left-hand side by χ meanwhile took place. Since the rational fraction produced in the left-hand side of Eq. (1.1065) is irregular, one should proceed to preliminary polynomial division to get

$$\int \left(-\frac{1}{\Theta} + \frac{\dfrac{\Omega}{\Theta}}{\Omega - \Theta\chi}\right) d\chi = \int d\zeta; \tag{1.1066}$$

hence, direct application of the rules of integration unfolds

$$\zeta = \kappa - \frac{1}{\Theta}\chi - \frac{\Omega}{\Theta}\frac{\ln\{\Omega - \Theta\chi\}}{\Theta}, \tag{1.1067}$$

with κ denoting an integration constant. Compatibility between Eqs. (1.1064) and (1.1067) implies

$$0 = \kappa - \frac{1}{\Theta}\left(1 + \frac{\Omega}{\Theta}\ln\{\Omega - \Theta\}\right), \tag{1.1068}$$

along with factoring out of $-1/\Theta$; and isolation of κ promptly yields

$$\kappa = \frac{1}{\Theta}\left(1 + \frac{\Omega}{\Theta}\ln\{\Omega - \Theta\}\right). \tag{1.1069}$$

Insertion of Eq. (1.1069) turns Eq. (1.1067) to

$$\zeta = \frac{1}{\Theta}\left(1-\chi-\frac{\Omega}{\Theta}\ln\frac{\Omega-\Theta\chi}{\Omega-\Theta}\right) \tag{1.1070}$$

– together with the operational rules of a logarithm, complemented by lumping of terms alike; Eqs. (1.1059)–(1.1062) may now be retrieved to reformulate Eq. (1.1070) as

$$\boxed{\frac{h_a A_f T_{f,0}}{\lambda_f m_{w,0}}t = \frac{1}{1-\dfrac{T_\infty}{T_{f,0}}}\left(\frac{m_i}{m_{w,0}} - \frac{\dfrac{\alpha_w X_{s,0}}{T_{f,0}}}{1-\dfrac{T_\infty}{T_{f,0}}}\ln\frac{\dfrac{\alpha_w X_{s,0}}{T_{f,0}}-\left(1-\dfrac{T_\infty}{T_{f,0}}\right)\left(1-\dfrac{m_i}{m_{w,0}}\right)}{\dfrac{\alpha_w X_{s,0}}{T_{f,0}}-\left(1-\dfrac{T_\infty}{T_{f,0}}\right)}\right)}, \tag{1.1071}$$

which entails an implicit relationship between dependent and independent variables. A plot of normalized mass of ice, $m_i/m_{w,0}$ versus dimensionless time, $h_aA_fT_{f,0}t/\lambda_fm_{w,0}$, is made available as Fig. 1.88ii. Note the increase of fraction of ice formed with time, t, departing from zero at t=0 as expected; a linear variation is associated to $1-T_\infty/T_{f,0}$ as proportionality constant – reflecting absence of freezing point depression, described by Eq. (1.1071) when the germane parameter $\alpha_wX_{s,0}/T_{f,0}$ is nil. The curves experience a stronger and stronger convex distortion as the said parameter takes higher values – and thus move gradually away (and down) from the aforementioned linear trend. The lower freezing extent when $\alpha_wX_{s,0}/T_{f,0}$ increases means that ice forms slower when solute molality in the initial solution, $X_{s,0}$, is higher, or degree of freezing point depression, measured by α_w, is more pronounced; in either case, a higher drop in freezing temperature will occur relative to pure water for a given amount of ice formed, thus hampering heat transfer out due to a poorer driving force, i.e. a smaller T_f-T_∞ in Eq. (1.1053).

The curves plotted in Fig. 3.58 of *Food Proc. Eng.: thermal & chemical operations* can be related to those in Fig. 1.88ii, upon careful conversion. Parameter $(T_{f,0}-T_r)/(-a)w_j$ should indeed be approximately replaced by the reciprocal of its analog $\alpha_wX_{s,0}/T_{f,0}$, with concomitant reversal of the order of labeling; whereas dimensionless time $UA(t-t_f)/m_jc_{Pj}$ used c_{Pj} as normalizing factor, while $\lambda_f>>c_{Pj}$ (with $h_a\approx U$) was instead used as normalizing factor in $h_aA_fT_{f,0}/\lambda_fm_{w,0}$ – so the time scale shrunk. One further realizes that $\dfrac{w_j}{w_l}=\dfrac{m_{s,j}}{m_{s,j}+m_{w,j}}\Big/\dfrac{m_{s,l}}{m_{s,l}+m_{w,l}}=\dfrac{m_s}{m_s+m_{w,j}}\Big/\dfrac{m_s}{m_s+m_{w,l}}=\dfrac{m_s+m_{w,l}}{m_s+m_{w,j}}$, based on the definition of mass fraction, followed by cancellation of common factors between numerator and denominator – with $m_{s,l}=m_{s,j}=m_s$ denoting mass of solids in liquor and (initial) juice, respectively (which remain constant because only water freezes out), and $m_{w,l}$ and $m_{w,j}$ denoting mass of water in liquor and juice, respectively. The ratio w_j/w_l further degenerates to $w_j/w_l\approx m_{w,l}/m_{w,j}\approx(m_{w,0}-m_i)/m_{w,0}=1-m_i/m_{w,0}$, since $m_s<<m_{w,l},m_{w,j}$ and $m_{w,j}\approx m_{w,0}$ usually; therefore, the curves in Fig. 3.58 of *Food Proc. Eng.: thermal & chemical operations* are to be flipped vertically – and the essential coincidence of the result (especially at larger $\lambda_f/(-a)c_{Pj}w_j$) confirms reasonability of neglecting sensible heat effects.

When the variation of T_f with X_s entertains a quadratic relationship, then one should resort to Eq. (1.1042) to get $T_f\equiv T_f\{m_i\}$ – accordingly revisited as

$$T_f = T_{f,0}-\frac{\beta_w X_{s,0}^2}{4}\left(\left(\frac{\alpha_w}{\beta_w X_{s,0}}+\frac{2}{1-\dfrac{m_i}{m_{w,0}}}\right)^2-\left(\frac{\alpha_w}{\beta_w X_{s,0}}\right)^2\right), \tag{1.1072}$$

toward transformation of Eq. (1.1053) to

$$\frac{dm_i}{dt}=\frac{h_aA_f}{\lambda_f}\left(T_{f,0}-T_\infty+\frac{\beta_w X_{s,0}^2}{4}\left(\frac{\alpha_w}{\beta_w X_{s,0}}\right)^2-\frac{\beta_w X_{s,0}^2}{4}\left(\frac{\alpha_w}{\beta_w X_{s,0}}+\frac{2}{1-\dfrac{m_i}{m_{w,0}}}\right)^2\right), \tag{1.1073}$$

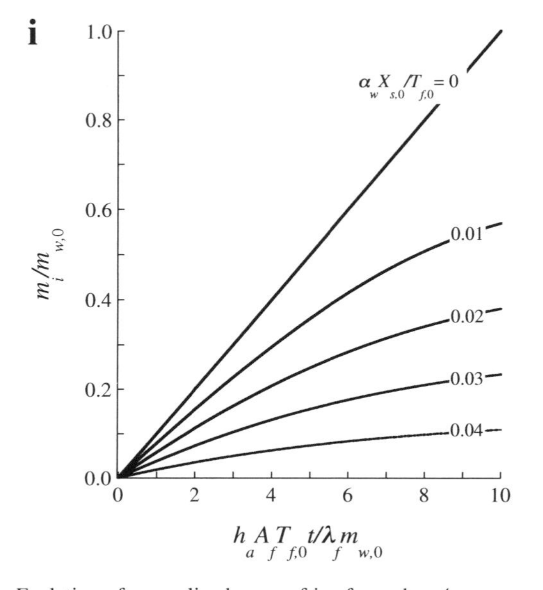

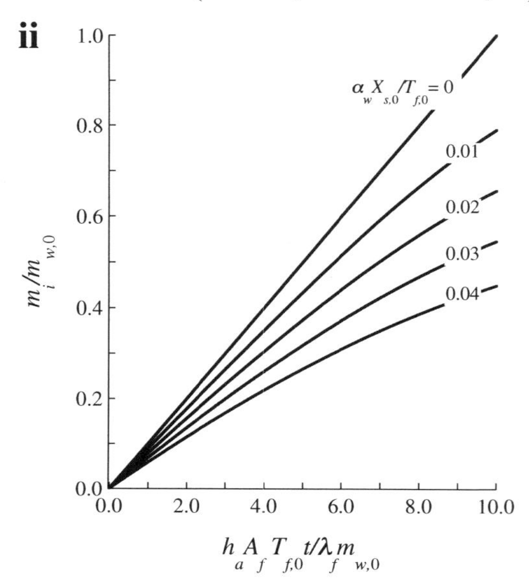

FIGURE 1.88 Evolution of normalized mass of ice formed, $m_i/m_{w,0}$, versus dimensionless time, $h_aA_fT_{f,0}t/\lambda_fm_{w,0}$ – assuming (i) quadratic freezing point depression for $\alpha_w/\beta_wX_{s,0}=1$, or (ii) linear freezing point depression with regard to solute molality, for $T_\infty/T_{f,0}=0.9$ and selected values of dimensionless parameter $\alpha_wX_{s,0}/T_{f,0}$, throughout slow freezing.

along with division of both sides by λ_f; further division of both numerator and denominator of the left-hand side by (constant) $m_{w,0}$, and then of both sides by (constant) $h_a A_f T_{f,0}/\lambda_f$ yield

$$\frac{d\left(\frac{m_i}{m_{w,0}}\right)}{d\left(\frac{h_a A_f T_{f,0}}{\lambda_f m_{w,0}}t\right)} = \left(1-\frac{T_\infty}{T_{f,0}}\right) + \frac{1}{4}\frac{\beta_w X_{s,0}^2}{T_{f,0}}\left(\frac{\alpha_w}{\beta_w X_{s,0}}\right)^2 - \frac{1}{4}\frac{\beta_w X_{s,0}^2}{T_{f,0}}\left(\frac{\alpha_w}{\beta_w X_{s,0}} + \frac{2}{1-\frac{m_i}{m_{w,0}}}\right)^2. \quad (1.1074)$$

Equation (1.1074) may be rewritten as

$$\boxed{\frac{d\xi}{d(\Lambda_1 t)} = \frac{\Lambda_3}{4}\left(\Lambda_4 + \frac{2}{\xi}\right)^2 - \left(\Lambda_2 + \frac{\Lambda_3}{4}\Lambda_4^2\right)} \quad (1.1075)$$

in simplified form – provided that auxiliary (dimensionless) parameters are defined as

$$\Lambda_1 \equiv \frac{h_a A_f T_{f,0}}{\lambda_f m_{w,0}}, \quad (1.1076)$$

$$\Lambda_2 \equiv 1 - \frac{T_\infty}{T_{f,0}}, \quad (1.1077)$$

$$\Lambda_3 \equiv \frac{\beta_w X_{s,0}^2}{T_{f,0}} \quad (1.1078)$$

and

$$\Lambda_4 \equiv \frac{\alpha_w}{\beta_w X_{s,0}}, \quad (1.1079)$$

and an auxiliary variable is defined as

$$\xi \equiv 1 - \frac{m_i}{m_{w,0}}; \quad (1.1080)$$

the negatives of both sides were meanwhile taken, for convenience. Note that

$$\left.\frac{d\xi}{d(\Lambda_1 t)}\right|_{\Lambda_3=\Lambda_4=0} = -\Lambda_2, \quad (1.1081)$$

will hold, as long as both Λ_3 and Λ_4 are nil – to yield a constant rate of disappearance of liquid water, should freezing point depression be absent. Integration of Eq. (1.1075) may proceed via separation of variables, according to

$$\int \frac{d\xi}{\frac{\Lambda_3}{4}\left(\Lambda_4 + \frac{2}{\xi}\right)^2 - \left(\Lambda_2 + \frac{\Lambda_3}{4}\Lambda_4^2\right)} = \int d(\Lambda_1 t), \quad (1.1082)$$

where multiplication of both numerator and denominator in the left-hand side by ξ^2 generates

$$\int \frac{\xi^2 d\xi}{\frac{\Lambda_3}{4}(\Lambda_4\xi + 2)^2 - \left(\Lambda_2 + \frac{\Lambda_3}{4}\Lambda_4^2\right)\xi^2} = \int d(\Lambda_1 t); \quad (1.1083)$$

expansion of the denominator via Newton's binomial theorem gives then rise to

$$\int \frac{\xi^2 d\xi}{\frac{\Lambda_3}{4}\Lambda_4^2\xi^2 + \frac{\Lambda_3}{4}4\Lambda_4\xi + \frac{\Lambda_3}{4}4 - \left(\Lambda_2 + \frac{\Lambda_3}{4}\Lambda_4^2\right)\xi^2} = \int d(\Lambda_1 t), \quad (1.1084)$$

where algebraic regrouping of terms by power of ξ yields

$$\int \frac{\xi^2 d\xi}{\left(\frac{\Lambda_3}{4}\Lambda_4^2 - \Lambda_2 - \frac{\Lambda_3}{4}\Lambda_4^2\right)\xi^2 + \Lambda_3\Lambda_4\xi + \Lambda_3} = \int d(\Lambda_1 t). \quad (1.1085)$$

Equation (1.1085) simplifies to

$$-\int \frac{\xi^2 d\xi}{\Lambda_2\xi^2 - \Lambda_3\Lambda_4\xi - \Lambda_3} = \int d(\Lambda_1 t) \quad (1.1086)$$

upon lumping of similar terms in denominator, followed by factoring out of –1. Since the numerator in the left-hand side is quadratic on ξ, and the corresponding denominator is quadratic as well, polynomial division is to be carried out before proceeding any further – and leads to

$$-\int\left(\frac{1}{\Lambda_2} + \frac{\frac{\Lambda_3\Lambda_4}{\Lambda_2}\xi + \frac{\Lambda_3}{\Lambda_2}}{\Lambda_2\xi^2 - \Lambda_3\Lambda_4\xi - \Lambda_3}\right)d\xi = \int d(\Lambda_1 t), \quad (1.1087)$$

where $1/\Lambda_2$ can be taken off the integral as

$$-\frac{1}{\Lambda_2}\int\left(1 + \frac{\Lambda_3\Lambda_4\xi + \Lambda_3}{\Lambda_2\xi^2 - \Lambda_3\Lambda_4\xi - \Lambda_3}\right)d\xi = \int d(\Lambda_1 t); \quad (1.1088)$$

the second fraction in the kernel of the integral in the left-hand side is now to be expanded in partial fractions – which demands previous knowledge of its poles, accessible through

$$\Lambda_2\xi^2 - \Lambda_3\Lambda_4\xi - \Lambda_3 = 0. \quad (1.1089)$$

Application of the solving formula for a quadratic equation to Eq. (1.1089) unfolds

$$\xi_1, \xi_2 = \frac{\Lambda_3\Lambda_4 \pm \sqrt{(\Lambda_3\Lambda_4)^2 + 4\Lambda_2\Lambda_3}}{2\Lambda_2} \quad (1.1090)$$

– where ξ_1 and ξ_2 denote the two solutions (both real, and distinct from each other because $\sqrt{(\Lambda_3\Lambda_4)^2 + 4\Lambda_2\Lambda_3}$ exceeds $\sqrt{(\Lambda_3\Lambda_4)^2} = \Lambda_3\Lambda_4 > 0$), corresponding to either the minus or the plus sign preceding the square root; hence, one may factorize the denominator of the (regular) rational fraction in Eq. (1.1088) as

$$\frac{\Lambda_3\Lambda_4\xi+\Lambda_3}{\Lambda_2\xi^2-\Lambda_3\Lambda_4\xi-\Lambda_3}=\frac{\Lambda_3\Lambda_4\xi+\Lambda_3}{\Lambda_2\left(\xi-\xi_1\right)\left(\xi-\xi_2\right)}, \quad (1.1091)$$

or else

$$\frac{\Lambda_3\Lambda_4\xi+\Lambda_3}{\Lambda_2\xi^2-\Lambda_3\Lambda_4\xi-\Lambda_3}=\frac{\frac{\Lambda_3\Lambda_4}{\Lambda_2}\xi+\frac{\Lambda_3}{\Lambda_2}}{\left(\xi-\xi_1\right)\left(\xi-\xi_2\right)} \quad (1.1092)$$

upon lumping Λ_2 in denominator with the terms in numerator. Expansion of Eq. (1.1092) as partial fractions is finally feasible, viz.

$$\frac{\frac{\Lambda_3\Lambda_4}{\Lambda_2}\xi+\frac{\Lambda_3}{\Lambda_2}}{\left(\xi-\xi_1\right)\left(\xi-\xi_2\right)}=\frac{k_1}{\xi-\xi_1}+\frac{k_2}{\xi-\xi_2}, \quad (1.1093)$$

where k_1 and k_2 denote constants to be calculated. After collapsing fractions in the right-hand side, Eq. (1.1093) becomes

$$\frac{\frac{\Lambda_3\Lambda_4}{\Lambda_2}\xi+\frac{\Lambda_3}{\Lambda_2}}{\left(\xi-\xi_1\right)\left(\xi-\xi_2\right)}=\frac{k_1\left(\xi-\xi_2\right)+k_2\left(\xi-\xi_1\right)}{\left(\xi-\xi_1\right)\left(\xi-\xi_2\right)}, \quad (1.1094)$$

where equality of numerators of the two sides suffices in view of their common denominator, i.e.

$$\frac{\Lambda_3\Lambda_4}{\Lambda_2}\xi+\frac{\Lambda_3}{\Lambda_2}=k_1\left(\xi-\xi_2\right)+k_2\left(\xi-\xi_1\right); \quad (1.1095)$$

once parentheses are removed, Eq. (1.1095) turns to

$$\frac{\Lambda_3\Lambda_4}{\Lambda_2}\xi+\frac{\Lambda_3}{\Lambda_2}=k_1\xi-k_1\xi_2+k_2\xi-k_2\xi_1, \quad (1.1096)$$

which may be rephrased as

$$\frac{\Lambda_3\Lambda_4}{\Lambda_2}\xi+\frac{\Lambda_3}{\Lambda_2}=\left(k_1+k_2\right)\xi-\left(k_1\xi_2+k_2\xi_1\right) \quad (1.1097)$$

upon regrouping of similar terms. To guarantee validity of Eq. (1.1097), irrespective of the value taken by ξ, one must enforce

$$k_1+k_2=\frac{\Lambda_3\Lambda_4}{\Lambda_2}, \quad (1.1098)$$

coupled with

$$-\xi_2k_1-\xi_1k_2=\frac{\Lambda_3}{\Lambda_2} \quad (1.1099)$$

– as coefficients for the first and zero-th order terms in ξ, respectively, in both sides. Application of Cramer's rule to the set of Eqs. (1.1098) and (1.1099) is in order, according to

$$k_1=\frac{\begin{vmatrix}\frac{\Lambda_3\Lambda_4}{\Lambda_2} & 1\\ \frac{\Lambda_3}{\Lambda_2} & -\xi_1\end{vmatrix}}{\begin{vmatrix}1 & 1\\ -\xi_2 & -\xi_1\end{vmatrix}} \quad (1.1100)$$

and

$$k_2=\frac{\begin{vmatrix}1 & \frac{\Lambda_3\Lambda_4}{\Lambda_2}\\ -\xi_2 & \frac{\Lambda_3}{\Lambda_2}\end{vmatrix}}{\begin{vmatrix}1 & 1\\ -\xi_2 & -\xi_1\end{vmatrix}}. \quad (1.1101)$$

Recalling the definition of second-order determinant, Eq. (1.1000) may be rewritten as

$$k_1=\frac{\frac{\Lambda_3\Lambda_4}{\Lambda_2}\left(-\xi_1\right)-\frac{\Lambda_3}{\Lambda_2}1}{1\left(-\xi_1\right)-\left(-\xi_2\right)1}, \quad (1.1102)$$

which degenerates to

$$k_1=-\frac{\frac{\Lambda_3\Lambda_4}{\Lambda_2}\xi_1+\frac{\Lambda_3}{\Lambda_2}}{\xi_2-\xi_1}=-\frac{\Lambda_3}{\Lambda_2}\frac{\Lambda_4\xi_1+1}{\xi_2-\xi_1}; \quad (1.1103)$$

by the same token, Eq. (1.1101) gives rise to

$$k_2=\frac{1\frac{\Lambda_3}{\Lambda_2}-\left(-\xi_2\right)\frac{\Lambda_3\Lambda_4}{\Lambda_2}}{1\left(-\xi_1\right)-\left(-\xi_2\right)1}, \quad (1.1104)$$

where algebraic simplification leaves

$$k_2=\frac{\frac{\Lambda_3}{\Lambda_2}+\xi_2\frac{\Lambda_3\Lambda_4}{\Lambda_2}}{\xi_2-\xi_1}=\frac{\Lambda_3}{\Lambda_2}\frac{\Lambda_4\xi_2+1}{\xi_2-\xi_1}. \quad (1.1105)$$

One will now resort to Eqs. (1.1092) and (1.1093) to rewrite Eq. (1.1088) as

$$\int\left(1+\frac{k_1}{\xi-\xi_1}+\frac{k_2}{\xi-\xi_2}\right)d\xi=-\Lambda_2\int d\left(\Lambda_1t\right) \quad (1.1106)$$

after having multiplied both sides by $-\Lambda_2$, where the method of decomposition of an integral can be invoked to write

$$\int d\xi+k_1\int\frac{d\xi}{\xi-\xi_1}+k_2\int\frac{d\xi}{\xi-\xi_2}=-\Lambda_2\int d\left(\Lambda_1t\right); \quad (1.1107)$$

Eq. (1.1107) is equivalent to

$$\xi+k_1\ln\left\{\xi-\xi_1\right\}+k_2\ln\left\{\xi-\xi_2\right\}=\kappa-\Lambda_2\Lambda_1t, \quad (1.1108)$$

with κ serving as (arbitrary) integration constant. Equation (1.1108) cannot hold unless it satisfies

$$\boxed{\xi\big|_{\Lambda_1t=0}=1} \quad (1.1109)$$

as alias to Eq. (1.1064), written with the aid of Eq. (1.1080); this enforces

$$\kappa=1+k_1\ln\left\{1-\xi_1\right\}+k_2\ln\left\{1-\xi_2\right\} \quad (1.1110)$$

stemming from Eq. (1.1108). Insertion of Eq. (1.1110) transforms Eq. (1.1108) finally to

$$\Lambda_2\Lambda_1 t = 1-\xi + k_1 \ln\frac{1-\xi_1}{\xi-\xi_1} + k_2 \ln\frac{1-\xi_2}{\xi-\xi_2}, \qquad (1.1111)$$

where terms alike have meanwhile been collapsed; Eqs. (1.1090), (1.1103), and (1.1105) are now to be retrieved to eliminate k_1 and k_2, besides ξ_1 and ξ_2, at the expense of Λ_2, Λ_3, and Λ_4 – whereas the original notation can be recovered upon further combination with Eqs. (1.1076)–(1.1080). A graphical interpretation is conveyed by Fig. 1.88i; presence of a quadratic, rather than a linear dependence of freezing point depression upon solute molality produces a similar pattern for the $m_i/m_{w,0}$ vs. $h_aA_fT_{f,0}t/\lambda_f m_{w,0}$ curves – except with regard to their wider spread down when parameter $\alpha_w X_{s,0}/T_{f,0}$ increases, since $\alpha_w/\beta_w X_{s,0}$ magnifies the curvature (already) caused thereby.

In either case, the rate of heat withdrawal from the food, q_{frz}, may be obtained via the trivial enthalpic balance

$$q_{frz} = \lambda_f \frac{dm_i}{dt} \qquad (1.1112)$$

that mimics Eq. (1.935) – again after discarding sensible heat effects, for being negligible compared to latent heat in transit. One may, for convenience, rewrite Eq. (1.1112) as

$$q_{frz} = \frac{\lambda_f m_{w,0}}{\dfrac{\lambda_f m_{w,0}}{h_aA_fT_{f,0}}} \frac{d\left(\dfrac{m_i}{m_{w,0}}\right)}{d\left(\dfrac{h_aA_fT_{f,0}}{\lambda_f m_{w,0}}t\right)}, \qquad (1.1113)$$

applied after multiplying and dividing the numerator in the right-hand side by $m_{w,0}$; and the denominator likewise by $\lambda_f m_{w,0}/h_aA_fT_{f,0}$ – at the expense of the classical rules of differentiation. Equation (1.1113) reduces further to

$$q_{frz} = h_aA_fT_{f,0} \frac{d\left(\dfrac{m_i}{m_{w,0}}\right)}{d\left(\dfrac{h_aA_fT_{f,0}}{\lambda_f m_{w,0}}t\right)}, \qquad (1.1114)$$

or else

$$\frac{q_{frz}}{h_aA_fT_{f,0}} = \frac{d\left(\dfrac{m_i}{m_{w,0}}\right)}{d\left(\dfrac{h_aA_fT_{f,0}}{\lambda_f m_{w,0}}t\right)} \qquad (1.1115)$$

upon division of both sides by $h_aA_fT_{f,0}$. In the case of linear freezing point depression, one may resort to Eq. (1.1071), viewed as $\Phi\{m_i/m_{w,0}, h_aA_fT_{f,0}t/\lambda_f m_{w,0}\} = 0$, where Φ represents several algebraic operations, to obtain

$$\frac{d\left(\dfrac{m_i}{m_{w,0}}\right)}{d\left(\dfrac{h_aA_fT_{f,0}}{\lambda_f m_{w,0}}t\right)} = -\frac{-1}{\dfrac{1}{1-\dfrac{T_\infty}{T_{f,0}}}\left(1-\dfrac{\dfrac{\alpha_w X_{s,0}}{T_{f,0}} - \left(1-\dfrac{T_\infty}{T_{f,0}}\right)(-1)}{1-\dfrac{T_\infty}{T_{f,0}}\dfrac{\alpha_w X_{s,0}}{T_{f,0}} - \left(1-\dfrac{T_\infty}{T_{f,0}}\right)\left(1-\dfrac{m_i}{m_{w,0}}\right)}\right)} \qquad (1.1116)$$

upon implicit differentiation, since $d(m_i/m_{w,0})/d(h_aA_fT_{f,0}t/\lambda_f m_{w,0})$ equals the negative of the ratio of $\partial\Phi/\partial(h_aA_fT_{f,0}t/\lambda_f m_{w,0})$ to $\partial\Phi/\partial(m_i/m_{w,0})$; insertion in Eq. (1.1115), followed by elementary algebraic rearrangement, unfold

$$\frac{q_{frz}}{h_aA_fT_{f,0}} = \frac{1-\dfrac{T_\infty}{T_{f,0}}}{1-\dfrac{\dfrac{\alpha_w X_{s,0}}{T_{f,0}}}{\dfrac{\alpha_w X_{s,0}}{T_{f,0}} - \left(1-\dfrac{T_\infty}{T_{f,0}}\right)\left(1-\dfrac{m_i}{m_{w,0}}\right)}}, \qquad (1.1117)$$

as illustrated in Fig. 1.89ii. Note that multiplication of both numerator and denominator of the right-hand side of Eq. (1.1117) by $\alpha_w X_{s,0}/T_{f,0}-(1-T_\infty/T_{f,0})(1-m_i/m_{w,0})$, followed by cancellation of $\alpha_w X_{s,0}/T_{f,0}$ with its negative in denominator, and then of $1-T_\infty/T_{f,0}$ between numerator and denominator will eventually retrieve the negative of Eq. (1.1058) – as expected, since $d(1-m_i/m_{w,0}) = -dm_i$. In the absence of freezing point depression, i.e. $\alpha_w=0$, the rate of heat withdrawn remains obviously constant in time, as per the constancy of T_f in Eq. (1.1053); it decreases when $\alpha_w X_{s,0}/T_{f,0}$ increases, due to the larger impact of a change in solute molality upon the driving force for heat transfer. Note that q_{frz} decreases as time elapses; this arises from reduction in the driving force for heat transfer, see again Eq. (1.1053), that goes along with an increasing molality of solute in the food as ice separates out.

One may instead reformulate Eq. (1.1115) to

$$\frac{q_{frz}}{h_aA_fT_{f,0}} = \frac{d(1-\xi)}{d(\Lambda_1 t)} = -\frac{d\xi}{d(\Lambda_1 t)}, \qquad (1.1118)$$

with the aid of Eqs. (1.1076) and (1.1080); upon insertion of Eq. (1.1075), one gets

$$\frac{q_{frz}}{h_aA_fT_{f,0}} = \Lambda_2 + \frac{\Lambda_3\Lambda_4}{4}\Lambda_4 - \frac{\Lambda_3\Lambda_4}{4}\Lambda_4\left(1+\frac{2}{\Lambda_4\xi}\right)^2. \qquad (1.1119)$$

Here Λ_4^2 was factored out to make lumped parameter $\Lambda_3\Lambda_4 = \alpha_w X_{s,0}/T_{f,0}$, as per Eqs. (1.1078) and (1.1079), appear as such in Eq. (1.1119); with ξ obtainable, in turn, from Eq. (1.1111) – and applicable in the presence of quadratic freezing point depression. Equation (1.1119) is depicted in Fig. 1.89i. Note again the horizontal line (exceptionally) associated to $\alpha_w/\beta_w X_{s,0} \equiv \Lambda_4 = 0$, on account

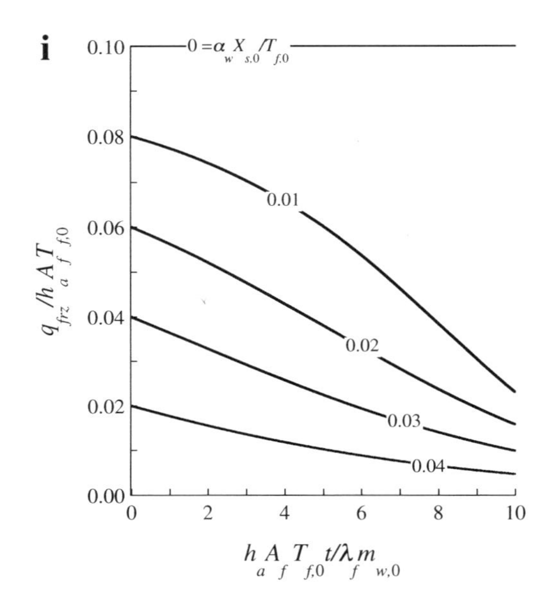

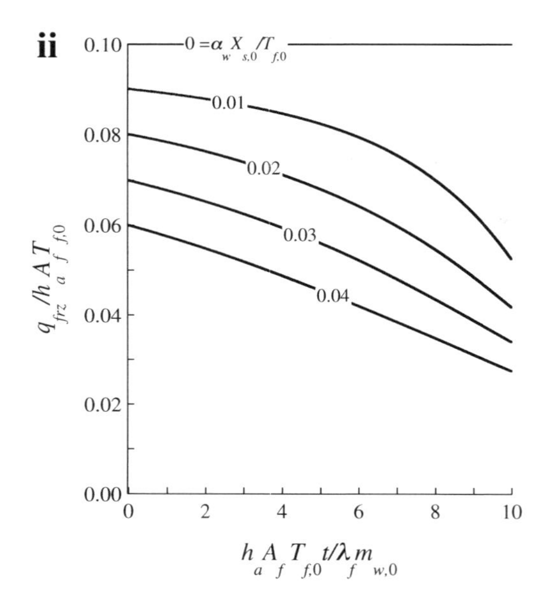

FIGURE 1.89 Evolution of dimensionless rate of heat withdrawn, $q_{frz}/h_aA_fT_{f,0}$, versus dimensionless time, $h_aA_fT_{f,0}t/\lambda_f m_{w,0}$ – assuming (i) quadratic freezing point depression for $\alpha_w/\beta_w X_{s,0}=1$ or (ii) linear freezing point depression with regard to solute molality, for $T_\infty/T_{f,0}=0.9$ and selected values of dimensionless parameter $\alpha_w X_{s,0}/T_{f,0}$, throughout slow freezing.

of $\alpha_w X_{s,0}/T_{f,0}=0$ – along with $\alpha_w/\beta_w X_{s,0}=1$ for hypothesis, in agreement with the unknown quantity arising from $\alpha_w=\beta_w=0$; coupled with an enhanced variation of heat withdrawn versus parameter $\alpha_w X_{s,0}/T_{f,0}$ at a given time, since $\alpha_w/\beta_w X_{s,0}\neq 0$ potentiates the effect of the former parameter.

The cumulative heat withdrawn, Q_{frz}, may likewise be calculated via

$$Q_{frz}=\lambda_f m_i \tag{1.1120}$$

– obtainable from Eq. (1.1112) via plain integration; division of both sides by $\lambda_f m_{w,0}$ leads promptly to

$$\frac{q_{frz}}{\lambda_f m_{w,0}}=\frac{m_i}{m_{w,0}}. \tag{1.1121}$$

If the freezing temperature decreases linearly with solute molality, then Eq. (1.1071) may be retrieved to transform Eq. (1.1121) to

$$\boxed{\Phi\left\{\frac{h_aA_fT_{f,0}}{\lambda_f m_{w,0}}t,\frac{m_i}{m_{w,0}};\frac{T_\infty}{T_{f,0}},\frac{\alpha_w X_{s,0}}{T_{f,0}}\right\}=0,} \tag{1.1122}$$

where the implicit nature of $\frac{m_i}{m_{w,0}}\equiv\frac{m_i}{m_{w,0}}\left\{\frac{h_aA_fT_{f,0}}{\lambda_f m_{w,0}}t\right\}$, through parameters $T_\infty/T_{f,0}$ and $\alpha_w X_{s,0}/T_{f,0}$, was again highlighted; a graphical interpretation of Eq. (1.1122) is provided in Fig. 1.90ii. As expected, the curves in this figure coincide with those in Fig. 1.88ii, since the heat received was hypothesized to be used in full to produce ice – as long as the vertical axis in the former uses $Q_{frz}/\lambda_f m_{w,0}$ as normalized coordinate, *in lieu* of $m_i/m_{w,0}$ as employed in the latter and duly linked by Eq. (1.1121).

If the full form of Eq. (1.874) is to be retained, then one should to advantage rephrase Eq. (1.1121) as

$$\frac{Q_{frz}}{\lambda_f m_{w,0}}=1-\left(1-\frac{m_i}{m_{w,0}}\right), \tag{1.1123}$$

which is equivalent to

$$\boxed{\frac{Q_{frz}}{\lambda_f m_{w,0}}=1-\xi} \tag{1.1124}$$

as per Eq. (1.1080); $\xi\{\Lambda_1 t\}$ is then to be recalled from Eq. (1.1111) – as plotted in Fig. 1.90i. Once again, one retrieves the curves already plotted in Fig. 1.88i, because combination of Eqs. (1.1080) and (1.1124) yields $Q_{frz}/\lambda_f m_{w,0}=1-(1-(m_i/m_{w,0}))=m_i/m_{w,0}$.

The behavior during thawing may be modeled via essentially the same mathematical approach – on the hypothesis that freezing can be reversed; this requirement is likely to be fulfilled, as chemical equilibrium was assumed to prevail during freezing since the very beginning. Under this type of process, m_i should be replaced by $m_{w,0}-m_i$, on account of $m_i|_{t=0}=m_{w,0}$ serving as initial condition instead of Eq. (1.1054), where m_i (still) represents the amount of (remaining) ice; and $T_{f,0}-T_f$ by $T_f-T_{f,0}$, with $T_{f,0}$ denoting original temperature of frozen food. Therefore, T_f will increase as m_i decreases, in agreement with Eq. (1.1042) for a quadratic change of freezing point with solute molality, or Eq. (1.1045) in the case of a linear relationship; and Fig. 1.87 will accordingly retain its usefulness. By the same token, one may resort to Eqs. (1.1071) or (1.1111), in attempts to relate fraction of remaining ice with current (equilibrium) temperature – for the linear and quadratic versions of Eq. (1.874), respectively; remember that temperature is fixed by solute molality in the newly formed aqueous solution, in each case. However, $T_\infty>T_{f,0}$ in this process, so as to assure heat transfer into the food – meaning that $T_\infty/T_{f,0}>1$, and thus $\Lambda_2\equiv 1-T_\infty/T_{f,0}<0$ as per Eq. (1.1077). This implies a decreasing m_i with elapsing t, as expected – consistent with Eq. (1.1071). By the same token, the vertical axis in Fig. 1.88 should now represent $1-m_i/m_{w,0}$ rather than $m_i/m_{w,0}$; and Figs. 1.89 and 1.90 will depict evolution with time of rate of heat transfer in, and cumulative heat supplied to the food, respectively.

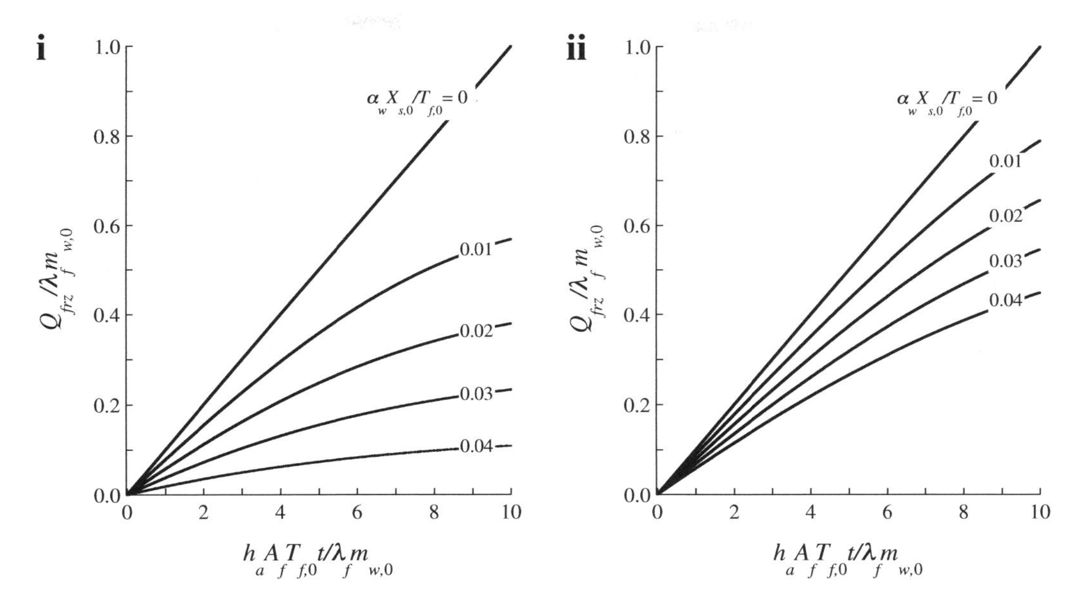

FIGURE 1.90 Evolution of normalized heat withdrawn, $Q_{frz}/\lambda_f m_{w,0}$, versus dimensionless time, $h_a A_f T_{f,0} t/\lambda_f m_{w,0}$ – assuming (i) quadratic freezing point depression for $\alpha_w/\beta_w X_{s,0}=1$, or (ii) linear freezing point depression with regard to solute molality, for $T_\infty/T_{f,0}=0.9$ and selected values of dimensionless parameter $\alpha_w X_{s,0}/T_{f,0}$, throughout slow freezing.

If the food matrix is prone to build up a temperature profile, then an enthalpy balance is to be applied to a layer of differential thickness dx and surface area A_f, located at position x, according to

$$-k_f A_f \left.\frac{\partial T_f}{\partial x}\right|_x + \rho_f A_f dx \lambda_f \chi_f \frac{\partial}{\partial t}\left(\frac{m_i}{m_{w,0}}\right) = -k_f A_f \left.\frac{\partial T_f}{\partial x}\right|_{x+dx} + \rho_f c_{P,f} A_f dx \frac{\partial T_f}{\partial t}. \tag{1.1125}$$

The first term in either side of Eq. (1.1125) represents rate of heat transfer by conduction – in or out, respectively; the second term in the left-hand side represents heat of freezing released – which plays the role of heat source inside the food matrix; and the last term represents rate of accumulation of enthalpy as sensible heat. Here k_f denotes thermal conductivity – assumed constant within and between unfrozen and frozen phases (for simplicity), and λ_f denotes (constant) latent heat of freezing per unit mass of ice formed; while χ_f denotes mass fraction of water in food – in either ice or liquid form (constant, when food is seen as a whole), ρ_f denotes (constant) mass density of food, $c_{P,f}$ denotes (constant) specific heat capacity of food, m_i denotes mass of ice formed, and t denotes time. Should Duhring's (linear) rule apply to describe freezing temperature vs. solute molality, then one may revisit Eq. (1.1045) as

$$\frac{m_i}{m_{w,0}} = 1 - \frac{\alpha_w X_{s,0}}{T_{f,0} - T_f}; \tag{1.1126}$$

Eq. (1.1126) conveys a direct relationship of mass fraction of ice, $m_i/m_{w,0}$, formed so far in each elementary volume, to freezing temperature prevailing therein, T_f, set by the local solute molality; no migration of solute throughout the food (via Fick's law) is postulated as well – even though $m_i/m_{w,0}$ will be a function of x (besides t). The time derivative of $m_i/m_{w,0}$ then looks like

$$\frac{\partial}{\partial t}\left(\frac{m_i}{m_{w,0}}\right) = -\frac{\alpha_w X_{s,0}}{\left(T_{f,0} - T_f\right)^2}\frac{\partial T_f}{\partial t}, \tag{1.1127}$$

departing from Eq. (1.1126); insertion of Eq. (1.1127) transforms Eq. (1.1125) to

$$-k_f \left.\frac{\partial T_f}{\partial x}\right|_x A_f - \rho_f \lambda_f \chi_f \frac{\alpha_w X_{s,0}}{\left(T_{f,0} - T_f\right)^2}\frac{\partial T_f}{\partial t} A_f dx = -k_f \left.\frac{\partial T_f}{\partial x}\right|_{x+dx} A_f + \rho_f c_{P,f} \frac{\partial T_f}{\partial t} A_f dx, \tag{1.1128}$$

where terms alike can be merged as

$$k_f\left(\left.\frac{\partial T_f}{\partial x}\right|_{x+dx} - \left.\frac{\partial T_f}{\partial x}\right|_x\right) A_f - \frac{\rho_f \lambda_f \chi_f \alpha_w X_{s,0}}{\left(T_{f,0} - T_f\right)^2}\frac{\partial T_f}{\partial t} A_f dx = \rho_f c_{P,f} \frac{\partial T_f}{\partial t} A_f dx. \tag{1.1129}$$

Division of both sides by $\rho_f c_{P,f} A_f dx$ converts Eq. (1.1129) to

$$\frac{k_f}{\rho_f c_{P,f}} \frac{\left.\frac{\partial T_f}{\partial x}\right|_{x+dx} - \left.\frac{\partial T_f}{\partial x}\right|_x}{dx} = \frac{\lambda_f \chi_f \alpha_w X_{s,0}}{c_{P,f}\left(T_{f,0} - T_f\right)^2}\frac{\partial T_f}{\partial t} + \frac{\partial T_f}{\partial t}; \tag{1.1130}$$

where the definition of (second-order) derivative, coupled with factoring out of $\partial T_f/\partial t$ support further transformation to

$$\boxed{\frac{k_f}{\rho_f c_{P,f}} \frac{\partial^2 T_f}{\partial x^2} = \left(1 + \frac{\lambda_f \chi_f \alpha_w X_{s,0}}{c_{P,f}\left(T_{f,0} - T_f\right)^2}\right)\frac{\partial T_f}{\partial t}.} \tag{1.1131}$$

Equation (1.1131) is (usually) subjected to

$$\left. T_f \right|_{t=0} = \left. T_f \right|_{X_s=X_{s,0}} = T_{in}, \tag{1.1132}$$

as initial condition – so $\alpha_w X_{s,0}$ in Eq. (1.1131) is sometimes replaced by $T_{f,0}-T_{in}$, see Eq. (1.1043) for $X_s=X_{s,0}$ after coupling with Eq. (1.1132); while

$$h_a\left(\left. T_f \right|_{x=L} - T_\infty\right) = -k_f \left.\frac{\partial T_f}{\partial x}\right|_{x=L}, \tag{1.1133}$$

with h_a denoting heat transfer coefficient of cold air flowing along the outer surface of the food, and

$$\left.\frac{\partial T_f}{\partial x}\right|_{x=0} = 0 \tag{1.1134}$$

serve (normally) as boundary conditions. Equation (1.1133) reflects identical rates of heat transfer out of the food (i.e. $k_f \partial T_f / \partial x$) and into the cooling utility (i.e. $h_a(T_f - T_\infty)$) on the top surface of the former (i.e. $x=L$); whereas insulation of the tray (i.e. $\partial T_f / \partial x = 0$), as per Fourier's law, should be satisfied on the bottom surface of the food (i.e. $x=0$). As done previously, it is convenient to reduce overparameterization – so Eqs. (3.1158) and (3.1159) of *Food Proc. Eng.: thermal & chemical operations* will be retrieved as definition of x^* and t^*, respectively; while an alias of Eq. (1.884), viz.

$$T_f^* \equiv \frac{T_f - T_\infty}{T_{in} - T_\infty}, \tag{1.1135}$$

is to be used as normalized freezing temperature. Under these circumstances, Eq. (1.1131) may be rewritten as

$$\frac{\partial^2 T_f^*}{\partial x^{*2}} = \left(1 + \frac{\dfrac{\alpha_w X_{s,0}}{T_{f,0}}}{\dfrac{c_{P,f} T_{f,0}}{\lambda_f \chi_f}\left(1 - \dfrac{T_f}{T_{f,0}}\right)^2}\right)\frac{\partial T_f^*}{\partial t^*}, \tag{1.1136}$$

where $T_{f,0}^2$ was factored out in denominator. After isolating T_f from Eq. (1.1135), one gets

$$T_f = T_\infty + \left(T_{in} - T_\infty\right) T_f^* \tag{1.1137}$$

– where negatives can be taken of both sides, and $T_{f,0}$ added thereto to obtain

$$T_{f,0} - T_f = T_{f,0} - T_\infty - \left(T_{in} - T_\infty\right) T_f^*; \tag{1.1138}$$

a final division of both sides by $T_{f,0}$ unfolds

$$1 - \frac{T_f}{T_{f,0}} = \frac{T_{f,0} - T_f}{T_{f,0}} = \frac{T_{f,0} - T_\infty - \left(T_{in} - T_\infty\right) T_f^*}{T_{f,0}}$$

$$= 1 - \frac{T_\infty}{T_{f,0}} - \left(\frac{T_{in}}{T_{f,0}} - \frac{T_\infty}{T_{f,0}}\right) T_f^*, \tag{1.1139}$$

which supports conversion of Eq. (1.1136) to

$$\frac{\partial^2 T_f^*}{\partial x^{*2}} = \left(1 + \frac{\dfrac{\alpha_w X_{s,0}}{T_{f,0}}}{\dfrac{c_{P,f} T_{f,0}}{\lambda_f \chi_f}\left(1 - \dfrac{T_\infty}{T_{f,0}} - \left(\dfrac{T_{in}}{T_{f,0}} - \dfrac{T_\infty}{T_{f,0}}\right) T_f^*\right)^2}\right)\frac{\partial T_f^*}{\partial t^*}. \tag{1.1140}$$

Besides $\alpha_w X_{s,0}/T_{f,0}$ and $T_\infty/T_{f,0}$ – ubiquitous in the previous analyses, two new dimensionless parameters appear, namely, $c_{P,f} T_{f,0}/\lambda_f \chi_f$ (or Stefan's number) as the ratio of sensible heat to latent heat (corrected by χ_f for portion that undergoes phase change); and $T_{in}/T_{f,0}$ as ratio of initial temperature to freezing point of plain water. By the same token, Eq. (1.1133) may be rephrased as

$$\frac{h_a L}{k_f} \left. T_f^* \right|_{x^*=1} = -\left.\frac{\partial T_f^*}{\partial x^*}\right|_{x^*=1} \tag{1.1141}$$

with the aid of Eq. (1.1135), complemented by Eq. (3.1158) in *Food Proc. Eng.: thermal & chemical operations* – where $h_a L/k_f$ represents Biot's number for heat transfer; and Eq. (1.1134) likewise becomes

$$\left.\frac{\partial T_f^*}{\partial x^*}\right|_{x^*=0} = 0. \tag{1.1142}$$

Finally, Eq. (1.1132) takes the form

$$\left. T_f^* \right|_{t^*=0} = 1, \tag{1.1143}$$

at the expense of Eq. (1.1135), coupled with Eq. (3.1159) of *Food Proc. Eng.: thermal & chemical operations*. Although sufficiently accurate, Eq. (1.1136) is a nonlinear, partial differential equation that cannot unfortunately be integrated – unless via some powerful numerical method as tool; the complexity of Eq. (1.1141), standing as boundary condition, also contributes to make the problem harder to solve.

For slow freezing in the presence of a non-uniform temperature profile, the rate of heat transfer is given by

$$q_{frz} = -k_f A_f \left.\frac{\partial T_f}{\partial x}\right|_{x=L}; \tag{1.1144}$$

it can be redone to

$$q_{frz} = -k_f A_f \frac{T_{in} - T_\infty}{L} \left.\frac{\partial T_f^*}{\partial x^*}\right|_{x^*=1} \tag{1.1145}$$

with the aid of Eq. (1.1135), as well as Eq. (3.1158) of *Food Proc. Eng.: thermal & chemical operations* – or, equivalently,

$$\frac{q_{frz}}{k_f A_f \dfrac{T_{in} - T_\infty}{L}} = -\left.\frac{\partial T_f^*}{\partial x^*}\right|_{x^*=1}, \tag{1.1146}$$

after dividing both sides by $k_f A_f (T_{in} - T_\infty)/L$. In view of Eq. (1.1133), q_{frz} may alternatively be calculated via

$$q_{frz} = h_a A_f \left(\left. T_f \right|_{x=L} - T_\infty\right) \tag{1.1147}$$

– where division of both sides by $h_aA_f(T_{in}-T_\infty)$ gives rise to

$$\frac{q_{frz}}{h_aA_f\left(T_{in}-T_\infty\right)}=T_f^*\Big|_{x^*=1}, \tag{1.1148}$$

at the expense again of Eq. (1.1135), complemented by Eq. (3.1158) of *Food Proc. Eng.: thermal & chemical operations*; the cumulative amount of heat transferred becomes then accessible as

$$Q_{frz}=\int_0^t q_{frz}\,d\tilde{t}, \tag{1.1149}$$

or else

$$\boxed{\frac{Q_{frz}}{\rho_f c_{P,f} LA_f\left(T_{in}-T_\infty\right)}=\int_0^{\frac{k_f t}{\rho_f c_{P,f}L^2}}\frac{q_{frz}}{k_fA_f\dfrac{T_{in}-T_\infty}{L}}\,d\left(\frac{k_f}{\rho_f c_{P,f}L^2}\tilde{t}\right)} \tag{1.1150}$$

coupled with Eq. (1.1146), which conveys $\dfrac{q_{frz}}{k_fA_f\dfrac{T_{in}-T_\infty}{L}}$ as a function of $k_f t/\rho_f c_{P,f}L^2$ suitable for integration – and after multiplication of both sides by $k_f/\rho_f c_{P,f}L^2$, followed by division by $k_fA_f(T_{in}-T_\infty)/L$. Unlike done in Eq. (1.944) pertaining to fast freezing, or Eq. (1.1124) pertaining to slow freezing, λ_f cannot here replace $c_{P,f}(T_{in}-T_\infty)$ as normalizing factor for Q_{frz} – because both latent heat and sensible heat (rather than just the former) are at stake.

If λ_f were nil, then Eq. (1.1131) would degenerate to Eq. (2.525) coupled to Eq. (1.7), both from *Food Proc. Eng.: thermal & chemical operations* – valid for cooling of a sab-shaped food in the absence of phase change; the term $\lambda_f\chi_f\alpha_w X_{s,0}/c_{P,f}(T_{f,0}-T_f)^2>0$ accordingly acts as a magnifying factor for the time derivative in the right-hand side of Eq. (1.1131), so temperature does not drop so fast as in the absence of freezing – due to the continuous, internal generation of heat derived from freezing of otherwise liquid water.

As previously discussed, thawing may be regarded as the reverse of freezing – so the same enthalpic balance, i.e. Eq. (1.1140), applies, along with similar boundary conditions, i.e. Eqs. (1.1141) and (1.1142), and initial condition, i.e. Eq. (1.1143). However, the relative magnitude of the temperatures is reversed, i.e. $T_\infty>T_f>T_{in}$ (with $T_\infty>T_{f,0}>T_f$ usually) – with $T_f^*\equiv(T_\infty-T_f)/(T_\infty-T_{in})$ being mathematically equivalent to Eq. (1.1135), but physically more sound for being the ratio of two positive quantities. One may similarly resort to Eq. (1.1146) as a tool to estimate the rate of heat transfer throughout thawing – using $T_\infty-T_{in}$ in the left-hand side, so as to get a positive heating rate; as well as Eq. (1.1150) to ascertain cumulative heat of thawing absorbed, again with $T_\infty-T_{in}$ in the denominator of both left-hand side and kernel of integral in right-hand side.

PROBLEM 1.20

Thawing at a given melting temperature entails a phase transition of water from solid to liquid form – which (as explored before) may be described by a partial differential equation in the case of a nonuniform temperature profile; such a phase transition should, in general, produce an interface that moves with time, and separates two homogeneous media. From a mathematical point of view, a phase is merely a region of space in which the solution of a partial differential equation is continuous and differentiable up to its order; in view of the moving boundary (or interface) as an infinitesimally thin surface that separates adjacent phases, solution of the underlying partial differential equation (and its derivatives) may exhibit a discontinuity across the interface. Therefore, the aforementioned partial differential equation is not valid at an interface consubstantiating a phase change – and an additional condition, also known as Stefan's condition, is needed to obtain closure. The said condition expresses the local velocity of the moving boundary as a function of quantities evaluated at both its sides; and is usually derived from a physical constraint – which, in the case of heat transfer, is but the equation of conservation of enthalpy, with such a velocity depending on the magnitude of the heat flux discontinuity at the interface.

a) Describe mathematically the slow, isothermal thawing of a semi-infinite, slab-shaped food in the presence of a melting interface, using a continuous approach in spatial coordinate x and time coordinate t – and ignoring any volume change upon melting.
b) After introducing a similarity variable ξ, defined as

$$\xi\equiv\frac{x}{\sqrt{t}}, \tag{1.20.1}$$

seek a solution of the form

$$T^*\equiv\Xi\{\xi\}, \tag{1.20.2}$$

and then recover the original notation.

Solution

a) The enthalpy balance, in terms of sensible heat, within the liquid region characterized by $0\le x<s\{t\}$ – where $s\{t\}$ denotes location of thawing front, may be set as

$$\frac{\partial T}{\partial t}=\frac{k}{\rho c_P}\frac{\partial^2 T}{\partial x^2} \tag{1.20.3}$$

in parallel to Eqs. (1.7) and (2.525) in *Food Proc. Eng.: thermal & chemical operations*; insertion of

$$T^*\equiv\frac{T-T_i}{T^\sigma-T_i} \tag{1.20.4}$$

permits transformation of Eq. (1.20.3) to

$$\frac{\partial T^*}{\partial t} = \frac{k}{\rho c_P} \frac{\partial^2 T^*}{\partial x^2}, \tag{1.20.5}$$

since $d((T-T_i)/(T^\sigma-T_i)) = d(T-T_i)/(T^\sigma-T_i) = dT/(T^\sigma-T_i)$ and $d^2((T-T_i)/(T^\sigma-T_i)) = d(d((T-T_i)/(T^\sigma-T_i))) = d(dT/(T^\sigma-T_i)) = d^2T/(T^\sigma-T_i)$, thus allowing the reciprocal of $T^\sigma-T_i$ drop off both sides; here T^σ denotes melting point and T_i denotes initial temperature of the frozen food, k denotes thermal conductivity of the thawed food, and ρ denotes mass density thereof. A reasonable boundary condition reads

$$T^*\Big|_{x=0} = T_0^*, \tag{1.20.6}$$

with $T_0^* > 1$ so as to provide a driving force for heat transfer associated with melting; while the initial condition will necessarily be phrased as

$$T^*\Big|_{t=0} = 0. \tag{1.20.7}$$

The other boundary condition, valid at $x=s\{t\}$, takes the form

$$\lambda \rho A \frac{ds}{dt} = -kA\left(T^\sigma - T_i\right)\frac{\partial T^*}{\partial x}\bigg|_{x=s\{t\}} \tag{1.20.8}$$

– as per Stefan's condition and consistent with Eq. (1.20.4), where A denotes cross-sectional area of slab and λ denotes latent heat of melting; after division of both sides by $A(T^\sigma-T_i)$, Eq. (1.20.8) becomes

$$\frac{\lambda \rho}{T^\sigma - T_i} \frac{ds}{dt} = -k \frac{\partial T^*}{\partial x}\bigg|_{x=s\{t\}}. \tag{1.20.9}$$

The initial position of the melting interface abides to

$$s\Big|_{t=0} = 0, \tag{1.20.10}$$

whereas Dirichlet's condition at the interface proper reads

$$T^*\Big|_{x=s\{t\}} = 1 \tag{1.20.11}$$

in view of the freezing temperature, T^σ, prevailing thereon. With regard to the solid region, corresponding to $s\{t\} < x < \infty$, one is faced with

$$T^* = 0 \tag{1.20.12}$$

due to prevalence of the initial temperature T_i, consistent with the semi-infinite assumption – so only the thawed portion of the food will hereafter be considered, given the trivial temperature profile prevailing in its frozen portion.

b) Partial differentiation of both sides of Eq. (1.20.2), with regard to t, unfolds

$$\frac{\partial T^*}{\partial t} = \frac{d\Xi}{d\xi} \frac{\partial \xi}{\partial t} \tag{1.20.13}$$

via the chain rule, which becomes

$$\frac{\partial T^*}{\partial t} = \frac{d\Xi}{d\xi} x \left(-\frac{1}{2}\right) t^{-\frac{3}{2}} = -\frac{x}{2t\sqrt{t}} \frac{d\Xi}{d\xi} \tag{1.20.14}$$

after recalling Eq. (1.20.1). By the same token, one obtains

$$\frac{\partial T^*}{\partial x} = \frac{d\Xi}{d\xi} \frac{\partial \xi}{\partial x} \tag{1.20.15}$$

from Eq. (1.20.2), upon partial differentiation with regard to x – where Eq. (1.20.1) again supports transformation to

$$\frac{\partial T^*}{\partial x} = \frac{1}{\sqrt{t}} \frac{d\Xi}{d\xi}; \tag{1.20.16}$$

a further differentiation of Eq. (1.20.16), with regard to x, gives rise to

$$\frac{\partial^2 T^*}{\partial x^2} \equiv \frac{\partial}{\partial x}\left(\frac{1}{\sqrt{t}} \frac{d\Xi}{d\xi}\right) = \frac{1}{\sqrt{t}} \frac{d}{d\xi}\left(\frac{d\Xi}{d\xi}\right)\frac{\partial \xi}{\partial x}, \tag{1.20.17}$$

once more with the aid of the chain differentiation rule – where Eq. (1.20.1) may be invoked once more to write

$$\frac{\partial^2 T^*}{\partial x^2} = \frac{1}{\sqrt{t}} \frac{d^2\Xi}{d\xi^2} \frac{1}{\sqrt{t}} = \frac{1}{t} \frac{d^2\Xi}{d\xi^2}, \tag{1.20.18}$$

along with condensation of factors alike. Insertion of Eqs. (1.20.14) and (1.20.18) transforms Eq. (1.20.5) to

$$-\frac{x}{2t\sqrt{t}} \frac{d\Xi}{d\xi} = \frac{k}{\rho c_P} \frac{1}{t} \frac{d^2\Xi}{d\xi^2}, \tag{1.20.19}$$

which reduces to

$$2\frac{k}{\rho c_P} \frac{d^2\Xi}{d\xi^2} + \frac{x}{\sqrt{t}} \frac{d\Xi}{d\xi} = 0 \tag{1.20.20}$$

after multiplying both sides by $2t$; insertion of Eq. (1.20.1) supports simplification to

$$\frac{d^2\Xi}{d\xi^2} + \frac{\xi}{\eta} \frac{d\Xi}{d\xi} = 0, \tag{1.20.21}$$

along with division of both sides by η – defined as

$$\eta \equiv 2\frac{k}{\rho c_P}. \tag{1.20.22}$$

An attempt to solve Eq. (1.20.21) may proceed via definition of an integrating factor, $I\{\xi\}$, abiding to

$$I\{\xi\} \equiv \exp\left\{\int_0^\xi \frac{\zeta}{\eta} d\zeta\right\} = \exp\left\{\frac{1}{\eta}\int_0^\xi \zeta \, d\zeta\right\}, \tag{1.20.23}$$

where application of the fundamental theorem of integral calculus unfolds

$$I\{\xi\} = \exp\left\{k_0 + \frac{1}{\eta} \frac{\zeta^2}{2}\bigg|_0^\xi\right\} = \exp\left\{k_0 + \frac{1}{\eta}\frac{\xi^2}{2}\right\}$$

$$= e^{k_0} \exp\left\{\frac{\xi^2}{2\eta}\right\} = k_1 \exp\left\{\frac{\xi^2}{2\eta}\right\}, \tag{1.20.24}$$

with k_0 and $k_1 \equiv e^{k_0}$ denoting integration constants. Multiplication of Eq. (1.20.21) by $I\{\xi\}$ produces

$$I\{\xi\}\frac{d^2\Xi}{d\xi^2}+\frac{\xi}{\eta}I\{\xi\}\frac{d\Xi}{d\xi}=0 \tag{1.20.25}$$

– where

$$I\{\xi\}\frac{d^2\Xi}{d\xi^2}+\frac{\xi}{\eta}I\{\xi\}\frac{d\Xi}{d\xi}=\frac{d}{d\xi}\left(I\{\xi\}\frac{d\Xi}{d\xi}\right) \tag{1.20.26}$$

is to be enforced, so as to justify definition of $I\{\xi\}$ in the first place; under these circumstances, one can rewrite Eq. (1.20.25) as

$$\frac{d}{d\xi}\left(I\{\xi\}\frac{d\Xi}{d\xi}\right)=0. \tag{1.20.27}$$

Application of the rule of differentiation of a product converts Eq. (1.20.26) to

$$I\frac{d^2\Xi}{d\xi^2}+\frac{\xi}{\eta}I\frac{d\Xi}{d\xi}=\frac{dI}{d\xi}\frac{d\Xi}{d\xi}+I\frac{d^2\Xi}{d\xi^2}, \tag{1.20.28}$$

where cancellation of terms alike between sides permits simplification to

$$\frac{\xi}{\eta}I\frac{d\Xi}{d\xi}=\frac{dI}{d\xi}\frac{d\Xi}{d\xi}; \tag{1.20.29}$$

a further dropping of similar factors from both sides permits extra simplification to

$$\frac{\xi}{\eta}I=\frac{dI}{d\xi}. \tag{1.20.30}$$

Integration via separation of variables is then feasible – and converts Eq. (1.20.30) to

$$\int \xi d\xi=\eta\int\frac{dI}{I}, \tag{1.20.31}$$

or else

$$\frac{\xi^2}{2}=k_2+\eta\ln I \tag{1.20.32}$$

– with k_2 denoting another integration constant; isolation of $I\{\xi\}$ in Eq. (1.20.32) yields

$$I\{\xi\}=\exp\left\{\frac{\frac{\xi^2}{2}-k_2}{\eta}\right\}=\exp\left\{-\frac{k_2}{\eta}\right\}\exp\left\{\frac{\xi^2}{2\eta}\right\}\equiv k_3\exp\left\{\frac{\xi^2}{2\eta}\right\}, \tag{1.20.33}$$

where k_3 denotes still another constant, meant to be used *in lieu* of k_2 and equal to $e^{-k_2/\eta}$. Equation (1.20.27) may now be (trivially) integrated to

$$I\{\xi\}\frac{d\Xi}{d\xi}=k_4, \tag{1.20.34}$$

provided that k_4 denotes again a constant; a second integration of Eq. (1.20.34) gives rise to

$$\int d\Xi=k_4\int\frac{d\xi}{I\{\xi\}}, \tag{1.20.35}$$

which is equivalent to

$$\Xi=k_5+k_4\int_0^{\xi}\frac{d\zeta}{I\{\zeta\}} \tag{1.20.36}$$

– where constant k_5 hinges upon the lower limit of integration displayed, arbitrarily (yet conveniently) selected in Eq. (1.20.36). Upon insertion of Eq. (1.20.33), one obtains

$$\Xi=k_5+k_4\int_0^{\xi}\frac{d\zeta}{k_3\exp\left\{\frac{\zeta^2}{2\eta}\right\}}=k_5+\frac{k_4}{k_3}\int_0^{\xi}\frac{d\zeta}{\exp\left\{\left(\frac{\zeta}{\sqrt{2\eta}}\right)^2\right\}} \tag{1.20.37}$$

from Eq. (1.20.36), which may be rewritten as

$$\Xi\{\xi\}=k_5+k_6\frac{2}{\sqrt{\pi}}\int_0^{\xi}\exp\left\{-\left(\frac{\zeta}{\sqrt{2\eta}}\right)^2\right\}d\zeta \tag{1.20.38}$$

– with k_6 denoting another constant, given by the ratio of k_4 to k_3 and further multiplied by $\sqrt{\pi}/2$; Eq. (1.20.38) will finally be reformulated to

$$\begin{aligned}T^*\{\xi\}&=k_5+k_6\sqrt{2\eta}\,\frac{2}{\sqrt{\pi}}\int_0^{\frac{\xi}{\sqrt{2\eta}}}\exp\left\{-\left(\frac{\zeta}{\sqrt{2\eta}}\right)^2\right\}d\left(\frac{\zeta}{\sqrt{2\eta}}\right)\\&=k_5+k_7\,\mathrm{erf}\left\{\frac{\xi}{\sqrt{2\eta}}\right\}\end{aligned} \tag{1.20.39}$$

in view of Eq. (1.202), and after multiplying and dividing the kernel by $\sqrt{2\eta}$ – with erf denoting error function, and Eq. (1.20.22) turning $k_7 \equiv \sqrt{2\eta}k_6$ into another constant. Equation (1.20.6) may now be recoined as

$$T^*\big|_{\xi=0}=T_0^*, \tag{1.20.40}$$

after bringing Eq. (19.20.1) on board; compatibility between Eqs. (1.20.39) and (1.20.40) enforces

$$T_0^*=k_5+k_7\,\mathrm{erf}\{0\}, \tag{1.20.41}$$

which breaks down to merely

$$k_5=T_0^* \tag{1.20.42}$$

because of the nil value of erf{0} – so Eq. (1.20.39) becomes

$$T^*\{\xi\} = T_0^* + k_7 \operatorname{erf}\left\{\frac{\xi}{\sqrt{2\eta}}\right\}, \tag{1.20.43}$$

following combination with Eq. (1.20.42). By the same token, Eq. (1.20.11) may be rewritten as

$$T^*\Big|_{\xi=\frac{s\{t\}}{\sqrt{t}}} = 0, \tag{1.20.44}$$

again at the expense of Eq. (1.20.1); insertion of Eq. (1.20.44) transforms Eq. (1.20.43) to

$$0 = T_0^* + k_7 \operatorname{erf}\left\{\frac{\frac{s\{t\}}{\sqrt{t}}}{\sqrt{2\eta}}\right\}, \tag{1.20.45}$$

where k_7 can be isolated as

$$k_7 = -\frac{T_0^*}{\operatorname{erf}\{\chi\}} \tag{1.20.46}$$

with χ defined as

$$\chi \equiv \frac{s\{t\}}{\sqrt{2\eta t}}. \tag{1.20.47}$$

Since k_7 in Eq. (1.20.39) was defined as a constant, χ must also be a constant by virtue of Eq. (1.20.46) – so one may rewrite Eq. (1.20.47) as

$$s\{t\} = \chi\sqrt{2\eta t}; \tag{1.20.48}$$

the derivative of $s\{t\}$ with regard to t will accordingly look like

$$\frac{ds}{dt} = \chi\frac{d\sqrt{2\eta t}}{dt} = \chi\frac{1}{2}(2\eta t)^{-\frac{1}{2}}2\eta = \frac{2\eta\chi}{2\sqrt{2\eta t}} = \chi\sqrt{\frac{\eta}{2t}}, \tag{1.20.49}$$

at the expense of the chain differentiation rule – valid as long as the rate of progression of the melting front is slow compared to the rate of heat transfer, i.e. assuming a pseudo-steady state. In view of Eqs. (1.20.46) and (1.20.47), one will redo Eq. (1.20.43) to

$$T^*\{\xi,t\} = T_0^* - \frac{T_0^*}{\operatorname{erf}\left\{\frac{s\{t\}}{\sqrt{2\eta t}}\right\}}\operatorname{erf}\left\{\frac{\xi}{\sqrt{2\eta}}\right\} = T_0^*\left(1 - \frac{\operatorname{erf}\left\{\frac{\xi}{\sqrt{2\eta}}\right\}}{\operatorname{erf}\left\{\frac{s\{t\}}{\sqrt{2\eta t}}\right\}}\right), \tag{1.20.50}$$

where T_0^* was meanwhile factored out; the derivative of both sides of Eq. (1.20.50), with regard to x, reads

$$\frac{\partial T^*}{\partial x} = -\frac{T_0^*}{\operatorname{erf}\left\{\frac{s\{t\}}{\sqrt{2\eta t}}\right\}}\frac{\partial}{\partial x}\left(\frac{2}{\sqrt{\pi}}\int_0^{\frac{\xi}{\sqrt{2\eta}}} e^{-\varphi^2}\,d\varphi\right) = -\frac{T_0^*}{\operatorname{erf}\{\chi\}}\frac{2}{\sqrt{\pi}}\frac{d}{d\xi}\left(\int_0^{\frac{\xi}{\sqrt{2\eta}}} e^{-\varphi^2}\,d\varphi\right)\frac{\partial\xi}{\partial x} \tag{1.20.51}$$

at the expense of the rules of differentiation of a product and of a composite function, as well as the definition of error function, complemented by Eqs. (1.20.39) and (1.20.47). The rule of differentiation of an integral with regard to a variable that only appears as argument in the upper limit of integration gives rise to

$$\frac{\partial T^*}{\partial x} = -\frac{2T_0^*}{\sqrt{\pi}\operatorname{erf}\{\chi\}}\exp\left\{-\left(\frac{\xi}{\sqrt{2\eta}}\right)^2\right\}\frac{1}{\sqrt{2\eta}}\frac{1}{\sqrt{t}} = -\frac{2T_0^*\exp\left\{-\left(\frac{\frac{x}{\sqrt{t}}}{\sqrt{2\eta}}\right)^2\right\}}{\sqrt{2\pi\eta t}\operatorname{erf}\{\chi\}} = -\sqrt{\frac{2}{\pi\eta t}}T_0^*\frac{\exp\left\{-\left(\frac{x}{\sqrt{2\eta t}}\right)^2\right\}}{\operatorname{erf}\{\chi\}}, \tag{1.20.52}$$

stemming from Eq. (1.20.51) – also with the aid of Eq. (1.20.1). At $x=s\{t\}$, Eq. (1.20.52) becomes

$$\left.\frac{\partial T^*}{\partial x}\right|_{x=s\{t\}} = -\sqrt{\frac{2}{\pi\eta t}}T_0^*\frac{\exp\left\{-\left(\frac{s\{t\}}{\sqrt{2\eta t}}\right)^2\right\}}{\operatorname{erf}\{\chi\}} = -\sqrt{\frac{2}{\pi\eta t}}T_0^*\frac{e^{-\chi^2}}{\operatorname{erf}\{\chi\}}, \tag{1.20.53}$$

with the aid again of Eq. (1.20.47); insertion of Eqs. (1.20.49) and (1.20.53) transforms Eq. (1.20.9) to

$$\frac{\lambda\rho}{T^\sigma - T_i}\chi\sqrt{\frac{\eta}{2t}} = k\sqrt{\frac{2}{\pi\eta t}}T_0^*\frac{e^{-\chi^2}}{\operatorname{erf}\{\chi\}}, \tag{1.20.54}$$

where condensation of factors alike and isolation of factors in χ unfold

$$\chi e^{\chi^2}\operatorname{erf}\{\chi\} = \frac{2}{\sqrt{\pi}}\frac{k\left(T^\sigma - T_i\right)T_0^*}{\lambda\rho\eta}; \tag{1.20.55}$$

based on Eq. (1.20.22), one will redo Eq. (1.20.55) to

$$\chi e^{\chi^2} \operatorname{erf}\{\chi\} = \frac{2}{\sqrt{\pi}} \frac{k\left(T^{\sigma}-T_i\right)T_0^*}{\lambda\rho 2\dfrac{k}{\rho c_P}}, \quad (1.20.56)$$

where common factors may drop off numerator and denominator to yield

$$\chi e^{\chi^2} \operatorname{erf}\{\chi\} = \frac{1}{\sqrt{\pi}} \frac{c_P\left(T^{\sigma}-T_i\right)T_0^*}{\lambda}. \quad (1.20.57)$$

Equation (1.20.57) can appear in a more condensed manner as

$$\chi e^{\chi^2} \operatorname{erf}\{\chi\} = \frac{StT_0^*}{\sqrt{\pi}}, \quad (1.20.58)$$

with St denoting Stefan's number – defined as

$$St \equiv \frac{c_P\left(T^{\sigma}-T_i\right)}{\lambda}, \quad (1.20.59)$$

i.e. the ratio of sensible to latent heat. Once (algebraic, yet transcendental) Eq. (1.20.58) is solved for (constant) χ, one may obtain $s\{t\}$ via Eqs. (1.20.22) and (1.20.48), i.e.

$$s\{t\} = \chi\sqrt{2\frac{2k}{\rho c_P}t} = 2\chi\sqrt{\frac{k}{\rho c_P}}\sqrt{t}; \quad (1.20.60)$$

Eq. (1.20.60) indicates that the thawing front progresses with the square root of time. Once in possession of Eq. (1.20.60), one will obtain T^* as

$$T^*\{x,t\} = T_0^*\left(1-\frac{\operatorname{erf}\left\{\dfrac{\dfrac{x}{\sqrt{t}}}{2\sqrt{\dfrac{k}{\rho c_P}}}\right\}}{\operatorname{erf}\{\chi\}}\right) = T_0^*\left(1-\frac{\operatorname{erf}\left\{\dfrac{x}{2}\sqrt{\dfrac{\rho c_P}{kt}}\right\}}{\operatorname{erf}\{\chi\}}\right) \quad (1.20.61)$$

from Eq. (1.20.50), with the aid again of Eqs. (1.20.1), (1.20.22), and (1.20.47) – as descriptor of the temperature profile, in dimensioness form, prevailing in the portion of food already thawed; and valid for $0 \leq x \leq s\{t\}$, with $s\{t\}$ sliding in time as per Eq. (1.20.60).

1.5 Irradiation

1.5.1 Introduction

Ionizing radiation – namely γ-rays from isotopes, or from electrons or X-rays to a much lesser extent, can be used to destroy pathogenic or spoilage microorganisms, and disinfest raw materials; as well as slow down germination, ripening, and sprouting. Hence, it can be employed to preserve foods, in alternative to classical heat-based techniques – and without the need for physical contact with the source. Irradiation is obviously unable to revert spoiled or over-ripened materials to a fresher condition, destroy (most) microbial toxins once synthesized/released, or repair texture, color, or taste of a food.

The radiation source supplies energetic particles, or waves (depending on the mechanistic point of view); as they pass through the food matrix, they collide with its molecules, so chemical bonds are broken around the sites of such collisions to eventually generate charged species – thus justifying being called ionizing radiation. The associated energies are of the order of MeV – with 1 eV defined as the energy gained by a single electron accelerating across an electrical potential difference of 1 V; this is equivalent to 1.602176×10^{-9} J.

HISTORICAL FRAMEWORK

Rationalization of ionizing radiation got its start in 1895, when Wilhelm C. Rontgen discovered that a portion of light generated in a vacuum, or Crook's tube, was able to penetrate through an opaque screen – and eventually react with a barium solution beneath. After taking a photo of his wife's hand skeletal structure, he coined the term X-rays temporarily as a designation of something unknown – yet the said name stuck ever after. In early 1896, H. L. Smith came up with an X-ray machine, but a fully functioning unit was not introduced until the 1904 World's Fair, by the hand of Clarence Dally.

Still in 1896, Henri Becquerel found that uranium salts gave off similar rays spontaneously; however, it was his doctoral student, Marie Curie, who is credited for the term radioactivity. She would indeed develop most pioneering work with radioactive materials – for which she shared the Nobel Prize in Physics with H. Becquerel; she was awarded years later also with the Nobel Prize in Chemistry, for her discovery specifically of radium and polonium. The suggestion to use ionizing energy to destroy pathogenic and spoilage microorganisms in food was conveyed by a German medical journal, soon after Becquerel's report; and Samuel Prescott described the bactericidal effects of radiation in 1904. The first patents came up in the early 1900s in the United Kingdom and the United States – namely, use of radioactive isotopes for irradiation in a flowing bed in 1906 by Appleby and Banks, and use of X-rays for food preservation in 1918 by Gillet, respectively.

The modern era of food irradiation began in 1951, with the wide research and development program coordinated by the US Atomic Energy Commission – after spent fuel rods from nuclear reactors started being generated. The first country to grant a clearance for human consumption of irradiated foods was the (former) Soviet Union in 1958 – viz. irradiated potatoes to prevent sprouting, and irradiated grain to prevent insect infestation one year later; irradiation of wheat flour was legalized in the United States by 1963. At present, food irradiation is legally permitted in more than 60 countries – not all of which possess processing plants, though; more than 250 food products are commercially available in irradiated form in more than 30 countries, adding up to ca. 500,000 $ton.yr^{-1}$ in total.

The rules that dictate how food should be irradiated vary greatly from country to country. The EU authorizes irradiation of only herbs and spices, seven categories of foods are allowed in the United Kingdom, a wide range of products in the Netherlands, France, and Belgium (including freshwater shrimp, frog legs, and deboned chicken), and all foods at any dose in Brazil. Irradiation is permitted in the United States for disinfestation and treatment of fresh produce and minced beef, chiefly to help reduce incidence of *E. coli* O157:H7.

Generation of a stream of electrons resorts to acceleration via an electric voltage or radio frequency wave, to a velocity close to that of light; upon interaction with the atoms of molecules in food, high-energy electrons produce a cascade of electrons moving through the target material – where they are free to interact with the molecules of the material, eject (other) electrons, and generate free radicals. Because electrons possess mass and electric charge, penetration into the food product cannot extend beyond a few centimeters – depending on product density.

Conversely, γ-rays possess a high penetration capability, and are thus the most widespread in industrial applications. Photons within this wavelength range, see Fig. 2.6 in *Food Proc. Eng.: thermal & chemical operations*, are also able to eject electrons, and thus generate electrically charged ions or neutral free radicals – known as Compton's effect; together with ejected electrons, they further react with food atoms in a process termed radiolysis.

Following acceleration via linear induction, electrons may be targeted at a heavy metal converter plate, thus generating X-rays; unlike electron beams, X-rays can penetrate up to 60–400 cm, depending on the actual photon energy. The recent development of an electronic switch – able to repeatedly deliver ultra-short pulses of power, at frequencies up to 1000 Hz, in the gigawatt power range, and using voltages in the megavolt range, has made it possible to generate pulsed X-rays. This pattern of delivery of energy appears much more effective against microorganisms than impinging the food with a uniform (average) intensity, for a given period of time; the underlying mechanism remains, however, to be fully understood.

The aforementioned reactions are the direct cause of destruction of microorganisms, insects, and parasites at large during food irradiation – immediately, due to changes in structure of the cell membrane and metabolic enzyme activity; and, on a longer run, via breaks in DNA that prevent unwinding of its double helix – and thus hamper expression of metabolically relevant enzymes, besides halting DNA replication and thus cell division. The rate of destruction of individual cells correlates directly to the rate at which ions are produced and inter-react with DNA – with reduction in cell numbers depending on total dose of radiation received thereby; as with thermal processing, a logarithmic reduction in microorganism viable numbers is typically found, with kinetics dependent on the species/strain at stake. Illustrative *D*-values for microorganisms relevant to foods, when subjected to irradiation, are included in Fig. 1.91. As a heuristic rule, the smaller and simpler the organism, the higher the radiation dose required for its destruction; commercial levels of irradiation hardly affect viruses – while species able to form spores (e.g. *Clostridium botulinum* and *Bacillus cereus*) or repair damaged DNA rapidly (e.g. *Deionococcus radiodurans*) are more resistant than vegetative cells and nonspore forming bacteria, and $Gram^-$ bacteria (e.g. *Salmonella* and *Shigella* spp.) are normally more sensitive than their $Gram^+$ counterparts.

Although radiation is able to penetrate packaging materials – and thus reduce risk of postprocessing contamination and facilitate product handling, packaging materials are themselves subjected to changes induced by radiation. Such radiolysis products as hydrocarbons, alcohols, ketones, and carboxylic acids can migrate from the package into the food itself, thus bringing unacceptable flavor taints with them – especiallyin the case of fatty foods (e.g. meats, nuts, and oils). This shortcoming can be addressed by careful choice of packaging materials and adhesives, additives, and printing materials; for instance, polystyrene does not withstand more than 5 MGy and polyethylene more than 1 MGy – while polyvinylchloride undergoes browning and releases HCl, and paper and board lose mechanical strength above 100 kGy, polypropylene becomes brittle above 25 kGy, and glass becomes brownish above 10 kGy.

Irradiation-mediated radicals also account for subtle alterations in the chemical structure of food; they cause chemical changes by bonding with and/or stripping particles from nearby molecules, namely contributed by the package itself. Excessive amounts of free radicals can lead to cell injury, and eventually to death thereof – which are the basis of many diseases, and ultimately the cause of aging; their being strong oxidizers account for such features. In high-moisture foods, water becomes ionized by irradiation, with formation of H_2O^+ and release of an electron; the latter can then combine with another water molecule to form H_2O^- – whereas H_2O^+ may degenerate to $OH^\cdot$ and H^+, and H_2O^- to $H^\cdot$ and OH^-. These products may recombine afterward as stable entities, viz. $H^\cdot$ with another $H^\cdot$ to give H_2, or $OH^\cdot$ with

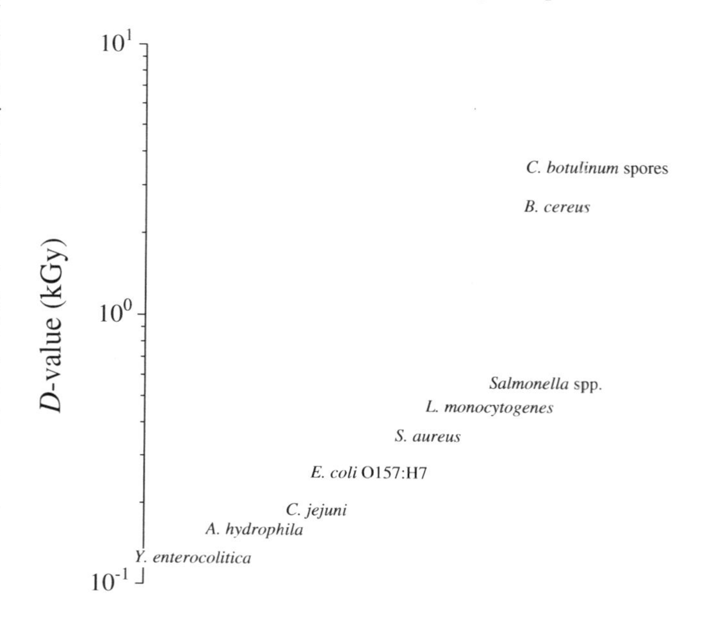

FIGURE 1.91 Resistance of selected pathogens or spoilage microorganisms upon exposure to irradiation, expressed as decimal reduction dose, *D*.

another OH· to give H_2O_2, or even H· with OH· to give back H_2O; conversely, new (and even more active) free radicals may form, namely, OH· (besides H_2) from H· and H_2O, or HO_2· (besides H_2O) from OH· and H_2O_2. Hydroxyl radicals (OH·) are powerful oxidizing agents, and react easily with unsaturated compounds – whereas expelled electrons react with aromatic compounds (especially ketones, aldehydes, and carboxylic acids). Despite their short lives (not exceeding 10^{-5} s), such radicals promptly disrupt the viability of microbial cells – but their molecular mobility is strongly hampered by absence of free liquid water (as happens in dry or frozen foods). Presence of oxygen in food enhances the radiolytic effects referred to above; such products as ozone, hydroperoxy radicals, and superoxide anions, as well as H_2O_2 afterward may form – all sharing a strong oxidizing power. This is why dairy products should not be irradiated (otherwise unacceptable rancid flavors would develop), and fat-containing meats should only be irradiated in vacuum packs. Formation of the aforementioned radicals is not unique to irradiated foods; they may also appear as a consequence of lipoxygenase or peroxidase action, UV-induced oxidation of fats and oils, and degradation of fat-soluble vitamins and pigments.

Irradiation at large improves microbial safety and reduces risk of foodborne illnesses – without resorting to chemicals; hence, occupational safety and environmental benefits are enhanced, while reducing chemical contamination of foods. Since heating is a minor side effect, the nutritional and sensory properties of the original food are essentially kept; energy requirements are low, as well as associated processing costs; and packaged foods may be treated and immediately released for shipping, thus minimizing stockholding – which is compatible with just-in-time manufacture.

Conversely, selective elimination of spoilage microorganisms relative to pathogens may compromise indicators of wholesomeness of a food – besides toxins released by the latter being normally not affected; consumption of radiolytic products (e.g. free radicals) may also pose adverse health effects. Wide acceptance of irradiated products has been further hampered by public resistance, due to concerns with nuclear industry – despite studies on subchronic and chronic changes in metabolism, histopathology, effects upon reproductive system, growth, teratogenicity, and mutagenicity having supported safety of irradiated foods for consumption at large. The capital costs incurred in when building an irradiation plant (2–7 M€ investment in premises, able to effect from low- to high-dose treatments; hardware – irradiator, totes and conveyors, control systems, and warehouse; and at least an extra 1.5 ha of secured land) also raise practical limitations to this processing technique. The costs of irradiation are lowest in the case of electron beams, increase in the case of X-rays, and further increase for γ-rays; typical values are 1.6, 5.1, and 8.2 €/m^{-3}, respectively, for a reference throughput rate of 125,000 m^3.yr^{-1} – with ^{60}Co costing ca. 1.5 €.Ci^{-1} (remember that 1 Ci represents 3.7×10^{10} disintegrations per second).

Current applications of food irradiation are schematized in Fig. 1.92 – classified by dose, i.e. low, medium, and high; an alternative classification resorts to effect, i.e. radappertization, radicidation, and radurization. Recall that gray (Gy) is the SI unit for absorption of radiation – corresponding to absorption of 1 J of energy per kg of food; given the specific heat capacity of water of 4.184 kJ.kg^{-1}.K^{-1}, this means that 4 kGy will eventually raise the temperature of 1 kg-food product by ca. 1°C.

Irradiation inhibits hormone production, and discontinues cell division and growth specifically in plant tissues; therefore, the ripening of fresh products is slowed down. Examples entail sprouting of potatoes (e.g. 150 Gy has been routinely used in Japan since 1973), garlic, onions, root ginger, and yam; low-level radiation has also been applied to arugula, ash gourd, bamboo shoots, cauliflower, coriander, parsley, spinach, and watercress. Doses between 0.25 and 1 kGy stop growth of parasitic protozoa and helminths (e.g. *Toxoplasma gondii* and *Trichinella spiralis*) in pork meat and fresh fish; as well as prevent development of insects (e.g. arthropods, butterflies, fruit flies, mealybugs, mites, moths) in such dried foods as fruits, grains, flours, and cocoa beans. The latter has overcome the need for quarantine of tropical fruit imports, and banned fumigants (e.g. ethylene dibromide, -dichloride, -oxide) and pesticides – known for their chemical toxicity.

Food shelf-life can be extended via irradiation (or radurization) using medium doses – sufficient to destroy yeasts, fungi, and nonspore-forming bacteria; spore-forming bacteria become more susceptible to subsequent thermal treatment, anyway. For instance, the shelf-life of strawberries, tomatoes, or mushrooms is doubled (or even tripled) via doses of 2–3 kGy; and synergisms can be gained upon combination with modified atmosphere packaging. Parasites can also be destroyed in meats, molds controlled in fresh fruits, and viability of contaminating microorganisms (with pathogeninc or spoilage role) reduced in meat, raw and frozen poultry, shrimps, and spices.

Irradiation of red meat has become legal in the United States since 1999, and became a commercial technique since 2002 in attempts to reduce contaminating pathogens (radicidation) – chiefly in ground/minced meat for burger patties. Fresh poultry carcasses become virtually free of *Salmonella* upon 2.5 kGy; and higher doses can be applied to (frozen) poultry or shellfish to get rid of viable *Campylobacter* spp. and *Escherichia coli*

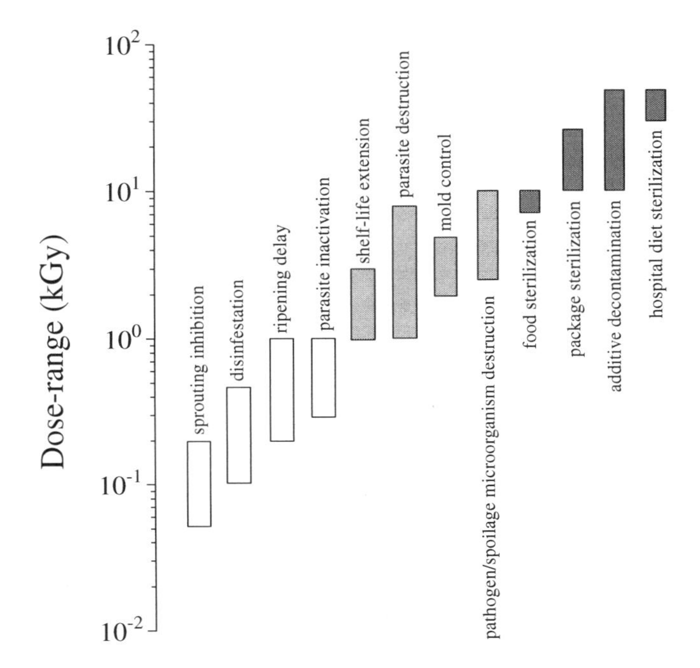

FIGURE 1.92 Irradiation ranges in selected applications encompassing foods – classified as low dose (□), medium dose (▒), and high dose (■).

FIGURE 1.93 International version of Radura logo, a labeling requirement/suggestion for irradiated foods.

O157:H7, as well as *Vibrio cholerae*, *V. parahaemolyticus*, and *V. vulnificus* – in the absence of perceptible organoleptic defects.

Although technically feasible to sterilize meats and the like, irradiation doses required would be of the order of 50 kGy for a 12 *D*-reduction of *Clostridium botulinum* – thus compromising acceptability of the final product by consumers. There is indeed little commercial interest in using irradiation for sterilizing purposes (radappertization) – except in the case of herbs and spices, which are often contaminated by heat-resistant, spore-forming bacteria. This procedure has indeed replaced chemical sterilization via ethylene oxide – prohibited in the EU since 1991, for concerns regarding worker safety and residue ingestion by final consumers. High irradiation doses have also been useful in decontaminating enzyme preparations and natural gums, as well as sterilizing wine corks and ready-meals for hospital patients.

The maximum allowable dose of irradiation for foods is 15 kGy, so as to ensure safety and wholesomeness – with average dosing not to exceed 10 kGy, as per the *Codex Alimentarius*. When irradiation of foods is permitted, labeling regulations require manufacturers to indicate so – with regard to the food itself or any listed ingredient, using the statement "Treated with/by irradiation"; in addition, wholesale foods are required to be labeled with "Do not irradiate again." In either case, the international Radura logo, see Fig. 1.93, should be visibly included (even though the EU States do not yet legally enforce it). The aforementioned symbol is usually green, and resembles a plant in a circle – the top half of which is dashed so as to mimic the discrete nature of photon-based radiation; Ulmann's original proposal (when acting as Director of the pilot plant for food irradiation at Wageningen, Netherlands) intended to be a symbol of quality, and thus a marketing tool for food processed by ionizing radiation. This logo circumvents compulsory use of international warning symbols for radiation- or biohazard – since irradiated food is not expected to pose afterward any radiological or biological hazard whatsoever. The word *Radura* is derived from radurization – in itself a portmanteau combining the initial letters of *radiation*, with stem *durus* as Latin word for hard (or lasting).

1.5.2 Product and process overview

For stable atoms with fewer than 20 protons, the number of neutrons in their nuclei equals the number of protons; nuclear stability beyond that atomic number requires, however, more neutrons than protons, due to the increasing difficulty of keeping similar charges from repelling each other when confined to the minute space occupied by the nucleus. A systematic imbalance eventually occurs when the atomic number exceeds 82 (corresponding to lead) – so Bi and all other elements beyond exhibit at least one unstable isotope.

Unstable isotopes may experience radioactive decay, thus releasing α-, β-, and/or γ-radiation; the first is restricted to the heavier elements in the periodic table (after Tl) – and releases a characteristic ^{4}He, or α-particle, containing two protons and two neutrons. Three alternatives modes exist for β-decay: electron (e^-) emission (or β^--decay), electron capture (or β^+-decay), and positron (e^+) emission (or p-decay); e^- emission entails ejection of a single electron from the nucleus (together with an antineutrino), so the charge of the latter increases by one unit – and is found within the range of lightest to heaviest elements, including the first step of decay of either ^{60}Co or ^{137}Cs. Nuclei can also decay by capturing one of the inner electrons surrounding their nucleus (and concomitantly emit an electron neutrino) – thus implying a decrease of their charge by one unit; this phenomenon is accompanied by release of an X-ray photon. The third possibility entails release a positron, or antimatter equivalent to an electron – i.e. with the same mass, but opposite charge (besides an electron neutrino once again); a daughter nuclide, with one less positive charge in the nucleus than its parent, concomitantly results. Positrons have a rather short lifetime – and rapidly lose kinetic energy as they pass through matter; upon coming to rest, they combine with an electron to generate two γ-ray photons – in a matter/antimatter annihilation event. An illustration of initial and final nuclide forms created by radioactivity is provided as Fig. 1.94. The order along the horizontal axis mimics Mendeléev's periodic table of elements – whereas abscissa and ordinate add up to the mass number of the nuclide under scrutiny.

The above forms of β-decay correspond to transformation of a neutron into a proton with release of e^- (the most likely to occur, in the case of neutron-rich radioactive isotopes); or

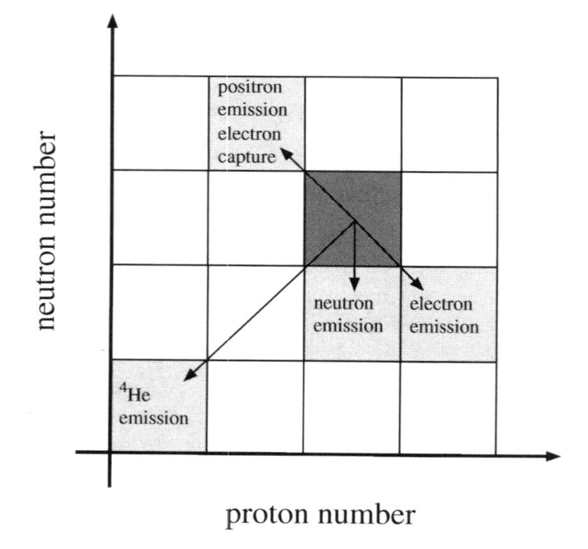

FIGURE 1.94 Schematic representation of transition from parent radionuclide (■) to daughter nuclide(s) (□), through emission of ^{4}He (α-decay), electron (β^--decay), positron (p-decay), or neutron (n-decay), or capture of electron (β^+-decay).

else conversion of a neutron into a proton via emission of e^- or capture of e^+. The daughter nuclides, produced by either α- or β-decay, are often in an excited state – and such an excess energy is promptly released *a posteriori* (within ca. 10^{-12} s) in the γ-portion of the electromagnetic spectrum. However, γ-decay is sometimes delayed – and a short-lived, or metastable nuclide meanwhile forms.

The remaining form of radioactive decay is known as spontaneous fission – and is usually restricted to nuclides with atomic numbers not below 90, with parent nucleus splitting into a set of two smaller nuclei; this process (i.e. n-decay) is accompanied by ejection of one or more neutrons, see Fig. 1.94.

Due to their distinct intrinsic energy and charge/size ratio, the particles produced by radioactive decay have disparate requirements for protection therefrom; α-particles are completely stopped by even a sheet of paper, and β-particles by aluminum shielding – while γ-rays are substantially reduced only by a thick layer of lead and/or concrete.

Only γ-radiation and accelerated electrons (which may be converted to X-rays) can be employed in food processing, since other forms of mass-based radiation can induce radioactivity; electron energy is indeed limited to a legal maximum of 10 MeV, and X- and γ-rays to 5 MeV for food irradiation – not enough to induce changes in the nucleus of the food atoms. When such quanta/particles hit the target materials, they may free other highly energetic particles; however, this effect ends shortly once exposure is over – in much the same way objects stop reflecting light when its source is turned off. Two radioisotopes abide to the above limits – besides exhibiting acceptable rates of decay, and being commercially available at reasonable cost: ^{60}Co and ^{137}Cs, with photon emission illustrated in Fig. 1.95. Note the β^--decay of ^{60}Co to (energized) ^{60}Ni* in a first step, or instead to (less energized) ^{60}Ni**; and degeneration of the former to the latter, and of the latter to ^{60}Ni – both accompanied by emission of γ-photons, but bearing distint energies. In the case of cesium, ^{137}Cs decays to either (energized) ^{137}Ba* or ^{137}Ba, along with β^--radiation; while decay of ^{137}Ba* to ^{137}Ba involves emission of a γ-photon.

Cobalt-60 is a synthetic radioactive isotope of cobalt, bearing a half-life of 5.2714 yr; it is artificially bred in specifically designed nuclear reactors, via deliberate neutron bombardment of bulk samples of monoisotopic and mononuclidic ^{59}Co – even though it is regularly generated as byproduct in a typical nuclear power plant. As emphasized in Fig. 1.95i, it decays via the β^--path – characterized by an energy of 0.3100 MeV (99.9% probability) to (metastable) ^{60}Ni*, or 1.4832 MeV (0.1% probability) to metastable ^{60}Ni**; further decay of ^{60}Ni* occurs to ^{60}Ni**, with emission of a γ-photon of 1.1732 MeV, which in turn decays to stable ^{60}Ni with emission of a γ-photon of 1.3325 MeV – see Fig. 1.95i. Note that an intermediate electron level exists (not shown for the sake of clarity), so a total of 6 distinct γ-photons may in fact be released – even though the overriding ones are by far the two mentioned above. The main advantage of this radioactive isotope lies on its high intensity of emission, coupled with deep penetration of the resulting major γ-rays – besides the low, and thus easily shielded β^--decay.

Cesium-137 (or radiocesium) is a radioactive isotope of cesium, and is one of the more common fission products of ^{235}U and other fissionable isotopes; it is a less costly alternative for being a byproduct of processing spent nuclear fuel, yet its availability is more restricted. Unfortunately, it poses environmental issues, due to the high solubility of its salts in natural water pools; and handling problems, owing to its strong chemical reactivity. With a half-life of 30.170 yr, ^{137}Cs decays to (metastable) ^{137}Ba* via emission of an electron with 0.5120 MeV of energy (94.6% probability), which decays, in turn, to stable, or ground-state ^{137}Ba, with emission of a γ-photon with 0.6617 MeV – or else directly to ^{137}Ba via emission of an electron with 1.1737 MeV (5.4% probability), as per Fig. 1.95ii.

Commercial irradiation equipment consists of an unstable isotope that produces γ-rays – or, to a lesser extent, a machine source that generates a high-energy electron beam, under a high vacuum, high-voltage electric fied. The said beam is to be used as such, or in the form of X-rays after hitting a tungsten anode (in a process known as *bremsstrahlung*, or conversion in German); the latter process is relatively inefficient, because only ca. 12%

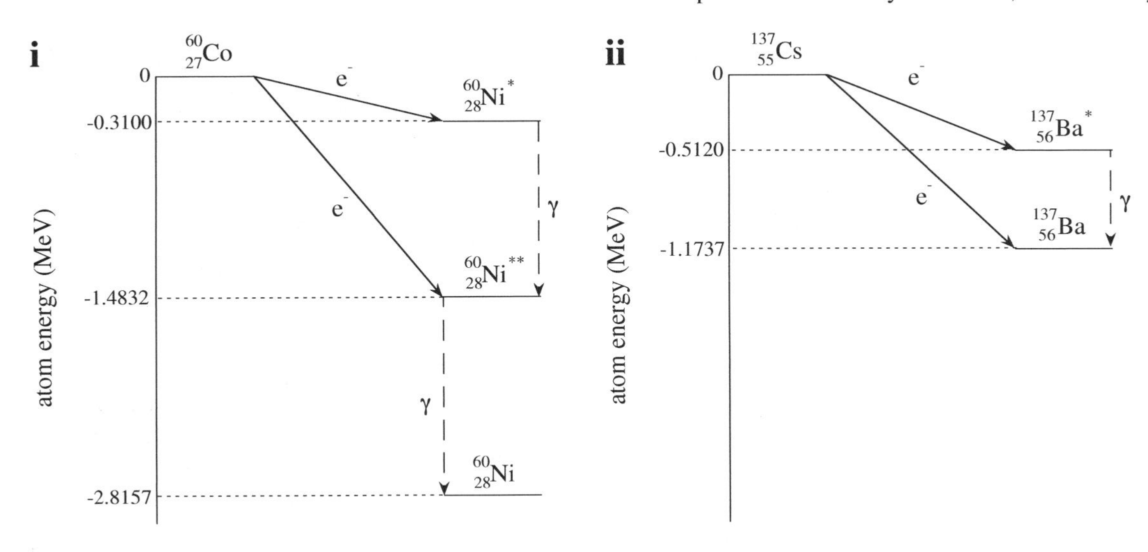

FIGURE 1.95 Schematic representation of most likely routes for radioactive decay of (i) cobalt-60 and (ii) cesium-137, involving electron (e^-) or γ-photon (γ) release, and associated energies.

of the input energy is converted to X-rays (with the remainder being dissipated as heat) – yet it is the standard technique available for commercial production of X-rays. Examples of machine and isotope sources are given in Fig. 1.96. The layout of product affects directly the dose distribution received by a food; good plant designs minimize nonuniformity of dosing throughout the food material, while maximizing dose efficiency – i.e. fraction of radiation absorbed relative to radiation emitted. High-energy electrons are directed over the food, yet they experience lower penetration than γ-rays (as mentioned before) – and are thus not suitable for bulky foods; they are normally used for surface treatments of unpackaged food, or food packaged with a thin film.

Most commercial plants resort to ^{60}Co, with nominal activities within the range $(220–370) \times 10^{13}$ $Bq.kg^{-1}$ – although ^{137}Cs may be used, also around 10^{13} $Bq.kg^{-1}$ (where Bq stands as SI unit for radioactivity, with 1 Bq = 2.703×10^{-11} Ci); in either case, typical dose rates are of the order of $kGy.hr^{-1}$. Since they can be highly focused via electromagnetic lenses acting upon their intrinsically negative charge, the dose rates of electron beams lie within the $kGy.s^{-1}$ range; due to the underlying electric generator of the machine source, they are rated as beam power – usually between 25 and 50 kW.

The main advantages of machine sources (see Fig. 1.96i and ii) is that they can be switched on and off; in the particular case of electron beams, they can be directed over the packaged food to ensure an even dose distribution. A powered roller conveyor normally carries packaged foods through two sequential radiation beams; however, this type of equipment is expensive and relatively inefficient in producing radiation. At processing speeds of up to ca. 10 $ton.hr^{-1}$ under 4 kGy, X-rays can be used with source energy of ca. 5 MeV and penetration depth of 80–100 cm – thus leading to high dose rate and homogeneity; in the case of sequential, in- or at-line processing of primary/secondary packaged products, irradiation is typically accomplished via electron beams with source energies of 5–10 MeV and penetration depth of 8–10 cm – thus assuring low-dose homogeneity, but high-dose rate. Bulk irradiation of large boxes or palletized products in shipping containers, at throughput rates above 10 $ton.hr^{-1}$, normally resorts to γ-rays with source energy of 1.33 MeV and penetration depth within 80–100 cm; this guarantees a low-dose rate and a high-dose homogeneity.

Radiation is contained within the processing chamber via concrete walls of a minimum of 3 m in thickness, complemented by lead shielding; opening for entry of products or personnel must be carefully constructed to prevent leakage – since a dose of 5 Gy is fatal to the human body. This is achieved at the expense of independently operated mechanical, electrical, and hydraulic interlocks – designed so as to avoid raising the radioactive source when staff is present, or accidental entry of staff during processing. Typically, a 10-ton plug door is used as formal entrance of personnel – with entry via the conveyor system being prohibited. Since the isotope source (see Fig. 1.96vi) cannot be switched on and off (unlike electron beams and X-rays), it must be shielded within a 6 m-deep water-filled storage pool (see Fig. 1.96vii), located below the processing area, when not in use; this procedure nullifies the effects of radiation, via absorption thereof without becoming radioactive. Economic arguments call for continuous operation – since ca. 10% of the radioactivity of ^{60}Co is lost every year, irrespective of time spent online. By the same token, special precautions are to be taken in transportation of radioactive isotopes – including sturdy metal containers (see Fig. 1.96ix), duly labelled on the outside (see Fig. 1.96viii). Packaged foods are loaded onto automatic conveyors (see Fig. 1.96iii), and transported through the

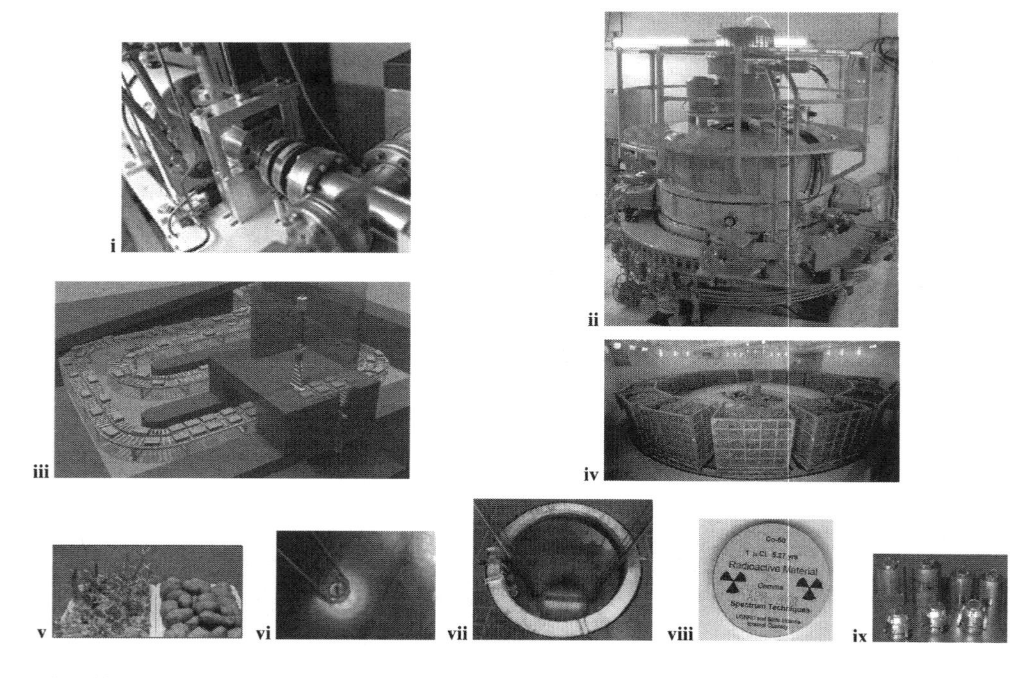

FIGURE 1.96 Examples of food irradiation equipment: (i) electron beam irradiator (courtesy of Steris, Mentor, OH), (ii) X-ray beam irradiator (courtesy of IBA Industrial, Louvain-la-Neuve, Belgium), and (iii) γ-ray beam irradiation plant (courtesy of TITANSCAN, Sioux City, IA) – with detail of (iv) irradiation chamber of potatoes (courtesy of Hokkaido Kato-gun, Shihoro, Japan) with (v) comparison between raw (left) and irradiated (right) potatoes, (vi) radioactive source (courtesy of Nordion, Laval, QC, Canada), (vii) radioisotope storage pool (courtesy of ANSTO, Lucas Heights, Australia), and (viii) supplier label and (ix) containers for transportation of radioisotopes (IZOTOP, Budapest, Hungary).

radiation field (see Fig. 1.96iv) – thus exposing both sides to the source, while ensuring a relatively uniform distribution of radiation; the effect of irradiation upon sprouting of potatoes is apparent from inspection of Fig. 1.96v. Packages suspended from an overhead rail are preferred to powered roller conveyors, because the latter use up more space in the most effective part of the irradiation chamber; pneumatic/hydraulic systems are employed for transport, with supporting cylinders located outside the chamber to facilitate maintenance – while preventing radiation-mediated damage to oil and seals. Programmable logic controllers are normally used to control the whole process, from goods-in to final distribution; the space time of the food is indeed determined by dose required, mass density of food, and power output of the source (which gradually decays over time).

At recommended doses, ^{60}Co and ^{137}Cs have insufficient emission energies to induce radioactivity in foods – unlike happens with machine sources of electrons or X-rays; however, realistic processing via the latter does not normally induce more than 0.0001% radiation (with an upper threshold of 2%, in the worst case). Ionizing radiation has also little or no effect upon the macronutrients in foods, if employed at low-dose levels; cleavage of sulfhydryl groups from the corresponding amino acid residues in proteins may, however, occur at high dosage – and are responsible for off-flavors. Polysaccharides may be hydrolyzed and oxidized to simpler compounds, more prone to enzyme-mediated decay – thus affecting viscosity and texture; while hydroperoxides may appear in irradiated lipids, in much the same manner autooxidation occurs – which may produce unacceptable changes in food taste and odor. Thiamin in meat and poultry products is the vitamin most susceptible to irradiation; ascorbic acid, pyridoxine, riboflavin, folic acid, cobalamin, and nicotinic acid follow, sorted by decreasing sensitivity. Furthermore, vitamin E is more susceptible than carotene, and carotene > vitamin A > vitamin D > vitamin K.

Illegally irradiated foods being sold without clearance – once a major concern, has no longer been a problem since the advent of dosimetry. Dosimeters are devices formed into pellets, films, or cylinders that produce a quantifiable and reproducible physicochemical change upon exposure to radiation – which correlates directly to dose received. Examples include Ala and ethanol-chlorobenzene (suitable for dose range 10^{-2}–10^{2} kGy), other amino acids (10^{-2}–10^{1} kGy), various dyes (10^{-2}–10^{0} kGy), lithium borate/fluoride (10^{-2}–10^{-1} kGy), dyed polymethylmethacrylate, PMMA (10^{-1}–10^{0} kGy), ceric/cerous sulfate and plain PMMA (10^{0}–10^{2}), ferrous cupric sulfate (10^{0}–10^{1} kGy), and cellulose triacetate (10^{1}–10^{2} kGy). For instance, free radicals are produced when crystalline Ala is irradiated, at a rate proportional to radiation dose absorbed by the crystal; their amount is assayed for afterward, via electron spin resonance spectroscopy.

In the absence of dosimeters, one may seek for minute changes in physical, chemical, or biological properties – in attempts to trace irradiation of a given food, since no perceptible alterations are at stake (as discussed previously); however, no single method is appropriate for all irradiated foods.

In the case of pistachio nuts or paprika, foods with bones, shells or seeds, and foods containing cellulose or crystalline sugars, one may resort to electron spin resonance spectroscopy – because free radicals produced by irradiation are relatively stable in solid and dry components of food.

Thermoluminescence consists of emission of light when electrons, previously ejected during irradiation and meanwhile trapped inside crystalline lattices, are released upon heating to ca. 400°C. This is an assay technique suitable for mineral-containing foods – as is the case of contaminating dust on spices, minerals from seabeds in the intestine of shellfish, and mineral salts in fruits and vegetables. Despite its wide applicability and unequivocal results, this method is quite laborious – as organic matter-free mineral samples, at mg-level, are to be prepared in advance, and strict procedures are necessary to prevent laboratory-borne contamination by dust. The said limitations have been essentially overcome via use of pulsed infrared light, instead of high-temperature heat – which does not require preliminary isolation of minerals, can resort to smaller samples, and provides results within a few minutes.

Radiolytic products are produced from fats in foods during irradiation – which, despite appearing as a consequence of other forms of food processing and to similar levels, exhibit a unique distribution pattern; gas chromatography (or tandem liquid- and gas chromatographies), coupled with detection by mass spectrometry of 2-alkylcyclobutanones serve well in the case of (irradiated) beef, lamb, pork, or poultry meat – and also for liquid whole egg, and such fruits as avocado, mango, or papaya.

Detection of DNA fragmentation, via microgel electrophoresis of single cells or nuclei, has proven useful – yet it is restricted to foods that have not undergone heat treatment (as the latter also causes such a fragmentation); when appropriate, ELISA tests may detect dihydrothymidine – a breakdown product of DNA. In alternative, one may search for trapped hydrogen or carbon monoxide produced by irradiation – as successfully used to identify irradiated cod, oysters, and shrimps. Finally, one may resort to a *Limulus* amebocyte lysate test – where an aqueous extract of blood cells from Atlantic horseshoe crab is reacted with endotoxin lipopolysaccharide of bacterial membranes; or the direct epifluorescence filter technique – where cells captured in a polycarbonate membrane are stained with acridine, and upon exposure to UV light fluoresce orange if viable or green if dead. Both of them estimate reduction in viability of microorganisms in a food following irradiation; the former resorts to Gram^{-} bacterium count as reference, whereas the latter uses aerobic plate count for the same purpose. These two biological methods have been successfully used to detect irradiated poultry meat – yet the second method provides false negatives if initial contamination is low, or low-dose levels have been employed.

1.5.3 Mathematical simulation

1.5.3.1 Radioactive decay

Radioactive decay is a stochastic phenomenon at the level of single atoms; hence, quantum theory claims that it is impossible to predict when a particular atom will decay, irrespective of how long the atom has existed as such in Eq. (1.1119).

A number of postulates may, however, be put forward regarding this process: the decay events within any partition of a time interval are mutually independent – so the probability of simultaneous events equals the product of probabilities of each single event; the distribution of events by subinterval is the same for all of them; the probability, ΔP_1, for one decay event within a subinterval of width Δt is essentially proportional to the said width, i.e.

$$\Delta P_1 \approx k\Delta t, \tag{1.1151}$$

with k denoting a positive constant – which, in the limit, leads to

$$\lim_{\Delta t \to 0} \frac{\Delta P_1}{\Delta t} \equiv \frac{dP_1}{dt} = k, \tag{1.1152}$$

according to the definition of derivative; and the probability for two, three, or, in general, N decay events within any subinterval of width Δt is negligible when compared to ΔP_1, viz.

$$\left. \left(\lim_{\Delta t \to 0} \frac{\Delta P_N}{\Delta t} \equiv \frac{dP_N}{dt} = 0 \right) \right|_{N \geq 2}, \tag{1.1153}$$

using Eq. (1.1152) as template. Here $P_N\{t\}$ denotes probability of occurrence of N decay events, within time interval $[0,t]$, so $P_N\{t+\Delta t\}$ will denote probability of occurrence of the very same N events within interval $[0,t+\Delta t]$; the cases of $N=0$ and $N>0$ will to advantage be treated separately. The former case, i.e. lack of events within $[0,t+\Delta t]$, necessarily implies lack of events within $[0,t]$; since the events are independent, one readily realizes that

$$P_0\{t+\Delta t\} = P_0\{t\}\Delta P_0. \tag{1.1154}$$

On the other hand, the probability, ΔP_0 of no events within $[t,t+\Delta t]$ is thus given by

$$\Delta P_0 = 1 - \Delta P_1 - \Delta P_2 - \ldots = 1 - \Delta P_1 - \sum_{i=2}^{\infty} \Delta P_i \tag{1.1155}$$

– because the probabilities must, by definition, add up to unity; insertion of Eq. (1.1155) transforms Eq. (1.1154) to

$$P_0\{t+\Delta t\} = P_0\{t\}\left(1 - \Delta P_1 - \sum_{i=2}^{\infty} \Delta P_i \right), \tag{1.1156}$$

where $P_0\{t\}$ may be moved to the left-hand side, after applying the distributive property, to obtain

$$P_0\{t+\Delta t\} - P_0\{t\} = -P_0\{t\}\left(\Delta P_1 + \sum_{i=2}^{\infty} \Delta P_i \right). \tag{1.1157}$$

Upon division of both sides by Δt, Eq. (1.1157) becomes

$$\frac{P_0\{t+\Delta t\} - P_0\{t\}}{\Delta t} = -P_0\{t\}\left(\frac{\Delta P_1}{\Delta t} + \sum_{i=2}^{\infty} \frac{\Delta P_i}{\Delta t} \right) \tag{1.1158}$$

– so a limit may be imposed, when Δt approaches zero, as

$$\lim_{\Delta t \to 0} \frac{P_0\{t+\Delta t\} - P_0\{t\}}{\Delta t} = -P_0\{t\}\left(\lim_{\Delta t \to 0} \frac{\Delta P_1}{\Delta t} + \sum_{i=2}^{\infty} \lim_{\Delta t \to 0} \frac{\Delta P_i}{\Delta t} \right), \tag{1.1159}$$

in agreement with the classical theorems on limits; Eq. (1.1159) can be rewritten, in a more condensed fashion, as

$$\frac{dP_0}{dt} = -P_0\{t\}\left(\frac{dP_1}{dt} + \sum_{i=2}^{\infty} \frac{dP_i}{dt} \right), \tag{1.1160}$$

at the expense of the definition of derivative – so combination with Eqs. (1.1152) and (1.1153) unfolds

$$\frac{dP_0}{dt} = -P_0\{t\}\left(k + \sum_{i=2}^{\infty} 0 \right) = -P_0\{t\}(k+0), \tag{1.1161}$$

or merely

$$\boxed{\frac{dP_0}{dt} = -kP_0.} \tag{1.1162}$$

When $N>0$, the events within $[0,t+\Delta t]$ should distribute between the consecutive subintervals $[0,t]$ and $[t,t+\Delta t]$; for instance, if $N=3$, then the possible distributions of events between such subintervals are (3,0), (2,1), (1,2), and (0,3) – meaning that

$$P_3\{t+\Delta t\} = P_3\{t\}\Delta P_0 + P_2\{t\}\Delta P_1 + P_1\{t\}\Delta P_2 + P_0\{t\}\Delta P_3, \tag{1.1163}$$

in view of the postulated independence of all such events. After retrieving ΔP_0 from Eq. (1.1155), one may transform Eq. (1.1163) to

$$\begin{aligned} P_3\{t+\Delta t\} = {} & P_3\{t\}\left(1 - \Delta P_1 - \sum_{i=2}^{\infty} \Delta P_i \right) \\ & + P_2\{t\}\Delta P_1 + P_1\{t\}\Delta P_2 + P_0\{t\}\Delta P_3; \end{aligned} \tag{1.1164}$$

once $P_3\{t\}$ is factored in and $-P_3\{t\}$ is added to both sides, Eq. (1.1164) becomes

$$\begin{aligned} P_3\{t+\Delta t\} - P_3\{t\} = {} & -P_3\{t\}\Delta P_1 - P_3\{t\}\sum_{i=2}^{\infty} \Delta P_i \\ & + P_2\{t\}\Delta P_1 + P_1\{t\}\Delta P_2 + P_0\{t\}\Delta P_3. \end{aligned} \tag{1.1165}$$

Following division of both sides by Δt, Eq. (1.1165) turns to

$$\begin{aligned} \frac{P_3\{t+\Delta t\} - P_3\{t\}}{\Delta t} = {} & -P_3\{t\}\frac{\Delta P_1}{\Delta t} - P_3\{t\}\sum_{i=2}^{\infty} \frac{\Delta P_i}{\Delta t} \\ & + P_2\{t\}\frac{\Delta P_1}{\Delta t} + P_1\{t\}\frac{\Delta P_2}{\Delta t} + P_0\{t\}\frac{\Delta P_3}{\Delta t}, \end{aligned} \tag{1.1166}$$

where limits are again to be taken of both sides when $\Delta t \to 0$ as

$$\begin{aligned} \lim_{\Delta t \to 0} \frac{P_3\{t+\Delta t\} - P_3\{t\}}{\Delta t} = {} & -P_3\{t\}\lim_{\Delta t \to 0} \frac{\Delta P_1}{\Delta t} - P_3\{t\}\sum_{i=2}^{\infty} \lim_{\Delta t \to 0} \frac{\Delta P_i}{\Delta t} \\ & + P_2\{t\}\lim_{\Delta t \to 0} \frac{\Delta P_1}{\Delta t} + P_1\{t\}\lim_{\Delta t \to 0} \frac{\Delta P_2}{\Delta t} + P_0\{t\}\lim_{\Delta t \to 0} \frac{\Delta P_3}{\Delta t}; \end{aligned} \tag{1.1167}$$

Eq. (1.1167) is equivalent to

$$\frac{dP_3}{dt} = -P_3\{t\}\frac{dP_1}{dt} - P_3\{t\}\sum_{i=2}^{\infty}\frac{dP_i}{dt} + P_2\{t\}\frac{dP_1}{dt} + P_1\{t\}\frac{dP_2}{dt} + P_0\{t\}\frac{dP_3}{dt}, \tag{1.1168}$$

based once more on the definition of derivative. Upon combination with Eqs. (1.1152) and (1.1153), it is possible to redo Eq. (1.1168) as

$$\frac{dP_3}{dt} = -P_3\{t\}k - P_3\{t\}\sum_{i=2}^{\infty}0 + P_2\{t\}k + P_1\{t\}0 + P_0\{t\}0, \tag{1.1169}$$

which breaks down to

$$\frac{dP_3}{dt} = -kP_3\{t\} + kP_2\{t\} = -k\left(P_3\{t\} - P_2\{t\}\right) \tag{1.1170}$$

after dropping nil terms and factoring $-k$ out; Eq. (1.1170) and its supporting derivation will then be generalized to any positive (integer) N as

$$\boxed{\frac{dP_N}{dt} = -k\left(P_N - P_{N-1}\right);\quad N \geq 1.} \tag{1.1171}$$

Equations (1.1162) and (1.1171) hold

$$\boxed{P_N = \frac{(kt)^N \mathrm{e}^{-kt}}{N!}} \tag{1.1172}$$

as general solution – known as Poisson's probability density function. In fact, Eq. (1.1172) readily implies

$$\frac{dP_N}{dt} = \frac{d}{dt}\left(\frac{(kt)^N \mathrm{e}^{-kt}}{N!}\right) = \frac{1}{N!}\left(N(kt)^{N-1}k\mathrm{e}^{-kt} + (kt)^N \mathrm{e}^{-kt}(-k)\right) \tag{1.1173}$$

upon straightforward application of the classical rules of differentiation, which reduces further to

$$\frac{dP_N}{dt} = -k\left(\frac{(kt)^N \mathrm{e}^{-kt}}{N!} - \frac{(kt)^{N-1}\mathrm{e}^{-kt}}{(N-1)!}\right) \tag{1.1174}$$

after factoring $N!$ in and $-k$ out; hence, Eq. (1.1171) is retrieved – in view of Eq. (1.1172) applied to N and $N-1$, thus confirming that Eq. (1.1172) is the solution to the differential equation labeled as Eq. (1.1171). When N is set equal to zero, Eq. (1.1172) simplifies to

$$P_0 = \frac{(kt)^0 \mathrm{e}^{-kt}}{0!} = \frac{1\mathrm{e}^{-kt}}{1} = \mathrm{e}^{-kt}; \tag{1.1175}$$

differentiation of both sides with regard to t then produces

$$\frac{dP_0}{dt} = -k\mathrm{e}^{-kt} \tag{1.1176}$$

– which coincides with Eq. (1.1162), upon combination with Eq. (1.1175). A graphical representation of Eq. (1.1172) is provided in Fig. 1.97. Except for absence of decay events – characterized by a monotonically decreasing probability with time, all other curves go through a maximum; such a stationary point is located later in terms of abscissa and lower in terms of ordinate as N increases, while the curve exhibits a larger and larger dispersion around its mean. If Eq. (1.1173) is rephrased as

$$\frac{dP_N}{d(kt)} = \frac{k\mathrm{e}^{-kt}}{N!}\left(N(kt)^{N-1} - (kt)^N\right) \tag{1.1177}$$

upon factoring out ke^{-kt}, then one may set $dP_N/d(kt)$ equal to zero at $kt = kt_{\max}$ to get

$$\frac{k\mathrm{e}^{-kt_{max}}}{N!}\left(N\left(kt_{\max}\right)^{N-1} - \left(kt_{\max}\right)^N\right) = 0; \tag{1.1178}$$

Eq. (1.1178) degenerates to

$$N\left(kt_{\max}\right)^{N-1} = \left(kt_{\max}\right)^N \tag{1.1179}$$

as per $kt^{-kt_{\max}}/N! \neq 0$, where cancellation of $(kt_{\max})^{N-1}$ between sides supports simplification to

$$\boxed{kt_{\max} = N.} \tag{1.1180}$$

Equation (1.1180) indicates that the maxima under scrutiny, also known as modes, coincide with the corresponding value of N – and thus translate rightward as N increases, as grasped in Fig. 1.97. In other words, the most likely number of events is N, should the time period extend from zero up to N/k; furthermore, insertion of Eq. (1.1180) transforms Eq. (1.1172) to

$$P_{N,\max} \equiv P_N\Big|_{kt=kt_{\max}} = \frac{N^N \mathrm{e}^{-N}}{N!} \tag{1.1181}$$

– which describes the ordinates of the said maxima.

A more useful description of the shape of the curves in Fig. 1.97 should resort to their expected value, μ, and variance, σ^2 – both accessible via the moment-generating function, $G\{\tau\}$, defined as

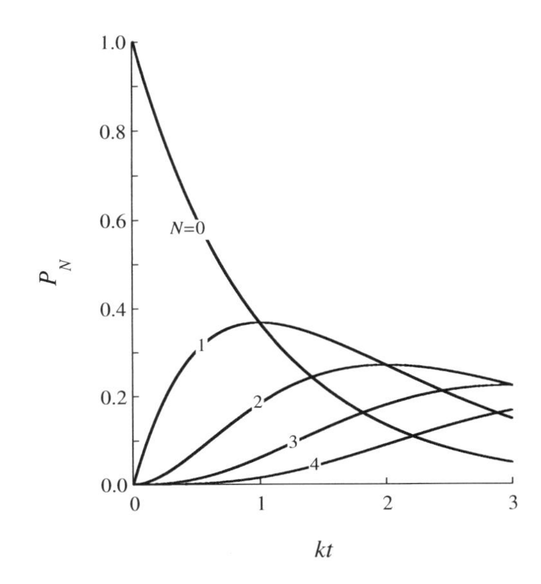

FIGURE 1.97 Variation of probability of N decay events, P_N, during interval of dimensionless time with magnitude kt, versus kt itself, for selected values of N.

$$\boxed{G\{\tau\} \equiv E\left\{e^{\tau N}\right\};} \tag{1.1182}$$

here N denotes a random variable, distributed according to Eq. (1.1172), whereas n will hereafter denote every specific value taken thereby. Recalling the definition of expectation function E, one may rewrite Eq. (1.1182) as

$$G\{\tau\} \equiv \sum_{n=0}^{\infty} e^{\tau n} P_n = \sum_{n=0}^{\infty} e^{\tau n} \frac{(kt)^n e^{-kt}}{n!} \tag{1.1183}$$

after bringing Eq. (1.1172) on board – where e^{-kt} may be factored out, and the other exponential function lumped with the power function to get

$$G\{\tau\} = e^{-kt} \sum_{n=0}^{\infty} \frac{\left(kt\,e^{\tau}\right)^n}{n!}; \tag{1.1184}$$

the summation in Eq. (1.1184) is but Taylor's expansion, about zero, of the exponential of kte^{τ}, so simplification ensues to

$$\boxed{G\{\tau\} = e^{-kt} e^{kt\,e^{\tau}}} \tag{1.1185}$$

– since the aforementioned expansion holds an infinite radius of convergence. Remember that

$$G\{\tau\} \equiv E\left\{e^{\tau Y}\right\} \tag{1.1186}$$

holds at large, when applied to a generic, random, discrete variable Y – after using Eq. (1.182) as template; Eq. (1.1186) supports

$$\frac{dG}{d\tau} \equiv \frac{d}{d\tau}\left(E\left\{e^{\tau Y}\right\}\right) \equiv \frac{d}{d\tau}\left(\sum_{y} e^{\tau y} P_y\right) = \sum_{y} \frac{d}{d\tau}\left(e^{\tau y} P_y\right)$$
$$= \sum_{y} \frac{d e^{\tau y}}{d\tau} P_y \equiv E\left\{\frac{d e^{\tau Y}}{d\tau}\right\} = E\left\{Y e^{\tau Y}\right\}, \tag{1.1187}$$

due to linearity of the differential operator coupled with the rule of differentiation of an exponential function. When $\tau=0$, Eq. (1.1187) reduces to

$$\left.\frac{dG}{d\tau}\right|_{\tau=0} = E\left\{Ye^{0}\right\} = E\{Y\} \equiv \mu, \tag{1.1188}$$

where the mean, μ, is accordingly termed first-order absolute moment of Y. By the same token, the second derivative of $G\{\tau\}$ reads

$$\frac{d^2G}{d\tau^2} \equiv \frac{d}{d\tau}\left(\frac{dG}{d\tau}\right) \equiv \frac{d}{d\tau}\left(E\left\{Ye^{\tau Y}\right\}\right) = E\left\{\frac{d}{d\tau}\left(Ye^{\tau Y}\right)\right\}$$
$$= E\left\{Ye^{\tau Y}Y\right\} = E\left\{Y^2 e^{\tau Y}\right\} \tag{1.1189}$$

with the aid of Eq. (1.1187) – thus leaving

$$\left.\frac{d^2G}{d\tau^2}\right|_{\tau=0} = E\left\{Y^2 e^{0}\right\} = E\left\{Y^2\right\} \equiv \mu_2 \tag{1.1190}$$

when τ is nil – with μ_2 denoting second-order absolute moment of Y; therefore, one may claim

$$\mu \equiv \sum_{n=0}^{\infty} nP_n = \left.\frac{dG}{d\tau}\right|_{\tau=0} \tag{1.1191}$$

as per Eq. (1.1188), and

$$\mu_2 \equiv \sum_{n=0}^{\infty} n^2 P_n = \left.\frac{d^2G}{d\tau^2}\right|_{\tau=0} \tag{1.1192}$$

as per Eq. (1.1190), both applying to the above Poisson's (discrete) distribution of N as a particular case of Y. Chain differentiation of Eq. (1.1185), with regard to τ, then unfolds

$$\frac{dG}{d\tau} = e^{-kt} e^{kt\,e^{\tau}} kt\,e^{\tau} = kt\,e^{-kt} e^{\tau + kt\,e^{\tau}} \tag{1.1193}$$

together with lumping of two of the three exponential functions; hence, insertion in Eq. (1.1191) permits conversion to

$$\mu = kt\,e^{-kt} e^{\tau + kt\,e^{\tau}}\Big|_{\tau=0} = kt\,e^{-kt} e^{0 + kt\,e^{0}} = kt\,e^{-kt} e^{kt}, \tag{1.1194}$$

or merely

$$\boxed{\mu = kt,} \tag{1.1195}$$

after the outstanding exponential functions are collapsed. When t is set equal to N/k, one obtains

$$\mu\Big|_{t=\frac{N}{k}} = k\frac{N}{k} = N \tag{1.1196}$$

from Eq. (1.1195) – which, in view of Eq. (1.1180), guarantees that the expected value (or mean) coincides with the most likely value (or mode). The second derivative of G with regard to τ may now depart from Eq. (1.1193) as

$$\frac{d^2G}{d\tau^2} \equiv \frac{d}{d\tau}\left(\frac{dG}{d\tau}\right) = \frac{d}{d\tau}\left(kt\,e^{-kt} e^{\tau + kt\,e^{\tau}}\right), \tag{1.1197}$$

where application of the classical rules of differentiation gives rise to

$$\frac{d^2G}{d\tau^2} = kt\,e^{-kt} e^{\tau + kt\,e^{\tau}}\left(1 + kt\,e^{\tau}\right); \tag{1.1198}$$

insertion of Eq. (1.1198) transforms Eq. (1.1192) to

$$\mu_2 = kt\,e^{-kt} e^{\tau + kt\,e^{\tau}}\left(1 + kt\,e^{\tau}\right)\Big|_{\tau=0} = kt\,e^{-kt} e^{0 + kt\,e^{0}}\left(1 + kt\,e^{0}\right)$$
$$= kt\,e^{-kt} e^{kt}\left(1 + kt\right), \tag{1.1199}$$

which breaks down to merely

$$\mu_2 = kt(1 + kt) \tag{1.1200}$$

upon lumping exponential functions with symmetrical arguments. On the other hand, the variance is defined by

$$\sigma^2 \equiv \sum_{n=0}^{\infty} (n - \mu)^2 P_n, \tag{1.1201}$$

so application of Newton's binomial, followed by splitting of the summation and factoring out of constants unfold

$$\sigma^2 = \sum_{n=0}^{\infty}\left(n^2 - 2n\mu + \mu^2\right)P_n = \sum_{n=0}^{\infty} n^2 P_n - 2\mu\sum_{n=0}^{\infty} nP_n + \mu^2\sum_{n=0}^{\infty} P_n; \tag{1.1202}$$

recalling Eqs. (1.1191) and (1.1192), one may convert Eq. (1.1202) to

$$\sigma^2 = \mu_2 - 2\mu\mu + \mu^2, \tag{1.1203}$$

since the sum of all P_n's, i.e. $\sum_{n=0}^{\infty} P_n$, is necessarily unity (given the definition of probability). Condensation of terms alike allows simplification of Eq. (1.1203) to

$$\sigma^2 = \mu_2 - 2\mu^2 + \mu^2 = \mu_2 - \mu^2, \tag{1.1204}$$

where insertion of Eqs. (1.1195) and (1.1200) gives rise to

$$\sigma^2 = kt(1+kt) - (kt)^2 = kt + (kt)^2 - (kt)^2 \tag{1.1205}$$

along with elimination of the first parenthesis, or simply

$$\boxed{\sigma^2 = kt} \tag{1.1206}$$

upon dropping symmetrical terms – so dispersion increases exactly as the expected value does, see Eq. (1.1195), as can be grasped in Fig. 1.97; after setting t equal to N/k once again, Eq. (1.1206) becomes

$$\sigma^2\Big|_{t=\frac{N}{k}} = k\frac{N}{k} = N \tag{1.1207}$$

that retrieves Eq. (1.1196).

The result conveyed by Eq. (1.1196) is of particular relevance – because it states that one decay event is expected to occur at every interval of amplitude $1/k$ (and so N events will likely occur within an interval of amplitude N/k, or N-fold $1/k$). This conclusion is ultimately a consequence of Eq. (1.1152) that may, for convenience, be rewritten as

$$\frac{\frac{-dM}{M}}{dt} = k, \tag{1.1208}$$

since probability (of a single event) may be seen as

$$dP_1 \equiv \frac{-dM}{M} \tag{1.1209}$$

when the number of entities is very large – i.e. the number of actual atoms undergoing radioactive decay, $-dM$, divided by the total number of atoms still in their original form, M. Equation (1.1208) usually appears as

$$\boxed{\frac{dM}{dt} = -kM,} \tag{1.1210}$$

where M indistinctly represents number or mass concentration of isotope atoms in the sample remaining in native form; in fact, they are proportional to each other via their atomic weight, and the said proportionality constant appears as such in, and thus drops off both sides of the equation. The differential equation labeled as Eq. (1.1210) is subjected to

$$\boxed{M\big|_{t=0} = M_0} \tag{1.1211}$$

as initial condition, with M_0 denoting number/mass of native isotope atoms in the starting sample. The above rate of decay is independent of the physicochemical state of the isotope; therefore, it has been used for atomic dating of archeological artefacts and fossil specimens. For instance, ^{14}C is produced in the atmosphere at a more or less constant rate, due to bombardment by cosmic rays – and circulates between atmosphere, oceans, and living organisms through their metabolic pathways at a rate much faster than its decay; however, it is no longer picked up by living organisms once they die, so the ratio of radioactivity of a dead sample to that of a currently living tissue can be used to estimate how long ago death took place. Integration of Eq. (1.1210) may proceed with the aid of Eq. (1.1211) to give

$$\int_{M_0}^{M} \frac{d\tilde{M}}{\tilde{M}} = -k\int_0^t d\tilde{t}, \tag{1.1212}$$

which is equivalent to

$$\ln \tilde{M}\Big|_{M_0}^{M} = -k\tilde{t}\Big|_0^t; \tag{1.1213}$$

Eq. (1.1213) degenerates to

$$\ln\frac{M}{M_0} = -kt, \tag{1.1214}$$

where application of exponentials to both sides finally leaves

$$\boxed{\frac{M}{M_0} = e^{-kt}} \tag{1.1215}$$

– as sketched in Fig. 1.98. A typical exponential decay is thus obtained – departing from unity, and approaching the horizontal axis in an asymptotic fashion as time grows unbounded. The rate of decay conveyed by k is also termed activity – and is expressed in becquerel (Bq) as SI unit (as pointed out before), corresponding to one disintegration per second. Nevertheless, the unit curie (Ci), originally defined as the number of disintegrations per second in 1 g of ^{226}Ra, is still widely used – with 1 Ci being equal to 36.99 GBq; a power of 14.8 kW is, in turn, associated to 1 MCi.

The half-life for decay, $t_{½}$, of a radioactive nuclide is the time it takes for half of the nuclei in the original sample undergo disintegration; if Eq. (1.1210) were integrated still with the aid of Eq. (1.1211), complemented by

$$t_{1/2} \equiv t\Big|_{M=\frac{M_0}{2}} \tag{1.1216}$$

– according to

$$\int_{M_0}^{\frac{M_0}{2}} \frac{dM}{M} = -k\int_0^{t_{1/2}} dt, \tag{1.1217}$$

then straight application of the fundamental theorem of integral calculus would unfold

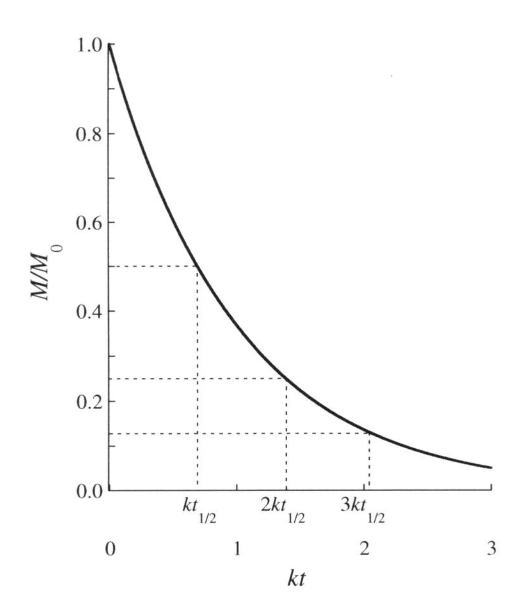

FIGURE 1.98 Evolution of normalized amount of original atoms, M/M_0, versus dimensionless time, kt – with indication of dimensionless half-life, $kt_{½}$.

$$\ln M\Big|_{M_0}^{\frac{M_0}{2}} = -kt\Big|_0^{t_{1/2}}, \tag{1.1218}$$

or else

$$\ln\frac{\frac{M_0}{2}}{M_0} = \ln\frac{1}{2} = -\ln 2 = -kt_{1/2} \tag{1.1219}$$

upon elementary algebraic manipulation, complemented by the operational features of a logarithmic function. One may finally isolate $t_{½}$ from Eq. (1.1219) as

$$\boxed{t_{1/2} = \frac{\ln 2}{k};} \tag{1.1220}$$

in view of the underlying first-order kinetics, the decay of one-half of the remaining isotope atoms always takes $t_{½}$, irrespective of the starting M. The said parameter is also represented in Fig. 1.98, in dimensionless form – as $\ln 2 = kt_{1/2}$, and corresponding first integer multiples. Remember that the half-life of ^{60}Co is 5.3 years, whereas ^{137}Cs possesses a half-life of 30.2 yr; this means that the former must be replaced more frequently as source in an irradiation plant – while the latter poses a tougher hazard, in terms of radioactivity of its disposed-off residues. The (average) activity of 1 g of ^{60}Co during the first half-life is 44 TBq, whereas its ^{137}Cs counterpart is characterized by 0.0032 TBq. There is no known upper limit for the half-lives of radioactive atoms – which span a time range of amplitude above 55 orders of magnitude, from nearly instantaneous to far longer the age of the Universe.

PROBLEM 1.21

Consider a radioactive decay process, following first-order kinetics characterized by constant k – with the useful life of the radioisotope denoted as τ.

a) Prove that such a rate of decay is compatible with a binomial probability distribution, when the number of atoms of the radioactive source is very large.
b) To damp changes in irradiation power throughout time, one may resort to two halves, or three thirds, ... of the regular mass of radioisotope, with identical overall initial specification – purchased and put to operation at uniformly spaced intervals, but combined to generate a single beam. Plot the evolution of the fraction of undecayed atoms with time for one single, two halves, and three-thirds – with identical time lags in between (as appropriate), and conclude on the periodicity of their sum.
c) Based on the conclusions reached in b), calculate the average fraction of undecayed material in those situations – and discuss the result found.
d) Derive an expression for the average difference in fraction of undecayed material relative to the corresponding grand average – taking again the observation in b) into account, and for a portion of radioisotope splitted as N equal fractions.
e) Conclude that stable irradiation power will eventually be attained with a very many (small) portions of radioactive source, based on the expression obtained in d).

Solution

a) During a time interval of amplitude t, a given atom will either undergo or not radioactive decay; consider that the probability of decay, p, remains unchanged with elapsing time, and that the probability of decay of an atom is independent of the corresponding probability of any other atom. When this set of postulates hold, the associated entities consubstantiate a Bernoulli's system – so the number N of atoms undergoing decay during time t, in such a system containing M_0 atoms in total, will follow a binomial distribution. To calculate an expression for the aforementioned probability density function, one should enumerate all possibilities of decay (denoted hereafter as 1) or no decay (denoted hereafter as 0) for all atoms in the system. Take $M_0=4$ as an illustration; if all atoms decay, then only sequence 1111 applies, corresponding to $1=\binom{4}{4}$ possibility and thus to an overall probability $P=p^4=p^4(1-p)^0=p^4(1-p)^{4-4}$ – in view of the individual probability p of decay, and the mutual independence of all decays. If there are three decaying atoms, then one may have 1110, 1101, 1011, or 0111 as sequences, thus adding up to $4=\binom{4}{3}$ possibilities associated to an overall probability $P=p^3(1-p)=p^3(1-p)^{4-3}$ – since absence of decay will exhibit the complementary probability $1-p$ of individual occurrence. If there are two decaying atoms, then one may get 1100, 1010, 1001, 0110, 0101, or 0011 as sequences, thus adding up to $6=\binom{4}{2}$ possibilities with an overall probability $P=p^2(1-p)^2=p^2(1-p)^{4-2}$. If there is one decaying atom, then one will obtain 1000, 0100, 0010, or 0001 as feasible sequences, with a total of $4=\binom{4}{1}$ possibilities entailing an overall probability $P=p(1-p)^3=p^1(1-p)^{4-1}$. Finally, no decay at all would require 0000 as only sequence, i.e. $1=\binom{4}{0}$ possibility with an overall probability $P=(1-p)^4=p^0(1-p)^4(1-p)^{4-0}$. Therefore, the overall probability of N atoms decaying out of M_0 atoms will, in general, look like

$$P_N=\binom{M_0}{N}p^N(1-p)^{M_0-N}, \tag{1.21.1}$$

for every N comprised between 0 and M_0; recalling Pascal's expression for the binomial coefficient, i.e.

$$\binom{M_0}{N}=\frac{M_0!}{N!(M_0-N)!}, \tag{1.21.2}$$

and the definition of factorial, one may expand the numerator of Eq. (1.21.2) to get

$$\binom{M_0}{N}=\frac{M_0(M_0-1)\cdots(M_0-N+1)(M_0-N)!}{N!(M_0-N)!}, \tag{1.21.3}$$

which reduces to

$$\binom{M_0}{N}=\frac{(M_0-0)(M_0-1)\cdots(M_0-(N-1))}{N!} \tag{1.21.4}$$

after canceling $(M_0-N)!$ between numerator and denominator – where N different factors appear in numerator, as many as those appearing in denominator. Insertion of Eq. (1.21.4) transforms Eq. (1.21.1) to

$$P_N=\frac{1}{N!}\frac{(M_0-0)(M_0-1)\cdots(M_0-(N-1))}{{M_0}^N}\times {M_0}^N p^N(1-p)^{M_0-N}, \tag{1.21.5}$$

where multiplication and division by ${M_0}^N$ were meanwhile effected (for convenience); upon splitting the overall fraction and lumping powers with identical exponents, Eq. (1.21.5) becomes

$$P_N=\frac{1}{N!}\frac{M_0-0}{M_0}\frac{M_0-1}{M_0}\cdots\frac{M_0-(N-1)}{M_0}(M_0p)^N(1-p)^{M_0-N}, \tag{1.21.6}$$

or else

$$P_N=\frac{1}{N!}\left(1-\frac{0}{M_0}\right)\left(1-\frac{1}{M_0}\right)\cdots\left(1-\frac{N-1}{M_0}\right)(M_0p)^N(1-p)^{M_0-N} \tag{1.21.7}$$

after splitting each fraction once more. Since M_0p represents the total number of atoms expected to undergo decay within time interval t – and in view of the definition of kinetic constant, k, as expected number of atoms decaying per unit time, one concludes that

$$M_0p\equiv kt; \tag{1.21.8}$$

Eq. (1.21.8) permits simplification of Eq. (1.21.7) to

$$P_N=\frac{1}{N!}\left(1-\frac{0}{M_0}\right)\left(1-\frac{1}{M_0}\right)\cdots\left(1-\frac{N-1}{M_0}\right)(kt)^N\left(1-\frac{kt}{M_0}\right)^{M_0-N} \tag{1.21.9}$$

where the last power may be conveniently split to give

$$P_N=\frac{(kt)^N}{N!}\left(1-\frac{kt}{M_0}\right)^{M_0}\frac{\left(1-\frac{0}{M_0}\right)\left(1-\frac{1}{M_0}\right)\cdots\left(1-\frac{N-1}{M_0}\right)}{\left(1-\frac{kt}{M_0}\right)^N}. \tag{1.21.10}$$

When the total number of atoms is large – as (always) happens in industrial practice, one should take the limit of Eq. (1.21.10) when $M_0\to\infty$, viz.

$$\lim_{M_0\to\infty}P_N=\frac{(kt)^N}{N!}\lim_{M_0\to\infty}\left(1-\frac{kt}{M_0}\right)^{M_0}\lim_{M_0\to\infty}\frac{\left(1-\frac{0}{M_0}\right)\left(1-\frac{1}{M_0}\right)\cdots\left(1-\frac{N-1}{M_0}\right)}{\left(1-\frac{kt}{M_0}\right)^N}, \tag{1.21.11}$$

together with the classical theorems on limits; Eq. (1.21.11) reduces to

$$\lim_{M_0\to\infty}P_N=\frac{(kt)^N}{N!}\lim_{M_0\to\infty}\left(\left(1+\left(-\frac{kt}{M_0}\right)\right)^{\frac{M_0}{kt}}\right)^{kt}\frac{(1-0)(1-0)\cdots(1-0)}{(1-0)^N}, \tag{1.21.12}$$

after raising the second factor to kt and taking the corresponding root – since $N << M_0$ implies $N-1 << M_0$, and thus $i/M_0 \leq (N-1)/M_0 << 1$ for $i \leq N-1$, coupled with very large values of M_0 and finite N. Equation Eq. (1.21.12) may be rewritten as

$$\lim_{M_0 \to \infty} P_N = \frac{(kt)^N}{N!} \left(\lim_{\frac{M_0}{kt} \to \infty} \left(1 + \frac{-1}{\frac{M_0}{kt}} \right)^{\frac{M_0}{kt}} \right)^{kt} \frac{1^N}{1^N}, \quad (1.21.13)$$

upon exchange of power and limit operators as permitted by their operational features; and because $M_0/kt \to \infty$ when $M_0 \to \infty$, under finite kt. Equation (1.21.13) degenerates to

$$\lim_{M_0 \to \infty} P_N = \frac{(kt)^N}{N!} \left(e^{-1} \right)^{kt} = \frac{(kt)^N}{N!} e^{-kt}, \quad (1.21.14)$$

upon cancellation of common factors between numerator and denominator – complemented by the definition of Neper's number, i.e.

$$e^{\alpha} \equiv \lim_{\xi \to \infty} \left(1 + \frac{\alpha}{\xi} \right)^{\xi} \quad (1.21.15)$$

– or, after replacing (dummy) variable ξ by M_0/kt and (dummy) parameter α by -1,

$$\lim_{\frac{M_0}{kt} \to \infty} \left(1 + \frac{-1}{\frac{M_0}{kt}} \right)^{\frac{M_0}{kt}} = e^{-1}. \quad (1.21.16)$$

Inspection of Eq. (1.21.14) retrieves Poisson's distribution, as given by Eq. (1.1172) – meaning that these two distributions coincide, when the size of the population is very large.

b) If the overall amount of radioactive source, M_0, is used as such, or instead subdivided two ways or three ways in identical portions, then the contribution of each portion to the total fraction of undecayed material is represented below as solid thin lines – satisfying Eq. (1.1215); while the combined (additive) contribution, on the hypothesis of a constant phase shift between start of operation of each said fraction relative to the previous one, is conveyed by the solid thick lines.

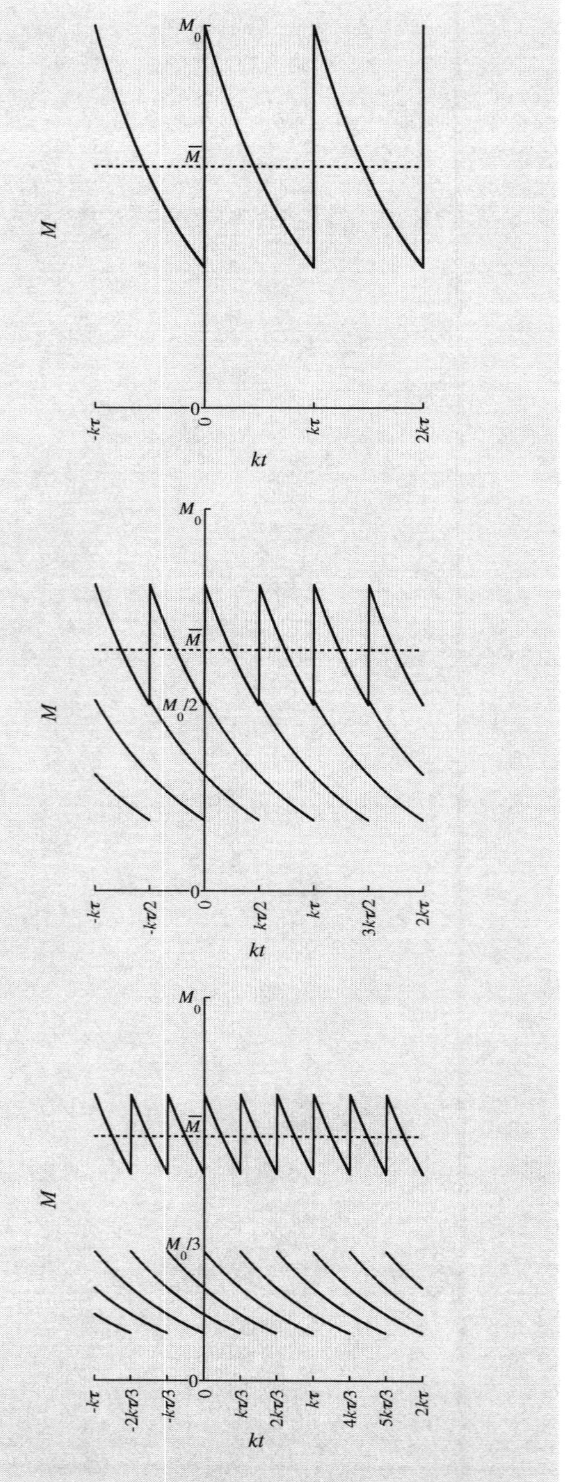

The (grand) average, indicated as dashed line, is the same in all cases – when computed over τ, as useful life of the radioactive material; however, dispersion around the said average is reduced when M_0 is subdivided in (more than one) identical fractions. The plots above unfold a periodic pattern, of period τ in the case of a single portion, $\tau/2$ in the case of two halves, and $\tau/3$ in the case of three thirds; in general, one easily concludes on a period τ/N, in the case of N fractions – each with a weight equal to $1/N$ of the required overall weight.

c) In view of the periodicity of the fraction of undecayed radioisotope, with period τ/N – as grasped in b), the average irradiation power becomes proportional to the average value of the said fraction, $\overline{(M/M_0)}$, calculated over its specific period – to be obtained via

$$\overline{\left(\frac{M}{M_0}\right)} = \frac{\bar{M}}{M_0} \equiv \frac{\sum_{i=1}^{N} \int_{(i-1)\frac{\tau}{N}}^{i\frac{\tau}{N}} \frac{1}{N} \frac{M}{M_0} dt}{\int_0^{\frac{\tau}{N}} dt}; \quad (1.21.17)$$

here $1/N$ accounts for the fractional weight of each of the N (equal) portions of radioisotope utilized, whereas the distinct sets of lower and upper limits of integration result from the different stages of decay thereof. Equation (1.1215) allows transformation of Eq. (1.21.17) to

$$\frac{\bar{M}}{M_0} = \frac{\frac{1}{N}\sum_{i=1}^{N} \int_{(i-1)\frac{\tau}{N}}^{i\frac{\tau}{N}} e^{-kt} dt}{\int_0^{\frac{\tau}{N}} dt}, \quad (1.21.18)$$

where the constancy of N was already taken advantage of in its factoring out; application of the fundamental theorem of integral calculus gives then rise to

$$\frac{\bar{M}}{M_0} = \frac{\frac{1}{N}\sum_{i=1}^{N} \left.\frac{e^{-kt}}{(-k)}\right|_{(i-1)\frac{\tau}{N}}^{i\frac{\tau}{N}}}{\left. t \right|_0^{\frac{\tau}{N}}} \quad (1.21.19)$$

– which degenerates to

$$\frac{\bar{M}}{M_0} = \frac{\frac{1}{N}\sum_{i=1}^{N} \frac{e^{-k(i-1)\frac{\tau}{N}} - e^{-ki\frac{\tau}{N}}}{k}}{\frac{\tau}{N} - 0} = \frac{\sum_{i=1}^{N}\left(e^{-(i-1)\frac{k\tau}{N}} - e^{-i\frac{k\tau}{N}} \right)}{k\tau}, \quad (1.21.20)$$

upon factoring out k for being constant, and dropping $1/N$ from both numerator and denominator afterward. The telescopic nature of the outstanding series in Eq. (1.21.20) allows cancellation of consecutive terms, so it reduces to just the first exponential function of the first term and the second exponential function of the last term, according to

$$\frac{\bar{M}}{M_0} = \frac{e^{-(1-1)\frac{k\tau}{N}} - e^{-N\frac{k\tau}{N}}}{k\tau} = \frac{1-e^{-k\tau}}{k\tau}; \quad (1.21.21)$$

therefore, the average fraction of radiosiotopes in native form is independent of their partitioning into N portions – as anticipated and graphically observed in b). This is so because of the independence in decay between atoms in the same portion, which extends immediately to atoms in identically-sized, yet distinct portions.

d) As emphasized in c), analysis of performance over a single period suffices – since behavior in every other turns identical; hence, one may again focus on t varying within $[0,\tau/N]$. In this regard, it is instructive to note that

$$\frac{1}{N}\sum_{i=1}^{N} \int_{(i-1)\frac{\tau}{N}}^{(i-1)\frac{\tau}{N}+t} e^{-k\tilde{t}} d\tilde{t} = \frac{1}{N}\sum_{i=1}^{N} \left.\frac{e^{-k\tilde{t}}}{(-k)}\right|_{(i-1)\frac{\tau}{N}}^{(i-1)\frac{\tau}{N}+t} \quad (1.21.22)$$

– where the right-hand side would, in general, replace the numerator of Eq. (1.21.18) as descriptor of the cumulative value of disintegrations over period $[(i-1)\tau/N,(i-1)\tau/N+t]$ if only the initial portion of a cycle were of interest, i.e. when $0<t<\tau/N=(i-1)\tau/N+\tau/N-(i-1)\tau/N=i\tau/N-(i-1)\tau/N$. Application of the fundamental theorem of integral calculus to Eq. (1.21.22) may now be finalized as

$$\frac{1}{N}\sum_{i=1}^{N} \int_{(i-1)\frac{\tau}{N}}^{(i-1)\frac{\tau}{N}+t} e^{-k\tilde{t}} d\tilde{t} = \frac{1}{Nk}\sum_{i=1}^{N}\left(e^{-k(i-1)\frac{\tau}{N}} - e^{-k\left((i-1)\frac{\tau}{N}+t\right)} \right), \quad (1.21.23)$$

where $e^{-k(i-1)\frac{\tau}{N}}$ can still be factored out as

$$\frac{1}{N}\sum_{i=1}^{N} \int_{(i-1)\frac{\tau}{N}}^{(i-1)\frac{\tau}{N}+t} e^{-k\tilde{t}} d\tilde{t} = \frac{1}{Nk}\sum_{i=1}^{N} e^{-k(i-1)\frac{\tau}{N}}\left(1-e^{-kt}\right). \quad (1.21.24)$$

The right-hand side of Eq. (1.21.24) may, in turn, be rewritten as

$$\frac{1}{N}\sum_{i=1}^{N} \int_{(i-1)\frac{\tau}{N}}^{(i-1)\frac{\tau}{N}+t} e^{-k\tilde{t}} d\tilde{t} = \frac{1}{N}\sum_{i=1}^{N} e^{-k(i-1)\frac{\tau}{N}} \left.\frac{e^{-k\tilde{t}}}{(-k)}\right|_0^t, \quad (1.21.25)$$

which is equivalent to

$$\frac{1}{N}\sum_{i=1}^{N} \int_{(i-1)\frac{\tau}{N}}^{(i-1)\frac{\tau}{N}+t} e^{-k\tilde{t}} d\tilde{t} = \frac{1}{N}\sum_{i=1}^{N} e^{-k(i-1)\frac{\tau}{N}} \int_0^t e^{-k\tilde{t}} d\tilde{t} \quad (1.21.26)$$

after reversed application of the fundamental theorem of integral calculus; $e^{-k(i-1)\frac{\tau}{N}}$ will finally be moved back into the kernel to give

$$\frac{1}{N}\sum_{i=1}^{N}\int_{(i-1)\frac{\tau}{N}}^{(i-1)\frac{\tau}{N}+t} e^{-k\tilde{t}}d\tilde{t} = \frac{1}{N}\sum_{i=1}^{N}\int_{0}^{t} e^{-k(i-1)\frac{\tau}{N}} e^{-k\tilde{t}}d\tilde{t} \quad (1.21.27)$$

– where summation and integral operators can finally be swapped due to their linearity, and exponential functions lumped as

$$\frac{1}{N}\sum_{i=1}^{N}\int_{(i-1)\frac{\tau}{N}}^{(i-1)\frac{\tau}{N}+t} e^{-k\tilde{t}}d\tilde{t} = \int_{0}^{t}\frac{1}{N}\sum_{i=1}^{N} e^{-k\left((i-1)\frac{\tau}{N}+\tilde{t}\right)}d\tilde{t}. \quad (1.21.28)$$

One may now search for the time, $t=t_0$, when the grand average over a full cycle, $\overline{M}/M_0$, is crossed by the instantaneous value, M/M_0, of the ratio of undecayed to native material, i.e.

$$\left.\frac{M}{M_0}\right|_{t=t_0} = \frac{1}{N}\sum_{i=1}^{N} e^{-k\left((i-1)\frac{\tau}{N}+t_0\right)} = \frac{1-e^{-k\tau}}{k\tau} = \frac{\overline{M}}{M_0}, \quad (1.21.29)$$

written with the aid of Eq. (1.21.21) – where only the kernel of the integral in the right-hand side of Eq. (1.21.28) was obviously retrieved. After factoring out $1/N$ and e^{-kt_0}, Eq. (1.21.29) becomes

$$\frac{1}{N}e^{-kt_0}\sum_{i=1}^{N} e^{-(i-1)\frac{k\tau}{N}} = \frac{1-e^{-k\tau}}{k\tau}, \quad (1.21.30)$$

or else

$$\sum_{j=0}^{N-1} e^{-j\frac{k\tau}{N}} = \frac{1-e^{-k\tau}}{\frac{k\tau}{N}}e^{kt_0} \quad (1.21.31)$$

upon isolation of the summation and decrease of its counting variable by one unit; a geometric series has meanwhile arisen in the left-hand side – with 1 as first term, and $e^{-k\tau/N}$ for ratio between consecutive terms. One may, therefore, retrieve the formula for the sum of the first $N-1$ terms of such a geometric series, and accordingly reformulate the left-hand side of Eq. (1.21.31) to read

$$\frac{1-e^{-((N-1)+1)\frac{k\tau}{N}}}{1-e^{-\frac{k\tau}{N}}} = \frac{1-e^{-N\frac{k\tau}{N}}}{1-e^{-\frac{k\tau}{N}}} = \frac{1-e^{-k\tau}}{1-e^{-\frac{k\tau}{N}}} = \frac{1-e^{-k\tau}}{\frac{k\tau}{N}}e^{kt_0}, \quad (1.21.32)$$

which reduces to

$$e^{kt_0} = \frac{\frac{k\tau}{N}}{1-e^{-k\tau}}\frac{1-e^{-k\tau}}{1-e^{-\frac{k\tau}{N}}} \quad (1.21.33)$$

after solving for e^{kt_0}; cancellation of $1-e^{-k\tau}$ between numerator and denominator is now in order, viz.

$$e^{kt_0} = \frac{\frac{k\tau}{N}}{1-e^{-\frac{k\tau}{N}}} \quad (1.21.34)$$

– while logarithms may be taken of both sides, viz.

$$kt_0 = \ln\frac{\frac{k\tau}{N}}{1-e^{-\frac{k\tau}{N}}}. \quad (1.21.35)$$

In view of the decreasing, concave shape of the curves within every (consecutive) cycle of amplitude τ/N, the abscissa of the first (and representative) intercepts of the solid thick curves with the horizontal dashed line is lower than $\tau/2N$ (as would happen in the case of a decreasing straight line); this is confirmed via calculation of t_0 using Eq. (1.21.35). The maximum average (positive) deviation, $\left(\overline{\Delta M}/M_0\right)_{max}$, between M/M_0 and $\overline{M}/M_0$ – given by

$$\left(\frac{\overline{\Delta M}}{M_0}\right)_{max} \equiv \frac{\int_0^{t_0}\left(\frac{M}{M_0}-\frac{\overline{M}}{M_0}\right)dt}{\int_0^{t_0}dt} \quad (1.21.36)$$

as per definition, will then look like

$$\left(\frac{\overline{\Delta M}}{M_0}\right)_{max} = \frac{\int_0^{t_0}\left(\frac{1}{N}\sum_{i=1}^{N} e^{-k\left((i-1)\frac{\tau}{N}+t\right)} - \frac{1-e^{-k\tau}}{k\tau}\right)dt}{\int_0^{t_0}dt} \quad (1.21.37)$$

after recalling the left-hand side of Eq. (1.21.29) and the right-hand side of Eq. (1.21.21); note that only the integral between 0 and t_0, conveying a positive value, is of interest here – knowing that the corresponding integral between t_0 and τ/N will be exactly its negative, otherwise $\overline{M}/M_0$ would not represent the average of M/M_0 over $[0,\tau/N]$. The integral in numerator of Eq. (1.21.37) may, in turn, be decomposed as

$$\left(\frac{\overline{\Delta M}}{M_0}\right)_{max} = \frac{\sum_{i=1}^{N}\int_0^{kt_0} e^{-k\left((i-1)\frac{\tau}{N}+t\right)}d(kt)}{N\int_0^{kt_0}d(kt)} - \frac{\frac{1-e^{-k\tau}}{k\tau}\int_0^{t_0}dt}{\int_0^{t_0}dt}, \quad (1.21.38)$$

with both numerator and denominator of the first fraction meanwhile multiplied by k, $1/N$ taken off the kernel, and summation and integral operators exchanged; after taking $e^{-k(i-1)\tau/N}$ off the kernel in the first term, and dropping identical integrals from numerator and denominator in the second term, Eq. (1.21.38) simplifies to

$$\left(\frac{\overline{\Delta M}}{M_0}\right)_{\max} = \frac{\sum_{i=1}^{N} e^{-(i-1)\frac{k\tau}{N}} \int_0^{kt_0} e^{-kt} d(kt)}{N \int_0^{kt_0} d(kt)} - \frac{1-e^{-k\tau}}{k\tau}. \quad (1.21.39)$$

The fundamental theorem of integral calculus can be invoked again to obtain

$$\left(\frac{\overline{\Delta M}}{M_0}\right)_{\max} = \frac{\sum_{i=1}^{N} e^{-(i-1)\frac{k\tau}{N}} \left.\frac{e^{-kt}}{-1}\right|_0^{kt_0}}{Nkt\Big|_0^{kt_0}} - \frac{1-e^{-k\tau}}{k\tau} \quad (1.21.40)$$

from Eq. (1.21.39), which readily becomes

$$\left(\frac{\overline{\Delta M}}{M_0}\right)_{\max} = \frac{\sum_{i=1}^{N} e^{-(i-1)\frac{k\tau}{N}} \left(1-e^{-kt_0}\right)}{Nkt_0} - \frac{1-e^{-k\tau}}{k\tau}. \quad (1.21.41)$$

After factoring out $1-e^{-kt_0}$ and decreasing by one unit the counting variable in the summation, Eq. (1.21.41) turns to

$$\left(\frac{\overline{\Delta M}}{M_0}\right)_{\max} = \frac{1-e^{-kt_0}}{Nkt_0} \sum_{j=0}^{N-1} e^{-j\frac{k\tau}{N}} - \frac{1-e^{-k\tau}}{k\tau} \quad (1.21.42)$$

– where the formula for the sum of the first terms of a geometric series may be recalled once more, thus allowing conversion to

$$\left(\frac{\overline{\Delta M}}{M_0}\right)_{\max} = \frac{1-e^{-kt_0}}{Nkt_0} \frac{1-e^{-N\frac{k\tau}{N}}}{1-e^{-\frac{k\tau}{N}}} - \frac{1-e^{-k\tau}}{k\tau} \quad (1.21.43)$$

since the first term is 1 and the ratio between consecutive terms is $e^{-k\tau/N}$. Algebraic lumping of common factors in the exponential functions of Eq. (1.21.43) unfolds

$$\left(\frac{\overline{\Delta M}}{M_0}\right)_{\max} = \frac{1-e^{-kt_0}}{Nkt_0} \frac{1-e^{-k\tau}}{1-e^{-\frac{k\tau}{N}}} - \frac{1-e^{-k\tau}}{k\tau}, \quad (1.21.44)$$

which turns equivalent to

$$\left(\frac{\overline{\Delta M}}{M_0}\right)_{\max} = \frac{1-e^{-k\tau}}{k\tau} \left(\frac{\frac{k\tau}{N}}{kt_0} \frac{1-e^{-kt_0}}{1-e^{-\frac{k\tau}{N}}} - 1 \right) \quad (1.21.45)$$

after factoring $1-e^{-k\tau}$ and $k\tau$ out, besides multiplying and dividing the first term in parenthesis by k. Meanwhile, Eq. (1.21.34) will be algebraically rearranged to read

$$1-e^{-kt_0} = 1 - \frac{1-e^{-\frac{k\tau}{N}}}{\frac{k\tau}{N}} = \frac{\frac{k\tau}{N} - 1 + e^{-\frac{k\tau}{N}}}{\frac{k\tau}{N}} \quad (1.21.46)$$

– after adding unity to the negatives of the reciprocals of both sides, and reducing the outcome to the same denominator; insertion of Eqs. (1.21.35) and (1.21.46) transforms Eq. (1.21.45) to

$$\left(\frac{\overline{\Delta M}}{M_0}\right)_{\max} = \frac{1-e^{-k\tau}}{k\tau} \left(\frac{\frac{k\tau}{N}}{\ln \frac{\frac{k\tau}{N}}{1-e^{-\frac{k\tau}{N}}}} \frac{\frac{\frac{k\tau}{N} - 1 + e^{-\frac{k\tau}{N}}}{\frac{k\tau}{N}}}{1-e^{-\frac{k\tau}{N}}} - 1 \right). \quad (1.21.47)$$

Upon dropping of $k\tau/N$ from both numerator and denominator, and division of both sides by $(1-e^{-k\tau})/k\tau$, Eq. (1.21.47) becomes

$$\frac{\left(\frac{\overline{\Delta M}}{M_0}\right)_{\max}}{\frac{1-e^{-k\tau}}{k\tau}} = \frac{\frac{k\tau}{N} - \left(1-e^{-\frac{k\tau}{N}}\right)}{\left(1-e^{-\frac{k\tau}{N}}\right) \ln \frac{\frac{k\tau}{N}}{1-e^{-\frac{k\tau}{N}}}} - 1, \quad (1.21.48)$$

or else

$$\frac{k\tau}{1-e^{-k\tau}} \left(\frac{\overline{\Delta M}}{M_0}\right)_{\max} = \frac{\frac{\frac{k\tau}{N}}{1-e^{-\frac{k\tau}{N}}} - 1}{\ln \frac{\frac{k\tau}{N}}{1-e^{-\frac{k\tau}{N}}}} - 1 \quad (1.21.49)$$

along with division of both numerator and denominator of the first term in the right-hand side by $1-e^{-k\tau/N}$; Eq. (1.21.49) is plotted below.

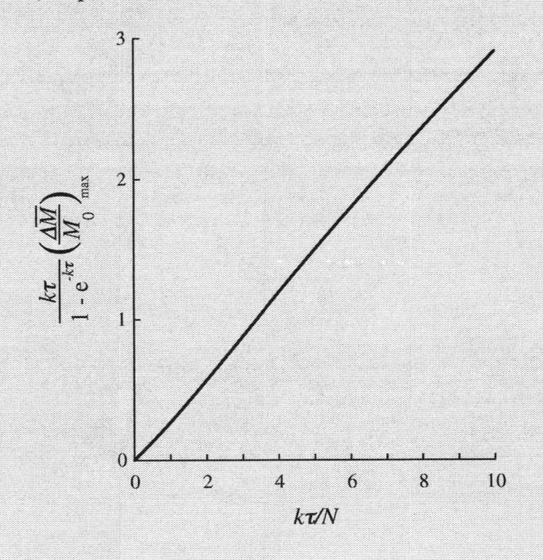

Note that $\frac{k\tau}{1-e^{-k\tau}}\left(\frac{\overline{\Delta M}}{M_0}\right)_{max}+1$ in the left-hand side of Eq. (1.21.49) must, in fact, be positive – because $k\tau/N>1-e^{-k\tau/N}$ implies $\frac{k\tau/N}{1-e^{-k\tau/N}}-1>0$ in numerator, and also $\ln\frac{k\tau/N}{1-e^{-k\tau/N}}>0$ in denominator; while $k\tau/N<1-e^{-k\tau/N}$ implies $\frac{k\tau/N}{1-e^{-k\tau/N}}-1<0$ in numerator, and thus $\ln\frac{k\tau/N}{1-e^{-k\tau/N}}<0$ in denominator. The curve described by Eq. (1.21.49) is almost a straight line – with behavior at large $k\tau/N$ given (in principle) by

$$\lim_{\frac{k\tau}{N}\to\infty}\frac{k\tau}{1-e^{-k\tau}}\left(\frac{\overline{\Delta M}}{M_0}\right)_{max}=\lim_{k\tau\to\infty}\frac{\dfrac{\frac{k\tau}{N}}{1-e^{-\frac{k\tau}{N}}}-1}{\ln\dfrac{\frac{k\tau}{N}}{1-e^{-\frac{k\tau}{N}}}}-1$$

$$=\lim_{\frac{k\tau}{N}\to\infty}\frac{\frac{k\tau}{N}-1}{\ln\frac{k\tau}{N}}-1=\frac{\infty}{\infty}-1, \tag{1.21.50}$$

since $e^{-\infty}<<1$ when $k\tau/N$ grows unbounded. In view of the unknown quantity found, of the ∞/∞ type, one should resort to l'Hôpital's rule to get

$$\lim_{\frac{k\tau}{N}\to\infty}\frac{k\tau}{1-e^{-k\tau}}\left(\frac{\overline{\Delta M}}{M_0}\right)_{max}=\lim_{\frac{k\tau}{N}\to\infty}\frac{\dfrac{1-e^{-\frac{k\tau}{N}}-\frac{k\tau}{N}e^{-\frac{k\tau}{N}}}{\left(1-e^{-\frac{k\tau}{N}}\right)^2}}{\dfrac{\dfrac{1-e^{-\frac{k\tau}{N}}-\frac{k\tau}{N}e^{-\frac{k\tau}{N}}}{\left(1-e^{-\frac{k\tau}{N}}\right)^2}}{\dfrac{\frac{k\tau}{N}}{1-e^{-\frac{k\tau}{N}}}}}-1 \tag{1.21.51}$$

from Eq. (1.21.50) instead – where factors alike cancel out between numerator and denominator, and the number of fraction levels reduces as

$$\lim_{\frac{k\tau}{N}\to\infty}\frac{k\tau}{1-e^{-k\tau}}\left(\frac{\overline{\Delta M}}{M_0}\right)_{max}=\lim_{\frac{k\tau}{N}\to\infty}\frac{\frac{k\tau}{N}}{1-e^{-\frac{k\tau}{N}}}-1$$

$$=\frac{\infty}{1-e^{-\infty}}-1=\frac{\infty}{1}-1=\infty; \tag{1.21.52}$$

hence, $k\tau\left(\overline{\Delta M}/M_0\right)_{max}/\left(1-e^{-k\tau}\right)$ grows unbounded when $k\tau/N$ does. Conversely, the decrease of $k\tau/(1-e^{-k\tau})$ as $k\tau/N$ decreases, coupled with the decrease in $k\tau\left(\overline{\Delta M}/M_0\right)_{max}/\left(1-e^{-k\tau}\right)$ as per the plot above imply a monotonic decrease of $\left(\overline{\Delta M}/M_0\right)_{max}$ itself when N increases. Therefore, breaking the supply of radioisotope into equal portions, and (ordering and putting) the said portions in operation at a given (constant) pace reduce the variability with elapsing time of irradiation power relative to its average – besides diluting costs throughout operation, and avoiding stops for full recharge.

e) When N is large, one realizes that

$$\lim_{N\to\infty}\frac{k\tau}{N}=0, \tag{1.21.53}$$

and consequently

$$\lim_{N\to\infty}\frac{\frac{k\tau}{N}}{1-e^{-\frac{k\tau}{N}}}=\lim_{\frac{k\tau}{N}\to 0}\frac{\frac{k\tau}{N}}{1-e^{-\frac{k\tau}{N}}}=\frac{0}{1-e^0}=\frac{0}{1-1}=\frac{0}{0}; \tag{1.21.54}$$

one should retrieve again l'Hôpital's rule to circumvent this unknown quantity – and accordingly get

$$\lim_{N\to\infty}\frac{\frac{k\tau}{N}}{1-e^{-\frac{k\tau}{N}}}=\lim_{\frac{k\tau}{N}\to 0}\frac{1}{-e^{-\frac{k\tau}{N}}(-1)}=\lim_{\frac{k\tau}{N}\to 0}\frac{1}{e^{-\frac{k\tau}{N}}}=\frac{1}{e^0}=\frac{1}{1}=1. \tag{1.21.55}$$

Once in possession of Eq. (1.21.55), one may proceed to

$$\lim_{N\to\infty}\frac{k\tau}{1-e^{-k\tau}}\left(\frac{\overline{\Delta M}}{M_0}\right)_{max}=\lim_{N\to\infty}\frac{\dfrac{\frac{k\tau}{N}}{1-e^{-\frac{k\tau}{N}}}-1}{\ln\dfrac{\frac{k\tau}{N}}{1-e^{-\frac{k\tau}{N}}}}-1 \tag{1.21.56}$$

based on Eq. (1.21.49) – which, to facilitate algebraic manipulation, is to be redone as

$$\lim_{N\to\infty}\frac{k\tau}{1-e^{-k\tau}}\left(\frac{\overline{\Delta M}}{M_0}\right)_{max}=\lim_{\Lambda\to 1}\frac{\Lambda-1}{\ln\Lambda}-1=\frac{1-1}{\ln 1}-1=\frac{0}{0}-1 \quad (1.21.57)$$

in view of Eq. (1.21.55), provided that

$$\Lambda\equiv\frac{\frac{k\tau}{N}}{1-e^{-\frac{k\tau}{N}}}. \quad (1.21.58)$$

One should differentiate numerator and denominator of Eq. (1.21.57) separately, with regard to Λ as per l'Hôpital's rule, to get rid of the emerging unknown quantity, viz.

$$\lim_{N\to\infty}\frac{k\tau}{1-e^{-k\tau}}\left(\frac{\overline{\Delta M}}{M_0}\right)_{max}=\lim_{\Lambda\to 1}\frac{1}{\frac{1}{\Lambda}}-1=\lim_{\Lambda\to 1}\Lambda-1=1-1=0; \quad (1.21.59)$$

since $k\tau/(1-e^{-k\tau})>0$ is a constant, one concludes that $\left(\overline{\Delta M}/M_0\right)_{max}\to 0$, as per Eq. (1.21.59), when N grows without limit. Under these circumstances, M/M_0 will essentially coincide with $\bar{M}/M_0$; since irradiation intensity is proportional to $-dM/dt$ – which is, in turn, proportional to M as per Eq. (1.1210), one confirms that constancy of M (or M/M_0, for that matter) when $N\to\infty$ guarantees stability of irradiation intensity over time.

1.5.3.2 Penetration depth

The mechanism of absorption of photons by matter is termed photoelectric effect; in particular, an incident γ-photon – with slightly more energy than the binding energy of an orbital electron it collides with, will strike that electron and concomitantly eject it from its original orbital. This phenomenon may or may not entail rupture of a covalent bond – depending on whether a molecular or an atomic orbital is at stake; but will, in principle, originate a species with an unpaired electron (e.g. cation, free radical). In either case, the photon ceases to exist as a result of the said interaction, with all its energy imparted to the orbital electron. Most energy is required to overcome the binding (potential) energy of the orbital electron as a result of attraction thereof to the nucleus – with the remainder being carried away (as kinetic energy) by the electron upon ejection; such a translational kinetic energy acquired contributes itself to heating, since ejection occurs at a random direction. The ejected electron normally travels a short distance – due to its extremely high charge density; and will eventually give up some of its energy, in the form of a photon with relatively low energy – oftentimes within the infrared range. This photon contributes, in turn, to directly raise temperature of a target atom/molecule, owing to the enhanced vibration/oscillation upon capture thereof. Electrons from outer orbitals eventually drop inward to fill the void left by the ejected electron, so the valence orbital will likely acquire another electron – and form (or not) another covalent bond, thus permitting the atom/molecule to recover its normal chemical stability. The aforementioned set of phenomena justify the ionizing effect of γ-photons, as well as their (minor) heating effect.

As already emphasized, photoelectric interaction is most likely to happen when energy of the incident photon exceeds, but is relatively close to the binding energy of the electron struck off. If less, not enough energy would be available to eject the orbital electron; however, it cannot be much higher as well – and it has been found that the magnitude of the photoelectric effect in this case would be inversely proportional to the third power of the photon energy. Remember, in this regard, that all forms of electromagnetic radiation (γ-rays obviously included) possess a sinusoidal waveform – characterized by frequency, ν, and wavelength, λ, and propagate through a (transparent) medium with velocity v that is upper bounded by c, i.e. the velocity of light in vacuum. The energy of each photon, E, satisfies, in turn,

$$\boxed{E=k_p\nu,} \quad (1.1221)$$

where k_p denotes Plank's constant; while wavelength and frequency are inversely proportional to each other, according to

$$\lambda\nu=v\le c, \quad (1.1222)$$

in agreement with Eq. (2.109) of *Food Proc. Eng.: thermal & chemical operations*.

The depth of penetration of electrons, γ-rays and X-rays in an irradiated food is a direct consequence of the probability of collision with atoms/molecules found along their trajectory inside the target matrix; hence, the degree of penetration is expected to decrease when food density, ρ, increases – since ρ measures concentration, and thus number of atomic/molecular entities per unit volume. However, it also depends on energy, E, of those quanta – since energies excessively above the binding energies of most orbital electrons fail to produce any photoelectric effect (as discussed above), and thus travel longer before being absorbed. In any case, the outer parts of a food are expected to receive a higher dose of radiation than their inner counterparts. The underlying relationship is known as Beer and Lambert's law – and pertains to attenuation of radiation caused by the intrinsic properties of the material struck; it states that the intensity, $I\{z\}$, of an electromagnetic wave emerging from a slice of thickness dz, measured normally to the incident beam with intensity I_0, decreases proportionally to its own intensity – according to

$$dI=-\mu I\,dz, \quad (1.1223)$$

where $\mu\equiv\mu\{\rho,E\}$ denotes attenuation coefficient – subjected to

$$I\big|_{z=0}=I_0 \quad (1.1224)$$

as boundary condition. Integration of Eq. (1.1223), via separation of variables, gives rise to

$$\int_{I_0}^{I}\frac{d\tilde{I}}{\tilde{I}}=-\mu\int_0^z d\tilde{z} \quad (1.1225)$$

with the aid of Eq. (1.1224) – and assuming a uniform, isotropic medium; Eq. (1.1225) degenerates to

$$\ln\tilde{I}\Big|_{I_0}^{I}=-\mu\,\tilde{z}\Big|_0^z \quad (1.1226)$$

at the expense of the fundamental theorem of integral calculus, or else

$$\ln\frac{I}{I_0}=-\mu z. \tag{1.1227}$$

If exponentials are taken of both sides, then Eq. (1.1227) becomes

$$\boxed{I=I_0 e^{-\mu z},} \tag{1.1228}$$

together with multiplication of both sides by I_0 afterward. The most common form of this law refects, however, the functional form of Eq. (1.1227), viz.

$$T=\alpha z, \tag{1.1229}$$

where T denotes transmittance – defined as

$$T\equiv\log_{10}\frac{I_0}{I}, \tag{1.1230}$$

with z referred to as optical path; while α, satisfying

$$\alpha\equiv\frac{\mu}{\ln 10} \tag{1.1231}$$

in view of $\log_{10}\frac{I_0}{I}=\frac{\ln I_0/I}{\ln 10}$, is termed absorbance – and is widely used as operational parameter in the applied literature.

PROBLEM 1.22

Consider a parallelepipedal food formed by N adjacent layers – with attenuation coefficient μ_1, μ_2, ..., μ_N, and thickness z_1, z_2, ..., z_N, respectively, exposed to radiation impinging normally on its top surface.

a) Obtain an expression for intensity of radiation received at the bottom surface of the aforementioned layered food.
b) Prove that Beer and Lambert's law remains valid if the food is subdivided into N adjacent layers, irrespective of the thickness of each one.
c) How does radiation density, D, vary with radius r in the case of a spherically-shaped food? Assume again that radiation strikes normally and uniformly on its outer surface, of radius R – and produce a normalized expression for D.
d) Find the location of (a putative) cold spot in the food described in c).

Solution

a) Beer and Lambert's law, as per Eq. (1.1228), can be applied to a generic i-th layer of food ($i=1, 2, ..., N$) – provided that I is replaced by I_i and I_0 is replaced by I_{i-1}, while μ is to be replaced by μ_i; one therefore obtains

$$\begin{aligned} I_1 &= I_0 e^{-\mu_1 z_1} \\ I_2 &= I_1 e^{-\mu_2 z_2} \\ &\ldots \\ I_N &= I_{N-1} e^{-\mu_N z_N} \end{aligned} \tag{1.22.1}$$

for N consecutive layers – where insertion of the i-th expression on the next one gives rise to

$$\begin{aligned} I_1 &= I_0 e^{-\mu_1 z_1} \\ I_2 &= I_0 e^{-\mu_1 z_1} e^{-\mu_2 z_2} \\ &\ldots \\ I_N &= I_0 e^{-\mu_1 z_1} e^{-\mu_2 z_2} \ldots e^{-\mu_N z_N}, \end{aligned} \tag{1.22.2}$$

since the incident radiation on the top surface of every layer coincides with the leaving radiation at the bottom surface of the previous layer. In view of the operational features of an exponential function, one may lump said functions in Eq. (1.22.2) to get

$$I_N = I_0 e^{-(\mu_1 z_1+\mu_2 z_2+\ldots+\mu_N z_N)} \tag{1.22.3}$$

– thus materializing a descriptor for radiation, I_N, received at the bottom surface of the food.

b) Partition of a food as N consecutive layers – of thickness z_1, z_2, ..., z_N, allows retrieval of Eq. (1.22.3) as

$$I_N = I_0 e^{-(\mu z_1+\mu z_2+\ldots+\mu z_N)} \tag{1.22.4}$$

– as long as μ denotes attenuation coefficient of the (uniform) material of the food; after factoring μ out in the argument of the exponential function, Eq. (1.22.4) turns to

$$I_N = I_0 e^{-\mu(z_1+z_2+\ldots+z_N)} = I_0\exp\left\{-\mu\sum_{i=1}^{N} z_i\right\}. \tag{1.22.5}$$

Since the whole thickness, z, of the food under scrutiny abides to

$$z=\sum_{i=1}^{N} z_i, \tag{1.22.6}$$

Eq. (1.22.5) can be rewritten as

$$I_N = I_0 e^{-\mu z} \tag{1.22.7}$$

– which mimics Eq. (1.1228), since I_N denotes intensity of radiation received at depth z within the food.

c) The radiation density, D, received by a food with area A is defined by

$$D\equiv\frac{I}{A}, \tag{1.22.8}$$

whereas the area of a sphere or radius r satisfies

$$A = 4\pi r^2; \tag{1.22.9}$$

insertion of Eqs. (1.1228) and (1.22.9) transforms Eq. (1.22.8) to

$$D = \frac{I_0 e^{-\mu(R-r)}}{4\pi r^2}, \tag{1.22.10}$$

where distance to surface, z, was replaced by

$$z = R - r. \tag{1.22.11}$$

The radiation density on the surface of the food, D_0, accordingly reads

$$D_0 \equiv D\big|_{r=R} = \frac{I_0 e^{-\mu(R-R)}}{4\pi R^2} = \frac{I_0 e^0}{4\pi R^2} = \frac{I_0}{4\pi R^2}, \tag{1.22.12}$$

obtained from Eq. (1.22.10) after setting $r=R$; ordered division of Eq. (1.22.10) by Eq. (1.22.12) gives rise to

$$\frac{D}{D_0} = \frac{\dfrac{I_0 e^{-\mu(R-r)}}{4\pi r^2}}{\dfrac{I_0}{4\pi R^2}}, \tag{1.22.13}$$

where lumping of factors alike between numerator and denominator generates

$$\frac{D}{D_0} = \frac{e^{-\mu(R-r)}}{\left(\dfrac{r}{R}\right)^2} = \left(\frac{r}{R}\right)^{-2} e^{-\mu R\left(1-\frac{r}{R}\right)} \tag{1.22.14}$$

along with factoring out of R in the argument of the exponential function – thus yielding (normalized radiation density) D/D_0 vs. (normalized radius) r/R as sought.

d) Inspection of Eq. (1.22.14) unfolds two opposing tendencies with increasing r/R – a decreasing tendency associated with $(r/R)^{-2}$ and an increasing tendency associated with $e^{-\mu R(1-r/R)}$; hence, a critical (stationary) point is likely to exist, putatively described by

$$\frac{d\left(\dfrac{D}{D_0}\right)}{d\left(\dfrac{r}{R}\right)} = 0 \tag{1.22.15}$$

as necessary condition. Upon insertion of Eq. (1.22.14), one obtains

$$\frac{d}{d\left(\dfrac{r}{R}\right)}\left(\left(\frac{r}{R}\right)^{-2} e^{-\mu R\left(1-\frac{r}{R}\right)}\right) = 0 \tag{1.22.16}$$

from Eq. (1.22.15), where application of the classical rules of differentiation leads to

$$-2\left(\frac{r}{R}\right)^{-3} e^{-\mu R\left(1-\frac{r}{R}\right)} + \left(\frac{r}{R}\right)^{-2} e^{-\mu R\left(1-\frac{r}{R}\right)} \mu R = 0; \tag{1.22.17}$$

after factoring $e^{-\mu R(1-r/R)}$ and $(r/R)^{-3}$ out, Eq. (1.22.17) becomes

$$e^{-\mu R\left(1-\frac{r}{R}\right)}\left(\frac{r}{R}\right)^{-3}\left(\mu R\frac{r}{R} - 2\right) = 0 \tag{1.22.18}$$

– which readily implies

$$\mu R\frac{r}{R} = 2, \tag{1.22.19}$$

since the (exponential) factor cannot be nil and r/R takes only finite values. Upon dropping R off both numerator and denominator, Eq. (1.22.19) will look like

$$r\mu = 2, \tag{1.22.20}$$

or else

$$r_{\min} = \frac{2}{\mu}; \quad \mu > \frac{2}{R} \tag{1.22.21}$$

after solving for r – valid only if μ exceeds $2/R$, otherwise $r_{\min} > R$ and thus devoid of practical interest. The nature of $r_{\min}$ (when, and if it exists) can be ascertained via the sign of its second derivative, defined as

$$\frac{d^2\left(\dfrac{D}{D_0}\right)}{d\left(\dfrac{r}{R}\right)^2} \equiv \frac{d}{d\left(\dfrac{r}{R}\right)}\left(\frac{d}{d\left(\dfrac{r}{R}\right)}\left(\left(\frac{r}{R}\right)^{-2} e^{-\mu R\left(1-\frac{r}{R}\right)}\right)\right) \tag{1.22.22}$$

– where insertion of the left-hand side of Eq. (1.22.18), as per its equivalence to the left-hand side of Eq. (1.22.16), gives rise to

$$\frac{d^2\left(\dfrac{D}{D_0}\right)}{d\left(\dfrac{r}{R}\right)^2} = \frac{d}{d\left(\dfrac{r}{R}\right)}\left(e^{-\mu R\left(1-\frac{r}{R}\right)}\left(\frac{r}{R}\right)^{-3}\left(\mu R\frac{r}{R} - 2\right)\right). \tag{1.22.23}$$

The rules of differentiation of a multiple product, a power, and an exponential support transformation of Eq. (1.22.23) to

$$\begin{aligned}\frac{d^2\left(\dfrac{D}{D_0}\right)}{d\left(\dfrac{r}{R}\right)^2} &= \mu R\, e^{-\mu R\left(1-\frac{r}{R}\right)}\left(\frac{r}{R}\right)^{-3}\left(\mu R\frac{r}{R} - 2\right) \\ &\quad - 3e^{-\mu R\left(1-\frac{r}{R}\right)}\left(\frac{r}{R}\right)^{-4}\left(\mu R\frac{r}{R} - 2\right) \\ &\quad + \mu R\, e^{-\mu R\left(1-\frac{r}{R}\right)}\left(\frac{r}{R}\right)^{-3},\end{aligned} \tag{1.22.24}$$

where $e^{-\mu R\left(1-\frac{r}{R}\right)}$ and $(r/R)^{-4}$ may, in turn, be factored out to get

$$\frac{d^2\left(\frac{D}{D_0}\right)}{d\left(\frac{r}{R}\right)^2} = e^{-\mu R\left(1-\frac{r}{R}\right)}\left(\frac{r}{R}\right)^{-4}\left(\begin{array}{c}\mu R\frac{r}{R}\left(\mu R\frac{r}{R}-2\right)\\ -3\left(\mu R\frac{r}{R}-2\right)+\mu R\frac{r}{R}\end{array}\right); \tag{1.22.25}$$

a further factoring out of $\mu R(r/R)-2$ unfolds

$$\frac{d^2\left(\frac{D}{D_0}\right)}{d\left(\frac{r}{R}\right)^2} = e^{-\mu R\left(1-\frac{r}{R}\right)}\left(\frac{r}{R}\right)^{-4}\left(\left(\mu R\frac{r}{R}-3\right)\left(\mu R\frac{r}{R}-2\right)+\mu R\frac{r}{R}\right). \tag{1.22.26}$$

In view of Eq. (1.22.19), one obtains

$$\left.\frac{d^2\left(\frac{D}{D_0}\right)}{d\left(\frac{r}{R}\right)^2}\right|_{r=r_{min}} = e^{-\mu R\left(1-\frac{2}{\mu R}\right)}\left(\frac{2}{\mu R}\right)^{-4}\left((2-3)(2-2)+2\right) \tag{1.22.27}$$

from Eq. (1.22.26), which breaks down to

$$\left.\frac{d^2\left(\frac{D}{D_0}\right)}{d\left(\frac{r}{R}\right)^2}\right|_{r=r_{min}} = 2e^{-(\mu R-2)}\left(\frac{\mu R}{2}\right)^4 > 0 \tag{1.22.28}$$

after factoring in μR at the argument of the exponential function; therefore, r_{min} as per Eq. (1.22.21) corresponds indeed to the radial location where the minimum radiation density is received – or cold spot.

The penetration depth, δ_p, is defined as the value of z at which I falls to 1/e of its original value, i.e.

$$\boxed{\delta_p \equiv z\Big|_{\frac{I}{I_0}=\frac{1}{e}}} \tag{1.1232}$$

– which, in view of Eq. (1.1227), leads to

$$-\mu\delta_p = \ln\frac{1}{e} = -\ln e = -1 \tag{1.1233}$$

or, equivalently, to

$$\delta_p \equiv \frac{1}{\mu}; \tag{1.1234}$$

Eq. (1.1230) may instead be taken advantage of to rewrite Eq. (1.1232) as

$$\delta_p \equiv z\Big|_{\ln\frac{I}{I_0}=-1} = z\Big|_{\frac{\log_{10}\frac{I}{I_0}}{\log_{10}e}=-1} = z\Big|_{\log_{10}\frac{I}{I_0}=-\log_{10}e} = z\Big|_{T=0.434}, \tag{1.1235}$$

since $\log_{10}e = -0.434$.

In view of previous discussion, the penetration depth should be somehow proportional to individual quantum energy, measured by E, owing to the lower probability of undergoing absorption by matter; but inversely proportional to concentration of absorbing atoms, as measured by ρ, due to the higher probability of collision with matter. It has been experimentally found that a more accurate relationship should include an additive correction to the former, according to

$$\boxed{\delta_p = \frac{0.5240E-0.1337}{\rho};} \tag{1.1236}$$

Eq. (1.1236) is valid for condensed phases and irradiation energies above 0.1337/0.524 = 0.255 MeV, with E expressed in MeV, ρ in g.cm^{-3}, and δ_p in cm. For instance, with a source energy of 10 MeV, the penetration depth in air is 3800 cm (ρ = 1 kg.m^{-3}); it reduces dramatically to 5.1 cm in the case of water (ρ = 1000 kg.m^{-3}), 4.2 cm for plastic (ρ = 1200 kg.m^{-3}), 2.1 cm for glass (ρ = 2400 kg.m^{-3}), and 1.8 cm for aluminum (ρ = 2700 kg.m^{-3}). If the source energy is reduced to 8 MeV, then δ_p reduces to 4.0 cm in the case of water – but it increases to 6.1 cm when the source energy is rated at 12 MeV.

Since radiation does not penetrate the food matrix in a uniform manner, the degree of irradiation, Δ, ranges between a minimum value, Δ_{min} – attained at the innermost portion, and required to attain the desired sterility degree; and a maximum value, Δ_{max} – attained on the surface, and entailing acceptable sensory changes; hence, the ratio $\Delta_{max}/\Delta_{min}$ measures the uniformity of dose distribution within the food. Such an overdose ratio is useful in attempts to appropriately design and economically operate an irradiation process; for instance, a sensitive food as chicken does not withstand values above 1.5, while a resistant food as onions can go up to 3.0 in terms of overdose ratio.

2

Complementary Operations

Food processing has classically encompassed operations under the scope of the secondary sector, i.e. of a transformation nature; and implemented in industrial settings. However, it should be emphasized that most foods are manufactured from raw materials supplied by plants or animals in the first place (primary sector), and will eventually be delivered to consumers in a direct or indirect mode (tertiary sector). Therefore, transformed foods bearing high quality and safety at the moment of ingestion require raw materials abiding to high standards, as well as final products maintained under appropriate conditions of transportation and storage. This calls for preprocessing operations of materials handling be carried out with great care – e.g. cleaning, sorting, and peeling of raw materials; complemented by processing operations of materials handling – e.g. in-plant transportation of intermediate products between pieces of equipment, and temporary storage; and finalizing with postprocessing operations of materials handling – e.g. storage and distribution of final products.

A finer analysis of the aforementioned complementary operations further unfolds the critical impact of equipment design – bearing hygienic features, and targeted at safe food processing; as well as of criterion-based (and periodic) operations of cleaning, disinfection, and rinsing of the said equipment – prior to and between effective use(s). When cleaning-in-place systems are at stake, spraying or sprinkling the cleaning solution *in situ* has to be dealt with; whereas soaking proves germane for recycling of bottles toward reuse as drink reservoirs. Another relevant issue is supply of freshwater – a ubiquitous element of cleaning (besides playing the frequent role of ingredient), and subsequent treatment of water effluents (and similar wastes) after cleaning/disinfection/rinsing have taken place – with an emphasis on reduction, reutilization, and recycling whenever possible. After the final product is ready, and before entering the distribution chain, food packaging is normally required as well.

Packaging possesses indeed a unique relevance in food processing, which justifies separate (and in-depth) treatment; this is so not because other industrial products do not need packages for convenient handling and storage – but because the food package plays germane roles of protection from physicochemical damage and microbiological contamination, while accounting for a major fraction of the overall unit cost. A wide variety of packaging materials are commercially available – ranging from natural fibers obtained almost directly from plants, to polymers produced via chemical synthesis that often depart from petrol fractions; a hybrid possibility is materialized by paper and the like, where natural fibers (i.e. cellulose from trees) are extensively reprocessed to acquire tailor-made functional features. All sorts of flexible films, as well as rigid and semirigid containers can be manufactured via extrusion/molding; when only rigid containers are intended, metals or glass become an option as packaging material. A key issue for selection of a packaging material comes from the need of transparency – because surface appearance of the food may serve as a strong factor of appeal; or instead from the need of opacity – so as to protect the food from light-induced decay.

Despite sealed packages being the rule – in attempts to minimize environmental contamination of food (and food loss, for that matter), the degree of tightness is an important issue; when the food matrix is still respiring, as frequently happens with fresh produce, controlled permeability to (only) gases contributes to extend its shelf-life – or, equivalently, provision of a modified atmosphere inside the package itself may prove useful. On the other hand, the basic goal of microbiological integrity pursued by food packaging may be extrapolated to protection on site via automatic (yet controlled) delivery of some suitable chemical compound with antimicrobial or antioxidant features; this is the case of active packaging – provided that the said chemical can be loaded as part of the original packaging material. One may go one step further and have the packaging serve as a vehicle of reliable (dynamic) information on storage conditions along the food chain; intelligent packaging serves this purpose, and may resort to time/temperature indicators incorporated in the package.

Between packaging materials and the final packaged product comes a number of filling and sealing processes – complemented by the required labeling that bears (static) information on composition and nutritional features; the strategy followed depends, however, on a variety of factors of economic and environmental nature.

In view of more and more stringent rules pertaining to the environmental impact associated with disposal of packages, biodegradability has arisen as an alternative solution – which can encompass edibility of the packaging film itself; other options are offered by the promising field of nanotechnology, in that packaging materials will be manufactured with target features.

2.1 Handling, cleaning, disinfection, rinsing, and effluent treatment

2.1.1 Introduction

Correct manipulation of foods, ingredients, and packaging materials – from supplier, through the whole production and distribution chains, until reaching the consumer, is essential for optimal product quality and safety; it also helps reduce costs, namely, those incurred in fixing *a posteriori* whatever could have been done properly in the first place. In fact, improvement in handling technologies at all stages of the process – as is the case of harvest, transportation, and storage of raw materials, preparation procedures and movement inside the factory, collection and disposal of wastes, collation of packaged foods and movement to finished product warehouses, distribution to wholesale and retailers, and presentation of products for sale; complemented by correct equipment design and sanitation, have permitted substantial increases in production efficiency over the last few decades.

Most raw materials are likely to carry physicochemical/microbiological contaminants, including inedible materials; and may exhibit (intrinsically) large variability of such physical characteristics as shape, size, and color. High-quality final food products obviously require standardized raw materials – so cleaning, sorting, grading, and peeling are frequently performed prior to nuclear processing, or before sale in the fresh market sector.

Cleaning, disinfection, and rinsing are, in general, aimed at eliminating contamination and destroying microorganisms that may be present in food-processing equipment – or in storage containers where the food is to be kept prior to, or between operations. Cleaning aims primarily at removing foreign materials (e.g. soil, dirt, animal fragments), eliminating chemical contaminants (e.g. agrochemicals), and preventing accumulation of residues of biological origin or otherwise. Disinfection (or sanitization) is targeted chiefly at microbial contaminants, and resorts to chemical aids or other methods able to reduce initial microbial load on the surface of raw materials or food processing equipment. Finally, rinsing is required to remove the said cleaning and disinfection agents, otherwise they would themselves play the role of chemical contaminants in the final food. Efficient sanitization measures also include approved materials for construction; adequate light; proper rate of air renewal and air velocity pattern; separation between raw material reception, food processing, and final product storage areas; sufficient space for operation and movement; approved plumbing; adequate water supply and draining; appropriate sewage disposal system and waste treatment facilities; and suitable soil conditions and surrounding environment.

HISTORICAL FRAMEWORK

Genuine knowledge and rationalization of human diseases was not available until the 19th century – yet suspicion had for long fallen on the harmful action of animalcules (or small living organisms). In the 1st century BC, Terentius Varro claimed that a disinfectant was effective if it had an apparent corrosive, suffocating, or toxic effect upon (visible) living creatures. The oldest reference to disinfection with a chemical agent, however, dates back to the 8th century BC, with Homer in *Book XII* of the *Odyssey* – where the hero, having killed his rivals, demanded that sulfur be burnt in the house they had occupied. The purifying effect of sulfur dioxide fumes was prescribed as well by Susruta, around the 4th century BC in India, to be carried out in rooms prior to surgical operations. Use of mercurial compounds in medicine was developed by the Arabs, who eventually disseminated such a know-how to Europeans – as recorded by Mathaeus Platerius in 1140. However, it was not until the work by Robert Koch, in the late 19th century, that a conclusive demonstration on the efficacy of said corrosive sublimate upon *in vitro* cultures of microorganisms was provided – thus paving the way for merthiolate and mercurochrome as disinfecting agents. On the other hand, seamen had for long been aware that algae and fungi are unable to grow on the hulls of boats sheathed in copper; this realization may have inspired the practice of applying an aqueous solution of Bordeaux mixture (containing copper sulfate) against mildew by French wine growers.

The first alkalis, meant for neutralization of microorganisms and viruses, were probably derived from lime; the striking detergent effect of sodium hydroxide and the white traces left by lime enabled, from early times, straightforward check of correct application of the said cleaning/disinfectant agent (alkali was actually coined, in 1509, from the Arab word for soda, *al-quali*). In 1715, Giovanni Lancisi, physician and private chamberlain to Pope Innocent XII, recommended use of concentrated soda lime to wash fountains, vessels, and drinking troughs. On the other hand, the corrosive action of strong acids upon hard substances (e.g. stone, metal), coupled with the preserving effect of vinegar on fruits and vegetables led embalmers propose use of acids as disinfectants. In his *Natural History*, published in 1625, Francis Bacon advised on disinfecting drinking water with small amounts of oil of vitriol (alias of sulfuric acid) to keep it fresh. However, only in 1676 did van Leeuwenhoek offer the first scientific proof of action of acids upon animalcules invisible to the human eye – which he had discovered, using the microscope of his own invention; when bacteria, previously collected from tooth surfaces, were covered with wine vinegar, such "very small objects moving with a swift motion, like eels, ceased their activity."

Disinfection by physical methods has also been empirically practiced from very early times; temperature rise, fumigation, drying, and filtration were routinely applied throughout Greco-Roman Antiquity. Soldiers of Alexander, the Great, followed the advice of Aristotle and boiled their drinking water; but Spalanzani was the first to experimentally demonstrate, in 1776, that spontaneous generation of microorganisms becomes impossible once the fluid they have lived in has been boiled for one hour. Fumigation was in use for centuries to purify the air – perhaps due to observation that smoke chases away insects, which were for long implicated with human diseases; in 429 BC, fumigation by burning odoriferous herbs was indeed recommended by Hippocrates to control a human epidemic in Athens. Drying often involved combined action of heat and UV light obtained via exposure to sunlight; it was recommended, around the 7th century BC, in *Avesta Vandidad*– the code of doctrine by Zarathustra, to purify surfaces where cadavers had been lying. Drying was also used in ancient Egypt to complete embalming of corpses, after soaking them in a salt solution – stemming from observation that cadavers spontaneously mummify upon drying in the desert. Although ancient Egyptians resorted to filtration to clarify grape juice by passing it through fabrics, it was not until the early 2nd century that Sayyid Jorjani, in Persia, recorded that filtered water took longer to go stale; filtration was eventually enforced in 1757 by the British Navy for purification of water aboard, via passing it through sand and charcoal.

Besides assuring safety, cleaning and disinfection are relevant in the food industry for sensory reasons; the quality of the final product is frequently influenced by unwanted tastes due to untimely contact with extraneous chemicals or untimely microbial development. Microbial growth may indeed occur on residues of products left in the processing equipment after completion of the previous operation – or else on fouling that takes place during regular processing; and existence of such residues as crusts of dried (or even spoiled) products, or presence of insects and their larvae, or even droppings of rodents will have a catastrophic effect upon safety – and thus eventual consumer acceptance. On the other hand, if a food product is not appropriately packaged upon processing, reinfection may occur – thus compromising the very effect of disinfection in the first place, upon extension of shelf-life and guarantee of safety.

2.1.2 Product and process overview

Efficient handling of materials consists of the organized movement of materials in correct quantities, from and to the correct places – effected as fast and safely as possible, and with as low as possible expenditure of labor and generation of waste. Appropriate procedures encompass moving materials only when necessary, and preferably in unit or bulk form to avoid cross-contamination; placing related operations as close as possible to each other, so as to reduce need for manipulation – using all layers along height of plant buildings to reduce floor space needs, providing correct storage conditions, and taking advantage of gravitational flow whenever possible; minimizing manual handling, at the expense of mechanized (and preferably automated) continuous handling – via equipment adaptable to distinct applications, in sufficient number and with convenient power; and optimizing material flow, via reduction of path length – to facilitate supply in the correct time and proper condition, and reduce fuel consumption and driver's time, as well as via establishment of a correct sequence to avoid shortages and bottlenecks – with logistics aided by a (rational) systems approach. By doing so, savings are expected in storage and operating space – further to improvements in stock control, product quality, processing time, production costs, and working conditions (including lower risk of accidents), as well as reduction of environmental impact.

Equipment (and facility) hygienic design, combined with cleaning routines therefor constitute a seminal portion of GMP codes and HACCP – as discussed in *Food Proc. Eng.: basics & mechanical operations*; commercial software exists to assist managers in devising adequate cleaning schedules. A basic issue may, however, arise because hygiene practices may conflict with safety precautions – namely due to the toxic nature of some detergents/disinfectants utilized; a preliminary risk assessment is advised in these cases.

2.1.2.1 Preprocess handling

The three most important steps prior to processing at a food plant (which may not all be present, or present at all) are cleaning of raw materials, sorting of particulate foods, and peeling of fruits; they assume special relevance in the case of fruits and vegetables.

Cleaning is primarily intended not only at removal of contaminating materials from raw foods, but also at preliminary separation thereof – so as to leave them suitable for sale in the fresh market sector, or else undergo further processing; this often extends to descaling of fish, peeling of fruits and vegetables, and skinning of meat. Contaminants found in raw foods arise from irrigation water, manure traces, and contact with skin, face, and feet of animals themselves, as well as incorrect application or overuse of agrochemicals – prior to harvest or slaughter proper; mechanized handling and processing equipment – during harvest or slaughter; and use of unclean washing water or equipment, as well as cross-contamination – following harvest or slaughter. Common examples include: animals and fragments thereof – e.g. insects and larvae, or blood, bone, excreta and hair, respectively; plant portions – e.g. leaves, twigs, weed seeds, pods, and skins; microbial cells – e.g. soft rots, and fungal and yeast colonies; microbial metabolites – e.g. toxins, odors, tastes, and colors; chemicals – e.g. fertilizers, herbicides, and pesticides (viz. insecticides, fungicides); minerals – e.g. soil, engine oil, grease, and stones; and ferrous and nonferrous metals – e.g. fillings, nuts, and bolts.

In order to fulfill its basic role in HACCP systems – besides reducing food wastage and improving process economics, cleaning should take place as early as possible along the process; this not only prevents damage to subsequent equipment (in the case of stones, bone, or metal debris), but also avoids loss of remaining bulk due to unwanted microbial growth, and spending valuable resources to remove contaminants *a posteriori* (with eventual discarding of product itself). Presence of contaminants in processed foods is in fact a major cause of legal prosecution of food companies; such contaminants normally take the form of extraneous matter, or else chemical or microbiological entities – further to being a consequence of lack of hygiene at large. Composition, labeling, and presentation come next; with food quality per se and other offenses coming last in the list of causes for lawsuits. Therefore, retail buyers have enforced detailed specifications on the maximum tolerable levels of contaminants.

Cleaning operations are categorized as wet or dry; the former include soaking and spraying – while dry procedures include air-, magnetic-, and mechanical-mediated separations. Selection of the right procedure hinges critically on the nature of product, and the qualitative and quantitative (known or suspected) profiles of contaminant(s); a combination of methods may actually prove the best option. A compromise ought to be reached between sufficiently extensive cleaning (enforced by food buyers) and least expensive cleaning (enforced by food processors).

Wet cleaning is inherently more effective than its dry counterpart in attempts to remove soil from root crops, or dust and pesticide residues from fruits and vegetables; various physical processing conditions and chemical additives may even improve the efficiency of that mode of cleaning. For instance, water warming and detergent addition are helpful when mineral oil is present – but cold water is preferred to reduce operational costs and texture damages imparted to, besides lowering microbial-mediated decay in food. Conversely, wet procedures require large amounts of freshwater, and thus produce large amounts of effluents – often bearing high concentrations of suspended solids, as well as dissolved solids; the latter have classically been measured by their chemical oxygen demand (COD) and biological oxygen demand (BOD). To reduce costs of purchasing freshwater and treating effluent water, filtration followed by recirculation of water have been employed – usually after chlorination treatment (up to 100–200 ppm) to prevent microorganism buildup.

Examples of wet-cleaning equipment include soaking tanks, spray and brush washers, drum and rod washers, ultrasonic cleaners, and flotation tanks; a few industrial examples are conveyed in Fig. 2.1.

Soaking, as illustrated in Fig. 2.1i, is a preliminary operation aimed at (partially) removing soil and stones from food items; it is frequently required prior to cleaning of heavily contaminated root crops. Some level of stirring may be provided by paddles to facilitate detachment of impurities from the food surface; in the case of delicate crops (e.g. asparagus, strawberries) or a major tendency for internal accumulation of dirt (e.g. celery), stirring may be achieved via air bubbling. Small, not too dense fruits and vegetables (e.g. beans, peas) may also undergo fluming, thus being carried by water in troughs over a number of weirs – while being simultaneously displaced, and thus brought to the next processing step. Dewatering screens are normally provided at the end, to permit removal of most wash water from the clean product.

Soaking is also employed to clean recycled glass bottles before filling them; they are sequentially injected with a solution of detergent, and then subjected to up to three hot water rinses and one final cold water rinse. Hot water from the rinse section may be advantageously employed as pre-rinse, thus warming bottles before they enter the bath. Since a portion of washing solution is carried away by the bottles as they leave the bath, makeup with fresh solution has to be performed on a continuous basis – so as to keep the volume of the soaking bath essentially constant. Transport of detergent from the main bath into the rinsing zone leads to loss of valuable cleaning solution, besides increasing *pH* of the warm water rinse – due to the alkaline nature of the detergent, with such drawbacks as scale formation and waste of rinse water. This is why the warm rinse water in most plants is neutralized with sulfuric acid or carbon dioxide, or even completely distilled via the heating steam of the main bath; if the warm rinse water is properly neutralized, the cold rinse water is needed chiefly to cool down the bottles. Electrolysis of rinse water has also been proposed – able not only to decrease *pH*, but also to promote recovery of the caustic agent. Besides soaking proper, cleaning of bottles may be brought about via discontinuous jets. Containers of plastic or metal are in general used only once, so they undergo a single rinse with hot water possessing a cleaning agent; the same applies to single-use glass bottles, or to the first use of (to-be-recycled) glass bottles.

Spray washing, as per Fig. 2.1ii, is used for many crops; drum or belt washers are regularly employed, as well as rotating baskets. The efficiency of cleaning increases with pressure and temperature – as long as food texture is sufficiently robust; it also depends on amount of water used, and corresponding contact time. Cleaning (and concomitant disinfection) of chicken carcasses with jets of vapor from organic liquids has also been attempted with success. Brushes, see Fig. 2.1iii, or flexible rubber disks, see Fig. 2.1iv, may be provided as part of the equipment to gently clean the food surface – thus decreasing processing time lo less than 30 s; or else ultrasound devices, to vigorously vibrate the food matrix and debris, see Fig. 2.1v. To assure uniform cleaning of their whole surface, larger foods should be rotated. Foams have also been employed to enhance cleaning – namely in the case of dressed poultry or mechanically harvested tomatoes; in the latter case, the foam acts as a cushion when the food pieces are dropped into large transport bins, thus reducing mechanical damage. As the foam drains, it also removes soil – besides destroying bacteria, which would otherwise start the spoilage process already during transportation to the processing plant.

Flotation washing, see Fig. 2.1vi, takes advantage of density differences between items that typically float (i.e. fruits and crops) and contaminating soil, stones, or rotten portions thereof that tend to sink in water due to their higher density. Better cleaning is achieved when foods (e.g. corn, lima beans, peas) are dipped in an oil/detergent emulsion before being poured into

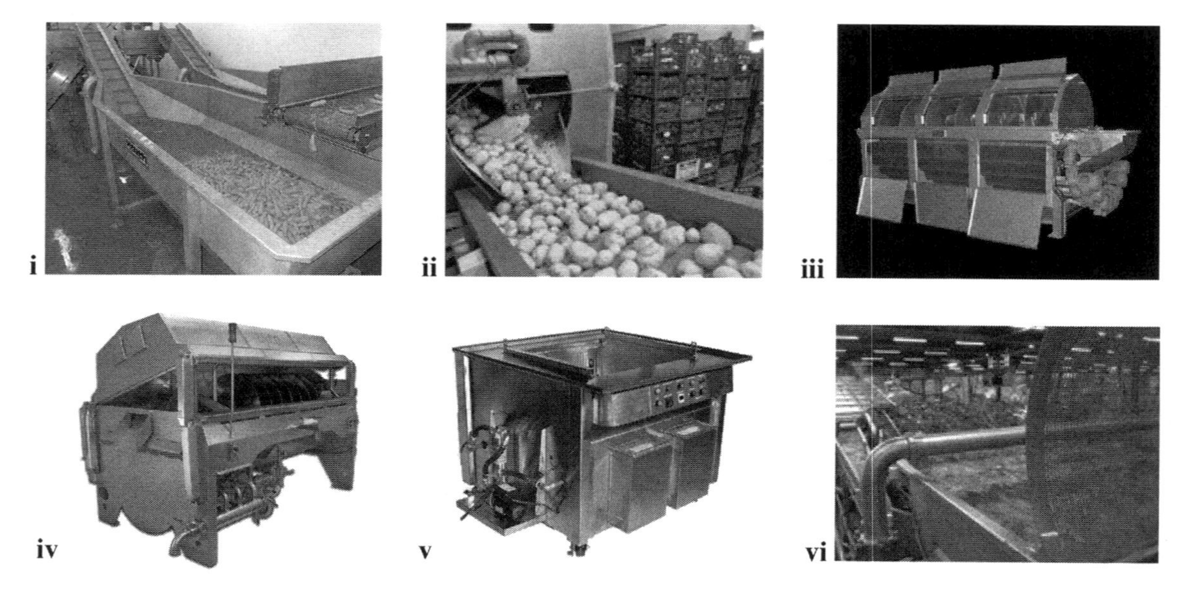

FIGURE 2.1 Examples of wet cleaning equipment: (i) soaking tank (courtesy of Tickhill Engineering, Doncaster, UK), (ii) spray washer (courtesy of Cam Spray, Iowa Falls, IA), (iii) brush washer (courtesy of Tummers Machinebowv, Hoogerheide, The Netherlands), (iv) rod washer (courtesy of Lyco Manufacturing, Columbus, WI), (v) ultrasonic cleaner (courtesy of Sharpertek, Pontiac, MI), and (vi) flotation tank (courtesy of FTNON, Almelo, The Netherlands).

water, through which air is continuously bubbled; the foam thus formed washes away contaminating material more easily than plain water – and will be rinsed via a final spray wash.

The main equipment utilized for dry cleaning of food includes air classifiers, and magnetic or electrostatic separators; industrial examples are shown in Fig. 2.2.

Air classifiers (also known as aspiration cleaners), see Fig. 2.2i, employ a fast-moving stream of air with velocity above the terminal velocity of impurities less dense than food particles, or below the terminal velocity of impurities denser than food particles; hence, they take advantage of differences in density (besides projected area of food particle, typically larger than that of impurities) to effect cleaning via mechanical separation – with contaminants flown upward or downward, respectively. They have met with success in harvesting machines to get rid of leaves, stalks, and husks, or else soil and stones, respectively, from grains and legumes; and also in cleaning eggs from animal feathers. Since both food and impurities remain dry upon separation from each other, no further surface drying is required – while microbial decay is also delayed; disposal of concentrated solid wastes is cheaper as well. However, this mode of cleaning is less effective than its wet counterpart(s) – and measures to protect from recontamination by dust, and further to prevent explosion hazards may be necessary.

Electrostatic cleaning, exemplified in Fig. 2.2ii, can be utilized whenever the surface charge on a raw material differs from that of its contaminants; it has found application in cleaning grains from seeds possessing similar geometries but disparate surface charges, as well as in cleaning of tea leaves. Usually the food is conveyed on a charged belt – with food or contaminant particles removed by attraction to the oppositely charged electrode.

Screen shape sorters take the form of rotary drums, or flatbeds as in Fig. 2.2iii, and can separate contaminants by size – being common in grain and sugar industries; leaves and stalks may accordingly be removed from smaller food particles (a process termed scalping), or else sand or dust removed from larger food particles (a process termed sifting). The efficiency of screening depends directly on the degree of openness of screen apertures – and is enhanced by screen vibration, or movement of abrasive disks or brushes over the screen. Mechanical separation is also possible, should food particles exhibit a regular, well-defined shape (e.g. blackcurrants, peas, rapeseed) unlike their impurities; when placed in an inclined conveyor belt moving upward, said food particles roll uniformly down – while weed seeds (or even snails and the like) will be carried up the conveyor, and be eventually recovered at its top end. An alternative design is the spiral conveyor – again meant to separate nonround, lurking materials (e.g. chaff, leaves, seeds) from round seeds (e.g. mustard, peas, soybeans). Movement along two spiraled (concentric) channels is driven by gravity, with no need for engines or moving parts of any sort – with round particles traveling down faster than nonround impurities. The former will ultimately gain sufficient momentum to run over the edge of the inner spiral, thus dropping over to the outer spiral – and being discharged at the bottom of the machine; while nonround material will remain on the inner flights, thus sliding down to a separate chute at the bottom.

Optical and machine vision systems – originally developed to sort and grade foods, may also be utilized to get rid of food contaminants, as illustrated in Fig. 2.2iv. Small-particulate foods may indeed be checked for contaminants via microprocessor-controlled color sorters; the food is well illuminated, and light

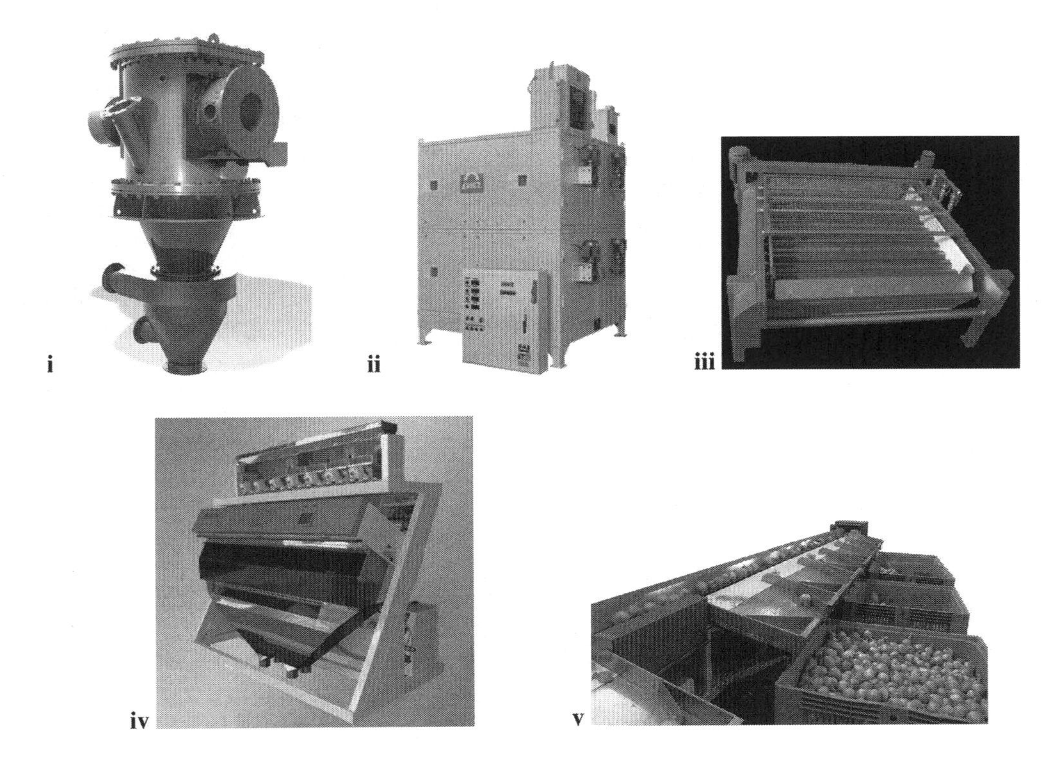

FIGURE 2.2 Examples of dry cleaning equipment: (i) air classifier (courtesy of Hosokawa Micro Powder Systems, Summit, NJ), (ii) electrostatic separator (courtesy of Eriez, Erie, PA), (iii) screen shape sorter (courtesy of Tummers Machinebowv, Hoogerheide, The Netherlands), (iv) color sorter (courtesy of Buhler Wijiete Color Sorting, Anhui, China), and (v) weight sorter (courtesy of Futura, Cesena, Italy).

reflected thereby is compared to a preset standard – with immediate rejection of contaminants exhibiting a disparate color. Light in the infrared range (from 700 nm to 1 mm wavelength) is effective in detecting nutshells, stones, and pits in nuts, fruits, and vegetables; light in the visible range (i.e. 400–700 nm wavelength) can be used to detect stones and stalks, carried in loose form by fruits or vegetables; and light in the ultraviolet range (i.e. 1–400 nm wavelength) has met with success in identifying fat, sinews, stones, and pits in meat, fruit, and vegetables. Smart cameras and use of laser light sources enable separation of contaminants with similar color, but different shape (e.g. green stalks from green beans); their use has been on the rise, due to an increasing societal awareness of food allergens (e.g. peanuts, shellfish, soybean, tree nuts).

Presence of metal fragments or loosened bolts from machinery constitutes a potential hazard in foods; since most such pieces are of a ferrous nature, magnetized drums and conveyor belts, or else magnets placed above conveyors or filters in pipework can be efficiently utilized for cleaning. Although permanent magnets are less expensive, they require periodic inspection to avoid excessive buildup of debris; in fact, these may start being lost back into the food (especially when in overload), thus causing gross (and avoidable) recontamination. Electromagnets are more expensive, but also easier to clean – since a mere switching off of the power supply will do. Magnetic separators have been classically constructed from ceramic magnets – yet rare, earth-based materials (e.g. Ne + Fe + Bo, or Sa + Co alloys) produce much more powerful magnets. Since this type of contamination may occur wherever along a process – as most pipework and equipment are made of metallic items, magnets are to be placed at the entry point, as well as before and after processing units; and further at the end of the processing line. This strategy also helps trace the cause of a contamination, besides preventing early failure of equipment due to (otherwise preventable) wear. Nonferrous metals do not respond to magnetic fields, so magnetic cleaners become useless in these cases; metal detectors may be employed instead – at intermediate stages to prevent equipment damage, and at the final packaging stage to eventually protect the consumer.

Magnetic techniques are frequent in detection of metals in loose and packaged foods. The most common detector resorts to a balanced coil system – made of a coil of wire conducting a high voltage that, in turn, generates a high-frequency magnetic field; two receiver coils are then placed on either side. The voltages induced in the receiver coils are tuned to exactly cancel each other in the case of an undisturbed magnetic field; the conductivity of the target food is taken into account in the said tuning. When an electrical conductor (e.g. a ferrous or nonferrous metal contaminant) passes through the detector, it changes the amplitude or phase of the electrical signal induced in the coils; such a change is electronically detected, and an alarm is set off or a rejection mechanism is started. For checking of aluminum containers or foil, a ferrous-only metal detector is frequently employed – where coils of wire are wrapped around a former that contains several magnets; when a ferrous piece passes through the magnetic field, a voltage is generated in the coil windings – and promptly used to trigger removal of such an impurity.

X-rays (<1 nm wavelength) may be used to detect metal pieces, as well as stones, bone fragments, seafood shell debris, glass, ceramics, concrete, and even rubber and plastics – in both raw materials and packaged foods. Detection is achieved via an incident beam as food passes on a conveyor, with a visible image produced that is transmitted to an image intensifier – for computerized treatment, or else yielding signals generated by solid-state, X-ray sensitive elements. This technique has proven quite accurate, with discrimination being feasible between container wall and contaminant itself – besides detection of contaminants by size, shape, density, and texture.

Microwaves (within wavelength range 1–100 mm) have been employed to detect pits in fruits. Surface-penetrating radar, also resorting to microwaves, is effective in detecting metal foreign bodies, as well as small (i.e. ≤ 1 mm in diameter) pieces of stone, glass, stainless steel, or plastic appearing as contaminants in wet, homogeneous foods; however, products in metallic or foil-wrapped containers are not eligible for handling via this technique. Microwave reflectance holography takes advantage of reflected or back-scattered radiation, and holds several advantages; in fact, many contaminants differ from foods in their specific microwave impedance – and even very small contaminants may be detected in three dimensions.

Nuclear magnetic resonance (1–10 mm wavelength) and magnetic resonance imaging have proven useful in detecting pits and stones in fruits and vegetables – yet they still suffer from the relatively long acquisition times of the large datasets required to generate a three-dimensional image with sufficient detail.

Novel imaging techniques include ultrasound – based again on realization that foreign bodies possess distinct acoustic impedance from that of food materials; hence, they can be detected and identified by changes in reflection, refraction, or scattering of ultrasound waves as they pass through the food – as is notably the case of stones in potatoes. A number of other promising methods are currently undergoing investigation – chiefly based on electromagnetic phenomena; these include capacitive systems, impedance spectroscopy, and electrical resistance tomography.

Sorting of foods entails separation into categories, on the basis of a measurable physical property – namely weight (see Fig. 2.2v), size, shape, color, or consistency; in this regard, such an operation is not essentially distinct from the physical separation treated under dry cleaning. Although sometimes confounded with grading, it should be emphasized that the latter refers specifically to assessment of overall quality of a food based on a number of prespecified attributes – oftentimes including sensory analysis.

Sorting and grading should take place as soon as possible along a food process – as this guarantees more uniform raw materials for subsequent handling; besides reducing expenditure of materials used in processing of foods that will eventually be discarded due to nonconformities. Effectiveness of a sorting procedure, η, has classically resorted to

$$\eta = \eta_a \eta_r = \frac{P x_P}{F x_F} \frac{R\left(1 - x_R\right)}{F\left(1 - x_F\right)} \tag{2.1}$$

as quantitative descriptor; F, P, and R denote mass flow rate of feed, product, and rejected food, respectively, and x_F, x_P, and x_R denote mass fraction of desired material in feed, product, and rejected food, respectively. Inspection of Eq. (2.1) unfolds an efficiency of acceptance, η_a – i.e. amount of desired material present in (accepted) product, Px_P, relative to total amount of desired material in feed, Fx_F; multiplied by an efficiency of rejection, η_r – i.e. amount of undesired material present in rejected product, $R(1-x_R)$, relative to total amount of undesired material in feed,

$F(1-x_F)$. Hence, the two relevant factors affecting performance of a sorter – i.e. η_a and η_r as actual efficiencies dependent, in turn, on P/F and R/F as relative throughput rates, are taken into account; note that $1-\eta_a$ quantifies false reject, whereas $1-\eta_r$ quantifies false accept. Application of Eq. (2.1) becomes more complex when products exhibit more than a single type of defect; oftentimes, the magnitude of P/F becomes the limiting factor – so relief of acceptance/rejection criteria may be in order. Rejection systems resort to air blast, conveyor stop, or pusher arm for items up to ca. 50 kg per unit; or to retracting sections of conveyor that allow unwanted product to fall into a collection bin underneath.

Suitability of several foods for (further) processing sometimes depends on their shape – namely in terms of retail value in the fresh-pack sector; for instance, potatoes should be uniformly ellipsoidal with no protuberances or cavities for easier peeling, cucumbers and gherkins are better when straight to facilitate packaging, and spherical shapes of cherry tomatoes and unique shapes of pears or bananas increase their market value. Furthermore, retailers frequently specify size ranges of products to be sold as fresh items (in bulk, or individually packaged form), in response to specific market demand. On the other hand, size and shape of individual (solid) foods affect the rate of heat and mass transfer – so preoptimized processes may lead to under- or overprocessing, if significant differences occur in raw materials relative to geometrical specifications of the original design. In the case of solid ingredients in powdered form (e.g. colorants, thickeners), correct size distribution is also relevant to attain uniform products upon mixing and blending operations; screening is thus in order – a nuclear operation dealing with solid mechanics (and discussed in *Food Proc. Eng.: basics & mechanical operations* to some length).

When solid foods have similar round shapes, sorting may be accomplished via drum screens – where holes with selected diameters have been drilled on plates, able to separate at various levels by size, see Fig. 2.3ii; however, machine vision systems exhibit better performance, and assure sorting even under some variability in those features, check Fig. 2.3i. Drum screens are appropriate for size-sorting of small particulate, mechanically robust foods (e.g. beans, nuts, peas); they possess a few advantages over the flatbed screen designs regularly used for grain

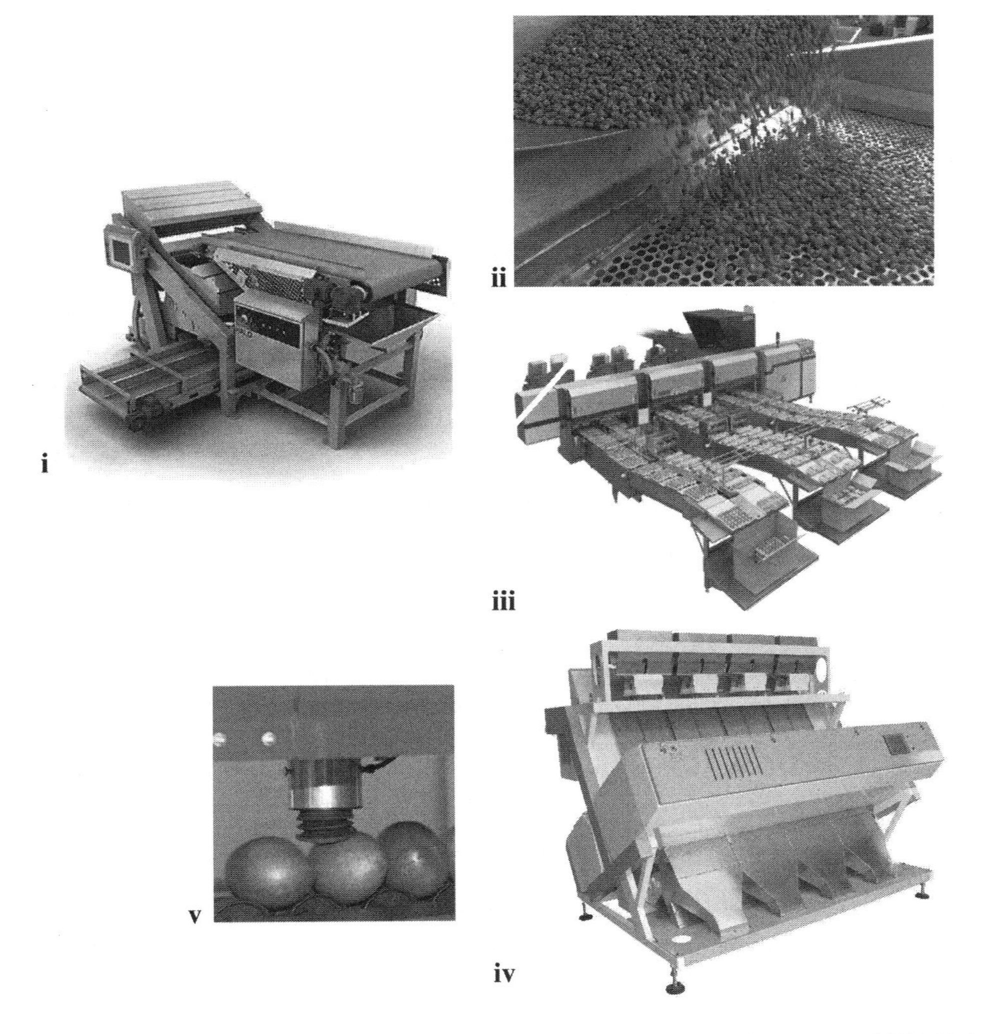

FIGURE 2.3 Examples of sorting equipment: by (i) size/shape (courtesy of TOMRA, Asker, Norway), (ii) size (courtesy of KCB, Lieshout, The Netherlands, (iii) weight (courtesy of MOBA, Barneveld, The Netherlands), (iv) color (courtesy of Buhler, Uzwil, Switzerland), and (v) firmness (courtesy of Aweta, Nootdorp, The Netherlands).

sieving (although characterized by much smaller apertures) – namely, higher handling capacity and lower susceptibility to blinding, especially if fitted with brushes. They are almost horizontal – and may be built concentrically (i.e. one inside another), in parallel (i.e. outlet of one screen coinciding with inlet of another), or in series (i.e. single drum constructed with sections of differently sized apertures). Variable aperture screens possess either a continuously or a stepwise diverging aperture; both types handle foods more gently than regular drum screens, and should consequently be employed for easily damaged (larger) fruits. Continuous variable screens are usually built as a set of diverging rollers, cables, or conveyor belts; inside an expanding roller sorter, foods move along until clearance between rollers or belts becomes sufficiently large for them to pass through. Fruit rotation becomes possible by driving such belts or rollers at different speeds – thus aligning the fruit, so as to expose the smallest dimension to the aperture. Stepwise increases in aperture are produced by adjusting the gap between driven rollers and an inclined conveyor belt; the food rotates, so the same dimension serves as criterion for sorting.

Weight sorting is more accurate than other methods of sorting – so it is preferred when handling more valuable foods (e.g. cut meats, eggs, some tropical fruits); operation rates may exceed 10,000 units per hour. The most typical case is indeed eggs – first graded by candling, and then sorted into five to nine categories, with differential increments as small as 0.5 g (see Fig. 2.3iii). The equipment consists of a conveyor, which transports eggs to a cascade of spring-loaded, strain gauge or electronic weighing devices – incorporated into the conveyor undergoing intermittent operation. When stationary, tipping or compressed air-driven devices remove heavier eggs – discharged into a padded chute; lighter eggs are replaced on the conveyor, and travel to the next weighing device – where the procedure is iterated.

Aspiration (as discussed previously for dry cleaning of foods) and flotation (also addressed as technique of wet cleaning) both exploit differences in food density; hence, they may likewise be used to sort foods – as is notably the case of grains, nuts, and pulses via aspiration, and vegetables via floating in brines with densities of 1.12–1.14 $g.cm^{-3}$. This strategy is useful whenever food density correlates directly to some intrinsic food feature – as happens with tenderness or sweetness in the case of peas; denser particles are indeed richer in starch, and thus supposedly more mature – so they are recovered after sinking in the brine. Denser potatoes are also characterized by higher solids content; sinking potatoes are accordingly removed for further processing into crisp items.

A final example of weight-based sorting pertains to collation of individual fish fillets (or the like) into packs, meant to exhibit a preset nominal weight; since individual units may vary considerably in weight, a laborious manual task was for long a must – where pieces from a large pool were collated, by trial-and-error, into packs supposed to approach (but not fall below) the weight specification. Nowadays, automatic collation has taken over, with many pieces individually weighed and placed in a magazine – with weights stored in the computer memory; the best combination is then calculated by the computer, which sends information for collation of the right pieces into each pack. This process holds economic relevance, because compliance with fill-weight legislation normally entails some give-away – which will be minimized if the said automatic procedure is put into place.

Depending on the size range of food pieces, color-mediated sorters resort to photodetectors, see Fig. 2.3iv – designed to sort small particulate foods (e.g. diced carrots, maize kernels, navy beans, peanuts, rice, small fruits, snack foods); or to video cameras – meant to sort larger foods (e.g. bakery products; such fruits as avocados, kiwis, mangoes, peaches; such vegetables as bell peppers, tomatoes). In both situations, sorting rates may reach 15 $ton.hr^{-1}$. In the former case, particles are fed into the chute – with angle, shape, and lining material chosen/tuned to permit one-at-a-time feeding, and thus generating an individual signal as they pass through a photodetector. Color of background and type and intensity of incident light are selected, so as to perfectly match each product – since photodetectors measure the color reflected by each piece, and compare it with a set of standards; defective pieces are pushed away by a short blast of compressed air.

Color sorting of fruits and vegetables classically operates under free-fall, and scanning at frequencies of the order of 1 kHz is made as they leave the conveyor belt. Concentrated He-Ne discharge light or laser light beams are employed, along with a high-speed rotating mirror; the machine detects differences in reflectivity between desired products and products off-specification. Another configuration consists of a sensor, located above the conveyor belt, which views products as they pass beneath; a rejection mechanism is put in action whenever a preselected color (e.g. burnt area on bakery products) or intensity of fluorescent light (e.g. green light emitted by chlorophyll in beans, peas, or spinach) is detected – yet different colored foods may be separated for distinct treatment afterward. For instance, color-based sorting of bread, crackers, and cookies is able to remove misshapen products, and thus avoid jamming of automatic packaging machines downstream. In the case of larger sized pieces, each item may be gently picked up by a gripper, rotated as necessary in front of the camera, and then transported to the right batch in a robotized manner. The flexibility and discriminatory ability of laser light have been taken advantage of in sorting potatoes for French fries after peeling (to check for remaining peel), before cutting (to check for rotten potatoes that would damage the knives), and upon freezing (to check for discoloration). If near-infrared detectors are employed, then sugar and water content, acidity, and degree of ripeness can be estimated – in a quite accurate and noninvasive fashion. The light that passes through the food is accordingly collected by a sensor placed opposite thereto; subjected to Fourier-transform analysis of absorption degrees throughout the spectrum of incident light; and eventually translated into useful indicators. Some machines resort instead to X-ray-based vision systems – able to detect internal flaws, or physiological conditions of the food not accessible via mere scanning of its outer surface.

Recent developments in machine vision technology encompass high-resolution digital cameras – fitted with telecentric optics, light-sensitive cells, and complementary metal-oxide semiconductor (or CMOS) logic circuits, able to scan finer line widths and areas of product; high computing power, based on 64-bit processing – able to rapidly handle the data outputs of the said cameras, and to run advanced image processing algorithms and complex decision-making mechanisms based on neural networks and fuzzy logic (incrementally trained by providing examples to the system, rather than resorting to exhaustive programming); and sophisticated lamps with specific spectral outputs – based on light-emitting diode (LED) technology, and custom-designed for each food product. Future developments will likely include wireless cameras and fiber-optic cabling – able to generate high-resolution images from multiple units, and carry them over long distances.

Acoustic firmness sensing, see Fig. 2.3v, enables online, non-destructive measurement of firmness and related characteristics of whole fruits; the grader apparatus gently taps the product, and listens to the vibration pattern thus generated. The acoustic signal, or resonance-attenuated vibration, relates directly to overall firmness – as well as juice content, freshness, and even internal product structure (including degree of tissue breakdown or level of dehydration). Upon calibration with more than one firmness index, it is possible to produce an accurate, reliable, and reproducible estimate of ripeness degree.

One of the best examples of grading is examination of meat carcasses by veterinarians and sanitarian/food inspectors – looking for evidence of disease, besides distribution of fat, bone/flesh ratio, and carcass size and shape. Cheese and tea are also graded by sensory panels, in terms of flavor, aroma, and color; eggs can be visually inspected by operators, over tungsten light (candling), aimed at removal of fertilized or malformed ones, as well as blood spot- or rot-containing ones; and wheat flour is assessed for color and presence of insects, in complement to (instrumental assay) for protein and moisture contents, and dough extensibility. With the advent of sophisticated machine vision systems, several of the aforementioned attributes may now be instrumentally ascertained, e.g. presence and extent of bruising, or skin color of chicken meat, or even existence of cracks in eggshells; hence, distinction between grading and sorting has become increasingly blurred. This trend has also been favored by the requirement of uniform fresh-pack products raised by retailers; the need for product tracking and traceability enforced by food processors, to more rapidly respond to liability claims; the increased labor costs of inspectors and operators; and the improved performance of automatic systems relative to humans – especially when routine operations are concerned, with loss of visual acuity or discriminatory ability as working time elapses being no longer a problem.

Removal of unwanted or inedible material from fruits and vegetables often calls for peeling; in addition, this unit operation improves overall appearance of the final product. Flash steam, knife, and abrasion account for the most important techniques to bring about peeling. The aim of this unit operation is to take off as little of the raw material as possible – since its value will be reduced by a weight decrease (although not linearly); while leaving the food surface clean and undamaged – and keeping energy, labor, and material costs as low as possible.

In the case of steam peeling, exemplified in Fig. 2.4i, batches of root crops are fed into a vessel containing high-pressure steam (ca. 1.5 MPa); uniform exposure of food to steam is promoted via slow rotation of the said vessel (ca. 5 rpm). The high temperature of the steam causes rapid heating of the surface layer – a process that takes typically no longer than 30 s; the relatively low thermal conductivity of the food bulk material prevents extensive heat penetration – otherwise the food would undergo cooking, and thus lose its characteristic texture and color. Upon sudden decrease to atmospheric pressure, water vapor readily forms under the skin – so the surface of the food flashes off; most resulting material is discharged with the steam in suspended form, while complementary water spraying guarantees removal of any residuals left. The final product holds a good appearance, which may be monitored via continuous scanning – able to adjust operating parameters in response to fluctuations in raw material. The associated water consumption and product loss are typically low, and throughput rates as high as 5 ton.hr^{-1} are feasible.

Dedicated knife-based peeling machines have been developed to handle specific foods – as is the case of fruits, onions, or shrimps, see Fig. 2.4ii. In fruit peeling, stationary blades are pressed against the surface of rotating foods, or instead rotate against stationary foods to remove their skin; the latter approach is typically elected for citrus fruits. In the case of onions – which exhibit diameters between 4 and 11 cm, a blade slits the onion, while the outer skin is gently removed via compressed air; hourly rates up to 4,000 units have been attained. Shrimp peelers can operate up to 5,000 units per hour – by first picking up the animal with a clamp, and carrying it through a centering system where a second clamp grips the shell; the tail segment of the shell is broken by a first cutter, while a second cutter clamp accurately splits the shell and brushes to remove the vein thereafter – until a fork finally pulls the shrimp meat off the shell (with the shell being discharged separately).

Such root crops as potatoes, carrots, celeriac, and beets may be peeled via abrasion – against rollers with their outer surface covered in silicon carbide (also termed carborundum), as per Fig. 2.4iii; they may instead be placed in a rotating bowl, lined again with carborundum on its inner surface. The abrasive surface removes the skin, which is then washed away by water spraying; in the case of onions, rates up to 2.5 ton.hr^{-1} have been recorded. Despite the good final appearance of the food surface – due to the polishing action of the abrasive surface (unable to cause significant thermal decay, though), and the low costs of equipment and operation, larger product losses (which may attain 25%(w/w)) and production of larger volume

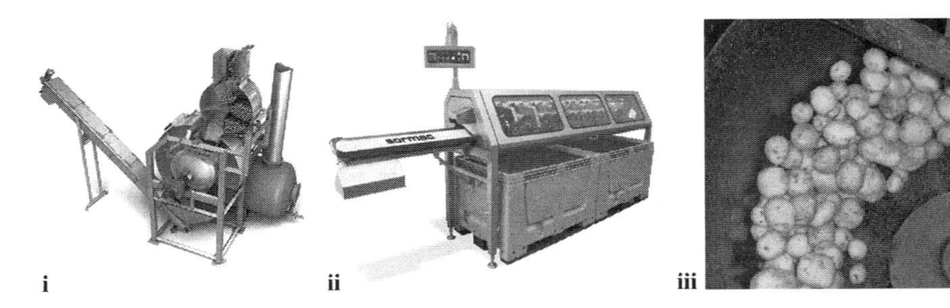

FIGURE 2.4 Examples of peeling equipment: (i) flash steam peeler (courtesy of TOMRA, Asker, Norway), (ii) knife peeler (courtesy of North Star Engineered Products, Perrysburg, OH), and (iii) abrasion peeler (courtesy of Tomanus, St. Louis, MI).

of more dilute waste (and thus more expensive to treat) than in steam peeling are found. Furthermore, all pieces of food require contact with the abrasive surface(s), so throughputs are relatively low – while irregular and concave surfaces (e.g. potato eyes) require hand finishing afterward.

Older techniques like caustic peeling and flame peeling have meanwhile been discontinued as large-scale operations. In any case, the latter leads to rather low product losses in the case of onions (less than 10%(w/w)), which are carried by a conveyor belt through a furnace kept at ca. 1000°C – a temperature sufficiently high to burn off the outer shell and root hairs; with the (charred) skin being finally taken off via water spraying at high pressure. Caustic peeling resorts instead to a bath of 1–2%(w/v) sodium hydroxide, at 100–120°C; the skin softens after the food is submerged in the said bath, so it is easily removed afterward via high-pressure water sprays. Higher concentrations, ca. 10%(w/w), of sodium hydroxide may be employed – in which case rubber disks or rollers will suffice toward removal of the softened skin. Besides the relatively high losses in food material – approaching 20%(w/w), and effluents carrying suspended solids that require specific treatment, an extra need arises of (chemical) neutralization of the solution prior to its disposal in the environment.

2.1.2.2 Hygienic design of equipment

All equipment intended for food handling should be designed for protection against physical, chemical, and biological hazards – besides allowing easy and effective cleaning and disinfection; for instance, a cylindrical tank possessing a vertical axis is much easier to mechanically clean than a horizontal tank (unlike happens with manual cleaning). Furthermore, dead spaces, liquid-level tubes, pressure gauges, pumps, gate and three-way valves, heat exchanger joints, pipe tops, and cover spatters complicate cleaning and sanitation procedures. Hygienic design is critical in apparatuses used to process foods that pose a higher risk to consumers – as is the case of cold fill cook-in sauce lines, cooked meat slicers, liquid dairy product filling, salad and fresh produce conveyor systems, short shelf-life chilled and cook-chill food handling devices, and sandwich-making equipment. Best-practice guidelines for hygienic design are available, and subjected to periodic updating by the European Hygiene Equipment Design Group in EU countries, or the Food and Drug Administration and the Food Safety and Inspection Service in the United States.

Surfaces of food processing equipment may obviously be in direct contact with foods – so residues thereof are expected to drip, drain, or diffuse through them; if contaminated, the said surfaces will serve as vehicle for contamination of further food. Surfaces that do not contact directly with foods (e.g. equipment legs, housings, supports) are also a target of sanitary design, owing to the possibility of indirect contamination. The former surfaces should simultaneously be smooth, impervious, and free of sharp corners and sides, as well as cracks and crevices; non-porous, non-absorbent, and unable to transfer odors and taints to the product(s); non-reactive, non-toxic, and resistant to corrosion; and durable, maintenance-free, accessible for cleaning and inspection, and easily disassembled.

Due to the aforementioned requirements, stainless steel is the default choice as material for surfaces of processing equipment directly exposed to foods – namely, because of its high resistance to corrosion (dependent on Cr content) and mechanical durability (dependent on Ni content); one of the most popular examples is series 300 stainless steel 18/8 (i.e. containing 18%(w/w) Cr and 8%(w/w) Ni), or series 316 sanitary standard steel 18/10. Stainless steel alloys containing Ti are employed in equipment surfaces designed for contact with high-acid or -salt content foods (e.g. citrus juice, tomato paste); whereas copper is instead used for brewing and distilling apparatuses handling low-acid foods – due to its higher thermal conductivity, coupled with ability to sequester compounds responsible for off-flavors. The exposed surface of stainless steel normally undergoes a previous treatment with nitric acid (or another strong oxidant) for passivation – i.e. creation of a very thin, chemically inert oxide film; ground or polished stainless steel surfaces are further required to meet a prespecified smoothness or finish – characterized by a quite small rugosity factor. Cast Fe is a sturdy material, commonly found in frying and baking equipment. Cooking utensils intended for manual handling are preferably built with Al for its low density, and consequent low overall weight per tool; however, poor resistance to corrosion causes pitting and cracking upon contact with cleaning and sanitizing chemicals – so it is normally coated with polytetrafluoroethylene (Teflon™), which also dramatically reduces adhesion of food residues (thus facilitating cleaning after use). Gaskets and membranes intended for food contact are typically manufactured with thermally resistant plastic or rubber materials – the low durability of which calls for frequent checks of deterioration. Other food contact materials of nonmetal nature are ceramics (e.g. in membrane filtration systems), as well as glass – tempered to resist mechanical and thermal shock (e.g. Pyrex™) in transparent vessels and sight-windows.

With regard to shape, dead spaces or bends in pipework are to be avoided – since they permit accumulation of food product; pipework should also slope to a drain, if not routinely disassembled – while pipes, gauges, and probes should not possess dead ends or contain open zones inaccessible to cleaning. Threaded bolts, screws, or rivets should not be present in or above food contact surfaces; and welded joints should be of butt-type, continuous, and flush – besides ground to a smooth finish. Standard specifications require predefined radii for internal body and top rim angles; for instance, an internal angle of 135° entails a minimum radius of 6.35 mm – so that no stagnant film forms that might be susceptible to chemical or microbial contamination. Vessels and tanks are to be self-draining by gravity; and their openings should be lipped and covered with overhanging lids. Shafts, bearings, and seals should be self- or product-lubricated; they should resort to food-grade lubricants, but the food contact area is still to be sealed against contamination thereby. Nonproduct contact surfaces are to be cleanable, corrosion-resistant, and maintenance-free. For instance, equipment legs should be sealed at their base, and not be hollow so as to avoid harboring microorganisms, insects, or rodents; bolts or studs on the surface are to be avoided; and attachments should be welded to the surface, rather than attached through drilled holes. Finally, keypads and touchscreens, as human/machine interfaces in the food process environment, are typically difficult to clean due to their susceptibility to chemical attack by cleaning/disinfection liquids containing moisture, oil, or organic solvents – yet they constitute an ubiquitous source of contamination carried by operators.

The layout of equipment at industrial plants should allow for sufficient space for cleaning, while avoiding holes that may harbor insects or rodents; unless fitted to the wall itself, equipment should keep a safe distance of at least 10 cm from the closest wall to permit cleaning in between – whereas floor-mounted equipment should sit above 15 cm therefrom (or else be sealed to the floor or a pedestal). The same principles apply to equipment sitting on a table – where sealing thereto, or provision of a ≥10 cm-clearance underneath is to be ensured.

2.1.2.3 *Cleaning, disinfection, and rinsing of equipment*

Two distinct stages exist in a typical operation of equipment cleaning: removal of soil-mediated contaminants, followed by disinfection proper – complemented with rinsing to eliminate traces of cleaning and sanitizing solutions. The reason for this sequence derives from realization that the ability of a disinfectant to inactivate microorganisms is reduced by presence of soil or contaminants alike. It should be emphasized that regular processing (including necessarily some form of heat treatment) of an unclean food makes eventual cleaning of both food and contacting equipment more difficult; for instance, sugar components of soil on a food or equipment surface may caramelize, fatty components polymerize, proteinaceous components denature, and polyvalent salts become complexed – with the resulting products sticking more strongly to the surface than their precursors, and thus becoming harder to remove.

The mechanism of bonding between soil and a surface relies on adhesion phenomena at solid/solid and solid/liquid interfaces – including electrostatic interactions, preferential wetting by a nonaqueous phase, and multilayer adsorption phenomena. Cleaning is intended to disrupt such interactions – which is normally achieved via progressive soaking and dispersion of soil and residues, rather than plain dissolution; note that the phenomena involved in removing dirt and contaminants from either food outer surface or equipment inner surfaces are in essence similar to each other.

The simplest approach to clean equipment is to use water in the form of sufficiently energetic jets; the mechanical energy to be imparted may be considerably reduced if a cleaning aid (i.e. a detergent) is added thereto. A detergent should possess good wetting and penetration properties, i.e. be able to reduce surface tension between the two adjacent phases – so as to promote intimate contact of water with the soil to be removed; its specific properties depend largely on the orientation of molecules at the aqueous interface – which, in turn, hinges critically on the ability of the detergent molecules to simultaneous bind to polar and nonpolar molecules. Remember that molecules of water are attracted to each other via (relatively strong) hydrogen bonds; hence, a significant amount of energy (ca. 0.072 $J.m^{-2}$) must be supplied to increase the area of contact between water and another phase, e.g. air or oil. In the case of slightly polar organic liquids and air, attraction between like molecules is due only to (much weaker) van der Waals' forces (ca. 0.025 $J.m^{-2}$). When surfactants are present, they tend to accumulate at the interface – thus turning it less polar, and therefore more similar to a nonpolar surface, with consequent reduction in surface tension.

In view of the above-mentioned considerations, one concludes that detergent molecules should be amphipathic in character, i.e. possess a hydrophobic region – normally a long-chain fatty acid; linked to a hydrophilic region – usually a sodium salt of a carboxylic acid (in the case of soapy, or cationic detergents), or a sodium salt of an alkyl sulfate, or even an alkyl or aryl sulfonate moiety (in the case of anionic detergents). The latter are not affected by hard water, whereas soapy detergents form a scum therein; nonionic detergents, on the contrary, produce little foam, so they are easily rinsed off – with alcohols, esters, or ethers (namely condensates of ethylene oxide with alkylphenols or fatty alcohols) accounting for their hydrophilic portion. Soils are accordingly displaced from the surface of a piece of equipment via saponification and emulsification – in the case of fatty components, via hydrolysis in the case of proteinaceous components, or merely by solubilization in the case of soluble minerals; enzymes may be added to bring about hydrolysis of fats and proteins, since their constitutive residues are more soluble in free form. Finally, the soil itself is dispersed in water – and eventual redeposition is prevented via suitable rinsing.

The *pH* of the cleaning solution is often adjusted to match the target component of soil: sugar is water-soluble, and thus easy to remove at neutral *pH*; fat is water-insoluble, but alkali-soluble – being difficult to remove around neutrality; fibrous proteins are water-insoluble, but alkali- and acid-soluble – being particularly difficult to get rid of; monovalent salts are water- and acid-soluble, so they are easy to take away via neutral media; and polyvalent salts are water-insoluble, but acid-soluble – so they are once more difficult to remove. The quality of water also plays a role upon cleaning effectiveness; if it contains suspended Mg or Ca salts, it is said to be hard – and cleaning agents become less effective under such circumstances, besides leading to formation of surface deposits on otherwise clean equipment. If water contains soluble Fe or Mn salts, then colored deposits may instead be left.

To maximize the cleaning action of detergents in aqueous solutions, namely the mechanistic steps of cleaning listed above, several compounds are added to the cleaning liquids for specific functions: deflocculating agents – to break up soil flocs on surfaces, thus facilitating removal thereof; dispersing agents – to flocculate mineral films, and prevent redeposition afterward; emulsifiers – to stabilize fat droplets or solid powders in suspensions; peptidizing agents – to breakdown and disperse proteins; rinsing agents – to detach solids from the surface, when flushed with freshwater; saponifiers – to transform fats into soaps; sequestering agents – to avoid deposition of mineral salts; solubilizers – to aid in formation of a solution of organic or inorganic solids; and wetting agents – to lower surface tension, thus facilitating penetration of water into the soil matrix.

Chemically speaking, the aforementioned agents fall into one of five major categories: acids – with sequestering and solubilizing roles; alkalis – with emulsifying, peptidizing, and saponifying roles; chelators – with dispersing, peptidizing, and sequestering roles; complex phosphates – with emulsifying, deflocculating, dispersing, peptidizing, rinsing, and sequestering roles; and surfactants – with dispersing and wetting roles. Acids produce $pH < 2.5$, and typically consist of blends of such organic acids as acetic, lactic, hydroxyacetic, citric, levulinic, and tartaric acids, or such inorganic acids (or acid salts, for that mater) as sulfuric, nitric, and phosphoric. Alkalis generate a *pH* of 11–12, and normally consist of caustic soda (plain or chlorinated), trisodium phosphate, sodium metasilicate, or hypochlorides. Chelating agents fix alkaline earth cations in the form of nonionizable complexes; polyphosphates (or salts of partial anhydrides of phosphoric acid, also able to prevent precipitation), ethylenediaminetetraacetic acid (EDTA), and gluconates or glucoheptonates are used most often. Complex phosphates are accounted for by sodium hexameta- and tetra-phosphate – yet a reduction in their levels has been enforced, as per stricter environmental legislation in this regard. Surfactants include (anionic) alkyl- and aryl-sulfonates, or sulfated alcohols – or else (cationic) quaternary ammonium compounds and ethoxylated amines.

Foam cleaning is one of the most common approaches to wet cleaning – whereby a layer of foam is formed and allowed to spread over the equipment, thus extending contact time of detergent solution with its surface. The said foam is produced by introducing air into the detergent solution as it is sprayed; if high pressure is utilized, then an extra mechanical force becomes available for removal of soil and the like – often complemented by high temperature.

Cleaning of pipes is best effected by pumping the germane solution so as to produce sufficient turbulence; the cleaning effect, C_{ef}, has been classically considered as proportional to shear stress on the inner wall – which correlates, in turn, with pressure drop, ΔP. A more accurate (yet empirical) correlation unfolds a proportionality of C_{ef} to $\Delta P^{1.5}$ – which implies that the ratio of cleaning action to pumping costs actually decreases with increasing power of pumping equipment.

Use of air bubbles dispersed in the cleaning water has been suggested – in that it decreases the amount of detergent needed; unfortunately, air will eventually separate off somewhere inside the equipment – and thus form pockets that may hamper an effective cleaning action. Furthermore, the yield of cleaning action over pumping costs fails to increase when air bubbling is employed, at relatively low flow rates. Conversely, annular flow develops at water linear velocities, v, above 1 m.s^{-1}, and air fractions within the range 0.01–0.1 kg$_{air}$.kg$_{water}$$^{-1}$; under such conditions, ΔP becomes approximately proportional to $Q^{1.4}$, where Q denotes volumetric flow rate – which compares with Q in true laminar flow, and to Q^2 in fully developed turbulent flow. The rate of momentum transfer in the annular liquid layer becomes quite intense, thus justifying a stronger cleaning action with limited pumping costs – yet it is obviously restricted to equipment parts possessing cylindrical symmetry (e.g. tube bundles in an evaporator). Recall the definition of Reynolds' number, i.e.

$$Re \equiv \frac{\rho R v}{\mu} \tag{2.2}$$

in parallel to Eq. (2.92) of *Food Proc. Eng.: basics & mechanical operations*; and the relationship between volumetric flow rate and velocity under cylindrical geometry, viz.

$$Q = \pi R^2 v, \tag{2.3}$$

with R denoting radius. Elimination of v from Eq. (2.2) via Eq. (2.3) unfolds

$$Re = \frac{\rho R \dfrac{Q}{\pi R^2}}{\mu} = \frac{\rho Q}{\pi \mu} \frac{1}{R}, \tag{2.4}$$

so Re is inversely proportional to R for a given Q – knowing that transition from laminar to turbulent flow will eventually occur at large enough Re. In other words, any enlargement of equipment (including passing from a pipe to an actual piece of cylindrical equipment) implies a reduction in turbulence – and thus a poorer scrubbing action; hence, satisfactory cleaning results will be harder to reach. In the specific case of three-way valves in a circuit, sterilization is recommended – while turning those valves between their extreme positions, due to possible existence of localized holdup volumes.

Effective cleaning of large tanks and similar pieces of equipment cannot, in general, be effected by plain circulation of cleaning solution (as emphasized above) – because the volumes needed would be too high, while turbulence would be too low, see Eq. (2.4). Therefore, some mechanical action is in order – namely brushing, either manually or with the aid of a scrubbing machine; mechanical action will accordingly complement chemical action in cleaning. Direct manual brushing is heavily dependent on the human factor – and should thus be avoided wherever possible; although manual jets constitute an improvement, the human factor is still relevant therein – with hot jets being more effective than room temperature ones. The said jets may become of a purely mechanical nature if resorting to high-pressure flow – as happens with jets forced through nozzles turned by impulse reaction, or proceeding from perforated balls/cylinders or mechanically driven swivel nozzles, see Fig. 2.5i–iv. This approach requires the jet-producing apparatus be physically present inside the equipment being cleaned; if temporarily introduced therein, sufficient time should be allowed for it to move thoroughly inside the equipment – which turns a hassle, if sequential steps of cleaning (followed by disinfection and rinsing) are to take place.

The modern tendency is toward having such apparatuses – consisting of either perforated balls or cylinders, or even rotating systems, built in the large-scale equipment itself (together with associated pipework, fittings, and instrumentation), where they produce jets or sprays; this constitutes the so-called cleaning-in-place (CIP) devices, see Fig. 2.5v pertaining to an overall integrated system. Such a strategy is quite convenient, as it speeds up cleaning (with concomitantly shorter downtimes) and reduces human labor requirement (with consequently lower costs) – further to being prone to standardization; such permanently mounted systems spray indeed the cleaning solution more or less uniformly in all directions. Cleaning-in-place may also be brought about via a jet of foam; a cleaning product is accordingly blended with a strong foaming agent – which allows a layer of foam spread over the equipment, and act for a desired time prior to rinsing.

In spray cleaning, the droplets of the cleaning solution are projected toward the inner wall of the equipment; it is thus reasonable to assume that cleaning action is proportional to kinetic energy of the droplets during perpendicular impact, and proportional to momentum of the droplets during tangential motion – with droplet velocity playing a decisive role in either case. Neglecting evaporation – as expected in a closed vessel full of moisture-saturated air, the droplets decelerate as they move toward the wall, due to mechanical resistance offered by the air medium. For a given jet velocity at nozzle exit (and thus similar pumping costs per unit volume of cleaning solution), the kinetic energy of individual droplets is higher for larger sizes – so a larger droplet produces a stronger cleaning action upon impact; the nozzle aperture should be consistently fixed. Note that clusters of droplets and air form at nozzle outlet – with overall diameters easily exceeding 10 cm, and which possess a mass density intermediate between those of liquid water and plain air. This realization is to be taken into account when estimating deceleration – especially if compressed air is employed to disperse and propel the droplets, since clumping expands the radius of the fluid particles and viscous forces increase proportionally as per Stokes' law. An innovative approach takes advantage of

FIGURE 2.5 Examples of CIP devices: (i) fixed spraying nozzles, of various sizes and orifice patterns (courtesy of Nocado GmbH, Delmenhorst, Germany), operated via (ii) wide-angle flat spray, (iii) multiple orientation spraying, or (iv) atomizing pattern (courtesy of H. Iheuchi & Co., Osaka, Japan); and (v) complete automated system (courtesy of Sani-Matic, Madison, WI).

the velocity of sound being ca. 1500 $m.s^{-1}$ in liquid water, and ca. 340 $m.s^{-1}$ in plain air at atmospheric pressure – but a mere ca. 30 $m.s^{-1}$ in water droplets suspended in air; therefore, sonic disruption will occur when air and water are mixed prior to spraying – complemented by ejection velocities up to 30 $m.s^{-1}$. This process proves quite economical and efficient, as even small droplets can reach the wall of large containers with considerable velocity.

In CIP of interior surfaces of tanks and pipelines, the cleaning solution is normally recirculated after filtering, and returned to an external reservoir for reuse. Operation of these systems typically resorts to programmable logic controllers that set pumps on and off, and open and close valves along a predefined period and according to a predefined pattern; a representative sequence starts with plain water flushing twice (with the first effluent discarded, as it is richer in suspended matter), cleaning with alkali, rinsing with water, cleaning with acid, and rinsing again with water – with final drying via an air heater and fan.

If dismantling is possible or necessary, then cleaning-out-of-place (COP) may be in order – where parts are soaked in a stirred dedicated tank, containing a hot cleaning solution. Power ultrasound (20–50 kHz) may be supplied to improve the effectiveness of cleaning – based on microbubble formation due to cavitation, in a uniform manner, including otherwise inaccessible spots; this technique is most appropriate for hard materials (e.g. ceramics, glass, metals, plastics) that reflect sound.

Food may to advantage undergo prior cleaning, in attempts to reduce the cleaning duty for the equipment itself – namely, with the aid of detergent-mediated foam; draining foam removes soil sediments and destroys bacteria that might otherwise spoil fresh items – and eventually contaminate the equipment. Other modes of cleaning of food products just before processing encompass rotating baskets subjected to jets or sprays of air, water, detergents or bactericidal solutions; or else vapors from boiling organic liquids, as happens with chicken carcasses.

Once equipment cleaning is over, the next step is sanitizing; the corresponding *modus operandi* includes dipping and soaking for smaller pieces, and (high-pressure) spraying for large apparatuses. Oftentimes, the detergents used previously also play a sanitizing role; this is notably the case of strong alkalis, e.g. caustic soda (NaOH) – which possesses sporicidal and virucidal activities, further to detergency. Strong acids – e.g. nitric, hydrochloric, and phosphoric acids, are also excellent detergents and bactericides; passive sulfuric acid exhibits, in turn, a strong bactericidal action, even against some spores. Conversely, such weak acids as gluconic, citric, or sulfamic acids possess a cleaning action inferior to that of strong acids – and an almost negligible bactericidal action; the same holds with regard to alkaline salts, with microbicidal action hardly comparable to that of caustic soda. When the equipment at stake permits their use, there is no reason to resort to extra (specific) disinfectants when cleaning agents already performed that role; this philosophy also contributes to reduce processing costs (in terms of chemicals, labor, and time) and environmental impacts. It should be stressed that effectiveness of disinfection depends critically on both qualitative and quantitative microbial profiles prevailing on the equipment surface at startup; while the amount of preexisting organic and mineral deposits is germane, because the sanitizer combines preferentially with (particulate) organic matter deposited than with (much smaller) microorganisms in biofilms.

Use of a dedicated microbicidal agent for disinfection, following application of a detergent for cleaning, will therefore be justified only when the equipment cannot stand strong acids or alkalis; or else if it is intended to complete the disinfection process, or reinforce it in attempts to minimize the risk of reinfection afterward. Compounds utilized specifically for their sanitizing capacity belong to two major groups – halogens (i.e. bromine, chlorine, iodine) to be used at $pH > 4.5$, and quaternary ammonium compounds to be employed at $pH < 4.5$; such compounds have been approved by regulatory authorities at large, and are active at ppm-concentration levels. The cocrystallization product of Na_3PO_4 and NaClO, containing ca. 3% active chlorine in a (relatively) stable form, has also been employed; as well as Ag^+ ions, formaldehyde, 3,5,4′-tribromosalicylanalide, O_3, SO_2, and sulfites – besides H_2O_2 and peracetic acid (despite their instability and mutagenic features). Chlorine and hypochlorites are sometimes complemented with hypobromite added to small levels – as the bactericidal activity of either one is significantly enhanced; the former compounds may instead be replaced by chloramines or chlorocyanuric acid, characterized by higher stability – all containing –NCl as part of their chemical structure. This group reacts with HCl to produce –NH, and simultaneously release Cl_2 as active chlorine form. Phenols have been gradually discontinued for their risk of tainting the food matrix – whereas ozone has been increasingly selected, for both plant sanitization and water treatment.

With regard to sanitizing efficiency, *pH*, temperature, and duration of treatment play a role, further to the intrinsic germicidal action of the disinfectant; for instance, chlorine is less effective at higher temperature and *pH*. Although an Arrhenius-type relationship would support an increasing germicidal action with temperature that involves some form of elementary chemical reaction, a lower equilibrium concentration of dissolved Cl_2 will be attained when temperature is increased – as per Henry's law, thus justifying the above (somewhat unexpected) observation. Due to a fugacious nature and reactivity, chlorine and its derivatives can bring about significant corrosion of the equipment material; type 18/8 stainless steel is indeed particularly sensitive thereto, and may undergo pitting – thus developing a surface porosity that severely hampers effectiveness of equipment cleaning procedures afterward. In addition, Cl_2 is incompatible with surfactants – and will react therewith during storage, if included in cleaning/sanitizing formulations; this justifies alternative use of I_2 with nonionic surface-active agents, termed iodophores – which exhibit bactericidal, and even sporicidal action at concentrations of the order of ppm and $pH < 4.5$. By the same token, one may resort to ICl_3 – due to its compatibility with anionic surface-active agents; complemented by its good stability, low tendency for corrosion, and bactericidal effect – even in the neighborhood of neutrality.

Disinfection of equipment surfaces is feasible also via heating; however, a steam jet may not suffice to kill bacteria – because the actual temperature attained on the said surface can favor incubation, rather than lethality. Sterilization via elevation of temperature requires specifically designed apparatuses for use of very hot water or condensing steam as heating medium – with judiciously placed purge valves, meant to deal with air and condensed steam.

The bactericidal action of a chemical agent often results from the (relatively nonselective) modification of permeability undergone by bacterial cell membranes, or even rupture thereof; or from reaction with certain enzymes and other cellular proteins – namely, oxidation, reduction, or hydrolysis that compromise activity of vital constituents; or from interference with specific substrates instead. For instance, Cl_2 reacts with oxidizable noncellular groups, and also with cellular constituents; and Ag reacts with –SH groups, in both exo- and endocellular compounds. In view of the decrease in concentration of a sanitizing agent with time – due to chemical decay, and consequent reduction in the rate of destruction of bacterium viability, an excess is regularly utilized relative to the expected concentration of organic material; hence, some residual concentration of disinfectant normally persists. A reference scale has classically been followed to grade sanitizing agents, the phenol coefficient – with a higher value indicating a stronger sanitizer; it is defined as the highest dilution of sanitizer that kills essentially all microorganisms within 10 min, divided by the highest dilution of phenol in water that would produce an equivalent effect.

When chemical disinfection of a surface is at stake, the interactions of putative biofilms thereon with the surface material are to be taken into account – as lethality may be reduced relative to the very same cells if in planktonic form; the death kinetics is expected to retain its order on active agent concentration – although described by a different kinetic constant. Disinfection of deep deposits, porous materials, and insterstices at joints, valves, cracks, and crevices is usually difficult with such highly reactive compounds as chlorine; in fact, the concentration of sanitizing agent decreases rapidly, even at small depth. On the other hand, such compound as formaldehyde may harden proteinaceous deposits on the surface, thus rendering their removal more difficult; hence, its concentration in the interior of the said deposit may never attain a bactericidal threshold.

The nature of the surface also plays a relevant role in disinfection efficiency; a reduction of 10^{-6} in viable numbers by a bactericidal agent on a surface of (polished) stainless steel typically worsens to a mere 10^{-3}–10^{-2} on an aluminum surface, or a metal surface lined with a plastic material. The chemical nature (namely its corrosive potential) of the disinfecting agent is often so critical that the nature of the equipment material is determined thereby, rather than by the food to be processed. This is the case of equipment used in the dairy industry, systematically built with stainless steel that is resistant to cheap and widely available agents, e.g. caustic soda or nitric acid; however, aluminum would do as well for contact with milk, further to being lighter and less expensive. A list of construction materials, including their resistance to various disinfectants, is provided in Table 2.1. Finally, note that surface finishing has an effect on disinfection as well; polished surfaces are easier to disinfect and less prone to establishment of biofilms – yet unpolished stainless steel is more resistant to corrosion.

Food products themselves may also be disinfected, so as to alleviate the level of disinfection of processing equipment afterward. Examples include use of ethylene oxide, mixed with an inert gas; as well as sulfites and sulfur dioxide, in the case of large-scale production of dried fruits and wine. Lye-mediated peeling also reduces spore numbers on fruit products, and so does washing with chlorinated water; whereas ultrasonic action appears to reinforce bactericidal action, namely, in the case of hydrogen peroxide.

2.1.2.4 Process handling

As discussed previously, fresh crops and other raw materials are routinely subjected to mechanized handling – in order to obtain washed and graded crops, as demanded by both processors and retailers; representative examples are combine harvesters, pea viners, mobile crop washers, destoners, grading equipment, gentle-flow box tippers (able to transport and unload crops with marginal damage), and automatic cascade fillers (for large and intermediate bulk containers of various sizes, e.g. boxes, combo

TABLE 2.1

Chemical resistance of selected materials used for equipment construction/lining, against selected cleaning/disinfection agents.

completely stable	sufficiently stable for regular use	sufficiently stable for sporadic contact	insufficiently stable for normal use	unstable

Cleaning/disinfectant agent	Metal					Polymer					
				steel							
	aluminium	copper	monel alloys	mild	stainless (18/8)	epoxy	phenolformaldeyde	polyester	polyethylene	poyvinylchloride	rubber
acetic acid	completely stable	unstable	sufficiently stable for sporadic contact	unstable	sufficiently stable for regular use	completely stable	completely stable	sufficiently stable for regular use	completely stable	completely stable	unstable
ammonia	completely stable	unstable	unstable	sufficiently stable for regular use	sufficiently stable for regular use	sufficiently stable for regular use	unstable	unstable	completely stable	insufficiently stable for normal use	sufficiently stable for sporadic contact
citric acid	completely stable	sufficiently stable for sporadic contact	sufficiently stable for sporadic contact	unstable	sufficiently stable for sporadic contact	completely stable	completely stable	completely stable	completely stable	completely stable	sufficiently stable for regular use
ethanol	completely stable	sufficiently stable for regular use	sufficiently stable for regular use	sufficiently stable for regular use	sufficiently stable for regular use	completely stable	completely stable	completely stable	insufficiently stable for normal use	sufficiently stable for regular use	completely stable
fatty acids (C_{12}-C_{18})	completely stable	unstable	sufficiently stable for sporadic contact	unstable	sufficiently stable for regular use	completely stable	completely stable	completely stable	unstable	insufficiently stable for normal use	unstable
glycerol	completely stable	insufficiently stable for normal use	sufficiently stable for regular use	sufficiently stable for regular use	sufficiently stable for regular use	completely stable	completely stable	completely stable	completely stable	completely stable	completely stable
hydrochloric acid	unstable	sufficiently stable for regular use	unstable	unstable	unstable	completely stable	completely stable	completely stable	completely stable	completely stable	sufficiently stable for regular use
hydrogen peroxide	completely stable	unstable	sufficiently stable for regular use	unstable	sufficiently stable for regular use	insufficiently stable for normal use	sufficiently stable for sporadic contact	sufficiently stable for regular use	sufficiently stable for regular use	completely stable	sufficiently stable for sporadic contact
nitric acid	insufficiently stable for normal use	unstable	unstable	unstable	completely stable	unstable	insufficiently stable for normal use	sufficiently stable for sporadic contact	completely stable	completely stable	unstable
phosphoric acid	insufficiently stable for normal use	insufficiently stable for normal use	unstable	unstable	sufficiently stable for regular use	completely stable	completely stable	completely stable	completely stable	completely stable	completely stable
sodium chloride	sufficiently stable for regular use	sufficiently stable for regular use	sufficiently stable for regular use	insufficiently stable for normal use	insufficiently stable for normal use	completely stable	completely stable	completely stable	completely stable	completely stable	completely stable
sodium hydroxide	unstable	insufficiently stable for normal use	completely stable	sufficiently stable for regular use	sufficiently stable for regular use	sufficiently stable for regular use	unstable	unstable	completely stable	completely stable	completely stable
sodium hypochlorite	unstable	unstable	unstable	unstable	unstable	sufficiently stable for sporadic contact	sufficiently stable for sporadic contact	sufficiently stable for sporadic contact	completely stable	completely stable	sufficiently stable for regular use
sulfuric acid	unstable	unstable	insufficiently stable for normal use	unstable	insufficiently stable for normal use	completely stable	completely stable	completely stable	completely stable	completely stable	completely stable
sulfurous acid	insufficiently stable for normal use	insufficiently stable for normal use	unstable	unstable	unstable	completely stable	completely stable	completely stable	sufficiently stable for regular use	sufficiently stable for sporadic contact	completely stable

bins, bags). Other well-established handling practices, pertaining to particulate and powdered solid foods and liquid foods (and ingredients), include bulk movement by road or rail tanker, and storage thereof in large tanks or silos – where fill-level, humidity, and temperature are continuously monitored; such sensors are able to detect weight loss as the tank/silo is emptied, thus permitting calculation of weight of product being used. Intermediate bulk containers have also been increasingly utilized to ship ingredients, displace foods within a production line, and move incompletely processed foods between production sites. In the case of mixing vessels, weight sensors have been utilized to detect an increase in weight as the ingredients are added – and flow sensors to estimate amount of ingredient supplied within a given time, according to preprogrammed recipes and formulations; they are simultaneously used to control operation of pumps, Archimedes' screws, and conveyors – while recording data for stock control and production costing.

Continuous handling equipment is a basic part of continuous, large-scale industrial processes – but it also contributes favorably to increase efficiency when operating batchwise. In the case of solid ingredients and foods, conveyors and elevators suffice for most handling operations; chutes, cranes and hoists, pneumatic devices, water flumes, and trucks are employed to a lesser extent.

Conveyors and elevators are best suited for movement of large quantities along a fixed direction of flow – as relatively constant amounts moved per unit time; when their holdup volumes are significant, they may also serve as reservoir of work-in-progress. Conveyors displace solids within and between operations, and very many designs are available for continuous operation on a permanent basis: horizontal and incline-up and -down configurations, able to serve point and path locations; positioned at overhead, working, floor level, and underfloor heights; and suitable for handling packed, bulk, and solid materials. Selected examples are given in Fig. 2.6.

One of the most common configurations is belt conveyor, see Fig. 2.6i – an endless belt, held under tension between two rollers, one of which is actively driven by a motor. Construction materials range from stainless steel mesh or wire to synthetic rubber – through composite materials including canvas, steel, polyester, or polyurethane. Small pieces and particulate materials are moved via trough-shaped fixed belts – check Fig. 2.6iii;

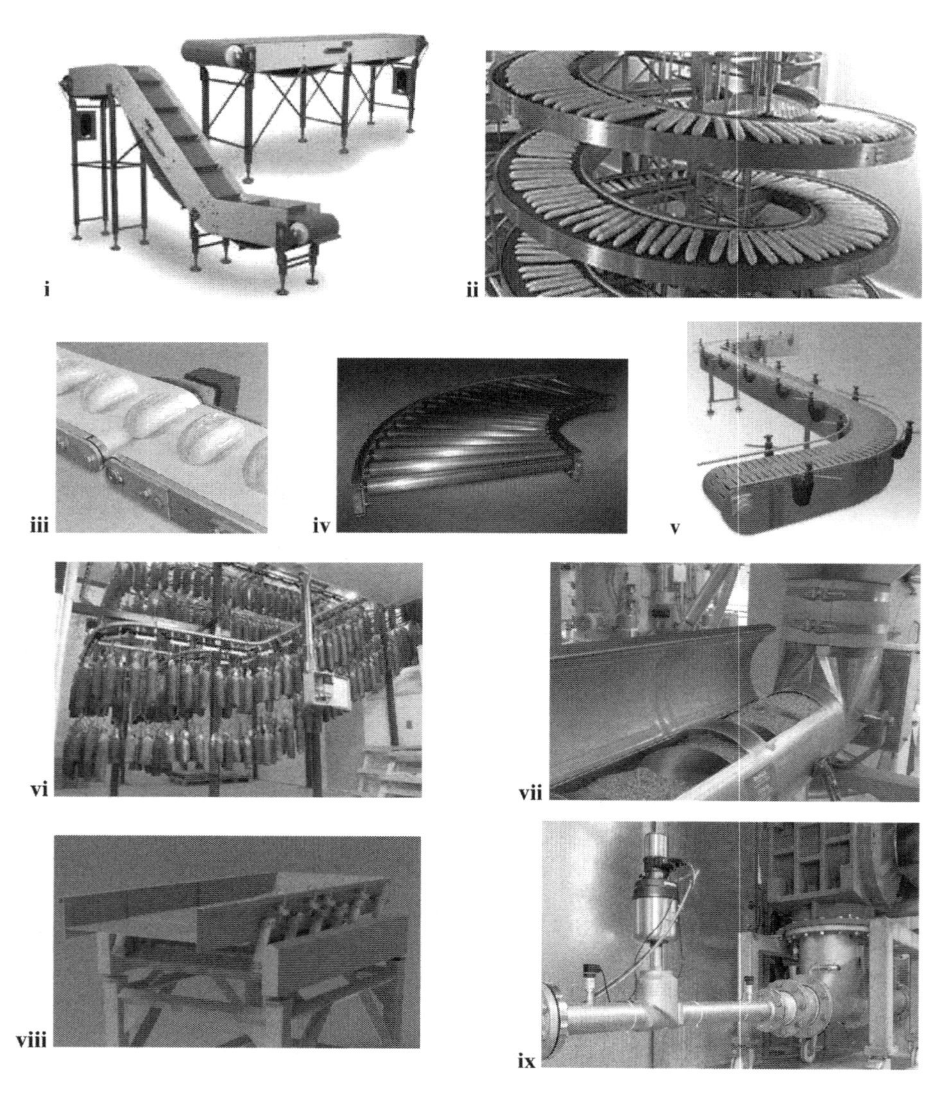

FIGURE 2.6 Examples of process handling equipment, of conveyor type: (i) belt conveyor, in inclined and horizontal designs (courtesy of DynaClean, Muskegon, MI), (ii) spiral conveyor (courtesy of Jonge Poerink Conveyors, Borne, The Netherlands), (iii) trough-shaped belt (courtesy of Teknik Konveyor Bant, Istanbul, Turkey), (iv) roller conveyor (courtesy of Triton Innovation, Alpena, MI), (v) chain conveyor (courtesy of Automac Engineering, Manukau, New Zealand), (vi) monorail conveyor (courtesy of Pacline, Mississauga, ON, Canada), (vii) screw conveyor (courtesy of Daxner, Wels, Austria), (viii) vibratory conveyor (courtesy of Dodman, Norfolk, UK), and (ix) pneumatic conveyor (courtesy of Burkert Fluid Control Systems, Thiruvanmiyur, India).

whereas flat belts are used to carry packed foods; inclination may go up to 45° – provided that they are fitted with cross-slats or raised chevrons, to avoid slipping back of transported material. Spiral conveyors constitute a handy mode of raising or lowering products continuously, as illustrated in Fig. 2.6ii.

Roller conveyors (as per Fig. 2.6iv) and skate wheel conveyors are usually unpowered – even though the former may be. Their rollers or wheels are placed either horizontally – to push packed foods along; or slightly inclined (2–3°) – to allow packs rollover under the sole effect of gravity. When heavier loads are at stake, rollers are the preferred choice for being stronger; conversely, their larger inertia makes startup and shutdown harder – and operation around corners is also more difficult. An ingenious way to circumvent these difficulties is to carry materials on a film of air, blown into tubular trough sections with the aid of a fan – in much the same way a hovercraft ship floats over water or land.

Chain conveyors, as depicted in Fig. 2.6v, are used to move barrels, churns, crates, and similar bulk containers placed directly over a powered chain – possessing protruding lugs at floor level. Displacement of meat and poultry carcasses often resorts to a monorail conveyor, mounted on an overhead track – as detailed in Fig. 2.6vi. Screw conveyors, represented in Fig. 2.6vii, comprise a rotating Archimedes' screw inside a metal trough; they are suitable to move grain, flour, and sugar, as well as small particulate fruits and vegetables (e.g. peas). A uniform, easily controlled flowrate, a compact cross-section without need of a return conveyor, and the possibility of total enclosure (that keeps risk of contamination to a minimum) constitute their main advantages. Both horizontal and inclined apparatuses are found in industrial settings – yet with lengths not exceeding ca. 6 m; above this threshold, friction forces result in excessive power consumption.

Vibratory conveyors, see Fig. 2.6viii, effect small vertical movements of particulate foods or powders – which raises them a small distance (of the order of millimeter) above the conveyor; coupled with a forward movement that carries said particles along the conveyor. The speed and direction of movement may be accurately controlled by tuning the amplitude and frequency of vibration – so these machines end up being useful as feeders of processing equipment.

Pneumatic conveyors consist of a system of pipes through which powders or small particulate foods are suspended in moving air, and transported at velocities ranging from 25 to 35 $m.s^{-1}$ – as illustrated in Fig. 2.6ix. If operated vertically, the velocity of air as carrier should (clearly) exceed the terminal velocity of the food particles in air, at the same pressure and temperature – to prevent particles from settling down (which would block the pipe); if operated horizontally, the particles readily attain the carrier air velocity – yet their tendency for settling in the vertical direction must be overcome by deliberate induction of eddies, formed at the expense of a sufficiently turbulent flow pattern. In either case, air velocity turns out critical: if too low, unwanted settling will take place; if too high, a risk of damage to the food (and the inner wall of the pipe) arises due to abrasion. A major shortcoming of their *modus operandi* is generation of static electricity, arising from friction of solid particles against the wall – which may lead to dust explosion in the case of powders; to minimize this hazard, the equipment must be duly earthed and vented – yet measures for explosion containment and suppression are still to be taken. Pneumatic conveyors will hardly get overloaded, and possess few moving parts – since a supply of high-velocity air essentially suffices as driving force; hence, maintenance costs can be kept low. If particle size is quite uniform and traveling distance is relatively short, then air velocity can be reduced to just above the terminal velocity of the particles – thus leading to dense-phase pneumatic conveyors; smaller turbulence and reduced attrition entail less abrasive wear on equipment, as well as lower overall energy consumption.

Magnetic conveyors have been used to a moderate extent, and only when the food is transported in metal-canned form; the local magnetic field produced holds cans in place with minimal noise – and may, in particular, be used to invert empty cans for washing.

Other conveyor designs used to date encompass singulators, sorters, diverters, descramblers, and accumulators; the first separate product along their path, while the last hold product in place until a given signal is received that triggers release thereof (e.g. biscuit units to be filled in constant-weight packs). Recent improvements include belts made of materials with antimicrobial features; and intelligent conveyors – able to divide a process into intelligent zones that know when they should operate, thus avoiding the hassles of turning on and off the whole machine. Information on the position of the product is generated via photoelectric sensors, and fed to a microprocessor that controls conveyor movement via servo-motors. In each zone, the said microprocessor collates information on products upstream and downstream, and decides at each time whether to run, accumulate, or idle – using inputs from neighboring zones.

Despite their many possible designs, the most common elevator is of the bucket type – as illustrated in Fig. 2.7i. The said elevator consists of metal (or plastic) buckets fixed between two endless chains, meant for permanent continuous operation at some point location; it is particularly appropriate for bulk movement of free-flowing powders and particulate foods at large, either vertically up and down or inclined-up and -down at overhead height – even though packed items can also be handled. The flow rate depends on spacing between consecutive buckets and their individual capacity, as well as conveyor speed – which typically ranges from 15 up to 100 $m.min^{-1}$.

Cranes and hoists are exemplified in Figs. 2.7ii and iii, respectively; by construction, these machines are intended to operate in the vertical direction, both up- and downward – on an intermittent basis, under temporary or permanent service. They serve a limited area – defined by the surface spanned by their supporting (sliding) arm(s); and work overhead and at floor level, handling all types of solid materials – from bulk to packed form.

Trucks are the working horses of food transport outside plant factories – and may actually be viewed as horizontal elevators; they deliver both raw materials to the food plant, and finished products therefrom to gross and retail warehouses, see Fig. 2.7iv. Their dominant direction of work is indeed horizontal – eventhough they can climb and descend hills; and their mode of operation is intermittent, and necessarily on a temporary basis – due to regulations on maximum working time of drivers and limited capacity of fuel tank. Unlike other types of handling apparatuses, unlimited locations may, in theory, be served by trucks – as long as roads are available that drive thereto. With working and floor level heights of operation, they can carry all sorts of products, including liquids, besides solids in bulk and packaged forms – depending on the type of installed chassis.

Further to pneumatic transport of particulate foods, compressed air may be used for a whole range of industrial applications – namely operation of valves and springs, as well as hydraulic presses (e.g. in corking bottles, or in releasing cooled chocolate off the mold); examples are provided in Figs. 2.7v and vi. Hence, local electric work can be avoided – owing to the high cost of electricity required when small engines are used (characterized by poor efficiency, along with high maintenance costs), or else to safety considerations (in terms of electrocution); mechanical work, in the form of a pressurized fluid generated elsewhere, is employed to advantage. Remember that solids transmit forces, whereas fluids transmit pressures; therefore, supply of air at some pressure to a piston with sufficiently large surface area, moving in a sealed (hollow) cylinder, can generate a very large force, even when a low-power compressor is utilized – provided that enough air is supplied. Pneumatic equipment can produce movement in all directions of space, and will work continuously on a permanent basis – as long as compressed air remains available; both point and path locations are feasible, at heights overhead, working and underfloor – being particularly appropriate to handle bulk solids and liquids.

Washing and simultaneous transport of small particulate foods can be brought about by shallow inclined troughs (or flumes), as grasped in Fig. 2.7vii, as well as by pipes; if water flows under gravity – as may happen in food plants located on hillsides, energy consumption is quite low. Besides the obvious preference for operation inclined down, water flumes may operate at working, floor level and underfloor heights, intermittently or permanently; they normally handle solid units, in both point and path locations. To reduce operation costs, water may be recirculated – following filtration and treatment with chlorine, so as to prevent growth of microorganisms and buildup of biofilms.

FIGURE 2.7 Examples of process handling equipment, of elevator type (and the like): (i) plain elevator (courtesy of NuPac Industries, Mentone, Australia), (ii) crane (courtesy of Konecranes, Hyvinkaa, Finland), and (iii) hoist (courtesy of DavidRound, Streetsboro, OH); (iv) transportation truck (courtesy of Domenico Transportation, Denver, CO); pneumatic equipment for (v) valve control (courtesy of AccroSeal, Vicksburgh, MI) and (vi) bottling (courtesy of Melegari Manghi, Alberi di Vigatto, Italy); and (vii) water flume (courtesy of Hughes Equipment, Columbus, WI).

Liquids in food processing environments are usually handled via pipework, and associated valves and pumps; this covers both food matrices and cleaning fluids. Such devices have been discussed to some length, when addressing fluid dynamics in *Food Proc. Eng.: basics & mechanical operations* – so a brief coverage ensues next.

Selection of a particular pump hinges on the type of product to be manipulated, namely, viscosity and shear sensitivity thereof; product flow rate, coupled with suction and discharge pressure; and mode of operation, i.e. continuous versus intermittent. Centrifugal pumps possess a rotating impeller inside a stationary casing; they are normally selected against a moderate downstream overpressure, and when variable flow rates are pursued – attained by either adjusting speed of the impeller, or throttling flow via an adjustable valve in the discharge duct. Conversely, positive displacement pumps have a cavity that is alternately expanded and contracted – with liquid admitted into the suction side at the former stage, and expelled from the discharge side at the latter stage; common designs include rotary lobe or piston pumps, as well as gear and diaphragm pumps. Unlike centrifugal pumps, the same flow rate is produced irrespective of discharge pressure; positive displacement pumps are thus suitable for viscous or shear-sensitive products, large downstream pressures, and accurate flow control.

Very many types of valves exist commercially, for use in pipelines and the like – meant for manual or mechanized operation; most can be fitted with proximity switches to detect and transmit the position of their splindle, and with pneumatic or electro/pneumatic actuators to accurately position the said spindle. In applications involving very high pressures or large diameter pipes, the force required to close a conventional valve is too high – so one usually resorts to a two-seat design; this entails two valve plugs on a common spindle, coupled with two valve seats. Industrial valves are grouped as two- and three-port valves; the former are meant to restrict flow of fluid through them, whereas the latter are used to mix or divert fluid streams.

Two-port valves with a linear spindle include slide and plug valves – whereas a rotary spindle is present in ball and butterfly valves. The spindle moves up and down along a frame in slide valves, whereas plug valves take advantage of a wedge gate; both can be used to isolate fluid, since they permit leakage-proof shutoff. Ball valves comprise a spherical ball bearing a hole – which permits passage of fluid through, is located between two sealing rings in the valve body; hence, a 90°-rotation suffices to move between on and off position, with the latter being sufficiently tight to withstand high-pressure steam. Butterfly valves are a weaker (but also less expensive) version of ball valves – and consist of a disc that rotates on bearings inside the valve; in the open position, the disk remains parallel to the pipe wall, thus allowing full flow therethrough – whereas rotation to the closed position, perpendicular to the pipe wall and against a seat, discontinues flow.

Three basic types of three-port valves exist in the market – viz. piston, plug, and rotating shoe; selected examples are provided in Fig. 2.8. Piston valves (see Fig. 2.8i) possess a hollow

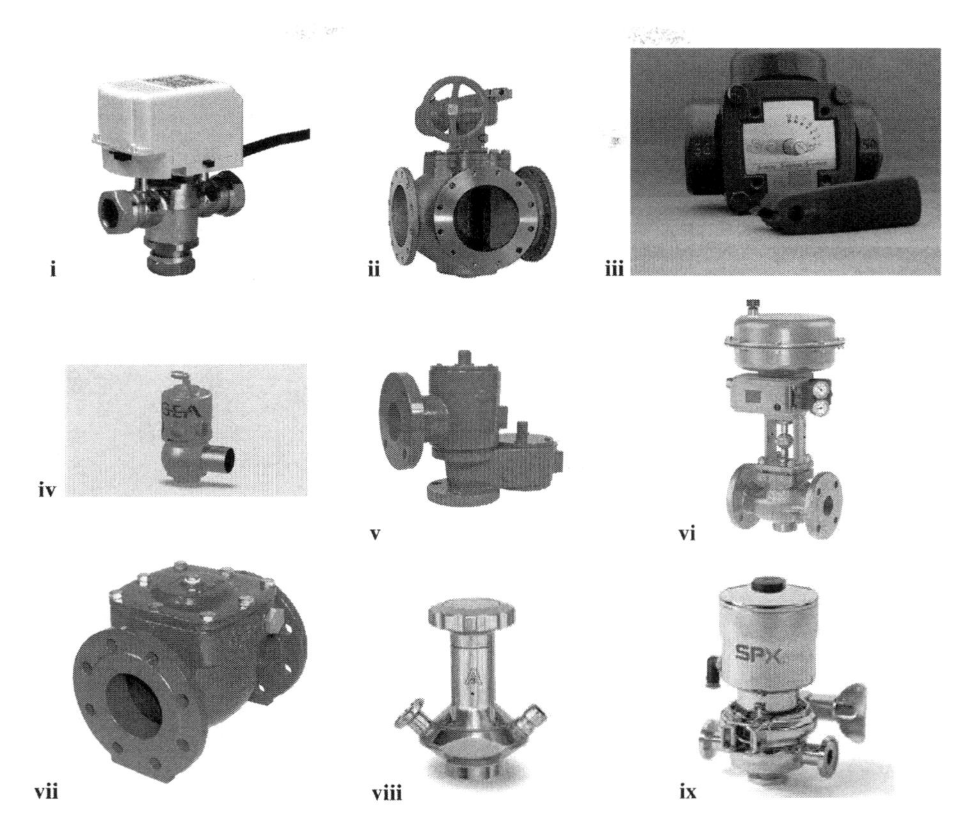

FIGURE 2.8 Examples of three-port valves: (i) piston-type (courtesy of Drayton Controls, Plymouth, UK), (ii) plug-type (courtesy of Orbinox, Anoeta, Spain), and (iii) rotating shoe-type (courtesy of Danfoss, Nordborg, Denmark); and examples of special-purpose valves: (iv) safety valve (courtesy of GEA, Dusseldorf, Germany), (v) vacuum valve (courtesy of Emerson Electric, St. Louis, MO), (vi) modulating valve (courtesy of Samson Controls, Baytown, TX), (vii) nonreturn valve (courtesy of Johnson Valves, Southampton, UK), (viii) sampling valve (courtesy of AlfaLaval, Lund, Sweden), and (ix) diaphragm valve (courtesy of SPX Flow Technology, Crawley, UK).

piston that moves up and down when driven by an actuator – thus covering, and concomitantly uncovering two ports; consequently, the cumulative cross-sectional area remains the same. The actuator pushes instead a disc, or pair of plugs between two seats in the case of plug-type valves (see Fig. 2.8ii); this action accordingly increases or decreases flow through the two ports, in a corresponding fashion. In the last type of three-port valves (see Fig. 2.8iii), a rotating shoe can be positioned for mixing applications – as it permits admission of distinct proportions of fluids, as supplied by the two inlet pipes into the same space; the mixture thereof then exits through the third port.

Other types of valves exist that are conceived for specific roles – namely safety valves (see Fig. 2.8iv) to prevent excess pressure in pressurized vessels; vacuum valves (see Fig. 2.8v) to protect vessels or tanks from collapse under unwanted vacuum; modulating valves (see Fig. 2.8vi) that permit exact control of product throughput; nonreturn valves (see Fig. 2.8vii) to prevent reversal of flow, namely, during process shutdown; sampling valves (see Fig. 2.8viii) that allow microbiologically safe collection of samples from the production line, without running the risk of contaminating the product; and diaphragm valves (see Fig. 2.8ix) consisting of a polymer membrane, or stainless steel bellows that keeps product away from contact with the valve shaft – and is convenient for applications where sterility is a must.

2.1.2.5 Postprocess handling

The batch nature of most channels of supply of raw materials and distribution of finished product requires intermediate buffering at the interface with processing operations – either batch or (more often) continuous; such a buffering is normally provided by some form of storage – the management of which constitutes the focus of logistics.

The amount of stored ingredients or products has classically been kept to a minimum for financial reasons, unavoidable decay in quality, and risk of pilferage. In fact, large amounts of raw materials are unfavorable for cashflow – as money is tied up in materials that have been already paid for, or in final products that have already incurred in costs of production; warehousing space is also expensive per se, in terms of investment reintegration (associated with usage) and regular maintenance. On the other hand, (bio)chemical changes take place continuously in foods, even when stored under low temperatures or in lyophilized form – which reduce their quality and market value as the prospective shelf-life is approached, to eventually render then unusable at all. Finally, the possibility of pilferage is not negligible, especially when premium-priced products are at stake. On the other hand, the intrinsic seasonality of (most) agricultural and (some) animal products serving as raw materials, or the seasonal demand by the market call for stocks of ingredients and final products – to be reduced as much as possible anyway, but without affecting overall processing fluidity. Although the just-in-time approach pioneered by Japanese companies has met with success in assemblage industries (e.g. car making), the same does not hold with regard to food manufacturing in general; an exception applies, however, to some retailing operations.

Raw materials, work-in-progress, and stored goods account for the whole sets of stocks; decisions on the relative size of their inventories depend on buying and storage costs. Three classes have accordingly been set, focused on usage value – defined as rate of usage multiplied by individual value: class A, accounting for ca. 20% high-value materials and ca. 80% total usage value; class B, accounting for ca. 30% intermediate-value materials and ca. 10% usage value; and class C, accounting for ca. 50% low-value materials and ca. 10% usage value. Therefore, class A products are given preference, in terms of inventory, relative to class B ones, and these over their class C counterparts. Stored goods encompass buffer (or safety) inventory – meant to compensate for uncertainties in supply or demand; cycle inventory – when batch production is greater than immediate demand; anticipation inventory – should seasonal demand (e.g. Easter eggs) or supply fluctuations (e.g. seasonal fruits) be significant, yet predictable; and pipeline (or in-transit) inventory – materialized in products undergoing movement between point of supply and point of demand.

When building storerooms and warehouses, special care is to be exercised to prevent access thereto by rodents, insects, and birds. Specific measures include: doors fitted with screens or air curtains; windows screened against insects; ceilings and walls designed to prevent entry and nesting of insects, rodents, and birds; floors covered with vinyl-based coatings, to prevent cracks that would otherwise harbor insects (besides microorganisms); rooms equipped with electric insect killers; and drainage channeling and power cable ducting fitted with antirodent devices. Inspection on a regular basis is essential to ensure that preventative measures remain in place, and have been effective so far.

Regular storage requires cool rooms, with good ventilation and protected from direct sunlight; fresh foods stored for short periods of time require high humidity to avoid wilting and drying out; and such hygroscopic ingredients as sugar, salt, and powdered flavorings or colorants need relatively dry air – i.e. carrying a relative humidity below their characteristic equilibrium relative humidity (e.g. 60% for sucrose). Fats and oils, on the one hand – for their being susceptible to pick up odors; and spices, on the other – for their releasing strong odors, require storage separate from other food items as general rule.

To minimize handling, packages are grouped together into larger (or unitized) loads – with pallets being frequently used to move unitized loads of cases or sacks, via fork-lifter or stacker trucks. Fiberboard slipsheets reduce the volume occupied by pallets in vehicles and warehouses – with products secured to the pallet or slipsheet by shrink- or stretch-film. Stacking should not exceed the recommended height for each material, otherwise crushing can occur – or collapse, with a risk of injuries to the operators. Lighting should be bright, and placed at a high level to reduce shadowing by stacked pallets; periodic cleaning routines are necessary to prevent accumulation of dust or spilled foods – which attract insects and rodents. Monitoring of material movements into and out of the stores – necessary for checking stock levels and ascertaining rotation turnovers, as well as degree of usage of raw materials and delivery of final products have been the focus of dedicated management systems; these resort, in turn, to optical barcode and radiofrequency identification tagging. Larger warehouses use computerized truck-routing systems, as well as robotic handling and automated guided vehicles (with fixed or optimized routes) – namely conveyor- and lift-deck, counterbalance, and aisle narrow machines, able to outrigger, reach, and tug/tow.

Remember that the food chain spreads from harvesting and production, through processing, to purchase and use by the final customer; besides storage at the interface of the process with upstream supply and downstream selling, distribution is germane – and constitutes a part of logistics as well. Efficient distribution means not only providing the customer with the product at the right place and time, to qualitatively and quantitatively acceptable levels – but also keeping the associated cost to a minimum, since it often does not add value to the product itself. The increasing importance of distribution has been driven by consumer pressure for a greater variety of food products – with higher quality and freshness, and availability year-round.

The sharp increase in competitiveness among large food-retail businesses has accelerated development of global value chains; and the associated growth in buying power has been pushing down the prices paid to food processors. Development of information, communication, and automation technologies has played a central role toward globalization of distribution – while reducing the need for highly-skilled, and thus highly-paid workforce. Hence, food companies have increasingly moved their operations to developing countries – where unskilled, and thus underpaid workers are available; while food production has been coordinated between distant sites – with suppliers now called upon to transfer goods across the world on short notice.

Storage and distribution of foods along the cold chain, in particular, have meanwhile undergone major changes. Products from a given food manufacturer were once transported to a relatively large number of small distribution depots – each one handling a single product, because small delivery volumes made it uneconomical to operate every day; along with separate vehicles, bearing dedicated refrigeration facilities, and subjected to their own contractors' distribution policies and delivery schedules. The associated increase in distribution costs and reduction in product quality caused by the said practices urged development of mathematically optimized routines – resorting to combination of distribution streams of various suppliers, transport of fresh, chilled, and frozen foods in the same cargo, alterations in method and frequency of ordering, and redesign and reorganization of warehouses. Nowadays, fewer but larger regional distribution centers handle a wider variety of food products – with occupied floor area up to 25,000 m^2, and separate zones for ambient temperature (ca. 10°C), cooling temperature (ca. 5°C), chilling temperature (ca. 0°C), and freezing temperature (ca. –25°C); together with (insulated) delivery vehicles, fitted with movable bulkheads and refrigeration units (able to create at least three of those temperature zones) – which permit throughputs of up to 30 million food cases yearly, and can service up to 50 retail outlets.

The aforementioned primary distribution, from manufacturer to regional distribution centers, is followed by secondary distribution to retail stores. Short-lived products are normally received during the late afternoon or evening, and shipped to retail stores early in the next morning; this is known as first-wave delivery. Conversely, longer shelf-life products are taken from stock, and formed into orders for each retail store over a 24-hr period – thus accounting for the second-wave delivery, encompassing regular working hours. Subcontracting logistics to a third party also constitutes an expanding option – namely, via introduction of factory-gate pricing, where extra transport costs are to be borne by retailers; subcontracted haulers – with highly efficient logistics able to keep transportation costs to a minimum, have accordingly operated both ways, in a process known as backhauling.

2.1.2.6 Water supply and effluent treatment

The quality of water supply plays an important role upon quality of the final food products, further to its effect upon raw materials and processing conditions. Depending on location of the food plant, coupled to costs of water itself and treatment thereof, this utility may be obtained from the public mains supply, from surface water collectors, or even from private wells. The former is appropriate for food processing at large in most countries, since potable drinking water standards are regularly enforced – via previous treatment and continuous monitoring, in microbiological, chemical, and radiological terms.

Four major uses exist for water in industrial facilities: general purpose, process, cooling, and boiler. The first type entails washing, cleaning, and sanitizing operations of raw materials and equipment; the only intramural treatment effected is chlorination. Process water is used for blanching, as cooking medium, or added directly to food; no significant alteration of product quality is expected, both in taste and texture, so it should exhibit sufficiently high quality – including lack of dissolved materials that would make it excessively hard. Calcium and magnesium bicarbonates are indeed frequently present in water, and account for temporary hardness – as they precipitate upon heating (to form scales); conversely, calcium and magnesium chlorides, nitrates, and sulfates are responsible for permanent hardness – and may harden vegetables during blanching and canning. Hard water also leads to scum or deposits when added with alkaline detergents; methods for hardness correction (or softening) include precipitation, cation exchange, reverse osmosis, and demineralization – see Fig. 2.9. Cooling and boiler waters do not contact food, or equipment surfaces directly exposed to food – so they do not need to be potable; however, some treatment to prevent accumulation of scale in pipes and equipment may be required – namely, removal of hardness, or total demineralization in the case of high-pressure boilers. When cooling thermally processed, low-acid foods in hermetically sealed containers, cooling water should be essentially devoid of chlorine and other sanitizers – to avoid chemical attack whatsoever on the packaging material, likely to compromise long-term storage.

FIGURE 2.9 Example of freshwater treatment unit (courtesy of Osmonic, Minnetonka, MN).

Should water be obtained from wells or surface sources, several impurities are expected – as is the case of suspended solids, dissolved gases, microorganisms, organic matter, and other contaminants responsible for off-colors, -odors, and -tastes; this calls for treatment prior to industrial use, so that acceptable thresholds are not exceeded.

Screening, sedimentation, coagulation, and flocculation – and eventually filtration, are the most common processes to remove suspended solids; they should be used in this sequence, as required by the decreasing size and density of solid particles in suspension.

Such dissolved gases as CO_2, O_2, N_2, and H_2S are naturally found in water, but may cause corrosion (with subsequent reduction in thermal conductivity of the material) or Fe pick-up. They can be removed via boiling – thus taking advantage of Henry's law, as its proportionality constant decreases with increasing temperature; other possibilities are addition of oxygen scavengers (in the case of dissolved O_2), aeration followed by chlorination (in the case of dissolved H_2S), or plain venting with air (in the case of CO_2 or H_2S).

Methods for disinfection of water resort to chemicals, heat, ultraviolet radiation, or ultrasound; one of these methods (or a combination thereof) will normally get rid of (viable) bacteria, viruses, protozoan cysts, and even worms. The least expensive, and thus more often utilized technique is chlorination – with treatment levels normally expressed in ppm (or mg.L^{-1}), and residual chlorine assayed for with *N,N*-diethyl-phenylenediamine test kits; reuse of water is permitted upon chlorination.

Organic compounds can be removed via coagulation, followed by settling or filtration; as alternative, one may use adsorption on activated carbon – where advantage is taken of its outstandingly high specific surface area (based on large porosity and small pore size) and chemical affinity (associated with its carbon atoms).

Due to global warming and associated climate change – including droughts inland and concomitant excessive precipitation over the sea, the sources of freshwater have been steadily decreasing; hence, saving water will become a more and more important issue in the near future. The simplest way to do so is employing good manufacturing practices (GMPs) as starting point – including improved housekeeping (e.g. repair of leaking pipes, turning off of water cooling when not in use); the range of savings obtainable, and the costs incurred in are summarized in Fig. 2.10. If further savings are sought, then one may implement the *R*-trilogy – reduction, reuse, and recycling of water; enough savings may not be attainable, however, unless extensive redesign of the process itself is carried out – yet at the expense of much higher costs.

Although water supply is of the utmost relevance in food processes, aqueous effluents (and other wastes) are also of concern – especially in view of increasingly strict legal requirements; with the notable exception of extrusion or baking, most food processes produce large amounts of effluents – as qualitatively illustrated in Fig. 2.11i, and quantitatively depicted in Fig. 2.11ii. Effluents arise from cleaning and preparation of raw materials, spillages, cleaning of equipment and floors, emptying of vessels, and changing over to different products; and are to be

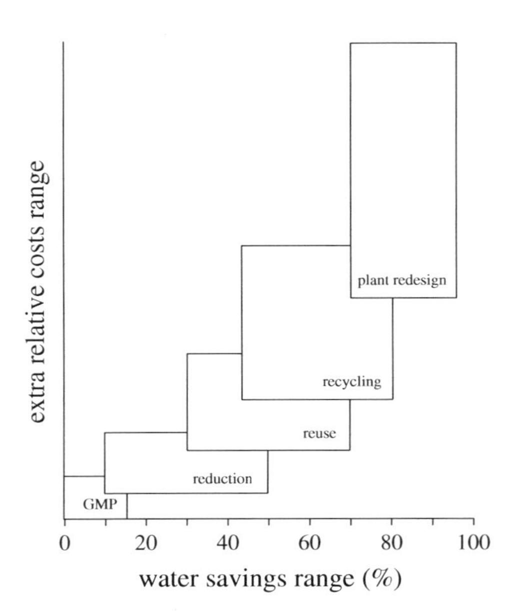

FIGURE 2.10 Strategies for reduction of (fresh) water consumption in food plants – expressed as ranges (▭) of extra relative costs and water savings.

complemented by solid wastes, e.g. discarded packaging materials and office paper. Fruit and vegetable processing effluents exhibit high concentrations of sugars, starch, and solid matter – namely associated with peelings; higher proportions of fat and proteins are instead typical of animal products, e.g. meat and dairy.

In large processing plants, or those located in unpopulated areas, purpose-built facilities exist for effluent treatment on-site – see Fig. 2.12. Besides lagoons and ponds with activated sludge, approaches include trickling filters, low-pressure membranes, oxidation ditches, ozone treatment, spray irrigation, and anaerobic digestion or fermentation. In all other cases, pretreatments to remove solids (including fats and oils) or adjust *pH* are normally required, followed by treatment by municipal authorities or private suppliers of water utilities.

The cost of effluent treatment arises from a combination of volumetric flow rate and polluting load – the latter measured by COD and/or BOD, coupled with the amount of suspended solids (see Fig. 2.11i). The former is a measure of the amount of (dissolved) oxygen taken up by a given oxidation reaction, and expressed as $mg_{oxygen}.L^{-1}$ (or ppm); in the case of organic compounds, of general chemical formula $C_aH_bO_cN_d$, complete oxidation to carbon dioxide, water and ammonia is envisaged, according to

$$C_aH_bO_cN_d + \left(a + \frac{b}{4} - \frac{c}{2} - \frac{3d}{4}\right)O_2 \rightarrow aCO_2 + \left(\frac{b}{2} - \frac{3d}{2}\right)H_2O + dNH_3. \tag{2.5}$$

The above reaction has traditionally been effected by dichromate – a strong oxidizing agent under acidic conditions, with general stoichiometry described by

$$C_aH_bO_cN_d + eCr_2O_7^{2-} + (d+8e)H^+ \rightarrow aCO_2 + \left(\frac{b}{2} - \frac{3d}{2} + \frac{8e}{2}\right)H_2O + dNH_4^+ + 2eCr^{3+}. \tag{2.6}$$

The amount spent of $Cr_2O_7^{2-}$, say n_d, is quantified by titration, and then converted to the equivalent molar amount of O_2, $n_o = n_d$, for the same chemical fate of C, H, O, and N in the organic compound taken as reference – thus avoiding the need to know *a*, *b*, *c*, and *d* exactly; this in turn becomes COD, after referring to solution volume. Note that

$$e = \frac{2a}{3} + \frac{b}{6} - \frac{c}{3} - \frac{d}{2} \tag{2.7}$$

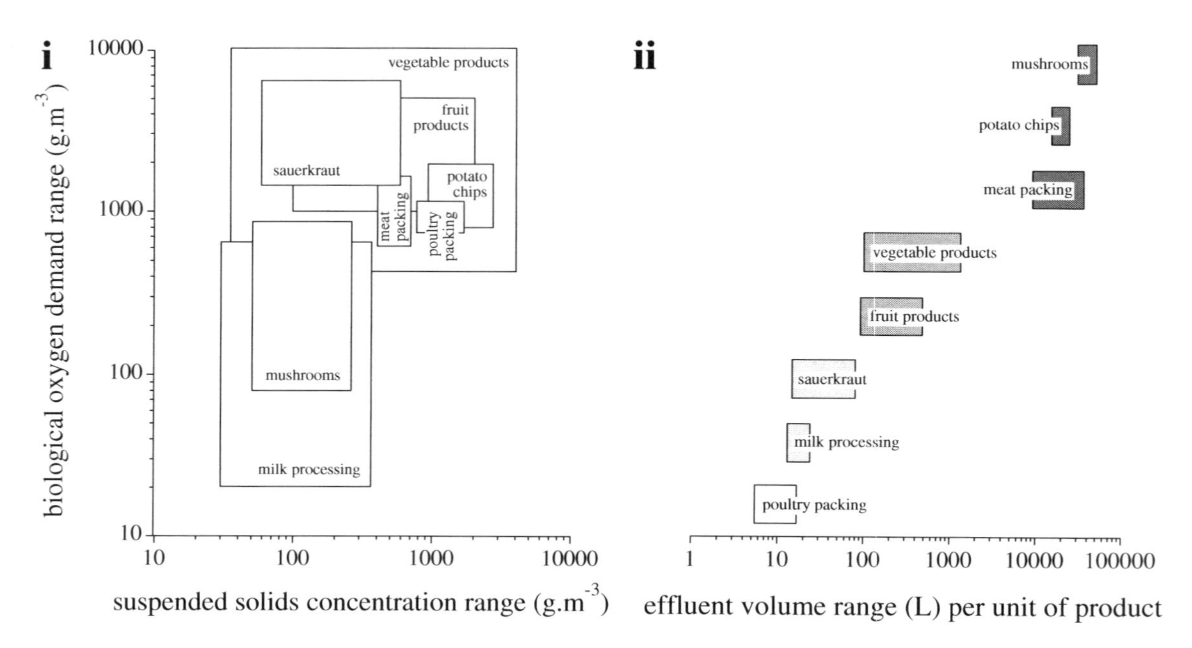

FIGURE 2.11 Ranges of effluent features in selected food processes – expressed as (i) biological oxygen demand and suspended solids concentration (▭), and (ii) effluent volumes generated per unit of product, using carcass (▭), L or kg (▭), case (▭), or ton (■) for unit; data on fruits include typically apples, apricots, citrus, pumpkin, and tomatoes, whereas data on vegetables include typically beans, carrots, corn, peas, peppers, and spinach.

FIGURE 2.12 Example of wastewater treatment unit (courtesy of SUEZ, Paris, France).

applies to guarantee balancing of oxygen atoms in Eq. (2.6) – a formula useful if a single organic species, i.e. $C_aH_bO_cN_d$, were present in the effluent. In view of the above, the oxygen demand of nitrification, i.e.

$$NH_3 + 2O_2 \rightarrow NO_3^- + H_2O + H^+ \tag{2.8}$$

representing oxidation of ammonia to nitrate, is not included in the COD index. On the other hand, BOD is the amount of dissolved oxygen demanded by (micro)organisms to aerobically breakdown organic material – at a given temperature and over a given period; it is again expressed as $mg_{oxygen}.L^{-1}$ (or ppm), with 20°C and 5 days being (by default) taken as assay conditions. The relevance of either of these indicators of pollution is that high values thereof may compromise survival of fish and other heterotrophic aquatic beings that require O_2 for respiration – if enough extra oxygen cannot dissolve within the timeframe required for the said oxidation to occur. As emphasized above, food industry effluents are often characterized by large concentrations of sugars, starches, and oils – characterized by a very strong polluting potential, due to COD within the range 500–4000 ppm; whereas domestic sewage ranges within merely 200–400 ppm. The cost of effluent treatment per unit volume, C_w, may then be roughly estimated via

$$C_w = R + V + B\frac{O_t}{O_s} + S\frac{S_t}{S_s}, \tag{2.9}$$

known as Mogden's formula; here R denotes reception and transport charges per unit volume; V denotes bulk and physicochemical/biological primary treatment charges per unit volume, or else treatment and disposal charges per unit volume at designated or nondesignated sea outfalls; B denotes biological oxidation charges per unit volume of settled sewage, O_t denotes COD of effluent after 1 h-settlement, and O_s denotes mean COD of settled sewage at treatment works; and S denotes treatment and disposal charges per unit volume of primary sludge, S_t denotes suspended solids, and S_s denotes mean suspended solids at treatment works.

Strategies to reduce treatment costs include separation of concentrated (e.g. processing) from diluted (e.g. washing) effluent streams – or else storage of concentrated effluents, and gradual blending thereof with dilute effluents to get a moderately dilute overall effluent in the end; removal of suspended solids via screening, sedimentation, or centrifugation, or chemically induced flocculation (e.g. addition of lime or ferrous sulfate); and dispersion of fats/oils by aeration flotation, followed by disposal of the solid or liquid waste for composting or the like. Some form of mechanical brushing/sweeping on the pipe inside (known as pigging), to both clean the pipe and recover product before product change-over – namely, by sending a rubber block that fits closely therein, is also employed. Solid wastes (including packaging and office materials) are to be separately collected, and eventually transported by municipal authorities for dumping in landfill sites; or increasingly by private recycling companies, in the case of paper, metal, and plastic. Structured management approaches to control and reduce industrial wastage and pollution – complemented by reduction of energy consumption, have meanwhile led to the concept of environment management systems (EMS, for short), required to fully meet with the corresponding international legislation (as set by *ISO 14001*).

2.1.3 Mathematical simulation

2.1.3.1 Cleaning

Cleaning and disinfection of kettles and tanks cannot, in general, be achieved by circulation of plain water – due to the paramount volumes that would be needed, and the insufficient stress that would develop on the wall (as discussed previously). The cleaning solution should indeed contain some detergent to aid in detaching dirt and contaminants, and dissolve or suspend them as micelles in water. Although some mechanical action can be produced with the aid of a brush, such a method is still strongly dependent on the human factor – and thus meant to be abolished. A better alternative is introduction of a jet-producing apparatus into the reservoir, for a sufficiently long period of time so as to span the whole inner surface of the equipment; this movement may, however, be annoying, and proceed along a chaotic (and thus unpredictable) trajectory.

In the simplest form, the amount of (unwanted) solids left, m_S, evolves in time, t, according to first-order kinetics, i.e.

$$\boxed{\frac{dm_S}{dt} = -k_S m_S,} \tag{2.10}$$

where k_S denotes a first-order kinetic constant;

$$\boxed{m_S\big|_{t=0} = m_{S,0}} \tag{2.11}$$

plays the role of initial condition, as long as $m_{S,0}$ denotes initial load of extraneous material. The aforementioned k_S depends on various factors: chemical nature and state of dirt; type of inner surface of the equipment wall – for instance, rough or plastic surfaces are characterized by lower k_S than smoother ones, e.g. polished stainless steel; and qualitative composition and quantitative concentration, C_{ca}, of cleaning aid utilized – where

$$k_S = k_0 C_{ca} \tag{2.12}$$

is often found to hold, with $k_0 > 0$ denoting a proportionality constant, even though such a relationship may not fully extend to true detergents. The operating (absolute) temperature affects k_S via

$$k_S = k_{1,0} \exp\left\{-\frac{k_{1,1}}{\mathcal{R}T}\right\} \tag{2.13}$$

as per a classical Arrhenius' law, where $k_{1,0}$ and $k_{1,1}$ denote a pre-exponential factor and a pseudo activation energy; where Reynolds' number – as a measure of the intensity of convection, plays a role quantified by

$$k_S = k_{2,0} + k_{2,1} \ln Re, \tag{2.14}$$

where $k_{2,0}$ and $k_{2,1}$ denote adjustable parameters. Integration of Eq. (2.10) via separation of variables, with the aid of Eq. (2.11) – i.e.

$$\int_{m_{S,0}}^{m_S} \frac{d\tilde{m}_S}{\tilde{m}_S} = -k_S \int_0^t d\tilde{t}, \tag{2.15}$$

readily leads to

$$\ln \tilde{m}_S \Big|_{m_{S,0}}^{m_S} = -k_S \tilde{t} \Big|_0^t \tag{2.16}$$

at the expense of the fundamental theorem of integral calculus – which is equivalent to writing

$$\ln \frac{m_S}{m_{S,0}} = -k_S t; \tag{2.17}$$

Eq. (2.17) is valid for isothermal processes, in which C_{ca} remains essentially constant (as is normally the case). After taking exponentials of both sides, Eq. (2.17) becomes

$$\boxed{\frac{m_S}{m_{S,0}} = \mathrm{e}^{-k_S t}}, \tag{2.18}$$

as plotted in Fig. 2.13. An exponential decay in the amount of dirt present on the surface accordingly results – implying that every logarithmic cycle of cleaning will take the same amount of time; in other words, the time taken to clean from 10% to 1% dirt is the same to clean from 1% to 0.1% dirt, the same to clean from 0.1% to 0.01% dirt, and so on.

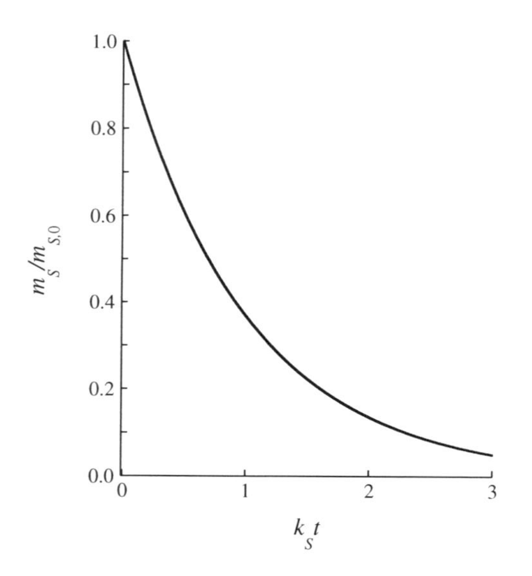

FIGURE 2.13 Evolution of normalized amount of dirt, $m_S/m_{S,0}$, remaining on equipment surface, versus dimensionless time of cleaning, $k_S t$.

Mechanical action overrides in cleaning when $Re > 10{,}000$; this is a consequence of the high stress experienced in the wall vicinity, caused by flow of fluid at high velocity – and may be calculated via Eq. (2.1) of *Food Proc. Eng.: basics & mechanical operations*, applied to $y=0$, i.e. on the surface wall – provided that the actual velocity profile can be ascertained in advance. Since (the average of) τ_{yx} correlates with pressure drop, ΔP, in the x-direction due to viscous flow, it is more useful to resort to Eq. (2.86) in *Food Proc. Eng.: basics & mechanical operations* for an estimate of ΔP, and Fig. 2.9 in the same book for an estimate of Fanning's friction factor, f. Alternatively – and in attempts to cover the intermediate range of Re, one may proceed directly to Eq. (2.18) and combine it with Eq. (2.14) to get

$$\frac{m_S}{m_{S,0}} = \mathrm{e}^{-(k_{2,0}+k_{2,1}\ln Re)t}, \tag{2.19}$$

which can be algebraically rearranged as

$$\frac{m_S}{m_{S,0}} = \mathrm{e}^{-k_{2,0}t}\mathrm{e}^{-k_{2,1}t\ln Re} \tag{2.20}$$

after splitting the exponential function – or else

$$\frac{m_S}{m_{S,0}} = \mathrm{e}^{-k_{2,0}t} \exp\left\{\ln Re^{-k_{2,1}t}\right\}, \tag{2.21}$$

in view of the operational features of a logarithmic function; composition of the exponential function with its inverse produces merely

$$\frac{m_S}{m_{S,0}} = \mathrm{e}^{-k_{2,0}t} Re^{-k_{2,1}t}, \tag{2.22}$$

valid at large Re. The aforementioned mechanical action is influenced by presence of foam, as it likely prevents direct contact between target surface and cleaning solution; conversely, the underlying abrasive action can be enhanced by dispersing air in the said solution. The same principle justifies flowing a suspension of solid or elastic particulate matter (e.g. rubber balls) through a pipe, due to the associated scrubbing action.

Material removal is normally faster at startup, but more and more difficult toward the end of a cleaning process – thus justifying the decrease of k_S with time that is frequently observed; in fact, earlier removal is affected chiefly by adhesion of target material to similar material as clumps, whereas the (stronger) forces prevailing between wall itself and deposited material dominate as the cleaning process progresses.

Since cleaning oftentimes resorts to surface-active agents, consideration of the underlying thermodynamic equilibrium is in order; one should consequently retrieve Eq. (3.35) of *Food Proc. Eng.: basics & mechanical operations* as

$$\boxed{dG = VdP - SdT + \sigma dA + \sum_{i=1}^{N} \mu_i dn_i} \tag{2.23}$$

– i.e. a more general form of the fundamental relationship in thermodynamics, which takes into account also the contribution of surface energy (or energy associated with an interfacial surface area) to overall Gibbs' energy – with

$$\boxed{\sigma \equiv \left(\frac{\partial G}{\partial A}\right)_{P,T,n}} \tag{2.24}$$

complementing Eqs. (3.31), (3.32), and (3.36) in *Food Proc. Eng.: basics & mechanical operations*. By the same token,

$$\boxed{G = \sigma A + \sum_{i=1}^{N} \mu_i n_i} \tag{2.25}$$

should be used *in lieu* of Eq. (3.59) in *Food Proc. Eng.: basics & mechanical operations* – since Gibbs' energy is accordingly carried not just by every i-th component of the N-component system (each with extensity measured by n_i), but also by the interface of that system (of overall extensity measured by A); upon differentiation of both sides, Eq. (2.25) gives rise to

$$dG = Ad\sigma + \sigma dA + \sum_{i=1}^{N} \left(n_i d\mu_i + \mu_i dn_i\right), \tag{2.26}$$

as per the rules of differentiation of a sum and of a product of functions, or else

$$dG = Ad\sigma + \sigma dA + \sum_{i=1}^{N} n_i d\mu_i + \sum_{i=1}^{N} \mu_i dn_i \tag{2.27}$$

after splitting the summation. Insertion of Eq. (2.23) transforms Eq. (2.27) to

$$VdP - SdT + \sigma dA + \sum_{i=1}^{N} \mu_i dn_i = Ad\sigma + \sigma dA + \sum_{i=1}^{N} n_i d\mu_i + \sum_{i=1}^{N} \mu_i dn_i, \tag{2.28}$$

where cancellation of common terms between sides unfolds

$$\boxed{Ad\sigma + \sum_{i=1}^{N} n_i d\mu_i = VdP - SdT;} \tag{2.29}$$

Eq. (2.29) is known as generalized Gibbs and Duhem's relationship, in view of Eq. (3.65) of *Food Proc. Eng.: basics & mechanical operations*. Under constant temperature and pressure – as happens in most cleaning situations of practical interest, $dT=0$ and $dP=0$, so Eq. (2.29) breaks down to

$$\left(Ad\sigma + \sum_{i=1}^{N} n_i d\mu_i = 0\right)_{P,T}; \tag{2.30}$$

in the case of a binary solution of surfactant, referred to by subscript 2, in water as solvent, referred to by subscript 1, Eq. (2.30) further reduces to

$$Ad\sigma + n_1 d\mu_1 + n_2 d\mu_2 = 0. \tag{2.31}$$

In the bulk of the said solution, no surface phenomena are (obviously) present; therefore, one realizes that

$$n_{1,b} d\mu_1 + n_{2,b} d\mu_2 = 0 \tag{2.32}$$

using Eq. (2.31) as template – with subscript b referring to bulk. In view of Eq. (3.103) in *Food Proc. Eng.: basics & mechanical operations*, the interface (as molecular separation between two contacting phases) must be at equilibrium with the bulk phase in terms of potential of solvent; hence, μ_1 and μ_2, in Eq. (2.31) and (2.32), are shared by bulk and interface. Upon multiplication of both sides by $n_1/n_{1,b}$, Eq. (2.32) becomes

$$n_1 d\mu_1 + \frac{n_1}{n_{1,b}} n_{2,b} d\mu_2 = 0 \tag{2.33}$$

– with ordered subtraction of Eq. (2.33) from Eq. (2.31) giving then rise to

$$Ad\sigma + n_2 d\mu_2 - \frac{n_1}{n_{1,b}} n_{2,b} d\mu_2 = 0; \tag{2.34}$$

division of both sides by $Ad\mu_2$, followed by factoring out of $1/A$ convert Eq. (2.34) to

$$\frac{d\sigma}{d\mu_2} + \frac{1}{A}\left(n_2 - \frac{n_1}{n_{1,b}} n_{2,b}\right) = 0, \tag{2.35}$$

where n_1 may be factored out as

$$-\frac{d\sigma}{d\mu_2} = \frac{n_1}{A}\left(\frac{n_2}{n_1} - \frac{n_{2,b}}{n_{1,b}}\right). \tag{2.36}$$

Equation (3.155) of *Food Proc. Eng.: basics & mechanical operations* – pertaining to a single-phase solution in the absence of surface effects, may now be retrieved as

$$\mu_2 = \mu_2^{\theta} + \mathcal{R}T \ln a_2, \tag{2.37}$$

which reads

$$\left(d\mu_2 = \mathcal{R}Td \ln a_2\right)_T \tag{2.38}$$

in differential form under constant temperature; Eq. (2.38) permits reformulation of Eq. (2.36) to

$$-\frac{d\sigma}{\mathcal{R}Td \ln a_2} = \frac{n_1}{A}\left(\frac{n_2}{n_1} - \frac{n_{2,b}}{n_{1,b}}\right). \tag{2.39}$$

The differential of the logarithm of a function is but the differential of the function divided by the function itself, so Eq. (2.39) may instead appear as

$$-\frac{a_2}{\mathcal{R}T} \frac{d\sigma}{da_2} = \frac{n_1}{A}\left(\frac{n_2}{n_1} - \frac{n_{2,b}}{n_{1,b}}\right); \tag{2.40}$$

definition of surface concentration excess, Γ, as

$$\boxed{\Gamma \equiv \frac{n_1}{A}\left(\frac{n_2}{n_1} - \frac{n_{2,b}}{n_{1,b}}\right)} \tag{2.41}$$

with n_1/A denoting areal concentration of solvent, supports transformation of Eq. (2.40) to

$$\Gamma = -\frac{a_2}{\mathcal{R}T} \frac{d\sigma}{da_2} \tag{2.42}$$

– where $n_2/n_1 - n_{2,b}/n_{1,b}$ represents dimensionless surface concentration excess of surfactant, or difference between the ratio of concentrations of surfactant to water at the interface, to the said ratio in the bulk. If the activity coefficient of surfactant, γ_2, does not considerably change within the concentration range of interest (as happens far from saturation at the interface), then Eq. (2.42) may be reformulated as

$$\Gamma = -\frac{\gamma_2 x_2}{\mathcal{R}T}\frac{d\sigma}{\gamma_2 dx_2}, \tag{2.43}$$

at the expense of Eq. (3.156) in *Food Proc. Eng.: basics & mechanical operations*; γ_2 can be dropped from both numerator and denominator to give

$$\Gamma = -\frac{x_2}{\mathcal{R}T}\frac{d\sigma}{dx_2}, \tag{2.44}$$

with x_2 denoting molar fraction of surfactant. Equation (3.44) of *Food Proc. Eng.: basics & mechanical operations* prompts transformation of Eq. (2.44) to

$$\Gamma = -\frac{\frac{n_2}{n}}{\mathcal{R}T}\frac{d\sigma}{d\left(\frac{n_2}{n}\right)}, \tag{2.45}$$

where the total number of moles, n, may be taken off the differential sign for being essentially constant in dilute solutions of surfactant, and then canceled out with that in the numerator to yield

$$\Gamma = -\frac{n_2}{\mathcal{R}T}\frac{d\sigma}{dn_2}; \tag{2.46}$$

definition of areal concentration of surfactant as

$$C_2 \equiv \frac{n_2}{A}, \tag{2.47}$$

using Eq. (3.165) of *Food Proc. Eng.: basics & mechanical operations* for template, further transforms Eq. (2.46) to

$$\boxed{\Gamma = \frac{C_2}{\mathcal{R}T}\left(-\frac{d\sigma}{dC_2}\right)}, \tag{2.48}$$

with the constancy of A taken advantage of. Inspection of Eq. (2.48) indicates that surface concentration excess is proportional to decrease in surface tension (or surface energy), $d\sigma$, with logarithmic concentration of surfactant at the interface, $d\ln C_2 = dC_2/C_2$. When a solution of surfactant is exposed to air, its amphiphilic molecules orient themselves gradually at the interface – thus causing a decrease in surface tension with time, while C_2 increases: this realization should be taken into account when implementing a cleaning process.

Equation (2.48) may alternatively be formulated to

$$\frac{n_1}{A}\left(\frac{n_2}{n_1} - \frac{n_{2,b}}{n_{1,b}}\right) = \frac{C_2}{\mathcal{R}T}\left(-\frac{d\sigma}{dC_2}\right), \tag{2.49}$$

upon combination with Eq. (2.41); after multiplying both sides by $\mathcal{R}T/C_2$, Eq. (2.49) becomes

$$\mathcal{R}T\frac{n_1}{AC_2}\left(\frac{n_2}{n_1} - \frac{n_{2,b}}{n_{1,b}}\right) = -\frac{d\sigma}{dC_2} \tag{2.50}$$

– where Eq. (2.47) supports complementary transformation to

$$\frac{d\sigma}{dC_2} = -\mathcal{R}T\frac{n_1}{n_2}\left(\frac{n_2}{n_1} - \frac{n_{2,b}}{n_{1,b}}\right). \tag{2.51}$$

If n_1/n_2 is factored in, then Eq. (2.51) yields

$$\frac{d\sigma}{dC_2} = -\mathcal{R}T\left(\frac{n_1}{n_2}\frac{n_2}{n_1} - \frac{n_1}{n_2}\frac{n_{2,b}}{n_{1,b}}\right), \tag{2.52}$$

which simplifies to

$$\boxed{\frac{d\sigma}{dC_2} = -\mathcal{R}T\left(1 - \frac{\frac{n_{2,b}}{n_{1,b}}}{\frac{n_2}{n_1}}\right)} \tag{2.53}$$

upon straightforward algebraic rearrangement. Experimental evidence indicates that the ratio of $n_{2,b}/n_{1,b}$ to n_2/n_1 remains approximately constant over most of the concentration range; in much the same way the ratio y_2/y_1 pertaining to mole fractions in the vapor phase, divided by the ratio x_2/x_1 pertaining to mole fractions in the liquid phase, for a binary mixture of two volatiles, 1 and 2, is approximately constant – known as relative volatility (and quite useful when simulating distillation). One therefore concludes from Eq. (2.53) that a straight line will likely represent the variation of σ with C_2, at a given temperature and within the lower concentration range – as illustrated in Fig. 2.14, and in full adherence to experimental evidence; this holds until saturation is attained at the interface. Such an

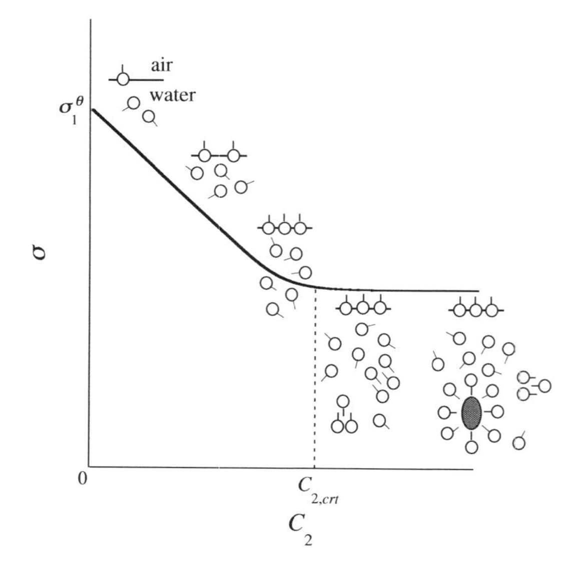

FIGURE 2.14 Variation of surface tension, σ, of binary liquid solution in contact with air, versus areal concentration of surfactant, C_2, in water – characterized by surface tension σ_1^θ when pure; with dissolved surfactant (, constituted by hydrophilic head, O, and hydrophobic tail, I) distributed between (random) solution and (ordered) interface, and as aggregates and eventually micelles enrobing dirt particles () – characterized by critical areal concentration at interface $C_{2,crt}$.

almost linear variation of σ departs from σ_1^θ (as surface tension of pure water at $C_2=0$), and decreases with increasing (areal) concentration of surfactant at the interface, C_2. When saturation of the interface with surfactant, at concentration $C_{2,crt}$, is approached, molecular crowding is no longer consistent with (an essentially) constant value of γ_2 – and a relatively sharp shoulder is observed in practice; the corresponding volumetric concentration of surfactant in the bulk solution, $C_{2,cmc}$, is termed critical micellar concentration. If more surfactant is dissolved in water, its excess relative to saturation will engage in aggregates of surfactant molecules, with their polar heads pointing outward (see Fig. 2.14) – either as such and stabilized by the close proximity of their hydrophobic tails, or surrounding other chemicals they have been meanwhile exposed to, e.g. (hydrophobic) dirt particles, thus producing a colloidal mixture. Since micelles exhibit dimensions at molecular level (say, $1–2\times10^{-8}$ m), such supramolecular aggregates become a suspension – and behave hydrodynamically (but not thermodynamically) as if they were dissolved in water. Therefore, actual detergency (including both penetration of cleaning agent and removal of dirt thereby) only manifests itself at concentrations above the aforementioned critical micellar concentration – which may be as low as 0.1%(w/w) in the case of alkyl aryl sulfonates. Note that temperature plays an important role upon $C_{2,cmc}$; Eq. (2.53) cannot, however, be directly used to ascertain such a role because it was derived under the hypothesis of a constant temperature, see Eq. (2.38). On the other hand, presence of salts may strongly diminish $C_{2,cmc}$, thus enhancing detergent action.

PROBLEM 2.1

Besides the effect of surfactant concentration upon surface tension, σ, temperature, T, plays a role – which may be estimated from thermodynamic considerations; on the other hand, contact of the aqueous cleaning solution with the solid surface of an equipment, in addition to contact with air, brings three surface tensions into play, pertaining to solid/liquid ($\sigma_{s,l}$), solid/gas ($\sigma_{s,g}$), and liquid/gas ($\sigma_{l,g}$) interfaces – which are not necessarily independent of each other.

a) Find a thermodynamic relationship describing the effect of T upon σ, resorting to surface entropy – assuming a homogeneous interface, and a system of constant composition under isobaric conditions.
b) Estimate surface enthalpy, departing from the result in a); assume that dependence of σ on T is known in advance.
c) Derive a relationship between $\sigma_{l,g}$, $\sigma_{s,l}$, and $\sigma_{s,g}$, given the angle of contact between liquid and gas, θ.

Solution

a) Since isobaric conditions and invariant composition were set by hypothesis, only two degrees of freedom – i.e. T and area, A, remain of relevance; the partial derivative of σ with regard to T may thus be obtained from Eq. (2.24) as

$$\left(\frac{\partial\sigma}{\partial T}\right)_A = \left(\frac{\partial}{\partial T}\left(\frac{\partial G}{\partial A}\right)\right)_A . \tag{2.1.1}$$

Differentiation of both sides of Eq. (3.24) in *Food Proc. Eng.: basics & mechanical operations* unfolds, in turn,

$$\left(\frac{\partial G}{\partial A}\right)_T = \left(\frac{\partial H}{\partial A}\right)_T - T\left(\frac{\partial S}{\partial A}\right)_T ; \tag{2.1.2}$$

Eq. (2.1.2) supports transformation of Eq. (2.1.1) to

$$\left(\frac{\partial\sigma}{\partial T}\right)_A = \left(\frac{\partial}{\partial T}\left(\left(\frac{\partial H}{\partial A}\right)_T - T\left(\frac{\partial S}{\partial A}\right)_T\right)\right)_A \tag{2.1.3}$$

– or merely

$$\left(\frac{\partial\sigma}{\partial T}\right)_A = -\left(\frac{\partial S}{\partial A}\right)_T , \tag{2.1.4}$$

because H and S being (obviously) proportional to A as measure of extensity enforce the constancy of $\left(\partial H/\partial A\right)_T$ and $\left(\partial S/\partial A\right)_T$, respectively – which, in turn, imply a nil value for their partial derivatives with regard to T. The above consideration, pertaining to a homogeneous interface, may be mathematically expressed as

$$\frac{\partial S}{\partial A} = \frac{S}{A}, \tag{2.1.5}$$

as applied in the case of entropy – with either side being equal to the underlying proportionality constant; this result allows simplification of Eq. (2.1.4) to

$$\left(\frac{\partial\sigma}{\partial T}\right)_A = -\frac{S}{A}, \tag{2.1.6}$$

which conveys the dependence of σ on T sought in the first place.

b) Following algebraic rearrangement of Eq. (2.1.2) to

$$\left(\frac{\partial H}{\partial A}\right)_T = \left(\frac{\partial G}{\partial A}\right)_T + T\left(\frac{\partial S}{\partial A}\right)_T , \tag{2.1.7}$$

one may retrieve Eqs. (2.24) and (2.1.4) to write

$$\left(\frac{\partial H}{\partial A}\right)_T = \sigma - T\left(\frac{\partial\sigma}{\partial T}\right)_A ; \tag{2.1.8}$$

as mentioned before with regard to interfacial entropy, it is expected that interfacial enthalpy varies proportionally to the amount of interfacial area in a homogeneous interface, i.e.

$$\frac{\partial H}{\partial T} = \frac{H}{A} \tag{2.1.9}$$

in parallel to Eq. (2.1.5) – thus allowing transformation of Eq. (2.1.8) to

$$\frac{H}{A} = \sigma - T\left(\frac{\partial \sigma}{\partial T}\right)_A . \quad (2.1.10)$$

Hence, a mode of estimation of surface enthalpy, given $\sigma \equiv \sigma\{T\}$ that supports calculation of $d\sigma/dT$, has been obtained – classically known as Kelvin's equation.

c) Consider, for simplicity, a liquid intruding a capillary pore of solid walls, with inner radius R – at the expense of extruding gas previously therein; if volumetric work on the gas and gravitational work on the liquid are both negligible (as found at relatively low pressures in horizontal setups), then

$$đW_l = 2\pi R\sigma_{s,l} dL \quad (2.1.11)$$

should describe the elementary surface work performed when liquid progresses elementary length dL inside the pore. Simultaneously, a resistive surface work will be experienced by the gas, according to

$$đW_g = -2\pi R\sigma_{s,g} dL; \quad (2.1.12)$$

ordered addition of Eqs. (2.1.11) and (2.1.12) unfolds

$$đW_{surf} \equiv đW_l + đW_g = 2\pi R\sigma_{s,l} dL - 2\pi R\sigma_{s,g} dL, \quad (2.1.13)$$

with $đW_{surf}$ denoting net surface work – where $2\pi RdL$ may be factored out to generate

$$đW_{surf} = 2\pi R\left(\sigma_{s,l} - \sigma_{s,g}\right) dL. \quad (2.1.14)$$

On the other hand, the aforementioned work can be calculated as net work performed by contact forces, associated with pressures P_{in} and P_{out} prevailing at inlet and outlet, respectively, of the pore – according to

$$đW_{in} = -\pi R^2 P_{in} dL \quad (2.1.15)$$

representing work input, and

$$đW_{out} = \pi R^2 P_{out} dL \quad (2.1.16)$$

describing work output; ordered addition of Eqs. (2.1.15) and (2.1.16) generates, in turn,

$$đW_{bulk} \equiv đW_{in} + đW_{out} = -\pi R^2 P_{in} dL + \pi R^2 P_{out} dL, \quad (2.1.17)$$

where $đW_{bulk}$ denotes net bulk work – equivalent to

$$đW_{bulk} = -\pi R^2 \left(P_{in} - P_{out}\right) dL, \quad (2.1.18)$$

once $\pi R^2 dL$ is factored out. Physicochemical equivalence between the above two reasonings implies

$$đW_{surf} = đW_{bulk}, \quad (2.1.19)$$

and thus

$$2\pi R\left(\sigma_{s,l} - \sigma_{s,g}\right) dL = -\pi R^2 \left(P_{in} - P_{out}\right) dL \quad (2.1.20)$$

at the expense of Eqs. (2.1.14) and (2.1.18); cancellation of πRdL between sides permits simplification to

$$2\left(\sigma_{s,l} - \sigma_{s,g}\right) = -R\left(P_{in} - P_{out}\right), \quad (2.1.21)$$

or else

$$\frac{2}{R}\left(\sigma_{s,g} - \sigma_{s,l}\right) = P_{in} - P_{out}. \quad (2.1.22)$$

Since the surface tension that drives the pressure differential between liquid and gas phases in a pore refers specifically to the liquid/gas interface, one may retrieve Young and Laplace's equation, labeled as Eq. (4.183) in *Food Proc. Eng.: basics & mechanical operations*, i.e.

$$P_{in} - P_{out} = \frac{2\sigma_{l,g}\cos\theta}{R} \quad (2.1.23)$$

– because the circular cross-section of the cylindrical pore implies equal radii of curvature, taken at any normal direction relative to each other; combining Eqs. (2.1.22) and (2.1.23) gives then rise to

$$\frac{2}{R}\left(\sigma_{s,g} - \sigma_{s,l}\right) = \frac{2}{R}\sigma_{l,g}\cos\theta, \quad (2.1.24)$$

where $2/R$ may drop off both sides, and the result be divided by $\sigma_{l,g}$ to get

$$\frac{\sigma_{s,g} - \sigma_{s,l}}{\sigma_{l,g}} = \cos\theta. \quad (2.1.25)$$

Equation (2.1.25) – traditionally known as Yang and Dupré's law, conveys (the physicochemical) constraint to be satisfied by $\sigma_{s,g}$, $\sigma_{s,l}$, and $\sigma_{l,g}$.

2.1.3.2 Spraying

As emphasized previously, the droplets of cleaning solution are thrown toward the walls of the equipment as part of spraying; the efficacy of the associated cleaning is thus expected to be proportional to the kinetic energy of the said droplets upon normal impact, and to their linear momentum throughout tangential motion. Therefore, estimation of the actual velocity in the radial direction is in order. Toward this goal, one may neglect evaporation – because a closed vessel is at stake, and the air inside is saturated with water vapor that precludes mass transfer within the gas phase as per Fick's law; creeping flow is also an acceptable hypothesis, in view of the minute sizes of typical droplets (as observed in most spargers), along with the assumption of a solid sphere behavior therefor. Under these conditions, Stokes' law as per Eq. (4.509) in *Food Proc. Eng.: basics & mechanical operations* applies to the radial component of droplet velocity, v_r, i.e.

$$F_r = -6\pi\mu_{air} R v_r \quad (2.54)$$

– meaning that the radial force, F_r, experienced by a droplet of radius R is (essentially) proportional to v_r, where μ_{air} denotes viscosity of air. On the other hand, Newton's second law of mechanics may be invoked to write

$$\frac{4}{3}\pi R^3\rho_W\frac{dv_r}{dt}=-6\pi\mu_{air}Rv_r, \tag{2.55}$$

where ρ_W denotes mass density of aqueous droplet – with $4\pi R^3/3$ representing its volume, and dv_r/dt describing its radial acceleration; upon cancellation of common factors between sides, Eq. (2.55) simplifies to

$$\frac{dv_r}{dt}=-\frac{9}{2}\frac{\mu_{air}}{\rho_W R^2}v_r, \tag{2.56}$$

satisfying

$$v_r\Big|_{t=0}=v_0\sin\phi \tag{2.57}$$

as initial condition – where ϕ denotes angle of inclination of nozzle mouth relative to vessel axis, and v_0 denotes (spatially uniform) speed of fluid at said outlet. Integration of Eq. (2.56) is feasible, via separation of variables, according to

$$\int_{v_0\sin\phi}^{v_r}\frac{d\tilde{v}_r}{\tilde{v}_r}=-\frac{9}{2}\frac{\mu_{air}}{\rho_W R^2}\int_0^t d\tilde{t}, \tag{2.58}$$

with the aid of Eq. (2.57) and assumption of constant physicochemical properties – which breaks down to

$$\ln\tilde{v}_r\Big|_{v_0\sin\phi}^{v_r}=-\frac{9}{2}\frac{\mu_{air}}{\rho_W R^2}\tilde{t}\Big|_0^t, \tag{2.59}$$

upon application of the fundamental theorem of integral calculus; Eq. (2.59) degenerates to

$$\ln\frac{v_r}{v_0\sin\phi}=-\frac{9}{2}\frac{\mu_{air}}{\rho_W R^2}t \tag{2.60}$$

– or, after taking exponentials of both sides and isolating v_r/v_0,

$$\frac{v_r}{v_0}=\exp\left\{-\frac{9}{2}\frac{\mu_{air}}{\rho_W R^2}t\right\}\sin\phi \tag{2.61}$$

that yields $v_r\equiv v_r\{t\}$.

The definition of radial velocity (i.e. $v_r\equiv dr/dt$) allows Eq. (2.61) be rewritten as

$$\frac{dr}{dt}=v_0\exp\left\{-\frac{9}{2}\frac{\mu_{air}}{\rho_W R^2}t\right\}\sin\phi, \tag{2.62}$$

together with multiplication of both sides by v_0 – which readily integrates to

$$\int_0^r d\tilde{r}=v_0\sin\phi\int_0^t\exp\left\{-\frac{9}{2}\frac{\mu_{air}}{\rho_W R^2}\tilde{t}\right\}d\tilde{t}, \tag{2.63}$$

at the expense of

$$r\Big|_{t=0}=0 \tag{2.64}$$

for initial condition; Eq. (2.63) leads to

$$\tilde{r}\Big|_0^r=v_0\sin\phi\left.\frac{\exp\left\{-\frac{9}{2}\frac{\mu_{air}}{\rho_W R^2}\tilde{t}\right\}}{-\frac{9}{2}\frac{\mu_{air}}{\rho_W R^2}}\right|_0^t, \tag{2.65}$$

or else

$$r=\frac{2}{9}\frac{\rho_W R^2 v_0}{\mu_{air}}\sin\phi\left(1-\exp\left\{-\frac{9}{2}\frac{\mu_{air}}{\rho_W R^2}t\right\}\right) \tag{2.66}$$

after swapping integration limits in the right-hand side – which conveys $r\equiv r\{t\}$. If radius and height of the (cylindrical) vessel are denoted by R_v and H_v, respectively, then the maximum linear distance, L_{max}, of a droplet when impinging on the inner wall of the vessel reads

$$L_{max}=\sqrt{R_v^{\,2}+H_v^{\,2}} \tag{2.67}$$

– assuming the sprayer is placed at the axis and on the top of the vessel. Both sides of Eq. (2.66) may be conveniently divided by $R_v=R_v^2/R_v$, and the argument of the exponential function multiplied and divided by $v_0R_v^2$ to yield

$$\boxed{\begin{aligned}\frac{r}{R_v}=\frac{2}{9}\frac{\rho_W R_v v_0}{\mu_{air}}\left(\frac{R}{R_v}\right)^2\sin\phi\\ \times\left(1-\exp\left\{-\frac{9}{2}\frac{1}{\frac{\rho_W R_v v_0}{\mu_{air}}\left(\frac{R}{R_v}\right)^2}\frac{t}{\frac{R_v}{v_0}}\right\}\right).\end{aligned}} \tag{2.68}$$

Equation (2.68) entails normalized radial distance, r/R_v, as a function of normalized droplet radius, R/R_v, ejection angle, ϕ, normalized time, $t/(R_v/v_0)$, and a system/process dimensionless parameter, $\rho_W R_v v_0/\mu_{air}$; after redoing Eq. (2.61) to

$$\boxed{\frac{v_r}{v_0}=\exp\left\{-\frac{9}{2}\frac{1}{\frac{\rho_W R_v v_0}{\mu_{air}}\left(\frac{R}{R_v}\right)^2}\frac{t}{\frac{R_v}{v_0}}\right\}\sin\phi,} \tag{2.69}$$

one again realizes that the dimensionless radial velocity, v_r/v_0, is a function solely of R/R_v, ϕ, $t/(R_v/v_0)$, and $\rho_W R_v v_0/\mu_{air}$ – so v_r/v_0 ends up depending on r/R_v via parameters R/R_v, $\rho_W R_v v_0/\mu_{air}$, and ϕ, following elimination of $t/(R_v/v_0)$ as intermediate (parametric) variable between Eqs. (2.68) and (2.69). Since only the viscous force raised by motion is present in the horizontal direction, the droplet undergoes continuous deceleration – consubstantiated by the decaying exponential in Eq. (2.69). At short times, the exponential function in Eq. (2.68) may be well approximated by its Taylor's expansion around $\frac{9}{2}\frac{1}{\frac{\rho_W R_v v_0}{\mu_{air}}\left(\frac{R}{R_v}\right)^2}\frac{t}{\frac{R_v}{v_0}}=0$, truncated after the linear term, viz.

$$\lim_{\frac{9}{2}\frac{1}{\frac{\rho_W R_v v_0}{\mu_{air}}\left(\frac{R}{R_v}\right)^2}\frac{t}{\frac{R_v}{v_0}}\to 0}\frac{r}{R_v}=\frac{2}{9}\frac{\rho_W R_v v_0}{\mu_{air}}\left(\frac{R}{R_v}\right)^2$$

$$\times\left(1-\left(1+\left(-\frac{9}{2}\frac{1}{\frac{\rho_W R_v v_0}{\mu_{air}}\left(\frac{R}{R_v}\right)^2}\frac{t}{\frac{R_v}{v_0}}\right)\right)\right)\sin\phi \quad (2.70)$$

$$=\frac{2}{9}\left(\frac{R}{R_v}\right)^2\frac{\rho_W R_v v_0}{\mu_{air}}\frac{9}{2}\frac{1}{\frac{\rho_W R_v v_0}{\mu_{air}}\left(\frac{R}{R_v}\right)^2}\frac{t}{\frac{R_v}{v_0}}\sin\phi$$

that breaks down to merely

$$\lim_{t\to 0}\frac{r}{t}=v_0\sin\phi \quad (2.71)$$

upon cancellation of like factors between numerator and denominator, and further dividing both sides by t/R_v; this result is consistent with the initial radial velocity, as conveyed by Eq. (2.57).

With regard to the vertical component of droplet velocity, v_x (oriented downward), one may similarly write

$$F_x=\frac{4}{3}\pi R^3\left(\rho_W-\rho_{air}\right)g-6\pi\mu_{air}Rv_x \quad (2.72)$$

– where drag force, accounted for by the second term in Eq. (2.72), now competes with gravitational force, $4\pi R^3\rho_W g/3$, but is reinforced by buoyancy force exerted by air as a continuous phase, $-4\pi R^3\rho_{air}g/3$; hence, the longitudinal force experienced by the droplet will vary linearly, but not proportionally to v_x – unlike observed with $F_r\equiv F_r\{v_r\}$ with regard to v_r, as per Eq. (2.54). After recalling once again Newton's second law of mechanics, Eq. (2.72) will be redone to

$$\frac{4}{3}\pi R^3\rho_W\frac{dv_x}{dt}=\frac{4}{3}\pi R^3\left(\rho_W-\rho_{air}\right)g-6\pi\mu_{air}Rv_x, \quad (2.73)$$

where both sides can be divided by $4\pi\rho_W R^3/3$ to get

$$\frac{dv_x}{dt}=\left(1-\frac{\rho_{air}}{\rho_W}\right)g-\frac{9}{2}\frac{\mu_{air}}{\rho_W R^2}v_x. \quad (2.74)$$

The initial condition looks like

$$v_x\big|_{t=0}=v_0\cos\phi, \quad (2.75)$$

consistent with Eq. (2.57); after having redone Eq. (2.74) to

$$\frac{dv_x}{dt}=-\frac{9}{2}\frac{\mu_{air}}{\rho_W R^2}\hat{v}_x \quad (2.76)$$

with the aid of auxiliary variable $\hat{v}_x$, defined by

$$\hat{v}_x\equiv v_x-\frac{\left(1-\frac{\rho_{air}}{\rho_W}\right)g}{\frac{9}{2}\frac{\mu_{air}}{\rho_W R^2}}=v_x-\frac{2}{9}\frac{\left(\rho_W-\rho_{air}\right)gR^2}{\mu_{air}}, \quad (2.77)$$

one may further rewrite Eq. (2.76) as

$$\frac{d\hat{v}_x}{dt}=-\frac{9}{2}\frac{\mu_{air}}{\rho_W R^2}\hat{v}_x \quad (2.78)$$

– based on the nil derivative of an additive constant. Integration of Eq. (2.78) via separation of variables is now in order, according to

$$\int_{v_0\cos\phi-\frac{2(\rho_W-\rho_{air})gR^2}{9\mu_{air}}}^{v_x-\frac{2(\rho_W-\rho_{air})gR^2}{9\mu_{air}}}\frac{d\hat{v}_x}{\hat{v}_x}=-\frac{9}{2}\frac{\mu_{air}}{\rho_W R^2}\int_0^t d\tilde{t} \quad (2.79)$$

where an analog to Eq. (2.75), i.e.

$$\hat{v}_x\big|_{t=0}=v_0\cos\phi-\frac{2}{9}\frac{\left(\rho_W-\rho_{air}\right)gR^2}{\mu_{air}} \quad (2.80)$$

after Eq. (2.77), was duly considered as initial condition; Eq. (2.79) is equivalent to

$$\ln\hat{v}_x\Big|_{v_0\cos\phi-\frac{2(\rho_W-\rho_{air})gR^2}{9\mu_{air}}}^{v_x-\frac{2(\rho_W-\rho_{air})gR^2}{9\mu_{air}}}=-\frac{9}{2}\frac{\mu_{air}}{\rho_W R^2}\tilde{t}\Big|_0^t. \quad (2.81)$$

Upon replacement by upper and lower limits of integration, Eq. (2.81) becomes

$$\ln\frac{v_x-\frac{2}{9}\frac{\left(\rho_W-\rho_{air}\right)gR^2}{\mu_{air}}}{v_0\cos\phi-\frac{2}{9}\frac{\left(\rho_W-\rho_{air}\right)gR^2}{\mu_{air}}}=-\frac{9}{2}\frac{\mu_{air}}{\rho_W R^2}t \quad (2.82)$$

– or, after taking exponentials of both sides,

$$\frac{v_x-\frac{2}{9}\frac{\left(\rho_W-\rho_{air}\right)gR^2}{\mu_{air}}}{v_0\cos\phi-\frac{2}{9}\frac{\left(\rho_W-\rho_{air}\right)gR^2}{\mu_{air}}}=\exp\left\{-\frac{9}{2}\frac{\mu_{air}}{\rho_W R^2}t\right\}; \quad (2.83)$$

Eq. (2.83) can be algebraically rearranged to read

$$\frac{v_x}{v_0}=\frac{2}{9}\frac{\left(\rho_W-\rho_{air}\right)gR^2}{\mu_{air}v_0}+\left(\cos\phi-\frac{2}{9}\frac{\left(\rho_W-\rho_{air}\right)gR^2}{\mu_{air}v_0}\right)\exp\left\{-\frac{9}{2}\frac{\mu_{air}}{\rho_W R^2}t\right\} \quad (2.84)$$

– where a functionality of the type $v_x\equiv v_x\{t\}$ is apparent. Inspection of Eq. (2.84) indicates that a constant velocity, $2/(\rho_W-\rho_{air})gR^2/9\mu_{air}$, will be asymptotically reached after a period of time long enough to permit the exponential function decay essentially to zero – corresponding to a steady state (characterized by nil acceleration), and termed terminal velocity; this is the velocity that raises a viscous force able to exactly balance the net force associated with gravity and buoyancy.

The axial velocity (i.e. $v_x\equiv dx/dt$) may, in turn, be formulated in differential form as

$$\frac{dx}{dt}=\frac{2}{9}\frac{\left(\rho_W-\rho_{air}\right)gR^2}{\mu_{air}}+\left(v_0\cos\phi-\frac{2}{9}\frac{\left(\rho_W-\rho_{air}\right)gR^2}{\mu_{air}}\right)\exp\left\{-\frac{9}{2}\frac{\mu_{air}}{\rho_W R^2}t\right\} \quad (2.85)$$

using Eq. (2.84) as starting point, and upon multiplication of both sides by v_0: separation of variables and integration promptly yields

$$\int_0^x d\tilde{x} = \frac{2}{9}\frac{(\rho_W-\rho_{air})gR^2}{\mu_{air}}\int_0^t d\tilde{t} + \left(v_0\cos\phi - \frac{2}{9}\frac{(\rho_W-\rho_{air})gR^2}{\mu_{air}}\right)\int_0^t \exp\left\{-\frac{9}{2}\frac{\mu_{air}}{\rho_W R^2}\tilde{t}\right\}d\tilde{t}, \tag{2.86}$$

with the aid of the decomposition rule for integrals, together with

$$x\big|_{t=0} = 0 \tag{2.87}$$

serving as initial condition. Equation (2.86) breaks down to

$$\tilde{x}\Big|_0^x = \frac{2}{9}\frac{(\rho_W-\rho_{air})gR^2}{\mu_{air}}\tilde{t}\Big|_0^t - \left(v_0\cos\phi - \frac{2}{9}\frac{(\rho_W-\rho_{air})gR^2}{\mu_{air}}\right)\frac{\exp\left\{-\frac{9}{2}\frac{\mu_{air}}{\rho_W R^2}\tilde{t}\right\}}{\frac{9}{2}\frac{\mu_{air}}{\rho_W R^2}}\Bigg|_0^t \tag{2.88}$$

or, upon condensation of factors alike, coupled with replacement by upper and lower limits of integration,

$$x = \frac{2}{9}\frac{(\rho_W-\rho_{air})gR^2}{\mu_{air}}t + \left(\frac{2}{9}\frac{\rho_W R^2 v_0}{\mu_{air}}\cos\phi - \frac{4}{81}\frac{\rho_W(\rho_W-\rho_{air})gR^4}{\mu_{air}^2}\right)\left(1-\exp\left\{-\frac{9}{2}\frac{\mu_{air}}{\rho_W R^2}t\right\}\right); \tag{2.89}$$

division of both sides by R_v, complemented by multiplication and division of R^2 by R_v^2 then yield

$$\frac{x}{R_v} = \frac{2}{9}\left(\frac{R}{R_v}\right)^2\frac{(\rho_W-\rho_{air})gR_v}{\mu_{air}}t + \left(\frac{2}{9}\left(\frac{R}{R_v}\right)^2\frac{\rho_W R_v v_0}{\mu_{air}}\cos\phi - \frac{4}{81}\left(\frac{R}{R_v}\right)^4\frac{\rho_W(\rho_W-\rho_{air})gR_v^3}{\mu_{air}^2}\right) \times \left(1-\exp\left\{-\frac{9}{2}\left(\frac{R_v}{R}\right)^2\frac{\mu_{air}}{\rho_W R_v^2}t\right\}\right). \tag{2.90}$$

After sequentially multiplying and dividing t by R_v/v_0, factoring out $2(R/R_v)^2/9$ in the second term, and multiplying and dividing the first term by ρ_W/μ_{air}, Eq. (2.90) becomes

$$\frac{x}{R_v} = \frac{2}{9}\frac{\frac{\rho_W(\rho_W-\rho_{air})gR_v^3}{\mu_{air}^2}}{\frac{\rho_W R_v v_0}{\mu_{air}}}\left(\frac{R}{R_v}\right)^2\frac{t}{\frac{R_v}{v_0}} + \frac{2}{9}\left(\frac{\rho_W R_v v_0}{\mu_{air}}\cos\phi - \frac{2}{9}\frac{\rho_W(\rho_W-\rho_{air})gR_v^3}{\mu_{air}^2}\left(\frac{R}{R_v}\right)^2\right)\left(\frac{R}{R_v}\right)^2 \times \left(1-\exp\left\{-\frac{9}{2}\frac{1}{\frac{\rho_W R_v v_0}{\mu_{air}}\left(\frac{R}{R_v}\right)^2}\frac{t}{\frac{R_v}{v_0}}\right\}\right). \tag{2.91}$$

Upon inspection of Eq. (2.91), one realizes that the normalized longitudinal distance, x/R_v, appears as a function of R/R_v, ϕ, $t/(R_v/v_0)$, and two system/process dimensionless parameters – $\rho_W R_v v_0/\mu_{air}$ that resembles Reynolds' number as per Eq. (2.92) in *Food Proc. Eng.: basics & mechanical operations*, and $\rho_W(\rho_W-\rho_{air})gR_v^3/\mu_{air}^2$ that resembles Galileo's number as per Eq. (5.630) of *Food Proc. Eng.: thermal & chemical operations*; one may accordingly rewrite Eq. (2.84) as

$$\frac{v_x}{v_0} = \frac{2}{9}\frac{\frac{\rho_W(\rho_W-\rho_{air})gR_v^3}{\mu_{air}^2}}{\frac{\rho_W R_v v_0}{\mu_{air}}}\left(\frac{R}{R_v}\right)^2 + \left(\cos\phi - \frac{\frac{2}{9}}{\frac{\rho_W R_v v_0}{\mu_{air}}}\frac{\rho_W(\rho_W-\rho_{air})gR_v^3}{\mu_{air}^2}\left(\frac{R}{R_v}\right)^2\right) \times \exp\left\{-\frac{9}{2}\frac{1}{\frac{\rho_W R_v v_0}{\mu_{air}}\left(\frac{R}{R_v}\right)^2}\frac{t}{\frac{R_v}{v_0}}\right\}, \tag{2.92}$$

upon multiplication and division by $\rho_W R_v^3$, and of the argument of the exponential function by R_v/v_0. Hence, the dimensionless axial velocity, v_x/v_0, becomes also a function of just R/R_v, ϕ, $t/(R_v/v_0)$, $\rho_W R_v v_0/\mu_{air}$, and $\rho_W(\rho_W-\rho_{air})gR_v^3/\mu_{air}^2$; this means that v_x/v_0 depends on x/R_v via R/R_v, ϕ, $\rho_W R_v v_0/\mu_{air}$, and $\rho_W(\rho_W-\rho_{air})gR_v^3/\mu_{air}^2$, following elimination of $t/(R_v/v_0)$ between Eqs. (2.91) and (2.92). As before, Eq. (2.90) is driven by the linear version of the corresponding Taylor's series, viz.

$$\lim_{\frac{9}{2}\left(\frac{R_v}{R}\right)^2 \frac{\mu_{air}}{\rho_W R_v^2} t \to 0} \frac{x}{R_v} = \frac{2}{9}\left(\frac{R}{R_v}\right)^2 \frac{(\rho_W - \rho_{air}) g R_v}{\mu_{air}} t + \left(\begin{array}{c} \frac{2}{9}\left(\frac{R}{R_v}\right)^2 \frac{\rho_W R_v v_0}{\mu_{air}} \cos\phi \\ -\frac{4}{81}\left(\frac{R}{R_v}\right)^4 \frac{\rho_W (\rho_W - \rho_{air}) g R_v^3}{\mu_{air}^2} \end{array} \right) \frac{9}{2}\left(\frac{R_v}{R}\right)^2 \frac{\mu_{air}}{\rho_W R_v^2} t \quad (2.93)$$

in the vicinity of $\frac{9}{2}\left(\frac{R_v}{R}\right)^2 \frac{\mu_{air}}{\rho_W R_v^2} t = 0$, where the second term in the right-hand side may be condensed to

$$\lim_{\frac{9}{2}\left(\frac{R_v}{R}\right)^2 \frac{\mu_{air}}{\rho_W R_v^2} t \to 0} \frac{x}{R_v} = \frac{2}{9}\left(\frac{R}{R_v}\right)^2 \frac{(\rho_W - \rho_{air}) g R_v}{\mu_{air}} t + \left(\frac{v_0}{R_v} \cos\phi - \frac{2}{9}\left(\frac{R}{R_v}\right)^2 \frac{(\rho_W - \rho_{air}) g R_v}{\mu_{air}} \right) t; \quad (2.94)$$

after canceling symmetrical terms, Eq. (2.94) turns to

$$\lim_{t \to 0} \frac{x}{R_v} = \frac{v_0}{R_v} t \cos\phi \quad (2.95)$$

in condensed notation – where multiplication of both sides by R_v/t finally gives

$$\lim_{t \to 0} \frac{x}{t} = v_0 \cos\phi, \quad (2.96)$$

in full agreement with Eq. (2.75) pertaining to the initial longitudinal velocity.

The distance of the droplet to its originating nozzle, z, is accessible through

$$z = \sqrt{r^2 + x^2}, \quad (2.97)$$

as per Pythagoras' theorem; the trajectories followed by the droplets can be ascertained via Eq. (2.68) in terms of radial position, coupled to Eq. (2.91) pertaining to longitudinal position – as done in Fig. 2.15. The maximum possible linear distance of a droplet of cleaning liquid, L_{max}, was already given by Eq. (2.67) – and corresponds to the light dashed line in Fig. 2.15; note that the actual distance traveled by a droplet along each curve is larger than the linear distance to the originating nozzle, because of the curvature of its trajectory. A portion of the whole set of droplets will be caught by the side wall of the vessel (e.g. for $\phi = \pi/4$ rad), while the remainder will fall onto the bottom thereof (e.g. for $\phi = \pi/6$ rad). As expected, the trajectories curve down as flight time elapses – due to the effect of gravity that increases v_x, concomitant with the effect of air drag that decreases v_r (besides decreasing v_x); the eventual point of impact with the vessel side wall moves downward as the angle of ejection with the vessel axis decreases, due to the associated lower momentum in the r-direction. Changes of R_v and v_0, such that their ratio R_v/v_0 (representing time taken by a droplet to reach side wall at a sustained ejection velocity) remains constant, do not lead to the same flight

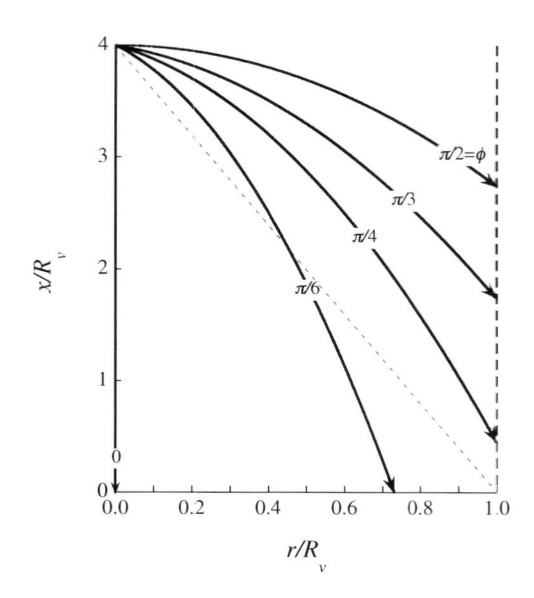

FIGURE 2.15 Trajectories (——) of droplets of cleaning (aqueous) solution, sprayed by a fixed nozzle (as part of a CIP system) located at $r=0$ and $x=H_v=4R_v$, in terms of normalized radial coordinate, r/R_v, and dimensionless axial coordinate, x/R_v, before reaching the side wall (---) of a cylindrical vessel with height H_v and radius R_v = 1 m, ejected at angle ϕ with the vertical axis – for droplet radius R=0.001 m, throwing velocity v_0=2 m.s^{-1}, air viscosity μ_{air}=0.00002 kg.m^{-1}.s^{-1}, air mass density ρ_{air}= 1.2 kg.m^{-3}, water mass density ρ_W= 1000 kg.m^{-3}, and acceleration of gravity g=9.8 m.s^{-2}, with indication of maximum (linear) distance to nozzle (- - -).

time in normalized form, nor to the same shape of the droplet trajectory – because the germane (dimensionless) parameters are independently affected by R_v and v_0, rather than by the said ratio.

The variation of radial component of velocity with radial position is illustrated in Fig. 2.16, for selected droplet sizes – based on Eq. (2.61), after elimination of t at the expense of Eq. (2.68). It is remarkable that the radial velocity decreases almost linearly with radial position – as a consequence of the small magnitude of the proportionality constant in Eq. (2.56); and that an increase in droplet size promotes an increase in radial velocity at any given distance. For the example worked out, a 0.5 mm-droplet hits the vessel wall with ca. 11% of its ejection velocity, whereas a 1 mm-droplet retains ca. 66% of its initial velocity by the time of hitting the side wall; therefore, a twofold increase in droplet radius, corresponding to an eightfold increase in mass, produces a sixfold increase in velocity, and thus a 36-fold increase in square velocity – which translates to a 48-fold increase in linear momentum (relevant for tangential impact) and a 288-fold increase in kinetic energy (relevant for normal impact) – which would translate to 6- and 36-fold increases, respectively, per unit volume of cleaning solution. This observation is a consequence of linear momentum, $V\rho v$, being proportional to velocity v, as well as to R^3 through volume V; and kinetic energy, $V\rho v^2/2$, being proportional to v^2, as well as R^3 through volume V – for a given mass density of liquid ρ. When droplets are too small, both their linear momentum and their kinetic energy at the inner wall become negligible; hence, their cleaning effect will prove poor.

Note that several simplifying assumptions were made above to facilitate mathematical simulation – namely, motionless air; however, this not exactly true, because spraying involves producing

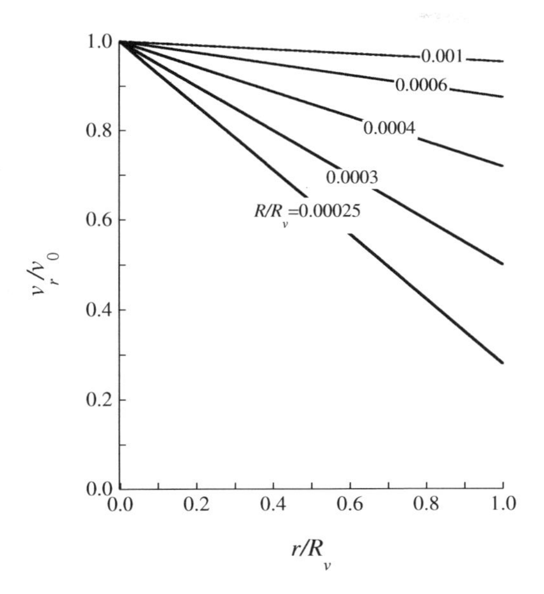

FIGURE 2.16 Variation of normalized radial velocity, v_r/v_0, versus normalized radial distance, r/R_v, of droplets of cleaning (aqueous) solution, sprayed with selected values of normalized radius, R/R_v, at ejection angle $\phi=\pi/2$ rad with the vertical axis, by a fixed nozzle (as part of a CIP system) located at $r=0$ and $x=H_v=4R_v$ in a cylindrical vessel with height H_v and radius $R_v=1$ m – for throwing velocity $v_0=2$ m.s^{-1}, air viscosity $\mu_{air}=0.00002$ kg.m^{-1}.s^{-1}, air mass density $\rho_{air}=1.2$ kg.m^{-3}, water mass density $\rho_W=1000$ kg.m^{-3}, and acceleration of gravity $g=9.8$ m.s^{-2}.

droplets that are carried by air. Furthermore, the droplets are normally small at nozzle exit, so clusters of droplets and air, characterized by relatively larger radii and density – intermediate between air and liquid water, should instead be used to calculate viscous deceleration. On the other hand, the assumption of creeping flow that supported use of Stokes' law in the first place may not hold – since the nonsliding condition fails to exactly apply on their surface for being an interface between water and air, and some turbulence of carrier air is likely present (as compressed air is used to disperse and propel the cleaning solution). Nevertheless, introduction of corrections that account for such nonidealities will hardly change the overall trends – and reasonable accuracy can be claimed for the model derived above. It should be emphasized that application of CIP may be hampered by equipment shape – with cylindrical apparatuses being usually more suitable; as well as existence of such devices as liquid-level tubes, dead spaces, pump connections, heat exchanger joints, valves and fittings, pressure gauges, and spatter on the cover.

PROBLEM 2.2

Consider a cleaning system, based on spraying of an aqueous solution of detergent at velocity v_0, from a center nozzle at angle ϕ, toward the wall of a cylindrically-shaped apparatus.

a) Prove that the reciprocal of the tangent of the inclination angle, Φ, of velocity vector $\boldsymbol{v}$ of each droplet – characterized by v_r and v_x as radial and axial components, respectively, and by R as radius, reduces to a function of a dimensionless time, $t^* \equiv \frac{t}{\frac{R}{v_0}}$, including Reynolds' and Froude's numbers, as well as ϕ and ρ_{air}/ρ_W as dimensionless parameters.
b) Find an expression relating normalized distance in radial direction traveled by a droplet, $r^* \equiv r/R_v$, to a function solely of Reynolds' number, a normalized droplet radius, R^*, a dimensionless time, t^*, a ratio of mass densities, and the ejection angle at nozzle mouth, ϕ.
c) Using the result in b), express the corresponding normalized distance traveled by the droplet in the longitudinal direction, $x^* \equiv x/R_v$, as a function of Reynolds' and Galileo's numbers, besides R^*, t^*, ϕ, and ρ_{air}/ρ_W.

Solution

a) After recalling the definition of trigonometric tangent,

$$\frac{1}{\tan\Phi} \equiv \frac{v_x}{v_r} \tag{2.2.1}$$

– where $\Phi \equiv \Phi\{t\}$ denotes angle formed by vector $\boldsymbol{v}(v_x,v_y)$ with the vertical axis, one may proceed to ordered division of Eq. (2.84) by Eq. (2.61) to get

$$\frac{1}{\tan\Phi} = \frac{1}{\tan\phi} + \frac{2}{9}\frac{(\rho_W-\rho_{air})gR^2}{\mu_{air}v_0\sin\phi}\left(\exp\left\{\frac{9}{2}\frac{\mu_{air}}{\rho_W R^2}t\right\}-1\right), \tag{2.2.2}$$

with $2(\rho_W-\rho_{air})gR^2/9\mu_{air}v_0$ meanwhile factored out, and with the aid of the definition of tangent of ϕ in terms of sine and cosine of ϕ. Following multiplication and division of the second term by $\rho_W\rho_{air}v_0$ and the argument of the exponential function by $\rho_{air}v_0$, Eq. (2.2.2) may be rewritten as

$$\frac{1}{\tan\Phi} = \frac{1}{\tan\phi} + \frac{\frac{2}{9}}{\sin\phi}\frac{1-\frac{\rho_{air}}{\rho_W}}{\frac{\rho_{air}}{\rho_W}}\frac{\rho_{air}Rv_0}{\mu_{air}}\frac{gR}{v_0^2}\left(\exp\left\{\frac{9}{2}\frac{\frac{\rho_{air}}{\rho_W}}{\frac{\rho_{air}Rv_0}{\mu_{air}}}\frac{t}{\frac{R}{v_0}}\right\}-1\right). \tag{2.2.3}$$

As expected, $1/\tan\Phi$ coincides with $1/\tan\phi$ at $t=0$, as per Eq. (2.2.3) – with Φ deviating further and further from ϕ as time elapses. In view of the definition of Reynolds' number, $Re \equiv \rho_{air}Rv_0/\mu_{air}$, and Froude's number, $Fr \equiv v_0^2/gR$, one can reformulate Eq. (2.2.3) to

$$\frac{1}{\tan\Phi} = \frac{1}{\tan\phi} + \frac{\frac{2}{9}}{\sin\phi}\frac{1-\frac{\rho_{air}}{\rho_W}}{\frac{\rho_{air}}{\rho_W}}\frac{Re}{Fr}\left(\exp\left\{\frac{9}{2}\frac{\frac{\rho_{air}}{\rho_W}}{Re}t^*\right\}-1\right) \tag{2.2.4}$$

– where Re, Fr, ϕ, and ρ_{air}/ρ_W appear as parameters in the right-hand side, as part of the intended functional dependence of Φ on t^*.

b) After multiplying and dividing ρ_W in the right-hand side by ρ_{air}, Eq. (2.68) becomes

$$\frac{r}{R_v}=\frac{2}{9}\frac{\rho_W}{\rho_{air}}\frac{\rho_{air}R_v v_0}{\mu_{air}}\left(\frac{R}{R_v}\right)^2\sin\phi \times\left(1-\exp\left\{-\frac{9}{2}\frac{1}{\frac{\rho_W}{\rho_{air}}\frac{\rho_{air}R_v v_0}{\mu_{air}}\left(\frac{R}{R_v}\right)^2}\frac{t}{\frac{R_v}{v_0}}\right\}\right). \quad (2.2.5)$$

Reynolds' number, Re, should hereby be defined as $\rho_{air}R_v v_0/\mu_{air}$, pertaining to air as fluid and using R_v for reference lenth – so Eq. (2.2.5) will be rewritten as

$$r^*=\frac{2}{9}\frac{\rho_W}{\rho_{air}}ReR^{*2}\sin\phi\left(1-\exp\left\{-\frac{9}{2}\frac{1}{\frac{\rho_W}{\mu_{air}}Re}\frac{t^*}{R^{*2}}\right\}\right); \quad (2.2.6)$$

R^* looks like R/R_v, while t^* appears as the ratio of t to the time taken by a droplet to travel the maximum radial distance under its initial velocity (i.e. R_v/v_0) – with the remaining (dimensionless) parameters accounted for by ρ_W/ρ_{air} as ratio of mass densities, and ϕ as angle of nozzle mouth.

c) By the same token – and after multiplying and dividing the right-hand side and the argument of the exponential function by ρ_{air}, Eq. (2.91) takes the form

$$\frac{x}{R_v}=\frac{2}{9}\frac{\frac{\rho_{air}(\rho_W-\rho_{air})gR_v^3}{\mu_{air}^2}}{\frac{\rho_{air}R_v v_0}{\mu_{air}}}\left(\frac{R}{R_v}\right)^2\frac{t}{\frac{R_v}{v_0}} +\frac{2}{9}\left(\frac{\frac{\rho_W}{\rho_{air}}\frac{\rho_{air}R_v v_0}{\mu_{air}}\cos\phi}{-\frac{2}{9}\frac{\rho_W}{\rho_{air}}\frac{\rho_{air}(\rho_W-\rho_{air})gR_v^3}{\mu_{air}^2}\left(\frac{R}{R_v}\right)^2}\right)\left(\frac{R}{R_v}\right)^2 \times\left(1-\exp\left\{-\frac{9}{2}\frac{1}{\frac{\rho_W}{\rho_{air}}\frac{\rho_{air}R_v v_0}{\mu_{air}}\left(\frac{R}{R_v}\right)^2}\frac{t}{\frac{R_v}{v_0}}\right\}\right); \quad (2.2.7)$$

if the previous definitions for r^*, R^*, and t^* are retrieved, and $2R^{*2}/9$ is factored out, then Eq. (2.2.7) will appear as

$$x^*=\frac{2}{9}\frac{Ga}{Re}R^{*2}t^* +\frac{2}{9}\frac{R^{*2}}{\frac{\rho_{air}}{\rho_W}}\left(Re\cos\phi-\frac{2}{9}GaR^{*2}\right)\times\left(1-\exp\left\{-\frac{\rho_{air}}{\frac{9}{2}\frac{\rho_W}{Re}\frac{t^*}{R^{*2}}}\right\}\right) \quad (2.2.8)$$

– with functionalities of x^* explicit on Re, Ga, R^*, t^*, ϕ, and ρ_{air}/ρ_W – provided that Galileo's number reads $\rho_{air}(\rho_W-\rho_{air})gR_v^3/\mu_{air}^2$.

2.1.3.3 Sprinkling

A sprinkler (or hydraulic tourniquet) – used to deliver a jet of liquid solution for cleaning/disinfection purposes, moves around its axis owing to the mere action of bulk liquid leaving at relatively high velocity from its tip(s).

In attempts to simulate this device, one should assess the thrust on the pipe bend, and corresponding ending nozzle – as illustrated in Fig. 2.17. Change in both magnitude and direction of fluid velocity at either tip (2), relative to that prevailing at their common feed (1), requires application of an external force by the corresponding sprinkler arm; in agreement with Newton's law of action and reaction, an opposite force is simultaneously exerted by flowing fluid upon the said sprinkler arm. Since two opposing forces, colinear and of identical magnitude, apply at some distance from each other, a torque develops – which is taken advantage of to set the sprinkler in (sustained) rotational motion.

Based on the nomenclature utilized in Fig. 2.17, one may set a balance to overall volume between planes 1 and 2 via the continuity equation, according to

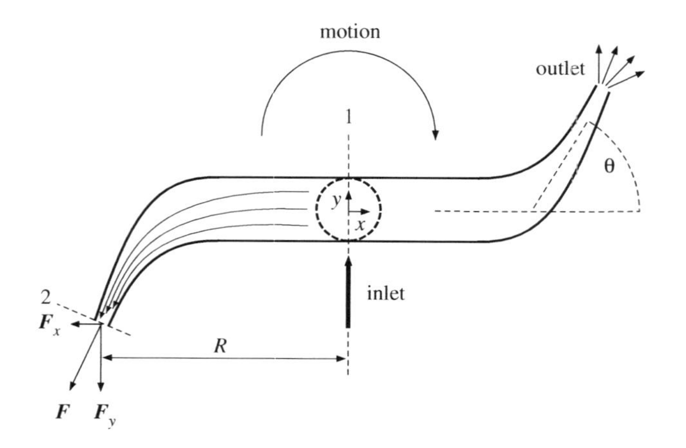

FIGURE 2.17 Schematic representation of rotating sprinkler, characterized by angle θ between outlet and inlet horizontal axes, fed with liquid at center (plane 1) in the x-direction, and delivering it equally through its lateral arms until leaving at their tips (plane 2) at inclined, yet opposite directions – with indication of force $\boldsymbol{F}$ exerted by device upon fluid at distance R from inlet pipe, split as horizontal, $\boldsymbol{F}_x$, and vertical, $\boldsymbol{F}_y$, projections, and causing an opposite reaction upon the device that accounts for rotational motion.

$$\boxed{\frac{Q_{ov}}{2} \equiv Q = S_1 v_1 = S_2 v_2} \tag{2.98}$$

that reflects Eq. (2.212) of *Food Proc. Eng.: basics & mechanical operations* – where Q_{ov} denotes overall volumetric flow rate and Q denotes volumetric flow rate through each sprinkler arm, S_1 and S_2 denote cross-sectional areas of inlet and outlet streams, respectively, and v_1 and v_2 denote associated (average) velocities. A relationship between v_1 and v_2, and (more useful) relationships thereof with volumetric flow rate through the sprinkler have consequently been found, and made available as Eq. (2.98).

A mechanical energy balance applied to the same control volume reads, in turn,

$$\boxed{\frac{1}{2}\left(v_2^2 - v_1^2\right) + \frac{P_2 - P_1}{\rho} + \frac{1}{2}K_h v_2^2 = 0} \tag{2.99}$$

as per Eq. (2.207) of *Food Proc. Eng.: basics & mechanical operations*, in the absence of shaft work as per hypothesis – where P_1 and P_2 denote pressure at inlet and outlet, respectively, ρ denotes mass density of liquid, and K_h denotes (localized) friction loss factor of nozzle; sprinklers usually rotate horizontally, so no change in potential energy of position within the gravitational field needs to be considered. In view of the smooth shape of the constriction, one may resort to the straight line (labeled as sudden constriction) in Fig. 2.11, or to Eq. (2.4.18) for that matter – both of *Food Proc. Eng.: basics & mechanical operations*, provided that a somewhat lower proportionality factor is used, say $0.4 = 2/5 < 0.45$. Equation (2.99) accordingly becomes

$$\frac{1}{2}v_2^2 - \frac{1}{2}v_1^2 + \frac{P_2 - P_1}{\rho} + \frac{1}{2}\frac{2}{5}\left(1 - \frac{S_2}{S_1}\right)v_2^2 = 0, \tag{2.100}$$

where lumping of terms in v_2^2 unfolds

$$\left(\frac{1}{2} + \frac{1}{5}\left(1 - \frac{S_2}{S_1}\right)\right)v_2^2 - \frac{1}{2}v_1^2 + \frac{P_2 - P_1}{\rho} = 0; \tag{2.101}$$

friction loss arising from the linear portion of the pipe itself was neglected (as usual), for being marginal when compared to that associated with the constriction. After multiplying and dividing the first term by S_2^2 and the second term similarly by S_1^2, Eq. (2.101) turns to

$$\left(\frac{1}{2} + \frac{1}{5}\left(1 - \frac{S_2}{S_1}\right)\right)\frac{\left(S_2 v_2\right)^2}{S_2^2} - \frac{1}{2}\frac{\left(S_1 v_1\right)^2}{S_1^2} + \frac{P_2 - P_1}{\rho} = 0 \tag{2.102}$$

– or else

$$\left(\frac{1}{2} + \frac{1}{5}\left(1 - \frac{S_2}{S_1}\right)\right)\frac{Q^2}{S_2^2} - \frac{1}{2}\frac{Q^2}{S_1^2} + \frac{P_2 - P_1}{\rho} = 0 \tag{2.103}$$

in view of Eq. (2.98); upon multiplication of both sides by S_1^2, Eq. (2.103) yields

$$\left(\frac{1}{2} + \frac{1}{5}\left(1 - \frac{S_2}{S_1}\right)\right)\frac{Q^2}{\frac{S_2^2}{S_1^2}} - \frac{1}{2}Q^2 + S_1^2\frac{P_2 - P_1}{\rho} = 0, \tag{2.104}$$

where Q^2 may, in turn, be factored out as

$$\left(\frac{\frac{1}{2} + \frac{1}{5}\left(1 - \frac{S_2}{S_1}\right)}{\left(\frac{S_2}{S_1}\right)^2} - \frac{1}{2}\right)Q^2 + S_1^2\frac{P_2 - P_1}{\rho} = 0. \tag{2.105}$$

An extra multiplication of both sides by $\rho/S_1^2P_2$, coupled with elimination of outer parentheses convert Eq. (2.105) to

$$\frac{\frac{1}{2} + \frac{1}{5}\left(1 - \frac{S_2}{S_1}\right) - \frac{1}{2}\left(\frac{S_2}{S_1}\right)^2}{\left(\frac{S_2}{S_1}\right)^2}\frac{\rho Q^2}{S_1^2 P_2} + 1 - \frac{P_1}{P_2} = 0, \tag{2.106}$$

where all terms independent of Q will be moved to the right-hand side as

$$\frac{\frac{1}{5}\left(1 - \frac{S_2}{S_1}\right) + \frac{1}{2}\left(1 - \left(\frac{S_2}{S_1}\right)^2\right)}{\left(\frac{S_2}{S_1}\right)^2}\frac{\rho Q^2}{S_1^2 P_2} = \frac{P_1}{P_2} - 1 \tag{2.107}$$

– with ½ meanwhile factored out in the numerator of the first term; once $1 - S_2/S_1$ itself is factored out, Eq. (2.107) becomes

$$\frac{\left(1 - \frac{S_2}{S_1}\right)\left(\frac{1}{5} + \frac{1}{2}\left(1 + \frac{S_2}{S_1}\right)\right)}{\left(\frac{S_2}{S_1}\right)^2}\frac{\rho Q^2}{S_1^2 P_2} = \frac{P_1}{P_2} - 1 \tag{2.108}$$

or, equivalently,

$$\frac{\left(1 - \frac{S_2}{S_1}\right)\left(2 + 5\left(1 + \frac{S_2}{S_1}\right)\right)}{10\left(\frac{S_2}{S_1}\right)^2}\frac{\rho Q^2}{S_1^2 P_2} = \frac{P_1}{P_2} - 1 \tag{2.109}$$

after multiplying both numerator and denominator of the left-hand side by 10. One may finally isolate $\rho Q^2/S_1^2P_2$ from Eq. (2.109) to get

$$\boxed{\frac{\rho Q^2}{S_1^2 P_2} = \frac{10\left(\frac{S_2}{S_1}\right)^2\left(\frac{P_1}{P_2} - 1\right)}{\left(1 - \frac{S_2}{S_1}\right)\left(7 + 5\frac{S_2}{S_1}\right)},} \tag{2.110}$$

along with lumping of constants in denominator; a graphical representation of such a dimensionless squared volumetric flow rate, as a function of S_2/S_1 and P_1/P_2, is made available in Fig. 2.18. The volumetric flow rate increases with the ratio of outlet tip to inlet feed cross-section, justified by the lower resistance raised against flow – since the loss term, appearing third in Eq. (2.99), becomes smaller; complemented by more gauge pressure available, materialized by a larger second term in Eq. (2.99) – needed

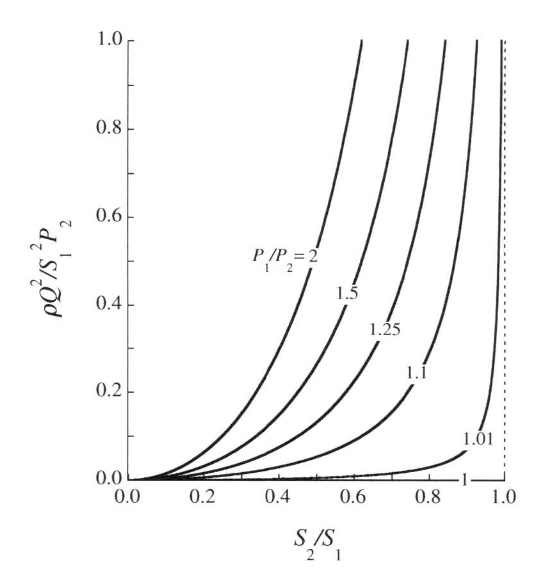

FIGURE 2.18 Variation of dimensionless squared volumetric flow rate of solution through sprinkler, $\rho Q^2/S_1^2P_2$, versus normalized cross-sectional area of nozzle, S_2/S_1, for selected values of dimensionless pressure inlet, P_1/P_2.

to balance the first term representing lower mechanical energy used up for acceleration of liquid. A higher pressure imposed upstream also contributes to an increase in volumetric flow rate, as expected; when $P_1=P_2$, fluid will not flow for lack of driving force – see line labeled as $P_1/P_2=1$ in Fig. 2.18. Estimation of volumetric flow rate in the vicinity of $S_2=S_1$ gives

$$\lim_{\frac{S_2}{S_1}\to 1}\frac{\rho Q^2}{S_1^{\,2}P_2}=\frac{10\left(\frac{P_1}{P_2}-1\right)}{(1-1)(7+5)}=\frac{\frac{5}{6}\left(\frac{P_1}{P_2}-1\right)}{0}=\infty, \tag{2.111}$$

which departs from Eq. (2.110) and is consistent with the vertical asymptote (marked as dashed line) in Fig. 2.18 at $S_2/S_1=1$; however, this mathematical result may be physically misleading. In fact, $S_2=S_1$ implies $v_2=v_1$ as per Eq. (2.98) – so Eq. (2.99) would degenerate to

$$\frac{P_2-P_1}{\rho}=0, \tag{2.112}$$

because

$$\lim_{\frac{S_2}{S_1}\to 1}K_h=\lim_{\frac{S_2}{S_1}\to 1}\frac{2}{5}\left(1-\frac{S_2}{S_1}\right)=0 \tag{2.113}$$

as assumed in Eq. (2.100); hence, the only line valid at that singularity would be the one labeled $P_2/P_1=1$ – associated with nil volumetric flow rate, rather than the vertical asymptote proper. This contradictory realization arises because all losses have been consubstantiated specifically in the constriction, and thus described solely by $K_h v_2^2/2$ in Eq. (2.99); however, this rationale holds only while there is flow at a finite rate through the nozzle. When $S_2\to S_1$, Eq. (2.99) is no longer valid, because the term of localized mechanical energy loss would disappear, and viscous dissipation along the tourniquet arms would have to be taken into account; this case is devoid of practical interest, anyway, as per the very definition of nozzle.

A momentum balance, between planes 1 and 2, is now in order – see Eq. (2.101) in *Food Proc. Eng.: basics & mechanical operations*, where the x- and y-components should be handled separately; the former looks like

$$\boxed{(S_1v_1)\rho v_{1,x}+S_{1,x}P_1+F_x=(S_2v_2)\rho v_{2,x}+S_{2,x}P_2,} \tag{2.114}$$

where $S_{1,x}$ and $S_{2,x}$ denote the projections of S_1 and S_2, respectively, normal to the x-direction of original flow, $v_{1,x}$ and $v_{2,x}$ denote the x-components of velocity at inlet and outlet, respectively, and F_x denotes the x-component of the external force applied onto the liquid by the sprinkler arm wall. Note that the balance above pertains to linear momentum, i.e. a vector quantity – which, referred to a unit volume, reads $\rho v_{1,x}$ and $\rho v_{2,x}$ in the x-direction, at inlet and outlet, respectively; whereas volumetric flow rate is a scalar quantity – which appears as S_1v_1 and S_2v_2, respectively, consistent with Eq. (2.98). By the same token, one may write

$$\boxed{(S_1v_1)\rho v_{1,y}+S_{1,y}P_1+F_y=(S_2v_2)\rho v_{2,y}+S_{2,y}P_2} \tag{2.115}$$

pertaining to the y-direction – where $S_{1,y}$ and $S_{2,y}$ denote the projections of S_1 and S_2 normal to the y-direction, while $v_{1,y}$ and $v_{2,y}$ denote the y-projections of $\boldsymbol{v}_1$ and $\boldsymbol{v}_2$, respectively, onto the same y-direction, and F_y denotes the y-component of $\boldsymbol{F}$. The underlying geometry of the problem indicates that

$$S_{1,y}=0 \tag{2.116}$$

and

$$v_{1,y}=0; \tag{2.117}$$

therefore, Eq. (2.115) simplifies to

$$F_y=\rho S_2v_2v_{2,y}+S_{2,y}P_2, \tag{2.118}$$

upon combination with Eqs. (2.116) and (2.117), followed by elimination of parenthesis. Further geometrical considerations have it that

$$S_{2,y}=S_2\sin\theta \tag{2.119}$$

and

$$v_{2,y}=v_2\sin\theta; \tag{2.120}$$

Eqs. (2.119) and (2.120) justify further transformation of Eq. (2.118) to

$$F_y=\rho S_2v_2(v_2\sin\theta)+(S_2\sin\theta)P_2, \tag{2.121}$$

which becomes

$$F_y=\rho\frac{(S_2v_2)^2}{S_2}\sin\theta+S_2P_2\sin\theta \tag{2.122}$$

after multiplication and division of the first term in the right-hand side by S_2. In view of Eq. (2.98), one may redo Eq. (2.122) as

$$F_y=\frac{\rho}{S_2}Q^2\sin\theta+S_2P_2\sin\theta, \tag{2.123}$$

while division of both sides by S_1 (i.e. the cross-sectional area of the feed inlet) and P_2 (usually atmospheric pressure) unfolds

$$\frac{F_y}{S_1P_2} = \frac{\rho Q^2}{S_1S_2P_2}\sin\theta + \frac{S_2}{S_1}\sin\theta; \tag{2.124}$$

further multiplication and division of the first term in the right-hand side of Eq. (2.124) by S_1, complemented by factoring out of sin θ give then rise to

$$\frac{F_y}{S_1P_2} = \left(\frac{\rho Q^2}{S_1^{\,2}P_2}\frac{S_1}{S_2} + \frac{S_2}{S_1}\right)\sin\theta. \tag{2.125}$$

Insertion of Eq. (2.110) converts Eq. (2.125) to

$$\frac{F_y}{S_1P_2} = \left(\frac{10\left(\frac{S_2}{S_1}\right)^2\left(\frac{P_1}{P_2}-1\right)}{\left(1-\frac{S_2}{S_1}\right)\left(7+5\frac{S_2}{S_1}\right)}\frac{S_1}{S_2} + \frac{S_2}{S_1}\right)\sin\theta, \tag{2.126}$$

or else

$$\frac{F_y}{S_1P_2} = \left(\frac{10\frac{S_2}{S_1}\left(\frac{P_1}{P_2}-1\right)}{\left(1-\frac{S_2}{S_1}\right)\left(7+5\frac{S_2}{S_1}\right)} + \frac{S_2}{S_1}\right)\sin\theta \tag{2.127}$$

upon lumping factors alike; a final factoring out of S_2/S_1 leaves

$$\frac{F_y}{S_1P_2} = \frac{S_2}{S_1}\left(1+\frac{10\left(\frac{P_1}{P_2}-1\right)}{\left(1-\frac{S_2}{S_1}\right)\left(7+5\frac{S_2}{S_1}\right)}\right)\sin\theta. \tag{2.128}$$

According to Newton's law of action/reaction, the force exerted by the fluid upon the sprinkler wall, in the vertical direction, $F_{y,s}$, looks like

$$\boxed{F_{y,s} = -F_y,} \tag{2.129}$$

so combination of Eqs. (2.128) and (2.129) gives rise to

$$\boxed{\frac{-F_{y,s}}{S_1P_2} = \frac{S_2}{S_1}\left(1+\frac{10\left(\frac{P_1}{P_2}-1\right)}{\left(1-\frac{S_2}{S_1}\right)\left(7+5\frac{S_2}{S_1}\right)}\right)\sin\theta;} \tag{2.130}$$

as expected, $-F_{y,s}/S_1P_2 \to 0$ when $\theta \to 0$, since sin $0=0$ – consistent with absence of y-momentum when the sprinkler arm is straight, in which case no rotational motion would be generated at all. A graphical representation of Eq. (2.130) is conveyed in Fig. 2.19, as dimensionless vertical force, $-F_{y,s}/S_1P_2$, versus normalized nozzle radius, S_2/S_1, and dimensionless pressure, P_1/P_2. Inspection of this figure unfolds an increase in $F_{y,s}$ that favors sprinkler rotational motion as S_2 increases; the reason for this realization comes from more pressure gauge available, as less and less is used up to accelerate liquid at the nozzle tip. The apparent vertical asymptote at $S_2/S_1=1$ should (again) be analyzed with care, owing to reasons presented before when discussing a similar issue pertaining to Q^2 – see Eqs. (2.111) and (2.112). On the other end of the scale, no effect would be felt in the absence of outlet flow – as per

$$\lim_{\frac{S_2}{S_1}\to 0}\frac{-F_{y,s}}{S_1P_2} = 0\left(1+\frac{10\left(\frac{P_1}{P_2}-1\right)}{(1-0)(7+0)}\right)\sin\theta = 0 \tag{2.131}$$

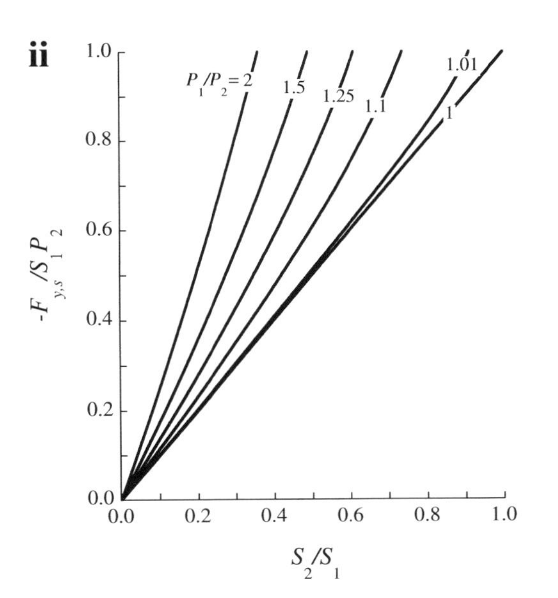

FIGURE 2.19 Variation of negative of dimensionless y-component of force exerted by solution upon sprinkler arm, $-F_{y,s}/S_1P_2$, versus normalized cross-sectional area of nozzle, S_2/S_1, for selected values of dimensionless inlet pressure, P_1/P_2, and inclination outflow angle, (i) $\theta = 45°$ or (ii) $\theta = 90°$.

stemming from Eq. (2.130); a hydrostatic (rather than hydrodynamic) situation would arise in this case, so no rotational movement would be produced. When inlet pressure essentially equals outlet (atmospheric) pressure, it is still possible to operate the sprinkler – as long as a change in direction of liquid between inlet and outlet occurs; the associated asymptote would read

$$\lim_{\frac{P_1}{P_2}\to 1}\frac{-F_{y,s}}{S_1P_2}=\frac{S_2}{S_1}\left(1+\frac{10(1-1)}{\left(1-\frac{S_2}{S_1}\right)\left(7+5\frac{S_2}{S_1}\right)}\right)\sin\theta=\frac{S_2}{S_1}\sin\theta, \quad (2.132)$$

based on Eq. (2.130) once more – which would lead to

$$\lim_{\frac{P_1}{P_2}\to 1}\left.\frac{-F_{y,s}}{S_1P_2}\right|_{\theta=0}=\frac{S_2}{S_1}\sin 0=0, \quad (2.133)$$

and thus no rotation whatsoever in the absence of bending (because of the nil $F_{y,s}$). The effect of the outlet angle can be better appreciated when going from Fig. 2.19i to ii; there is an increase in $F_{y,s}$ due to the increase in sin θ as per Eq. (2.130), due to the larger fraction of y-momentum generated. The tips of a typical sprinkler, however, do usually bend at an angle below 90°; this prevents an unnecessarily large turning rate for the marginal improvement obtained in cleaning efficiency – which would also bring about higher loss of mechanical energy at the inlet bearings.

A similar approach may be followed with regard to the x-component of $\boldsymbol{F}$ – in which case Eq. (2.114) is to be revisited as

$$\rho S_1v_1(v_1)+(S_1)P_1+F_x=\rho S_2v_2(v_2\cos\theta)+(S_2\cos\theta)P_2 \quad (2.134)$$

– since geometrical considerations support

$$S_{1,x}=S_1 \quad (2.135)$$

and

$$S_{2,x}=S_2\cos\theta, \quad (2.136)$$

encompassing the projections of cross sections S_1 and S_2, respectively, normally to the x-direction; complemented by

$$v_{1,x}=v_1 \quad (2.137)$$

and

$$v_{2,x}=v_2\cos\theta, \quad (2.138)$$

pertaining to the projections of $\boldsymbol{v}_1$ and $\boldsymbol{v}_2$, respectively, onto the x-direction. After pooling together identical factors, Eq. (2.134) becomes

$$\rho S_1v_1^2+S_1P_1+F_x=\rho S_2v_2^2\cos\theta+S_2P_2\cos\theta. \quad (2.139)$$

Combination with the second equality in Eq. (2.98) transforms Eq. (2.139) to

$$\rho S_1\left(\frac{S_2}{S_1}\right)^2v_2^2+S_1P_1+F_x=\rho S_2v_2^2\cos\theta+S_2P_2\cos\theta, \quad (2.140)$$

along with multiplication and division of the first term by $S_1=S_1^2/S_1$ – where ρv_2^2 can be factored out as

$$F_x=\rho\left(S_2\cos\theta-S_1\left(\frac{S_2}{S_1}\right)^2\right)v_2^2+S_2P_2\cos\theta-S_1P_1, \quad (2.141)$$

together with isolation of F_x; insertion of the equality between left- and right-hand side in Eq. (2.98) gives, in turn, rise to

$$F_x=\rho\left(S_2\cos\theta-S_1\left(\frac{S_2}{S_1}\right)^2\right)\left(\frac{Q}{S_2}\right)^2+S_2P_2\cos\theta-S_1P_1, \quad (2.142)$$

while multiplication and division by $S_1^2P_2$ generate

$$F_x=\frac{S_1^2}{S_2^2}P_2\left(S_2\cos\theta-S_1\left(\frac{S_2}{S_1}\right)^2\right)\frac{\rho Q^2}{S_1^2P_2}+S_2P_2\cos\theta-S_1P_1. \quad (2.143)$$

Upon insertion of Eq. (2.110), one obtains

$$F_x=\left(\frac{S_1}{S_2}\right)^2P_2\left(S_2\cos\theta-S_1\left(\frac{S_2}{S_1}\right)^2\right)\frac{10\left(\frac{S_2}{S_1}\right)^2\left(\frac{P_1}{P_2}-1\right)}{\left(1-\frac{S_2}{S_1}\right)\left(7+5\frac{S_2}{S_1}\right)} +S_2P_2\cos\theta-S_1P_1 \quad (2.144)$$

from Eq. (2.143) – where cancellation of the square of S_1/S_2 with its reciprocal permits simplification to

$$F_x=P_2\left(S_2\cos\theta-S_1\left(\frac{S_2}{S_1}\right)^2\right)\frac{10\left(\frac{P_1}{P_2}-1\right)}{\left(1-\frac{S_2}{S_1}\right)\left(7+5\frac{S_2}{S_1}\right)} +S_2P_2\cos\theta-S_1P_1; \quad (2.145)$$

division of both sides by S_1P_2 then produces

$$\frac{F_x}{S_1P_2}=\frac{10\left(\frac{S_2}{S_1}\cos\theta-\left(\frac{S_2}{S_1}\right)^2\right)\left(\frac{P_1}{P_2}-1\right)}{\left(1-\frac{S_2}{S_1}\right)\left(7+5\frac{S_2}{S_1}\right)}+\frac{S_2}{S_1}\cos\theta-\frac{P_1}{P_2}. \quad (2.146)$$

Once S_2/S_1 has been factored out in the numerator of the first term of the right-hand side, Eq. (2.146) takes the form

$$\frac{F_x}{S_1P_2}=\frac{10\frac{S_2}{S_1}\left(\cos\theta-\frac{S_2}{S_1}\right)\left(\frac{P_1}{P_2}-1\right)}{\left(1-\frac{S_2}{S_1}\right)\left(7+5\frac{S_2}{S_1}\right)}+\frac{S_2}{S_1}\cos\theta-\frac{P_1}{P_2}; \quad (2.147)$$

S_2/S_1 may be factored out once more, between first and second terms in the right-hand side, to get

$$\frac{F_x}{S_1P_2}=\frac{S_2}{S_1}\left(\cos\theta+\frac{10\left(\cos\theta-\frac{S_2}{S_1}\right)\left(\frac{P_1}{P_2}-1\right)}{\left(1-\frac{S_2}{S_1}\right)\left(7+5\frac{S_2}{S_1}\right)}\right)-\frac{P_1}{P_2}. \quad (2.148)$$

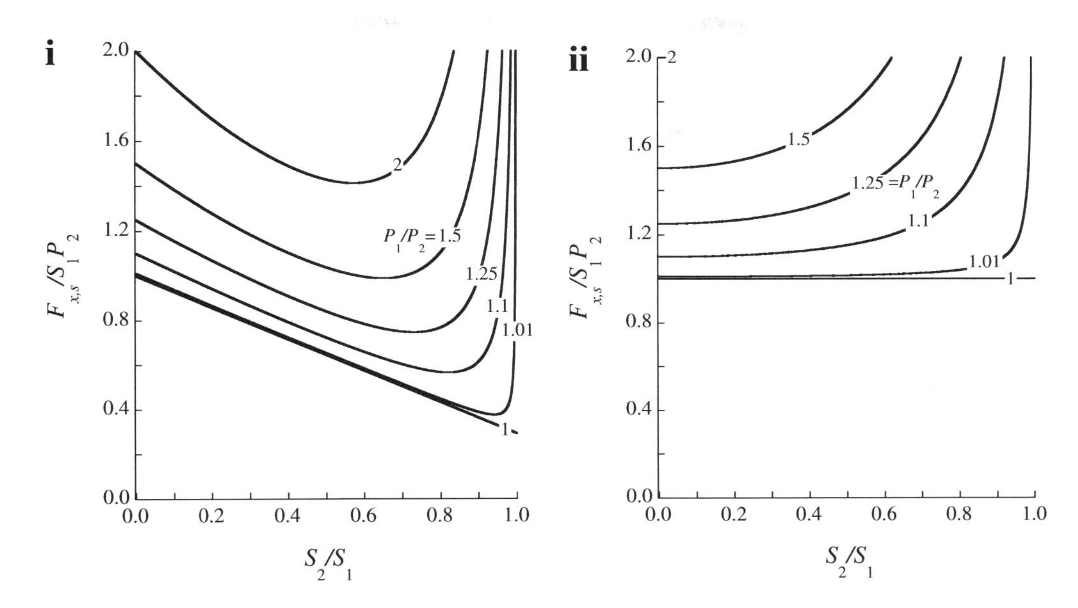

FIGURE 2.20 Variation of dimensionless x-component of force exerted by solution upon sprinkler arm, $F_{x,s}/S_1P_2$, versus normalized cross-sectional area of nozzle, S_2/S_1, for selected values of dimensionless inlet pressure, P_1/P_2, and inclination outflow angle, (i) $\theta = 45°$ or (ii) $\theta = 90°$.

In view of Newton's principle of action and reaction applied along the x-direction, one finds that

$$\boxed{F_{x,s} = -F_x,} \tag{2.149}$$

where $F_{x,s}$ pertains to the x-component of force $\boldsymbol{F}_s$ exerted by flowing liquid onto the sprinkler arm; insertion of Eq. (2.148) finally gives

$$\boxed{\frac{F_{x,s}}{S_1P_2} = \frac{P_1}{P_2} - \frac{S_2}{S_1}\left(\cos\theta + \frac{10\left(\cos\theta - \frac{S_2}{S_1}\right)\left(\frac{P_1}{P_2} - 1\right)}{\left(1 - \frac{S_2}{S_1}\right)\left(7 + 5\frac{S_2}{S_1}\right)}\right)} \tag{2.150}$$

when departing from Eq. (2.149) – which describes magnitude of the horizontal component of $\boldsymbol{F}_s$. A graphical interpretation of Eq. (2.150) is provided as Fig. 2.20. The curves start from P_1/P_2, in agreement with

$$\lim_{\frac{S_2}{S_1}\to 0} \frac{F_{x,s}}{S_1P_2} = \frac{P_1}{P_2} - 0\left(\cos\theta + \frac{10(\cos\theta - 0)\left(\frac{P_1}{P_2} - 1\right)}{(1-0)(7+0)}\right) = \frac{P_1}{P_2} \tag{2.151}$$

stemming directly from Eq. (2.150); and then decrease almost linearly until reaching a minimum – after which they increase hyperbolically toward infinity. The curves described by Eq. (2.150) approach a straight line, of slope $-\cos\theta$ and unit vertical intercept, when $P_1 \rightarrow P_2$, viz.

$$\lim_{\frac{P_1}{P_2}\to 1} \frac{F_{x,s}}{S_1P_2} = 1 - \frac{S_2}{S_1}\left(\cos\theta + \frac{10\left(\cos\theta - \frac{S_2}{S_1}\right)(1-1)}{\left(1 - \frac{S_2}{S_1}\right)\left(7 + 5\frac{S_2}{S_1}\right)}\right) = 1 - \frac{S_2}{S_1}\cos\theta \tag{2.152}$$

– which thus drives the asymptotic behavior associated with no pressure drop between inlet and outlet; when $\theta=90°$, the said line degenerates to a horizontal line (since cos 90° is nil), see Fig. 2.20ii. This special case implies coincidence of $F_{x,s}/S_1P_2$ with F/S_1P_1 – where S_1P_1 as single source of momentum must be exactly overcome by F, because no x-momentum exists at the nozzle tip. In fact, a plain hydrostatic problem would result, since $P_2=P_1$ enforces $Q=0$ as per Eq. (2.110), and thus $v_2^2=0$ and $v_1^2=0$, irrespective of S_2/S_1; hence, the effect of S_2/S_1 would become immaterial. The $F_{x,s}$s prevailing at nozzle tips coincide in magnitude – and lie on the same straight line, despite being opposite to each other; therefore, no torque is produced nor is there any contribution to motion – as they balance each other collinearly.

Operation around the minimum of $F_{x,s}$ is *a priori* recommended, since static stresses imposed on the material in the x-direction would become as low as feasible; this deed is enhanced by a lower P_1/P_2, as expected (see Fig. 2.20) – while a smaller angle θ causes such a minimum of $F_{x,s}$ fall below those recorded for $\theta=90°$, see Fig. 2.20i *vis-à-vis* with Fig. 2.20ii. Therefore, another justification emerges for use of an inclination angle for the nozzle tip below 90°.

The forces in the y-direction do, in practice, counterbalance frictional resistance to rotational motion – so the sprinkler will rapidly attain a constant angular frequency of operation.

PROBLEM 2.3

Consider a regular sprinkler, employed for disinfection in place of a cylindrical vessel used in food processing.

a) Calculate the distance to be traveled by the disinfecting solution when the sprinkler arm bends at 90° and 0°, and conclude on the convenience of avoiding the former (despite yielding the maximum rotational thrust).

b) Find the necessary condition describing the minima in Fig. 2.20, and discuss the effect of both P_1/P_2 and θ on such optimum values for S_2/S_1.

Solution

a) Consider the two extreme situations of angle suggested, i.e. 90° and 0°, made by the sprinkler tip with its arm, as illustrated below

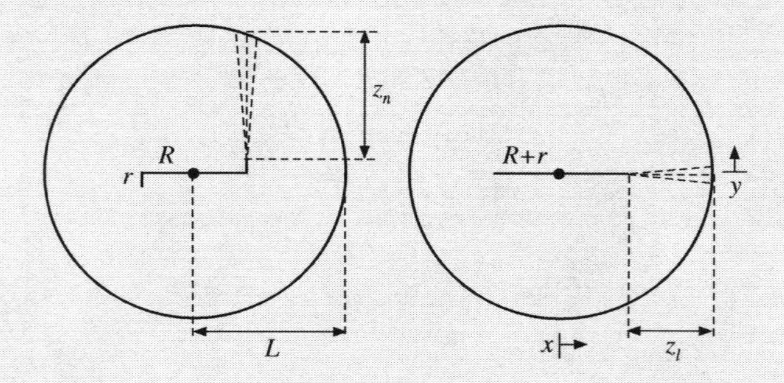

– with vessel inner radius equal to L, and sprinkler arm length and tip length equal to R and r, respectively. The distance to be traveled by the jet (considered to essentially keep the cross-section of the nozzle tip) until impinging on the vessel wall is denoted by z_n and z_l, for the normal and longitudinal configuration, respectively. If the vessel is cylindrical, and its axis coincides with the center of the cross-section circle, then the x- and y-coordinates of the corresponding circumference must satisfy

$$x^2 + y^2 = L^2, \tag{2.3.1}$$

which may be solved with regard to y as

$$y = \sqrt{L^2 - x^2}. \tag{2.3.2}$$

In the normal configuration, the nozzle tip is described by (x,y) coordinates (R,r), so the point of impact of the corresponding jet with the wall will be characterized by $(R,r+z_n)$ on the aforementioned circumference – where z_n satisfies

$$r + z_n = \sqrt{L^2 - R^2}, \tag{2.3.3}$$

as per Eq. (2.3.2); hence, one readily concludes that

$$z_n = \sqrt{L^2 - R^2} - r \tag{2.3.4}$$

from Eq. (2.3.3). In the longitudinal configuration, the nozzle tip is characterized by $(R+r,0)$ as coordinates, whereas the point of impact of its jet with the inner wall abides to $(L,0)$; the distance traveled by the jet, in this case, is simply given by

$$z_l = L - (R + r) = (L - R) - r. \tag{2.3.5}$$

The difference between the two germane distances may be obtained via ordered subtraction of Eq. (2.3.5) from Eq. (2.3.4), according to

$$z_n - z_l = \sqrt{L^2 - R^2} - r - (L - R) + r \tag{2.3.6}$$

that breaks down to

$$z_n - z_l = \sqrt{L^2 - R^2} - (L - R) \tag{2.3.7}$$

after dropping symmetrical terms. On the other hand, the difference of two squares coincides with the product of the conjugated binomials of their bases, so Eq. (2.3.7) may undergo reformulation to

$$z_n - z_l = \sqrt{(L - R)(L + R)} - \sqrt{(L - R)^2}, \tag{2.3.8}$$

along with replacement of $L - R$ by the square root of its square power. After factoring $\sqrt{L - R}$ out, Eq. (2.3.8) becomes

$$z_n - z_l = \sqrt{L - R}(\sqrt{L + R} - \sqrt{L - R}), \tag{2.3.9}$$

with $L > R$ by construction implying $L - R > 0$ and $\sqrt{L - R} > 0$; in addition, $L+R > L > L-R$ implies $\sqrt{L + R} > \sqrt{L - R} > 0$. Therefore, Eq. (2.3.9) supports

$$z_n - z_l > 0 \tag{2.3.10}$$

– meaning that z_n exceeds z_l; in other words, a direct jet needs to span a smaller distance until its impact point – so it should have a higher efficacy of cleaning. Unfortunately, a straight nozzle would produce no thrust, as required for rotational motion of the sprinkler; cleaning would thus be confined to a minor portion of the vessel inner wall, and be essentially useless.

b) The necessary condition for a minimum of $F_{x,s}/S_1P_2$, with regard to S_2/S_1, reads

$$\frac{\partial}{\partial x}\left(\frac{F_{x,s}}{S_1P_2}\right) = \frac{\partial}{\partial x}\left(a - x\left(b + \frac{10(a-1)(b-x)}{(1-x)(7+5x)}\right)\right) = 0 \tag{2.3.11}$$

based on Eq. (2.150) – as long as auxiliary parameters a and b are defined as

$$a \equiv \frac{P_1}{P_2} \tag{2.3.12}$$

and

$$b \equiv \cos\theta, \tag{2.3.13}$$

respectively; and auxiliary variable x is defined as

$$x \equiv \frac{S_2}{S_1}. \tag{2.3.14}$$

Application of the classical rules of differentiation to Eq. (2.3.11) gives rise to

$$-\left(b + \frac{10(a-1)(b-x)}{(1-x)(7+5x)}\right) + 10(a-1)x\frac{(1-x)(7+5x) + (b-x)\big(5(1-x) - (7+5x)\big)}{(1-x)^2(7+5x)^2} = 0, \tag{2.3.15}$$

where conversion of both terms in the left hand side to a similar denominator entails

$$10(a-1)x\frac{(1-x)(7+5x)+(b-x)(5-5x-7-5x)}{(1-x)^2(7+5x)^2}$$
$$=\frac{b(1-x)^2(7+5x)^2+10(a-1)(b-x)(1-x)(7+5x)}{(1-x)^2(7+5x)^2} \tag{2.3.16}$$

– along with elimination of inner parenthesis (where appropriate); Eq. (2.3.16) reduces to an equality of numerators of the two sides in view of their common denominator, i.e.

$$10(a-1)x\big((1-x)(7+5x)-(b-x)(2+10x)\big)$$
$$=b(1-x)^2(7+5x)^2+10(a-1)(b-x)(1-x)(7+5x) \tag{2.3.17}$$

– where $10(a-1)$ may be isolated to give

$$10(a-1)$$
$$=\frac{b(1-x)^2(7+5x)^2}{x\big((1-x)(7+5x)-(b-x)(2+10x)\big)-(b-x)(1-x)(7+5x)}. \tag{2.3.18}$$

Upon removal of the outer parenthesis in the first term in denominator, Eq. (2.3.18) becomes

$$10(a-1)$$
$$=\frac{b(1-x)^2(7+5x)^2}{x(1-x)(7+5x)-x(b-x)(2+10x)-(b-x)(1-x)(7+5x)}, \tag{2.3.19}$$

whereas factoring out of $(1-x)(7+5x)$ or 2 (for convenience) leads to

$$10(a-1)=\frac{b(1-x)^2(7+5x)^2}{\big(x-(b-x)\big)(1-x)(7+5x)-2x(b-x)(1+5x)}; \tag{2.3.20}$$

lumping of similar terms produces

$$10(a-1)=\frac{b(1-x)^2(7+5x)^2}{(1-x)(2x-b)(7+5x)-2x(b-x)(1+5x)}, \tag{2.3.21}$$

while isolation of a eventually yields

$$a=$$
$$1+\frac{\frac{b}{10}\left(1-x_{opt}\right)^2\left(7+5x_{opt}\right)^2}{\left(1-x_{opt}\right)\left(2x_{opt}-b\right)\left(7+5x_{opt}\right)-2x_{opt}\left(b-x_{opt}\right)\left(1+5x_{opt}\right)}. \tag{2.3.22}$$

The loci of optima, $x_{opt} \equiv x_{opt}\{a\}$, are plotted below for several values of b associated with the corresponding inclination angles θ via Eq. (2.3.13), after using $a \equiv a\{x_{opt}\}$ reversewise as per Eq. (2.3.22) – and recalling Eqs. (2.3.12) and (2.3.14).

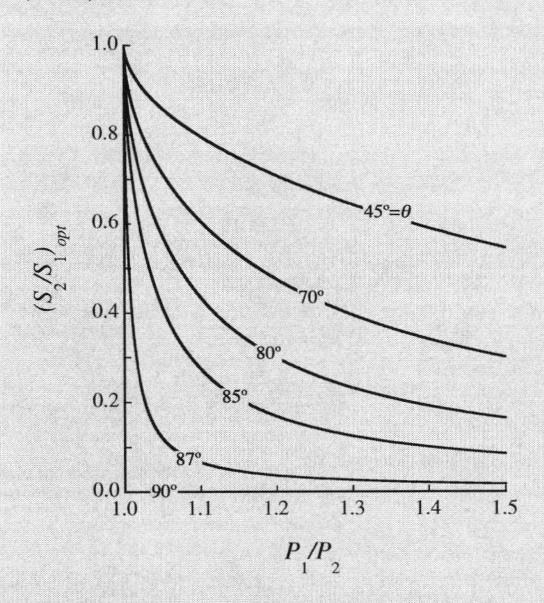

As P_1 increases for a given P_2, the minimum occurs for a nozzle characterized by a smaller and smaller cross-sectional area S_2 – as also apparent in Fig. 2.20i; note the asymptotic tendency for the loci to coincide with the vertical axis at unity as $P_1/P_2 \rightarrow 1$, and with the horizontal axis when $\theta \rightarrow 90°$ – consistent with the monotonically increasing pattern exhibited by the $F_{x,s}/S_1P_2$-vs.-S_2/S_1 curves in Fig. 2.20ii.

The overall force acting upon each arm of the sprinkler, $\boldsymbol{F}_s$, may now be ascertained from $F_{x,s}$ and $F_{y,s}$, according to

$$|\boldsymbol{F}_s|=\sqrt{F_{x,s}^{\ 2}+F_{y,s}^{\ 2}} \tag{2.153}$$

in terms of magnitude, and

$$\Theta=\tan^{-1}\frac{F_y}{F_x} \tag{2.154}$$

for inclination angle, Θ – with $F_{x,s}$ and $F_{y,s}$ given by Eqs. (2.150) and (2.130), respectively.

As emphasized before, the forces along the x-direction, i.e. $F_{x,s}$ and $-F_{x,s}$, pertaining to the left and right nozzle tips, respectively, balance each other – because they have the same magnitude, but opposite directions along the same line passing through the axis of the sprinkler. However, the same does not happen with regard to the forces in the y-direction, i.e. $-F_{y,s}$ and $F_{y,s}$, respectively; despite also sharing the same magnitude and opposite directions, they have application points separated by distance $2R\neq 0$, see again Fig. 2.17. The torque developed, $\mathcal{T}$, associated to each force may thus be calculated as

$$\boxed{\mathcal{T}_{y,s} \equiv F_{y,s}R,} \tag{2.155}$$

which gives

$$\boxed{\frac{\mathcal{T}_{y,s}}{RS_1P_2} = \frac{S_2}{S_1}\left(1+\frac{10\left(\dfrac{P_1}{P_2}-1\right)}{\left(1-\dfrac{S_2}{S_1}\right)\left(7+5\dfrac{S_2}{S_1}\right)}\right)\sin\theta} \tag{2.156}$$

in dimensionless form – after taking Eq. (2.130) into account; in view of the linear increase of $\mathcal{T}_{y,s}$ with R, a long arm should be sought for the sprinkler so as to maximize thrust – which, however, would come at the expense of a larger weight, implying higher inertia and thus a tendentially slower operation. Therefore, some compromise is to be found in practice.

2.1.3.4 Rinsing

Rinsing consists on elimination of cleaning (and disinfecting) solutions – using typically freshwater or a convenient solvent, and (possibly) original liquid food afterward; the overall concentration of agent(s) undergoing removal decreases with time, as tendentially sketched in Fig. 2.21ii – along with a decrease in solution volume, as per Fig. 2.21i. Similarly to cleaning, the rate of rinsing depends on the agent(s) to be taken away – namely, solubility in water, whether foam develops, initial concentration, and flow rate of rinsing water utilized; the geometry of equipment, surface morphology, and operating temperature also play a role.

The three basic (and essentially sequential) mechanisms germane to rinsing are film (laminar) flow, boundary layer mixing, and boundary layer diffusion. The initial emptying period entails a (variable) power-type decrease in volume – followed by a logarithmic decrease in boundary layer volume during the mixing period, and a final constant volume for the boundary layer during the diffusion period – see Fig. 2.21i. Conversely, a constant concentration of solute prevails during the emptying period; a logarithmic decay in concentration appears as characteristic of the mixing period – described by a negative slope κ_1, see Fig. 2.21ii; and a second logarithmic decay is observed throughout the diffusion period – eventually characterized by a negative slope κ_2, such that $|\kappa_2| < |\kappa_1|$. Mixing is promoted by high flow rates of rinsing water, whereas diffusion requires less water but a longer period to take place; foaming by the cleaning agent interferes with both mechanisms. Rinsing is oftentimes improved by pulsatile flow – as it promotes formation of small-scale eddies that contribute directly to (micro)mixing. The problem of rinsing optimization is normally to be solved as a compromise between total amount of rinsing water and whole rinsing time needed; this will be feasible only if a water cost-equivalent of rinsing time, or instead a rinsing time-equivalent of wasted water can be pre-established.

The preliminary step of rinsing is obviously emptying of the apparatus under scrutiny; the rate of displacement of bulk liquid abides to Eq. (2.247), derived previously in *Food Proc. Eng.: basics & mechanical operations*. However, a liquid film always remains stuck onto its inner wall – the thicker the film, the higher the liquid viscosity. In view of its small thickness compared to the interfacial surface, a parallelepipedal geometry can safely be claimed – of constant width Y in the y-direction (equal to $2\pi R_v$, in the case of a cylinder of inner radius R_v), length Z in the z-direction (equal to H_v, in the case of a cylinder of height H_v), and variable thickness $X<<Y,Z$ in the x-direction. An overall balance to the volume of liquid film along the vessel wall – being drained in the z-direction at (average) velocity, $\bar{v}_z$, takes the form

$$X\,\bar{v}_z\big|_z\,Ydt = X\,\bar{v}_z\big|_{z+dz}\,Ydt + YdXdz; \tag{2.157}$$

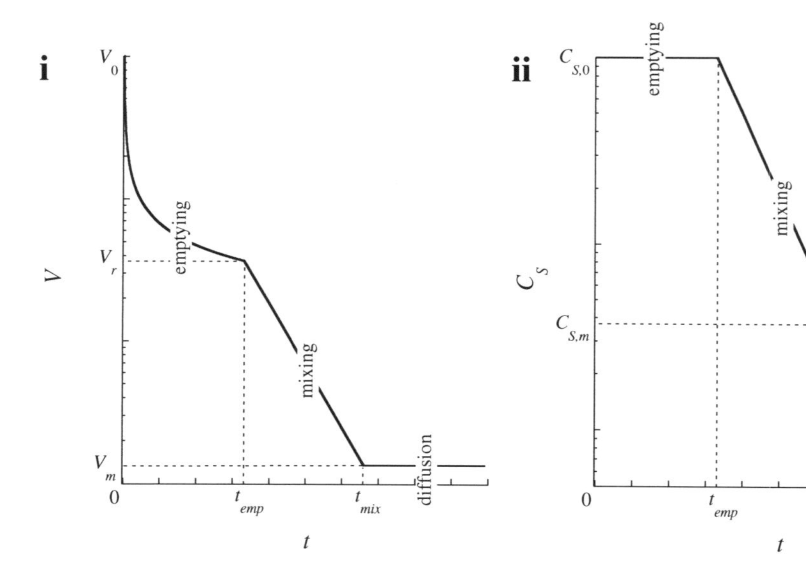

FIGURE 2.21 Evolution of (i) volume, V, and (ii) concentration in solute, C_S, of residual film of solution sticking onto the inner surface of the vessel wall, versus rinsing time, t, for three consecutive stages of: emptying, between time t equal to 0 and t_{emp} – with variation of V from V_0 down to V_r, and C_S remaining at $C_{S,0}$; mixing, between t equal to t_{emp} and t_{mix} – with variation of V from V_r down to V_m, and variation of C_S from $C_{S,0}$ down to $C_{S,m}$; and diffusion, from t equal to t_{mix} on – with V remaining at V_m, and variation of C_S from $C_{S,m}$ down.

here XY accounts for area of cross-section of flow, and dz denotes length of elementary control volume along the z-direction, with $X \equiv X\{z,t\}$. The left-hand side and the first term in the right-hand side of Eq. (2.157) represent inlet and outlet volume, respectively, by convection during elementary time period dt; and the outstanding term represents variation of volume during that period. After condensing terms alike and factoring Ydt out, one obtains

$$Y\left(X\bar{v}_z\big|_{z+dz} - X\bar{v}_z\big|_z \right)dt + YdXdz = 0 \tag{2.158}$$

from Eq. (2.157), where division of both sides by $Ydzdt$ yields

$$\boxed{\frac{\partial}{\partial z}\left(X\bar{v}_z \right) + \frac{\partial X}{\partial t} = 0.} \tag{2.159}$$

Integration of Eq. (2.159) is not possible unless $\bar{v}_z \equiv \bar{v}_z\{z\}$ or $\bar{v}_z \equiv \bar{v}_z\{X\}$ is known in advance; hence, one should resort to Navier and Stokes' equation, in Cartesian coordinates, describing gravitational flow of a parallelepipedal layer of liquid down a solid vertical wall under steady state conditions, see Eq. (2.122) in *Food Proc. Eng.: basics & mechanical operations* – after realizing that $v_x = v_y = 0$ and $g_x = g_y = 0$, and specifically imposing $\partial\mathbf{v}/\partial t = 0$. Under these circumstances, the x-component of the said equation reduces to

$$\boxed{\frac{\partial P}{\partial x} = 0,} \tag{2.160}$$

because $v_x\partial v_x/\partial x = v_y\partial v_x/\partial y = v_z\partial v_x/dz = \partial v_x/\partial t = \partial^2 v_x/\partial x^2 = \partial^2 v_x/\partial y^2 = \partial^2 v_x/\partial z^2 = 0$ – meaning that no pressure change occurs in the x-direction. The y-component similarly becomes

$$\boxed{\frac{\partial P}{\partial y} = 0,} \tag{2.161}$$

since $v_x\partial v_y/\partial x = v_y\partial v_y/\partial y = v_z\partial v_y/\partial z = \partial v_y/\partial t = \partial^2 v_y/\partial x^2 = \partial^2 v_y/\partial y^2 = \partial^2 v_y/\partial z^2 = 0$ – thus indicating that pressure does not depend on y either. One may finally resort to the z-component of the equation of motion, viz.

$$\rho v_z \frac{\partial v_z}{\partial z} = \rho g - \frac{\partial P}{\partial z} + \mu\left(\frac{\partial^2 v_z}{\partial x^2} + \frac{\partial^2 v_z}{\partial z^2} \right), \tag{2.162}$$

after realizing that $v_x\partial v_z/\partial x = v_y\partial v_z/\partial y = \partial v_z/\partial t = 0$ and $g_z = g$ for gravitational pull acting vertically – and parallelepipedal symmetry enforcing $\partial v_z/\partial y = 0$, and thus $\partial^2 v_z/\partial y^2 = 0$. On the other hand, the equation of continuity, as per Eq. (2.121) of *Food Proc. Eng.: basics & mechanical operations*, has it that

$$\boxed{\frac{\partial v_z}{\partial z} = 0,} \tag{2.163}$$

in view of $\partial v_x/\partial_x = \partial v_y/\partial_y = 0$; insertion of Eq. (2.163) permits simplification of Eq. (2.162) to

$$\rho g - \frac{\partial P}{\partial z} + \mu \frac{\partial^2 v_z}{\partial x^2} = 0, \tag{2.164}$$

since Eq. (2.163) also implies $\partial^2 v_z/\partial z^2 = \partial(\partial v_z/\partial z)/\partial z = 0$. The flow under scrutiny is free flow – exposed to the open atmosphere along the whole outer surface of liquid, so no change in pressure is allowed inside the fluid; this enforces $\partial P/\partial z = 0$ in particular, so Eq. (2.164) breaks down to just

$$\boxed{\frac{d}{dx}\left(\frac{dv_z}{dx} \right) = -\frac{\rho g}{\mu}} \tag{2.165}$$

– where both sides were meanwhile divided by μ, and the definition of second derivative was taken advantage of (besides $v_z \equiv v_z\{x\}$ allowing replacement of $\partial^2 v_z/\partial x^2 \equiv \partial(\partial v_z/\partial x)/\partial x$ by $d(dv_z/dx)/dx$). Integration of Eq. (2.165) is feasible, via separation of variables, as

$$\int d\left(\frac{dv_z}{dx} \right) = -\frac{\rho g}{\mu}\int dx, \tag{2.166}$$

since ρ and μ are essentially constant for an isothermal liquid – which is equivalent to writing

$$\frac{dv_z}{dx} = k_1 - \frac{\rho g}{\mu}x; \tag{2.167}$$

here k_1 denotes an (arbitrary) integration constant. A second step of integration of Eq. (2.167), again by separation of variables, is in order, viz.

$$\int dv_z = k_1\int dx - \frac{\rho g}{\mu}\int xdx \tag{2.168}$$

– where the linearity of this operator and the constancy of $\rho g/\mu$ were taken advantage of; Eq. (2.168) turns to

$$v_z = k_2 + k_1 x - \frac{\rho g}{2\mu}x^2, \tag{2.169}$$

with the aid of a second integration constant, k_2. Boundary conditions may be conveniently formulated as

$$\boxed{v_z\big|_{x=0} = 0,} \tag{2.170}$$

as a consequence of the nonslip condition on the inner (solid) wall; coupled with

$$\tau_{xz}\big|_{x=X} = 0, \tag{2.171}$$

describing the practical impossibility of overhead air resist shearing at the interface with a (much more viscous) liquid – or, after recalling Newton's law,

$$\boxed{-\mu\frac{dv_z}{dx}\bigg|_{x=X} = 0.} \tag{2.172}$$

Combination of Eqs. (2.169) and (2.170) implies that

$$\left(k_2 + k_1 x - \frac{\rho g}{2\mu}x^2 = 0 \right)\bigg|_{x=0}, \tag{2.173}$$

so one readily concludes that

$$k_2 = 0; \tag{2.174}$$

Eq. (2.169) accordingly simplifies to

$$v_z = k_1 x - \frac{\rho g}{2\mu} x^2. \tag{2.175}$$

One may now eliminate dv_z/dx between Eqs. (2.167) and (2.172) as

$$\left.\left(k_1 - \frac{\rho g}{\mu} x = 0\right)\right|_{x=X}, \tag{2.176}$$

or else

$$k_1 = \frac{\rho g X}{\mu}; \tag{2.177}$$

in view of Eq. (2.177), one can redo Eq. (2.175) to

$$v_z = \frac{\rho g X}{\mu} x - \frac{\rho g}{2\mu} x^2 \tag{2.178}$$

– where factoring out of $\rho g x/2\mu$ unfolds

$$\boxed{v_z\{x\} = \frac{\rho g}{2\mu} x(2X - x).} \tag{2.179}$$

Once in possession of the functional dependence of v_z on x as per Eq. (2.179) – valid for steady state, calculation of the average flow velocity over a cross-section of the film flowing down ensues as

$$\bar{v}_z \equiv \frac{\int_0^Y \int_0^X v_z\{x\} dx dy}{\int_0^Y \int_0^X dx dy}; \tag{2.180}$$

application of Fubini's (weak) theorem allows conversion of the composite integrals to plain multiplicative integrals, i.e.

$$\bar{v}_z = \frac{\int_0^Y dy \int_0^X v_z dx}{\int_0^Y dy \int_0^X dx}, \tag{2.181}$$

since the integration limits do not depend on either integration variable. Upon insertion of Eq. (2.179) and realization that $\rho g/2\mu$ is constant, Eq. (2.181) turns to

$$\bar{v}_z = \frac{Y \int_0^X \frac{\rho g}{2\mu} x(2X - x) dx}{Y \int_0^X dx} = \frac{\frac{\rho g}{2\mu} \int_0^X (2Xx - x^2) dx}{\int_0^X dx} \tag{2.182}$$

– where similar factors were meanwhile dropped from numerator and denominator, and x was factored in. Application of the fundamental theorem of integral calculus transforms Eq. (2.182) to

$$\bar{v}_z = \frac{\rho g}{2\mu} \frac{\left.\left(2X \frac{x^2}{2} - \frac{x^3}{3}\right)\right|_0^X}{x\big|_0^X} = \frac{\rho g}{2\mu} \frac{X x^2\big|_0^X - \frac{1}{3} x^3\big|_0^X}{x\big|_0^X}; \tag{2.183}$$

straightforward algebraic manipulation gives then rise to

$$\bar{v}_z = \frac{\rho g}{2\mu} \frac{X(X^2 - 0^2) - \frac{1}{3}(X^3 - 0^3)}{X}. \tag{2.184}$$

Condensation of factors in numerator converts Eq. (2.184) to

$$\bar{v}_z = \frac{\rho g}{2\mu} \frac{X^3 - \frac{X^3}{3}}{X}, \tag{2.185}$$

whereas lumping of similar terms in numerator, and dropping of $2X$ between numerator and denominator afterward yield

$$\boxed{\bar{v}_z\{X\} = \frac{\rho g X^2}{3\mu};} \tag{2.186}$$

recall that Eq. (2.186) applies under steady state conditions, and so considers fully developed laminar flow at every point in the z-direction along fluid path – for a constant thickness of film X.

On the hypothesis of a pseudo steady state for the equipment emptying problem – i.e. assuming that fully developed laminar flow establishes very fast compared to the rate of flow proper (as experimentally confirmed), one may resort to Eq. (2.186) to convert Eq. (2.159) to

$$\frac{\partial}{\partial z}\left(\frac{\rho g X^3}{3\mu}\right) + \frac{\partial X}{\partial t} = 0, \tag{2.187}$$

where factors in X were already lumped; based on the rule of differentiation of a power, Eq. (2.187) becomes

$$\frac{\rho g}{\mu} X^2 \frac{\partial X}{\partial z} + \frac{\partial X}{\partial t} = 0, \tag{2.188}$$

along with cancelling of 3 between numerator and denominator. A solution of the general form

$$X \equiv X\left\{\frac{z}{t}\right\} \tag{2.189}$$

may hereafter be postulated for the partial differential equation labeled as Eq. (2.188); its partial derivative with regard to z looks indeed like

$$\frac{\partial X}{\partial z} = \frac{dX}{d\left(\frac{z}{t}\right)} \frac{\partial\left(\frac{z}{t}\right)}{\partial z}, \tag{2.190}$$

and likewise

$$\frac{\partial X}{\partial t} = \frac{dX}{d\left(\frac{z}{t}\right)} \frac{\partial\left(\frac{z}{t}\right)}{\partial t} \tag{2.191}$$

for the partial derivative of X with regard to t – both supported by the chain differentiation rule. After rewriting Eqs. (2.190) and (2.191) as

$$\frac{\partial X}{\partial z} = \frac{1}{t} \frac{dX}{d\left(\frac{z}{t}\right)} \tag{2.192}$$

and

$$\frac{\partial X}{\partial t}=-\frac{z}{t^2}\frac{dX}{d\left(\frac{z}{t}\right)}, \tag{2.193}$$

respectively, one may resort to Eqs. (2.192) and (2.193) to transform Eq. (2.188) to

$$\frac{\rho g}{\mu}X^2\frac{1}{t}\frac{dX}{d\left(\frac{z}{t}\right)}-\frac{z}{t^2}\frac{dX}{d\left(\frac{z}{t}\right)}=0; \tag{2.194}$$

if both sides are now multiplied by t and divided by $dX/d(z/t)$, then Eq. (2.194) breaks down to

$$\frac{\rho g}{\mu}X^2-\frac{z}{t}=0 \tag{2.195}$$

– from where X can be isolated as

$$\boxed{X\{z,t\}=\sqrt{\frac{\mu}{\rho g}}\sqrt{\frac{z}{t}}.} \tag{2.196}$$

Therefore, the film thickness of liquid on the surface of the vessel wall depends on ratio z/t, rather than on z and t separately; note that the functional form of Eq. (2.196) is consistent with the form postulated via Eq. (2.189). A more convenient approach – which also overcomes overparameterization, is possible upon division of both sides of Eq. (2.196) by Z, viz.

$$\frac{X}{Z}=\frac{1}{Z}\sqrt{\frac{\mu}{\rho g}}\sqrt{\frac{z}{t}}; \tag{2.197}$$

straightforward algebraic rearrangement then produces

$$\boxed{\frac{X}{Z}=\sqrt{\frac{\frac{z}{Z}}{\frac{\rho g Z}{\mu}t}},} \tag{2.198}$$

as plotted in Fig. 2.22i – whereas a graphical account of the evolution in time of the vertical cross-section of the liquid film sticking to the container wall, as per Eq. (2.196), is plotted in Fig. 2.22ii. For $z=0$, the lumped variable z/t is also nil – and inspection of Fig. 2.22ii indicates that $X=0$ as well (except for $t=0$), i.e. no liquid remains exactly at the level of its original free surface. Specifically for $t=0$, one realizes that $z/t \rightarrow \infty$ for a finite z; under these circumstances, Eq. (2.196) would predict an infinite thickness – which violates the departing assumption of a sufficiently thin residual film, as required for the slab geometry be valid. Nevertheless, liquid filling the vessel up to the initial level is compatible with the line labelled as $t_0=0$ in Fig. 2.22ii – including $z/t \rightarrow 0/0$ (as unknown quantity) at $z=0$.

The total residual (volumetric) amount of liquid, V, along length Z of container inner wall, may be calculated via

$$V\{t\}\equiv\int_0^Z\int_0^Y X\{z,t\}dydz \tag{2.199}$$

– where independence of X on y as enforced by Eq. (2.189), and Z and Y being given constants allow Fubini's (weak) theorem be again invoked to get

$$V=\int_0^Z Xdz\int_0^Y dy; \tag{2.200}$$

insertion of Eq. (2.196) transforms Eq. (2.200) to

$$V=\sqrt{\frac{\mu}{\rho g t}}\int_0^Z \sqrt{z}\,dz\int_0^Y dy \tag{2.201}$$

– with advantage gained from $\mu/\rho g t$ being constant with regard to integration in z. Equation (2.201) readily becomes

$$V=\sqrt{\frac{\mu}{\rho g t}}\left.\frac{z^{\frac{3}{2}}}{\frac{3}{2}}\right|_0^Z y\Big|_0^Y \tag{2.202}$$

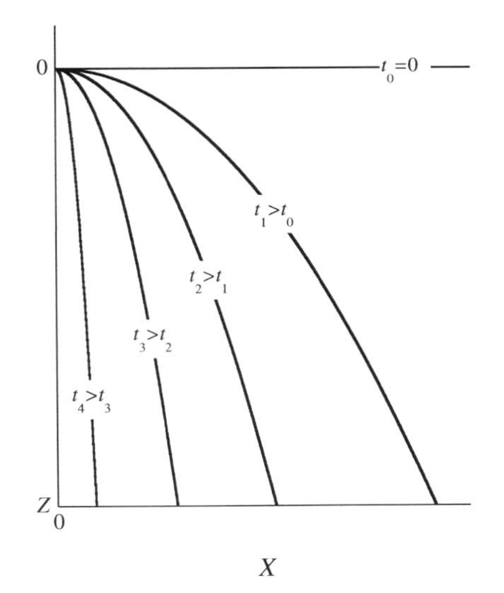

FIGURE 2.22 Variation of (i) normalized thickness, X/Z, of residual film of solution sticking onto the inner surface of the vessel wall, versus dimensionless lumped variable, $\mu z/\rho g Z^2 t$, and (ii) thickness, X, versus vertical position, z, for selected values of time, t_0, t_1, t_2, t_3, and t_4, throughout the emptying stage of rinsing.

as per the fundamental theorem of integral calculus, which gives rise to

$$\boxed{V = \frac{2}{3}\frac{Y\sqrt{\frac{\mu Z^3}{\rho g}}}{\sqrt{t}}};\tag{2.203}$$

for consistency with the rationale above, one should divide both sides of Eq. (2.203) by a reference volume, YZ^2, to get

$$\frac{V}{YZ^2} = \frac{2}{3}\sqrt{\frac{\mu}{\rho g}}\sqrt{\frac{1}{Zt}},\tag{2.204}$$

which may be further rearranged as

$$\boxed{\frac{V}{YZ^2} = \frac{2}{3}\sqrt{\frac{1}{\frac{\rho g Z}{\mu}t}}}.\tag{2.205}$$

A graphical representation of Eq. (2.205) is provided in Fig. 2.23; whereas Eq. (2.203) describes the curve labeled as emptying stage in Fig. 2.21i – with V arbitrarily set equal to V_0 at $t=0$ so as to meet with physical evidence, since Eq. (2.198) holds only when a film forms from the initial bulk liquid. A distorted hyperbolic evolution of V/YZ^2 with $\rho gZt/\mu$ can be grasped in Fig. 2.23, consistent with the functional form of Eq. (2.205) – corresponding to a thinner and thinner film lying on the inner wall of the apparatus. Since plain emptying is at stake, the concentration of solute in the residual film remains constant – in agreement with Fig. 2.21ii.

The volume of liquid still remaining at the end of the emptying stage, by time t_{emp}, will hereafter be denoted simply as V_r, i.e.

$$V_r \equiv V\Big|_{t=t_{emp}},\tag{2.206}$$

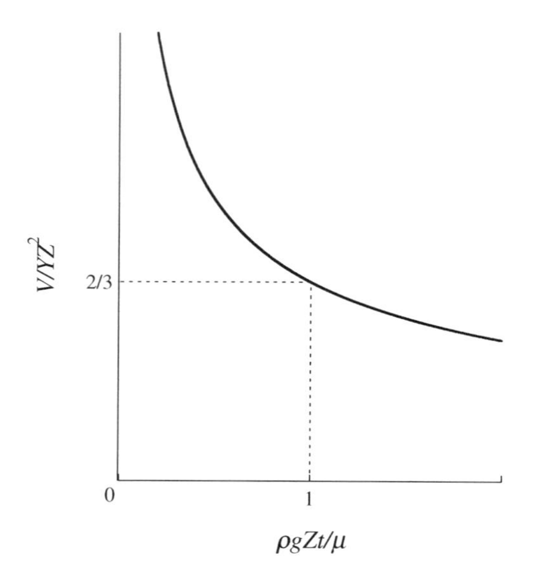

FIGURE 2.23 Evolution of dimensionless volume, V/YZ^2, of residual film of solution sticking onto the inner surface of the vessel wall, versus dimensionless time, $\rho gZt/\mu$, throughout the emptying stage of rinsing.

consistent with Fig. 2.21i; Eq. (2.203) will thus support

$$\boxed{V_r = \frac{2}{3}Y\sqrt{\frac{\mu Z^3}{\rho g t_{emp}}}},\tag{2.207}$$

upon combination with Eq. (2.206). In practice, the average film thickness, $\bar{X}$, is of higher relevance during emptying than its exact profile; it can readily be defined as

$$\bar{X} \equiv \frac{V_r}{YZ},\tag{2.208}$$

so one may resort to Eq. (2.203) to obtain

$$\boxed{\bar{X}\{t\} = \frac{2}{3}\frac{\sqrt{\frac{\mu Z}{\rho g}}}{\sqrt{t}}}\tag{2.209}$$

after dropping Y and taking Z into the square root – meaning that $\bar{X}$ varies with $t^{-1/2}$. One may likewise define $\bar{X}_r$ as

$$\bar{X}_r \equiv \bar{X}\Big|_{t=t_{emp}},\tag{2.210}$$

in agreement with Eq. (2.206) – where insertion of Eq. (2.209) leads to

$$\bar{X}_r = \frac{2}{3}\sqrt{\frac{\mu Z}{\rho g t_{emp}}};\tag{2.211}$$

this estimate of average film thickness will serve as departure point for simulation of the next stages of rinsing.

The following steps of (actual) rinsing resort normally to freshwater. During the first one, good mixing will be assumed of fresh (plain) water – supplied on a continuous basis at volumetric flow rate Q, and occupying (constant) volume V_f inside the equipment, with solute concentration $C_{S,f}\{t\}$; together with a residual layer, with (decreasing) volume $V\{t\}$, of original liquid containing solute at concentration $C_S\{t\}$. The underlying mass balance to solute over the whole system accordingly reads

$$\boxed{\frac{d}{dt}\left(VC_S + V_f C_{S,f}\right) + QC_{S,f} = 0},\tag{2.212}$$

due to absence of solute in the inlet stream – where $C_{S,f}$ also denotes (varying) concentration of solute in outlet rinsing stream, as per the perfect mixing conditions (implicitly) hypothesized to prevail inside the vessel. Equation (2.212) is subjected to

$$\boxed{C_S\Big|_{t=t_{emp}} = C_{S,0}}\tag{2.213}$$

– with $C_{S,0}$ fixed by the liquid food previously occupying the vessel; and which remained constant during the initial emptying period (as highlighted in Fig. 2.21ii). The rate of dissolution of the residual layer containing solute may be taken as proportional to the amount still left, i.e.

$$\boxed{\frac{dV}{dt} = -k_d V},\tag{2.214}$$

with k_d denoting a constant – consistent with Eq. (2.10), yet expected to increase with *Re*, as also postulated in Eq. (2.14); this model implies that it turns more and more difficult to carry the film away as its thickness diminishes, due to a lesser mobility thereof (as experimentally observed). The associated boundary condition reads

$$\boxed{V\big|_{t=t_{emp}} = V_r,} \tag{2.215}$$

where V_r was given by Eqs. (2.206) and (2.207). Integration of Eq. (2.214) can proceed by parts, viz.

$$\int_{V_r}^{V} \frac{d\tilde{V}}{\tilde{V}} = -k_d \int_{t_{emp}}^{t} d\tilde{t}, \tag{2.216}$$

with the aid of Eq. (2.215); the fundamental theorem of integral calculus may be invoked to write

$$\ln \tilde{V}\Big|_{V_r}^{V} = -k_d\, \tilde{t}\Big|_{t_{emp}}^{t}, \tag{2.217}$$

which eventually produces

$$\ln \frac{V}{V_r} = -k_d\left(t - t_{emp}\right) \tag{2.218}$$

after bringing upper and lowers limits on board – or else

$$\boxed{V\{t\} = V_r \mathrm{e}^{-k_d\left(t - t_{emp}\right)},} \tag{2.219}$$

once exponentials are taken of both sides, followed by multiplication thereof by V_r. A graphical account of Eq. (2.219) is provided in Fig. 2.21i – more specifically, its intermediate portion; an exponential decrease is apparent, as per the straight line in the logarithmic scale elected for the vertical axis.

In view of the rules of differentiation of a sum and of a product of functions, Eq. (2.212) may be converted to

$$C_S \frac{dV}{dt} + V \frac{dC_S}{dt} + V_f \frac{dC_{S,f}}{dt} + QC_{S,f} = 0, \tag{2.220}$$

where the constancy of V_f was already taken advantage of. On the other hand, the strong turbulence undergone by the rinsing water stream, coupled with the turbulence induced on the residual film lying on the equipment wall justify a very thin boundary layer in between; this means that the rates of mass transfer between such adjacent phases should be sufficiently high for thermodynamic equilibrium be readily attained. Under these circumstances, one may hypothesize that

$$\boxed{C_{S,f} = \kappa C_S,} \tag{2.221}$$

with κ denoting a partition coefficient – usually (but not necessarily) close to unity, since water serves as solvent in both phases. Upon insertion of Eq. (2.221), one gets

$$C_S \frac{dV}{dt} + V \frac{dC_S}{dt} + V_f \frac{d\left(\kappa C_S\right)}{dt} + Q\kappa C_S = 0 \tag{2.222}$$

from Eq. (2.220), which breaks down to

$$\kappa Q C_S + C_S \frac{dV}{dt} + V \frac{dC_S}{dt} + \kappa V_f \frac{dC_S}{dt} = 0; \tag{2.223}$$

insertion of Eq. (2.219) supports further transformation to

$$\kappa Q C_S + C_S V_r \mathrm{e}^{-k_d\left(t - t_{emp}\right)}\left(-k_d\right) + V_r \mathrm{e}^{-k_d\left(t - t_{emp}\right)} \frac{dC_S}{dt} + \kappa V_f \frac{dC_S}{dt} = 0, \tag{2.224}$$

where C_S or dC_S/dt (as appropriate) can be factored out as

$$\left(\kappa V_f + V_r \mathrm{e}^{-k_d\left(t - t_{emp}\right)}\right)\frac{dC_S}{dt} = -\left(\kappa Q - k_d V_r \mathrm{e}^{-k_d\left(t - t_{emp}\right)}\right)C_S. \tag{2.225}$$

Integration of Eq. (2.225) may be carried out via separation of variables, according to

$$\int_{C_{S,0}}^{C_S} \frac{d\tilde{C}_S}{\tilde{C}_S} = -\int_{t_{emp}}^{t} \frac{\kappa Q - k_d V_r \mathrm{e}^{-k_d\left(\tilde{t} - t_{emp}\right)}}{\kappa V_f + V_r \mathrm{e}^{-k_d\left(\tilde{t} - t_{emp}\right)}}\, d\tilde{t} \tag{2.226}$$

with the aid of Eq. (2.213). An auxiliary variable ξ will, to advantage, be defined as

$$\xi \equiv \mathrm{e}^{-k_d\left(t - t_{emp}\right)}, \tag{2.227}$$

thus implying

$$d\xi = -k_d \mathrm{e}^{-k_d\left(t - t_{emp}\right)} dt \tag{2.228}$$

upon differentiation of both sides – or else

$$dt = -\frac{d\xi}{k_d \mathrm{e}^{-k_d\left(t - t_{emp}\right)}} = -\frac{d\xi}{k_d \xi}, \tag{2.229}$$

after isolation of dt and combination back with Eq. (2.227);

$$\xi\big|_{t=t_{emp}} = \mathrm{e}^{-k_d\left(t_{emp} - t_{emp}\right)} = \mathrm{e}^0 = 1 \tag{2.230}$$

will accordingly stand as asymptotic case of Eq. (2.227). Insertion of Eqs. (2.227), (2.229), and (2.230) supports reformulation of Eq. (2.226) to

$$\int_{C_{S,0}}^{C_S} \frac{d\tilde{C}_S}{\tilde{C}_S} = \int_1^{\xi} \frac{\kappa Q - k_d V_r \tilde{\xi}}{\kappa V_f + V_r \tilde{\xi}} \frac{d\tilde{\xi}}{k_d \tilde{\xi}} = \int_1^{\xi} \frac{\kappa Q - k_d V_r \tilde{\xi}}{k_d \tilde{\xi}\left(\kappa V_f + V_r \tilde{\xi}\right)} d\tilde{\xi}, \tag{2.231}$$

where the right-hand side may be rewritten as

$$\int_{C_{S,0}}^{C_S} \frac{d\tilde{C}_S}{\tilde{C}_S} = \frac{1}{k_d}\int_1^{\xi}\left(\frac{k_1}{\tilde{\xi}} + \frac{k_2}{\kappa V_f + V_r \tilde{\xi}}\right) d\tilde{\xi} \tag{2.232}$$

at the expense of k_d being constant – and following Heaviside's expansion of the kernel in partial fractions, with k_1 and k_2 denoting constants to be determined. The fundamental theorem of integral calculus then permits transformation of Eq. (2.232) to

$$\ln \tilde{C}_S\Big|_{C_{S,0}}^{C_S} = \frac{1}{k_d}\left(k_1 \ln \tilde{\xi} + k_2 \frac{\ln\left\{\kappa V_f + V_r \tilde{\xi}\right\}}{V_r}\right)\Bigg|_1^{\xi} \tag{2.233}$$

which leaves

$$\ln \frac{C_S}{C_{S,0}} = \frac{1}{k_d}\left(k_1 \ln \frac{\xi}{1} + \frac{k_2}{V_r} \ln \frac{\kappa V_f + V_r \xi}{\kappa V_f + V_r 1}\right) \tag{2.234}$$

after replacing dummy integration variables by upper and lower limits of integration. Equation (2.234) further simplifies to

$$\ln\frac{C_S}{C_{S,0}}=\frac{k_1}{k_d}\ln\xi+\frac{k_2}{k_dV_r}\ln\frac{1+\dfrac{V_r}{\kappa V_f}\xi}{1+\dfrac{V_r}{\kappa V_f}} \tag{2.235}$$

upon elimination of parenthesis, complemented by division of both numerator and denominator of the argument of the last logarithmic function by κV_f – or else

$$\ln\frac{C_S}{C_{S,0}}=\ln\xi^{\frac{k_1}{k_d}}+\ln\left(\frac{1+\dfrac{V_r}{\kappa V_f}\xi}{1+\dfrac{V_r}{\kappa V_f}}\right)^{\frac{k_2}{k_dV_r}} \tag{2.236}$$

as per the operational features of logarithmic functions; if exponentials of both sides are considered, then Eq. (2.236) gains the form

$$\frac{C_S}{C_{S,0}}=\xi^{\frac{k_1}{k_d}}\left(\frac{1+\dfrac{V_r}{\kappa V_f}\xi}{1+\dfrac{V_r}{\kappa V_f}}\right)^{\frac{k_2}{k_dV_r}}. \tag{2.237}$$

Equation (2.227) permits recovery of the original notation in Eq. (2.237) as

$$\frac{C_S}{C_{S,0}}=\left(e^{-k_d\left(t-t_{emp}\right)}\right)^{\frac{k_1}{k_d}}\left(\frac{1+\dfrac{V_r}{\kappa V_f}e^{-k_d\left(t-t_{emp}\right)}}{1+\dfrac{V_r}{\kappa V_f}}\right)^{\frac{k_2}{k_dV_r}}; \tag{2.238}$$

lumping of powers yields, in turn,

$$\frac{C_S}{C_{S,0}}=e^{-\frac{k_1}{k_d}\left(k_dt-k_dt_{emp}\right)}\left(\frac{1+\dfrac{V_r}{\kappa V_f}e^{-\left(k_dt-k_dt_{emp}\right)}}{1+\dfrac{V_r}{\kappa V_f}}\right)^{\frac{k_2}{k_dV_r}}. \tag{2.239}$$

One should now formally retrieve Heaviside's expansion in attempts as to compute k_1 and k_2, i.e.

$$\frac{\kappa Q-k_dV_r\xi}{\xi\left(\kappa V_f+V_r\xi\right)}=\frac{k_1}{\xi}+\frac{k_2}{\kappa V_f+V_r\xi}, \tag{2.240}$$

which has supported conversion between Eqs. (2.231) and (2.232) – to eventually get

$$\kappa Q-k_dV_r\xi=k_1\left(\kappa V_f+V_r\xi\right)+k_2\xi \tag{2.241}$$

upon elimination of denominators; after setting $\xi=0$, Eq. (2.241) becomes

$$\kappa Q=k_1\kappa V_f, \tag{2.242}$$

which breaks down to

$$\frac{k_1}{k_d}=\frac{Q}{k_dV_f} \tag{2.243}$$

following cancellation of κ between sides, and division thereof by k_dV_f. By the same token, one may set $\xi=-\kappa V_f/V_r$ – with Eq. (2.241) reducing, in this case, to

$$\kappa Q-k_dV_r\left(-\frac{\kappa V_f}{V_r}\right)=k_2\left(-\frac{\kappa V_f}{V_r}\right), \tag{2.244}$$

or else

$$\kappa Q+k_d\kappa V_f=-k_2\frac{\kappa V_f}{V_r}; \tag{2.245}$$

isolation of κ_2 then unfolds

$$k_2=-\frac{\kappa Q+k_d\kappa V_f}{\dfrac{\kappa V_f}{V_r}}, \tag{2.246}$$

where straightforward algebraic rearrangement gives rise to

$$k_2=-V_r\left(k_d+\frac{Q}{V_f}\right) \tag{2.247}$$

– or, equivalently,

$$\frac{k_2}{k_dV_r}=-\left(1+\frac{Q}{k_dV_f}\right) \tag{2.248}$$

after dividing both sides by k_dV_r. In view of Eqs. (2.243) and (2.248), one can reformulate Eq. (2.239) as

$$\boxed{\frac{C_S}{C_{S,0}}=e^{-\frac{Q}{k_dV_f}\left(k_dt-k_dt_{emp}\right)}\left(\frac{1+\dfrac{V_r}{\kappa V_f}e^{-\left(k_dt-k_dt_{emp}\right)}}{1+\dfrac{V_r}{\kappa V_f}}\right)^{-\left(1+\frac{Q}{k_dV_f}\right)}} \tag{2.249}$$

– as depicted in Figs. 2.24ii and 2.24iii, using $C_S/C_{S,0}$ versus $k_dt-k_dt_{emp}$, for selected values of (dimensionless) parameters Q/k_dV_f and $V_r/\kappa V_f$. An exponential-type decrease in solute concentration takes place as time elapses, due to the (dominant) first-order nature of the mixing/dilution process. Such a decay occurs faster when (dimensionless) parameter Q/k_dV_f is larger – because the space time of bulk washing water, i.e. V_f/Q, becomes smaller compared to the time taken by the film to dissolve, measured by $1/k_d$; although this means that convection becomes the rate-controlling step, the larger Q is able to carry more solute away from the film per unit time – while a smaller k_d also distorts the curves horizontally leftward. The second (dimensionless) parameter, $V_r/\kappa V_f$, affects both spread and sigmoidicity of the curves; wider spread and more sigmoidal curves are indeed observed when $V_r/\kappa V_f$ is higher. A V_r larger than V_f makes it more difficult for the solute be removed, in view of its larger initial inventory compared to the bulk solvent currently available – thus pushing the curves upward, as grasped when going from Fig. 2.24i to ii. Remember that κ is expected to be close to unity, by virtue of the similar aqueous solvent in both film and rinsing water; while larger volumes of rinsing solution, V_f, being normally utilized than of film on the vessel wall, V_r, account for the represented values of $V_r/\kappa V_f$ lying below unity.

To avoid overparameterization, one may redo Eq. (2.219) as

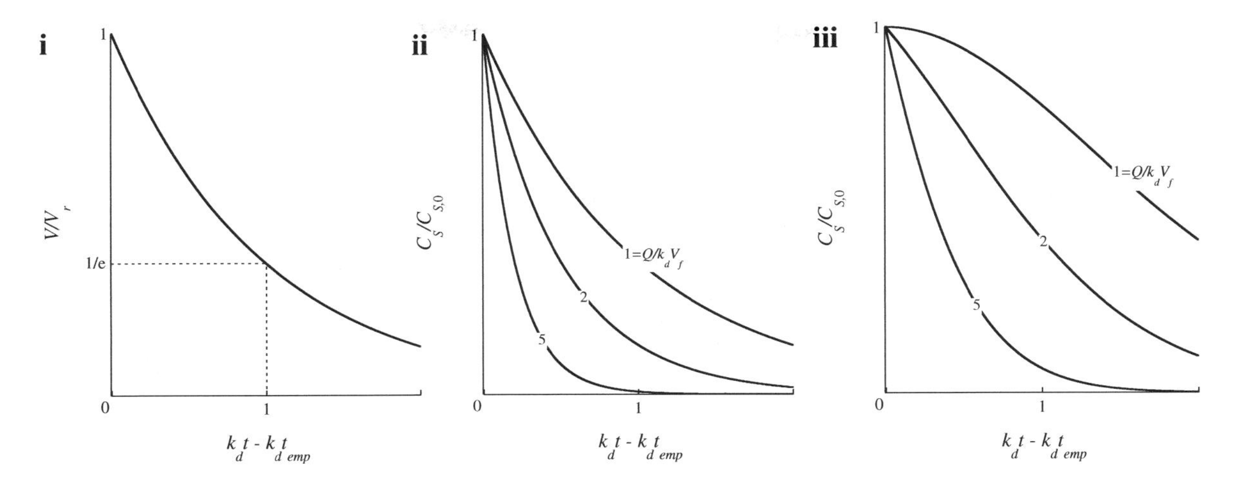

FIGURE 2.24 Evolution of (i) dimensionless volume, V/V_r, and (ii, iii) dimensionless concentration, $C_S/C_{S,0}$, of residual film of solution sticking onto the inner surface of the vessel wall, versus dimensionless time, $k_dt - k_dt_{emp}$, for selected values of parameters Q/k_dV_f, and (ii) $V_r/\kappa V_f = 0.01$ and (iii) $V_r/\kappa V_f = 1$, throughout the mixing stage of rinsing.

$$\boxed{\frac{V}{V_r} = \mathrm{e}^{-\left(k_dt - k_dt_{emp}\right)}} \tag{2.250}$$

– with (normalized) volume of film, V/V_r, evolving with (dimensionless) rinsing time, $k_dt - k_dt_{emp}$, as depicted in Fig. 2.24i; note again the exponential decay – a direct consequence of the first-order nature of the dissolution process, as already apparent in Eq. (2.219). After redoing Eq. (2.249) to

$$\boxed{C_S = C_{S,0}\mathrm{e}^{-\frac{Q}{V_f}\left(t - t_{emp}\right)}\left(\frac{1 + \dfrac{V_r}{\kappa V_f}\mathrm{e}^{-k_d\left(t - t_{emp}\right)}}{1 + \dfrac{V_r}{\kappa V_f}}\right)^{-\left(1 + \frac{Q}{k_dV_f}\right)}} \tag{2.251}$$

via multiplication of both sides by $C_{S,0}$ and factoring k_d out, one obtains the graphical evolution depicted in the middle portion of Fig. 2.21ii – corresponding to the mixing period; a linear decrease is apparent due to the overriding role of the first (exponential) factor in Eq. (2.251), relative to the second (power) factor – since the effect of the exponential function below unity in the latter is damped by addition thereof to unity.

Although this intermediate stage of rinsing may progress for a long time, the remaining volume will eventually degenerate to a thin layer strongly adherent onto the equipment surface – with thickness X_m so small that it can hardly be reduced anymore; hence, it will play the role of (an essentially stationary) boundary layer. The volume of residual liquid by the end of the mixing stage, V_m, viz.

$$V_m \equiv V\Big|_{t = t_{mix}}, \tag{2.252}$$

is obtainable from Eq. (2.219) after setting $t = t_{mix}$; in other words,

$$\boxed{V_m = V_r\mathrm{e}^{-k_d\left(t_{mix} - t_{emp}\right)}.} \tag{2.253}$$

The time t_{mix} when the said thickness – defined as

$$X_m \equiv \frac{V_m}{YZ}, \tag{2.254}$$

is reached may thus be ascertained upon combination of Eqs. (2.253) and (2.254), viz.

$$\boxed{X_m = \frac{V_r}{YZ}\mathrm{e}^{-k_d\left(t_{mix} - t_{emp}\right)}.} \tag{2.255}$$

By the same token, one may calculate the concentration of solute(s), $C_{S,m}$, by the end of the mixing stage, i.e.

$$C_{S,m} \equiv C_S\Big|_{t = t_{mix}}, \tag{2.256}$$

by resorting to Eq. (2.251) as

$$\boxed{C_{S,m} = C_{S,0}\mathrm{e}^{-\frac{Q}{V_f}\left(t_{mix} - t_{emp}\right)}\left(\frac{1 + \dfrac{V_r}{\kappa V_f}\mathrm{e}^{-k_d\left(t_{mix} - t_{emp}\right)}}{1 + \dfrac{V_r}{\kappa V_f}}\right)^{-\left(1 + \frac{Q}{k_dV_f}\right)},} \tag{2.257}$$

where $t = t_{mix}$ was taken again.

The final stage of rinsing entails plain diffusion of the remaining solute through the aforementioned boundary layer of constant volume, toward the fresh stream of water flowing over its surface – which is expected to undergo a negligible increase in concentration, as per the residual amount of solute in the film itself. Under these conditions, a mass balance to solute within a slice of cross-sectional area YZ and differential thickness dx will read

$$\boxed{-YZ\mathcal{D}_S\left.\frac{\partial C_S}{\partial x}\right|_x = -YZ\mathcal{D}_S\left.\frac{\partial C_S}{\partial x}\right|_{x+dx} + YZdx\frac{\partial C_S}{\partial t}} \tag{2.258}$$

– subjected to

$$\boxed{C_S\Big|_{x = X} = 0} \tag{2.259}$$

and

$$\boxed{\left.\frac{\partial C_S}{\partial x}\right|_{x=0} = 0} \tag{2.260}$$

as boundary conditions, complemented by

$$\boxed{C_S\big|_{t=t_{mix}} = C_{S,m}} \tag{2.261}$$

to serve as initial condition. The left-hand side and the first term in the right-hand side of Eq. (2.258) represent inlet and outlet, respectively, flow rate of solute(s) as per Fick's law of diffusion – characterized by diffusivity $\mathcal{D}_S$; whereas the last term represents accumulation of solute(s) in the elementary control volume. A uniform concentration $C_{S,m}$ is hypothesized at startup – see Eq. (2.261), and previous postulates applying to the mixing stage; whereas the surface concentration of solute (i.e. at $x=X$) is kept nil due to sufficient circulation of freshwater thereover, see Eq. (2.259) – while no mass transfer is allowed on the equipment surface proper, thus justifying the nil concentration gradient in Eq. (2.260) at $x=0$. Upon factoring out $YZ\mathcal{D}_S$ and dividing both sides by dx afterward, Eq. (2.258) becomes

$$YZ\mathcal{D}_S \frac{\left.\frac{\partial C_S}{\partial x}\right|_{x+dx} - \left.\frac{\partial C_S}{\partial x}\right|_{x}}{dx} = YZ\frac{\partial C_S}{\partial t} \tag{2.262}$$

– where the definition of second-order derivative permits notation be simplified to

$$\mathcal{D}_S \frac{\partial^2 C_S}{\partial x^2} = \frac{\partial C_S}{\partial t}, \tag{2.263}$$

along with dropping of YZ from both sides; multiplication and division of the left-hand side by X^2, and division of both sides by $C_{S,m}$ then transform Eq. (2.263) to

$$\frac{\mathcal{D}_S}{X^2} \frac{\partial^2\left(\frac{C_S}{C_{S,m}}\right)}{\partial\left(\frac{x}{X}\right)^2} = \frac{\partial\left(\frac{C_S}{C_{S,m}}\right)}{\partial\left(t-t_{mix}\right)} \tag{2.264}$$

– together with realization that $d(t-t_{mix})=dt$. Definition of auxiliary variables as

$$C_S^* \equiv \frac{C_S}{C_{S,m}} \tag{2.265}$$

pertaining to normalized concentration of solute, as well as

$$x^* \equiv \frac{X-x}{X} \tag{2.266}$$

encompassing normalized distance to layer surface and

$$t^* \equiv \frac{\mathcal{D}_S\left(t-t_{mix}\right)}{X^2} \tag{2.267}$$

referring to dimensionless time elapsed in current stage of rinsing, permit reformulation of Eq. (2.264) to

$$\boxed{\frac{\partial^2 C_S^*}{\partial x^{*2}} = \frac{\partial C_S^*}{\partial t^*}} \tag{2.268}$$

– because $d(X-x)=-dx$, and thus $d(X-x)^2=(d(X-x))^2=d(X-x)d(X-x)=-dx(-dx)=(dx)^2=dx^2$, complemented by division of both sides by (constant) $\mathcal{D}_S/X^2$. In view of Eqs. (2.265)–(2.267), one can redo Eq. (2.261) as

$$\boxed{C_S^*\big|_{t^*=0} = 1;} \tag{2.269}$$

and likewise Eqs. (2.259) and (2.260) to

$$\boxed{C_S^*\big|_{x^*=0} = 0} \tag{2.270}$$

and

$$\boxed{\left.\frac{\partial C_S^*}{\partial x^*}\right|_{x^*=1} = 0,} \tag{2.271}$$

respectively. Inspection of Eqs. (2.268)–(2.271), *vis-à-vis* with Eqs. (2.551)–(2.553) and (2.605) of *Food Proc. Eng.: thermal & chemical operations,* unfolds exact coincidence of both differential balance and limiting conditions – as long as T^* is replaced by C_S^*; hence, one may jump immediately to

$$\boxed{C_S^*\left\{x^*,t^*\right\} = \frac{4}{\pi}\sum_{i=0}^{\infty} \frac{\sin\frac{(2i+1)\pi x^*}{2}}{2i+1} e^{-\frac{(2i+1)^2\pi^2 t^*}{4}},} \tag{2.272}$$

as analog to Eq. (2.627) in *Food Proc. Eng.: thermal & chemical operations.* The evolution of C_S^* versus $1-x^*$ and t^*, as per Eq. (2.272), after bringing Eqs. (2.265)–(2.267) back on board, is plotted in Fig. 2.25i. The solute concentration profiles are consistent with a nil concentration on the surface of the film (i.e. $x^*=0$, or equivalently $1-x^*=1$) as per Eq. (2.270); while the horizontal tangent thereto at $x^*=1$ (or else $1-x^*=0$) agrees with no loss of solute through the equipment surface – see Eq. (2.271). As time elapses, less and less solute remains in the stagnant film – so the rate of loss thereof, measured by the slope of the curves at $1-x^*=1$ in agreement with Fick's law, becomes smaller and smaller.

Once again, the average concentration of solute(s), $\overline{C_S^*}$, in the aforementioned boundary layer – rather than its exact profile, is of higher practical interest; one may accordingly integrate C_S^* between $x^*=0$ and $x^*=1$, viz.

$$\overline{C_S^*} \equiv \frac{\int_0^1 C_S^*\,dx^*}{\int_0^1 dx^*} = \int_0^1 C_S^*\,dx^* \tag{2.273}$$

– where advantage was obtained from the parallelepipedal symmetry of the film, and from x^* being a normalized variable, see Eq. (2.266). Insertion of Eq. (2.272) transforms Eq. (2.273) to

$$\overline{C_S^*} = \frac{4}{\pi}\sum_{i=0}^{\infty} \frac{e^{-\frac{(2i+1)^2\pi^2 t^*}{4}}}{2i+1} \int_0^1 \sin\frac{(2i+1)\pi x^*}{2}\,dx^*, \tag{2.274}$$

with the aid of the intrinsic linearity of the integral operator. Direct application of the fundamental theorem of integral calculus transforms Eq. (2.274) to

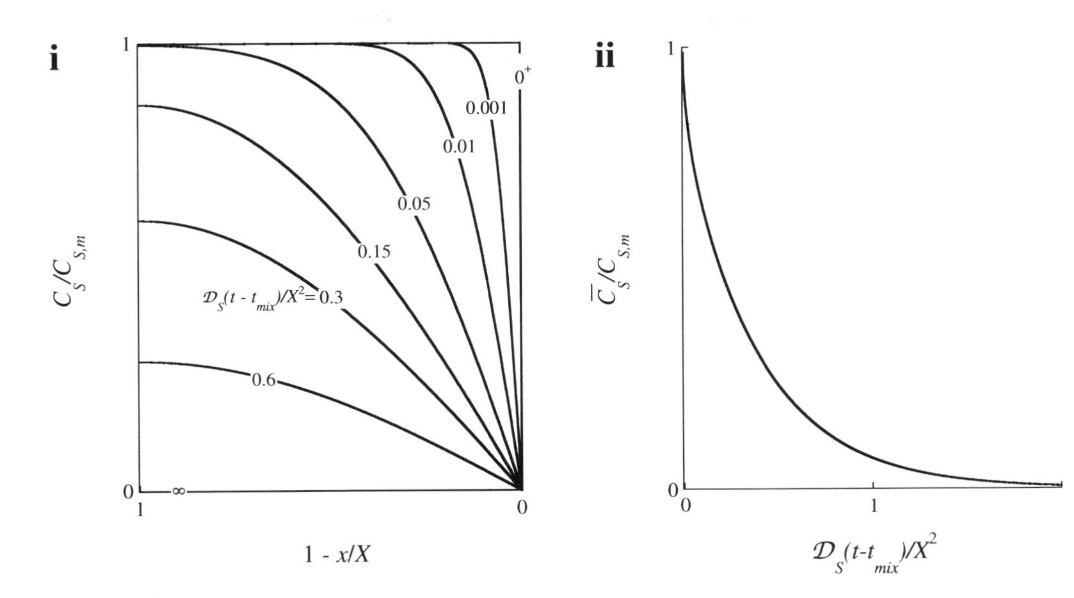

FIGURE 2.25 Variation of (i) normalized concentration of solute, $C_S/C_{S,m}$, in residual film of solution sticking onto the inner surface of the vessel wall, versus normalized location, $1-x/X$, for selected dimensionless times, $\mathcal{D}_S(t-t_{mix})/X^2$; and evolution of (ii) normalized average concentration, $\bar{C}_S / C_{S,m}$, versus $\mathcal{D}_S(t-t_{mix})/X^2$, throughout the diffusion stage of rinsing.

$$\overline{C_S^*} = \frac{4}{\pi}\sum_{i=0}^{\infty} \frac{e^{-\frac{(2i+1)^2\pi^2 t^*}{4}}}{2i+1} \left. \frac{-\cos\frac{(2i+1)\pi x^*}{2}}{\frac{(2i+1)\pi}{2}} \right|_0^1, \tag{2.275}$$

which degenerates to

$$\overline{C_S^*} = \frac{4}{\pi}\sum_{i=0}^{\infty} \frac{e^{-\frac{(2i+1)^2\pi^2 t^*}{4}}}{(2i+1)\frac{(2i+1)\pi}{2}} \left. \cos\frac{(2i+1)\pi x^*}{2} \right|_1^0 \tag{2.276}$$

after swapping integration limits, on account of taking the negative of the right-hand side; Eq. (2.276) is equivalent to

$$\overline{C_S^*} = \frac{4}{\pi}\sum_{i=0}^{\infty} \frac{\frac{2}{\pi} e^{-\frac{(2i+1)^2\pi^2 t^*}{4}}}{(2i+1)^2} \left(\cos 0 - \cos\frac{(2i+1)\pi}{2} \right), \tag{2.277}$$

or merely

$$\overline{C_S^*} = \frac{8}{\pi^2}\sum_{i=0}^{\infty} \frac{e^{-\frac{(2i+1)^2\pi^2 t^*}{4}}}{(2i+1)^2}, \tag{2.278}$$

– since cos 0 is unity, while the cosine of an odd integer-multiple of $\pi/2$ rad is nil. Finally, one may retrieve the original notation by inserting Eqs. (2.265) and (2.267) in Eq. (2.278), viz.

$$\frac{\bar{C}_S}{C_{S,m}} = \frac{8}{\pi^2}\sum_{i=0}^{\infty} \frac{\exp\left\{-\frac{(2i+1)^2\pi^2\mathcal{D}_S\left(t-t_{mix}\right)}{4X^2}\right\}}{(2i+1)^2}; \tag{2.279}$$

the third portion of Fig. 2.21ii is described by Eq. (2.279), phenomenologically corresponding to diffusion of solute. The evolution of $\bar{C}_S/C_{S,m}$ in time is plotted in Fig. 2.25ii – and reflects the (anticipated) lower and lower concentration of residual solute in the film as time elapses. After an initial transient, an essentially straight line results – which, due to the logarithmic scale used for the vertical axis, unfolds an exponential decrease, compatible with the overriding role of the first term of the summation in Eq. (2.279) – for carrying a decaying exponential function with the smallest absolute argument; the slope of such a line, however, is typically smaller in magnitude than that observed during the mixing phase – since convection is more efficient than plain diffusion toward solute removal. As emphasized previously, the volume of film remains constant during the diffusion period – thus justifying the horizontal line in Fig. 2.21i with V_m for vertical intercept, in agreement with Eq. (2.252).

One should again emphasize that the above discussion applies to a vessel initially containing a solution of some solute in solvent (often water) – so it can be used in the case of a liquid food containing dissolved substrate(s), or liquid food itself undergoing removal from said vessel via rinsing water; as well as in the case of a cleaning solution containing some detergent as solute, or else a disinfecting solution containing some antiseptic agent as solute.

PROBLEM 2.4

Both volume and composition of the solution containing detergent/disinfectant (or residual food or dirt), throughout the various stages of rinsing after emptying are of concern, as long as a fully clean equipment is the goal – i.e. devoid not only of the original produt, and of chemical and microbial contaminants, but also of detergent/disinfectant itself.

a) Develop an expression for the volume of solution remaining on the inner wall of a cylindrical container – with inner radius R_c and height L, using an exit pipe of radius $R_e << R_c$, after regular emptying.
b) The typical finishing used on equipment inner surfaces, coupled with some chemicals classically employed for cleaning/disinfection often lead to formation of puddles of residual (or spilled) liquid

on horizontal (or very shallow) surfaces – as illustrated below for spontaneously wettable surfaces.

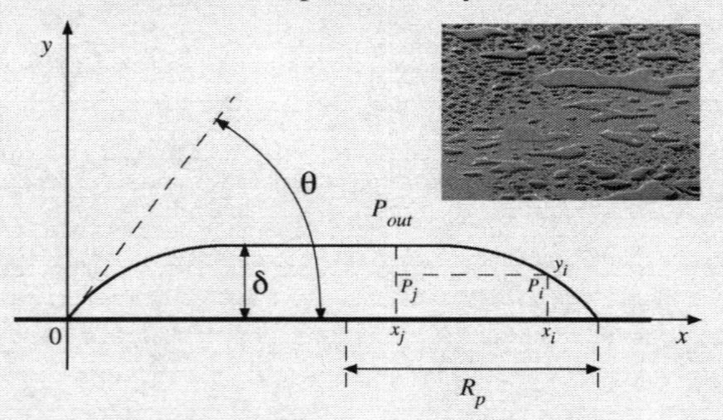

For a contact angle θ at the edge of a sufficiently large, circular puddle described by radius R_p, a maximum thickness, δ, is attained not far from the edge; estimate the said thickness, as a function of θ, acceleration of gravity, g, and physicochemical properties of liquid, i.e. surface tension, σ, and mass density, ρ.

Solution

a) After revisiting Eq. (2.261) of *Food Proc. Eng.: basics & mechanical operations* as

$$t_{emp} = \frac{\sqrt{\dfrac{L}{g}}}{\dfrac{A_e}{A_c}\sqrt{\dfrac{10}{29-9\dfrac{A_e}{A_c}}}} \approx \frac{\sqrt{\dfrac{29L}{10g}}}{\dfrac{A_e}{A_c}} \tag{2.4.1}$$

– since $R_e << R_c$ by hypothesis implies $A_e = \pi R_e^2 <<< \pi R_c^2$, and thus $9A_e/A_c <<< 29$; one may thus further proceed to

$$t_{emp} = \frac{\sqrt{\dfrac{29L}{10g}}}{\dfrac{\pi R_e^2}{\pi R_c^2}} = \sqrt{\frac{29}{10}}\left(\frac{R_c}{R_e}\right)^2\sqrt{\frac{L}{g}}, \tag{2.4.2}$$

after recalling the expression for circular cross-sectional area, and algebraically rearranging afterward. Insertion of Eq. (2.4.2) then transforms Eqs. (2.203) and (2.206) to

$$V_r\Big|_{t_{emp}} = \frac{2}{3}2\pi R_c\sqrt{\frac{\mu}{\rho g}}\sqrt{\frac{L^3}{\sqrt{\dfrac{29}{10}}\left(\dfrac{R_c}{R_e}\right)^2\sqrt{\dfrac{L}{g}}}}, \tag{2.4.3}$$

along with the definition of container perimeter (i.e. $Y = 2\pi R_c$) and height (i.e. $Z = L$) – where algebraic simplification leaves

$$V_r\Big|_{t_{emp}} = \frac{4\pi}{3}\left(\frac{10}{29}\right)^{\frac{1}{4}}\frac{\mu^{\frac{1}{2}}}{\rho^{\frac{1}{2}}g^{\frac{1}{4}}}R_e L^{\frac{5}{4}} \tag{2.4.4}$$

as expression for the residual volume at emptying time; V_r accordingly appears as a function of geometrical parameters of the container (i.e. L, but not R_c) and emptying tube (i.e. R_e), gravitational field (i.e. g), and physicochemical properties of fluid (i.e. ρ and μ).

b) Consider first the curve $y \equiv y\{x\}$ plotted below.

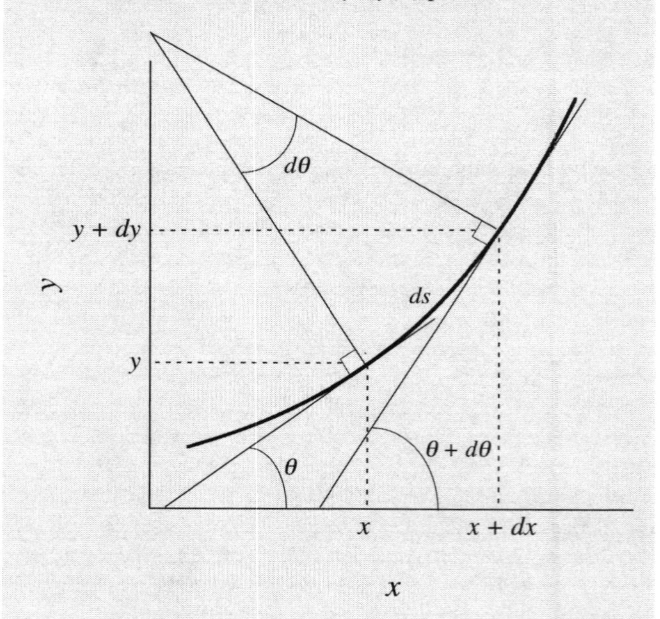

Curvature, κ, is defined as the variation in direction of angle θ, between two close tangents to said curve. Hence, the larger the increment in arc length ds required for a given variation $d\theta$, the smaller κ – or, in a more quantitative manner,

$$\kappa \equiv \left|\frac{d\theta}{ds}\right|, \tag{2.4.5}$$

where θ is defined as

$$\theta \equiv \tan^{-1}\frac{dy}{dx}; \tag{2.4.6}$$

hence, one may write

$$\frac{d\theta}{dx} = \frac{\dfrac{d}{dx}\left(\dfrac{dy}{dx}\right)}{1+\left(\dfrac{dy}{dx}\right)^2} = \frac{\dfrac{d^2y}{dx^2}}{1+\left(\dfrac{dy}{dx}\right)^2} \tag{2.4.7}$$

based on Eq. (2.4.6) – coupled with the rules of differentiation of a composite function and of the derivative of the inverse tangent itself, complemented by the definition of second-order derivative. The chain differentiation rule may again be invoked to obtain

$$\frac{d\theta}{ds} = \frac{d\theta}{dx}\frac{dx}{ds}, \tag{2.4.8}$$

which may instead appear as

$$\frac{d\theta}{ds} = \frac{\dfrac{d\theta}{dx}}{\dfrac{ds}{dx}} \tag{2.4.9}$$

owing to the rule of differentiation of the inverse function. On the other hand, the differential length of a curve abides to

$$ds = \sqrt{1 + \left(\frac{dy}{dx}\right)^2}\, dx, \tag{2.4.10}$$

so insertion of Eqs. (2.4.7) and (2.4.10) transforms Eq. (2.4.9) to

$$\frac{d\theta}{ds} = \frac{\dfrac{\dfrac{d^2y}{dx^2}}{1 + \left(\dfrac{dy}{dx}\right)^2}}{\sqrt{1 + \left(\dfrac{dy}{dx}\right)^2}}; \tag{2.4.11}$$

Eq. (2.4.11) degenerates to

$$\frac{d\theta}{ds} = \frac{\dfrac{d^2y}{dx^2}}{\left(1 + \left(\dfrac{dy}{dx}\right)^2\right)^{\frac{3}{2}}} \tag{2.4.12}$$

after lumping factors alike – and Eq. (2.4.5) ends up as

$$\kappa = \frac{\left|\dfrac{d^2y}{dx^2}\right|}{\left(1 + \left(\dfrac{dy}{dx}\right)^2\right)^{\frac{3}{2}}}. \tag{2.4.13}$$

A circle of radius R, centered at the origin of coordinates (0,0), is specifically described by

$$x^2 + y^2 = R^2; \tag{2.4.14}$$

differentiation of both sides unfolds

$$2x dx + 2y dy = 0 \tag{2.4.15}$$

– where division of both sides by $2dx$, followed by isolation of dy/dx yield

$$\frac{dy}{dx} = -\frac{x}{y}. \tag{2.4.16}$$

A further differentiation of both sides, with regard to x, transforms Eq. (2.4.16) to

$$\frac{d}{dx}\left(\frac{dy}{dx}\right) = \frac{d}{dx}\left(-\frac{x}{y}\right) = -\frac{y - x\dfrac{dy}{dx}}{y^2}, \tag{2.4.17}$$

or else

$$\frac{d^2y}{dx^2} = \frac{x\dfrac{dy}{dx} - y}{y^2}; \tag{2.4.18}$$

combination with Eq. (2.4.16) then permits simplification to

$$\frac{d^2y}{dx^2} = \frac{x\left(-\dfrac{x}{y}\right) - y}{y^2} = -\frac{\dfrac{x^2}{y} + y}{y^2} = -\frac{x^2 + y^2}{y^3}, \tag{2.4.19}$$

via elementary algebraic rearrangement. Insertion of Eqs. (2.4.16) and (2.4.19) converts Eq. (2.4.13) to

$$\kappa = \frac{\left|-\dfrac{x^2 + y^2}{y^3}\right|}{\left(1 + \left(-\dfrac{x}{y}\right)^2\right)^{\frac{3}{2}}}, \tag{2.4.20}$$

which may be manipulated to read

$$\kappa = \frac{\left(1 + \dfrac{x^2}{y^2}\right)\left|-\dfrac{1}{y}\right|}{\left(1 + \dfrac{x^2}{y^2}\right)^{\frac{3}{2}}} = \frac{\left|\dfrac{1}{y}\right|}{\sqrt{1 + \dfrac{x^2}{y^2}}}; \tag{2.4.21}$$

a final multiplication of both numerator and denominator by $|y|$ produces

$$\kappa = \frac{1}{|y|\sqrt{1 + \dfrac{x^2}{y^2}}} = \frac{1}{\sqrt{x^2 + y^2}}, \tag{2.4.22}$$

or else

$$\kappa = \frac{1}{\sqrt{R^2}} = \frac{1}{|R|} = \frac{1}{R} \tag{2.4.23}$$

in view of Eq. (2.4.14) and $R>0$ as per definition of radius. The fact that curvature coincides with the reciprocal of the radius of a circumference has prompted generalization of the concept of curvature to any (plane) curve – so the radius of curvature, R_c, is, in general, defined as

$$R_c \equiv \frac{1}{\kappa}, \tag{2.4.24}$$

inspired by Eq. (2.4.23). Consider now the relatively large, and essentially circular puddle referred to above, with contour curve $y \equiv y\{x\}$; its two radii of curvature, R_c (along the cross-section at the edge) and R_p (along the vertical projection), satisfy

$$R_p >> R_c = \frac{1}{\kappa} = \frac{\left(1 + \left(\dfrac{dy}{dx}\right)^2\right)^{\frac{3}{2}}}{\left|\dfrac{d^2y}{dx^2}\right|} \approx \frac{1}{\left|\dfrac{d^2y}{dx^2}\right|} \tag{2.4.25}$$

as per hypothesis, coupled with Eqs. (2.4.13) and (2.4.24) – besides realization that $dy/dx < 1$ when θ is small, as happens in the case of a wetting liquid spilled on a surface, and thus $(dy/dx)^2 << 1$. Young and Laplace's equation, as per Eq. (4.47) in *Food Proc. Eng.: basics & mechanical operations*, has it that

$$P_i - P_{out} = \sigma\left(\frac{1}{R_P} + \frac{1}{R_c}\right), \tag{2.4.26}$$

where P_i denotes pressure of liquid on the inner side of the surface at point of coordinates (x_i, y_i) – when pressure on the outer side (uniform over the whole surface) equals $P_{out} < P_i$; in view of Eq. (2.4.25), one promptly obtains

$$P_i - P_{out} \approx \frac{\sigma}{R_c} = \sigma\left|\frac{d^2y}{dx^2}\right|, \tag{2.4.27}$$

because $1/R_p << 1/R_c$ when $R_p >> R_c$. Plain inspection of the shape of the curve describing the cross-sectional profile near the edge of the puddle indicates that

$$\frac{d^2y}{dx^2} < 0; \tag{2.4.28}$$

hence, Eq. (2.4.27) may be rewritten as

$$\sigma\left.\frac{d^2y}{dx^2}\right|_{y_i} = P_{out} - P_i, \tag{2.4.29}$$

based on the definition of absolute value of a function. On the other hand, essentially no curvature exists for the surface at abscissa x_j well appart from x_i on the edge – so pressure P_j, at point of coordinates (x_j, y_i), may be related to P_{out} via

$$P_j = P_{out} + \rho g(\delta - y_i), \tag{2.4.30}$$

stemming from the fundamental law of hydrostatics; since

$$P_i = P_j \tag{2.4.31}$$

again based on the same law of hydrostatics – owing to the identical ordinate within a given (homogeneous liquid) phase placed horizontally, one may eliminate P_i from Eq. (2.4.29) as

$$\sigma\left.\frac{d^2y}{dx^2}\right|_{y_i} = P_{out} - \left(P_{out} + \rho g(\delta - y_i)\right), \tag{2.4.32}$$

with the aid of Eqs. (2.4.30) and (2.4.31); simplification is then in order, viz.

$$\sigma\left.\frac{d^2y}{dx^2}\right|_{y_i} = \rho g y_i - \rho g \delta, \tag{2.4.33}$$

following removal of parentheses and cancellation of symmetrical terms. Division of both sides by σ, and replacement of (dummy) variable y_i by generic variable y transform Eq. (2.4.33) to

$$\frac{d^2y}{dx^2} - \frac{\rho g}{\sigma}y = -\frac{\rho g \delta}{\sigma} \tag{2.4.34}$$

– as underlying differential equation describing the shape of the curve that serves as contour for the cross-section of the (circular) puddle; suitable boundary conditions read

$$y\big|_{x=0} = 0 \tag{2.4.35}$$

and

$$\lim_{x\to\infty} y = \delta, \tag{2.4.36}$$

with ∞ meaning sufficiently away from the edge(s). In view of being nonhomogeneous, one should first solve the homogeneous counterpart of Eq. (2.4.34), i.e.

$$\frac{d^2y}{dx^2} - \frac{\rho g}{\sigma}y = 0; \tag{2.4.37}$$

the characteristic (polynomial) equation associated therewith takes the form

$$m^2 - \frac{\rho g}{\sigma} = 0 \tag{2.4.38}$$

– so two eigenvalues, m_1 and m_2, result as solutions, viz.

$$m_1, m_2 = \pm\sqrt{\frac{\rho g}{\sigma}}. \tag{2.4.39}$$

Therefore, the general solution of Eq. (2.4.37) reads

$$y = b_1 \exp\left\{-\sqrt{\frac{\rho g}{\sigma}}x\right\} + b_2 \exp\left\{\sqrt{\frac{\rho g}{\sigma}}x\right\}, \tag{2.4.40}$$

where constants b_1 and b_2 accompany the first and second terms, respectively – which thus serve as eigenvectors. Because of the constant nature of the right-hand side of Eq. (2.4.34), its general solution may be coined as

$$y = b_1 \exp\left\{-\sqrt{\frac{\rho g}{\sigma}}x\right\} + b_2 \exp\left\{\sqrt{\frac{\rho g}{\sigma}}x\right\} + b_3, \tag{2.4.41}$$

using Eq. (2.4.40) for template; here (added) b_3 denotes a third constant to be determined as specific integral. In fact, the first two terms in the right-hand side of Eq. (2.4.40) satisfy Eq. (2.4.37) by hypothesis; further to

$$\frac{d^2b_3}{dx^2} = \frac{db_3}{dx} = 0, \tag{2.4.42}$$

due to the constancy of b_3 – so Eq. (2.4.34) reduces to

$$-\frac{\rho g}{\sigma}b_3 = -\frac{\rho g}{\sigma}\delta, \tag{2.4.43}$$

after replacing y by b_3. Equation (2.4.43) implies

$$b_3 = \delta, \tag{2.4.44}$$

once $-\rho g/\sigma$ is dropped off both sides; which converts, in turn, Eq. (2.4.41) to

$$y = \delta + b_1 \exp\left\{-\sqrt{\frac{\rho g}{\sigma}}x\right\} + b_2 \exp\left\{\sqrt{\frac{\rho g}{\sigma}}x\right\}. \quad (2.4.45)$$

Equation (2.4.45) is asymptotically driven by

$$\lim_{x\to\infty} y = \lim_{x\to\infty} b_2 \exp\left\{\sqrt{\frac{\rho g}{\sigma}}x\right\} = b_2 e^{\infty} = b_2\infty, \quad (2.4.46)$$

as the last term overrides any constant or decaying exponential function; hence, the finite nature of δ in Eq. (2.4.36) cannot be accommodated unless

$$b_2 = 0, \quad (2.4.47)$$

so Eq. (2.4.45) simplifies further to

$$y = \delta + b_1 \exp\left\{-\sqrt{\frac{\rho g}{\sigma}}x\right\}. \quad (2.4.48)$$

When $x=0$, Eq. (2.4.48) becomes

$$y\big|_{x=0} = \delta + b_1 e^0 = \delta + b_1; \quad (2.4.49)$$

elimination of $y\big|_{x=0}$ between Eqs. (2.4.35) and (2.4.49) gives rise to

$$0 = \delta + b_1, \quad (2.4.50)$$

which readily unfolds

$$b_1 = -\delta \quad (2.4.51)$$

– meaning that Eq. (2.4.48) will take the form

$$y = \delta - \delta \exp\left\{-\sqrt{\frac{\rho g}{\sigma}}x\right\}, \quad (2.4.52)$$

or simply

$$y = \delta\left(1 - \exp\left\{-\sqrt{\frac{\rho g}{\sigma}}x\right\}\right) \quad (2.4.53)$$

after having δ factored out. Although Eq. (2.4.53) describes the shape of the contour curve of the puddle, δ still remains to be calculated; one should accordingly realize that

$$\left.\frac{dy}{dx}\right|_{x=0} = \tan\theta \quad (2.4.54)$$

provides a supplementary boundary condition, which may be used to eliminate the outstanding degree of freedom. Differentiation of both sides of Eq. (2.4.53) produces, in turn,

$$\frac{dy}{dx} = -\delta \exp\left\{-\sqrt{\frac{\rho g}{\sigma}}x\right\}\left(-\sqrt{\frac{\rho g}{\sigma}}\right), \quad (2.4.55)$$

so one gets

$$\left.\frac{dy}{dx}\right|_{x=0} = \sqrt{\frac{\rho g}{\sigma}}\delta e^0 = \sqrt{\frac{\rho g}{\sigma}}\delta \quad (2.4.56)$$

after setting $x=0$; comparative inspection of Eqs. (2.4.54) and (2.4.56) enforces

$$\tan\theta = \sqrt{\frac{\rho g}{\sigma}}\delta, \quad (2.4.57)$$

which leads to

$$\delta = \sqrt{\frac{\sigma}{\rho g}}\tan\theta \quad (2.4.58)$$

as estimate for δ – dependent only on θ and (lumped parameter) $\rho g/\sigma$, as intended.

2.1.3.5 Soaking

Bottle washing has attracted a renewed interest in recent times, owing to the strong environmental impact of single-use containers. Reusable bottles are accordingly prerinsed, and then soaked in an aqueous solution of NaOH – usually formulated with wetting and antifoaming agents, as well as chelating agents to aid in rinsing afterward; typical temperatures of operation range within 70–80°C.

A mass balance to dirt in a (well-stirred) soaking solution may accordingly be setup as

$$\boxed{V_B \frac{d\rho_d}{dt} = \dot{N}_b m_{d,b} - \dot{N}_b V_b \rho_d} \quad (2.280)$$

– where V_B and V_b denote (constant) volumes of soaking bath and individual bottle, respectively, ρ_d denotes mass concentration of dirt in solution, $m_{d,b}$ denotes mass of dirt carried (on average) by each bottle, and $\dot{N}_b$ denotes number of bottles fed per unit time; V_B remains constant via (continuous) addition of volumetric flow rate $\dot{N}_b V_b$ of fresh solution, to account for the volume of washing solution transported by bottles leaving the equipment. Equation (2.280) assumes no mass transfer limitations, and thus a thermodynamically-controlled process – where all solid dirt is immediately solubilized/suspended in the soaking liquid; it may be rewritten as

$$V_B \frac{d}{dt}\left(\rho_d - \frac{m_{d,b}}{V_b}\right) = -\dot{N}_b V_b\left(\rho_d - \frac{m_{d,b}}{V_b}\right) \quad (2.281)$$

after factoring out $-\dot{N}_b V_b$ in the right-hand side, and taking advantage of the nil derivative of (constant) $-m_{d,b}/V_b$. A suitable initial condition reads

$$\boxed{\rho_d\big|_{t=0} = 0} \quad (2.282)$$

or, after subtracting $m_{d,b}/V_b$ from both sides,

$$\left.\left(\rho_d - \frac{m_{d,b}}{V_b}\right)\right|_{t=0} = -\frac{m_{d,b}}{V_b}. \quad (2.283)$$

Upon separation of variables and integration, Eq. (2.281) becomes

$$\int_{-\frac{m_{d,b}}{V_b}}^{\rho_d-\frac{m_{d,b}}{V_b}} \frac{d\left(\tilde{\rho}_d-\frac{m_{d,b}}{V_b}\right)}{\tilde{\rho}_d-\frac{m_{d,b}}{V_b}}=-\frac{\dot{N}_b V_b}{V_B}\int_0^t d\tilde{t}, \tag{2.284}$$

also with the aid of Eq. (2.283) – where $\dot{N}_b$ was assumed constant, as normally happens in practice; Eq. (2.284) degenerates to

$$\ln\left\{\tilde{\rho}_d-\frac{m_{d,b}}{V_b}\right\}\Bigg|_{-\frac{m_{d,b}}{V_b}}^{\rho_d-\frac{m_{d,b}}{V_b}}=-\frac{\dot{N}_b V_b}{V_B}\tilde{t}\,\Big|_0^t, \tag{2.285}$$

upon application of the fundamental theorem of integral calculus. When upper and lower limits of integration are taken into account, Eq. (2.285) yields

$$\ln\frac{\rho_d-\frac{m_{d,b}}{V_b}}{-\frac{m_{d,b}}{V_b}}=\ln\frac{\frac{m_{d,b}}{V_b}-\rho_d}{\frac{m_{d,b}}{V_b}}=-\frac{\dot{N}_b V_b}{V_B}(t-0)=-\frac{\dot{N}_b V_b}{V_B}t \tag{2.286}$$

– where negatives of both numerator and denominator of the argument of the logarithm were considered for convenience. Exponentials may now be taken of both sides of Eq. (2.286) to get

$$\frac{\frac{m_{d,b}}{V_b}-\rho_d}{\frac{m_{d,b}}{V_b}}=\exp\left\{-\frac{\dot{N}_b V_b}{V_B}t\right\} \tag{2.287}$$

or, equivalently,

$$1-\frac{V_b\rho_d}{m_{d,b}}=\exp\left\{-\frac{\dot{N}_b V_b}{V_B}t\right\} \tag{2.288}$$

upon multiplication of both numerator and denominator of the left-hand side by $V_b/m_{d,b}$; isolation of $V_b\rho_d/m_{d,b}$ finally yields

$$\boxed{\frac{V_b}{m_{d,b}}\rho_d=1-\exp\left\{-\frac{\dot{N}_b V_b}{V_B}t\right\}}, \tag{2.289}$$

as illustrated in Fig. 2.26. The rectangular-hyperbolic pattern observed is typical of a first-order process – so the mass concentration of dirt increases fast at early times, due chiefly to the input of dirt carried by already (used) bottles versus the output of dirt as a dilute solution; the said increase turns, however, smoother and smoother as time elapses, because the washing solution becomes progressively more concentrated – so the amount of dirt exiting the vessel in liquid form approaches that entering it in solid form. At startup (i.e. $t=0$), one realizes that $\exp\left\{-\dot{N}_b V_b t/V_B\right\}\to 1$, so $V_b\rho_d/m_{d,b}=0$ as per Eq. (2.289), or else $\rho_d=0$ (since $V_b/m_{d,b}\neq 0$) – in full agreement with Eq. (2.282); conversely, large times lead to

$$\boxed{\lim_{\frac{\dot{N}_b V_b}{V_B}t\to\infty}\frac{V_b}{m_{d,b}}\rho_d=1} \tag{2.290}$$

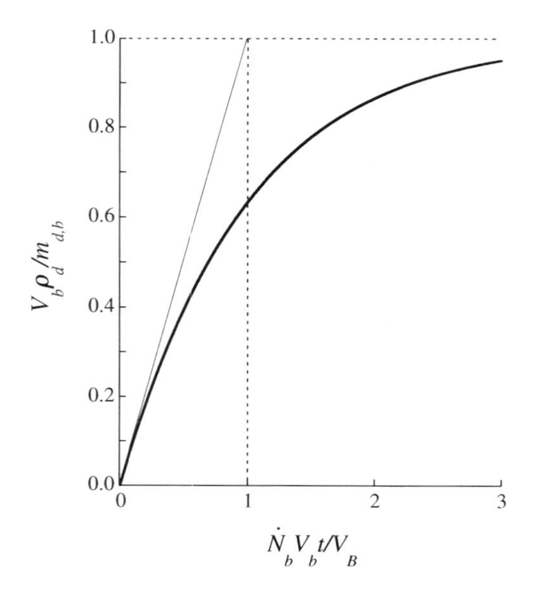

FIGURE 2.26 Evolution of normalized mass concentration of dirt in washing solution, $V_b\rho_d/m_{d,b}$(—), versus dimensionless time, $\dot{N}_b V_b t/V_B$ – with indication of initial tangent (—), throughout soaking of reusable bottles.

as the exponential function in Eq. (2.289) turns negligible – which describes the horizontal asymptote in Fig. 2.26, associated with steady state. One also finds that

$$\frac{d\left(\frac{V_b\rho_d}{m_{d,b}}\right)}{d\left(\frac{\dot{N}_b V_b t}{V_B}\right)}=-\exp\left\{-\frac{\dot{N}_b V_b}{V_B}t\right\}(-1)=\exp\left\{-\frac{\dot{N}_b V_b}{V_B}t\right\}, \tag{2.291}$$

obtained upon differentiation of Eq. (2.289); after setting $t=0$, Eq. (2.291) unfolds

$$\left.\frac{d\left(\frac{V_b\rho_d}{m_{d,b}}\right)}{d\left(\frac{\dot{N}_b V_b t}{V_B}\right)}\right|_{\frac{\dot{N}_b V_b t}{V_B}=0}=e^0=1 \tag{2.292}$$

Eq. (2.353) conveys the initial slope of the tangent to the $V_b\rho_d/m_{d,b}$ vs. $\dot{N}_b V_b t/V_B$ curve plotted in Fig. 2.26. In practice, steady state is normally reached after 200,000–300,000 bottles have been soaked per cubic meter of solution – although the bath has classically been renewed after 20,000–30,000, for safety reasons. The current trend, however, is toward operation of bottle washers at steady state – along with continuous purification of the bath, via settling, filtration, ultrafiltration, or chemical treatment.

PROBLEM 2.5

Consider a soaking operation, followed by a rinsing operation – aimed as a whole at washing reusable bottles in a brewery.

a) Should soaking be carried out under steady state, investigate whether any difference will arise if the soaking apparatus is split into two identical apparatuses – each one with half the total volume, but operated in parallel at the same overall throughput rates of washing solution and bottles.
b) If two similar apparatuses were connected in series, would there be any difference in performance under steady state operation? Assume that all washing water is fed to the first unit.
c) Consider now a rinsing unit of the same volume of the soaking unit, placed after its outlet – operated under unsteady state, and initially filled with freshwater; plot the evolution in concentration of dirt at the outlet stream, ρ_r, in dimensionless form, and discuss the shape of the curve produced.

Solution

a) Under steady state conditions, the mass balance to dirt in a single unit reduces to

$$\dot{N}_b m_{d,b} - \dot{N}_b V_b \rho_d = 0 \tag{2.5.1}$$

in lieu of Eq. (2.280) – since the accumulation term, $d\rho_d/dt$, turns immaterial; after dropping $\dot{N}_b$ from both sides and isolating ρ_d, Eq. (2.5.1) breaks down to

$$\rho_d = \frac{m_{d,b}}{V_b}, \tag{2.5.2}$$

consistent with Eq. (2.290) – as expected, for pertaining to the steady state that will eventually be attained. When two smaller soaking units are operated in parallel, the mass balance to either unit reads

$$\frac{\dot{N}_b}{2} m_{d,b} - \frac{\dot{N}_b}{2} V_b \rho_{d,i,par} = 0; \quad i = 1,2 \tag{2.5.3}$$

– with $\rho_{d,i,par}$ denoting mass concentration of dirt at the outlet of either apparatus of the parallel network; the above reasoning reflects the hypothesis that the same overall rate of supply of bottles is kept, but equally split by the identical units. The mass of dirt carried by each bottle at the entrance still reads $m_{d,b}$; and the volume of washing solution carried by each bottle at exit is also given by V_b. Cancellation of $\dot{N}_b/2$ between sides in Eq. (2.5.3) leads to

$$\rho_{d,1,par} = \frac{m_{d,b}}{V_b} = \rho_{d,2,par}, \tag{2.5.4}$$

upon isolation of $\rho_{d,1,par}$ or $\rho_{d,2,par}$. One promptly concludes that

$$\rho_{d,1,par} = \rho_{d,2,par} = \rho_d, \tag{2.5.5}$$

as per comparative inspection of Eqs. (2.5.2) and (2.5.4); therefore, no difference exists in steady-state performance between a single soaking unit, and two units with half the total capacity, each connected in parallel – because similar conditions prevail in the two configurations, namely identical space times.

b) The mass balance to dirt in the first unit, under steady-state, now takes the form

$$\dot{N}_b m_{d,b} - \dot{N}_b V_b \rho_{d,1,ser} - \dot{N}_b m_{d,b,1} = 0, \tag{2.5.6}$$

and likewise

$$\dot{N}_b m_{d,b,1} + \dot{N}_b V_b \rho_{d,1,ser} - \dot{N}_b V_b \rho_{d,2,ser} = 0 \tag{2.5.7}$$

pertaining to the second unit; here $m_{d,b,1} < m_{d,b}$ denotes mass of dirt carried by every bottle at the intermediate stage, due to (putatively) incomplete washing in the first unit – which will be carried on in the second unit. Note that the volume of freshwater supplied to the first unit is supposed to make up for $\dot{N}_b V_b$ carried away by the bottles leaving that unit, whereas that very same volume of (previously used) washing solution enters and leaves the second unit using the bottles as vehicles – because no makeup with fresh washing solution occurs in the latter unit. Ordered addition of Eqs. (2.5.6) and (2.5.7) gives rise to

$$\dot{N}_b m_{d,b} - \dot{N}_b V_b \rho_{d,2,ser} = 0, \tag{2.5.8}$$

since $-\dot{N}_b V_b \rho_{d,1,ser}$ and $-\dot{N}_b m_{d,b,1}$ in the first equation cancel out with $\dot{N}_b V_b \rho_{d,1,ser}$ and $\dot{N}_b m_{d,b,1}$, respectively, in the second equation; a further dropping of $\dot{N}_b$ from both sides reduces Eq. (2.5.8) to

$$m_{d,b} - V_b \rho_{d,2,ser} = 0, \tag{2.5.9}$$

thus immediately leading to

$$\rho_{d,2,ser} = \frac{m_{d,b}}{V_b}. \tag{2.5.10}$$

In view of Eq. (2.5.2), one concludes from Eq. (2.5.10) that

$$\rho_{d,2,ser} = \rho_d \tag{2.5.11}$$

– so, once again, no improvement will result from operating two similar units in series instead of a single one, in a sustained manner. The second unit actually proves useless, because it operates with amount of dirt inside each bottle and bulk washing solution imported directly from (and supplied only by) the previous unit – where equilibrium with each other prevailed, in the first place. The first unit mimicking the single unit in a), in terms of throughput rates, enforces, in fact, $\rho_{d,1,ser} = \rho_d$, since $m_{d,b,1}$ would necessarily be nil – see Eq. (2.5.6) *vis-à-vis* with Eq. (2.5.1); the said $\rho_{d,1,ser} = \rho_d$ coupled with Eq. (2.5.11) imply $\rho_{d,1,ser} = \rho_{d,2,ser}$ – which, together with the aforementioned $m_{d,b,1} = 0$, transform indeed Eq. (2.5.7) to a universal (and thus redundant) condition.

c) In the case of an (extra) rinsing unit, freshwater is supplied further to the unclean water carried by the bottles leaving the soaking unit – so the mass balance to dirt will take the form

$$V_B \frac{d\rho_r}{dt} = \dot{N}_b V_b \rho_d - \dot{N}_b V_b \rho_r, \tag{2.5.12}$$

using Eq. (2.280) as template and denoting mass concentration of dirt in the rinsing unit by ρ_r;

$$\rho_r \big|_{t=0} = 0 \tag{2.5.13}$$

may as well serve as initial condition, in parallel to Eq. (2.282). After dividing both sides by V_B, Eq. (2.5.12) becomes

$$\frac{d\rho_r}{dt} = \frac{\dot{N}_b V_b}{V_B}\rho_d - \frac{\dot{N}_b V_b}{V_B}\rho_r; \tag{2.5.14}$$

Eq. (2.289) may now be revisited as

$$\rho_d = \frac{m_{d,b}}{V_b}\left(1 - \exp\left\{-\frac{\dot{N}_b V_b}{V_B}t\right\}\right) \tag{2.5.15}$$

as descriptor of the inlet stream to the rinsing unit – which transforms Eq. (2.5.14) to

$$\frac{d\rho_r}{dt} = \frac{\dot{N}_b V_b}{V_B}\frac{m_{d,b}}{V_b}\left(1 - \exp\left\{-\frac{\dot{N}_b V_b}{V_B}t\right\}\right) - \frac{\dot{N}_b V_b}{V_B}\rho_r \tag{2.5.16}$$

or, equivalently,

$$\frac{d\rho_r}{dt} = \frac{\dot{N}_b m_{d,b}}{V_B}\left(1 - \exp\left\{-\frac{\dot{N}_b V_b}{V_B}t\right\}\right) - \frac{\dot{N}_b V_b}{V_B}\rho_r \tag{2.5.17}$$

after cancelling V_b between numerator and denominator. The solution to the homogeneous differential equation associated with Eq. (2.5.17), i.e.

$$\frac{d\rho_r}{dt} = -\frac{\dot{N}_b V_b}{V_B}\rho_r, \tag{2.5.18}$$

takes the form

$$\int \frac{d\rho_r}{\rho_r} = -\frac{\dot{N}_b V_b}{V_B}\int dt \tag{2.5.19}$$

upon resorting to integration via separation of variables; Eq. (2.5.19) supports

$$\ln \rho_r = k_0 - \frac{\dot{N}_b V_b}{V_B}t, \tag{2.5.20}$$

where exponentials may be taken of both sides to give

$$\rho_r = k_1 \exp\left\{-\frac{\dot{N}_b V_b}{V_B}t\right\} \tag{2.5.21}$$

– with k_0 and $k_1 \equiv e^{k_0}$ denoting (integration) constants. One can then apply the method of variation of the constant – by postulating $k_1 \equiv k_1\{t\}$; this justifies differentiation of Eq. (2.5.21) as

$$\frac{d\rho_r}{dt} = \frac{dk_1}{dt}\exp\left\{-\frac{\dot{N}_b V_b}{V_B}t\right\} + k_1 \exp\left\{-\frac{\dot{N}_b V_b}{V_B}t\right\}\left(-\frac{\dot{N}_b V_b}{V_B}\right). \tag{2.5.22}$$

Upon recalling Eq. (2.5.17), one can replace the left-hand side of Eq. (2.5.22) to get

$$\frac{\dot{N}_b m_{d,b}}{V_B}\left(1 - \exp\left\{-\frac{\dot{N}_b V_b}{V_B}t\right\}\right) - \frac{\dot{N}_b V_b}{V_B}\rho_r = \frac{dk_1}{dt}\exp\left\{-\frac{\dot{N}_b V_b}{V_B}t\right\} - k_1\frac{\dot{N}_b V_b}{V_B}\exp\left\{-\frac{\dot{N}_b V_b}{V_B}t\right\}, \tag{2.5.23}$$

whereas combination with Eq. (2.5.21) gives rise to

$$\frac{\dot{N}_b m_{d,b}}{V_B}\left(1 - \exp\left\{-\frac{\dot{N}_b V_b}{V_B}t\right\}\right) - k_1\frac{\dot{N}_b V_b}{V_B}\exp\left\{-\frac{\dot{N}_b V_b}{V_B}t\right\} = \frac{dk_1}{dt}\exp\left\{-\frac{\dot{N}_b V_b}{V_B}t\right\} - k_1\frac{\dot{N}_b V_b}{V_B}\exp\left\{-\frac{\dot{N}_b V_b}{V_B}t\right\}. \tag{2.5.24}$$

Cancellation of common terms between sides permits simplification of Eq. (2.5.24) to

$$\frac{\dot{N}_b m_{d,b}}{V_B}\left(1 - \exp\left\{-\frac{\dot{N}_b V_b}{V_B}t\right\}\right) = \frac{dk_1}{dt}\exp\left\{-\frac{\dot{N}_b V_b}{V_B}t\right\}, \tag{2.5.25}$$

where dk_1/dt is, in turn, to be isolated as

$$\frac{dk_1}{dt} = \frac{\dot{N}_b m_{d,b}}{V_B}\left(\exp\left\{\frac{\dot{N}_b V_b}{V_B}t\right\} - 1\right); \tag{2.5.26}$$

integration will again resort to separation of variables, viz.

$$\int dk_1 = \frac{\dot{N}_b m_{d,b}}{V_B}\int \left(\exp\left\{\frac{\dot{N}_b V_b}{V_B}t\right\} - 1\right)dt, \tag{2.5.27}$$

which yields

$$k_1\{t\} = k_2 + \frac{\dot{N}_b m_{d,b}}{V_B}\left(\frac{\exp\left\{\frac{\dot{N}_b V_b}{V_B}t\right\}}{\frac{\dot{N}_b V_b}{V_B}} - t\right) \tag{2.5.28}$$

– where k_2 is now a true (arbitrary) constant. Insertion of Eq. (2.5.28) transforms Eq. (2.5.21) to

$$\rho_r = \left(k_2 + \frac{\dot{N}_b m_{d,b}}{V_B}\left(\frac{\exp\left\{\frac{\dot{N}_b V_b}{V_B}t\right\}}{\frac{\dot{N}_b V_b}{V_B}} - t\right)\right)\exp\left\{-\frac{\dot{N}_b V_b}{V_B}t\right\}, \tag{2.5.29}$$

where $\exp\{-\dot{N}_b V_b t/V_B\}$ and $\dot{N}_b m_{d,b}/V_B$ can be factored in to produce

$$\rho_r = k_2 \exp\left\{-\frac{\dot{N}_b V_b}{V_B} t\right\} + \frac{m_{d,b}}{V_b} - \frac{\dot{N}_b m_{d,b}}{V_B} t \exp\left\{-\frac{\dot{N}_b V_b}{V_B} t\right\}; \tag{2.5.30}$$

when $t=0$, Eq. (2.5.30) becomes merely

$$\rho_r\big|_{t=0} = k_2 + \frac{m_{d,b}}{V_b}, \tag{2.5.31}$$

so elimination of $\rho_r\big|_{t=0}$ between Eqs. (2.5.13) and (2.5.31) leaves

$$k_2 = -\frac{m_{d,b}}{V_b}. \tag{2.5.32}$$

In view of Eq. (2.5.32), one may finally rephrase Eq. (2.5.30) as

$$\rho_r = -\frac{m_{d,b}}{V_b} \exp\left\{-\frac{\dot{N}_b V_b}{V_B} t\right\} + \frac{m_{d,b}}{V_b} - \frac{\dot{N}_b m_{d,b}}{V_B} t \exp\left\{-\frac{\dot{N}_b V_b}{V_B} t\right\}, \tag{2.5.33}$$

which will undergo algebraic rearrangement to read

$$\frac{V_b}{m_{d,b}} \rho_r = 1 - \exp\left\{-\frac{\dot{N}_b V_b}{V_B} t\right\} - \frac{\dot{N}_b V_b}{V_B} t \exp\left\{-\frac{\dot{N}_b V_b}{V_B} t\right\} \tag{2.5.34}$$

after multiplying both sides by $V_b/m_{d,b}$; Eq. (2.5.34) is plotted below, as $V_b \rho_r / m_{d,b}$ versus $\dot{N}_b V_b t / V_B$.

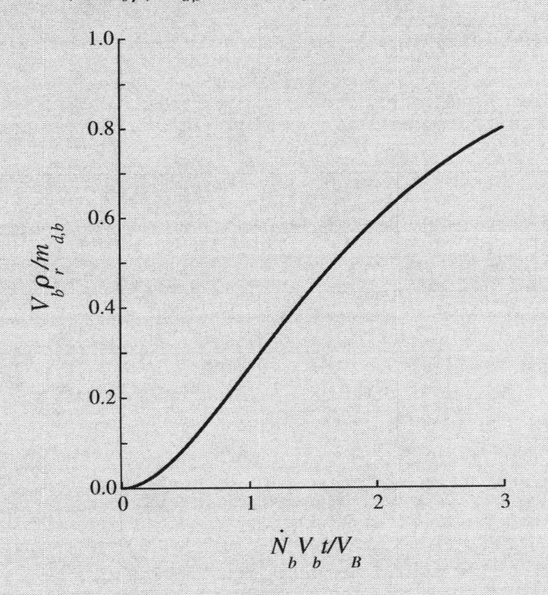

The curve generated possesses a sigmoidal shape – unlike the rectangular hyperbola in Fig. 2.26; this is typical of a second-order system, obtained here via cascading of two first-order systems, i.e. the soaking unit and the rinsing unit. On the other hand, ρ_r takes longer to buildup than ρ_d, see Fig. 2.26 – an anticipated feature, because extra freshwater is added along the process, when the rinsing unit is connected to the outlet of the soaking unit.

2.1.3.6 Disinfection

It has been known for more that one century that the rate of chemical elimination of bacteria is dependent on disinfectant concentration and temperature. Postulation that disinfection is, in essence, a chemical reaction has served as the basis for Chick's mechanistic hypothesis – the most widespread theory of disinfection.

Chemical destruction of microorganisms, at biomass concentration C_X, may accordingly be considered to follow first-order kinetics – as happens with thermal death thereof, i.e.

$$\boxed{\frac{dC_X}{dt} = -k_d\{C_B\} C_X,} \tag{2.293}$$

where the death kinetic constant, k_d, depends on target microorganism, besides microbicide at stake and corresponding concentration, C_B; temperature and medium composition are implicitly hypothesized to also affect k_d – the former somehow described by Arrhenius' law. A suitable initial condition reads

$$\boxed{C_X\big|_{t=0} = C_{X,0},} \tag{2.294}$$

where $C_{X,0}$ denotes starting biomass concentration of microorganism under scrutiny. It is usually found that

$$\boxed{k_d\{C_B\} = k_b C_B{}^n,} \tag{2.295}$$

with k_b denoting an n-th order, concentration-independent rate constant, and n denoting a dilution coefficient – again dependent on type of microorganism, nature and concentration of microbiocide, medium composition, and temperature; Eqs. (2.293) and (2.295) together constitute the so-called Chick and Watson's model of disinfection, an improvement built upon Chick's hypothesis – with typical values for n tabulated in Table 2.2. A wide range of values for n is indeed found – always above unity, thus supporting an increasing sensitivity of microorganism to biocide as the concentration of the latter increases. As stressed above, the said values depend on both chemical nature of biocide and target microorganism; note the particularly strong effect of phenol, second only to thymol (in the case of *Staph. aureus*, at least).

Integration of Eq. (2.293), via separation of variables, gives rise to

$$\int_{C_{X_0}}^{C_X} \frac{d\tilde{C}_X}{\tilde{C}_X} = -k_d \int_0^t d\tilde{t}, \tag{2.296}$$

TABLE 2.2

Values for parameter n in Chick and Watson's model, for selected disinfectants and microorganisms.

Disinfectant	Microorganism	n
C_{12}-Quaternary ammonium	*Staphylococcus aureus*	1.53
C_{14}-Quaternary ammonium	*Staph. aureus*	2.18
Chlorocresol	*Staph. aureus*	5.16
3-Chlorophenol	*Staph. aureus*	6.90
m-Cresol	*Staph. aureus*	7.59
EmpigenBB	*Staph. aureus*	2.53
Hydrogen peroxide	*Pseudomonas aeruginosa*	1.28
	Staph. aureus	1.27
	Staph. aureus	8.48
Phenol	*Salmonella* spp.	3.98
Sodium hypochlorite	*Staphylococcus aureus*	2.24
Silver nitrate	*Staph. aureus*	1.04
Thymol	*Staph. aureus*	9.32

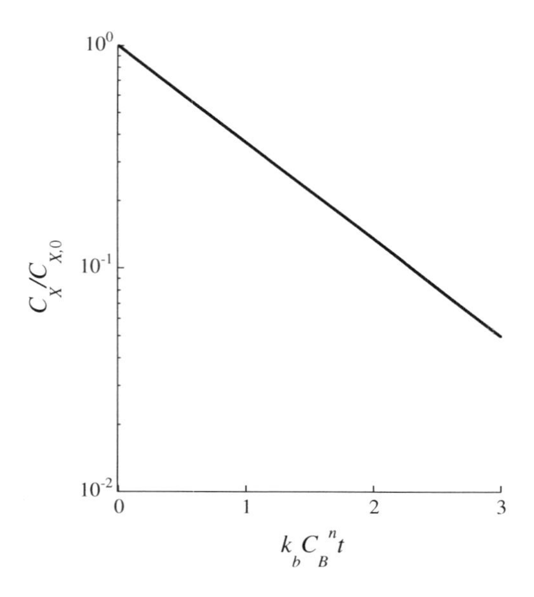

FIGURE 2.27 Evolution of normalized concentration of viable biomass, $C_X/C_{X,0}$, versus dimensionless time, $k_bC_B^nt$ – throughout disinfection following Chick and Watson's model.

with Eq. (2.294) providing the lower limit of integration – under the assumption that k_d does not vary with C_X or t; application of the fundamental theorem of integral calculus ensues as

$$\ln \tilde{C}_X \Big|_{C_{X,0}}^{C_X} = -k_d \tilde{t}\Big|_0^t, \tag{2.297}$$

or else

$$\ln \frac{C_X}{C_{X,0}} = -k_d t. \tag{2.298}$$

If exponentials are taken of both sides, then Eq. (2.298) becomes

$$\frac{C_X}{C_{X,0}} = e^{-k_d t} \tag{2.299}$$

– where Eq. (2.295) may be recalled to finally write

$$\boxed{\frac{C_X}{C_{X,0}} = e^{-k_b C_B^n t}}, \tag{2.300}$$

on the hypothesis that C_B does not significantly vary in time; Eq. (2.300) is plotted in Fig. 2.27, as (normalized) viable biomass, $C_X/C_{X,0}$, versus (dimensionless) time $k_bC_B^nt$. A single straight line results in the log/lin scale utilized, as anticipated; this corresponds to a constant specific rate of death of microorganisms, k_d (or $k_bC_B^n$, for that matter) as time elapses, along with decreasing C_X for an essentially constant C_B. The probabilistic nature of this process makes the said rate be proportional to the actual concentration of biomass, at each time.

For a given degree of destruction of microorganisms, $(C_X/C_{X,0})_{std}$, Eq. (2.298) may be revisited as

$$\ln\left(\frac{C_X}{C_{X,0}}\right)_{std} = -k_b C_B^n t_{std} \tag{2.301}$$

with the aid of Eq. (2.295), where isolation of t_{std} unfolds

$$t_{std} = \frac{\ln\left(\frac{C_{X,0}}{C_X}\right)_{std}}{k_b} \frac{1}{C_B^n}; \tag{2.302}$$

therefore, the time required for a standard degree of lethality, t_{std}, is inversely proportional to C_B^n – with (given) $k_b^{-1}\ln\left(C_{X,0}/C_X\right)_{std}$ serving as proportionality constant.

The above model is unable to simulate the whole distribution of resistance to chemical disinfection exhibited by several microbial populations, though – normally characterized by a tail of the logarithm of surviving viable biomass with disinfection time; nor does it provide accurate results when the microbiocide is not in vast excess. For instance, an aqueous solution of 0.01%(w/w) quaternary ammonium compound in a 10^8 cell.mL^{-1}-bacterial suspension contains roughly 2×10^9 molecules of biocide per cell; in view of its mechanism of action, via dissolution of the cell lipidic membrane – and recalling the average of 2.0×10^7 lipid molecules and 1.2×10^6 lipopolysaccharide molecules per cell of a typical bacterium, this would unfold an apparent 100-fold excess of surfactant biocide molecules over target microbial molecules. Nevertheless, a nonlinearity arises when survival data versus time are plotted in a log/lin scale; besides the possibility that disruption of functionality of each target molecule in the cell requires more than a single molecule of biocide, inherent instability of the biocide in aqueous media (as proven otherwise for e.g. ozone), or microbial-mediated quenching thereof may account for such an observation.

Useful refinements of the base model have accordingly been proposed – tentatively rationalized by a variable population resistance (alias vitalistic theory) or, alternatively, by a physicochemical route for microbiocide decay (alias mechanistic theory); in either case, k_d is hypothesized to vary with time. One may, in fact, consider that disinfectant concentration obeys

$$\boxed{\frac{dC_B}{dt} = -k_i C_B}, \tag{2.303}$$

as per Lambert and Johnston's model (also known as intrinsic quenching model) – where k_l denotes a first-order kinetic constant, with

$$\boxed{C_B\big|_{t=0} = C_{B,0}} \tag{2.304}$$

serving as initial condition; selected values for k_l are tabulated in Table 2.3. Quite a variability of values for k_l is again encountered – ranging within 10^{-4}–10^{-1} min^{-1}, with notably low values (and thus high stabilities) of thymol and 3-chlorophenol; this corresponds to half-lives of biocide from 115 hr down to 7 min.

Integration of Eq. (2.303) may proceed via separation of variables, viz.

$$\int_{C_{B,0}}^{C_B} \frac{d\tilde{C}_B}{\tilde{C}_B} = -k_l \int_0^t d\tilde{t} \tag{2.305}$$

at the expense of Eq. (2.304) – thus giving rise to

$$\ln \tilde{C}_B \Big|_{C_{B,0}}^{C_B} = -k_l \tilde{t}\Big|_0^t, \tag{2.306}$$

in view of the fundamental theorem of integral calculus; plain replacement of functions in both sides by their upper and lower limits yields

$$\ln \frac{C_B}{C_{B,0}} = -k_l t \tag{2.307}$$

– so exponentials may to advantage be taken of both sides to produce

$$\boxed{C_B = C_{B,0} e^{-k_l t},} \tag{2.308}$$

complemented by multiplication of both sides by $C_{B,0}$. Insertion of Eq. (2.308) transforms Eq. (2.295) to

$$k_d = k_b \left(C_{B,0} e^{-k_l t} \right)^n = k_b C_{B,0}{}^n e^{-nk_l t}, \tag{2.309}$$

where composite powers were meanwhile split; Eq. (2.309) may, in turn, be inserted in Eq. (2.293) to get

TABLE 2.3

Values for parameter k_l in Lambert and Johnston's model, for selected disinfectants and microorganisms.

Disinfectant	Microorganism	k_l (min^{-1})
C_{12}-Quaternary ammonium	*Staphylococcus aureus*	0.0817
C_{14}-Quaternary ammonium	*Staph. aureus*	0.0261
Chlorocresol	*Staph. aureus*	0.000678
3-Chlorophenol	*Staph. aureus*	0.000289
m-Cresol	*Staph. aureus*	0.00184
EmpigenBB	*Staph. aureus*	0.0482
Hydrogen peroxide	*Pseudomonas aeruginosa*	0.0314
	Staph. aureus	0.00126
Phenol	*Staph. aureus*	0.000590
Sodium hypochlorite	*Staph. aureus*	0.00625
Silver nitrate	*Staph. aureus*	0.364
Thymol	*Staph. aureus*	0.000172

$$\frac{dC_X}{dt} = -k_b C_{B,0}{}^n e^{-nk_l t} C_X. \tag{2.310}$$

Integration of Eq. (2.310) is again feasible via separation of variables, according to

$$\int_{C_{X,0}}^{C_X} \frac{d\tilde{C}_X}{\tilde{C}_X} = -k_b C_{B,0}{}^n \int_0^t e^{-nk_l \tilde{t}} \, d\tilde{t}, \tag{2.311}$$

with the contribution of Eq. (2.294). One may invoke the fundamental theorem of integral calculus to transform Eq. (2.311) to

$$\ln \tilde{C}_X \Big|_{C_{X,0}}^{C_X} = -k_b C_{B,0}{}^n \left. \frac{e^{-nk_l \tilde{t}}}{-nk_l} \right|_0^t, \tag{2.312}$$

which readily generates

$$\ln \frac{C_X}{C_{X,0}} = \frac{k_b C_{B,0}{}^n}{nk_l} \left(e^{-nk_l t} - e^0 \right); \tag{2.313}$$

once exponentials are taken of both sides, Eq. (2.313) becomes

$$\boxed{\frac{C_X}{C_{X,0}} = \exp\left\{ -\frac{k_b C_{B,0}{}^n}{nk_l} \left(1 - e^{-nk_l t} \right) \right\}.} \tag{2.314}$$

A graphical account of the relationship, conveyed by Eq. (2.314), between normalized dependent variable, $C_X/C_{X,0}$, and dimensionless independent variable, $nk_l t$, via a single dimensionless parameter, $k_b C_{B,0}{}^n/nk_l$, is provided in Fig. 2.28. The reduction in C_X, relative to its departing value $C_{X,0}$, is not linear in the log/lin scale selected, unlike observed in Fig. 2.27; the curves tail off in a more pronounced fashion at smaller values of parameter $k_b C_{B,0}{}^n/nk_l$ – as a consequence of a faster decay in biocide integrity associated to a larger k_l as per Eq. (2.303), further enhanced by larger k_b and n in view of Eq. (2.295).

When k_l is small, a slow decay in disinfectant activity will be observed as per Eq. (2.303), so the loss of microbial viability will occur fast – see curves associated with large $k_b C_{B,0}{}^n/nk_l$ in Fig. 2.28. In fact,

$$\lim_{k_l \to 0} \frac{C_X}{C_{X,0}} = \exp\left\{ -\frac{k_b C_{B,0}{}^n}{nk_l} \left(1 - \left(1 - nk_l t \right) \right) \right\} \tag{2.315}$$

holds exactly – as obtained upon Taylor's expansion, about $nk_l t = 0$, of the inner exponential function in Eq. (2.314), truncated after its linear term; Eq. (2.315) reduces to

$$\lim_{k_l \to 0} \frac{C_X}{C_{X,0}} = \exp\left\{ -\frac{k_b C_{B,0}{}^n}{nk_l} nk_l t \right\} \tag{2.316}$$

– or, after taking logarithms of both sides, exchanging logarithm and limit operators, and canceling common factors between numerator and denominator of the argument of the outstanding exponential function,

$$\ln\left\{ \lim_{k_l \to 0} \frac{C_X}{C_{X,0}} \right\} = \lim_{k_l \to 0} \left\{ \ln \frac{C_X}{C_{X,0}} \right\} = -k_b C_{B,0}{}^n t. \tag{2.317}$$

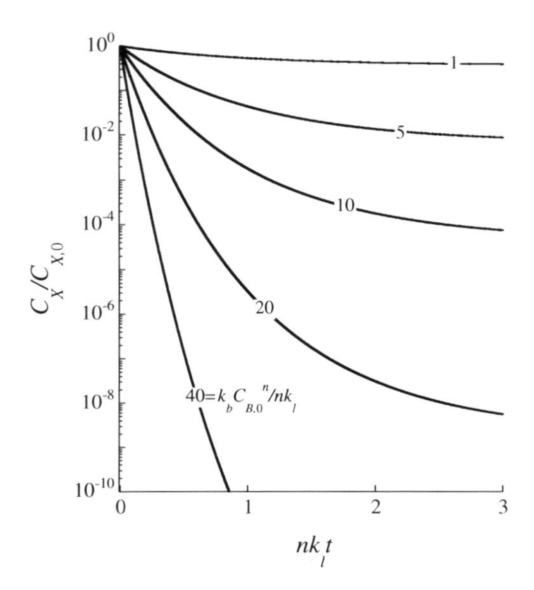

FIGURE 2.28 Evolution of normalized concentration of viable biomass, $C_X/C_{X,0}$, versus dimensionless time, nk_lt, for selected values of (dimensionless) parameter $k_bC_{B,0}{}^n/nk_l$, throughout disinfection following Lambert and Johnston's model.

Equation (2.317) not only confirms the tendency for a linear behavior in the log/lin scale elected for Fig. 2.28, under a large $k_bC_{B,0}{}^n/nk_l$; but also retrieves Eq. (2.300) after logarithms are taken of both its sides – since $C_B \approx C_{B,0}$ for a stable biocide (i.e. when $k_l \rightarrow 0$).

PROBLEM 2.6

Hom's (empirical) model for disinfection states that

$$\frac{dC_X}{dt} = -k_h C_B{}^m t^{h-1} C_X, \tag{2.6.1}$$

where k_h, m, and $h<1$ denote adjustable parameters – somehow corresponding to k_b, n, and k_l, respectively, in Lambert and Johnston's model; if disinfectant exists in great excess, i.e. $C_{B,0} >> C_{X,0}$, then $C_B \approx C_{B,0}$ – which supports reformulation of Eq. (2.6.1) to

$$\frac{dC_X}{dt} \approx -k_h C_{B,0}{}^m t^{h-1} C_X. \tag{2.6.2}$$

a) Show that Hom's model, in integrated form, coincides with Chick and Watson's model for $h=1$.
b) If $k_b = k_h/h$ and $n=m$, prove that Hom's model becomes equivalent to Lambert and Johnston's model when nk_l is small, and t lies in the neighborhood of unity.
c) Produce a linearization technique suitable to estimate the three parameters in Hom's model.

Solution

a) Integration of Eq. (2.6.2) may be achieved via separation of variables, viz.

$$\int_{C_{X,0}}^{C_X} \frac{d\tilde{C}_X}{\tilde{C}_X} = -k_h C_{B,0}{}^m \int_0^t \tilde{t}^{h-1} d\tilde{t}, \tag{2.6.3}$$

should Eq. (2.294) still be used for initial condition; the fundamental theorem of integral calculus allows transformation of Eq. (2.6.3) to

$$\ln \tilde{C}_X \Big|_{C_{X,0}}^{C_X} = -k_h C_{B,0}{}^m \frac{\tilde{t}^h}{h}\Big|_0^t, \tag{2.6.4}$$

or else

$$\ln \frac{C_X}{C_{X,0}} = -\frac{k_h C_{B,0}{}^m}{h} t^h \tag{2.6.5}$$

– which degenerates to

$$\frac{C_X}{C_{X,0}} = \exp\left\{-\frac{k_h C_{B,0}{}^m}{h} t^h\right\}, \tag{2.6.6}$$

after taking exponentials of both sides. If $h=1$, then Eq. (2.6.6) reduces to

$$\frac{C_X}{C_{X,0}} = e^{-k_h C_{B,0}{}^m t}, \tag{2.6.7}$$

which coincides with Eq. (2.300) after renaming k_h and m as k_b and n, respectively – and recalling that $C_B \approx C_{B,0}$ when no significant decay is postulated for the microbiocide; hence, Chick and Watson's model proves a particular case of Hom's model.

b) Taylor's expansion of the inner exponential in Eq. (2.314), about $t=1$, gives rise to

$$\frac{C_X}{C_{X,0}} \approx \exp\left\{-\frac{k_b C_{B,0}{}^n}{nk_l}\left(1-\left(e^{-nk_l} - nk_l e^{-nk_l}(t-1)\right)\right)\right\}, \tag{2.6.8}$$

as quadratic and higher-order terms can be safely discarded if $t \approx 1$ – which is equivalent

$$\frac{C_X}{C_{X,0}} = \exp\left\{-k_b C_{B,0}{}^n\left(\frac{1-e^{-nk_l}}{nk_l} + e^{-nk_l}(t-1)\right)\right\}, \tag{2.6.9}$$

once nk_l is factored in; by the same token, the integrated form of Hom's model as per Eq. (2.6.6) may be approximated by

$$\frac{C_X}{C_{X,0}} \approx \exp\left\{-\frac{k_h}{h} C_{B,0}{}^m \left(1+h(t-1)\right)\right\}, \tag{2.6.10}$$

via similar Taylor's expansion of t^h. Equations (2.6.9) and (2.6.10) will coincide when

$$\frac{1-e^{-nk_l}}{nk_l} = 1 \tag{2.6.11}$$

and

$$e^{-nk_l} = h, \tag{2.6.12}$$

irrespective of the actual value taken by t – since $k_b=k_h/h$ and $n=m$, by hypothesis; insertion of Eq. (2.6.12) transforms Eq. (2.6.11) to

$$\frac{1-h}{nk_l}=1 \tag{2.6.13}$$

that gives rise to

$$1-nk_l=h \tag{2.6.14}$$

– meant to replace Eq. (2.6.11). Compatibility between Eqs. (2.6.12) and (2.6.14) is observed when nk_l is small – since Taylor's expansion of e^{-nk_l}, around $nk_l=0$, reduces essentially to $1-nk_l$ under these circumstances; Hom's model and Lambert and Johnston's model accordingly become equivalent.

c) Linearization of Eq. (2.6.6) is possible by applying logarithms to its version labeled as Eq. (2.6.5) – after taking negatives of both sides, according to

$$\ln\left\{\ln\frac{C_{X,0}}{C_X}\right\}=\ln\left\{\frac{k_hC_{B,0}{}^m}{h}t^h\right\}, \tag{2.6.15}$$

where coincidence of the negative of the logarithm of an argument with the logarithm of the reciprocal of that argument was taken advantage of; a second step of algebraic manipulation of Eq. (2.6.15) gives rise to

$$\ln\left\{\ln\frac{C_{X,0}}{C_X}\right\}=\ln\frac{k_hC_{B,0}{}^m}{h}+h\ln t. \tag{2.6.16}$$

Equation (2.6.16) allows estimation of parameter h as slope, and (lumped) parameter $k_hC_{B,0}{}^m/h$ as exponential of the vertical intercept of a linear fit to a dataset of the form $\left(\ln t,\ln\left\{\ln C_{X,0}/C_X\right\}\right)$. If the dataset is extended so as to encompass $\left(\ln t,\ln C_{B,0},\ln\left\{\ln C_{X,0}/C_X\right\}\right)$, then Eq. (2.6.16) should to advantage be rewritten as

$$\ln\left\{\ln\frac{C_{X,0}}{C_X}\right\}=\ln\frac{k_h}{h}+m\ln C_{B,0}+h\ln t. \tag{2.6.17}$$

The bilinear equation labeled as Eq. (2.6.17) provides h directly as slope of $\ln\left\{\ln C_{X,0}/C_X\right\}$ vs. ln t, and m as slope of $\ln\left\{\ln C_{X,0}/C_X\right\}$ vs. ln $C_{B,0}$; meanwhile, k_h/h can be calculated as exponential of the corresponding vertical intercept – and the (already available) estimate for h may finally be used to get k_h from the (current) estimate for k_h/h.

2.2 Packaging

2.2.1 Introduction

The major goal of packaging is to properly contain foods, and protect them against several hazards throughout production, storage, and distribution. Furthermore, relevant information on nutritional and safety issues, as well as on culinary preparation is normally included on the label that accompanies a regular package – which materializes a communicational/educational role.

A package is indeed critical in most food products, namely for convenience purposes (e.g. easy transport and storage, easy manipulation, opening, dispensing, resealing, disposal, recycling, reuse) and securing purposes (e.g. leakage-proof features) related to handling – all of which consubstantiate a containment role. This calls for mechanically sturdy materials – able to resist such damages as fractures, tears, or dents during filling/closing, loading/unloading, and transportation/storage.

However, a package serves chiefly as a protective barrier from unwanted interaction with the environment (e.g. chemical and microbiological contamination, radiation-induced decay, volatile loss, and moisture loss or pickup) – thus entailing a preservation role. For instance, many foods cannot be kept crisp and fresh unless their package provides a barrier to moisture; in addition, oxidation-mediated rancidity can be kept to a minimum by blocking light and access of environmental oxygen via an appropriate opaque and nonporous package material; and specific materials may even be designed that offer a barrier to a few aromas.

All in all, proper packaging is crucial to extend the shelf-life of a food; however, a package should minimize unfavorable interaction with the food product itself, so it tendentially resorts to chemically inert materials. Migration of toxic compounds from the packaging material to the food matrix should be avoided at all cost; whereas conditions prevailing in the package, in terms of vacuum or modified atmosphere, should not create a selective pressure toward growth of undesirable microorganisms.

Unlike happens with most bulk and fine chemicals (parfums excluded), the package represents a major fraction of the final cost of a (processed) food; and food packages account for most nonorganic garbage produced in households, and are likely to be damped in landfills. Moreover, the package often constitutes the first interaction of the consumer with a food product – so it should be attractive and esthetically pleasing, besides possessing functional size and shape; it accordingly helps selling the product, thus playing a vital marketing function. On the other hand, periodic changes in size and design of packages contribute to product flexibility and innovation – and should appropriately address queries by retailers and final consumers.

The above considerations justify why packaging is currently an integral part of the food sector, and why it has undergone substantial developments in both materials and processes over the years. In fact, an impetus from Nature has marked its ineffaceable presence, either by pushing foods at time of consumption to be as close as possible to their sources (via minimal processing) or by reducing their environmental impact as much as possible (via reduction, reuse, and recycling of materials).

HISTORICAL FRAMEWORK

Foods have forever been kept stored in some form of container – normally wood in various shapes, e.g. wooden barrels (and later glass bottles), and using some form of lid, e.g. corks for wine and beer. However, a true packaging industry, based on dedicated devices, did not appear as such until the 19th century – urged by the persistent problem felt by the soft drink industry of lacking an effective seal for bottles. Carbonated drink bottles are indeed exposed to high overpressure (with a risk of explosion), so industrialists actively sought an efficient and safe way to prevent carbon dioxide from escaping – in attempts to preserve the unique bubbling features of such drinks. In 1870, Hiram Codd invented a bottling method suitable for the so-called Codd-neck bottle – designed to enclose a marble and a rubber washer in the neck. The bottles were filled upside down, and pressure of gas in the bottle forced the marble against the washer – thus sealing it during carbonation; the bottle neck was pinched into a special shape, so as to provide a chamber into which the marble was pushed to open the bottle – thus preventing it from blocking the neck as the drink was poured out. In 1892, the crown cork bottle seal was patented by William Painter – as the first closure able to successfully keep the bubbles inside the bottle. Furthermore, the first patent covering a glass-blowing machine for automatic production of bottles was granted to Michael Owens in 1899 – which avoided the cumbersome, manual work of blowing each bottle individually.

In the early 20th century, sales of bottled soda increased exponentially – and home-packs were made available in the 1920s, as six-pack cartons manufactured from cardboard; vending machines were also launched during this decade. In 1941, Rex Whinfield and James Dickson developed polyethylene terephthalate (PET); and Reynolds Metals Co. used surplus aluminum from World War II to make aluminum foil – traded as *Reynolds Wrap*. With the boom of domestic appliances in the 1950s, C. A. Swanson & Sons introduced the first TV dinner in 1954 – roast turkey with stuffing and gravy, as well as sweet potatoes and peas, coming in an aluminum tray for direct processing in the oven; this apparently came after an idea by its executive officer, Gerald Thomas, to use tons of turkey leftover from Thanksgiving. Aluminum cans were first used on a commercial basis in 1960 for foods and beverages; Coca Cola introduced the 12-oz can in 1961. In this very year, boiling bags – i.e. frozen plastic packages of food that could be safely dropped in boiling water for heating before serving, entered the general market. In 1963, Ermal Fraze revolutionized the beverage industry with his invention of pull-tab openers for cans – promptly sold to Alcoa; meanwhile, Schlitz Brewing Co. introduced the first pop-top beer can. Plastic milk containers appeared in 1964; and an easy-open, front pull-tab made its way to the SPAM (or SPiced hAM) can through Hormel Foods in 1967. In the next year, Alexander Liepa invented *Pringles* – a form of potato- and wheat-based stackable snack chips (now owned by Kellogg's); this analog to classical chips was packaged in a tubular can, with a foil-coated interior and a resealable plastic lid. In 1973, Nathaniel Weyth received a patent for PET beverage bottles – after revisiting the earlier development of PET, and taking advantage of its being chemically safe and strong enough to hold carbonated beverages without bursting; the first commercial soda bottle was created by Dupont two years later. On June 26 of 1974, a 10-pack of *Wrigley's Juicy Fruit* was run through a handmade laser scanner at a Marsh Supermarket in Troy, OH; by the end of the year, more than 1,000 food and beverage companies already registered, and had bar codes assigned thereto. LaBatt Brewing Co. introduced the twist-off cap on a refillable bottle in 1984 – and modified atmosphere packaging reached the marketplace in the following year; lunchables were proposed by Kraft Foods in 1988.

An increasing need to legislate for consumer's protection eventually led to the *Nutrition Labeling and Education Act*, put into action in 1990; it required all packaged food to bear nutritional labeling, and set the terms of use of all health claims. The *Food Allergy Labeling and Consumer Protection Act* of 2004 enforced specific mention in the label to any food containing protein(s) derived from peanuts, soybeans, cow's milk, eggs, fish, crustacean shellfish, tree nuts, and wheat – the most common source of food-borne allergens.

Environmental concerns were meanwhile on the rise – echoing a more educated population, with worldwide awareness enhanced by Internet-mediated communication channels and the like. Coca Cola introduced the first bottle made of recycled plastic in 1991; and Walkers (i.e. Pepsico's British chips brand) became, in 2007, the first major brand to display a carbon footprint reduction logo on its packs.

In classical food processing, the final packaged food is subjected to some form of treatment lethal to contaminating microorganisms (e.g. heat, pressure, irradiation). This treatment accordingly affects not only the food matrix, but also the packaging material – namely, its (moisture and oxygen) barrier features; it may also induce migration of packaging material into the food – e.g. monomers released by radiation- or heat-induced depolymerization. Hence, a careful choice of packaging material is essential for overall success of a food processing operation, and full safety of the final food product.

Instead of sterilizing the packaged food product – since this poses limitations arising from the nature of packaging materials and the patterns followed by heat transfer, one may proceed to aseptic processing and packaging; this technique involves sterilization of food product and container separately – followed by filling the latter with processed food, and sealing it in a sterile environment that guarantees aseptic conditions.

HISTORICAL FRAMEWORK

The pioneering efforts of Olin Ball in 1927 in the United States led to development of HCF, or heat–cool–fill processing – departing from the basic concept of sterilizing empty containers and food material separately, and filling the packages afterward with food under a steam environment; this was indeed the forerunner of aseptic technology – even though economic constraints hampered its immediate success. Avoset's process was based on the very same idea, but resorted to an air stream – previously sterilized by UV light, around the filling area; complemented by a positive air pressure, to avoid spurious contamination *in situ*.

A major leap took place in the 1940s, with Dole's process designed for the canning industry (e.g. milk products, purées, specialty sauces, split-pea soups) – using superheated steam to sterilize metal containers and covers, flash heating and cooling of liquid food in a tube-and-shell heat exchanger, aseptic filling of cold sterile product into sterile containers, and final sealing thereof under a steam atmosphere. Commercial development of a 55-gallon drum filler for tomato paste in 1958 represented the next milestone. It was followed in the 1960s by use of hydrogen peroxide by TetraPak – coupled to a vertical form–fill–seal unit (currently known as brickpak filler), using laminated packaging material tailor-made to serve as seal and barrier to air and light; it required below-ambient processing temperatures, and thus the said hydrogen peroxide for asepsis (provided that its residual level would not exceed 0.1 ppm, for the sake of consumer safety). Later on, Scholle Co. developed the bag-in-box concept, with 5–2000 L capacity range – thus allowing replacement of metal cans for the institutional distribution market.

A notorious growth in market share of aseptic food products took place in the 1980s, and was accounted for by fruit juices and juice drinks in individual serving containers; complemented by UHT-sterilized milk in family-sized containers, as well as soups, puddings, and baby foods to a lesser degree. However, challenges still remain concerning application of this technology to low-acid, particulate foods (e.g. fruits in carrier liquids, meatballs in gravy) – due to unmatched heat capacities and thermal conductivities of solid particles versus those of surrounding liquid; the latter is usually overprocessed to ensure adequate thermal processing of the suspended particles.

The nuclear components of an aseptic processing and packaging system are: deaerator, heating/cooling section, holding tube, packaging system, and pumps and flow control. The food product is exposed, on a continuous basis, to a vacuum during deaeration; removal of air (and thus oxygen) is a must, so as to minimize the chance for oxidative reactions afterward – which may compromise product quality and shorten its shelf-life. Heating is normally brought about by classical tubular, plate, or scraped-surface heat exchangers; direct heat exchangers, using steam injection in the liquid food itself, may also be employed – followed by flash cooling, to remove the extra water added due to steam condensation (as discussed previously). This heating section *per se* is often unable to achieve the intended lethality, especially when highly viscous liquids, heat-resistant enzymes, or large suspended particles are at stake; hence, a holding tube is normally included – designed to achieve a uniform temperature for a sufficiently long time. Critical features thereof include: upward slope, of at least 2 $cm.m^{-1}$ (to eliminate air pockets, and facilitate draining at cleaning stage); avoidance of extra heating along tube path (to minimize overprocessing); easy assemblage/disassemblage (for inspection purposes); insulation from condensate or cold air drafts (to avoid temperature drops); inside pressure above vapor pressure of liquid food (to prevent flashing/boiling); and inclusion of adequate temperature sensors at entrance and exit (for supplementary control). The packaging section is variable – depending on nature of food, as well as size, shape, and material of package; careful design of the space where the product is inserted into the package is of the utmost relevance, so as to permit full sterilization *in situ* – and thus prevent postprocessing contamination. A metering pump (usually of the positive displacement type) is located upstream of the holding tube – and guarantees the desired throughput rate; for cleaning-in-place purposes, however, a centrifugal pump suffices.

Active packaging refers to packaging systems used with foods (besides pharmaceuticals), specifically designed to extend shelf-life and improve safety; such a concept encompasses active functions played by the package itself, beyond passive containment and protection of food product. One of the oldest examples is inclusion of a desiccant (e.g. silica gel), designed to control moisture inside a package of a moisture-sensitive food; it is usually a hygroscopic substance, contained in a porous pouch or sachet, and placed inside a sealed package. On the other hand, trace transition metals in foods (e.g. Fe) can catalyze oxidative degradation of many food components, chiefly lipids – and thus compromise the quality of those products by the time of consumption; suitable active packaging materials can be manufactured via immobilization of metal-chelating agents onto the packaging material – which will scavenge the said transition metals, and thus enhance oxidative stability of the product. A similar approach can be developed with (natural or synthetic) antioxidants or oxygen scavengers – which react with oxygen right on the surface of the food (where it is more likely to show up), thus avoiding formation of free radicals on the long run.

Besides moisture control and oxygen scavenging, sustained and/or controlled release packaging constitutes a (commercially) successful example in the emerging area of active food packaging – aimed at extending duration and decreasing level of delivery of an active compound, along with a much more predictable and reproducible release pattern. Should antimicrobial ingredientes be at stake, longer shelf lives of food products, still holding reasonable quality descriptors, will indeed become possible – based on inhibition or retarding of growth of unwanted microorganisms. Theoretical prediction of the said release is of the utmost importance; too fast a release will cause depletion of the antimicrobial compound within a short time – after which the minimal concentration required to prevent microbial growth cannot be maintained on the food surface. Conversely, spoilage reactions may occur on the food surface, should too slow a release of the said compound from the package take place.

Remember that antimicrobial agents have traditionally been a part of food formulation in the first place; however, direct addition all at once may decrease the concentration of antimicrobial agent on the food surface owing to diffusion into the bulk – while its effect in the bulk can meanwhile (or eventually) be neutralized by complex interactions with the other food components. In either case, the minimum concentration required to inhibit growth may not be attained at all. Moreover, direct addition brings about utilization of excessive amounts of antimicrobial agent – which may compromise taste of food, further to incurring in higher costs. Controlled release of antimicrobial agents offers an alternative to circumvent these shortcomings; a reduction in the level of some antimicrobial compound in the food itself, at any given moment, will indeed be achieved – while the typical taste will likely remain unchanged.

A few approaches exist at present that assure a controlled diffusion of an antimicrobial, or active agent at large from the packaging material; for instance, incorporation of the compound into/onto the packaging material itself – including encapsulation thereof in the constitutive polymer material or via extraneous encapsulating agents, or development of polymer blends with tailor-made morphologies. Changes in polymer structure – namely porosity, have met with success in controlled release of lysozyme; this is the case of cinnamaldehyde-crosslinked gliadin films – characterized by slower release due to decreased water swelling, as well as genipin-crosslinked gelatin and cellulose acetate films. In the case of coatings, one has proceeded via changes in such polymer features as permeability and diffusivity; this includes polylactic acid films, containing silver coated with beeswax, for delayed antimicrobial release. Changes in film morphology become also accessible upon polymer blending; this happens with plasticized starch/poly(butylene succinate co-butylene adipate) films with increased tortuosity – as well as zein-oleic acid films with lysozyme, meant to induce lysis of contaminant microorganisms; or else cellulose acetate multilayer films, with potassium sorbate bearing enhanced barrier features.

More recent developments have encompassed intelligent packaging – and one of the best examples is time-temperature indicators; this is a family of devices or smart labels, which somehow display the temperature history of a food product. Although not yet a generalized practice, such devices have been successfully employed in pharmaceutical and medical items to indicate exposure to excessively high temperatures, or for too long periods at a temperature higher than expected; use in foods is anticipated to become routine practice in the near future, provided that their unit price keeps dropping (as observed so far). Conversely, temperature datalogers measure and record temperatures for a specified period – and are normally reusable, owing to their high unit price; hence, they cannot accompany food down to its final destination. When required, the stored data can be downloaded in digital form though, and analyzed at will. They accordingly help identify the time period when out-of-tolerance temperatures prevailed, or aid in estimation of the remaining shelf-life, or qualitatively ascertain the likelihood of spoilage. The said small recorders are also employed to identify time (and thus location) of a shipment when a problem has been found – thus permitting corrective actions to taken, within a useful time and at the right spot.

Intelligent packaging accordingly remains a field of intensive research; innovation is underway into thin-film devices that produce audio and/or visual information – either in response to motion or touch, or to some form of activation (e.g. scanning), thus permitting direct communication with the customer. This and several other concepts alike unfortunately share the same drawback – it is not (yet) clear how applicable and affordable they will eventually become. In fact, commercial success depends heavily upon acceptance by the packaging industry, food manufacturers, and consumers at large – where the former already face increasing environmental concerns over lack of reusability/recyclability of food packages; too complex packages, in terms of building materials involved, will surely worsen that issue. Furthermore, a clearly perceived improvement in terms of quality/safety is required by food manufacturers and retailers – sufficient to justify the significant extra costs involved. Besides cost of materials themselves, the sophistication of the technologies under scrutiny will require more advanced apparatuses, as well as much more expensive technicians to maintain and repair them – due to their (unavoidable) advanced skills and training. On the other hand, putative unreliability of indicator devices within their operating range – e.g. inaccuracy as per showing food to be safe when it is not (i.e. false positive), or else lack of reproducibility at large may result in expensive liability and lawsuits; this will hardly contribute to overcome the existing barriers against wider acceptance of intelligent packaging. The technologies underlying intelligent, miniaturized devices will also have to comply with food safety regulations – namely, concerning potential migration of components from complex packaging materials into the food products. All these arguments will surely slowdown routine implementation of intelligent packaging approaches – except for some specialty or gourmet foods, sold at premium prices and designed for niche markets.

2.2.2 Product and process overview

Selection of a packaging material hinges on a number of issues related to mechanical strength, interaction with light, response to temperature, grease, and moisture/gases, and lability to chemical/biological contamination – in terms of their effects upon intended display or protection of the food product, and environmental impacts (envisaged or anticipated) throughout its distribution and storage.

Packaged foods are regularly subjected to a number of handling operations and steps that may cause damage: crushing – caused by stacking in warehouses and vehicles; abrasion – caused by rubbing against equipment and throughout handling; puncturing and fracturing – caused by impacts during handling; and vibration – caused chiefly by transportation. The ability of a package to effectively protect food from the aforementioned forms of damage derives directly from its mechanical strength. When foods are particularly fragile (e.g. biscuits, cakes, eggs, fresh fruits), a higher degree of protection conveyed by the package is a *sine qua non*; strategies include cushioning with tissue paper, or use of containers shaped for individual pieces, and manufactured from paper pulp or foamed sheets (e.g. egg cartons, fruit trays). Protection may also be assured by a rigid container; and/or restricted movement, either via shrink- or stretch-wrapping, or by resorting to plastic film tightly formed around the product. Polymer pots, trays, and multilayered cartons bring about mechanical protection to specific foods; while stronger materials, e.g. metal cans, glass, or polyethylene therephthalate bottles are required to withstand pressure in the case of carbonated beverages – which increases with temperature, due to both decrease of gas solubility in the liquid phase as per Henry's law, and more energetic molecular collisions against the walls. In terms of shipping of bulk or multiple individual units of food, good mechanical protection is provided by metal crates, barrels, and drums

– thus justifying their classical election for shipping containers; due to the associated high cost and weight, however, a trend has been observed toward gradual replacement by (cheaper) composite intermediate bulk containers – usually made from fiberboard and polypropylene.

As detailed previously in *Food Proc. Eng.: basics & mechanical operations* when addressing textural characterization of foods, mechanical strength of a packaging material can be ascertained via measurement of its elongation (or strain) caused by application of a force (or stress). An elastic behavior is typically observed at low stress/strain – as per Hooke's law, with proportionality constant termed Young's modulus (as already discussed); if stress is released, the original shape and size will be retrieved – so the mechanical work initially applied has meanwhile been converted to internal elastic potential energy, which may be recovered in full afterward. A threshold will eventually be attained – termed yield point, after which strain will go on increasing under a constant stress, thus giving rise to fluid-like behavior. The extra work performed on the system is thus converted to internal plastic potential energy – meaning that irreversible changes in molecular structure and orientation have necessarily occurred, which prevent recovery as work when the external forces cease to be applied. After sufficient elongation, the material will reach a breaking point – with concomitant failure, i.e. the material will eventually be discontinued.

Young's modulus accordingly entails a measure of stiffness; its value is often the same, irrespective of direction, in the case of glass and metals. Nevertheless, anisotropy is frequent in polymer materials, owing to the specific orientation of their macromolecules. This causes the material to be stronger in the machining direction, but weaker in the transverse direction; if molecules are oriented uniformly in both directions, then the overall behavior becomes more uniform – as happens with polyethylene, polypropylene, polyethylene terephthalate, and polystyrene. Examples of mechanical strength in the machine direction are provided in Fig. 2.29; for instance, the tensile strength decreases from 145–200 $MN.m^{-2}$ in the machine direction, to a mere 0.4–0.6 $MN.m^{-2}$ in the transverse direction for the case of uniaxially-oriented polypropylene.

Although transparent packaging is instrumental in displaying contents and enhancing appearance of foods at large – which constitute important marketing strategies, several foods do not withstand light (or, at least, some ranges of the spectrum); this may be due to susceptibility to rancidity development (due to lipid oxidation), nutrient decay (as in vitamin loss), or color compromise (when natural pigments undergo chemical transformation). The fraction of light absorbed or transmitted varies with the packaging material and the spectrum of incident light; such materials as low-density polyethylene transmit both visible and UV light essentially to the same degree, whereas the former is preferentially transmitted and the latter preferentially absorbed in the case of polyvinylidene chloride. Total light transmission can be as low as 0.5–3% in the case of metalized polypropylene, and as high as 87–88% in the case of polyester – or even reach 90%, in the case of polyvinylidene chloride or cellulose. Gloss is also a major feature associated with light effects; it describes the fraction of incident light reflected by the surface, at an angle identical to the angle of incidence (usually 45°) – and accounts for a pleasing sparkle. Specific data in this regard are conveyed by Fig. 2.30, using gloss of standard polyvinylidene chloride as reference; note the improvement in gloss of this polymer when combined with cellulose, and the further improvement when its metallized version is considered. Remember that light can also be reflected after it enters a matrix, and become diffuse due to scattering; this phenomenon accounts for the colors and patterns perceived by vision. To modulate the quantity and quality of light that enters a packaged food product, pigments have been incorporated in glass containers or polymer films themselves; as an alternative, packages may be overwrapped with opaque labels or printed – whereas containment in fiberboard cartons can be taken advantage of, during distribution and storage.

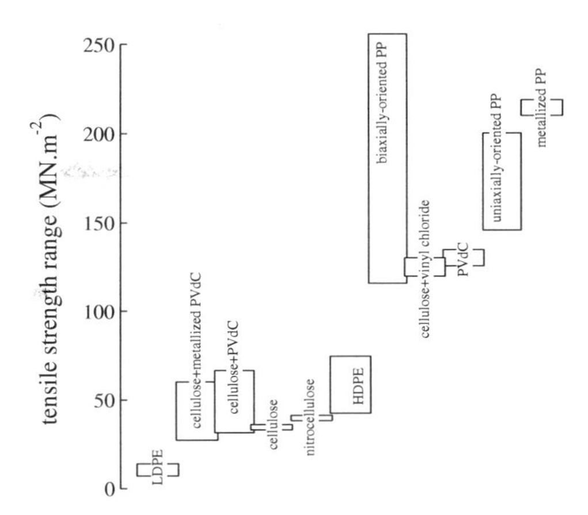

FIGURE 2.29 Ranges of tensile strength along machine direction (▭), for selected polymeric films used in food packaging.

The insulating effect of a package is determined by its thermal conductivity – as per Fourier's law; coupled with its reflectivity – as per Maxwell and Boltzmann's law. Low thermal conductivities (e.g. paperboard, polystyrene, polyurethane foams) reduce heat transfer by conduction, whereas reflective materials (e.g. aluminum foil, metallized films) reduce heat transfer by radiation. In any case, control of environmental temperature is far more relevant than the package insulation features themselves. When the food is designed to undergo heating in packaged form – as is the case of sterilization of canned foods in an industrial plant, or preparation of microwaveable, ready-meals at home, the packaging matrix is expected to withstand the applied processing conditions without undergoing damage. Unintended physicochemical interactions with the food are indeed to be avoided – e.g. melting of the package and subsequent sticking to the food surface, or release of monomers via depolymerization of the packaging material that may end up contaminating food, in the case of plastic packages. This issue is of a particular relevance, since food packages are often sealed via heating to a high temperature – see Fig. 2.31. To avoid thermal shock, and thus minimize risk of breakage, glass containers are to be heated and cooled more slowly than their metal or plastic counterparts. By the same token, the flexibility of the packaging material should be retained in the case of foods intended for freezing and storage in frozen state – otherwise the package may crack along the cold network, thus exposing the food to putative contamination.

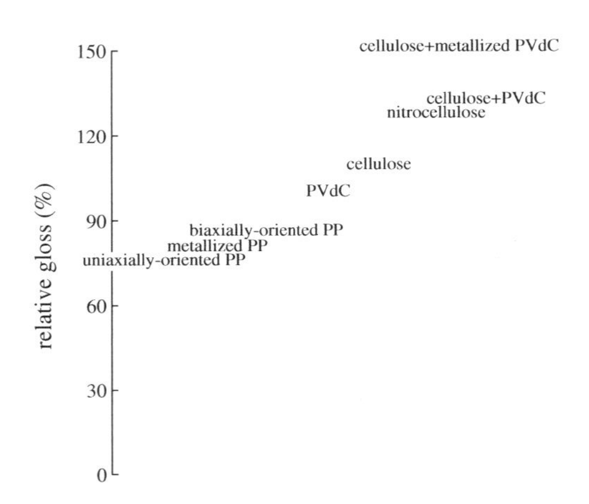

FIGURE 2.30 Relative gloss for selected polymeric films used in food packaging – using gloss of PVdC as reference (100%).

Leakage of oils and fats causes product loss, besides compromising appearance of the package – and even making it unsuitable for further manipulation; this justifies packaging of cooking oils in glass/plastic bottles or metal cans – and cooking fats in grease-proof paper, aluminum foil, or plastic tubs. Such dry fatty foods as chocolate are packaged in foil or plastic films – while such wet fatty foods as meat and fish resort to laminated, or somehow treated papers and films.

The shelf-life of foods is frequently controlled by the rate of moisture uptake or loss – the former for increasing the chance of microbial growth or enzyme-mediated decay, and the latter for leading to unacceptable sensory quality of food (e.g. drying out of fresh foods, freezer burn of frozen foods). In fact, a microclimate develops inside a package – determined by the vapor pressure of moisture at the food storage temperature, coupled with the permeability of the packaging material to the said moisture. Higher permeabilities of the packaging material are required in the case of fresh vegetables and bread – to prevent moisture condensation on the inside of the package, as it is prone to mold growth. To prevent fogging of the display area (should a change in temperature occur), movement of water vapor out of the pack should be allowed in the case of chilled foods. Conversely, low permeability packaging is a must in the case of hygroscopic items, as happens with such dried foods as biscuits, icing sugar, custard/gravy powder, or such extruded foods as snacks – otherwise their intended crispness will be disrupted, or caking may occur due to absorption of moisture from the outer atmosphere; microbial growth may also occur if a_W exceeds a certain threshold. Special care is to be exercised with regard to nature and amount of inks, adhesives, and solvents utilized, to avoid odor pick-up by the food through the packaging material; as well as to such defects as splits, pinholes, and inadequately formed seals in flexible films – since complete isolation of the food from its environment will not be guaranteed.

Food containing significant amounts of lipids may undergo spoilage if the package raises an inadequate barrier to oxygen; while loss of ethylene (as ripening hormone) by climacteric fruits will induce shriveling. Conversely, fresh foods require a high degree of permeability to permit exchange, with the atmosphere, of O_2 and CO_2 associated to respiration; while fresh red meats require oxygen to maintain the red color of hemoglobin. By the same token, modified atmosphere packaging (of e.g. cheeses, cooked meats, egg powder) requires materials impermeable to N_2 and/or CO_2, otherwise their shelf-life will be compromised; and similar barrier properties are required to prevent loss of volatile compounds associated with desirable odors (e.g. coffee), or avoid pick-up of compounds associated with off-flavors (e.g. powders, fatty foods). Relevant data pertaining to the permeability of various polymeric materials to the most important gases are conveyed by Fig. 2.32. Transmission rate of water vapor varies from less than 1 $mL.m^{-2}.d^{-1}$ in low permeability films up to above 10^3 $mL.m^{-2}.d^{-1}$ in highly permeable films; and of gases varies from ca. 1 $mL.m^{-2}.d^{-1}$ for oxygen and nitrogen, to 10^4 or 10^3 $mL.m^{-2}.d^{-1}$, respectively. Carbon dioxide permeates typical polymeric films at rates within $10–10^3$ $mL.m^{-2}.d^{-1}$. Both food constituents and environmental composition affect permeability, however; for instance, plasticizers and pigments, often used in the formulation of packaging films, facilitate gas permeation for loosening their structure – whereas permeability of cellulose, nylon, and polyvinyl alcohol films changes with outer humidity levels, due to interaction of water molecules with those fibers via hydrogen bonding. Furthermore, permeability increases exponentially with temperature – not only due to an increase in the average pore size of the packaging material itself due to expansion; but also because gas molecules possess a higher average kinetic energy, and thus undergo more intense collisions with each other and with the molecules of the container film. In both cases, warmer packages enhance plain diffusion – based on more frequent and energetic collisions of gas molecules with one another; and Knudsen-type diffusion – based on more frequent and energetic collisions of gas molecules with molecules of the packaging material along its pore walls.

Packs that are folded, stapled, or twist-wrapped are not truly sealed, so they are somewhat susceptible to microbial

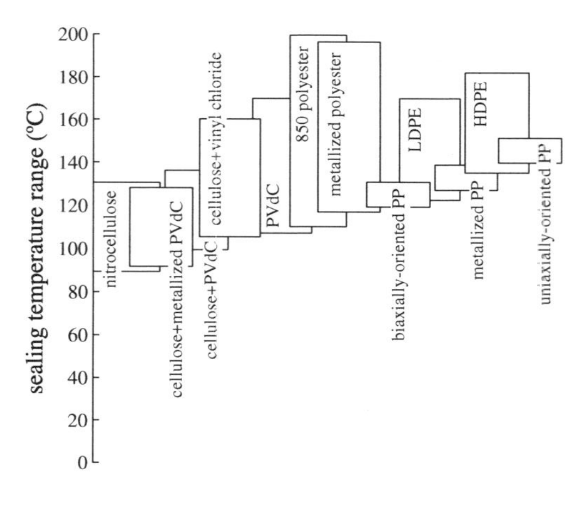

FIGURE 2.31 Ranges of sealing temperature (▭), for selected polymeric films used in food packaging.

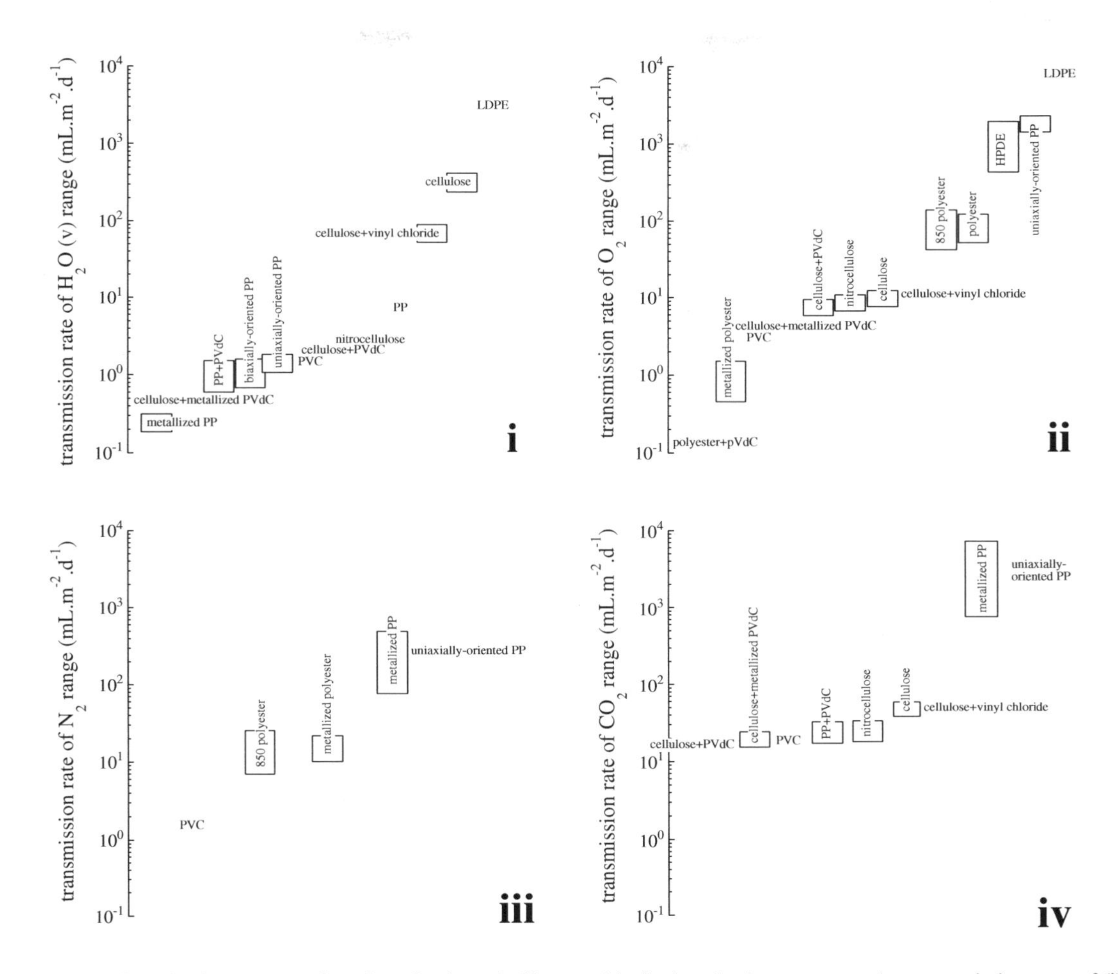

FIGURE 2.32 Ranges of barrier features (▭), for selected polymeric films used in food packaging – expressed as transmission rates of (i) water vapor (25°C, 85% relative humidity, RH), (ii) oxygen (25°C, 45% RH), (iii) nitrogen (25°C, 0% RH), and (iv) carbon dioxide (25°C, 0% RH).

contamination; by the same token, impermeability of metal, glass, and polymer packaging materials to microorganisms will not serve its purpose unless all seals have been properly made. The chief causes of microbial contamination through the package material are: contaminated air or water, which gain access to food through pinholes, tears, or creases of otherwise hermetically sealed containers – namely, driven by the headspace vacuum formed during cooling upon sterilization, when water vapor returns to its (much more condensed) liquid form; and inadequate heat seals in polymer films – namely, due to contamination of seal with product, or inadequate temperature or duration of the heat sealing process. If specific hurdles (e.g. low *pH*, low a_w, low storage temperature, presence of native or deliberately added preservatives) are applied to the food throughout its useful lifetime, then slight failure of package in its barrier role will not be critical. This is not the case, however, when thermal treatment or irradiation has taken place at a specific time in processing – and maintenance of sterility up to consumption stage requires a strict and sustained barrier thereafter, for lack of any extra hurdle. In either case, the package is expected to protect the food against contamination by soil and dust; while protection against insects, rodents, and birds will be absolute only when intact glass or metal containers are employed.

Interactions between food matrix and food packaging material are typically undesirable, unless the latter is of an edible nature; this general concern arises not only due to potentially toxic effects of the packaging material – but also because the shelf-life of food may be reduced, and its sensory quality will be at risk. Of particular concern is migration of oil from food into a plastic material (namely, flexible films) serving as package; besides reducing its barrier features, the said oil may serve as solvent, and thus a vehicle for liquid-mediated diffusional transport of residual monomers from the polymerization process, as well as all sorts of additives used in its formulation (e.g. antifogging agents, fillers, nucleating agents, pigments, plasticizers, stabilizers) into the food. Some packaging materials may also contain compounds that cause tainting of food – if absorbed thereby, or adsorbed thereon; these include solvents used to make polymer films or containers in the first place – and such additives as wax coating on papers, lacquers and sealing compounds in cans and closures, and printing inks and adhesives in labels. Acids, anthocyanins, and sulfur compounds present in food may specifically interact with metal containers; previous application of lacquers and coatings on steel, tin, or aluminum normally avoid this problem. However, failure in the lacquer of a tinplate container may allow reaction of food acids with the tin coating on the steel material, thus releasing hydrogen – which, in extreme situations, may cause swelling of the can (sometimes mistaken for *Clostridium botulinum* contamination). Although glass is an inert material – meaning that no interaction results from direct contact with food, the same does not often apply to the closure of containers made from glass.

Packaging materials are of two basic types: shipping containers – with no marketing function, and bearing concise information for the carrier about destination and special handling/storage requirements; and retail containers (or consumer units) – conveying convenient quantities for retail sale and home storage/consumption, and able to protect and effectively advertise the product. Selected examples of the latter are provided in Table 2.4. Bakery products include bread, cakes, and pies; beverages include beers, carbonated sodas, juices, and wines; dairy products include milk and yogurt, as well as butter and cheese; dried foods include cereals, coffee, fruits, and spices; pastes and purées include peanut butter and tomato/garlic paste; preserves include chutneys, jams, pickles, and sauces; snack foods include fried and extruded items; and sterilized foods include canned foods and UHT milk. Package features range from permeable films, susceptible to gas exchange in the case of respiring fresh fruits, and relying on mechanical flexibility; to sturdy, strictly sealed containers, designed for longtime preservation of low-acid sterilized foods, and bearing mechanical robustness and integrity.

Shipping containers come in the form of barrels, combo bins, crates, drums, corrugated fiberboard and plastic/metal cases, sacks, shrink- or stretch-wrapped trays, wooden structures, and woven plastic-fabric bags; and may be single-use or reusable, depending on their structure and cost. Retail containers include collapsible tubes, flexible plastic bags, glass or plastic bottles and jars, metal cans, overwraps, paperboard cartons, rigid and semi-rigid plastic tubs and trays, and sachets.

Evolution in handling and distribution methods have brought along a higher complexity in materials and modes of packaging – with fewer foods resorting to a single packaging material, or to a dedicated shipping container; combined packaging systems have indeed become the rule. One illustrative example is display cartons – containing multiple packs of food packaged in flexible film; which are, in turn, collated and stretch-wrapped, or else placed in corrugated board shipping containers. Other systems resort to shrink-wrapped trays of cans or glass containers, duly palletized and stretch- or shrink-wrapped. A third representative example is bag-in-box packages – consisting of a laminated (or coextruded) film bag, fitted with an integral plastic tap, and contained within a solid or corrugated fiberboard display case. These are, in turn, packed into fiberboard shipping containers. The bag collapses evenly as liquid is withdrawn, so trapping of food product in the folds of the bag is prevented, as well as oxidation thereof by atmospheric air; with weight reduction up to 70% (relative to glass bottles), this type of container is secure for liquid foods (e.g. cooking oils, dairy products, fruit juices, liquid egg, milk, purées, sauces, syrups, wines) and fully recyclable. Intermediate bulk containers (e.g. combo bins, large bags), made from woven plastic fabric – with capacities ranging from 1 m^3-containers (with integral pallet and bottom discharge valve) up to bulk road tankers, were introduced for enhanced handling efficiency; as a consequence, wooden crates and cases have been abandoned to a large degree. These packages may carry up to 20% more product than drums for the same space occupied; while their being flat when empty calls for up to 80% less storage space.

Remember that the shelf-life of packaged foods is controlled by food properties – playing the role of intrinsic factors; as well as by environmental conditions and package features – known as extrinsic factors. Water activity, *pH*, susceptibility to chemical/enzymatic/microbiological decay, and requirement of/sensitivity to O_2, CO_2, light, and moisture constitute the major intrinsic factors. Ultraviolet radiation, relative moisture, oxygen content, temperature, contact with microorganisms/insects/soil, and mechanical forces (arising from impact, vibration, compression, and abrasion) account for the most important environmental factors – felt by the food through its outer package; besides deliberate malevolous actions, such as pilferage, tampering, and adulteration.

2.2.2.1 Natural fibers

Textile containers exhibit poor moisture and gas barrier features – further to being labile to contamination by microorganisms and insects; they also have a poorer appearance than plastic ones, and are unsuitable for high-speed filling. Hence, their use has been restricted to shipping containers for dried foods, as well as overwraps of other packages – as a distinctive mark for some niche markets. The most common is woven jute (or burlap), used as plain weave, single yarn – chemically treated to preclude rotting and reduce flammability; it allows safe stacking, resists tearing, does not undergo significant extension, and possesses good durability. Other materials used with food products include tarpaulin (double weave) and twill – as illustrated in Fig. 2.33. These textile materials are still used to transport several bulk foods, e.g. flour, grain, salt, sugar – even though they have been gradually replaced by multiwalled paper or PP-reinforced sacks.

Wooden shipping containers have traditionally been used for a range of solid and liquid foods – as is the case of beers, fruits and vegetables, and spirits and wines; examples are included in Fig. 2.34. Wood offers good mechanical protection, further to good stacking characteristics – including a remarkable vertical compression strength-to-weight ratio. However, PP and PE drums, crates, and boxes are less expensive – and have accordingly replaced wood in many applications. Wood continues to be used for storage and maturation of many wines and spirits; in fact, transfer of flavor compounds from the wood of the barrels contributes favorably to the quality of the final drink – by bringing unique flavor keynotes. Wooden tea chests are manufactured more cheaply than other containers in tea-producing regions, so they are still widely utilized as well.

2.2.2.2 Paper and board

The raw material for paper is paper pulp – produced via preliminary grinding of wood chips (mostly spruce and eucalyptus), followed by digesting the pulp via hydrolysis under alkaline or acidic conditions; lignin, carbohydrates, resins, and gums in the pulp are accordingly dissolved, and eventually removed by washing – thus leaving chiefly cellulose fibers.

Digestion as per the alkaline (also known as Kraft's) process takes several hours, and resorts to sodium hydroxide and sulfate as active agents; this process gives higher yields, while byproducts are more completely and economically recovered

TABLE 2.4

Examples of packages, utilized for foods characterized by short ( ) or medium/long ( ) shelf-lives.

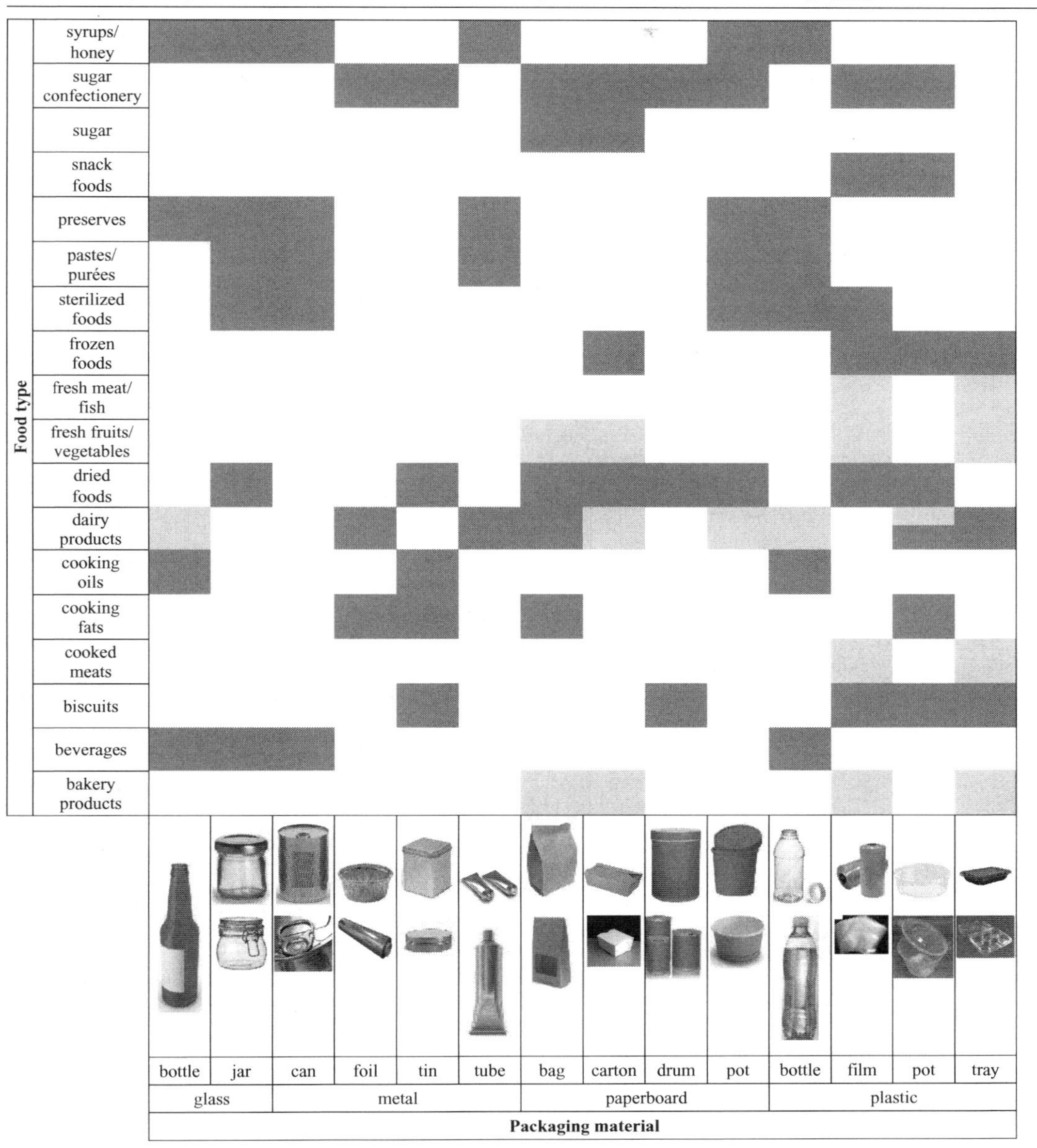

than via its acid counterpart. Kraft (or sulfate) paper is strong, and thus used in 25–50 kg (multiwalled) sacks for powders (e.g. flour, fruits, sugar, vegetables), as well as liner for corrugated boards; several layers (or plies) are normally employed, so as to attain the intended mechanical strength. It can be used as such, bearing a typical brown color – or instead undergo bleaching to become white, or even printed.

In the acid (also known as sulfite) process, sulfur dioxide and calcium bisulfite are heated with pulp to 140°C; washing and bleaching with calcium hypochlorite ensues, thus yielding quite pure cellulose fibers. Sulfite papers are lighter and weaker than their sulfate counterparts; hence, they are suitable for the manufacture of grocery bags and pouches, waxed papers, confectionery wrappers, and labels – and they may as well serve as inner liner for biscuit packs, or be laid on foil laminates.

Following digestion, either type of pulp undergoes beating – aimed at splitting individual cellulose fibers, in the longitudinal direction; a mass of thin fibrils is consequently obtained, which bind together more strongly – and so permit an increase in burst, tensile, and tear strengths. While extent of heating and thickness of fibers determine final paper strength, specific compounds may be added to the pulp to impart particular features: fillers, or loading agents (e.g. china clay) – to increase opacity and brightness, while improving surface smoothness and printability; binders (e.g. gums, starches, synthetic resins) – to improve mechanical strength; sizing agents (e.g. resins, wax emulsions) – to reduce penetration by water or printing inks; pigments – to change original color; and antifoaming agents – to assist in the manufacturing process proper.

Once in possession of paper pulp, one of two methods is normally followed to obtain paper. In the first one, fibers are

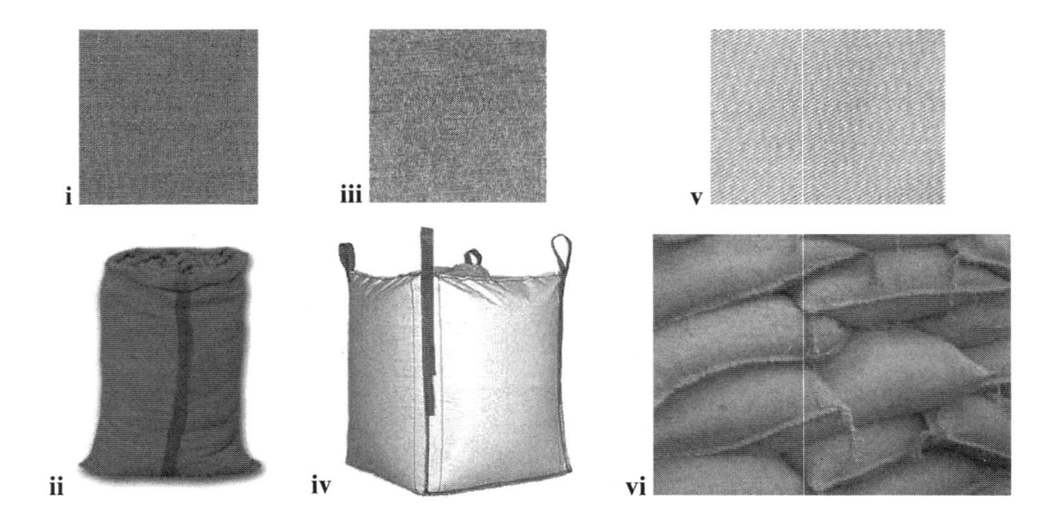

FIGURE 2.33 Examples of textile packaging for foods: burlap, as (i) detail of cloth and (ii) sack form; tarpaulin, as (iii) detail of cloth and (iv) intermediate bulk container; and twill, as (v) detail of cloth and (vi) sack form.

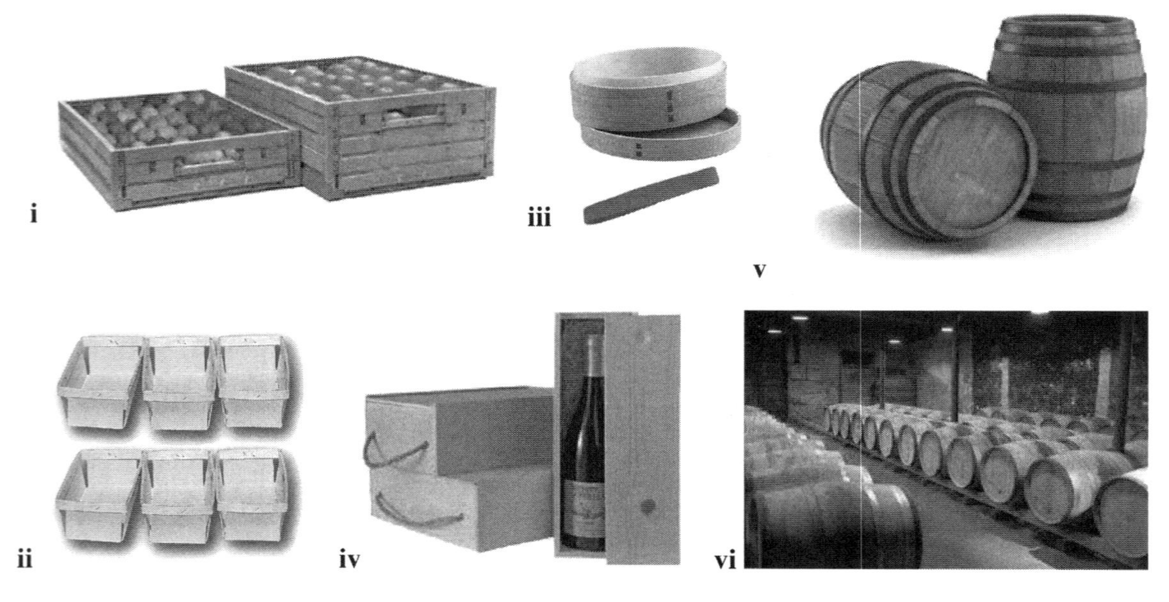

FIGURE 2.34 Examples of wooden packaging for foods: open cases for (i) transport of large quantities and (ii) display of small quantities of fresh fruits; (iii) closed boxes for gourmet products and (iv) outer packages for wine bottles; and barrels, as (v) detail of belted structure and (vi) collective view in cellar storage.

suspended in water, and then transferred to a finely woven mesh belt in a Fourdrinier machine; vacuum filtration is then brought about, to reduce moisture content of fibers down to 75–80%(w/w) – followed by pressing and drying, to attain just 4–8%(w/w) in the final paper. In the second method, 6 (or more) wire mesh cylinders, mounted in series, are partly submerged in the pulp suspension; as they rotate, fibers are picked up and deposited onto a moving felt belt. Water is accordingly absorbed from the paper, while pressing reduces the moisture content of fibers down to ca. 60%(w/w); the paper is finally dried via heated cylinders.

Many types of papers can be obtained – with characteristic weight and tensile strength ranges as indicated in Fig. 2.35. High-gloss machine-glazed (MG) or machine-finished (MF) papers are obtained by passing them between a cascade of highly polished cylinders (or calendar stack); only one roller is driven in each set, while the other ends up moving by friction with the paper – thus eventually generating a smooth surface.

Vegetable parchment is produced from sulfate pulp, after passing it through a bath of concentrated sulfuric acid – meant to swell, and partially dissolve the cellulose fibers afterward, which brings about their plasticization; the pores become accordingly closed, and voids are filled within the fiber network. The surface becomes more intact than typical kraft paper – so it is more resistant to grease and oils, and also mechanically stronger when wet; this makes it particularly suitable to wrap or line boxes used for butter, cheese, and fresh fish or meat.

When glazed, sulfite papers experience an improvement in wet strength and oil resistance; truly greaseproof paper is indeed obtained from sulfite paper pulp, following a more thorough beating that creates a closer structure. It is resistant to oils and fats, yet such a feature is essentially lost upon wetting; nevertheless, it is widely used to wrap fish and meat, as well as bakery and (fatty) dairy products.

Glassine is similar to greaseproof paper, but undergoes additional calendaring – to increase density, and generate a close-knit structure and a high-gloss finish. It is relatively resistant to water when dry – but loses it once wet; its water-resistance is substantially improved upon wax-coating, though. It is appropriate for odor-resistant, greaseproof bags, and wrappers or liners for boxes; it is frequently found in packages of cake mixes, coffee, dried soups, dry cereals, potato crisps, and sugar.

Finally, a mention is deserved by tissue paper – a soft, non-resilient paper; it is used, namely, to protect bread and fruits against dust and bruising.

Paper holds a number of advantages when used for food packaging: besides the many grades available (as described above), it is quite flexible to package shaping; it is recyclable and biodegradable; and it is easily combined with other materials, to make coated or laminated packaging. Selected examples are provided in Fig. 2.36. All types of paper protect food from dust and soil; however, they exhibit, in general, a rather poor performance as a barrier to water vapor (or liquid water, for that mater) and gases – and cannot be heat-sealed. Coated papers, in particular, may lose moisture from one face, and thus become curled; if too smooth, they will in turn tend to block when pressed in a stack. Therefore, the optimum storage conditions for papers are ca. 50%(RH) and 20°C – even though papers are poorly affected by temperature.

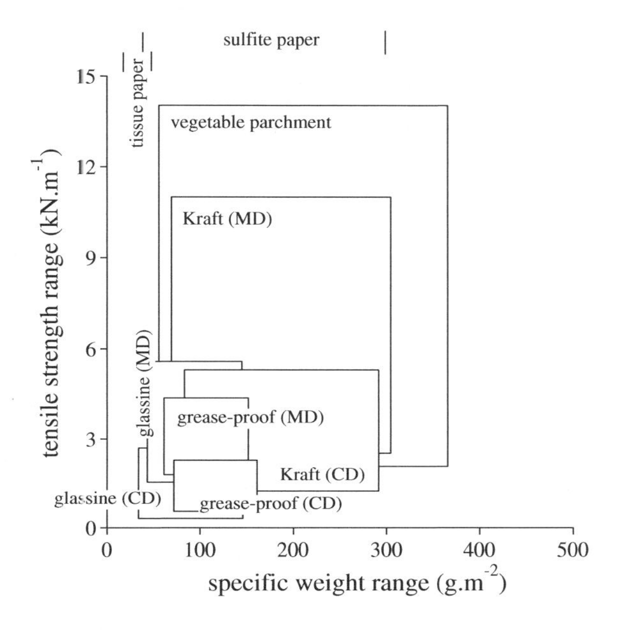

FIGURE 2.35 Ranges of physical and mechanical features (▭) for selected papers used in food packaging – expressed as tensile strengths and specific weights (MD: machine direction; CD: cross direction).

Many papers undergo treatment with wax – due to its moisture barrier features, while bringing about the possibility for sealing of paper matrices; waxed papers are utilized as bread wrappers and inner liners for cereal cartons, among other applications. There are three basic modes of wax application: wax sizing – in which wax is added during pulp preparation; dry waxing – with wax applied while the paper is still hot; and wax coating – which consists of laying wax after the paper has cooled to ambient temperature. In the latter case, easy damage by (repeated) folding or (attritional) contact with abrasive foods can be avoided by laminating wax between layers of paper and/or polyethylene.

The standard thickness of some plastic films – as required for a given degree of protection, is less than can be handled by some filling and forming machines; when the said films are relatively expensive, producing a thicker film is hardly an alternative – yet coating onto (a less expensive) paper substrate provides the intended strength and required handling features, besides turning paper greaseproof. Coatings can be applied from aqueous solutions (e.g. cellulose ethers, polyvinyl alcohol), or organic solvent solutions of lacquers or aqueous dispersions (PVdC); or else be applied as hot melts (e.g. microcrystalline wax, PE, and copolymers of ethylene and vinyl acetate) or via extrusion (e.g. polyethylene) – to increase durability, gloss, scuff, and crease resistance.

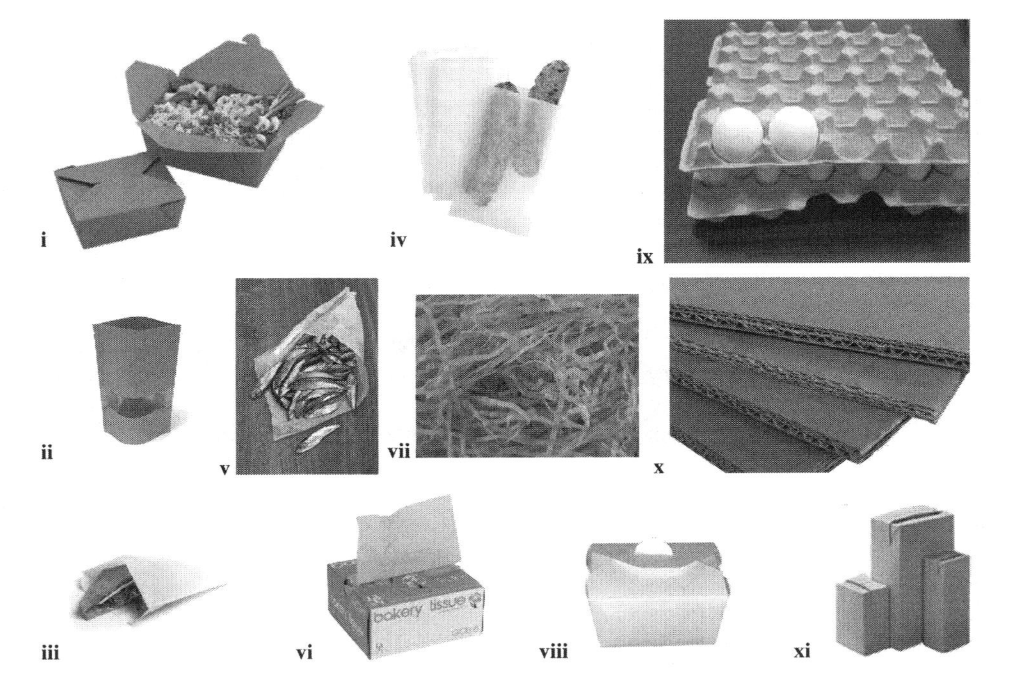

FIGURE 2.36 Examples of paper and board packaging for foods: (i) closed boxes and (ii) pouch, made of kraft paper; (iii) pouch made of greaseproof paper; (iv) pouch made of glassine paper; (v) wrap made of vegetable parchment; tissue paper in (vi) napkin and (vii) shredded form; (viii) whiteboard container; (ix) molded paper-pulp container; (x) corrugated board plate; and (xi) laminated paperboard packages.

When referring to boxboard, chipboard, and (corrugated or solid) fiberboard, the generic term paperboard has classically been employed. Most paper board possesses fourfold structure – with plies glued together with hot melt or aqueous adhesives: a top ply of bleached pulp – meant to provide mechanical strength to the surface, as well as printability; middle plie(s) made of a lower grade material – meant to create bulk volume; an underliner of white pulp – meant to reduce the tendency for the grey/brown color of middle plies show through; and a back ply – of either lower grade paper pulp when intended for plain stacking, or higher grade paper pulp when mechanical strength or printability are a must. Boards are manufactured in a way similar to paper – but are thicker, and chiefly designed to protect foods from mechanical damage; mechanical features sought include thickness, stiffness, and ability to crease without cracking – while visual features include surface smoothness, level of whiteness, and suitability for printing. Four major types can be pinpointed: whiteboard, fiberboard, corrugated board, and laminated paperboard. When compared to metal or glass containers, these types of cartons are stronger and unlikely to fracture, in mechanical terms; require no additional capping or labeling – and bold graphics on the sides may even create a billboard effect, when in display shelves; their manufacture is less energy-intensive; and permit substantial savings in weight – relevant to reduce transportation and handling costs, as well as in volume – relevant to save space throughout distribution, storage, and display.

Whiteboard is suitable to directly contact food in the form of cartons (e.g. chocolate, ice cream, frozen items) – and is often coated with PE, PVC, or wax to bring about heat sealability; conversely, chipboard is manufactured from recycled paper, and used for outer cartons not in direct contact with foods (e.g. breakfast cereals, tea) – being sometimes lined with whiteboard, for better appearance and higher strength. Duplex board cartons (e.g. biscuits, frozen foods) are typically 0.3–1 mm thick, and possesses two layers; the inner one is manufactured from bleached wood pulp, whereas the outer one is unbleached.

Fiberboard – meant to resist compression (and impact, to a lesser degree) and bearing a thickness above 0.1 mm, is available in either solid or corrugated form; the former consists of an outer layer of kraft paper and an inner layer of bleached board. Small fiberboard cylinders, also known as composite tubs or cans (used for e.g. cocoa powder, confectionery, frozen juice concentrates, nuts, salt, snack foods, spices), are manufactured from single ply-board – and may include an extra aluminum foil or LDPE layer. After spirally wound around a mandrel (in a helical pattern) and adhesive-bonded, they are cut to the intended can length – and flanges are formed at each end; they are eventually fitted with plastic or metal caps, with an easy-open or a pouring mechanism. Larger drums may also be manufactured from fiberboard – with volumes up to 400 L, and constitute a less expensive alternative to metal drums for powders and other dry foods (e.g. frozen fruits and vegetables). They may be lined or laminated with PE, to handle greasy or otherwise liquid foods (e.g. cooking fats, peanut butter, sauces, wine); or be obtained from single-ply board, with a moisture-proof membrane below the surface for transport cases exposed to moisture (e.g. chilled foods) – since they are lightweight, resist compression, and may be made waterproof for outside storage.

Corrugated board consists of an outer and an inner lining of kraft paper, complemented by a layer of corrugated (or fluted) material in between; this composite material possesses enhanced mechanical resistance to impact, abrasion, and compression – thus making it particularly appropriate for shipping containers of bottled, canned, or plastic-packaged foods. There are four commercial types; coarse (with 4.5–4.7 mm height flutes, at a pace of 104–125 flute.m^{-1}, and a minimum flat crush of 140 N.m^{-2}); fine (2.1–2.9 mm, 150–184 flute.m^{-1}, and 180 N.m^{-2}); medium (3.5–3.7 mm, 120–145 flute.m^{-1}, and 165 N.m^{-2}); and very fine (1.15–1.65 mm, 275–310 flute.m^{-1}, and 485 N.m^{-2}). Higher rigidity is associated with smaller and more frequent corrugations; larger corrugations, or double or triple walls convey better resistance to damage by impact; and twin-ply fluting, with a strengthening agent between layers, has the same stacking strength but half the weight of solid board – so it permits space savings of ca. 30%, *vis-à-vis* with double-corrugated boards of comparable strength. Corrugation is accomplished via softening the fluting material with steam, and then passing it over corrugating rollers; the liners are then applied onto each side, upon spreading a suitable adhesive. The board is formed into cut-outs, which are finally assembled into cases along the filling line. For being based on (hygroscopic) paper, boards are to be stored in a dry atmosphere – otherwise they risk losing their strength, and may even undergo delamination of the corrugated material. Packaging of wet foods (e.g. chilled bulk meat) requires previous lining of corrugated board with PE – aimed at preventing moisture migration and tainting; as an alternative, a laminate of greaseproof paper, coated with microcrystalline wax and PE, can serve as liner (suitable for e.g. fresh fruit and vegetables, frozen food, dairy products, meats).

As referred to before, paper may also be combined with aluminum, LDPE, polyamides, or PVdC to manufacture laminated paperboard cartons – meant to aseptically package sterilized foods; and two basic approaches exist to do so. In the first one, the laminated material is supplied as individual, preformed, collapsed sleeves – with the side seam already provided by the manufacturer; the cartons are erected at the filling line in the food plant, filled up, and sealed – with the top seal formed above the food, with generation of a headspace that aids in mixing of the food product upon shaking, while reducing the risk of spillage once opened. The laminate is supplied as a roll in the second approach – with lower space requirements for storage and prone to easier handling; cartons are formed therefrom, in form–fill–seal equipment.

Paper and fiberboard containers are safe and cheap, and are mechanically sturdy; they resort to natural and (synthetic) water-based adhesives, also used for corrugated board – and are useful for labeling of cans and bottles, seams of bags, and winding of paper tubes. In the case of carton and case closing, tube winding, and paperboard lamination and labeling, starch pastes and dextrins are elected. For carton making, cold-seal coatings for confectionery wrappers, and water-resistant labeling of drums, cans, and bottles, latex is blended with acrylic resins. Hot melt adhesives contain no water or solvent, and are heated until they become sufficiently fluid for application; upon cooling, they set rapidly. The most frequently employed is EVA copolymer – but low molecular weight PE is also suitable for carton and bag sealing.

Molded paper-pulp containers are designed to absorb shocks, by distorting and even crushing the matrix at the point of impact – thus avoiding transmission of such shocks to the packaged food product; they are lightweight, with typical 2.5 mm-thick walls – and employed in bottle sleeves, as well as in trays for eggs, fish, fruits, and meats. Paper pulp is used as starting material, and is subjected to molding via either pressure injection or suction injection. In the former, air at 480°C is used to form the pack that leaves at ca. 50%(w/w) moisture; the product is finally dried. The (more expensive) suction injection resorts to a perforated mold, with the pack leaving at ca. 85%(w/w) moisture prior to drying. When compared to competing shock-absorbing, lightweight materials, molded paper pulp exhibits lower fire risk and static problems than expanded polystyrene; and allows more complex final shapes be obtained than with corrugated board.

2.2.2.3 Synthetic polymers

A polymeric material may, in a sense, be viewed as an aggregate of wriggling worms – each one accounting for a long molecule, bearing an intrinsic flexibility high enough to support almost unrestricted shaping. Spacing between said worm-like molecules brings about porosity, i.e. supports a network of (tiny) pores within the material that permit solute movement; while wriggling arises from large chain segments sliding over each other, due to thermal agitation. Despite representing distinct features of a polymeric material, permeability (as measure of resistance to flow through) and porosity (as measure of fraction of voids) are closely related to each other; a fluid will indeed permeate the solid structure by moving along its constitutive pores – and the greater the number and size of the said pores in a given mass of solid, the easier the fluid will pass through. In other words, a higher porosity in a material is normally accompanied by a higher permeability.

The long molecules that makeup plastics are formed via addition reactions (e.g. polyethylene from ethylene monomers) or condensation reactions (e.g. polyethylene terephthalate from ethylene glycol and terephthalic acid, with concomitant elimination of water). Thermoplastic materials can undergo repeated softening upon heating, and subsequent hardening upon cooling. The aforementioned long molecules will become cross-linked upon heating, or treatment with specific chemicals, in the case of thermosetting plastics; under these circumstances, they will no longer be able to resoften. Two major forms are available for food packaging purposes – flexible films, and rigid or semirigid plastic containers; they serve distinct purposes, in terms of nature of food and intended mode of preservation, as well as of associated extent of shelf-life.

2.2.2.3.1 *Flexible films*

Flexible packaging refers to any type of material that is intrinsically not rigid – even though this term has classically been applied to nonfibrous plastic polymers, with thickness not exceeding 0.25 mm; examples provided in Fig. 2.37 illustrate their ability to meet the packaging requirements of a multitude of foods. Flexible films exhibit a wide range of barrier properties against moisture and gases, further to a wide range of tensile and impact strengths (under wet and dry forms) – and are, in general, nonexpensive; these features have made them quite popular. Besides being heat-sealable, and thus able to prevent leakage of contents *a posteriori*, polymeric films can be laminated to paper, aluminum, or other polymers – thus giving rise to composite materials with unique characteristics. From a processing point of view, they are compatible with high-speed filling, and can easily be handled and printed; they add little weight to the food product; and their fitting closely to the food shape minimizes wasting space throughout storage and distribution.

Variation in film thickness, orientation of polymer molecules, qualitative and quantitative profiles of additives, and type and thickness of coatings permit modulation of barrier, mechanical, optical, and thermal properties of flexible films. Selected data on thickness and yield ranges are displayed in Fig. 2.38. Some films (e.g. polyamide, polyester, PE, PP) are oriented by stretching softened material – thus bringing about alignment of their molecules, in either a single direction (uniaxial orientation) or two perpendicular directions (biaxial orientation); besides increasing their strength, transparency, and flexibility, this process is likely to increase moisture and gas barrier features – relative to random (unoriented) films. A number of additives have meanwhile been employed for specific purposes – as is the case of antioxidants, pigments, plasticizers, slip agents, and stabilizers. Inclusion of pigments is prone to decrease the need for extended printing areas; while plasticizers contribute to soften the film and make it more flexible – an important feature for use in cold climates or with frozen foods. Films are currently utilized in single form, coated with another polymer or aluminum, or as multilayered laminates or coextrudates.

Manufacture of most polymer films usually resorts to extrusion; pellets of the polymer are first melted and then pumped, under pressure, through a nozzle – thus generating a continuous sheet or tube, which is finally cut at will; alternative processes encompass coextrusion, calendaring, and casting. Coextrusion consists

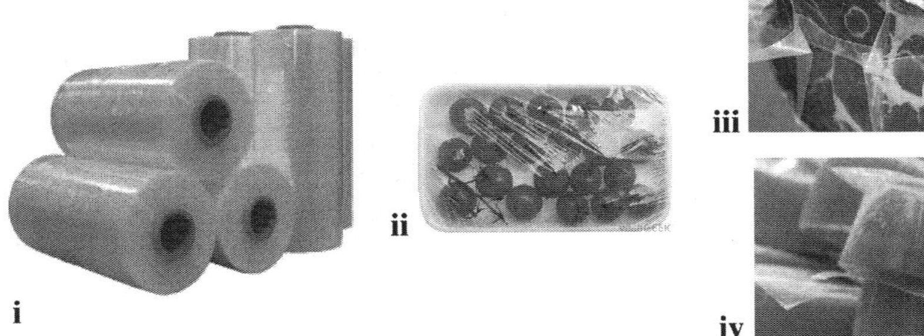

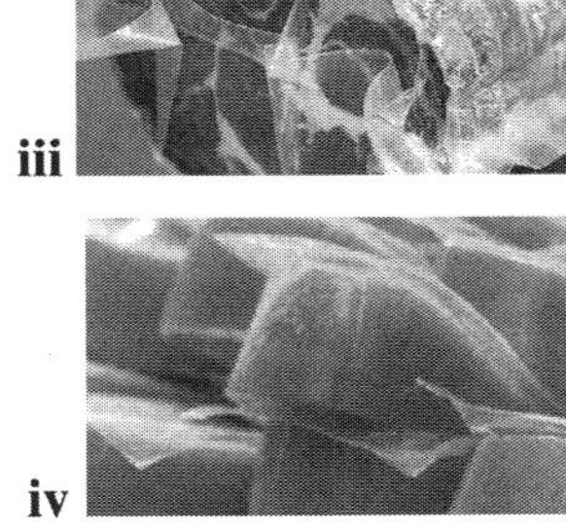

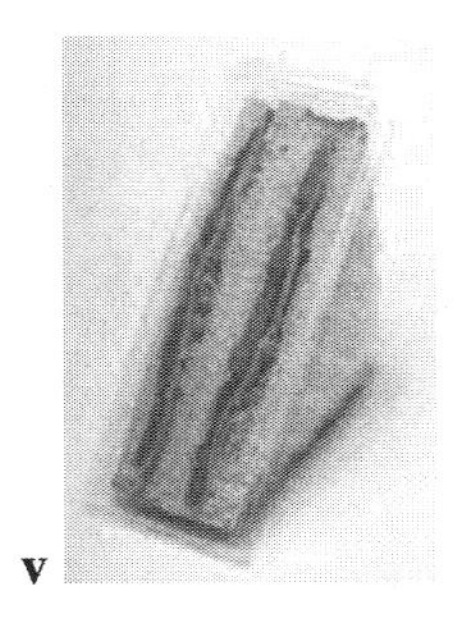

FIGURE 2.37 Examples of flexible polymeric packaging for foods: (i) wrapping film proper; and wraps of various thicknesses, enclosing such fresh vegetables as (ii) cherry tomatoes, and such transformed foods as (iii) processed meat, (iv) cheese, and (v) ready-to-eat sandwich.

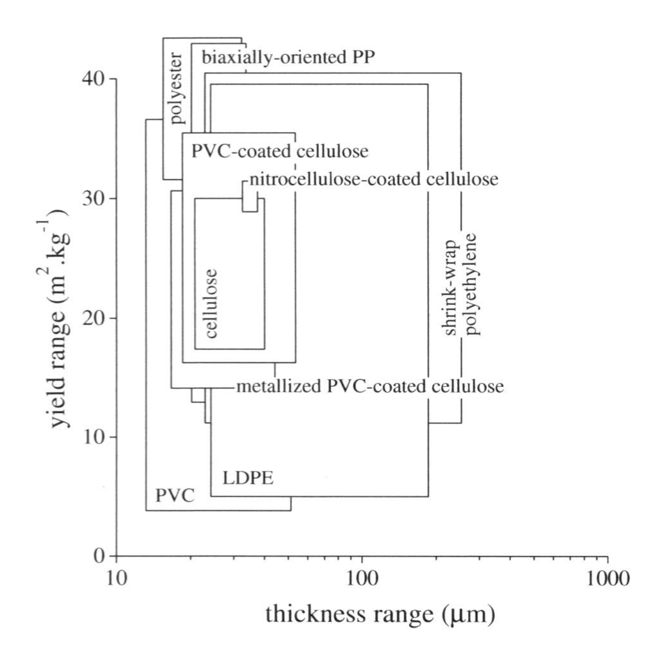

FIGURE 2.38 Ranges of physical and mechanical features (▭) for selected polymeric films used in food packaging – expressed as yields and thicknesses.

of simultaneously extruding two (or more) polymers – which will eventually fuse together, to yield a single film. In calendaring, a polymer (e.g. ethylene vinyl acetate, polyvinyl chloride) is passed through heated rollers, until reaching the intended thickness. A polymer solution (e.g. cellulose acetate), together with suitable additives may instead undergo casting through a slot onto a stainless steel belt; the solvent is then evaporated by heat, thus leaving a transparent, sparkling film. Film surfaces may be treated *a posteriori* with ion beams or cold gaseous plasmas; or undergo flame- or corona-treatments – aimed at improving sealability, adhesion, printability, or barrier properties.

The most important types of film used in food packaging are: cellulose, polyethylene (PE for short) in low- (LDPE for short) and high-density (HDPE for short) versions, ethylene vinyl-acetate (EVA for short) or -alcohol (EVOH for short), polyamides (PA, nylon for short), polyethylene terephthalate (PET for short), polypropylene (PP for short), regular- (PS for short) and high impact-polystyrene (HIPS for short), polyvinyl chloride (PVC for short), polyvinylidene chloride (PVdC for short) and rubber hydrochloride. Using PET as reference for cost, EVOH is fourfold more expensive and nylon is twofold; whereas PP (0.85-fold), HIPS (0.82-fold), HDPE (0.75-fold), and LDPE (0.70-fold) appear as the least expensive alternatives.

Cellulose films are obtained by first dissolving sulfite paper pulp in caustic soda; following a ripening period of 2–3 days aimed at reducing length of the polymer chains, sodium cellulose is formed – which is then converted to cellulose xanthate via treatment with carbon disulfide, and ripened for 4–5 days to produce viscose. Cellulose (in the form of cellulose hydrate) is finally regenerated, via extrusion or casting into an acid-salt bath – where glycerol is frequently added as softener; the film undergoes final drying on heated rollers. Increasing the level of softener and the space time spent in the bath lead to more flexible and permeable films. A wide portfolio of cellulose films is commercially available: plain, and thus non-moisture proof (code P), semi-moisture-proof (Q), or moisture-proof (M); transparent (T), colored (C), or opaque (B); anchored or lacquer-coated (A), or coated on one side only (D); copolymer coated from aqueous dispersion (/A) or solvent (/S); copolymer coated on one side (X) or both sides (XX); heat-sealable (S); and meant for such specific applications as twist-wrapping (F), adhesive tape (U), or low temperature storage (W). Plain cellulose generates glossy transparent, odorless, tasteless, and greaseproof films; and is susceptible to biodegradation within 100 days. It is tough and puncture-resistant, despite being easily torn off; its low-slip and dead-folding features, coupled with its being unaffected by static buildup make it appropriate for twist-wrapping. Unfortunately, it is not heat-sealable *per se*; and increasing environmental humidity expands it, thus enhancing permeability to a considerable extent. Therefore, it should be used only to package foods that do not require complete gas or moisture barrier (e.g. fresh bread, some sugar confectionery). Cellulose acetate requires plasticizers for film manufacture; it is a clear, glossy transparent, and sparkling film – through which vapor, odors, and gases permeate easily. It is chiefly used as window material in paperboard cartons – and is specifically characterized by good printability, rigidity, and dimensional stability, as well as toughness and resistance to puncture; however, it is not tearing-proof.

LDPE film is odorless and chemically inert, and shrinks upon heating; it is heat-sealable, and suitable for shrink- or stretch-wrapping – with the latter normally resorting to thinner films. It constitutes a good barrier to moisture, yet it is relatively permeable to gases; furthermore, it is sensitive to contact with oils, and resists poorly to pre-existing odors. Both low- and high-slip features can be imparted to the film – the former aimed at safer stacking, and the latter meant for easier filling of packs into an outer container. It is less expensive than most films, and thus widely applied in bags, coating papers or boards, and pouches – as well as a component of laminates. Linear LDPE (or LLDPE for short) consists of a highly linear arrangement of molecules, with lower average molecular weight than LDPE; it exhibits greater strength and higher restraining force. The cling properties of both LDPE and LLDPE may be biased on one side – thus maximizing adhesion between film layers, while minimizing adhesion between adjacent packages. HDPE is stronger, thicker, less flexible, and more brittle than LDPE – and possesses lower permeability to moisture and gases. In view of its higher softening temperature (ca. 120°C), it can be heat-sterilized or used for boil-in-bag applications; however, it is still suitable for shrink-wrapping. Sacks made from this material display high tear and tensile strengths, penetration resistance, and seal robustness; they are also waterproof and chemically inert – and have thus been increasingly utilized in shipping containers, instead of multiwalled paper sacks. A foamed HDPE film is thicker and stiffer than conventional films, and has dead-folding properties; it has met with success in the case of edible fats – and, when perforated (with up to 80 hole.cm^{-2}), it becomes suitable for use with fresh bakery products (and the like).

EVA consists of LDPE polymerized with vinyl acetate – which imparts high mechanical strength and flexibility, experienced even at low temperatures; this copolymerization brings about extra flexibility and permeability to gases and vapors relative to LDPE, as well as higher resilience than PVC. Three major types are available, depending on the range of content in vinyl acetate (VA for short): <5% VA film is used for deep-freeze applications;

6–10% VA is suitable for bag-in-box applications and milk pouches; and >10% VA is employed as hot-melt adhesive.

EVOH presents a higher barrier to O_2 – of the same order of magnitude of PVdC, despite its being more expensive; its hydrophilicity makes it more permeable to moisture – which comes along with a higher sealing temperature (ca. 185°C). It is used mainly as laminate, with either PP or PE – thus conveying the required moisture barrier properties, together with sealability.

Nylon and PA are clear, mechanically strong films, over a temperature range as wide as from –60 up to 200°C; they are greaseproof, and exhibit low permeability to gases – even though permeability to water vapor depends on the type of film. The major shortcomings are their high cost of manufacture and high sealing temperature (above 140°C), and the effect of environmental humidity upon permeability thereof. They are often coextruded, coated, or laminated with other polymers – in attempts to make the former heat-sealable at lower temperatures, and improve their barrier properties; major applications include packaging of cheese and meat.

PET is a very strong and glossy transparent film, bearing good gas and moisture barrier properties; moreover, it maintains flexibility within a temperature range spanning from –70 to 135°C – and undergoes little shrinkage/expansion upon variations in temperature or humidity. Two types exist – amorphous (APET for short) and crystalline (CPET for short). The former is clear and biaxially-oriented, and possesses the high tensile strength required by boil-in-bag packages or bottles for carbonated drinks. Conversely, CPET is opaque – and common in microwave trays, or such semirigid containers as tubs.

PP is a strong film, yet it becomes brittle at low temperatures; furthermore, it exhibits low permeability to water vapor and gases. To lower its sealing temperature (ca. 170°C), it is tendentially coated or laminated with PE or PVdC/PVC; its applications are shared with LDPE. Oriented PP (or OPP for short) is a clear, glossy transparent film, possessing good optical features – besides high tensile strength and puncture resistance (even at low temperatures); it holds moderate permeability to gases and odor-associated volatiles, and raises a high barrier to water vapor – which is hardly affected by environmental humidity. In view of its thermoplasticity, it stretches upon heating – although less than PE; and exhibits low friction, so static buildup is kept to a minimum – while permitting high-speed filling. The coated or laminated forms are used in a large number of applications – e.g. packs for biscuits, cheese, coffee, and meat. Biaxially-oriented polypropylene (or BOPP for short) behaves in much the same manner as PP – but is quite stronger; both PP and BOPP are employed to manufacture biscuit wrappers, bottles, crisp packets, and jars.

PS is a brittle, clear sparkling film – which possesses good gas permeability, and is relatively inexpensive; its biaxially-oriented version (BOPS for short) exhibits better gas barrier properties and mechanical strength – yet a high permeability to gases is still observed. As a film, it can be used to wrap fresh produce; its foam version is used in cartons or trays for transport and display of eggs, fresh fruits, and takeaway meals. It is often coextruded with EVOH or PVdC/PVC, to make semirigid containers and blow-molded bottles. HIPS is used to make semirigid containers and trays – known for their being able to withstand freezing temperatures; however, they are still not suitable for use in microwave and conventional ovens, or even in modified atmosphere packaging.

PVC is a clear, transparent, brittle, and grease-proof film – manufactured via either extrusion or calendaring. Its flexibility is improved via addition of plasticizers – thus making it suitable for stretch-wrapping and use as a cling film; plasticizers also aid in modulating permeability to water vapor, gases, and volatiles.

PVdC is stiff and brittle; its unique mechanical strength permits utilization as a single film. Despite being poorly permeable to water vapor and gases, it is heat-sealable and -shrinkable; it can be used in freezer-to-oven foods, since it resists contact with fats and does not melt in contact therewith (even when hot). It is normally used as a copolymer with PVC – which possesses even greater strength and barrier properties, while remaining heat-shrinkable; hence, it is widely utilized for shrink-wrapping poultry and meats. PVdC is frequently found as a component of laminates, and as a coating for film and bottles aimed at improving their barrier performance.

Finally, rubber hydrochloride is similar to PVC – but becomes brittle when exposed to ultraviolet light; in addition, it is unable to resist oils at low temperature. Therefore, its utilization with food items is quite limited.

Polymer films may be coated with other polymers or aluminum – in attempts to improve their barrier properties, or impart heat-sealability (as mentioned before); an example is nitrocellulose with added waxes and resins – coated on one side of a cellulose film to enhance moisture and gas barrier features, or on both sides to further improve barrier features to moisture, gases, and odor volatiles, while supporting heat seals. An alternative to improve heat sealability and barrier properties is application of PVdC/PVC coating on one side (coded as MXD cellulose), or on both sides of a cellulose film – from either an aqueous dispersion (MXXT/A cellulose) or an organic solvent (MXXT/S cellulose). If coating is performed with vinyl acetate, a stiffer film results showing intermediate permeability – useful to package meats prior to smoking or cooking; sleeves of this material are tough, stretchable, and permeable to air, moisture, and smoke. Metallization consists of producing a thin coat of aluminum – via deposition of vaporized Al particles onto the surface of cellulose, PP, or polyester film under vacuum; this process improves barrier features against moisture, odor volatiles, and gases – besides oils and light. Since such metallized films are less expensive and more flexible than foil laminates, they are suitable for high-speed filling in form–fill–seal apparatuses.

The most versatile method of lamination consists of applying an adhesive onto the surface of each film, followed by drying; the two films are then brought into contact, and undergo pressure-aided bonding by passing them between rollers (a technique known as dry bonding). Adhesives are of synthetic origin – typically aqueous dispersions (or suspensions) of polyvinyl acetate with other compounds (e.g. polyvinyl alcohol, 2-hydroxycellulose ether); urethane adhesives – consisting of a polyester- or ether-resin, added with an isocyanate for cross-linking, are employed to a lesser extent. Besides cost, human toxicity and environmental regulations dictate the type of adhesive to be employed. Not all films, however, can be successfully laminated; similar characteristics are normally a *sine qua non*. Film tension, adhesive application, and drying conditions must be accurately controlled – otherwise bulk smooth unwinding becomes a likely event, edge curling can develop, and even layer separation may take place. Lamination improves appearance, barrier properties, and mechanical strength – by coupling the unique properties brought about by each film; for instance, the strength of a nylon film can be combined with the barrier properties of a PVdC film – coupled, in turn, to the heat sealability of an LDPE film. Copolymerized vinyl acetate, and ethylene- or acrylic-esters improve adhesion; they are also used for case sealing, spiral tube winding, pressure-sensitive coatings, and plastic bottle labeling.

Typical examples of double laminates include (listed from outside to inside of package): metallized PET film with PP film – for retort pouches; metallized polyester film with PE film – for bag-in-box packaging, coffee, dried milk, frozen foods, and potato flakes; PE film with nylon film – for vacuum packs of bulk fresh cheese and meat; PP film with EVA film – for modified atmosphere packaging of bacon, cheese, and cooked meats; PVC film with LDPE film – for modified atmosphere packaging of respiring fresh produce; PVdC-coated PP film with PE film – for bakery products, cheese, confectionery, dried fruits, and frozen vegetables; and PVdC-coated PP film with another PVdC-coated PP film – for biscuits (regular and chocolate-containing), confectionery, crisps, ice cream, and snack foods. Common triple laminates include: BOPP film with nylon film with PE film – for retort pouches; cellulose film with PE film with cellulose film – for bacon, cheese, coffee, cooked meats, crusty bread, and pies; nylon film with PVdC film with LDPE film, or nylon film with EVOH film with LDPE film – for modified atmosphere packaging of nonrespiring fresh produce; and PE film with foil with paper – for chocolate, dried soup, and dried vegetables. Quadruple laminates commercially available include: cellulose acetate film with paper with foil with PE film – for dried soups; and nylon film with LDPE film with butene copolymer – for boil-in-bag packaging. A quintuple laminate, also accessible in the market, is nylon film with PVdC film with PE film with foil with PE film – for bag-in-box packaging.

Coextrusion consists of simultaneous extrusion of two (or more) layers of distinct polymers, so as to form a collapsed final film; blown film and flat-sheet coextrusion are the two major processes for manufacture thereof. Coextruded films possess three major advantages: notoriously high barrier features – similar to those of multilayered laminates or wax-coated paper, yet obtained at a lower cost; small thickness, which facilitates utilization in forming and filling equipment – since they are thinner than laminates, and close to mono-layered films; and virtual impossibility of layer separation. Once again, strong adhesion hinges critically upon similarity in chemical nature of adjacent films; similar hydrodynamic patterns (requiring, in turn, close viscosities) are also required throughout the extrusion process. The three major groups of polymers eligible for coextrusion are olefins – i.e. LDPE, HDPE, and PP; styrenes – i.e. PS and acrylonitrile/butadiene/styrene (or ABS, for short); and PVC polymers. All materials within each group, as well as ABS and PVC adhere spontaneously to each other; other combinations need bonding through EVA. Blown coextruded films are thinner and suitable for high-speed form–fill–seal machines and pouch/sachet equipment. Triple co-extruded films are the most common type – and used for confectionery, cereals, dry mixes, and snack foods: the outermost layer constitutes the presentation layer, so it possesses high gloss and printability; the middle (usually thicker) layer ensures such mechanical features as stiffness, strength, and split resistance; and the innermost layer permits heat-sealing. A typical double coextruded material is HIPS film with PET film – for packaging of butter and margarine. A triple coextruded material is PP film with saran film with PP film – for retortable trays of (sterilized) foods; another example is PS film with EVA film with PE film – for modified atmosphere packaging of fruits and meats. Quadruple coextruded materials include PS film with another PS film with PVC film with PS film – for juices, meats, and milk products, chosen for its good barrier features against UV light and odors; similar features are shown by co-extrudates of PS film with another PS film with PVdC film with PE film – appropriate for butter, cheese, coffee, margarine, mayonnaise, and sauces. Metallized polyester is often replaced by quintuple coextruded (blown) films for bag-in-box applications. Flat-sheet coextrusion produces 0.075–3 mm films – which may be formed into pots, tubs, or trays.

2.2.2.3.2 Rigid and semirigid containers

Single or coextruded polymers are often utilized to manufacture trays, cups, tubs, bottles, and jars – i.e. food containers that are intrinsically rigid or semirigid. Several advantages exist relative to similar containers made of metal or glass, namely: lower weight – thus permitting savings up to 40% in transportation costs; possibility of tapering of cups, tubs, and trays (via a rim wider than its base) – thus allowing more compact stacking, and thus more effective storage; lower manufacturing temperature – and so smaller energy costs; susceptibility to precise molding into a wider variety of shapes than glass – besides being tough, unbreakable when subjected to impact and overpressure, and easier to seal; stronger chemical resistance than metal; and easier coloring for esthetic appeal and protection from UV light. Commercial examples are provided in Fig. 2.39. Note that reuse of plastic containers is more difficult than of glass containers; whereas the former exhibit lower thermal resistance and rigidity than their glass or metal counterparts.

Seven distinct methods exist to manufacture plastic containers – thermoforming, blow molding, injection molding, injection blow molding, extrusion blow molding, stretch blow molding, and multilayer blow molding.

A film of polymer – usually PVC, PS, PP, PVC-PVDC, or PVC-PE-PVdC, is thermally softened over a mold in the case of thermoforming; pressure is then applied over the film against the mold, or vacuum is instead applied between film and mold – in both cases aimed at final shaping. Examples include trays or punnets for chocolates, eggs, or soft fruits, as well as cups and tubs for dairy products, dried foods, ice cream, and margarine; such containers are relatively weak, from a mechanical point of view, owing to their thin walls.

Blow molding resembles the blow processes used in glassmaking; bottles, jars, and pots are manufactured this way, through a single- or double-stage process – and eventually used as packages for beverages, cooking oils, sauces, and vinegar. Recent improvements have permitted extension of this technique to PVC matrices – namely to obtain bottles suitable for wine, cooking oils, and fruit juices; since thermal degradation of PVC occurs at a temperature slightly above its melting point, accurate control of temperature is a must – to avoid compromising mechanical and chemical integrity of the container.

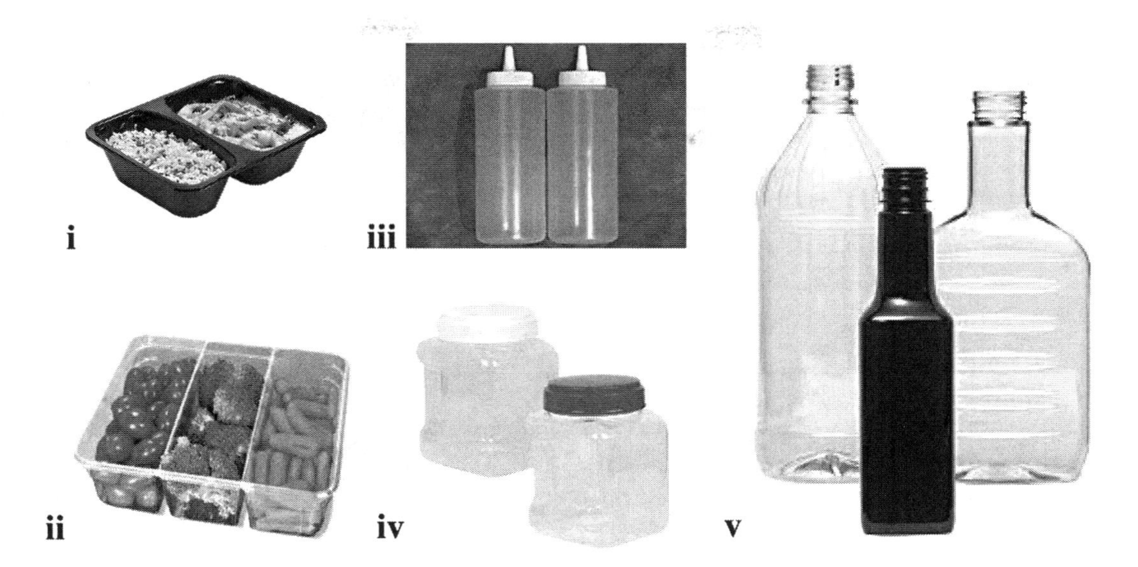

FIGURE 2.39 Examples of rigid/semirigid polymeric packaging for foods: (i) trays for ready-meals, (ii) tubs for fresh produce, (iii) squeezing containers for sauces, (iv) jars for jams or jellies, and (v) bottles for cooking oils or olive oil, fruit juices, sauces, vinegar, or water.

In injection molding, grains of a polymer are mixed and heated by a rotating screw – which also serves to propel the mix along the molding machine; and injected, under high pressure, into a cool mold. This process is normally employed to manufacture such wide-mouthed containers as jars and tubs, besides lids thereof.

As an alternative, the polymer (usually HDPE, PP, or PS) may be injection-molded around a blowing stick – and transferred to the blowing mold while still molten; compressed air is then admitted to produce the intended shape for the container – and thus finalize the injection blow molding process. Accurate control of container weight is possible, as well as precise neck finish; this technique is also more efficient than extrusion blow molding, and preferred for <0.5 L-bottles.

A continuously extruded tube of softened polymer is trapped between the two halves of a mold, and both ends are sealed as the mold closes – in the case of extrusion blow molding; the trapped portion is then inflated by compressed air to the shape of the mold. This process is common in the manufacture of >0.2 L-bottles – but such high-capacity plastic containers as 4.5 m^3-tanks can also be manufactured in this way. Another advantage arises from the possibility of forming handles and offset necks; once again, uniform thickness of the container wall is a must – which calls for tight control.

Stretch blow molding resorts to a pre-form (or parison) of PET (or PVC or PP, to a lesser extent) – made in advance by injection, injection blow, or extrusion blow molding of PET; after bringing it to the correct temperature, the molten material is rapidly stretched, and then cooled in both directions via compressed air. The biaxial orientation of the molecules allows a clear container be obtained – with increased stiffness, tensile strength, surface gloss, impact resistance, barrier properties (to moisture and gases), and stability over a wide range of temperatures. This approach is traditionally employed to make 0.45–1.8 L-bottles.

Multilayer blow molding has traditionally been confined to EVOH – which is used as a thin layer sandwiched (via an adhesive) between two PE or PP layers.

ABS and HIPS are widely employed to produce thermoformed trays, tubs, and cups – designed to contain cheeses, desserts, ice cream, margarine, spreads, and yogurt; despite the original translucency of the polymeric material, such containers can be easily colored.

HDPE is the most common material for bottles and jars – designed to handle milk, syrups, and vinegar; as well as for drums – conceived to manipulate bulk fruit juices and salt.

High nitrile resins (e.g. acrylonitrile-methyl acrylate, acrylonitrile-styrene copolymers) are molded into containers when supreme barrier features are sought – and used to package cheese, margarine, peanut butter, and processed meat.

PET is the polymer of election for carbonated beverages – as it can withstand high pressures, without compromising its barrier properties; in addition, it has a glossy transparency – particularly desired in the case of mineral water, or strongly colored fruit juices (e.g. orange juice). Multi-chamber PET trays are also known for their hygienic, smooth white finish; their being fat-resistant, heat-sealable, and lightweight support choice for chilled or frozen ready-meals – where the cover is left on during either microwaving or conventional cooking, and eventually peeled off to expose an attractive table dish.

PP is more expensive than HIPS or PVC, and thus less utilized for rigid and semirigid containers; however, it can bear a wider temperature range – besides providing a good barrier to water vapor and oxygen. PP is often coextruded, with EVA copolymer serving as central barrier material; its being shatterproof, resistant to moisture and oxygen, squeezable, and susceptible to hot filling make it the container material of election for jams, mayonnaise, mustard, tomato ketchup, and other sauces. A coextruded, five layer-sheet of PP (or polycarbonate), with PVdC or EVA layers in between, has been used in the manufacture of heat-sterilizable pots and trays – by either thermoforming, blow molding, or injection molding. Plastic cans are made from similar materials – with can body obtained again via thermoforming or blow injection; sealing is normally carried out via easy-open aluminum ends. If processed throughout existing canning lines, the associated noise level is substantially reduced.

PVC devices are produced via extrusion- or injection-stretch blow molding. High toughness, transparency, and colorability, good oil resistance, and low gas permeability of this material make it particularly indicated for bottles of cooking oils, fruit juices and concentrates, and squashes; as well as trays for chocolates and meat products, and tubs for jams and margarine.

2.2.2.4 Metals

The major advantages of metal cans over food containers built with competitor materials arise from their withstanding both high-temperature processing and low-temperature storage. In addition, they are impermeable to light, moisture, odors, and microorganisms if properly sealed – thus guaranteeing full protection of the food inside; and they are inherently tamper-resistant, and also susceptible to recycling. However, the high cost of metal as commodity, coupled to the relatively high manufacturing costs make metal cans more expensive than containers of similar capacity made of plastic (or even glass); they also incur in higher transportation costs, for being heavier than other materials (except glass) for a given capacity. The types of metal employed are tinplate, electrolytic chromium-coated steel (also known as tin-free steel), and aluminum; examples are given in Fig. 2.40.

Hermetically sealed, three-piece sanitary cans are produced from tinplate or tin-free steel – and consist of a can body and two end pieces; they are employed as package for such heat-sterilized foods as meat and fish products, some fruit preserves, precooked vegetables, and syrups – besides high-added value liquids, e.g. seasoning, and cooking oils or fats. Tinplate consists of low-carbon mild steel, with such minor constituents as Mn, P, or Cu; the strength of the final material depends indeed on the (relative) amounts of those minor constituents, further to material thickness and method of production.

Single (or cold) reduction and double reduction are the most widespread methods of manufacture of steel cans; in both cases, steel is first rolled to a strip 1.8 mm-thick, and then dipped into hot dilute sulfuric acid to activate the surface. The single reduction method proceeds through cold-rolls that strip to ca. 0.50 mm thickness, and then tamper-rolls to ca. 0.17 mm; the double reduction method proceeds through a sequence of two cold-rolling stages, so the product obtained is stiffer – and accordingly allows an even thinner sheet be obtained, with ca. 0.15 mm in thickness.

Tin coating is finally applied via electrolytic plating; depending on the intended final use, the same or distinct coating weights may be generated on the two sides – for instance, more acidic foods require a higher coating weight on the inner surface of the can. To prevent sulfide staining when in the presence of sulfur-containing foods (e.g. meat, fish, vegetables), the coating may instead be produced with zinc oxide, or else aluminum powder. A dull surface results from electroplating – due to the intrinsically porous finish; surface brightness can be improved via electrical induction heating – which slightly melts the tin, while simultaneously improving its resistance to corrosion. To protect the tinplate from scratching throughout can-making afterward, a monolayer of edible oil is normally applied.

To obtain an inert surface suitable for direct contact with food, the tin coating is, in turn, to be coated with a variety of lacquers: epoxy-phenolic compounds – resistant to heat and acids, and bearing good flexibility (e.g. canned beer, fish, fruit, meat, pasta, vegetables); vinyl chloride/acetate copolymers – resistant to acids and alkalis but not to heat, and with good adhesion and flexibility (e.g. canned beer, carbonated beverages, fruit juices, wines); phenolic compounds – resistant to acids and sulfide compounds (e.g. canned fish, fruits, meat, soups, vegetables); butadiene – resistant to heat, and further able to prevent discoloration (e.g. beer, soft drinks, vegetables); acrylic compounds – bearing a characteristic

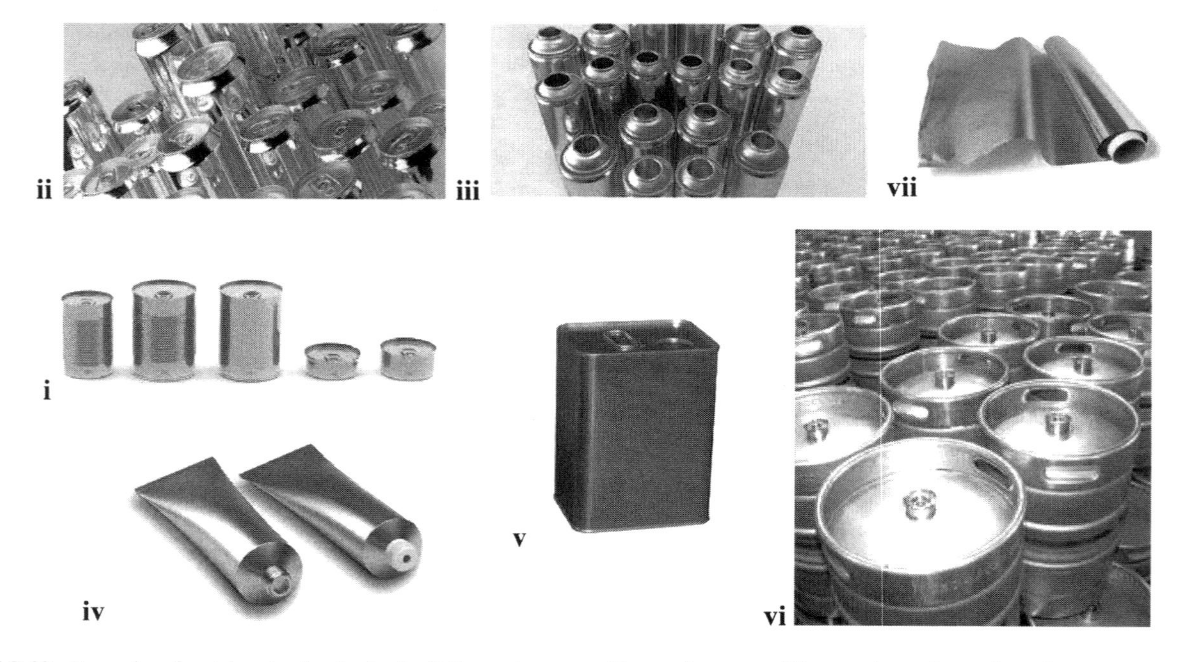

FIGURE 2.40 Examples of metal packaging for foods: (i) three-piece cans, (ii) two-piece cans, (iii) aerosol containers, (iv) collapsible tubes, (v) oil bin, (vi) pressurized barrels, and (vii) aluminum foil.

white color, and suitable for both internal and external coating (e.g. canned fruits and vegetables); epoxy amine – resistant to heat and abrasion, off-flavor-free, and possessing good adhesion features, despite being expensive (e.g. beer, dairy products, fish, meat, soft drinks); alkyl compounds – not expensive, and often used externally as varnish over inks, but unsuitable for internal use due to risk of off-flavor development; and oleoresinous compounds – not expensive and exhibiting good flexibility, and frequently employed due to their golden color (e.g. fruit drinks, meats, soups, vegetables). Once applied, the lacquer is cured by heating at 150–205°C for ca. 10 min – with thickness control afterward via optical interpherometry. Labeling and decoration of tinplate usually resort to external lithography – with ink cured in an oven upon application, and further protected by a varnish applied thereover.

To generate the body of a cylindrical can, the steel sheet is first slit by a set of revolving cutters – to a width equal to the diameter of the can; a second set of cutters then cuts strips at right angles to the first cut, to a width coinciding with the height of the can. The flat body bank is then rolled into an open-ended cylinder – with the two edges held together. A small overlap is left that serves as basis for a welded seam – obtained through melting, brought about by electrically heating with a copper wire; or else through bonding, upon application of a thermoplastic polyamide adhesive. This type of seam exhibits better integrity, along with better appearance than classical soldering; in view of its being the most critical portion of the can, a sidestripe of lacquer is applied externally and/or internally for extra protection. The ends of the body are then curled outward, thus producing a flange that is used to obtain a double seam; can strength is often improved via beading, i.e. production of corrugations in the metal around the can body. Finally, one can end is stamped out from the tinplate sheet, and double-seamed onto the can body; filling, followed by sealing are the final steps.

Tin-free plate is manufactured via a process similar to tinplate – except that tin coating is replaced by chromium–chromium oxide coating, at a nominal rating of ca. 0.15 $g.m^{-2}$, via electrolytic deposition; a lacquer is eventually applied, to minimize external or internal corrosion. It should be emphasized that use of lacquers at large has been gradually reduced, following claims that bisphenol A – an ingredient thereof, serves as endocrine disrupter (for being an analog to female estrogen). This has led to development of bilayer PE films, to be heat-laminated onto tin-free steel – thus providing an equivalent protection against corrosion.

Aluminum possesses three inherent advantages over steel: higher abundancy – Al is indeed the third-most abundant element in Earth crust, with bauxite (46–60%(w/w) hydrated aluminum oxide) permitting the most economical recovery; lower density; and lower susceptibility to corrosion. Aluminum alloys suitable for manufacture of two-piece cans contain 1.5–5.0%(w/w) Mg; they may undergo either the draw-and-wall-iron process, or the draw-and-redraw process of manufacture. The former produces thinner walls, more appropriate for carbonated beverages – where advantage is taken of the inner overpressure to avoid inward-bound mechanical collapse; cans obtained via the latter process are more sturdy, and thus able to withstand the headspace vacuum produced during cooling after sterilization. In both processes, the can end is applied by double seaming – followed by application of epoxy, phenolic, or vinyl-based lacquers to avoid food/metal interaction. Two-piece cans are in general characterized by a greater integrity than their three-piece counterparts – despite being less demanding in terms of metal material; they are also susceptible of more uniform lacquer coverage – with the blank first printed with abrasion-resistant inks.

In the draw-and-wall-iron process, a disc-shaped blank – with typical thickness within 0.3–0.4 mm, is cut and drawn into a cup with the final can diameter. The cup is then rammed through a series of rings, with internal surfaces coated with tungsten carbide, aimed at thinning the plate to ca. 0.1 mm in thickness – and concomitantly increasing height. The heat generated by friction is sufficient to soften the aluminum material; in addition, it aids in stretching the originally imprinted ink to the finished design. After forming the can body, the uneven top edge is trimmed to the intended height, and then flanged to accept the can end – to be fitted after filling up the can; beading may again be sought, for enhanced mechanical strength. Modifications to such basic two-piece design include: reduced diameter of the neck – for improved appearance, along with enhanced ability of can stacking; and ring-pull tabs or full-aperture easy-open ends – for greater convenience at consumption point. The draw-and-redraw process is similar to the above one in the first stages – although more commonly used to manufacture smaller cans; however, metal is pushed from the base of the container toward its wall, by reducing the diameter of the container – instead of ironing to uniformly reduce wall thickness.

Aerosol cans are two- or three-piece cans – duly lacquered, but fitted with a valve designed to dispense the product; the propellant gas is kept inside the can – but is either mixed with the food product, or kept separate by a plastic bag or a piston device. The can should withstand at least 1.5-fold the vapor pressure of the filled aerosol at 55°C – with a minimum of 1 MPa. In the case of UHT-sterilized cream, nitrous oxide is employed as propellant; such other gases as Ar, CO_2, and N_2 have been cleared as well for use with foods – namely cheese spreads for sandwich filling, and oil sprays for baking pans.

Aluminum in foil form is used for wrappers, lids, cups and trays, laminated pouches, and collapsible tubes; and in (thicker) sheet form for barrels of pressurized liquid and closures. Foil is usually manufactured by a cold reduction process – with >99.5% pure aluminum passed through rollers, designed to decrease its thickness to <0.152 mm; it is then annealed, i.e. heated for ductility control – thus bringing about dead-folding features.

The abovementioned uses of foil are materialized in wraps with nominal thickness of 0.009 mm, bottle caps with 0.05 mm thickness, and trays for frozen and ready meals with 0.05–0.1 mm thickness; such metal films are regularly coated with nitrocellulose, if intended for contact with acid or salty foods. A number of advantages have indeed been ascribed to aluminum foil: good looking, with high-quality surface suitable for decorating and printing; no need for application of lacquers, since a protective layer of aluminum oxide spontaneously forms on the surface as soon as it contacts (atmospheric) air; lack of inherent odor or taint compounds; impermeability to moisture, gases, volatiles, radiant energy, and microorganisms; good weight:strength ratio; suitability for lamination with paper or plastic; and compatibility with a wide range of sealing resins and coatings – as required by several types of closure. Unfortunately, aluminum foil is incompatible with microwave ovens; although no back effect has been recorded with modern magnetrons, arcing between foil and oven wall does sometimes occur.

Collapsible tubes are also manufactured from aluminum foil, but are supplied preformed – with an epoxy-phenolic or acrylic lacquer applied on the internal surface, a sealed nozzle, and an open-end ready for filling. They are normally utilized with viscous food products (e.g. garlic paste, tomato purée); and preferred to PE tubes in food applications, because they collapse permanently upon squeezing (unlike plastic tubes) – thus preventing air, and airborne (potential) contaminants be drawn into contact with the product left in the tube.

2.2.2.5 Glass

Glass is generally described as an inorganic solid below its glass transition temperature, ca. 570°C – even though it shares some structural features of a typical supercooled liquid; mass density and heat capacity lie in the vicinity of 2500 $kg.m^{-3}$ and 500 $J.kg^{-1}.K^{-1}$, respectively. Glass devices and containers are manufactured from a mixture of sand – normally constituted by 70–74% SiO_2, 12–16% Na_2O, 5–11% CaO, 1–4% Al_2O_3, 1–3% MgO, ca. 0.3% K_2O, ca. 0.2% SO_3, ca. 0.04% Fe_2O_3, and ca. 0.01%(w/w) TiO_2; with 30–50%(w/w) broken glass (for recycling, or cullet), heated to 1350–1600°C. To enhance chemical durability, the content of Al_2O_3 can be expanded; while reduction of temperature and time required for melting – along with removal of gas bubbles, become possible at the expense of addition of refining agents. Increased strength can be achieved upon surface treatment of glass with Ti, Al, or Zr compounds – so the resulting containers will be lighter for a given mechanical specification. Typical coloring includes green (via addition of chromic oxide), amber (via addition of iron, sulfur, and carbon), or blue (via addition of cobalt oxide); whereas masking of unwanted color derived from trace impurities (e.g. Fe) resorts to Ni and Co as additives, thus leading to clear glass (or flint).

A wide variety of glass containers are available – and a few examples are included in Fig. 2.41. From a physical point of view, glass containers raise a total barrier to microorganisms, as well as to moisture, volatile odors, and gases; and are completely inert from a chemical point of view, i.e. their components do not react with, or else migrate into the food product. Their intrinsic rigidity provides good vertical strength, and thus allows stacking without damaging the container; to minimize contact between glass surfaces susceptible of causing mutual damage, a protruding shoulder on the lateral surface is often included. Sharp corners and surface abrasion weaken glass devices, so cylindrical shapes with surface polishing are preferred – for the sake of durability; the relative strength of a cylindrical container is ca. tenfold that of a paralellepipedal container with sharp corners, ca. fivefold that of a 2:1 (prolate) ellipsoidal container, and twofold that of a parallellepipedal container with round corners.

In what concerns processing in the food plant, the filling speeds of glass containers match those of metallic cans – and such vessels are suitable for thermal treatment (if hermetically sealed), besides being transparent to microwaves. In marketing terms, glass containers can display their contents due to an

FIGURE 2.41 Examples of glass packaging for foods: (i) threaded jar, (ii) glass-lidded pot, (iii) regular and (iv) pressurized wine bottles, and (v) juice or sauce containers.

inherent transparency, and can be decorated or molded into a wide variety of shapes and colors; in fact, glass is perceived by customers as a clue to a high-value product – and this helps justify why good-quality wines, vintage Ports, liqueurs, and spirits at large always resort to glass bottles. From an environmental point of view, glass containers are reusable and recyclable. Their being resealable, and possessing a smooth internal surface easy to clean and sterilize justify straightforward reuse of glass containers (e.g. beer, milk); while plain reheating until molten, followed by container reforming justify their prompt recycling – without loss of container quality, or release of dangerous byproducts (as can happen with plastic materials).

The main disadvantages of glass containers stem from their intrinsically high weight – with consequent larger transportation costs; coupled to their susceptibility to thermal shock, fracturing, and breakage – thus releasing splinters or fragments, which may constitute serious hazards if accidentally ingested.

Two basic modes exist to manufacture glass containers – blow-and-blow and press-and-blow processes; the former is used for narrow-neck containers, while wide-neck containers are obtained via the latter. In the blow-and-blow process, a gob of glass enters a parison mold at ca. 1000°C, and undergoes blow-settling to form the finish (i.e. the part supporting the closure); the parison is then inverted, and the body gets formed via injection of compressed air into the mold. Conversely, the gob is shaped into a parison mold, and the finish is molded by upward action of a plunger – followed by blow-molding via compressed air, when the press-and-blow method is employed. Note that glassmaking is an energy-intensive process, owing to the very high temperatures necessary – alleviated by up to 40% when recycled glass is used, though.

When bottles leave the mold in either process, at ca. 450°C, they are allowed to cool on their own under carefully controlled conditions. If forced cooling were employed, the interior would cool more slowly than the outside due to the lower thermal conductivity of its more liquid-type nature – so internal stresses would develop due to distinct rates of contraction, likely to turn the glass unstable. In practice, glass is annealed at 540–570°C to remove stresses, and then passed through a long tunnel (ca. 30 m) to prevent distortion and fracturing; improvements in glass-making include plasma-arc crucibles to melt the raw ingredients, followed by coextrusion of the molten glass in a way similar to plastic container manufacture. Critical faults incurred in production include broken, cracked, or chipped glass, strands of glass stretched across the inside of containers, or air bubbles in the glass material (that make it thinner, and thus more fragile); the seriousness of potential faults requires inspection of every single container, as routinely accomplished via automated equipment.

2.2.2.6 Modified atmosphere packaging

Modified atmosphere packaging (or MAP, for short) refers to the introduction of an atmosphere other than air into retail packages, without subsequent monitoring or modification thereof; this contrasts with controlled atmosphere packaging (or CAP, for short) – where continuous monitoring and control of gas composition takes place in bulk containers. Either approach is meant to extend the shelf-life of food products – typically by between 50 and 400%. Therefore, additional time will become available for storage by processors, grocers, and retailers – and longer transportation distances (with less frequent trips) will be possible for distributors, without significantly compromising food freshness, or quality at large; while decreasing (or even eliminating) the need for use of chemical preservatives, facilitating separation of sliced foods, and conveying better visual appearance.

MAP is used for fresh foods, and an increasing number of mildly processed foods (e.g. bakery products, coffee, fish, fresh pasta, cheese, partially baked bread croissants, peeled fruits, pizzas, potato crisps, poultry, prepared salads with dressings, raw and cooked meats, sandwiches, seafood, *sous-vide* foods, tea, vegetables) – and is rapidly gaining additional popularity. These advantages come, however, with a number of shortcomings: added cost of gases – with variable composition; control of temperature and initial degree of contamination required for distinct food products; special packaging equipment necessary; increased pack volume, with unfavorable impact upon costs of transportation and storage/display; and loss of MAP benefits, once the package is opened or some form of leaking has occurred. In addition, product safety is still to be fully established for a number of food products.

Distinct foods do indeed respond in different (and, sometimes, hardly predictable) ways to modified atmospheres. Therefore, assessment of responses, by each type of product, to microbial growth and activity – as well as to such physicochemical parameters as moisture content and pH, and to such sensory features as texture, flavor, and color are deemed necessary. On the other hand, composition of the modified atmosphere may change for physical reasons – e.g. mass transport through the packaging material, due to its non-nil permeability; or to metabolic reasons – e.g. food tissue respiration, or activity of (contaminant or deliberately added) microflora.

The longer shelf-life possible for a number of foods, relative to storage in the open atmosphere, under the recommended composition for MAP, is illustrated in Fig. 2.42; the extension in shelf-life of coffee by almost three orders of magnitude should be emphasized. In fresh fruits and vegetables, decay is usually avoided by 15–20%(v/v) CO_2 – yet most products cannot tolerate such high levels of that gas; notorious exceptions include broccoli, spinach, and strawberries. Red meat joints and ground beef may experience an extension in shelf-life of up to 18 and 10 days, respectively, at 0–2°C under ca. 80%(v/v) O_2, with the remainder accounted for by CO_2.

The three major gases used for MAP are N_2, O_2, and CO_2; nitrogen, in particular, is inert and tasteless, and exhibits low solubility in both water and fats – being used to replace O_2, and thus inhibit chemical oxidation and growth of aerobic microorganisms. In the specific case of fresh produce, the aim of MAP is minimizing the rate of respiration and senescence – without harming the metabolic activity of plant tissues themselves, which might compromise the quality of the final product. However, the effects of low concentration of O_2 and high concentration of CO_2 upon respiration are cumulative and temperature-dependent; and the closed nature of the packaged food enforces a continuous change in composition of the modified atmosphere inside, as respiration evolves. In practice, the concentration of CO_2 is originally increased by gas flushing prior to sealing – although immediate packaging of bread after oven baking may take advantage of CO_2 (previously generated by fermentative or chemical leavening) still being released; a film permeable to O_2 and CO_2 is normally used for packaging of fresh produce, as it allows respiration to proceed.

In batch operation, preformed bags are filled, evacuated, gas-flushed, and heat-sealed; continuous operation encompasses one of three approaches – thermoformed trays covered with film (e.g. meats), pillow pouches (e.g. fresh salads), or flowpacks (e.g. baked products). Single or coextruded films or laminates of EVA, PP, PE, polyester, PET, and PVC – and even

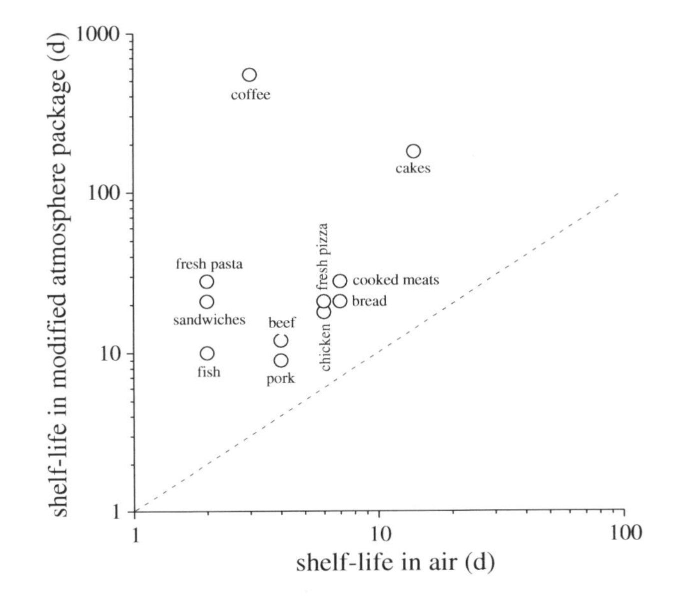

FIGURE 2.42 Extension of shelf-life of selected foods, under modified atmosphere packaging with recommended composition, relative to plain air.

polyamides/nylons (despite their relatively poor barrier features), are frequently employed. Films are normally coated on the inside of the pack with an antifogging agent (e.g. silicone, stearate), so as to disperse droplets formed upon condensation of moisture; this allows the food remain fully visible. Some of those films can actually change their permeability to water vapor and gases with the temperature elevation arising from respiratory metabolism – and are thus specifically designed to match the anticipated respiration rate of the fresh produce at stake.

The rate of gas exchange with the (outer) atmosphere depends obviously on the relative permeability to each gaseous compound offered by the film serving as interface (which is often affected by inside/outside relative humidity), as well as its surface area; while provision should be made for a balance between rate of metabolic production/consumption and rate of exchange through the packaging film, so as to avoid collapse or burst thereof. On the other hand, dissolution of CO_2 in water and fat components of a food should be taken into account – especially in the case of chilled products, since the said solubility increases when temperature is decreased as per Henry's law; if provisions are not taken, namely, addition of N_2 as filler gas, package collapse may occur (advantageous in hard cheeses, however). Examples of minimum and maximum levels of O_2 and CO_2 for MAP of selected foods are provided in Fig. 2.43; this piece of information, pertaining to the atmosphere inside the package, complements the data conveyed previously on optimum composition range for the overhead atmosphere inside the container during chilled storage, see Fig. 1.59. Oxygen at high concentration is used to maintain the red color of oxyhemoglobin – characteristic of (unprocessed) meats; however, its level is reduced for other applications, in attempts to prevent growth of spoilage microorganisms, or development of oxidative rancidity. For cured meats (e.g. bacon), appearance of off-colors is minimized by resorting to 35%(v/v) O_2 and 65%(v/v) CO_2, or else 70%(v/v) O_2, 20%(v/v) CO_2, and 10%(v/v) N_2; anaerobic bacteria are severely inhibited by such a high oxygen concentration, in either case. Pork, poultry, and cooked meats have no oxygen requirement to maintain color, so concentrations of CO_2 up to 90%(v/v) become possible – and support extension of shelf-life up to 11 d. For other processed, and thus non-respiring foods, atmospheres should be as low as possible in O_2 – and concomitantly be as high as possible in CO_2. For instance, a high CO_2 concentration prevents mold growth, thus allowing shelf-life expansion by up to 3–4 wk in the case of (hamburger) buns, or even 3–6 mo in the case of (bakery) cakes – along with retention of water vapor, which in turn assures product softness; whereas oxidation of ground coffee is prevented under a mixture of only N_2 and CO_2 for MAP, or else vacuum.

Reducing the concentration of O_2 inhibits development of several spoilage microorganisms, especially *Pseudomonas* spp.; and even those that are facultative anaerobes (e.g. lactic acid bacteria) grow slower when exposed to lower partial pressures of oxygen. The major risk arises from the fresh look exhibited by a food item – otherwise indicative as safe for consumption, when absence of the regular competing aerobic microflora has meanwhile permitted excessive growth of anaerobic pathogens (e.g. *Aeromonas hydrophila*, *Clostridium botulinum*, *Salmonella* spp., *Yersinia enterocolitica*). To reduce the probability of occurrence of such a hazard, supplementary hurdles are advisable – e.g. reduction of a_W to below 0.92, pH to below 4.5, or temperature to below 3°C, or addition of a chemical preservative (e.g. $NaNO_3$); treatment by heat, UV light, γ-radiation, or ozone are suitable alternatives.

The barrier properties of a packaging material can be expressed in terms of permeability, or ability to allow (or restrain) passage of fluids through it; in quantitative terms, it consists of the rate of transport of gas or vapor per unit area and per unit driving force – thus providing a measure of how well a certain gas (or vapor) can permeate the packaging material. In the case of diffusional (or molecular) mass transfer, the driving force is a gradient of concentration, or partial pressure – which dominates in the case of regular MAP; for convective (or bulk) mass transfer, a gradient in total pressure plays the germane role. Mass transport through a polymeric material occurs via a sequence of steps – as sketched in Fig. 2.44. The gas or vapor molecules start by dissolving in the polymeric material – acting, in this regard, as a very viscous liquid solvent, on the film side exposed to the higher concentration; those molecules then diffuse through the polymeric material, toward the side of the film exposed to the lower concentration; the final step encompasses desolvation of the gaseous molecules from the surface of the film.

According to their barrier properties against oxygen, usually measured at 90%RH and 25°C, materials are classified as: low barrier, i.e. >300 $mL.m^{-2}.d^{-1}$ – meant for overwraps on fresh meats, and a few other applications where transmission of O_2 is desirable; medium barrier, i.e. 50–300 $mL.m^{-2}.d^{-1}$; high barrier, i.e. 10–50 $mL.m^{-2}.d^{-1}$; and ultrahigh barrier, i.e. <10 $mL.m^{-2}.d^{-1}$ – which protects product from oxygen. Examples of ultrahigh barrier materials include PVdC, or else aluminum foil – the latter with the obvious inconvenient of blocking visual inspection of food, for being opaque. Recent developments in the field of transparent films exhibiting high-barrier features against O_2 include glass-coated microwaveable pouches, silicon oxide-coated films (<1 $mL.m^{-2}.d^{-1}$), aluminum oxide coatings, and nylon-based coextrudates (ca. 0.5 $mL.m^{-2}.d^{-1}$) – while SiO_x/PET films, trays, and bottles, produced by plasma-enhanced chemical vapor deposition, appear quite promising.

2.2.2.7 Frozen food packaging

A special mention is deserved by packaging of frozen foods – designed not only to protect from contamination and damage throughout storage, and especially during transit from the manufacturer to the consumer; but also to preserve the intrinsic value associated with their original texture, color, and flavor – despite the typically long storage period. The packaging materials are in general meant to be moisture- and air-proof – so air (including its constitutive oxygen) should to advantage be evacuated prior to packaging; in addition, they should not become brittle at low temperatures, so as to prevent cracking; and should be leakage-proof (a feature particularly relevant during thawing afterward) and resealable (if continuing frozen storage of the remainder of the food is intended).

There are typically three kinds of packages utilized with frozen foods: primary – in direct contact with food, and kept until final use; secondary – i.e. some form of multiple packaging, used to handle individual packages together for sale; and tertiary – used for bulk transportation of sale packages. Primary packaging resorts to a range of rigid containers – e.g. glass, plastic, tin, and heavily

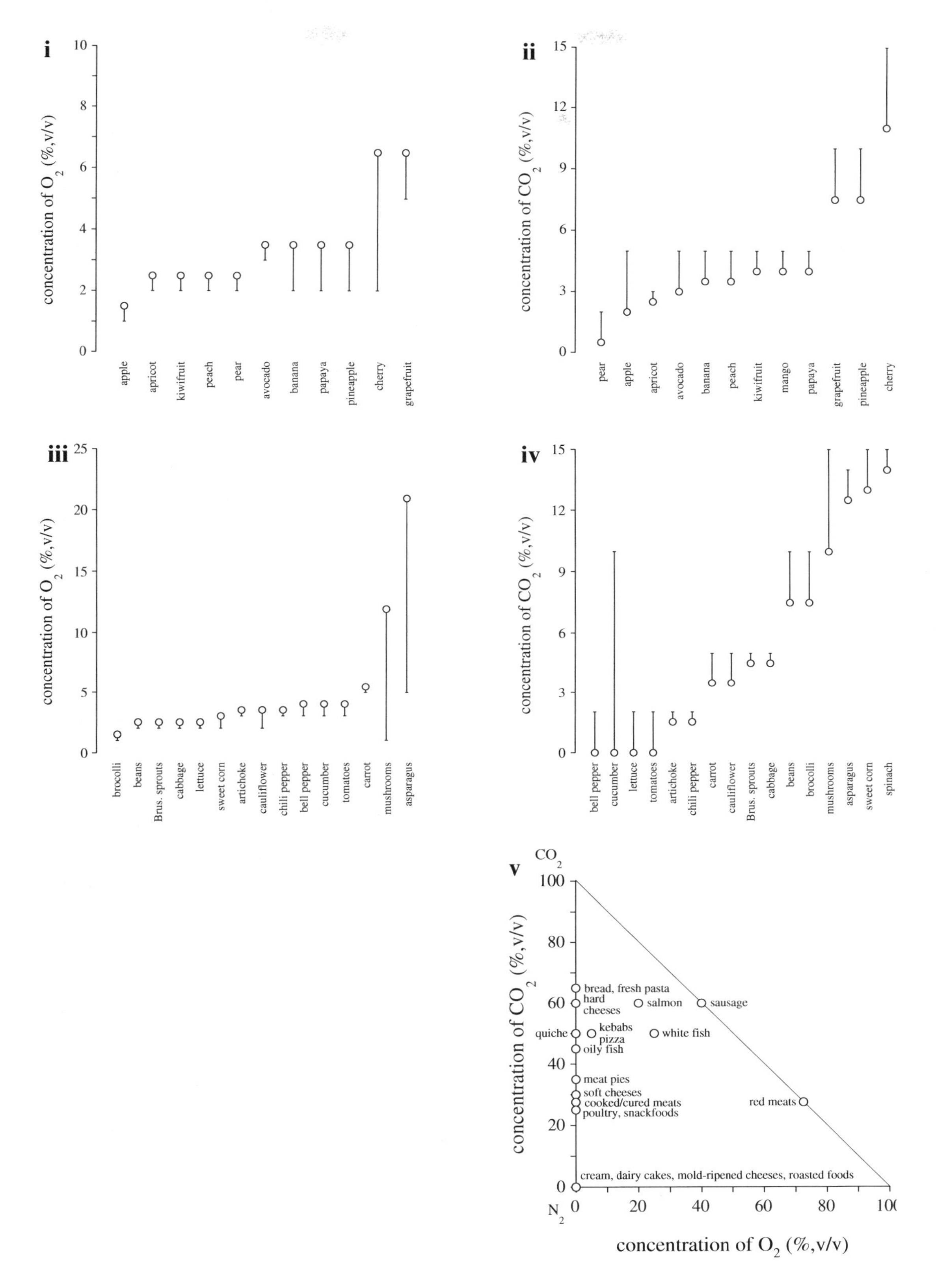

FIGURE 2.43 Recommended composition of modified atmosphere packaging for selected foods, namely, (i, ii) fruits, (iii, iv) vegetables, and (v) processed foods – expressed as average (○), and (i, iii) minimum (⊥) or (ii, iv) maximum (⊤) concentration of (i, iii, v) oxygen and (ii, iv, v) carbon dioxide.

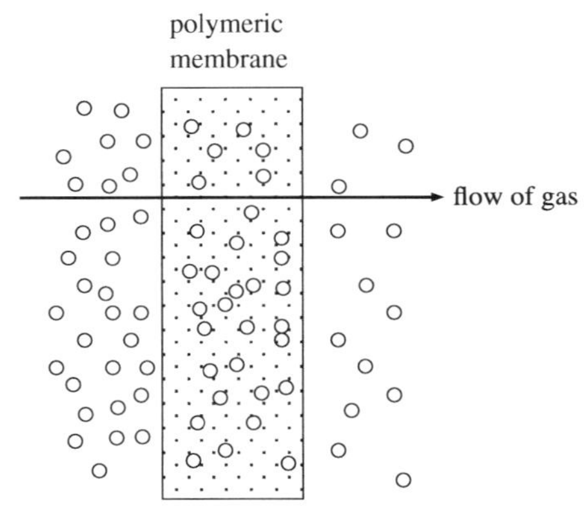

FIGURE 2.44 Schematic representation of molecular transport of gaseous solute, through a porous polymeric membrane used as packaging material.

waxed cardboard, especially conceived for otherwise liquid food products; as well as nonrigid containers, which include bags and sheets made of heavy aluminum foil or PE, or laminated paper instead. Cardboard cartons are often used for secondary packaging – to protect the primary package from tearing. Shape and size are also to be taken into account when selecting packages; serving size may vary from product to product (usually meal-portion), and it directly affects the size of primary packaging. In terms of shape, the main driver is optimization of freezer space use – so rigid containers with flat tops/bottoms and lateral sides that stack well are preferred to round containers.

Packaging of frozen fruits demands special precautions, in view of their susceptibility to sensory decay – and for responding to an ever-expanding market demand. Air should be excluded from the package to prevent oxygen-mediated browning (as already outlined) – either via vacuum, or replacement by nitrogen or other inert gas, dipping in a sugar solution, or addition (directly on the surface, or via controlled release from the package) of glucose oxidase that takes up molecular oxygen. Fruit packs designed specifically for fruit freezing include syrup pack, sugar pack, tray pack, sugar-substitute pack, and unsweetened pack. Syrup packs contain the fruit in a wet state, submerged in a sugar solution – 40%(w/w) syrup for most fruits. Lighter syrups are recommended in the case of mild-flavored fruits, to avoid masking their original flavor; while heavier syrups are recommended for very sour fruits – with pectin often included to reduce sugar content. In the preparation of a sugar pack, sugar powder is first sprinkled over the fruit; the container is then gently agitated, until the juice is drawn out and the sugar dissolves therein. This approach is common with soft sliced (or even whole) fruits, e.g. cherries, peaches, plums, strawberries. Unsweetened packs can be prepared in a number of ways – ranging from dry pack, through coverage with water containing ascorbic acid, to dipping in unsweetened juice. These packs typically yield lower quality products than sugar packs – except in the case of (scalded) apples, blueberries, cranberries, currants, gooseberries, and raspberries. Unsweetened packs are normally prepared using tray packs – where a single layer of fruit is spread on a shallow tray, frozen, and promptly packaged in a freezer bag afterward; the fruit sections thus remain loose, without significant clumping together. Artificial sweeteners may be used as sugar substitutes – yet such intrinsic features as color protection and thickness associated with regular sugar will hardly be replaced; sugar substitutes possess the advantage of reducing the rate of both freezing and thawing.

Several factors are to be considered when packaging frozen vegetables – including protection from atmospheric oxygen, prevention of moisture loss, retention of flavor, and enhanced rate of heat transfer through the package material; hence, two basic packaging methods are common – dry pack and tray pack.

In the former type of package, blanched and drained vegetables are placed into meal-sized freezer bags, and packed tightly so as to cut down the residual amount of air inside the package; a headspace of ca. 2 cm is, however, left at the top of rigid containers before closing – whereas a larger headspace is left in the case of freezer bags. Provision for headspace is not needed for such foods as broccoli, asparagus, and Brussels sprouts, as they do not pack tightly in containers.

In the tray pack method, chilled, well-drained vegetables are placed as a single layer on shallow trays or pans; these are, in turn, placed in a freezer until the vegetables become firm – and then removed and filled into containers. Tray-packed foods do not freeze as a block, but remain loosely distributed; hence, the amount needed at every single use by the final consumer can be easily poured from the container, and the package reclosed afterward.

2.2.2.8 Edibility/biodegradability of packaging materials

Widespread awareness, and consequent growing concern by consumers about the environmental effects of (food) packages manufactured from petroleum-based plastics have urged development of polymers from renewable sources (also known as bioplastics) – which are biodegradable and compostable. The said biodegradation is carried out by microorganisms – which may also be involved in food spoilage; therefore, the major challenge is to produce bioplastic materials that are sufficiently stable and perform sufficiently well, throughout the whole expected shelf-life of a food (i.e. able to maintain their intrinsic mechanical and/or barrier properties) – while degrading fast, once disposed off. Environmental conditions conducive to biodegradation should obviously be avoided during package manufacture and utilization; whereas optimum conditions for biodegradation of the already-used package are to be provided thereafter. Furthermore, economic feasibility requires such packaging materials to be of a cost essentially equivalent to that of regular packaging materials – while being compatible with conventional filling and sealing equipment.

According to their method of production, biodegradable materials have traditionally been classified as polymers directly extracted or removed from natural materials (e.g. starch from barley, maize, oats, potato, rice, wheat; cellulose from cotton, wood; gums such as alginates, carrageenan, guar, locust bean, pectins, or chitosan/chitin – as polysaccharides; casein, collagen, gelatin, whey – as animal proteins; gluten, soy protein, zein – as plant proteins; and cross-linked triglycerides as lipids); polymers produced through chemical synthesis from renewable monomers (e.g. polylactate from lactic acid); and polymers produced by microbial action upon fermentable substrates (e.g. bacterial cellulose, curdlan, polyhydroxyalkanoates, pullulan, xanthan). The biodegradable materials with the largest commercial expression are starch, cellulose, polylactates, and polyhydroxyalkanoates.

Starch is widely available from multiple sources (with maize being the largest one) – and is, at present, economically competitive with petroleum as raw material. The bioplastics obtained therefrom are, however, brittle – and thus unable to form films bearing the mechanical features of flexibility, elongation, and tensile strength required by packaging of many foods. Flexibility may be improved via addition of plasticizers (e.g. glycerol, low-molecular-weight polyhydroxy compounds, polyethers, urea); the oxygen-barrier properties of the resulting films are acceptable, yet the hygroscopic nature of starch prevents its utilization with high-moisture foods. Upon blending with hydrophobic polymers, semirigid and rigid containers can be manufactured – via injection- or blow-molding. Despite the aforementioned constraints, starch films have been used for wrapping, laminating, and coating paperboard and LDPE, and producing foodservice containers from biodegradable foam laminate, as well as thermoformed (or injection-molded) cups and egg trays.

Cellulose is the most abundant, and thus inexpensive natural polymer; despite being formed by the same glucose units as starch, the β-1,4-glycosidic linkages between such building blocks account for stronger intermolecular hydrogen bonding. Its side chains contribute to the highly crystalline structure that justifies brittleness, low flexibility, and poor tensile strength of the resulting films – along with weak moisture barrier features. Such cellulose derivatives as cellulose acetate possess noteworthy extrusion properties – suitable for films aimed at food wrapping, as well as injection- and blow-molding containers; however, large-scale production is still not economically competitive.

Polylactates (or polylactic acids, PLA for short) are thermoplastic polyesters chemically derived from lactic acid – which is, in turn, normally produced via *Lactobacillus*-mediated fermentation of starch or molasses; their commercial expression is still limited, due to the associated high production costs. The features shown by polylactates depend on the ratio of D- to L-monomers as residues; for instance, a 0:100 ratio leads to a quite crystal-like matrix – accordingly characterized by a high melting point; whereas a 90:10 ratio yields a polymer that melts more easily, and is thus suitable for packaging films (glass transition temperature of ca. 60°C, melting temperature around 150°C). PLA exhibits good aroma barrier properties when compared with LDPE, PET, PS, and PP. The films obtained therefrom – similar in appearance and properties to oriented PS films, exhibit good barrier features against moisture, oxygen, and (odor) volatiles; and their surface is suitable for printing. Cast films have been produced for wrapping bakery and confectionery products, cast sheets for thermoforming, and extruded films for coating paperboard. Injection molding, and blow- and vacuum-forming processes are also feasible; common examples are disposable containers and service tableware.

Polyhydroxyalkanoates are linear polyesters produced by e.g. *Alcaligenes eutrophus*- or *Ralstonia eutrophus*-mediated fermentation of sugars or lipids. Although the most common is polyhydroxybutyrate (PHB for short), there are 100+ monomers (dependent on microorganism and carbon source used) that can be combined to produce materials exhibiting a wide range of mechanical and barrier features. PHB has a melting temperature (175–180°C) similar to PP, so it can withstand retorting – but is stiffer and more brittle; it also possesses excellent moisture and aroma barrier features, but offers poor resistance to organic solvents when compared to PET. Due to the higher production costs than alternative petroleum-based plastics, its widespread use in food packaging is limited. Other polymers of the same family have lower glass transition and melting temperatures, and a (low) moisture permeability comparable to LDPE; they have found application as cheese coatings – and appear also adequate to manufacture bottles and trays.

Novel biodegradable polymers include thermoplastic polyesters and polyamides, polyurethanes from castor oil, polycaprolactone (PCL for short), and polymethyl-valerolactone; as well as copolymers of lactams and lactones, and laminated chitosan/cellulose/polycaprolactone.

An increased interest has meanwhile arisen for nonpetroleum-based, edible protective surface layers (EPSL for short); besides being inherently biodegradable, they may be ingested together with the food – without the need for unwrapping at all. Such coatings are applied directly onto the surface of a food, and are aimed at preventing loss of quality – as they consubstantiate an extra hurdle against contamination by environmental microorganisms, as well as against transfer of oxygen and moisture. They also enhance mechanical robustness, thus reducing susceptibility to handling; and may further control the rate of respiration, by acting as selective barriers to O_2 and/or CO_2 – relevant, in particular, for packaging of fresh fruits and vegetables. One of the first examples was collagen casing for meat products; gelatine (derived from collagen) can indeed be formed into films and light foams, despite its high sensitivity to moisture. Several EPSLs – characterized by hydrophilic character, further to good mechanical flexibility, and resistance to breakage and abrasion, have meanwhile resorted to hydrocolloids (e.g. alginates, pectin) and other carbohydrates (e.g. chitosan) – besides proteins (e.g. corn zein, wheat gluten, milk caseins, and proteins from cottonseed, milk whey, peanut, soybean). Currently, the most relevant commercial applications encompass corn zein – used to produce films via casting or extrusion; and chitosan – widely employed as coating, due not only to its good barrier features, but also to its inherent antimicrobial properties.

Barrier features, to oxygen and carbon dioxide, of selected biodegradable polymers are depicted in Fig. 2.45. The particularly low and high transmission rates of zein and wheat gluten, respectively, are to be outlined; however, gas transmission rates are affected by prevailing relative humidity (note the increase, by two orders of magnitude, when RH varies from 56% to 91% in the case of gluten/beeswax) and temperature.

The number of applications of biodegradable polymers is already considerable – and relevant examples will be mentioned next, sorted by major features exhibited as packaging material.

In terms of plain containment, powdered starch foam has been used in ready-to-eat meal containers for chicken, French fries, and hamburgers.

Regarding oxygen barrier, corn zein, hydroxypropyl cellulose, and whey protein isolates are a suitable option to coat roasted peanuts.

Acetylated monoglycerides and whey protein have been utilized as edible coating for fish – for their oxygen and moisture barrier, coupled to their antioxidant roles; paper coated with PLA or starch has been employed in cups for miscellaneous

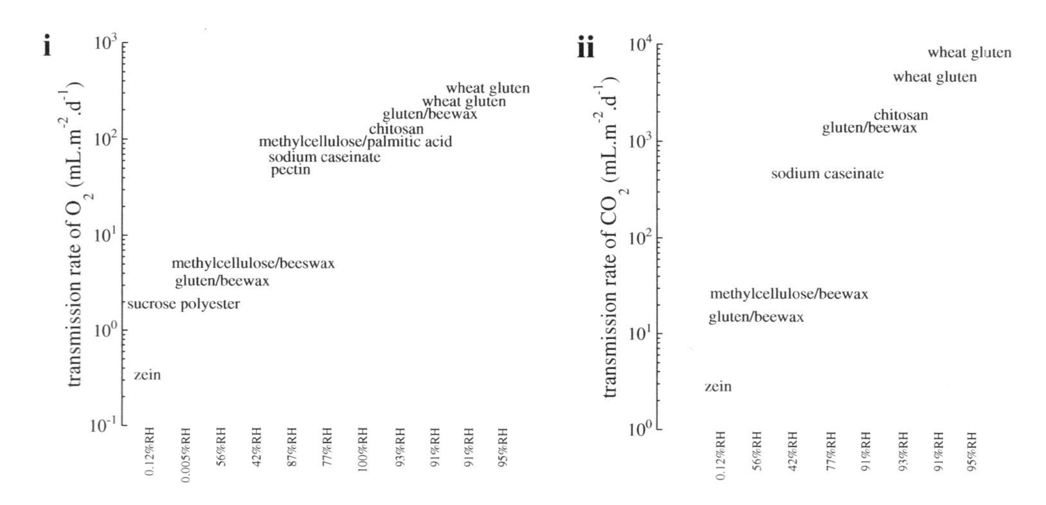

FIGURE 2.45 Barrier features, at 25°C, for selected biodegradable polymeric films used in food packaging – expressed as transmission rate of (i) oxygen and (ii) carbon dioxide, under the indicated relative humidity, %RH.

applications; PLA or starch have been used to coat paper or serve as window therein; and starch- or PHB-coated paperboard trays, or starch-overwrapped bags have been employed with fresh tomatoes.

In the case of oxygen and moisture barrier features, alginates serve appropriately as edible coating for mushrooms; starch-PE films have been used with ground beef; sodium or calcium caseinate-acetylated monoglyceride and sucrose-fatty acid esters have been employed as edible coating for fresh apples – also sought after for their ability to improve gloss and appearance, and to act as a vehicle for antioxidant or preservative agents (meant to delay browning and microbial growth, respectively).

With regard to mechanical protection, further to oxygen and moisture barrier features: alginate, carrageenan, caseins, lipids, and whey proteins have served as edible coatings for frozen fish and shrimps – also chosen for their batter adhesion, and possibility to carry antioxidant or antimicrobial agents.

In terms of CO_2 barrier, further to moisture and oxygen barrier: corn zein has been selected to coat fresh pears.

In what concerns mechanical protection, further to oxygen, moisture, and CO_2 barrier: cellulose laminated with chitosan, polycaprolactone, and protein have proven feasible as edible coatings for cut and whole fresh broccoli, cabbage, lettuce, and tomatoes.

Regarding grease barriers, further to oxygen, moisture, and CO_2 barriers, and mechanical protection: alginate, carrageenan, cellulose, gelatin, and soy protein have played the role of edible coatings and casings for fresh, cooked, and cured meats – also elected for their adhesion to batters, and ability to inhibit microbial growth.

Should light barrier be of interest, further to oxygen and moisture barrier: methylcellulose laminated with corn zein and stearic/palmitic acid have been utilized to pack potato chips; and PLA and PHB bottles or paperboard cartons, coated with PLA and/or PHB, have been chosen for contact with milk.

In terms of grease barrier, further to moisture and light barrier: PLA plus copolyester, PCL, polyamide or starch have been an option for pots of butter, margarine, and yogurt – also selected for their antifogging features.

Finally, examples of high oxygen permeability, along with low moisture permeability, include paper pulp, cellulose acetate or protein film, PLA and/or PHB trays with PLA top lids, and starch used as coating and pack for fresh meat portions – also convenient for their ability to absorb meat drip.

The expected time for composting of the aforementioned polymers is given in Fig. 2.46 – under mechanical turning and active aeration. Typical periods range from less than 1 mo – in the case of (biodegradable) thermoplastic starch and blends thereof, PHB, and proteins; such low figures compare with ca. 6 mo in the case of (synthetic) polyester amides and cellulose diacetate.

Remember that composting recycles the biopolymers (otherwise considered as waste products), and concomitantly produces a soil conditioner or fertilizer (known as compost). At the simplest level, it requires making a heap of organic matter or green waste (e.g. leaves, grass, food scraps), and waiting for the materials to breakdown into humus within a period of months. The microbial

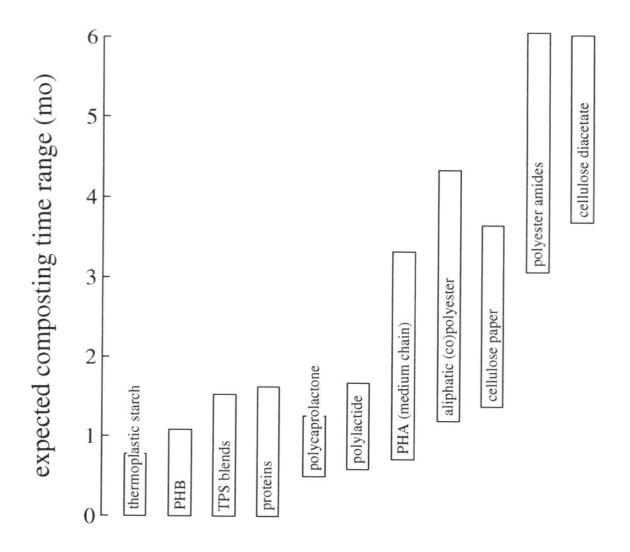

FIGURE 2.46 Ranges of expected composting time (▭), required by selected bioplastics and synthetic polymers used in food packaging.

decomposition process is aided by shredding the plant matter, adding water, ensuring proper aeration, and regularly turning the mixture; earthworms and mushrooms further break up the material.

2.2.2.9 Nanotechnology in packaging materials

Natural polymers exhibit inherent limitations – namely, low mechanical strength and poor moisture barrier properties; these may be circumvented by composites obtained via nanotechnology, a novel area dealing with characterization and manipulation of materials with dimensions within the range 1–100 nm. The basic approaches are either improving existing materials or developing new materials with tailored features; the former resorts to polymer/inorganic nanocomposites (e.g. polymer/silicate, polymer/organoclay), where ultrasmall inorganic particles are used to change the nature of a polymeric material. The result is composites bearing higher heat resistance, increased flexibility, lower gas permeability, and better dimensional stability – along with good surface appearance. A unique advantage of this approach is that improvement in material properties is possible independently of each other; in other words, stronger barrier properties do not rule out flexibility or transparency, while higher stiffness does not compromise ductility or lightweight.

The most common silicate additive is montmorillonite, a layered smectite clay – which breaks down to nano-sized platelets, with thickness of the order of 1 nm and length within 100–1000 nm. Due to its intrinsic high hydrophilicity, it is first modified via replacement of sodium ions by organic ammonium ions – thus unfolding organoclay complexes; this derivatization facilitates dispersion into conventional (hydrophobic) polymers. Nanocomposites can be prepared by dissolving both polymer and complex in a solvent, or instead shearing a polymer melt to assist in dispersion; one example is nylon-6 nanocomposites – with oxygen transmission rates four times lower than their unfilled counterpart, while retaining the original transparency. Besides increased mechanical strength, lower gas permeabilities, and higher moisture barrier properties, bioactive components (e.g. antimicrobial agents) can be added during formulation.

Another approach to produce nanocomposites is chemical vapor deposition – which permits generation of highly accurate and consistent layer coatings (ca. 1 nm-thick) of, namely, SiO_2; the resulting product offers ultrahigh gas barrier features, crack resistance, and light transmission. Unfortunately, permeability to moisture remains high, and gas barrier features become compromised at high RH – thus calling for an extra coating, with a component characterized by low water vapor permeability.

2.2.2.10 Active packaging

Further to the basic functions played by a regular package, active prevention of food spoilage, preservation of integrity throughout shelf-life, and enhancement of product attributes become possible via active packaging; the package accordingly senses (or anticipates) changes in storage conditions, and responds automatically by altering some property or releasing some chemical. From the early widget in cans of beer that released pressurized gas toward formation of a foamy head on the liquid once the can was opened, active packaging has experienced extensive developments; these have been focused on gas control, antimicrobial release, and temperature record, second only to moisture control. Release of specific compounds from the package, in a gradual manner, has indeed accounted for one of the most efficient forms of active packaging; this includes antimicrobials (some with flavoring effects), as well as antioxidants, ethanol, CO_2, SO_2, and even pesticides.

In solid foods, deteriorative reactions brought about by spoilage or pathogenic microorganisms occur preferentially on their surface – due to the high resistance offered by their bulk matrix against transport of substrates and growth of cells. Hence, lower amounts of antimicrobial compounds will be necessary to inhibit pathogens if incorporated only in the packaging material – rather than added to the whole bulk food itself; in this way, the compounds of interest are released when they are needed the most, and where the highest probability of microbial contamination exists. Slow (and controlled) release thereof makes available just the appropriate amount, while providing protection over an extended period of time – a deed particularly useful if long shelf-lives are sought; since the preservative is meant for the product surface only, the amount required is quite lower. Examples pertaining to bread, cheese, fish, fruits and vegetables, and meat and poultry include bacteriocins (e.g. nisin and pediocin), and fungicides and antibiotics to a lesser extent; organic acids (e.g. benzoates, propionates, sorbates); spices and other natural extracts (e.g. cloves, cinnamon, grapefruit, horseradish) – which may also serve as sensory agents, thus improving current flavor or masking developing off-flavors; silver zeolites; allylisothiocyanate; and lysozyme. Many more compounds appear promising, but are not widely employed at present due to high cost or regulatory constraints. Active EPSLs have also been produced that bear antimicrobial features – via incorporation of cinnamaldehyde from cinnamon, or else sorbic acid.

The same rationale essentially applies to oxidative decay – more probable on the food surface, as unwanted air admitted through the packaging material, or oxygen produced by residual photosynthesis within a plant food matrix tends to concentrate in the overhead space between food surface and outer package. Examples of antioxidante additives include BHA, BHT, rosemary extract, and tocopherol – or else carvacrol from oregano, in EPSLs; they are intended to inhibit fat oxidation, and successful in dried and fatty foods.

Ethanol possesses antimicrobial properties, especially against molds and yeasts – and is thus useful toward growth inhibition thereof, in such foods as bakery products, cheeses, and dried fish. A common delivery method is via sachets – made from a thin film highly permeable to ethanol (e.g. EVA), inside which ethanol/water is adsorbed onto SiO_2 powder.

Carbon dioxide is able to inhibit Gram$^-$ bacteria and molds in fish, fruits, meat, poultry, and vegetables; sachets of sodium hydrogen carbonate and ascorbate have been utilized in this framework to generate that gas.

Sachets containing sodium metabisulfite, in a microporous material, have proven favorable in attempts to inhibit molds in packaged fruits – including grapes, in particular; this salt reacts *in situ* with food acids, already present in the food material or meanwhile leached from the packaging film (e.g. sorbic acid) – and concomitantly releases SO_2.

Pesticides may, in turn, be adsorbed onto the inner layer of shipping containers or sacks of dried foods (e.g. cereal grains, flour), thus preventing growth of pests, molds, and bacteria – not only at the interface right away, but also in the food bulk as they slowly diffuse in.

Another field where active packaging is relevant is in controlled removal of specific compounds from the food and its overhead space – namely CO_2, O_2, moisture, C_2H_4, and flavor-related chemicals; as well as in enzyme immobilization – in or on the film, to bring about one or more of the said processes.

Sachets of polyacrylate sheets or propylene glycol films – containing, namely, $Ca(OH)_2$, NaOH, or KOH, are able to absorb CO_2 generated by such foods as (packaged) coffee, dried beef, or poultry products; calcium hydroxide is readily converted to $CaCO_3$ and water, upon contact with CO_2. Note that freshly roasted coffee still emits carbon dioxide, which may lead to bursting of sealed pouches where it has been packaged; whereas meat matrices still undergo some degree of aerobic respiration following animal slaughter and sectioning. Dual-action scavengers and emitters of CO_2 have been used with cakes, nuts, and snack foods. Scavenging of CO_2 may occur simultaneously with O_2; this is the case of a sachet system containing iron powder and calcium hydroxide, able to scavenge both gases. Extension of shelf-life of packaged ground coffee up to threefold has accordingly proven feasible.

Among packages with gas control capability, oxygen scavenging is currently the most developed subsector in commercial terms. Levels below 0.01%(v/v) of O_2 are easily attainable, thus entailing a better performance than modified atmosphere packaging – which can reach (at most) 0.3%(v/v); oftentimes, the economically most feasible approach is to use both, in a synergistic manner. Oxygen scavengers minimize oxidative damage to flavors, oils, pigments, and vitamins; hence, they prevent development of off-flavors, rancidity, and off-colors, and loss of nutritional value, respectively – in such foods as bakery products, cheeses, coffee, meat products, milk powder, ready-to-eat foods, and nuts. In addition, growth of insects, molds, and strictly aerobic bacteria is halted for lack of oxygen – thus circumventing the need for fumigation, or addition of chemicals with antifungal or -bacterial features for that matter. The earliest oxygen scavenging systems resorted to sachets with Fe-based powders and catalysts, able to speed up their reaction with molecular oxygen to form iron oxides. Modern approaches employ a polymer (e.g. PET) as packaging material, blended with an oxygen scavenger/absorber (e.g. nylon polymer MXD6) and a catalyst (e.g. cobalt salt); this type of system is stable up to 2 years – and has proven successful in beer, fruit juice, mayonnaise, and wine packaging. Another system is a multilayered polymer tray – encompassing an outside protective layer, an intermediate EVOH oxygen barrier layer, another (adjacent) intermediate oxygen absorbing layer, and finally an inside heat-sealable layer; oxygen-scavenging adhesive labels have also been applied to baked products, dried foods (e.g. coffee), and pizzas. Oxygen scavenging features are often combined with antimicrobial action in sachets, labels, films, or corks. This is notably the case of a polymeric film containing ascorbic acid – bearing an antimicrobial role; or an enzyme, normally alcohol oxidase or glucose oxidase, attached to its inner surface – able to catalyze oxidation of alcohol or glucose, respectively, at the expense of molecular oxygen. The resulting H_2O_2 also plays an antimicrobial role (e.g. in fresh fish) – further to the lower pH brought about by ascorbic acid itself.

Desiccant sachets or cartridges have been commercially available for a long time; a representative example is a sachet system that undergoes a fast increase in its capacity to absorb moisture as the dewpoint temperature is approached. Hence, formation of water droplets on the product is avoided – which are prone to microbial growth, as per the associated (local) increase in a_W. Clay sheets, polyacrylate sheets, propylene glycol films, and silica gel have also met with success in controlling excess moisture in packaged bakery products, cut fruits and vegetables, fish, meat, and poultry – thus delaying quality and safety deterioration. For instance, propylene glycol or diatomaceous earth have been entrapped in polymeric films or mats, and placed in contact with the surface of fresh fish or meat for drip absorption – thus constraining growth of spoilage bacteria at the interface, where it is most likely to occur.

Ethylene scavenging (via oxidation) is possible at the expense of sachets containing activated carbon, aluminum oxide, clay, potassium permanganate, silica gel, or zeolites; it is useful to reduce (or even prevent) ripening of climacteric fruits, e.g. apples, apricots, avocados, bananas, mangoes, tomatoes.

Sachets containing citric acid or ferrous salts have been employed to improve flavor of biscuits, cereal products, and potato crisps (and other fatty foods); they are, in particular, able to reduce bitterness of grapefruit juice.

When removal is to be effected via a (bio)chemical pathway, a possibility is immobilization in the packaging film of an enzyme that catalyzes it (as already discussed with regard to molecular oxygen); further examples are cholesterol reductase for removal of atherosclerosis-causing cholesterol in dairy foods, lactase for removal of lactose in dairy products for lactose-intolerant people, and naringinase for removal of naringin responsible for bitterness in grapefruit juice.

Active packaging may as well entail protection from light; for instance, films containing an UV-absorbing agent (e.g. nylon-6) may be used to coat PET bottles, and thus reduce light-induced oxidation of beer or wine therein.

Finally, active packaging can play a role in heating control – namely by conveying localized heating under a microwave field; or else heat addition or removal right upon container opening, or even control of respiratory metabolism.

The first effect resorts to susceptor films – e.g. PET lightly metallized with aluminum that supports application of a high-temperature treatment, and duly laminated to a paperboard substrate. Note that most microwavable packs are unable to heat up significantly (by themselves) when subjected to a microwave field; conversely, susceptors are able to absorb a fraction of the microwave energy in the form of thermal energy – thus reaching temperatures as high as 220°C, which directly alter the rate (and pattern) of heating of the food matrix inside. The bottom line here is to impart crispness to, or bring about browning and drying of the surface of some (microwaveable) foods – e.g. popcorn, French fries, pizzas, pies and other baked goods, and ready-to-eat foods at large. If metallization is etched off during manufacture, heat may instead be directed to specific areas of the pack; this is useful to avoid melting of, or volatile release by carton glues – which would otherwise contaminate the food product. Self-venting microwave packs possess, in addition, a vent – which opens after a preset temperature is reached, and recloses upon cooling.

A self-heating pack is but a double-walled metal container, which warms up upon reaction of CaO with H_2O; it has been used for heating and/or cooking such foods as coffee, soup, tea, or ready-to-eat meals. Despite its potential, it is still far from being a commercial reality.

A similar goal is pursued by self-cooling containers, e.g. for beer or (noncarbonated) soft drinks. One approach resorts to the latent heat of evaporation of water to produce the cooling effect; the water is accordingly bound in a gel layer that coats a separate container within the beverage can. When the base of the can is twisted, a valve is opened – which exposes the water to a desiccant held in a separated, evacuated chamber; this causes water to evaporate at room temperature – thus allowing ca. 300 mL of beverage be cooled from room temperature down ca. 17°C within 3 min. Another approach is a double-walled metal container – cooled by ammonium chloride or nitrate from one compartment contacting and (endothermically) dissolving in water supplied by the other compartment.

Finally, temperature-sensitive films – obtained with fillers in regular polymeric films, may control gas permeability at different temperatures; since temperature in freshly packaged fruits and vegetables rises spontaneously due to respiration, exchange between inner and outer atmospheres at early times may prevent buildup of anaerobic conditions.

2.2.2.11 Intelligent packaging

Intelligent (or smart) packaging refers to packaging systems able to actively respond to changes in product or environmental conditions, monitor freshness, display information on quality or product history, confirm product authenticity, enhance convenience, and counteract tampering or theft. They usually involve the ability to sense or measure an attribute of the product (e.g. composition of inner atmosphere) or of its surroundings (e.g. shipping environment). Therefore, the package may switch on and off in response to some varying internal or external condition; and either trigger active packaging function(s), or merely communicate its current status to the retailer or the consumer – with the possibility for connection of signals delivered to control mechanisms of home appliances, or the storage system itself.

A first possibility is supplying some form of relevant information – which encompasses, for miscellaneous products, type of food and degree of ripeness, or safety, quality and nutritional attributes; or inventory data instead, relevant for tracing and tracking foods throughout the distribution chain. Besides the quite widespread barcodes, novel methods include radio frequency identification tags (RFIDs for short), and magnetic strips and electronic article surveillance tags (EASTs for short). RFIDs are smart labels that permit a package or unit load be tracked or traced – with the possibility of simultaneously recording some elementary history; they contain an electronic chip, a power source, and an antenna. Data can be read from the tag, or sent to the tag without having a line of sight thereto – while a large number of tags can be simultaneously handled (i.e. identified and/or addressed); these constitute the chief advantages associated with such a technology. EASTs are security tags, materialized as miniature electromagnetic devices – able to activate alarms in shops, if not previously deactivated at cashier/checkout; they allow storage of more extensive data, thus conveying stronger protection from fraud than plain barcodes. The price of some of these devices is still relatively high – so they are utilized on reusable distribution containers, as an integral part of the management routines of logistic chains.

One major opportunity for intelligent packaging arises in MAP, as gas concentration probes or control devices. The former are relevant to monitor storage conditions, current gas composition, and even leakage (as nondestructive check of package integrity); they encompass dyes that change color when a *pH* or E_h threshold is reached, as long as it somehow correlates to the actual levels of O_2 and CO_2 – sometimes through reactions catalyzed by specific enzymes. For instance, an oxygen-sensitive ink and an indicator have been developed that change from pink to blue when oxygen level rises from <0.1%(v/v) to >0.5%(v/v). An outstanding response causing an impact well beyond mere informational content is, however, possible with breathable intelligent polymer films – able to cope with high respiration rates, and already in commercial use for freshly cut vegetables and fruits. The said films are acrylic side-chain polymers that undergo a reversible phase change at a specified temperature; when the side-chain components melt, gas permeation increases dramatically, so the rate of exchange of O_2 and CO_2 with the outer environment will eventually be regulated as a function of temperature – and the packaging material can be designed to adjust permeation ratios to distinct products. An optimum atmosphere composition may accordingly be maintained around the food product, throughout the whole storage and distribution network – thus permitting extension of shelf-life, with product remaining above minimum acceptable levels of freshness and quality; and allowing for some degree of reoxygenation of the food, able to constrain growth of (anaerobic) pathogenic bacteria.

Freshness monitoring and control are also possible via microbial growth indicators – for instance, *pH* dyes (e.g. bromocresol green, cresol red, methyl red, and xylenol blue) that react with specific metabolites (e.g. CO_2, SO_2, NH_3, H_2S, organic acids), or chemical/immunochemical moieties that selectively attach to toxins; spoilage bacteria, and even such a harmful pathogen as *Escherichia coli* O157:H7 may accordingly be detected in fresh fish, meats, or poultry. One example of gas detection-mediated sensing is an adhesive label to be posted on the outside of a packaging film – and tailored to probe freshness of seafood products, via development of color in a tag; a barb on its reverse penetrates the film, and volatile amines – generated by spoilage bacteria, can reach a chemical sensor that turns progressively bright pink as seafood ages. Another example of the same type is a metmyoglobin-based indicator, which changes its color from brown to red when in the presence of sufficient H_2S produced by spoilage bacteria in packaged poultry. With regard to detection of the enterotoxin produced by *Escherichia coli* O157:H7, one may resort to a cross-linked polymerized polydiacetylene sensor, incorporated into plastic – which develops a deep blue color in its presence.

Another form of intelligent packages encompass mechanical damage indicators – namely physical shock and tamper attempt.

Shock detectors have been available for quite some time; they are classically based on overload devices, e.g. spring/mass systems, magnets, or drops of dye, attached to the package or the product itself – and convey some form of perceptible information if an excessively violent shock has been experienced. More recent devices include optically variable films – e.g. piezoelectric polymers in packaging materials that change color at a certain stress threshold; or gas-sensing dyes that undergo an irreversible modification when exposed to some form of mechanical impact – indicative of poor handling.

Self-bruising closures on bottles or jars, designed for a variety of foods, are probably the simplest way to indicate that deliberate

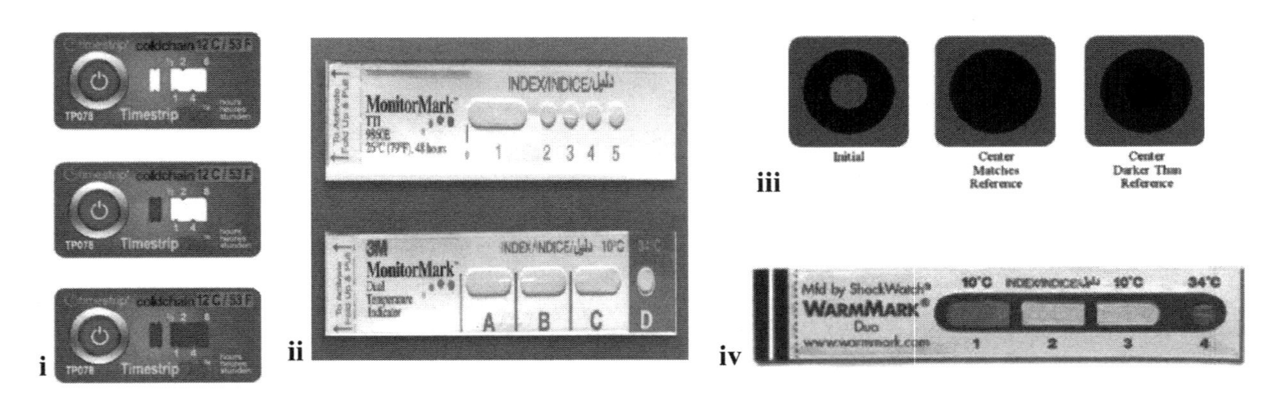

FIGURE 2.47 Examples of time-temperature indicators: (i) Timestrip™, (ii) MonitorMark™, (iii) MRE ShelfLife™, and (iv) WarmMark™.

attempts have been made to open them; an alternative is tamper-evident labels – which change color when removed, or leave a message on the pack that cannot be hidden.

Finally, temperature control has open wide opportunities for application of intelligent packaging; the two most important forms are critical temperature indicators (CTIs for short) – which alert when a prespecified (maximum) temperature has been exceeded, and time-temperature indicators (TTIs for short) – which signal a critical accumulation of temperature deviation over time, also known as time-temperature history. To assure improved reliability, the mechanism of indicator decay in the latter case should match as far as possible the mechanism of decay of the critical ingredient (or overall product, for that matter). Commercial examples of these indicators are presented in Fig. 2.47. Either type is particularly relevant in the case of chilled and frozen foods – as per their ability to indicate whether a food product has been held at the correct temperature throughout storage, thus abiding to the expected shelf-life; or if temperature abuse has taken place – in which case the food at stake must undergo faster movement forward along the cold chain (if still feasible).

While CTIs indicate whether a product has been exposed to temperatures above a reference temperature for sufficiently long a time to bring about significant changes in its quality/safety, they neither convey information on how long such an exposure lasted – nor for how much the reference temperature was exceeded; hence, they are appropriate for foods that undergo irreversible damage above or below a given temperature – including foods susceptible to growth of pathogen(s) above a critical temperature. Conversely, TTIs integrate over time the temperature of the product they are attached to; and are formally classified as one of two types – critical (or partial history) and full history time/temperature indicators. The former show the cumulative time/temperature exposure above a reference (critical) temperature – and are thus particularly useful in the case of foods that experience chemical, enzymatic, or microbial transformations significantly only above a given temperature. Full history TTIs reflect the whole temperature history of the food or, at least, of its surface (usually the location most prone to alteration) and are expected to give a clear, accurate, and unambiguous indication of product quality, safety, and remaining shelf-life; they accordingly materialize general-purpose devices.

The aforementioned indicators usually rely on irreversible mechanical, chemical, or enzymatic mechanisms, e.g. melting, thermochromic ink, polymerization, electrochemical corrosion, or acid/base indicator. Representative examples include: wax that melts, thus releasing a colored dye upon unacceptable increase in temperature; liquid crystal coatings that change color with storage temperature; a label with an outer ring printed with a stable reference color, and a bull's eye at the center printed with diacetylene – with the latter undergoing progressive, predictable, and irreversible color change, until reaching the outer ring color (indicating that actual shelf-life has expired); or enzyme-mediated reaction that produces/consumes a carboxylic acid, thus changing color of a pH indicator. An integrated system has also been proposed that can be applied to a pack as the product is dispatched – with a barcode containing information on product identity, date of manufacture, and batch number (among other data relevant for unique identification of container); a second code identifying reactivity of the TTI chosen; and a third section containing the indicator material itself. When the barcode and the color/shape of the TTI are scanned, a hand-held microcomputer immediately displays the current quality status of the product – and prompts a variety of straightforward, preprogrammed messages.

CTIs may also be taken advantage of as convenience aid, rather than quality/safety control; a representative example is an intelligent self-heating or -cooling container (e.g. beer bottle labels), with thermochromic ink dots to inform the consumer that it has reached the correct temperature – following microwave heating or regular refrigeration.

2.2.2.12 *Filling, sealing, and labeling*

Accurate filling of food containers is of practical relevance – not only to ensure compliance with fill-weight legislation, but also to minimize product give-away caused by overfilling. Furthermore, food preservation requires adequate sealing of the said containers – to avoid physicochemical and microbial contamination that would jeopardize the expected shelf-life of the product. However, seals are normally the weakest part of a container – since they are not produced as a continuum, unlike normally happens with the remainder of the package. Finally, information regarding identification of product and manufacturer, as well as nutritional content, amount, storage conditions, and shelf-life, and (optionally) preparation/serving suggestions is to be included on top of the product – besides some form of coding for computerized handling, and specific information about the previous manufacture process to facilitate tracking upstream (should some problem arise with the product); such a type of information should accordingly be made available somewhere on the package.

Metal and glass containers are industrially supplied as palletized loads, duly wrapped in shrink/stretch film – and usually in a status known as commercially clean (but insufficient to assure food safety). Upon receipt, they are depalletized, and inverted over steam or water sprays for additional cleaning, and kept in that inverted position until filling thereof is in order – thus minimizing the chance for contamination, by any agent contributed by the outer environment and driven by gravity. Wide-mouthed plastic pots or tubs are normally supplied in stacks, fitted one inside the other for the sake of spacekeeping, and contained in fiberboard cases or shrink film. Prior to use, they are cleaned by moist air; if intended for aseptic packaging, they are instead sterilized with hydrogen peroxide. Laminated paperboard cartons are, in turn, supplied either as a continuous reel or as partly formed flat containers; once again, sterilization is via H_2O_2, prior to filling with products subjected to UHT treatment in advance.

2.2.2.12.1 Filling

Filling equipment, at the industrial level, is normally required to be accurate to ±1% of the target volume or weight – and without spillage over, or contamination of the surrounding area; provision for no-container/no-fill capability is also a must, besides the possibility to accommodate containers of different sizes and shapes. Selection of a liquid filler hinges upon physicochemical characteristics of the product (e.g. viscosity, foaming features, particle size if in suspended form) and operating conditions (e.g. temperature, throughput rate); the most important types are overflow, servo pump, peristaltic, gravity, piston, and net-weight fillers, as illustrated in Fig. 2.48. Filling heads may be arranged in line, or as a carousel – thus permitting longitudinal or rotary motion; the latter type has been extensively used in automatic filling (and sealing afterward) of glass bottles and plastic cups (with e.g. beer, coffee, jam, juice, milk, mineral water, yogurt), especially because their throughputs may easily go over 20,000 bottle.hr^{-1}.

Overflow, or fill-to-level fillers (see Fig. 2.48i), often handle low- to intermediate-viscosity, foamy liquids (e.g. some milk products, mineral water) – including suspensions with particle size not exceeding ca. 1.5 mm; filling takes place until a target level in the container is attained, rather than a preset volume.

Servo pump fillers (see Fig. 2.48ii) are appropriate for liquids spanning the whole range of viscosities – including particulate liquids (e.g. sauces); each nozzle is connected to a dedicated servo-controlled pump of the positive-displacement type, so their cost is relatively high.

Small-volume filling of high-value liquids (e.g. essential oils) is normally handled by peristaltic fillers (see Fig. 2.48iii), due to the need of extra accuracy in volume delivered. The rollers of rotary pumps make intermittent contact with, and thus squeeze the outside of a flexible tube – so the liquid is propelled in the direction of their movement, with no chance for leakage or product contamination; the said tube is often disposable.

The most economical liquid fillers resort to gravity (see Fig. 2.48iv), yet they are restricted to nonfoamy, low-viscosity liquids (e.g. alcoholic spirits, mineral water) – otherwise the driving force, in terms of hydrostatic pressure, will not suffice for the flow rates desired. Flow control for delivery of the target volume resorts to valves, thus avoiding the need for pumps.

The oldest, and still most reliable apparatuses are piston fillers (see Fig. 2.48v) – suitable for viscous products (e.g. creams, food pastes, heavy sauces, salad dressings). They are quite accurate and non-expensive, despite being unsuitable for low-viscosity liquids (due to the possibility of leakage between piston and cylinder); but hardly flexible, in terms of volumes delivered. A

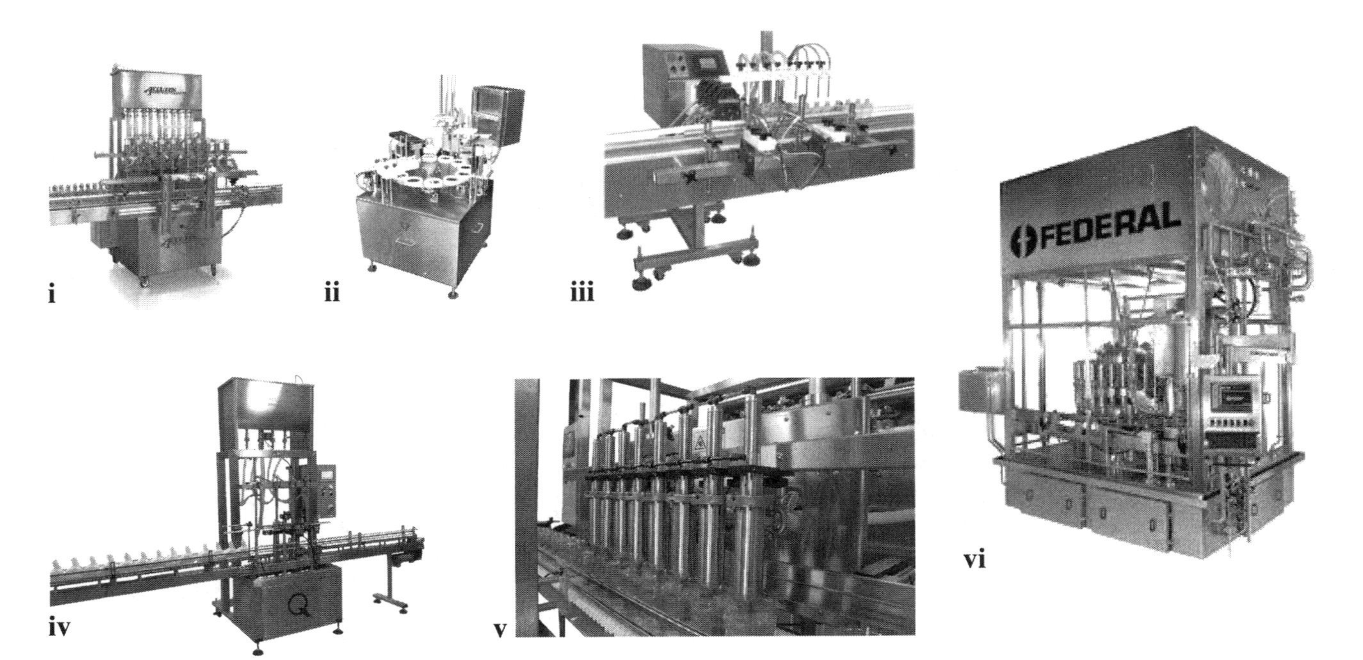

FIGURE 2.48 Examples of liquid filling equipment: (i) overflow filler (courtesy of Accutek, Irving, TX), (ii) servo pump filler (courtesy of Volumetric Technologies, Cannon Falls, MN), (iii) peristaltic filler (courtesy of Inine Filling Systems, Venice, FL), (iv) gravity filler (courtesy of Liquid Packaging Solutions, La Porte, IN), (v) piston filler (courtesy of Apacks, La Porte, IN), and (vi) net-weight filler (courtesy of Berks Plant Design and Maintenance, Shoemakersville, PA).

piston accordingly draws product from a hopper into a cylinder as it slides, and then a (rotary) valve changes position before the piston reverses its movement – so that product will be delivered, through a nozzle, into the containers passing below.

When liquids are to be filled in bulk amounts, or smaller amounts to be sold by weight, then net weight fillers are in order (see Fig. 2.48vi). The product is pumped to a holding tank with a built-in manometer, connected to a pneumatically operated valve at its bottom outlet; the net weight of product remaining in the container is monitored in real-time, via change of (hydrostatic) pressure – and the valve is shut when the target weight delivered has been attained.

Small particulate foods (e.g. rice and other grains, powdered foods) may be filled using equipment similar to liquid fillers or form–fill–seal apparatuses – since their flow somehow mimics the flow pattern of a true fluid; examples are depicted in Fig. 2.49. Larger foods (e.g. confectionery products, fruits) can be filled into rigid containers with the aid of photoelectric devices – able to count and ascertain size/weight of individual items; hence, their operation is similar to that of food sorters. In alternative, a disc fitted with a number of recesses may be employed – able to hold individual items, as it rotates below a holding container; such recesses become consecutively filled with food pieces, until the required number is reached – when the said disc pours its contents into a pack. To produce mixed salads, nuts, or confectionery items (e.g. biscuits, cookies), multihead weighers are used instead; they are designed to separately handle distinct products, and fill a given container with a more or less random (yet balanced) selection thereof.

Containers can be filled by weight using a net-weight or gross-weight system; the difference between the two lies on whether the product is weighed before, or after filling and sealing the container – in which case the package and seal weights will be added to the food weight proper. For more accurate filling, a bulk feeder should be employed first – able to quickly fill the pack to ca. 90% of its capacity; after weighing the exact amount filled so far, a controller estimates the amount still left to fill – and promptly activates a fine feeder that tops up the package, prior to final weighing as doublecheck. Commercial systems currently available can perform up to 200 weighing.min^{-1}, with a precision of the order of $\pm 0.5\%$; in view of the intrinsic variability when the food is made up of large, irregular pieces, a statistical record of fill weight history is generated – useful for periodic calibration, thus minimizing the chance for giveaway product.

Fresh fruits and vegetables can be filled into distribution cartons or bins using a number of devices; for such fruits as citrus, kiwifruits, and tomatoes, as well as such vegetables as onions, the most cost-effective mode of packing is volume filling or loose filling. Equipment configurations range from manual to semiautomatic systems – including rotary pack- and belt-tables, with padded vinyl surfaces designed for protection of food items from bruising and mechanical damage. The former facilitates hand-placement packing, for continuous supply of fruits within a wide size range; while the latter allows access of multiple packers to a given size/grade of fruit. High-volume, gentle filling (of e.g. presized apples, avocados, kiwifruits, pears) often resorts to tray-filling conveyors – while delicate products are filled into retail bags via air bagging heads; bin fillers are employed to evenly and gently fill shipment bins with fruits.

When heat-sterilized foods are packed into glass or metal containers, these are not filled to full volume; a 5–10%(v/v) expansion space (or ullage) above the food is indeed necessary to create a partial vacuum – which not only reduces oxidative deterioration of food owing to absence of atmospheric oxygen, but also buffers pressure changes inside the container. To avoid air being trapped in food, and thus eventually reduce headspace vacuum, (viscous) sauces and gravies are filled before solid pieces of food are dropped in; air might otherwise be caught in the interstitial spacing between the (essentially stationary) solids – which would hardly be occupied by the liquid within the processing period, due to kinetic limitations on its gravity-driven flow and possibility of air bubble imprisonment.

2.2.2.12.2 Sealing

The finish of glass containers – i.e. the part of the bottle or jar that holds the lid or cap, is the most critical part with regard to sealing; it possesses lugs or threads to secure the cap, along with a smooth surface meant to form a seal with the closure.

Pressure seals, regular seals, and vacuum seals are the three types of bottle closures regularly employed. The former have traditionally been used with carbonated beverages, owing to the overatmospheric pressure of CO_2 inside the vessel, and include: screw-in-screw-out (also known as internal screw), or screw-on-screw-off (or external screw); crimp-on-lever-off, crimp-on-screw-off, or crimp-on-pull-off; and roll-on (or spin-on) or screw-off (or roll-on-pilfer-proof). Typical examples encompass cork, injection-molded PE stopper, screw cap, crown cap or pressed

FIGURE 2.49 Examples of solid filling equipment: (i) cereal filler (courtesy of Shenzhen Penglai Industrial Corporation, Shenzen, China), (ii) small fruit filler (courtesy of Automation Techniques, Tauranga, New Zealand), and (iii) large fruit filler (courtesy of Burg Machinefabriek, Kruiningen, The Netherlands).

tinplate lined with PVC , aluminum roll-on screw cap, and lug cap – as exemplified in Figs. 2.50i–vi.

Regular seals are utilized in noncarbonated beverages, where overpressure is not an issue – and include one- or two-piece prethreaded, screw-on-screw-off; lug-type screw-on-twist-off; spin-on-screw-off (or roll-on-screw-off); press-on-prise-off; crimp-on-prise-off, or crimp-on-screw-off; and push-in-pull-out, or push-on-pull-off. The most representative examples are cork or synthetic cork stoppers fitted with tinned lead, PE, or aluminum capsules, metal or plastic caps, and aluminum foil lids – see Figs. 2.50vii and viii.

Finally, vacuum seals are employed for hermetically sealed containers, and preserves or paste jars – and include screw-off-twist-off; press-on-prise-off, or press-on-twist-off; two-piece screw-on-screw-off, or roll-on-screw-off; and crimp-on-prise-off. This is the case of lug or twist cap (steel cap with a few inward protrusions from the side of the cap), plastisol-lined continuous thread cap (metal cap with a knurled threaded side), and press-on-twist-off cap, see Figs. 2.50ix and x.

Sealing of wide-mouthed rigid and semirigid plastic pots and tubs may be effected via push-on, snap-on or clip-on lids, or push-on, or crimp-on metal or plastic caps. Plastic jars and bottles are sealed with closures that can be tamper-evident (for customers' protection) and recloseable (for customers' convenience). They may instead possess an aperture or pouring spout designed for easy dispensing of contents; this includes push-on or screw-threaded caps for squeezable bottles (of e.g. cream, mayonnaise, sauces, syrups) – usually possessing a profiled pin that cleans the aperture on reclosing, and is aimed at preventing microbial growth. After filling, thermoformed pots or trays are lidded with a polymer film or laminated foil – duly heat-sealed to the top flanges; a single machine is used to form, fill, and seal small containers used for UHT-sterilized products (e.g. jam, honey, milk). More rigid containers (including glass) may be used for dry products – since the detachable contents can be directly poured upon inversion of, or shaken from the pack when the lid is opened; supplementary configurations include disc- or snap-top closures.

Generally speaking, the seal is formed when a resilient material is pressed against the rim of the container; the pressure should thus be evenly distributed – and be of a sustained nature, so as to produce a uniform seal around the whole cushioning material in direct contact with the rim. Glass bottles and jars have a narrow round sealing edge, while plastic bottles possess flat sealing edges. Such a resilient material is often constructed in EVA, PE, or PVC, or else obtained from composite cork or pulpboard sheet as bulk material – lined with one of these (essentially inert) polymers, aimed at protecting it from direct contact with the food contents. A preselected torque guarantees fitting tightness of cap to container; in alternative, pressure directly applied during the sealing process ensures such a tightness, as is the case with roll-on, crimped-on, and pressed-on caps. Should caps resort to a rotary action for effective sealing, their thread engagement and pitch are then to be carefully designed. The former pertains to the number of turns of the cap – from first engagement between cap and rim, to point where the liner is sufficiently engaged with the rim; at least one full turn is needed, otherwise uniform engagement of liner with rim will be difficult to attain. Thread pitch refers to steepness of thread; the steeper the

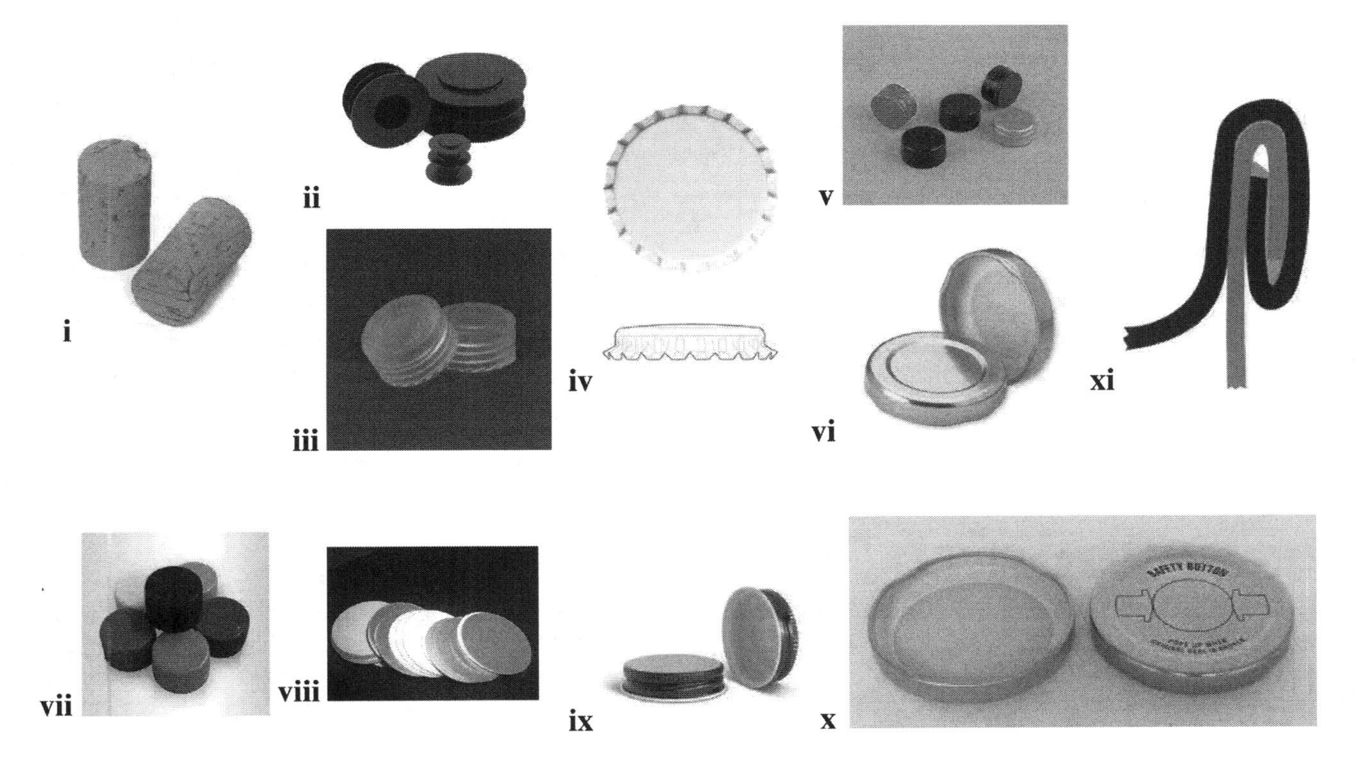

FIGURE 2.50 Examples of closures for rigid and semirigid food containers: pressure seal, as (i) cork, (ii) PE stopper, (iii) screw cap, (iv) crown cap, (v) aluminum roll-on screw cap, and (vi) lug cap; regular seal, as (vii) plastic cap and (viii) aluminum foil lid; vacuum seal, as (ix) plastisol-lined continuous thread cap and (x) press-on-twist-off cap; and (xi) detail of double seam in can lid, made from terminus of can top lid (■) and terminus of can side wall (■), with sealing material (■) in between.

slope, the lower the number of turns required to bring liner and rim into contact with each other (i.e. the more rapidly the cap will screw on and off) – but the higher the torque required for effective sealing.

A seaming machine is employed to seal can lids, typically via a double seam (see Fig. 2.50xi) produced by sequential operation of two rollers; the first rolls the cover hook around the body hook, whereas the second tightens the two hooks to produce the aforementioned double seam. A thermoplastic sealing compound is placed between metal sheets during this process, to fill the spaces inside the seam; it melts during retorting, thus providing an additional barrier against contaminants. Various types of easy-open ends may be fitted to cans, depending on the product at stake; for instance, ring-pull closures are used for two-piece aluminum beverage cans, while full-aperture ring pull closures are preferred for meat products, nuts, and snack foods – with both types manufactured by scoring the metal lid, and coating it with an internal lacquer. Collapsible tubes made of aluminum are instead sealed by folding and crimping the open end of the tube after filling – whereas tubes made of PE or laminated plastic are sealed using a heat sealer.

In the case of paper-based containers, plain or corrugated cases are manufactured as a flat blank – which is cut, creased, and folded to form the intended case (or carton); to minimize wastage, as many blanks as possible should be fitted to each sheet of paperboard. The board is printed and stacked into multiple layers for blanks to be cut out using a guillotine – either at the food premises, or already upstream by the case supplier; each blank is then precisely creased and formed into a carton – with shape held in position via gluing or stapling.

A thermoplastic film forms the inner layer of rigid laminated paperboard cartons for aseptically processed foods. Two alternative production systems exist for carton manufacture: either a continuous roll of laminated material is aseptically formed–filled–sealed; or pre-formed cartons are erected, filled in, and then sealed in an aseptic filler. To avoid the need of using scissors to open UHT cartons (as happened with earlier products), a peelable (tamper-evident) foil strip is now incorporated that reveals a dispensing hole upon removal. More recently, a range of screw caps with various sizes have been utilized – with the further advantage of permitting multiple usage, and thus intermittent consumption of the food. In either case, a hole is to be cut in the original cartons, ready to receive the cap – which is applied by the filling machine, after the carton has been filled; it is finally sealed in position, by resorting usually to hot air – which melts its constitutive polymer directly, or the glue meanwhile applied.

Although heat-seals are used for most flexible films, cold (or adhesive) seals are required for packaging of heat-sensitive products (e.g. chocolate, chocolate-coated or -filled foods, ice cream); remember that thermoplastic materials or coatings become fluid as they are heated, but resolidify upon cooling. To seal flexible films, the surfaces of the two films (or webs) are heated until their interface vanishes – with pressure then applied thereon, to complete/reinforce the fusion process; hence, temperature and pressure, as well as processing time will directly determine final seal strength. Bead seals, fin seals, and lap seals are the most common types of seal used by the food industry. A bead seal consists of a narrow weld at the end of the pack, whereas opposite surfaces are sealed in a lap seal – meaning that both are to show thermoplastic features. A fin seal has the same surface of a sheet sealed – but only one side of the film, or one component of a laminate needs to be thermoplastic; a major advantage arises from its protruding from the pack, as the food does not experience extra pressure during the sealing process – so heat-sensitive foods, or foods labile to deformation (e.g. biscuits, soft bakery) may be safely handled.

Sealing of films can occur simultaneously with cutting thereof – in which case a metal wire is heated to red, and used to form a bead-type seal for a single film; in the case of two films, a hot bar (or jaw sealer) can be utilized, which holds the said films in place between heated jaws until the seal forms. Films may instead be clamped between two cold jaws as in impulse sealers; fusion will occur only when those jaws are suddenly heated – and they remain in place until the seal cools and sets, so as to avoid film shrinkage or wrinkling that would jeopardize final appearance. Since heat applied locally to form a seal dissipates throughout the remainder of the film, special care is to be exercised – with regard to operating temperature on the one hand, and thickness and melting temperature of film on the other; otherwise, some form of thermal decay will likely occur, including generation of a poor seal.

Higher filling speeds require rotary (or band) sealers; stationary shoes accordingly heat the centers of metal belts, while their edges support the unsoftened film. The mouth of a package passes between the belt, and the two films are then welded together; passage through cooling belts ensues, which clamp the seal in position until proper setting. Seal heating may also be brought about via application of a high frequency (1–50 MHz) electromagnetic field, or ultrasound (>20 kHz); sealing is a consequence of increased molecular vibration, brought about by either electromagnetic or mechanical energy received by the film(s) – with the former requiring a high loss factor to work out.

Form–fill–seal equipment has indeed constituted a breakthrough in food packaging – in view of the several advantages associated thereto, namely reduced transport, handling, and storage costs of materials when compared to preformed containers. Furthermore, package production *in situ* is simple and nonexpensive, requires little human labor, and permits much higher throughputs be attained. Two major types of vertical form–fill–seal machines exist – transwrap and flow pack; and one horizontal – flow wrap (or pillow pack).

A film is intermittently pulled over a forming shoulder in the transwrap equipment, see Fig. 2.51i, via the vertical movement of a set of sealing jaws; films should thus possess good slip properties, as well as resistance to creasing and cracking – so as to permit trouble-free passes over the said shoulder. The bottom is meanwhile sealed by jaw sealers, and the product is filled in the resulting top-open package; a high melt strength is accordingly required for the film, in order to support the product on the bottom seal while it remains hot. A second seal is then produced that closes the top of the package – while simultaneously forming the bottom seal for the next package. This type of filling equipment is suitable for powders and granular products, besides sticky items – and can go up to 120 pack.min^{-1}. Vertical form–fill–seal machines may also be used to package liquids and high-viscosity pastes (e.g. butter, cream cheese, marzipan, sausage meat) in so-called chub packs; the film is again formed into a tube, and filled with product – yet sealing resorts to metal or plastic clips at each

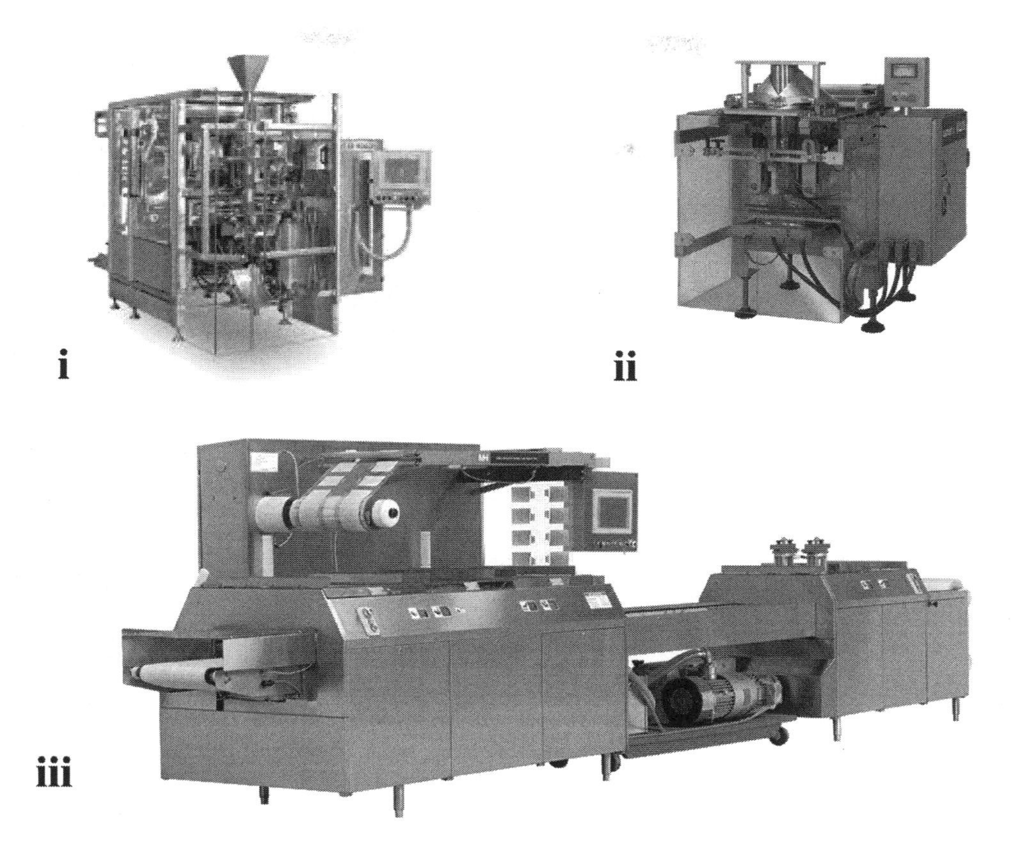

FIGURE 2.51 Examples of form–fill–seal equipment: (i) transwrap type (courtesy of Bosch, Waiblingen, Germany), (ii) flow pack type (courtesy of Ilapak, Newtown, PA), and (iii) flow wrap type (courtesy of Ossid, Battleboro, NC).

end, rather than seal/cutting of individual packs through heating. The film may instead be folded over a triangular shoulder, thus forming two side seams; the sachets are then separated, opened by a jet of compressed air, filled, and heat-sealed across the top.

The flow pack equipment, illustrated in Fig. 2.51ii and suitable for up to 600 pack.min^{-1}, differs from its transwrap counterpart because a forming shoulder is not employed – as two sheets of film are used; the film(s) are thus subjected to less stress. The thermoforming machine preheats the said plastic films, and forms side seams that create a continuous sleeve – via heaters and crimp rollers; the strips are sealed at the bottom, filled with product, and finally sealed at the top – with strips finally cut into individual packages (or sale units of 5–15 portions).

The flow wrap equipment displayed in Fig. 2.51iii somehow mimics a flow pack equipment, except that it works horizontally; products are thus pushed into the sleeve of the film as it is formed, and transverse seals are produced by rotary sealers – which also cut the packs separate. These apparatuses are widely used to form and fill sachets with powders and granules (e.g. coffee, salt, sweeteners) or liquids (e.g. cream, ketchup, salad cream, sauces). They will attain throughputs of 1000 pack.min^{-1}, as long as a strong seal can be produced within a short heating time; this calls for films sufficiently thin for quick melting, but possessing a high melting strength – while such sachets may be automatically cartoned at the end. Further to allowing greater throughputs, flow wrap equipment has become popular for its greater flexibility – ranging from single round pieces to irregularly shaped foods and multiple pieces meant for packaging, besides aseptic packaging of laminated cartons.

Wrapping by polymeric films applied onto foods normally resorts to either shrinking or stretching; other flexible films may require plain mechanical reshaping, as is the case of aluminum foil (for such unusually shaped products as Easter eggs or bunnies) or twist-wrapped cellulose film (for confectionery). Shrinking ratios in machine and transverse directions are normally provided when films are intended for shrinking – with performance rated as fully balanced (characterized by percent shrinkage to ca. 50:50%), preferentially balanced (ca. 50:20%), and low balanced (ca. 10:10%); film tightening requires some degree of shrinkage, which increases substantially if contoured packages are meant. LDPE has been extensively used to bring about shrink-wrapping, since the biaxial orientation of its molecular chains produces uniform shrinking in the two normal directions. Shrinking is attained in a continuous, gradual manner by passing the film through a hot air tunnel (with heating power of the order of 20–30 kW), or below radiant heaters; intermittent pulses of hot air may instead be fired against the package, when it passes beneath the heat gun – with substantial energy savings. Stretch-wrapping resorts to LDPE, LLDPE, or PVC films, challenged with tension around collated packages; much lower energies are required in this case (with mechanical power within 1–6 kW). Shrink-wrapping reduces the amount of film area by 5–10% (at least), while stretch-wrapping actually elongates the film by ca. 2–5% – which confers economic and technological advantages to the latter approach.

To facilitate opening of flexible packs (especially when made of strong films), tear tape can be applied longitudinally or transversely to the pack; or slits and perforations may be produced mechanically by the wrapping machine itself, or by a laser-cutting device afterward. On the other hand, closures that enable intermittent consumption of a food product have been on the rise, owing to their obvious convenience – which imply tamper-evident and -proof features be included in such closures. These are important measures to deter prospective consumers from breaking the closure of a food package for tasting, and then returning it to the retail shelf; or, in the worst case, from deliberate poisoning of foods, underlying extorsion attempts *a posteriori*.

Although total protection is obviously impossible, tamper-resistant closures delay violation of package, and leave a perceptible clue of tampering that cannot be reversed or easily disguised. Several examples of tamper-proof closures are provided in Fig. 2.52. The most common features in tamper-proof closures of bottles and jars are: foil or membrane seals (see Fig. 2.52i) for wide-mouthed containers; heat-shrinkable sleeves for bottlenecks (see Fig. 2.52v), or bands or wrappers placed over lids; breakable caps (see Fig. 2.52ii) – in the form of rings or bridges that join the cap or lid to a lower section, and are irreversibly broken when opened; roll-on-pilfer-proof (or ROPP for short) caps – where a tamper-evident ring in the cap locks onto a special bead in the neck or the jar or bottle, thus producing a seal that breaks upon opening and drops slightly; a safety button in press-on-twist-off closures for heat-sterilized jars and bottles (see Fig. 2.52iv), bearing a concave section in the lid previously formed by headspace vacuum when residual steam condenses upon cooling to room temperature – which becomes quickly convex upon opening, and simultaneously generates a typical click noise; and a breakable plastic strip that conveys a visual signal if the jar has been previously opened (see Fig. 2.52iii). Aluminum or plastic tubes may carry a membrane over their mouth, which needs to be punctured prior to use. Three- and two-piece cans (see Fig. 2.52vii), as well as aerosol containers are inherently tamper-resistant; in the case of composite containers, the ends may be joined to the walls – and thus cannot be pulled apart without notice.

With regard to flexible film wrappers, these must be cut or torn, and laminated plastic/foil pouches (see Fig. 2.52vi) have also to be cut to gain access; hence, tampering becomes readily apparent. Blister or bubble packs give, in turn, visible evidence of backing material separating from blister – as each compartment must be broken, cut, or torn to access its contents.

A special mention is deserved by modified atmosphere packaging – where fin seals are the rule, and perform well as barriers against transfer of gas and moisture. Heat-forming of semirigid and flexible, single-piece containers has been used for bakery, fish (cooked and raw), meat, and poultry – or else of composite structures of carton blank and plastic tray with in-line lidding, in the case of bakery products, cheese, fish, meat, poultry, prepared meals, and salads and vegetables; vacuum application, followed by gas flushing is used at the end in both cases. Cheese, coffee, fish, fruits and vegetables, meat, nuts, and salads have resorted to single flexible web forming–filling–sealing, with gas flushing by lance or tube followed by venting to the atmosphere. Fish, meat, nuts, and prepared meals have also been packed in HDPE, PET, and HIPS trays or preformed plastic bags, with vacuum followed by gas flushing; while bulk meat, cheese, dried powders, fruit, nuts, and poultry in barrier and nonbarrier bags have successfully been packed into corrugated or solid board case – again with vacuum and gas flushing. Vacuum skin packs have been employed for meat and fish – and are made from a multilayered film top web shrunk over the product, contained in easy-peel film forming the bottom web; or from two similar film webs, with the upper heated and shrunk over the product placed on the lower web;

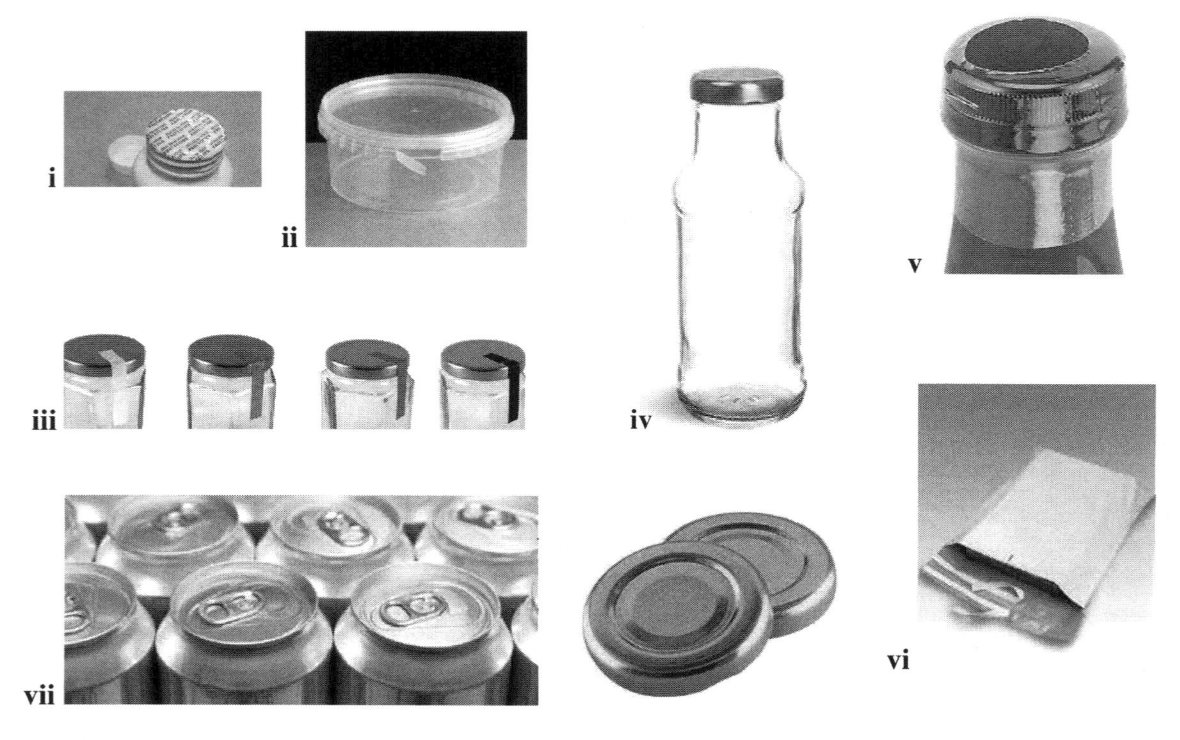

FIGURE 2.52 Examples of tamper-evident food packages: (i) foil seal, (ii) breakable cap, (iii) breakable plastic strips, (iv) press-on-twist-off closure applied onto bottle (top) and detail of safety button therein (bottom), (v) heat-shrinkable sleeve, (vi) laminated pouch, and (vii) two-piece cans.

or even from laminated top web shrunk over the product on a laminated board/film bottom. Finally, bag-in-carton approaches have been employed for food powders and granular products, using lined carton eventually flushed with desired gas.

2.2.2.12.3 Labeling

The first contact of a customer with a food is normally via its visual appearance – with label playing a minor role; however, the label is the first (and most important) means of contact between processor and customer, and thus a nuclear part of the marketing strategy associated with the product. Conversely, when choosing between competing food products, the information included in the corresponding labels acquires a major relevance for consumer's final decision – since watching follows seeing.

The legislation of most countries requires the following minimum information be included in the label: name of product; list of ingredients (normally in descending order of weight fraction, or nutritional relevance); name and contact of manufacturer; net weight or volume per package; a "use by," "best before," or "sell by" date notice – accompanied by permanent markings (often inscribed by noncontact laser tools, applicable to every type of surface nature and shape) specific to each manufacturer, and identifying factory, production line, and manufacture batch (useful for traceability and tracking, if legally necessary); instructions on best storage conditions, prior and posterior to opening; and a product barcode. Special instructions for preparation of food may be included, although not strictly required.

To standardize product description (and manufacturer identification), a universal product code (UPC for short) has been developed in the United States, and extended afterward via European article number (or EAN for short) to such other countries as Canada, most European States, Japan, Australia, and New Zealand. It consists of 12 numeric digits in its default UPC-A version, and the same plus one (or two, or even three digits) preceding it as identification of country of origin in the default EAN-13 version (for instance, 45 stands for Japan, while 0 stands for standard UPC) – uniquely assigned to each (food) item and manufacturer. This labeling system is particularly handy and useful for scanning at sale point – as it consists of a strip of black bars and white spaces, which can be accurately and reproducibly scanned with an X-shaped laser beam, above a sequence of 12 (or 13) numerical digits; examples are provided in Figs. 2.53i and iii. No letters, characters, or content of any other kind are allowed in the said barcode; and there is a one-to-one correspondence between the 12-digit number and the strip of black bars and white spaces.

FIGURE 2.53 Examples of (i) UPC-A, (ii) UPC-E, (iii) EAN-13, and (iv) EAN-8 barcodes.

The scannable area of every UPC-A barcode is of the form $sL_1L_2L_3L_4L_5L_6mR_1R_2R_3R_4R_5R_6e$ – where s denotes start, m denotes middle, and e denotes end guard patterns; $L_1L_2L_3L_4L_5L_6$ on the left, and $R_1R_2R_3R_4R_5R_6$ likewise on the right account for the unique set of 12 digits in all. The first digit of the left subgroup, L_1, indicates the particular number system followed by the following digits; whereas the last digit of the right subgroup, R_6, is an error-detection check digit. The UPC-E code – see Fig. 2.53ii, is a compressed barcode intended for use only on physically small items; compression consists of squeezing extra zeroes out of the barcode, and then automatically reinserting them upon scanning (so only barcodes containing zeroes are eligible thereto). A similar rationale applies to the compressed version of EAN-13, termed EAN-8 – see Fig. 2.53iv, containing 8 digits only.

The UPC barcodes can be printed at various densities, so as to accommodate a variety of printing and scanning processes. The significant dimensional parameter is the x-dimension, or nominal width (0.33 mm) of a single module element – complemented by the nominal symbol height (25.9 mm); the width of each bar or space is determined by multiplying such an x-dimension by the specific module width of the said bar or space – while the bars corresponding to s, m, and e are extended downward by fivefold the x-dimension (with a resulting overall symbol height of 27.6 mm). A quiet zone, bearing a width of (at least) ninefold the x-dimension, must be present on each side of the scannable area – with L_1 and R_6 placed outside the symbol in UPC-A barcodes to clearly indicate the quiet zones necessary for proper scanning, but replaced by country code and >, respectively, in EAN-13 barcodes. Reduction or magnification of the whole barcode is allowed, but only within the range 80–200%.

Each digit in a UPC-A barcode is uniquely represented by a strip of two bars alternating with two spaces; the guard patterns include two bars each. Hence, all such barcodes consist of $(3 \times 2) + (12 \times 2) = 30$ bars – of which 6 represent guard patterns and 24 represent numerical digits. The bars and spaces are, however, of variable width – i.e. 1, 2, 3, or 4 modules wide; since the total width for each digit is always 7 modules, the whole barcode requires $7 \times 12 = 84$ modules – with encoding rules specified in Fig. 2.54. A complete UPC-A barcode is 95 modules wide – 84 modules for the digits (as seen above), plus 11 modules for the set of s, m, and e guard patterns; s and e are 3 modules wide each, and utilize the pattern bar-space-bar (all with unit width) – whereas the m is 5 modules wide, and uses the pattern space-bar-space-bar-space (all also with unit width). The $L_1L_2L_3L_4L_5L_6$ subgroup possesses odd parity, i.e. the total width of the bars is an odd number; conversely, the $R_1R_2R_3R_4R_5R_6$ subgroup possesses even parity. Hence, a scanner can immediately tell whether the symbol is read in the regular position or upside down (since the s and e symbols are coincident) – and thus trigger an alarm requesting correct positioning, or automatically correct for the right sequence. Furthermore, the right-hand side digits are the optical inverse of the same digits if placed in the left-hand side (i.e. black bars and white spaces are exchanged); hence, the

FIGURE 2.54 Numerical and graphical encoding syntax in UPC barcode system.

scanner can reconstruct the full code, as long as the barcode center marker is included.

In view of the above features, the total number of combinations in UPC-A (or EAN-13 for each country) is $10^6 \times 10^5 = 10^{11}$, corresponding to 10 possible digits in 6 left positions and in 5 right positions; in the case of UPC-E, the number of combinations is reduced to $10^6 \times 2 = 2 \times 10^6$, corresponding to 10 possible digits in 6 positions and 2 possible parity patterns – and likewise with regard to EAN-8, with 10 possible digits in 8 positions instead. The $L_2L_3L_4L_5L_6$ subgroup (with digits 0, 1, 6, 7, 8, 9) is a manufacturer code, whereas the $R_1R_2R_3R_4R_5$ subgroup (with digits 0, 1, 6, 7, 8, 9) is a product code – thus covering the majority of situations. In the case of a food item sold by variable weight (e.g. fresh fruits, meats, vegetables) and packaged at the store/warehouse, digit 2 may be used – with $L_2L_3L_4L_5L_6$ representing item number and $R_1R_2R_3R_4R_5$ representing either weight ($R_1=0$) or price. Digit 3 is reserved for (pharmaceutical) drugs; digit 4 is reserved for local use (usually loyalty cards, or store coupons); and digit 5 is reserved for coupons at large – with $L_2L_3L_4L_5L_6$ representing manufacturer code, $R_1R_2R_3$ representing family code set by manufacturer, and R_4R_5 representing coupon code (that determines specific amount of discount). Formally, R_6 satisfies

$$R_6 = 10 - \mathrm{mod}_{10}\left\{\begin{matrix} 3\left(L_1 + L_3 + L_5 + R_1 + R_3 + R_5\right) \\ +L_2 + L_4 + L_6 + R_2 + R_4 \end{matrix}\right\}, \tag{2.318}$$

and stands as an error check digit – where mod_{10} denotes remainder of division by 10; if $\mathrm{mod}_{10}=0$, then $R_6=0$. This last digit is useful, especially when the integrity of the graphical barcode has somehow been disrupted, or misplaced on the product label – thus compromising proper scanning; in this situation, the number code has to be entered manually by the cashier. Almost all typing (or scanning, for that matter) errors correspond to either a single mistyped digit or two consecutive digits inadvertently swapped; this error check digit can detect 100% errors in the former case, and 90% in the latter case.

The UPC code is printed on consumer packs, and is intended for cross laser-aided reading at retail checkout; it does not contain price information, yet it circumvents the need for individual price labeling – while allowing automatic generation of itemized bills for customers in real-time. In fact, the scanner sends a signal to a central computer system, which identifies the UPC number and returns the current price of that item; hence, no human labor is required to tag (and retag) products with price (or even memorize prices, in the first place) – as it suffices to enter this information once into the computer. At the same time, the computer automatically deducts the item from the store inventory – thus dramatically facilitating stock management, namely, adjustment of shelf space allotment to specific items, and generation of information on sales as a result of specific marketing strategies; automatic reordering is also permitted, as well as detection of pilferage. Shipping containers are also barcoded to inform the carrier about their destination.

Labels are made of preprinted paper, plastic film, foil, or laminated materials – but may be absent if printing is performed on the package surface, as in the case of some glass bottles. Due to the sturdy and crystalline nature of glass, direct printing thereon is indeed possible, and yields quite good results in terms of quality and durability – thus circumventing printout on a separate film, followed by gluing onto the glass surface. A similar approach is printing before or during plastic bottle manufacture, as long as stretchable ink is utilized.

Modes of inclusion of a label in a food product include: glued-on labels (e.g. cans, glass bottles) – using pre-glued labels wetted at the time of application, or applying some sort of adhesive at that time; thermo-sensitive labels (e.g. biscuits, bread) – often serving as closure, and stuck via application of heat; pressure-sensitive labels – pre-coated with adhesive and protected with waxy paper, mounted on a roll, and requiring removal of the said paper just prior to application followed by (slight) pressing; insert labels – to be included, together with the food items, inside transparent packages; heat-transfer labels – where the design is preprinted onto paper or polyester substrate, and will be transferred via application of heat; and in-mold labels – where container and label are simultaneously thermoformed, with a printed paper label (bearing a heat-activated coating on its reverse side) placed in the mold before inserting the parison, and activated when hot air is injected to blow the package shape.

In alternative, a shrink sleeve can be used for cylindrical containers; the former is made of axially oriented PP or UV-curable urethane-acrylate – with inner diameter slightly larger than the package so as to enrobe it, being heat-shrunk to fit in the end. An LDPE sleeve with inner diameter slightly smaller than that of the package can also be utilized – in which case it is mechanically stretched to fit. Besides being less expensive than direct printing, such sleeves have met with success in glass bottles, as well as PVC and PET bottles – which would otherwise not be eligible for direct surface printing. The sleeve material is normally designed for good printability with flexographic inks, good adhesion, good resistance to scratching and wrinkling, and good clarity upon shrinking or stretching.

Printing inks for films and papers used in labeling consist of a dye dispersed in a (blend of) solvent(s), and a resin that forms a varnish; the ink should be as nonexpensive as possible, but compatible with the nature of the film due for printing – so as to attain a high bond strength. Solvents must be removed following ink application, otherwise contamination of the product by their odor can occur – or film transparency may be compromised during use *a posteriori*. Printing on the outside of the packaging material is simple to perform, and safe from a food (chemical) contamination point of view; however, it must have a high gloss and be scuff-resistant – otherwise it will be rapidly rubbed off throughout handling. Conversely, printing on the inside (or reverse printing) readily permits a high-gloss finish – at the expense of a critical requirement for negligible odor, however. It may also be applied between two layers of a laminated material – in which case reverse printing on one material is followed by lamination with the other material.

Printing on films and papers resorts to one of five alternative approaches: flexographic, rotogravure, offset lithography, screen, and ink-jet.

Flexographic printing (or letterpress) is the oldest technique – and uses a flexible rubber plate with raised characters. Such characters are previously pressed against an inked roller, to cover the raised portions with ink – which passes the ink onto the film surface afterward, again via pressing, followed by fast dryout. This mode of printing is appropriate for lines and blocks of (up to six different) colors.

Rotogravure (or intaglio) resorts to an engraved chromium-plated roller – where the printing surfaces are instead recessed in the metal; hence, ink is applied to the roller, with excess being wiped off the whole surface except the said recesses – thus making it available for transfer onto the film. Rotogravure is appropriate for high-quality, detailed, and realistic printing.

Offset (or planographic) lithography takes advantage of the intrinsic immiscibility of grease with water – with greasy ink being repelled by premoistened portions of a printing plate, while remaining on the hydrophobic parts carrying the intended design. The associated printing quality mimics that of rotogravure – and this technique can be applied to rougher materials that would be hardly suitable for rotogravure.

In the case of screen printing, the ink is applied onto a porous surface – but passes through only where the screen has not been blocked by some sizing agent (e.g. modified starch, polyvinyl alcohol, carboxymethylcellulose).

Ink-jet printing relies on spraying of droplets of ink, via thermal or pressure waves created behind the ink nozzle by a vibrating rod – taking advantage of Rayleigh and Plateau's instability. Such droplets are tiny, but exhibit predictable sizes; and acquire an electrostatic charge as they leave the nozzle. In view of the said charge, the path of each droplet (and thus its ultimate position on the label) is controlled by its charge-to-mass ratio, as it passes through an electrostatic field – modulated by a computer as it reads through each pixel; the resulting minute dots are perceived as a continuous layout, when observed by the human eye from a distance. Elimination of satellite droplets – a lurking phenomenon (not predicted by Rayleigh's linear approach) is nevertheless an ongoing technical problem.

Laser coders have been in use for some time now, and operate by firing a dot-matrix laser to build up each character of the UPC code; such a laser beam either removes colored ink from a label to reveal a white surface beneath, or etches the surface to leave a mark. A recent development in printing includes online laser decoration, in which a laser beam, connected to an ultrafast 3D-displacement engine, produces photographic decoration on polymer materials – as long as they possess special pigments in their original formulation, able to change color only when struck by laser light; besides zero printing residues, a wide flexibility is possible – since no prepared plates are necessary in advance for printing. A major improvement underway, pertaining to labeling, includes a microdot on the pack – containing all required label information (as well as nutritional composition and instructions for use) in a number of languages; hence, the main area of the label will be left for graphic design and branding. Another possibility is radiofrequency identification tag chips also on the pack, containing the UPC number – to be made available when challenged by a radiofrequency field generated in a tunnel at retail cashier (resembling electronic tolls at highways); this will avoid the current need to take products out of the shopping cart, place them on the cashier stand or belt, and have them manually scanned by the cashier on a one-by-one basis, before eventually returning the products to the cart.

2.2.2.13 Economic and environmental impact

Unlike most goods that begin their useful life upon purchase, packaging materials typically cease their usefulness at this stage – apart from a limited period of home storage. In more developed countries, almost two-thirds of all packaging is utilized by the food industry, and typical households use (and eventually dispose off) more than 200 types of packages somewhat related to food; in Europe alone, about half of all packaged foods resort to plastic – which accounts for ca. 11% of overall municipal waste. Therefore, reduction, reuse, and recycling are three keywords to bear in mind, in attempts to keep the ecological footprint of (food) packaging at an acceptable level.

Reusable bottles remain common in some countries for milk or beer; this is feasible whenever the consumers are relatively close to the manufacturer – or a closed system is in operation, with delivery vehicles of fresh products collecting also empty bottles. However, reusable bottles are normally thicker (and thus heavier) than single-use bottles, so as to withstand multiple uses; hence, extra energy is required for bottle production and transportation, besides cleaning.

An alternative is to remanufacture bottles using glass from single-use bottles upon recycling; more than 50% incorporation of recycled glass in new containers is already in action in Europe and Japan. Such an effort has come along with domestic separation of glass, as well as metal (ca. 40% Al and 45% Sn), plastic (>20%), wood (>15%), and paper (>60%) – which are separately collected by municipal services, and sent to recycling facilities. Recycling to higher extents has, however, been hampered by existence of specialty packages – namely, multilayered structures designed for specific barrier properties, of disparate nature and thus difficult to separate for recycling. Furthermore, the suitability of plastics for recycling depends on their inertness – as only those plastics, e.g. RVC and PET, unable to absorb significant amounts of food chemicals (seen as contaminants here) are suitable for recycling; this essentially rules out most PS and HDPE. Although modern sorting technologies can provide recycled plastic that contains almost a single polymer type, recycled plastics are normally blended with new polymer(s) from 5 up to 50%; as alternative, the recycled plastic may be used as core material in multilayered coextruded packages – protected from direct contact with the food material by a layer (or functional barrier) made of fresh polymer. Shortage of supply of oil, and accompanying higher prices thereof – complemented with pollution taxes, will likely enhance the relative benefits of recyclable materials versus new materials in the coming future.

While incineration of disposed packages can recover part of the original energy required for their manufacture, this comes at the expense of releasing greenhouse gases – so composting of (biodegradable) plastics has been on the rise, also motivated by the increasing environmental awareness of consumers at large. On the other hand, a range of organic chemicals and heavy metals – besides effluent waters with suspended solids, result from manufacture of many packaging materials – including aluminum, board, glass, paper, plastics, and even steel. Retail packs have also received severe criticism due to the excessive resources required for their production, and major contribution to litter and landfill – while unnecessarily adding to the final cost of the food product (as a trade-off for convenience).

Besides costs associated with waste handling and effluent treatment, packaging costs on the manufacturer side encompass procurement of packaging materials themselves, shipping of containers, depreciation of packaging machinery, and labor requirements – as well as energy used in the process of production of packaging material from its raw material(s), in the first place. Another issue to take into account is yield; specific information in this regard was provided in Fig. 2.38, for the case of polymeric films. Note that a weight basis was used for comparability of data, yet packaging often resorts to an areal or a volumetric basis; for instance, 1 kg of plastic material produces much larger an area of plastic film than 1 kg of tin produces of can sheet.

In attempts to reduce costs of packaging, food manufacturers have increasingly moved away from glass and metal items – owing to their high energy demands; and resorted to plastics. Unfortunately, most plastics are still produced from (nonrenewable) fossil sources – while glass and metals are recyclable to a large degree. A recent trend is use of paper and board packaging, as these materials are sourced from (renewable) forests; nevertheless, there are limitations in functionality of cellulose-based materials, chiefly because of their hygroscopic nature. Measures taken to reduce cost of delivery of raw materials include bulk handling, instead of utilization of small containers – as is the case of tankers and large combo bins, tanks, or bags; film rolls for form–fill–seal machines instead of preformed packs, or stackable pots rather than cans and jars; lighter materials – already accounting for 80% less in the case of plastics compared to the 1980s, or 50% less tinplate and 30% less paper than in the 1930s; and development of improved barrier materials – e.g. PET instead of glass bottles, coextruded polymer containers instead of glass jars with metal lid, and polymer trays and films instead of metal and glass packages. The energy used for transportation of packaging materials is also relevant – with trucks filled before they reach their weight limit in the case of plastic and paper, or reaching their weight limit before they are full in the case of glass; the average consumption for roadway transport of ca. 1 $MJ.ton^{-1}.km^{-1}$ compares with ca. 0.6 $MJ.ton^{-1}.km^{-1}$ for railway transport, and ca. 0.1 $MJ.ton^{-1}.km^{-1}$ for (surface) sea transport.

2.2.3 Mathematical simulation

2.2.3.1 Transparent packaging

The fate of incident light power, $P_{i,p}$, over a package abides, in general, to

$$\boxed{P_{i,p} = P_{a,p} + P_{r,p} + P_{t,p},} \tag{2.319}$$

where $P_{a,p}$, $P_{r,p}$, and $P_{t,p}$ denote light power absorbed, reflected, and transmitted, respectively, by the package; the concepts of absorptivity of package, α_p, i.e.

$$\boxed{\alpha_p \equiv \frac{P_{a,p}}{P_{i,p}},} \tag{2.320}$$

reflectivity of package, ρ_p, i.e.

$$\boxed{\rho_p \equiv \frac{P_{r,p}}{P_{i,p}},} \tag{2.321}$$

and transmissivity of package, τ_p, i.e.

$$\boxed{\tau_p \equiv \frac{P_{t,p}}{P_{i,p}},} \tag{2.322}$$

may then be invoked to write

$$P_{i,p} = \alpha_p P_{i,p} + \rho_p P_{i,p} + \tau_p P_{i,p} \tag{2.323}$$

– which would retrieve Eq. (2.116) of *Food Proc. Eng.: thermal & chemical operations*, if both sides were divided by $P_{i,p}$. Beer and Lambert's law, as per Eq. (2.503) in the same book, may now be recalled to write

$$P_{i,f} = \left(P_{i,p} - P_{r,p}\right)e^{-\mu_p \delta_p} \tag{2.324}$$

– where $P_{i,f}$ denotes light power transmitted at depth δ_p equal to thickness of outer package, and μ_p denotes attenuation coefficient of the said packaging material. Equation (2.324) therefore represents the incident light that reaches the surface of the food proper; combination with Eq. (2.321) supports transformation of Eq. (2.324) to

$$P_{i,f} = \left(P_{i,p} - \rho_p P_{i,p}\right)e^{-\mu_p \delta_p}, \tag{2.325}$$

which simplifies to

$$\boxed{P_{i,f} = \left(1 - \rho_p\right)e^{-\mu_p \delta_p} P_{i,p}} \tag{2.326}$$

after factoring $P_{i,p}$ out. The chemical nature of the packaging material affects μ_p considerably – both quantitatively and qualitatively; for instance, LDPE transmits visible and ultraviolet light to similar degrees, whereas PVC transmits visible but not UV light (as referred to previously). Pigments can be incorporated to prevent transmission of certain wavelengths, though – as happens in the case of wine or beer bottles, to block UV light that speeds up oxidative decay (via formation of free radicals).

If, for simplicity, the food is taken as slab-shaped with thickness δ_f, and is characterized by attenuation coefficient μ_f, then a rationale similar to that supporting Eq. (2.326) can be followed to write

$$P_{o,f}=\left(1-\rho_f\right)e^{-\mu_f\delta_f}P_{i,f} \tag{2.327}$$

– pertaining to the light power still available on the other side of the food, $P_{o,f}$; insertion of Eq. (2.326) allows conversion of Eq. (2.327) to

$$P_{o,f}=\left(1-\rho_f\right)e^{-\mu_f\delta_f}\left(1-\rho_p\right)e^{-\mu_p\delta_p}P_{i,p}, \tag{2.328}$$

where lumping of exponential functions finally gives

$$\boxed{P_{o,f}=\left(1-\rho_f\right)\left(1-\rho_p\right)e^{-\left(\mu_f\delta_f+\mu_p\delta_p\right)}P_{i,p}.} \tag{2.329}$$

If food packages are piled up consecutively, with negligible space in between, then light has still to go through the package on the bottom side of the food product before reaching the next food item. Therefore, the light power coming out of the food package, $P_{u,p}$, should be given by

$$P_{u,p}=\left(1-\rho_p\right)e^{-\mu_p\delta_p}P_{o,f}; \tag{2.330}$$

combination with Eq. (2.329) generates

$$P_{u,p}=\left(1-\rho_p\right)e^{-\mu_p\delta_p}\left(1-\rho_f\right)\left(1-\rho_p\right)e^{-\left(\mu_f\delta_f+\mu_p\delta_p\right)}P_{i,p}, \tag{2.331}$$

which simplifies to

$$\boxed{P_{u,p}=\left(1-\rho_f\right)\left(1-\rho_p\right)^2e^{-\left(\mu_f\delta_f+2\mu_p\delta_p\right)}P_{i,p},} \tag{2.332}$$

after lumping factors alike. Equation (2.332) pertains to the whole set of a food item and its outer package; any extra packaged food placed adjacent thereto will obviously behave the same way, with available light power decreasing in a cumulative manner. Therefore, one may iterate Eq. (2.332) as

$$P_{u,f,N}=\prod_{j=1}^{N}\left(1-\rho_{f,j}\right)\left(1-\rho_{p,j}\right)^2e^{-\left(\mu_{f,j}\delta_{f,j}+2\mu_{p,j}\delta_{p,j}\right)}P_{i,p}, \tag{2.333}$$

corresponding to the light power still available at the bottom of a pile of N food packages placed next to each other – where the operational features of exponential functions support transformation to

$$\boxed{P_{u,f,N}=\left(\prod_{j=1}^{N}\left(1-\rho_{f,j}\right)\right)\left(\prod_{j=1}^{N}\left(1-\rho_{p,j}\right)^2\right)e^{-\sum_{j=1}^{n}\left(\mu_{f,j}\delta_{f,j}+2\mu_{p,j}\delta_{p,j}\right)}P_{i,p}.} \tag{2.334}$$

In most practical situations, it is expectable that

$$\rho_{f,j}=\rho_f;\quad j=1,2,\ldots,N, \tag{2.335}$$

$$\rho_{p,j}=\rho_p;\quad j=1,2,\ldots,N, \tag{2.336}$$

$$\mu_{f,j}=\mu_f;\quad j=1,2,\ldots,N, \tag{2.337}$$

$$\mu_{p,j}=\mu_p;\quad j=1,2,\ldots,N, \tag{2.338}$$

$$\delta_{f,j}=\delta_f;\quad j=1,2,\ldots,N, \tag{2.339}$$

and

$$\delta_{p,j}=\delta_p;\quad j=1,2,\ldots,N; \tag{2.340}$$

under these circumstances, Eq. (2.334) degenerates to

$$P_{u,f,N}=\left(\prod_{j=1}^{N}\left(1-\rho_f\right)\right)\left(\prod_{j=1}^{N}\left(1-\rho_p\right)^2\right)e^{-\sum_{j=1}^{N}\left(\mu_f\delta_f+2\mu_p\delta_p\right)}P_{i,p}. \tag{2.341}$$

The (mathematical) definitions of power and product permit simplification of Eq. (2.341) to

$$\boxed{\frac{P_{u,f,N}}{P_{i,p}}=\left(1-\rho_f\right)^N\left(1-\rho_p\right)^{2N}e^{-N\left(\mu_f\delta_f+2\mu_p\delta_p\right)},} \tag{2.342}$$

along with division of both sides by $P_{i,p}$; here $P_{u,f,N}/P_{i,p}$ represents fraction of the original incident light that reaches the outer package side of the lowermost (or innermost) food product in the storage stack.

The simplest case of

$$\rho_f=\rho_p=\rho, \tag{2.343}$$

coupled with

$$\mu_f=\mu_p=\mu \tag{2.344}$$

and

$$\delta_f=\delta_p=\delta \tag{2.345}$$

further reduces Eq. (2.342) to

$$\frac{P_{u,f,N}}{P_{i,p}}=(1-\rho)^N(1-\rho)^{2N}e^{-N(\mu\delta+2\mu\delta)} \tag{2.346}$$

– or else

$$\frac{P_{u,f,N}}{P_{i,p}}=(1-\rho)^{3N}e^{-3N\mu\delta}. \tag{2.347}$$

In this particular case, the overall fraction of incident light either reflected by, or absorbed through the pile of food packs is given by

$$\boxed{1-\frac{P_{u,f,N}}{P_{i,p}}=1-(1-\rho)^{3N}e^{-3N\mu\delta}} \tag{2.348}$$

– as depicted in Fig. 2.55, for illustration purposes. The overall fraction of incident light unable to reach the bottom of the food pile increases with overall number of units of the said pile, N; as well as with attenuation coefficient, μ, and thickness, δ, of each unit (with $3N\delta$ serving as optical path) – in agreement with their definition. The increase caused by a higher reflectivity – apparent when going from Figs. 2.55i to 2.55ii, is due not to absorption of light by the packaged foods making up the pile, but to reflection (and, consequently, irreversible loss) of light thereby. As typically happens with an exponential decay, the increase in amount of light absorbed/reflected by a set of two adjacent packaged foods relative to a single packaged food is much higher than from a set

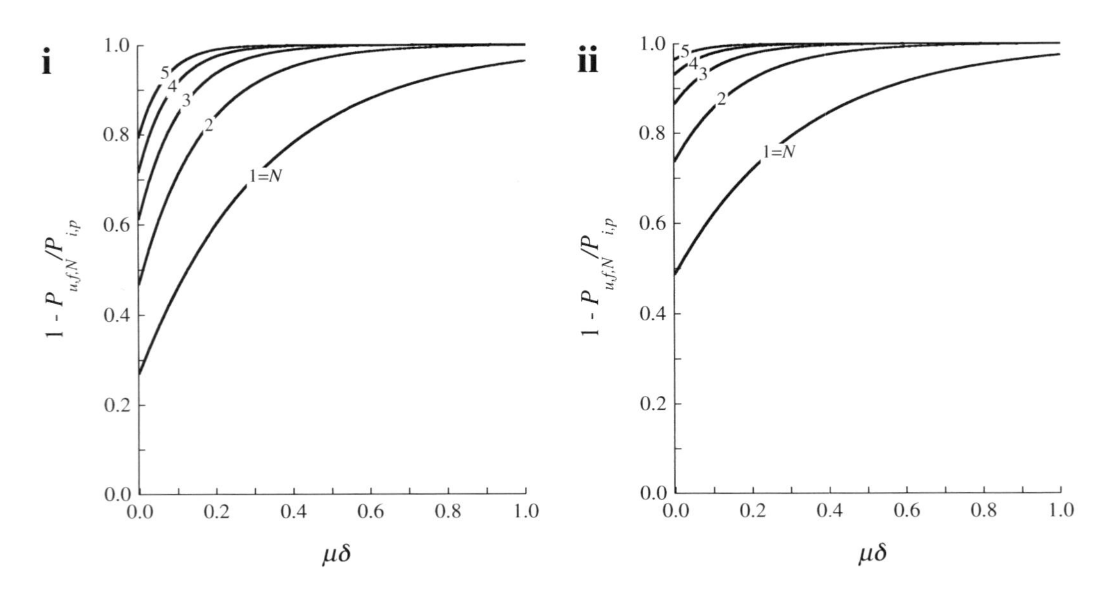

FIGURE 2.55 Variation of dimensionless light power reflected or absorbed by N packaged foods piled next to each other, $1-P_{u,f,N}/P_{i,p}$, versus dimensionless attenuation coefficient, $\mu\delta$, for selected values of N and reflectivity, (i) $\rho=0.1$ and (ii) $\rho=0.2$.

of two to a set of three packaged foods, and so on; this realization can be easily grasped in Fig. 2.55. An increase in reflectivity magnifies the vertical intercept of the curves – since less light will be available, in the first place, for absorption/transmission; while a horizontal asymptote exists, described by

$$\lim_{\mu\delta\to\infty}\left(1-\frac{P_{u,f,N}}{P_{i,p}}\right)=1-(1-\rho)^{3N}\mathrm{e}^{-\infty}=1-(1-\rho)^{3N}0=1 \qquad (2.349)$$

and stemming from Eq. (2.348) – which bears a unit vertical intercept, corresponding to full extinction of light. A similar finding arises when N grows unbounded, i.e.

$$\lim_{N\to\infty}\left(1-\frac{P_{u,f,N}}{P_{i,p}}\right)=1-\lim_{N\to\infty}\frac{(1-\rho)^{3N}}{\mathrm{e}^{3N\mu\delta}}=1-\frac{0}{\infty}=1, \qquad (2.350)$$

again based on Eq. (2.348) – since $(1-\rho)^{3N}\to 0$ because $0<1-\rho<1$, as a consequence, in turn, of $0<\rho<1$ by definition; while $\mathrm{e}^{3N\mu\delta}\to\mathrm{e}^{\infty}=\infty$.

In view of Eq. (2.334), the loss in transmitted energy, when going from the (i–1)-th to the i-th food pack, can be calculated via

$$P_{u,f,i}=\left(1-\rho_{f,i}\right)\left(1-\rho_{p,i}\right)^2\mathrm{e}^{-\left(\mu_{f,i}\delta_{f,i}+2\mu_{p,i}\delta_{p,i}\right)}P_{u,f,i-1} \qquad (2.351)$$

– which resembles Eq. (2.332) pertaining to a single unit; hence, the light power absorbed/reflected by the i-th unit reads

$$P_{u,f,i-1}-P_{u,f,i}$$
$$=P_{u,f,i-1}-\left(1-\rho_{f,i}\right)\left(1-\rho_{p,i}\right)^2\mathrm{e}^{-\left(\mu_{f,i}\delta_{f,i}+2\mu_{p,i}\delta_{p,i}\right)}P_{u,f,i-1}. \qquad (2.352)$$

Normalization by the light power entering the i-th unit, $P_{u,f,i-1}$, transforms Eq. (2.352) to

$$1-\frac{P_{u,f,i}}{P_{u,f,i-1}}=1-\left(1-\rho_{f,i}\right)\left(1-\rho_{p,i}\right)^2\mathrm{e}^{-\left(\mu_{f,i}\delta_{f,i}+2\mu_{p,i}\delta_{p,i}\right)}; \qquad (2.353)$$

Eq. (2.353) conveys the incremental variation between consecutive units, rather than between the last unit relative to the incident light on the first unit – as per the simple case described by Eq. (2.348).

PROBLEM 2.7

When estimating the amount of light available after a stack of semitransparent liquid foods packaged in semitransparent material, described by fully similar physicochemical and geometrical features, the light reflected was taken into account as diffusively lost to the surroundings.

a) Obtain the upper (and thus conservative) estimate for the fraction of light absorbed by a pile of food packs, constituted by N units – and provide a graphical interpretation.
b) Although most light power leaving a food pack and impinging on the outer package surface of the next food pack will be scattered and eventually lost, the light power leaving the inner surface of the package and hitting the upper surface of the food will probably undergo minor reflection; and the light power leaving the lower surface of the food and striking the inner surface of its package will likely be reflected back into the liquid. Derive an estimate of the overall light power absorbed/reflected, based on this hypothesis – and produce the corresponding plot of $1-P_{u,f,N}/P_{i,p}$ versus $\mu\delta$, for selected values of N.

Solution

a) The conservative upper estimate sought considers that no light whatsoever is reflected back – which implies $\rho=0$; under these circumstances, Eq. (2.348) will reduce to

$$\left.\left(1-\frac{P_{u,f,N}}{P_{i,p}}\right)\right|_{\rho=0}=1-(1-0)^{3N}\,\mathrm{e}^{-3N\mu\delta}=1-\mathrm{e}^{-3N\mu\delta} \tag{2.7.1}$$

– as plotted below.

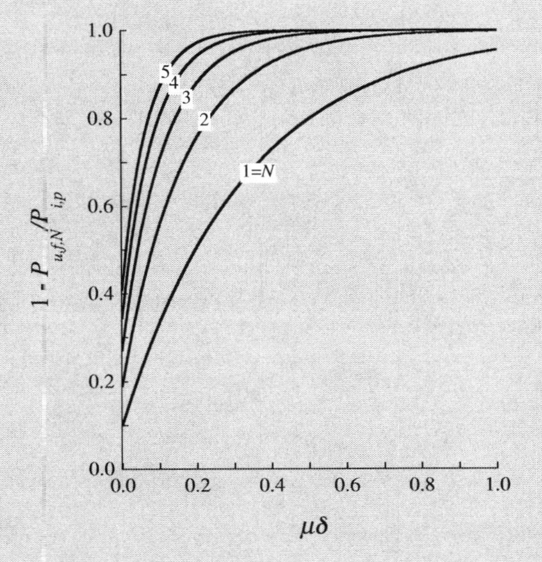

As expected, the plot above represents the limiting form of the plots labeled as Fig. 2.55, when $\rho\to 0$.

b) The situations of poor reflection, or reflection back into the liquid are essentially equivalent to nil reflectivity at the package/food upper and lower interfaces; therefore, Eq. (2.332) should simplify to

$$P_{u,p}=(1-\rho_p)\mathrm{e}^{-(\mu_f\delta_f+2\mu_p\delta_p)}P_{i,p} \tag{2.7.2}$$

pertaining to a single pack, and likewise

$$P_{u,f,N}=\left(\prod_{j=1}^{N}(1-\rho_{p,j})\right)\mathrm{e}^{-\sum_{j=1}^{N}(\mu_{f,j}\delta_{f,j}+2\mu_{p,j}\delta_{p,j})}P_{i,p} \tag{2.7.3}$$

pertaining to the whole stack – to be used *in lieu* of Eq. (2.334). If similar features of package and liquid were again considered, then the analog to Eq. (2.348) would look like

$$1-\frac{P_{u,f,N}}{P_{i,p}}=1-(1-\rho)^N\mathrm{e}^{-3N\mu\delta}, \tag{2.7.4}$$

as plotted next for $\rho=0.1$.

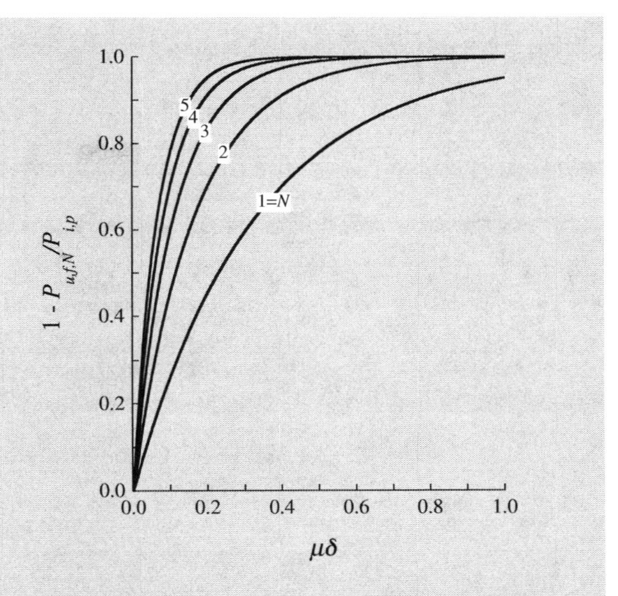

Since $0<\rho<1$ implies $1>1-\rho>0$, and consequently $0<(1-\rho)^{3N}<(1-\rho)^N<1$, one concludes that the situation under scrutiny here lies between Eqs. (2.348) and (2.7.1), since $\mathrm{e}^{-3N\mu\delta}$ is shared as factor by all three.

2.2.3.2 Permeable packaging

In the case of a packaging material sufficiently porous to permit mobility of a fluid through it, the germane mass balance to a gaseous species characterized by diffusivity $\mathcal{D}_g$ looks like

$$-\mathcal{D}_g A\left.\frac{dC_g}{dx}\right|_x=-\mathcal{D}_g A\left.\frac{dC_g}{dx}\right|_{x+dx}, \tag{2.354}$$

as per Fick's law applied under steady-state conditions; clear dominance of cross-sectional area A over thickness δ justifies use of a parallelepipedal geometry, irrespective of the actual film shape – as long as a convex configuration (or, at most, characterized by smooth folding and twisting) is at stake. Since A is independent of space variable x under these circumstances, it can be dropped off both sides in Eq. (2.354) to get

$$\begin{aligned}\mathcal{D}_g\left.\frac{dC_g}{dx}\right|_{x+dx}-\mathcal{D}_g\left.\frac{dC_g}{dx}\right|_x&=\mathcal{D}_g\left(\left.\frac{dC_g}{dx}\right|_{x+dx}-\left.\frac{dC_g}{dx}\right|_x\right)\\&=\mathcal{D}_g d\left(\frac{dC_g}{dx}\right)=0\end{aligned} \tag{2.355}$$

– where $\mathcal{D}_g$ was meanwhile factored out, and the concept of differential recalled; further division of both sides by $\mathcal{D}_g dx$ unfolds

$$\frac{d}{dx}\left(\frac{dC_g}{dx}\right)=0, \tag{2.356}$$

or else

$$\boxed{\frac{d^2C_g}{dx^2}=0} \tag{2.357}$$

in view of the definition of second-order derivative. Suitable boundary conditions look like

$$\boxed{C_g\big|_{x=0} = C^{\sigma}_{g,in}} \tag{2.358}$$

and

$$\boxed{C_g\big|_{x=\delta} = C^{\sigma}_{g,out}} \tag{2.359}$$

– with C^{σ}_{g} denoting solubility (or equilibrium concentration) of the gas/vapor of interest in the packaging material, and subscripts *in* and *out* referring to inside and outside surface, respectively, of the package, respectively. Integration of Eq. (2.356) is now in order, via separation of variables, viz.

$$\int d\left(\frac{dC_g}{dx}\right) = \int 0\,dx, \tag{2.360}$$

which is equivalent to writing

$$\frac{dC_g}{dx} = k_1 \tag{2.361}$$

as long as k_1 denotes an arbitrary (integration) constant. A second integration step, again via separation of variables, unfolds

$$\int dC_g = k_1 \int dx \tag{2.362}$$

from Eq. (2.361) or, equivalently,

$$C_g = k_0 + k_1 x \tag{2.363}$$

– with k_0 denoting a second arbitrary constant. Equations (2.358) and (2.359) may finally be combined with Eq. (2.363) to yield

$$C^{\sigma}_{g,in} = k_0 + k_1 0 \tag{2.364}$$

and

$$C^{\sigma}_{g,out} = k_0 + k_1\delta, \tag{2.365}$$

respectively; Eq. (2.364) breaks down to

$$k_0 = C^{\sigma}_{g,in}, \tag{2.366}$$

so insertion of Eq. (2.366) transforms Eq. (2.365) to

$$C^{\sigma}_{g,out} = C^{\sigma}_{g,in} + k_1\delta \tag{2.367}$$

– where isolation of k_1 yields, in turn,

$$k_1 = \frac{C^{\sigma}_{g,out} - C^{\sigma}_{g,in}}{\delta}. \tag{2.368}$$

Based on Eqs. (2.366) and (2.368), one can rewrite Eq. (2.363) as

$$C_g = C^{\sigma}_{g,in} + \frac{C^{\sigma}_{g,out} - C^{\sigma}_{g,in}}{\delta} x, \tag{2.369}$$

which entails a linear concentration profile for the gaseous species diffusing through the porous film. This result is anticipated because no accumulation term was considered, along with a parallelepipedal geometry (characterized by constant cross-section) – meaning that the rate of transport, F_g, must be identical irrespective of location, i.e.

$$F_g = -\mathcal{D}_g A \frac{dC_g}{dx} = -\mathcal{D}_g A \frac{C^{\sigma}_{g,out} - C^{\sigma}_{g,in}}{\delta} \tag{2.370}$$

again in agreement with Fick's law; combined with the differentiated form of Eq. (2.369) or, in alternative, Eq. (2.361) coupled with Eq. (2.368).

When a gaseous species is under scrutiny, the equilibrium concentration prevailing at an interface is usually related to its partial pressure in the bulk phase, p_g, according to

$$\boxed{C^{\sigma}_{g} = He_g p_g,} \tag{2.371}$$

where He_g denotes the associated Henry's constant; this applies whenever the packaging material behaves as a (very) viscous liquid, as in a polymeric film – able to dissolve the gas. In the case of porous materials with truly solid behavior, no dissolution of gas takes place therein; hence, plain gas occupies its pores – and He_g breaks down to just $1/\mathcal{R}T$, which retrieves the ideal gas equation upon combination with Eq. (2.371). Insertion of Eq. (2.371) transforms Eq. (2.370) to

$$-F_g = \mathcal{D}_g A \frac{He_g p_{g,out} - He_g p_{g,in}}{\delta} = \mathcal{D}_g He_g A \frac{p_{g,out} - p_{g,in}}{\delta}, \tag{2.372}$$

where He_g was meanwhile factored out and negatives taken of both sides. The quantity $\mathcal{D}_g He_g$ is commonly referred to as permeability coefficient, $\mathcal{P}_g$, i.e.

$$\boxed{\mathcal{P}_g \equiv \mathcal{D}_g He_g;} \tag{2.373}$$

hence, Eq. (2.372) may reappear as

$$\boxed{-F_g = \mathcal{P}_g A \frac{p_{g,out} - p_{g,in}}{\delta}}, \tag{2.374}$$

in simpler form. Values of permeability for selected polymeric materials and the three gases and the one vapor of practical interest – i.e. O_2 and N_2 for being the chief constituents of the regular atmosphere, and CO_2 and H_2O for being the principal products of respiration, are provided in Table 2.5. Note the wide diversity in performance – with permeabilities ranging from 10^{-10} up to 10^{0} m^2.s^{-1}.Pa^{-1}. Remember that cellulose is commonly traded under the name Cellophane™, poly(ethylene terephthalate) under PET, Terylene™, or Dacron™, low- and high-density polyethylene as LDPE and HDPE, respectively, poly(imino(1-oxohexamethylene)) and -(-oxoundecamethylene) under Nylon™ 6 and 11, respectively, poly(vinyl chloride) under PVC, and poly(vinylidene chloride) under Saran™.

In view of its definition as per Eq. (2.373), one may postulate

$$\mathcal{P}_g = \frac{\kappa_g T}{\mu\{T\}} \exp\left\{-\frac{\Delta g_{sol,g}}{\mathcal{R}T}\right\} \tag{2.375}$$

for the dependence of permeability on absolute temperature, T – after recalling Eq. (3.205) of *Food Proc. Eng.: basics & mechanical operations* describing variation of diffusivity with temperature as per Stoke's and Einstein's equation, and Eq. (1.105) relating value of an equilibrium constant to temperature; here $\Delta g_{sol,g}$ denotes molar Gibbs' energy of solubilization, κ_g denotes a proportionality constant, $\mathcal{R}$ denotes ideal gas constant, and $\mu\{T\}$ denotes temperature-dependent viscosity. If Guzmán and Andrade's model applies, see Table 2.1 of *Food Proc. Eng.:*

TABLE 2.5

Average values of permeability coefficient, $\mathcal{P}_g$ (in m^2.s^{-1}.Pa^{-1}), of most common gases and vapor at 25°C, through selected polymeric materials used in food packaging.

Packaging material		Gases/vapor			
		O_2	N_2	CO_2	H_2O
Cellulose	Plain	1.58×10^{-10}	2.40×10^{-10}	3.53×10^{-10}	1.43×10^{-4}
	Acetate	5.85×10^{-8}	2.10×10^{-8}	1.70×10^{-6}	4.13×10^{-4}
Polyethylene	Low density	2.16×10^{-7}	7.27×10^{-8}	9.45×10^{-7}	6.75×10^{-6}
	High density	3.02×10^{-8}	1.07×10^{-8}	1.27×10^{-7}	9.02×10^{-7}
Polypropylene		2.30×10^{-8}	1.73×10^{-7}	3.33×10^{-8}	6.90×10^{-7}
Poly(imino(1-oxohexamethylene))		3.80×10^{-2}	2.85×10^{-9}	7.13×10^{-10}	6.60×10^{-9}
Poly(imino(1-oxoundecamethylene))				7.51×10^{-8}	
Polyethyene terephthalate		3.50×10^{-2}	2.63×10^{-9}	4.88×10^{-10}	1.28×10^{-8}
Polyvinyl acetate		5.05×10^{-1}	3.75×10^{-8}	2.48×10^{-8}	8.98×10^{-10}
Polyvinyl alcohol		8.90×10^{-3}	6.68×10^{-10}	7.50×10^{-11}	1.13×10^{-9}
Polyvinyl chloride		4.53×10^{-2}	3.39×10^{-9}	8.85×10^{-10}	1.18×10^{-8}
Polyvinylidene chloride		5.30×10^{-3}	3.98×10^{-10}	7.05×10^{-11}	2.25×10^{-9}

basics & mechanical operations, then Eq. (2.375) can be reformulated as

$$\mathcal{P}_g = \frac{\kappa_g T}{B\exp\left\{\frac{\kappa}{T}\right\}}\exp\left\{-\frac{\Delta g_{sol,g}}{\mathcal{R}T}\right\} = K_g T\exp\left\{-\left(\kappa + \frac{\Delta g_{sol,g}}{\mathcal{R}}\right)\frac{1}{T}\right\} \quad (2.376)$$

– where $K_g \equiv \kappa_g/B$ denotes a new (lumped) proportionality constant. For some materials, there is a break in permeability at a critical temperature; the material becomes indeed much more permeable when temperature exceeds the said threshold. Such a critical temperature is ca. 30°C for polyvinyl acetate, and ca. 80°C for PS; it normally matches the corresponding glass transition temperature, since the material is glassy below but rubbery above this temperature – when it acquires a fluid-like behavior that offers a much lower resistance to molecular transport.

Equation (2.374) is, in general, applicable to transfer of any gaseous substance through a packaging material – and, in particular, water vapor. In view of the effect of water content of a food item upon its water activity (critical for microbial activity and textural quality, as already discussed), estimation of shelf-life often resorts to a mass balance to moisture inside a packaged food – according to

$$\boxed{V_f\frac{dC_w}{dt} = \mathcal{P}_w\frac{A}{\delta}\left(p_{w,out} - p_w^\sigma\right)} \quad (2.377)$$

that takes advantage of Eq. (2.374). Here V_f denotes volume of food, C_w denotes water content of food, and $\mathcal{P}_w$ denotes permeability of package to moisture; $p_{w,out}$ denotes partial pressure of water vapor in atmosphere surrounding the package – since partial pressure of water vapor in the overhead space inside the package is ultimately fixed by the moisture content of the food, while p_w^σ denotes equilibrium pressure of water vapor with food. After recalling the definition of a_w as per Eq. (1.147), one may write

$$\boxed{p_w^\sigma = a_w p_{w,0}^\sigma} \quad (2.378)$$

– where $p_{w,0}^\sigma$ denotes vapor pressure of pure water; Eq. (2.378) allows reformulation of Eq. (2.377) to

$$V_f\frac{dC_w}{dt} = \mathcal{P}_w\frac{A}{\delta}\left(p_{w,out} - a_w p_{w,0}^\sigma\right). \quad (2.379)$$

The water activity of a food relates, in turn, to its moisture concentration via

$$a_w = \gamma_w x_w = \gamma_w\frac{C_w}{C_{tot}}, \quad (2.380)$$

upon sequential application of Eq. (3.156), and Eqs. (3.43) and (3.165) of *Food Proc. Eng.: basics & mechanical operations* – where γ_w denotes activity coefficient and C_{tot} denotes total (molar) concentration of food components; after setting

$$\boxed{\Gamma_w \equiv \frac{\gamma_w p_{w,0}^\sigma}{C_{tot}}} \quad (2.381)$$

as auxiliary constant, Eq. (2.379) may be reformulated to

$$V_f\frac{dC_w}{dt} = \mathcal{P}_w\frac{A}{\delta}\left(p_{w,out} - \Gamma_w C_w\right), \quad (2.382)$$

or else

$$V_f\frac{dC_w}{dt} = \mathcal{P}_w\Gamma_w\frac{A}{\delta}\left(\frac{p_{w,out}}{\Gamma_w} - C_w\right) \quad (2.383)$$

upon factoring out Γ_w. On the reasonable hypothesis that environmental $p_{w,out}$ remains essentially constant throughout food storage, and the same happens with storage temperature – thus implying a constant $p_{w,0}^\sigma$ as per Antoine's equation, and thus also a constant $\mathcal{P}_g$ as per Eq. (2.376), one may redo Eq. (2.383) to

$$\frac{d}{dt}\left(C_w - \frac{p_{w,out}}{\Gamma_w}\right) = -\frac{\mathcal{P}_w\Gamma_w A}{\delta V_f}\left(C_w - \frac{p_{w,out}}{\Gamma_w}\right), \quad (2.384)$$

along with division of both sides by V_f and factoring out of –1 in the right-hand side; validity of Eq. (2.384) requires, in addition, that C_{tot} and γ_w are rather weak functions of composition (at least within the range of changes expected during storage), so as to guarantee constancy of Γ_w as per Eq. (2.381). A suitable boundary condition for Eq. (2.384) reads

$$\left. C_w \right|_{t=0} = C_{w,0} \tag{2.385}$$

– which readily implies

$$\left. \left(C_w - \frac{p_{w,out}}{\Gamma_w} \right) \right|_{t=0} = C_{w,0} - \frac{p_{w,out}}{\Gamma_w}. \tag{2.386}$$

Integration of Eq. (2.384) will now proceed via separation of variables, viz.

$$\int_{C_{w,0}-\frac{p_{w,out}}{\Gamma_w}}^{C_w-\frac{p_{w,out}}{\Gamma_w}} \frac{d\left(\tilde{C}_w - \frac{p_{w,out}}{\Gamma_w} \right)}{\tilde{C}_w - \frac{p_{w,out}}{\Gamma_w}} = -\frac{\mathcal{P}_w \Gamma_w A}{\delta V_f} \int_0^t d\tilde{t}, \tag{2.387}$$

with the aid of Eq. (2.386); the fundamental theorem of integral calculus then leads to

$$\ln \left\{ \tilde{C}_w - \frac{p_{w,out}}{\Gamma_w} \right\} \Bigg|_{C_{w,0}-\frac{p_{w,out}}{\Gamma_w}}^{C_w-\frac{p_{w,out}}{\Gamma_w}} = -\frac{\mathcal{P}_w \Gamma_w A}{\delta V_f} \tilde{t} \Big|_0^t, \tag{2.388}$$

or else

$$\ln \frac{C_w - \frac{p_{w,out}}{\Gamma_w}}{C_{w,0} - \frac{p_{w,out}}{\Gamma_w}} = -\frac{\mathcal{P}_w \Gamma_w A}{\delta V_f} t \tag{2.389}$$

that breaks down to

$$\ln \frac{\Gamma_w C_w - p_{w,out}}{\Gamma_w C_{w,0} - p_{w,out}} = -\frac{\mathcal{P}_w \Gamma_w A}{\delta V_f} t \tag{2.390}$$

after multiplying numerator and denominator of the argument of the logarithmic function by Γ_w. The shelf-life may be defined as time, t_{sl}, when C_w reaches a critical moisture content, $C_{w,crt} < C_{w,0}$, i.e.

$$\left. C_w \right|_{t=t_{sl}} = C_{w,crt}; \tag{2.391}$$

insertion of Eq. (2.391) transforms Eq. (2.390) to

$$\frac{\mathcal{P}_w \Gamma_w A}{\delta V_f} t_{sl} = \ln \frac{\Gamma_w C_{w,0} - p_{w,out}}{\Gamma_w C_{w,crt} - p_{w,out}}, \tag{2.392}$$

since the symmetrical of a logarithm equals the logarithm of the reciprocal of its argument – as plotted in Fig. 2.56. If the critical moisture concentration, $C_{w,crt}$, in food decreases, or its initial moisture concentration, $C_{w,0}$, increases, then $(\Gamma_w C_{w,0} - p_{w,out})/(\Gamma_w C_{w,crt} - p_{w,out})$ increases – so a longer shelf-life results, as per inspection of Fig. 2.56; these conditions do indeed delay attainment of unacceptable degree of dehydration. A similar reasoning may be laid out when absorption of moisture from the surroundings (rather than loss thereof to the surroundings) is an issue; in this case, Eq. (2.392) should be more appropriately rewritten as

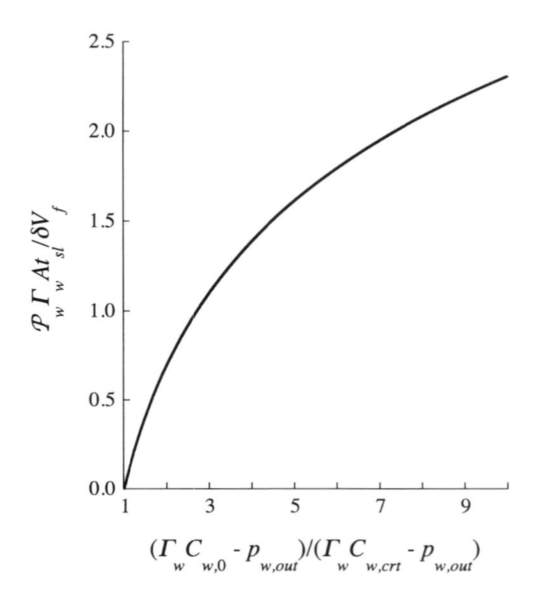

FIGURE 2.56 Variation of dimensionless shelf-life, $\mathcal{P}_w \Gamma_w At_{sl}/\delta V_f$, versus dimensionless critical moisture concentration, $(\Gamma_w C_{w,0} - p_{w,out})/(\Gamma_w C_{w,crt} - p_{w,out})$.

$$\frac{\mathcal{P}_w \Gamma_w A}{\delta V_f} t_{sl} = \ln \frac{p_{w,out} - \Gamma_w C_{w,0}}{p_{w,out} - \Gamma_w C_{w,crt}} \tag{2.393}$$

– after negatives are taken of both numerator and denominator of the argument of the logarithmic function, where $C_{w,crt} > C_{w,0}$.

PROBLEM 2.8

Consider a closed storage room, where N identical food products are kept – with circulation, but not renewal of air, and an initial partial pressure $p_{w,out,0}$ of water vapor. Estimate the shelf-life of those foods, and compare with the situation of continuous renewal of air kept at $p_{w,out,0}$.

Solution

Besides Eq. (2.382) – serving as mass balance to water in each individual food, an extra mass balance to water in the overhead atmosphere is in order, according to

$$V_{out} \frac{dC_{w,out}}{dt} = -N\mathcal{P}_w \frac{A}{\delta} \left(p_{w,out} - \Gamma_w C_w \right); \tag{2.8.1}$$

here V_{out} denotes volume of air in storage room and $C_{w,out}$ denotes (molar) concentration of moisture therein. Assuming ideal gas behavior for the said atmosphere, one may retrieve the corresponding state equation, i.e. Eq. (3.132) of *Food Proc. Eng.: basics & mechanical operations* – reformulated as

$$p_{w,out} V_{out} = n_{w,out} \mathcal{R} T, \tag{2.8.2}$$

where $n_{w,out}$ denotes number of moles of water vapor in overhead atmosphere inside the room, and T denotes (absolute) temperature thereof; after dividing both sides by V_{out}, Eq. (2.8.2) becomes

$$p_{w,out} = \frac{n_{w,out}}{V_{out}} \mathcal{R}T \equiv \mathcal{R}TC_{w,out}, \quad (2.8.3)$$

along with the definition of (mole) concentration of water as ratio of $n_{w,out}$ to V_{out}. Upon insertion of Eq. (2.8.3), one gets

$$V_{out} \frac{dC_{w,out}}{dt} = -N\mathcal{P}_w \frac{A}{\delta}\left(\mathcal{R}TC_{w,out} - \Gamma_w C_w\right) \quad (2.8.4)$$

from Eq. (2.8.1); the associated boundary condition will, by hypothesis, be set as

$$C_{w,out}\big|_{t=0} = C_{w,out,0}, \quad (2.8.5)$$

related to $p_{w,out,0}$ via an analogue to Eq. (2.8.3). Ordered addition of Eq. (2.382), after multiplying both sides by N, to Eq. (2.8.1) leads to merely

$$NV_f \frac{dC_w}{dt} + V_{out} \frac{dC_{w,out}}{dt} = 0 \quad (2.8.6)$$

– where dt may be dropped from both sides to obtain

$$V_{out} dC_{w,out} = -NV_f dC_w; \quad (2.8.7)$$

integration of Eq. (2.8.7) can now proceed as

$$\int_{C_{w,out,0}}^{C_{w,out}} d\tilde{C}_{w,out} = -\frac{NV_f}{V_{out}} \int_{C_{w,0}}^{C_w} d\tilde{C}_w, \quad (2.8.8)$$

with the aid of Eqs. (2.385) and (2.8.5). Application of the fundamental theorem of integral calculus transforms Eq. (2.8.8) to

$$\tilde{C}_{w,out}\Big|_{C_{w,0}}^{C_w} = -\frac{NV_f}{V_{out}} \tilde{C}_w\Big|_{C_{w,0}}^{C_w}, \quad (2.8.9)$$

or else

$$C_{w,out} - C_{w,out,0} = -\frac{NV_f}{V_{out}}\left(C_w - C_{w,0}\right) \quad (2.8.10)$$

after replacing by upper and lower limits of integration; isolation of $C_{w,out}$ finally gives

$$C_{w,out} = C_{w,out,0} - \frac{NV_f}{V_{out}}\left(C_w - C_{w,0}\right), \quad (2.8.11)$$

with $C_{w,out} \equiv C_{w,out}\{C_w\}$ appearing in the form of a linear relationship. Insertion of Eq. (2.8.3) converts Eq. (2.382) to

$$V_f \frac{dC_w}{dt} = \mathcal{P}_w \frac{A}{\delta}\left(\mathcal{R}TC_{w,out} - \Gamma_w C_w\right), \quad (2.8.12)$$

whereas combination with Eq. (2.8.11) gives rise to

$$\frac{dC_w}{dt} = \mathcal{P}_w \frac{A}{\delta}\left(\frac{\mathcal{R}TC_{w,out,0}}{V_f} - \frac{\mathcal{R}NT}{V_{out}}\left(C_w - C_{w,0}\right) - \frac{\Gamma_w}{V_f} C_w\right) \quad (2.8.13)$$

– where both sides were meanwhile divided by V_f; the notation in Eq. (2.8.13) may be condensed to

$$\frac{dC_w}{dt} = \alpha - \beta C_w, \quad (2.8.14)$$

provided that auxiliary constants α and β are defined as

$$\alpha \equiv \mathcal{P}_w \frac{A}{\delta} \mathcal{R}T\left(\frac{C_{w,out,0}}{V_f} + \frac{N}{V_{out}} C_{w,0}\right) \quad (2.8.15)$$

and

$$\beta \equiv \mathcal{P}_w \frac{A}{\delta}\left(\frac{\mathcal{R}NT}{V_{out}} + \frac{\Gamma_w}{V_f}\right), \quad (2.8.16)$$

respectively. Integration of Eq. (2.8.14) is feasible via separation of variables, viz.

$$\int_{C_{w,0}}^{C_w} \frac{d\tilde{C}_w}{\alpha - \beta \tilde{C}_w} = \int_0^t d\tilde{t}, \quad (2.8.17)$$

where Eq. (2.385) was again taken on board; the fundamental theorem of integral calculus is then to be invoked to get

$$\left.\frac{\ln\left\{\alpha - \beta \tilde{C}_w\right\}}{-\beta}\right|_{C_{w,0}}^{C_w} = \tilde{t}\Big|_0^t \quad (2.8.18)$$

or, equivalently,

$$-\frac{1}{\beta} \ln \frac{\alpha - \beta C_w}{\alpha - \beta C_{w,0}} = t. \quad (2.8.19)$$

The minus sign in the left-hand side of Eq. (2.8.19) can be removed, provided that the reciprocal of the argument of the logarithm function is simultaneously taken, i.e.

$$t = \frac{1}{\beta} \ln \frac{\alpha - \beta C_{w,0}}{\alpha - \beta C_w} \quad (2.8.20)$$

– while multiplication of both sides by β leaves

$$\beta t = \ln \frac{\alpha - \beta C_{w,0}}{\alpha - \beta C_w}; \quad (2.8.21)$$

recalling Eq. (2.391), one will be able to write

$$\beta t_{sl} = \ln \frac{\alpha - \beta C_{w,0}}{\alpha - \beta C_{w,crt}}. \quad (2.8.22)$$

The original notation can finally be recovered in Eq. (2.8.22), with the aid of Eqs. (2.8.15) and (2.8.16) – according to

$$\mathcal{P}_w \frac{A}{\delta}\left(\frac{\mathcal{R}NT}{V_{out}}+\frac{\Gamma_w}{V_f}\right)t_{sl} = \ln \frac{\mathcal{P}_w \frac{A}{\delta}\mathcal{R}T\left(\frac{C_{w,out,0}}{V_f}+\frac{N}{V_{out}}C_{w,0}\right)-\mathcal{P}_w \frac{A}{\delta}\left(\frac{\mathcal{R}NT}{V_{out}}+\frac{\Gamma_w}{V_f}\right)C_{w,0}}{\mathcal{P}_w \frac{A}{\delta}\mathcal{R}T\left(\frac{C_{w,out,0}}{V_f}+\frac{N}{V_{out}}C_{w,0}\right)-\mathcal{P}_w \frac{A}{\delta}\left(\frac{\mathcal{R}NT}{V_{out}}+\frac{\Gamma_w}{V_f}\right)C_{w,crt}}, \tag{2.8.23}$$

where $\mathcal{P}_w A/\delta$ may be dropped off both numerator and denominator, and parentheses removed in the right-hand side of Eq. (2.8.23) to get

$$\mathcal{P}_w \frac{A}{\delta}\left(\frac{\mathcal{R}NT}{V_{out}}+\frac{\Gamma_w}{V_f}\right)t_{sl} = \ln \frac{\frac{\mathcal{R}T}{V_f}C_{w,out,0}+\frac{\mathcal{R}NT}{V_{out}}C_{w,0}-\frac{\mathcal{R}NT}{V_{out}}C_{w,0}-\frac{\Gamma_w}{V_f}C_{w,0}}{\frac{\mathcal{R}T}{V_f}C_{w,out,0}+\frac{\mathcal{R}NT}{V_{out}}C_{w,0}-\frac{\mathcal{R}NT}{V_{out}}C_{w,crt}-\frac{\Gamma_w}{V_f}C_{w,crt}}; \tag{2.8.24}$$

upon cancellation of terms with their negatives in numerator of the argument of the logarithm function, complemented by factoring out of $\mathcal{R}NT/V_{out}$ in denominator, Eq. (2.8.24) becomes

$$\mathcal{P}_w \frac{A}{\delta}\left(\frac{\mathcal{R}NT}{V_{out}}+\frac{\Gamma_w}{V_f}\right)t_{sl} = \ln \frac{\frac{\mathcal{R}T}{V_f}C_{w,out,0}-\frac{\Gamma_w}{V_f}C_{w,0}}{\frac{\mathcal{R}T}{V_f}C_{w,out,0}-\frac{\Gamma_w}{V_f}C_{w,crt}+\frac{\mathcal{R}NT}{V_{out}}\left(C_{w,0}-C_{w,crt}\right)}. \tag{2.8.25}$$

Addition and subtraction of $\Gamma_w C_{w,crt}/V_f$ to the numerator transforms Eq. (2.8.25) to

$$\mathcal{P}_w \frac{A}{\delta}\left(\frac{\mathcal{R}NT}{V_{out}}+\frac{\Gamma_w}{V_f}\right)t_{sl} = \ln \frac{\left(\frac{\mathcal{R}T}{V_f}C_{w,out,0}-\frac{\Gamma_w}{V_f}C_{w,crt}\right)-\frac{\Gamma_w}{V_f}\left(C_{w,0}-C_{w,crt}\right)}{\left(\frac{\mathcal{R}T}{V_f}C_{w,out,0}-\frac{\Gamma_w}{V_f}C_{w,crt}\right)+\frac{\mathcal{R}NT}{V_{out}}\left(C_{w,0}-C_{w,crt}\right)}, \tag{2.8.26}$$

along with convenient factoring out of Γ_w/V_f; division of both numerator and denominator in the right-hand side by $\mathcal{R}TC_{w,out,0}/V_f - \Gamma_w C_{w,crt}/V_f$ then yields

$$\mathcal{P}_w \frac{A}{\delta}\left(\frac{\mathcal{R}NT}{V_{out}}+\frac{\Gamma_w}{V_f}\right)t_{sl} = \ln \frac{1-\dfrac{\frac{\Gamma_w}{V_f}\left(C_{w,0}-C_{w,crt}\right)}{\frac{\mathcal{R}T}{V_f}C_{w,out,0}-\frac{\Gamma_w}{V_f}C_{w,crt}}}{1+\dfrac{\frac{\mathcal{R}NT}{V_{out}}\left(C_{w,0}-C_{w,crt}\right)}{\frac{\mathcal{R}T}{V_f}C_{w,out,0}-\frac{\Gamma_w}{V_f}C_{w,crt}}} \tag{2.8.27}$$

– and a further division of both numerator and denominator of the outstanding fractions by Γ_w/V_f, followed by elementary algebraic manipulation eventually generate

$$\mathcal{P}_w \frac{A}{\delta}\left(\frac{\mathcal{R}NT}{V_{out}}+\frac{\Gamma_w}{V_f}\right)t_{sl} = \ln \frac{1+\dfrac{C_{w,0}-C_{w,crt}}{C_{w,crt}-\frac{\mathcal{R}T}{\Gamma_w}C_{w,out,0}}}{1-\dfrac{\mathcal{R}NTV_f}{V_{out}\Gamma_w}\dfrac{C_{w,0}-C_{w,crt}}{C_{w,crt}-\frac{\mathcal{R}T}{\Gamma_w}C_{w,out,0}}}. \tag{2.8.28}$$

Inspection of Eq. (2.8.28) unfolds a germane (dimensionless) parameter, $\mathcal{R}NTV_f/V_{out}\Gamma_w$; an illustration of its effect is provided below – via a plot of (dimensionless) $\mathcal{P}_w A(\mathcal{R}NT/V_{out}+\Gamma_w/V_f)t_{sl}/\delta$ as dependent variable, versus (dimensionless) $(C_{w,0}-C_{w,crt})/(C_{w,crt}-\mathcal{R}TC_{w,out,0}/\Gamma_w)>0$ for a drying effect upon food.

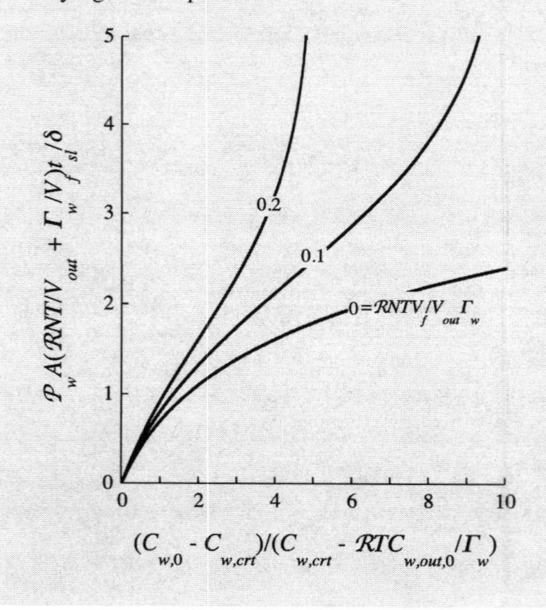

An increase in characteristic parameter as a consequence of an increase in fraction of storage room volume occupied by food products, NV_f/V_{out}, permits a longer shelf-life; this is so because saturation of the overhead atmosphere is attainable at the expense of a smaller degree of dehydration of the food inventory. The lower and lower driving force for mass transfer of moisture out of the food, as moisture is withdrawn toward the overhead atmosphere present in the vessel, justifies the sigmoidal curves – and unfolds a cooperative effect between those two (concomitant) events. Note that $\mathcal{R}NTV_f/V_{out}\Gamma_w = (\mathcal{R}T/\Gamma_w)NV_f/V_{out} \rightarrow 0$, and thus $NV_f/V_{out} \rightarrow 0$ as per finite $\mathcal{R}$, T, and Γ_w would imply $dC_{w,out}/dt=0$ as per Eq. (2.8.6), since dC_w/dt is also expected to remain finite; consequently, $C_{w,out} \approx C_{w,out,0}$ – and the simpler case of constant $C_{w,out}$, described by Fig. 2.56, would be retrieved as the lowest curve in the plot above. Under such circumstances, one finds

$$\lim_{\substack{\mathcal{R}NTV_f \\ \overline{V_{out}\Gamma_w}} \rightarrow 0} \mathcal{P}_w \frac{A}{\delta}\left(\frac{\mathcal{R}NT}{V_{out}} + \frac{\Gamma_w}{V_f}\right) t_{sl} = \ln \frac{1 + \dfrac{C_{w,0} - C_{w,crt}}{C_{w,crt} - \dfrac{\mathcal{R}T}{\Gamma_w} C_{w,out}}}{1 - 0 \dfrac{C_{w,0} - C_{w,crt}}{C_{w,crt} - \dfrac{\mathcal{R}T}{\Gamma_w} C_{w,out}}} \quad (2.8.29)$$

based on Eq. (2.8.28), which degenerates to

$$\lim_{\substack{\mathcal{R}NTV_f \\ \overline{V_{out}\Gamma_w}} \rightarrow 0} \mathcal{P}_w \frac{A}{\delta} \frac{\Gamma_w}{V_f} t_{sl} = \ln \left\{ 1 + \frac{C_{w,0} - C_{w,crt}}{C_{w,crt} - \dfrac{\mathcal{R}T}{\Gamma_w} C_{w,out}} \right\} \quad (2.8.30)$$

– in view of $\mathcal{R}NTV_f/V_{out}\Gamma_w = (V_f/\Gamma_w)(\mathcal{R}NT/V_{out}) \rightarrow 0$ also implying $\mathcal{R}NT/V_{out} \rightarrow 0$, as long as V_f and Γ_w are finite. Elimination of brackets in Eq. (2.8.30) gives rise to

$$\lim_{\substack{\mathcal{R}NTV_f \\ \overline{V_{out}\Gamma_w}} \rightarrow 0} \frac{\mathcal{P}_w \Gamma_w A}{\delta V_f} t_{sl} = \ln \frac{C_{w,crt} - \dfrac{\mathcal{R}T}{\Gamma_w} C_{w,out} + C_{w,0} - C_{w,crt}}{C_{w,crt} - \dfrac{\mathcal{R}T}{\Gamma_w} C_{w,out}}, \quad (2.8.31)$$

where cancellation of symmetrical terms supports simplification to

$$\lim_{\substack{\mathcal{R}NTV_f \\ \overline{V_{out}\Gamma_w}} \rightarrow 0} \frac{\mathcal{P}_w \Gamma_w A}{\delta V_f} t_{sl} = \ln \frac{C_{w,0} - \dfrac{\mathcal{R}T}{\Gamma_w} C_{w,out}}{C_{w,crt} - \dfrac{\mathcal{R}T}{\Gamma_w} C_{w,out}}; \quad (2.8.32)$$

multiplication of both numerator and denominator by Γ_w then yields

$$\lim_{\substack{\mathcal{R}NTV_f \\ \overline{V_{out}\Gamma_w}} \rightarrow 0} \frac{\mathcal{P}_w \Gamma_w A}{\delta V_f} t_{sl} = \ln \frac{\Gamma_w C_{w,0} - \mathcal{R}TC_{w,out}}{\Gamma_w C_{w,crt} - \mathcal{R}TC_{w,out}}. \quad (2.8.33)$$

Based on Eq. (2.8.3), one may reformulate Eq. (2.8.33) to

$$\lim_{\substack{\mathcal{R}NTV_f \\ \overline{V_{out}\Gamma_w}} \rightarrow 0} \frac{\mathcal{P}_w \Gamma_w A}{\delta V_f} t_{sl} = \ln \frac{\Gamma_w C_{w,0} - p_{w,out}}{\Gamma_{w,crt} C_{w,crt} - p_{w,out}} \quad (2.8.34)$$

– which, as expected, recovers Eq. (2.392). In fact, $\mathcal{R}NTV_f/V_{out}\Gamma_w \rightarrow 0$ also means $V_{out} \rightarrow \infty$ – in which case the storage room would operate as in an open atmosphere, equivalent in turn to an infinite rate of circulation of fresh air; hence, $C_{w,out}$ and thus $p_{w,out}$ would remain constant in time, and equal to $C_{w,out,0}$ and $P_{w,out,0}$, respectively.

2.2.3.3 *Active packaging*

Consider a spherically shaped food (e.g. a fruit piece), of outer radius R_f, wrapped with some film containing an active component in its formulation – able to gradually diffuse away into the food matrix. The mass balance to the said component, under transient conditions and using an elementary shell within the food as control volume, may be written as

$$-\mathcal{D}_c A \left.\frac{\partial C_c}{\partial r}\right|_r = -\mathcal{D}_c A \left.\frac{\partial C_c}{\partial r}\right|_{r+dr} + A dr \frac{\partial C_c}{\partial t}; \quad (2.394)$$

here $\mathcal{D}_c$ denotes effective diffusivity of active component within the porous food matrix, A denotes area of hypothetical sphere with radius r – given by

$$A = 4\pi r^2, \quad (2.395)$$

and C_c denotes molar concentration of active component at time t and radial location r. After bringing Eq. (2.395) on board and moving the space derivative at $r+dr$ to the left-hand side, Eq. (2.394) becomes

$$\mathcal{D}_c 4\pi (r+dr)^2 \left.\frac{\partial C_c}{\partial r}\right|_{r+dr} - \mathcal{D}_c 4\pi r^2 \left.\frac{\partial C_c}{\partial r}\right|_r = 4\pi r^2 dr \frac{\partial C_c}{\partial t} \quad (2.396)$$

– where $\mathcal{D}_c$ (taken as constant) may be factored out in the left-hand side, and both sides divided by $4\pi dr$ afterward to produce

$$\mathcal{D}_c \frac{(r+dr)^2 \left.\dfrac{\partial C_c}{\partial r}\right|_{r+dr} - r^2 \left.\dfrac{\partial C_c}{\partial r}\right|_r}{dr} = r^2 \frac{\partial C_c}{\partial t}; \quad (2.397)$$

in view of the concept of (partial) derivative, the notation in Eq. (2.397) will be simplified to

$$\boxed{\mathcal{D}_c \frac{1}{r^2} \frac{\partial}{\partial r}\left(r^2 \frac{\partial C_c}{\partial r}\right) = \frac{\partial C_c}{\partial t},} \quad (2.398)$$

with both sides meanwhile divided by r^2. The rule of differentiation of a product, coupled with the definition of second-order derivative convey transformation of Eq. (2.398) to

$$\mathcal{D}_c \frac{1}{r^2}\left(2r\frac{\partial C_c}{\partial r}+r^2\frac{\partial^2 C_c}{\partial r^2}\right)=\frac{\partial C_c}{\partial t}, \tag{2.399}$$

or else

$$\mathcal{D}_c\left(\frac{2}{r}\frac{\partial C_c}{\partial r}+\frac{\partial^2 C_c}{\partial r^2}\right)=\frac{\partial C_c}{\partial t} \tag{2.400}$$

after factoring in r^2; a suitable initial condition reads

$$\boxed{C_c\big|_{t=0}=0,} \tag{2.401}$$

with one (consistent) boundary condition looking like

$$\boxed{C_c\big|_{r=0}=0} \tag{2.402}$$

– whereas the other boundary condition may take the form

$$\boxed{C_c\big|_{r=R_f}=C_{c,0}^{\sigma},} \tag{2.403}$$

with $C_{c,0}^{\sigma}$ denoting equilibrium concentration of active compound on the food matrix side, with its concentration (assumed constant) prevailing in the outer wrapping film, $C_{c,0}$.

In attempts to solve Eq. (2.398), it is useful to define an auxiliary variable as

$$\xi \equiv rC_c, \tag{2.404}$$

or else

$$C_c=\frac{\xi}{r}; \tag{2.405}$$

Eq. (2.405) produces

$$\frac{\partial C_c}{\partial r}=-\frac{1}{r^2}\xi+\frac{1}{r}\frac{\partial \xi}{\partial r} \tag{2.406}$$

upon differentiation with regard to r, and thus

$$\begin{aligned}\frac{\partial^2 C_c}{\partial r^2}&=\frac{2}{r^3}\xi-\frac{1}{r^2}\frac{\partial \xi}{\partial r}-\frac{1}{r^2}\frac{\partial \xi}{\partial r}+\frac{1}{r}\frac{\partial^2 \xi}{\partial r^2}\\&=\frac{2}{r^3}\xi-\frac{2}{r^2}\frac{\partial \xi}{\partial r}+\frac{1}{r}\frac{\partial^2 \xi}{\partial r^2}\end{aligned} \tag{2.407}$$

after a second differentiation with regard to r, followed by condensation of terms alike. By the same token, differentiation of Eq. (2.405) with regard to t yields just

$$\frac{\partial C_c}{\partial t}=\frac{1}{r}\frac{\partial \xi}{\partial t}. \tag{2.408}$$

Insertion of Eqs. (2.406)–(2.408) transforms Eq. (2.400) to

$$\mathcal{D}_c\left(\frac{2}{r}\left(-\frac{1}{r^2}\xi+\frac{1}{r}\frac{\partial \xi}{\partial r}\right)+\frac{2}{r^3}\xi-\frac{2}{r^2}\frac{\partial \xi}{\partial r}+\frac{1}{r}\frac{\partial^2 \xi}{\partial r^2}\right)=\frac{1}{r}\frac{\partial \xi}{\partial t}, \tag{2.409}$$

where elimination of parenthesis leaves

$$\mathcal{D}_c\left(-\frac{2}{r^3}\xi+\frac{2}{r^2}\frac{\partial \xi}{\partial r}+\frac{2}{r^3}\xi-\frac{2}{r^2}\frac{\partial \xi}{\partial r}+\frac{1}{r}\frac{\partial^2 \xi}{\partial r^2}\right)=\frac{1}{r}\frac{\partial \xi}{\partial t}; \tag{2.410}$$

after collapsing similar terms, Eq. (2.410) becomes merely

$$\mathcal{D}_c\frac{1}{r}\frac{\partial^2 \xi}{\partial r^2}=\frac{1}{r}\frac{\partial \xi}{\partial t}, \tag{2.411}$$

which further reduces to

$$\boxed{\mathcal{D}_c\frac{\partial^2 \xi}{\partial r^2}=\frac{\partial \xi}{\partial t}} \tag{2.412}$$

upon multiplication of both sides by r. Equations (2.401)–(2.403) can likewise be redone to

$$\boxed{\xi\big|_{t=0}=0,} \tag{2.413}$$

as well as

$$\boxed{\xi\big|_{r=0}=0} \tag{2. 414}$$

and

$$\boxed{\xi\big|_{r=R_f}=R_f C_{c,0}^{\sigma},} \tag{2.415}$$

respectively – with the aid of Eq. (2.404). In view of the constant differential coefficients of Eq. (2.412) – unlike happened with Eq. (2.400), one may apply Laplace's transforms (with regard to time) to get

$$\mathcal{D}_c\frac{d^2\bar{\xi}}{dr^2}=s\bar{\xi}-\xi\big|_{t=0}, \tag{2.416}$$

where $\bar{\xi}\equiv\mathcal{L}(\xi)$ – thus generating an ordinary differential equation in r; Eq. (2.413) permits simplification of Eq. (2.416) to

$$\mathcal{D}_c\frac{d^2\bar{\xi}}{dr^2}-s\bar{\xi}=0 \tag{2.417}$$

or, equivalently,

$$\frac{d^2\bar{\xi}}{dr^2}-\frac{s}{\mathcal{D}_c}\bar{\xi}=0 \tag{2.418}$$

after dividing both sides by $\mathcal{D}_c$. The constancy of the differential coefficients in Eq. (2.418) justifies its (polynomial) characteristic equation looking like

$$m^2-\frac{s}{\mathcal{D}_c}=0; \tag{2.419}$$

Eq. (2.419) accepts

$$m=\pm\sqrt{\frac{s}{\mathcal{D}_c}} \tag{2.420}$$

as solutions – so the general form for the solution to Eq. (2.418) may be coined as

$$\bar{\xi}=k_1\exp\left\{-\sqrt{\frac{s}{\mathcal{D}_c}}r\right\}+k_2\exp\left\{\sqrt{\frac{s}{\mathcal{D}_c}}r\right\}, \tag{2.421}$$

where k_1 and k_2 denote arbitrary constants (still to be determined). If Eq. (2.414) holds, then the center will never experience any change in ξ; should $\hat{r}$ denote distance to the food surface, i.e.

$$\hat{r} \equiv R_f - r, \tag{2.422}$$

then

$$\xi\big|_{\hat{r}\to R_f} = 0 \tag{2.423}$$

would to advantage replace Eq. (2.414). If the original food is thick enough, and diffusion is a relatively slow phenomenon compared to the time framework of its shelf-life, then Eq. (2.423) will become essentially equivalent to

$$\xi\big|_{\hat{r}\to\infty} = 0 \tag{2.424}$$

– or, in Laplace's domain,

$$\bar{\xi}\big|_{\hat{r}\to\infty} = 0. \tag{2.425}$$

Recalling Eq. (2.422), one may rewrite Eq. (2.421) as

$$\bar{\xi} = k_1 \exp\left\{-\sqrt{\frac{s}{\mathcal{D}_c}}\left(R_f - \hat{r}\right)\right\} + k_2 \exp\left\{\sqrt{\frac{s}{\mathcal{D}_c}}\left(R_f - \hat{r}\right)\right\}; \tag{2.426}$$

elimination of ξ between Eq. (2.425), and Eq. (2.426) for $\hat{r}\to\infty$ gives rise to

$$0 = k_1 \exp\left\{-\sqrt{\frac{s}{\mathcal{D}_c}}(-\infty)\right\} + k_2 \exp\left\{\sqrt{\frac{s}{\mathcal{D}}}(-\infty)\right\} \tag{2.427}$$

$$= k_1 e^{\infty} + k_2 e^{-\infty} = k_1\infty + k_2 0 = k_1\infty,$$

because $R_f - \hat{r}\to -\infty$ when $\hat{r}\to\infty$; Eq. (2.427) will lead to an impossible condition unless

$$k_1 = 0 \tag{2.428}$$

– since only $0\cdot\infty$, as unknown quantity, can accomodate the nil value for the product at stake. In view of Eqs. (2.422) and (2.428), one may simply Eq. (2.421) to

$$\bar{\xi} = k_2 \exp\left\{\sqrt{\frac{s}{\mathcal{D}_c}}\left(R_f - \hat{r}\right)\right\}; \tag{2.429}$$

application of Eq. (2.415) as remaining boundary condition – upon combining with Eq. (2.422) once more and converting to Laplace's domain afterward, i.e.

$$\bar{\xi}\big|_{\hat{r}=0} = \frac{R_f C_{c,0}^{\sigma}}{s}, \tag{2.430}$$

permits transformation of Eq. (2.429) to

$$\frac{R_f C_{c,0}^{\sigma}}{s} = k_2 \exp\left\{\sqrt{\frac{s}{\mathcal{D}_c}} R_f\right\}. \tag{2.431}$$

Isolation of k_2 in Eq. (2.431) unfolds

$$k_2 = \frac{R_f C_{c,0}^{\sigma}}{s} \exp\left\{-\sqrt{\frac{s}{\mathcal{D}_c}} R_f\right\}, \tag{2.432}$$

which, in turn, supports transformation of Eq. (2.429) to

$$\bar{\xi} = \frac{R_f C_{c,0}^{\sigma}}{s} \exp\left\{-\sqrt{\frac{s}{\mathcal{D}_c}} R_f\right\} \exp\left\{\sqrt{\frac{s}{\mathcal{D}_c}}\left(R_f - \hat{r}\right)\right\}; \tag{2.433}$$

after pooling exponential functions, Eq. (2.433) simplifies to

$$\bar{\xi} = \frac{R_f C_{c,0}^{\sigma}}{s} \exp\left\{-\sqrt{\frac{s}{\mathcal{D}_c}}\hat{r}\right\}. \tag{2.434}$$

Inverse Laplace's transforms, applied to both sides of Eq. (2.434), produce

$$\xi \equiv \mathcal{L}^{-1}\left(R_f C_{c,0}^{\sigma} \frac{\exp\left\{-\frac{\hat{r}}{\sqrt{\mathcal{D}_c}}\sqrt{s}\right\}}{s}\right) = R_f C_{c,0}^{\sigma} \mathcal{L}^{-1}\left(\frac{\exp\left\{-\frac{\hat{r}}{\sqrt{\mathcal{D}_c}}\sqrt{s}\right\}}{s}\right)$$

$$= R_f C_{c,0}^{\sigma}\,\mathrm{erfc}\left\{\frac{1}{2}\frac{\hat{r}}{\sqrt{\mathcal{D}_c}}\frac{1}{\sqrt{t}}\right\} = R_f C_{c,0}^{\sigma}\left(1 - \mathrm{erf}\left\{\frac{\hat{r}}{2\sqrt{\mathcal{D}_c t}}\right\}\right) \tag{2.435}$$

in view of the intrinsic linearity of such an integral operator – coupled to the definition of error function, erf, and complementary error function, erfc, as well as straightforward algebraic rearrangement. The original notation can now be retrieved with the aid of Eqs. (2.404) and (2.422), viz.

$$rC_c = R_f C_{c,0}^{\sigma}\left(1 - \mathrm{erf}\left\{\frac{R_f - r}{2\sqrt{\mathcal{D}_c t}}\right\}\right); \tag{2.436}$$

upon isolation of C_c, Eq. (2.436) gives rise to

$$C_c = R_f C_{c,0}^{\sigma} \frac{1 - \mathrm{erf}\left\{\frac{R_f - r}{2\sqrt{\mathcal{D}_c t}}\right\}}{r}. \tag{2.437}$$

If r is converted to ratio r/R_f, and both sides are divided by $C_{c,0}^{\sigma}$, then Eq. (2.437) will read

$$\boxed{\frac{C_c}{C_{c,0}^{\sigma}} = \frac{1 - \mathrm{erf}\left\{\frac{1 - \frac{r}{R_f}}{2\sqrt{\frac{\mathcal{D}_c t}{R_f^2}}}\right\}}{\frac{r}{R_f}}}; \tag{2.438}$$

$C_c/C_{c,0}^{\sigma}$ represents normalized concentration of active compound – while r/R_f represents normalized radial distance to the center of the food, and $\mathcal{D}_c t/R_f^2 = t/(R_f^2/\mathcal{D}_c)$ represents time elapsed (i.e. t) made dimensionless by the time scale associated with transport by molecular diffusion (i.e. $R_f^2/\mathcal{D}_c$). The evolution in the concentration profile of active compound, as conveyed by Eq. (2.438), is depicted in Fig. 2.57. At startup, Eq. (2.438) degenerates to

$$\lim_{\frac{\mathcal{D}_c t}{R_f^2}\to 0}\frac{C_c}{C_{c,0}^\sigma}=\frac{1-\operatorname{erf}\left\{\frac{1-\frac{r}{R_f}}{2\sqrt{0}}\right\}}{\frac{r}{R_f}}=\frac{1-\operatorname{erf}\{\infty\}}{\frac{r}{R_f}}=\frac{1-1}{\frac{r}{R_f}}=\frac{0}{\frac{r}{R_f}}=0 \tag{2.439}$$

– corresponding to the horizontal axis serving as representative curve, except for the singularity at $r=R_f$ that is required to satisfy Eq. (2.403) as boundary condition; as time elapses, the concentration profile becomes smoother and smoother in shape, yet it remains equal to unity at $r=R_f$, i.e.

$$\lim_{\frac{r}{R_f}\to 1}\frac{C_c}{C_{c,0}^\sigma}=\frac{1-\operatorname{erf}\left\{\frac{1-1}{2\sqrt{\frac{\mathcal{D}_c t}{R_f^2}}}\right\}}{1}=1-\operatorname{erf}\{0\}=1-0=1, \tag{2.440}$$

based again on Eq. (2.438). Conversely, if $\hat{r}$ were made to grow toward ∞ or, equivalently, $r\to-\infty$, then Eq. (2.338) would be driven by

$$\lim_{\frac{r}{R_f}\to -\infty}\frac{C_c}{C_{c,0}^\sigma}=\frac{1-\operatorname{erf}\left\{\frac{1-(-\infty)}{2\sqrt{\frac{\mathcal{D}_c t}{R_f^2}}}\right\}}{-\infty}=\frac{1-\operatorname{erf}\{\infty\}}{-\infty} \tag{2.441}$$

$$\equiv -\frac{1-1}{\infty}=-\frac{0}{\infty}=\frac{0}{\infty}=0$$

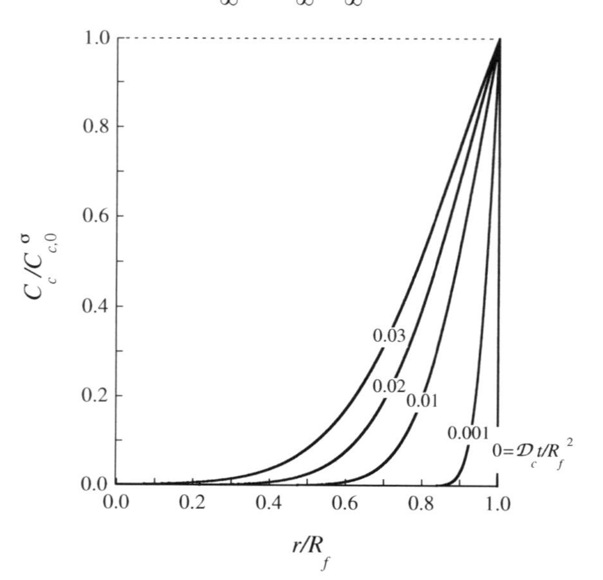

FIGURE 2.57 Variation of normalized concentration of active compound, $C_c/C_{c,0}^\sigma$, versus normalized radial position, r/R_f, for selected dimensionless times, $\mathcal{D}_c t/R_f^2$.

– which indeed coincides with Eq. (2.425), after taking Eq. (2.422) on board; and is compatible with Eq. (2.402), since $C_c|_{-\infty<r\leq 0}=0$ per hypothesis. The solution conveyed by Eq. (2.438) holds while the center of the food does not experience a significant increase in concentration – otherwise Eq. (2.424) will no longer be equivalent to Eq. (2.414) as boundary condition.

PROBLEM 2.9

An Edam cheese, with approximately spherical shape and 15 cm for outer diameter, is wrapped in a cellulose acetate film – prepared by a dry-phase inversion technique, and formulated with potassium sorbate (E202) as antifungal agent; as it dissolves, the concentration front gradually progresses within the cheese, but is not expected to reach its center within the anticipated shelf-life. The solubility of potassium sorbate (with molecular weight $M_c=150.2$ g.mol^{-1}) in cheese curd is $w_c^\sigma=15.5$ g.L^{-1}; and its diffusivity in the cheese matrix is 4.85×10^{-11} m^2.s^{-1}. Calculate the time required for its concentration reach $w_c=0.385$ g.L^{-1} at the median of the cheese.

Solution

The equilibrium concentration of potassium sorbate, on the cheese side at the interface of cheese with wrapping film, can be calculated as

$$C_{c,o}^\sigma=\frac{w_c^\sigma}{M_c}=\frac{15.5\times10^3}{150.2}=103.2\ \text{mol.m}^{-3}, \tag{2.9.1}$$

based on the ratio of $w_c^\sigma=15.5\times10^3$ g.m^{-3} to $M_c=150.2$ g.mol^{-1}; the target concentration may be expressed in normalized form as

$$\frac{C_c}{C_{c,0}^\sigma}=\frac{\frac{w_c}{M_c}}{C_{c,0}^\sigma}=\frac{\frac{0.385}{0.1502}}{103.2}=0.02484. \tag{2.9.2}$$

On the other hand, the median of the cheese radius corresponds to

$$\frac{r}{R_f}=0.5; \tag{2.9.3}$$

the point with abscissa given by Eq. (2.9.3) and ordinate abiding to Eq. (2.9.2) lies approximately on the curve labeled as 0.02 in Fig. 2.57, so

$$\frac{\mathcal{D}_c t}{R_f^2}\approx 0.02; \tag{2.9.4}$$

isolation of t gives then rise to

$$t=0.02\frac{R_f^2}{\mathcal{D}_c}=0.02\frac{0.075^2}{4.85\times10^{-11}}=2.32\times10^6\ \text{s}, \tag{2.9.5}$$

since $R_f=0.075$ m per hypothesis. This corresponds to ca. 38 wk – compatible with the regular shelf-life of this type of cheese.

2.2.3.4 Intelligent packaging

When searching for an integral time-temperature indicator, for instance built in a food label, one should first identify the parameter, say Q, that best describes the decay in quality of the target food; a chemical, enzymatic, or microbial reaction may suit this purpose – and is often associated with some representative compound that evolves in concentration as time elapses.

Since a given packaged food may be regarded as a batch system, one may proceed to a mass balance to the compound of interest as

$$\boxed{\frac{dC_Q}{dt} = -k_Q\{T\}C_Q^{m_Q}} \tag{2.442}$$

– where subscript Q refers to the said quality parameter, while C denotes concentration, k denotes kinetic constant, m denotes reaction order (usually between 0 and 2), and T denotes (absolute) temperature; the compound chosen for indicator should as well undergo depletion with time, according to

$$\boxed{\frac{dC_I}{dt} = -k_I\{T\}C_I^{m_I}} \tag{2.443}$$

that follows Eq. (2.442) as template – with subscript I now referring to indicator. Equation (2.442) will abide to

$$\boxed{C_Q\big|_{t=0} = C_{Q,0}} \tag{2.444}$$

as initial condition, and Eq. (2.443) will likewise satisfy

$$\boxed{C_I\big|_{t=0} = C_{I,0};} \tag{2.445}$$

here subscript 0 refers obviously to initial conditions. Arrhenius' law may be retrieved, see Eq. (2.2) of *Food Proc. Eng.: thermal & chemical operations*, so as to rewrite $k_Q\{T\}$ as

$$\boxed{k_Q\{T\} = k_{0,Q}\exp\left\{-\frac{E_{act,Q}}{\mathcal{R}T}\right\},} \tag{2.446}$$

and $k_I\{T\}$ similarly as

$$\boxed{k_I\{T\} = k_{0,I}\exp\left\{-\frac{E_{act,I}}{\mathcal{R}T}\right\}} \tag{2.447}$$

– where $k_\bullet$ denotes pre-exponential factor and E_{act} denotes activation energy. Insertion of Eq. (2.447) transforms Eq. (2.443) to

$$\frac{dC_I}{dt} = -k_{0,I}\exp\left\{-\frac{E_{act,I}}{\mathcal{R}T}\right\}C_I^{m_I}, \tag{2.448}$$

where algebraic rearrangement unfolds

$$\frac{1}{k_{0,I}C_I^{m_I}}\left(-\frac{dC_I}{dt}\right) = \exp\left\{-\frac{E_{act,I}}{\mathcal{R}T}\right\}; \tag{2.449}$$

after having taken logarithms of both sides, Eq. (2.449) becomes

$$-\frac{E_{act,I}}{\mathcal{R}T} = \ln\frac{-\dfrac{dC_I}{dt}}{k_{0,I}C_I^{m_I}} \tag{2.450}$$

– where division of both sides by $E_{act,I}$ yields, in turn,

$$-\frac{1}{\mathcal{R}T} = \frac{1}{E_{act,I}}\ln\frac{-\dfrac{dC_I}{dt}}{k_{0,I}C_I^{m_I}}. \tag{2.451}$$

Equation (2.451) may now be combined with Eq. (2.446) to get

$$k_Q\{T\} = k_{0,Q}\exp\left\{\frac{E_{act,Q}}{E_{act,I}}\ln\frac{-\dfrac{dC_I}{dt}}{k_{0,I}C_I^{m_I}}\right\} \tag{2.452}$$

– which may, in turn, be used to transform Eq. (2.442) to

$$-\frac{dC_Q}{dt} = k_{0,Q}\exp\left\{\frac{E_{act,Q}}{E_{act,I}}\ln\frac{-\dfrac{dC_I}{dt}}{k_{0,I}C_I^{m_I}}\right\}C_Q^{m_Q}. \tag{2.453}$$

The exponential function in Eq. (2.453) can be isolated as

$$\frac{-\dfrac{dC_Q}{dt}}{k_{0,Q}C_Q^{m_Q}} = \exp\left\{\frac{E_{act,Q}}{E_{act,I}}\ln\frac{-\dfrac{dC_I}{dt}}{k_{0,I}C_I^{m_I}}\right\}, \tag{2.454}$$

where logarithms are to be taken off both sides to obtain

$$\ln\frac{-\dfrac{dC_Q}{dt}}{k_{0,Q}C_Q^{m_Q}} = \frac{E_{act,Q}}{E_{act,I}}\ln\frac{-\dfrac{dC_I}{dt}}{k_{0,I}C_I^{m_I}}; \tag{2.455}$$

the operational features of a logarithm support reformulation of Eq. (2.455) to

$$\ln\frac{-\dfrac{dC_Q}{dt}}{k_{0,Q}C_Q^{m_Q}} = \ln\left(\frac{-\dfrac{dC_I}{dt}}{k_{0,I}C_I^{m_I}}\right)^{\frac{E_{act,Q}}{E_{act,I}}}, \tag{2.456}$$

which simplifies to

$$\frac{-\dfrac{dC_Q}{dt}}{k_{0,Q}C_Q^{m_Q}} = \left(\frac{-\dfrac{dC_I}{dt}}{k_{0,I}C_I^{m_I}}\right)^{\frac{E_{act,Q}}{E_{act,I}}} \tag{2.457}$$

upon application of exponentials to both sides – or, after taking $E_{act,Q}$-th roots of both sides,

$$\boxed{\left(\frac{-\dfrac{dC_Q}{dt}}{k_{0,Q}C_Q^{m_Q}}\right)^{\frac{1}{E_{act,Q}}} = \left(\frac{-\dfrac{dC_I}{dt}}{k_{0,I}C_I^{m_I}}\right)^{\frac{1}{E_{act,I}}}.} \tag{2.458}$$

If $C_I \equiv C_I\{t\}$ is known, then dC_I/dt will promptly be obtained – such that

$$\boxed{-\frac{\frac{dC_I}{dt}}{k_{0,I}C_I^{m_I}} \equiv \psi_I\{t\}} \tag{2.459}$$

will serve as base for the power in the right-hand side of Eq. (2.458); insertion of Eq. (2.459) transforms Eq. (2.457) to

$$-\frac{\frac{dC_Q}{dt}}{k_{0,Q}C_Q^{m_Q}} = \psi_I^{\frac{E_{act,Q}}{E_{act,I}}}, \tag{2.460}$$

where integration ensues, via separation of variables, as

$$-\int_{C_{Q,0}}^{C_Q} \tilde{C}_Q^{-m_Q} d\tilde{C}_Q = k_{0,Q}\int_0^t \psi_I^{\frac{E_{act,Q}}{E_{act,I}}}\{\tilde{t}\}d\tilde{t} \tag{2.461}$$

with the aid of Eq. (2.444). Application of the fundamental theorem of integral calculus supports transformation of Eq. (2.461) to

$$-\left.\frac{\tilde{C}_Q^{1-m_Q}}{1-m_Q}\right|_{C_{Q,0}}^{C_Q} = k_{0,Q}I_I\{t\} \tag{2.462}$$

should $m_Q \neq 1$, where $I_I\{t\}$ satisfies

$$\boxed{I_I\{t\} \equiv \int_0^t \psi_I^{\frac{E_{act,Q}}{E_{act,I}}}\{\tilde{t}\}d\tilde{t};} \tag{2.463}$$

Eq. (2.462) degenerates to

$$\frac{C_{Q,0}^{1-m_Q} - C_Q^{1-m_Q}}{1-m_Q} = k_{0,Q}I_I\{t\} \tag{2.464}$$

– so $C_Q^{1-m_Q}$ may be isolated as

$$C_Q^{1-m_Q} = C_{Q,0}^{1-m_Q} - k_{0,Q}\left(1-m_Q\right)I_I\{t\}, \tag{2.465}$$

and both sides raised to $1/(1-m_Q)$ to finally get

$$\boxed{C_Q = \sqrt[1-m_Q]{C_{Q,0}^{1-m_Q} - k_{0,Q}\left(1-m_Q\right)I_I\{t\}}.} \tag{2.466}$$

In the particular case of $m_Q = 1$, one should instead resort to

$$-\int_{C_{Q,0}}^{C_Q} \frac{d\tilde{C}_Q}{\tilde{C}_Q} = k_{0,Q}\int_0^t \psi_I^{\frac{E_{act,Q}}{E_{act,I}}}\{\tilde{t}\}d\tilde{t} \tag{2.467}$$

in lieu of Eq. (2.461), which breaks down to

$$-\ln\tilde{C}_Q\Big|_{C_{Q,0}}^{C_Q} = k_{0,Q}I_I\{t\} \tag{2.468}$$

with $I_I\{t\}$ still abiding to Eq. (2.463); reformulation of Eq. (2.468) unfolds

$$\ln\frac{C_{Q,0}}{C_Q} = k_{0,Q}I_I\{t\}, \tag{2.469}$$

where exponentials are to be taken of both sides to finally obtain

$$\boxed{C_Q = C_{Q,0}e^{-k_{0,Q}I_I\{t\}}.} \tag{2.470}$$

For either Eq. (2.466) or Eq. (2.470), the pending problem lies on calculation of $I_I\{t\}$, as given by Eq. (2.463); after redoing Eq. (2.459) to

$$\psi_I\{t\} = \frac{1}{k_{0,I}}\frac{d}{dt}\left(-\frac{C_I^{1-m_I}}{1-m_I}\right) \tag{2.471}$$

on the hypothesis that $m_I \neq 1$, one will postulate existence of a series of the type

$$\boxed{-\frac{C_I^{1-m_I}\{t\}}{1-m_I} = a_0 + a_1t + a_2t^2 + a_3t^3 + \ldots} \tag{2.472}$$

– with coefficients a_0, a_1, a_2, a_3, … to be obtained (via some sort of fitting) in each specific case. By the same token, one should resort to

$$\psi_I\{t\} = \frac{1}{k_{0,I}}\frac{d}{dt}\left(-\ln C_I\right) \tag{2.473}$$

as alias to Eq. (2.459) if m_I were unity – in which case

$$\boxed{-\ln C_I\{t\} = a_0 + a_1t + a_2t^2 + a_3t^3 + \ldots} \tag{2.474}$$

should replace Eq. (2.472). Upon differentiation of either Eq. (2.472) or Eq. (2.474) with regard to t, one gets

$$-\left.\frac{d\ln C_I}{dt}\right|_{m_I=1} = -\left.\frac{d}{dt}\left(\frac{C_I^{1-m_I}}{1-m_I}\right)\right|_{m_I\neq 1} = a_1 + 2a_2t + 3a_3t^2 + \ldots, \tag{2.475}$$

thus yielding

$$\psi_I\{t\} = \frac{a_1 + 2a_2t + 3a_3t^2 + \ldots}{k_{0,I}} \tag{2.476}$$

from either Eq. (2.471) or Eq. (2.473); one may now use Eq. (2.476) to transform Eq. (2.463) to

$$I_I\{t\} = k_{0,I}^{-\frac{E_{act,Q}}{E_{act,I}}}\int_0^t \left(a_1 + 2a_2\tilde{t} + 3a_3\tilde{t}^2 + \ldots\right)^{\frac{E_{act,Q}}{E_{act,I}}} d\tilde{t}. \tag{2.477}$$

Taylor's expansion around zero can be invoked to reformulate the kernel in Eq. (2.477) as

$$\begin{aligned}
\left(a_1 + 2a_2\tilde{t} + 3a_3\tilde{t}^2 + \ldots\right)^{\frac{E_{act,Q}}{E_{act,I}}} &= \left.\left(a_1 + 2a_2\tilde{t} + 3a_3\tilde{t}^2 + \ldots\right)^{\frac{E_{act,Q}}{E_{act,I}}}\right|_{\tilde{t}=0} \\
&+ \frac{E_{act,Q}}{E_{act,I}}\left.\left(a_1 + 2a_2\tilde{t} + 3a_3\tilde{t}^2 + \ldots\right)^{\frac{E_{act,Q}}{E_{act,I}}-1}\left(2a_2 + 6a_3\tilde{t} + \ldots\right)\right|_{\tilde{t}=0}\tilde{t} \\
&+ \frac{E_{act,Q}}{E_{act,I}}\left(\frac{E_{act,Q}}{E_{act,I}} - 1\right)\left.\left(a_1 + 2a_2\tilde{t} + 3a_3\tilde{t}^2 + \ldots\right)^{\frac{E_{act,Q}}{E_{act,I}}-2}\left(2a_2 + 6a_3\tilde{t} + \ldots\right)^2\right|_{\tilde{t}=0}\frac{\tilde{t}^2}{2} \\
&+ \frac{E_{act,Q}}{E_{act,I}}\left.\left(a_1 + 2a_2\tilde{t} + 3a_3\tilde{t}^2 + \ldots\right)^{\frac{E_{act,Q}}{E_{act,I}}-1}\left(6a_3 + \ldots\right)\right|_{\tilde{t}=0}\frac{\tilde{t}^2}{2} + \ldots;
\end{aligned} \tag{2.478}$$

after setting $\tilde{t}=0$ as indicated, Eq. (2.478) reduces to

$$\left(a_1+2a_2\tilde{t}+3a_3\tilde{t}^2+\ldots\right)^{\frac{E_{act,Q}}{E_{act,I}}}=a_1^{\frac{E_{act,Q}}{E_{act,I}}}+2\frac{E_{act,Q}}{E_{act,I}}a_2a_1^{\frac{E_{act,Q}}{E_{act,I}}-1}\tilde{t}$$
$$+2\frac{E_{act,Q}}{E_{act,I}}\left(\frac{E_{act,Q}}{E_{act,I}}-1\right)a_2^2a_1^{\frac{E_{act,Q}}{E_{act,I}}-2}\tilde{t}^2+3\frac{E_{act,Q}}{E_{act,I}}a_3a_1^{\frac{E_{act,Q}}{E_{act,I}}-1}\tilde{t}^2+\ldots. \tag{2.479}$$

Integration of Eq. (2.479), between 0 and t, will then give rise to

$$\int_0^t\left(a_1+2a_2\tilde{t}+3a_3\tilde{t}^2+\ldots\right)^{\frac{E_{act,Q}}{E_{act,I}}}d\tilde{t}$$
$$=a_1^{\frac{E_{act,Q}}{E_{act,I}}}\tilde{t}\Big|_0^t+2\frac{E_{act,Q}}{E_{act,I}}a_2a_1^{\frac{E_{act,Q}}{E_{act,I}}-1}\frac{\tilde{t}^2}{2}\Bigg|_0^t$$
$$+2\frac{E_{act,Q}}{E_{act,I}}\left(\frac{E_{act,Q}}{E_{act,I}}-1\right)a_2^2a_1^{\frac{E_{act,Q}}{E_{act,I}}-2}\frac{\tilde{t}^3}{3}\Bigg|_0^t \tag{2.480}$$
$$+3\frac{E_{act,Q}}{E_{act,I}}a_3a_1^{\frac{E_{act,Q}}{E_{act,I}}-1}\frac{\tilde{t}^3}{3}\Bigg|_0^t+\ldots,$$

where replacement by upper and lower limits, followed by insertion in Eq. (2.477) generate

$$I_I\{t\}=k_{0,I}^{-\frac{E_{act,Q}}{E_{act,I}}}\left(\begin{array}{l}a_1^{\frac{E_{act,Q}}{E_{act,I}}}t+\frac{E_{act,Q}}{E_{act,I}}a_2a_1^{\frac{E_{act,Q}}{E_{act,I}}-1}t^2\\+\left(\begin{array}{l}\frac{2}{3}\frac{E_{act,Q}}{E_{act,I}}\left(\frac{E_{act,Q}}{E_{act,I}}-1\right)a_2^2a_1^{\frac{E_{act,Q}}{E_{act,I}}-2}\\+\frac{E_{act,Q}}{E_{act,I}}a_3a_1^{\frac{E_{act,Q}}{E_{act,I}}-1}\end{array}\right)t^3+\ldots\end{array}\right); \tag{2.481}$$

after factoring $a_1^{\frac{E_{act,Q}}{E_{act,I}}}$ and t out, Eq. (2.481) becomes

$$\boxed{I_I\{t\}=\left(\frac{a_1}{k_{0,I}}\right)^{\frac{E_{act,Q}}{E_{act,I}}}\left(\begin{array}{l}1+\frac{E_{act,Q}}{E_{act,I}}\frac{a_2}{a_1}t+\frac{E_{act,Q}}{E_{act,I}}\frac{a_3}{a_1}t^2\\+\frac{2}{3}\frac{E_{act,Q}}{E_{act,I}}\left(\frac{E_{act,Q}}{E_{act,I}}-1\right)\left(\frac{a_2}{a_1}\right)^2t^2+\ldots\end{array}\right)t} \tag{2.482}$$

– to be inserted in Eq. (2.466) or (2.470) as appropriate, and consequently produce the evolution in time of C_Q as quality indicator. This deed is feasible with the aid of a computer, but hardly practical when standing in front of a supermarket shelf – and about to decide whether to buy some food product under scrutiny.

Nevertheless, if $E_{act,Q}$ coincides with $E_{act,I}$, then Eq. (2.458) simplifies dramatically to

$$\frac{-\frac{dC_Q}{dt}}{k_{0,Q}C_Q^{m_Q}}=\frac{-\frac{dC_I}{dt}}{k_{0,I}C_I^{m_I}} \tag{2.483}$$

– once both sides are raised to $E_{act,Q}$ (or $E_{act,I}$, for that matter). Cancellation of dt between sides permits further simplification of Eq. (2.483) to

$$-\frac{dC_Q}{k_{0,Q}C_Q^{m_Q}}=-\frac{dC_I}{k_{0,I}C_I^{m_I}}, \tag{2.484}$$

where integration of both sides follows with the aid of Eqs. (2.444) and (2.445), i.e.

$$-\frac{1}{k_{0,Q}}\int_{C_{Q,0}}^{C_Q}\frac{d\tilde{C}_Q}{\tilde{C}_Q^{m_Q}}=-\frac{1}{k_{0,I}}\int_{C_{I,0}}^{C_I}\frac{d\tilde{C}_I}{\tilde{C}_I^{m_I}}; \tag{2.485}$$

the fundamental theorem of integral calculus, applied to Eq. (2.485), has it that

$$\frac{1}{k_{0,Q}\left(1-m_Q\right)}\left(C_{Q,0}^{1-m_Q}-C_Q^{1-m_Q}\right)=\frac{1}{k_{0,I}\left(1-m_I\right)}\left(C_{I,0}^{1-m_I}-C_I^{1-m_I}\right), \tag{2.486}$$

in much the same way the left-hand side of Eq. (2.464) was obtained from the left-hand side of Eq. (2.461) – valid for $m_Q\neq 1$ and $m_I\neq 1$;

$$\frac{1}{k_{0,Q}}\ln\frac{C_{Q,0}}{C_Q}=\frac{1}{k_{0,I}\left(1-m_I\right)}\left(C_{I,0}^{1-m_I}-C_I^{1-m_I}\right), \tag{2.487}$$

in parallel to the rationale that supported transformation of the left-hand side of Eq. (2.467) to the left-hand side of Eq. (2.469) – and valid for $m_Q=1$ and $m_I\neq 1$;

$$\frac{1}{k_{0,Q}\left(1-m_Q\right)}\left(C_{Q,0}^{1-m_Q}-C_Q^{1-m_Q}\right)=\frac{1}{k_{0,I}}\ln\frac{C_{I,0}}{C_I}, \tag{2.488}$$

valid for $m_Q\neq 1$ and $m_I=1$; or, finally,

$$\frac{1}{k_{0,Q}}\ln\frac{C_{Q,0}}{C_Q}=\frac{1}{k_{0,I}}\ln\frac{C_{I,0}}{C_I}, \tag{2.489}$$

valid for $m_Q=1$ and $m_I=1$. For simplicity, the indicator reaction is normally sought such that it is described by the same kinetic order as decay of the elected quality parameter – meaning that one ends up, in practice, with

$$\left(C_{Q,0}^{1-m}-C_Q^{1-m}\right)=\frac{k_{0,Q}}{k_{0,I}}\left(C_{I,0}^{1-m}-C_I^{1-m}\right), \tag{2.490}$$

based on Eq. (2.486) after setting $m_I=m_Q=m\neq 1$, where division of both sides by $C_{Q,0}^{1-m}$, complemented by multiplication and division of the right-hand side by $C_{I,0}^{1-m}$ give rise to

$$\boxed{\left(\left(\frac{C_Q}{C_{Q,0}}\right)^{1-m}=1+\frac{k_{0,Q}}{k_{0,I}}\left(\frac{C_{I,0}}{C_{Q,0}}\right)^{1-m}\left(\left(\frac{C_I}{C_{I,0}}\right)^{1-m}-1\right)\right)\Bigg|_{\substack{E_{act,I}=E_{act,Q}\\ m_I=m_Q=m\neq 1}}}; \tag{2.491}$$

or instead

$$\ln\frac{C_Q}{C_{Q,0}}=\frac{k_{0,Q}}{k_{0,I}}\ln\frac{C_{I,0}}{C_I}, \tag{2.492}$$

departing from Eq. (2.489) – where exponentials may be taken of both sides to yield

$$\exp\left\{\ln\frac{C_{Q,0}}{C_Q}\right\}=\exp\left\{\frac{k_{0,Q}}{k_{0,I}}\ln\frac{C_{I,0}}{C_I}\right\}=\exp\left\{\ln\left(\frac{C_{I,0}}{C_I}\right)^{\frac{k_{0,Q}}{k_{0,I}}}\right\} \tag{2.493}$$

together with the operational rules of logarithms, or merely

$$\boxed{\left.\left(\frac{C_Q}{C_{Q,0}}=\left(\frac{C_I}{C_{I,0}}\right)^{\frac{k_{0,Q}}{k_{0,I}}}\right)\right|_{\substack{E_{act,I}=E_{act,Q}\\ m_I=m_Q=m=1}}} \tag{2.494}$$

upon taking reciprocals of both sides, and because the logarithm is the inverse function of the corresponding exponential. When the indicator property is easily measured, and responds to log C_I (e.g. color of pH indicator, or absorbance of solution), then one obtains the plot labeled as Fig. 2.58 – valid for first-order decay of both Q and I. The mathematical basis of this figure is Eq. (2.494), after having taken decimal logarithms of both sides, i.e.

$$\log_{10}\frac{C_Q}{C_{Q,0}}=\frac{k_{0,Q}}{k_{0,I}}\log_{10}\frac{C_I}{C_{I,0}}; \tag{2.495}$$

this promptly justifies the slope being given by $k_{0,Q}/k_{0,I}$ – the constancy of which allows straightforward (linear) calibration of the time/temperature history indicator. All lines obviously intercept at (1,1) – corresponding to food at startup conditions, satisfying Eqs. (2.444) and (2.445); and evolve with elapsing time as indicated by the arrows. If a relationship of the type conveyed by Eq. (2.491) is anticipated, then one obtains the plot

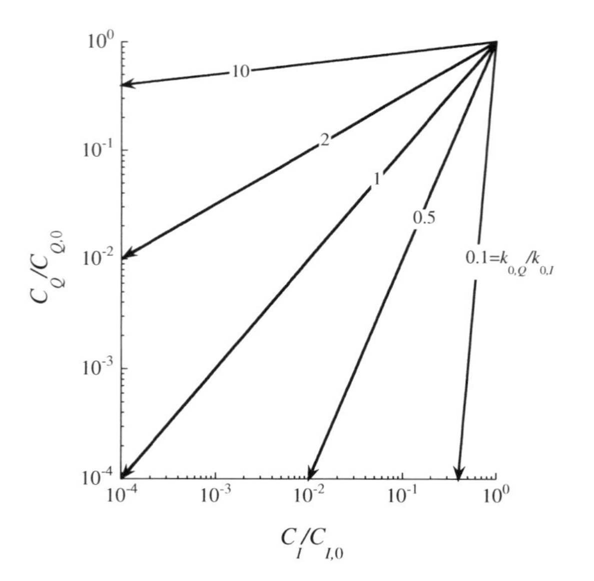

FIGURE 2.58 Variation of fractional decay in quality compound, $C_Q/C_{Q,0}$, versus fractional decay in indicator compound, $C_I/C_{I,0}$, for first-order response and selected values of dimensionless parameter $k_{0,Q}/k_{0,I}$.

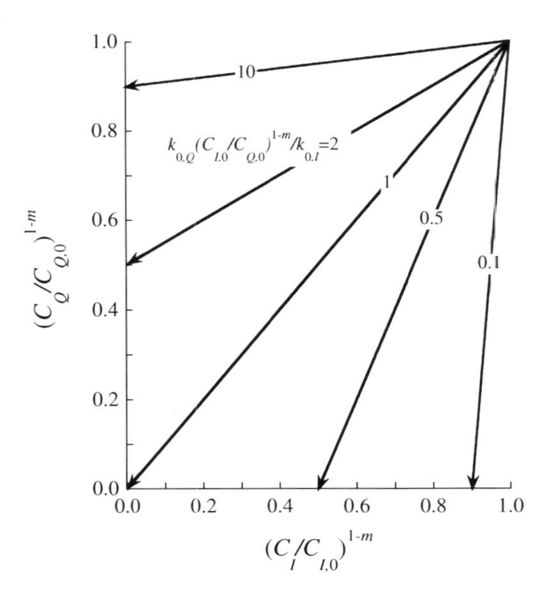

FIGURE 2.59 Variation of dimensionless decay in quality compound, $(C_Q/C_{Q,0})^{1-m}$, versus dimensionless decay in indicator compound, $(C_I/C_{I,0})^{1-m}$, for m-th order response and selected values of dimensionless parameter $k_{0,Q}(C_{I,0}/C_{Q,0})^{1-m}/k_{0,I}$.

labeled as Fig. 2.59. In this case, the characteristic parameter, $k_{0,Q}(C_{I,0}/C_{Q,0})^{1-m}/k_{0,I}$, depends not only on $k_{0,Q}/k_{0,I}$ as in Fig. 2.58, but also on the initial levels of both quality and indicator compounds; and requires *a priori* knowledge of the specific value of m. A linear pattern can be grasped in Fig. 2.58, similar to that prevailing in Fig. 2.59 – as long as the logarithmic scale is swapped with a power scale.

PROBLEM 2.10

Consider a food product, on top of which a full-history time-temperature indicator is affixed as part of its label – also containing the UPC-A barcode depicted in Fig. 2.53i.

a) Such a product, characterized by mass density ρ, isobaric specific heat capacity c_P, volume V, and exposed surface area A, is initially at temperature T_0, and will be subjected to chilling for a certain period of time – until reaching some target temperature below T_0. Most resistance to heat transfer is expected to lie on the film of (cold) air over the product – characterized by overall heat transfer coefficient U, under constant (uniform) bulk temperature T_∞. Estimate the evolution in time, t, of (uniform) food temperature, T.
b) If the indicator undergoes a reaction following first-order kinetics, with activation energy $E_{act,I}$ and pre-exponential factor $k_{0,I}$, find an approximate expression for the variation of its concentration with time – along the evolution in temperature found in a).
c) Should the quality descriptor of interest follow also first-order kinetics, with activation energy $E_{act,Q}$ and

pre-exponential factor $k_{0,Q}$, estimate the fraction of quality remaining as a function of fraction of intact indicator remaining – dependent on $E_{act,Q}/\mathcal{R}T_0$, $E_{act,I}/\mathcal{R}T_0$, T_∞/T_0, $UA/\rho c_P V k_{0,Q}$ and $UA/\rho c_P V k_{0,I}$ as dimensionless parameters.

d) Confirm that $R_6=4$ in the proposed label.

e) Prove that the final digit in an UPC-A barcode, intended for error checking, is in general able to detect all single-digit errors.

f) What is the expected percentage of transposition errors, between adjacent digits, remaining undetected by the check digit of an UPC-A bar code?

g) Repeat f), but for any type of transposition errors.

Solution

a) An enthalpy balance to the food item takes the form

$$\rho c_P V \frac{dT}{dt} = -UA\left(T - T_\infty\right), \tag{2.10.1}$$

consisting solely of accumulation term materialized by sensible heat, and outlet conductive term of heat transfer through the gaseous film – with no relevant temperature gradients expected inside the food matrix, since most thermal resistance is, by hypothesis, accounted for by the gaseous film thereon; Eq. (2.10.1) is to satisfy

$$T\big|_{t=0} = T_0, \tag{2.10.2}$$

as initial condition. Since $dT_\infty=0$ due to the constancy of T_∞, one may redo Eq. (2.10.1) to

$$\frac{d}{dt}\left(T - T_\infty\right) = -\frac{UA}{\rho c_P V}\left(T - T_\infty\right), \tag{2.10.3}$$

along with division of both sides by $\rho c_P V$. Integration of Eq. (2.10.3) may now proceed via separation of variables, viz.

$$\int_{T_0 - T_\infty}^{T - T_\infty} \frac{d\left(\tilde{T} - T_\infty\right)}{\tilde{T} - T_\infty} = -\frac{UA}{\rho c_P V}\int_0^t d\tilde{t}, \tag{2.10.4}$$

with the aid of Eq. (2.10.2); the fundamental theorem of integral calculus can be invoked to write

$$\ln\left\{\tilde{T} - T_\infty\right\}\Big|_{T_0 - T_\infty}^{T - T_\infty} = -\frac{UA}{\rho c_P V}\tilde{t}\Big|_0^t, \tag{2.10.5}$$

which degenerates to

$$\ln\frac{T - T_\infty}{T_0 - T_\infty} = -\frac{UA}{\rho c_P V}t. \tag{2.10.6}$$

After taking exponentials of both sides, Eq. (2.10.6) becomes

$$\frac{T - T_\infty}{T_0 - T_\infty} = \exp\left\{-\frac{UA}{\rho c_P V}t\right\}, \tag{2.10.7}$$

while isolation of T unfolds

$$T = T_\infty + \left(T_0 - T_\infty\right)\exp\left\{-\frac{UA}{\rho c_P V}t\right\}; \tag{2.10.8}$$

Eq. (2.10.8) accordingly describes the (exponential) decay in temperature with time that accompanies chilling of the food product at stake.

b) The first-order kinetics of transformation of species I supports

$$\frac{dC_I}{dt} = -k_I\{T\}C_I \tag{2.10.9}$$

as mass balance thereto – consistent with Eq. (2.443), and to be complemented by

$$C_I\big|_{t=0} = C_{I,0} \tag{2.10.10}$$

for initial condition in parallel to Eq. (2.445). The kinetic constant is, in turn, anticipated to vary with temperature as per Arrhenius' relationship, viz.

$$k_I\{T\} = k_{0,I}\exp\left\{-\frac{E_{act,I}}{\mathcal{R}T}\right\}, \tag{2.10.11}$$

in agreement with Eq. (2.447). In view of Eq. (2.10.8), one finds that

$$-\frac{E_{act,I}}{\mathcal{R}T} = -\frac{E_{act,I}}{\mathcal{R}\left(T_\infty + \left(T_0 - T_\infty\right)\exp\left\{-\frac{UA}{\rho c_P V}t\right\}\right)} \tag{2.10.12}$$

prevails in the situation under scrutiny – where Taylor's expansion of the right-hand side, about $t=0$ and truncated after the linear term, gives rise to

$$-\frac{E_{act,I}}{\mathcal{R}T} \approx -\frac{E_{act,I}}{\mathcal{R}\left(T_\infty + \left(T_0 - T_\infty\right)\exp\left\{-\frac{UA}{\rho c_P V}t\right\}\right)}\Bigg|_{t=0} + \frac{E_{act,I}\left(T_0 - T_\infty\right)\exp\left\{-\frac{UA}{\rho c_P V}t\right\}\left(-\frac{UA}{\rho c_P V}\right)}{\mathcal{R}\left(T_\infty + \left(T_0 - T_\infty\right)\exp\left\{-\frac{UA}{\rho c_P V}t\right\}\right)^2}\Bigg|_{t=0} t \tag{2.10.13}$$

– assumed to convey a reasonable approximation to the evolution of $-E_{act,I}/\mathcal{R}T$ within time interval $[0,t]$. Equation (2.10.13) simplifies to

$$-\frac{E_{act,I}}{\mathcal{R}T} = -\frac{E_{act,I}}{\mathcal{R}\left(T_\infty + T_0 - T_\infty\right)} - \frac{E_{act,I}\left(T_0 - T_\infty\right)\frac{UA}{\rho c_P V}}{\mathcal{R}\left(T_\infty + T_0 - T_\infty\right)^2}t \tag{2.10.14}$$

after setting $t=0$ (as indicated), and thus realizing that the pending exponential functions reduce to unity – or else

$$-\frac{E_{act,I}}{\mathcal{R}T}=-\frac{E_{act,I}}{\mathcal{R}T_0}-\frac{E_{act,I}}{\mathcal{R}T_0}\left(1-\frac{T_\infty}{T_0}\right)\frac{UA}{\rho c_P V}t, \quad (2.10.15)$$

upon cancellation of symmetrical terms and factoring in of T_0 afterward. Insertion of Eq. (2.10.15) converts Eq. (2.10.11) to

$$k_I\{T\}\approx k_{0,I}\exp\left\{-\frac{E_{act,I}}{\mathcal{R}T_0}-\frac{E_{act,I}}{\mathcal{R}T_0}\left(1-\frac{T_\infty}{T_0}\right)\frac{UA}{\rho c_P V}t\right\}, \quad (2.10.16)$$

which, in turn, condenses to

$$k_I\{T\}=\alpha\,\mathrm{e}^{-\beta t} \quad (2.10.17)$$

– provided that auxiliary parameters α and β are defined as

$$\alpha\equiv k_{0,I}\exp\left\{-\frac{E_{act,I}}{\mathcal{R}T_0}\right\} \quad (2.10.18)$$

and

$$\beta\equiv\frac{E_{act,I}}{\mathcal{R}T_0}\left(1-\frac{T_\infty}{T_0}\right)\frac{UA}{\rho c_P V}, \quad (2.10.19)$$

respectively. Combination of Eqs. (2.10.9) and (2.10.17) gives rise to

$$\frac{dC_I}{dt}\approx-\alpha\mathrm{e}^{-\beta t}C_I, \quad (2.10.20)$$

where integration via separation of variables takes place as

$$\int_{C_{I,0}}^{C_I}\frac{d\tilde{C}_I}{\tilde{C}_I}=-\alpha\int_0^t \mathrm{e}^{-\beta\tilde{t}}\,d\tilde{t} \quad (2.10.21)$$

– at the expense of Eq. (2.10.10); the fundamental theorem of integral calculus then supports transformation to

$$\ln\tilde{C}_I\Big|_{C_{I,0}}^{C_I}=-\alpha\left.\frac{\mathrm{e}^{-\beta\tilde{t}}}{-\beta}\right|_0^t, \quad (2.10.22)$$

which is equivalent to

$$\ln\frac{C_I}{C_{I,0}}=-\frac{\alpha}{\beta}\left(1-\mathrm{e}^{-\beta t}\right). \quad (2.10.23)$$

Equation (2.10.23) may be rewritten as

$$\ln C_I=\ln C_{I,0}-\frac{\alpha}{\beta}\left(1-\mathrm{e}^{-\beta t}\right), \quad (2.10.24)$$

while exponentials are to be taken of both sides to get

$$C_I=C_{I,0}\exp\left\{-\frac{\alpha}{\beta}\left(1-\mathrm{e}^{-\beta t}\right)\right\} \quad (2.10.25)$$

– which unfolds the evolution in time of indicator concentration.

c) Since $m_I=m_Q=1$ per hypothesis, Eq. (2.458) simplifies to

$$\left(\frac{-\frac{dC_Q}{dt}}{k_{0,Q}C_Q}\right)^{\frac{1}{E_{act,Q}}}=\left(\frac{-\frac{dC_I}{dt}}{k_{0,I}C_I}\right)^{\frac{1}{E_{act,I}}}; \quad (2.10.26)$$

the concentration in each denominator may then be lumped with its differential in the corresponding numerator as

$$\left(-\frac{1}{k_{0,Q}}\frac{d\ln C_Q}{dt}\right)^{\frac{1}{E_{act,Q}}}=\left(-\frac{1}{k_{0,I}}\frac{d\ln C_I}{dt}\right)^{\frac{1}{E_{act,I}}}, \quad (2.10.27)$$

or else

$$-\frac{1}{k_{0,Q}}\frac{d\ln C_Q}{dt}=\left(-\frac{1}{k_{0,I}}\frac{d\ln C_I}{dt}\right)^{\frac{E_{act,Q}}{E_{act,I}}} \quad (2.10.28)$$

after raising both sides to $E_{act,Q}$. On the other hand, differentiation of both sides with regard to t converts Eq. (2.10.24) to

$$\frac{d\ln C_I}{dt}=\frac{\alpha}{\beta}\mathrm{e}^{-\beta t}(-\beta)=-\alpha\,\mathrm{e}^{-\beta t} \quad (2.10.29)$$

– which can be used to transform Eq. (2.10.28) to

$$-\frac{1}{k_{0,Q}}\frac{d\ln C_Q}{dt}=\left(\frac{\alpha\,\mathrm{e}^{-\beta t}}{k_{0,I}}\right)^{\frac{E_{act,Q}}{E_{act,I}}}; \quad (2.10.30)$$

after setting

$$\ln C_Q\big|_{t=0}=\ln C_{Q,0} \quad (2.10.31)$$

as initial condition pertaining to (known) departing quality factor – in parallel to Eq. (2.444), one may proceed to integration of Eq. (2.10.30) as

$$-\frac{1}{k_{0,Q}}\int_{\ln C_{Q0}}^{\ln C_Q}d\ln\tilde{C}_Q=\left(\frac{\alpha}{k_{0,I}}\right)^{\frac{E_{act,Q}}{E_{act,I}}}\int_0^t \mathrm{e}^{-\frac{E_{act,Q}}{E_{act,I}}\beta\tilde{t}}\,d\tilde{t}. \quad (2.10.32)$$

The fundamental theorem of integral calculus supports conversion of Eq. (2.10.32) to

$$-\frac{1}{k_{0,Q}}\ln\tilde{C}_Q\Big|_{\ln C_{Q,0}}^{\ln C_Q}=\left(\frac{\alpha}{k_{0,I}}\right)^{\frac{E_{act,Q}}{E_{act,I}}}\left.\frac{\mathrm{e}^{-\frac{E_{act,Q}}{E_{act,I}}\beta\tilde{t}}}{-\frac{E_{act,Q}}{E_{act,I}}\beta}\right|_0^t, \quad (2.10.33)$$

which breaks down to

$$-\frac{1}{k_{0,Q}}\left(\ln C_Q-\ln C_{Q,0}\right)=\frac{1}{\beta}\frac{E_{act,I}}{E_{act,Q}}\left(\frac{\alpha}{k_{0,I}}\right)^{\frac{E_{act,Q}}{E_{act,I}}}\left(1-e^{-\frac{E_{act,Q}}{E_{act,I}}\beta t}\right); \tag{2.10.34}$$

after multiplying both sides by $-k_{0,Q}$, lumping the two logarithms, and redoing the outstanding exponential to a power, Eq. (2.10.34) becomes

$$\ln\frac{C_Q}{C_{Q,0}}=-\frac{k_{0,Q}}{\beta}\frac{E_{act,I}}{E_{act,Q}}\left(\frac{\alpha}{k_{0,I}}\right)^{\frac{E_{act,Q}}{E_{act,I}}}\left(1-\left(e^{-\beta t}\right)^{\frac{E_{act,Q}}{E_{act,I}}}\right), \tag{2.10.35}$$

which yields

$$\frac{C_Q}{C_{Q,0}}=\exp\left\{-\frac{k_{0,Q}}{\beta}\frac{E_{act,I}}{E_{act,Q}}\left(\frac{\alpha}{k_{0,I}}\right)^{\frac{E_{act,Q}}{E_{act,I}}}\left(1-\left(e^{-\beta t}\right)^{\frac{E_{act,Q}}{E_{act,I}}}\right)\right\} \tag{2.10.36}$$

once exponentials are taken of both sides. Equation (2.10.23) may now be retrieved, upon isolation of $e^{-\beta t}$, to convert Eq. (2.10.36) to

$$\frac{C_Q}{C_{Q,0}}=\exp\left\{-\frac{k_{0,Q}}{\beta}\frac{E_{act,I}}{E_{act,Q}}\left(\frac{\alpha}{k_{0,I}}\right)^{\frac{E_{act,Q}}{E_{act,I}}}\left(1-\left(1+\frac{\beta}{\alpha}\ln\frac{C_I}{C_{I,0}}\right)^{\frac{E_{act,Q}}{E_{act,I}}}\right)\right\}, \tag{2.10.37}$$

where insertion of Eqs. (2.10.18) and (2.10.19) permits further transformation to

$$\frac{C_Q}{C_{Q,0}}=\exp\left\{\begin{array}{l}-\dfrac{k_{0,Q}}{\dfrac{E_{act,I}}{\mathcal{R}T_0}\left(1-\dfrac{T_\infty}{T_0}\right)\dfrac{UA}{\rho c_P V}}\dfrac{E_{act,I}}{E_{act,Q}}\left(\dfrac{k_{0,I}\exp\left\{-\dfrac{E_{act,I}}{\mathcal{R}T_0}\right\}}{k_{0,I}}\right)^{\frac{E_{act,Q}}{E_{act,I}}}\\ \times\left(1-\left(1+\dfrac{\dfrac{E_{act,I}}{\mathcal{R}T_0}\left(1-\dfrac{T_\infty}{T_0}\right)\dfrac{UA}{\rho c_P V}}{k_{0,I}\exp\left\{-\dfrac{E_{act,I}}{\mathcal{R}T_0}\right\}}\ln\dfrac{C_I}{C_{I,0}}\right)^{\frac{E_{act,Q}}{E_{act,I}}}\right)\end{array}\right\} \tag{2.10.38}$$

– which simplifies to

$$\frac{C_Q}{C_{Q,0}}=\exp\left\{\begin{array}{l}-\dfrac{\exp\left\{-\dfrac{E_{act,Q}}{\mathcal{R}T_0}\right\}}{\dfrac{E_{act,Q}}{\mathcal{R}T_0}\left(1-\dfrac{T_\infty}{T_0}\right)\dfrac{UA}{\rho c_P V k_{0,Q}}}\\ \times\left(1-\left(1+\dfrac{\dfrac{E_{act,I}}{\mathcal{R}T_0}\left(1-\dfrac{T_\infty}{T_0}\right)\dfrac{UA}{\rho c_P V k_{0,I}}}{\exp\left\{-\dfrac{E_{act,I}}{\mathcal{R}T_0}\right\}}\ln\dfrac{C_I}{C_{I,0}}\right)^{\frac{E_{act,Q}}{E_{act,I}}}\right)\end{array}\right\} \tag{2.10.39}$$

upon canceling common factors between numerator and denominator, complemented by lumping powers and constants. Equation (2.10.39) can be reformulated, in condensed notation, to just

$$\frac{C_Q}{C_{Q,0}}=\exp\left\{-\Psi\left\{k_{0,Q},E_{act,Q}\right\}\left(1-\left(1+\frac{\ln\dfrac{C_I}{C_{I,0}}}{\Psi\left\{k_{0,I},E_{act,I}\right\}}\right)^{\frac{E_{act,Q}}{E_{act,I}}}\right)\right\}, \tag{2.10.40}$$

as long as auxiliary function Ψ is defined as

$$\Psi\left\{k_0,E_{act}\right\}\equiv\frac{\exp\left\{-\dfrac{E_{act}}{\mathcal{R}T_0}\right\}}{\dfrac{E_{act}}{\mathcal{R}T_0}\left(1-\dfrac{T_\infty}{T_0}\right)\dfrac{UA}{\rho c_P V k_0}}; \tag{2.10.41}$$

Ψ applies when $E_{act}=E_{act,Q}$ and $k_0=k_{0,Q}$, and likewise when $E_{act}=E_{act,I}$ and $k_0=k_{0,I}$. Equation (2.10.40) relates fraction of remaining quality, $C_Q/C_{Q,0}$, to fraction of remaining (intact) indicator, $C_I/C_{I,0}$ – via $E_{act,Q}/\mathcal{R}T_0$, $E_{act,I}/\mathcal{R}T_0$, T_∞/T_0, $UA/\rho c_P V k_{0,Q}$, and $UA/\rho c_P V k_{0,I}$ as germane dimensionless parameters – as originally intended, since $E_{act,Q}/E_{act,I}$ as exponent may still be expressed as the ratio of $E_{act,Q}/\mathcal{R}T_0$ to $E_{act,I}/\mathcal{R}T_0$.

d) Inspection of Fig. 2.53i indicates that $L_1=L_5=L_6=R_1=R_2=0$, $L_4=1$, $L_3=R_4=2$, $L_2=4$, $R_3=5$, and $R_5=6$; hence, Eq. (2.318) becomes

$$R_6=10-\text{mod}_{10}\left\{\begin{array}{l}3(0+2+0+0+5+6)\\+4+1+0+0+2\end{array}\right\}. \tag{2.10.42}$$

After performing the stated operations, Eq. (2.10.42) simplifies to

$$\begin{aligned}R_6&=10-\text{mod}_{10}\{3\cdot 13+7\}=10-\text{mod}_{10}\{46\}\\&=10-6=4\end{aligned} \tag{2.10.43}$$

– thus confirming 4 as last digit in Fig. 2.53i.

e) To facilitate algebraic handling hereafter, one will define an auxiliary summation S as

$$S \equiv 3\left(x_1 + x_3 + x_5 + x_7 + x_9 + x_{11}\right) + x_2 + x_4 + x_6 + x_8 + x_{10}, \tag{2.10.44}$$

where x_i denotes digit (i.e. 0, 1, 2, 3, 4, 5, 6, 7, 8, or 9) in i-th position (i = 1, 2, 3, 4, 5, 6, 7, 8, 9, 10, 11) of the barcode; under these circumstances, Eq. (2.318) may be rewritten as

$$x_{12} = 10 - \text{mod}_{10}\{S\}, \tag{2.10.45}$$

where x_{12} denotes check digit located last in the barcode (corresponding to R_6). All possibilities of a mistaken digit in the i-th (generic) position of the barcode, due to faulty reading/typing of said code, are considered below – and the associated changes in the summation defined by Eq. (2.10.44), i.e. ΔS, are tabulated for each case; when i is odd, one accordingly obtains

Original digit at odd i-th position	Mistaken digit at odd i-th position									
	0	1	2	3	4	5	6	7	8	9
0		+3	+6	+9	+12	+15	+18	+21	+24	+27
1	−3		+3	+6	+9	+12	+15	+18	+21	+24
2	−6	−3		+3	+6	+9	+12	+15	+18	+21
3	−9	−6	−3		+3	+6	+9	+12	+15	+18
4	−12	−9	−6	−3		+3	+6	+9	+12	+15
5	−15	−12	−9	−6	−3		+3	+6	+9	+12
6	−18	−15	−12	−9	−6	−3		+3	+6	+9
7	−21	−18	−15	−12	−9	−6	−3		+3	+6
8	−24	−21	−18	−15	−12	−9	−6	−3		+3
9	−27	−24	−21	−18	−15	−12	−9	−6	−3	

whereas an even i leads to

Original digit at even i-th position	Mistaken digit at even i-th position									
	0	1	2	3	4	5	6	7	8	9
0		+1	+2	+3	+4	+5	+6	+7	+8	+9
1	−1		+1	+2	+3	+4	+5	+6	+7	+8
2	−2	−1		+1	+2	+3	+4	+5	+6	+7
3	−3	−2	−1		+1	+2	+3	+4	+5	+6
4	−4	−3	−2	−1		+1	+2	+3	+4	+5
5	−5	−4	−3	−2	−1		+1	+2	+3	+4
6	−6	−5	−4	−3	−2	−1		+1	+2	+3
7	−7	−6	−5	−4	−3	−2	−1		+1	+2
8	−8	−7	−6	−5	−4	−3	−2	−1		+1
9	−9	−8	−7	−6	−5	−4	−3	−2	−1	

The above values found for ΔS produce, in turn, a change in the corresponding modulo 10, i.e. $\text{mod}_{10}\{S+\Delta S\}$ (left), and in the digit placed at x_{12} to a new digit, ξ_{12} (right, in bold), abiding to

$$\xi_{12} = 10 - \text{mod}_{10}\{S + \Delta S\}; \tag{2.10.46}$$

these are listed in the next tables – the first covering the range from −9 to+9 via a stepwise unit increase, i.e.

	x_{12}									
ΔS	**0**	**1**	**2**	**3**	**4**	**5**	**6**	**7**	**8**	**9**
−9	1 **9**	0 **0**	9 **1**	8 **2**	7 **3**	6 **4**	5 **5**	4 **6**	3 **7**	2 **8**
−8	2 **8**	1 **9**	0 **0**	9 **1**	8 **2**	7 **3**	6 **4**	5 **5**	4 **6**	3 **7**
−7	3 **7**	2 **8**	1 **9**	0 **0**	9 **1**	8 **2**	7 **3**	6 **4**	5 **5**	4 **6**
−6	4 **6**	3 **7**	2 **8**	1 **9**	0 **0**	9 **1**	8 **2**	7 **3**	6 **4**	5 **5**
−5	5 **5**	4 **6**	3 **7**	2 **8**	1 **9**	0 **0**	9 **1**	8 **2**	7 **3**	6 **4**
−4	6 **4**	5 **5**	4 **6**	3 **7**	2 **8**	1 **9**	0 **0**	9 **1**	8 **2**	7 **3**
−3	7 **3**	6 **4**	5 **5**	4 **6**	3 **7**	2 **8**	1 **9**	0 **0**	9 **1**	8 **2**
−2	8 **2**	7 **3**	6 **4**	5 **5**	4 **6**	3 **7**	2 **8**	1 **9**	0 **0**	9 **1**
−1	9 **1**	8 **2**	7 **3**	6 **4**	5 **5**	4 **6**	3 **7**	2 **8**	1 **9**	0 **0**
+1	1 **9**	0 **0**	9 **1**	8 **2**	7 **3**	6 **4**	5 **5**	4 **6**	3 **7**	2 **8**
+2	2 **8**	1 **9**	0 **0**	9 **1**	8 **2**	7 **3**	6 **4**	5 **5**	4 **6**	3 **7**
+3	3 **7**	2 **8**	1 **9**	0 **0**	9 **1**	8 **2**	7 **3**	6 **4**	5 **5**	4 **6**
+4	4 **6**	3 **7**	2 **8**	1 **9**	0 **0**	9 **1**	8 **2**	7 **3**	6 **4**	5 **5**
+5	5 **5**	4 **6**	3 **7**	2 **8**	1 **9**	0 **0**	9 **1**	8 **2**	7 **3**	6 **4**
+6	6 **4**	5 **5**	4 **6**	3 **7**	2 **8**	1 **9**	0 **0**	9 **1**	8 **2**	7 **3**
+7	7 **3**	6 **4**	5 **5**	4 **6**	3 **7**	2 **8**	1 **9**	0 **0**	9 **1**	8 **2**
+8	8 **2**	7 **3**	6 **4**	5 **5**	4 **6**	3 **7**	2 **8**	1 **9**	0 **0**	9 **1**
+9	9 **1**	8 **2**	7 **3**	6 **4**	5 **5**	4 **6**	3 **7**	2 **8**	1 **9**	0 **0**

and the second covering the range from −27 to + 27 via jumps by multiples of 3, viz.

	x_{12}									
ΔS	**0**	**1**	**2**	**3**	**4**	**5**	**6**	**7**	**8**	**9**
−27	3 **7**	2 **8**	1 **9**	0 **0**	9 **1**	8 **2**	7 **3**	6 **4**	5 **5**	4 **6**
−24	6 **4**	5 **5**	4 **6**	3 **7**	2 **8**	1 **9**	0 **0**	9 **1**	8 **2**	7 **3**
−21	9 **1**	8 **2**	7 **3**	6 **4**	5 **5**	4 **6**	3 **7**	2 **8**	1 **9**	0 **0**
−18	2 **8**	1 **9**	0 **0**	9 **1**	8 **2**	7 **3**	6 **4**	5 **5**	4 **6**	3 **7**
−15	5 **5**	4 **6**	3 **7**	2 **8**	1 **9**	0 **0**	9 **1**	8 **2**	7 **3**	6 **4**
−12	8 **2**	7 **3**	6 **4**	5 **5**	4 **6**	3 **7**	2 **8**	1 **9**	0 **0**	9 **1**
−9	1 **9**	0 **0**	9 **1**	8 **2**	7 **3**	6 **4**	5 **5**	4 **6**	3 **7**	2 **8**
−6	4 **6**	3 **7**	2 **8**	1 **9**	0 **0**	9 **1**	8 **2**	7 **3**	6 **4**	5 **5**
−3	7 **3**	6 **4**	5 **5**	4 **6**	3 **7**	2 **8**	1 **9**	0 **0**	9 **1**	8 **2**
+3	3 **7**	2 **8**	1 **9**	0 **0**	9 **1**	8 **2**	7 **3**	6 **4**	5 **5**	4 **6**
+6	6 **4**	5 **5**	4 **6**	3 **7**	2 **8**	1 **9**	0 **0**	9 **1**	8 **2**	7 **3**
+9	9 **1**	8 **2**	7 **3**	6 **4**	5 **5**	4 **6**	3 **7**	2 **8**	1 **9**	0 **0**
+12	2 **8**	1 **9**	0 **0**	9 **1**	8 **2**	7 **3**	6 **4**	5 **5**	4 **6**	3 **7**
+15	5 **5**	4 **6**	3 **7**	2 **8**	1 **9**	0 **0**	9 **1**	8 **2**	7 **3**	6 **4**
+18	8 **2**	7 **3**	6 **4**	5 **5**	4 **6**	3 **7**	2 **8**	1 **9**	0 **0**	9 **1**
+21	1 **9**	0 **0**	9 **1**	8 **2**	7 **3**	6 **4**	5 **5**	4 **6**	3 **7**	2 **8**
+24	4 **6**	3 **7**	2 **8**	1 **9**	0 **0**	9 **1**	8 **2**	7 **3**	6 **4**	5 **5**
+27	7 **3**	6 **4**	5 **5**	4 **6**	3 **7**	2 **8**	1 **9**	0 **0**	9 **1**	8 **2**

Inspection of the digits in bold in each column (ξ_{12}), in both tables above, indicates no coincidence whatsoever between ξ_{12} and x_{12} – so the last digit of UPC-A will indeed detect all single-digit errors.

f) For the case of transposition errors between adjacent digits, consider that the digit in some position differs by δ from the next digit, i.e.

$$x_{2i} = x_{2i-1} + \delta; \quad i = 1,2,3,4,5 \tag{2.10.47}$$

when a digit in an odd position is exchanged with the next digit, or

$$x_{2i+1} = x_{2i} + \delta; \quad i = 1,2,3,4,5 \tag{2.10.48}$$

when a digit in an even position is exchanged with the next digit; here $\delta = -9, -8, -7, -6, -5, -4, -3, -2, -1, +1, +2, +3, +4, +5, +6, +7, +8, +9$ for occurrence of an error. When two consecutive digits are swapped, the change in S will thus look like

$$\Delta S = \left(x_{2i-1} + 3x_{2i}\right) - \left(3x_{2i-1} + x_{2i}\right) \tag{2.10.49}$$

for swapping digits between an odd and its next position, or

$$\Delta S = \left(3x_{2i} + x_{2i+1}\right) - \left(x_{2i} + 3x_{2i+1}\right) \tag{2.10.50}$$

for swapping digits between an even and its next position – with both cases being consistent with Eq. (2.10.44). Insertion of Eq. (2.10.47) transforms Eq. (2.10.49) to

$$\Delta S = x_{2i-1} + 3\left(x_{2i-1} + \delta\right) - 3x_{2i-1} - \left(x_{2i-1} + \delta\right), \tag{2.10.51}$$

which readily degenerates to

$$\Delta S = x_{2i-1} + 3x_{2i-1} + 3\delta - 3x_{2i-1} - x_{2i-1} - \delta = 2\delta \tag{2.10.52}$$

upon elimination of parentheses followed by lumping of terms alike; by the same token, combination of Eqs. (2.10.48) and (2.10.50) gives rise to

$$\Delta S = 3x_{2i} + \left(x_{2i} + \delta\right) - x_{2i} - 3\left(x_{2i} + \delta\right), \tag{2.10.53}$$

which breaks down to

$$\Delta S = 3x_{2i} + x_{2i} + \delta - x_{2i} - 3x_{2i} - 3\delta = -2\delta \tag{2.10.54}$$

following removal of parentheses and condensation of similar terms. All possible values of ΔS associated with Eqs. (2.10.52) and (2.10.54) are depicted in the second column, and $\mathrm{mod}_{10}\{S + \Delta S\}$ for each value of x_{12} appears as entry to the first column of each set of two columns in the table below; whereas the bold figure in the second column of each set represents the corresponding ξ_{12}.

		x_{12}																			
δ	ΔS	0		1		2		3		4		5		6		7		8		9	
−9	−18	2	**8**	1	**9**	0	**0**	9	**1**	8	**2**	7	**3**	6	**4**	5	**5**	4	**6**	3	**7**
−8	−16	4	**6**	3	**7**	2	**8**	1	**9**	0	**0**	9	**1**	8	**2**	7	**3**	6	**4**	5	**5**
−7	−14	6	**4**	5	**5**	4	**6**	3	**7**	2	**8**	1	**9**	0	**0**	9	**1**	8	**2**	7	**3**
−6	−12	8	**2**	7	**3**	6	**4**	5	**5**	4	**6**	3	**7**	2	**8**	1	**9**	0	**0**	9	**1**
−5	−10	0	**0**	9	**1**	8	**2**	7	**3**	6	**4**	5	**5**	4	**6**	3	**7**	2	**8**	1	**9**
−4	−8	2	**8**	1	**9**	0	**0**	9	**1**	8	**2**	7	**3**	6	**4**	5	**5**	4	**6**	3	**7**
−3	−6	4	**6**	3	**7**	2	**8**	1	**9**	0	**0**	9	**1**	8	**2**	7	**3**	6	**4**	5	**5**
−2	−4	6	**4**	5	**5**	4	**6**	3	**7**	2	**8**	1	**9**	0	**0**	9	**1**	8	**2**	7	**3**
−1	−2	8	**2**	7	**3**	6	**4**	5	**5**	4	**6**	3	**7**	2	**8**	1	**9**	0	**0**	9	**1**
+1	+2	2	**8**	1	**9**	0	**0**	9	**1**	8	**2**	7	**3**	6	**4**	5	**5**	4	**6**	3	**7**
+2	+4	4	**6**	3	**7**	2	**8**	1	**9**	0	**0**	9	**1**	8	**2**	7	**3**	6	**4**	5	**5**
+3	+6	6	**4**	5	**5**	4	**6**	3	**7**	2	**8**	1	**9**	0	**0**	9	**1**	8	**2**	7	**3**
+4	+8	8	**2**	7	**3**	6	**4**	5	**5**	4	**6**	3	**7**	2	**8**	1	**9**	0	**0**	9	**1**
+5	+10	0	**0**	9	**1**	8	**2**	7	**3**	6	**4**	5	**5**	4	**6**	3	**7**	2	**8**	1	**9**
+6	+12	2	**8**	1	**9**	0	**0**	9	**1**	8	**2**	7	**3**	6	**4**	5	**5**	4	**6**	3	**7**
+7	+14	4	**6**	3	**7**	2	**8**	1	**9**	0	**0**	9	**1**	8	**2**	7	**3**	6	**4**	5	**5**
+8	+16	6	**4**	5	**5**	4	**6**	3	**7**	2	**8**	1	**9**	0	**0**	9	**1**	8	**2**	7	**3**
+9	+18	8	**2**	7	**3**	6	**4**	5	**5**	4	**6**	3	**7**	2	**8**	1	**9**	0	**0**	9	**1**

Note that the second column entails all possibilities for ΔS as twice the actual value of δ appearing in the first column, even though each δ may produce two distinct values for ΔS – one being the negative of the other. Since both negative and positive values are considered for δ, the corresponding values of ΔS given by Eq. (2.10.52) suffice – as they will coincide with the corresponding values of ΔS given by Eq. (2.10.54). Inspection of the remainder of the table above indicates that adjacent digits differing by 5 units (up or down) cannot be detected by the (final) check digit – since ξ_{12} (marked in engraved bold) coincides with x_{12} (used as heading for the corresponding column). This entails $2 \times 10 = 20$ possibilities, out of $18 \times 10 = 180$ total possibilities of adjacent distinct digits – thus yielding an (average) error of 11.1%; and unfolding a relatively low probability for such a type of frequent error going undetected.

g) The analysis presented in f) may somehow be simplified, because the behavior of positive values for ΔS mimics that of negative differences – although in reverse order. Therefore, it suffices to only consider positive deviations $\delta = 1, 2, 3, 4, 5, 6, 7, 8, 9$; and realize that the only faulty case of distinct digits, swapped between adjacent positions, corresponds to $\delta = 5$ – i.e. 1 case out of 9 total cases, with a probability, e, given by

$$e = \frac{1}{9} = 11.1\%, \tag{2.10.55}$$

as found before. If erroneous swapping of distinct digits occurs between any two positions, then two possibilities are to be analyzed: either the digit in an even position is swapped with the digit in an odd position – with ΔS given by

$$\Delta S\big|_{odd \leftrightarrow even} = 2\delta, \tag{2.10.56}$$

using Eq. (2.10.52) as a template; or the digit in an even position is swapped with the digit in another even position, or the digit in an odd position is swapped with the digit in another odd position – thus leading to

$$\Delta S\big|_{even \leftrightarrow even} = 0 \tag{2.10.57}$$

and

$$\Delta S\big|_{odd \leftrightarrow odd} = 0, \tag{2.10.58}$$

respectively. The total number of possible transpositions, N, is given by

$$N = \sum_{i=1}^{10} N_{x_i}, \tag{2.10.59}$$

where N_{x_i} denotes number of transpositions between the i-th position and every position afterward; since the order of transposition is immaterial, backward transposition from position x_j should not be considered, as it will be already found for some $x_{k>j}$. Hence, Eq. (2.10.59) may be redone to

$$N=\sum_{i=1}^{10}(11-i)=\sum_{i=1}^{10}11-\sum_{i=1}^{10}i, \qquad (2.10.60)$$

along with splitting of the summation – which applies to all eleven positions, each one occupied by one out of ten distinct digits; definition of multiplication and sum of the first terms of an arithmetic progression support

$$N=10\cdot 11-\frac{10(10+1)}{2}=110-\frac{110}{2}=55. \qquad (2.10.61)$$

Using the same rationale, the number of transpositions between alike positions, $N_{=}$, is given by

$$\begin{aligned}N_{=} &= N_{odd\leftrightarrow odd,1}+N_{even\leftrightarrow even,2}+N_{odd\leftrightarrow odd,3}\\ &+N_{even\leftrightarrow even,4}+N_{odd\leftrightarrow odd,5}+N_{even\leftrightarrow even,6}\\ &+N_{odd\leftrightarrow odd,7}+N_{even\leftrightarrow even,8}+N_{odd\leftrightarrow odd,9},\end{aligned} \qquad (2.10.62)$$

which corresponds to

$$N_{=}=5+4+4+3+3+2+2+1+1=25 \qquad (2.10.63)$$

– with five positions (i.e. 3, 5, 7, 9, and 11) available after position 1 as $N_{odd\leftrightarrow odd,1}$, four positions (i.e. 4, 6, 8, and 10) available after position 2 as $N_{even\leftrightarrow even,2}$, four positions (i.e. 5, 7, 9, and 11) available after position 3 as $N_{odd\leftrightarrow odd,3}$, and so on. The remainder, $N_{\neq}$, corresponds to changes between an odd and an even position, i.e.

$$\begin{aligned}N_{\neq} &= N_{odd\leftrightarrow even,1}+N_{even\leftrightarrow odd,2}+N_{odd\leftrightarrow even,3}\\ &+N_{even\leftrightarrow odd,4}+N_{odd\leftrightarrow even,5}+N_{even\leftrightarrow odd,6}\\ &+N_{odd\leftrightarrow even,7}+N_{even\leftrightarrow odd,8}+N_{odd\leftrightarrow even,9}\\ &+N_{even\leftrightarrow odd,10},\end{aligned} \qquad (2.10.64)$$

accounted for by $N_{odd\leftrightarrow even,1}=5$ as per five positions available (i.e. 2, 4, 6, 8, and 10) after position 1, $N_{even\leftrightarrow odd,2}=5$ as five position available (i.e. 3, 5, 7, 9, and 11) after position 2, $N_{odd\leftrightarrow even,3}=4$ as per four positions available (i.e. 4, 6, 8, 10) after position 3, and so on; all cases taken together accordingly lead to

$$N_{\neq}=5+5+4+4+3+3+2+2+1+1=30, \qquad (2.10.65)$$

with

$$N=N_{=}+N_{\neq} \qquad (2.10.66)$$

consistent with Eqs. (2.10.61), (2.10.63), and (2.10.65). Since there are 10 possible digits, the number of distinctive swaps, n_s, reads

$$n_s=10(10-1)=90; \qquad (2.10.67)$$

therefore, the total number of swaps undetected by the check digit, E, should be given by

$$E=\frac{N_{=}n_S+N_{\neq}en_s}{Nn_s}=\frac{N_{=}+N_{\neq}e}{N}, \qquad (2.10.68)$$

being consequently independent of n_s – as it drops off both numerator and denominator; insertion of Eqs. (2.10.55), (2.10.61), (2.10.63), and (2.10.65) transforms Eq. (2.10.68) finally to

$$E=\frac{25+30\frac{1}{9}}{55}=51.5\%. \qquad (2.10.69)$$

The large possibility that this type of error goes undetected, via the check digit, is to be emphasized; however, this is still acceptable in practice, because the intrinsic probability of occurrence of the underlying reading/typing mistake is much lower than swapping two adjacent digits.

3

Industrial Utilities

Despite the unit operation approach followed for the relevant physicochemical transformations undergone by food (effected as part of processing or handling/storage), strong interaction exists between them in actual plants; they are aimed at saving energy, as well as minimizing wastes and effluents, both with a positive impact upon global environment – besides improving economic feasibility of the process as a whole. On the other hand, industrial operation *per se* is rationally based on existence of economic gains following scale-up, relative to mere craftsmanship or local, small-scale manufacture. In fact, cost of equipment typically correlates with its size/capacity via a power relationship – with characteristic exponent between 0 and 1, usually around 0.6. Together with quantity discounts due on purchase of raw materials and fuels, the aforementioned concave (power) relationship makes processing costs per unit amount of final product decrease consistently with increasing throughput rate and equipment/plant size – and also justify mergers that merely seek an extra economic gain. This is why a plant usually possesses a single, large, and dedicated facility to produce each of a number of utilities – rather than having many small units scattered all over the premises, closer to where they will ultimately be utilized. The aforementioned strategy proves more cost-effective in that larger quantities are obtained at prefixed costs; the utilities are then to be delivered to wherever needed in the plant, while also facilitating recycle and reuse thereof in view of the underlying integrated approach.

The most common utilities include fuel combustion for production of steam – to be used directly for heating purposes; or to be fed to thermal engines – where enthalpy of steam is, in turn, (partially) converted to mechanical power via Rankine's cycle. The latter process cannot proceed to completion, due to restrictions on overall entropy variation imposed by the second law of thermodynamics; the theoretical (or Carnot's) efficiency sets indeed an upper limit for performance of thermal engines. Conversely, the first law of thermodynamics supports full conversion of work to heat – unfortunately too expensive a strategy for mere heating puposes. The second law of thermodynamics also forbids spontaneous flow of heat from a lower to a higher temperature – since work has to be applied to overcome the unfavorable Gibbs' energy variation accompanying the said process. This calls for supply of external work to produce cold and cryogenic fluids – which normally takes place via (single) vapor compression cycles, sequentially applied in the case of gas liquefaction. In alternative, a purely thermal process may be utilized to remove heat below ambient temperature, thus producing cold through a vapor absorption cycle.

When local production of mechanical power is necessary, it is normally more convenient to resort to electric motors – since they are quite versatile in size, performance, and control, and electric current may be easily transported everywhere. To minimize heat losses, such a transport should be carried out at high voltage and low current intensity, for a given electrical power; conversely, applications resort normally to low-voltage mains for safety reasons. Matching between those two types of circuits is possible via utilization of transformers. Since Joule's effect is ubiquitously present, one should anticipate thermal outcomes of electric currents flowing in cables; their mode of dissipation hinges critically upon thermal resistance thereof. Hence, electric cables tend to be used as such, without outer protective insulation – which calls for aerial layouts, and thus (periodically) spaced supports at a height. The issue of mechanical resistance then arises, for a matter of user safety – especially when their weight per unit length is not negligible. On the other hand, a given (and the same) electric network is often to be simultaneously accessible to a variety of resistors – so the pattern followed to connect them to the mains should be carefully designed, in attempts to optimize use of the power made available through electric grids.

Irrespective of whether steam-based work or electrical work is at stake, knowledge of its magnitude is of the utmost importance to design equipment and enhance operating conditions. This may be done by attaching a suitable meter to the shaft of an engine, and measuring how much work is performed thereby – as product of torque (measured with a dynamometer) by rotational speed (measured with a frequency meter). Such a black-box approach does not obviously require knowledge of what happens to a putative working fluid inside the engine casing – as only interaction of such a system with its surroundings is considered. It is, nevertheless, mechanistically more informative to ascertain the said work based on behavior of the fluid itself, and its evolution between states inside the engine – which calls for arguments stemming from fluid mechanics. This fundamental approach will consequently be pursued here – because it allows a physical meaning be gained, and entitles some degree of extrapolation.

Finally, a need exists for process control – from sensing a system state, through automatically taking a decision in terms of response to a change relative to a given set point, to eventual manipulation of an actuator (usually a valve), pertaining to either an individual apparatus or an integrated subsystem in a plant. Although water or oil can be used as transmission fluid, one often resorts to compressed air for this purpose – since it is readily available as raw material, and may be safely released in need. This utility is also required whenever drying processes or carbonation of drinks are at stake – in which case large (or batteries of) compressors, coupled to storage vessels to avoid pressure fluctuations should be equated.

3.1 Introduction

The energy requirements for food processing are met in a variety of ways; traditional energy sources are indeed utilized at large

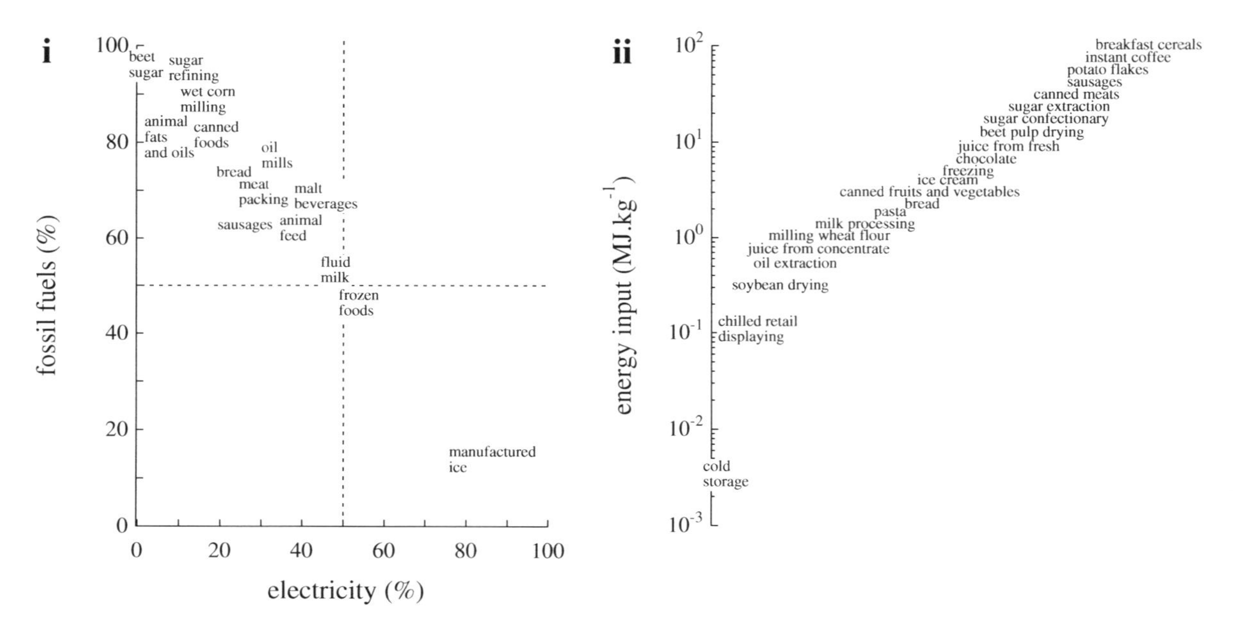

FIGURE 3.1 Energy requirement/utilization by some of the most energy-intensive industrial food processes, expressed as (i) approximate percents of total energy utilized as fossil fuel burning and electricity, and (ii) average amount of energy utilized per unit mass of selected foods.

in food processing plants, to generate steam and provide other utilities. Brief information on the most energy-demanding food industries is presented in Fig. 3.1i; and on the energy used per unit amount of final product in Fig. 3.1ii. Fossil fuels include natural gas, petroleum products, and coal; most food industries rely heavily thereon – with the major exception of manufactured ice, due to the dominant work-intensive freezing component. With regard to individual product requirements, breakfast cereals dominate the energy input – owing to extensive costs of extrusion-mediated pumping/cooking/drying; while cold storage lies at the bottom, in the case of well-insulated networks.

The (chemical) energy associated with fossil fuels will not become available unless burning takes place. A burner is thus the primary component of a system capable of extracting energy from natural gas or petroleum products; it is designed to admit fuel and air into the combustion chamber, so as to maximize efficiency of heat generation, and concomitantly produce hot gases as required to generate steam from liquid water.

HISTORICAL FRAMEWORK

Archeologists found evidence of surface mining and household usage of coal in China, dating back to 3490 BC; small mining operations spread throughout Europe much later, during the Middle Ages – to supply forges, smithies, and breweries. The growing popularity of coal soon exhausted it on the surface – so coal miners began digging, and eventually created coal mines; these would frequently fill with water as they were dug out – which provided the impetus for Thomas Newcomen create a steam engine to pump it out. During the first half of the 1800s, the Industrial Revolution spread to the United States, so the use of coal boomed; it eventually replaced low-energy firewood as leading source of energy in steam production.

According to Herodotus, asphalt obtained from oil pits near Ardericca was used in the construction of walls and towers in Babylon. A pitch spring apparently existed in the Ionian Islands in Greece, and great quantities were found on the banks of river Issus (one of the tributaries to river Euphrates). Although utilization of petroleum in raw state dates back to ancient China – namely, in I Ching city, its application as fuel did not become a routine practice until the 4th century. The earliest known oil wells were drilled also in China in 347 BC, using bits attached to bamboo poles; oil was then burned to evaporate brine and obtain solid salt. By the 10th century, extensive bamboo pipelines connected oil wells with salt springs. Distillation of petroleum was pioneered by Persian alchemist Muhammad ibn Zakariya Razi; distillation was then disseminated in Western Europe through Islamic Spain. Edwin Drake drilled the first rock oil well in Pennsylvania in 1859, which constituted the dawn of the petroleum age; its first major impact was replacement of whale oil in lighting. Development of drilling technology in the mid-1800s permitted mass-exploration and consumption of petroleum as a fuel – which surpassed coal by the mid-1950s.

The famous Oracle at Delphi, on Mount Parnassus in ancient Greece, was built where natural gas seeped from the ground in a flame – ca. 1,000 BC. Five centuries later, the Chinese used crude bamboo pipelines to transport natural gas that also seeped to the surface, and used it to boil seawater into drinkable water. In 1626, French explorers reported that natives ignited gases that were seeping into and around Lake Erie. However, natural gas was commercialized much later, starting in 1785 in Britain to light houses; and the streets of Baltimore, MA were lighted with natural gas from 1816 on. In 1821, William Hart dug the first successful natural gas well in Fredonia, NY – and eventually established Fredonia Gas Light Co., the pioneer company in natural gas distribution.

Nowadays, industry at large is quite dependent on fossil fuels; between 1967 and 2007, the worldwide consumption increased from 3.8 to 11.1 billion tons of oil-equivalent – and the rate of growth has not decreased ever since.

A heat engine is a cyclically operated device, which exchanges only heat and work with the surroundings; the cycle is closed in thermodynamic terms – meaning that no material is gained or lost overall. The simplest form of heat engine is a pair of thermocouples, accounting for junctions J_h and J_l – with heat flown into J_h from the surroundings at higher temperature T_h, and flown out from J_l to the surroundings at lower temperature T_l. This set of thermocouples generates electric power from the said heat exchange – which can then be used to drive an electric motor, and thus be converted to mechanical power *tout court*. The applicability of this device is, however, restricted to bench scale, due to the low currents generated.

In less constrained terms, a heat engine receives heat at a high temperature and loses heat at a low temperature; the difference between these two quantities appears in the form of work produced thereby, as per the first law of thermodynamics. At industrial scale – and for economic and convenience reasons, the low-temperature sink is some fluid at ambient temperature (usually atmospheric air, or river/lake water), while (water) steam plays the role of high-temperature source. Liquid water supplied by the feed pump is evaporated at the steam generator by providing heat at high temperature – and the resulting high-pressure, high-temperature steam drives a turbine that eventually produces (mechanical) work. The steam leaves the said turbine at lower temperature and pressure – and eventually undergoes condensation to liquid water via the room-temperature coolant; it is subsequently returned to the pump, thus closing the cycle. Since the working fluid is separated from the heat source, combustion products of any fuel, as well as solar, nuclear, or geo/volcanic thermal energies directly can be used to obtain the high temperature in the aforementioned fluid.

HISTORICAL FRAMEWORK

The aeolipile – described by Hero of Alexandria (1st century BC), rotates due to escaping steam from two nozzles that serve as arms and face opposite directions. Although no practical use was made of this effect by that time, it represents the first recorded steam engine in History. Jerónimo de Ayanz y Beaumont obtained a patent for a rudimentary steam-powered water pump in 1606; however, the first practical steam engine, working at atmospheric pressure, is credited to Thomas Savery in 1698 – without piston or moving parts, just taps. Thomas Newcomen developed, in 1712, the first commercially successful piston-based steam engine – resorting to steam condensation in a cylinder, which caused a subatmospheric pressure able to push a piston and concomitantly generate mechanical work. An improvement by James Watt, patented in 1781, originated the first steam engine able to produce sustained rotary motion; for his invention – particularly suited for large-scale machinery operated continuouswise, he has been denominated the father of Industrial Revolution.

If a standard thermodynamic cycle is operated in reverse, then a refrigerator results; a heat engine is again at stake, except that heat is withdrawn at low temperature, with the aid of externally supplied work – to produce more heat (equal to the sum of low-temperature heat and work received) that is released at high temperature, again satisfying the first law of thermodynamics. In this case, the high-temperature sink is normally provided by a fluid at ambient temperature, again for economic considerations – whereas the low-temperature source turns to be a refrigerant below ambient temperature. Hence, water is not so useful as circulating fluid – owing to the possibility of formation of ice that cannot flow (for being a solid), and its high boiling point (well above the working range of interest). One has resorted with advantage to refrigerants possessing much lower boiling points, so that a vapor phase will exist at a sufficiently low temperature. A wide variety of such refrigerants are available – so choice should reach a compromise between such other features as safety, unit cost, viscosity, vapor pressure, latent heat of vaporization, and boiling and critical points.

HISTORICAL FRAMEWORK

The first cooling systems resorted to ice – yet artificial cold was not available until the work of William Cullen in 1755. He basically used a pump to create a partial vacuum over a container with a volatile compound inside – which boiled up, thus converting sensible heat from the surroundings to latent heat of vaporization, and simultaneously creating a minute amount of ice. In 1805, Oliver Evans described a closed vapor-compression refrigeration cycle to manufacture ice, based on application of a vacuum over ether. However, the first working vapor-compression refrigeration system came from the hands of Jacob Perkins, in 1834. The oldest gas absorption refrigeration system was, in turn, developed by Ferdinand Carré in 1859 – and resorted to dissolution of gaseous ammonia in water.

Carl von Linde patented, in 1876, a process of liquefying gases – which permitted use of NH_3, SO_2, and CH_3Cl as refrigerants until the 1920s. He accordingly took advantage of Joule and Thomson's effect, which causes a (minor) drop in air temperature when it is throttled adiabatically; such an effect becomes cumulative if operation is continuously iterated, in a cyclic manner. Oxygen was liquefied for the first time in 1877, by Louis Cailletet and Raoult Pictet independently – departing, in essence, from von Linde's work. In 1902, George Claude came up with a competing mode of liquefying air – via an isentropic process, more efficient than the existing isenthalpic process. This permitted production of industrial quantities of liquid nitrogen, besides oxygen and argon. Another crucial invention in this regard was meanwhile made by James Dewar in 1892 – the double-walled, vacuum-insulated vessel; this device permits convenient storage and transportation of cryogenic fluids, and has ever since been used in industrial practice. Finally, Heike K. Onnes succeeded in liquefying helium in 1908 – which led him, a few years later, to the discovery of superconductivity at very low temperatures; in 1913, he received the Nobel Prize for his work.

A compressor converts power – made available by an electric engine or otherwise (e.g. a gas/diesel-driven thermal engine), into internal energy of a gas, stored at higher pressure; usually a compressor forces more and more gas into a reservoir, and shuts off when its pressure reaches a preset threshold. High pressure – which is an elastic form of potential energy, may then be used to propel the gas for a variety of final uses, e.g. bubbling CO_2 for soda manufacture, foaming foods with air, chilling and freezing with cryogenic fluids, regenerating steam between evaporation stages, and even operating final control elements with air (e.g. pneumatic valves). Air compressors are used in a number of industries, beyond the food sector – including construction industry and tooling machine operation; earlier compressors were much less versatile indeed.

A compressor is to be differentiated from a gas pump – as the latter is meant to overcome pressure drop due to viscous flow; hence, it does not require a damping reservoir downstream – and operates, in general, much quieter, besides being slower and less expensive. Positive displacement compressors work by decreasing the volume of a chamber, prior to gas admission to the pressurized reservoir. Conversely, dynamic displacement compressors entail a rotating component that imparts its kinetic energy to the gas – which will eventually be brought to a higher pressure level upon conversion of such a kinetic energy.

HISTORICAL FRAMEWORK

The earliest air compressor known is actually the human lung – with breath, characterized by a mere 0.02–0.08 bar overpressure, having been instrumental to stoke fires at the dawn of civilization, despite its unfavorable content of CO_2. This crude practice faded away by 3000 BC, as metallurgy rose – since melting down of such materials as gold and copper required much higher temperatures to be kept in a sustained fashion. A new type of air compressor, called bellows, was put forward by 1500 AD – a hand-held, and later foot-controlled flexible bag, able to produce a blast of air as necessary to keep high-temperature fires. A water-wheel-driven blowing cylinder was invented in 1762 by John Smeaton, which gradually replaced the bellows. Despite its efficiency, it was in turn replaced by the blasting machine, due to John Winkinson in 1776 – which eventually became the archetype for modern air compressors.

Besides metal working, air compressors soon began to be used for ventilation of underground areas, namely, in mining and tunnel construction. Air compressors were used to transmit energy by the 1800s. The first compression plant was built in Paris in 1888, under the guidance of Viktor Popp – with an original power of 1,500 kW that soon grew up to 18,000 kW. The next milestone was use of electricity to drive compressors. Today, there is a wide variety of air compressors commercially available, ranging from reciprocating units to turbo-machinery.

As mentioned before, the most common types of air compressors are positive displacement units – based on reciprocating or rotational motion, coupled with strict sealing to prevent leakage; and turbo-compressors, with axial or radial flow. The latter may be regarded as a reaction turbine operated in reverse – from the point of view of exchange of energy between fluid and rotor.

Unlike happens with the positive displacement compressor, there is only a limited range of flow rates, at any given frequency of rotation, within which an axial turbocompressor can operate; if flow rate is excessively reduced, the flow pattern breaks down – and the compressor is said to surge. This type of apparatuses are accordingly appropriate to compress large quantities of gas, to comparatively low overpressures. Their efficiency changes considerably within a narrow range of flow rates; it can be kept at a high level by varying the angle of the blades during operation, but this calls for expensive designs.

The aforementioned lack of flexibility has prompted development of large radial turbocompressors. Their basic design consists of an impeller – i.e. a disk on which one or more rings of radial or curved blades are mounted, which is rotated at high speed. Fluid is admitted into the center (or eye) of the impeller, and leaves from its periphery at high velocity. The acquired kinetic energy is then converted to pressure increase – in a ring of stationary diffusing blades, or a free vortex diffuser surrounding the impeller. The pressure ratio of each stage is normally restricted to ca. 1.6; overall ratios greater than this (up to 20) are handled by mounting several stages on a common shaft – with vanes in the outer casing being used to guide the gas inward from one stage to the next. However, cooling is strictly required when pressure ratios exceed ca. 5 – and is usually achieved by water-jacketing the ducts, or passing the gas through intercoolers. Typical flow rates range from 3 to 15 $m^3.s^{-1}$, but power consumption is ca. 5% greater than their axial counterparts – because of a more tortuous path between stages. Efficiency increases with load, so size of a radial turbocompressor may reach threefold that of the corresponding axial machine; being less expensive, more robust, more stable, and able to deliver overpressures in excess of 9 bar constitute its major advantages over an axial turbocompressor.

If a compressor is operated in reverse, then a vacuum machine results – for which a storage reservoir makes no sense. These devices can be broadly categorized as per their underlying mode of operation: positive displacement – where a sealed cavity is repeatedly expanded to allow gas in, which is then exhausted to the atmosphere; momentum transfer – which resorts to high-speed jets of dense fluid, or high-speed rotating blades to knock gas molecules off the chamber; and entrapment pumps – able to capture gases via adsorption on a solid matrix.

HISTORICAL FRAMEWORK

The predecessor of the vacuum pump was the suction pump, already known to Ancient Romans; dual-action machines were indeed found in the remains of Pompeii. Suction pumps were afterward described in the 13th century by Arab Al Jazari, who claimed them to be larger versions of Byzantine siphons used to discharge Greek fire during battles; such apparatuses reappeared in Europe throughout the 15th century. By the 17th century, water pump designs had improved to the point of producing measurable degrees of vacuum – even though their applicability was not immediately realized; heuristic rules published in 1635 had it that water could not be pulled up beyond 18 Florentine yards, ca. 10 m (known today as the water head equivalent to 1 atm). This limit was actually a concern for irrigation projects, mine drainage, and decorative water fountains – so the Duke of Tuscany commissioned Galileo Galilei to help solve the problem, with the aid of his peers. Gaspar Berti came up with the first water barometer in 1639, which gave rise to the first mercury barometer by Evangelista Torricelli in 1643; he also put forward a convincing argument that the overhead space over the mercury column was a vacuum.

The first modern vacuum pump was invented soon after, by Otto von Guericke in 1654, and was widely publicized through his famous Magdeburg experiment – showing that two hemispheres duly sealed to each other, enclosing a vacuum, could not be separated by even several horses pulling in opposite directions. Improvements of such an apparatus came by the hand of Robert Boyle and Robert Hooke – including development of an air pump useful to produce vacuum. In 1855, Heinrich Geissler invented the mercury displacement pump – which was able to reach as low a pressure as 10 Pa. In the 19th century, Nikola Tesla designed an apparatus resorting to a Sprengel pump – which could attain an even higher degree of exhaustion.

Electric devices are crucial to any modern food process, namely because electrical work delivered in the form of an electric voltage/current can be readily (and fully) converted to all forms of mechanical work, or to heat as most degraded form of energy; electrical energy is also quite easy to transport. These advantages overcome its disadvantages, associated chiefy with difficulty of storage and risk of electrocution. Electric apparatuses include (active) electric motors – which convert electrical energy to mechanical work; and (passive) resistors – which convert electrical energy to heat, or such related forms as light (radiant energy) or electrolysis (chemical energy). Associated devices encompass transformers that change electric voltage (and concomitantly electric current intensity) – thus adding to the flexibility of use of electrical energy, by adapting its features to the type of final user's apparatus.

As stressed above, an electric motor is an engine that transforms electrical energy to mechanical energy; the reverse is carried out by an electric generator – which has indeed much in common with an electric motor. This type of device typically operates through the interaction between a magnetic field and a winding current to generate a force. Its applications are as diverse as industrial fans, blowers, pumps, and compressors; general-purpose motors – of standardized dimensions and features, provide on their own convenient mechanical power for industrial use. Electric motors are employed to produce linear or, more frequently, rotary force (or torque), so they should be distinguished from such devices as actuators (e.g. magnetic solenoids) and transducers (e.g. loudspeakers); despite converting also electricity into motion, the latter are unable to generate usable mechanical power.

HISTORICAL FRAMEWORK

The pioneer electric motors were probably the simple electrostatic devices created by monk Andrew Gordon in the 1740s; soon after, André-Marie Ampère described the interaction between electric current and magnetic field. In 1821, Michael Faraday demonstrated, in turn, conversion of electrical to mechanical energy via electromagnetic means; a free-hanging wire was dipped into a pool of mercury, on which a permanent magnet was placed – and the wire rotated around the magnet, when a current was passed through the former. An early refinement of Faraday's demonstration was Barlow's wheel – but this type of homopolar motor was still devoid of practical interest. In 1824, François Arago formulated the existence of rotating magnetic fields – which was taken advantage of by Walter Baily to develop the first induction motor, by manually turning switches on and off.

Ànyos Jedlik meanwhile solved the technical problem of continuous rotation in direct current (DC) motors, with his invention of the commutator; in 1828, he made a public demonstration of the first device containing the three basic components of modern motors, i.e. stator, rotor, and commutator – with no need for permanent magnets, as the magnetic fields of both stationary and revolving components were generated solely by currents flowing through their windings. The first commutator DC electric motor was put forward by William Sturgeon in 1832 – yet the first actually rotating electric motor was invented only in 1834, by Moritz von Jacobi. Four years later, he developed a motor able to drive a boat with 14 people across a wide river.

Due to the high cost of primary battery power, however, electric motors were commercially unsuccessful until electricity distribution became available. A major turning point in DC motors took place in 1864, when Antonio Pacinotti described the first ring armature, with symmetrically grouped coils closed upon themselves – and connected to the bars of a commutator; the brushes of the latter delivered essentially nonfluctuating current. In 1886, Frank Sprague invented the first practical DC motor – a nonsparking motor, able to maintain a relatively constant speed under variable loads.

Meanwhile, a collective research effort was on the move to develop workable alternate current (AC) motors – owing to the advantages of AC for long-distance, high-voltage-mediated transmission; Galileo Ferraris and Nikola Tesla were (independently) the first in the field of commutatorless induction motors – in 1885 and 1887, respectively. A shorted-winding rotor induction motor was included in one of Tesla's patents, bought by George Westinghouse afterward; his engineers adapted it to power a mining operation in Colorado, in 1891. The first three-phase cage-rotor induction motor came in 1889, by the hand of Mikhail Dolivo-Dobrovolsky – which is now used for many industrial applications; the impetus for his invention came from the two-phase pulsatory behavior of Tesla's motor, which rendered it hardly practical. General Electric and Westinghouse companies signed, in 1896, a cross-licensing agreement for the bar-winding-rotor design – called squirrel-cage rotor afterward, which eventually became widespread for countless applications.

The major electric motors operate on direct current (DC for short) or alternate current (AC for short) – with the former increasingly overridden by the latter; they appear as self- or externally commutated designs – the former resorting to either mechanical or electronic commutators, and including universal AC/DC motors, besides brushed and brushless DC motors, switched reluctance motors, and electrically excited DC motors. Externally commutated engines can be either asynchronous or synchronous; once started, the latter require synchronism with the magnetic field own frequency for delivery of a regular torque. Asynchronous machines entail squirrel-cage induction motor (SCIM for short) and three-phase wound-rotor induction motor; whereas synchronous configurations encompass AC wound-rotor synchronous motor (WRSM for short), hysteresis motor, and synchronous reluctance motor.

A transformer is a device that transfers electrical energy between two (or more) circuits through electromagnetic induction; it basically consists of two windings (or coils) of wire, around a core of ferrous material – since Fe possesses electromagnetic features. A varying current in one coil generates a varying magnetic field – which, in turn, induces a varying voltage in the other coil; hence, electric power is transferred between such coils, without the need of a metallic connection between them.

HISTORICAL FRAMEWORK

The principle underlying operation of a transformer – electromagnetic induction, was discovered independently by Michael Faraday in 1831, and Joseph Henry in the following year; early experiments included winding a pair of coils around an iron ring, thus giving rise to the first toroidal, closed-core transformer. However, practical application of this principle came only by the hand of Nicholas Callan – who, in 1836, found that more turns in the secondary winding relative to the number of turns in the primary winding induce a larger voltage in the latter.

Batteries produce DC rather than AC, so induction coils by these times resorted to vibrating electrical contacts – able to interrupt electric current in the primary coil, at regular time intervals; generators able to efficiently produce AC without resorting to interrupters became available only in the 1870s. Pavel Yablochkov invented a lighting system in 1876 relying on a set of induction coils, with the primary winding connected to an AC source; the secondary winding was then connected to arc lamps – and large-scale manufacture of such type of lighting systems was started by Ganz factory in Hungary a few years later. In 1882, Lucien Gaulard and John D. Gibbs first exhibited an apparatus with an open iron core – later sold to Westinghouse, headquartered in the United States.

Induction coils with open magnetic circuits are, however, inefficient at transferring power to loads, and thus incapable of reliably regulating electric voltage; this shortcoming was overcome by Károly Zipernowsky, Ottó Bláthy, and Miksa Déri in 1884 – who described, in their joint patent, copper wire windings as per two alternative designs, i.e. either wound around or surrounded by an iron wire ring core. These so-called ZBD transformers (acronym coined with the initials of their family names) – the first of which had a power of 1.4 kW and converted 120 to 72 V at 40 Hz, were instrumental for the eventual triumph of AC distribution systems over their DC counterparts. Besides being three- to fourfold more efficient than Gaulard and Gibbs' open-core bipolar devices, the said transformers were complemented with parallel-, instead of series-connected utilization loads; and were also built with very high turn ratios, thus allowing conversion of high-voltage power at transmission to low-voltage power at utilization.

Westinghouse, Stanley, and Associates soon developed an easier way to manufacture the core – consisting of a stack of thin, E-shaped iron plates, insulated by thin sheets of paper; this permitted prewound copper coils to be slid into place, and straight iron plates laid to create a closed magnetic circuit. In 1889, Mikhail Dolivo-Dobrovolsky developed the first three-phase transformer – and two years later, Nikola Tesla was able to generate very high voltages, at high frequency, via his unique coil design; audio-frequency transformers were meanwhile used by experimenters as early developments of telephone.

Transmission of electrical power through wires over long distances is subjected to Joule's heating effect – which takes place at a rate proportional to the square of electric current intensity, with resistance serving as proportionality constant. Since electrical power is given by product of voltage difference by current intensity, conversion of a given power to a higher voltage and a concomitant lower intensity produces a quadratic decrease in heating loss; this provides the major (technical and economic) justification for use of transformers in current practice – and has permitted points of generation be remotely located relative to points of demand.

HISTORICAL FRAMEWORK

Benjamin Franklin has become famous for tying a key to a kite string during a thunderstorm in 1752 – and accordingly proving that a discharge of static electricity was the reason for lighting; however, controlled conversion of electricity to light actually required intermediate generation of heat, as per Joule's heating effect. This laid the grounds for Thomas Edison's invention of a practical light bulb in 1879 – able to last for a sufficiently long time before burning out; the critical issue was finding a material strong enough for the filament

– and an ordinary (sewing) cotton tread, following careful carbonization, ended up playing this role satisfactorily.

On the other hand, the problem of transmission of electrical power comes along with the problem of generation thereof in the first place. Alessandro Volta was responsible for a breakthrough in this field: after soaking paper in salt water, and then placing zinc and copper plates on opposite sides of the paper, he observed occurrence of a chemical reaction that produced an electric current. This constituted the first electric cell; by connecting many of these cells together, he obtained a battery – i.e. the first safe and dependable source of electricity. However, supply at large scale is not compatible with batteries. Such a production became feasible only upon invention of the electric motor; in fact when operated in reverse, it allows conversion of externally supplied mechanical power to electric current.

The major practical application of transformers is to increase or decrease (alternating) voltages, so as to make them appropriate for distinct electrical power applications. Industrial equipment and household appliances use indeed electrical power at relatively low voltages (especially if electronic components and circuits are at stake) – to minimize the risk of electrocution, so transformers are needed to deliver the proper high voltage upstream of the AC mains suitable for transport, and the proper low voltage at inlet of the equipment proper. Note that a transformer effectively insulates an apparatus (and thus its operation) from the mains supply – because electric current transfer is not direct, but instead occurs via induction at a distance.

Finally, the issue arises of controlling each processing unit – and also a food process as a whole; the bottom line is guaranteeing that the techno-economic requirements imposed by their designers, as well as environmental and social constraints are met in full – despite presence of ever-changing, spurious external disturbances, or else deliberately altered set-points. Representative examples include product and process specifications, operational and safety constraints, environmental regulations, and economic performance. Variability in final product specification is thus to be avoided as much as possible – so as to fully abide to market requirements; while all safety and environmental regulations are to be enforced – for ethical and legal reasons; and utilization of resources is to be kept to a minimum – to assure overall competitiveness.

HISTORICAL FRAMEWORK

Although controls are considered an intrinsic part of any industrial process since the 1800s, early Greeks and Arabs pioneered in this regard; for instance, they already resorted to float-valve regulators in such devices as water clocks, oil lamps, wine dispensers, and water tanks. One of the first feedback control devices was the ancient water clock of Ktesibios in Alexandria, Egypt, ca. 250 BC; its accuracy was so high that was only matched by that of the pendulum clock invented by Christiaan Huygens in the 17th century. In 1620, Cornelis Drebbel designed a closed loop control system to operate a furnace – thus giving rise to the first thermostat. His idea departed from the possibility of measuring temperature via expansion of a liquid from a vessel into a U-shaped tube containing mercury; a float in the mercury operated an arm that, in turn, controlled the draft to a furnace via a mechanical linkage. A mechanism, patented by Edmund Lee in 1745, was used to tent the sails of windmills in order to control the gap between grain grinding stones driven by the rotating sails; this concept ultimately led to one of the most significant control developments of the 18th century – the steam engine governor.

Relay logic was introduced along with factory electrification, and quickly adopted by the 1920s. This is, in essence, a means to represent a manufacturing program (or other logic) of on/off type, in a form normally used for relay; the underlying concept was eventually incorporated in programmable logic controllers. Central control rooms meanwhile became a common part of power plants of many factories in the 1920s. This period closed with the advent of the communications boom, as wired (and wireless) systems emerged – which are able to pass information over a distance, combined with advancements in control; the stage was thus set for most control applications as are known today.

Some form of control has always been present in a food process – starting with several ways of human-mediated optical or audio reading; complemented by decision taken by an operator based on the said reading and their own skill and judgement, and followed by manual action thereby upon a valve or the like. Escalating labor costs, along with increasingly stringent regulations for standardized and safe foods have provided an impetus for a shift to automatic control systems – which take full advantage of technological advances (e.g. electric/pneumatic operation), including versatility of microelectronics toward data generation, handling, and recording. Nowadays, extensive reporting is also a must for an increasing number of food products, for traceability purposes – in case an inconformity is found *a posteriori*.

The aforementioned developments have been implemented throughout the food plant, and at all stages of processing, namely: ordering and supplying of raw materials (including just-in-time and material resource planning approaches); planning and management of production (namely, processing of orders, recipes, and batches); assuring processing conditions and overviewing flow product throughout the process; collation and evaluation of process and product data, for record-keeping and optimization; and implementation of cleaning-in-place procedures, packaging, and warehouse storage and distribution. Notwithstanding their unique advantages in terms of efficiency and reliability of process control (especially when performance of routine actions is concerned), automation has brought about social drawbacks in terms of reduction of employment – along with higher setup and maintenance costs, and increased risks, delays, and costs in the case of system failure.

Three general classes of needs have classically been addressed by a control system: suppress influence of external disturbances, ensure stability of operation, and optimize performance. Since disturbances denote the effect of surroundings upon the equipment/process, they are normally out of reach of a human operator;

hence, control mechanisms are in order, available and operative on a continuous basis – which will make proper changes to processing conditions aimed at cancelling (or, at least, minimizing) the unfavorable impact of the said disturbances. This strategy will bring about stability, as long as operation is kept within narrow limits of prespecified set points; such set points often reflect previous optimization of some objective function, laid on technical and/or economic grounds. Typically, a variable that (somehow) characterizes the state of a system is measured and compared to a given set point; any deviation will trigger manipulation of another variable in the process, so as to keep deviation as small and short as possible – usually via a programmable logic controller. By default, this active device responds proportionally to the current value of the said deviation – but its response may be refined by taking past evolution into account, and even current tendency of evolution; a proportional (P for short), proportional-integral (PI for short), or proportional-integral-derivative (PID for short) control, respectively, will accordingly do.

3.2 Product and process overview

3.2.1 Steam engine

There are two fundamental components of a steam engine – the boiler, or steam generator, and the engine itself, usually a turbine; for safety or efficiency reasons, they may be installed in premises some distance apart from each other. Extra auxiliary components are present as well: pump (to supply water to the boiler), condenser (to recirculate water and recover latent heat of vaporization), and superheater (to raise steam temperature above its point of saturation).

The first step to produce steam is generation of hot gases via an appropriate burner; examples are given in Fig. 3.2. Should a liquid fuel be used, it must first undergo atomization into a fine spray – usually by forcing it under pressure through a nozzle; this spray is often ignited by an electric spark. The air is forced through around the said nozzle, at the end of the blast tube, via a fan driven by the fuel burner engine. Orientation of the doors in the air register provides the turbulence necessary to bring about thorough mixing of fuel and air, so as to generate a (desirably) short flame. In the case of coal (and other solid fuels), a mechanical device is required to previously pulverize it down to the size of fines – suitable to feed a flame, in a controlled manner. This coal burner is normally operated by an engine – fueled, in turn, with oil; ignition often occurs via the oil burning igniter. The steam generated can be supplied for direct heating purposes – via heat exchangers; or to propel turbines that, in turn, produce mechanical work (namely in electric form) as typical heat engines.

Safety is the dominant concern in operating any system involving combustion – so ignition of a burner should occur at a location close to the burner, and at a much higher air flow than actually required; special precautions are to be taken also when shutting down the system, or varying load of fuel (or fuel type itself). It should be stressed that natural gas explodes when mixed with air to proportions within 4.7–15%, with an ignition temperature range of 482–632°C; propane explodes within 2.2–9.6%, with ignition range of 493–604°C; butane explodes within 1.9–8.5%, with ignition range of 482–538°C; and hydrogen explodes within 4–75%, with ignition temperature of 500°C. Burners that resort to natural gas as fuel are fed directly therewith, under pressure and upon mixing with air – at air ratios safely above those explosion thresholds.

Since the ultimate goal of boilers is to generate steam at high pressure, a high temperature should be generated; examples of boilers are conveyed by Fig. 3.3. As discussed above, the heat required to boil water, and concomitantly supply steam as utility can be obtained from a number of sources – most often burning of combustible materials with an appropriate supply of air in a closed space, i.e. the burner (also known as combustion chamber or firebox). Boilers, as high-temperature-heat sources, are pressurized vessels containing water to be boiled – together with some kind of device to transfer heat thereto. Although a plain, closed tank containing water can be employed with hot gases flown through an outer jacket (see Fig. 3.3i), the most common layouts are the water/tube boiler (see Fig. 3.3ii) – with water running through one or several tubes, surrounded by hot gases; and the fire/tube boiler (see Fig. 3.3iii) – where water partially fills a vessel, with hot gases passed through submerged tubes. Steam boilers in food processing are normally of the water/tube type – owing to faster heat transfer (as water is pumped through the pipes under turbulent flow), larger capacity, higher pressure of final steam, greater flexibility of operation, and ease of cleaning (as water can be flown at higher velocity, along a linear path, to remove wall-sticking films and precipitates via shearing); in addition, they are safer – since steam is generated in small, independent tubes, rather than in a large boiler vessel altogether.

i

ii

FIGURE 3.2 Examples of burners: (i) gas type (courtesy of Webster, Winfield, KS) and (ii) oil type (courtesy of Accuterm International, Mulgrave, Australia).

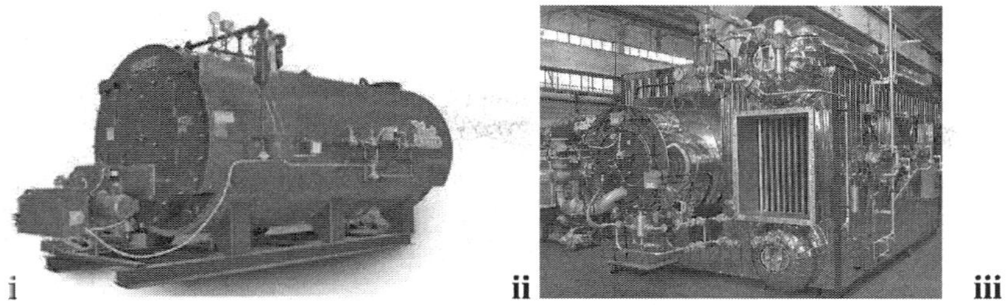

FIGURE 3.3 Examples of boilers: (i) tank boiler (courtesy of Hurst Boiler & Welding, Coolidge, GA), (ii) water/tube boiler (courtesy of Performance Heating, Alpharetta, GA), and (iii) fire/tube boiler (courtesy of Dallol Energy, York, UK).

The second stage of a heat engine is a machine, normally operated under steady state flow conditions, which will accordingly deliver (or receive) work at a constant rate; most said machines are either of a rotary nature – e.g. centrifugal pump, fan, turbine or compressor, or of a positive displacement nature – e.g. a piston-in-cylinder apparatus. They are employed to produce work from a high-pressure fluid (usually steam), or in reverse to compress a fluid by input of work. Despite being essentially equivalent in thermodynamic terms, two basic types have classically been addressed in a separate manner: turbines and reciprocating apparatuses. The former are present in most heat engines – and are schematically represented in Fig. 3.4; the latter are more often operated in reverse mode, typically as reciprocating compressors (and will thus be discussed later on their own).

A turbine consists of a series of banks of blades, fixed alternately to the casing (termed stators, or static disks), and to the rotor or drive shaft (termed rotating disks) – see Figs. 3.4i, iii, iv; they are shaped in such a way that the rotor turns and produces (shaft) work when fluid flows through the turbine – as apparent in Fig. 3.4ii.

Steam turbines are generally more efficient than their reciprocating piston counterparts, have fewer moving parts, and provide torque directly rather than through a connecting rod; illustrative examples are provided in Fig. 3.5. The turbine takes a supply of steam at high pressure and temperature, and gives out a supply of steam at lower pressure and temperature; the associated difference in internal energy is converted to mechanical work. The rotors follow a propeller-like arrangement of blades at the outer edge; steam acts upon these blades as emphasized in Fig. 3.4ii, thus producing rotary motion. The stator consists of a similar, yet fixed series of blades – aimed at redirecting the steam flow onto the next rotor stage. The stages of a steam turbine are typically arranged to extract the maximum work from steam with a given velocity and pressure, thus giving rise to variable-sized, high- and low-pressure stages. Turbines are more efficient when they rotate at higher speed, so reduction gearing is convenient if intended to directly drive lower speed applications (e.g. food mixers).

Each fixed plus moving pair of banks of blades constitutes a stage; the fixed blades of the stator form a converging channel (usually referred to as nozzle) – so that the fluid flowing through is accelerated, and driven by the pressure drop across it. Therefore, the emerging fluid possesses a higher kinetic energy, but a lower pressure than the ingoing fluid – in general agreement with Bernoulli's law. This accelerated fluid impinges on the blades attached to the rotor – thus driving it around, with an associated torque that leads overall to generation of shaft work; since kinetic energy is transmitted to the moving blade, it is concomitantly lost by the fluid left behind.

The simplest version is the (ideal) impulse turbine – where all kinetic energy gained in flowing through the fixed blades is given

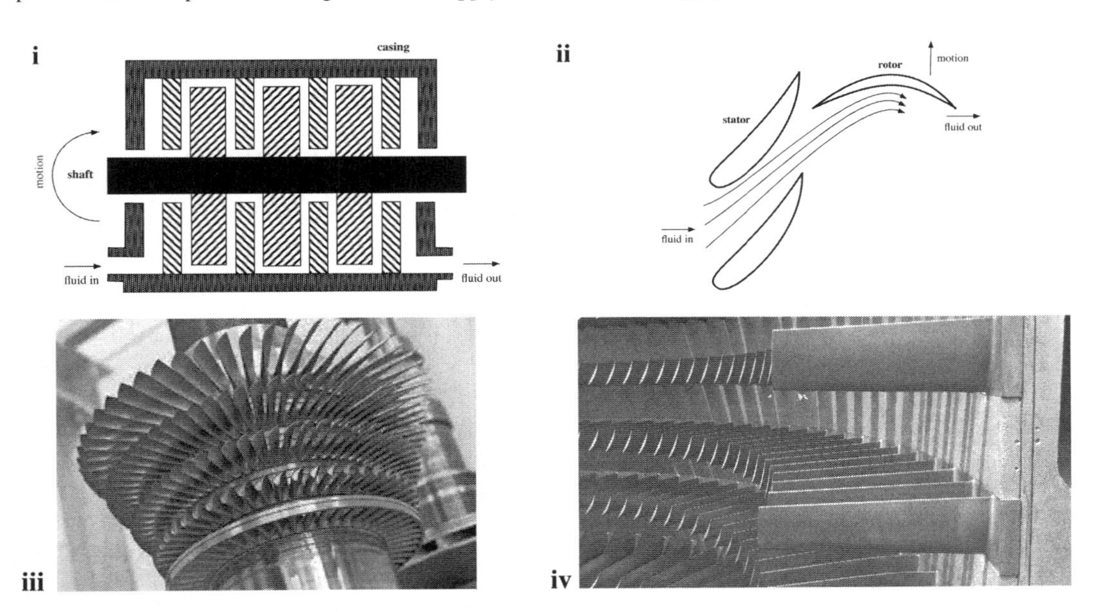

FIGURE 3.4 Schematic representation of steam turbine, as (i) overall cross-section and (ii) detail of blade arrangement in stator and rotor relative to fluid flow; and examples of (iii) rotor (courtesy of Sulzer, Winterthur, Switzerland) and (iv) stator (courtesy of Universal Dry Ice Blasting, Apple Valley, CA) as major constitutive parts.

FIGURE 3.5 Examples of steam turbines: (i) general outer view (courtesy of GE Power, Boston, MA), (ii) inner view (courtesy of Maxwatt, Karnataka, India), and (iii) detail of mobile shaft (courtesy of Alstom, Saint-Ouen, France).

up to the moving blades immediately after; and the fluid experiences no pressure drop through the moving blades, i.e. all pressure drop occurs in the fixed bank of blades owing to integral conversion of kinetic energy to potential energy. In the absence of frictional losses and considering the flow as unidimensional (despite the channel being curved), one can calculate the kinetic energy gained by the fluid as it passes through the fixed blades – which is then transferred in full to the moving ones; this exercise was developed previously in *Food Proc. Eng.: thermal & chemical operations* – when describing the performance of a steam compressor between evaporators. An alternative design is the reaction turbine – where expansion of gas to a low pressure occurs via passage through the stator, instead of the ring of nozzles.

Finally, a cold sink is required to absorb heat released at lower temperature; the simplest is a steam vent to the environment – yet water from a local river (or the like) may be utilized for the same purpose. In order to recirculate water, or top up boiler water in a continuous fashion, a pump is normally present – usually of a multistage, centrifugal type. For safety reasons, nearly all steam engines are equipped with devices to monitor the boiler – viz. a pressure gauge to avoid explosion, and sight glass to monitor water level; the engines themselves are often fitted with a governor to regulate speed of shaft rotation, coupled to a variable steam cutoff.

3.2.2 Refrigerator

The vapor compression cycle is, by far, the most commonly utilized to produce cold in both household- and industrial-scale refrigerating apparatuses; it resorts to a liquid/vapor system – and takes advantage of the high volatility at low temperature, and the large latent heat of vaporization of a given refrigerant.

A wide variety of refrigerants are available, consisting of pure compounds; it is also possible to use azeotropic mixtures, e.g. 73.8%(w/w) dichlorodifluoromethane with 26.2%(w/w) 1,1-difluoroethane, since such a mixture does not change composition when undergoing phase change – yet it exhibits features distinct from those of its constitutive components when pure. To avoid the confusion caused by different manufacturers supplying the same compound under a different trade name, an international agreement has been reached on an alphanumeric code; for instance, the aforementioned azeotrope is labeled as R500, whereas its components are referred to as R12 and R152a, respectively – while R507 denotes azeotropic mixture of R125 and R143a. Inorganic refrigerants like ammonia, carbon dioxide, and water are labeled as R717, R744, and R718, respectively. Organic refrigerants encompass such hydrocarbons as ethane (R170), propane (R290), butane (R600), isobutene (R600a), and propylene (R1270); such hydrochlorofluorocarbons as dichlorodifluoromethane (R12) and monochlorodifluoromethane (R22), 1,1-dichloro-2,2,2-trifluoroethane (R123), 1-chloro-1,2,2,2-tetrafluoroethane (R124) and 1,1-dichloro-1-fluoroethane (R141b); and such hydrofluorocarbons as difluoromethane (R32), pentafluoroethane (R125), 1,1,1,2-tetrafluoroethane (R134a), 1,1,1-trifluoroethane (R143a), and 1,1-difluoroethane (R152a).

Selection of a specific refrigerant from the above list for a vapor compression system is based on several performance characteristics, namely: (i) latent heat of vaporization – a high value is preferred because a smaller amount of refrigerant needs to be circulated per unit time for a given refrigeration load; (ii) condensing pressure – an excessively high value requires heavy construction and piping, so it should be avoided; (iii) freezing temperature – which should be lower than the intended evaporator temperature, otherwise no refrigeration effect will be produced; (iv) critical temperature – since refrigerant cannot be liquefied above its critical point, the critical temperature should be as high as possible; (v) density – a dense vapor reduces volumetric ratio between entrance and exit of the compressor, and thus contributes to a smaller size of the said equipment; (v) toxicity – small scale (namely domestic) uses require nontoxic compounds, especially because they are normally placed in closed rooms; (vi) flammability – a nonflammable refrigerant is preferred, for safety precautions; (vii) corrosiveness – the refrigerant should be compatible with piping and equipment materials, besides being insoluble in the oil used to lubricate the compressor; (viii) chemical stability – refrigerant performance will decay in time if it is not sufficiently stable; (ix) leak detection – a portion of refrigerant escaping should be immediately detectable for prompt fixing; (x) environmental impact – putative leaks should not have a major effect upon the environment, namely, integrity of ozone upper layer in the upper atmosphere; and (ix) cost – a low cost helps support economically feasible processes. In terms of safety, refrigerants are classified as A – for which toxicity has not been found at concentrations below 400 mg.kg^{-1}, and B otherwise; whereas flammability features rank them as either class 1 (not burning in air at 21°C and 1 atm), class 2 (with lower flammability limit above 0.1 kg.m^{-3} at 21°C and 1 atm, and heat of combustion below 19 kJ.kg^{-1}), or class 3 (with lower flammability limit below 0.1 kg.m^{-3} at 21°C and 1 atm, or heat of combustion above 19 kJ.kg^{-1}).

Among the physicochemical properties of the aforementioned refrigerants, the most important is probably its saturated vapor pressure – as it establishes the range of operating pressure for the compressor, as the most expensive device in the refrigeration system; examples of vapor pressure behavior of common refrigerants with temperature are given in Fig. 3.6. Water was included in this plot for comparative purposes only – since it is hardly suitable for most refrigeration duties, in view (also) of its high

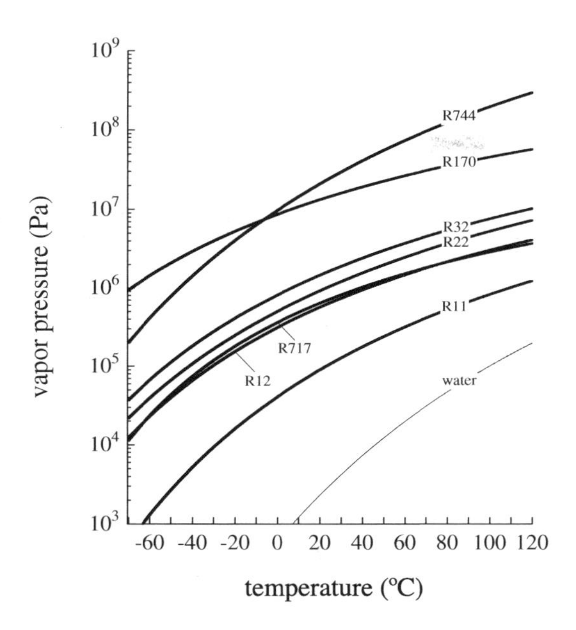

FIGURE 3.6 Variation of vapor pressure of selected refrigerants versus temperature.

solidification point. Besides delivering a relatively high vapor pressure, ammonia (R717) offers an exceptionally high latent heat of vaporization – 1314 kJ.kg^{-1}, versus 162 and 217 kJ.kg^{-1} for R12 and R22, respectively; it is noncorrosive to iron and steel (but not to copper, brass, and bronze); and it is immiscible with oil. Nevertheless, it is irritating to the mucous membranes and eyes, and will become toxic at concentrations as low as 0.5%(v/v) in air. Therefore, it is not suitable for domestic use, even though it constitutes an appropriate solution for industrial scale applications – usually setup in large (or open) spaces, where a leak may be easily detected by smell, while venting will occur spontaneously. Carbon dioxide (R744) is nonflammable and nontoxic, but can cause asphyxiation even at relatively low concentrations in air; it is commonly used in refrigerated ships, yet it requires considerable higher operating pressures than ammonia. It was once customary to choose a refrigerant so that the pressure prevailing in the evaporator is slightly above atmospheric pressure, to prevent ingress of air; however, use of nonflammable refrigerants, development of efficient purge systems, and availability of hermetically sealed compressors have meanwhile permitted operation at lower evaporator pressures.

Note that several refrigerants belong to the chlorofluorocarbon family – all nontoxic and nonflammable; they further exhibit good heat transfer properties, and are less expensive than other refrigerants. They are particularly stable in the lower atmosphere when (and if) released – but will eventually migrate to the stratosphere, in either vapor or liquid droplet form. The chlorine portion of their molecule will then be split off by solar UV radiation, and eventually react with ozone – thus contributing to deplete the ozone layer that protects Earth from excess UV radiation. Use of such refrigerants has thus been gradually phased out worldwide, under the auspices of the *Montreal Protocol* (1989). Replacement by refrigerants bearing a much lower ozone-depleting potential has meanwhile been proposed, as is the notable case of hydrochlorofluorocarbons – introduced as temporary replacements for chlorofluorocarbons; and hydrofluorocarbons – which contain weaker C–H bonds that make them more labile to cleavage, meaning that they will entertain shorter lives in the atmosphere. Unfortunately, both these types of compounds possess greenhouse effects, i.e. are impermeable to IR radiation emitted by the surface of land and sea back to space, but permeable to incident sunlight with shorter and longer wavelengths that will eventually produce thermal effects.

The vapor compression system basically requires a compressor, due to increase pressure of the vapor phase – with concomitant increase in temperature, as per Joule and Thomson's effect; enthalpy is then extracted back in both sensible and latent heat forms, via a condenser that delivers a saturated liquid at its exit – still under high pressure. When pressure is released – usually through an expansion (or throttling) valve, the high vapor pressure of the refrigerant drives its (flash) vaporization; the required latent heat is then obtained at the expense of sensible heat of the refrigerant itself, since the valve operates under essentially adiabatic conditions – and a lower temperature will accordingly be attained. The low-pressure, low-temperature liquid/vapor mixture is then passed to the evaporator – where heat contributed by the intended food commodity is absorbed, owing to a favorable temperature difference that acts as driving force, and becomes a low-pressure vapor again. In summary, the four major components of a standard mechanical refrigeration system are compressor, condenser, expansion valve, and evaporator – as illustrated in Fig. 3.7. The only active apparatus is the compressor, which supplies external work – none of which is, however, recoverable, because the (passive) expansion valve, found later along the cycle, operates irreversibly; extended areas of contact of refrigerant with outer atmosphere (at the condenser) and with food/fluid undergoing refrigeration (at the evaporator) are normally pursued for faster heat transfer – which justify finned surfaces, or plate-shaped tubing. As an alternative (or in addition), a blowing fan can be used to displace air (or water) at room temperature along the outer surface of the condenser or the inner surface of the evaporator as a form of forced convection; this approach reduces thickness of the boundary (stagnant) layer of fluid adjacent to either metallic surface, thus promoting heat transfer due to lower overall thermal resistance.

Depending on the required throughput rate and intended final pressure, compressors for refrigeration systems may be of the reciprocating or screw types – both being positive displacement machines, as illustrated in Fig. 3.8. When the cooling action is to be delivered at a distance, the cold refrigerant should be transported via well-insulated pipes to the location of the evaporator – as happens with chillers, with a convenient reservoir to account for volume holdup under intermittent, thermostatted operation.

Remember that a refrigerant is meant to absorb heat at a lower temperature in the evaporator, and release heat at a higher temperature in the condenser, both referred to ambient temperature – concomitant with supply of work in between, since the whole phenomenon is nonspontaneous. Therefore, foods (or other bodies/fluids) undergoing refrigeration are placed in the evaporator, or else in a closed room where a fluid is circulated after having contacted the evaporator – while the condenser is exposed directly to room conditions, usually atmosphere or a natural supply of water (for their negligible cost). This type of system may be operated in reverse as a heat pump – by exposing the evaporator to outside conditions, and moving the condenser to the container of interest. The latter configuration brings about

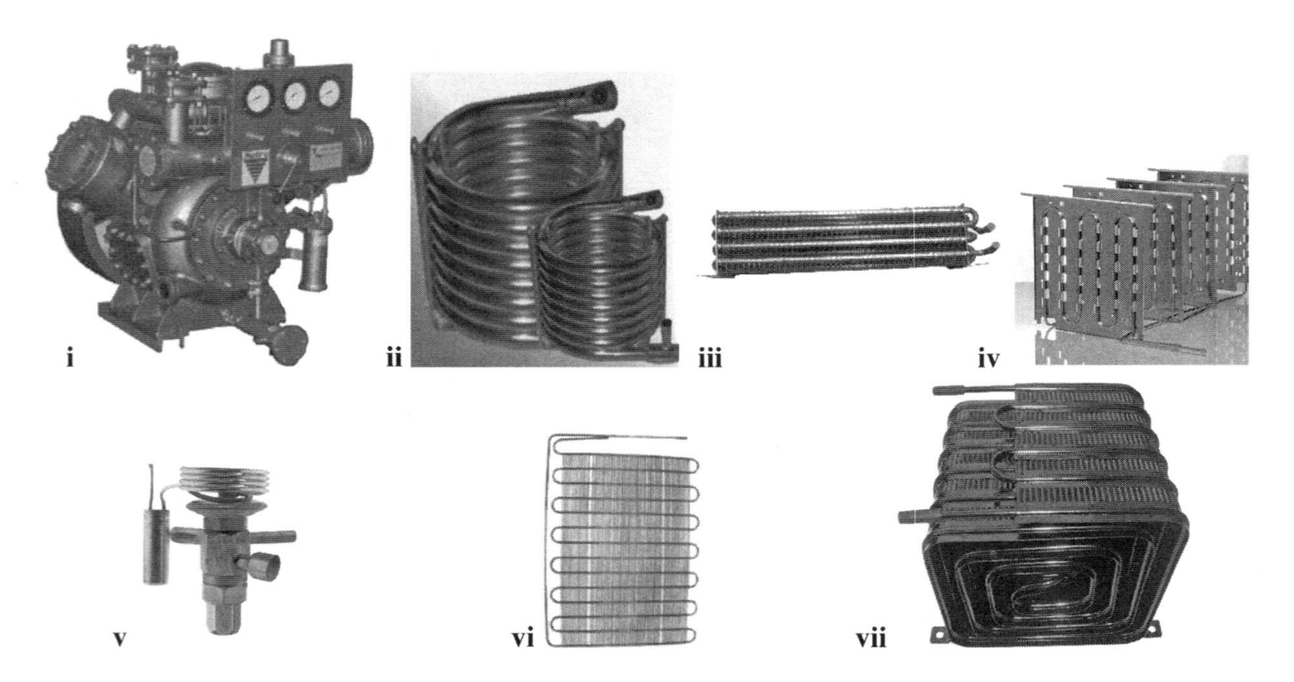

FIGURE 3.7 Examples of major constitutive parts of vapor compression refrigeration system: (i) compressor (courtesy of Metro Refrigeration Industries, Uttar Pradesh, India); evaporator, as (ii) bare pipe (courtesy of Hengyue Refrigeration Equipment, Shangai, China), (iii) finned tube (courtesy of Jinan Retek Industries, Shandong, China), or (iv) plate (courtesy of Arotubi Metais, Curitiba, Brazil); (v) expansion valve (courtesy of Parker Sporlan, Washington, DC); and condenser, in (vi) coiled form (courtesy of Flygrow Refrigeration, Xinxiang, China) or (vii) spiral tube-and-plate form (courtesy of Retekool, Jinan, China).

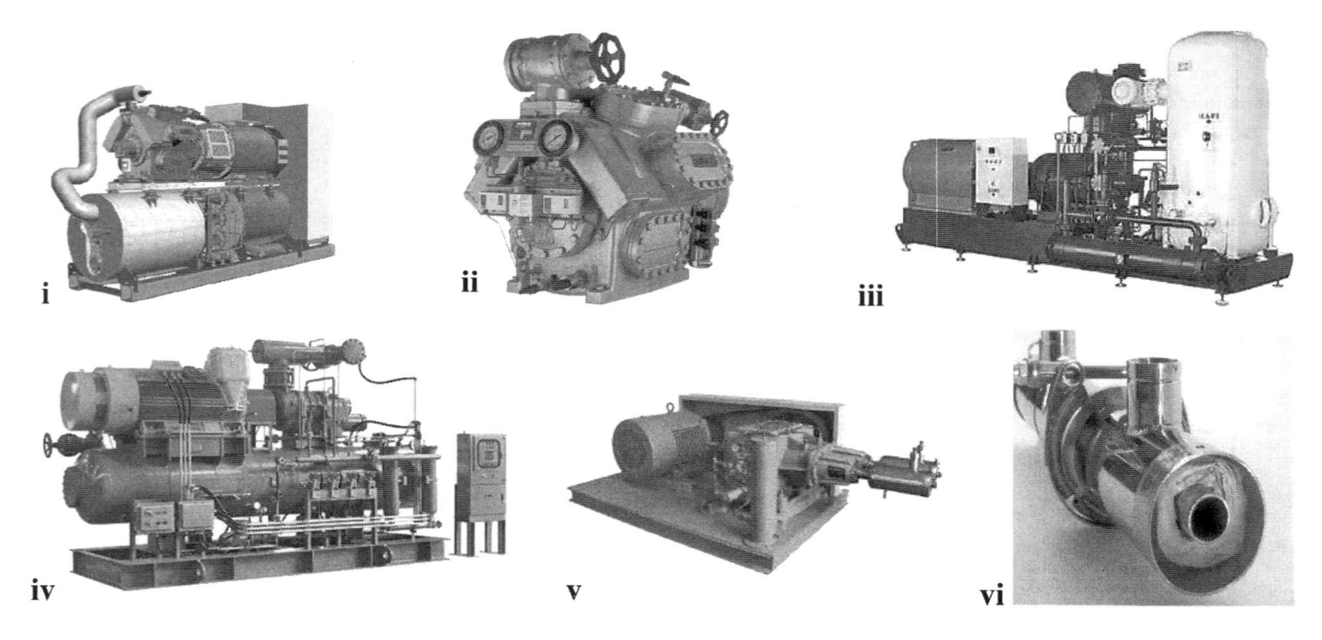

FIGURE 3.8 Examples of compressors for vapor compression refrigeration system: (i) chiller (courtesy of Johnson Controls, Milwaukee, WI), (ii) reciprocating compressor (courtesy of M&M Refrigeration, Federalsburg, MD), (iii) screw compressor (courtesy of HAFI Engineering and Consulting, Feldkirch, Austria), (iv) heat pump (courtesy of Kaishhan, Quzhou City, China), and (v) pump for cryogenic fluids (courtesy of Tai Lian Cryogenic Equipment, Hang Zhou, China); and piping for liquid nitrogen (courtesy of Airgases Projects, Kolkata, India).

heating of the said container and corresponding contents – at the expense of heat removed from the surroundings; however, this approach is more expensive than using direct (or indirect) heating via fuel burning. This is so because the cost of electricity, i.e. a nondegraded form of energy, required to drive the compressor is (still) quite higher than the cost of an equivalent amount of fuel – which generates a degraded form of energy. Heat pumps are, nevertheless, versatile apparatuses in small-scale air conditioning – because they can bring about refrigeration of rooms during warm weather, and the very same equipment can be operated in reverse for heating those rooms during cold weather.

To make refrigeration transients shorter, one may resort to single-use cryogenic fluids; in other words, a very cold liquid (e.g. liquid nitrogen) or solid (e.g. powdered carbon dioxide) is generated elsewhere by some refrigeration system, duly stored in a well-insulated, pressurized container, and then made available as needed for refrigeration of foods – where it is used only once, with the resulting gas being directly released (and thus lost) to the atmosphere that contains identical constituents. In view of the much lower temperatures at stake than those attained with fluid refrigerants in single-cycle vapor compression systems, pumps with special features are required to properly (and safely) handle cryogenic fluids; this entails not only particularly efficient insulation to minimize unwanted heat input, but also moving parts that do not become fragile at very low temperatures – as fractures would rapidly compromise operation.

In the early 20th century, the vapor absorption cycle resorting to water-ammonia systems became quite popular – namely because heat, produced by fuel burning, can be directly used, without need to generate work in between. With invention of the vapor-compression cycle and wide availability of electricity, such a chemical cycle lost most of its importance, though – except for some recreational vehicles that carry liquefied propane gas, or industrial environments where plentiful waste heat overcomes its inherent inefficiency; in fact, the characteristic coefficient of performance is of the order of only one-fifth that of a regular vapor compression system. The absorption system is not conceptually different from its compression counterpart, in terms of requirement of (external) supply of work; however, the hypothetical compressor of the latter is replaced by an absorber (i.e. water, or lithium bromide) that dissolves the refrigerant (i.e. ammonia, or water, respectively), a liquid pump that raises pressure of the resulting mixture, and a generator – which, upon absorbing heat from the food (or some fluid meant to contact the food afterward), drives off the refrigerant gas/vapor from the high-pressure liquid. The work necessary to operate the said pump is much lower than that required by the compressor of a regular vapor compression system anyway – since liquid pumping, rather than vapor compression are at stake. The condensed form and almost incompressible nature of the liquid entails, in fact, a dramatically lower volume displacement between any given pressure limits than if a gas/vapor were at stake – because the product of volume by pressure gauge represents work to be supplied.

When the working fluid in a compression system is a gas (often air) – rather than a liquid/vapor system, the refrigeration cycle is termed gas cycle. Since this approach does not include any change of phase – as operation takes place above critical temperature (as per definition of a gas), hot and cold gas-to-gas heat exchangers are used *in lieu* of evaporator and condenser, respectively. The underlying gas cycle is less efficient than a vapor compression cycle, since only sensible heat is exchanged – which is accordingly received and rejected at varying temperatures. The refrigeration effect is, in this case, measured by the product of specific heat capacity of gas by temperature rise thereof in the evaporator, instead of its specific heat of vaporization; hence, a given cooling load requires a much larger mass flow rate – and the underlying system is bulkier than an equivalent vapor compression system. Air cycle coolers are seldom used in terrestrial cooling devices, yet they are quite common in gas turbine-powered jet aircraft – as long as compressed air is readily available from the engine compressor sections; besides contributing to keep food catering cold onboard, such units are also employed to pressurize the aircraft cabin.

Cooling can instead resort to the thermoelectric, or Peltier's effect – which creates a heat flux at the junction of two different types of metals, should a (direct) difference of potential be applied. Due to its low capacity and high cost, utilization of this system is normally restricted to camping and portable coolers, or else for topical cooling of electronic and small analytical instruments.

Magnetic refrigeration – or adiabatic demagnetization, is a cooling technology based on the magnetocaloric effect exhibited by intrinsically magnetic solids. The refrigerant of choice is usually a paramagnetic salt (e.g. cesium-magnesium nitrate); and the active magnetic dipoles are accounted for by the electron shells of paramagnetic atoms. When a strong magnetic field is applied to the refrigerant, the magnetic dipoles are forced to align with each other – thus leading to a lower state of entropy; a heat sink absorbs the heat thereby released by the refrigerant. After removal of the said heat sink, along with insulation of the system, the magnetic field is switched off – which increases the heat capacity of the refrigerant, and so decreases its temperature to below that of the heat sink; hence, the refrigerant will be available to absorb energy by conduction, due to a favorable temperature difference. Since very few materials exhibit the needed features at room temperature (or nearby), applications of this mode of refrigeration have been limited to bench scale, or specific, high-valued cryogenic fluids.

3.2.3 Gas compressor

Pressures of industrial interest in food processing may range from high vacuum (as in freeze-drying) to hundreds of megapascal (as in hyperbaric sterilization); the compressors required thereby may be classified as dynamic or positive displacement.

In dynamic devices, pressure is increased by imparting a high velocity to the gas – which is converted afterward to pressure in a diffuser; this occurs in (centrifugal) radial or axial flow machines – the operation of which essentially mimics that of centrifugal pumps, or power turbines operated in reverse. Both types are collectively known as turbomachines – and resort to a multibladed rotor or impeller to increase velocity of the gas. An alternative is jet compressors – with no moving parts; here a high-velocity jet of water, steam, or air is used to compress gas. These types have previously been addressed to some extent, namely in *Food Proc. Eng.: thermal & chemical operations*.

In the case of positive displacement devices, gas is prevented from flowing back down the pressure gradient by a solid boundary. Such machines include reciprocating compressor – in which the volume change is effected by a piston in a cylinder; diaphragm compressor – in which the volume change is brought about by elastic deflection of a membrane; and rotary compressor – which employs either a single rotor or twin rotors. Examples are provided in Fig. 3.9.

A reciprocating compressor converts shaft work, available in the form of rotary motion with an associated torque as provided by a flywheel; to a shuttle-type motion, driven by a crankshaft. This type of apparatus basically consists of a piston moving back and forth inside a cylinder – see Fig. 3.9i, and includes spring-loaded inlet and outlet valves. The intake gas enters the suction manifold; it then flows into the compression cylinder, where it undergoes a pressure increase brought about by the reciprocating piston; and is finally discharged. The vessel connecting to outlet pipe is meant for damping pressure fluctuations that would otherwise necessarily develop – but may, in some cases, reduce

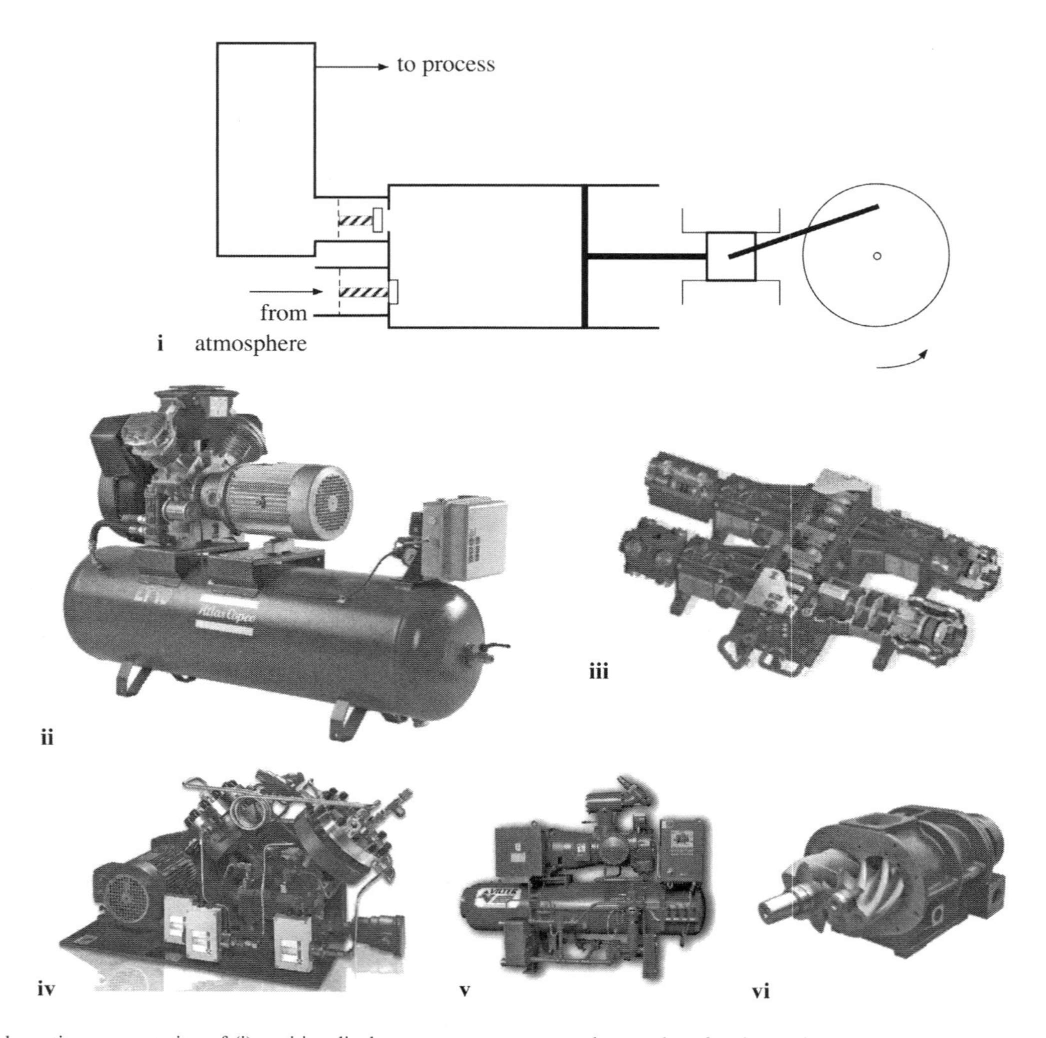

FIGURE 3.9 Schematic representation of (i) positive displacement compressor; and examples of reciprocating compressors, as (ii) overall view, including storage reservoir (courtesy of Atlas Copco, Nacka, Sweden) and (iii) cross-sectional view (courtesy of Ariel Co., Mount Vernon, OH); (iv) diaphragm compressor (courtesy of SERA Gmbh, Immenhausen, Germany); and rotary compressors, with (v) single screw (courtesy of Emerson, St. Louis, MI) or (vi) twin screw (courtesy of Kaeser Compressors, Fredericksburg, VA).

to mere lengths of tubing. The overall performance of the cylinder plus pulsation dampeners produces essentially steady flow compression, even though compression of fluid is carried out batchwise – provided that the frequency of compression strokes is high enough. Should the role of the moving piston be played by an elastic membrane – alternating between convex and concave shapes, then a diaphragm compressor would result; this configuration is, however, far less common than the reciprocating piston-based machine.

Rotary compressors are smaller than reciprocating machines, for comparable flow rates – since their continuous rotary action enables greater speeds be employed; they are generally uncooled, so their operation may be taken as approximately adiabatic. There is no need for suction and discharge valves – and clearance volume is small; hence, their throughput rate is determined directly by pressures prevailing up- and downstream. They may contain a single or a twin rotor, or screw; if the (single) screw is mounted eccentrically in a casing and slotted to house vanes, then a sliding vane machine results – characterized by a decrease in volume trapped between rotor and casing, from suction to discharge port. Root's compressor possesses two 8-shaped rotors – each one fitting one of its convex portions to one of the concave portions of the others.

Regular application of each family of compressors to different operating conditions is depicted in Fig. 3.10. The main field of application of reciprocating compressors if for not too high volumetric flow rates, up to relatively high pressures – as happens with carbonation of sodas in tall containers, or performance of hyperbaric treatments; their efficiency is particularly high, yet power savings are normally offset by high maintenance costs stemming from intermittent operation. Difficulty in lubricating piston rings, and prevention of valve and piping failure caused by mechanical vibration induced by pressure pulsations are among the most common (and difficult) problems faced in practice. At very high pressures, there is no alternative to the piston compressor, though; in other situations, many duties have been taken over by centrifugal machines under radial operation (suitable for relatively high pressures, but relatively low flow rates), or axial operation (suitable for lower pressures, yet higher flow rates) – despite their exhibiting much lower efficiencies.

Machines with intake pressure below atmospheric pressure and discharge pressure at atmospheric pressure are specifically

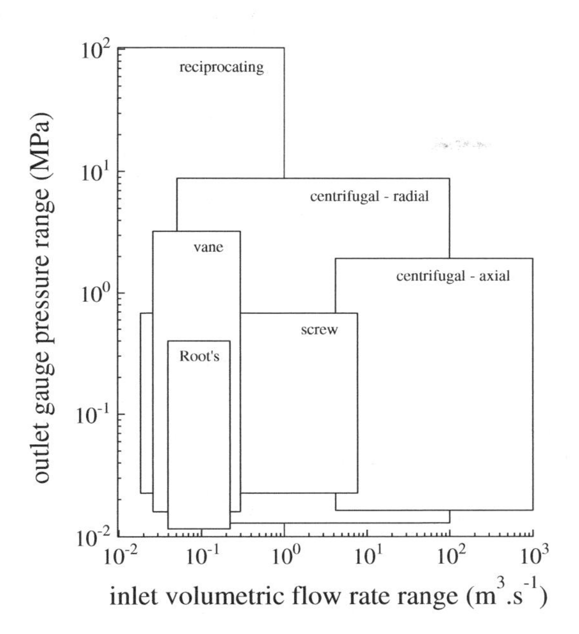

FIGURE 3.10 Ranges of duties (▭) for selected compressor types – expressed as pressure increases between inlet and outlet, and inlet flow rates.

termed vacuum pumps, or exhausters; however, the working principle is the same as true compressors – in view of the outlet stream being delivered at a higher pressure than its inlet counterpart. Fans can be used to overcome resistance to flow, but solely around atmospheric pressure; actual recirculating compressors are due when operation is meant at elevated negative pressure (gauges).

3.2.4 Electric engines and circuitry

Electrical power is crucial in modern food plants – thus justifying ubiquitous existence of electric power generators for backup, should the primary supply be disrupted; its versatility in terms of final utilization and intermediate transport (but not storage) essentialy justify that status. With costs falling due to larger production from renewable sources (e.g. hydraulic turbines in river dams, eolic generators powered by wind, photovoltaic cells for sunlight collection), electric power is likely to increase its share over the total energy industrially used in the coming future.

In its very essence, electricity consists of an ordered flow of electrons from atom to atom, through an electrical conductor – when subjected to an external difference of potential (expressed in V, as SI unit); remember that electrons are spontaneously and randomly exchanged between metal atoms at all times, especially at high temperatures – which justifies their high electrical (and thermal) conductivities. Therefore, a higher current intensity (in A, as SI unit) means a higher number of electrons involved (6.06×10^{18} e^- per second account for 1 A); while resistance, or reciprocal of conductance, measures how difficult transport is of such electrons. If an electric current flows always in the same direction, it is termed direct current – as mentioned before; if polarity is (periodically) reversed, then alternate current is said to be present.

Electric motors operate on either of three distinct physical principles: magnetic, electrostatic, or piezoelectric; however, the first type is the most common by far. Electrostatic motors resort to attraction and repulsion between electric charges, and typically require a high-voltage power supply – unlike magnetic motors, which rely on magnetic attraction and repulsion, so high current at low voltage suffices. The former find frequent use in micro-electromechanical systems, since charged plates are far easier to fabricate than coils and iron cores. Conversely, a piezoelectric motor is based upon change in shape of a piezoelectric material, under the influence of an electric field.

Magnetic-based motors are the default choice for food processing, and most rely on three basic concepts: electromagnet – formed by winding insulated wire around a soft iron core; electromagnetic induction – i.e. electric current that is induced in a circuit as it moves through a magnetic force field; and alternating current – i.e. electric current that periodically changes its direction of flow (or AC). The typical frequency of change of electric flow direction in AC is 50 Hz in Europe and 60 Hz in the United States. When current flows through the said wire, it produces a magnetic field in the iron core – with orientation dependent on current direction. This induced electric current produces, in turn, a voltage in the circuit – with magnitude dependent on magnetic field strength, speed of relative movement of circuit through field, and number of conductor circuits. The above concepts are operationalized in an electric motor via basically a rotor and a stator – as illustrated in Fig. 3.11i; the former is the moving part, while the stator is the stationary one. Both rotor and stator consist of electromagnets mounted on their surface – in such a way that rotor poles face stator poles.

In its simplest version, a stator (also known as armature winding) consists of a casing with two iron cores, or many thin metal sheets (in attempts to reduce energy losses), each one wound with insulated wire – made of copper for its highest electrical conductivity; they are placed opposite to each other, and the leads from the windings are connected to the AC power mains, see Fig. 3.11ia. This configuration creates an electromagnet, which reverses polarity as electric current alternates. The rotor is, in turn, constituted by a (rotating) drum of iron with copper bars – and is placed between the two poles of the stator, see Figs. 3.11ib and c; it is actually mounted on the motor shaft. An electric current is supplied to the said copper bars – and this current creates, in turn, magnetic poles (*S*/*N*); their response to the magnetic field created by the stator produces displacement, in the form of a rotational movement (since the angular coordinate accounts for the only degree of freedom mechanically allowed).

One (or both) of the aforementioned magnetic fields must be made to change with rotation of the motor; this is done by switching the poles on and off at the right time, or varying the strength of the pole. The two magnetic fields at stake produce antiparallel forces, and thus a torque on the motor shaft; the product of such a torque by its frequency of rotation gives the power of the motor. Although rotational speed should, in principle, be consistent with the AC frequency (e.g. $50\ \text{Hz} \times 60\ \text{s.min}^{-1} = 3000$ rpm), smaller speeds are attained in practice; fixed-speed AC motors are provided with direct online, or soft-start starters. Variable speed motors are instead provided with a power inverter, and a variable-frequency driver or an electronic commutator.

The rotor is supported by bearings that permit turning of the shaft around its axis; the bearings are, in turn, supported by the motor housing. The shaft extends through those bearings to the outside of the motor – where the load is applied. The distance

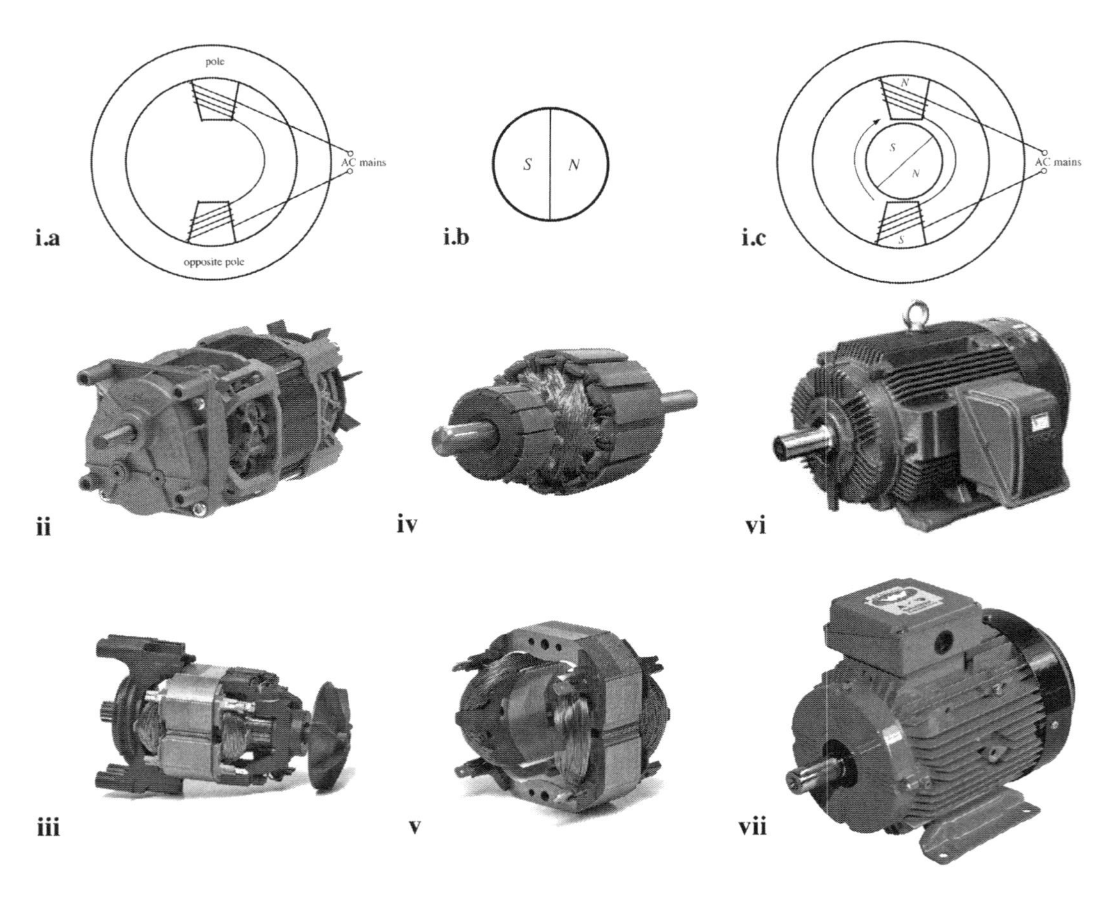

FIGURE 3.11 Schematic representation of (i) magnetic-based electric motor, with detail of (a) stator, (b) rotor, and (c) whole ensemble; and examples of universal electric motor, as (ii) overall view (courtesy of EBMPAPST Zeitlauf, Pegnitz, Germany), and detail of (iii) wiring (courtesy of Industria Motori Elettrici, Manerbio, Italy), (iv) rotor (courtesy of Lutron Industrial, New Taipei City, Taiwan), and (v) stator (courtesy of Foshan Shymould, Guangdong, China); (vi) squirrel-caged induction motor (courtesy of Xtreme Engineering Equipment, Pune, India); and (viii) wound-rotor synchronous, three-phase motor (courtesy of Brook Crompton, Huddersfield, UK).

between rotor and stator is termed air gap – and is supposed to be as small as possible for maximum performance. In fact, a larger air gap increases the magnetizing current necessary – while very small air gaps raise mechanical problems of friction upon thermal expansion, further to noise.

Windings are wires laid as coils, wrapped around a solid or laminated soft iron magnetic core – meant to produce magnetic poles, when supplied with an electric current. Two basic configurations exist: salient-pole – where winding is done around the pole below its face; and nonsalient-pole – where winding is distributed in pole face slots. Some motors also have conductors consisting of thicker metal – usually copper, or aluminum to a lesser extent, in the form of bars or sheets; these are powered by magnetic induction.

Finally, a commutator is a device used to switch the input of most DC machines and several AC machines; it consists of slip ring segments, insulated from each other and from the shaft. The armature current is supplied through stationary brushes in contact with the revolving commutator, thus causing (the required) current reversal – while applying power to the machine in an optimal manner, as the rotor is displaced from pole to pole. It should be emphasized that absence of current reversal would make the motor simply stop. Electromechanical commutation has been increasingly displaced by external commutated-induction and permanent-magnet motors.

Remember that the frequency of AC is measured by the number of times the voltage (and thus the electric current) changes direction per unit time; such a frequency, ω, is normally constant in the industrial power supply. The current and voltage may to advantage be represented by a phasor, i.e. a vector that rotates at (constant) frequency ω, and characterized by a magnitude proportional thereto; its vertical projection as time, t, elapses generates a sinusoidal representation of instantaneous magnitude of current, $I\{t\}$, in the case of phasor $\boldsymbol{I}$, and voltage, $V\{t\}$, in the case of phasor $\boldsymbol{V}$. The angle between $\boldsymbol{I}$ and $\boldsymbol{V}$ (often termed power angle), ϕ, depends on the specific features of the circuit at stake – namely whether it is resistive, capacitive, or inductive. Resistive circuits are the simplest – because $\boldsymbol{I}$ and $\boldsymbol{V}$ are in phase, i.e. their power angle is nil; the instantaneous power delivered by the source to the load equals the product of instantaneous voltage by instantaneous current, so one may write

$$\mathcal{S} = \|\boldsymbol{V}\|\|\boldsymbol{I}\| = VI \tag{3.1}$$

pertaining to resistive circuits – where $\mathcal{S}$ denotes apparent power. In a purely capacitive circuit, current leads voltage by 90° – while a purely inductive one is characterized by current lagging voltage by 90°. In general, ϕ depends on resistance, capacitance, and inductance of the load. When both capacitive and inductive elements are present, the circuit is termed reactive – and an active power, $\mathcal{P}$, may be defined by

$$\mathcal{P} = \|\mathbf{V}\| \|\mathbf{I}\| \cos\phi = VI\cos\phi, \tag{3.2}$$

whereas a reactive power, $\mathcal{Q}$, may likewise be defined by

$$\mathcal{Q} = \|\mathbf{V}\| \|\mathbf{I}\| \sin\phi = VI\sin\phi. \tag{3.3}$$

The active power of a circuit denotes a real energy flow; in the case of purely resistive circuits, $\phi=0°$, so cos $\phi=1$ – and $\mathcal{P}$ coincides with $\mathcal{S}$, see Eq. (3.2) *vis-à-vis* with Eq. (3.1), with power being dissipated in full as heat. However, $\mathcal{P}$ will instead appear in the form of work, or mechanical energy, in an electric engine. Equations (3.1)–(3.3) support a right triangle – with $\mathcal{S}$ serving as hypothenuse, $\mathcal{P}$ serving as horizontal cathetus, and $\mathcal{Q}$ serving as vertical cathetus.

Several types of electric motors of the magnetic type exist, with differences related primarily to their mode of starting; the three most important configurations, in food plant settings, are the universal motor and the squirrel-caged induction motor (both single-phase motors), as well as the wound-rotor synchronous motor (three-phase motor) – see Fig. 3.11. Choice of one over another should consider efficiency in conversion of electrical to mechanical energy, power supply available, period and mode of utilization, power load intended, and free room space.

Single-phase refers to type of electric current generated by a single set of windings in an electric generator – designed to convert some form of mechanical power to electric voltage; it uses two wires for transport (phase and neutral). A generator is but an electric motor operated in reverse; hence, its rotor is a magnet that produces magnetic lines as it rotates – and such lines generate, in turn, a voltage in the iron frame (or stator) that holds the windings. The voltage produced becomes the driving force for AC.

When three sets of windings are present in the stator, three-phase electric current is generated – which may be transported using three wires as well; the active power per phase still abides to Eq. (3.2). Three AC voltages being simultaneously generated reduces oscillations in the overall voltage – so transport becomes more economical, since only one-third of conductor material is used to transmit the same amount of electric power. Even more uniform power transfer, by cancelling phase currents, would in theory be possible with any number (greater than three) of phases; however, more that three phases unnecessarily complicates the infrastructure, in view of the marginal gain thus obtained – whereas two-phase power results in a less smooth (or pulsating) torque in a generator or motor.

As mentioned above, one AC current is obtained from each of the three coils in a three-phase generator – and the corresponding windings are arranged in such a way that the said currents vary in a sinusoidal manner, with the same frequency; furthermore, the peaks and troughs of their waves offset each other, via relative phase separation of one-third of a cycle (or $2\pi/3$ rad). The two basic configurations are schematized in Figs. 3.12i and ii. The Δ-, or delta-configuration requires only three wires for transmission, while the Y-, or star-configuration, may have an (optional) fourth wire; when present, the latter is provided as a neutral – and is normally grounded, thus conducting electricity should the source or the load become unbalanced. The four-wire system, holding symmetrical voltages between phase and neutral, is obtained when the neutral is connected to the common star point of all supply windings; under such circumstances, the three phases will share the same voltage magnitude relative to the neutral. The four-wire Y-system is employed when a combination of single- and three-phase loads are to be served, such as mixed lighting and motor loads.

The polarity of stator poles is periodically reversed, so that their combined magnetic field rotates within the core; since unlike poles attract each other, the moving magnetic field of the rotor is associated with the opposite polarity of the stator – and torque is consequently generated on the rotor (as emphasized before). The said torque moves the shaft mounted on its rotor; this association between moving magnetic field and moving rotor is termed synchronization – and the speed at which rotor and stator synchronize with each other is called synchronous speed. A motor performs at its maximum capability when the moving magnetic field generated by the stator is used solely to rotate (the

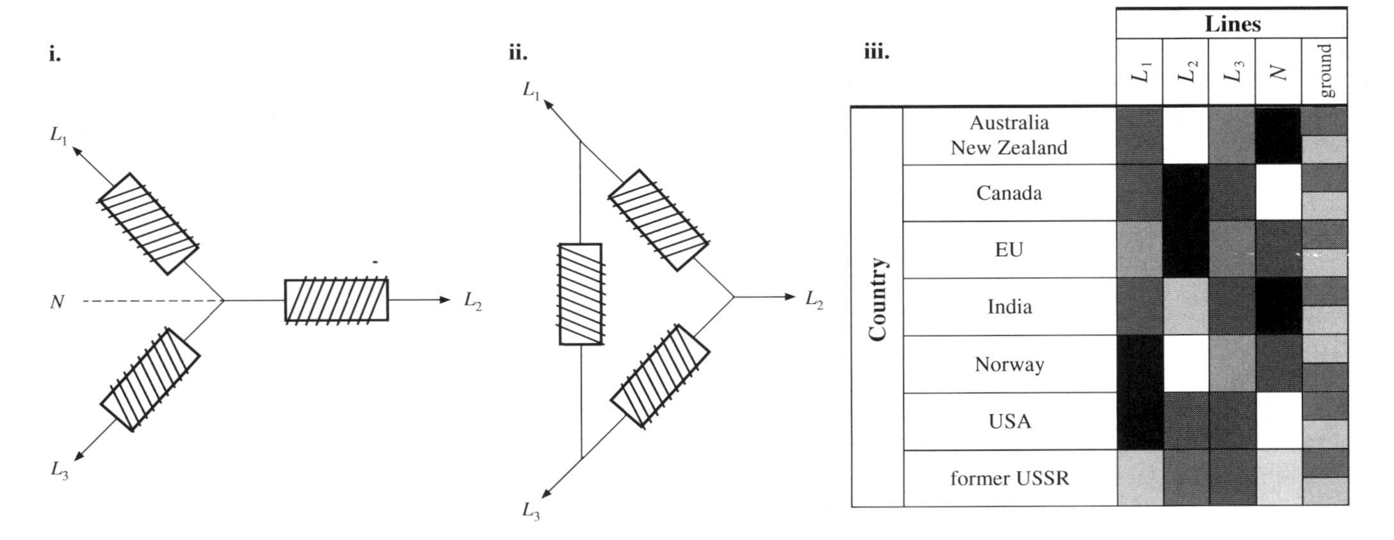

FIGURE 3.12 Schematic representation of (i) star configuration and (ii) delta configuration for three-phase electric generator; and (iii) color codes utilized for the corresponding wires (energized, L_1–L_3, or neutral, N) in selected countries.

shaft of) the rotor; in other words, the synchronous speed of stator and rotor equals the actual speed of the motor shaft – with zero slip. A more realistic scenario, however, corresponds to rotor speed lagging stator magnetic field speed – which produces slip between rotor and stator; when it increases, so do electric losses (termed slip losses) – thus calling for a higher current to feed the motor for an intended output. Besides incurring in a larger consumption of energy, slip losses increase physical burden upon stator coils – with consequent stator overheating; this contributes to premature failure of the electric motor. Besides slip losses, electric motors are also susceptible to Joule's losses – due to dissipation of electrical current as heat, in both stator and rotor; eddy and circulating current losses in windings – due to parasitic currents induced therein; iron losses – mainly in the stator, due to hysteresis in stator laminations; parasitic losses – due to induced currents in all metallic components (e.g. bolts, frame), friction and windage, cooling vents and bearings; and exogenous losses – arising from auxiliary equipment, namely excitation and lubrication oil pumps.

A commutated, electrically excited series- or parallel-wound motor is referred to as universal motor – as it can be designed to operate on either AC or DC. It can use the former because the current in both the field and armature coils, and thus the resulting magnetic field alternates in synchronism (or reverses polarity); hence, the resulting force is generated in a constant direction of rotation. This type of motors entail a wide application in both industrial devices and home appliances – and often deliver up to 1 kW of power. One major advantage is that AC supplies may be used together with features typical of DC motors – specifically high starting torque, compact design, and high running speeds; conversely, maintenance costs caused primarily by the commutator (namely, its brushes), besides acoustic noisiness constitute major shortcomings. They are appropriate for such devices as food mixers and power tools – in view of their intermittent use, along with requirement for a high starting torque; they also lend themselves to electronic speed control, and can operate in reverse by switching the field winding with respect to the armature. Unlike squirrel-caged induction motors (to be discussed next) – which cannot turn the shaft faster than allowed by the power line frequency, universal motors can run at higher speeds, up to 30,000 rpm in e.g. miniature food grinders.

A squirrel-caged induction motor is an asynchronous AC motor, in which power is transferred to the rotor by electromagnetic induction – much like what happens in an electric transformer; in fact, this motor resembles a rotating transformer, because the stator is essentially the primary side of the transformer, and the rotor is its secondary side. It possesses a heavy winding made up of solid bars (usually of aluminum or copper), joined by rings and the ends of the rotor; when the said bars and rings are considered as a whole, they resemble a pet animal's rotating exercise cage (thus justifying its name). Currents induced into the winding create the rotor magnetic field; and the shape of the rotor bars determines the speed-torque features. When operated at low speeds, the current induced in the squirrel cage is nearly at line frequency – and tends to materialize in the outer parts of the rotor cage; as the motor accelerates, the slip frequency becomes lower – and more current resides in the winding inside. By shaping the bars to change resistance of the winding portions in the inner and outer parts of the cage, a variable resistance may effectively be inserted in the rotor circuit.

The wound-rotor synchronous motor is the most common example of a three-phase induction motor in industrial settings; although more costly, it bears a simple design, a high efficiency, and an inherently high starting torque – further to its ability to operate at fixed or variable speed. It is also more compact, and undergoes less vibration as a consequence of a more stable voltage – so it is less noisy and lasts longer than its single-phase equivalent; hence, it is typically the choice when power demand exceeds 5 kW. Its rotor winding is made of many turns of insulated wire, and connected to slip rings on the motor shaft; an external resistor, or other speed control device can be connected in the rotor circuit – even though the former may entail significant power dissipation. This motor is appropriate to start a high inertia load, or a load that requires a very high starting torque across the full speed range; its speed can be changed because the torque curve is modified by the amount of resistance connected to the rotor circuit – with an increasing resistance causing a decrease in speed at maximum torque. When employed with a load for which torque increases with speed, the motor will operate at the speed where motor and load torque coincide; reducing load will cause the motor to speed up, while increasing load will cause the motor to slow down – until the torques become equal once again. If operated in this manner, the slip losses are dissipated in the secondary resistors – but may be quite significant; a converter can instead be fed from the rotor circuit, to return the slip-frequency power (which would otherwise be wasted) back into the power system – through an inverter, or separate power-generator.

Since the invention of the first constant-potential transformer, these apparatuses have become essential for transmission, distribution, and utilization of AC. The simplest representation of a transformer is provided in Fig. 3.13i; it basically consists of one core and two windings – with primary voltage applied to the primary winding, and secondary voltage delivered by the secondary winding. Closed-core transformers are constructed in core-form (see Figs. 3.13iv and v) or shell-form (see Figs. 3.13vi and vii); in the former type, winding surrounds the core, while winding is surrounded by the core in the latter. Shell-form designs are more prevalent for distribution transformer applications – due to the relative ease of stacking the core around winding coils; conversely, core-form designs are more economical for high voltage power transformer applications.

Transformer cores are typically manufactured with high permeability silicon steel; this material possesses a permeability many times that of free space, so the core serves to greatly reduce the magnetizing current – while confining the flux to a path that closely couples the windings. Such a material is normally supplied in laminated form – with each metal layer separated from its neighbor(s) by a thin, nonconductive layer of insulation.

Windings are often arranged concentrically – so as to keep flux leakage to a minimum. Power transformers use multiple-stranded conductors, so as to avoid nonuniform distribution of current; each strand is insulated individually – and strands are arranged so that each portion occupies different relative positions in the complete conductor, at certain points in the winding or throughout the whole winding. This transposition equalizes current flowing

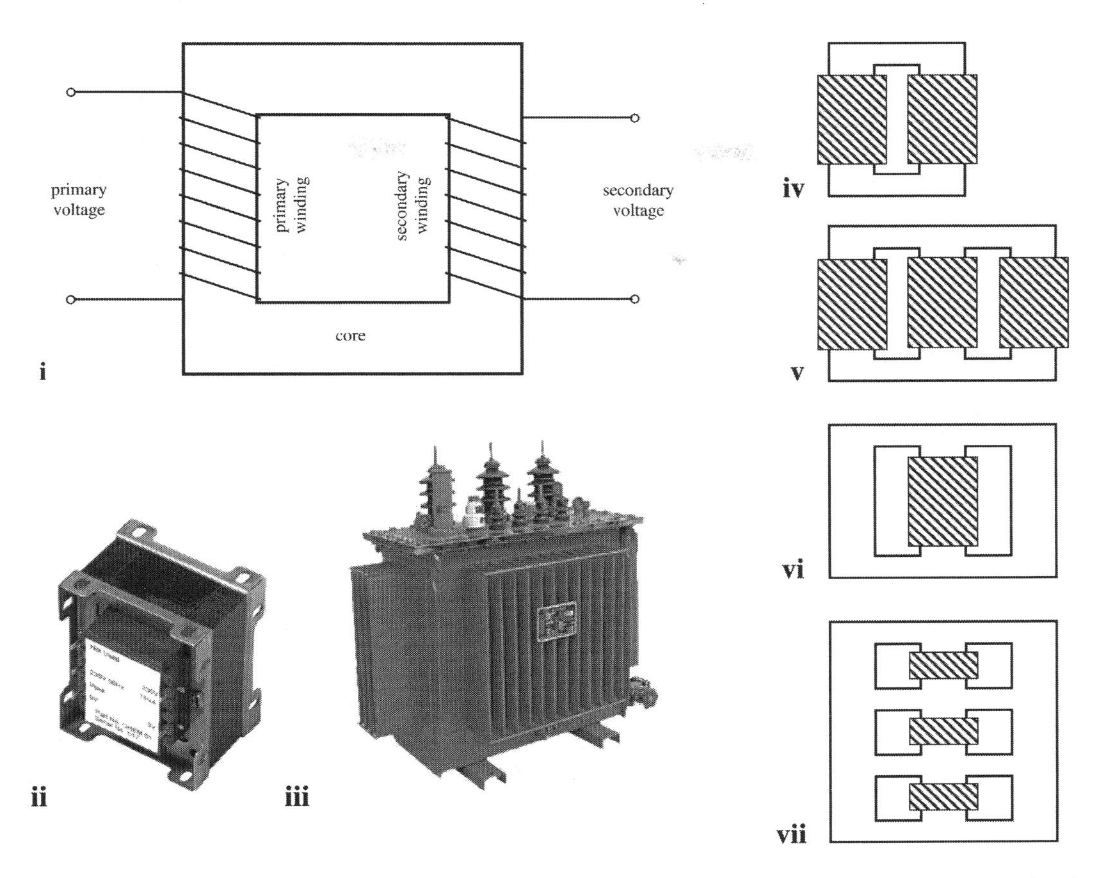

FIGURE 3.13 Schematic representation of (i) plain electric transformer, and (iv, v) core-type or (vi, vii) shell-type transformers – designed to work with (iv, vi) single-phase or (v, vii) three-phase AC; and examples of (ii) small size, single-phase, open-frame transformer (courtesy of Majestic Transformer, Dorset, UK) and (iii) large size, oil-bathed, finned three-phase transformer (courtesy of ProtekG Power Electronics, Ahmedabad, India).

in each strand of the conductor, and reduces eddy current losses in the winding itself; the stranded conductor is also more flexible than a solid conductor of similar size, thus making its manufacture easier. The conducting material used for windings depends on the application foreseen; in any case, individual turns must be electrically insulated from each other, so as to ensure that current travels throughout every turn rather than engaging in short-circuits. In small power transformers – where currents are relatively low and potential difference between adjacent turns is small, coils are often wound using enameled magnet wire; larger power transformers, designed for operation at high voltages, are preferably wound with copper rectangular strip conductors – insulated by oil-impregnated paper, and blocks of pressboard.

As a rule of thumb, the life expectancy of electrical insulation is halved for every 10°C increase in regular operating temperature; therefore, care is to be exercised to avoid excessive heating, especially during long-run operation – to prevent fast decay in insulation features, and eventual (catastrophic) transformer failure. Small dry-type and liquid-immersed transformers are self-cooled by natural convection and radiation; as power ratings increase, one needs to resort to forced-air, -oil or -water cooling (or combinations thereof). Larger transformers are filled with a highly refined mineral oil, which serves a dual purpose – insulation and cooling of windings; the storage tank often has fins on its outer walls to promote natural convection, yet electric fans for forced air-cooling or heat exchangers for forced water-cooling may still be needed. Polychlorinated biphenyls exhibit properties that once favored their application as dielectric coolant – but rising concerns over their environmental persistence led to a widespread ban on their use; they were forbidden in all new equipment built after 1981. Nontoxic, stable silicone-based oils may be employed – where fire-resistance of the liquid offsets additional construction costs of a transformer vault.

As emphasized before, all magnetic flux generated by passage of an electric current in the primary winding of an ideal transformer is transferred to the secondary winding – where it will generate another electric current; in real situations, however, some flux transverses paths that take outside the windings – thus giving rise to leakage flux. This phenomenon entails energy being alternately stored in and discharged from the magnetic fields, within each cycle of power supply. Unless it intercepts nearby conductive materials, such as the transformer support structure (in which case dissipation as heat will occur), leakage flux does not represent an actual power loss; however, it hampers strict voltage regulation – so the secondary voltage may not be directly proportional to its primary counterpart, particularly under heavy loads. Other unwanted losses include: hysteresis – due to nonlinear application of voltage in the core; eddy current losses – due to Joule's heating in the core; and heating of both windings – due also to Joule's effect, as a result of their inherent resistances. Furthermore, parasitic capacitance and self-resonance, owing to electric field distribution, may also occur; the former results from capacitance between adjacent turns in a

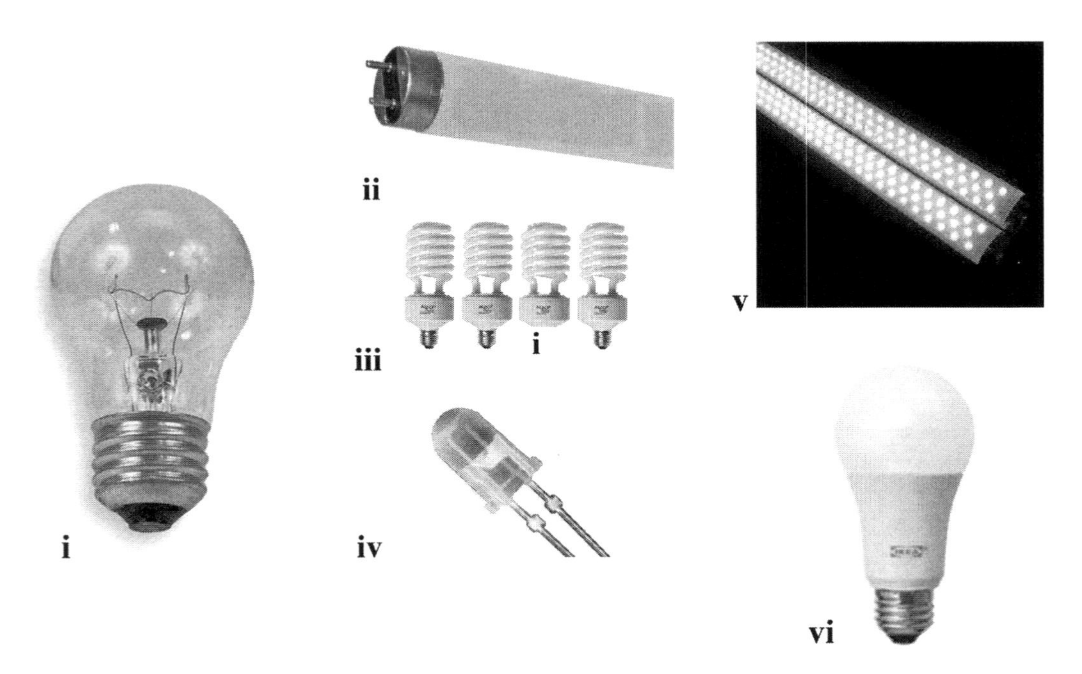

FIGURE 3.14 Examples of lighting devices: (i) incandescent lamp (courtesy of General Electric, Boston, MA); fluorescent lamps, in (ii) tube form (courtesy of Osram Sylvania, Wilmington, MA) and (iii) compact form (courtesy of ALZO Digital, Bethel, CT); and light-emitting diodes, in (iv) unit form (courtesy of Alarmtec, Devon, UK), (v) combined form (courtesy of GetTesting, Sheffield, UK), and (vi) bulb-type shape (courtesy of RYET/IKEA, Leiden, The Netherlands).

given layer, between adjacent layers, or between core and layers adjacent thereto. Still another effect accompanying transformer operation is audible noise (or hum) – especially objectionable in transformers supplied at power frequencies; it results from slight physical expansion and contraction of the transformer core (as happens with any ferromagnetic material) along each cycle of the magnetic field – a phenomenon known as magnetostriction.

Hysteresis and eddy current losses are constant at all load levels, and so dominate overwhelmingly without load; hysteresis dissipates electrical power as heat, at a rate proportional to $\omega B_{max}{}^{1.6}$ as per Steinmetz's formula – where ω denotes current frequency, and B_{max} denotes maximum flux density. Eddy current losses are produced in the metal core and bring about heating – being a complex function of ω^2 and L^2, where L denotes core thickness; they may be reduced by constructing the core as a stack of juxtaposed metal plates, electrically insulated from each other – rather than a solid block (as already explained). The effect of laminations is to confine the eddy currents to highly elliptical paths that enclose little flux – thus reducing the magnitude of such losses; thinner laminations are more effective, but are also more laborious, and thus more expensive to build. Joule's losses increasingly dominate as load increases; a higher frequency, coupled with proximity and skin effects also contribute to increase this type of losses. In any case, transformer efficiency normally improves with increasing capacity.

Electric power is frequently used to provide illumination of work spaces in food processing plants; efficiency and comfort of workers indeed hinge on proper lighting – i.e. characterized by sufficient (but not excessive) power and uniform distribution. Furthermore, the light source should be properly supported, as well as easily repaired or serviced. The intensity of light is measured in lux – corresponding to the magnitude of illumination at 1 m distance from a standard candle; while the amount of light is measured in lumen – corresponding to the magnitude of illumination received by 1 m^2 of surface when light intensity is 1 lux.

Three major types of light sources are present in food processing plants: incandescent lamps, fluorescent lamps, and light-emitting diodes (LED) – as illustrated in Fig. 3.14. Incandescent lamps typically resort to a tungsten filament as resistance to electric current; the high resistance of the wire produces a notable increase in temperature, arising from Joule's effect – high enough to produce light within the visible range, as per Planck's law of emissive power, see Eq. (2.120) of *Food Process Engineering: thermal & chemical operations*, but below the melting temperature of tungsten (i.e. 3422°C). The filament accordingly glows – but efficiency of conversion to visible light is poor, ca. 20 lumen.W^{-1}.

A fluorescence lamp uses an inductance coil to create a current discharge within a tube; the associated heat removes electrons from mercury vapor inside the said tube – and spontaneous return of such electrons to their fundamental state releases ultraviolet photons. These photons stimulate, in turn, phosphor crystals on the inner tube surface to produce light – a phenomenon termed fluorescence. These lamps are two- to threefold more efficient than incandescent lamps of the same (nominal) power – and their useful life is much longer, usually between 6,000 and 15,000 hr. In order to make them handier – and also compatible with most fixtures designed for incandescent bulbs, fluorescent lights are currently available in compact formats – with integral ballasts and screw bases.

A LED is a two-lead semiconductor light source, of small size (<1 mm^2), consisting of a specialized *p*–*n* junction diode. As soon as a suitable voltage is applied to the leads, holes from the *p*-type region and electrons from the *n*-type region enter the aforementioned junction, and recombine like in a regular diode – thus enabling flow of an electric current; some of the associated energy is then released in the form of photons characterized by frequencies within the visible range – a phenomenon termed electroluminescence. The color of light corresponds to the energy of the photon(s) at stake – and is determined by the energy band gap of the semiconductor; integrated optical components can be used to shape the radiation pattern – and bunches of diodes are to be combined for higher intensities or use with regular fixtures.

Choice among the above alternatives depends on number of supports necessary/feasible, coefficient of utilization of space (accounting for room size and shape, and recommended light intensity), light loss factor (accounting for room surface and lamp dust, as well as lumen depreciation), replacement cost (accounting for bulb individual cost and lifespan), and efficiency of conversion of electric power to lighting.

3.2.5 Automatic control

A food process as a whole, or each food operation in particular may, in its simplest form, be schematically viewed as a system with two inputs – one susceptible of change due to interaction with the environment (or disturbance variable, x_{dt}), and the other deliberately changed by the operator (or manipulated variable, x_{mp}); this material process is depicted in Fig. 3.15i, in block form.

Automatic control is required to keep the aforementioned system under steady state operation – previously set so as to optimize some objective function; hence, a virtual process should be added, connected via some interface, which receives information from the material system and consequently triggers a change thereof – as depicted in Fig. 3.15ii, encompassing the classical situation of feedback control. This entails an initial measure (e.g. temperature by thermocouple – see Fig. 3.15iii, pressure by manometer, level by photoelectric cell), taken/transmitted as an electric signal, z_{ms}, via a transducer (or measuring device) – which correlates with process conditions, and provides information on current system status.

A comprehensive list of sensors is provided in Table 3.1. Sensors provide primary measurements (e.g. flowrate, pressure, temperature, weight), comparative measurements (e.g. density), inferred measurements (e.g. hardness), or calculated measurements (e.g. biomass growth in fermenter). Solid-state electronic sensors have overridden chemical types, and also mechanical ones (yet to a lesser extent); this is notably the case of capacitance, near-infrared, or ultrasound cells. Irrespective of their nature or working principle, however, sensors in food processing must be: exempt of contaminants (of both chemical and microbial nature); robust to thermal treatment, and

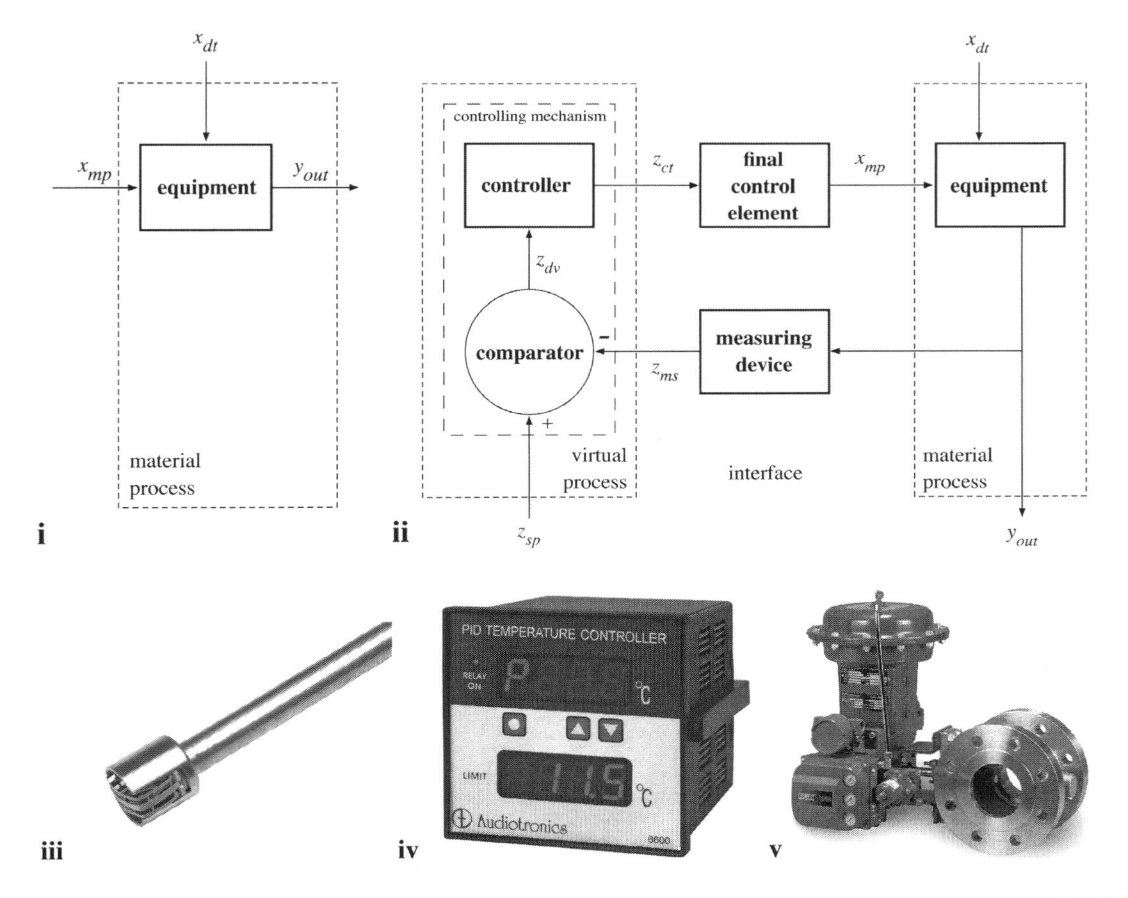

FIGURE 3.15 Schematic representation of material process in (i) uncontrolled and (ii) feedback-controlled configuration – encompassing (i, ii) equipment with two inlets, disturbance (x_{dt}) and manipulation (x_{mp}), and one outlet, product (y_{out}); (ii) interface – encompassing measuring device with one inlet (y_{out}) and one outlet, measurement (z_{ms}), and final control element with one inlet, control action (z_{ct}), and one outlet, manipulated variable (x_{mp}); and (ii) virtual process – encompassing comparator with two inlets, measurement (z_{ms}) and setpoint (z_{sp}), and one outlet, deviation (z_{dv}), and controller with one inlet, deviation (z_{dv}), and one outlet, control signal (z_{ct}); and examples of (iii) measuring device, temperature probe (courtesy of Fluke, Everett, WA), (iv) programmable logic controller, PID (courtesy of Audiotronics, North Hollywood, CA), and (v) final control element, valve (courtesy of Emerson, St. Louis, MO).

TABLE 3.1

List of selected sensors, with corresponding physicochemical measurement(s) and applicability in food processing.

Sensor	Measurement	Application
Absorbance meter	Turbidity	Food fermentations
Acoustic emission	Powder flow	Blending, dehydration
Capacitance	Conductivity	Beverage manufacture, cleaning solution strength, liquid processing
	Level	Filling
	Humidity	Freezing
Electric potential	*pH*	Various liquid foods
Electromagnetic induction	Foreign body detection	Various processes
Diaphragm sensor	Pressure, vacuum	Canning, extrusion, evaporation
Enzyme-coupled sensor	Specific sugars, alcohols	Spoilage of high-risk foods
Fiber-optic sensor	Temperature	Various processes
γ-rays	Density	Liquid foods
Hygrometer	Humidity	Drying, chilled storage
Immunosensor	Specific microorganisms	Pathogens in foods
	Specific toxins	High-risk foods
Mechanical float	Level	Automatic vessel filling
Mechanical resonance	Density	Solid foods
	Level	Filling
	Viscosity	Blending, dairy products
Microwave	Fat, protein, carbohydrate	Various foods
Near infrared	Fat, protein, carbohydrate	Various foods
	Caffeine	Coffee
	Headspace volatiles	Canning, modified atmosphere packaging
	Temperature	Various processes
	Thickness	Laminate packaging
	Water content	Baking, drying
Nuclear magnetic resonance	Solid/liquid ratio	Fresh produce
Proximity switch	Valve position	Various processes
Radiowave	Bulk density	Granules, powders
	Particle size, shape	Dehydration
	Salt content	Pickle brines
Refractometer	Polarized light rotation	Preserves, sugar processing
Resistance thermometer	Temperature	Various processes
Static manometer	Level	Automatic filling of vessels
Strain gauge	Level	Automatic filling of vessels
	Pressure, vacuum	Canning, extrusion, evaporation
	Weight	Checkweight, tank content
Tachometer	Pump/motor speed	Various processes
Thermocouple	Temperature	Various processes
Ultrasound	Counting	Packaging
	Dispersed bubbles	Foam/liquid consistency, interface location
	Dispersed droplets	Emulsion stability
	Level	Filling
	Suspended solids	Wastewater streams
Ultraviolet light	Shape and size	Optical imaging for identification/measurement
Venturi meter	Flow rate	Various processes
Visible light	Color	Color sorting
	Counting	Various foods
Vortex meter	Flow rate	Various processes
X-rays	Foreign body detection	Various processes

high- or low-pressure operation, and resistant to food components (in fluid and particulate forms), cleaning-in-place chemicals, and effluents; inexpensive; and reliable and reproducible (for accurate and consistent information, even when exposed to moisture, steam, dust, fouling, mechanical vibration, or electromagnetic interference). In addition, they should bear easily replaced, disposable heads, require low maintenance, and be unable to carry foreign bodies into the food (e.g. fragile glass components that may constitute safety hazards for the consumer).

Sensors may be positioned in-line (i.e. contacting directly or indirectly with the main flow), on-line (i.e. contacting by-passed flow), at-line (i.e. providing frequent sampling for immediate analysis), or off-line (i.e. conveying periodic sampling, to be processed *a posteriori* by an analytical laboratory). On- and in-line sensors are quite popular, owing to their rapid response; whereas noncontact, at-line sensors exhibit several advantages derived from their resorting to electromagnetic waves (e.g. visible light, infrared radiation, microwave, radiofrequency), ultrasound, or γ-rays.

Current research efforts will likely lead, in the near future, to more widespread use of inexpensive sensors, namely, based on nuclear magnetic, electron spin, and plasmon resonance – as well as time-resolved, diffuse, near-infrared reflectance spectroscopy and antibody-based sensors. Furthermore, electronic noses are available that detect flavor compounds, as well as electronic tongues (once based on mass spectrometers and gas chromatographs) that detect tastes – both resorting to arrays of simple sensors able to measure changes in voltage or frequency, and linked to neural network-based software for pattern recognition. These integrated devices bear a great potential, since they mimic actual sensory responses by the human body.

The aforementioned measure is to be compared with a setpoint, z_{sp} – which constitutes a new input variable (see Fig. 3.15ii); and the difference, z_{dv}, is normally fed to a programmable logical controller (see Fig. 3.15iv) – able to trigger a correction, z_{ct}, using such criteria as proportional, integral or derivative of deviation relative to setpoint. Such devices operate using machine logic, in a way similar to the decision-making logic underlying human thought; their expanding use arises from flexibility in operation and ability to log data for subsequent calculations – besides easy reprogramming to accommodate new products or process changes. These simple, flexible, inexpensive, easy-to-install, and reliable microcomputer-type controllers have typically a fixed program, stored as two modes in their memory: teach mode – allowing instructions be programmed in a straightforward manner by an operator, using a keyboard or touch screen; and run mode – when the program is executed automatically, in response to data received from the sensor(s). Manipulation of a wide range of products that require changes in product formulation and operating conditions can be easily handled via batching – with a code assigned to each set of instructions duly stored in memory, and easily retrievable when needed.

If complex, nonlinear relationships (or lack of knowledge thereof) exist between a measured variable and a process or product, then expert systems or neural networks have been able to address the severe challenges posed by automation. These systems are able to deduce complex relationships (not easily transcribed as analytical expressions), and to quickly learn from experience; hence, they have to be trained prior to application to process control. Their wide versatility sponsors ubiquitous applications – ranging from crop inspection (i.e. check of fruits for such defects as mis-shapes, poor color, undersized items, or foreign bodies), through manufacture of ready-to-eat foods (i.e. monitoring of correct positioning and proportion of ingredients in pizzas), to packaging (i.e. correct placement of labels, and inspection for damage of container or leakage of content).

Another successful approach to control is having computers reason like humans do – but much faster, via fuzzy logic control; these systems do not need precise inputs to generate usable outputs, i.e. such blurred terms as "little", "not so large" or "larger" may be handled as well as strict terms "on/off" or "yes/no". Other advantages include their empirical basis, which circumvents requirement for mathematical models; and the possibility of modification and fine-tuning during operation. Fuzzy logic resorts to simple Boolean algorithms; for instance, if "$x=A$ and $y=B$, then $z=C$", where x and y denote input (measured) variables, z denotes output (manipulated) variable, and A, B, and C denote nonrigid linguistic variables; a rule matrix is a *sine qua non*, containing information (previously gathered and organized) between said inputs and corresponding linguistic output values.

Advances in neural networks, complemented with vision systems, pressure-sensitive grippers, and laser-guidance systems have been widely applied in robotics; the nuclear components of a robotic machine are indeed a (microelectronic) processor that uses information, made available by a sensing device, to recognize a product or perform inspection, and to find its orientation – and servo motors to implement a mechanical action, e.g. through arms and grippers moved so as to appropriately manipulate the product. Modern robots in food processing are capable of maintaining hygienic surfaces, and thus avoid food contamination by e.g. grease, lubricants; at present, they are most frequently used to pick up and replace items, for meat deboning and cutting, and for cake decoration – as well as for primary (or single portion) and secondary (or multiple portion) packaging, carton erection, and palletizing.

Finally, controllers are also expected to survey a process for faults (i.e. self-diagnosis), and provide management information. When a fault is detected or a required condition is not met, adequate alarms or interlocks are to be activated to discontinue processing – with automatic restart, once the malfunction has been fixed. Monitoring also permits collation and storage of data, which are made available at preset intervals (e.g. end of shift, end of day); this piece of information is useful to backup cost analysis and maintenance schedules.

The correction generated by the controller, delivered as a continuous or a discrete on/off signal, is transmitted via electric means to an actuator, or final control element (see Figs. 3.15ii and v) – where some action, x_{mp}, is taken to counteract the original disturbance, and accordingly bring the process/product back to original specifications. Several devices are available for the above actions, such as: magnetic relays – which use a coil to produce electromagnetism that mechanically moves a contact and closes the primary circuit; motors – which provide mechanical motion to implement some action; timing apparatuses – which take advantage of a clock mechanism to mechanically bring two points into contact, and thus complete the electric circuit; or (more frequently) valves – which open or close some aperture, thus controlling flow.

The classical, and most robust form of control resorts to feedback, for possessing two major intrinsic advantages: it is quite flexible – since it responds independently of the cause of disturbance, and requires minimal knowledge of the behavior of the system; and it will keep acting until the deviation (or, at least, most of it) has been eliminated. Its major shortcoming is the need for a disturbance have already been felt by a system before a control action can be triggered – meaning that some product will already be off-specification, and thus likely to be rejected.

To avoid the above shortcoming, one may implement feedforward control, as illustrated in Fig. 3.16. In this case, the disturbance, x_{dt}, is measured before it has a chance to affect the system – and the result, z_{ms}, is duly compared with the setpoint,

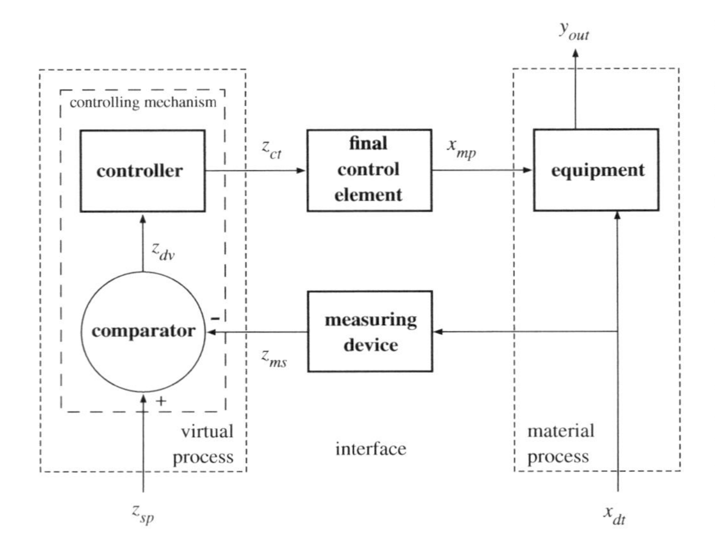

FIGURE 3.16 Schematic representation of material process in feedforward-controlled configuration – encompassing equipment with two inlets, disturbance (x_{dt}) and manipulation (x_{mp}), and one outlet, product (y_{out}); interface – encompassing measuring device with one inlet (x_{dt}) and one outlet, measurement (z_{ms}), and final control element with one inlet, control action (z_{ct}), and one outlet, manipulated variable (x_{mp}); and virtual process – encompassing comparator with two inlets, measurement (z_{ms}) and setpoint (z_{sp}), and one outlet, deviation (z_{dv}), and controller with one inlet, deviation (z_{dv}), and one outlet, control signal (z_{ct}).

z_{sp}; the difference, z_{dv}, is then fed to the controller that delivers a signal, z_{ct}. This signal determines the response of the final control element, which causes a change in x_{mp}, the manipulated variable, aimed at completely eliminating the effect of the disturbance – so that the output of the system, y_{out}, remains unchanged at all times. Despite its perfect performance in theory, feedforward control requires a detailed knowledge of the behavior of the system, including accurate model form and parameter estimates – so it is quite demanding; and can only respond to disturbance(s) accounted for and measured in advance, meaning that spurious changes of all possible inputs have to be considered in the model in terms of their effects upon the controlled variable(s) – so it is poorly flexible. Furthermore, it requires sophisticated, computer-based hard- and software to handle the germane mathematical model, and find the right control action at each time. All in all, a feedforward control rarely stands on its own – being typically coupled to a (more robust and flexible) feedback control.

Extensive developments in both hardware and software have enabled connection of programmable logic controllers to larger computers in supra-control systems; therefore, dedicated control systems, ensuring control of a single unit and communicating with a central control panel, can be made to exchange information with systems alike and a mainframe computer, toward centralized, overall control aimed at maximum operating efficiency. Despite the advantages of accessing an overall picture to optimize use of raw materials and energy, while minimizing generation of wastes and effluents, putative failure of the central computer will cause whole plant shutdown – unless a backup computer, of similar capacity (and price), in on place. This justifies why distributed systems are preferred in food processing – where the dedicated controls of a given portion of the process are interlocked via a medium-sized computer, for which some degree of redundancy is more reasonable from an economic standpoint. Although more expensive overall in capital and programming costs than a centralized system, the said intermediate approach is more flexible to the frequent changes in processing conditions and raw material/product specifications, and does not suffer from the vulnerability of total process shutdown should one component fail. A hybrid approach, i.e. integration of control systems, is possible – meaning that the distributed small computers, used for control in various sections of the process, are interconnected to form a larger management information system; these are linked, in turn, to the central computer, via a communications network. Mass data storage, sophisticated data manipulation, and communication capacities with other management computers are thus utilized to advantage, as they provide supporting information for marketing, quality assurance, and plant maintenance – further to process control.

In terms of software developed specifically for control, a special mention is deserved by Supervisory Control And Data Acquisition (SCADA) – which collects data from programmable logic controllers in charge of a piece of equipment or a part of a process, and displays them to plant operators in real-time as animated graphics; such graphics are interactive, so the operator can easily adjust most process parameters and variables. Analysis of trends and recall of historical data, in spreadsheet form, are possible via Dynamic Data Exchange (DDE) – linked to office computer systems; the latter can thus be used to adjust recipes and schedule batches in real time, as well as produce management reports. Communication between different software systems may resort to Open Data-Base Connectivity (ODBC) and Object-linking and -embedding for Process Control (OPC) standards – thus permitting plant status be incorporated into company business systems. More recently, Common Object Resource-Based Architecture (CORBA) has supported development of Manufacturing Execution Systems (MES) – which act as an information broker between SCADA, OPC, and office business systems; besides being able to exchange information also with barcode readers, checkweighing apparatuses, formulation programs, and laboratory systems. This information can be used to increase productivity and reduce downtime – as well as track ingredients and products throughout the whole manufacturing process, thus generating a product genealogy. If a problem occurs, such a genealogy can be taken advantage of to quickly traceback, through each stage of the process, until establishing the likely cause – and ascertain which products may have been affected; this factual information is crucial to rationally decide on the extent of product recall.

Complementary management systems include Enterprise Resource Planning (ERP) and Materials Requirement Planning (MRP) – which aid in coordinating decisions on ordering, stock levels, work-in-progress, storage, and distribution of final products. The latter has meanwhile expanded so as to encompass other parts of the business, thus giving rise to Manufacture Resource Planning version II (MRPII) – a single integrated system, with a database accessible by engineering and process management, sales and marketing, and finance and accounting departments. Manufacture control systems allow one to keep track of movement of each ingredient, work-in-progress, and final product during production and into stores – with production date and time being recorded for each unit of product, and each pallet being labeled with a barcode or a radiofrequency identification tag;

hence, accurate inventories can be kept at all times, while ensuring first-in-first-out rotation of both ingredients and final products.

3.3 Mathematical simulation

3.3.1 Fuel combustion

The major components of fossil fuels, possessing an enthalpic interest toward burning, are of the generic chemical form $C_mH_nO_q$ – where m, n, and q denote integer numbers; for instance, $m=1$, $n=4$, and $q=0$ in the case of methane (of chemical formula CH_4), or $n=2$ and $m=q=0$ in the case of hydrogen (of molecular formula H_2).

The basic chemical equation for complete combustion eventually leads to quantitative conversion of carbon to CO_2 and hydrogen to H_2O, according to

$$\boxed{C_mH_nO_q + rO_2 \rightarrow sCO_2 + tH_2O} \tag{3.4}$$

– where r, s, and t denote stoichiometric coefficients; an elemental mass balance requires that

$$m = s \tag{3.5}$$

pertaining to carbon, and

$$n = 2t \tag{3.6}$$

encompassing hydrogen – coupled with

$$q + 2r = 2s + t, \tag{3.7}$$

associated with oxygen. After retrieving Eq. (3.5) as

$$s = m \tag{3.8}$$

for known m, and isolating t in Eq. (3.6) as

$$t = \frac{n}{2} \tag{3.9}$$

related to given n, one may solve Eq. (3.7) for r as

$$r = \frac{2s + t - q}{2}; \tag{3.10}$$

combination with Eqs. (3.8) and (3.9) transforms Eq. (3.10) to

$$r = \frac{2m + \frac{n}{2} - q}{2}, \tag{3.11}$$

or else

$$\boxed{r = m + \frac{n}{4} - \frac{q}{2}} \tag{3.12}$$

as minimum molar ratio of molecular oxygen to original fuel able to assure quantitative combustion. Since air is composed of ca. 21%(mol/mol) O_2 and 79%(mol/mol) N_2 – corresponding to a ratio $\nu_{N:O}$ equal to 0.79/0.21 = 3.762, then nitrogen must also be taken into account (as inert) in the atmospheric mixture fed to the burner.

The first step in modeling of burner operation is computing the enthalpy of reaction, $\Delta h_{c,C_mH_nO_q}\big|_{T=T_{in}}$, at inlet temperature, T_{in} – knowing that enthalpy is a function of state. One should accordingly bring reactants and inerts from T_{in} to a reference temperature T^θ (usually 15°C) – as manifestation of sensible heat, with $c_{P,C_mH_nO_q}$, c_{P,O_2} and c_{P,N_2} denoting (isobaric) heat capacity of $C_mH_nO_q$, O_2, and N_2, respectively; perform combustion at T^θ, characterized by $\Delta h^\theta_{c,C_mH_nO_q} < 0$ as molar latent heat of reaction – yielding water in the liquid state; then vaporize water at T^θ, with $\Delta h^\theta_{v,H_2O}$ as molar latent heat of phase change; and finally bring products (including water vapor) and inerts back to T_{in} in a second manifestation of sensible heat – with c_{P,CO_2} and c_{P,H_2O} denoting (isobaric) molar heat capacity of CO_2 and (vapor) H_2O, respectively. In mathematical terms, this reasoning looks like

$$\begin{aligned}\Delta h_{c,C_mH_nO_q}\Big|_{T_{in}} &= \left(c_{P,C_mH_nO_q} + rc_{P,O_2} + \nu_{N:O}rc_{P,N_2}\right)\left(T^\theta - T_{in}\right)\\ &+ \Delta h^\theta_{c,C_mH_nO_q} + t\Delta h^\theta_{v,H_2O}\\ &+ \left(sc_{P,CO_2} + tc_{P,H_2O} + \nu_{N:O}rc_{P,N_2}\right)\left(T_{in} - T^\theta\right),\end{aligned} \tag{3.13}$$

where $c_{P,i}$ is assumed to remain essentially constant within $[T^\theta, T_{in}]$.

The second step is an enthalpy balance to the burner itself, assumed to operate adiabatically, and to utilize the stoichiometric amount of molecular oxygen (provided by atmospheric air) for full combustion – using inlet temperature, T_{in}, as reference temperature to estimate the enthalpy associated with reactants, inerts, and products; this leads to

$$\begin{aligned}&F_{C_mH_nO_q}\left(c_{P,C_mH_nO_q} + rc_{P,O_2} + \nu_{N:O}rc_{P,N_2}\right)\left(T_{in} - T_{in}\right)\\ &+ F_{C_mH_nO_q}\left(-\Delta h_{c,C_mH_nO_q}\Big|_{T_{in}}\right)\\ &= F_{C_mH_nO_q}\left(sc_{P,CO_2} + tc_{P,H_2O} + \nu_{N:O}rc_{P,N_2}\right)\left(T_{out} - T_{in}\right),\end{aligned} \tag{3.14}$$

where $F_{C_mH_nO_q}$ denotes molar flow rate of burning fuel – and where the first term in the left-hand side and the right-hand side represent rate of enthalpy in and out, respectively, by convection, and the remaining term represents rate of enthalpy generation via chemical reaction. After dropping the first term for being nil, and dividing both sides by $F_{C_mH_nO_q}$, Eq. (3.14) reduces to

$$-\Delta h_{c,C_mH_nO_q}\Big|_{T_{in}} = \left(sc_{P,CO_2} + tc_{P,H_2O} + \nu_{N:O}rc_{P,N_2}\right)\left(T_{out} - T_{in}\right) \tag{3.15}$$

– where insertion of Eq. (3.13) unfolds

$$\begin{aligned}&\left(c_{P,C_mH_nO_q} + rc_{P,O_2} + \nu_{N:O}rc_{P,N_2}\right)\left(T_{in} - T^\theta\right) - \Delta h^\theta_{c,C_mH_nO_q} - t\Delta h^\theta_{v,H_2O}\\ &- \left(sc_{P,CO_2} + tc_{P,H_2O} + \nu_{N:O}rc_{P,N_2}\right)\left(T_{in} - T^\theta\right)\\ &= \left(sc_{P,CO_2} + tc_{P,H_2O} + \nu_{N:O}rc_{P,N_2}\right)\left(T_{out} - T_{in}\right);\end{aligned} \tag{3.16}$$

upon condensation of terms alike, Eq. (3.16) gives rise to

$$\left(sc_{P,CO_2} + tc_{P,H_2O} + \nu_{N:O} rc_{P,N_2}\right)\left(T_{out} - T^\theta\right)$$
$$= \left(c_{P,C_mH_nO_q} + r\left(c_{P,O_2} + \nu_{N:O} c_{P,N_2}\right)\right)\left(T_{in} - T^\theta\right), \qquad (3.17)$$
$$+ \left(-\Delta h^\theta_{c,C_mH_nO_q}\right) - t\Delta h^\theta_{v,H_2O}$$

along with convenient isolation of the term in $T_{out} - T^\theta$ in the left-hand side and factoring out of r in the right-hand side. Upon solution for $T_{out} - T^\theta$ itself, complemented by multiplication of both sides by $c_{P,O_2}/\Delta h^\theta_{v,H_2O}$, Eq. (3.17) becomes

$$\frac{c_{P,O_2}\left(T_{out} - T^\theta\right)}{\Delta h^\theta_{v,H_2O}} = \frac{c_{P,C_mH_nO_q} + r\left(c_{P,O_2} + \nu_{N:O} c_{P,N_2}\right)}{sc_{P,CO_2} + tc_{P,H_2O} + \nu_{N:O} rc_{P,N_2}} \frac{c_{P,O_2}\left(T_{in} - T^\theta\right)}{\Delta h^\theta_{v,H_2O}} + \frac{c_{P,O_2}\left(\dfrac{-\Delta h^\theta_{c,C_mH_nO_q}}{\Delta h^\theta_{v,H_2O}} - t\right)}{sc_{P,CO_2} + tc_{P,H_2O} + \nu_{N:O} rc_{P,N_2}}; \qquad (3.18)$$

Eqs. (3.8), (3.9), and (3.12) may now be inserted to eventually obtain

$$\frac{c_{P,O_2}\left(T_{out} - T^\theta\right)}{\Delta h^\theta_{v,H_2O}} = \frac{c_{P,C_mH_nO_q} + \left(m + \dfrac{n}{4} - \dfrac{q}{2}\right)\left(c_{P,O_2} + \nu_{N:O} c_{P,N_2}\right)}{mc_{P,CO_2} + \dfrac{n}{2} c_{P,H_2O} + \nu_{N:O}\left(m + \dfrac{n}{4} - \dfrac{q}{2}\right) c_{P,N_2}} \frac{c_{P,O_2}\left(T_{in} - T^\theta\right)}{\Delta h^\theta_{v,H_2O}} + \frac{c_{P,O_2}\left(\dfrac{-\Delta h^\theta_{c,C_mH_nO_q}}{\Delta h^\theta_{v,H_2O}} - \dfrac{n}{2}\right)}{mc_{P,CO_2} + \dfrac{n}{2} c_{P,H_2O} + \nu_{N:O}\left(m + \dfrac{n}{4} - \dfrac{q}{2}\right) c_{P,N_2}} \qquad (3.19)$$

in dimensionless form – where m, n, and q, and obviously $c_{P,C_mH_nO_q}$ and $\Delta h^\theta_{c,C_mH_nO_q}$, appear as characteristic parameters of the compound used as fuel. Since $\Delta h^\theta_{v,H_2O} = 4.442 \times 10^4$ J.mol^{-1}, $c_{P,O_2} = 14.67$ J.mol^{-1}.K^{-1}, $c_{P,N_2} = 14.55$ J.mol^{-1}.K^{-1}, $c_{P,CO_2} = 18.32$ J.mol^{-1}.K^{-1}, $c_{P,H_2O} = 33.51$ J.mol^{-1}.K^{-1}, and $\nu_{N:O} = 3.762$ (as seen before), one may redo Eq. (3.19) to

$$\frac{14.67\left(T_{out} - T^\theta\right)}{44420} = \frac{c_{P,C_mH_nO_q} + \left(m + \dfrac{n}{4} - \dfrac{q}{2}\right)(14.67 + 3.762 \cdot 14.55)}{m18.32 + \dfrac{n}{2}33.51 + 3.762\left(m + \dfrac{n}{4} - \dfrac{q}{2}\right)14.55} \frac{14.67\left(T_{in} - T^\theta\right)}{44420} + \frac{14.67\left(\dfrac{-\Delta h^\theta_{c,C_mH_nO_q}}{44420} - \dfrac{n}{2}\right)}{m18.32 + \dfrac{n}{2}33.51 + 3.762\left(m + \dfrac{n}{4} - \dfrac{q}{2}\right)14.55} \qquad (3.20)$$

that breaks down to

$$3.303 \times 10^{-4}\left(T_{out} - T^\theta\right) = \frac{c_{P,C_mH_nO_q} + 69.41m + 17.35n - 34.71}{73.06m + 30.44n - 27.37q} 3.303 \times 10^{-4}\left(T_{in} - T^\theta\right) + \frac{3.303 \times 10^{-4}\left(-\Delta h^\theta_{c,C_mH_nO_q}\right) - 7.335n}{73.06m + 30.44n - 27.37q} \qquad (3.21)$$

as simpler version – quite handy for industrial practice (and valid in *SI*); examples of flame temperature, or T_{out}, calculated under the postulated adiabatic conditions are tabulated in Table 3.2. Each entry (except the last two) corresponds to a fuel constituted by a pure compound; when a mixture of compounds is present (as is often the case), r, and consequently s and t must be recalculated so as to take into account the different requirements of each combustible compound for molecular oxygen – and averages of c_P and Δh^θ_c pertaining to the said fuel have to be taken, weighed by the corresponding mole fractions. Note that the flame temperature is higher when pure oxygen is used rather than air – because no nitrogen is present; hence, the whole heat of reaction released will be used to heat up only the reaction products CO_2 and H_2O, without interference by (inert) N_2 that dilutes sensible heat effects.

If T_{in} coincides with T^θ, then Eq. (3.19) simplifies dramatically to

$$T_{out}\Big|_{T_{in}=T^\theta} = T_{in} + \frac{-\Delta h^\theta_{c,C_mH_nO_q} - \dfrac{n}{2}\Delta h^\theta_{v,H_2O}}{mc_{P,CO_2} + \dfrac{n}{2}c_{P,H_2O} + \nu_{N:O}\left(m + \dfrac{n}{4} - \dfrac{q}{2}\right)c_{P,N_2}} \qquad (3.22)$$

– where multiplication of both sides by $\Delta h^\theta_{v,H_2O}/c_{P,O_2}$ meanwhile took place; oftentimes $T_{in} \approx T^\theta$, so the approximation conveyed by Eq. (3.22) proves sufficiently accurate in practice.

TABLE 3.2

Flame temperature, T_{out}, of selected fuels when burned with oxygen/air – assuming $T^\theta = T_{in} = 15$°C and stoichiometric proportions of oxygen.

Fuel/oxidant	T_{out} (°C)
Acetylene/air	2550
/oxygen	3100
Butane/air	1970
Coal/air	1900
Cyanogen/air	4525
Dicyanoacetylene/oxygen	4982
Hydrogen/air	2111
Methane/air	1950
Propane/air	1980
/oxygen	2800
Natural gas/air	ca. 1970
Wood/air	ca. 1980

TABLE 3.3

Standard specific heat of combustion, Δh_c^θ, at T^θ = 15°C, of selected fuels when burned with oxygen.

Fuel		$-\Delta h_c^\theta$ (J.kg^{-1})
Butane		4.95×10^7
Coal	Anthracite	3.25×10^7
	Lignite	1.50×10^7
Diesel		4.48×10^7
Ethane		5.19×10^7
Ethanol		2.97×10^7
Gasoline		4.70×10^7
Hydrogen		1.42×10^8
Kerosene		4.62×10^7
Methane		5.55×10^7
Natural gas		5.40×10^7
Pentane		4.86×10^7
Propane		5.04×10^7
Turf	Dry	1.50×10^7
	Damp	6.00×10^6
Wood		$1.50–2.17\times 10^6$

Standard heats of combustion (per unit mass), pertaining to 15°C as reference temperature, are made available in Table 3.3; conversion to their molar counterparts ensues, via multiplication by the corresponding (average) molecular weight. The aforementioned heats of combustion are all negative in the case of organic compounds, because the double bond in molecular oxygen is weaker than the double bonds in CO_2 and the single bonds in H_2O, and the bonds C–O and C–H in organic compounds are typically much weaker than those in CO_2 and H_2O. One may, in general, resort to

$$(-\Delta \tilde{h})^{\theta}_{c,C_mH_nO_q} \approx 4.18\times 10^5 (m + 0.3n - 0.5q)\ \text{J.mol}^{-1} \quad (3.23)$$

as empiric rule to estimate (within ±3% error) the molar heat of combustion of an organic compound of chemical formula $C_mH_nO_q$; it is interesting that the content of the parenthesis resembles the right-hand side of Eq. (3.12) – whereas the proportionality factor is approximately equal to one half the standard molar heat of combustion of methane.

In order to assure complete combustion, up to 10% excess air is normally utilized. Without sufficient oxygen, combustion risks being incomplete – with likely production of (poisonous) carbon monoxide, which entails safety hazards to human operators in processing plants. The above upper threshold is set so as to minimize the dilution effect discussed previously for excess oxygen (acting as inert) – while reducing the risk of formation of CO to an acceptable level.

PROBLEM 3.1

Standard heats of reaction (including combustion) are normally tabulated at a single (reference) temperature T^θ, say 288 or 298 K. Consider, in this regard, a reaction with R (gaseous) reactants and P (gaseous) products, with chemical equation reading

$$\sum_{i=1}^{R} \nu_i S_i + \sum_{j=R+1}^{R+P} \nu_j S_j = 0 \quad (3.1.1)$$

– where ν_i denotes stoichiometric algebraic coefficient of i-th substrate S_i; assume that the specific (molar) heat capacity, $c_{P,i}$, is known for all compounds involved.

a) Calculate the enthalpy of reaction, at atmospheric pressure and any temperature T, assuming $c_{P,i}$ to remain essentially constant within the temperature range of interest.
b) repeat *a*), assuming that a second-order empirical correlation of the type

$$c_{P,i} = a_i + b_i T + c_i T^2 \quad (3.1.2)$$

– derived as Taylor's expansion of $c_{P,i}\{T\}$ around T = 0, describes the variation of $c_{P,i}$ with (absolute) temperature.

Solution

a) Since enthalpy is a state function, its change between two states depends only on those states – being thus independent of the path followed between them. Therefore, the enthalpy change accompanying reaction at temperature T, denoted hereafter as Δh_{rxn}, may instead be calculated as the sum of enthalpy change accompanying transfer of sensible heat undergone by reactants between temperatures T and T^θ, or Δh_r, with enthalpy change accompanying reaction at temperature T^θ, or Δh^θ_{rxn}, with enthalpy change accompanying transfer of sensible heat undergone by products between temperatures T^θ and T, or Δh_p; this can be mathematically expressed as

$$\Delta h_{rxn} = \Delta h_r + \Delta h^\theta_{rxn} + \Delta h_p. \quad (3.1.3)$$

Equation (3.1.3) parallels the rationale underlying Eq. (3.13); however, there is no need to account for latent heat of phase change now, since all reactants/products exist in gaseous form per hypothesis. If (isobaric) specific heat capacities are constant (as also postulated), then one may write

$$\Delta h_r = \left(T^\theta - T\right) \sum_{i=1}^{R} (-\nu_i) c_{P,i} \quad (3.1.4)$$

pertaining to reactants brought from T to T^θ – referred to 1 mol of reaction, as working basis; and likewise

$$\Delta h_p = \left(T - T^\theta\right) \sum_{j=R+1}^{R+P} \nu_j c_{P,j} \quad (3.1.5)$$

encompassing products brought from T^θ to T, on the same molar basis. In both cases, a stoichiometric combination of $c_{P,i}$s was taken, since $|\nu_i|$ moles of each reactant and ν_j moles of each product are at stake for each mole of reaction; ν_i was thus converted to its negative in the case of reactants (because $\nu_i<0$, and thus $|\nu_i| = -\nu_i$), for the sake of physical consistency. Insertion of Eqs. (3.1.4) and (3.1.5) converts Eq. (3.1.3) to

$$\Delta h_{rxn} = \left(T^\theta - T\right) \sum_{i=1}^{R} (-\nu_i) c_{P,i} + \Delta h^\theta_{rxn} + \left(T - T^\theta\right) \sum_{j=R+1}^{R+P} \nu_j c_{P,j}, \quad (3.1.6)$$

or else

$$\Delta h_{rxn} = \Delta h_{rxn}^{\theta} + \left(T - T^{\theta}\right)\left(\sum_{i=1}^{R} \nu_i c_{P,i} + \sum_{j=R+1}^{R+P} \nu_j c_{P,j}\right) \tag{3.1.7}$$

after factoring $T-T^{\theta}$ out. The notation in Eq. (3.1.7) may be condensed to

$$\Delta h_{rxn} = \Delta h_{rxn}^{\theta} + \left(T - T^{\theta}\right)\sum_{i=1}^{R+P} \nu_i c_{P,i}, \tag{3.1.8}$$

after having collapsed summations sharing the same functional form for their kernel, and consecutive upper and lower limits – or else

$$\Delta h_{rxn} = \Delta h_{rxn}^{\theta} + \left(T - T^{\theta}\right)\Delta c_P, \tag{3.1.9}$$

in view of the definition of operator $\Delta\xi$, i.e.

$$\Delta\xi \equiv \sum_{i=1}^{R+P} \nu_i \xi_i \tag{3.1.10}$$

in parallel to Eqs. (1.31) or (1.32). The equality labeled as Eq. (3.1.9) is classically known as Kirchhoff's law of thermochemistry – and holds irrespective of the magnitude of T relative to T^{θ}, as long as the $C_{P,i}$s are constant within the temperature range of interest.

b) If (isobaric) heat capacities vary considerably with temperature, then one should resort to

$$\Delta h_r = \int_{T}^{T^{\theta}} \sum_{i=1}^{R} (-\nu_i) c_{P,i}\left\{\tilde{T}\right\} d\tilde{T} \tag{3.1.11}$$

and similarly

$$\Delta h_p = \int_{T^{\theta}}^{T} \sum_{j=R+1}^{R+P} \nu_j c_{P,j}\left\{\tilde{T}\right\} d\tilde{T}, \tag{3.1.12}$$

in lieu of Eqs. (3.1.4) and (3.1.5), respectively; note that constant $c_{P,i}$s would permit $\sum_{i=1}^{R}(-\nu_i)c_{P,i}$ be taken off the kernel in Eq. (3.1.11), with the remaining integral reduced to $\int_{T}^{T^{\theta}} d\tilde{T} = \tilde{T}\Big|_{T}^{T^{\theta}} = T^{\theta} - T$ that retrieves Eq. (3.1.4) pertaining to reactants – and a similar rationale would apply to products. Due to the intrinsic linearity of integral and summation operators, Eq. (3.1.11) may be rephrased as

$$\Delta h_r = \sum_{i=1}^{R} (-\nu_i) \int_{T}^{T^{\theta}} c_{P,i}\{\tilde{T}\} d\tilde{T} \tag{3.1.13}$$

or, after exchanging order of integration at the expense of the preceding minus sign,

$$\Delta h_r = \sum_{i=1}^{R} \nu_i \int_{T^{\theta}}^{T} c_{P,i}\{\tilde{T}\} d\tilde{T}; \tag{3.1.14}$$

Eq. (3.1.12) likewise becomes

$$\Delta h_p = \sum_{j=R+1}^{R+P} \nu_j \int_{T^{\theta}}^{T} c_{P,j}\{\tilde{T}\} d\tilde{T}. \tag{3.1.15}$$

Upon insertion of Eqs. (3.1.14) and (3.1.15), one obtains

$$\Delta h_{rxn} = \sum_{i=1}^{R} \nu_i \int_{T^{\theta}}^{T} c_{P,i}\{\tilde{T}\} d\tilde{T} + \Delta h_{rxn}^{\theta} + \sum_{j=R+1}^{R+P} \nu_j \int_{T^{\theta}}^{T} c_{P,j}\{\tilde{T}\} d\tilde{T} \tag{3.1.16}$$

from Eq. (3.1.3) – where similarity of their kernels, and consecutiveness of upper limit of first summation and lower limit of second summation permit their lumping as

$$\Delta h_{rxn} = \Delta h_{rxn}^{\theta} + \sum_{i=1}^{R+P} \nu_i \int_{T^{\theta}}^{T} c_{P,i}\{\tilde{T}\} d\tilde{T}; \tag{3.1.17}$$

combination with Eq (3.1.2) is finally in order, to get

$$\Delta h_{rxn} = \Delta h_{rxn}^{\theta} + \sum_{i=1}^{R+P} \nu_i \int_{T^{\theta}}^{T} \left(a_i + b_i\tilde{T} + c_i\tilde{T}^2\right) d\tilde{T}, \tag{3.1.18}$$

where the fundamental theorem of integral calculus supports further transformation to

$$\Delta h_{rxn} = \Delta h_{rxn}^{\theta} + \sum_{i=1}^{R+P} \nu_i \left(a_i\tilde{T} + b_i\frac{\tilde{T}^2}{2} + c_i\frac{\tilde{T}^3}{3}\right)\Bigg|_{T^{\theta}}^{T}. \tag{3.1.19}$$

Straightforward algebraic manipulation of Eq. (3.1.19) unfolds

$$\Delta h_{rxn} = \Delta h_{rxn}^{\theta} + \sum_{i=1}^{R+P} \nu_i \left(a_i\left(T - T^{\theta}\right) + \frac{b_i}{2}\left(T^2 - T^{\theta^2}\right) + \frac{c_i}{3}\left(T^3 - T^{\theta^3}\right)\right), \tag{3.1.20}$$

where ν_i can be factored in, $T-T^{\theta}$, $\left(T^2 - T^{\theta^2}\right)/2$, or $\left(T^3 - T^{\theta^3}\right)/3$ (as appropriate) factored out, and the summation split to produce

$$\Delta h_{rxn} = \Delta h_{rxn}^{\theta} + \left(T - T^{\theta}\right)\sum_{i=1}^{R+P} \nu_i a_i + \frac{1}{2}\left(T^2 - T^{\theta^2}\right)\sum_{i=1}^{R+P} \nu_i b_i + \frac{1}{3}\left(T^3 - T^{\theta^3}\right)\sum_{i=1}^{R+P} \nu_i c_i; \tag{3.1.21}$$

after recalling Eq (3.1.10), notation will be condensed in Eq. (3.1.21) as

$$\Delta h_{rxn} = \Delta h_{rxn}^{\theta} + \left(T - T^{\theta}\right)\Delta a + \left(T^2 - T^{\theta 2}\right)\frac{\Delta b}{2} + \left(T^3 - T^{\theta 3}\right)\frac{\Delta c}{3}. \tag{3.1.22}$$

As expected, a constant (isobaric) specific heat capacity would imply $b_i = c_i = 0$ in Eq. (3.1.21), and thus $\Delta b = \Delta c = 0$ – in which case Eq. (3.1.22) would reduce to Eq. (3.1.9), as per Δc_P being equivalent to Δa in this case, see Eq. (3.1.2).

3.3.2 Carnot's thermal engine

Carnot's heat engine represents a particularly simple design for a device able to (partially) transform heat into work. It is efficient from a conceptual point of view, yet difficult to implement in practice; nevertheless, it consubstantiates the standard against which all industrially relevant designs of power production, via (direct or indirect) fuel burning, are judged.

The basis of the said engine encompasses a sequence of four steps – known as Carnot's cycle, and sketched in Fig. 3.17 using either temperature and volume, temperature and entropy, or pressure and volume as coordinates; an ideal gas is normally hypothesized as working fluid, under steady state operation. Steps $1 \rightarrow 2$ and $3 \rightarrow 4$ are isothermal, whereas steps $2 \rightarrow 3$ and $4 \rightarrow 1$ are adiabatic; all such steps are taken as reversible, which implies the latter also being isentropic.

Carnot's engine accordingly entails four distinct devices – as detailed in Fig. 3.18. The devices labeled as expanders can

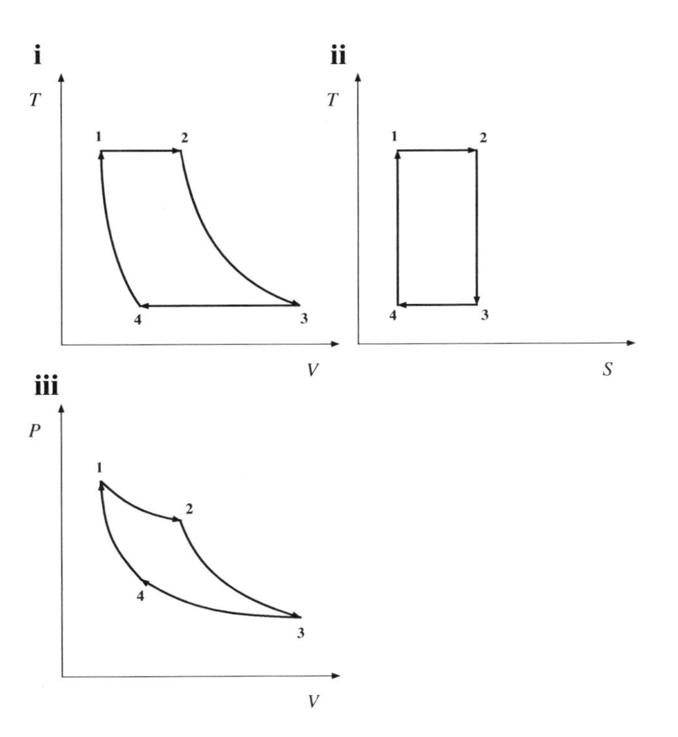

FIGURE 3.17 Schematic representation of Carnot's cycle, pertaining to a heat engine based on an ideal gas – encompassing two isothermal steps, between states **1** and **2**, and **3** and **4**, alternated with two adiabatic steps, between states **2** and **3**, and **4** and **1**, laid out as (i) temperature, *T*, versus volume, *V*, (ii) *T* versus entropy, *S*, and (iii) pressure, *P*, versus *V*.

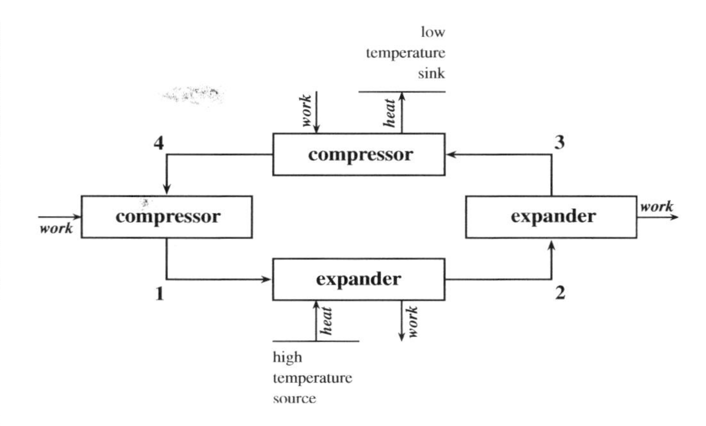

FIGURE 3.18 Schematic representation of Carnot's heat engine, comprising two compressors and two expanders, operated along cycle $\mathbf{1 \rightarrow 2 \rightarrow 3 \rightarrow 4 \rightarrow 1}$ – and exchanging work with its surroundings, as well as heat from a high-temperature source or to a low-temperature sink.

be viewed as expansion turbines – designed to produce work, at the expense of an expanding fluid; conversely, compressors generate a high-pressure fluid, following input of external work. Moreover, adiabatic operation of devices in steps $2 \rightarrow 3$ and $4 \rightarrow 1$ is attainable with the aid of appropriate thermal insulation, so that only work is exchanged with the surroundings; whereas devices operated in steps $1 \rightarrow 2$ and $3 \rightarrow 4$ require heat input or output, respectively – necessary to balance work performed or received, respectively, by the system, for compatibility with isothermal operation. Practical implementation would profit from mechanical linking of all such devices – namely by having them share a common shaft, with net work produced by the engine considered as a whole.

When going from 1 to 2, gas undergoes (reversible) isothermal expansion; energy is lost to the surroundings in the form of work, so the accompanying drop in temperature is to be matched by a continuous supply of energy in the form of heat. The work, W_{12}, involved in this step is given by

$$W_{12} \equiv \int_1^2 đW = \int_{P_1}^{P_2} V dP, \tag{3.24}$$

in the case of a system operated continuously under steady state; remember that a closed system should resort to $-\int_{V_1}^{V_2} P dV$, instead of $\int_{P_1}^{P_2} V dP$. Recalling Clausius' definition of entropy, one may write

$$Q_{12} \equiv \int_1^2 đQ = \int_{S_1}^{S_2} T dS, \tag{3.25}$$

pertaining to heat involved in step $1 \rightarrow 2$. The next step runs from 2 to 3 – and entails reversible adiabatic expansion; by the same token, the heat involved satisfies

$$Q_{23} \equiv \int_2^3 đQ = \int_{S_2}^{S_3} T dS, \tag{3.26}$$

which degenerates to

$$Q_{23} = \int_{S_2}^{S_2} T dS = 0 \quad (3.27)$$

for the process being adiabatic – which, combined with reversibility, assures indeed that $S_3 = S_2$. The associated work satisfies

$$W_{23} \equiv \int_2^3 đW = \int_{P_2}^{P_3} V dP; \quad (3.28)$$

a temperature drop will be noticed in this case, because $P_3 < P_2$ as $W_{23} < 0$ by hypothesis. The following step is (qualitatively speaking) a reversal of the first one; the surroundings now perform work on the gas, which accordingly increases its pressure from P_3 to P_4 – and experiences an equivalent loss of energy in the form of heat. The associated work is described by

$$W_{34} \equiv \int_3^4 đW = \int_{P_3}^{P_4} V dP, \quad (3.29)$$

using Eq. (3.24) as template; while heat is given by

$$Q_{34} \equiv \int_3^4 đQ = \int_{S_3}^{S_4} T dS, \quad (3.30)$$

in parallel to Eq. (3.25). The final stage encompasses adiabatic compression of gas, from P_4 back to initial pressure, P_1; the associated work reads

$$W_{41} \equiv \int_4^1 đW = \int_{P_4}^{P_1} V dP, \quad (3.31)$$

while heat abides to

$$Q_{41} \equiv \int_4^1 đQ = \int_{S_4}^{S_1} T dS \quad (3.32)$$

– equivalent to merely

$$Q_{41} = \int_{S_4}^{S_4} T dS = 0, \quad (3.33)$$

due to prevalence of adiabatic and reversible conditions implying $S_1 = S_4$.

The net result of operating the aforementioned heat engine, throughout one complete cycle, must be nil in terms of enthalpy change, ΔH, i.e.

$$\Delta H = 0, \quad (3.34)$$

as per the final state coinciding with the initial one – couped with enthalpy being a state function; the first law of thermodynamics enforces

$$dH = đW + đQ \quad (3.35)$$

for systems operated continuously, so integration over a full cycle yields

$$\Delta H \equiv \oint dH = \int_1^2 đW + \int_2^3 đW + \int_3^4 đW + \int_4^1 đW + \int_1^2 đQ + \int_2^3 đQ + \int_3^4 đQ + \int_4^1 đQ, \quad (3.36)$$

since W and Q may be split in their corresponding components in every step. Upon insertion of Eqs. (3.24), (3.25), (3.27)–(3.31), (3.33), and (3.36), one obtains

$$0 = W_{12} + W_{23} + W_{34} + W_{41} + Q_{12} + Q_{34} \quad (3.37)$$

from Eq. (3.34); the overall work actually generated, W_c – defined as

$$W_c \equiv \oint đW = W_{12} + W_{23} + W_{34} + W_{41} = \oint V dP < 0, \quad (3.38)$$

is graphically described by the (negative of the) area in Fig. 3.19i. Equation (3.37) may be combined with Eq. (3.38) to produce

$$Q_c \equiv \oint đQ = Q_{12} + Q_{34} = \oint T dS = -W_c > 0 \quad (3.39)$$

– meaning that $Q_{12} + Q_{34}$ must be supplied as net heat to the engine, in order for it be able to generate W_c; therefore, the shaded area in Fig. 3.19i also represents the net heat required. In alternative, one may resort to Fig. 3.19ii – where the area at stake represents $Q_{12} + Q_{34}$ directly. Note that $\oint V\,dP \neq V \oint dP = 0$ and $\oint T\,dS \neq T \oint dS = 0$, since $V \equiv V\{P\}$ and $T \equiv T\{S\}$ – further to P and S being state functions (i.e. $\oint dP = \oint dS = 0$); hence, evolution of V and T depend on the path followed within the cycle.

It should be emphasized that $Q_{12} > 0$, while $Q_{34} < 0$ as per Fig. 3.18; however, $Q_{12} = |Q_{12}| > |Q_{34}| = -Q_{34}$, so $Q_{12} + Q_{34} > 0$ and more heat is supplied to gas at high temperature than extracted therefrom at low temperature. The cost of heat supplied at high temperature is normally much higher – since it involves some form of fuel burning upstream, or the like; whereas heat is typically withdrawn toward some heat sink, namely air or water at ambient temperature – available at essentially nil cost. Therefore, the efficiency of Carnot's engine, η_c, aimed at producing work at the expense of (high-temperature) heat has traditionally been defined as

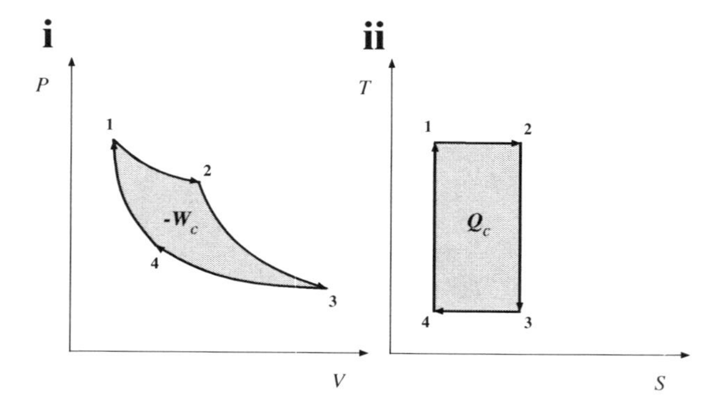

FIGURE 3.19 Schematic representation of Carnot's cycle, pertaining to a heat engine based on an ideal gas – encompassing two isothermal steps, between states **1** and **2**, and **3** and **4**, alternated with two adiabatic steps, between states **2** and **3**, and **4** and **1**, laid out as (i) pressure, P, versus volume, V, with overall work produced, $-W_c$, given by shaded area ( ), and (ii) temperature, T, versus entropy, S, with overall heat supplied, Q_c, given by shaded area ( ).

$$\boxed{\eta_c \equiv -\frac{W_{12} + W_{23} + W_{34} + W_{41}}{Q_{12}};} \tag{3.40}$$

in view of Eq. (3.37), one finds that

$$\eta_c = \frac{Q_{12} + Q_{34}}{Q_{12}} = 1 - \frac{-Q_{34}}{Q_{12}} \tag{3.41}$$

from Eq. (3.40) – and consequently

$$\boxed{\eta_c < 1,} \tag{3.42}$$

because $-Q_{34} > 0$ and $|Q_{34}| < Q_{12}$ as stressed above. This point is of the utmost importance, since it indicates that a given amount of high-temperature heat cannot be fully converted to work; a portion thereof must be lost to the surroundings, at a lower temperature – otherwise the engine will not operate at all, in a sustained mode. On the other hand, η_c proves a univariate, monotonically decreasing function of $|Q_{34}|/Q_{12}$ – meaning that a smaller amount of heat lost to the low-temperature sink, or a larger amount of heat absorbed from the high-temperature source favor thermodynamic efficiency (as both imply a larger amount of work produced).

Consider now a set of devices as in Fig. 3.20, able to perform the same four operations as Carnot's engine originally designed to produce power – but in reverse order. Note the change in role from expansion to compression, and *vice versa* – accompanying reversal of direction of flow of both heat and work. In this case, there will be a net heat flow from the heat source, at lower temperature, toward the heat sink, at higher temperature – at the expense of net work supplied to the heat engine operated reversewise, hereafter termed refrigerator.

Operation of the aforementioned system may again be described by a cycle, laid on an appropriate thermodynamic diagram – as done in Fig. 3.21 using three alternatives, each one characterized by a distinct set of two degrees of freedom. The first step, $1 \rightarrow 4$, consists of a reversible adiabatic expansion, and is followed by a reversible isothermal expansion, $4 \rightarrow 3$; a reversible adiabatic compression, $3 \rightarrow 2$, ensues, and the cycle comes to completion through a reversible isothermal compression, $2 \rightarrow 1$.

A thermodynamic analysis of this refrigeration cycle can be put forward, along the same lines as done for the power cycle presented before – except that

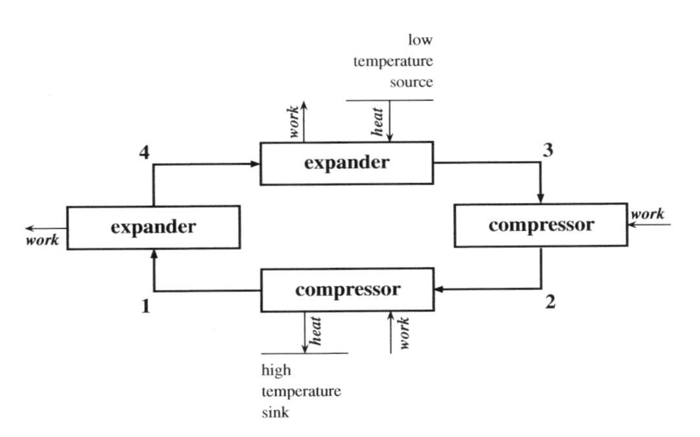

FIGURE 3.20 Schematic representation of Carnot's refrigerator, comprising two compressors and two expanders, operated along cycle **1→2→3→4→1** – and exchanging work with its surroundings, as well as heat from a low-temperature source or to a high-temperature sink.

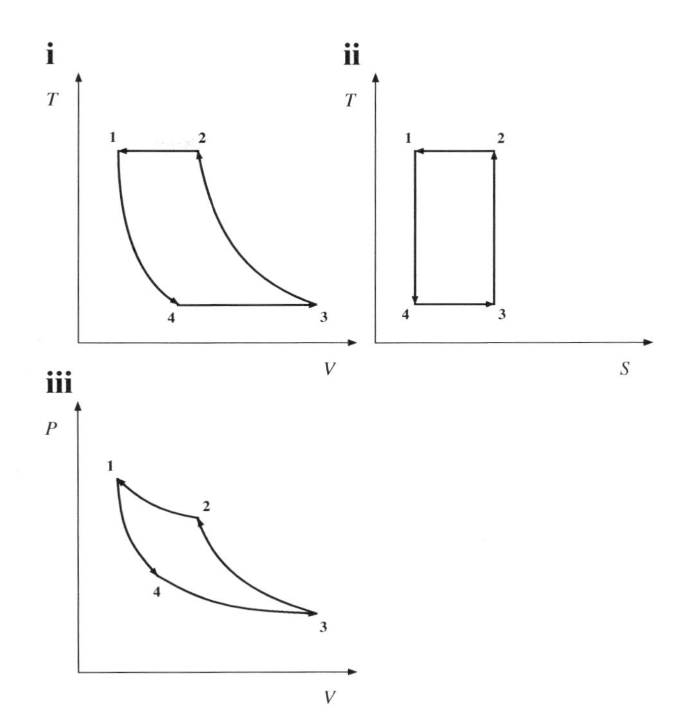

FIGURE 3.21 Schematic representation of Carnot's cycle, pertaining to a refrigerator based on an ideal gas – encompassing two isothermal steps, between states **2** and **1**, and **4** and **3**, alternated with two adiabatic steps, between states **3** and **2**, and **1** and **4**, laid out as (i) temperature, *T*, versus volume, *V*, (ii) *T* versus entropy, *S*, and (iii) pressure, *P*, versus *V*.

$$W_c \equiv \oint đW = W_{14} + W_{43} + W_{32} + W_{21} = \oint V\,dP > 0 \tag{3.43}$$

should now replace Eq. (3.38), complemented by

$$Q_c \equiv \oint đQ = Q_{21} + Q_{43} = \oint T\,dS < 0 \tag{3.44}$$

in lieu of Eq. (3.39) – along with $Q_{14} = Q_{32} = 0$; consequently,

$$Q_c = -W_c < 0, \tag{3.45}$$

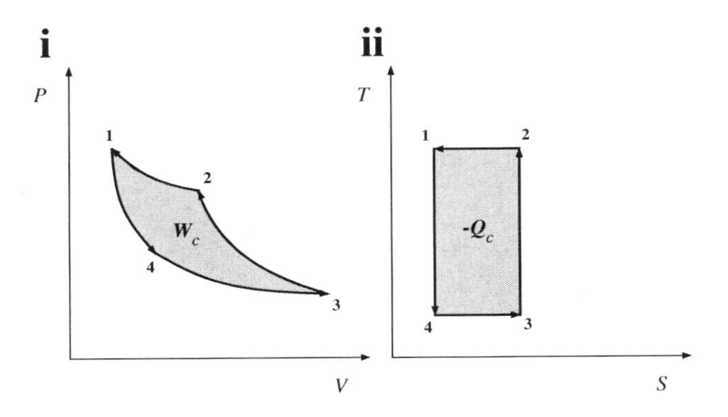

FIGURE 3.22 Schematic representation of Carnot's cycle, pertaining to a refrigerator based on an ideal gas – encompassing two isothermal steps, between states **2** and **1**, and **4** and **3**, alternated with two adiabatic steps, between states **3** and **2**, and **1** and **4**, laid out as (i) pressure, *P*, versus volume, *V*, with overall work received, W_c, given by shaded area (▒), and (ii) temperature, *T*, versus entropy, *S*, with overall heat withdrawn, $-Q_c$, given by shaded area (▒).

in view of

$$0 = W_{14} + W_{43} + W_{32} + W_{21} + Q_{21} + Q_{43} \tag{3.46}$$

as analog to Eq. (3.37) pertaining to a (closed) cycle. An illustration of Eq. (3.43) is provided in Fig. 3.22i, whereas Eq. (3.44) is depicted in Fig. 3.22ii. The shaded area in the former figure represents the work input to Carnot's refrigerator, while the shaded area in Fig. 3.22ii accounts for the heat concomitantly released to the surroundings; these are equivalent in absolute value, as enforced by Eq. (3.45).

A refrigerator is expected to extract heat, Q_{43}, from a low temperature system; this heat, together with the net work input thereto, is then rejected as $-Q_{21} = |Q_{21}| > |Q_{43}| = Q_{43}$ to (higher temperature) surroundings. Hence, the objective here is to remove heat at low temperature – while the major cost arises from work required to do so; once again, heat is rejected to some (very large) reservoir in the surroundings at essentially no cost. Therefore, a suitable measure of the efficiency of a refrigerator – normally coined as coefficient of performance, β_c, looks like

$$\boxed{\beta_c \equiv \frac{Q_{43}}{W_{14} + W_{43} + W_{32} + W_{21}};} \tag{3.47}$$

Eq. (3.46) can then be invoked to rewrite Eq. (3.47) as

$$\beta_c = \frac{Q_{43}}{-Q_{21} - Q_{43}} = \frac{-Q_{34}}{Q_{12} + Q_{34}} = \frac{|Q_{34}|}{Q_{12} + Q_{34}}, \tag{3.48}$$

together with realization that $Q_{43}=-Q_{34}$ and $Q_{21}=-Q_{12}$ as per the equivalence between (reversible) Carnot's heat engine and refrigerator, see Fig. 3.17 *vis-à-vis* with Fig. 3.21. One may now retrieve Eq. (3.41) to get

$$\beta_c = \frac{|Q_{34}|}{Q_{12}} \frac{1}{\eta_c}, \tag{3.49}$$

where $Q_{12} > |Q_{34}| > 0$ (as seen above) implies $|Q_{34}| / Q_{12} < 1$; this is often (but not necessarily) sufficient to override, and eventually support $\beta_c > 1$, once $|Q_{43}| / Q_{12} < 1$ is multiplied by $1/\eta_c > 1$. On the other hand, β_c is given by the ratio of $Q_{43} / |Q_{21}|$ to $1 - Q_{43} / |Q_{21}|$, upon division of both numerator and denominator of Eq. (3.48) by $|Q_{21}| = -Q_{21}$, meaning that $\beta_c \equiv \beta_c\{Q_{43} / |Q_{21}|\}$; this functional form entails a monotonically increasing behavior of β_c with $Q_{43} / |Q_{21}|$, since $Q_{43} / |Q_{21}| > 0$ appears as such in numerator and its negative as term in denominator. Therefore, a larger amount of heat absorbed from the low-temperature source, or a smaller amount of heat lost to the high-temperature sink (for implying a smaller work required) contribute favorably to the coefficient of performance.

Common experience indicates that it is impossible to continuously extract heat from a body at lower temperature, T_l, and give that heat in full to a body at a higher temperature, $T_h > T_l$ – despite the first law of thermodynamics guaranteeing (quantitative) equivalence between such heats; or else completely (and cyclically) use heat at a higher temperature to quantitatively produce work, without any other effect upon the surroundings – which would again run in parallel to the first law of thermodynamics. The former experimental realization actually implies

$$|Q_{21}| \geq |Q_{43}| \tag{3.50}$$

in the case of a refrigerator, see Fig. 3.20 – where the equal sign is reserved to reversible operation (characterized, namely, by $T_l \approx T_h$); whereas the latter realization can mathematically be expressed as

$$|Q_{12}| \geq |W_{12} + W_{23} + W_{34} + W_{41}| \tag{3.51}$$

pertaining to a power generator, see Fig. 3.18 – again with the equality restricted to a reversible process (with $T_l \approx T_h$). Although a mathematical proof of such impossibilities cannot be produced, experimental evidence has invariably confirmed so; they constitute the basis of Clausius' statement, or Kelvin and Planck's statement, respectively, of the second law of thermodynamics. This law is concerned with quality of energy – i.e. (ordered, macroscopic) work versus (random, macroscopic) heat; rather than with equivalence between quantity of energy.

One has focused so far on a reversible Carnot's engine, operated as either a power generator or a refrigerator; however, most heat engines do actually operate in an irreversible manner – because of mechanical and viscous dissipation of work conveyed by a fluid flowing through some machine, or imperfect insulation. Consider, in this context, a generic heat engine – either reversible or irreversible; and Carnot's engine, operating (reversibly) between the same higher temperature, T_h, and lower temperature, $T_l < T_h$ – such that both engines handle the same amount of work per cycle, W, either as output if operated as power generator or as input if operated as refrigerator. These two systems are depicted in Fig. 3.23i. If Carnot's heat engine is reversed – and

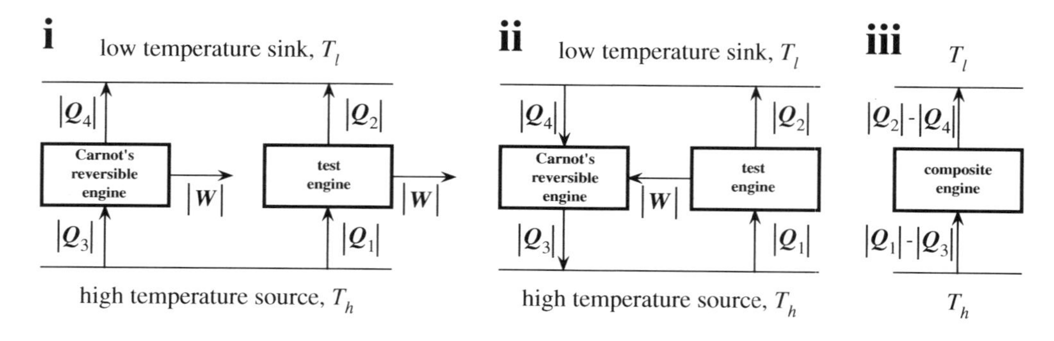

FIGURE 3.23 Schematic representation of Carnot's reversible heat engine and a generic test heat engine, operated between a heat reservoir at higher temperature T_h and a heat reservoir at lower temperature $T_l < T_h$ – either (i) separately and producing work $|W|$ at the expense of heat $|Q_3|$ received at T_h concomitant with heat $|Q_4|$ released at T_l, or heat $|Q_1|$ received at T_h concomitant with heat $|Q_2|$ released at T_l, respectively; or with the reversible heat engine operated reversewise as refrigerator, either (ii) as two devices connected to each other or (iii) one single, composite device with heat $|Q_1|-|Q_3|$ absorbed at T_h concomitant with heat $|Q_2|-|Q_4|$ released at T_l.

thus employed as a refrigerator, then the magnitudes of Q_3, Q_4, and W are kept owing to definition of a reversible process; however, their signs become opposite due to flow in the opposite direction (as already taken advantage of). Therefore, the work output of the generic engine may be used to drive Carnot's refrigerator – as highlighted in Fig. 3.23ii. Since there is no reason why the set of the two engines should be considered as separate any longer, a composite heat engine can be conceived, as done in Fig. 3.23iii – with exchanges of only heat with the surroundings, since exchange of work has now become an internal process.

The first law of thermodynamics has it that

$$|Q_1| - |Q_3| = |Q_2| - |Q_4| \tag{3.52}$$

– for constancy of overall energy in the universe, constituted by composite engine and surroundings. The second law of thermodynamics, as per Clausius' statement, requires, in turn,

$$|Q_1| \geq |Q_3| \tag{3.53}$$

referring to Fig. 3.23iii – otherwise net heat flow from a cold to a hot reservoir, or $|Q_1| - |Q_3| < 0$, would take place in the absence of any other interaction with the surroundings; remember that the first law of thermodynamics would enforce $|Q_2| < |Q_4|$ if $|Q_1| < |Q_3|$, consistent again with Fig. 3.23iii. Equation (3.53) may be rephrased as

$$\frac{1}{|Q_1|} \leq \frac{1}{|Q_3|} \tag{3.54}$$

after taking reciprocals of both (nonnegative) sides, or else

$$\frac{|W|}{|Q_1|} \leq \frac{|W|}{|Q_3|} \tag{3.55}$$

once both sides are multiplied by $|W| \geq 0$; in view of Eq. (3.40) with $|W| = -(W_{12} + W_{23} + W_{34} + W_{41})$, and $Q_{12} = |Q_1|$ or $Q_{12} = |Q_3|$ as per Fig. 3.23i, one may rewrite Eq. (3.55) as

$$\boxed{\eta_{irr} \leq \eta_c} \tag{3.56}$$

– i.e. the efficiency of any irreversible (test) heat engine, η_{irr}, cannot exceed that of (reversible) Carnot's engine.

Consider now that the test engine is also reversible, see Fig. 3.24i; in this case, its direction of flow may be changed toward operation as a refrigerator – which could, in turn, be driven via supply of work by Carnot's (reversible) engine, as illustrated in Fig. 3.24ii for separate engines or in Fig. 3.24iii as a composite (lumped) engine. The first law of thermodynamics applied to Fig. 3.24iii enforces

$$|Q_3| - |Q_1| = |Q_4| - |Q_2|, \tag{3.57}$$

whereas the second law of thermodynamics assures that

$$|Q_3| \geq |Q_1| \tag{3.58}$$

based on Fig. 3.24iii – again at the expense of Clausius' formulation; after taking reciprocals of both (nonnegative) sides, Eq. (3.58) becomes

$$\frac{1}{|Q_3|} \leq \frac{1}{|Q_1|}, \tag{3.59}$$

or else

$$\frac{|W|}{|Q_3|} \leq \frac{|W|}{|Q_1|} \tag{3.60}$$

upon multiplying both sides by $|W| \geq 0$. On the other hand, the reasoning followed previously still applies to Carnot's and test reversible engines – so Eq. (3.55) remains valid. Equation (3.60) is incompatible with Eq. (3.55) unless

$$\frac{|W|}{|Q_1|} = \frac{|W|}{|Q_3|}, \tag{3.61}$$

meaning that

$$\boxed{\eta_{rev} = \eta_c} \tag{3.62}$$

– written with the aid of Eq. (3.40), coupled with equivalence of $|W|$, $|Q_1|$, and $|Q_3|$ to $-(W_{12}+W_{23}+W_{34}+W_{41})$, Q_{12}, and (again) Q_{12}, respectively; this is usually known as Carnot's theorem. Therefore, the efficiency of any reversible engine coincides necessarily with that of Carnot's engine. Equations (3.56) and (3.62) justify the practical relevance of Carnot's engine as standard for comparative purposes – namely in setting the upper threshold for efficiency of an actual heat engine, despite its nature and mode of operation.

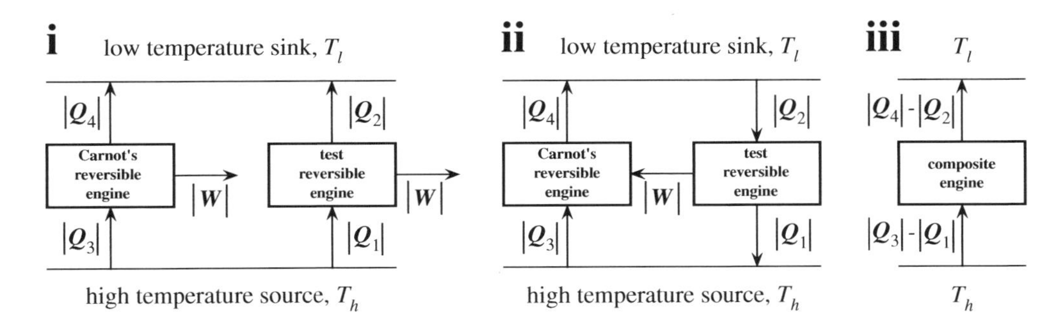

FIGURE 3.24 Schematic representation of two reversible heat engines, Carnot's engine and a test heat engine, operated between a heat reservoir at higher temperature T_h and a heat reservoir at lower temperature $T_l < T_h$ – either (i) separately and producing work $|W|$ at the expense of heat $|Q_3|$ received at T_h and heat $|Q_4|$ released at T_l, or heat $|Q_1|$ received at T_h and heat $|Q_2|$ released at T_l, respectively; or with the reversible test heat engine operated reversewise as refrigerator, either (ii) as two devices connected to each other or (iii) one single, composite device with heat $|Q_3|-|Q_1|$ absorbed at T_h concomitant with heat $|Q_4|-|Q_2|$ released at T_l.

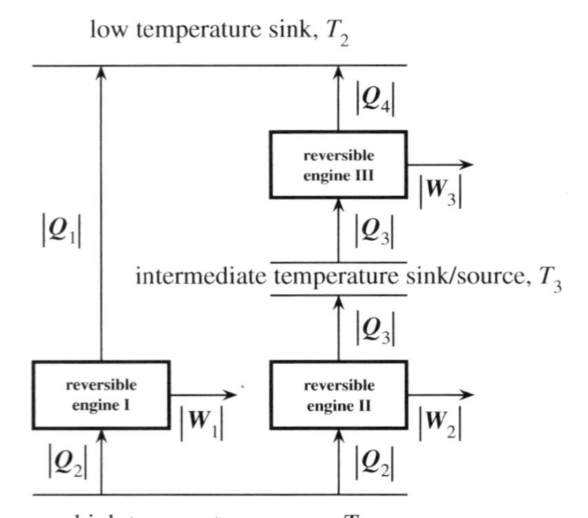

FIGURE 3.25 Schematic representation of three reversible heat engines – engine I operated between a heat reservoir at highest temperature T_1 and a heat reservoir at lowest temperature $T_2 < T_1$, engine II operated between the heat reservoir at T_1 and a heat reservoir at intermediate temperature $T_3 < T_1$, and engine III operated between the heat reservoir at T_3 and the heat reservoir at $T_2 < T_3$; with engine I producing work $|W_1|$ at the expense of heat $|Q_2|$ received at T_1 and heat $|Q_1|$ released at T_2, engine II producing work $|W_2|$ at the expense of heat $|Q_2|$ received at T_1 and heat $|Q_3|$ released at T_3, and engine III producing work $|W_3|$ at the expense of heat $|Q_3|$ received at T_3 and heat $|Q_4|$ released at T_2.

No restrictive assumptions whatsoever were imposed along the above reasoning upon Carnot's heat engine in terms of fluid, or pressure (or volume) for that matter; furthermore, nothing was said on how the efficiency of the said engine varies with the temperatures prevailing in the two reservoirs – which, in general, is expected to abide to

$$\eta_c = \phi\{T_h, T_l\}, \tag{3.63}$$

with ϕ denoting a bivariate function (still to be determined). To further address this issue, consider the arrangement of heat engines depicted in Fig. 3.25. Carnot's engine on the left operates between T_1 and $T_2 < T_1$ – whereas two Carnot's engines are connected in series on the right, sharing by hypothesis the overall heat, $|Q_2|$, received at T_1; the sizes of the latter engines are such as to avoid any net heat gain or loss by reservoir at temperature T_3, comprised between T_2 and T_1 – thus justifying $|Q_3|$ entering and leaving it.

The efficiency of engine II, when viewed separately, looks like

$$\eta_{c,II} = \frac{|W_2|}{|Q_2|} = \phi\{T_1, T_3\}, \tag{3.64}$$

in agreement with both Eqs. (3.40) and (3.63); the first law of thermodynamics supports

$$|W_2| = |Q_2| - |Q_3|, \tag{3.65}$$

which may be used to convert Eq. (3.64) to

$$\frac{|Q_2| - |Q_3|}{|Q_2|} = \phi\{T_1, T_3\}. \tag{3.66}$$

After splitting the fraction in the left-hand side, Eq. (3.66) turns to

$$\frac{|Q_2|}{|Q_2|} - \frac{|Q_3|}{|Q_2|} = 1 - \frac{|Q_3|}{|Q_2|} = \phi\{T_1, T_3\}, \tag{3.67}$$

which can, in turn, be rephrased as

$$\frac{|Q_3|}{|Q_2|} = 1 - \phi\{T_1, T_3\} \equiv \Phi\{T_1, T_3\} \tag{3.68}$$

– where Φ denotes another bivariate function of T_1 and T_3. By the same token, one finds

$$\eta_{c,III} = \frac{|W_3|}{|Q_3|} = \phi\{T_3, T_2\} \tag{3.69}$$

pertaining to engine III seen also on its own, and reflecting again Eqs. (3.40) and (3.63) – where the first law of thermodynamics unfolds

$$|W_3| = |Q_3| - |Q_4|; \tag{3.70}$$

insertion of Eq. (3.70) transforms Eq. (3.69) to

$$\frac{|Q_3| - |Q_4|}{|Q_3|} = \phi\{T_3, T_2\} \tag{3.71}$$

that degenerates to

$$\frac{|Q_3|}{|Q_3|} - \frac{|Q_4|}{|Q_3|} = 1 - \frac{|Q_4|}{|Q_3|} = \phi\{T_3, T_2\} \tag{3.72}$$

upon straightforward algebraic rearrangement, or else

$$\frac{|Q_4|}{|Q_3|} = 1 - \phi\{T_3, T_2\} \equiv \Phi\{T_3, T_2\} \tag{3.73}$$

– with the functional form $\Phi \equiv \Phi\{T_h, T_l\}$ being obviously shared by Eqs. (3.68) and (3.73). Engines II and III may now be connected in series, with heat input $|Q_2|$ at temperature T_1, heat output $|Q_4|$ at temperature T_2, and total work output $|W_2| + |W_3|$; Eq. (3.40) further supports

$$\eta_{c,I} = \frac{|W_1|}{|Q_2|}, \tag{3.74}$$

when applied to the single heat engine on the left of Fig. 3.25. In view of Eq. (3.62),

$$\eta_{c,II \times III} = \eta_{c,I} \tag{3.75}$$

because engine I is reversible, and combination of reversible engines II and III must also be reversible; the efficiency of the composite machine accordingly reads

$$\eta_{c,I \times III} = \frac{|W_2| + |W_3|}{|Q_2|}. \tag{3.76}$$

Insertion of Eqs. (3.74) and (3.76) transforms Eq. (3.75) to

$$\frac{|W_1|}{|Q_2|} = \frac{|W_2| + |W_3|}{|Q_2|}, \tag{3.77}$$

where cancellation of $|Q_2| \neq 0$ between sides promptly yields

$$|W_1| = |W_2| + |W_3|. \tag{3.78}$$

On the other hand, the first law of thermodynamics enforces

$$|Q_2| = |W_1| + |Q_1| \tag{3.79}$$

applying to engine I, as well as

$$|Q_2| = |W_2| + |W_3| + |Q_4| \tag{3.80}$$

encompassing composite engine II×III; elimination of $|Q_2|$ between Eqs. (3.79) and (3.80) leaves

$$|W_1| + |Q_1| = |W_2| + |W_3| + |Q_4| \tag{3.81}$$

– while insertion of Eq. (3.78) permits simplification to

$$|Q_1| = |Q_4|, \tag{3.82}$$

after dropping $|W_1| = |W_2| + |W_3|$ from both sides. Based on Eq. (3.82), one can redo Eq. (3.73) to

$$\frac{|Q_1|}{|Q_3|} = \Phi\{T_3, T_2\}; \tag{3.83}$$

ordered multiplication of Eqs. (3.68) and (3.83) gives rise to

$$\frac{|Q_3|}{|Q_2|}\frac{|Q_1|}{|Q_3|} = \Phi\{T_1, T_3\}\Phi\{T_3, T_2\} \tag{3.84}$$

that breaks down to just

$$\frac{|Q_1|}{|Q_2|} = \Phi\{T_1, T_3\}\Phi\{T_3, T_2\} \tag{3.85}$$

upon cancellation of $|Q_3|$ between numerator and denominator. On the other hand, the definition of Φ that supports Eqs. (3.68) and (3.73) similarly enforces

$$\frac{|Q_1|}{|Q_2|} = \Phi\{T_1, T_2\} \tag{3.86}$$

– see Fig. 3.25, where $|Q_2|$ enters the engine at T_1 and $|Q_1|$ leaves it at T_2; comparative inspection of Eqs. (3.85) and (3.86) indicates that the effect of T_3 upon the product of $\Phi\{T_1, T_3\}$ by $\Phi\{T_3, T_2\}$ must vanish so as to produce $\Phi\{T_1, T_2\}$ as result. Mathematically speaking, this is possible only if $\Phi\{T_h, T_l\}$ can be written as

$$\Phi\{T_h, T_l\} \equiv \frac{\Psi\{T_h\}}{\Psi\{T_l\}}, \tag{3.87}$$

with $\Psi\{T\}$ denoting a (still unknown) univariate function of T; therefore, Eq. (3.85) will become

$$\frac{|Q_1|}{|Q_2|} = \frac{\Psi\{T_1\}}{\Psi\{T_3\}}\frac{\Psi\{T_3\}}{\Psi\{T_2\}} \tag{3.88}$$

after applying Eq. (3.87) twice, compatible with

$$\frac{|Q_1|}{|Q_2|} = \frac{\Psi\{T_1\}}{\Psi\{T_2\}} \tag{3.89}$$

obtained upon cancelling $\Psi\{T_3\}$ in numerator with that in denominator. There are a very many functional forms possible for $\Psi\{T\}$ – yet scientists and engineers have almost exclusively resorted to

$$\Psi\{T\} \equiv \frac{\kappa}{T}, \tag{3.90}$$

where κ denotes an arbitrary constant; using this definition, Eq. (3.89) becomes

$$\frac{|Q_1|}{|Q_2|} = \frac{\dfrac{\kappa}{T_1}}{\dfrac{\kappa}{T_2}} = \frac{T_2}{T_1}. \tag{3.91}$$

After applying the first law of thermodynamics to engine I as

$$|W_1| = |Q_2| - |Q_1|, \tag{3.92}$$

Eq. (3.74) may be retrieved as

$$\eta_{c,I} = \frac{|Q_2| - |Q_1|}{|Q_2|} = 1 - \frac{|Q_1|}{|Q_2|}; \tag{3.93}$$

insertion of Eq. (3.91) gives then rise to

$$\boxed{\eta_{c,I} = 1 - \frac{T_2}{T_1}.} \tag{3.94}$$

If engine I is driven in reverse as a refrigerator, then the corresponding coefficient of performance, i.e.

$$\beta_{c,I} = \frac{|Q_1|}{|W_1|} \tag{3.95}$$

– consistent with Eq. (3.47), can be reformulated as

$$\beta_{c,I} = \frac{|Q_1|}{|Q_2| - |Q_1|} \tag{3.96}$$

with the aid of Eq. (3.92); after dividing both numerator and denominator by $|Q_2|$, Eq. (3.96) becomes

$$\beta_{c,I} = \frac{\dfrac{|Q_1|}{|Q_2|}}{1 - \dfrac{|Q_1|}{|Q_2|}} \tag{3.97}$$

– where insertion of Eq. (3.91) permits final transformation to

$$\boxed{\beta_{c,I} = \frac{\dfrac{T_2}{T_1}}{1 - \dfrac{T_2}{T_1}},} \tag{3.98}$$

i.e. $\beta_{c,I}$ ends up given by

$$\beta_{c,I} = \frac{\dfrac{T_2}{T_1}}{\eta_{c,I}} = \frac{\dfrac{Q_1}{Q_2}}{\eta_{c,I}} \tag{3.99}$$

after taking Eq. (3.94) into account, and in full agreement with Eq. (3.49).

PROBLEM 3.2

Prove that the total change in entropy, S, of an ideal gas undergoing Carnot's cycle is nil; recall that

$$dS \equiv \frac{đQ}{T} \tag{3.2.1}$$

stands as Clausius' definition of entropy – with Q denoting heat exchanged at constant (absolute) temperature T.

Solution

According to Eq. (3.2.1), change in entropy occurs as a consequence only of energy exchanged in the form of heat, but not in the form of (reversible) work. Equation (3.27), pertaining to an adiabatic expansion, coupled with Eq. (3.2.1) accordingly enforce

$$\Delta S_{23} \equiv \int_{S_2}^{S_3} dS = \int_{Q_{23}} \frac{đQ}{T} = 0, \tag{3.2.2}$$

and Eq. (3.33) similarly supports

$$\Delta S_{41} \equiv \int_{S_4}^{S_1} dS = \int_{Q_{41}} \frac{đQ}{T} = 0; \tag{3.2.3}$$

Eqs. (3.2.2) and (3.2.3) are consistent with the vertical paths in Fig. 3.17ii. The remaining two steps of Carnot's cycle are isothermal, by hypothesis – see horizontal paths in Fig. 3.17ii; during expansion in step $1 \rightarrow 2$, work produced is balanced by heat absorbed, i.e.

$$Q_{12} = -W_{12}, \tag{3.2.4}$$

whereas compression in step $3 \rightarrow 4$ requires release of heat to maintain temperature constant, such that

$$-Q_{34} = W_{34} \tag{3.2.5}$$

– both following the first law of thermodynamics. In view of Eq. (3.24), one may redo Eq. (3.2.4) to

$$Q_{12} = -\int_{P_1}^{P_2} V dP, \tag{3.2.6}$$

while Eq. (3.29) yields

$$Q_{34} = -\int_{P_3}^{P_4} V dP \tag{3.2.7}$$

departing from Eq. (3.2.5). By hypothesis, the working fluid is an ideal gas, so one may retrieve its equation of state, see Eq. (3.135) of *Food Proc. Eng.: basics & mechanical operations* – coined here as

$$V = \frac{n\mathcal{R}T}{P} \tag{3.2.8}$$

for pertaining to n moles, to transform Eqs. (3.2.6) and (3.2.7) to

$$Q_{12} = -\int_{P_1}^{P_2} \frac{n\mathcal{R}T_1}{P} dP \tag{3.2.9}$$

and

$$Q_{34} = -\int_{P_3}^{P_4} \frac{n\mathcal{R}T_3}{P} dP, \tag{3.2.10}$$

respectively; these represent isothermal steps that take place at temperatures T_1 and T_3, respectively. Since n and $\mathcal{R}$ are constant, and so is temperature during either path under scrutiny, they will be taken off the kernel to get

$$Q_{12} = -n\mathcal{R}T_1 \int_{P_1}^{P_2} \frac{dP}{P} \tag{3.2.11}$$

from Eq. (3.2.9), which integrates to

$$Q_{12} = -n\mathcal{R}T_1 \ln P \Big|_{P_1}^{P_2} = n\mathcal{R}T_1 \ln \frac{P_1}{P_2}; \tag{3.2.12}$$

by the same token, Eq. (3.2.10) becomes

$$Q_{34} = -n\mathcal{R}T_3 \int_{P_3}^{P_4} \frac{dP}{P} \tag{3.2.13}$$

that degenerates to

$$Q_{34} = -n\mathcal{R}T_3 \ln P \Big|_{P_3}^{P_4} = n\mathcal{R}T_3 \ln \frac{P_3}{P_4} \tag{3.2.14}$$

– in both cases with the preceding minus sign accounting for the reciprocal taken of the argument of the logarithm. When temperature remains constant at T_1, Eq. (3.2.1) integrates as

$$\Delta S_{12} \equiv \int_{S_1}^{S_2} dS = \int_{Q_{12}} \frac{đQ}{T_1} = \frac{1}{T_1} \int_{Q_{12}} đQ = \frac{Q_{12}}{T_1} \tag{3.2.15}$$

pertaining to step $1 \rightarrow 2$, and likewise

$$\Delta S_{34} \equiv \int_{S_3}^{S_4} dS = \int_{Q_{34}} \frac{đQ}{T_3} = \frac{1}{T_3} \int_{Q_{34}} đQ = \frac{Q_{34}}{T_3} \tag{3.2.16}$$

encompassing step $3 \rightarrow 4$, owing to its constant temperature T_3; combination of Eqs. (3.2.12) and (3.2.15) unfolds

$$\Delta S_{12} = \frac{n\mathcal{R}T_1}{T_1} \ln \frac{P_1}{P_2} = n\mathcal{R} \ln \frac{P_1}{P_2}, \tag{3.2.17}$$

and insertion of Eq. (3.2.14) likewise transforms Eq. (3.2.16) to

$$\Delta S_{34} = \frac{n\mathcal{R}T_3}{T_3} \ln \frac{P_3}{P_4} = n\mathcal{R} \ln \frac{P_3}{P_4} \tag{3.2.18}$$

– where similar factors were meanwhile dropped from numerator and denominator. The total variation of entropy in a full cycle is given by

$$\Delta S = \Delta S_{12} + \Delta S_{23} + \Delta S_{34} + \Delta S_{41}, \quad (3.2.19)$$

so combination with Eqs. (3.2.2), (3.2.3), (3.2.17), and (3.2.18) unfolds

$$\Delta S = n\mathcal{R}\ln\frac{P_1}{P_2} + 0 + n\mathcal{R}\ln\frac{P_3}{P_4} + 0 = n\mathcal{R}\left(\ln\frac{P_1}{P_2} + \ln\frac{P_3}{P_4}\right) \quad (3.2.20)$$

– with $n\mathcal{R}$ meanwhile factored out; the operational features of logarithms permit further algebraic rearrangement to

$$\Delta S = n\mathcal{R}\ln\frac{P_1P_3}{P_2P_4} = n\mathcal{R}\left(\ln\frac{P_3}{P_2} + \ln\frac{P_1}{P_4}\right). \quad (3.2.21)$$

In agreement with Eq. (3.136) of *Food Proc. Eng: thermal & chemical operations*,

$$\frac{T_3}{T_2} = \left(\frac{P_3}{P_2}\right)^{\frac{\gamma-1}{\gamma}} \quad (3.2.22)$$

and

$$\frac{T_1}{T_4} = \left(\frac{P_1}{P_4}\right)^{\frac{\gamma-1}{\gamma}}, \quad (3.2.23)$$

with γ denoting ratio of isobaric to isochoric heat capacities – since $2 \to 3$ and $4 \to 1$, respectively, are both adiabatic processes; once logarithms are taken of both sides, Eqs. (3.2.22) and (3.2.23) yield

$$\ln\frac{T_3}{T_2} = \frac{\gamma-1}{\gamma}\ln\frac{P_3}{P_2} \quad (3.2.24)$$

and

$$\ln\frac{T_1}{T_4} = \frac{\gamma-1}{\gamma}\ln\frac{P_1}{P_4}, \quad (3.2.25)$$

respectively. Insertion of Eqs. (3.2.24) and (3.2.25) converts Eq. (2.2.21) to

$$\Delta S = n\mathcal{R}\left(\frac{\gamma}{\gamma-1}\ln\frac{T_3}{T_2} + \frac{\gamma}{\gamma-1}\ln\frac{T_1}{T_4}\right) = n\mathcal{R}\frac{\gamma}{\gamma-1}\left(\ln\frac{T_3}{T_2} + \ln\frac{T_1}{T_4}\right), \quad (3.2.26)$$

along with factoring out of $\gamma/(\gamma - 1)$; Eq. (3.2.26) reduces further to

$$\Delta S = n\mathcal{R}\frac{\gamma}{\gamma-1}\ln\frac{T_3T_1}{T_2T_4} = n\mathcal{R}\frac{\gamma}{\gamma-1}\left(\ln\frac{T_1}{T_2} + \ln\frac{T_3}{T_4}\right) \quad (3.2.27)$$

via straightforward algebraic rearrangement. Since steps $1 \to 2$ and $3 \to 4$ are isothermal, they satisfy

$$T_2 = T_1 \quad (3.2.28)$$

and

$$T_4 = T_3, \quad (3.2.29)$$

respectively – so Eq. (3.2.27) degenerates to

$$\Delta S = n\mathcal{R}\frac{\gamma}{\gamma-1}(\ln 1 + \ln 1) = n\mathcal{R}\frac{\gamma}{\gamma-1}(0+0) = 0; \quad (3.2.30)$$

therefore, no overall change in entropy occurs in a full cycle of Carnot's engine – consistent not only with the inherent reversibility of the latter, but also with S being a state variable.

3.3.3 Rankine's power cycle

The food industry utilizes both mechanical power and heat – and it is regular practice, in larger plants, to have both produced on site, rather than purchasing electrical power from the grid, for being more economical. The classical procedure is to generate high-pressure steam from water – using as source of heat the hot products of combustion of conventional fossil fuels, or (preferably) of some waste product. The steam may then be used directly as process heat, or indirectly for production of mechanical power via passage through a turbine – aimed at driving e.g. compressors, vacuum pumps, liquid pumps, electric generators.

The most efficient engine for operation between two specified temperatures has previously been shown to be Carnot's heat engine; it absorbs heat at the higher temperature, and rejects a portion thereof at the lower (usually ambient) temperature – with the difference accounting for mechanical work generated thereby. However, it is impractical to use Carnot's cycle in steam-based power production, because the turbine would have to handle steam with low quality, and thus prone to cause erosion and wear in the turbine blades – besides being difficult to process two phases; such a hypothetical Carnot's cycle is tentatively laid out in Fig. 3.26ii on a *TS*-diagram similar to that conveyed by Fig. 3.17ii – versus

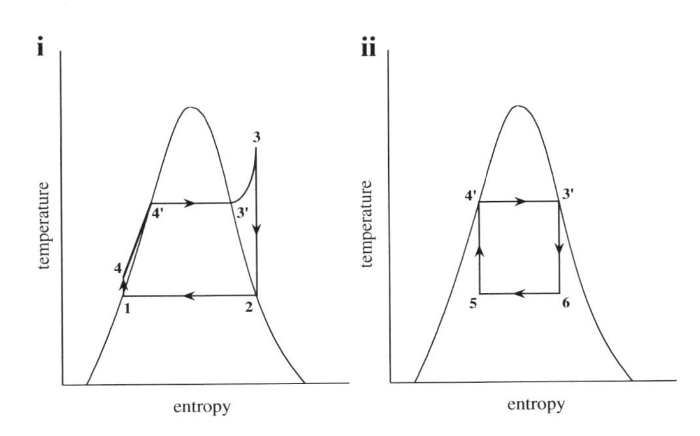

FIGURE 3.26 Thermodynamic diagram of water, as temperature versus entropy, after superimposing (i) a practical Rankine's vapor expansion/compression power cycle, $\mathbf{1} \to \mathbf{4} \to \mathbf{4'} \to \mathbf{3'} \to \mathbf{3} \to \mathbf{2} \to \mathbf{1}$, or (ii) an ideal Carnot's cycle operated between the same extreme temperatures, $\mathbf{5} \to \mathbf{4'} \to \mathbf{3'} \to \mathbf{6} \to \mathbf{5}$.

an actual operative cycle, as shown in Fig. 3.26i. Note the use of either saturated liquid (1) or saturated steam (2) in Fig. 3.26i – instead of a mixture of liquid and vapor at equilibrium (5, 6) in Fig. 3.26ii. In addition, it would be hard to control the condensation process so as to exactly end up in the desired point 3', on the biphasic equilibrium line – see again Fig. 3.26ii; instead of entering the superheated steam region, as in point 3 of Fig. 3.26i. On the other hand, it is difficult to carry out isentropic compression to extremely high pressures, and isothermal heat transfer at variable pressure – due to the sensible nature of heat transferred, coupled to the proportionality between pressure and temperature in the case of a given volume of an ideal gas/vapor; this justifies why point 3 is not far away from point 3' in Fig. 3.26i.

In view of the above arguments, Carnot's cycle would hardly be approximated in actual devices, so it does not represent a realistic model for steam power cycles; however, it may still be used as estimator of the maximum efficiency ever attainable – to which the efficiency of an actual apparatus is to be compared. In practice, one resorts to Rankine's cycle as sketched in Fig. 3.26i, implemented via the layout depicted in Fig. 3.27i in block diagram form. The said cycle takes advantage of phase change of water between liquid and vapor (i.e. boiling of liquid water that produces steam, and condensation of exhaust steam that produces liquid water again) to provide a practical heat/power conversion system; both such processes are isobaric, because of the equilibrium that holds during the given phase change.

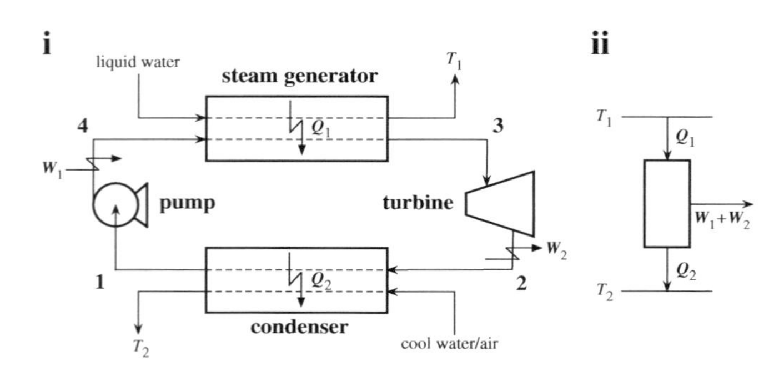

FIGURE 3.27 Schematic representation of (mechanical) power generation system, based on Rankine's vapor expansion/compression cycle (**1 → 4 → 3 → 2 → 1**) – with indication of (i) apparatuses and streams involved, and corresponding operating temperatures (T_1, $T_2 < T_1$), as well as (i, ii) energy exchanged in the form of heat ($Q_1 > 0$, $Q_2 < 0$) and work ($W_1 > 0$, $W_2 < 0$).

In the simple power-generation process displayed in Figs. 3.26i and 3.27i, high-pressure liquid water (4) enters the boiler of the steam generator, where it is converted to saturated steam (3'), and then to high-pressure, high-temperature supersaturated steam (3) in its superheater section – with heat transferred, at constant pressure, from the products of combustion of some suitable fuel; once flown through the turbine, such supersaturated steam performs work as it expands to condenser lower temperature and pressure (2); the exhaust (saturated) steam is then condensed at constant pressure (and temperature), by rejecting heat to cooling water (or air) – and the condensate constituted by saturated liquid (1) is returned to the boiler via a feed pump, thus closing the cycle after increasing its pressure to subcooled liquid state (4).

The aforementioned heat engine may be thermodynamically conceptualized as done in Fig. 3.27ii; heat $|Q_2|$ is rejected in the condenser at (lower) temperature T_2, while heat $|Q_1|$ is received in the steam generator at (higher) temperature T_1; the difference between $|Q_1|$ and $|Q_2|$ accounts for the net work $|W_2|-|W_1|$ generated by the machine, with $|W_1| \ll |W_2|$ together with $W_2 \ll -W_1 < 0$.

The ideal version of Rankine's cycle accordingly includes four consecutive steps: isentropic compression (step $1 \rightarrow 4$), isobaric heating (step $4 \rightarrow 3$), isentropic expansion (step $3 \rightarrow 2$), and final isobaric cooling (step $2 \rightarrow 1$) – see Fig. 3.26i; after recalling the thermodynamic diagram of water on a *PH*-plane, see Fig. 3.28i, one can superimpose the above steps so as to produce Fig. 3.28ii. The isobaric steps are now represented by horizontal straight lines – pointing rightward in the case of heat supply (see step $4 \rightarrow 3$), but leftward in the case of heat withdrawal (see step $2 \rightarrow 1$); pumping of liquid water (i.e. step $1 \rightarrow 4$) is an almost isochoric process, due to the essential incompressibility of liquid water – so it is depicted as a nearly vertical straight line pointing upward, because of the little enthalpy in transit; while adiabatic expansion of superheated steam (see step $3 \rightarrow 2$) accounts for a curved line pointing left- and downward, for being associated with decreasing pressure and enthalpy.

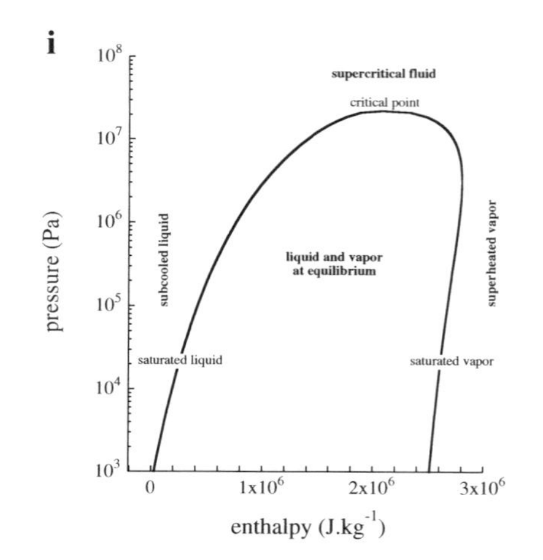

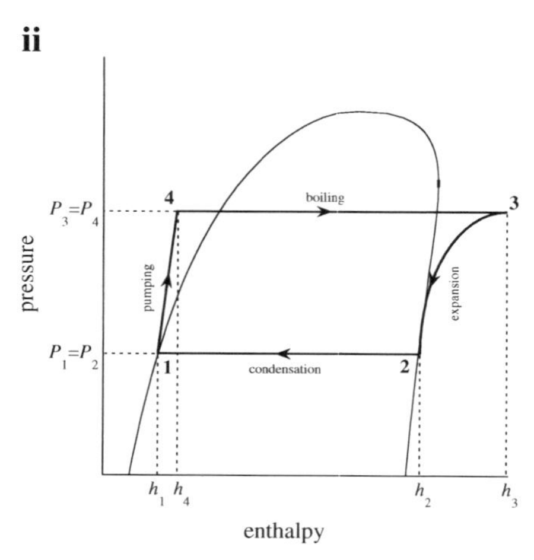

FIGURE 3.28 Thermodynamic diagram of water, as pressure versus enthalpy (using 1 J.kg^{-1} for saturated liquid at 0.01°C and 612 Pa as reference state), (i) with indication of critical point as maximum of binodal curve encompassing saturated liquid on the left (as separation between subcooled liquid and liquid/vapor at equilibrium), and saturated vapor on the right (as separation between liquid/vapor at equilibrium and superheated vapor), and (ii) after superimposing the power cycle (**1→4→3→2→1**) – corresponding to the sequence pumping, boiling, expansion, and condensation, with specification of germane specific enthalpies (h_1, h_2, h_3, h_4) and pressures (P_1, P_2, P_3, P_4).

Based on Eq. (3.40), one can express the efficiency of Rankine's cycle, η_r, as

$$\eta_r = -\frac{W_{14} + W_{32}}{Q_{43}}; \tag{3.100}$$

note that $Q_{14}=0$ and $Q_{32}=0$ since steps $1 \rightarrow 4$ and $3 \rightarrow 2$, respectively, are taken as adiabatic – complemented with $W_{43}=0$ and $W_{21}=0$ because steps $4 \rightarrow 3$ and $2 \rightarrow 1$ encompass solely exchange of heat between system and surroundings. In view of the first law of thermodynamics applied to a steady state, flow process, Eq. (3.100) will be reformulated as

$$\eta_r = -\frac{(h_4 - h_1) + (h_2 - h_3)}{h_3 - h_4}, \tag{3.101}$$

in agreement with Fig. 3.28ii – and because the extensity measure required by W in numerator cancels out with the very same measure required by Q in denominator (i.e. $W/Q = w/q$); here $h_4 - h_1 > 0$ represents the specific work of pumping received by the system in step $1 \rightarrow 4$, $h_2 - h_3 < 0$ represents the specific work of expansion performed by the turbine, and $h_3 - h_4$ represents the specific heat supplied to the system in the boiler/superheater. As stressed before, the essential incompressibility of liquid water supports

$$h_4 - h_1 \approx v_1 (P_4 - P_1) \approx 0 \tag{3.102}$$

for estimate of pumping work; since specific volume, v_1, remains almost constant and is small for being a liquid, one finds that $|h_4 - h_1|$ is normally negligible when compared to $|h_2 - h_3|$. In fact, only up to ca. 3% of the power generated at the turbine is actually consumed by the pump – so it is customary to simplify Eq. (3.101) to

$$\boxed{\eta_r = \frac{h_3 - h_2}{h_3 - h_4},} \tag{3.103}$$

The efficiency of Rankine's cycle depends on (and is limited by) the working fluid; the temperature range over which it can operate is normally quite narrow, so as to avoid pressure approaching critical levels – with concomitant reduction of difference in specific enthalpy between liquid and vapor, which would require larger flow rates of fluid for the same power produced. In steam turbines, anyway, the entry temperature is upper limited by ca. 560°C, for being the creep limit of stainless steel; whereas condenser temperatures are ca. 25°C – i.e. the regular ambient temperature of air or water. Recalling Eq. (3.94), one obtains $\eta_c = (560 - 25)/(560 + 273.2) = 0.64$ as the maximum efficiency ever attainable by Carnot's cycle – yet actual Rankine's cycles seldom go above 0.42. For a given mass flow rate of refrigerant, $\dot{m}$, the (work) power obtained from the turbine, $|\dot{W}_{32}|$, will look like

$$\boxed{|\dot{W}_{32}| = \dot{m}(h_3 - h_2),} \tag{3.104}$$

with $\dot{W}_{32} < 0$; by the same token, the (heat) power to be supplied in the steam generator, $\dot{Q}_{43}$, satisfies

$$\boxed{\dot{Q}_{43} = \dot{m}(h_3 - h_4)} \tag{3.105}$$

– with (heat) power lost to the cooling fluid, $|\dot{Q}_{21}|$, abiding to

$$|\dot{Q}_{21}| = \dot{m}(h_2 - h_1), \tag{3.106}$$

along with $\dot{Q}_{21} < 0$. After realizing that

$$\begin{aligned} h_3 - h_4 &= (h_3 - h_2) + (h_2 - h_1) + (h_1 - h_4) \\ &\approx (h_3 - h_2) + (h_2 - h_1) \end{aligned} \tag{3.107}$$

– following addition and subtraction of h_2 and h_1, and combination with Eq. (3.102), one may redo Eq. (3.105) as

$$\frac{\dot{Q}_{43}}{\dot{m}} = \frac{|\dot{W}_{32}|}{\dot{m}} + \frac{|\dot{Q}_{21}|}{\dot{m}}, \tag{3.108}$$

with the aid of Eqs. (3.104) and (3.106), and division of both sides by $\dot{m}$; Eq. (3.108) promptly justifies

$$\dot{Q}_{43} > |\dot{W}_{32}|, \tag{3.109}$$

consistent with the second law of thermodynamics – since $|\dot{Q}_{21}| > 0$.

An alternative representation of Rankine's cycle resorts to the *TS*-plane – with the process cycle superimposed on the binodal curve pertaining to water in Fig. 3.29ii (which essentially mimics Fig. 3.26i), and further detail of the binodal curve in Fig. 3.29i. Note the small increase in temperature associated with pumping ($1 \rightarrow 4$) – since a liquid is at stake, unlike what happens with expansion of superheated vapor ($3 \rightarrow 2$); the constant temperature, corresponding to horizontal lines during boiling proper of liquid and condensation of vapor ($2 \rightarrow 1$); and the vertical lines ($1 \rightarrow 4$, $3 \rightarrow 2$) representing reversible, adiabatic – and thus isentropic processes.

As emphasized before, the efficiency of current Rankine's machines is poor compared to the upper thermodynamic threshold conveyed by Carnot's cycle; improvements should, in principle, come at the extent of increasing the temperature at which heat is absorbed, or decreasing the temperature at which heat is rejected – as illustrated in Fig. 3.30.

Decrease in condenser temperature, see Fig. 3.30ii, below the thermal reservoir temperature (i.e. lake, river, atmosphere) requires a decrease in its operating pressure – since vapor pressure (of water, in this case) increases exponentially with temperature, as per Antoine's equation; this entails extra costs arising from vacuum generation. Furthermore, heat transfer is not instantaneous – with magnitude of driving force and area available thereto also playing a role. At least 10°C for the former is required in common practice, so the effective value for T_2 as per Fig. 3.29ii is actually higher than expected for a thermodynamically-controlled phenomenon; while economical size considerations for the condenser upper constrain the heat transfer area.

Decreasing the temperature of the superheated steam as it enters the turbine, see Fig. 3.30i, is undesirable – since it decreases not only the mean temperature at which heat is absorbed (and hence Carnot's efficiency of the cycle), but also worsens the quality of steam eventually exhausted to the condenser (for bearing a higher liquid moisture content after passing the turbine); conversely, increasing superheat temperature decreases specific steam consumption, for a given power output.

Raising pressure while keeping temperature of the superheated steam constant, see Fig. 3.30iii, also increases the mean temperature at which heat is absorbed – due to the increase in

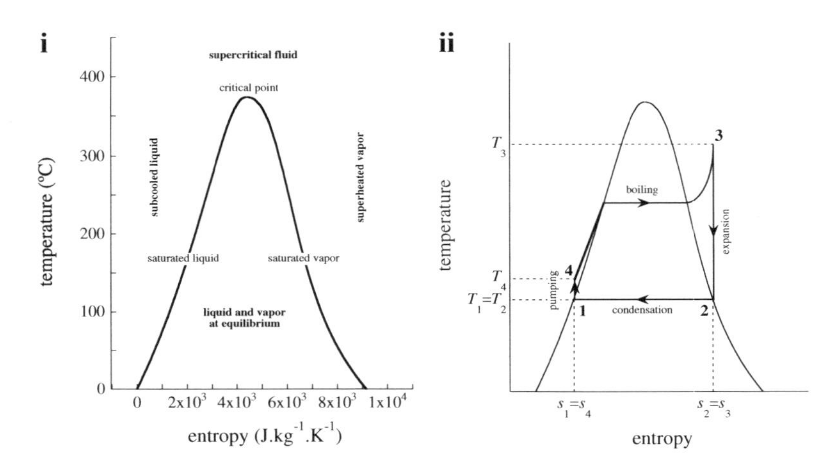

FIGURE 3.29 Thermodynamic diagram of water, as temperature versus entropy (using 0 J.kg^{-1}.K^{-1} for saturated liquid at 0.01°C and 612 Pa as reference state), (i) with indication of critical point as maximum of binodal curve encompassing saturated liquid on the left (as separation between subcooled liquid and liquid/vapor at equilibrium), and saturated vapor on the right (as separation between liquid/vapor at equilibrium and superheated vapor), and (ii) after superimposing the power cycle ($\mathbf{1 \rightarrow 4 \rightarrow 3 \rightarrow 2 \rightarrow 1}$) – corresponding to the sequence pumping, boiling, expansion, and condensation, with specification of germane specific entropies (s_1, s_2, s_3, s_4) and temperatures (T_1, T_2, T_3, T_4).

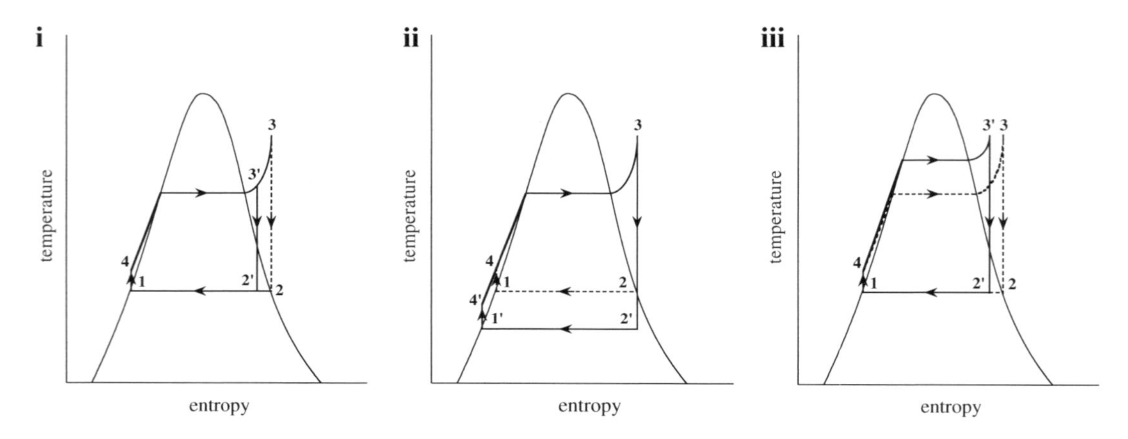

FIGURE 3.30 Thermodynamic diagram of water, as temperature versus entropy, after superimposing the power cycle – with elucidation of effects of (i) (lower) temperature of superheated steam, $\mathbf{1 \rightarrow 4 \rightarrow 3' \rightarrow 2' \rightarrow 1}$ versus $\mathbf{1 \rightarrow 4 \rightarrow 3 \rightarrow 2 \rightarrow 1}$, (ii) (lower) temperature of condensing steam, $\mathbf{1' \rightarrow 4' \rightarrow 3 \rightarrow 2' \rightarrow 1'}$ versus $\mathbf{1 \rightarrow 4 \rightarrow 3 \rightarrow 2 \rightarrow 1}$, and (iii) (higher) temperature of boiling liquid, $\mathbf{1 \rightarrow 4 \rightarrow 3' \rightarrow 2' \rightarrow 1}$ versus $\mathbf{1 \rightarrow 4 \rightarrow 3 \rightarrow 2 \rightarrow 1}$.

boiling temperature predicted by Antoine's equation; hence, cycle efficiency is improved. However, this effect vanishes as pressure approaches its critical value, since $\left(\partial T/\partial P\right)_\sigma$ decreases; hence, the positive outcome is partly offset by the accompanying decrease in quality of exhaust steam – see increase in liquid moisture content of steam as the cycle is shifted to the left, because entropy decreases with increasing pressure. This phenomenon affects unfavorably work output, and may even lead to erosion of turbine blades due to formation of liquid droplets.

To avoid the latter shortcoming, one may use two turbines – and reheat the steam leaving the higher pressure unit with flue gases, before it enters the lower pressure one; the corresponding layout is provided in Fig. 3.31i, and the underlying thermodynamic cycle is overlaid on a *TS*-diagram in Fig. 3.31ii. The overall heating cycle (i.e. $4' \rightarrow 6$) resembles a conventional cycle – except that a double peak of supersaturated steam arises, with regular heating ($4' \rightarrow 3$) followed by reheating ($5 \rightarrow 6$), as the single turbine is replaced by a set of two, usually with a common shaft and admission of hot steam in between; the first turbine accounts indeed for step $3 \rightarrow 5$, and the second turbine for step $6 \rightarrow 2'$.

The corresponding efficiency, $\eta_{r,r}$ is given by

$$\eta_{r,r} = -\frac{W_1 + W_2 + W_3}{Q_1 + Q_3} \equiv -\frac{W_{1'4'} + W_{35} + W_{62'}}{Q_{4'3} + Q_{56}}, \tag{3.110}$$

again consistent with Eq. (3.40) – but with work produced subdivided as W_2 and W_3, and high-temperature heat absorbed splitted as Q_1 and Q_3; correspondence between (specific) work or heat involved in each step to the corresponding enthalpy change permits reformulation of Eq. (3.110) to

$$\eta_{r,r} = -\frac{\left(h_{4'} - h_{1'}\right) + \left(h_5 - h_3\right) + \left(h_{2'} - h_6\right)}{\left(h_3 - h_{4'}\right) + \left(h_6 - h_5\right)} \tag{3.111}$$

that mimics Eq. (3.101) in functional form. Equation (3.110) implicitly assumes that $Q_{1'4'} = Q_{35} = Q_{62'} = 0$ – for representing

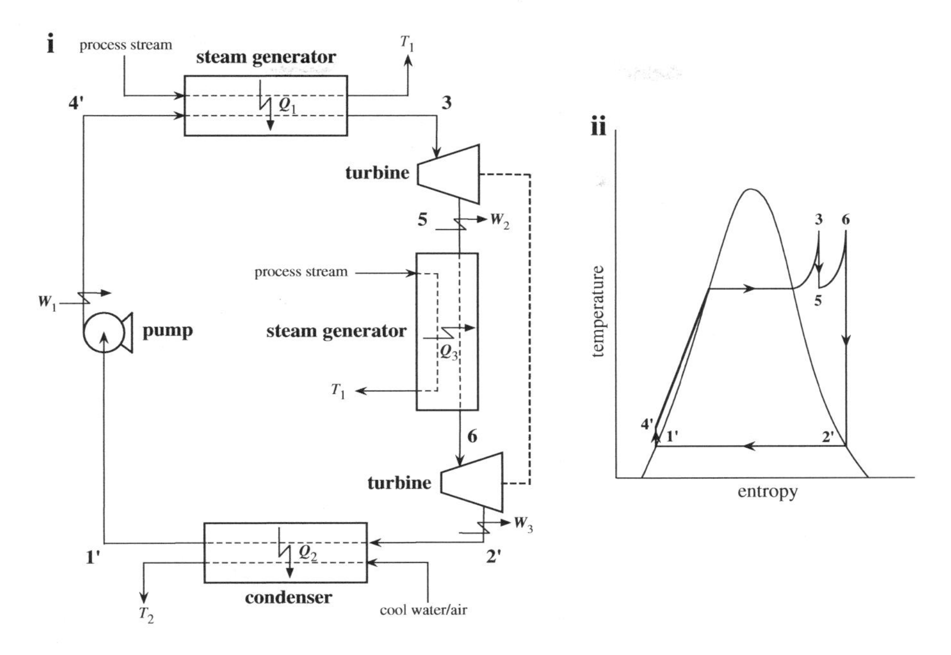

FIGURE 3.31 Schematic representation of (mechanical) power generation system, based on Rankine's vapor expansion/compression cycle, with intermediate reheating ($\mathbf{1'}\rightarrow\mathbf{4'}\rightarrow\mathbf{3}\rightarrow\mathbf{5}\rightarrow\mathbf{6}\rightarrow\mathbf{2'}\rightarrow\mathbf{1'}$) – with (i) indication of apparatuses and streams involved, and corresponding operating temperatures (T_1, $T_2<T_1$), as well as energy exchanged in the form of heat ($Q_1>0$, $Q_2<0$, $Q_3>0$) and work ($W_1>0$, $W_2<0$, $W_3<0$), and (ii) overlaid on thermodynamic diagram of water, as temperature versus entropy.

adiabatic steps, and $W_{4'3}=W_{56}=0$ – for being associated to plain heat transfer steps. In view of $h_{4'}\approx h_{1'}$, serving as alias to Eq. (3.102), simplification of Eq. (3.111) ensues as

$$\eta_{r,r}=\frac{h_3+h_6-h_{2'}-h_5}{h_3+h_6-h_{4'}-h_5}; \tag{3.112}$$

the temperature of the steam entering the first turbine frequently coincides with the temperature of steam entering the second turbine, i.e. $h_3=h_6$, so Eq. (3.112) simplifies further to

$$\eta_{r,r}=\frac{2h_3-h_{2'}-h_5}{2h_3-h_{4'}-h_5}. \tag{3.113}$$

Under these circumstances, there is an optimum reheat pressure between that prevailing in the boiler and that prevailing in the condenser – for which $\eta_{r,r}$ becomes maximum; such an optimum lies usually in the neighborhood of 20% the boiler pressure. Reheating provides a slight improvement in efficiency (ca. 5%), while considerably reducing specific steam consumption. In any case, the maximum temperature of steam is governed by mechanical strength of the materials used to build the parts subjected to the highest stresses, i.e. superheater tubes and turbine blades; creep-resistant stainless steels are quite expensive, yet required for operating temperatures above 600°C – so again economic considerations will override. One should emphasize again that irreversibilities during operation of the various components (e.g. fluid friction, heat loss) contribute also to deviate Rankine's actual efficiency from its ideal value.

As discussed above, increasing the temperature and/or pressure of steam, as it is admitted to the turbine(s), raises the mean temperature at which heat is absorbed – and thus increases efficiency of the associated cycle, as suggested by Eq. (3.94); conversely, the sensible heat needed to increase temperature of water through the pump and the latent heat needed to bring about boiling contribute unfavorably to that temperature rise, for a given amount of overall enthalpy supplied. Hence, another logical way of raising the mean temperature is to preheat the feed water internally within the cycle – via transfer of heat from the steam expanding in the turbine to the liquid water between pump and boiler; the associated process has been termed regenerative Rankine's cycle. The maximum improvement in efficiency would come from reversible transfer of heat – calling for an infinitesimally small temperature difference as driving force; this means passing the liquid water from the condenser, at constant pressure, through a theoretically infinite sequence of heat exchangers, each with infinite area and located between the stages of the turbine – constituted, in turn, by an infinite number of stages. This hypothetical strategy is obviously impractical – so one often extracts (or bleeds) steam from the turbine at a few intermediate pressures during the expansion process, and brings about transfer of heat therefrom to the feed water; this may occur through the wall of a heat exchanger (closed feedwater heating), or through direct mixing (open feedwater heat).

When heat is necessary for both heating proper (called process heat) and generation of mechanical power – while steam is available at 5–7 atm and 150–200°C, it makes sense, from an economic point of view, to resort to cogeneration; this is illustrated in Fig. 3.32. The major advantage of this combined approach is that steam may be mainly routed to the process heater (through a three-way valve, followed by an expansion valve) rather than the turbine, at times of high demand for process heat; otherwise, steam can be used chiefly to produce work – which, when in electrical form, can be injected into the electrical mains and get paid for, thus contributing to improve plant economics.

Gas turbine cycles operate at considerably higher temperatures than steam cycles; the maximum fluid temperature at inlet

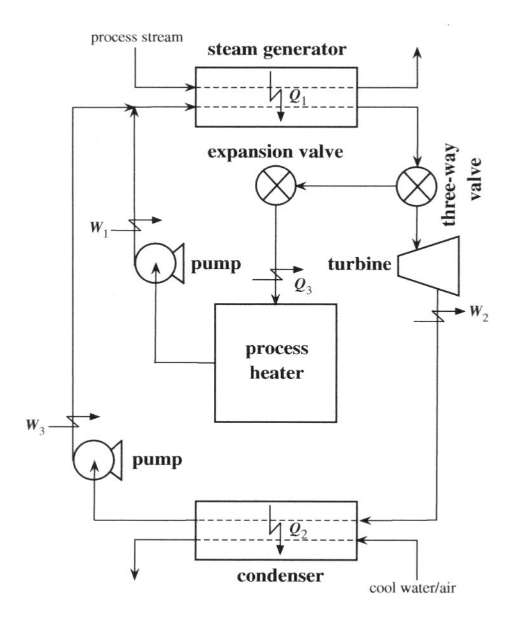

FIGURE 3.32 Schematic representation of cogeneration system, based on available steam ($Q_1>0$) – with adjustable loads between production of mechanical power ($W_2<0$) and partial rejection to low temperature ($Q_2<0$) followed by pumping ($W_3>0$), and utilization as process heat ($Q_3<0$) followed by pumping ($W_1>0$).

is ca. 1400°C for gas turbines, and over 1500°C at the burner exit of turbojet engines. Therefore, exhaust gases from gas turbines may be used as energy source for steam generation in regular power plants; this is called combined cycle. The major advantage is the high overall thermal efficiency attainable – which can approach 60%, knowing that the maximum predicted by Eq. (3.94) reads 1 – (15 + 273)/(1600 + 273) = 85%.

PROBLEM 3.3

Otto's cycle is an idealized thermodynamic path, able to describe functioning of an ignition piston engine – commonly found in portable electrical generators, besides internal combustion engines of automobiles and trucks. The system is defined as a portion of air drawn from the atmosphere, and mixed with fuel sprayed from a tank at temperature T_1; it is compressed adiabatically by the piston to volume V_2, thus causing a temperature rise to $T_2>T_1$, heated to temperature $T_3>T_2$ at constant volume by ignition of the added fuel (either spontaneous as in diesel engines, or triggered by a spark in regular gasoline engines), allowed to expand adiabatically to volume $V_1>V_2$ and temperature $T_4<T_3$ as it pushes the piston back, and finally cooled down to temperature $T_1<T_4$ at constant volume. The additional steps required by an actual 4-stroke engine – i.e. exhaust of waste heat and combustion products at constant pressure, and intake of ambient air and fuel droplets also at constant pressure, can be omitted to simplify thermodynamic analysis; while simultaneously guaranteeing a closed system, useful for the subsequent analysis. Consider, in addition, that: the system is unaffected in its physical properties by occurrence of combustion; compression and expansion steps are reversible (meaning that no work is dissipated as heat); useful work is extracted only by gas expansion in the cylinder – part of which is used to bring about gas compression (with net work used for propulsion, or to drive another engine); and only heat arising from combustion of fuel is added to the air/fuel mixture.

a) Sketch Otto's cycle in *TV*- and *PV*-diagrams.
b) Obtain an expression for efficiency, η_o, as a function of heat exchanged by system with surroundings.
c) Reformulate the result obtained in b) to the functional form $\eta_o \equiv \eta_o\{r_{cp},\gamma\}$, where r_{cp} denotes ratio of the higher to the lower volume, and γ denotes ratio of isobaric to isochoric heat capacities of fluid – on the hypothesis that the gaseous mixture behaves as an ideal gas.
d) Confirm that ideal Otto's engine is less efficient than ideal Carnot's heat engine, operating between the same temperature levels.
e) Confirm that, in practice, Otto's engine is less efficient than its ideal counterpart.

Solution

a) The isochoric steps, i.e. $2\rightarrow3$ and $4\rightarrow1$, are represented in both diagrams below as vertical arrows – pointing, however, in opposite directions; the remaining (adiabatic) steps, i.e. $1\rightarrow2$ and $3\rightarrow4$, correspond to curves that are steeper at higher temperature, and thus higher pressure.

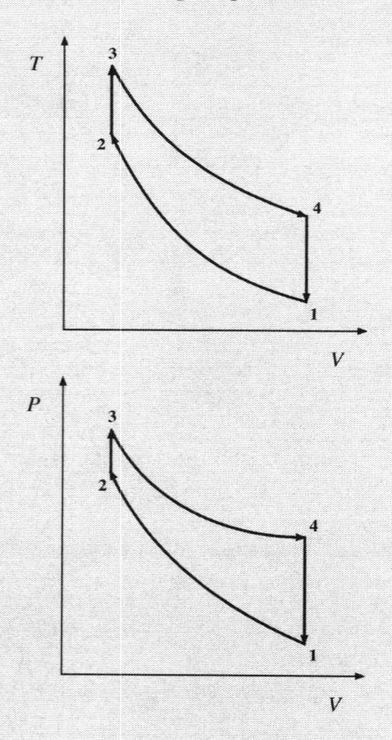

A compression ratio, r_{cp}, has classically been defined as

$$r_{cp} \equiv \frac{V_1}{V_2}; \tag{3.3.1}$$

in gasoline engines, $r_{cp}<10$ to avoid uncontrolled detonation (known as knocking), whereas $r_{cp}>14$ is required for spontaneous ignition in diesel engines – in view of Joule and Thomson's effect, i.e. $(\partial T/\partial P)_H > 0$, coupled with the underlying gas equation of state, i.e. $(\partial P/\partial V)_T < 0$.

b) The efficiency is given by the ratio of net work obtained to heat supplied at high temperature, i.e.

$$\eta_o \equiv -\frac{W_{12} + W_{23} + W_{34} + W_{41}}{Q_{23}}, \tag{3.3.2}$$

in agreement with Eq. (3.40) and the diagrams in a); for a full cycle, the final state coincides with the initial state – so the overall ΔU_{1234} satisfies

$$\Delta U \equiv \oint dU = 0, \tag{3.3.3}$$

because internal energy is a state function. On the other hand, the first law of thermodynamics, labeled as Eq. (3.16) of *Food Proc. Eng.: basics & mechanical operations*, supports

$$\begin{aligned}\Delta U &= \oint \left(đW + đQ \right) = \oint đW + \oint đQ \\ &= \int_1^2 đW + \int_2^3 đW + \int_3^4 đW + \int_4^1 đW \\ &\quad + \int_1^2 đQ + \int_2^3 đQ + \int_3^4 đQ + \int_4^1 đQ,\end{aligned} \tag{3.3.4}$$

which simplifies to

$$\Delta U = W_{12} + W_{34} + Q_{23} + Q_{41} \tag{3.3.5}$$

– at the expense of

$$Q_{12} = 0 \tag{3.3.6}$$

and

$$Q_{34} = 0, \tag{3.3.7}$$

for $1 \rightarrow 2$ and $3 \rightarrow 4$, respectively, being adiabatic steps by hypothesis; complemented with

$$W_{23} \equiv -\int_{V_2}^{V_3} P dV = -\int_{V_2}^{V_2} P dV = 0 \tag{3.3.8}$$

and

$$W_{41} \equiv -\int_{V_4}^{V_1} P dV = -\int_{V_1}^{V_1} P dV = 0 \tag{3.3.9}$$

for $2 \rightarrow 3$ and $4 \rightarrow 1$, respectively, taking place at constant volume in a closed system, see Eq. (3.17) of *Food Proc. Eng.: basics & mechanical operations*. Combination of Eqs. (3.3.3) and (3.3.5) gives rise to

$$W_{12} + W_{34} + Q_{23} + Q_{41} = 0, \tag{3.3.10}$$

or else

$$W_{12} + W_{34} = -\left(Q_{23} + Q_{41} \right); \tag{3.3.11}$$

insertion of Eqs. (3.3.8), (3.3.9), and (3.3.11) then transforms Eq. (3.3.2) to

$$\eta_o = \frac{Q_{23} + Q_{41}}{Q_{23}} = 1 - \frac{-Q_{41}}{Q_{23}}, \tag{3.3.12}$$

where $-Q_{41}$ represents heat released by system to surroundings throughout cooling, and Q_{23} denotes heat generated by combustion (analogous to heat received from an external source).

c) In view of Eqs. (3.16) of *Food Proc. Eng.: basics & mechanical operations*, complemented by Eqs. (3.3.8) and (3.3.9), one may redo Eq. (3.3.12) to

$$\eta_o = 1 - \frac{-\Delta U_{41}}{\Delta U_{23}}; \tag{3.3.13}$$

the definition of isochoric heat capacity, conveyed by Eq. (3.109) in *Food Proc. Eng.: thermal & chemical operations*, supports, in turn,

$$\Delta U_{41} = \int_{T_4}^{T_1} C_V \, dT \tag{3.3.14}$$

pertaining to step $4 \rightarrow 1$ implemented at constant volume V_1, and likewise

$$\Delta U_{23} = \int_{T_2}^{T_3} C_V dT \tag{3.3.15}$$

pertaining to step $2 \rightarrow 3$ carried out at constant volume V_2. For an ideal gas, C_V is independent of temperature, so Eqs. (3.3.14) and (3.3.15) can be algebraically rearranged to

$$-\Delta U_{41} = -C_V \int_{T_4}^{T_1} dT = C_V T \Big|_{T_1}^{T_4} = C_V \left(T_4 - T_1 \right) \tag{3.3.16}$$

and

$$\Delta U_{23} = C_V \int_{T_2}^{T_3} dT = C_V T \Big|_{T_2}^{T_3} = C_V \left(T_3 - T_2 \right), \tag{3.3.17}$$

respectively – after invoking the fundamental theorem of integral calculus; insertion of Eqs. (3.3.16) and (3.3.17) converts Eq. (3.3.13) to

$$\eta_o = 1 - \frac{C_V \left(T_4 - T_1 \right)}{C_V \left(T_3 - T_2 \right)} = 1 - \frac{T_4 - T_1}{T_3 - T_2}, \tag{3.3.18}$$

where C_V meanwhile dropped from both numerator and denominator. At this stage, it is convenient to apply Eq. (3.136) of *Food Proc. Eng.: thermal & chemical operations* to (adiabatic) step $1 \rightarrow 2$ to get

$$\frac{T_2}{T_1}=\left(\frac{P_2}{P_1}\right)^{\frac{\gamma-1}{\gamma}}, \qquad (3.3.19)$$

and likewise to (adiabatic) step $3 \rightarrow 4$ to obtain

$$\frac{T_4}{T_3}=\left(\frac{P_4}{P_3}\right)^{\frac{\gamma-1}{\gamma}} \qquad (3.3.20)$$

– where γ was defined in Eq. (3.126) of *Food Proc. Eng.: thermal & chemical operations*; the state equation for an ideal gas, i.e.

$$P=\frac{n\mathcal{R}T}{V} \qquad (3.3.21)$$

in parallel to Eq. (3.135) of *Food Proc. Eng: basics & mechanical operations*, will then be utilized to re-express P_i as a function of T_i and V_i in Eqs. (3.3.19) and (3.3.20), according to

$$\frac{T_2}{T_1}=\left(\frac{\frac{n\mathcal{R}T_2}{V_2}}{\frac{n\mathcal{R}T_1}{V_1}}\right)^{\frac{\gamma-1}{\gamma}}=\left(\frac{T_2}{T_1}\right)^{\frac{\gamma-1}{\gamma}}\left(\frac{V_1}{V_2}\right)^{\frac{\gamma-1}{\gamma}} \qquad (3.3.22)$$

and

$$\frac{T_4}{T_3}=\left(\frac{\frac{nRT_4}{V_4}}{\frac{nRT_3}{V_3}}\right)^{\frac{\gamma-1}{\gamma}}=\left(\frac{\frac{nRT_4}{V_1}}{\frac{nRT_3}{V_2}}\right)^{\frac{\gamma-1}{\gamma}}=\left(\frac{T_4}{T_3}\right)^{\frac{\gamma-1}{\gamma}}\left(\frac{V_2}{V_1}\right)^{\frac{\gamma-1}{\gamma}}, \qquad (3.3.23)$$

respectively – where common factors cancelled between numerator and denominator, with n being constant as per a closed system, and after realizing that $V_3=V_2$ and $V_4=V_1$. Upon lumping factors in T_2/T_1, Eq. (3.3.22) becomes

$$\left(\frac{T_2}{T_1}\right)^{\frac{1}{\gamma}}=\left(\frac{T_2}{T_1}\right)^{1-\frac{\gamma-1}{\gamma}}=\left(\frac{V_1}{V_2}\right)^{\frac{\gamma-1}{\gamma}}, \qquad (3.3.24)$$

and Eq. (3.3.23) similarly produces

$$\left(\frac{T_4}{T_3}\right)^{\frac{1}{\gamma}}=\left(\frac{T_4}{T_3}\right)^{1-\frac{\gamma-1}{\gamma}}=\left(\frac{V_2}{V_1}\right)^{\frac{\gamma-1}{\gamma}}; \qquad (3.3.25)$$

T_2/T_1 may then be calculated from Eq. (3.3.24) as

$$\frac{T_2}{T_1}=\left(\frac{V_1}{V_2}\right)^{\gamma-1}, \qquad (3.3.26)$$

while T_4/T_3 can be obtained as

$$\frac{T_4}{T_3}=\left(\frac{V_2}{V_1}\right)^{\gamma-1} \qquad (3.3.27)$$

from Eq. (3.3.25) – after raising both sides to γ. Therefore, Eq. (3.3.18) turns to

$$\eta_o=1-\frac{T_3\left(\frac{V_2}{V_1}\right)^{\gamma-1}-T_1}{T_3-T_1\left(\frac{V_1}{V_2}\right)^{\gamma-1}}, \qquad (3.3.28)$$

upon elimination of T_2 and T_4 with the aid of Eqs. (3.3.26) and (3.3.27), respectively; after factoring out $(V_2/V_1)^{\gamma-1}$, Eq. (3.2.28) yields

$$\eta_o=1-\frac{T_3-T_1\left(\frac{V_1}{V_2}\right)^{\gamma-1}}{T_3-T_1\left(\frac{V_1}{V_2}\right)^{\gamma-1}}\left(\frac{V_2}{V_1}\right)^{\gamma-1}, \qquad (3.3.29)$$

where the numerator coinciding with the denominator in the second term of the right-hand side justifies simplification to

$$\eta_o=1-\left(\frac{V_2}{V_1}\right)^{\gamma-1}=1-\left(\frac{1}{\frac{V_1}{V_2}}\right)^{\gamma-1}. \qquad (3.3.30)$$

Equation (3.3.30) can be reformulated to

$$\eta_o=1-\frac{1}{r_{cp}{}^{\gamma-1}} \qquad (3.3.31)$$

at the expense of Eq. (3.3.1) – which exhibits the (intended) functional dependence of η_o on r_{cp} and γ.

d) After reformulation of Eq. (3.3.18) to

$$\eta_o=1-\frac{T_1}{T_2}\frac{\frac{T_4}{T_1}-1}{\frac{T_3}{T_2}-1} \qquad (3.3.32)$$

via factoring out of T_1/T_2, one should proceed to ordered multiplication of Eqs. (3.3.26) and (3.3.27) to get

$$\frac{\frac{T_4}{T_1}}{\frac{T_3}{T_2}}=\frac{T_2}{T_1}\frac{T_4}{T_3}=\left(\frac{V_1}{V_2}\right)^{\gamma-1}\left(\frac{V_2}{V_1}\right)^{\gamma-1}=1 \qquad (3.3.33)$$

– or else

$$\frac{T_4}{T_1}=\frac{T_3}{T_2}; \qquad (3.3.34)$$

insertion of Eq. (3.3.33) allows simplification of Eq. (3.3.32) to

$$\eta_o=1-\frac{T_1}{T_2}, \qquad (3.3.35)$$

since the same factor T_4/T_1-1, or T_3/T_2-1 for that matter appears in both numerator and denominator. Inspection of the first plot in a) indicates that $T_3>T_2$, and thus $T_3/T_2>1$ – so Eq. (3.3.35) will support

$$\eta_o = 1 - \frac{T_3}{T_2}\frac{T_1}{T_3} < 1 - \frac{T_1}{T_3} = \eta_c, \tag{3.3.36}$$

upon multiplication and division by T_3 and the extra aid of Eq. (3.94); note that T_1 denotes the lower temperature (at which heat is lost) and T_3 denotes the higher temperature (at which heat is produced/received). Therefore, Otto's engine is less efficient than Carnot's engine if operated between the same upper and lower temperature levels.

e) Inspection of Eq. (3.3.31) *vis-à-vis* with Eq. (3.3.35) indicates that the temperature right before combustion, T_2, will not permit maximum thermodynamic performance, η_o, be attained unless

$$\frac{T_1}{T_2} = \frac{1}{r_{cp}^{\gamma-1}}; \tag{3.3.37}$$

the said temperature would thus be given by

$$T_2 = T_1 r_{cp}^{\gamma-1}. \tag{3.3.38}$$

If the fuel + air mixture departed from room temperature (say, 20°C), the best performant diesel engines known to date – with $r_{cp} \approx 23$, would then require

$$T_2 = 293 \cdot 23^{1.3-1} = 751\ \mathrm{K} \tag{3.3.39}$$

for temperature right before ignition (and using the typical value of $\gamma = 1.3$ pertaining exactly to plain air, and approximately to the combustion gases produced therefrom); this is far too high a temperature, knowing that autoignition of common diesel would occur at ca. 600 K. Therefore, the efficiency of actual Otto's cycles lies necessarily below that of ideal Otto's cycle, as claimed.

3.3.4 Vapor compression/expansion refrigeration cycle

It is difficult to implement, in practice, Carnot's cycle intended for refrigeration, using a superheated vapor or gas as refrigerant fluid – because sensible heat effects would preclude isothermal steps of absorption and release of heat. Consequently, most refrigeration systems in operation, at both industrial and domestic scales, resort to a vapor compression/expansion cycle, entailing continuous isothermal evaporation of a refrigerant in liquid form to bring about the refrigeration effect.

A comparison of the aforementioned cycle with a putative Carnot's cycle – operating between the same temperatures and entirely within the two-phase region, is provided in Figs. 3.33i and ii, respectively. Note the coincidence between states 3' (i.e. saturated vapor) and 4 (i.e. saturated liquid) between cycles; and the similarity between Figs. 3.26ii and 3.33ii, except for the reverse direction of operation (besides use of a distinct working fluid). The first main difference of a practical refrigeration cycle to Carnot's cycle is that compression takes place in the superheated region, see Fig. 3.33i – as it would be difficult to stop the evaporative process just prior to compression step 6→3' in Fig. 3.33ii; furthermore, it is hard to compress wet vapor efficiently,

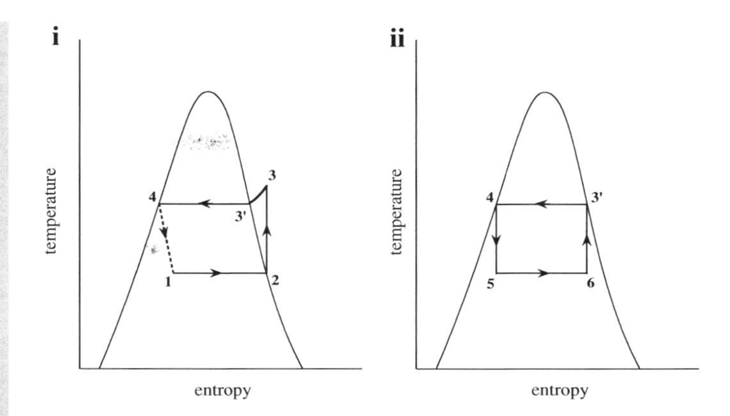

FIGURE 3.33 Thermodynamic diagram of ammonia (R717), as temperature versus entropy, after superimposing (i) a practical vapor compression/expansion refrigeration cycle, **1→2→3→3′→4→1**, or (ii) an ideal Carnot's cycle operated between the same temperatures, **5→6→3′→4→5**.

with discharge of solely saturated vapor at 3′. The second main-difference is that saturated liquid (4) is expanded isenthalpically through a valve (4→1), instead of isentropically in an engine or turbine as per Carnot's cycle (4→5); inclusion of an irreversible step in common refrigeration cycles is justified by the work that would ever be recovered by an expander being small compared to the work needed to compress the vapor – so the value of the work saved via use of an expander would easily be offset by the increased cost of the said expander relative to a valve (except in very large facilities).

If the ideal vapor compression/expansion cycle coincided with Carnot's cycle, all refrigerants would share the same (maximum) coefficient of performance for operation between given temperatures; therefore, practical choice of refrigerant should bear in mind how the underlying cycle deviates from Carnot's counterpart.

The layout most commonly elected to bring about a vapor compression/expansion cycle, for refrigeration purposes, is conveyed by Fig. 3.34. Inspection of Fig. 3.34i *vis-à-vis* with Fig. 3.33i indicates that saturated vapor (2) leaves the evaporator, and is compressed reversibly and adiabatically to superheated vapor (3); it is then cooled, at constant pressure, through saturated

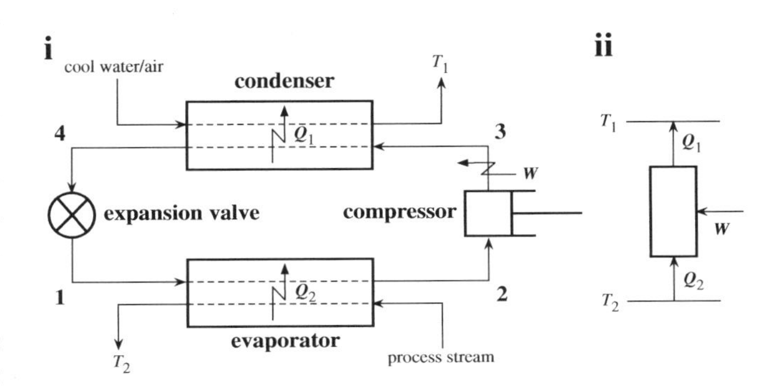

FIGURE 3.34 Schematic representation of (mechanical) refrigeration system, based on vapor compression/expansion cycle (**1→2→3→4→1**) – with indication of (i) apparatuses and streams involved, and corresponding operating temperatures (T_1, $T_2 < T_1$), as well as (i, ii) energy exchanged in the form of heat ($Q_1 < 0$, $Q_2 > 0$) and work ($W > 0$).

vapor until full condensation has occurred to saturated liquid (4); finally, an (irreversible) adiabatic, and thus isenthalpic expansion takes place through a valve, which conveys partial flashing toward a mixture of saturated liquid and vapor (1) – which closes the refrigeration cycle.

In fundamental terms, the cycle under scrutiny represents a heat engine operated backward – as grasped when Fig. 3.34ii is compared with Fig. 3.27ii. Heat Q_2 is removed in the evaporator at (lower) temperature T_2, as effected by work W supplied by the compressor; both end up as heat Q_1, which is eventually released in the condenser at (higher) temperature T_1.

Since pressure and enthalpy of refrigerant change as it passes through the sequence of components of the refrigeration system sketched in Fig. 3.34i, it is convenient to retrieve the characteristic thermodynamic diagram of such a refrigerant, and overlay the refrigeration cycle thereon – namely, using enthalpy and pressure as variables (in view of existence of two degrees of freedom, as per Gibbs' phase rule); this is done in Fig. 3.35ii. The plain *PH*-diagram in Fig. 3.35i is divided into four distinct regions, separated by the liquid- and vapor-saturation curves (each with a single degree of freedom, as two phases are present); the inner area, enclosed by such lines, contains a mixture of liquid and vapor at equilibrium with each other. The area to the left of the saturated liquid curve represents subcooled liquid, while the area to the right of the saturated vapor curve corresponds to superheated vapor; a point in either area requires two coordinates for full definition – as many as the associated thermodynamic degrees of freedom. Above the critical point, no distinction further exists between liquid and vapor – and a single (supercritical) fluid phase results.

In both evaporator and condenser, pressure of refrigerant remains constant while enthalpy changes – see Fig. 3.35ii; this is indicated by a rightward horizontal (isobaric) trajectory ($1 \rightarrow 2$) corresponding to evaporation, and likewise by a leftward horizontal trajectory ($3 \rightarrow 4$) corresponding to condensation at a higher pressure. A mixture of saturated liquid and vapor (1) accordingly passes through the evaporator – where refrigerant still in the liquid state evaporates (under reduced pressure) to become saturated vapor in full (2); in doing so, it absorbs latent heat of vaporization, and consequently cools the surroundings. When refrigerant passes through the condenser, after having gained enthalpy via compression, (cool) air or water flowing through its coils absorbs heat from the said hot refrigerant vapor (3) – thus causing it to condense in full (4); sensible heat is first released by the refrigerant until reaching the saturation line, and then latent heat of condensation is withdrawn, until the system becomes plain saturated liquid. During the compression step ($2 \rightarrow 3$), work is performed upon the refrigerant – departing from saturated vapor (2) and ending as superheated vapor (3), with an increase in enthalpy as a consequence of the increase in pressure – accompanied by an increase in temperature; this process is hypothesized as adiabatic and reversible, so exchange of heat and dissipation of work as thermal energy are neglected – and an isentropic path accordingly results, see Fig. 3.33i. The outlet pressure from the compressor must lie below the critical pressure of the refrigerant, but should be high enough to reach a temperature above that of the cooling medium, usually at ambient temperature; whereas the size of the compressor should be compatible with the intended flow rate and pressure increase. Conversely, expansion occurs at a valve via an essentially isenthalpic process – described by a downward vertical line $4 \rightarrow 1$ in Fig. 3.35ii, but an inclined line pointing down- and rightward in Fig. 3.33i as per the irreversible nature of this step; high-pressure liquid refrigerant (4) accordingly passes, at a controlled rate, to the low-pressure section of the refrigeration system – thus generating a mixture of saturated liquid and vapor (1).

The characteristic shape of the binodal curve in Fig. 3.35ii indicates that, for a fixed pressure difference P_4–P_2, the coefficient of performance falls off rapidy as critical conditions are approached. In fact, h_2–$h_1 \rightarrow 0$ in the vicinity of the point where the saturated liquid curve and the saturated vapor curve merge; hence, the ratio of h_2–h_1 as cooling load to (a given) h_3–h_2 as work load becomes smaller and smaller as $P \rightarrow P_{crt}$. Conversely, the power required to produce an intended refrigeration duty becomes quite insensitive to refrigerant choice when $P_2 \ll P_{crt}$ – as the two portions of the binodal curve keep an essentially constant distance between them, along with the similar pattern exhibited by isentropic lines in the superheated vapor region, see Fig. 3.36.

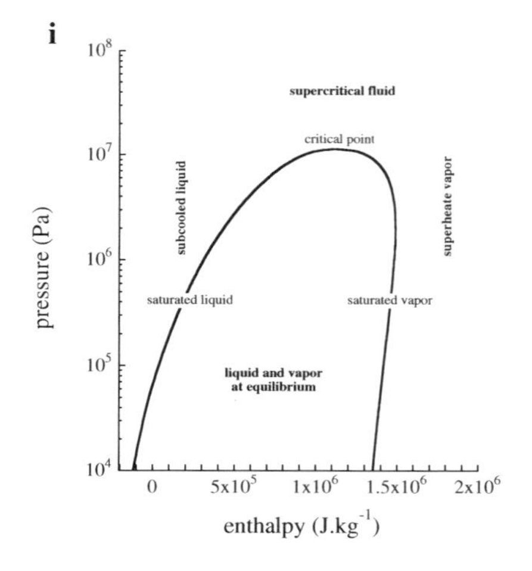

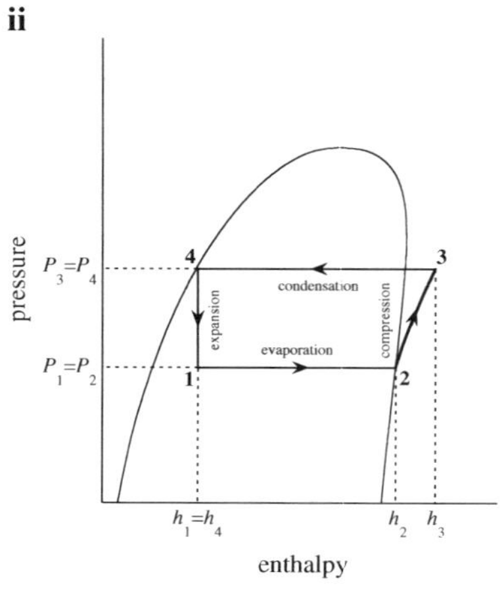

FIGURE 3.35 Thermodynamic diagram of ammonia (R717), as pressure versus enthalpy (using 2×10^5 J.kg^{-1} for saturated liquid at 0°C and 0.43 MPa as reference state), (i) with indication of critical point as maximum of binodal curve, encompassing saturated liquid on the left (as separation between subcooled liquid and liquid/vapor at equilibrium) and saturated vapor on the right (as separation between liquid/vapor at equilibrium and superheated vapor), and (ii) after superimposing the refrigeration cycle (**1**$\rightarrow$**2**$\rightarrow$**3**$\rightarrow$**4**$\rightarrow$**1**) – corresponding to the sequence evaporation, compression, condensation, and expansion, with specification of germane specific enthalpies (h_1, h_2, h_3, h_4) and pressures (P_1, P_2, P_3, P_4).

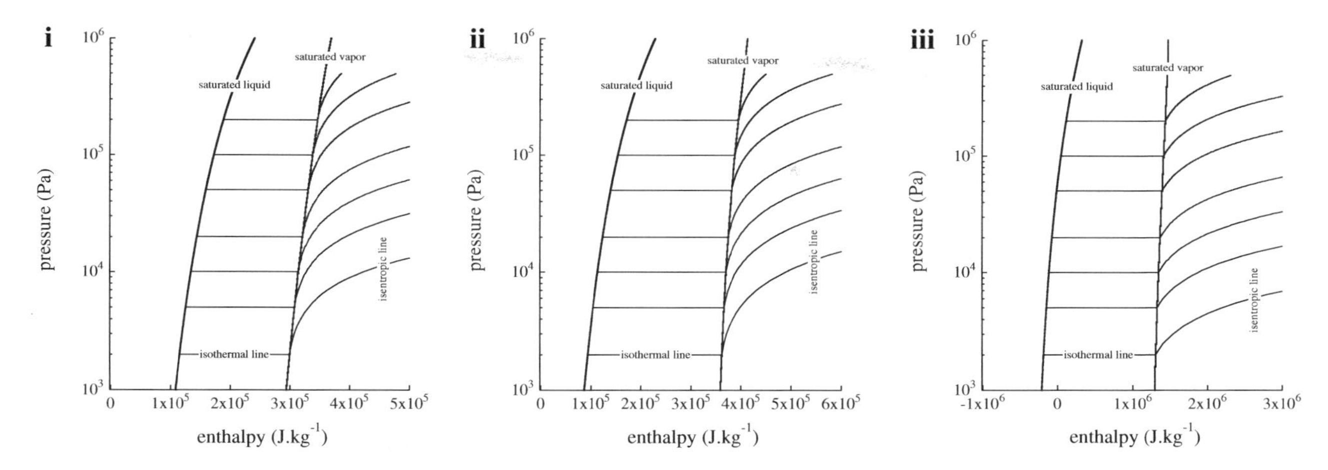

FIGURE 3.36 Thermodynamic diagrams of (i) R12, (ii) R22, and (iii) R717, as pressure versus enthalpy (using 2×10^5 J.kg^{-1} for saturated liquid at 0°C as reference state) – highlighting the relevant portions of the saturated liquid and vapor curves, as well as isothermal trajectories within the saturated liquid/vapor region, and isentropic trajectories within the superheated vapor region.

Unlike water – used as universal fluid in power generation cycles, a number of compounds can play the role of refrigerant; *PH*-diagrams of three common refrigerants are conveyed in Fig. 3.36, focused on the region of operational interest (i.e. below their critical point) – including isothermal and isentropic trajectories. Existence of two degrees of freedom in the superheated vapor region implies that isentropic lines (characterized by a fixed entropy) in such a region represent pressure as a function of enthalpy, for the given entropy; under the binodal curve, all trajectories are necessarily horizontal, because of the single degree of freedom available – i.e. pressure, must remain constant when enthalpy is varied, since temperature has itself been set as constant. The magnitude of the enthalpy change between liquid and vapor saturated phases narrows as the critical point is approached – as already stressed; this will eventually compromise performance of the refrigeration cycle, for requiring a large mass flow rate of refrigerant for a given heat load released at the temperature of the surroundings. Hence, the portion of the thermodynamic diagram in the vicinity of the critical point was ignored in Fig. 3.36, for practical reasons – even though the binodal curve in Fig. 3.36iii coincides with that plotted in Fig. 3.35i. Isobaric steps are obviously represented by horizontal lines in Fig. 3.36 – corresponding to constant pressure, as per the ordinate of the plots; whereas isothermal steps would be represented by almost vertical lines pointing upward in the (subcooled) liquid region (where enthalpy is a very weak function of increasing pressure), horizontal lines under the saturation curve, and lines curved up- and rightward in the (superheated) vapor region.

Using Eq. (3.47) as template, one may express the coefficient of performance of the refrigeration cycle as

$$\beta = \frac{Q_{12}}{W_{23}} \tag{3.114}$$

– where Q_{12} denotes heat absorbed by (liquid) refrigerant in the evaporator, or Q_2 in Fig. 3.34i, and W_{23} denotes work supplied by the compressor to the vapor refrigerant, or W in Fig. 3.34i. Inspection of Fig. 3.35ii supports transformation of Eq. (3.114) to

$$\boxed{\beta = \frac{h_2 - h_1}{h_3 - h_2},} \tag{3.115}$$

since the (specific) heat involved in step $1\rightarrow2$ coincides with the change in specific enthalpy between those states, h_2-h_1 – and the (specific) work involved in step $2\rightarrow3$ likewise entails a variation in specific enthalpy from h_2 to h_3. Note that β conveys an important measure of performance of a given refrigeration system; common systems are characterized by β lying between 3 and 6. Fluid friction, heat transfer loss, and component inefficiency prevent, however, actual refrigeration cycles from operating at their best ever, or Carnot's coefficient of performance. If the mass flow rate of refrigerant, $\dot{m}$, is known, then one may readily access the (work) power required for compression, $\dot{W}_{23}$, via

$$\boxed{\dot{W}_{23} = \dot{m}\left(h_3 - h_2\right);} \tag{3.116}$$

by the same token, the (heat) power removed in the condenser by the cooling utility would abide to

$$\left|\dot{Q}_{34}\right| = \dot{m}\left(h_3 - h_4\right), \tag{3.117}$$

while the (heat) power absorbed in the evaporator should look like

$$\boxed{\dot{Q}_{12} = \dot{m}\left(h_2 - h_1\right)} \tag{3.118}$$

– also known as refrigeration effect. In view of the hypothesized $h_1=h_4$, one gets

$$h_3 - h_4 = h_3 - h_1 = \left(h_3 - h_2\right) + \left(h_2 - h_1\right) \tag{3.119}$$

along with addition and subtraction of h_2; one thus finds that

$$h_3 - h_4 > h_2 - h_1, \tag{3.120}$$

since $h_3-h_2>0$. After multiplying both sides by $\dot{m}>0$, Eq. (3.120) yields

$$\dot{m}\left(h_3 - h_4\right) > \dot{m}\left(h_2 - h_1\right), \tag{3.121}$$

where insertion of Eqs. (3.117) and (3.118) permits simplification to

$$\left|\dot{Q}_{34}\right| > \dot{Q}_{12} \tag{3.122}$$

– in full agreement with the second law of thermodynamics, as per Clausius' statement.

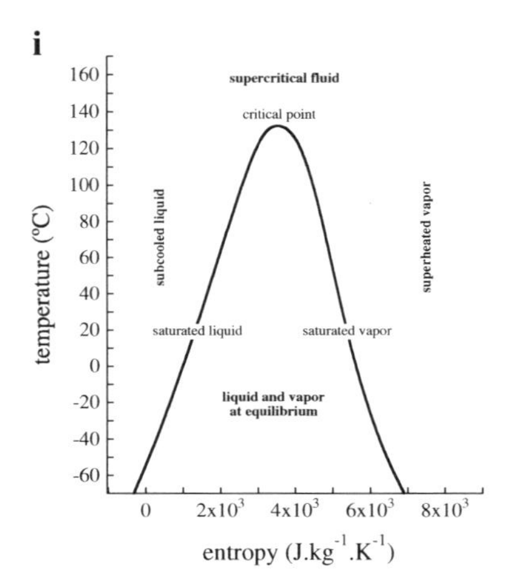

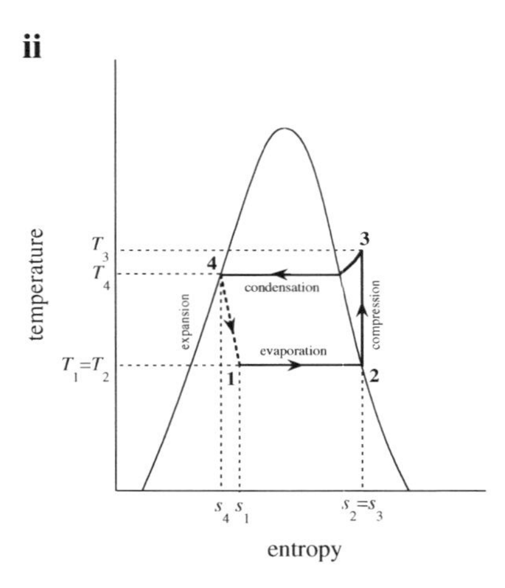

FIGURE 3.37 Thermodynamic diagram of ammonia (R717), as temperature versus entropy (using 1×10^3 J.kg^{-1}.K^{-1} for saturated liquid at 0°C and 0.43 MPa as reference state), (i) with indication of critical point as maximum of binodal curve, encompassing saturated liquid on the left (as separation between subcooled liquid and liquid/vapor at equilibrium) and saturated vapor on the right (as separation between liquid/vapor at equilibrium and superheated vapor), and (ii) after superimposing the refrigeration cycle (**1** → **2** → **3** → **4** → **1**) – corresponding to the sequence evaporation, compression, condensation, and expansion, with specification of germane specific entropies (s_1, s_2, s_3, s_4) and temperatures (T_1, T_2, T_3, T_4).

Despite the practical applicability of pressure vs. enthalpy charts in designing and simulating refrigeration systems, temperature and entropy are sometimes selected as alternative (independent) variables; an example is included in Fig. 3.37, pertaining again to R717 as refrigerant (but somehow more informative than Fig. 3.33). The saturation curve – representing liquid/vapor equilibrium, sections again the thermodynamic plane into four regions; a biphasic mixture of saturated liquid and saturated vapor accounts, in particular, for the area under the said curve. Furthermore, the area to the left of the saturated liquid curve corresponds to uniphasic subcooled liquid, whereas the area to the right of the saturated vapor curve representes uniphasic superheated vapor. The saturated-liquid and -vapor lines merge at the critical point – above which a uni-phase area appears again, constituted by supercritical fluid; as emphasized previously, this region is of no interest toward refrigeration purposes, since no phase change will be physically possible.

Recall that a single degree of freedom exists along a saturation curve – so temperature remains constant (since pressure is fixed) while entropy increases in the case of evaporation, with an accompanying horizontal line that points rightward (1 → 2), see Fig. 3.37ii. Condensation starts with cooling from the superheated region (3) down to the saturated vapor line, after which a constant temperature (T_4, above that prevailing during evaporation, T_2) is observed – represented thereafter by a leftward horizontal line (→4). Compression occurs isentropically, for having been assumed as adiabatic and reversible; hence, a vertical path pointing upward ensues (2 → 3), which crosses the superheated vapor region until a maximum temperature is reached (T_3). Conversely, expansion is intrinsically irreversible, so an (ill-defined) dashed line pointing down- and rightward (4 → 1) was used in Fig. 3.37ii; this representation reflects the lack of knowledge on the exact path tracked by the refrigerant throughout this step. Despite being thermodynamically equivalent to Fig. 3.35ii, the information required to compute the coefficient of performance cannot be directly retrieved from the independent variables used in Fig. 3.37ii – thus justifying the more universal utilization of the former in refrigeration design.

The type of compressor utilized depends on the temperature intended at the evaporator, T_2, and the refrigeration duty desired, Q_2 – see Fig. 3.34i; for powers not exceeding 500 kW and T_2=–20°C, reciprocating compressors (with operating pressure ratio below 9) are employed – whereas centrifugal compressors (with operating pressure ratio of ca. 5) are preferred for higher powers, say up to 20 MW. Remember that the said pressure ratio is given by the quotient of saturation pressure of refrigerant, at condensation temperature, T_1, to that at evaporation temperature, T_2. Should evaporator pressure be lower than atmospheric pressure, it will be convenient to employ a hermetically sealed compressor to prevent ingress of atmospheric air – an option feasible for powers below 75 or 750 kW, for reciprocating or centrifugal compressors, respectively.

PROBLEM 3.4

Consider Carnot's heat engine operated as refrigerator.

a) Obtain an expression for the differential work performed on a flow system, operated under steady state.
b) Calculate the overall work involved in either mode of operation of Carnot's refrigerator, and compare them.
c) Justify that the heat involved in every step of the said cycle is the same, irrespective of the refrigerator at stake being a closed or an open system.
d) Prove that the coefficient of performance of closed Carnot's refrigerator, β_c, coincides with the coefficient of performance of its open counterpart, β_s.

Solution

a) The elementary volumetric work, $\text{đ}W_c$, associated to a change of volume dV of a closed system at pressure P, has been given by Eq. (3.17) in *Food Prod. Eng.: basics & mechanical operations* – and reads

$$\text{đ}W_c = -PdV; \quad (3.4.1)$$

for a flow system, operated under steady state conditions, the work at stake, $\text{đ}W_s$, is the sum of the said volumetric work, with the work required to push the fluid into the elementary volume and the work needed to push the fluid out thereof – according to

$$\text{đ}W_s = -(PV)\big|_V + \text{đ}W_c + (PV)\big|_{V+dV} \quad (3.4.2)$$

In view of Eq. (3.4.1), coupled with definition of differential, Eq. (3.4.2) may be rewritten as

$$\begin{aligned} \text{đ}W_s &= (PV)\big|_{V+dV} - (PV)\big|_V - PdV \\ &= d(PV) - PdV \equiv \frac{d(PV)}{dV}dV - PdV; \end{aligned} \quad (3.4.3)$$

the rule of calculation of the derivative of a product supports further transformation to

$$\text{đ}W_s = \left(V\frac{dP}{dV} + P\right)dV - PdV. \quad (3.4.4)$$

Upon factoring in dV, and dropping common factors between numerator and denominator afterward, Eq. (3.4.4) becomes

$$\text{đ}W_s = V\frac{dP}{dV}dV + PdV - PdV = VdP + PdV - PdV, \quad (3.4.5)$$

with cancellation of PdV with its negative unfolding

$$\text{đ}W_s = VdP \quad (3.4.6)$$

– which differs from Eq. (3.4.1) owing to exchange of V with P, and dP with $-dV$.

b) In the case of a closed refrigerator, the total work involved, W_{1234}, abides to

$$W_{1234,c} \equiv W_{14,c} + W_{43,c} + W_{32,c} + W_{21,c}, \quad (3.4.7)$$

following the nomenclature in Fig. 3.21; one may then invoke Eq. (3.4.1) to write

$$W_{14,c} \equiv \int_1^4 \text{đ}W_c = -\int_{V_1}^{V_4} PdV \quad (3.4.8)$$

pertaining to step $1 \rightarrow 4$,

$$W_{43,c} \equiv \int_4^3 \text{đ}W_c = -\int_{V_4}^{V_3} PdV \quad (3.4.9)$$

pertaining to step $4 \rightarrow 3$,

$$W_{32,c} \equiv \int_3^2 \text{đ}W_c = -\int_{V_3}^{V_2} PdV \quad (3.4.10)$$

pertaining to step $3 \rightarrow 2$, and finally

$$W_{21,c} \equiv \int_2^1 \text{đ}W_c = -\int_{V_2}^{V_1} PdV \quad (3.4.11)$$

pertaining to step $2 \rightarrow 1$. The values of the partial works are graphically depicted below, on the PV plane – each as the area of the trapezoid between the horizontal, or V-axis, and the curve defining the corresponding step; while $W_{1234,c}$ coincides with the area bounded by the closed curve representing the cycle, after using the plus or minus sign (as explicitly indicated) to algebraically add the corresponding areas.

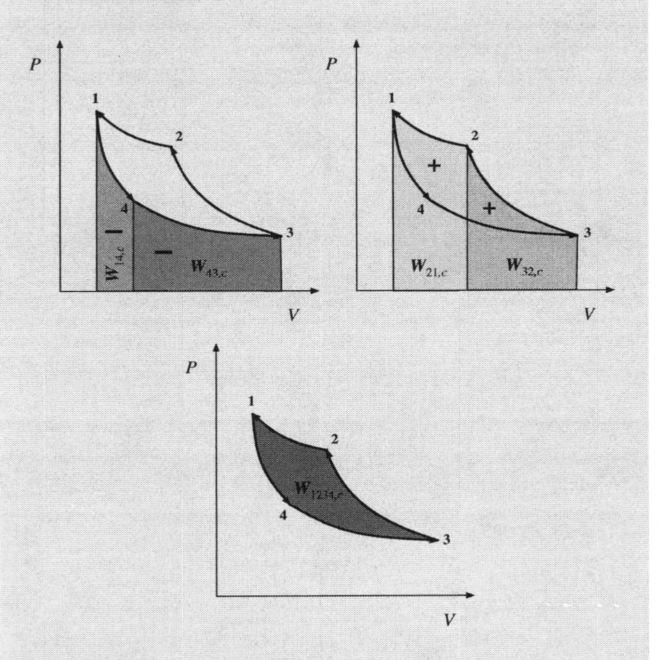

Remember that the area associated with a definite integral is considered positive when the kernel is negative (i.e. $-P < 0$) and the integration variable decreases (i.e. $dV < 0$) – as is the case of steps $3 \rightarrow 2$ and $2 \rightarrow 1$; and negative when the kernel is still negative, but the integration variable increases – as is the case of steps $1 \rightarrow 4$ and $4 \rightarrow 3$. By the same token, the total work involved in steady state operation of a continuous refrigerator, $W_{1234,s}$, is given by

$$W_{1234,s} \equiv W_{14,s} + W_{43,s} + W_{32,s} + W_{21,s}; \tag{3.4.12}$$

each term may now be calculated with the aid of Eq. (3.4.6), to eventually give

$$W_{14,s} \equiv \int_1^4 đW_s = \int_{P_1}^{P_4} VdP \tag{3.4.13}$$

in the case step $1 \rightarrow 4$,

$$W_{43,s} \equiv \int_4^3 đW_s = \int_{P_4}^{P_3} VdP \tag{3.4.14}$$

in the case of step $4 \rightarrow 3$,

$$W_{32,s} \equiv \int_3^2 đW_s = \int_{P_3}^{P_2} VdP \tag{3.4.15}$$

in the case of step $3 \rightarrow 2$, and finally

$$W_{21,s} \equiv \int_2^1 đW_s = \int_{P_2}^{P_1} VdP \tag{3.4.16}$$

in the case of step $2 \rightarrow 1$. The work associated with each step is represented below, as area of the trapezoid located between the vertical, or P-axis, and the corresponding curve; once again, $W_{1234,s}$ is accounted for by the area bounded by the closed curve representing the cycle, obtained after algebraic addition of all four areas under scrutiny.

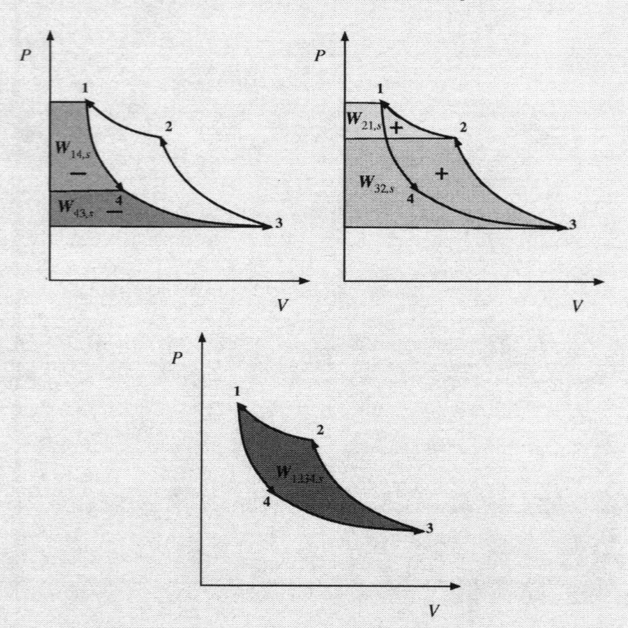

The area associated with the definite integral has now been taken as positive when the kernel is positive (i.e. $V > 0$) and the integration variable increases (i.e. $dP > 0$) – as is the case of steps $3 \rightarrow 2$ and $2 \rightarrow 1$; and negative when the kernel is again positive, along with a decrease of the integration variable – as is the case of steps $1 \rightarrow 4$ and $4 \rightarrow 3$. Note that $W_{14,c} \neq W_{14,s}$, $W_{43,c} \neq W_{43,s}$, $W_{32,c} \neq W_{32,s}$ and $W_{21,c} = W_{21,s}$, in general; however,

$$W_{1234,c} = W_{1234,s} \tag{3.4.17}$$

– since the very same plane surface, comprised within curves $1 \rightarrow 4$, $4 \rightarrow 3$, $3 \rightarrow 2$, and $2 \rightarrow 1$, accounts for the area representing $W_{1234,c}$ and $W_{1234,s}$.

c) Consider a generic step $i \rightarrow j$ of Carnot's cycle; the heat involved therein will hereafter be denoted as $Q_{ij,c}$ or $Q_{ij,s}$, for closed system or steady-state flow operation, respectively. Combination of Eqs. (3.16) and (3.17) of *Food Proc. Eng.: basics & mechanical operations*, pertaining to the differential formulation of the first law of thermodynamics, leads to

$$dU_{ij} = đQ_{ij,c} - PdV \tag{3.4.18}$$

applying to a closed system; whereas combinations of Eqs. (3.15) and (3.23) of *Food Proc. Eng.: basics & mechanical operations* gives rise to

$$dH_{ij} = đQ_{ij,s} + VdP \tag{3.4.19}$$

applying to an open system, operated under steady state. In view of Eq. (3.21) of *Food Proc. Eng.: basics & mechanical operations*, one may write

$$dH_{ij} = dU_{ij} + PdV + VdP, \tag{3.4.20}$$

so insertion of Eq. (3.4.18) yields

$$dH_{ij} = đQ_{ij,c} - PdV + PdV + VdP = đQ_{ij,c} + VdP \tag{3.4.21}$$

along with cancellation of symmetrical terms; elimination of dH_{ij} between Eqs. (3.4.19) and (3.4.21) leaves

$$đQ_{ij,c} + VdP = đQ_{ij,s} + VdP, \tag{3.4.22}$$

which reduces to

$$đQ_{ij,c} = đQ_{ij,s} \tag{3.4.23}$$

after dropping VdP from both sides. The aforementioned conclusion is a consequence of both internal energy, U, and enthalpy, H, being state functions; therefore, their values depend only on the current state, and not on the trajectory followed to reach it – namely, within a closed system or a steady state flow process. Integration of both sides Eq. (3.4.23) between any given limits then supports

$$Q_{ij,c} \equiv \int_i^j đQ_{ij,c} = \int_i^j đQ_{ij,s} \equiv Q_{ij,s} \tag{3.4.24}$$

– thus indicating that the same heat is exchanged with the surroundings, by either a closed or a (steady state) flow system operating between the same temperatures.

d) In view of Eq. (3.47), one has it that

$$\beta_c \equiv \frac{Q_{43,c}}{W_{1234,c}} \tag{3.4.25}$$

and likewise

$$\beta_s \equiv \frac{Q_{43,s}}{W_{1234,s}}. \tag{3.4.26}$$

Equation (3.4.24) implies, in particular, that

$$Q_{43,c} = Q_{43,s} \tag{3.4.27}$$

when subscript *ij* refers to step 4 → 3, which may be combined with Eq. (3.4.17) to conclude that

$$\beta_s = \beta_c \tag{3.4.28}$$

from Eqs. (3.4.25) and (3.4.26) – as originally claimed.

3.3.5 Vapor absorption/stripping refrigeration cycle

An absorption refrigerator resorts to two working substances – a refrigerant proper, usually NH_3, and an absorbent therefor, usually water (or $CaCl_2$ to a lesser extent, or even $SrCl_2$); water as refrigerant and concentrated lithium bromide as absorbent have also been applied in large plants.

The thermodynamic principle behind this mode of refrigeration does not differ substantially from a vapor compression/expansion process – except that the compressor of the latter is replaced by an absorber and a stripper as heat engine; its essential components are schematized in Fig. 3.38. Ammonia vapor forms via vaporization of liquid ammonia at low pressure (1), with the required latent heat Q_2 being contributed by the process stream of interest, at (low) temperature T_2; after leaving the evaporator (2), it dissolves in water in the absorber, thus releasing the corresponding enthalpy change of mixing – with Q_3 denoting heat lost here, to temperature $T_0 > T_2$ assured by some utility (air or water) at ambient temperature. The concentrated, cool solution is then pumped to a stripper, where it receives heat Q_1 from a high-temperature source – so dissolved ammonia is released, in agreement with Henry's law that predicts a lower solubility at a higher temperature; it then enters the condenser (3), where it condenses under high pressure, via loss of latent heat Q_4 again to room temperature, T_0. Finally, flash release of pressure will force partial vaporization of liquid ammonia – with the underlying latent heat being received at the expense of sensible heat made available by the prevailing adiabatic conditions, so the liquid/vapor mixture will attain a final lower temperature. The cycle comes to completion by returning the dilute aqueous solution of ammonia, generated in the stripper, to the absorber, through a heat exchanger – meant to reduce the heat input required in the stripper, with a portion of heat, Q, transferred from the hotter dilute solution to the colder concentrated solution *en route* to the stipper. To improve efficiency, a rectifier is sometimes placed between stripper and condenser – consisting of a few distillation plates, meant to remove most water vapor from the ammonia-rich vapor mixture leaving the stripper by taking advantage of differences in volatility; the water condensate is then returned to the absorber. Since the stripper portion of the circuit experiences a higher pressure than the absorber portion thereof, valves are required not only between 4 and 1, but also between 3 and 2. The work incurred in compression, W, delivered by a pump is, however, much less than in regular vapor compression systems – due to the lower energy input needed to pump a liquid, of specific volume v_{liq}, from P_1 toward a higher pressure P_2, viz. $\int_{P_1}^{P_2} v_{liq} dP$, when compared to that required to compress a vapor, of specific volume $v_{gas} >> v_{liq}$, across the same pressure difference, i.e. $\int_{P_1}^{P_2} v_{gas} dP$. Natural convection caused by concentration gradients may even suffice to move the refrigerant along the circuit – a process known as Electrolux™, still available in small refrigerators meant for camping and caravaning, where burning of bottled propane gas serves as heat source. Note that any available waste heat, at temperature $T_1 > T_0$, can play the role of heat source in the stripper – thus justifying why this process is attractive under such circumstances, in a food plant or otherwise.

Inspection of the vapor absorption/stripping cycle indicates that heat is transferred at three temperature levels – as emphasized in Fig. 3.38ii: room temperature, T_0, at which heat is rejected at both condenser and absorber; refrigeration temperature, T_2, at which heat is absorbed at evaporator; and heating temperature, T_1, at which heat is added to stripper. A portion of heat, Q, is also internally transferred between refrigerant and absorber streams – thus reducing the heat rejected at T_0 as Q_3, while adding to Q_1 to increase the heat supplied at T_1. The set absorber/stripper actually operates as a (reversible) heat engine – which receives heat Q_1 at temperature T_1, and rejects heat Q_3 at temperature $T_0 < T_1$. The work thereby produced, W_1, is then utilized to make up for pumping work W necessary to drive the refrigerant around the system; this produces a (net) useful work W_2, required to operate

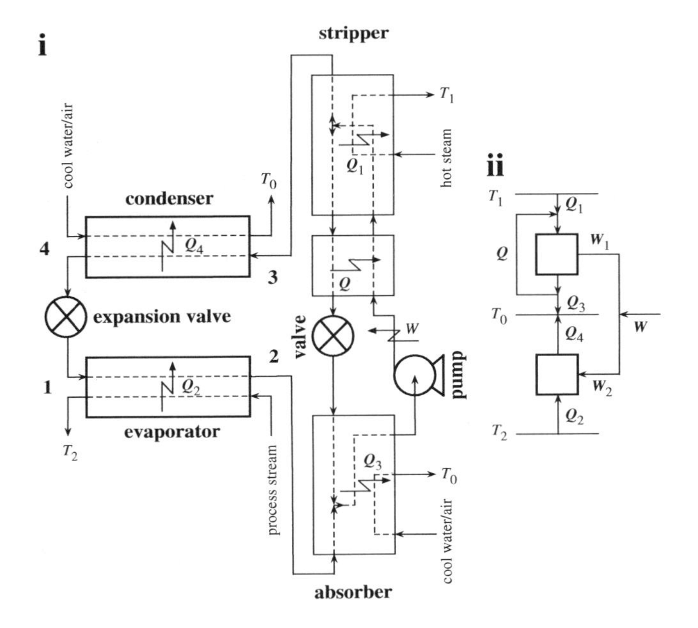

FIGURE 3.38 Schematic representation of (chemical) refrigeration system, based on vapor absorption/stripping cycle (**1→2→3→4→1**) – with indication of (i) apparatuses and streams involved, and corresponding operating temperatures (T_0, $T_1 > T_0$, $T_2 < T_0$), as well as (i, ii) energy exchanged in the form of heat (Q, $Q_1 > 0$, $Q_2 > 0$, $Q_3 < 0$, $Q_4 < 0$) and work ($W > 0$, W_1, W_2).

the refrigerator proper – meant to abstract heat Q_2 at low temperature T_2, and reject it as part of Q_4 at temperature $T_0 > T_2$.

In view of the description above, one may phrase the efficiency of the (auxiliary) heat engine, η_c, as

$$\eta_c \equiv -\frac{W_1}{Q_1} = 1 - \frac{T_0}{T_1}, \tag{3.123}$$

on the hypothesis of reversible operation as per Carnot's cycle – and based on Eqs. (3.40) and (3.94); whereas the coefficient of performance of the refrigerator, β_c, again hypothesized as Carnot's refrigerator, abides to

$$\beta_c \equiv \frac{Q_2}{W_2} = \frac{T_2}{T_0 - T_2} \tag{3.124}$$

on account of Eq. (3.47), coupled to Eq. (3.98) after getting rid of denominators. Ordered multiplication of Eqs. (3.123) and (3.124) then ensues as

$$\frac{-W_1}{Q_1}\frac{Q_2}{W_2} = \left(1 - \frac{T_0}{T_1}\right)\frac{T_2}{T_0 - T_2}; \tag{3.125}$$

since $W << |W_1|$, one may take

$$W_2 \approx -W_1 \tag{3.126}$$

as essentially valid – in which case Eq. (3.125) will be reformulated to

$$\beta = \frac{T_2\left(T_1 - T_0\right)}{T_1\left(T_0 - T_2\right)} \tag{3.127}$$

upon straightforward algebraic rearrangement; complemented by definition of a lumped coefficient of performance, β, as

$$\beta \equiv \frac{Q_2}{Q_1}, \tag{3.128}$$

i.e. target heat to be removed, Q_2, over expensive heat to be made available, Q_1. In practice, temperature of the cooling water/air, T_0, and refrigeration temperature intended, T_2, are fixed *a priori* – so only temperature of the heat source may be subjected to change; when T_1 is sufficiently large, one finds

$$\lim_{T_1 \to \infty} \beta = \frac{T_2 T_1}{T_1\left(T_0 - T_2\right)} = \frac{T_2}{T_0 - T_2} \equiv \beta_c \tag{3.129}$$

based on Eq. (3.127) – upon realization that $T_1 - T_0$ would become approximately equal to T_1, followed by cancellation of T_1 between numerator and denominator. Comparison with Eq. (3.124) justifies the identity in Eq. (3.129), meaning that β will approach Carnot's coefficient of performance for the refrigerator, β_c, should the heat source be very hot.

In current practice, the thermodynamic properties of the ammonia aqueous solution constrain Q_2/Q_1 to ca. 0.9, as per Eq. (3.128); this low value for the lumped coefficient of performance proves, however, unimportant when waste heat is already available, in view of its almost nil cost of opportunity. On the other hand, β as per Eq. (3.127) decreases slowly as $T_0 - T_2$ is increased – due to the buffering effect of the ratio of two absolute values (i.e. T_2/T_1) divided by the ratio of two differences (i.e. $(T_0 - T_2)/(T_1 - T_0)$).

PROBLEM 3.5

The actual portion of the refrigeration cycle that abstracts heat from the process stream and releases heat to cool water/air in the vapor compression/expansion refrigeration cycle coincides with that in the corresponding vapor absorption/stripping cycle.

a) Provide a conceptual comparison of the cycles with each other, using the same refrigerant, say ammonia.
b) In either case, the ideal cycle may be essentially represented in a *TS*-diagram, as done in 3.37ii. Sketch an actual cycle, and justify your options – knowing that: (i) the process is intrinsically irreversible; (ii) fluid friction cannot be avoided during flow; (iii) valve operation is not completely adiabatic; (iv) exact equilibrium conditions cannot be guaranteed; and (v) heat transfer requires a finite temperature difference.

Solution

a) For the two cycles under scrutiny, heat is withdrawn from the process stream by cold refrigerant in the evaporator (see step $1 \rightarrow 2$ in Figs. 3.34i and 3.38i), while heat is rejected to water/air at room temperature in the condenser (see step $3 \rightarrow 4$ also in Figs. 3.34i and 3.38i). In both cases, an expansion valve provides closure between the high-pressure circuit at condenser exit (i.e. $3 \rightarrow 4$) and the low-pressure circuit (i.e. $1 \rightarrow 2$) at evaporator entrance; and an inherent refrigeration effect, derived from promotion of a quick (adiabatic) pressure drop, accompanied by endothermic vaporization. Hence, the major difference between cycles lies on the mode of generation of (hot) refrigerant vapor at high pressure. In the vapor absorption/stripping process, hot refrigerant is obtained via release thereof from the hot liquid absorbent where it was previously dissolved – as a consequence of heat provided at high temperature; complemented by heat released by dissolution of refrigerant in cool liquid absorbent (for being an exothermic phenomenon), meanwhile lost to the surroundings. In the vapor compression/expansion process, hot refrigerant is instead generated via direct supply of work to compress the cool vapor refrigerant, so no need exists for such an extra set of heat source/sink.
b) The actual cycle may be sketched as done below in thick line type – superimposed on the ideal cycle, in thin line type for easier comparison.

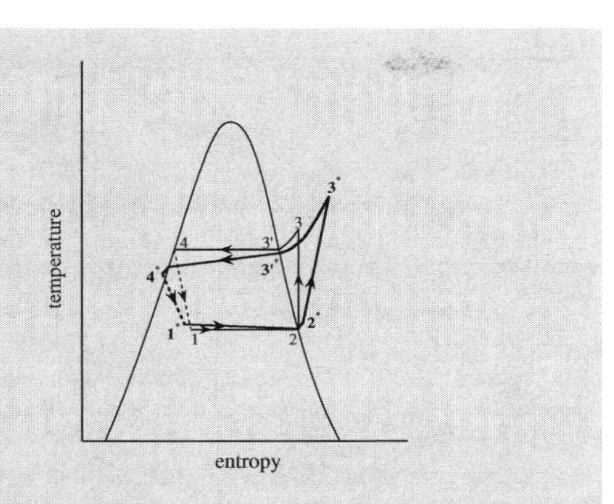

Presence of irreversibility, together with fluid friction imply that line [2*3*] is inclined instead of being vertical as line [23], and that the slightly inclined line [41] is to be replaced by a more inclined line [4*1*], aided by leakage of heat into the (otherwise adiabatic) valve. Fluid friction also causes a pressure drop between entrance to and exit from each apparatus, which produces slight decreases in temperature – so inclined lines [3'*4*] and [1*2*] result in the direction of flow, rather than horizontal lines [3'4] and [12]. Subcooling of condensate (4*) and extra superheating of vapor (3*) will occur because temperature of cooling water in the condenser has to be lower than that of the condensate, and work is not fully supplied in reversible form, respectively; while temperature of the process stream (2*) has to be higher than that prevailing in the evaporator (2) to drive heat transfer. Lack of a true chemical equilibrium also pushes the fluid away from the saturated curve, thus helping justify why points 2* and 4* lie somewhat off the binodal liquid/vapor boundary.

3.3.6 Chilled water production

Water is frequently employed in chilling, when target food temperatures do not fall below ca. 5°C. Direct refrigeration of water – by either conventional compression/expansion or absorption/stripping systems discussed previously, has been a common practice in industrial food processing; however, fluctuating needs are difficult to handle by either system.

On the other hand, advantage may be taken of the particularly high latent heat of vaporization of water toward its own cooling, provided that (some) loss of the original water is allowed in vapor form – as done in vacuum refrigeration, with typical duties up to 7 MW. Its chief disadvantage is the large specific volume of saturated vapor even at low temperatures, which calls for large compressors operated in reverse to produce a (partial) vacuum; reciprocating machines cannot respond to large duties, unlike centrifugal ones – yet steam-jet ejectors have been preferred, despite their low efficiency, due to their lower cost and maintenance requirements. An example of one such system is provided in Fig. 3.39. Feed water is sprayed into a thermally insulated flash chamber (1) at room temperature T_1, where the low pressure brought about by the first ejector (2), in the form of work W_1, forces its partial evaporation; the underlying latent heat is obtained at the expense of sensible heat of the remaining water, Q_2, so chilling will take place to

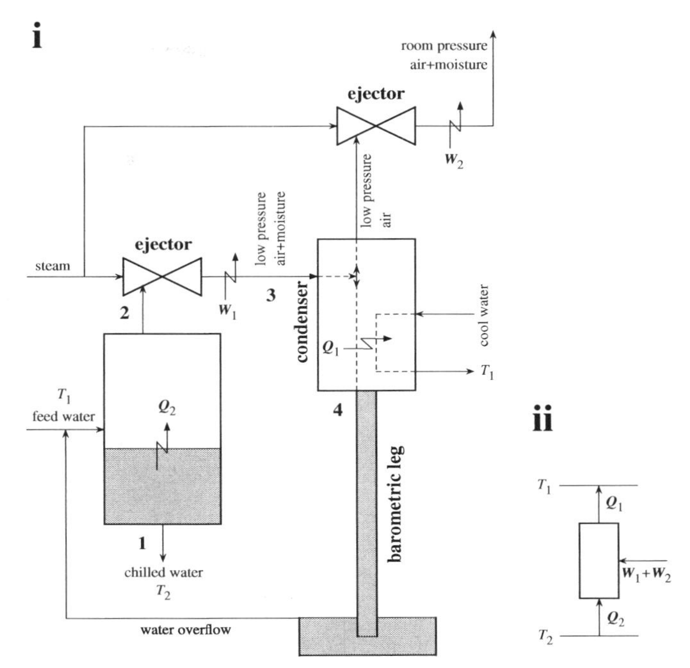

FIGURE 3.39 Schematic representation of (mechanical) refrigeration system for chilling of liquid water, based on vacuum-driven evaporation (**1→2→3→4→1**) – with indication of (i) apparatuses and streams involved, and corresponding operating temperatures (T_1, $T_2<T_1$), as well as (i, ii) energy exchanged in the form of heat ($Q_1<0$, $Q_2>0$) and work ($W_1>0$, $W_2>0$).

temperature $T_2<T_1$. The outlet stream from the ejector (3), containing steam with traces of air, enters a heat exchanger playing the role of condenser – where heat of condensation, Q_1, is removed by cool water at temperature T_1, which also sets the operation pressure; a barometric leg (4) is required to match the underpressure of the leaving liquid stream. The water condensate overflow may be added to the feed water stream, and thus used to partially replace the sprayed water eventually carried away as vapor. A second ejector is necessary, which delivers work W_2 and keeps the underpressure required by continuous operation of the condenser – besides removing air entrained in the vapor initially generated.

The system above is not closed, and so does not operate as a true cycle – yet it merits a thermodynamic analysis similar to that applied to the vapor compression/expansion refrigeration system, see Fig. 3.34ii. The two sources of vacuum-delivered work, W_1 and W_2, are used to drive heat Q_2 from the low-temperature source as $Q_1 = Q_2 + W_1 + W_2$ to the high-temperature sink – as schematized in Fig. 3.39ii. The barometric leg in Fig. 3.39i plays the role of expansion valve in Fig. 3.34i, while assuring closure of high- and low-pressure parts of the cycle; and the portion of feed water converted to chilled water is analogous to the process stream. The ejector(s) replace, in turn, the compressor – although operated backward to create a vacuum; while the cooling water mimics the cool water/air.

PROBLEM 3.6

Consider the simplified vacuum-based system sketched below – designed for production of chilled water at temperature T_1, from regular water at (ambient) temperature T_w, at a (given) mass flow rate Q_{cw}.

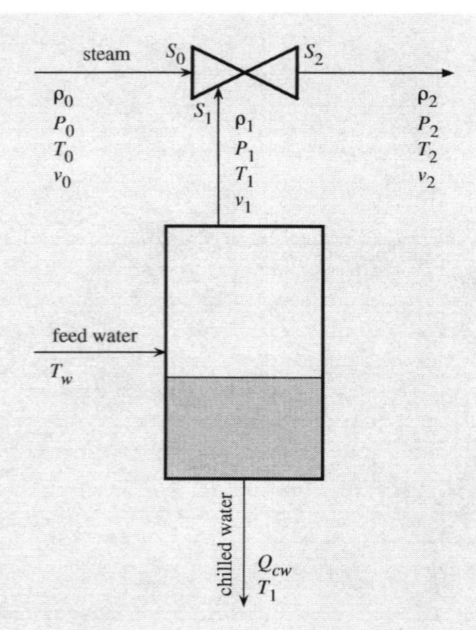

A jet nozzle, with cross-sectional area S_0, is employed to lower pressure in the vessel containing liquid water, thus producing vapor that enters the nozzle with inlet cross-sectional area S_1; the cross-sectional area S_2 for the lumped stream at the exit is but the sum of S_0 and S_1. Fresh superheated steam is supplied to the nozzle at velocity v_0, and is characterized by mass density ρ_0, (given) pressure P_0 and (given) temperature T_0; water evaporated from the vessel enters the nozzle at velocity v_1, being characterized by mass density ρ_1, pressure P_1, and temperature T_1; and lumped gaseous stream leaves the nozzle at velocity v_2, being characterized by mass density ρ_2, (atmospheric) pressure P_2 and temperature T_2. It can be reasonably hypothesized that: water steam approximates ideal gas behavior; water vessel is well isolated; vapor leaves vessel at equilibrium with liquid water left behind; and apparatus is operated under steady state. Devise a mathematical strategy to obtain the T_1 vs. v_0 profile.

Solution

Ideal gas behavior supports use of

$$\rho_i = \frac{P_i M_w}{\mathcal{R}T_i}; \quad i = 0,1,2 \tag{3.6.1}$$

for equation of state, in parallel to Eq. (3.135) of *Food Proc. Eng.: basics & mechanical operations* after replacing n/V by $(m/M_w)/V = (m/V)/M_w = \rho/M_w$ – where M_w denotes molecular weight of water; hence, the continuity equation applying to the jet nozzle, originally labeled as Eq. (3.193) of *Food Proc. Eng.: thermal & chemical operations*, will take the form

$$\frac{P_0 M_w}{\mathcal{R}T_0} S_0 v_0 + \frac{P_1 M_w}{\mathcal{R}T_1} S_1 v_1 = \frac{P_2 M_w}{\mathcal{R}T_2} S_2 v_2 \tag{3.6.2}$$

at the expense of Eq (3.6.1), or else

$$\frac{M_w S_0 P_0}{\mathcal{R}T_0} v_0 + \frac{M_w S_1}{\mathcal{R}} \frac{P_1 v_1}{T_1} = \frac{M_w S_2 P_2}{\mathcal{R}} \frac{v_2}{T_2} \tag{3.6.3}$$

upon algebraic rearrangement to highlight the true variables in the problem. By the same token, conservation of linear momentum within the nozzle requires

$$\frac{P_0 M_w}{\mathcal{R}T_0} S_0 v_0^2 + S_0 P_0 + \frac{P_1 M_w}{\mathcal{R}T_1} S_1 v_1^2 + S_1 P_1$$

$$= \frac{P_2 M_w}{\mathcal{R}T_2} S_2 v_2^2 + S_2 P_2, \tag{3.6.4}$$

obtained after inserting Eq. (3.6.1) in Eq. (3.194) of *Food Proc. Eng.: thermal & chemical operations* – which degenerates to

$$\frac{M_w S_0 P_0}{\mathcal{R}T_0} v_0^2 + S_0 P_0 + \frac{M_w S_1}{\mathcal{R}} \frac{P_1 v_1^2}{T_1} + S_1 P_1$$

$$= \frac{M_w S_2 P_2}{\mathcal{R}} \frac{v_2^2}{T_2} + S_2 P_2, \tag{3.6.5}$$

as a consequence of straightforward manipulation. Equations (3.200) and (3.201) in *Food Proc. Eng.: thermal & chemical operations*, consubstantiating the energy balance again to the nozzle, become, in turn,

$$\frac{P_0 M_w}{\mathcal{R}T_0} S_0 v_0 \left(\frac{1}{2} v_0^2 + c_{P,w,g} T_0 \right) + \frac{P_1 M_w}{\mathcal{R}T_1} S_1 v_1 \left(\frac{1}{2} v_1^2 + c_{P,w,g} T_1 \right)$$

$$= \frac{P_2 M_w}{\mathcal{R}T_2} S_2 v_2 \left(\frac{1}{2} v_2^2 + c_{P,w,g} T_2 \right) \tag{3.6.6}$$

following combination with Eq. (3.6.1) – or, equivalently,

$$\frac{M_w S_0 P_0}{\mathcal{R}} v_0 \left(\frac{1}{2T_0} v_0^2 + c_{P,w,g} \right) + \frac{M_w S_1}{\mathcal{R}} P_1 v_1 \left(\frac{1}{2} \frac{v_1^2}{T_1} + c_{P,w,g} \right)$$

$$= \frac{M_w S_2 P_2}{\mathcal{R}} v_2 \left(\frac{1}{2} \frac{v_2^2}{T_2} + c_{P,w,g} \right) \tag{3.6.7}$$

once factors have been shuffled, with $c_{P,w,g}$ denoting (isobaric) specific heat capacity of water vapor, and 0 K serving as reference temperature for enthalpy calculations associated with sensible heat. Liquid water and vapor, remaining by hypothesis in chemical equilibrium inside the vessel, can be handled as

$$P_1 = A_w \exp\left\{ -\frac{B_w}{C_w + T_1} \right\}, \tag{3.6.8}$$

as per Antoine's equation on vapor pressure – where A_w, B_w, and C_w denote parameters characteristic of water. On the other hand, an enthalpy balance to the vessel may take the form

$$\rho_1 S_1 v_1 \lambda_w + Q_{cw} c_{P,w,l} \left(T_1 - T_w\right) = 0, \tag{3.6.9}$$

on the hypothesis that no heat is exchanged by vessel contents with surroundings – or else

$$\frac{M_w S_1 \lambda_w}{\mathcal{R}} \frac{P_1 v_1}{T_1} + Q_{cw} c_{P,w,l} \left(T_1 - T_w\right) = 0, \tag{3.6.10}$$

upon combination with Eq. (3.6.1); here λ_w denotes enthalpy of vaporization of water, and $c_{P,w,l}$ denotes specific heat capacity of liquid water. The volumetric flow rate of (makeup) feed water is such that the volume of chilled water inside the vessel remains constant – for compatibility with steady state operation. Construction parameters S_0, S_1, and S_2 are accordingly known, as well as physical properties M_w, $c_{P,w,g}$, $c_{P,w,l}$, λ_w, A_w, B_w, and C_w, and operational conditions Q_{cw}, P_0, P_2, T_0, and T_w; hence, variables v_1, v_2, T_1, T_2, and P_1 can be numerically calculated with the aid of Eqs. (3.6.3), (3.6.5), (3.6.7), (3.6.8), and (3.6.10), for every value proposed of v_0. One efficient sequence is to first solve Eq. (3.6.8) for T_1, i.e.

$$T_1 = -C_w - \frac{B_w}{\ln \dfrac{P_1}{A_w}}, \tag{3.6.11}$$

thus generating

$$T_1 \equiv T_1\left\{P_1\right\}; \tag{3.6.12}$$

Eq. (3.6.12) may then be inserted in Eq. (3.6.10) to get

$$v_1 \equiv v_1\left\{P_1, T_1\right\} \equiv v_1\left\{P_1\right\}, \tag{3.6.13}$$

according to

$$v_1 = \frac{\mathcal{R} Q_{cw} c_{P,w,l}}{M_w S_1 \lambda_w} \frac{T_1\left(T_w - T_1\right)}{P_1}. \tag{3.6.14}$$

One will now solve Eq. (3.6.3) with regard to v_2/T_2, to obtain

$$\frac{v_2}{T_2} = \frac{S_0 P_0}{S_2 P_2 T_0} v_0 + \frac{S_1}{S_2 P_2} \frac{P_1 v_1}{T_1}, \tag{3.6.15}$$

which prompts

$$\frac{v_2}{T_2} \equiv \frac{v_2}{T_2}\left\{P_1, T_1, v_0, v_1\right\} \equiv \frac{v_2}{T_2}\left\{P_1, v_0\right\} \tag{3.6.16}$$

in view of Eqs. (3.6.12) and (3.6.13); and then solve Eq. (3.6.5) for v_2, viz.

$$v_2 = \frac{\dfrac{S_0 P_0}{S_2 P_2 T_0} {v_0}^2 + \dfrac{S_1}{S_2 P_2} \dfrac{P_1 {v_1}^2}{T_1} + \dfrac{\mathcal{R}}{M_w}\left(\dfrac{S_0 P_0}{S_2 P_2} + \dfrac{S_1}{S_2 P_2} P_1 - 1\right)}{\dfrac{v_2}{T_2}}, \tag{3.6.17}$$

which unfolds

$$v_2 \equiv v_2\left\{P_1, T_1, v_0, v_1, \frac{v_2}{T_2}\right\} \equiv v_2\left\{P_1, v_0\right\} \tag{3.6.18}$$

by virtue of Eqs. (3.6.12), (3.6.13), and (3.6.16). One may finally retrieve Eq. (3.6.7) as

$$\frac{M_w S_0 P_0}{\mathcal{R}} v_0 \left(\frac{1}{2T_0} {v_0}^2 + c_{P,w,g}\right) + \frac{M_w S_1}{\mathcal{R}} P_1 \left(\frac{1}{2} \frac{{v_1}^3}{T_1} + c_{P,wg} v_1\right)$$
$$= \frac{M_w S_2 P_2}{\mathcal{R}} \left(\frac{1}{2} {v_2}^2 \frac{v_2}{T_2} + c_{P,w,g} v_2\right), \tag{3.6.19}$$

which can, in turn, be rephrased as

$$\Phi\left\{v_0, P_1, T_1, v_1, v_2, \frac{v_2}{T_2}\right\} = 0 \tag{3.6.20}$$

in (implicit) functional form – with the aid of algebraic operator Φ; insertion of Eqs. (3.6.12), (3.6.13), (3.6.16), and (3.6.18) permits simplification of Eq. (3.6.20) to

$$\Phi\left\{v_0, P_1\right\} = 0, \tag{3.6.21}$$

so numerical solution follows to get

$$P_1 \equiv P_1\left\{v_0\right\}. \tag{3.6.22}$$

Once in possession of P_1, via Eq. (3.6.22), for an arbitrary v_0, one may resort to Eq. (3.6.18) to obtain v_2 from P_1 and v_0, to Eq. (3.6.16) to obtain v_2/T_2 from P_1 and v_0, to Eq. (3.6.13) to obtain v_1 from P_1, and to Eq. (3.6.12) to obtain T_1 from P_1 – so $T_1 \equiv T_1\{v_0\}$ will be found; iteration of the process for every other (arbitrary) v_0 will eventually generate the requested plot – useful to estimate the flow rate of fresh steam required by an intended extent of chilling of a given flow rate of water.

3.3.7 Liquid nitrogen production

A gas is, by definition, a gaseous phase at a temperature above the critical temperature (but below the critical pressure) of the compound at stake; its liquefaction therefore requires refrigeration to below the said temperature – unlike happens with vapors, which may turn liquid by mere compression. However, the critical temperature of most gases of practical interest is quite low, e.g. –146.9°C in the case of nitrogen; therefore, a refrigeration system with improved performance is necessary to carry out production of the corresponding liquid(s), for *a posteriori* use as cryogenic fluid(s).

Liquid nitrogen is industrially obtained by fractional distillation of liquid air (often as byproduct of oxygen production, for steelmaking or clinical purposes), or by mechanical separation from gaseous air – namely, via pressurized reverse osmosis or pressure-swing adsorption; although more expensive, the latter processes are more energy-efficient. Bulk nitrogen may also be chemically produced by treating ammonium chloride, in aqueous solution, with sodium nitrite – with release of gaseous N_2, which readily separates from byproduct NaCl that remains in solution; traces of NO and HNO_3 as impurities can be promptly removed, by passing the product gas through aqueous sulfuric acid in the presence of potassium dichromate.

Toward efficient liquefaction, the first step is to enhance the coefficients of performance of actual refrigeration cycles – which is relatively reduced compared to the theoretical (maximum) coefficient of performance of Carnot's ideal cycle, operating between the same upper and lower temperatures. This is typically due to superheating and expansion losses; their relative importance depends on the shape of the saturation line, and the critical temperature of the refrigerant versus its condensation temperature. Most losses are caused by irreversible expansion, in the case of halogenated hydrocarbons; hence, reduction in entropy across the valve is more likely to bring about an improvement in performance. In the case of such refrigerants as ammonia, large superheating losses are instead the major problem; these can be effectively reduced via incorporation of interstage coolers in the compressor – a process sometimes termed desuperheating.

As temperature in the evaporator is decreased, while keeping the condensation temperature constant, the required pressure ratio of the compressor increases – see Fig. 3.35ii as illustration; however, the said ratio cannot exceed 9 for common single-stage machines – so the refrigeration temperature of an ammonia-based cycle is limited to ca. –25°C, for a condensation temperature of 30°C (sufficiently above ambient temperature to permit spontaneous cooling). Lower temperatures – as needed for liquefaction of a gas, accordingly require multistage compression combined with interstage cooling; however, the approach differs from that employed in classical gas compression at room temperature, because temperature of the compressed vapor at condenser stage cannot be reduced any longer by ambient air/water-mediated cooling.

Production of liquid nitrogen, in particular, resorts to multistage compression, with one refrigerant in a given stage employed as coolant for the refrigerant in another stage. Three auxiliary refrigerants are normally used for this purpose, i.e. ammonia, ethylene, and methane – with normal boiling points of 239.7 K, 169.5 K, and 111.7 K, respectively, as illustrated in Fig. 3.40i. Inspection of this figure unfolds ammonia as refrigerant for the first nested stage of compression ($1 \rightarrow 2 \rightarrow 3 \rightarrow 4 \rightarrow 1$), ethylene as refrigerant for the second nested stage of compression ($5 \rightarrow 6 \rightarrow 7 \rightarrow 8 \rightarrow 5$), and methane as refrigerant for the third nested stage of compression ($9 \rightarrow 10 \rightarrow 11 \rightarrow 12 \rightarrow 9$); all these operate as regular refrigeration cycles, based on vapor compression/expansion. The outer cycle of refrigeration takes superheated gas nitrogen (20) following compression, and cools it down to water/air ambient temperature (21), then down to the evaporation temperature of ammonia (22), further down to the evaporation temperature of ethylene (13), an even further down to the evaporation temperature of methane (14) – before expanding it through a valve (15), and resorting to a final flash cooler ($16 \leftarrow 15 \rightarrow 17$) that releases liquid N_2 (16) and cold gaseous N_2 (17); makeup with gaseous nitrogen is now in order (18), so as to balance the liquid nitrogen left behind in the outlet stream – while the resulting stream (19) undergoes compression to superheated gas nitrogen (20), thus closing the cycle.

The nested cycles are laid on the corresponding thermodynamic diagrams in Fig. 3.40ii; advantage is accordingly taken of the distinct saturation curves, where the lower horizontal trajectory on each one coincides with the upper trajectory on the next – while the outer stream of nitrogen is forced to cool down sequentially to each such temperature. The outer stream of gaseous nitrogen supplied (18) is indeed mixed with the inner stream of cold nitrogen vapor (17), and the result as per the lever rule (19) undergoes a regular vapor compression/refrigeration cycle ($19 \rightarrow 20 \rightarrow 21 \rightarrow 22 \rightarrow 13 \rightarrow 14 \rightarrow 15 \rightarrow 17 \rightarrow 19$).

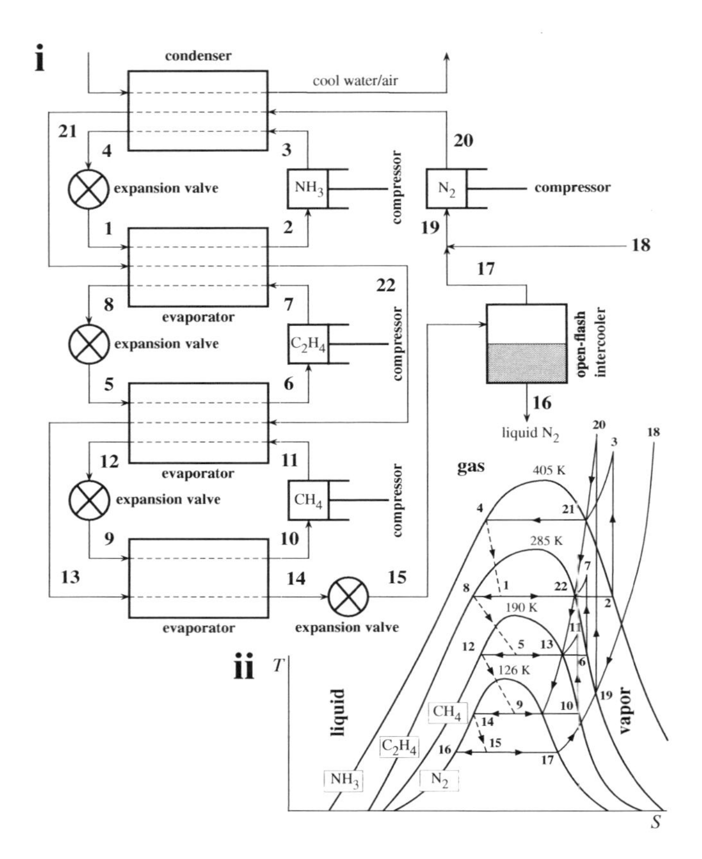

FIGURE 3.40 Schematic representation of (mechanical) refrigeration system for liquefaction of gaseous nitrogen (**18** → **16**), based on vapor compression/expansion cycle – with (i) indication of apparatuses and streams involved, and (ii) after superimposing the refrigeration cycles on the thermodynamic diagrams of ammonia (NH_4), ethylene (C_2H_4), methane (CH_4), and nitrogen (N_2) as temperature, *T*, versus entropy, *S*, with specification of critical temperatures.

The process sketched in Fig. 3.40 is more energy-efficient than several other available for the same purpose – with β_c=0.47; this compares with 0.310–0.346 via Claude's process, 0.312 via Heyland's process, 0.260 via Philips' machine, 0.250 via Bell and Coleman's cycle, 0.218 via two-stage throttling with precooling, 0.148 via plain two-stage throttling, and 0.073 via single-stage throttling. However, methane (the only readily available refrigerant to bridge the interval 100–170 K) may constitute a serious explosion hazard when in the presence of oxygen. Cascade refrigeration has also been employed to produce liquefied natural gas, consisting mainly of *n*-alkanes – among which methane exhibits the lowest boiling point.

PROBLEM 3.7

In theory, the lower temperature limit of a multistage cycle is fixed only by the triple point of the refrigerant at stake; however, difficulties arise in practice if pressure on the suction side of the compressor lies substantially below atmospheric pressure. The said limit may be lowered by changing to a different refrigerant with a lower triple point – but then pressure in the condenser will be high, should it be cooled by air/water at room temperature. Such a high condensation pressure can, in turn, be circumvented by resorting to a binary cascade cycle (or split-stage compression); two regular cycles, each using its own refrigerant, are thus to be connected via a heat exchanger – as shown below.

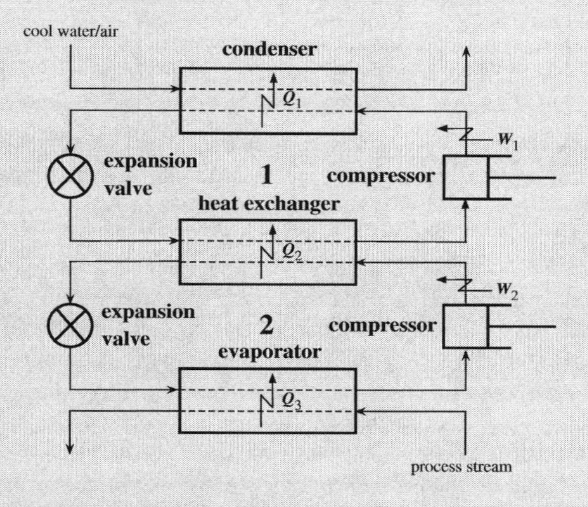

The refrigerant of higher vapor pressure, in cycle 2, is condensed by evaporating the other refrigerant – which is, in turn, condensed by cool air/water in cycle 1. Show that the overall coefficient of performance, β_{1+2}, describing this system can be related via

$$\left(1+\frac{1}{\beta_1}\right)\left(1+\frac{1}{\beta_2}\right)=1+\frac{1}{\beta_{1+2}} \tag{3.7.1}$$

to the coefficients of performance of the two component cycles, β_1 and β_2, if considered independently.

Solution

The coefficient of performance of cycle 1 can be coined as

$$\beta_1=\frac{|Q_2|}{W_1}, \tag{3.7.2}$$

using Eq. (3.47) as template – since the heat exchanger serves as low temperature source; modulus was used to circumvent the sign convention (since $Q_2 > 0$ in cycle 1, but $Q_2 < 0$ in cycle 2). The coefficient of performance of cycle 2 may likewise be phrased as

$$\beta_2=\frac{Q_3}{W_2}, \tag{3.7.3}$$

again stemming from Eq. (3.47). Once reciprocals are taken of, and unity is added to both sides, Eqs. (3.7.2) and (3.7.3) become

$$1+\frac{1}{\beta_1}=1+\frac{W_1}{|Q_2|} \tag{3.7.4}$$

and

$$1+\frac{1}{\beta_2}=1+\frac{W_2}{Q_3}, \tag{3.7.5}$$

respectively; ordered multiplication of Eqs. (3.7.4) and (3.7.5), i.e.

$$\left(1+\frac{1}{\beta_1}\right)\left(1+\frac{1}{\beta_2}\right)=\left(1+\frac{W_1}{|Q_2|}\right)\left(1+\frac{W_2}{Q_3}\right), \tag{3.7.6}$$

gives eventually rise to

$$\left(1+\frac{1}{\beta_1}\right)\left(1+\frac{1}{\beta_2}\right)=1+\frac{W_2}{Q_3}+\frac{W_1}{|Q_2|}+\frac{W_1W_2}{|Q_2|Q_3} \tag{3.7.7}$$

upon elimination of parentheses in the right-hand side. Addition and subtraction of W_1/Q_3 to the right-hand transforms Eq. (3.7.7) to

$$\left(1+\frac{1}{\beta_1}\right)\left(1+\frac{1}{\beta_2}\right)=1+\frac{W_2}{Q_3}+\frac{W_1}{Q_3}+\frac{W_1}{|Q_2|}-\frac{W_1}{Q_3}+\frac{W_1W_2}{|Q_2|Q_3}, \tag{3.7.8}$$

where W_1 may be factored out in the last three terms as

$$\left(1+\frac{1}{\beta_1}\right)\left(1+\frac{1}{\beta_2}\right)=1+\frac{W_2}{Q_3}+\frac{W_1}{Q_3}+W_1\left(\frac{1}{|Q_2|}-\frac{1}{Q_3}+\frac{W_2}{|Q_2|Q_3}\right); \tag{3.7.9}$$

elimination of the last parenthesis produces, in turn,

$$\left(1+\frac{1}{\beta_1}\right)\left(1+\frac{1}{\beta_2}\right)=1+\frac{W_1}{Q_3}+\frac{W_2}{Q_3}+W_1\frac{Q_3-|Q_2|+W_2}{|Q_2|Q_3}. \tag{3.7.10}$$

On the other hand, cycle 2 requires

$$Q_3-|Q_2|+W_2=0, \tag{3.7.11}$$

because its final state coinciding with its departing state enforces the same value for enthalpy as state function – and the first law of thermodynamics applies to this system operated continuously; Eq. (3.7.11) permits simplification of Eq. (3.7.10) to

$$\left(1+\frac{1}{\beta_1}\right)\left(1+\frac{1}{\beta_2}\right)=1+\frac{W_1}{Q_3}+\frac{W_2}{Q_3}. \tag{3.7.12}$$

After pooling together the last terms, Eq. (3.7.12) becomes

$$\left(1+\frac{1}{\beta_1}\right)\left(1+\frac{1}{\beta_2}\right)=1+\frac{W_1+W_2}{Q_3}, \tag{3.7.13}$$

which is algebraically equivalent to

$$\left(1+\frac{1}{\beta_1}\right)\left(1+\frac{1}{\beta_2}\right)=1+\frac{1}{\dfrac{Q_3}{W_1+W_2}}; \tag{3.7.14}$$

since the overall coefficient of performance satisfies

$$\beta_{1+2} = \frac{Q_3}{W_1 + W_2} \tag{3.7.15}$$

as the overall work supplied is split between cycles and Q_3 represents the refrigeration load at the lowest-temperature source, one promptly retrieves Eq. (3.7.1) – following insertion of Eq. (3.7.15) in Eq. (3.7.14).

3.3.8 Solid carbon dioxide production

Carbon dioxide can be obtained by distillation of air, yet this method is quite inefficient. Industrial supply relies predominantly on waste products, namely combustion of methane, petroleum distillates (e.g. gasoline, diesel, propane), coal, or wood. The aforementioned gas is also a byproduct of industrial production of hydrogen by steam reforming, or of the (so-called) gas shift reaction of CO with H_2O – which precedes Harber and Bosch's process for ammonia synthesis; as well as a byproduct of fermentation of sugar in brewing and manufacture of other alcoholic beverages, and in production of bulk bioethanol. Another large-scale source is thermal decomposition of limestone ($CaCO_3$), via calcination at ca. 850°C in the manufacture of quicklime (CaO); or reduction of iron from its oxides with coke, in blast furnaces.

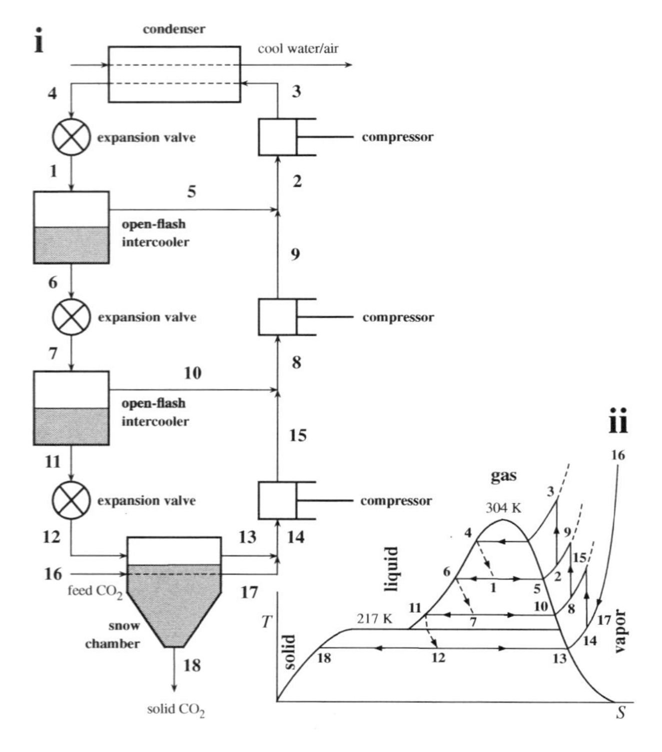

FIGURE 3.41 Schematic representation of (mechanical) refrigeration system for solidification of gaseous carbon dioxide (**16→18**), based on vapor compression/expansion cycle – with (i) indication of apparatuses and streams involved, and (ii) after superimposing the refrigeration cycles on the thermodynamic diagram of carbon dioxide, as temperature, T, versus entropy, S, with specification of critical and triple point temperatures.

Production of solid carbon dioxide requires temperature be lowered to below its triple point (217 K) – so that sublimation may take place; some form of multistage expansion/compression is therefore required, which normally resorts to a three-stage cascade as sketched in Fig. 3.41. This arrangement does not operate in a true (closed) cycle – in view of supply of gaseous CO_2 as inlet stream (16), and recovery of solid CO_2 as outlet stream (18); the stream of cool water/air is hereby meant to receive heat as high-temperature sink. Note the presence of open-flash devices containing CO_2 as refrigerant, under liquid/vapor equilibria; and a snow chamber, containing solid/vapor at equilibrium – where solid CO_2 is released as product. A first cycle of vapor compression/expansion ($4 \rightarrow 1 \rightarrow 5 \rightarrow 2 \rightarrow 3 \rightarrow 4$) lowers the temperature of (liquid) CO_2; which then undergoes a second cycle ($6 \rightarrow 7 \rightarrow 10 \rightarrow 8 \rightarrow 9$) that lowers the temperature of (liquid) CO_2 further; and finally goes through a third cycle ($11 \rightarrow 12 \rightarrow 13 \rightarrow 14 \rightarrow 15$) that lowers the temperature of (liquid, and then solid) CO_2 even further.

To permit a better insight to the fundamentals of the process under scrutiny, the three consecutive cycles have been overlaid on the thermodynamic diagram of carbon dioxide in Fig. 3.41ii – showing both liquid/vapor saturation line (between 217 and 304 K) and solid/vapor saturation line (below 217 K). Liquid CO_2 (4) from the condenser expands through a valve into an open-flash intercooler (1), with pressure identical to the inlet pressure in the first stage compressor; while the corresponding saturated vapor formed (5) gets mixed with vapor CO_2 leaving the second stage compressor (9) – with the resulting mixture (2) entering the first stage compressor. The compressed vapor CO_2 (3) is then admitted to the condenser, thus generating liquid CO_2 (4). The saturated liquid CO_2 remaining (6) flows through another expansion valve into a second open-flash intercooler (7) – with pressure equal to that prevailing at the inlet to the second stage compressor; the remaining saturated liquid CO_2 (11) flows, in turn, through another expansion valve (12) into the snow chamber, while the corresponding saturated vapor formed (10) gets mixed with vapor CO_2 leaving the third stage compressor (15) – with the resulting mixture (8) being fed to the second stage compressor. In the snow chamber, the residual liquid CO_2 flashes into a solid/vapor mixture – should the prevailing pressure (usually atmospheric) be lower than the pressure of the corresponding triple point; solid CO_2 (18) is then collected as product, while the overhead vapor (13) is passed to the third stage compressor, together with the fresh feed of gaseous CO_2 (16) precooled to below the critical temperature by passage through the snow chamber (17), to eventually produce a mixture (14).

The mass flow rates through the distinct parts of the cycle are not equal; in fact, the *TS*-diagram in Fig. 3.41ii shows only the thermodynamic state of the fluid at each point – so due allowance is required when interpreting areas of plane, enclosed by each closed cycle, as heat transferred (as done previously in Fig. 3.22ii). The change in slope of the irreversible expansion in the third valve ($11 \rightarrow 12$) indicates that the triple line has been crossed – which is characterized by coexistence of solid, liquid, and vapor phases; while each right- and upward translation of vertical paths from the vapor saturation line to the superheated region is the consequence of mixing processes (i.e. 14 from 13 + 17, 8 from 10 + 15, and 2 from 5 + 9).

The optimum interstage pressure depends on the adiabatic efficiencies of the compressors – and can be selected by iterative calculation, aimed at maximizing the overall coefficient of performance; if a simple, departing estimate is required, one may safely assume equal compression ratios for each stage. An improved coefficient of performance is observed for this arrangement, when compared to a single-stage cycle, operating between the same temperatures; this is due to reduction of work applied at the expense of interstage cooling, coupled with smaller amount of vapor compressed in the first stage.

PROBLEM 3.8

Refrigeration may be brought about by (the phase change-free) reverse Joule's cycle, also known as reverse Brayton's cycle – as per the diagram below.

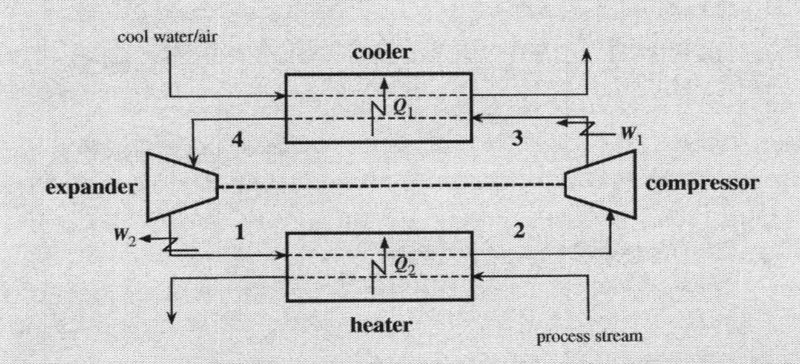

Following isentropic compression (2→3) accompanied by temperature increase, the gaseous refrigerant is cooled via isobaric heat transfer to the surroundings (3→4); isentropic expansion of the gas then takes place to the suction pressure of the compressor (4→1), concomitant with a drop in temperature – after which the gaseous refrigerant absorbs heat isobarically (1→2), until closing the cycle. The output work from the expander is actually used to help drive the compressor.

a) Show that, for an ideal gas – with (isobaric) heat capacity independent of temperature, the coefficient of performance, β_{rj}, is given by

$$\frac{1}{\beta_{rj}} = \left(\frac{P_3}{P_2}\right)^{\frac{\gamma-1}{\gamma}} - 1 \tag{3.8.1}$$

– where P_3/P_2 represents the pressure ratio prevailing at the compressor.

b) In classical reverse Joule's cycle, the difference between temperatures at which heat is absorbed and rejected is determined by the pressure ratio of the compressor. However, the said difference may be varied independently thereof, upon incorporation of an internal heat exchanger – with the resulting cycle being termed reverse regenerative Joule's cycle, or Bell and Coleman's cycle, as sketched below.

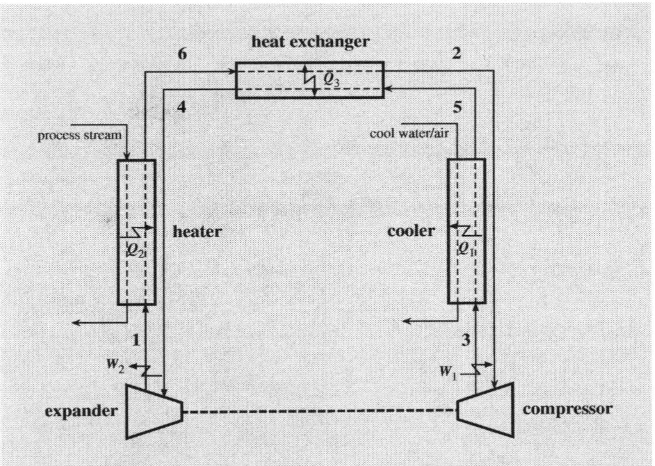

Assuming reversible heat transfer in the exchanger, i.e. driven by an infinitesimally small temperature difference between gas streams throughout the whole apparatus, prove that

$$\frac{1}{\beta_{rrj}} = \left(\frac{P_3}{P_2}\right)^{\frac{\gamma-1}{\gamma}} \frac{T_2}{T_4} - 1 \tag{3.8.2}$$

represents the (reciprocal of the) associated coefficient of performance, β_{rrj} – with T_2 denoting suction temperature to the compressor, and T_4 denoting inlet temperature to the expander; and investigate under what conditions an improvement will be possible in β_{rrj} relative to β_{rj}.

Solution

a) As usual, the coefficient of performance is calculated as the ratio of heat extracted at low temperature to (net) work required to do so, viz.

$$\beta_{rj} \equiv \frac{Q_2}{W_1 + W_2} \tag{3.8.3}$$

that mimics Eq. (3.47); in fact, no heat is exchanged with the surroundings during isentropic compression and expansion, while no work is performed or received during isobaric heating and cooling. Since the refrigerant retrieves its original state once a cycle is completed, the overall change in enthalpy is nil for its being a state function – so one may write

$$W_1 + W_2 + Q_1 + Q_2 = 0; \tag{3.8.4}$$

isolation of $W_1 + W_2$ then produces

$$W_1 + W_2 = -(Q_1 + Q_2). \tag{3.8.5}$$

Insertion of Eq. (3.8.5) converts Eq. (3.8.3) to

$$\beta_{rj} = -\frac{Q_2}{Q_1 + Q_2}, \tag{3.8.6}$$

where $Q_1 < 0$, $Q_2 > 0$, and $|Q_1| > Q_2$ (in agreement with the second law of thermodynamics) assure a positive value for

β_{rj}. Recalling the definition of isobaric heat capacity, C_P, as per Eq. (3.5) of *Food Proc. Eng.: thermal & chemical operations* in the absence of any phase change, it is possible to rewrite Eq. (3.8.6) as

$$\beta_{rj} = -\frac{C_P(T_2 - T_1)}{C_P(T_4 - T_3) + C_P(T_2 - T_1)}, \tag{3.8.7}$$

where C_P will be constant if ideal gas behavior holds; hence, one may drop C_P from numerator and denominator, and remove parentheses to get

$$\beta_{rj} = -\frac{T_2 - T_1}{T_4 - T_3 + T_2 - T_1}. \tag{3.8.8}$$

Equation (3.8.8) becomes

$$\frac{1}{\beta_{rj}} = \frac{(T_3 - T_4) - (T_2 - T_1)}{T_2 - T_1} \tag{3.8.9}$$

once reciprocals are taken of both sides, and the negative taken of its numerator afterward; and ultimately simplifies to

$$\frac{1}{\beta_{rj}} = \frac{T_3 - T_4}{T_2 - T_1} - 1, \tag{3.8.10}$$

after splitting the numerator. Since $2 \rightarrow 3$ and $4 \rightarrow 1$ are adiabatic steps, the germane pressures are related via

$$\frac{T_3}{T_2} = \left(\frac{P_3}{P_2}\right)^{\frac{\gamma-1}{\gamma}} \tag{3.8.11}$$

and

$$\frac{T_1}{T_4} = \left(\frac{P_1}{P_4}\right)^{\frac{\gamma-1}{\gamma}}, \tag{3.8.12}$$

respectively – as per Eq. (3.136) in *Food Proc. Eng.: thermal & chemical operations*, with γ denoting ratio of C_P to isochoric heat capacity, C_V; insertion of Eqs. (3.8.11) and (3.8.12) transforms Eq. (3.8.10) to

$$\frac{1}{\beta_{rj}} = \frac{T_2\left(\frac{P_3}{P_2}\right)^{\frac{\gamma-1}{\gamma}} - T_4}{T_2 - T_4\left(\frac{P_1}{P_4}\right)^{\frac{\gamma-1}{\gamma}}} - 1. \tag{3.8.13}$$

On the other hand, steps $1 \rightarrow 2$ and $3 \rightarrow 4$ are isobaric – thus satisfying

$$P_1 = P_2 \tag{3.8.14}$$

and

$$P_3 = P_4, \tag{3.8.15}$$

respectively; Eqs. (3.8.14) and (3.8.15) may then be taken advantage to transform Eq. (3.8.13) to

$$\frac{1}{\beta_{rj}} = \frac{T_2\left(\frac{P_3}{P_2}\right)^{\frac{\gamma-1}{\gamma}} - T_4}{T_2 - T_4\left(\frac{P_2}{P_3}\right)^{\frac{\gamma-1}{\gamma}}} - 1, \tag{3.8.16}$$

where multiplication of both numerator and denominator of the first term in the right-hand side by $\left(P_3/P_2\right)^{(\gamma-1)/\gamma}$ gives rise to

$$\frac{1}{\beta_{rj}} = \left(\frac{P_3}{P_2}\right)^{\frac{\gamma-1}{\gamma}} \frac{T_2\left(\frac{P_3}{P_2}\right)^{\frac{\gamma-1}{\gamma}} - T_4}{T_2\left(\frac{P_3}{P_2}\right)^{\frac{\gamma-1}{\gamma}} - T_4} - 1 \tag{3.8.17}$$

– which retrieves Eq. (3.8.1), because coincidence of numerator and denominator transform the outstanding fraction to unity.

b) Equation (3.8.3) may still be used to calculate the coefficient of performance of regenerative Joule's cycle, since only interactions between system and surroundings, in terms of energy exchange, are of relevance thereto; heat transfer within the heat exchanger is indeed a process internal to the system. For the same reason, the overall enthalpic balance labeled as Eq. (3.8.4), or Eq. (3.8.5) for that matter remains valid; consequently, one can revisit Eq. (3.8.6) with

$$Q_1 = C_P(T_5 - T_3) \tag{3.8.18}$$

and

$$Q_2 = C_P(T_6 - T_1) \tag{3.8.19}$$

pertaining to isobaric steps, to obtain

$$\beta_{rrj} = -\frac{C_P(T_6 - T_1)}{C_P(T_5 - T_3) + C_P(T_6 - T_1)}; \tag{3.8.20}$$

division of both numerator and denominator by C_P permits prompt simplification to

$$\beta_{rrj} = -\frac{T_6 - T_1}{T_5 - T_3 + T_6 - T_1}. \tag{3.8.21}$$

After taking reciprocals of both sides, and then the negative of the right-hand side, Eq. (3.8.21) becomes

$$\frac{1}{\beta_{rrj}} = \frac{(T_3 - T_5) - (T_6 - T_1)}{T_6 - T_1}, \tag{3.8.22}$$

which breaks down to

$$\frac{1}{\beta_{rrj}} = \frac{T_3 - T_5}{T_6 - T_1} - 1 \tag{3.8.23}$$

once the fraction is splitted. The adiabatic nature of steps $2 \rightarrow 3$ and $4 \rightarrow 1$ guarantee validity of Eqs. (3.8.11) and (3.8.12), respectively; therefore, Eq. (3.8.23) can be redone to

$$\frac{1}{\beta_{rrj}} = \frac{T_2\left(\frac{P_3}{P_2}\right)^{\frac{\gamma-1}{\gamma}} - T_5}{T_6 - T_4\left(\frac{P_1}{P_4}\right)^{\frac{\gamma-1}{\gamma}}} - 1. \quad (3.8.24)$$

Furthermore, Eqs. (3.8.14) and (3.8.15) apply, as well as their ordered quotient, $P_1/P_4 = P_2/P_3$, because steps $1 \rightarrow 6 \rightarrow 2$ and $3 \rightarrow 5 \rightarrow 4$, respectively, are isobaric transformations; this permits conversion of Eq. (3.8.24) to

$$\frac{1}{\beta_{rrj}} = \frac{T_2\left(\frac{P_3}{P_2}\right)^{\frac{\gamma-1}{\gamma}} - T_5}{T_6 - T_4\left(\frac{P_2}{P_3}\right)^{\frac{\gamma-1}{\gamma}}} - 1. \quad (3.8.25)$$

After multiplying and dividing the right hand side by $(P_3/P_2)^{(\gamma-1)/\gamma}$, Eq. (3.8.25) becomes

$$\frac{1}{\beta_{rrj}} = \left(\frac{P_3}{P_2}\right)^{\frac{\gamma-1}{\gamma}} \frac{T_2\left(\frac{P_3}{P_2}\right)^{\frac{\gamma-1}{\gamma}} - T_5}{T_6\left(\frac{P_3}{P_2}\right)^{\frac{\gamma-1}{\gamma}} - T_4} - 1; \quad (3.8.26)$$

T_2/T_4 can, in turn, be factored out to produce

$$\frac{1}{\beta_{rrj}} = \left(\frac{P_3}{P_2}\right)^{\frac{\gamma-1}{\gamma}} \frac{T_2}{T_4} \frac{\left(\frac{P_3}{P_2}\right)^{\frac{\gamma-1}{\gamma}} - \frac{T_5}{T_2}}{\frac{T_6}{T_4}\left(\frac{P_3}{P_2}\right)^{\frac{\gamma-1}{\gamma}} - 1} - 1. \quad (3.8.27)$$

The infinitesimally small temperature difference between gas streams that prevails (by hypothesis) along the heat exchanger implies, in particular,

$$\frac{T_5}{T_2} \approx \frac{T_4}{T_6} \approx 1 \quad (3.8.28)$$

– which may be taken advantage of to rewrite Eq. (3.8.27) as

$$\frac{1}{\beta_{rrj}} \approx \left(\frac{P_3}{P_2}\right)^{\frac{\gamma-1}{\gamma}} \frac{T_2}{T_4} \frac{\left(\frac{P_3}{P_2}\right)^{\frac{\gamma-1}{\gamma}} - 1}{1\left(\frac{P_3}{P_2}\right)^{\frac{\gamma-1}{\gamma}} - 1} - 1$$
$$= \left(\frac{P_3}{P_2}\right)^{\frac{\gamma-1}{\gamma}} \frac{T_2}{T_4} \frac{\left(\frac{P_3}{P_2}\right)^{\frac{\gamma-1}{\gamma}} - 1}{\left(\frac{P_3}{P_2}\right)^{\frac{\gamma-1}{\gamma}} - 1} - 1; \quad (3.8.29)$$

dropping of $(P_3/P_2)^{(\gamma-1)/\gamma} - 1$ from numerator and denominator finally retrieves Eq. (3.8.2). Note that a marginal driving force for heat transfer – compatible with $T_5 \approx T_2$ and $T_4 \approx T_6$ as per Eq. (3.8.28), and which remains essentially constant throughout the heat exchanger for being operated countercurrentwise, does not necessarily imply a negligible rate of heat transfer therein – which causes temperature to change from T_6 to T_2 in one stream, and from T_5 to T_4 in the other; hence, nothing can be concluded *a priori* on the magnitude of T_2/T_4. The proposed regenerative configuration will improve the coefficient of performance (i.e. $\beta_{rrj} > \beta_{rj}$, and thus $1/\beta_{rrj} < 1/\beta_{rj}$) as long as $T_2/T_4 < 1$, and thus $T_2 < T_4$; this implies upward direction for heat flow Q_3 in the diagram above, and consequently $T_6 < T_2$ and $T_5 > T_4$ due to the underlying enthalpy balance.

3.3.9 Gas compression

Gas compression normally resorts to a reciprocating compressor – regarded as steady state flow machine, to which work is supplied at an (essentially) uniform rate; the underlying cycle, with expansion and compression steps alternating with charge and discharge, is schematically represented in Fig. 3.42i. For simplicity, the fluid undergoing compression is hypothesized as an ideal gas, at constant state throughout suction (or charge) and release (or discharge) steps, and bearing negligible macroscopic kinetic energy; reversibility is also postulated for operation of the compressor. Under these circumstances, the underlying cycle can be sketched in a *PV*-diagram as done in Fig. 3.42ii, and in an *n*/*V*-diagram as done in Fig. 3.42iii.

In the first step, $1 \rightarrow 2$, n_2 moles of gas is compressed – with both suction and discharge valves closed, see Fig. 3.9i; this is considered to take place adiabatically, so

$$\boxed{P = \frac{K}{V^{\gamma}}} \quad (3.130)$$

applies in parallel to Eq. (3.131) of *Food Proc. Eng.: thermal & chemical operations*, with K and $\gamma > 1$ denoting constants. A steep curve results in Fig. 3.42ii, since $\gamma > 1$ – between an isotherm containing point 1 and another isotherm containing point 2, without heat lost to the surroundings; or instead a horizontal line corresponding to a constant number of moles, n_2, undergoing volume reduction, see Fig. 3.41iii. In step $2 \rightarrow 3$, n_2–n_4 moles of gas are displaced at constant pressure through the discharge valve – as indicated by the horizontal line in Fig. 3.42ii, and by the inclined straight line on the left in Fig. 3.42iii. The n_4 moles of gas trapped in the clearance volume, between end of cylinder and piston, undergo expansion in step $3 \rightarrow 4$ – with both valves closed, and abiding again to Eq. (3.130) of *Food Proc. Eng.: thermal & chemical operations* for being adiabatic, i.e. represented by a curve evolving between the isotherm containing point 3 and the isotherm containing point 4 in Fig. 3.42ii; a horizontal straight line results again in Fig. 3.42iii, pointing rightward as it accompanies a volume increase. Finally, n_2–n_4 moles of gas are drawn through the suction valve into the cylinder in step $4 \rightarrow 1$, at constant pressure and thus along the lower horizontal path

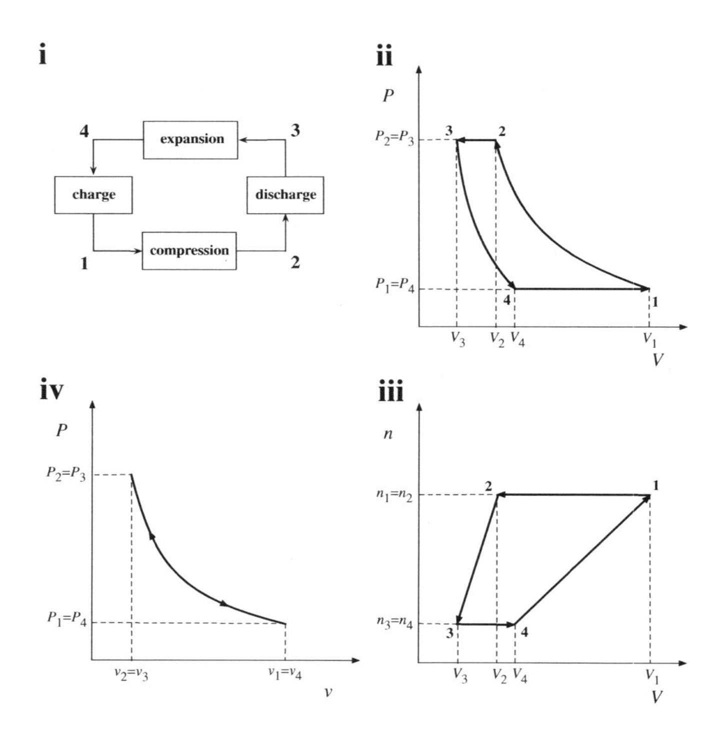

FIGURE 3.42 Schematic representation of reciprocating compressor, operated along cycle **1→2→3→4→1**, corresponding to (i) the sequence compression, discharge, expansion, and charge of an ideal gas, encompassing two isobaric, open system steps between states **2** and **3**, and **4** and **1**, alternated with two adiabatic, closed system steps between states **1** and **2**, and **3** and **4**, laid out as (ii) pressure, P, versus volume, V, (iii) number of moles, n, versus V, and (iv) P versus molar volume, v – with specification of germane pressures (P_1, P_2, P_3, P_4), volumes (V_1, V_2, V_3, V_4), number of moles (n_1, n_2, n_3, n_4), and molar volumes (v_1, v_2, v_3, v_4).

in Fig. 3.42ii – and they mix with the n_4 moles already present therein; this corresponds to the inclined straight line on the right in Fig. 3.42iii, with n increasing linearly with V. Note that the piston cannot displace all gas in a real reciprocating compressor – since some remains trapped in the ports leading to inlet and outlet valves, and some space is deliberately left, at construction stage, between piston and bottom of cylinder to avoid accidental collision during operation.

After revisiting Eq. (3.137) of *Food Proc. Eng.: basics & mechanical operations* as

$$V = \frac{\mathcal{R}T}{P}n, \tag{3.131}$$

one easily justifies the linear shape of lines describing $4\rightarrow1$ and $2\rightarrow3$ in Fig. 3.42iii – as long as temperature (besides pressure) remains constant during those strokes, toward expansion and compression, respectively. Increase from n_4 to n_2 in step $4\rightarrow1$, as well as a decrease from n_2 to n_4 in step $2\rightarrow3$ are directly proportional to piston displacement, but described by distinct slopes – since

$$\frac{1}{\mathcal{R}}\left(\frac{P}{T}\right)_{4\rightarrow1} < \frac{1}{\mathcal{R}}\left(\frac{P}{T}\right)_{2\rightarrow3} \tag{3.132}$$

in Eq. (3.131); the $4\rightarrow1$ trajectory is indeed shallower than its $2\rightarrow3$ counterpart. It should be stressed that the diagram in Fig. 3.42ii is not a true thermodynamic PV-diagram of gas along a compression/expansion cycle – because its number of moles does not remain constant. If molar volume were utilized instead, then the two curves in Fig. 3.42ii would collapse, as observed in Fig. 3.42iv – since points 2 and 3 coincide in pressure and temperature, according to

$$\boxed{v \equiv \frac{V}{n} = \frac{\mathcal{R}T}{P}}, \tag{3.133}$$

obtained from Eq. (3.131) after dividing both sides by n; the same reasoning justifies coincidence of points 1 and 4 in Fig. 3.42iv, yet lying below points 2 and 3 due to Eq. (3.132). The true thermodynamic process depicted in Fig. 3.42iv therefore appears as a single line – taken in the upward direction throughout compression involving n_4 moles, and in the downward direction during expansion involving n_2 moles; the net effect is thus compression and expansion of $n_2 - n_4$ moles between P_1 and $P_2 > P_1$, and vice versa. Equation (3.133) may be rephrased to $P/\mathcal{R}T = 1/v$, where insertion of Eq. (3.130) unfolds $P/\mathcal{R}T = \sqrt[\gamma]{P/K}$; therefore, $P/\mathcal{R}T$ increases when P increases, which prompty justifes the inequality labeled as Eq. (3.132) – since step $4\rightarrow1$ takes place at low pressure after an expansion, whereas step $2\rightarrow 3$ occurs at high pressure after a compression, see Fig. 3.42i.

The work involved in compression from 1 to 2 pertains to a closed, adiabatic system, characterized by a constant number of moles n_2; hence, one can resort to the first law of thermodynamics to write

$$W_{12} \equiv \int_1^2 đW = -\int_{V_1}^{V_2} P dV. \tag{3.134}$$

By the same token, the number of moles undergoing expansion between 3 and 4 entail a closed, adiabatic system as well, now containing a constant number of moles n_4; the accompanying work then looks like

$$W_{34} \equiv \int_3^4 đW = -\int_{V_3}^{V_4} P dV. \tag{3.135}$$

In the other two steps, the system is not closed, so one cannot resort directly to an equation similar to Eqs. (3.134) or (3.135); however, the work done on the gas is equal to the work performed by the piston – which exerts a force F, uniformly distributed as pressure P over its area A, and displaces its application point by a distance L to produce

$$W_{23} \equiv -F_2 L_{2\rightarrow 3} = -(P_2 A) L_{2\rightarrow 3} = -P_2 (A L_{2\rightarrow 3}) \tag{3.136}$$

in the case of step $2\rightarrow 3$, taken at constant pressure P_2; Eq. (3.136) breaks down to

$$W_{23} = -P_2 (V_3 - V_2), \tag{3.137}$$

after realizing that $A\Delta L$ coincides with ΔV. One may similarly write

$$W_{41} \equiv -F_1 L_{4\rightarrow 1} = -(P_4 A) L_{4\rightarrow 1} = -P_4 (A L_{4\rightarrow 1}) \tag{3.138}$$

in the case of step $4\rightarrow 1$, which occurs at constant pressure P_4; Eq. (3.128) becomes

$$W_{41} = -P_4 (V_1 - V_4), \tag{3.139}$$

upon straightforward algebraic rearrangement. Therefore, the overall work performed during a cycle, W_p – defined as

$$W_p \equiv \oint đW = W_{12} + W_{23} + W_{34} + W_{41}, \tag{3.140}$$

may be calculated upon insertion of Eqs. (3.134), (3.135), (3.137), and (3.139) as

$$W_p = -\int_{V_1}^{V_2} P dV - P_2 (V_3 - V_2) - \int_{V_3}^{V_4} P dV - P_4 (V_1 - V_4). \tag{3.141}$$

The fundamental theorem of integral calculus should now be invoked to transform Eq. (3.141) to

$$W_p = -\int_{V_1}^{V_2} P dV - P_2 \int_{V_2}^{V_3} dV - \int_{V_3}^{V_4} P dV - P_4 \int_{V_4}^{V_1} dV, \tag{3.142}$$

where P_2 and P_4 are to be taken into the corresponding kernels for being constant, viz.

$$W_p = -\left(\int_{V_1}^{V_2} P dV + \int_{V_2}^{V_3} P dV + \int_{V_3}^{V_4} P dV + \int_{V_4}^{V_1} P dV \right), \tag{3.143}$$

while –1 has been factored out; Eq. (3.143) can be condensed to merely

$$\boxed{W_p = -\oint P dV} \tag{3.144}$$

– meaning that the work done throughout an entire cycle is graphically given by the area of the PV plane enclosed by curve [1234] in Fig. 3.43. The shaded area in this figure results from subtraction of trapezoid areas below curve $3\rightarrow 4$ and below line $4\rightarrow 1$, corresponding to $W_{34}<0$ and $W_{41}<0$, respectively, from trapezoid areas below curve $1\rightarrow 2$ and below line $2\rightarrow 3$, corresponding to $W_{12}>0$ and $W_{23}>0$, respectively. In practice, the shaded trapezoid in Fig. 3.43 exhibits rounder corners – since the spring-loaded suction and delivery valves do not open instantaneously. Furthermore, pressure oscillates during those steps, due to valve bouncing – caused by the high initial pressure difference across the valve needed to overcome inertia of its moving parts; but even when valves are fully open, there is inevitably some pressure drop across them, due to fluid friction.

Based on Figs. 3.42iii and iv, combined with Eq. (3.133), one realizes that

$$V_1 = n_1 v_1 = n_2 v_4, \tag{3.145}$$

$$V_2 = n_2 v_2, \tag{3.146}$$

$$V_3 = n_3 v_3 = n_4 v_2, \tag{3.147}$$

and

$$V_4 = n_4 v_4; \tag{3.148}$$

insertion of Eqs. (3.145)–(3.148) then converts Eq. (3.141) to

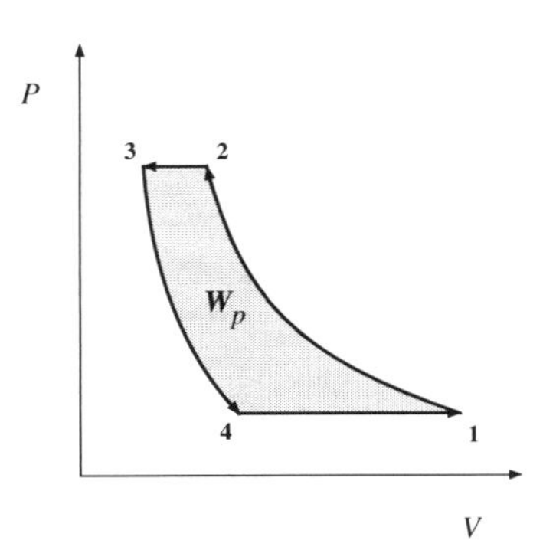

FIGURE 3.43 Schematic representation of reciprocating compressor cycle, based on an ideal gas – encompassing two isobaric, open-system steps between states **2** and **3**, and **4** and **1**, alternated with two adiabatic, closed-system steps between states **1** and **2**, and **3** and **4**, laid out as pressure, P, versus volume, V, with overall work received, W_p, given by shaded area ( ).

$$W_p = -n_2 \int_{v_4}^{v_2} P dv - P_2 \left(n_4 v_2 - n_2 v_2\right) - n_4 \int_{v_2}^{v_4} P dv - P_4 \left(n_2 v_4 - n_4 v_4\right) \tag{3.149}$$

– since $dV = d(n_2 v) = n_2 dv$ between 1 and 2, with n_2 constant, and likewise $dV = d(n_4 v) = n_4 dv$ between 3 and 4, with n_4 constant. The path followed by gas expansion coincides with that of gas compression, although spanned in opposite directions, see Fig. 3.42iv – as they are both adiabatic, and the ending state of either one coincides with the starting state of the other. Therefore, one is entitled to write

$$-\int_{v_4}^{v_2} P dv = \int_{v_2}^{v_4} P dv; \tag{3.150}$$

this is also supported by the status of P as exact differential, for being a state function – which justifies application of the fundamental theorem of integral calculus, and thus exchange of integration limits at the expense of taking the negative of the kernel. Equation (3.150) allows simplification of Eq. (3.149) to

$$W_p = -n_2 \int_{v_4}^{v_2} P dv - n_4 P_2 v_2 + n_2 P_2 v_2 + n_4 \int_{v_4}^{v_2} P dv - n_2 P_4 v_4 + n_4 P_4 v_4, \tag{3.151}$$

along with elimination of parentheses; after factoring the integral out, as well as $P_2 v_2$ and $P_4 v_4$ (as appropriate), Eq. (3.151) becomes

$$W_p = \left(n_4 - n_2\right) \int_{v_4}^{v_2} P dv + \left(n_2 - n_4\right) P_2 v_2 + \left(n_4 - n_2\right) P_4 v_4 \tag{3.152}$$

– where $n_2 - n_4$ may, in turn, be factored out to give

$$\boxed{W_p = \left(n_2 - n_4\right) \left(P_2 v_2 - P_4 v_4 - \int_{v_4}^{v_2} P dv \right).} \tag{3.153}$$

At this stage, it is instructive to recall the rule of differentiation of a product of functions, say x and y, viz.

$$d(xy) = xdy + ydx, \tag{3.154}$$

or else

$$ydx = d(xy) - xdy \tag{3.155}$$

upon solving for ydx; integration of both sides of Eq. (3.155), between extreme points 4 and 2, unfolds

$$\int_4^2 ydx = \int_4^2 d(xy) - \int_4^2 x dy, \tag{3.156}$$

which is equivalent to

$$\int_4^2 y dx = (xy)\Big|_4^2 - \int_4^2 x dy \tag{3.157}$$

– or else

$$\int_4^2 y dx = x_2 y_2 - x_4 y_4 - \int_4^2 x dy \tag{3.158}$$

that in essence retrieves the rule of (definite) integration by parts. If x is replaced by P, and y by v, then Eq. (3.158) will look like

$$\int_4^2 v dP = P_2 v_2 - P_4 v_4 - \int_4^2 P dv, \tag{3.159}$$

which is the same as writing

$$\int_{P_4}^{P_2} v dP = P_2 v_2 - P_4 v_4 - \int_{v_4}^{v_2} P dv \tag{3.160}$$

for consistency with the corresponding integration variables; comparative inspection of Eqs. (3.153) and (3.160) finally yields

$$\boxed{W_p = \left(n_2 - n_4\right) \int_{P_4}^{P_2} v dP.} \tag{3.161}$$

Therefore, a reciprocating compressor, with finite clearance, may be treated as a steady state flow device – where the work required to compress $n_2 - n_4$ moles of an (ideal) gas in each cycle/stroke, between pressures P_4 and $P_2 > P_4$, is obtainable at the expense solely of $\int_{P_4}^{P_2} vdP$ (multiplied obviously by $n_2 - n_4$). Upon multiplication and division of the right-hand side by (constant) molecular weight, M, of the gas at stake, Eq. (3.161) will instead appear as

$$W_p = \left(n_2 M - n_4 M\right) \int_{P_4}^{P_2} \frac{v}{M} dP. \tag{3.162}$$

Recall that

$$\rho \equiv \frac{m}{V} = \frac{\frac{m}{n}}{\frac{V}{n}} = \frac{M}{v}, \tag{3.163}$$

as per the definition of mass density, ρ, as m/V, molar volume, v, as per Eq. (3.133), and molecular weight, M, and m/n – along with division of both numerator and denominator by n; Eq. (3.162) may accordingly be redone to

$$W_p = \left(m_2 - m_4\right) \int_{P_4}^{P_2} \frac{dP}{\rho}, \tag{3.164}$$

which is equivalent to

$$w_p \equiv \frac{W_p}{m_2 - m_4} = \int_{P_4}^{P_2} \frac{dP}{\rho}, \tag{3.165}$$

following division of both sides by $m_2 - m_4$, to yield work performed per unit mass, w_p. It is remarkable that the very same expression obtained before for compression of steam using a turbine, see Eq. (3.147) in *Food Proc. Eng.: thermal & chemical operations*, is hereby retrieved. Therefore, one can insert Eq. (3.187) of *Food Proc. Eng.: thermal & chemical operations* to obtain

$$\frac{W_p}{m_2 - m_4} = \frac{\gamma}{\gamma - 1}\frac{P_4}{\rho_4}\left(\left(\frac{P_2}{P_4}\right)^{\frac{\gamma-1}{\gamma}} - 1\right), \tag{3.166}$$

since both compression and expansion took place adiabatically; to explicitly account for the piston/cylinder clearance occupied by gas mass m_4, Eq. (3.166) has classically been presented as

$$W_p = \frac{\gamma}{\gamma - 1}(m_2 - m_4)\frac{P_4}{\rho_4}\left(\left(\frac{P_2}{P_4}\right)^{\frac{\gamma-1}{\gamma}} - 1\right), \tag{3.167}$$

where multiplication and division of the right-hand side by M produces

$$W_p = \frac{\gamma}{\gamma - 1}\left(\frac{m_2}{M} - \frac{m_4}{M}\right)\frac{M}{\rho_4}P_4\left(\left(\frac{P_2}{P_4}\right)^{\frac{\gamma-1}{\gamma}} - 1\right) \tag{3.168}$$

– or simply

$$W_p = (n_2 - n_4)\frac{\gamma}{\gamma - 1}P_4 v_4\left(\left(\frac{P_2}{P_4}\right)^{\frac{\gamma-1}{\gamma}} - 1\right), \tag{3.169}$$

with the aid of Eq. (3.163). If Eqs. (3.145) and (3.148) are taken on board to eliminate n_2 and n_4, respectively, then Eq. (3.169) will acquire the form

$$W_p = \left(\frac{V_1}{v_4} - \frac{V_4}{v_4}\right)\frac{\gamma}{\gamma - 1}P_4 v_4\left(\left(\frac{P_2}{P_4}\right)^{\frac{\gamma-1}{\gamma}} - 1\right), \tag{3.170}$$

where factoring in of v_4 generates

$$W_p = P_4 V_{int}\frac{\gamma}{\gamma - 1}\left(\left(\frac{P_2}{P_4}\right)^{\frac{\gamma-1}{\gamma}} - 1\right); \tag{3.171}$$

here V_{int} is defined as

$$V_{int} \equiv V_1 - V_4, \tag{3.172}$$

i.e. intake volume of gas, at suction conditions. Since W_p in Eq. (3.171) is associated with a single cycle, the actual power required for compression, $\mathcal{P}_p$, abides to

$$\boxed{\mathcal{P}_p \equiv \Omega W_p,} \tag{3.173}$$

where Ω denotes reciprocating frequency, or number of cycles completed per unit time; insertion of Eq. (3.171) gives then rise to

$$\boxed{\mathcal{P}_p = P_1 V_{int}\Omega\frac{\gamma}{\gamma - 1}\left(\left(\frac{P_2}{P_1}\right)^{\frac{\gamma-1}{\gamma}} - 1\right),} \tag{3.174}$$

where P_4 coincides with P_1 (see Fig. 3.42ii).

The ratio of volume of gas actually compressed during each cycle, referred to suction conditions, V_{int}, to the volume swept by the piston along the cylinder, V_{swp} – given by

$$V_{swp} \equiv V_1 - V_3, \tag{3.175}$$

is termed volumetric efficiency of the compressor, ε_v, i.e.

$$\boxed{\varepsilon_v \equiv \frac{V_{int}}{V_{swp}};} \tag{3.176}$$

the clearance ratio, R_{clr}, has also been traditionally defined as

$$R_{clr} \equiv \frac{V_3}{V_{swp}}. \tag{3.177}$$

Equations (3.172) and (3.175) are graphically represented in Fig. 3.44. Note the coincidence of compression and expansion trajectories, should V_2 coincide with the clearance volume of the compressor device, V_3 – to be viewed as an asymptotic trend; complemented with

$$\lim_{P_2 \to P_1} V_{int} = V_{swp}, \tag{3.178}$$

serving as complementary asymptotic trend – which corresponds to coincidence of intake and swept volume when no compression takes place. Further inspection of Fig. 3.44

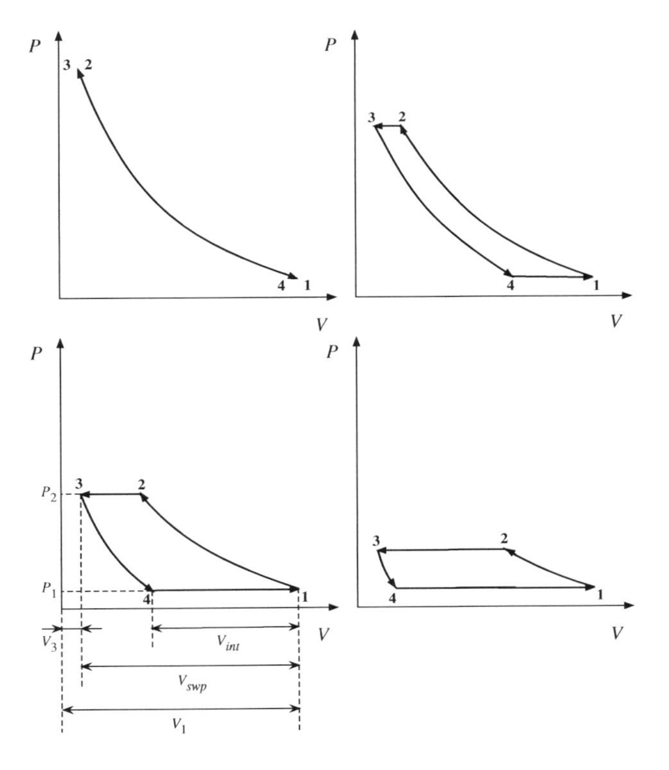

FIGURE 3.44 Schematic representation of reciprocating compressor operated along cycle **1→2→3→4→1** upon an ideal gas, laid out as pressure, P, versus volume, V – with elucidation of effect of pressure increase (from left to right, and then from top to bottom) from fixed P_1 up to variable P_2, at compression stage, upon intake volume, V_{int}, and swept volume, V_{swp}, of gas for fixed clearance volume, V_3, and fixed maximum stroke volume, V_1.

indicates that, for given P_1, V_1, and V_3, V_{int} decreases when P_2 increases; a more quantitative analysis follows, in terms of volumetric efficiency.

Using Eqs. (3.172) and (3.175), it is indeed possible to rewrite Eq. (3.176) as

$$\varepsilon_v = \frac{V_1 - V_4}{V_1 - V_3}, \tag{3.179}$$

where addition and subtraction of V_3 in numerator, fraction splitting afterward, and final multiplication and division by V_3 yield

$$\varepsilon_v = \frac{(V_1 - V_3) - (V_4 - V_3)}{V_1 - V_3} = 1 - \frac{V_4 - V_3}{V_1 - V_3} = 1 - \frac{V_4 - V_3}{V_3}\frac{V_3}{V_1 - V_3}; \tag{3.180}$$

insertion of Eqs. (3.175) and (3.177) gives then rise to

$$\varepsilon_v = 1 - R_{clr}\frac{V_4 - V_3}{V_3} = 1 - R_{clr}\left(\frac{V_4}{V_3} - 1\right). \tag{3.181}$$

Since process $3 \rightarrow 4$ occurs adiabatically (see Fig. 3.42ii) and with a constant number of moles (see Fig. 3.42iii), one may revisit Eq. (3.130) as

$$V^{\gamma} = \frac{\kappa}{P} \tag{3.182}$$

with $\kappa \equiv Kn_4^{\gamma}$ – or else

$$V = \left(\frac{\kappa}{P}\right)^{\frac{1}{\gamma}}, \tag{3.183}$$

after taking γ-th roots of both sides; Eq. (3.183) may now be applied to both states 3 and 4 as

$$V_3 = \left(\frac{\kappa}{P_3}\right)^{\frac{1}{\gamma}} \tag{3.184}$$

and

$$V_4 = \left(\frac{\kappa}{P_4}\right)^{\frac{1}{\gamma}}, \tag{3.185}$$

respectively. Ordered division of Eq. (3.184) by Eq. (3.185) unfolds

$$\frac{V_3}{V_4} = \frac{\left(\frac{\kappa}{P_3}\right)^{\frac{1}{\gamma}}}{\left(\frac{\kappa}{P_4}\right)^{\frac{1}{\gamma}}} \tag{3.186}$$

or, equivalently,

$$\frac{V_3}{V_4} = \left(\frac{P_4}{P_3}\right)^{\frac{1}{\gamma}} \tag{3.187}$$

after lumping powers, and then dropping κ between numerator and denominator. Combination of Eqs. (3.181) and (3.187) gives finally rise to

$$\varepsilon_v = 1 - R_{clr}\left(\left(\frac{P_3}{P_4}\right)^{\frac{1}{\gamma}} - 1\right), \tag{3.188}$$

or to its alias

$$\boxed{\varepsilon_v = 1 - R_{clr}\left(\left(\frac{P_2}{P_1}\right)^{\frac{1}{\gamma}} - 1\right)} \tag{3.189}$$

– owing to $P_3 = P_2$ and $P_4 = P_1$, as apparent in Fig. 3.42ii; inspection of Eq. (3.189) indicates that volumetric efficiency decreases as pressure ratio, P_2/P_1, increases. When P_2 approaches P_1, Eq. (3.189) enforces $\varepsilon_v \rightarrow 1$, in agreement with Eqs. (3.176) and (3.178) – thus confirming the graphical trend in Fig. 3.44.

PROBLEM 3.9

Consider a reciprocating compressor, meant to produce pressurized air – hypothesized to behave as an ideal gas.

a) Calculate the pressure ratio yielding minimum volumetric efficiency in the case of adiabatic operation – and provide a graphical/physical interpretation.
b) Obtain an expression for the power required for a given compression ratio P_2/P_1, as a function of γ, P_1V_{swp}, Ω, R_{clr}, and P_2/P_1.
c) Using the result obtained in b), derive the condition to be satisfied by pressure ratio if maximum power input is required.

Solution

a) The minimum volumetric efficiency corresponds to $\varepsilon_v = 0$ in Eq. (3.176) – in which case Eq. (3.189) becomes

$$1 - R_{clr}\left(\left(\frac{P_2}{P_1}\right)^{\frac{1}{\gamma}} - 1\right) = 0, \tag{3.9.1}$$

where the term containing pressures can be isolated as

$$\left(\frac{P_2}{P_1}\right)^{\frac{1}{\gamma}} = 1 + \frac{1}{R_{clr}}; \tag{3.9.2}$$

after taking the γ-th power of both sides, Eq. (3.9.2) turns to

$$\frac{P_2}{P_1} = \left(1 + \frac{1}{R_{clr}}\right)^{\gamma}, \tag{3.9.3}$$

or else

$$\left.\frac{P_2}{P_1}\right|_{\varepsilon_v=0} = \left(\frac{1+R_{clr}}{R_{clr}}\right)^{\gamma} \tag{3.9.4}$$

after merging terms in parenthesis. According to Fig. 3.44 and Eq. (3.179), the compression (upward) path will eventually coincide with the expansion (downward) path, i.e. $V_1 = V_4$ and thus $\varepsilon_v=0$, when the pressure ratio attains $\left(P_2/P_1\right)\big|_{\varepsilon_v=0}$; according to Eq. (3.172), V_{int} is nil under such circumstances, so no flow of gas will take place through the machine – which will repeatedly compress and expand the very same gas along isentrope $1 \leftrightarrow 2$.

b) After having eliminated V_{int} from Eq. (3.174) at the expense of Eq. (3.176), one obtains

$$\mathcal{P}_p = P_1 V_{swp}\varepsilon_v \Omega \frac{\gamma}{\gamma-1}\left(\left(\frac{P_2}{P_1}\right)^{\frac{\gamma-1}{\gamma}}-1\right); \tag{3.9.5}$$

Eq. (3.189) may now be retrieved to write

$$\mathcal{P}_p = k\left(1-R_{clr}\left(\xi^{\frac{1}{\gamma}}-1\right)\right)\left(\xi^{\frac{\gamma-1}{\gamma}}-1\right) \tag{3.9.6}$$

based on Eq. (3.9.5), which is of the form $\mathcal{P}_p \equiv \mathcal{P}_p\{k, R_{clr}, \xi, \gamma\}$ – where ξ denotes pressure ratio, abiding to

$$\xi \equiv \frac{P_2}{P_1}, \tag{3.9.7}$$

and k denotes a constant given by

$$k\{\gamma, P_1 V_{swp}\Omega\} \equiv \frac{\gamma}{\gamma-1} P_1 V_{swp}\Omega. \tag{3.9.8}$$

In view of Eqs. (3.9.6)–(3.9.8), one readily finds $\mathcal{P}_p \equiv \mathcal{P}_p\{\gamma, P_1 V_{swp}\Omega, R_{clr}, P_2/P_1\}$, as originally intended.

c) The necessary condition for the maximum $\mathcal{P}_p$, with regard to ξ as per Eq. (3.9.7), reads

$$\frac{d\mathcal{P}_p}{d\xi} = 0; \tag{3.9.9}$$

Eq. (3.9.9) becomes

$$-R_{clr}\frac{1}{\gamma}\xi^{\frac{1}{\gamma}-1}\left(\xi^{\frac{\gamma-1}{\gamma}}-1\right)+\left(1-R_{clr}\left(\xi^{\frac{1}{\gamma}}-1\right)\right)\frac{\gamma-1}{\gamma}\xi^{\frac{\gamma-1}{\gamma}-1} = 0, \tag{3.9.10}$$

upon application of derivatives to Eq. (3.9.6), and division of both sides by k afterward. Further division of both sides by $\xi^{1/\gamma-1}$ and $\xi^{(\gamma-1)/\gamma}$ transforms Eq. (3.9.10) to

$$\frac{\gamma-1}{\gamma}\left(\xi^{-\frac{1}{\gamma}}-R_{clr}\left(1-\xi^{-\frac{1}{\gamma}}\right)\right) = \frac{R_{clr}}{\gamma}\left(1-\xi^{\frac{1}{\gamma}-1}\right) \tag{3.9.11}$$

– where multiplication of both sides by γ/R_{clr}, followed by elimination of inner parenthesis produce

$$(\gamma-1)\left(\frac{\xi^{-\frac{1}{\gamma}}}{R_{clr}}-1+\xi^{-\frac{1}{\gamma}}\right) = 1-\xi^{\frac{1}{\gamma}-1}; \tag{3.9.12}$$

removal of parenthesis, complemented by factoring out of $\xi^{-1/\gamma}$ give rise to

$$(\gamma-1)\left(1+\frac{1}{R_{clr}}\right)\xi^{-\frac{1}{\gamma}}-\gamma+1 = 1-\xi^{\frac{1}{\gamma}-1}, \tag{3.9.13}$$

where unity may drop off both sides as

$$\frac{(\gamma-1)\left(1+R_{clr}\right)}{R_{clr}}\xi^{-\frac{1}{\gamma}}+\xi^{\frac{1}{\gamma}-1} = \gamma. \tag{3.9.14}$$

Equation (3.9.14) entails the optimality condition sought, and is to be numerically solved for ξ to produce the corresponding optimum ratio.

The volumetric efficiency of an actual compressor differs from that predicted by Eq. (3.189) for a number of reasons: the gas will likely exhibit nonideal behavior at high pressures, compression/expansion may not be exactly adiabatic, the state of gas at suction changes because of heat transferred, and some extent of leakage past valves and piston is unavoidable.

On the other hand, increasing the pressure ratio delivered by a compressor will also increase the temperature ratio, in agreement with Eq. (3.136) of *Food Proc. Eng.: thermal & chemical operations*, i.e.

$$\frac{T_2}{T_1} = \left(\frac{P_2}{P_1}\right)^{\frac{\gamma-1}{\gamma}} \tag{3.190}$$

– since $\gamma > 1$, as per Eq. (3.126) in the same book; for instance, in the case of nitrogen, characterized by $\gamma = 1.4$, an increase in pressure from 1 to 10 atm will increase its temperature from 20 to 293°C. Even if the cylinder of the reciprocating compressor were water-cooled, the temperature of the gas at discharge would hardly fall 10°C below the aforementioned 293°C; therefore, higher pressure ratios normally call for multistage compression – since it allows gas cooling between stages. By doing so, overall volumetric efficiency of the process is improved, maximum temperature attained will not be so high, and work of compression will be reduced – besides simplifying the problem of lubricating piston rings.

In attempts to optimize interstage pressure(s) – given inlet pressure, P_1, and (intended) outlet pressure, P_{N+1}, one may resort to minimum global work, W_{oa}, as objective function; once again, adiabatic (reversible) operation and ideal gas behavior will be assumed for simplicity of simulation, with γ independent of temperature – as well as gas cooling to the original suction temperature, after each compression stage. One should accordingly start by defining W_{oa} as

$$W_{oa} \equiv \sum_{i=1}^{N} W_{p,i}, \tag{3.191}$$

with work applied at i-th stage (out of a total of N stages) of compression denoted by $W_{p,i}$; insertion of Eq. (3.169) – after realizing that $n_4 \equiv n_i$, $n_2 \equiv n_{i+1}$, $P_4 \equiv P_i$, $P_2 \equiv P_{i+1}$, and $v_4 \equiv v_i$ in agreement with Figs. 3.42ii–iv, gives rise to

$$W_{oa} = \sum_{i=1}^{N} \left(n_{i+1} - n_i\right) \frac{\gamma}{\gamma - 1} P_i v_i \left(\left(\frac{P_{i+1}}{P_i} \right)^{\frac{\gamma-1}{\gamma}} - 1 \right). \tag{3.192}$$

The same amount of gas enters each compression stage, i.e.

$$n_{i+1} - n_i = n_2 - n_1; \quad i = 1,2,\ldots,N, \tag{3.193}$$

and the original temperature, T, is, by hypothesis, retrieved at the inlet of each stage, i.e.

$$P_i V_i = P_1 V_1 = \left(n_2 - n_1\right) \mathcal{R} T; \quad i = 1,2,\ldots,N; \tag{3.194}$$

based on reformulation of Eq. (3.132) of *Food Proc.Eng.: basics & mechanical operations*, Eq. (3.192) can be rewritten as

$$W_{oa} = \frac{n_2 - n_1}{\Gamma} P_1 v_1 \sum_{i=1}^{N} \left(\left(\frac{P_{i+1}}{P_i} \right)^{\Gamma} - 1 \right) \tag{3.195}$$

– where constant factors were taken off the summation sign, and auxiliary constant Γ was set as

$$\Gamma \equiv \frac{\gamma - 1}{\gamma}. \tag{3.196}$$

One may, for convenience, define an overall molar work, $\hat{w}_{oa}$, according to

$$\hat{w}_{oa} \equiv \frac{W_{oa}}{n_2 - n_1} \tag{3.197}$$

and using Eq. (3.165) for template – in that it allows further simplification of Eq. (3.195) to

$$\boxed{\hat{w}_{oa} = \frac{P_1 v_1}{\Gamma} \sum_{i=1}^{N} \left(\left(\frac{P_{i+1}}{P_i} \right)^{\Gamma} - 1 \right)}; \tag{3.198}$$

note that two extra constraints exist, since inlet pressure (i.e. P_1) and final outlet pressure (i.e. P_{N+1}) are fixed in advance, i.e.

$$\boxed{P_i\big|_{i=1} = P_1} \tag{3.199}$$

and

$$\boxed{P_i\big|_{i=N+1} = P_{N+1}}, \tag{3.200}$$

respectively. After recalling that

$$\boxed{\left(\frac{\partial \hat{w}_{oa}}{\partial P_i} \right)_{P_{j \neq i}} = 0; \quad i = 2,3,\ldots,N} \tag{3.201}$$

will serve as $N - 1$ necessary conditions for the optima under scrutiny, combination with Eq. (3.198) will give rise to

$$\frac{P_1 v_1}{\Gamma} \sum_{j=1}^{N} \frac{\partial}{\partial P_i} \left(\left(\frac{P_{j+1}}{P_j} \right)^{\Gamma} - 1 \right) = 0, \tag{3.202}$$

at the expense of the linearity of the differential operator; upon multiplication of both sides by Γ/P_1V_1, Eq. (3.202) reduces to

$$\frac{\partial}{\partial P_i} \left(\left(\left(\frac{P_i}{P_{i-1}} \right)^{\Gamma} - 1 \right) + \left(\left(\frac{P_{i+1}}{P_i} \right)^{\Gamma} - 1 \right) \right) = 0 \tag{3.203}$$

– since only two terms P_{j+1}/P_j of the summation contain P_i explicitly, or else

$$\frac{\partial}{\partial P_i} \left(\frac{P_i}{P_{i-1}} \right)^{\Gamma} + \frac{\partial}{\partial P_i} \left(\frac{P_{i+1}}{P_i} \right)^{\Gamma} = 0 \tag{3.204}$$

after getting rid of constants under the differential operator. Application of the rule of differentiation of a power to the two terms in Eq. (3.204) gives rise to

$$\Gamma \left(\frac{P_i}{P_{i-1}} \right)^{\Gamma - 1} \frac{1}{P_{i-1}} + \Gamma \left(\frac{P_{i+1}}{P_i} \right)^{\Gamma - 1} \left(-\frac{P_{i+1}}{P_i^2} \right) = 0, \tag{3.205}$$

where division of both sides by Γ, followed by lumping of factors alike generate

$$\frac{P_i^{\Gamma - 1}}{P_{i-1}^{\ \Gamma}} = \frac{P_{i+1}^{\ \Gamma}}{P_i^{\Gamma + 1}}; \tag{3.206}$$

upon elimination of denominators, Eq. (3.206) becomes

$$P_i^{2\Gamma} = P_{i-1}^{\ \Gamma} P_{i+1}^{\ \Gamma}, \tag{3.207}$$

where 2Γ-th roots should now be taken of both sides as

$$P_i = \sqrt{P_{i-1} P_{i+1}}. \tag{3.208}$$

Equation (3.208) indicates that the optimality condition is having pressure at each intake/delivery point equal to the geometric mean of previous and next stage pressures; after squaring both sides, Eq. (3.208) will appear as

$$P_i^2 = P_{i-1} P_{i+1}, \tag{3.209}$$

which may be further algebraically rearranged to

$$\boxed{\frac{P_i}{P_{i-1}} = \frac{P_{i+1}}{P_i}; \quad i = 2,3,\ldots,N} \tag{3.210}$$

– meaning that the pressure ratio, i.e. P_{i+1}/P_i for the i-th stage, should be the same for the $(i–1)$th stage, and thus for every stage. One may now rewrite P_{N+1}/P_1 as

$$\frac{P_{N+1}}{P_1} = \frac{P_2}{P_1} \frac{P_3}{P_2} \frac{P_4}{P_3} \cdots \frac{P_N}{P_{N-1}} \frac{P_{N+1}}{P_N} \tag{3.211}$$

– where P_{N+1}/P_1 is known *a priori*, see Eqs. (3.199) and (3.200); insertion of Eq. (3.210), rewritten as

$$\frac{P_i}{P_{i-1}} = \frac{P_2}{P_1} = \frac{P_{i+1}}{P_i} \tag{3.212}$$

and applicable to $i = 2, 3, \ldots, N$, transforms Eq. (3.211) to

$$\frac{P_2}{P_1}\frac{P_2}{P_1}\frac{P_2}{P_1}\cdots\frac{P_2}{P_1}\frac{P_2}{P_1} \equiv \left(\frac{P_2}{P_1}\right)^N = \frac{P_{N+1}}{P_1} \tag{3.213}$$

with the aid of the definition of power – where P_2/P_1 is to be isolated as

$$\frac{P_2}{P_1} = \left(\frac{P_{N+1}}{P_1}\right)^{\frac{1}{N}}, \tag{3.214}$$

provided that N-th roots are taken of both sides. Once Eq. (3.198) is redone as

$$\hat{w}_{oa} = \frac{P_1 v_1}{\Gamma}\left(\sum_{i=1}^{N}\left(\frac{P_{i+1}}{P_i}\right)^{\Gamma} - \sum_{i=1}^{N} 1\right) \tag{3.215}$$

after splitting the summation, one may insert Eq. (3.212) to get

$$\hat{w}_{oa} = \frac{P_1 v_1}{\Gamma}\left(\sum_{i=1}^{N}\left(\frac{P_2}{P_1}\right)^{\Gamma} - N\right) \equiv \frac{P_1 v_1}{\Gamma}\left(N\left(\frac{P_2}{P_1}\right)^{\Gamma} - N\right) \tag{3.216}$$

– now at the expense of the definition of multiplication; combination with Eq. (3.214) transforms Eq. (3.216) to

$$\hat{w}_{oa} = \frac{P_1 v_1}{\Gamma} N\left(\left(\left(\frac{P_{N+1}}{P_1}\right)^{\frac{1}{N}}\right)^{\Gamma} - 1\right) = \frac{N P_1 v_1}{\Gamma}\left(\left(\frac{P_{N+1}}{P_1}\right)^{\frac{\Gamma}{N}} - 1\right), \tag{3.217}$$

after having N previously factored out. The original notation will be recovered in Eq. (3.217) as

$$\boxed{\hat{w}_{oa} = N P_1 v_1 \frac{\gamma}{\gamma - 1}\left(\left(\frac{P_{N+1}}{P_1}\right)^{\frac{\gamma-1}{N\gamma}} - 1\right)}, \tag{3.218}$$

in view of Eq. (3.196) – thus yielding the minimum amount of average work required for multistage, adiabatic compression of an ideal gas between P_1 and P_{N+1}.

The question now remains as to which number of stages to use; inspection of Eq. (3.218), *vis-à-vis* with Eq. (5.714) of *Food Proc. Eng.: thermal & chemical operations* pertaining to a cascade of stirred tank reactors, unfolds a similar functional form with regard to N – as long as ${C^*_{L,N}}^{-1/N}$, is replaced by $\left(1/C^*_{L,N}\right)^{1/N}$, with $1/C^*_{L,N} > 1$ similar to $P_{N+1}/P_1 > 1$. Hence, an analog to Eq. (5.183) of *Food Proc. Eng.: thermal & chemical operations* applies, i.e.

$$\frac{d\hat{w}_{oa}}{dN} < 0 \tag{3.219}$$

using the current notation. Since $\hat{w}_{oa}$ is a monotonically decreasing function of N as per Eq. (3.219), its lowest bound will be described by

$$\lim_{N\to\infty} \hat{w}_{oa} = \lim_{N\to\infty} N P_1 v_1 \frac{\gamma}{\gamma - 1}\left(\left(\frac{P_{N+1}}{P_1}\right)^{\frac{\gamma-1}{N\gamma}} - 1\right), \tag{3.220}$$

in agreement with Eq. (3.218); direct application of the theorems on limits to Eq. (3.220) leads to an unknown quantity, i.e.

$$\begin{aligned}\lim_{N\to\infty} \hat{w}_{oa} &= \infty P_1 v_1 \frac{\gamma}{\gamma - 1}\left(\left(\frac{P_{N+1}}{P_1}\right)^{\frac{\gamma-1}{\infty}} - 1\right) = \infty\left(\left(\frac{P_{N+1}}{P_1}\right)^{0} - 1\right) \\ &= \infty(1-1) = 0\cdot\infty\end{aligned} \tag{3.221}$$

since $\gamma > 1$. To circumvent this problem, one should revisit Eq. (3.220) as

$$\lim_{N\to\infty} \hat{w}_{oa} = P_1 v_1 \frac{\gamma}{\gamma - 1} \lim_{N\to\infty} \frac{\left(\frac{P_{N+1}}{P_1}\right)^{\frac{\gamma-1}{N\gamma}} - 1}{\frac{1}{N}} = P_1 v_1 \frac{\gamma}{\gamma - 1}\frac{0}{0}, \tag{3.222}$$

which satisfies the conditions of validity for application of l'Hôpital's rule; hence, one may redo Eq. (3.222) to

$$\lim_{N\to\infty} \hat{w}_{oa} = P_1 v_1 \frac{\gamma}{\gamma - 1} \lim_{N\to\infty} \frac{\left(\frac{P_{N+1}}{P_1}\right)^{\frac{\gamma-1}{N\gamma}} \frac{\gamma-1}{\gamma}\left(-\frac{1}{N^2}\right)\ln\frac{P_{N+1}}{P_1}}{\left(-\frac{1}{N^2}\right)}, \tag{3.223}$$

upon independent differentiation of numerator and denominator with regard to N. After dropping common factors between numerator and denominator, Eq. (3.223) simplifies to

$$\lim_{N\to\infty} \hat{w}_{oa} = P_1 v_1 \ln\frac{P_{N+1}}{P_1} \lim_{N\to\infty}\left(\frac{P_{N+1}}{P_1}\right)^{\frac{\gamma-1}{N\gamma}}, \tag{3.224}$$

or else

$$\lim_{N\to\infty} \hat{w}_{oa} = P_1 v_1 \ln\frac{P_{N+1}}{P_1}\left(\frac{P_{N+1}}{P_1}\right)^{\frac{\gamma-1}{\infty}} = P_1 v_1 \ln\frac{P_{N+1}}{P_1}\left(\frac{P_{N+1}}{P_1}\right)^{0}; \tag{3.225}$$

one finally concludes that

$$\boxed{\lim_{N\to\infty} \hat{w}_{oa} = P_1 v_1 \ln\frac{P_{N+1}}{P_1}}, \tag{3.226}$$

representing the (finite) minimum molar work – attainable only via an infinite number of compression stages, though.

After defining

$$\hat{w}_T \equiv \frac{W_p}{n_2 - n_1} \tag{3.227}$$

in parallel to Eq. (3.197) – but now pertaining to an isothermal process describing compression within a single stage, one may revisit Eq. (3.161) as

$$\hat{w}_T = \int_{P_1}^{P_2} v\,dP, \tag{3.228}$$

again upon realization that n_4 and P_4 have hereby been renamed as n_1 and P_1, respectively; since the flow process is isothermal by hypothesis, application of Eq. (3.133) will transform Eq. (3.228) to

$$\hat{w}_T = \mathcal{R}T \int_{P_1}^{P_2} \frac{dP}{P}. \quad (3.229)$$

Once the fundamental theorem of integral calculus is invoked, Eq. (3.229) becomes

$$\hat{w}_T = \mathcal{R}T \ln P\Big|_{P_1}^{P_2} \quad (3.230)$$

that is equivalent to

$$\hat{w}_T = \mathcal{R}T \ln \frac{P_2}{P_1}; \quad (3.231)$$

on the other hand, $\mathcal{R}T$ being constant implies that Pv must be constant as well, in agreement again with Eq. (3.133) – which may instead be stated as

$$P_1 v_1 = P_2 v_2 = \mathcal{R}T. \quad (3.232)$$

Insertion of Eq. (3.232) converts Eq. (3.231) to

$$\hat{w}_T = P_1 v_1 \ln \frac{P_2}{P_1}, \quad (3.233)$$

or else

$$\hat{w}_{oaT} = P_1 v_1 \ln \frac{P_{N+1}}{P_1} \quad (3.234)$$

as alias encompassing overall molar work, $\hat{w}_{oaT}$, associated to isothermal compression directly from P_1 up to P_{N+1}; in the case of multistage isothermal compression, the condition labeled as Eq. (3.211) would lead to

$$\begin{aligned}\hat{w}_{oaT} &= P_1 v_1 \ln \frac{P_{N+1}}{P_N}\frac{P_N}{P_{N-1}}\ldots\frac{P_3}{P_2}\frac{P_2}{P_1} = P_1 v_1 \ln \left(\frac{P_2}{P_1}\right)^N \\ &= N P_1 v_1 \ln \frac{P_2}{P_1} = N\hat{w}_T\end{aligned} \quad (3.235)$$

– with the aid of the operational features of a logarithm, and after Eqs. (3.212) and (3.233) have been taken on board. One therefore concludes that

$$\boxed{\lim_{N\to\infty} \hat{w}_{oa} = \hat{w}_{oaT}} \quad (3.236)$$

upon comparative inspection of Eqs. (3.226) and Eq. (3.234); in other words, the minimum (molar) overall work associated with multistage adiabatic compression of a gas coincides with the overall work required for its single-stage, isothermal compression between the same overall pressure limits. This mathematical result is physically expected, because an infinite number of compression stages – with intermediate cooling to the original temperature, is equivalent to compressing under plain isothermal conditions all the way.

In the design of an actual multistage compressor network, it is necessary to allow for pressure drops through the interstage coolers – normally by increasing discharge pressure from each stage by one-half, and concomitantly decreasing suction pressure to the next stage by the other half. The (theoretical) optimum intermediate pressure ratios are not affected, but the cumulative power required to do the extra work should be added to the total compression power. Although substantial savings can be attained by going multistage rather than single-stage, the increased equipment costs incurred in seldom justify more than 4–5 stages, from an economic point of view. Extra savings arise from the need to build a smaller high-pressure cylinder, able to withstand the full (final) delivery pressure; and from the ease of balancing machine loads, provided that all stages are hooked on the same drive-shaft.

The classical mode of operation of a compressor, meant to handle ideal gases, entails adiabatic conditions, i.e. lack of exchange of heat with the surroundings. This mode of operation minimizes indeed the amount of work needed for a given pressure increase – since no portion whatsoever of work is lost to the surroundings in the form of heat. Remember that this assumption led to Eq. (3.131) in *Food Proc. Eng.: thermal & chemical operations* as defining condition for such a process – which may be alternatively formulated as

$$\boxed{\left(\frac{V_{in}}{V_{out}}\right)^{\gamma} = \frac{P_{out}}{P_{in}} = \left(\frac{T_{out}}{T_{in}}\right)^{\frac{\gamma}{\gamma-1}}}, \quad (3.237)$$

after algebraic rearrangement (and merging) of Eqs. (3.132) and (3.136) of *Food Proc. Eng.: thermal & chemical operations*; here γ is obtained from two properties of the gas at stake, i.e. its isobaric and isochoric specific heat capacities, following Eq. (3.126) from the same book. In practice, achieving exactly such operating conditions is hard, so γ in Eq. (3.237) may to advantage be replaced by an adjustable parameter ν – usually close, but not coincident with γ. Deviations between γ and ν will thus be due either to reversible but nonadiabatic compression, or irreversible but adiabatic compression (associated with internal friction and turbulence). Reciprocating compressors are more prone to the former mode of nonideality, should gas linear velocities be relatively small; if interstage cooling is provided, then $\nu<\gamma$. Turbocompressors tend to deviate from ideal behavior chiefly due to the latter case, since internal friction and turbulence are considerable – with $\nu>\gamma$. Once again, reversible and adiabatic conditions support $\nu=\gamma$, see Eq. (3.130); whereas $\nu=1$ describes the limiting case of isothermal compression, see Eq. (3.133).

3.3.10 Electric apparatuses

3.3.10.1 Motors

The seminal phenomenon in the vast majority of electric motors is electromagnetic induction of a relative movement between stator and rotor (with an air gap in between) – so that useful torque (or linear force) is generated.

Attempts to simulate the above phenomenon should consequently start by realizing that a particle of charge q, moving with velocity $\boldsymbol{v}$ in the presence of both an electric field $\boldsymbol{E}$ and a magnetic field $\boldsymbol{B}$, experiences a force $\boldsymbol{F}$ that satisfies

$$\boxed{\boldsymbol{F} = q(\boldsymbol{E} + \boldsymbol{v}\times\boldsymbol{B});} \quad (3.238)$$

this is usually known as Lorentz's law. Therefore, a positively charged particle will be accelerated in the same direction as $\boldsymbol{E}$, but will curve perpendicularly to both $\boldsymbol{v}$ and $\boldsymbol{B}$ – according to the right-hand rule; in other words, if the index finger of one's right hand is extended in the direction of $\boldsymbol{v}$, and the middle finger curled so as to point in the direction of $\boldsymbol{B}$, then the extended thumb will point in the direction of $\boldsymbol{F}$. The term $q\boldsymbol{E}$ in Eq. (3.238), sometimes called Lorentz's force, represents induced electromotive force – normally produced in a wire loop moving through an electric field; this is consubstantiated in a few electric motors/generators. Conversely, the term $q\boldsymbol{v}\times\boldsymbol{B}$, sometimes called Laplace's force, represents magnetic force acting on a current-carrying wire – and is responsible for motional electromotive force; this is the phenomenon underlying operation of most electric motors/generators, where conductors and magnets experience movement relative to each other. It should be emphasized that Lorentz's law complements the information carried by the four basic laws of electromagnetics, i.e. Eqs. (2.364), (2.370), (2.379), and (2.381) in *Food Proc. Eng.: thermal & chemical operations* – so Eq. (3.238) cannot be derived from such equations; the said laws basically indicate that electromotive fields are due to a given distribution of charges and currents – whereas Lorentz's law describes the response, in terms of motion, charges, and currents, to given electromotive fields.

A force, in general, is but the rate of creation/depletion of linear momentum, in agreement with Newton's second law of motion; Lorentz's force, in particular, describes the rate at which linear momentum is transferred from an electromagnetic field to a charged particle. If this (charged) particle travels at velocity $\boldsymbol{v}$ and experiences (electromagnetic) force $\boldsymbol{F}$, then the electromagnetic power, $\mathcal{P}$, transferred thereto reads

$$\boxed{\mathcal{P}\equiv\boldsymbol{v}\cdot\boldsymbol{F};} \tag{3.239}$$

combination with Eq. (3.238) unfolds

$$\mathcal{P}=\boldsymbol{v}\cdot q(\boldsymbol{E}+\boldsymbol{v}\times\boldsymbol{B}). \tag{3.240}$$

The mathematical properties of the scalar product between vectors, and of the product of a scalar by a vector support transformation of Eq. (3.240) to

$$\mathcal{P}=q(\boldsymbol{v}\cdot\boldsymbol{E}+\boldsymbol{v}\cdot(\boldsymbol{v}\times\boldsymbol{B})) \tag{3.241}$$

since, by definition of vector product, $\boldsymbol{v}\times\boldsymbol{B}$ is normal to $\boldsymbol{v}$, one promptly concludes that the scalar product between vectors $\boldsymbol{v}$ and $\boldsymbol{v}\times\boldsymbol{B}$ must be nil, as per the nil cosine of their angle in the defining algorithm – so Eq. (3.241) reduces to merely

$$\boxed{\mathcal{P}=q\boldsymbol{v}\cdot\boldsymbol{E}.} \tag{3.242}$$

Therefore, the magnetic field does not contribute to electrical power – in view of its being perpendicular to the electric field; in much the same way, the (vertical) weight of a body does not contribute to mechanical power, when the latter is displaced horizontally.

PROBLEM 3.10

The electric field, $\boldsymbol{E}$, and the magnetic field, $\boldsymbol{B}$, may be expressed as functions of an electrostatic (scalar) potential, ϕ, and a magnetic (vector) potential, $\boldsymbol{A}$, according to

$$\boldsymbol{E}=-\nabla\phi-\frac{\partial\boldsymbol{A}}{\partial t} \tag{3.10.1}$$

and

$$\boldsymbol{B}=\nabla\times\boldsymbol{A}, \tag{3.10.2}$$

respectively.

a) Express Lorentz's law in terms of ϕ and $\boldsymbol{A}$ (besides $\boldsymbol{v}$), resorting only to differential operators ∇ and d/dt.
b) Confirm that Eqs. (3.10.1) and (3.10.2) are consistent with Maxwell and Faraday's law of magnetism.
c) Check whether Eq. (3.10.2) is compatible with Gauss' second law of magnetism.
d) Although existence of ϕ and $\boldsymbol{A}$ is guaranteed by the results of b) and c) using Helmholtz's theorem, Eq. (3.10.2) does not define uniquely the magnetic (vector) potential – since curl-free components may be added thereto, without changing the magnetic field; one degree of freedom is thus available when choosing $\boldsymbol{A}$ – a realization termed gauge-invariance. Use Lorentz's gauge – according to which $\boldsymbol{A}$ is chosen so as to satisfy

$$\nabla\cdot\boldsymbol{A}+\frac{1}{c^2}\frac{\partial\phi}{\partial t}=0, \tag{3.10.3}$$

to obtain compact, similar forms for the remaining two fundamental laws of electromagnetism.

Solution

a) Insertion of Eqs. (3.10.1) and (3.10.2) transforms Eq. (3.238) to

$$\boldsymbol{F}=q\left(-\nabla\phi-\frac{\partial\boldsymbol{A}}{\partial t}+\boldsymbol{v}\times(\nabla\times\boldsymbol{A})\right), \tag{3.10.4}$$

where Lagrange's formula permits expansion of the triple vector product as

$$\boldsymbol{F}=q\left(-\nabla\phi-\frac{\partial\boldsymbol{A}}{\partial t}+\nabla(\boldsymbol{v}\cdot\boldsymbol{A})-(\boldsymbol{v}\cdot\nabla)\boldsymbol{A}\right). \tag{3.10.5}$$

On the other hand, the chain differentiation rule has it that

$$\frac{d\boldsymbol{A}}{dt}=\frac{\partial\boldsymbol{A}}{\partial t}+\left(\frac{\partial\boldsymbol{A}}{\partial x}\frac{dx}{dt}+\frac{\partial\boldsymbol{A}}{\partial y}\frac{dy}{dt}+\frac{\partial\boldsymbol{A}}{\partial z}\frac{dz}{dt}\right), \tag{3.10.6}$$

should $\boldsymbol{A}\equiv\boldsymbol{A}\{x,y,z,t\}$ – where definition of velocity vector and rectangular components thereof (i.e. v_x, v_y, and v_z), operator nabla, and scalar product allow simplification to

$$\frac{d\boldsymbol{A}}{dt}=\frac{\partial\boldsymbol{A}}{\partial t}+\left(v_x\frac{\partial\boldsymbol{A}}{\partial x}+v_y\frac{\partial\boldsymbol{A}}{\partial y}+v_z\frac{\partial\boldsymbol{A}}{\partial z}\right)=\frac{\partial\boldsymbol{A}}{\partial t}+(\boldsymbol{v}\cdot\nabla)\boldsymbol{A}; \tag{3.10.7}$$

isolation of the partial derivative with regard to time then unfolds

$$\frac{\partial \boldsymbol{A}}{\partial t}=\frac{d\boldsymbol{A}}{dt}-(\boldsymbol{v}\cdot\nabla)\boldsymbol{A}. \quad (3.10.8)$$

Insertion of Eq. (3.10.8) transforms Eq. (3.10.5) to

$$\boldsymbol{F}=q\left(-\nabla\phi-\frac{d\boldsymbol{A}}{dt}+(\boldsymbol{v}\cdot\nabla)\boldsymbol{A}+\nabla(\boldsymbol{v}\cdot\boldsymbol{A})-(\boldsymbol{v}\cdot\nabla)\boldsymbol{A}\right), \quad (3.10.9)$$

where $(\boldsymbol{v}\cdot\nabla)\boldsymbol{A}$ and its negative cancel out to give

$$\boldsymbol{F}=q\left(-\nabla\phi-\frac{d\boldsymbol{A}}{dt}+\nabla(\boldsymbol{v}\cdot\boldsymbol{A})\right); \quad (3.10.10)$$

a final factoring out of –1, and of the spatial differential operator afterward yield

$$\boldsymbol{F}=-q\left(\nabla(\phi-\boldsymbol{v}\cdot\boldsymbol{A})+\frac{d\boldsymbol{A}}{dt}\right) \quad (3.10.11)$$

as simplest form for the relationship sought.

b) Equation (3.10.1) supports

$$\nabla\times\boldsymbol{E}=\nabla\times\left(-\nabla\phi-\frac{\partial\boldsymbol{A}}{\partial t}\right)=-\left(\nabla\times(\nabla\phi)+\nabla\times\frac{\partial\boldsymbol{A}}{\partial t}\right)=-\left((\nabla\times\nabla)\phi+\nabla\times\frac{\partial\boldsymbol{A}}{\partial t}\right) \quad (3.10.12)$$

upon application of the curl operator to the electric field, where the distributive and associative properties of the vector product were taken advantage of. Definition of the latter supports, in turn,

$$\nabla\times\nabla\equiv\begin{vmatrix}\boldsymbol{i}_x & \boldsymbol{i}_y & \boldsymbol{i}_z\\ \frac{\partial}{\partial x} & \frac{\partial}{\partial y} & \frac{\partial}{\partial z}\\ \frac{\partial}{\partial x} & \frac{\partial}{\partial y} & \frac{\partial}{\partial z}\end{vmatrix}, \quad (3.10.13)$$

with $\boldsymbol{i}_x$, $\boldsymbol{i}_y$, and $\boldsymbol{i}_z$ denoting unit vectors in the x-, y-, and z-direction, respectively; Eq. (3.10.13) readily produces

$$\nabla\times\nabla=\boldsymbol{0}, \quad (3.10.14)$$

because the determinant of a matrix possessing two identical rows is nil. Equation (3.10.14) permits simplification of Eq. (3.10.12) to

$$\nabla\times\boldsymbol{E}=-\left(\boldsymbol{0}\phi+\nabla\times\frac{\partial\boldsymbol{A}}{\partial t}\right)=-\left(\boldsymbol{0}+\nabla\times\frac{\partial\boldsymbol{A}}{\partial t}\right)=-\nabla\times\frac{\partial\boldsymbol{A}}{\partial t}; \quad (3.10.15)$$

moreover, the time derivative being independent of space derivatives permits transformation of the second term in Eq. (3.10.15) to

$$\nabla\times\boldsymbol{E}=-\frac{\partial}{\partial t}(\nabla\times\boldsymbol{A}) \quad (3.10.16)$$

– so insertion of Eq. (3.10.2) finally yields

$$\nabla\times\boldsymbol{E}=-\frac{\partial\boldsymbol{B}}{\partial t}, \quad (3.10.17)$$

which permits retrieval of Eq. (2.379) of *Food Proc. Eng.: thermal & chemical operations* (as intended).

c) Equation (3.10.2) allows one to write

$$\nabla\cdot\boldsymbol{B}=\nabla\cdot(\nabla\times\boldsymbol{A}), \quad (3.10.18)$$

encompassing the divergent of the magnetic field; since vector $\nabla\times\boldsymbol{A}$ is normal to both vectors ∇ and $\boldsymbol{A}$ as per definition of vector product, the cosine of the angle formed thereby with ∇ is nil – so the defining algorithm of the scalar product implies that the right-hand side of Eq. (3.10.18) is nil as well, i.e.

$$\nabla\cdot\boldsymbol{B}=0, \quad (3.10.19)$$

– thus confirming validity of Eq. (2.375) of *Food Proc. Eng.: thermal & chemical operations.*

d) Combination with Eq. (2.370) transforms Eq. (2.364) of *Food Proc. Eng.: thermal & chemical operations* to

$$\nabla\cdot(\varepsilon\boldsymbol{E})=\rho_V, \quad (3.10.20)$$

whereas combination with Eqs. (2.370) and (2.376) turns Eq. (2.381) – all also from *Food Proc. Eng.: thermal & chemical operations*, to

$$\nabla\times\frac{\boldsymbol{B}}{\mu}=\frac{\partial}{\partial t}(\varepsilon\boldsymbol{E})+\boldsymbol{J}; \quad (3.10.21)$$

upon division of both sides by (constant) ε, Eq. (3.10.20) becomes

$$\nabla\cdot\boldsymbol{E}=\frac{\rho_V}{\varepsilon}, \quad (3.10.22)$$

or else

$$\nabla\cdot\left(\nabla\phi+\frac{\partial\boldsymbol{A}}{\partial t}\right)=-\frac{\rho_V}{\varepsilon} \quad (3.10.23)$$

once Eq. (3.10.1) is inserted and negatives taken of both sides. Upon multiplying both sides by μ and factoring out ε afterward for its constancy, Eq. (3.10.21) yields

$$\nabla\times\boldsymbol{B}=\mu\varepsilon\frac{\partial\boldsymbol{E}}{\partial t}+\mu\boldsymbol{J}, \quad (3.10.24)$$

which further becomes

$$\nabla \times (\nabla \times \boldsymbol{A}) = \mu\varepsilon \frac{\partial}{\partial t}\left(-\nabla \phi - \frac{\partial \boldsymbol{A}}{\partial t}\right) + \mu \boldsymbol{J} \quad (3.10.25)$$

– with $\boldsymbol{E}$ and $\boldsymbol{B}$ eliminated via Eqs. (3.10.1) and (3.10.2), respectively; after isolating $\mu \boldsymbol{J}$ and recalling Eq. (2.443) of *Food Proc. Eng.: thermal & chemical operations*, one can simplify Eq. (3.10.25) to

$$\nabla \times (\nabla \times \boldsymbol{A}) + \frac{1}{c^2}\frac{\partial}{\partial t}\left(\nabla \phi + \frac{\partial \boldsymbol{A}}{\partial t}\right) = \mu \boldsymbol{J}. \quad (3.10.26)$$

The distributive property of the scalar product can now be utilized to write

$$\nabla \cdot \nabla \phi + \nabla \cdot \frac{\partial \boldsymbol{A}}{\partial t} = -\frac{\rho_V}{\varepsilon} \quad (3.10.27)$$

based on Eq. (3.10.23), where a shorter notation is possible, i.e.

$$\nabla^2 \phi + \frac{\partial}{\partial t}(\nabla \cdot \boldsymbol{A}) = -\frac{\rho_V}{\varepsilon} \quad (3.10.28)$$

– with exchange of ∇ and $\partial/\partial t$ differential operators justified for their being mutually independent; Eq. (3.10.3) then permits transformation of Eq. (3.10.28) to

$$\nabla^2 \phi + \frac{\partial}{\partial t}\left(-\frac{1}{c^2}\frac{\partial \phi}{\partial t}\right) = -\frac{\rho_V}{\varepsilon}, \quad (3.10.29)$$

where $-c^2$ may be taken off the differential operator as

$$\nabla^2 \phi - \frac{1}{c^2}\frac{\partial^2 \phi}{\partial t^2} = -\frac{\rho_V}{\varepsilon} \quad (3.10.30)$$

– thus giving rise to the potential-based form of first Gauss' law of electromagnetism. The linearity of the differential operator $\partial/\partial t$ will, in turn, be taken advantage of to obtain

$$\nabla \times (\nabla \times \boldsymbol{A}) + \frac{1}{c^2}\left(\frac{\partial}{\partial t}\nabla \phi + \frac{\partial}{\partial t}\left(\frac{\partial \boldsymbol{A}}{\partial t}\right)\right) = \mu \boldsymbol{J} \quad (3.10.31)$$

from Eq. (3.10.26), where the triple product can be expanded as per Lagrange's formula, and the definition of second-order derivative can be recalled to write

$$\nabla(\nabla \cdot \boldsymbol{A}) - (\nabla \cdot \nabla)\boldsymbol{A} + \frac{1}{c^2}\left(\frac{\partial}{\partial t}(\nabla \phi) + \frac{\partial^2 \boldsymbol{A}}{\partial t^2}\right) = \mu \boldsymbol{J}; \quad (3.10.32)$$

insertion of Eq. (3.10.3) and simplification of notation in the composite differential operator then unfold

$$\nabla\left(-\frac{1}{c^2}\frac{\partial \phi}{\partial t}\right) - \nabla^2 \boldsymbol{A} + \frac{1}{c^2}\nabla \frac{\partial \phi}{\partial t} + \frac{1}{c^2}\frac{\partial^2 \boldsymbol{A}}{\partial t^2} = \mu \boldsymbol{J}, \quad (3.10.33)$$

again after swapping ∇ and $\partial/\partial t$ – or else

$$-\frac{1}{c^2}\nabla \frac{\partial \phi}{\partial t} - \nabla^2 \boldsymbol{A} + \frac{1}{c^2}\nabla \frac{\partial \phi}{\partial t} + \frac{1}{c^2}\frac{\partial^2 \boldsymbol{A}}{\partial t^2} = \mu \boldsymbol{J} \quad (3.10.34)$$

upon elimination of parenthesis, since c is constant. Cancelling out of symmetrical terms and taking negatives of both sides finally produce

$$\nabla^2 \boldsymbol{A} - \frac{1}{c^2}\frac{\partial^2 \boldsymbol{A}}{\partial t^2} = -\mu \boldsymbol{J} \quad (3.10.35)$$

as compact form of Maxwell and Ampère's law, in terms of electric and magnetic potentials. In operator form, it is still possible to rewrite Eqs. (3.10.30) and (3.10.35) as

$$\left(\nabla^2 - \frac{1}{c^2}\frac{\partial^2}{\partial t^2}\right)\phi = -\frac{\rho_V}{\varepsilon} \quad (3.10.36)$$

and

$$\left(\nabla^2 - \frac{1}{c^2}\frac{\partial^2}{\partial t^2}\right)\boldsymbol{A} = -\mu \boldsymbol{J}, \quad (3.10.37)$$

respectively. Note the analogy between Eqs. (3.10.36) and (3.10.37), which encompass a common differential operator; besides the similarity of treatment of time and spatial derivatives (i.e. via second-order derivatives) – thus justifying the success of Lorentz's gauge as compared to, namely, Coulomb's gauge.

The electric motors of interest in practice take advantage of an electric current of some given intensity, passed through a conductor freely suspended in a fixed magnetic field characterized by flux $\boldsymbol{B}$; this creates a force $\boldsymbol{F}$ that causes movement of the conductor through the field – as illustrated in Fig. 3.45i. The aforementioned electric current consists of a stream of (freely) moving charges with velocity $\boldsymbol{v}$ – so a continuous formulation is in order; such a magnetic field will exert Laplace's force thereon, of an infinitesimal magnitude $d\boldsymbol{F}$, according to

$$d\boldsymbol{F} = (\boldsymbol{v} \times \boldsymbol{B})dq \quad (3.243)$$

– based on Eq. (3.238), but restricted to the magnetic component of the force. In the case of electrons moving along a wire, the said charge is characterized by some linear density λ, so an elementary portion of charge dq will satisfy

$$dq = \lambda dl; \quad (3.244)$$

insertion of Eq. (3.244) then transforms Eq. (3.243) to

$$d\boldsymbol{F} = (\boldsymbol{v} \times \boldsymbol{B})\lambda dl. \quad (3.245)$$

A vector intensity of electric current, $\boldsymbol{I}$, may accordingly be defined as

$$\boxed{\boldsymbol{I} \equiv \lambda \boldsymbol{v},} \quad (3.246)$$

since intensity is characterized by both a magnitude and a direction; therefore, Eq. (3.245) can be rewritten as

$$d\boldsymbol{F} = \lambda(\boldsymbol{v} \times \boldsymbol{B})dl = (\lambda \boldsymbol{v}) \times \boldsymbol{B}dl = \boldsymbol{I} \times \boldsymbol{B}dl. \quad (3.247)$$

Integration of both sides of Eq. (3.247) – resorting to

$$\boldsymbol{F}\big|_{l=0} = \boldsymbol{0} \quad (3.248)$$

and

$$\boldsymbol{F}\big|_{l=L} = \boldsymbol{F}_L \quad (3.249)$$

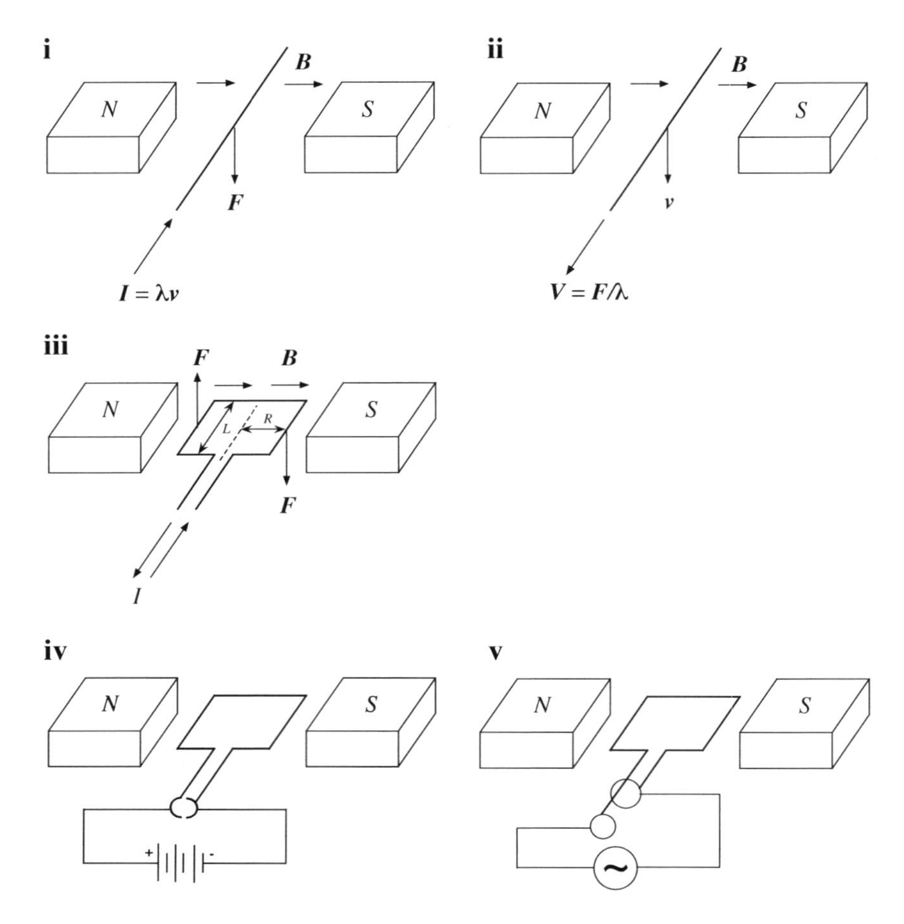

FIGURE 3.45 Schematic representation of: relationship of magnetic flux, $\boldsymbol{B}$, arising between two (fixed) electromagnet poles, N and S, to (i) intensity of current, $\boldsymbol{I}$, in and force, $\boldsymbol{F}$, on linear conductor as part of an electric motor, and (ii) velocity, $\boldsymbol{v}$, of and voltage difference, $\boldsymbol{V}$, in linear conductor as part of an electric generator; (iii) relationship between $\boldsymbol{B}$, I, and $\boldsymbol{F}$ in looped conductor, with length L and radius R; and detail of connection of looped, rotating conductor to (iv) DC or (v) AC source.

as boundary conditions, looks like

$$\int_0^{F_L} d\boldsymbol{F} = \int_0^L \boldsymbol{I} \times \boldsymbol{B} dl \tag{3.250}$$

– which breaks down to

$$\int_0^{F_L} d\boldsymbol{F} = \boldsymbol{I} \times \boldsymbol{B} \int_0^L dl, \tag{3.251}$$

should $\boldsymbol{I}$ and $\boldsymbol{B}$ remain constant along the l-coordinate; $\boldsymbol{F}_L$ represents here the macroscopic force, gradually accumulated by all moving charges along wire length L. Trivial application of the fundamental theorem of integral calculus permits transformation of Eq. (3.251) to

$$\boldsymbol{F}\Big|_0^{F_L} = \boldsymbol{I} \times \boldsymbol{B} l\Big|_0^L, \tag{3.252}$$

or else

$$\boldsymbol{F}_L = \boldsymbol{I} \times \boldsymbol{B} L = L\boldsymbol{I} \times \boldsymbol{B}; \tag{3.253}$$

a more general form reads

$$\boldsymbol{F} = \boldsymbol{J} \times \boldsymbol{B}, \tag{3.254}$$

applicable to any geometry – where $\boldsymbol{J}$ denotes current density (which reduces to $L\boldsymbol{I}$ in the linear case). For simplicity of notation, one often resorts to the scalar version of Eq. (3.253), i.e.

$$\boxed{F = BLI} \tag{3.255}$$

– knowing that the force, of magnitude F, applies normally to the magnetic field, of magnitude B and felt along a length L, in the presence of an electric current intensity of magnitude I, see Fig. 3.45i.

On the other hand, division of both sides of Eq. (3.243) by dq produces

$$\frac{d\boldsymbol{F}}{dq} = \boldsymbol{v} \times \boldsymbol{B}, \tag{3.256}$$

and combination with Eq. (3.244) gives rise, in turn, to

$$\frac{d\boldsymbol{F}}{\lambda dl} = \boldsymbol{v} \times \boldsymbol{B}; \tag{3.257}$$

integration of Eq. (3.257) ensues via separation of variables, viz.

$$\int_0^{F_L} \frac{d\boldsymbol{F}}{\lambda} = \int_0^L \boldsymbol{v} \times \boldsymbol{B} dl, \tag{3.258}$$

at the expense again of Eqs. (3.248) and (3.249). For constant λ and $\boldsymbol{v}\times\boldsymbol{B}$, Eq. (3.258) simplifies to

$$\frac{1}{\lambda}\int_{\mathbf{0}}^{\boldsymbol{F}_L} d\boldsymbol{F} = \boldsymbol{v}\times\boldsymbol{B}\int_0^L dl, \tag{3.259}$$

where the fundamental theorem of integral calculus may again be invoked to write

$$\frac{1}{\lambda}\boldsymbol{F}\Big|_{\mathbf{0}}^{\boldsymbol{F}_L} = \boldsymbol{v}\times\boldsymbol{B}\,l\Big|_0^L; \tag{3.260}$$

Eq. (3.260) ends up as

$$\boldsymbol{V} = \boldsymbol{v}\times\boldsymbol{B}L = L\boldsymbol{v}\times\boldsymbol{B} \tag{3.261}$$

– since $\boldsymbol{F}_L/\lambda$ coincides with electromotive force, *emf*, or electric potential difference, *V*, when electric current is marginally small in intensity, i.e.

$$\boxed{\boldsymbol{V} \equiv \frac{\boldsymbol{F}}{\lambda}.} \tag{3.262}$$

It should be emphasizes that *emf* is constant, and refers to energy provided by some generator per unit charge passing through it – whereas *V* is variable, and denotes amount of energy that the said unit charge uses to pass through. Therefore, $emf = V + \mathcal{R}I \geq V$ because some because some amount of energy is lost in overcoming the internal resistance of the source, $\mathcal{R}$; and because electrons possess extra kinetic energy when previously accelerated to flow as a current (further to kinetic energy associated to their chaotic movement, constant at a given temperature) – which increases with current intensity, *I*. As before, Eq. (3.261) normally takes the form

$$\boxed{V = BLv,} \tag{3.263}$$

applicable to linear geometries; the underlying situation is illustrated in Fig. 3.45ii, where *L* again denotes length along which the germane field is felt. Equations (3.255) and (3.263) constitute the set of fundamental equations describing transfer of power between regular electric and mechanical systems (as in electric motors), or the reverse (as in electric generators), respectively. The said interaction is two-way, i.e. if electric current generates mechanical force, then mechanical velocity generates (back) voltage difference; interaction is also power-continuous, i.e. power is transferred in full from one domain to another – with no power dissipation, or energy somehow stored for that matter. In fact, if Eq. (2.93) of *Food Proc. Eng.: thermal & chemical operations*, pertaining to electrical power, is revisited as

$$\mathcal{P} = I(BLv) \tag{3.264}$$

with the aid of Eq. (3.263), then the associative and commutative properties of product of scalars support reformulation to

$$\mathcal{P} = (BLI)v; \tag{3.265}$$

in view of Eq. (3.255), one is able to redo Eq. (3.265) to

$$\mathcal{P} = Fv \tag{3.266}$$

as scalar version of Eq. (3.239) – the latter applying in general, even if $\boldsymbol{F}$ and $\boldsymbol{v}$ do not share the same direction.

To make the above electromagnetic phenomena usable, the linear wire undergoing a force should be reshaped to a loop, and rotation around an axis be permitted – so as to generate a torque, as highlighted in Fig. 3.45iii. If *R* denotes distance between rotation axis and wire, then the associated torque, $\mathcal{T}$, reads

$$\mathcal{T} \equiv RF, \tag{3.267}$$

and Eq. (3.255) may come aboard to generate

$$\boxed{\mathcal{T} = RBLI} \tag{3.268}$$

– where *L* still denotes length of loop. In practice, coiled wires are used in multiple turns, say *N*, so current intensity in each one has to be multiplied by *N* to get overall current intensity; for a rotation frequency, Ω, the power delivered by an electric motor will thus be given by

$$\boxed{\mathcal{P} = NRBLI\Omega}, \tag{3.269}$$

or $N\mathcal{T}\Omega$ – with $\mathcal{T}$ given by Eq. (3.268), and Ω obviously equal to $\omega/2\pi$ should ω denote angular frequency. When $I\rightarrow 0$, Eq. (3.269) also permits calculation of electromotive force, ε, as $\mathcal{P}/I$, according to

$$\varepsilon = NRLB\Omega; \tag{3.270}$$

this is consistent with Eq. (2.93) in *Food Proc. Eng.: thermal & chemical operations*, since $\varepsilon = V$ under these circumstances. If DC is used, then the polarity of the current is to be changed at each half-turn – otherwise the coil will stop rotating; a slip commuter is utilized for this purpose, as depicted in Fig. 3.45iv. In the case of AC, inversion of polarity occurs on a periodic basis, yet electrical contact is to be provided permanently; this is done via some kind of brush sliding on concentric rings, which isolate the inlet and outlet wires from each other – see Fig. 3.45v.

3.3.10.2 Transformers

To simplify mathematical simulation, one will hereafter focus on an ideal transformer – conceptualized as a theoretical linear, lossless, perfectly coupled apparatus. Perfect coupling implies, in particular, infinitely high core magnetic permeability and winding inductance, as well as nil net magnetomotive force. A varying current in its primary winding accordingly creates a varying magnetic flux in the transformer core, and a varying magnetic field impinging on the secondary winding; the latter induces, in turn, a varying electromotive force (equal to voltage at negligible current intensity, as stressed above) in the secondary winding. If the primary and secondary windings are wrapped around a core of infinitely high magnetic permeability (as hypothesized above), then the whole magnetic flux will pass through both primary and secondary windings.

In attempts to model this type of device, the starting point is definition of Φ as magnetic flux, $\boldsymbol{B}$, integrated over circuit area, *S* – according to

$$\boxed{\Phi \equiv \iint_S \boldsymbol{B}\cdot d\tilde{S}.} \tag{3.271}$$

Consider now a change in magnetic flux, through a conducting loop, in the inertial frame of reference where the circuit is moving; when the loop moves from its initial position l_t to a new position l_{t+dt}, it generates a cylindrical volume in space subjected to magnetic field $\boldsymbol{B}$ – with side surface $\boldsymbol{\Sigma}$, further to two top surfaces, $\boldsymbol{S}\{t\}$ and $\boldsymbol{S}\{t+dt\}$. Therefore, one can write

$$\frac{d\Phi}{dt} \equiv \frac{\Phi\{t+dt\}-\Phi\{t+dt\}}{dt} = \frac{\iint_{S\{t+dt\}} \boldsymbol{B}\{t+dt\}\cdot d\tilde{\boldsymbol{S}} - \iint_{S\{t\}} \boldsymbol{B}\{t\}\cdot d\tilde{\boldsymbol{S}}}{dt} \tag{3.272}$$

based on the definition of derivative, coupled with Eq. (3.271); after noting that

$$\boxed{d\Phi = \left(\frac{\partial \Phi}{\partial t}\right)_S dt + \left(\frac{\partial \Phi}{\partial t}\right)_B dt} \tag{3.273}$$

as imposed by definition of a total differential with two degrees of freedom (i.e. $\boldsymbol{B}$ and $\boldsymbol{S}$), one may split Eq. (3.272) as

$$\frac{d\Phi}{dt} = \frac{\iint_{S\{t\}} \frac{\partial \boldsymbol{B}}{\partial t} dt \cdot d\tilde{\boldsymbol{S}}}{dt} + \frac{\iint_{S\{t+dt\}} \boldsymbol{B}\{t\}\cdot d\tilde{\boldsymbol{S}} - \iint_{S\{t\}} \boldsymbol{B}\{t\}\cdot d\tilde{\boldsymbol{S}}}{dt} \tag{3.274}$$

with the aid of Eq. (3.271) – which ends up separating the effects of t upon the kernel ($\boldsymbol{B}$) and the integration region ($\boldsymbol{S}$), respectively; dt will then be taken off the kernel in the first term of the right-hand side, and subsequently cancelled with dt in denominator to leave

$$\frac{d\Phi}{dt} = \iint_{S\{t\}} \frac{\partial \boldsymbol{B}}{\partial t} \cdot d\tilde{\boldsymbol{S}} + \frac{\iint_{S\{t+dt\}} \boldsymbol{B}\{t\}\cdot d\tilde{\boldsymbol{S}} - \iint_{S\{t\}} \boldsymbol{B}\{t\}\cdot d\tilde{\boldsymbol{S}}}{dt}, \tag{3.275}$$

since t is independent of (dummy) variable of integration $\tilde{\boldsymbol{S}}$. Equation (2.375) of *Food Proc. Eng.: thermal & chemical operations* is now to be revisited, in integrated form, as

$$\iiint_V (\nabla \cdot \boldsymbol{B})\, d\tilde{\mathcal{V}} = 0, \tag{3.276}$$

where $\mathcal{V}$ denotes volume generated by the plane of the loop when it rotates around itself; Gauss' divergence theorem supports transformation of Eq. (3.276) to

$$\iint_{S\{t\}+S\{t+dt\}+\Sigma} (\boldsymbol{B}\cdot \boldsymbol{i}_n)\, d\tilde{S} = 0, \tag{3.277}$$

where all bounding surfaces (in scalar form) were taken into account, i.e. cross sectional surface(s) S and side surface Σ – and where $\boldsymbol{i}_n$ denotes unit vector normal to the surface(s) at each point. After rewriting Eq. (3.277) as

$$\iint_{S\{t+dt\}+S\{t\}+\Sigma} \boldsymbol{B}\cdot d\tilde{\boldsymbol{S}} = 0 \tag{3.278}$$

once $\boldsymbol{i}_n$ is lumped with $\tilde{S}$ to yield $\tilde{\boldsymbol{S}}$, one may split its left-hand side to get

$$\iint_{S\{t+dt\}} \boldsymbol{B}\cdot d\tilde{\boldsymbol{S}} - \iint_{S\{t\}} \boldsymbol{B}\cdot d\tilde{\boldsymbol{S}} + \iint_{\Sigma} \boldsymbol{B}\cdot d\boldsymbol{S} = 0 \tag{3.279}$$

– bearing in mind the similar direction of vectors $\boldsymbol{B}$ and $d\tilde{\boldsymbol{S}}$ when spanning $\boldsymbol{S}\{t\}$ and their opposite direction when spanning $\boldsymbol{S}\{t+dt\}$; this justifies the plus sign preceding $\iint_{S\{t+dt\}} \boldsymbol{B}\cdot d\tilde{\boldsymbol{S}}$, and the minus sign preceding $\iint_{S\{t\}} \boldsymbol{B}\cdot d\tilde{\boldsymbol{S}}$, after the indicated scalar product has been performed. Equation (3.279) promptly gives rise to

$$\iint_{S\{t+dt\}} \boldsymbol{B}\cdot d\tilde{\boldsymbol{S}} - \iint_{S\{t\}} \boldsymbol{B}\cdot d\tilde{\boldsymbol{S}} = -\iint_{\Sigma} \boldsymbol{B}\cdot d\boldsymbol{S}, \tag{3.280}$$

after isolation of $-\iint_{\Sigma} \boldsymbol{B}\cdot d\boldsymbol{S}$; in view of Eq. (3.280), one can reduce Eq. (3.275) to

$$\frac{d\Phi}{dt} = \iint_{S\{t\}} \frac{\partial \boldsymbol{B}}{\partial t} \cdot d\tilde{\boldsymbol{S}} - \frac{\iint_{\Sigma} \boldsymbol{B}\cdot d\tilde{\boldsymbol{S}}}{dt}. \tag{3.281}$$

On the other hand, $d\boldsymbol{S}$ pertaining to side surface $\boldsymbol{\Sigma}$ may be rewritten as $d\tilde{\boldsymbol{l}} \times d\tilde{\boldsymbol{r}}$, so one may write

$$\iint_{\Sigma} \boldsymbol{B}\cdot d\tilde{\boldsymbol{S}} = \int_l \int_r \boldsymbol{B}\cdot \left(d\tilde{\boldsymbol{l}} \times d\tilde{\boldsymbol{r}}\right) \tag{3.282}$$

– where $\tilde{\boldsymbol{l}}$ denotes (vector) coordinate along the loop, while $\boldsymbol{l}$ denotes (vector) length of the said loop as (closed) contour of $\boldsymbol{\Sigma}$; $\tilde{\boldsymbol{r}}$ denotes, in turn, (vector) coordinate accompanying loop displacement, at velocity $\boldsymbol{v}$, knowing that

$$\boldsymbol{v} \equiv \frac{d\boldsymbol{r}}{dt}. \tag{3.283}$$

Following multiplication and division of the kernel of its right-hand side integral by dt, Eq. (3.282) becomes

$$\iint_{\Sigma} \boldsymbol{B}\cdot d\tilde{\boldsymbol{S}} = \int_l \int_r \boldsymbol{B}\cdot \left(d\tilde{\boldsymbol{l}} \times \frac{d\tilde{\boldsymbol{r}}}{dt} dt\right) = \oint_l \boldsymbol{B}\cdot \left(d\tilde{\boldsymbol{l}} \times \boldsymbol{v}\{\tilde{\boldsymbol{l}}\} dt\right) \tag{3.284}$$

with the aid of Eq. (3.283) – where dt will again be taken off the kernel as

$$\iint_{\Sigma} \boldsymbol{B}\cdot d\tilde{\boldsymbol{S}} = dt \oint_l \boldsymbol{B}\cdot (d\tilde{\boldsymbol{l}} \times \boldsymbol{v}); \tag{3.285}$$

hence, the original surface integral, $\iint_{\Sigma}$, has been transformed to a path (or line) integral along the wire loop, $\oint_l$. Since the scalar/vector triple product is invariant under a circular shift of its three operands, it is possible to rewrite Eq. (3.285) as

$$\iint_{\Sigma} \boldsymbol{B}\cdot d\tilde{\boldsymbol{S}} = dt \oint_l \boldsymbol{v}\cdot (\boldsymbol{B} \times d\tilde{\boldsymbol{l}}) = dt \oint_l d\tilde{\boldsymbol{l}} \cdot (\boldsymbol{v} \times \boldsymbol{B}), \tag{3.286}$$

where commutativity of the scalar product justifies, in turn,

$$\boxed{\iint_{\Sigma} \boldsymbol{B}\cdot d\tilde{\boldsymbol{S}} = dt \oint_l (\boldsymbol{v} \times \boldsymbol{B})\cdot d\tilde{\boldsymbol{l}};} \tag{3.287}$$

insertion of Eq. (3.287) converts Eq. (3.281) to

$$\frac{d\Phi}{dt} = \iint\limits_{S\{t\}} \frac{\partial \boldsymbol{B}}{\partial t} \cdot d\tilde{\boldsymbol{S}} - \frac{dt \oint_l (\boldsymbol{v} \times \boldsymbol{B}) \cdot d\tilde{\boldsymbol{l}}}{dt}, \tag{3.288}$$

which breaks down to

$$\frac{d\Phi}{dt} = \iint\limits_{S\{t\}} \frac{\partial \boldsymbol{B}}{\partial t} \cdot d\tilde{\boldsymbol{S}} - \oint\limits_l (\boldsymbol{v} \times \boldsymbol{B}) \cdot d\tilde{\boldsymbol{l}} \tag{3.289}$$

after dropping dt from both numerator and denominator. Equation (2.379) of *Food Proc. Eng.: thermal & chemical operations* can now be retrieved, upon integration over bounding surface $\boldsymbol{S}$, as

$$\iint\limits_{S\{t\}} (\nabla \times \boldsymbol{E}) \cdot d\tilde{\boldsymbol{S}} = -\iint\limits_{S\{t\}} \frac{\partial \boldsymbol{B}}{\partial t} \cdot d\tilde{\boldsymbol{S}}, \tag{3.290}$$

thus giving room to apply Kelvin and Stokes' theorem to the left-hand side to obtain

$$\boxed{\iint\limits_{S\{t\}} \frac{\partial \boldsymbol{B}}{\partial t} \cdot d\tilde{\boldsymbol{S}} = -\oint\limits_l \boldsymbol{E} \cdot d\tilde{\boldsymbol{l}};} \tag{3.291}$$

insertion of Eq. (3.291) then converts Eq. (3.289) to

$$\boxed{\frac{d\Phi}{dt} = -\oint\limits_l \boldsymbol{E} \cdot d\tilde{\boldsymbol{l}} - \oint\limits_l (\boldsymbol{v} \times \boldsymbol{B}) \cdot d\tilde{\boldsymbol{l}}.} \tag{3.292}$$

Comparative inspection of the right-hand sides of Eqs. (3.273) and (3.292) unfolds

$$\left(\frac{\partial \Phi}{\partial t}\right)_S \equiv \left(\frac{\partial \Phi}{\partial t}\right)_{\boldsymbol{v}=\boldsymbol{0}} = -\oint\limits_l \boldsymbol{E} \cdot d\tilde{\boldsymbol{l}}, \tag{3.293}$$

complemented by

$$\left(\frac{\partial \Phi}{\partial t}\right)_{\boldsymbol{B}} = -\oint\limits_l (\boldsymbol{v} \times \boldsymbol{B}) \cdot d\tilde{\boldsymbol{l}}. \tag{3.294}$$

If the line integrals, sharing the same integration region, are lumped with each other, then Eq. (3.292) will appear as

$$\frac{d\Phi}{dt} = -\oint\limits_l (\boldsymbol{E} + \boldsymbol{v} \times \boldsymbol{B}) \cdot d\tilde{\boldsymbol{l}}, \tag{3.295}$$

which gives rise to

$$\frac{d\Phi}{dt} = -\oint\limits_l \frac{\boldsymbol{F}}{q} \cdot d\tilde{\boldsymbol{l}} \tag{3.296}$$

once Lorentz's law is brought on board, see Eq. (3.238); note that Eq. (3.296) may be rewritten, in scalar form, as

$$\frac{d\Phi}{dt} = -\int_0^l \frac{F}{q}\, d\tilde{l} \tag{3.297}$$

upon leaving the fixed system of axes to ascertain coordinates, and instead adopting a moving axis along the loop itself. Equation (3.297) is equivalent to

$$\frac{d\Phi}{dt} = -\left(\int_0^\infty \frac{F}{q}\, dl + \int_\infty^l \frac{F}{q}\, d\tilde{l}\right), \tag{3.298}$$

since definite are additive if they share the same kernel, with the upper limit of integration of one coincident with the lower limit of the other; one may further transform Eq. (3.298) to

$$\frac{d\Phi}{dt} = \int_\infty^0 \frac{F}{q}\, dl - \int_\infty^l \frac{F}{q}\, d\tilde{l} = -\int_\infty^l \frac{F}{q}\, d\tilde{l} - \left(-\int_\infty^0 \frac{F}{q}\, dl\right) \tag{3.299}$$

– after taking in the minus sign, exchanging the limits of integration of the first integral at the expense of swapping the preceding sign, an splitting 1 as (–1)(–1). In view of Eqs. (2.82) and (2.83) from *Food Proc. Eng.: thermal & chemical operations*, reformulation of Eq. (3.299) is in order, viz.

$$\frac{d\Phi}{dt} = V\big|_l - V\big|_0 = \Delta V \tag{3.300}$$

– or, equivalently,

$$\boxed{\varepsilon \equiv \varepsilon_{transformer} + \varepsilon_{motional} = -\frac{d\Phi}{dt},} \tag{3.301}$$

since $\varepsilon = \Delta V|_{l \to 0}$ as emphasized before; here ε denotes electromotive force induced in a circuit, so it holds a sign opposite to that of ΔV (as a consequence of energy received or energy lost, respectively, per unit charge). Equation (3.301) has classically been known as Faraday's law of electromagnetic induction; consistency with Eqs. (3.273), (3.293), and (3.294) implies

$$\varepsilon_{transformer} \equiv -\left(\frac{\partial \Phi}{\partial t}\right)_{\boldsymbol{v}=\boldsymbol{0}} \tag{3.302}$$

and

$$\varepsilon_{motional} \equiv -\left(\frac{\partial \Phi}{\partial t}\right)_{\boldsymbol{B}}, \tag{3.303}$$

respectively – where Eq. (3.302) accounts for the so-called volta-electric induction, while Eq. (3.303) describes magneto-electric induction. Finally, it should be noted that Eq. (2.379) in *Food Proc. Eng.: thermal & chemical operations* is indeed a generalization of Eq. (3.301) – stating that a time-varying magnetic field always accompanies a spatially varying (nonconservative) electric field, and vice versa.

Once in possession of Eq. (3.301), one may write

$$\varepsilon_P = -N_P \frac{d\phi_P}{dt} \tag{3.304}$$

pertaining to the primary winding of an ideal transformer – where ε_P denotes global electromotive force, ϕ_P denotes magnetic flux per loop, and N_P denotes number of loops per unit area; by the same token,

$$\varepsilon_S = -N_S \frac{d\phi_S}{dt} \tag{3.305}$$

describes the secondary winding – with the corresponding global electromotive force, magnetic flux per loop, and number of loops per unit area denoted by ε_S, ϕ_S, and N_S, respectively. As emphasized previously, the same magnetic flux passes through the primary and secondary windings of an ideal transformer, so

$$\phi_P = \phi_S \tag{3.306}$$

that readily implies

$$\frac{d\phi_P}{dt} = \frac{d\phi_S}{dt} \tag{3.307}$$

following differentiation of both sides with regard to t; one may thus eliminate $d\phi_P/dt$ and $d\phi_S/dt$ between Eqs. (3.304) and (3.305) to get

$$-\frac{\varepsilon_P}{N_P} = -\frac{\varepsilon_S}{N_S} \tag{3.308}$$

or, in a more useful form,

$$\boxed{\frac{\Delta V_P}{\Delta V_S} = \frac{N_P}{N_S}} \tag{3.309}$$

– with the aid of Eq. (3.300) and Eq. (3.301), under negligible current intensity. Equation (3.309) entails one of the fundamental equations for an ideal transformer; in particular, it enforces $N_P/N_S > 1$ in step-down transformers, and likewise $N_P/N_S < 1$ in step-up transformers.

On the other hand, the law of conservation of energy applied to the windings of a transformer has it that

$$\mathcal{P}_P = \mathcal{P}_S, \tag{3.310}$$

as long as it is ideal – so one can retrieve Eq. (2.93) from *Food Proc. Eng.: thermal & chemical operations* to state that

$$I_P \Delta V_P = I_S \Delta V_S; \tag{3.311}$$

Eq. (3.311) will appear as

$$\frac{I_S}{I_P} = \frac{\Delta V_P}{\Delta V_S}, \tag{3.312}$$

upon division of both sides by $I_P \Delta V_S$. Insertion of Eq. (3.309) converts Eq. (3.312) to

$$\boxed{\frac{I_S}{I_P} = \frac{N_P}{N_S}}, \tag{3.313}$$

so current intensity is inversely proportional to number of loops; Eq. (3.313) completes the set of fundamental equations describing a transformer.

PROBLEM 3.11

Electrical impedance, $\mathcal{Z}$, measures the opposition offered by a circuit to flow of an electric current, when a voltage difference is applied thereto. It extends the concept of resistance, $\mathcal{R}$, to AC circuits – and is characterized by both magnitude and phase, i.e.

$$\mathcal{Z} = \mathcal{R} + \iota\mathcal{X} \tag{3.11.1}$$

in Cartesian complex form, or

$$\mathcal{Z} = |\mathcal{Z}| e^{\iota \arg \mathcal{Z}} \tag{3.11.2}$$

in polar (or Euler's) complex form; here $\mathcal{X}$ denotes reactance – with $|\mathcal{Z}| \equiv \sqrt{\mathcal{R}^2 + \mathcal{X}^2}$ and $\arg \mathcal{Z} \equiv \tan^{-1} \mathcal{X}/\mathcal{R}$.

a) Extend Ohm's law so as to obtain an expression for $\mathcal{Z}$ as a function of voltage difference applied, $\Delta V\{t\}$, and resulting current intensity, $I\{t\}$.
b) Calculate the apparent primary load impedance, given the impedance on the secondary load, and the number of loops in primary and secondary windings, for an ideal electric transformer.
c) In the case of a resistor the relationship between voltage difference and current intensity reads

$$\Delta V = \mathcal{R}I \tag{3.11.3}$$

– while an induction coil designed to store energy in magnetic form, and characterized by $\mathcal{I}$ as inductance, abides to

$$\Delta V = \mathcal{I}\frac{dI}{dt}; \tag{3.11.4}$$

for a capacitor designed to store energy in electrical form, and characterized by capacitance $\mathcal{C}$, the underlying relationship looks like

$$I = \mathcal{C}\frac{d\Delta V}{dt}. \tag{3.11.5}$$

Should these three elements be connected in series to produce a so-called tank circuit, calculate the overall impedance as a function of angular rate of oscillation of the AC voltage diference applied. Remember that, for a system described by transfer function $G\{s\}$ in Laplace's domain, the ratio of amplitude of the sustained (sinusoidal) response to the (applied) sinusoidal disturbance, characterized by angular frequency, ω (in rad.s^{-1}), equals the modulus of $G\{s\}$ after setting $s = \iota\omega$.

d) Prove that the resonance frequency of circuit in c) is simply the reciprocal of the geometric mean of $\mathcal{I}$ and $\mathcal{C}$.
e) Derive an expression for the root-mean squared voltage, V_{rms}, applied to the primary winding of a transformer – defined as the amount of AC power that produces the same heating effect as an equivalent DC power, for given Ω (in Hz, or s^{-1}) as supply frequency,

N_P as number of turns, and $\phi_{P,pk}$ as peak magnetic flux density per turn.

Solution

a) Due to the rotational mode of operation of regular electric generators, sinusoidal functions of time result for the AC generated – according to

$$\Delta V\{t\} = \Delta V_{\max} \, e^{\iota(\omega t+\phi_V)} \tag{3.11.6}$$

and

$$I\{t\} = I_{\max} \, e^{\iota(\omega t+\phi_I)}, \tag{3.11.7}$$

using Eq. (3.11.2) as template; here $\Delta V_{\max}$ and $I_{\max}$ denote nominal (peak) voltage difference and current intensity, respectively, ω denotes angular frequency of alternate current, and ϕ_V and ϕ_I denote phase shift of voltage difference and current intensity, respectively. The generalized form of Ohm's law should look like

$$Z\{t\} = \frac{\Delta V\{t\}}{I\{t\}}, \tag{3.11.8}$$

using Eq. (2.80) of *Food Proc. Eng.: thermal & chemical operations* as template – where insertion of Eqs. (3.11.6) and (3.11.7) produces

$$Z = \frac{\Delta V_{\max} e^{\iota(\omega t+\phi_V)}}{I_{\max} e^{\iota(\omega t+\phi_I)}}; \tag{3.11.9}$$

the operational rules involving exponential functions then allow one to write

$$Z = \frac{\Delta V_{\max}}{I_{\max}} e^{\iota(\omega t+\phi_V)-\iota(\omega t+\phi_I)} = \frac{\Delta V_{\max}}{I_{\max}} e^{\iota(\phi_V-\phi_I)}, \tag{3.11.10}$$

along with cancellation of symmetrical terms (where present). In other words, impedance is given by resistance, $\mathcal{R} = \Delta V_{\max}/I_{\max}$, as per classical Ohm's law – corrected by factor $e^{\iota(\phi_V-\phi_I)}$ that accounts for phase shift, $\phi_V - \phi_I$.

b) Since a transformer relies on electromagnetic induction and works necessarily with AC, one should resort to Eq. (3.11.8) – viz.

$$Z_S = \frac{\Delta V_S}{I_S} \tag{3.11.11}$$

as applied to its secondary winding, and likewise

$$Z_P = \frac{\Delta V_P}{I_P} \tag{3.11.12}$$

pertaining to its primary winding. Isolation of ΔV_P from Eq. (3.309) unfolds

$$\Delta V_P = \frac{N_P}{N_S}\Delta V_S, \tag{3.11.13}$$

and I_P will similarly be obtained from Eq. (3.313) as

$$I_P = \frac{I_S}{\dfrac{N_P}{N_S}}; \tag{3.11.14}$$

one may now proceed with reformulation of Eq. (3.11.12) to

$$Z_P = \frac{\dfrac{N_P}{N_S}\Delta V_S}{\dfrac{I_S}{\dfrac{N_P}{N_S}}}, \tag{3.11.15}$$

with the aid of Eqs. (3.11.13) and (3.11.14) – which is to be algebraically rearranged to read

$$Z_P = \left(\frac{N_P}{N_S}\right)^2 \frac{\Delta V_S}{I_S}. \tag{3.11.16}$$

Insertion of Eq. (3.11.11) in Eq. (3.11.16) finally gives

$$Z_P = \left(\frac{N_P}{N_S}\right)^2 Z_S; \tag{3.11.17}$$

hence, the apparent load impedance in the primary winding is proportional to that in its secondary counterpart, with the square of the ratio between turns serving as proportionality factor.

c) The differential equation describing response of $I\{t\}$ to $\Delta V\{t\}$ in the proposed circuit is more easily handled in Laplace's domain; toward this goal, one should first obtain the defining equations for each device, i.e.

$$\Delta\overline{V} = \mathcal{R}\overline{I} \tag{3.11.18}$$

based on Eq. (3.11.3) and pertaining to the resistor,

$$\Delta\overline{V} = \mathcal{I}\left(s\overline{I} - I\big|_{t=0}\right) \tag{3.11.19}$$

based on Eq. (3.11.4) and pertaining to the inductor, and

$$\overline{I} = \mathcal{C}\left(s\Delta\overline{V} - \Delta V\big|_{t=0}\right) \tag{3.11.20}$$

based on Eq. (3.11.5) and pertaining to the capacitor. After taking the usual hypothesis

$$I\big|_{t=0} = \Delta V\big|_{t=0} = 0 \tag{3.11.21}$$

applicable prior to closing the circuit, Eq. (3.11.19) can be redone to

$$G_I\{s\} = \frac{1}{\mathcal{I}s}, \tag{3.11.22}$$

and likewise Eq. (3.11.20) to

$$G_C\{s\} = \mathcal{C}s \tag{3.11.23}$$

– besides redoing Eq. (3.11.18) to

$$G_R\{s\} = \frac{1}{\mathcal{R}}; \tag{3.11.24}$$

here G_R, G_I, and G_C denote transfer functions, in Laplace's domain, pertaining to resistor, inductor, and capacitor, respectively – defined, in general, as

$$G\{s\} \equiv \frac{\bar{I}}{\Delta\bar{V}}. \tag{3.11.25}$$

Upon application of Kirchhoff's loop rule, see Eq. (3.315), to the circuit under scrutiny, one gets

$$\Delta V + (-\Delta V_R) + (-\Delta V_I) + (-\Delta V_C) = 0 \tag{3.11.26}$$

– where ΔV denotes voltage difference delivered by the electromotive force device as such, or taken up by each device in the circuit when appended with subscripts R, I, and C referring to resistor, inductor, and capacitor, respectively; remember that voltage difference, associated to every passive device, holds a sign opposite to that provided by the electromotive force supplied by the active device. After applying Laplace's transforms, Eq. (3.11.26) becomes

$$\Delta\bar{V} = \Delta\bar{V}_R + \Delta\bar{V}_I + \Delta\bar{V}_C, \tag{3.11.27}$$

where insertion of Eq. (3.11.25), pertaining to each device as appropriate, gives rise to

$$\Delta\bar{V} = \frac{\bar{I}}{G_R\{s\}} + \frac{\bar{I}}{G_I\{s\}} + \frac{\bar{I}}{G_C\{s\}}; \tag{3.11.28}$$

upon recalling Eqs. (3.11.22)–(3.11.24), complemented with factoring out of $\bar{I}$, it is possible to transform Eq. (3.11.28) to

$$\Delta\bar{V} = \frac{\bar{I}}{\frac{1}{\mathcal{R}}} + \frac{\bar{I}}{\frac{1}{\mathcal{I}s}} + \frac{\bar{I}}{\mathcal{C}s} = \left(\mathcal{R} + \mathcal{I}s + \frac{1}{\mathcal{C}s}\right)\bar{I}, \tag{3.11.29}$$

along with lumping of terms alike; $\bar{I}/\Delta\bar{V}$ may then be singled out as

$$\frac{\bar{I}}{\Delta\bar{V}} = \frac{1}{\mathcal{R} + \mathcal{I}s + \frac{1}{\mathcal{C}s}} = \frac{\mathcal{C}s}{1 + \mathcal{R}\mathcal{C}s + \mathcal{I}\mathcal{C}s^2}, \tag{3.11.30}$$

together with multiplication of both numerator and denominator by $\mathcal{C}s$. In view again of Eq. (3.11.25), one will finally reformulate Eq. (3.11.30) to

$$\frac{1}{G\{s\}} = \frac{1 + \mathcal{R}\mathcal{C}s + \mathcal{I}\mathcal{C}s^2}{\mathcal{C}s}, \tag{3.11.31}$$

where reciprocals of both sides were meanwhile taken for convenience. The information conveyed by the relationship between amplitudes of outlet to inlet sinusoidal events may be mathematically coined as

$$\alpha \equiv \lim_{t\to\infty} \frac{I_{max}}{\Delta V_{max}} = |G\{\iota\omega\}|, \tag{3.11.32}$$

consistent with Eq. (3.11.25) – and, consequently,

$$\mathcal{Z} = \frac{1}{\alpha} = \frac{1}{|G\{\iota\omega\}|} = \left|\frac{1}{G\{\iota\omega\}}\right| \tag{3.11.33}$$

based on Eq. (3.11.8) and the operational features of the modulus function; insertion of Eq. (3.11.31), after setting $s = \iota\omega$, transforms Eq. (3.11.33) to

$$\mathcal{Z} = \left|\frac{1 + \mathcal{R}\mathcal{C}s + \mathcal{I}\mathcal{C}s^2}{\mathcal{C}s}\right|_{s=\iota\omega} = \left|\frac{1 + \mathcal{R}\mathcal{C}\iota\omega + \mathcal{I}\mathcal{C}(\iota\omega)^2}{\mathcal{C}\iota\omega}\right|, \tag{3.11.34}$$

or else

$$\mathcal{Z} = \left|\frac{1 - \mathcal{I}\mathcal{C}\omega^2 + \iota\mathcal{R}\mathcal{C}\omega}{\iota\mathcal{C}\omega}\right| \tag{3.11.35}$$

since $\iota^2 = -1$. After multiplying both numerator and denominator of Eq. (3.11.35) by $-\iota$, one gets

$$\mathcal{Z} = \left|\frac{(-\iota)\left(1 - \mathcal{I}\mathcal{C}\omega^2 + \iota\mathcal{R}\mathcal{C}\omega\right)}{\mathcal{C}\omega}\right| \tag{3.11.36}$$

– where $-\iota$ may be factored in, and common factors dropped from both numerator and denominator to reach

$$\mathcal{Z} = \left|\frac{\mathcal{R}\mathcal{C}\omega}{\mathcal{C}\omega} - \iota\frac{1 - \mathcal{I}\mathcal{C}\omega^2}{\mathcal{C}\omega}\right| = \left|\mathcal{R} - \iota\frac{1 - \mathcal{I}\mathcal{C}\omega^2}{\mathcal{C}\omega}\right|; \tag{3.11.37}$$

definition of modulus of a complex number, consistent with Eqs. (3.11.1) and (3.11.2), then allows transformation of Eq. (3.11.37) to

$$\mathcal{Z} = \sqrt{\mathcal{R}^2 + \left(-\frac{1 - \mathcal{I}\mathcal{C}\omega^2}{\mathcal{C}\omega}\right)^2}, \tag{3.11.38}$$

or else

$$\mathcal{Z} = \sqrt{\mathcal{R}^2 + \left(\frac{1}{\mathcal{C}\omega} - \mathcal{I}\omega\right)^2} \tag{3.11.39}$$

after splitting the second term and swapping the base of the square with its negative – which conveys $\mathcal{Z} \equiv \mathcal{Z}\{\omega\}$.

d) By definition, the resonance frequency of a circuit, ω_{res}, is the one producing minimum impedance – which, in turn, corresponds to the minimum possible value for the kernel of the square root in Eq. (3.11.39); such a minimum is accordingly described by

$$\frac{1}{\mathcal{C}\omega_{res}} - \mathcal{I}\omega_{res} = 0 \tag{3.11.40}$$

for any given vaue of (ω-independent) $\mathcal{R}$, because the minimum value taken by a square is zero; isolation of ω_{res}^2, after multiplying both sides by ω_{res}, unfolds

$$\omega_{res}^2 = \frac{1}{IC}, \tag{3.11.41}$$

or simply

$$\omega_{res} = \frac{1}{\sqrt{IC}} \tag{3.11.42}$$

once square roots have been taken of both sides. Therefore, the resonance frequency equals the reciprocal of the geometric mean of I and C – and is, in particular, independent of $\mathcal{R}$; when $\omega = \omega_{res}$ – implying validity of Eq. (3.11.40), one finds that Eq. (3.11.39) promptly reduces to

$$Z_{res} \equiv Z\big|_{\omega=\omega_{res}} = \sqrt{\mathcal{R}^2}. \tag{3.11.43}$$

Equation (3.11.43) justifies applicability of the proposed circuit as band-pass filter in a radiotuner – because it raises the minimum impedance possible to any signal when its frequency equals $\omega_{res}/2\pi$; under these circumstances, only the intrinsic resistances connected in series, associated with the processor of such a signal (e.g. an ear-phone), will actually matter.

e) Based on Eqs. (3.300), (3.301), and (3.304), one may write

$$\Delta V_P = N_P \frac{d\phi_P}{dt} \tag{3.11.44}$$

– where the magnetic flux density per loop can, in turn, be expressed as

$$\phi_P = \phi_{P,pk} \sin 2\pi\Omega t, \tag{3.11.45}$$

given the sinusoidal pattern of the underlying function; the time derivative of ϕ_P accordingly reads

$$\frac{d\phi_P}{dt} = 2\pi\Omega\phi_{P,pk} \cos 2\pi\Omega t. \tag{3.11.46}$$

The root-mean square of $d\phi_P/dt$, spanning a full cycle of duration $1/\Omega$, will then ensue from Eq. (3.11.46), viz.

$$\left(\frac{d\phi_P}{dt}\right)_{rms} \equiv \sqrt{\frac{\int_0^{\frac{1}{\Omega}} \left(2\pi\Omega\phi_{P,pk} \cos 2\pi\Omega t\right)^2 dt}{\int_0^{\frac{1}{\Omega}} dt}}, \tag{3.11.47}$$

where constant factors may be taken off the kernel as

$$\left(\frac{d\phi_P}{dt}\right)_{rms} = \sqrt{\frac{\left(2\pi\Omega\phi_{P,pk}\right)^2 \int_0^{\frac{1}{\Omega}} \cos^2 2\pi\Omega t \, dt}{\int_0^{\frac{1}{\Omega}} dt}}; \tag{3.11.48}$$

Eq. (3.11.48) further simplifies to

$$\left(\frac{d\phi_P}{dt}\right)_{rms} = 2\pi\Omega\phi_{P,pk} \sqrt{\frac{\int_0^{\frac{1}{\Omega}} \cos^2 2\pi\Omega t \, dt}{\int_0^{\frac{1}{\Omega}} dt}}, \tag{3.11.49}$$

after taking the square power of the constant off the square root. The formula of duplication of cosine can then be invoked to write

$$\left(\frac{d\phi_P}{dt}\right)_{rms} = 2\pi\Omega\phi_{P,pk} \sqrt{\frac{\int_0^{\frac{1}{\Omega}} \frac{1 + \cos 4\pi\Omega t}{2} dt}{\int_0^{\frac{1}{\Omega}} dt}}, \tag{3.11.50}$$

whereas the fundamental theorem of integral calculus supports transformation to

$$\left(\frac{d\phi_P}{dt}\right)_{rms} = 2\pi\Omega\phi_{P,pk} \sqrt{\frac{\frac{1}{2}\left(t + \frac{\sin 4\pi\Omega t}{4\pi\Omega}\right)\Big|_0^{\frac{1}{\Omega}}}{t\Big|_0^{\frac{1}{\Omega}}}}. \tag{3.11.51}$$

Equation (3.11.51) becomes equivalent to

$$\left(\frac{d\phi_P}{dt}\right)_{rms} = 2\pi\Omega\phi_{P,pk} \sqrt{\frac{\frac{1}{2}\left(\frac{1}{\Omega} + \frac{\sin 4\pi}{4\pi\Omega}\right)}{\frac{1}{\Omega}}} \tag{3.11.52}$$

after replacing by lower and upper limits of integration, and realizing that sin 0 is nil; Eq. (3.11.52) readily yields

$$\left(\frac{d\phi_P}{dt}\right)_{rms} = 2\pi\Omega\phi_{P,pk} \sqrt{\frac{1}{2}}, \tag{3.11.53}$$

since $\sin 4\pi$ is also nil, and $1/\Omega$ can drop off both numerator and denominator. Equation (3.11.53) reduces to

$$\left(\frac{d\phi_P}{dt}\right)_{rms} = \sqrt{2}\pi\Omega\phi_{P,pk}, \tag{3.11.54}$$

once factors alike are lumped; using Eq. (3.11.44) for template, one may write

$$\Delta V_{P,rms} = N_P \left(\frac{d\phi_P}{dt}\right)_{rms} \tag{3.11.55}$$

– where insertion of Eq. (3.11.54) finally leaves

$$\Delta V_{P,rms} = \sqrt{2}\pi\Omega N_P \phi_{P,pk}, \tag{3.11.56}$$

describing the root-mean squared voltage that prevails in the primary winding of the transformer.

3.3.10.3 Cables

3.3.10.3.1 Electric resistance

A great many electric circuits of interest in industrial food processing consist of a network of resistors, which somehow connect

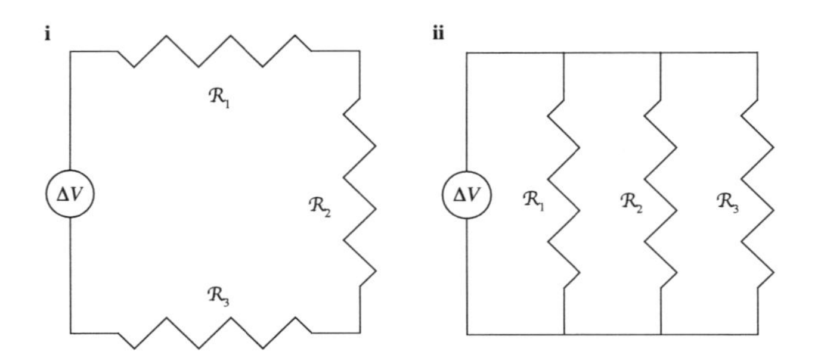

FIGURE 3.46 Schematic representation of electric network, comprising three resistors – characterized by resistances $\mathcal{R}_1$, $\mathcal{R}_2$, and $\mathcal{R}_3$, connected in (i) series or (ii) parallel, and submitted to a difference of potential ΔV, aimed at generating electric current(s) through them.

the two poles of an electrical power source; there are two major modes of connection thereof, namely, series and parallel – as illustrated in Fig. 3.46. Such resistors may, in general, be different from each other; however, they all undergo transients that are typically quite short – due to the very small inertia of electron/ion vehicles, so the associated electric current is normally modeled under a steady state.

Two fundamental laws, first proposed from empirical observation, relate differences in electrical potential, ΔV, to current intensity, I, in a node with N concurrent paths, or in a mesh with N connected resistors: Kirchhoff's first law, viz.

$$\boxed{\left.\left(\sum_{k=1}^{N} I_k = 0\right)\right|_{node}} \tag{3.314}$$

– also known as current law, or Kirchhoff's nodal rule; and Kirchhoff's second law, viz.

$$\boxed{\left.\left(\sum_{k=1}^{N} \Delta V_k = 0\right)\right|_{mesh}} \tag{3.315}$$

– also known as voltage law, or Kirchhoff's loop rule. Equation (3.314) stems from the principle of conservation of electric charge – since current intensity is but a measure of electrons passing through the wire cross section per unit time; it states that the sum of currents flowing into a node (or junction) must equal the sum of currents flowing out of that node, in any electric circuit. By the same token, Eq. (3.315) stems from the principles of conservation of charge and energy, since voltage difference is but a difference in amount of potential energy per unit charge; it states that the sum of potential differences around a closed mesh (or network) is zero, in any electric circuit. Both equations are valid at steady-state (or in the vicinity thereof) – as happens with the low frequencies available as AC, and thus of practical interest in food handling/processing. Application inside a capacitor is accordingly precluded – due to buildup of charge in its plates with time; as well as inside an inductor – because a significant magnetic flux passes through its coil, so closed paths running along said winding experience periodic voltage changes.

Both Kirchhoff's laws can be theoretically derived from the fundamental laws of electromagnetics – thus constituting logical corollaries thereof. For instance, when Eq. (2.379) of *Food Prod. Eng.: thermal & chemical operations* is integrated over a (vector) surface $\boldsymbol{S}$ – as done previously in Eq. (3.291), one ends up with

$$\oint_l \boldsymbol{E} \cdot d\tilde{\boldsymbol{l}} = -\iint_S \frac{\partial \boldsymbol{B}}{\partial t} \cdot d\tilde{\boldsymbol{S}} \tag{3.316}$$

over a wire of (vector) length $\boldsymbol{l}$, following application of Kelvin and Stokes' theorem. One may thus jump to Eq. (3.292), which reduces to

$$\boxed{\oint_l \boldsymbol{E} \cdot d\tilde{\boldsymbol{l}} = -\frac{d\Phi}{dt}} \tag{3.317}$$

when only an electric field is considered; since the integral of an electric field over a length is a voltage difference between its extremes, one may rewrite Eq. (3.317) as

$$\Delta V = -\frac{d\Phi}{dt} \tag{3.318}$$

– compatible with Eq. (3.300), except for the opposite sign due to the (potential) energy lost, rather than (potential) energy supplied. If the continuous, closed path is now subdivided into a series of discrete steps, then Eq. (3.318) acquires the form

$$\sum_k \Delta V_k = -\frac{d\Phi}{dt}; \tag{3.319}$$

should magnetic flux, as per Eq. (3.271), vary negligibly in time (as happens at low frequencies), then $d\Phi/dt \approx 0$, so Eq. (3.319) will retrieve Eq. (3.315) – thus confirming the intrinsic validity of Kirchhoff's loop rule, as well as consistency with the first principles of electrodynamics.

A similar rationale can be followed based on Eq. (2.381) of *Food Prod. Eng.: thermal & chemical operations* – starting from application of the nabla operator to both sides as

$$\nabla \cdot (\nabla \times \boldsymbol{H}) = \nabla \cdot \left(\frac{\partial \boldsymbol{D}}{\partial t} + \boldsymbol{J}\right) \tag{3.320}$$

– with advantage taken of the concept of scalar product for multiple, three-dimensional differentiation. The left-hand side of Eq. (3.320) can be redone as

$$\boldsymbol{H} \cdot (\nabla \times \nabla) = (\nabla \times \nabla) \cdot \boldsymbol{H} = \nabla \cdot \left(\frac{\partial \boldsymbol{D}}{\partial t} + \boldsymbol{J}\right), \tag{3.321}$$

in view of the underlying properties of the triple scalar/vector product and the scalar product – which becomes merely

$$0 = \boldsymbol{0} \cdot \boldsymbol{H} = \nabla \cdot \left(\frac{\partial \boldsymbol{D}}{\partial t} + \boldsymbol{J}\right), \tag{3.322}$$

owing to the definition of magnitude of vector product as product of magnitudes of two vectors by sine of the angle formed thereby (zero in this case, due to coincidence of vector ∇ with itself). Recalling the distributive property of scalar product, Eq. (3.322) becomes

$$\nabla \cdot \frac{\partial \boldsymbol{D}}{\partial t} + \nabla \cdot \boldsymbol{J} = 0, \tag{3.323}$$

where ∇ and $\partial/\partial t$ are allowed to exchange due to their mutual independence as operators – thus leaving

$$\frac{\partial}{\partial t}(\nabla \cdot \boldsymbol{D}) + \nabla \cdot \boldsymbol{J} = 0; \tag{3.324}$$

insertion of Eq. (2.364) from *Food Prod. Eng.: thermal & chemical operations* gives then rise to

$$\nabla \cdot \boldsymbol{J} = -\frac{\partial \rho_V}{\partial t}, \tag{3.325}$$

sometimes referred to as continuity equation. Integration of Eq. (3.325) over volume $\mathcal{V}$ of interest, i.e.

$$\iiint_{\mathcal{V}} \nabla \cdot \boldsymbol{J} \, d\tilde{\mathcal{V}} = -\iiint_{\mathcal{V}} \frac{\partial \rho_V}{\partial t} d\tilde{\mathcal{V}}, \tag{3.326}$$

permits application of Gauss' divergence theorem to get

$$\iint_S \boldsymbol{i}_n \cdot \boldsymbol{J} d\tilde{S} = \iint_S \boldsymbol{J} \cdot d\tilde{\boldsymbol{S}} = -\iiint_{\mathcal{V}} \frac{\partial \rho_V}{\partial t} d\tilde{\mathcal{V}} \tag{3.327}$$

– where S denotes surface serving as outer boundary to $\mathcal{V}$, and $\boldsymbol{S}$ denotes surface vector obtained as product of S by unit vector normal to the surface, $\boldsymbol{i}_n$, at each point. On the other hand, the partial derivative with regard to time may be taken off the triple integral in Eq. (3.327) as

$$\iint_S \boldsymbol{J} \cdot d\tilde{\boldsymbol{S}} = -\frac{\partial}{\partial t} \iiint_{\mathcal{V}} \rho_V d\tilde{\mathcal{V}}, \tag{3.328}$$

again because t and V are independent variables (and thus uncorrelated with each other). The volume integral of charge density is but total charge, Q, inside the volume under scrutiny – meaning that Eq. (3.328) is equivalent to

$$\boxed{\iint_S \boldsymbol{J} \cdot d\tilde{\boldsymbol{S}} = -\frac{\partial Q}{\partial t};} \tag{3.329}$$

furthermore, the surface integral of current density, $\boldsymbol{J}$, coincides with total current, I – so Eq. (3.329) degenerates further to

$$I = -\frac{\partial Q}{\partial t}. \tag{3.330}$$

The closed surface cuts through areas bearing current, e.g. conductors (or resistors, since they offer plain resistance); hence, I is actually the sum of currents in all conductors passing through the said surface – in which case Eq. (3.330) may be rephrased as

$$\sum_k I_k = -\frac{\partial Q}{\partial t}. \tag{3.331}$$

Note that (closed) surface normal vectors (or $d\tilde{\boldsymbol{S}}$) facing outward implies that their sum as in Eq. (3.331) assigns all current directions to be flowing away from the surface. Finally, if the total charge changes negligibly with time, i.e. $\partial Q/\partial t = 0$ (as happens at low frequencies), then Eq. (3.331) ends up coinciding with Eq. (3.314); therefore, validity by itself – further to consistency with the basic rules of electrodynamics, become also apparent in the case of Kirchhoff's nodal rule.

Return now to the resistor layouts of interest, and recall Ohm's law – initially labeled as Eq. (2.80) in *Food Prod. Eng.: thermal & chemical operations*, and revisited here as

$$\Delta V = \mathcal{R}I, \tag{3.332}$$

with $\mathcal{R}$ denoting electric resistance; Eq. (3.332) will then be applied to each resistor in Fig. 3.46i to get

$$\Delta V_1 = \mathcal{R}_1 I, \tag{3.333}$$

$$\Delta V_2 = \mathcal{R}_2 I, \tag{3.334}$$

and

$$\Delta V_3 = \mathcal{R}_3 I \tag{3.335}$$

– since I must be shared by all resistors, as implied by a steady state of operation along the linear circuit. Ordered addition of Eqs. (3.333)–(3.335) gives rise to

$$\Delta V_1 + \Delta V_2 + \Delta V_3 = \mathcal{R}_1 I + \mathcal{R}_2 I + \mathcal{R}_3 I, \tag{3.336}$$

where I can be factored out as

$$\Delta V_1 + \Delta V_2 + \Delta V_3 = \left(\mathcal{R}_1 + \mathcal{R}_2 + \mathcal{R}_3\right) I; \tag{3.337}$$

according to Eq. (3.315), the sequential drops in electric potential around the closed-loop add up to ΔV, i.e. the overall difference in potential imposed by the electromotive force device – so one may redo Eq. (3.337) as

$$\Delta V = \mathcal{R}_{eq} I. \tag{3.338}$$

The equivalent resistance, $\mathcal{R}_{eq}$, accordingly abides to

$$\mathcal{R}_{eq} = \mathcal{R}_1 + \mathcal{R}_2 + \mathcal{R}_3, \tag{3.339}$$

following comparative inspection of Eqs. (3.337) and (3.338); this rule will be generalized to N resistors as

$$\boxed{\mathcal{R}_{eq,ser} = \sum_{i=1}^{N} \mathcal{R}_i,} \tag{3.340}$$

with subscript *ser* referring to connection in series – also known as series resistance formula.

Consider instead the circuit laid out in Fig. 3.46ii; in this case, one should retrieve Eq. (3.332) as

$$I = \frac{\Delta V}{\mathcal{R}}. \tag{3.341}$$

Application of Eq. (3.341) to each resistor gives

$$I_1 = \frac{\Delta V}{\mathcal{R}_1}, \tag{3.342}$$

$$I_2 = \frac{\Delta V}{\mathcal{R}_2}, \tag{3.343}$$

and

$$I_3 = \frac{\Delta V}{\mathcal{R}_3}, \tag{3.344}$$

because all resistors are subjected to the same difference of potential, ΔV – for being directly connected to the source of electromotive force, and connected between them via resistanceless wires. Upon ordered addition, Eqs. (3.342)–(3.344) generate

$$I_1 + I_2 + I_3 = \frac{\Delta V}{\mathcal{R}_1} + \frac{\Delta V}{\mathcal{R}_2} + \frac{\Delta V}{\mathcal{R}_3}, \tag{3.345}$$

where factoring out of ΔV yields

$$I_1 + I_2 + I_3 = \left(\frac{1}{\mathcal{R}_1} + \frac{1}{\mathcal{R}_2} + \frac{1}{\mathcal{R}_3}\right)\Delta V. \tag{3.346}$$

Equation (3.346) preferentially appears as

$$I = \frac{1}{\mathcal{R}_{eq}}\Delta V; \tag{3.347}$$

this is so because overall current intensity, I, at the inlet to the upper node directly connected to the source of electromotive force (and thus supplied with electric current) must equal the sum of partial currents, I_1, I_2, and I_3, forming the outlet of the said node – in agreement with Eq. (3.314). The reciprocal of the equivalent resistance in Eq. (3.347) then reads

$$\frac{1}{\mathcal{R}_{eq}} = \frac{1}{\mathcal{R}_1} + \frac{1}{\mathcal{R}_2} + \frac{1}{\mathcal{R}_3}, \tag{3.348}$$

following inspection of Eq. (3.347) *vis-à-vis* with Eq. (3.346). A similar conclusion would be drawn from application of Kirchhoff's first law to the inlet and outlet of the lower node; in fact, such a node is also directly connected to the source of electromotive force that delivers a voltage difference to the circuit on the outlet side, and to all resistors on the inlet side – which, in turn, triggers a current of intensity I therein. Equation (3.348) appears, in a more general way, as

$$\boxed{\mathcal{R}_{eq,par} = \frac{1}{\sum_{i=1}^{N} \frac{1}{\mathcal{R}_i}}} \tag{3.349}$$

encompassing N resistors – after taking reciprocals of both sides, and using subscript *par* to refer to connection in parallel; this is the so-called parallel resistance formula.

PROBLEM 3.12

Consider an electrical network, composed of five resistors, $\mathcal{R}_1$–$\mathcal{R}_5$, and a source of electromotive force, *emf* – as laid out below.

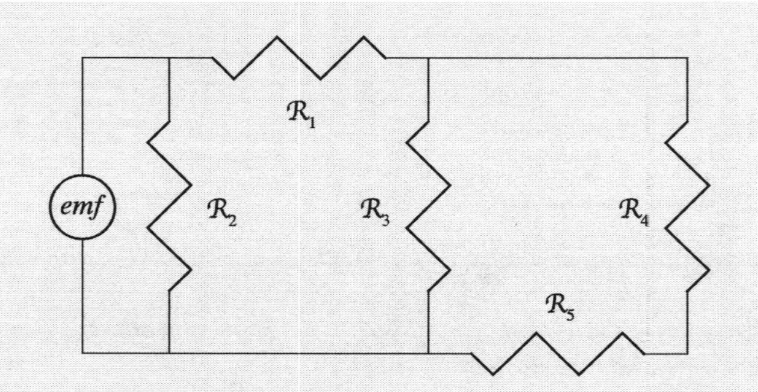

a) Calculate the overall electric resistance offered by the circuit.
b) If $\mathcal{R}_1 = \mathcal{R}_2$ and $\mathcal{R}_3 = \mathcal{R}_4 = \mathcal{R}_5$, prove that the sensitivity of the overall equivalent resistance to changes in $\mathcal{R}_1$ is larger than the negative of threefold the sensitivity with regard to changes in $\mathcal{R}_3$.
c) Consider replacing $\mathcal{R}_4$ by a second *emf* source, and eliminating $\mathcal{R}_2$ – as per the new layout below.

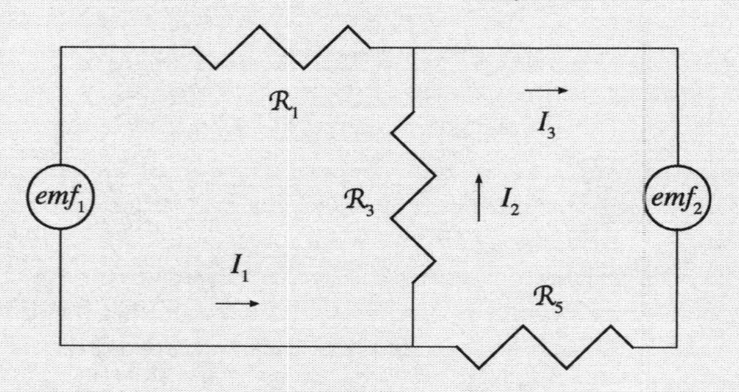

Calculate the intensities of the three electric currents, I_1–I_3, with putative directions as indicated – expressed as functions of the two electromotive forces, $\varepsilon_1 \equiv emf_1$ and $\varepsilon_2 \equiv emf_2$.

Solution

a) Since the formulae describing the equivalent resistance of a combination of resistors entail either series or parallel connection only, a hybrid network (as proposed) should be manipulated in a stepwise manner; one may thus start by replacing $\mathcal{R}_4$ and $\mathcal{R}_5$ by a single resistor $\mathcal{R}_{eq,45}$, via application of Eq. (3.340), viz.

$$\mathcal{R}_{eq,45} = \mathcal{R}_4 + \mathcal{R}_5, \tag{3.12.1}$$

so the circuit simplifies to

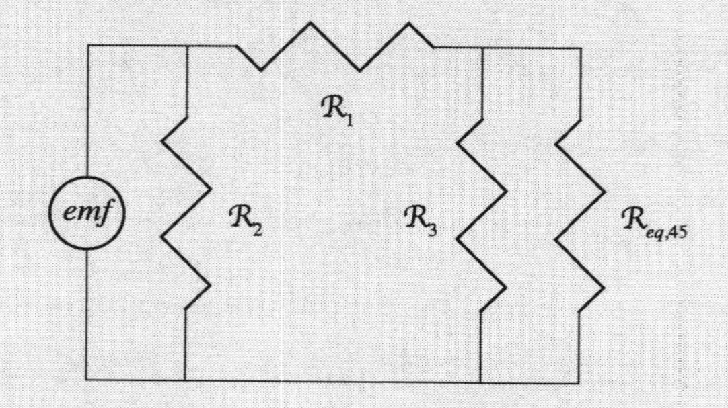

– with reduction of the overall inventory from five to four resistors. Resistors $\mathcal{R}_3$ and $\mathcal{R}_{eq,45}$ are now placed in parallel, so one may resort to Eq. (3.349) to write

$$\mathcal{R}_{eq,345} = \frac{1}{\frac{1}{\mathcal{R}_3} + \frac{1}{\mathcal{R}_{eq,45}}}; \tag{3.12.2}$$

hence, the circuit simplifies to

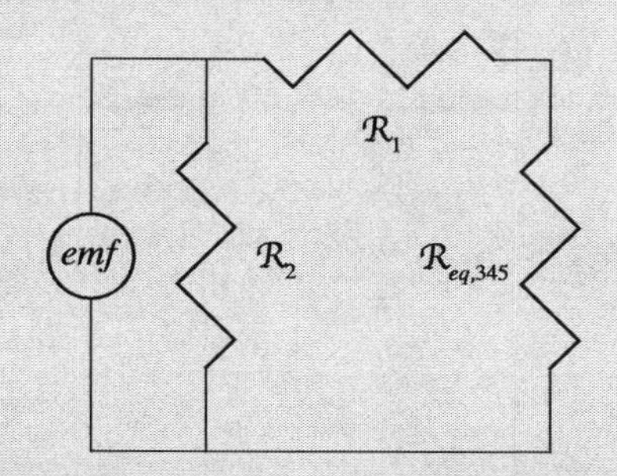

– with the number of actual resistors reduced to only three. A second application of Eq. (3.340), now to $\mathcal{R}_1$ and $\mathcal{R}_{eq,345}$ placed in series, unfolds

$$\mathcal{R}_{eq,1345} = \mathcal{R}_1 + \mathcal{R}_{eq,345}; \tag{3.12.3}$$

the associated simplified circuit looks like

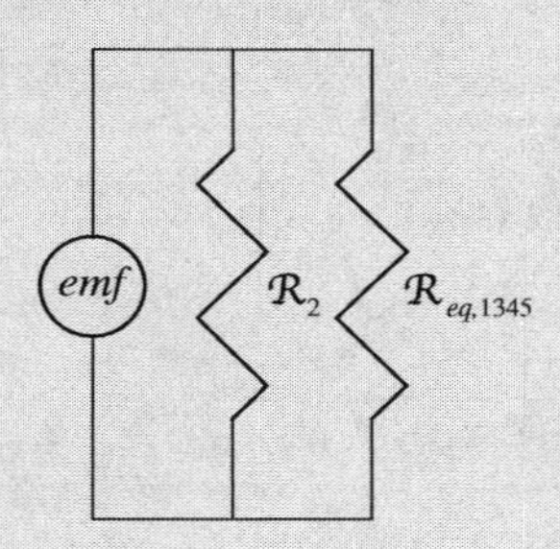

– containing only two resistors. Since the outstanding resistors $\mathcal{R}_2$ and $\mathcal{R}_{eq,1345}$ are connected in parallel, application of Eq. (3.349) is again in order, viz.

$$\mathcal{R}_{eq,12345} = \frac{1}{\frac{1}{\mathcal{R}_2} + \frac{1}{\mathcal{R}_{eq,1345}}}; \tag{3.12.4}$$

the associated simplified circuit reads merely

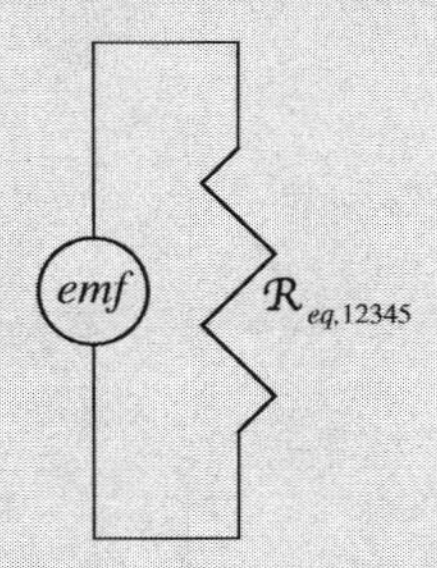

– meaning that the departing circuit was eventually reduced to just one resistor. In other words, $\mathcal{R}_{eq,12345}$ is equivalent to $\mathcal{R}_1$, $\mathcal{R}_2$, $\mathcal{R}_3$, $\mathcal{R}_4$, and $\mathcal{R}_5$ connected as in the original layout; combination of Eqs. (3.12.1)–(3.12.4) finally gives

$$\mathcal{R}_{eq,12345} = \frac{1}{\frac{1}{\mathcal{R}_2} + \frac{1}{\mathcal{R}_1 + \frac{1}{\frac{1}{\mathcal{R}_3} + \frac{1}{\mathcal{R}_4 + \mathcal{R}_5}}}}. \tag{3.12.5}$$

b) Upon lumping terms in the last denominator, Eq. (3.12.5) becomes

$$\mathcal{R}_{eq,12345} = \frac{1}{\frac{1}{\mathcal{R}_2} + \frac{1}{\mathcal{R}_1 + \frac{1}{\frac{\mathcal{R}_3 + \mathcal{R}_4 + \mathcal{R}_5}{\mathcal{R}_3(\mathcal{R}_4 + \mathcal{R}_5)}}}}, \tag{3.12.6}$$

which readily entails

$$\mathcal{R}_{eq,12345} = \frac{1}{\frac{1}{\mathcal{R}_2} + \frac{1}{\mathcal{R}_1 + \frac{\mathcal{R}_3(\mathcal{R}_4 + \mathcal{R}_5)}{\mathcal{R}_3 + \mathcal{R}_4 + \mathcal{R}_5}}}; \tag{3.12.7}$$

a similar procedure involving terms in the last denominator transforms Eq. (3.12.7) to

$$\mathcal{R}_{eq,12345} = \frac{1}{\frac{1}{\mathcal{R}_2} + \frac{1}{\frac{\mathcal{R}_1(\mathcal{R}_3 + \mathcal{R}_4 + \mathcal{R}_5) + \mathcal{R}_3(\mathcal{R}_4 + \mathcal{R}_5)}{\mathcal{R}_3 + \mathcal{R}_4 + \mathcal{R}_5}}} \tag{3.12.8}$$

that breaks down to

$$\mathcal{R}_{eq,12345} = \frac{1}{\frac{1}{\mathcal{R}_2} + \frac{\mathcal{R}_3 + \mathcal{R}_4 + \mathcal{R}_5}{\mathcal{R}_1(\mathcal{R}_3 + \mathcal{R}_4 + \mathcal{R}_5) + \mathcal{R}_3(\mathcal{R}_4 + \mathcal{R}_5)}}. \tag{3.12.9}$$

A final lumping of terms in the denominator of Eq. (3.12.9) gives rise to

$$\mathcal{R}_{eq,12345} = \frac{1}{\frac{\mathcal{R}_1(\mathcal{R}_3 + \mathcal{R}_4 + \mathcal{R}_5) + \mathcal{R}_3(\mathcal{R}_4 + \mathcal{R}_5) + \mathcal{R}_2(\mathcal{R}_3 + \mathcal{R}_4 + \mathcal{R}_5)}{\mathcal{R}_2(\mathcal{R}_1(\mathcal{R}_3 + \mathcal{R}_4 + \mathcal{R}_5) + \mathcal{R}_3(\mathcal{R}_4 + \mathcal{R}_5))}}, \tag{3.12.10}$$

or else

$$\mathcal{R}_{eq,12345} = \frac{\mathcal{R}_2(\mathcal{R}_3(\mathcal{R}_4 + \mathcal{R}_5) + \mathcal{R}_1(\mathcal{R}_3 + \mathcal{R}_4 + \mathcal{R}_5))}{\mathcal{R}_3(\mathcal{R}_4 + \mathcal{R}_5) + (\mathcal{R}_1 + \mathcal{R}_2)(\mathcal{R}_3 + \mathcal{R}_4 + \mathcal{R}_5)} \tag{3.12.11}$$

upon straightforward algebraic rearrangement – including reduction of fraction levels and condensation of terms alike. If $\mathcal{R}_1 = \mathcal{R}_2$ and $\mathcal{R}_3 = \mathcal{R}_4 = \mathcal{R}_5$ as suggested, then Eq. (3.12.11) becomes

$$\mathcal{R}_{eq,12345} = \frac{\mathcal{R}_1\left(\mathcal{R}_3\left(\mathcal{R}_3+\mathcal{R}_3\right)+\mathcal{R}_1\left(\mathcal{R}_3+\mathcal{R}_3+\mathcal{R}_3\right)\right)}{\mathcal{R}_3\left(\mathcal{R}_3+\mathcal{R}_3\right)+\left(\mathcal{R}_1+\mathcal{R}_1\right)\left(\mathcal{R}_3+\mathcal{R}_3+\mathcal{R}_3\right)}, \tag{3.12.12}$$

which breaks down to

$$\mathcal{R}_{eq,12345} = \frac{\mathcal{R}_1\left(2\mathcal{R}_3^2+3\mathcal{R}_1\mathcal{R}_3\right)}{2\mathcal{R}_3^2+6\mathcal{R}_1\mathcal{R}_3} = \frac{\mathcal{R}_1\left(3\mathcal{R}_1+2\mathcal{R}_3\right)}{6\mathcal{R}_1+2\mathcal{R}_3} \tag{3.12.13}$$

upon lumping terms alike, and dropping $\mathcal{R}_3$ from both numerator and denominator afterward. If addition and subtraction of $3\mathcal{R}_1$ in numerator is performed, then Eq. (3.12.13) turns to

$$\mathcal{R}_{eq,12345} = \frac{\mathcal{R}_1\left(\left(6\mathcal{R}_1+2\mathcal{R}_3\right)-3\mathcal{R}_1\right)}{6\mathcal{R}_1+2\mathcal{R}_3}, \tag{3.12.14}$$

which is equivalent to

$$\mathcal{R}_{eq,12345} = \mathcal{R}_1 - \frac{3\mathcal{R}_1^2}{6\mathcal{R}_1+2\mathcal{R}_3}. \tag{3.12.15}$$

The sensitivity of overall resistance to $\mathcal{R}_1$ may be ascertained as

$$\frac{\partial \mathcal{R}_{eq,12345}}{\partial \mathcal{R}_1} = 1 - 3\frac{2\mathcal{R}_1\left(6\mathcal{R}_1+2\mathcal{R}_3\right)-6\mathcal{R}_1^2}{\left(6\mathcal{R}_1+2\mathcal{R}_3\right)^2}, \tag{3.12.16}$$

stemming from Eq. (3.12.15) – and likewise with regard to $\mathcal{R}_3$, i.e.

$$\frac{\partial \mathcal{R}_{eq,12345}}{\partial \mathcal{R}_3} = \frac{6\mathcal{R}_1^2}{\left(6\mathcal{R}_1+2\mathcal{R}_3\right)^2}; \tag{3.12.17}$$

elimination of parenthesis, and then factoring out of $2\mathcal{R}_1$ permit further simplification of Eq. (3.12.16) to

$$\begin{aligned}\frac{\partial \mathcal{R}_{eq,12345}}{\partial \mathcal{R}_1} &= 1 - 3\frac{12\mathcal{R}_1^2+4\mathcal{R}_1\mathcal{R}_3-6\mathcal{R}_1^2}{\left(6\mathcal{R}_1+2\mathcal{R}_3\right)^2} \\ &= 1 - \frac{6\mathcal{R}_1\left(3\mathcal{R}_1+2\mathcal{R}_3\right)}{\left(6\mathcal{R}_1+2\mathcal{R}_3\right)^2},\end{aligned} \tag{3.12.18}$$

or else

$$\frac{\partial \mathcal{R}_{eq,12345}}{\partial \mathcal{R}_1} = 1 - \frac{12\mathcal{R}_1\mathcal{R}_3}{\left(6\mathcal{R}_1+2\mathcal{R}_3\right)^2} - 3\frac{6\mathcal{R}_1^2}{\left(6\mathcal{R}_1+2\mathcal{R}_3\right)^2} \tag{3.12.19}$$

after splitting the fraction. Insertion of Eq. (3.12.17) transforms Eq. (3.12.19) to

$$\frac{\partial \mathcal{R}_{eq,13345}}{\partial \mathcal{R}_1} = 1 - \frac{12\mathcal{R}_1\mathcal{R}_3}{\left(6\mathcal{R}_1+2\mathcal{R}_3\right)^2} - 3\frac{\partial \mathcal{R}_{eq,12345}}{\partial \mathcal{R}_3}, \tag{3.12.20}$$

while combination of the two first terms yields

$$\frac{\partial \mathcal{R}_{eq,12345}}{\partial \mathcal{R}_1} = \frac{\left(36\mathcal{R}_1^2+24\mathcal{R}_1\mathcal{R}_3+4\mathcal{R}_3^2\right)-12\mathcal{R}_1\mathcal{R}_3}{\left(6\mathcal{R}_1+2\mathcal{R}_3\right)^2} - 3\frac{\partial \mathcal{R}_{eq,12345}}{\partial \mathcal{R}_3} \tag{3.12.21}$$

along with expansion of the square of the binomial; after pooling similar terms, Eq. (3.12.21) becomes

$$\frac{\partial \mathcal{R}_{eq,12345}}{\partial \mathcal{R}_1} = \frac{4\left(9\mathcal{R}_1^2+3\mathcal{R}_1\mathcal{R}_3+\mathcal{R}_3^2\right)}{\left(6\mathcal{R}_1+2\mathcal{R}_3\right)^2} - 3\frac{\partial \mathcal{R}_{eq,12345}}{\partial \mathcal{R}_3}. \tag{3.12.22}$$

Since the first term in the right-hand side of Eq. (3.12.22) is always positive, one readily concludes that

$$\frac{\partial \mathcal{R}_{eq,12345}}{\partial \mathcal{R}_1} > -3\frac{\partial \mathcal{R}_{eq,12345}}{\partial \mathcal{R}_3} \tag{3.12.23}$$

as initially claimed.

c) According to Kirchhoff's first law, labeled as Eq. (3.314), one may write

$$I_1 - I_2 + I_3 = 0 \tag{3.12.24}$$

pertaining to either upper or lower node; Kirchhoff's second law, labeled as Eq. (3.315), can be applied to the left loop as

$$\varepsilon_1 - \mathcal{R}_3 I_2 - \mathcal{R}_1 I_1 = 0, \tag{3.12.25}$$

and similarly to the right loop as

$$\varepsilon_2 - \mathcal{R}_5 I_3 - \mathcal{R}_3 I_2 = 0. \tag{3.12.26}$$

Here $\varepsilon_1 > 0$ and $\varepsilon_2 > 0$ denote (given) electromotive forces applied to first and second loops, respectively, in the direction of I_1 and I_3, respectively; and Eq. (2.80) of *Food Proc. Eng.: thermal & chemical operations* was retrieved, after isolation of ΔV as product of the corresponding $\mathcal{R}$ and I. The set of three relationships, labeled as Eqs. (3.12.24)–(3.12.26), is represented in matrix form as

$$\begin{bmatrix} 1 & -1 & 1 \\ \mathcal{R}_1 & \mathcal{R}_3 & 0 \\ 0 & \mathcal{R}_3 & \mathcal{R}_5 \end{bmatrix} \begin{bmatrix} I_1 \\ I_2 \\ I_3 \end{bmatrix} = \begin{bmatrix} 0 \\ \varepsilon_1 \\ \varepsilon_2 \end{bmatrix}; \tag{3.12.27}$$

Cramer's rule can then be invoked to directly obtain the solutions sought for $[I_1\ I_2\ I_3]^T$ as vector of unknowns, i.e.

$$I_1 = \frac{\begin{vmatrix} 0 & -1 & 1 \\ \varepsilon_1 & \mathcal{R}_3 & 0 \\ \varepsilon_2 & \mathcal{R}_3 & \mathcal{R}_5 \end{vmatrix}}{\begin{vmatrix} 1 & -1 & 1 \\ \mathcal{R}_1 & \mathcal{R}_3 & 0 \\ 0 & \mathcal{R}_3 & \mathcal{R}_5 \end{vmatrix}} \tag{3.12.28}$$

pertaining to I_1,

$$I_2 = \frac{\begin{vmatrix} 1 & 0 & 1 \\ \mathcal{R}_1 & \varepsilon_1 & 0 \\ 0 & \varepsilon_2 & \mathcal{R}_5 \end{vmatrix}}{\begin{vmatrix} 1 & -1 & 1 \\ \mathcal{R}_1 & \mathcal{R}_3 & 0 \\ 0 & \mathcal{R}_3 & \mathcal{R}_5 \end{vmatrix}} \tag{3.12.29}$$

encompassing I_2, and

$$I_3 = \frac{\begin{vmatrix} 1 & -1 & 0 \\ \mathcal{R}_1 & \mathcal{R}_3 & \varepsilon_1 \\ 0 & \mathcal{R}_3 & \varepsilon_2 \end{vmatrix}}{\begin{vmatrix} 1 & -1 & 1 \\ \mathcal{R}_1 & \mathcal{R}_3 & 0 \\ 0 & \mathcal{R}_3 & \mathcal{R}_5 \end{vmatrix}} \tag{3.12.30}$$

underlying I_3. The (common) determinant in all three denominators of Eqs. (3.12.28)–(3.12.30) is given by

$$\begin{vmatrix} 1 & -1 & 1 \\ \mathcal{R}_1 & \mathcal{R}_3 & 0 \\ 0 & \mathcal{R}_3 & \mathcal{R}_5 \end{vmatrix} = -\begin{vmatrix} 1 & -1 & 1 \\ 0 & \mathcal{R}_3 & \mathcal{R}_5 \\ \mathcal{R}_1 & \mathcal{R}_3 & 0 \end{vmatrix} = -\begin{vmatrix} 1 & -1 & 1 \\ 0 & \mathcal{R}_3 & \mathcal{R}_5 \\ 0 & \mathcal{R}_1+\mathcal{R}_3 & -\mathcal{R}_1 \end{vmatrix}, \tag{3.12.31}$$

after swapping the second and third rows, and then replacing the last row by its sum with the first row upon multiplication by $-\mathcal{R}_1$; Laplace's theorem supports simplification to

$$\begin{vmatrix} 1 & -1 & 1 \\ \mathcal{R}_1 & \mathcal{R}_3 & 0 \\ 0 & \mathcal{R}_3 & \mathcal{R}_5 \end{vmatrix} = -\begin{vmatrix} \mathcal{R}_3 & \mathcal{R}_5 \\ \mathcal{R}_1+\mathcal{R}_3 & -\mathcal{R}_1 \end{vmatrix} = \mathcal{R}_1\mathcal{R}_3 + \mathcal{R}_5(\mathcal{R}_1+\mathcal{R}_3) \tag{3.12.32}$$

using the first column as pivot, coupled with definition of second-order determinant. By the same token, the determinant in numerator of Eq. (3.12.28) may be worked out as

$$\begin{vmatrix} 0 & -1 & 1 \\ \varepsilon_1 & \mathcal{R}_3 & 0 \\ \varepsilon_2 & \mathcal{R}_3 & \mathcal{R}_5 \end{vmatrix} = \begin{vmatrix} 0 & 0 & 1 \\ \varepsilon_1 & \mathcal{R}_3 & 0 \\ \varepsilon_2 & \mathcal{R}_3+\mathcal{R}_5 & \mathcal{R}_5 \end{vmatrix}, \tag{3.12.33}$$

following replacement of the second column by its sum with the third column; hence, application of Laplace's theorem through the first row ensues, viz.

$$\begin{vmatrix} 0 & -1 & 1 \\ \varepsilon_1 & \mathcal{R}_3 & 0 \\ \varepsilon_2 & \mathcal{R}_3 & \mathcal{R}_5 \end{vmatrix} = \begin{vmatrix} \varepsilon_1 & \mathcal{R}_3 \\ \varepsilon_2 & \mathcal{R}_3+\mathcal{R}_5 \end{vmatrix} = \varepsilon_1(\mathcal{R}_3+\mathcal{R}_5) - \varepsilon_2\mathcal{R}_3 \tag{3.12.34}$$

– complemented again by definition of second-order determinant. The determinant in numerator of Eq. (3.12.29) becomes

$$\begin{vmatrix} 1 & 0 & 1 \\ \mathcal{R}_1 & \varepsilon_1 & 0 \\ 0 & \varepsilon_2 & \mathcal{R}_5 \end{vmatrix} = \begin{vmatrix} 1 & 0 & 1 \\ 0 & \varepsilon_1 & -\mathcal{R}_1 \\ 0 & \varepsilon_2 & \mathcal{R}_5 \end{vmatrix}, \tag{3.12.35}$$

upon replacement of the second row by its sum with the first row premultiplied by $-\mathcal{R}_1$ – so Laplace's rule, applied through the first column, leads to

$$\begin{vmatrix} 1 & 0 & 1 \\ \mathcal{R}_1 & \varepsilon_1 & 0 \\ 0 & \varepsilon_2 & \mathcal{R}_5 \end{vmatrix} = \begin{vmatrix} \varepsilon_1 & -\mathcal{R}_1 \\ \varepsilon_2 & \mathcal{R}_5 \end{vmatrix} = \varepsilon_1\mathcal{R}_5 + \varepsilon_2\mathcal{R}_1, \tag{3.12.36}$$

with the aid again of definition of second-order determinant; finally, the determinant in numerator of Eq. (3.12.30) gives rise to

$$\begin{vmatrix} 1 & -1 & 0 \\ \mathcal{R}_1 & \mathcal{R}_3 & \varepsilon_1 \\ 0 & \mathcal{R}_3 & \varepsilon_2 \end{vmatrix} = \begin{vmatrix} 1 & 0 & 0 \\ \mathcal{R}_1 & \mathcal{R}_1+\mathcal{R}_3 & \varepsilon_1 \\ 0 & \mathcal{R}_3 & \varepsilon_2 \end{vmatrix}, \tag{3.12.37}$$

after replacement of the second column by its sum with the first column – thus allowing Laplace's theorem come aboard as

$$\begin{vmatrix} 1 & -1 & 0 \\ \mathcal{R}_1 & \mathcal{R}_3 & \varepsilon_1 \\ 0 & \mathcal{R}_3 & \varepsilon_2 \end{vmatrix} = \begin{vmatrix} \mathcal{R}_1+\mathcal{R}_3 & \varepsilon_1 \\ \mathcal{R}_3 & \varepsilon_2 \end{vmatrix} = \varepsilon_2(\mathcal{R}_1+\mathcal{R}_3) - \varepsilon_1\mathcal{R}_3, \tag{3.12.38}$$

using the first row as pivot and resorting once more to definition of second-order determinant. Insertion of Eqs. (3.12.32) and (3.12.34) transforms Eq. (3.12.28) to

$$I_1 = \frac{\varepsilon_1(\mathcal{R}_3+\mathcal{R}_5) - \varepsilon_2\mathcal{R}_3}{\mathcal{R}_1\mathcal{R}_3 + \mathcal{R}_5(\mathcal{R}_1+\mathcal{R}_3)}, \tag{3.12.39}$$

whereas insertion of Eqs. (3.12.32) and (3.12.36) converts Eq. (3.12.29) to

$$I_2 = \frac{\varepsilon_1\mathcal{R}_5 + \varepsilon_2\mathcal{R}_1}{\mathcal{R}_1\mathcal{R}_3 + \mathcal{R}_5(\mathcal{R}_1+\mathcal{R}_3)}; \tag{3.12.40}$$

one may similarly combine Eqs. (3.12.30), (3.12.32), and (3.12.38) to obtain

$$I_3 = \frac{\varepsilon_2(\mathcal{R}_1+\mathcal{R}_3) - \varepsilon_1\mathcal{R}_3}{\mathcal{R}_1\mathcal{R}_3 + \mathcal{R}_5(\mathcal{R}_1+\mathcal{R}_3)}. \tag{3.12.41}$$

Although the direction of I_2 is correct as arbitrated – as the right-hand side of Eq. (3.12.40) is always positive, one realizes that I_1 or I_3 will have to be reversed relative to the initial hypothesis if $\varepsilon_2\mathcal{R}_3 > \varepsilon_1(\mathcal{R}_3+\mathcal{R}_5)$ or $\varepsilon_1\mathcal{R}_3 > \varepsilon_2(\mathcal{R}_1+\mathcal{R}_3)$, respectively – since the (common) denominator of the right-hand sides of Eqs. (3.12.39) and (3.12.41) remains positive.

Once in possession of Eqs. (3.340) and (3.349), one realizes that the overall resistance of a circuit increases with addition of an extra resistor in series, while the overall conductance (or reciprocal resistance) increases with addition of extra resistor in parallel, respectively. Therefore, such an extra resistor decreases the overall electric current in series, but increases the overall electric current when mounted in parallel – for a given electromotive force. On the other hand, placing resistances in series runs the risk of breaking the whole circuit if at least one of them ceases to function – whereas all other resistances will remain in operation, and only the portion of the circuit containing a broken resistance will be discontinued if they are connected in parallel; this is why resistances are normally placed in parallel – except when a high voltage is at stake.

3.3.10.3.2 Thermal resistance

Electric currents are normally transported by metallic (usually copper) wires, of cylindrical shape – which behave as plain resistors; electrical power, $\mathcal{P}$, will thus be fully dissipated as thermal energy. One may therefore resort to Eqs. (2.81) and (2.95) in *Food Proc. Eng.: thermal & chemical operations* to write

$$\mathcal{P}=\frac{1}{\sigma}\frac{L}{A}I^2, \tag{3.350}$$

where σ denotes electric conductivity, L and A denote length and cross-section of wire, respectively, and I denotes electric current intensity therein; division of both sides by wire volume, i.e. AL, gives then rise to

$$\boxed{\hbar=\frac{1}{\sigma}\left(\frac{I}{A}\right)^2}, \tag{3.351}$$

where $\hbar$ denotes heating rate per unit volume, i.e.

$$\boxed{\hbar\equiv\frac{\mathcal{P}}{LA}}. \tag{3.352}$$

In view of the underlying cylindrical symmetry, an enthalpy balance can be set to an annulus of a wire, of elementary thickness, preferably using cylindrical coordinates – according to

$$-2\pi rLk\left.\frac{dT}{dr}\right|_r+2\pi rLdr\hbar=-2\pi(r+dr)Lk\left.\frac{dT}{dr}\right|_{r+dr}, \tag{3.353}$$

with k denoting thermal conductivity; here the first terms in both left- and right-hand sides represent areal rate of heat transmitted by thermal conduction at inlet (through side surface of area $2\pi rL$) and outlet (through side surface of area $2\pi(r+dr)L$), respectively. The second term in the left-hand side represents volumetric rate of production of heat, due to passage of the aforementioned electric current (through annular solid of volume $2\pi rLdr$). After factoring $2\pi kL$ out, Eq. (3.353) becomes

$$2\pi kL\left((r+dr)\left.\frac{dT}{dr}\right|_{r+dr}-r\left.\frac{dT}{dr}\right|_r\right)+2\pi\hbar Lrdr=0; \tag{3.354}$$

while division of both sides by $2\pi Ldr$ leaves

$$\boxed{k\frac{d}{dr}\left(r\frac{dT}{dr}\right)+\hbar r=0}, \tag{3.355}$$

as long as the definition of derivative is revisited. Integration of Eq. (3.355) via separation of variables, i.e.

$$k\int d\left(r\frac{dT}{dr}\right)=-\hbar\int rdr \tag{3.356}$$

on the hypothesis of constant k and $\hbar$, gives rise to

$$kr\frac{dT}{dr}=\kappa_1-\hbar\frac{r^2}{2}. \tag{3.357}$$

– with κ_1 denoting an arbitrary (integration) constant; division of both sides by r leads, in turn, to

$$k\frac{dT}{dr}=\frac{\kappa_1}{r}-\frac{1}{2}\hbar r. \tag{3.358}$$

The temperature gradient, responsible for (outward) transfer of heat as per Fourier's law, must remain finite at all locations – and, in particular, on the axis, according to

$$\left.\frac{dT}{dr}\right|_{r=0}=\text{finite}, \tag{3.359}$$

which may thus serve as first boundary condition; Eq. (3.359) enforces

$$\kappa_1=0, \tag{3.360}$$

otherwise $\kappa_1/r\to\infty$, and thus $dT/dr\to\infty$ in Eq. (3.358) when $r\to 0$. Equation (3.360) accordingly permits simplification of Eq. (3.358) to

$$k\frac{dT}{dr}=-\frac{1}{2}\hbar r. \tag{3.361}$$

Constant k and $\hbar$ permit integration of Eq. (3.361), again via separation of variables, as

$$k\int dT=-\frac{1}{2}\hbar\int rdr \tag{3.362}$$

– which is equivalent

$$kT=\kappa_2-\frac{1}{2}\hbar\frac{r^2}{2}, \tag{3.363}$$

provided that κ_2 denotes a second arbitrary constant; isolation of T then unfolds

$$T=\frac{\kappa_2}{k}-\frac{1}{4}\frac{\hbar}{k}r^2. \tag{3.364}$$

A second boundary condition may now be set as

$$\boxed{\left.T\right|_{r=R}=T_0}, \tag{3.365}$$

where R denotes wire outer radius and T_0 denotes room temperature; combination of Eqs. (3.364) and (3.365) leads to

$$T_0 = \frac{\kappa_2}{k} - \frac{1}{4}\frac{\hbar}{k}R^2, \tag{3.366}$$

which can be solved for κ_2/k as

$$\frac{\kappa_2}{k} = T_0 + \frac{1}{4}\frac{\hbar}{k}R^2. \tag{3.367}$$

Insertion of Eq. (3.367) transforms Eq. (3.364) to

$$T = T_0 + \frac{1}{4}\frac{\hbar}{k}R^2 - \frac{1}{4}\frac{\hbar}{k}r^2; \tag{3.368}$$

algebraic rearrangement of Eq. (3.368) yields

$$T = T_0 + \frac{1}{4}\frac{\hbar}{k}\left(R^2 - r^2\right) \tag{3.369}$$

via factoring out $\hbar/4k$, or else

$$\boxed{\frac{T}{T_0} = 1 + \frac{1}{4}\frac{\hbar R^2}{kT_0}\left(1 - \left(\frac{r}{R}\right)^2\right)} \tag{3.370}$$

after division of both sides by T_0, complemented by factoring out of R^2 in the last term – where combination with Eq. (3.351) produces

$$\frac{T}{T_0} = 1 + \frac{1}{4}\frac{\frac{1}{\sigma}\left(\frac{I}{A}\right)^2 R^2}{kT_0}\left(1 - \left(\frac{r}{R}\right)^2\right). \tag{3.371}$$

On the other hand, $\mathcal{R}$ can be eliminated between Eqs. (2.80) and (2.81) of *Food Proc. Eng.: thermal & chemical operations* to generate

$$\frac{1}{\sigma}\frac{L}{A} = \frac{\Delta V}{I} \tag{3.372}$$

or, after solving for I/A,

$$\frac{I}{A} = \frac{\sigma}{L}\Delta V. \tag{3.373}$$

Insertion of Eq. (3.373) transforms Eq. (3.371) to

$$\frac{T}{T_0} = 1 + \frac{1}{4}\frac{\frac{1}{\sigma}\left(\frac{\sigma}{L}\Delta V\right)^2 R^2}{kT_0}\left(1 - \left(\frac{r}{R}\right)^2\right), \tag{3.374}$$

where lumping of factors alike gives

$$\boxed{\frac{T}{T_0} = 1 + \frac{1}{4}\left(\frac{R}{L}\right)^2\frac{1}{Lz}\left(1 - \left(\frac{r}{R}\right)^2\right);} \tag{3.375}$$

here Lz denotes (dimensionless) Lorenz's number – defined as

$$\boxed{Lz \equiv \frac{kT_0}{\sigma \Delta V^2},} \tag{3.376}$$

while R/L denotes a characteristic geometric factor. The variation of dimensionless temperature with dimensionless radius, as per Eq. (3.375), is depicted in Fig. 3.47, for selected values of lumped parameter $(R/L)^2/Lz$. As expected, a uniform temperature profile results in the absence of ohmic heating – characterized by $Lz \to \infty$, as a consequence of either $\sigma \to 0$ in Eq. (3.376) or $\Delta V \to 0$ in Eq. (3.376). Larger heating rate per unit volume, $\hbar$, lower thermal conductivity, k, or larger outer radius, R, cause an increase in characteristic parameter, $(R/L)^2/Lz = \hbar R^2/kT_0$, as per Eqs. (3.370) and (3.375) – thus leading to the higher increases in bulk temperature apparent in Fig. 3.47, felt especially on the axis; this is a consequence of more heat produced per unit volume, higher difficulty to transfer heat outward by conduction, or smaller area available for heat transfer per unit volume, respectively.

Since T/T_0 is a monotonically decreasing function of r/R – due to the negative of its square appearing as sole functional dependence thereon in Eq. (3.375), the maximum temperature, T_{max}, will be attained at the axis of the wire, i.e.

$$\frac{T_{max}}{T_0} = \frac{T\big|_{r=0}}{T_0}. \tag{3.377}$$

Equation (3.377) degenerates to

$$\boxed{\frac{T_{max}}{T_0} = 1 + \frac{1}{4Lz}\left(\frac{R}{L}\right)^2,} \tag{3.378}$$

after taking Eq. (3.375) into account; Eq. (3.378) describes the loci of the vertical intercepts of the curves plotted in Fig. 3.47.

According to Wiedemann, Franz, and Lorenz's law,

$$\boxed{\frac{k}{\sigma T} = k_l,} \tag{3.379}$$

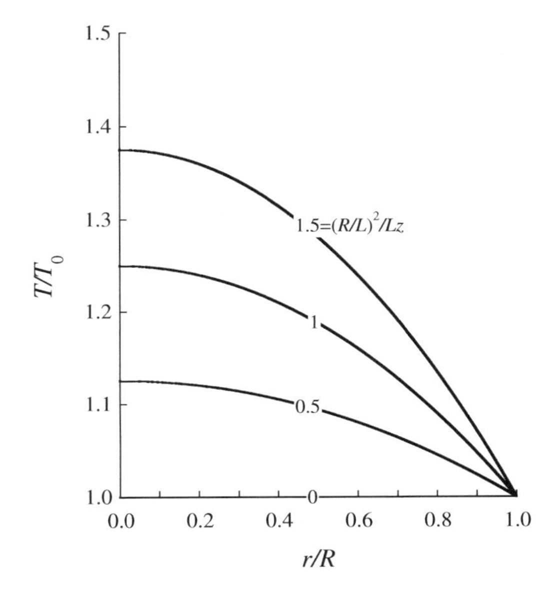

FIGURE 3.47 Variation of dimensionless temperature, T/T_0, of wire conducting electric current, versus normalized radial position, r/R, for selected values of dimensionless parameter $(R/L)^2/Lz$.

where k_l denotes Lorenz's constant (equal to 22–29 $\times 10^{-9}$ V$^2\cdot$K^{-2}) – valid for all pure metals, and accurate to 10% within a range in temperature of amplitude 1000°C. After rewriting Eq. (3.379) as

$$k = k_l T\sigma, \tag{3.380}$$

one realizes that thermal conductivity (i.e. k) is proportional to electric conductivity (i.e. σ), via a proportionality constant, $k_l T$, that increases linearly with temperature. The functional form of Eq. (3.380) is easily justified by collisions between free electrons accounting for the chief mechanism of transfer of both heat and electricity. After redoing Eq. (3.378) with the aid of Eq. (3.376), one gets

$$\frac{T_{max}}{T_0} = 1 + \frac{1}{4}\left(\frac{R}{L}\right)^2 \frac{\sigma \Delta V_{max}{}^2}{kT_0}, \tag{3.381}$$

where isolation of $\Delta V_{max}{}^2$ gives rise to

$$\Delta V_{max}{}^2 = \frac{4\dfrac{kT_0}{\sigma}\left(\dfrac{T_{max}}{T_0} - 1\right)}{\left(\dfrac{R}{L}\right)^2}; \tag{3.382}$$

once square roots are taken of both sides, Eq. (3.382) degenerates to

$$\boxed{\Delta V_{max} = 2\frac{L}{R}\sqrt{\frac{kT_0}{\sigma}\left(\frac{T_{max}}{T_0} - 1\right)}} \tag{3.383}$$

– which is useful to estimate the maximum voltage difference, ΔV_{max}, that can be carried by a wire, given L and R for geometric features, and k and σ for physical features thereof, with T_{max}/T_0 set as maximum temperature acceptable.

PROBLEM 3.13

Consider a cylindrical wire, of radius R, carrying an electric current under steady state operation.

a) Calculate the average (relative) temperature rise of the wire.
b) Estimate the total heat generated inside the wire.
c) If insulation is provided on the outer surface of the wire, characterized by heat transfer coefficient h, obtain the temperature profile therein.
d) For 100 m of copper wire, with 1 mm radius, calculate the maximum voltage difference that does not incur in temperature increases above 30°C relative to room temperature (taken as 20°C) – knowing that Lorenz's constant is 2.23 $\times 10^{-8}$ V^2.K^{-2} for copper at that temperature.
e) Repeat d), but now assuming that the said wire is wound 1000-fold to produce a transformer coil (to be seen as a cable with an equivalent radius) – and comment on the result obtained.
f) Discuss the analogy between Eq. (3.375) and the velocity profile for laminar flow inside a cylindrical pipe.

Solution

a) Departing from the definition of volume-averaged (relative) temperature rise, viz.

$$\overline{\left(\frac{T-T_0}{T_0}\right)} = \frac{\overline{T}-T_0}{T_0} = \frac{\overline{T}}{T_0} - 1 \equiv \frac{\displaystyle\int_0^R 2\pi r\left(\frac{T}{T_0}-1\right)dr}{\displaystyle\int_0^R 2\pi r dr}, \tag{3.13.1}$$

one may take 2π off the kernel and drop it from both numerator and denominator to get

$$\overline{\left(\frac{T-T_0}{T_0}\right)} = \frac{2\pi\displaystyle\int_0^R r\left(\frac{T}{T_0}-1\right)dr}{2\pi\displaystyle\int_0^R rdr} = \frac{\displaystyle\int_0^R \left(\frac{T}{T_0}-1\right)rdr}{\displaystyle\int_0^R rdr}; \tag{3.13.2}$$

Eq. (3.370) should now be retrieved to write

$$\overline{\left(\frac{T-T_0}{T_0}\right)} = \frac{\displaystyle\int_0^R \frac{1}{4}\frac{\hbar R^2}{kT_0}\left(1-\left(\frac{r}{R}\right)^2\right)rdr}{\displaystyle\int_0^R rdr} \tag{3.13.3}$$

– where dividing both numerator and denominator by R^2, and then taking constant factors off the kernel yield

$$\overline{\left(\frac{T-T_0}{T_0}\right)} = \frac{1}{4}\frac{\hbar R^2}{kT_0}\frac{\displaystyle\int_0^1 \left(1-\left(\frac{r}{R}\right)^2\right)\frac{r}{R}d\left(\frac{r}{R}\right)}{\displaystyle\int_0^1 \frac{r}{R}d\left(\frac{r}{R}\right)}, \tag{3.13.4}$$

or else

$$\overline{\left(\frac{T-T_0}{T_0}\right)} = \frac{1}{4}\frac{\hbar R^2}{kT_0}\frac{\displaystyle\int_0^1 \left(\frac{r}{R}-\left(\frac{r}{R}\right)^3\right)d\left(\frac{r}{R}\right)}{\displaystyle\int_0^1 \frac{r}{R}d\left(\frac{r}{R}\right)} \tag{3.13.5}$$

once r/R is factored in. The fundamental theorem of integral calculus can be invoked to rewrite Eq. (3.13.5) as

$$\overline{\left(\frac{T-T_0}{T_0}\right)} = \frac{1}{4}\frac{\hbar R^2}{kT_0}\frac{\left.\left(\dfrac{\left(\frac{r}{R}\right)^2}{2} - \dfrac{\left(\frac{r}{R}\right)^4}{4}\right)\right|_0^1}{\left.\dfrac{\left(\frac{r}{R}\right)^2}{2}\right|_0^1}, \tag{3.13.6}$$

which breaks down to

$$\overline{\left(\frac{T-T_0}{T_0}\right)} = \frac{1}{4}\frac{\hbar R^2}{kT_0}\frac{\left(\frac{1^2}{2}-\frac{1^4}{4}\right)-\left(\frac{0^2}{2}-\frac{0^4}{4}\right)}{\frac{1^2}{2}-\frac{0^2}{2}} = \frac{1}{4}\frac{\hbar R^2}{kT_0}\frac{\frac{1}{2}-\frac{1}{4}}{\frac{1}{2}} \tag{3.13.7}$$

after replacement by the indicated upper and lower limits of integration; Eq. (3.13.7) reduces to

$$\overline{\left(\frac{T-T_0}{T_0}\right)} = \frac{1}{8}\frac{\hbar R^2}{kT_0} \tag{3.13.8}$$

– so the average (relative) temperature rise over the cross-section, for a uniform behavior along length, is but one-half of the maximum temperature rise at the axis, see Eqs. (3.370) and (3.378).

b) Under steady state operation, all heat generated inside the wire, Q_{tot}, must leave via conduction through its outer side surface, $Q\big|_{r=R}$, i.e.

$$Q_{tot} = Q\big|_{r=R}; \tag{3.13.9}$$

Eq. (3.13.9) becomes

$$Q_{tot} = -k(2\pi rL)\big|_{r=R}\frac{dT}{dr}\bigg|_{r=R} = 2\pi kRL\left(-\frac{dT}{dr}\right)\bigg|_{r=R}, \tag{3.13.10}$$

in agreement with Fourier's law – while multiplication and division of the right-hand side by T_0, coupled with lumping of R with r leave

$$Q_{tot} = 2\pi kLT_0\left(-\frac{d\left(\frac{T}{T_0}\right)}{d\left(\frac{r}{R}\right)}\right)\Bigg|_{\frac{r}{R}=1}. \tag{3.13.11}$$

Differentiation of Eq. (3.370), with regard to r/R, gives rise to

$$\frac{d\left(\frac{T}{T_0}\right)}{d\left(\frac{r}{R}\right)} = \frac{1}{4}\frac{\hbar R^2}{kT_0}\left(-2\frac{r}{R}\right) = -\frac{1}{2}\frac{\hbar R^2}{kT_0}\frac{r}{R}, \tag{3.13.12}$$

which breaks down to

$$\left(-\frac{d\left(\frac{T}{T_0}\right)}{d\left(\frac{r}{R}\right)}\right)\Bigg|_{\frac{r}{R}=1} = \frac{1}{2}\frac{\hbar R^2}{kT_0} \tag{3.13.13}$$

at $r/R=1$; insertion of Eq. (3.13.13) then transforms Eq. (3.13.11) to

$$Q_{tot} = 2\pi kLT_0\frac{1}{2}\frac{\hbar R^2}{kT_0} = \pi\hbar LR^2, \tag{3.13.14}$$

along with cancellation of common factors between numerator and denominator. Note the linear increase of heat dissipated with wire length, but its quadratic increase with wire radius – with product of πR^2 by L unfolding volume V of cylinder; this is expected, since $\hbar$ refers to rate of production of heat per unit volume, and $Q_{tot} = \hbar V$.

c) If an insulation layer is provided on the outer surface of the wire, then

$$h\left(T\big|_{r=R^+} - T_0\right) = -k\frac{dT}{dr}\bigg|_{r=R^-} \tag{3.13.15}$$

should be used for second boundary condition, *in lieu* of Eq. (3.365) – where the same area and temperature apply on both sides of the wire/insulation interface. Hence, Eqs. (3.361) and (3.364) shoud be retrieved, and inserted in Eq. (3.13.15) to generate

$$h\left(\frac{\kappa_2}{k} - \frac{1}{4}\frac{\hbar}{k}r^2 - T_0\right)\bigg|_{r=R} = -k\left(-\frac{1}{2}\frac{\hbar}{k}r\right)\bigg|_{r=R}; \tag{3.13.16}$$

Eq. (3.13.16) reduces to

$$h\frac{\kappa_2}{k} - \frac{1}{4}\frac{h}{k}\hbar R^2 - hT_0 = \frac{1}{2}\hbar R \tag{3.13.17}$$

upon replacement of r by R, together with cancellation of common factors between numerator and denominator. Isolation of κ_2/k unfolds

$$\frac{\kappa_2}{k} = T_0 + \frac{1}{2}\frac{\hbar}{h}R + \frac{1}{4}\frac{\hbar}{k}R^2, \tag{3.13.18}$$

which permits reformulation of Eq. (3.364) to

$$T = T_0 + \frac{1}{2}\frac{\hbar}{h}R + \frac{1}{4}\frac{\hbar}{k}R^2 - \frac{1}{4}\frac{\hbar}{k}r^2; \tag{3.13.19}$$

after dividing both sides by T_0, complemented by factoring out of $\hbar R^2/4k$, Eq. (3.13.19) becomes

$$\frac{T}{T_0} = 1 + \frac{1}{4}\frac{\hbar R^2}{kT_0}\left(1-\left(\frac{r}{R}\right)^2\right) + \frac{1}{2}\frac{\hbar R}{hT_0}. \tag{3.13.20}$$

Inspection of Eq. (3.13.20) *vis-à-vis* with Eq. (3.370) unfolds a uniform increase, by $\hbar R/2hT_0$, of temperature along the whole axial position. This is why (electric) insulation for wiring is often chosen to be as thin as possible, so as to get the highest value possible for h – and thus minimize the impact of the last term in Eq. (3.13.29).

d) The maximum ΔV may be ascertained directly from Eq. (3.383), after recalling Eq. (3.380), i.e.

$$\Delta V_{max} = 2\frac{L}{R}\sqrt{k_l TT_0\left(\frac{T_{max}}{T_0}-1\right)} = 2\frac{L}{R}\sqrt{k_l T\left(T_{max}-T_0\right)}; \tag{3.13.21}$$

upon setting $L=100$ m, $R=0.001$ m, $k_l=2.23\times10^{-8}$ V^2.K^{-2}, $T\approx T_0=20+273=293$ K, and $T_{max}=(20+30)+273=323$ K, one obtains

$$\Delta V_{max}=2\frac{100}{0.001}\sqrt{2.23\times10^{-8}\cdot293(323-293)}=2{,}800V \tag{3.13.22}$$

from Eq. (3.13.21) – meaning that special care is to be exercised when intramural transport of high voltages in industrial premises is concerned, with regard to temperature increase in their carrying cable(s). Note that $T_0=293<T<323$, so $1\leq T/T_0\leq 323/293=1.102\approx1$ justifies the aforementioned approximation of setting $T\approx T_0$, to a relative error $\left(T-T_0\right)/T_0=T/T_0-1\leq1.102-1=10.2\%$.

e) If the original wire, or radius R_w, is wound N times upon itself, then the radius R_c of the equivalent cable in terms of heat transfer – assuming that cylindrical symmetry is retained, will abide to

$$N\pi R_w^{\,2}=\pi R_c^{\,2}; \tag{3.13.23}$$

upon isolation of R_c, one gets

$$R_c=\sqrt{N}R_w. \tag{3.13.24}$$

Note that length, L_c, associated with the winding layout still abides to

$$L_c=L_w, \tag{3.13.25}$$

with L_w denoting length of original wire; this is so because such a length pertains to the distance over which ΔV applies, see Eqs. (3.351) and (3.373), when serving as heat source as per Joule's law – which remains unaltered in the new layout, since the wire is wound upon itself (although in insulated form). One may then write

$$\Delta V_{max,c}=2\frac{L_c}{R_c}\sqrt{k_lT_0\left(T_{max}-T_0\right)} \tag{3.13.26}$$

pertaining to the maximum value of ΔV along the cable, using Eq. (3.13.21) as template with $T\approx T_0$ within the range of interest – where insertion of Eqs. (3.13.24) and (3.13.25) gives rise to

$$\Delta V_{max,c}=2\frac{L_w}{\sqrt{N}R_w}\sqrt{k_lT_0\left(T_{max}-T_0\right)}; \tag{3.13.27}$$

after rewriting Eq. (3.13.21) as

$$\Delta V_{max,w}=2\frac{L_w}{R_w}\sqrt{k_lT_0\left(T_{max}-T_0\right)} \tag{3.13.28}$$

encompassing the original unwound wire, Eq. (3.13.27) can be reformulated to

$$\Delta V_{max,c}=\frac{\Delta V_{max,w}}{\sqrt{N}}, \tag{3.13.29}$$

with $\Delta V_{max,c}$ denoting the maximum ΔV allowed for the wound wire. Therefore, winding the wire makes the system much more sensitive to voltage drops, in terms of temperature increase; for $N=1000$ as suggested, Eqs. (3.13.22) and (3.13.29) produce

$$\Delta V_{max,c}=\frac{2{,}800}{\sqrt{1000}}=89\text{ V}, \tag{3.13.30}$$

which is dramatically lower than without winding. This realization justifies why (liquid-mediated) cooling is used inside some electric machines, namely those encompassing large volumes of repeatedly coiled (cylindrical) wire – as is the case of large AC transformers; in attempts to extend their useful life – which hinges critically upon the thermal history of the insulation material used to enrob the said wire.

f) After redoing Eq. (2.45) in *Food Proc. Eng.: basics & mechanical operations* as

$$v_x=\frac{R^2\Delta P}{4\mu L}\left(1-\left(\frac{r}{R}\right)^2\right) \tag{3.13.31}$$

with R^2 factored out for convenience, both sides will be further multiplied by $\rho v_{x,max}/\Delta P$ to get

$$\frac{\rho v_{x,max}}{\Delta P}v_x=\frac{1}{4}\left(\frac{R}{L}\right)^2\frac{\rho Lv_{x,max}}{\mu}\left(1-\left(\frac{r}{R}\right)^2\right) \tag{3.13.32}$$

– along with multiplication of both numerator and denominator in the right-hand side by L; upon composing a square root with a square power in the left-hand side, and dividing both numerator and denominator thereof by 2, Eq (3.13.32) becomes

$$2\frac{\frac{1}{2}\left(\sqrt{v_{x,max}v_x}\right)^2}{\dfrac{\Delta P}{\rho}}=\frac{1}{4}\left(\frac{R}{L}\right)^2Re\left(1-\left(\frac{r}{R}\right)^2\right) \tag{3.13.33}$$

– where Reynolds' number is hereby defined as

$$Re\equiv\frac{\rho Lv_{x,max}}{\mu}, \tag{3.13.34}$$

in parallel to Eq. (2.92) in *Food Proc. Eng.: basics & mechanical operations*. If Re is replaced by the reciprocal of Lz, then Eq. (3.13.33) will coincide with Eq. (3.375) – while the left-hand side, equal to twice the kinetic energy term per unit mass, i.e. $\left(\sqrt{v_{x,max}v_x}\right)^2/2$ referred to the geometric mean of v_x and $v_{x,max}$, normalized by the dissipative work per unit mass, $\Delta P/\rho$, associated to viscous behavior, would play the role of dimensionless temperature, $\left(T-T_0\right)/T_0=T/T_0-1$. This noteworthy similarity arises from the analogy between momentum and heat transfer – where the contact (attrition) force associated with ΔP describes consumption of momentum per unit volume, in much the same way $\hbar$ describes production of heat per unit volume.

3.3.10.3.3 *Mechanical resistance*

In attempts to improve flexibility, electric circuitry resorts to cables rather than wires – as they can be more easily (and reversily) bent and folded for storage in coil form; however, such a flexibility also facilitates cables take a catenary-type shape when hanging between two fixation points – due to their own weight (and possibility of minute sliding of the individual, thinner wires relative to each other), as sketched in Fig. 3.48i. The freestanding, constant thickness cable under scrutiny takes indeed a (shallow) U-shape that resembles a parabolic arch – but is certainly not a true parabola. This shape is shared by suspension bridges, provided that weight of their walk/roadway is small compared to weight of the cable itself; if otherwise, then the latter would not hang freely, with weight being proportional to horizontal distance, rather than distance measured along the cable itself.

Mathematical simulation of the above system has classically departed from assumption of an idealized cable – so thin that it can be regarded as a curve, and so flexible that any force of tension exerted is tangential thereto; coupled with the hypothesis that the said cable behaves like a rigid body, once it has attained mechanical equilibrium. At the lowest point of the cable, termed vertex of the catenary, the slope dy/dx of the associated $y \equiv y\{x\}$ curve is accordingly nil; the tension developed is horizontal, with magnitude T_0, as highlighted in Fig. 3.48ii. When a finite portion of the cable, starting at the said minimum, is considered – as done in Fig. 3.48iii, then three forces act thereon: weight in the vertical direction, pointing downward and with magnitude W; tension at the lowest point, in the horizontal direction and magnitude T_0 (as seen above); and tension at the uppermost point, in tangential direction and with magnitude T. The weight is given by

$$W = \rho A s g, \tag{3.384}$$

where ρ denotes mass density of cable material, A and s denote (constant) cross-sectional area and (variable) length of cable, respectively, and g denotes acceleration of gravity. Tension $\boldsymbol{T}$ may, in turn, be decomposed as two normal components, see Fig. 3.48iv – one horizontal, of magnitude T_x given by

$$T_x \equiv T\cos\theta, \tag{3.385}$$

and another vertical, of magnitude T_y abiding to

$$T_y \equiv T\sin\theta; \tag{3.386}$$

here θ represents angle of the tangent to the curve with the horizontal plane, see Fig. 3.48iii. When the electric cable attains equilibrium, the sum of those forces must be nil, according to Newton's first law of mechanics – which implies

$$\boxed{T\cos\theta - T_0 = 0,} \tag{3.387}$$

valid for the horizontal direction and written with the aid of Eq. (3.385); and similarly

$$\boxed{T\sin\theta - \rho A g s = 0,} \tag{3.388}$$

valid for the vertical direction and resorting to Eqs. (3.384) and (3.386). After redoing Eqs. (3.387) and (3.388) as

$$\cos\theta = \frac{T_0}{T} \tag{3.389}$$

and

$$\sin\theta = \frac{\rho A g s}{T}, \tag{3.390}$$

respectively, one may pursue to ordered division of Eq. (3.390) by Eq. (3.389) to get

$$\frac{\sin\theta}{\cos\theta} = \frac{\dfrac{\rho A g s}{T}}{\dfrac{T_0}{T}}; \tag{3.391}$$

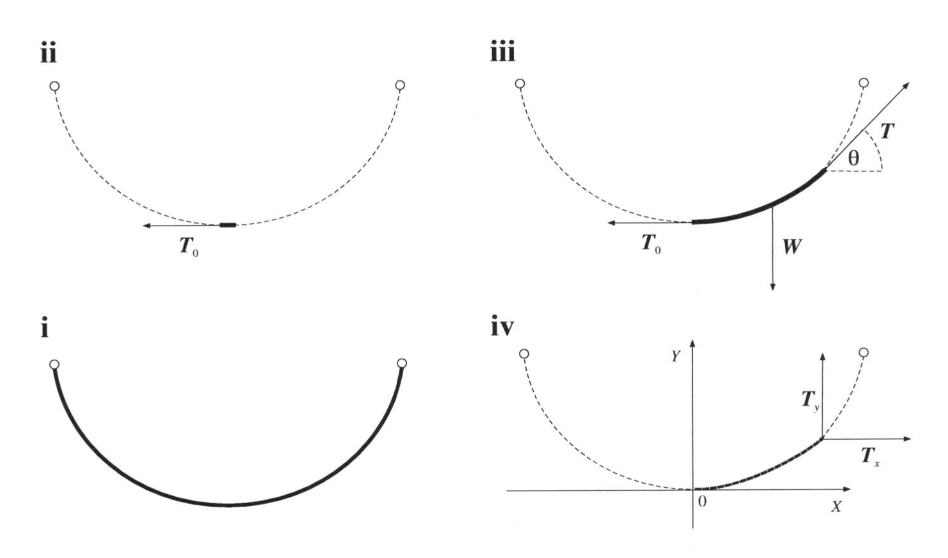

FIGURE 3.48 Schematic representation of: (i) full electric cable, suspended from two fixed points (○); detail of (ii) elementary segment at the vertex, subjected to horizontal tension T_0, and (iii) finite segment, with vertical weight $\boldsymbol{W}$, and subjected to $\boldsymbol{T}_0$ and inclined tension $\boldsymbol{T}$ forming an angle θ with the horizontal plane; and (iv) decomposition of $\boldsymbol{T}$ as Cartesian components, $\boldsymbol{T}_x$ in horizontal direction and $\boldsymbol{T}_y$ in vertical direction, with indication of reference set of Cartesian axes, $X0Y$, centered at the vertex.

Eq. (3.391) may instead appear as

$$\tan\theta = \Lambda s \tag{3.392}$$

– after dropping similar factors from numerator and denominator, defining auxiliary parameter Λ as

$$\boxed{\Lambda \equiv \frac{\rho A g}{T_0},} \tag{3.393}$$

and recalling the definition of trigonometric tangent. Note that the horizontal component of tension, T_x, is constant and equal to T_0, see Eqs. (3.385) and (3.387); while the vertical component thereof, T_y, is proportional (via factor ρAg) to the cable length considered so far starting at the vertex, i.e. s – see Eqs. (3.386) and (3.388). On the other hand,

$$\tan\theta \equiv \frac{dy}{dx}, \tag{3.394}$$

where the right-hand side represents slope of the straight line tangent to $y \equiv y\{x\}$ at the germane s; while the (differential) length of arch is given by

$$ds = \sqrt{1+\left(\frac{dy}{dx}\right)^2}\,dx. \tag{3.395}$$

Upon division of both sides by dx, followed by insertion of Eq. (3.394), one obtains

$$\frac{ds}{dx} = \sqrt{1+\tan^2\theta} \tag{3.396}$$

from Eq. (3.395) – where further combination with Eq. (3.392) unfolds

$$\frac{ds}{dx} = \sqrt{1+\left(\Lambda s\right)^2}; \tag{3.397}$$

after taking reciprocals of both sides and splitting the square power, Eq. (3.397) becomes

$$\boxed{\frac{dx}{ds} = \frac{1}{\sqrt{1+\Lambda^2 s^2}}.} \tag{3.398}$$

One may now rewrite dy/dx as

$$\frac{dy}{dx} = \frac{dy}{ds}\frac{ds}{dx}, \tag{3.399}$$

with the aid of the chain differentiation rule – using s as intermediate variable, and being aware of the univocal and monotonic relationship of y to s; where isolation of dy/ds yields

$$\frac{dy}{ds} = \frac{\dfrac{dy}{dx}}{\dfrac{ds}{dx}}. \tag{3.400}$$

Combination with Eqs. (3.392), (3.394), and (3.397) finally generates

$$\boxed{\frac{dy}{ds} = \frac{\Lambda s}{\sqrt{1+\Lambda^2 s^2}}} \tag{3.401}$$

from Eq. (3.400); note that Eqs. (3.398) and (3.401) represent a parametric version of dy/dx (or $y \equiv y\{x\}$, for that matter) as descriptor of cable shape.

Integration of Eq. (3.401) ensues via separation of variables, viz.

$$\int dy = \frac{1}{\Lambda}\int \frac{\Lambda s}{\sqrt{1+(\Lambda s)^2}}\,d(\Lambda s), \tag{3.402}$$

after multiplication and division of the right-hand side by (constant) Λ; Eq. (3.402) gives rise to

$$y = \frac{1}{\Lambda}\sqrt{1+(\Lambda s)^2} + \kappa_1, \tag{3.403}$$

where κ_1 denotes an (integration) arbitrary constant. Should the (vertical) y-axis be shifted by κ_1 in the y-direction, then a new variable Y would be in order, viz.

$$Y \equiv y - \kappa_1, \tag{3.404}$$

duly represented in Fig. 3.48iv; one would then rewrite Eq. (3.403) simply as

$$\boxed{Y = \frac{1}{\Lambda}\sqrt{1+(\Lambda s)^2}.} \tag{3.405}$$

The Y-axis is termed axis of the catenary curve – and contains its vertex, while serving as symmetry axis thereof.

Before proceeding with integration of the x-component of the curve as conveyed by Eq. (3.398), it is convenient to resort again to the chain-differentiation rule – besides the rules of differentiation of a logarithm and a square root, to get

$$\frac{d}{d(\kappa\zeta)}\left(\ln\left\{\kappa\zeta + \sqrt{1+(\kappa\zeta)^2}\right\}\right) = \frac{1+\dfrac{2(\kappa\zeta)}{2\sqrt{1+(\kappa\zeta)^2}}}{\kappa\zeta + \sqrt{1+(\kappa\zeta)^2}} = \frac{1+\dfrac{\kappa\zeta}{\sqrt{1+(\kappa\zeta)^2}}}{\kappa\zeta + \sqrt{1+(\kappa\zeta)^2}} \tag{3.406}$$

for derivative of $\ln\left\{\kappa\zeta + \sqrt{1+\left(\kappa\zeta\right)^2}\right\}$ with regard to $\kappa\zeta$, as part of an auxiliary calculation; here κ denotes some constant and ζ denotes a generic variable. Upon multiplication of both numerator and denominator by $\sqrt{1+(\kappa\zeta)^2}$, Eq. (3.406) becomes

$$\frac{d}{d\left(\kappa\zeta\right)}\left(\ln\left\{\kappa\zeta + \sqrt{1+(\kappa\zeta)^2}\right\}\right) = \frac{\kappa\zeta + \sqrt{1+(\kappa\zeta)^2}}{\sqrt{1+(\kappa\zeta)^2}\left(\kappa\zeta + \sqrt{1+(\kappa\zeta)^2}\right)}, \tag{3.407}$$

where cancellation of common factors between numerator and denominator leaves merely

$$\frac{d}{d(\kappa\zeta)}\left(\ln\left\{\kappa\zeta + \sqrt{1+(\kappa\zeta)^2}\right\}\right) = \frac{1}{\sqrt{1+(\kappa\zeta)^2}}. \tag{3.408}$$

One is now ready to proceed with integration of Eq. (3.398), via separation of variables, viz.

$$\int dx = \frac{1}{\Lambda}\int \frac{d(\Lambda s)}{\sqrt{1+(\Lambda s)^2}}, \tag{3.409}$$

again upon multiplication and division of the kernel by (constant) Λ; Eq. (3.409) readily leads to

$$x = \frac{1}{\Lambda}\ln\left\{\Lambda s + \sqrt{1+\left(\Lambda s\right)^2}\right\} + \kappa_2, \tag{3.410}$$

after combining with Eq. (3.408) with $\kappa \equiv \Lambda$ and $\zeta \equiv s$ – where κ_2 denotes another (integration) arbitrary constant.

Meanwhile, it is instructive to recall the inverse hyperbolic sine function, say,

$$\zeta \equiv \sinh^{-1}\xi, \tag{3.411}$$

with ζ and ξ denoting generic variables – where application of hyperbolic sine to both sides unfolds

$$\xi = \sinh\zeta; \tag{3.412}$$

recalling the fundamental relationship between hyperbolic functions, i.e.

$$\cosh^2\zeta - \sinh^2\zeta = 1, \tag{3.413}$$

one may insert Eq. (3.412) therein to get

$$\cosh^2\zeta - \xi^2 = 1 \tag{3.414}$$

– and, consequently,

$$\cosh\zeta = \sqrt{1+\xi^2} \tag{3.415}$$

upon isolation of the hyperbolic cosine (which only takes positive values). On the other hand, an exponential may always be expressed as a linear combination of hyperbolic sine and hyperbolic cosine of the same argument, according to

$$e^{\zeta} = \sinh\zeta + \cosh\zeta \tag{3.416}$$

– which will be reformulated to

$$e^{\zeta} = \xi + \sqrt{1+\xi^2}, \tag{3.417}$$

in view of Eqs. (3.412) and (3.415); after taking logarithms of both sides, Eq. (3.417) becomes

$$\zeta = \ln\left\{\xi + \sqrt{1+\xi^2}\right\} \tag{3.418}$$

or, after recalling Eq. (3.411),

$$\ln\left\{\xi + \sqrt{1+\xi^2}\right\} = \sinh^{-1}\xi. \tag{3.419}$$

Departing from the result conveyed by Eq. (3.419), it is possible to reformulate Eq. (3.410) to

$$x = \frac{1}{\Lambda}\sinh^{-1}\Lambda s + \kappa_2, \tag{3.420}$$

upon exchange of ξ and Λs; if the horizontal axis is shifted by κ_2 in the x-direction, with definition of new variable X as

$$X \equiv x - \kappa_2, \tag{3.421}$$

then Eq. (3.420) can be reformulated to

$$\boxed{X = \frac{1}{\Lambda}\sinh^{-1}\Lambda s} \tag{3.422}$$

– with the X-axis, called directrix of the catenary curve, containing its vertex (see Fig. 3.48iv). After rewriting Eq. (3.422) as

$$\sinh^{-1}\Lambda s = \Lambda X, \tag{3.423}$$

one may apply hyperbolic sine to both sides to get

$$\Lambda s = \sinh\Lambda X \tag{3.424}$$

as per (trivial) composition of a function with its inverse; Λs is now to be eliminated from Eq. (3.405) as

$$Y = \frac{1}{\Lambda}\sqrt{1+\sinh^2\Lambda X}, \tag{3.425}$$

at the expense of Eq. (3.424). On the other hand, Eq. (3.413) has it that

$$\cosh^2\zeta = 1 + \sinh^2\zeta, \tag{3.426}$$

which may be used to reformulate Eq. (3.425) to

$$Y = \frac{1}{\Lambda}\sqrt{\cosh^2\Lambda X} \tag{3.427}$$

after setting $\zeta \equiv \Lambda X$; Eq. (3.427) becomes merely

$$\boxed{\Lambda Y = \cosh\Lambda X,} \tag{3.428}$$

after composing square power with square root, and multiplying both sides by Λ. A graphical representation of the dimensionless

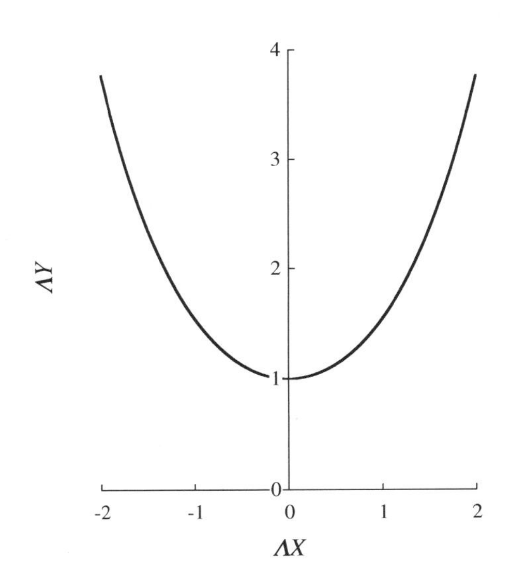

FIGURE 3.49 Shape acquired by electric cable, under mechanical equilibrium when suspended from its extremes, as variation of dimensionless vertical coordinate, ΛY, versus dimensionless horizontal coordinate, ΛX.

ordinate, as a function of the dimensionless abscissa is provided in Fig. 3.49. The exact shape of the catenary accordingly corresponds to a hyperbolic cosine; the surface of revolution of the catenary curve (or catenoid) is also a minimal surface, associated with the most stable mechanical configuration. The information conveyed by Fig. 3.49 is relevant, because it permits anticipation of the minimum vertical clearance required by (high-tension) cables – needed to define the height to the floor of the fixation points, so that operator safety is guaranteed in an industrial setting. Conversion to the original x- and y-coordinates then requires Eqs. (3.421) and (3.404), respectively, be taken into account – with constants κ_1 and κ_2 being fixed by height of vertex and half distance between the fixation points, respectively.

The magnitude of the tension experienced by the material of the cable is given by

$$T = \sqrt{T_x^2 + T_y^2}, \tag{3.429}$$

following definition of rectangular projections in Fig. 3.48iv, complemented by Pythagoras' theorem; insertion of Eqs. (3.385)–(3.388) gives then rise to

$$T = \sqrt{(T\cos\theta)^2 + (T\sin\theta)^2} = \sqrt{T_0^2 + (\rho Ags)^2}, \tag{3.430}$$

while division of both sides by T_0 unfolds

$$\boxed{\frac{T}{T_0} = \sqrt{1 + (\Lambda s)^2}} \tag{3.431}$$

with the aid of Eq. (3.393) – as plotted in Fig. 3.50, using cable length for abscissa. As expected, the overall tension is minimum at the bottom of the cable catenary, corresponding to only T_0, see Fig. 3.48ii and Eq. (3.431) for $s=0$; it then increases monotonicaly with s, to be eventually driven by a straight line with unit slope as $(\Lambda s)^2 \to \infty$, since $\sqrt{1+(\Lambda s)^2} \approx \sqrt{(\Lambda s)^2} = |\Lambda s| = \Lambda s$ under such circumstances – with cable weight overriding horizontal tension. A global maximum will accordingly be reached at either suspension point, abiding to

$$\boxed{\frac{T_{max}}{T_0} = \sqrt{1 + (\Lambda L)^2};} \tag{3.432}$$

T_{max} depends on both T_0 and full weight of cable, i.e. ρALg as per Eqs. (3.384) and (3.393), with L denoting full length.

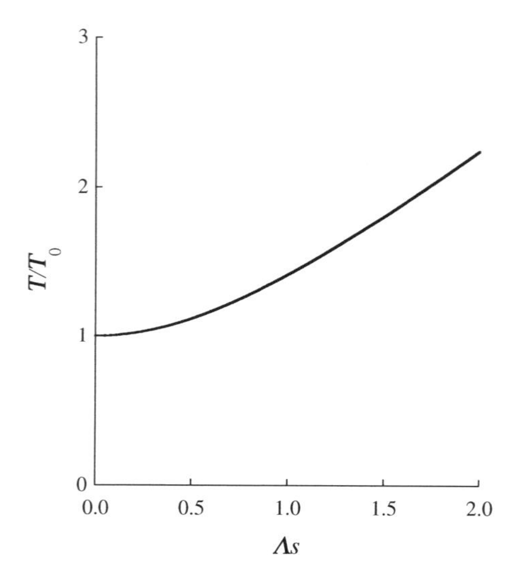

FIGURE 3.50 Variation of (tangential) dimensionless tension, T/T_0, in electric cable under mechanical equilibrium when suspended from its extremes, versus dimensionless distance along cable relative to vertex, Λs.

PROBLEM 3.14

Simulation of the static behavior of suspended electrical cables contributes to prevent safety problems in industrial environments – especially when high voltages are at stake.

a) Prove analytically that the maximum tangential tension occurs at the suspension points of an electrical cable.
b) Rewrite Eq. (3.428) as $y \equiv y\{x\}$, and comment on its symmetry.

Solution

a) The point(s) where magnitude of the tension force is maximum are described by

$$\frac{dT}{ds} = 0 \tag{3.14.1}$$

as necessary condition, using the s-coordinate along cable length for positioning; combination with Eq. (3.431) transforms Eq. (3.14.1) to

$$T_0 \frac{2(\Lambda s)\Lambda}{2\sqrt{1+(\Lambda s)^2}} = 0, \tag{3.14.2}$$

which readily simplifies to

$$\frac{s}{\sqrt{1+(\Lambda s)^2}} = 0 \tag{3.14.3}$$

after dropping 2 from both the numerator and denominator, and dividing both sides by $T_0\Lambda^2$. The solution of Eq. (3.14.3) is just

$$s = 0, \tag{3.14.4}$$

since $\sqrt{1+(\Lambda s)^2}$ remains always finite. The second-order derivative of T with regard to s reads, in turn,

$$\frac{\partial^2 T}{\partial s^2} \equiv \frac{\partial}{\partial s}\left(\frac{\partial T}{\partial s}\right) = \frac{\partial}{\partial s}\left(T_0 \frac{2(\Lambda s)\Lambda}{2\sqrt{1+(\Lambda s)^2}}\right) = \frac{\partial}{\partial s}\left(\frac{T_0\Lambda^2 s}{\sqrt{1+(\Lambda s)^2}}\right) \tag{3.14.5}$$

stemming from Eq. (3.14.2). Application of classical rules of differentiation converts Eq. (3.14.5) to

$$\frac{\partial^2 T}{\partial s^2}=T_0\Lambda^2\frac{\sqrt{1+(\Lambda s)^2}-s\dfrac{2(\Lambda s)\Lambda}{2\sqrt{1+(\Lambda s)^2}}}{\left(\sqrt{1+(\Lambda s)^2}\right)^2}\qquad(3.14.6)$$

$$=T_0\Lambda^2\frac{1+(\Lambda s)^2-(\Lambda s)^2}{\left(1+(\Lambda s)^2\right)\sqrt{1+(\Lambda s)^2}}$$

upon multiplication of both numerator and denominator by $\sqrt{1+(\Lambda s)^2}$; or else

$$\frac{\partial^2 T}{\partial s^2}=\frac{T_0\Lambda^2}{\left(1+(\Lambda s)^2\right)^{\frac{3}{2}}}>0,\qquad(3.14.7)$$

after cancelling out symmetrical terms in numerator and lumping similar factors in denominator. The positive sign of the left-hand side of Eq. (3.14.7) implies that a minimum for tension occurs at $s=0$ as stationary point, see Eq. (3.14.4); therefore, the maximum under scrutiny is of the local type, as (putatively) described by Eq. (3.14.1) – but simultaneously lies on a physical constraint (i.e. $s \geq 0$). After recalling that

$$\frac{\partial T}{\partial s}=T_0\frac{2(\Lambda s)\Lambda}{2\sqrt{1+(\Lambda s)^2}}=\frac{T_0\Lambda^2 s}{\sqrt{1+(\Lambda s)^2}}>0\qquad(3.14.8)$$

as per Eq. (3.14.2) again, one confirms that T increases monotonically with s (as mentioned previously) – so T will attain its (global) maximum value when s takes the uppermost value permitted, i.e. L; in other words, T/T_0 as per Eq. (3.431) will reach its (operational) maximum when $s=L$, in full agreement with Eq. (3.432).

b) Insertion of Eqs. (3.404) and (3.421) converts Eq. (3.428) to

$$\Lambda\left(y-\kappa_1\right)=\cosh\Lambda\left(x-\kappa_2\right);\qquad(3.14.9)$$

if the fixation points of the cable are described by (x_1,y_0) and (x_2,y_0) – corresponding to a common distance, y_0, to the floor, then

$$\Lambda\left(y_0-\kappa_1\right)=\cosh\Lambda\left(x_1-\kappa_2\right)\qquad(3.14.10)$$

and

$$\Lambda\left(y_0-\kappa_1\right)=\cosh\Lambda\left(x_2-\kappa_2\right)\qquad(3.14.11)$$

have to be enforced for compatibility with Eq. (3.14.9). Upon elimination of $\Lambda(y_0-\kappa_1)$ between Eqs. (3.14.10) and (3.14.11), one obtains

$$\cosh\Lambda\left(x_1-\kappa_2\right)=\cosh\Lambda\left(x_2-\kappa_2\right)\qquad(3.14.12)$$

that degenerates to

$$\Lambda\left(x_1-\kappa_2\right)=\Lambda\left(x_2-\kappa_2\right)\vee\Lambda\left(x_1-\kappa_2\right)=-\Lambda\left(x_2-\kappa_2\right)\qquad(3.14.13)$$

– because hyperbolic cosine is an even function; the first possibility is worthless, as it would imply coincidence of (distinct) fixation points. However, the second condition is to be retained – after rephrasing as

$$x_1-\kappa_2=\kappa_2-x_2\qquad(3.14.14)$$

via division by $\Lambda \neq 0$, consistent with Eq. (3.393); upon isolation of κ_2, Eq. (3.14.14) leads to

$$\kappa_2=\frac{x_1+x_2}{2}.\qquad(3.14.15)$$

Insertion of Eq. (3.14.15) transforms Eq. (3.14.10) to

$$\Lambda\left(y_0-\kappa_1\right)=\cosh\Lambda\left(x_1-\frac{x_1+x_2}{2}\right),\qquad(3.14.16)$$

where the two terms in parentheses can be lumped as

$$\Lambda\left(y_0-\kappa_1\right)=\cosh\Lambda\frac{2x_1-x_1-x_2}{2}=\cosh\Lambda\frac{x_1-x_2}{2};\qquad(3.14.17)$$

isolation of κ_1 then yields

$$\kappa_1=y_0-\frac{\cosh\Lambda\dfrac{x_1-x_2}{2}}{\Lambda}.\qquad(3.14.18)$$

In view of Eqs. (3.14.15) and (3.14.18), one may reformulate Eq. (3.14.9) to

$$\Lambda\left(y-\left(y_0-\frac{\cosh\Lambda\dfrac{x_1-x_2}{2}}{\Lambda}\right)\right)=\cosh\Lambda\left(x-\frac{x_1+x_2}{2}\right),\qquad(3.14.19)$$

which promptly simplifies to

$$\Lambda\left(y-y_0\right)+\cosh\Lambda\frac{x_1-x_2}{2}=\cosh\Lambda\left(x-\frac{x_1+x_2}{2}\right);\qquad(3.14.20)$$

isolation of $\Lambda(y-y_0)$ gives rise to

$$\Lambda\left(y-y_0\right)=\cosh\Lambda\left(x-\frac{x_1+x_2}{2}\right)-\cosh\Lambda\frac{x_1-x_2}{2},\qquad(3.14.21)$$

where solution for y unfolds

$$y=y_0+\frac{\cosh\Lambda\left(x-\dfrac{x_1+x_2}{2}\right)-\cosh\Lambda\dfrac{x_1-x_2}{2}}{\Lambda}.\qquad(3.14.22)$$

Recalling the universal mathematical relationship between diference of hyperbolic cosines and product of hyperbolic sines, viz.

$$\cosh\zeta - \cosh\xi = 2\sinh\frac{\zeta+\xi}{2}\sinh\frac{\zeta-\xi}{2}, \quad (3.14.23)$$

it becomes possible to rewrite Eq. (3.14.22) as

$$y = y_0 + \frac{2}{\Lambda}\sinh\frac{\Lambda\left(x - \frac{x_1+x_2}{2}\right) + \Lambda\frac{x_1-x_2}{2}}{2} \times\sinh\frac{\Lambda\left(x - \frac{x_1+x_2}{2}\right) - \Lambda\frac{x_1-x_2}{2}}{2} \quad (3.14.24)$$

– after replacing ζ by $\Lambda\left(x - \frac{x_1+x_2}{2}\right)$ and ξ by $\Lambda\frac{x_1-x_2}{2}$; elimination of parentheses and splitting of fractions then produce

$$y = y_0 + \frac{2}{\Lambda}\sinh\frac{\Lambda x - \Lambda\frac{x_1}{2} - \Lambda\frac{x_2}{2} + \Lambda\frac{x_1}{2} - \Lambda\frac{x_2}{2}}{2} \times\sinh\frac{\Lambda x - \Lambda\frac{x_1}{2} - \Lambda\frac{x_2}{2} - \Lambda\frac{x_1}{2} + \Lambda\frac{x_2}{2}}{2}, \quad (3.14.25)$$

where cancellation of symmetrical terms coupled with condensation of terms alike generate

$$y = y_0 + \frac{2}{\Lambda}\sinh\left\{\frac{\Lambda x - \Lambda x_2}{2}\right\}\sinh\left\{\frac{\Lambda x - \Lambda x_1}{2}\right\}. \quad (3.14.26)$$

A final factoring out of $\Lambda/2$ in the argument of the hyperbolic functions in Eq. (3.14.26) leaves

$$y = y_0 + \frac{2}{\Lambda}\sinh\left\{\frac{\Lambda}{2}(x - x_1)\right\}\sinh\left\{\frac{\Lambda}{2}(x - x_2)\right\} \quad (3.14.27)$$

that conveys y as a direct function of x, as requested – where the symmetrical behavior about the median of x_1 and x_2 is apparent, since y would remain unaltered should x_1 and x_2 be exchanged.

3.3.11 Sensors, controllers, and actuators

In its simplest form, a piece of equipment may be viewed as a black box – with an outlet response in time, $y_{out}\{t\}$, developing as a consequence of an inlet disturbance, $x_{dt}\{t\}$, and an inlet manipulated variable, $x_{mp}\{t\}$, in general agreement with Fig. 3.15i.

Consider, in this regard, a well-stirred tank – where a mixing/reaction process takes place at preset temperature T, above room temperature, and kept so by feeding saturated steam as needed, at temperature $T_{st} > T$; this process is susceptible to unexpected, spurious changes in temperature of its inlet stream, T_0. An enthalpy balance to the said equipment looks like

$$\boxed{Q\rho c_P\left(T_0 - T^\theta\right) + UA\left(T_{st} - T\right) = Q\rho c_P\left(T - T^\theta\right) + V\rho c_P\frac{dT}{dt},} \quad (3.433)$$

where Q denotes inlet/outlet volumetric flow rate of liquid, possessing mass density ρ and (isobaric) specific heat capacity c_P, T^θ denotes (arbitrary) temperature serving as a reference for measurement of enthalpy, V denotes holdup volume of tank, A denotes heat transfer area in tank, and U denotes overall heat transfer coefficient. After lumping similar terms, Eq. (3.433) becomes

$$UA\left(T_{st} - T\right) = Q\rho c_P\left(T - T_0\right) + V\rho c_P\frac{dT}{dt} \quad (3.434)$$

– where, as expected, T^θ drops out; division of both sides by $Q\rho c_P$ permits further simplification to

$$\frac{V}{Q}\frac{UA}{V\rho c_P}\left(T_{st} - T\right) = T - T_0 + \frac{V}{Q}\frac{dT}{dt}, \quad (3.435)$$

where the left-hand side was also multiplied and divided by V. After setting

$$\tau \equiv \frac{V}{Q}, \quad (3.436)$$

with τ denoting space-time as characteristic hydrodynamic parameter, coupled to

$$\kappa \equiv \frac{UA}{V\rho c_P}, \quad (3.437)$$

with κ denoting characteristic heat transfer parameter, Eq. (3.435) simplifies to

$$\tau\frac{dT}{dt} = \kappa\tau\left(T_{st} - T\right) + T_0 - T. \quad (3.438)$$

Under steady state conditions of operation, Eq. (3.438) degenerates to

$$0 = \kappa\tau\left(T_{st}^{ss} - T^{ss}\right) + T_0^{ss} - T^{ss}, \quad (3.439)$$

because $dT/dt = 0$ by definition of steady state – where superscript ss refers explicitly thereto; ordered subtraction of Eq. (3.439) from Eq. (3.438) gives rise to

$$\tau\frac{dT}{dt} = \kappa\tau\left(T_{st} - T\right) + T_0 - T - \kappa\tau\left(T_{st}^{ss} - T^{ss}\right) - \left(T_0^{ss} - T^{ss}\right), \quad (3.440)$$

which may be algebraically rearranged as

$$\tau\frac{dT}{dt} = \kappa\tau\left(\left(T_{st} - T_{st}^{ss}\right) - \left(T - T^{ss}\right)\right) + \left(T_0 - T_0^{ss}\right) - \left(T - T^{ss}\right) \quad (3.441)$$

after factoring out $\kappa\tau$ and grouping terms alike. Definition of deviation variables (relative to steady state) as

$$\hat{T} \equiv T - T^{ss}, \quad (3.442)$$

$$\hat{T}_{st} \equiv T_{st} - T_{st}^{ss}, \quad (3.443)$$

and

$$\hat{T}_0 \equiv T_0 - T_0^{ss} \quad (3.444)$$

permits simplification of notation in Eq. (3.441) to

$$\tau \frac{dT}{dt} = \kappa\tau\left(\hat{T}_{st} - \hat{T}\right) + \hat{T}_0 - \hat{T}; \quad (3.445)$$

on the other hand, differentiation of Eq. (3.442) produces

$$\frac{d\hat{T}}{dt} = \frac{dT}{dt}, \quad (3.446)$$

owing to constancy of T^{ss} – so Eq. (3.445) can be rewritten as

$$\tau \frac{d\hat{T}}{dt} = \kappa\tau\left(\hat{T}_{st} - \hat{T}\right) + \hat{T}_0 - \hat{T}. \quad (3.447)$$

Upon separate grouping of terms in $\hat{T}$, $\hat{T}_0$, and $\hat{T}_{st}$, Eq. (3.447) will take the form

$$\tau \frac{d\hat{T}}{dt} = \hat{T}_0 + \kappa\tau\hat{T}_{st} - (1+\kappa\tau)\hat{T}; \quad (3.448)$$

consequently, one ends up with a differential equation that relates $\hat{T}$ as product (outlet) variable (alias y_{out}) to $\hat{T}_0$ as disturbance (input) variable (alias x_{dt}), and also to $\hat{T}_{st}$ as manipulated (input) variable (alias x_{mp}); the alternative notation

$$\boxed{\tau \frac{dy_{out}}{dt} = x_{dt} + \kappa\tau x_{mp} - (1+\kappa\tau)y_{out}} \quad (3.449)$$

will thus be in order, as basic descriptor of transients in equipment-mediated processing. The control theory assumes, by default, that the system is initially operating under steady state, so any change in x_{dt} actually means a deviation from the said (intended) steady state – thus justifying the algebraic manipulation above, which resorted to deviation variables; the same obviously applies to x_{mp} and y_{out}. Note that a typical food process is normally described by some ordinary differential equation in time (also termed concentrated-parameter model), as is the case of Eq. (3.433) – or, at most, a partial differential equation, where terms in space coordinate(s) are also present (termed distributed-parameter model instead).

Since modeling of transients is nuclear in control, attempts to solve Eq. (3.449) ensue – so as to obtain $y_{out} \equiv y_{out}\{t\}$; this is easier to perform in Laplace's domain, according to

$$\tau\left(s\bar{y}_{out} - y_{out}\big|_{t=0}\right) = \bar{x}_{dt} + \kappa\tau\bar{x}_{mp} - (1+\kappa\tau)\bar{y}_{out}. \quad (3.450)$$

This approach is warranted because Laplace's transforms are integral transforms in time that start from $t=0$ – at which the (steady) state of the system is known; and eventually run unbounded in time – as would normally happen before a steady state is again attained. Since y_{out} represents a deviation variable relative to the state prevailing at $t=0$ as per Eq. (3.442), then Eq. (3.450) may be reduced to

$$\tau s\bar{y}_{out} = \bar{x}_{dt} + \kappa\tau\bar{x}_{mp} - (1+\kappa\tau)\bar{y}_{out} \quad (3.451)$$

as $y_{out}\big|_{t=0}$ is (obviously) nil. Remember that application of Laplace's transforms requires linearity of the departing (differential) equation – a requirement satisfied, in particular, by Eq. (3.449); in the general case, however, this may not hold, with Eq. (3.449) being replaced by

$$\frac{dY_{out}}{dt} = f\left\{X_{dt}, X_{mp}, Y_{out}\right\} \quad (3.452)$$

– where f denotes some nonlinear function of X_{dt}, X_{mp}, and Y_{out} as disturbance, manipulated, and controlled variables, respectively. In this case, one should first linearize the right-hand side of Eq. (3.452) using Taylor's multivariate expansion, around the initial steady state – according to

$$f\left\{X_{dt}, X_{mp}, Y_{out}\right\} \approx f\big|_{X_{dt}^{ss}, X_{mp}^{ss}, Y_{out}^{ss}} + \frac{\partial f}{\partial X_{dt}}\bigg|_{X_{dt}^{ss}, X_{mp}^{ss}, Y_{out}^{ss}}\left(X_{dt} - X_{dt}^{ss}\right) + \frac{\partial f}{\partial X_{mp}}\bigg|_{X_{dt}^{ss}, X_{mp}^{ss}, Y_{out}^{ss}}\left(X_{mp} - X_{mp}^{ss}\right) + \frac{\partial f}{\partial Y_{out}}\bigg|_{X_{dt}^{ss}, X_{mp}^{ss}, Y_{out}^{ss}}\left(Y_{out} - Y_{out}^{ss}\right); \quad (3.453)$$

after setting

$$x_{dt} \equiv X_{dt} - X_{dt}^{ss}, \quad (3.454)$$

$$x_{mp} \equiv X_{mp} - X_{mp}^{ss}, \quad (3.455)$$

and

$$y_{out} \equiv Y_{out} - Y_{out}^{ss}, \quad (3.456)$$

Eq. (3.453) becomes

$$f\left\{X_{dt}, X_{mp}, Y_{out}\right\} = f\big|_{X_{dt}^{ss}, X_{mp}^{ss}, Y_{out}^{ss}} + \frac{\partial f}{\partial X_{dt}}\bigg|_{X_{dt}^{ss}, X_{mp}^{ss}, Y_{out}^{ss}} x_{dt} + \frac{\partial f}{\partial X_{mp}}\bigg|_{X_{dt}^{ss}, X_{mp}^{ss}, Y_{out}^{ss}} x_{mp} + \frac{\partial f}{\partial Y_{out}}\bigg|_{X_{dt}^{ss}, X_{mp}^{ss}, Y_{out}^{ss}} y_{out}. \quad (3.457)$$

Therefore, Eq. (3.452) ought be replaced by

$$\frac{dy_{out}}{dt} \approx f\big|_{X_{dt}^{ss}, X_{mp}^{ss}, Y_{out}^{ss}} + \frac{\partial f}{\partial X_{dt}}\bigg|_{X_{dt}^{ss}, X_{mp}^{ss}, Y_{out}^{ss}} x_{dt} + \frac{\partial f}{\partial X_{mp}}\bigg|_{X_{dt}^{ss}, X_{mp}^{ss}, Y_{out}^{ss}} x_{mp} + \frac{\partial f}{\partial Y_{out}}\bigg|_{X_{dt}^{ss}, X_{mp}^{ss}, Y_{out}^{ss}} y_{out}, \quad (3.458)$$

upon combination with Eq. (3.457) – again because dy_{out} coincides with dY_{out}, following differentiation of Eq. (3.456); an ordinary linear differential equation has thus been produced – since $f|_{X_{dt}^{ss},X_{mp}^{ss},Y_{out}^{ss}}$, $(\partial f/\partial X_{dt})|_{X_{dt}^{ss},X_{mp}^{ss},Y_{out}^{ss}}$, $(\partial f/\partial X_{mp})|_{X_{dt}^{ss},X_{mp}^{ss},Y_{out}^{ss}}$, and $(\partial f/\partial Y_{out})|_{X_{dt}^{ss},X_{mp}^{ss},Y_{out}^{ss}}$ are all constants. At steady state – described by $dy_{out}/dt = 0$, Eq. (3.458) reduces to

$$0 = f|_{X_{dt}^{ss},X_{mp}^{ss},Y_{out}^{ss}} \tag{3.459}$$

since Eqs. (3.454)–(3.456) have it that $x_{dt}=x_{mp}=y_{out}=0$, in view of $X_{dt}=X_{dt}^{ss}$, $X_{mp}=X_{mp}^{ss}$, and $Y_{out}=Y_{out}^{ss}$; hence, Eq. (3.458) simplifies to

$$\frac{dy_{out}}{dt} = \left.\frac{\partial f}{\partial X_{dt}}\right|_{X_{dt}^{ss},X_{mp}^{ss},Y_{out}^{ss}} x_{dt} + \left.\frac{\partial f}{\partial X_{mp}}\right|_{X_{dt}^{ss},X_{mp}^{ss},Y_{out}^{ss}} x_{mp} + \left.\frac{\partial f}{\partial Y_{out}}\right|_{X_{dt}^{ss},X_{mp}^{ss},Y_{out}^{ss}} y_{out}. \tag{3.460}$$

Despite the inherent (and unavoidable) error associated with linearization of a function, it should be stressed that industrial control exists to avoid large deviations from an initial/intended steady state – so that the amount of product off-specification will be kept to a minimum; hence, Eq. (3.453) proves particularly appropriate for process modeling purposes, and to support simulation of efficient control devices – because it becomes more and more accurate as the expansion point of Taylor's series is approached.

In view of the similar functional forms of Eqs. (3.449) and (3.460) – as it suffices to replace $(\partial f/\partial X_{dt})|_{ss}$, $(\partial f/\partial X_{mp})|_{ss}$, and $(\partial f/\partial Y_{out})|_{ss}$ by $1/\tau$, κ, and $-(1+\kappa\tau)/\tau$, respectively, a single strategy of solution emerges; one should accordingly resort to Eq. (3.451), and lump like terms as

$$(1+\kappa\tau+\tau s)\overline{y}_{out} = \overline{x}_{dt} + \kappa\tau\overline{x}_{mp}. \tag{3.461}$$

Division of both sides of Eq. (3.461) by $1+\kappa\tau+\tau s$ unfolds

$$\overline{y}_{out} = \frac{1}{1+\kappa\tau+\tau s}\overline{x}_{dt} + \frac{\kappa\tau}{1+\kappa\tau+\tau s}\overline{x}_{mp}; \tag{3.462}$$

Eq. (3.462) appears more frequently as

$$\boxed{\overline{y}_{out} = G_{dt}\{s\}\overline{x}_{dt} + G_{mp}\{s\}\overline{x}_{mp},} \tag{3.463}$$

provided that

$$\boxed{G_{dt}\{s\} \equiv \left(\frac{\overline{y}_{out}}{\overline{x}_{dt}}\right)_{\overline{x}_{mp}=0}} \tag{3.464}$$

represents transfer function of the equipment relating output variabe to input disturbance, in general – with

$$\boxed{G_{dt}\{s\} = \frac{1}{1+\kappa\tau+\tau s}} \tag{3.465}$$

holding in the case under scrutiny, see Eq. (3.463) *vis-à-vis* with Eq. (3.462). By the same token,

$$\boxed{G_{mp}\{s\} \equiv \left(\frac{\overline{y}_{out}}{\overline{x}_{mp}}\right)_{\overline{x}_{dt}=0}} \tag{3.466}$$

represents transfer function of the equipment, relating (controlled) output variable to input actuated (or manipulated) variable at large – where

$$\boxed{G_{mp}\{s\} = \frac{\kappa\tau}{1+\kappa\tau+\tau s}} \tag{3.467}$$

represents the said transfer function in the current case, again based on comparative inspection of Eqs. (3.462) and (3.463); an illustration of Eq. (3.463), in block form and valid in Laplace's domain, is conveyed by Fig. 3.51i. Note the existence of material exchanges between process and surroundings; with two degrees of freedom as inlet variables, $\overline{x}_{dt}$ and $\overline{x}_{mp}$ – both affecting the (single) outlet variable, $\overline{y}_{out}$.

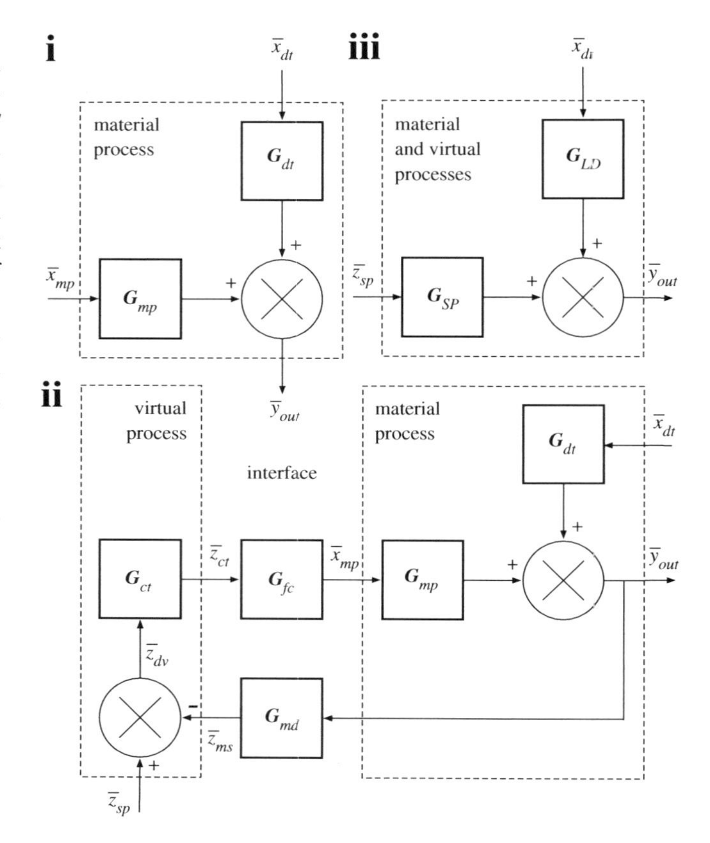

FIGURE 3.51 Schematic representation of material process in Laplace's domain – encompassing (i, ii, iii) equipment with two inlets, (disturbance) $\overline{x}_{dt}$ and (manipulation) $\overline{x}_{mp}$, and one outlet $\overline{y}_{out}$, related via transfer function G_{ct} and G_{mp}, respectively; and (ii) virtual process in controlled system – encompassing one material inlet, $\overline{y}_{out}$, and one virtual inlet, (setpoint) $\overline{z}_{sp}$, as well as one material outlet, $\overline{x}_{mp}$, besides one virtual inlet to, (measurement) $\overline{z}_{ms}$, one virtual outlet from, (control action) $\overline{z}_{ct}$, and one virtual transfer within, (deviation) $\overline{z}_{dv}$, making up the controlling mechanism, related via transfer function G_{ct} in the virtual process itself, and transfer functions G_{md} and G_{fc} at the interface between material and virtual processes; represented (ii) comprehensively, or (iii) briefly upon joining material and virtual processes, and defining lumped transfer functions G_{LD} and G_{SP}.

When a (feedback) control is connected to the aforementioned process, three extra elements are added – a measuring device (or sensor) and a final control element (or actuator), either of which plays the role of interface between (physical) process and (virtual) control; and the controller apparatus itself. The measuring device produces a signal $\bar{z}_{ms}$ from $\bar{y}_{out}$. The controller mechanism is constituted by a passive subunit (or comparator), which compares $\bar{z}_{ms}$ to a desired setpoint $\bar{z}_{sp}$ (obviously related to the true setpoint $\bar{y}_{sp}$, in the same form $\bar{z}_{ms}$ is related to $\bar{y}_{out}$); this originates a deviation response $\bar{z}_{dv}$. The comparator is coupled to an active subunit (or controller proper) that analyzes $\bar{z}_{dv}$, and accordingly produces a decision $\bar{z}_{ct}$. The actuator then takes up $\bar{z}_{ct}$, and converts it to a change in manipulated variable $\bar{x}_{mp}$. These apparatuses – each characterized by a transfer function in Laplace's domain, and the corresponding interactions are depicted in Fig. 3.51ii. It should be outlined that the degree of freedom originally associated with x_{mp} in the plain material process (or open-loop configuration), see Fig. 3.51i, has now been replaced by the degree of freedom accounted for by z_{sp} in the controlled process (or closed-loop configuration).

Two types of probems are normally considered: regulator problems – where $\bar{x}_{dt}$ varies, but $\bar{z}_{sp}$ is kept constant; and servo problems – where $\bar{z}_{sp}$ varies, but $\bar{x}_{dt}$ remains constant. The underlying approaches are not different in essence, yet attention will hereafter be focused on the former one – due to its overriding relevance in industrial practice.

The comparator subunit of the controlling mechanism operates in the time domain, and calculates the difference, Z_{dv}, between setpoint, Z_{sp}, and measurement, Z_{ms}, viz.

$$\boxed{Z_{dv}\{t\} = \left(Z_{sp}\{t\} - Z_{sp}^{ss}\right) - \left(Z_{ms}\{t\} - Z_{ms}^{ss}\right) + Z_{dv}^{ss}} \tag{3.468}$$

– written, for convenience, in deviation form. The baseline signal at steady state, Z_{dv}^{ss}, may, in turn, be decomposed as

$$Z_{dv}^{ss} \equiv Z_{sp}^{ss} - Z_{ms}^{ss} \tag{3.469}$$

– with Z_{sp}^{ss} and Z_{ms}^{ss} denoting steady state values for Z_{sp} and Z_{ms}, respectively; this permits expression of Z_{dv} as difference between Z_{sp} and Z_{ms}. More often, one resorts to

$$z_{dv} \equiv Z_{dv} - Z_{dv}^{ss} \tag{3.470}$$

in lieu of Z_{dv}, so Eq. (3.468) becomes analogous to

$$\boxed{z_{dv} \equiv z_{sp} - z_{ms}} \tag{3.471}$$

– provided that

$$z_{sp} \equiv Z_{sp} - Z_{sp}^{ss} \tag{3.472}$$

and

$$z_{ms} \equiv Z_{ms} - Z_{ms}^{ss} \tag{3.473}$$

are taken into account, together with Eq. (3.470); one accordingly resorts to

$$\boxed{\bar{z}_{dv}\{s\} \equiv \bar{z}_{sp} - \bar{z}_{ms}} \tag{3.474}$$

as alias of Eq. (3.471) in Laplace's domain.

The role of the controller itself is more complex; classically, three types have met with industrial success – proportional (P), proportional/integral (PI), and proportional/integral/derivative (PID), sorted by increasing degrees of effectiveness. All are easily implemented in industrial settings via programmable logical controllers, but differ in type of response triggered to the deviation signal received.

The proportional controller responds simply via $Z_{ct,P}$, given by

$$\boxed{Z_{ct,P}\{t\} = K_P z_{dv}\{t\} + Z_{ct}^{ss},} \tag{3.475}$$

where K_P denotes proportional gain; Z_{ct}^{ss} denotes bias signal of the controller – i.e. actuating signal when the processing unit is operated under steady state (in which case $z_{dv}=0$). The aforementioned proportionality constant is often accessed as proportional band, B_P, defined as

$$B_P \equiv \frac{100}{K_P}; \tag{3.476}$$

B_P characterizes the value by which the error must change, in order to drive the actuating signal of the controller over its full range. Therefore, the larger K_P (or the smaller B_P), the higher the sensitivity of the controller to z_{dv}; usual values for industrial operation lie within $1 \leq B_P \leq 500$. Subtraction of Z_{ct}^{ss} from both sides of Eq. (3.475) gives rise to

$$z_{ct,P} = K_P z_{dv} \tag{3.477}$$

– where z_{ct} denotes, in general, deviation of Z_{ct} relative to its steady state value, Z_{ct}^{ss}, i.e.

$$\boxed{z_{ct} \equiv Z_{ct} - Z_{ct}^{ss};} \tag{3.478}$$

the extra subscript P refers specifically to proportional action. After taking Laplace's transforms of both sides, Eq. (3.477) becomes

$$\bar{z}_{ct,P}\{s\} \equiv \mathcal{L}\left(z_{ct,P}\right) \equiv \mathcal{L}\left(Z_{ct,P} - Z_{ct}^{ss}\right) = K_P \bar{z}_{dv}\{s\} \tag{3.479}$$

also with the aid of Eq. (3.478), so the transfer function of a proportional (feedback) controller reads merely

$$\boxed{G_P\{s\} \equiv \frac{\bar{z}_{ct,P}}{\bar{z}_{dv}} = K_P,} \tag{3.480}$$

– as a particular case of

$$\boxed{G_{ct}\{s\} \equiv \frac{\bar{z}_{ct}}{\bar{z}_{dv}},} \tag{3.481}$$

applicable to any type of controller.

If integral action is added to pre-existing proportional action, then one obtains

$$\boxed{Z_{ct,PI}\{t\} = K_P\left(z_{dv}\{t\} + \frac{1}{\tau_I}\int_0^t z_{dv}\{\tilde{t}\}\,d\tilde{t}\right) + Z_{ct}^{ss}} \tag{3.482}$$

as design equation for a PI controller – where τ_I denotes integral time constant, or reset time; the most common range of operation is $5 \leq \tau_I \leq 3 \times 10^3$ s – yet its reciprocal, R_{PI} (or reset rate), viz.

$R_{PI} \equiv 1/\tau_I$, is again preferred for nominal reference in industrial practice. Should Laplace's transforms be taken of both sides of Eq. (3.482) – after moving Z_{ct}^{ss} to its left-hand side and using Eq. (3.478) as a template for definition of $z_{ct,PI}$, one gets

$$\overline{z}_{ct,PI}\{s\} \equiv \mathcal{L}\left(z_{ct,PI}\right) \equiv \mathcal{L}\left(Z_{ct,PI} - Z_{ct}^{ss}\right) = K_P\left(\overline{z}_{dv} + \frac{1}{\tau_I}\mathcal{L}\left(\int_0^t z_{dv}\, d\tilde{t}\right)\right); \tag{3.483}$$

advantage was meanwhile taken of the linearity of such an integral transform. Remember that Laplace's transform of an integral, between 0 and t, coincides with Laplace's transform of its kernel, divided by s afterward – so Eq. (3.483) may be rephrased as

$$\overline{z}_{ct,PI} = K_P\left(\overline{z}_{dv} + \frac{1}{\tau_I}\frac{1}{s}\mathcal{L}\left(z_{dv}\right)\right). \tag{3.484}$$

Upon standardizing notation and then factoring out $\overline{z}_{dv}$, one obtains

$$\overline{z}_{ct,PI} = K_P\left(1 + \frac{1}{\tau_I s}\right)\overline{z}_{dv} \tag{3.485}$$

from Eq. (3.484); hence, the transfer function of a PI (feedback) controller looks like

$$\boxed{G_{PI}\{s\} \equiv \frac{\overline{z}_{ct,PI}}{\overline{z}_{dv}} = K_P\left(1 + \frac{1}{\tau_I s}\right),} \tag{3.486}$$

consistent with Eq. (3.481). Integral action takes system history into account, and causes the controller output to change as long as a deviation from the setpoint remains; hence, PI controllers will be able to eventually eliminate even small errors. Conversely, if such errors do not vanish quickly and enough time elapses, they will generate larger and larger values for the integral term – which, in turn, keeps increasing the control action until saturation (attained when the actuator reaches its maximum, or minimum value possible). In other words, a PI controller usually needs special provisions to cope with the so-called integral windup.

If a derivative action is added to a PI, then a PID control arises – described by

$$\boxed{Z_{ct,PID}\{t\} = K_P\left(z_{dv}\{t\} + \frac{1}{\tau_I}\int_0^t z_{dv}\{\tilde{t}\}\, d\tilde{t} + \tau_D\frac{dz_{dv}\{t\}}{dt}\right) + Z_{ct}^{ss},} \tag{3.487}$$

where τ_D denotes derivative time constant; such a derivative term responds proportionally to the rate of variation of z_{dv}, so this type of controller is frequently called proportional-plus-reset-plus-rate in industrial settings. Recalling Eq. (3.478) – and after having abstracted Z_{ct}^{ss} from both sides, one may proceed to application of Laplace's transforms to Eq. (3.487), viz.

$$\begin{aligned}\overline{z}_{ct,PID}\{s\} &\equiv \mathcal{L}\left(z_{ct,PID}\right) \equiv \mathcal{L}\left(Z_{ct,PID} - Z_{ct}^{ss}\right)\\ &= K_P\left(\overline{z}_{dv} + \frac{1}{\tau_I}\mathcal{L}\left(\int_0^t z_{dv}\, d\tilde{t}\right) + \tau_D\mathcal{L}\left(\frac{dz_{dv}}{dt}\right)\right);\end{aligned} \tag{3.488}$$

once again, linearity of the said integral operator supports this reasoning. The mathematical property used to generate Eq. (3.485) from Eq. (3.484) can be recalled, and coupled with that associated with a derivative – as the product of s by Laplace's transform of the original function, to generate

$$\overline{z}_{ct,PID} = K_P\left(\overline{z}_{dv} + \frac{1}{\tau_I}\frac{1}{s}\mathcal{L}\left(z_{dv}\right) + \tau_D s\mathcal{L}\left(z_{dv}\right)\right) \tag{3.489}$$

from Eq. (3.488). After simplifying notation and factoring out $\overline{z}_{dv}$, Eq. (3.489) gives rise to

$$\overline{z}_{ct,PID} = K_P\left(1 + \frac{1}{\tau_I}\frac{1}{s} + \tau_D s\right)\overline{z}_{dv}; \tag{3.490}$$

division of both sides by $\overline{z}_{dv}$ further converts Eq. (3.490) to

$$\boxed{G_{PID}\{s\} \equiv \frac{\overline{z}_{ct,PID}}{\overline{z}_{dv}} = K_P\left(1 + \frac{1}{\tau_I s} + \tau_D s\right),} \tag{3.491}$$

representing the transfer function of a PID (feedback) controller – in agreement again with Eq. (3.481). Due to presence of a derivative term, a PID controller anticipates what the error will be in the immediate future – and accordingly applies a control action proportional to its current rate of change. The major drawbacks of this anticipatory feature are a lack of response in the presence of a constant, nonzero error – because of its nil derivative; and an unnecessarily large action for a noisy error of small amplitude – because its local derivative may be too large, and change sign very rapidly.

Besides the transfer function associated with the controller, descriptors can be provided, in Laplace's domain, for the remaining units at the interface between material and virtual processes – namely, the measuring device, viz.

$$\boxed{G_{md}\{s\} \equiv \frac{\overline{z}_{ms}}{\overline{y}_{out}};} \tag{3.492}$$

as well as the actuator, or final control element, viz.

$$\boxed{G_{fc}\{s\} \equiv \frac{\overline{x}_{mp}}{\overline{z}_{ct}}.} \tag{3.493}$$

The transfer functions defined by Eqs. (3.492) and (3.493) are sketched in Fig. 3.51ii.

Departing from the (reasonable) hypothesis that the dynamics of transmission lines is negligible, one will retrieve Eq. (3.463) – where insertion of Eq. (3.493) unfolds

$$\overline{y}_{out} = G_{dt}\overline{x}_{dt} + G_{mp}G_{fc}\overline{z}_{ct}; \tag{3.494}$$

Eq. (3.481) is now to be invoked to get

$$\overline{y}_{out} = G_{dt}\overline{x}_{dt} + G_{mp}G_{fc}G_{ct}\overline{z}_{dv}, \tag{3.495}$$

whereas Eq. (3.474) supports transformation to

$$\overline{y}_{out} = G_{dt}\overline{x}_{dt} + G_{mp}G_{fc}G_{ct}\left(\overline{z}_{sp} - \overline{z}_{ms}\right). \tag{3.496}$$

Equation (3.492) may then be revisited to produce

$$y_{out} = G_{dt}\bar{x}_{dt} + G_{mp}G_{fc}G_{ct}\left(\bar{z}_{sp} - G_{md}\bar{y}_{out}\right) \quad (3.497)$$

from Eq. (3.496) – or, upon elimination of parenthesis,

$$\bar{y}_{out} = G_{dt}\bar{x}_{dt} + G_{mp}G_{fc}G_{ct}\bar{z}_{sp} - G_{mp}G_{fc}G_{ct}G_{md}\bar{y}_{out}; \quad (3.498)$$

after factoring $\bar{y}_{out}$ out, Eq. (3.498) becomes

$$\left(1 + G_{md}G_{ct}G_{fc}G_{mp}\right)\bar{y}_{out} = G_{dt}\bar{x}_{dt} + G_{ct}G_{fc}G_{mp}\bar{z}_{sp}, \quad (3.499)$$

where division of both sides by $1 + G_{md}G_{ct}G_{fc}G_{mp}$ finally yields

$$\boxed{\bar{y}_{out} = G_{LD}\bar{x}_{dt} + G_{SP}\bar{z}_{sp}} \quad (3.500)$$

– as long as (lumped transfer function) G_{LD} is defined as

$$\boxed{G_{LD} \equiv \frac{G_{dt}}{1 + G_{md}G_{ct}G_{fc}G_{mp}},} \quad (3.501)$$

and (lumped transfer function) G_{SP} likewise abides to

$$\boxed{G_{SP} \equiv \frac{G_{ct}G_{fc}G_{mp}}{1 + G_{md}G_{ct}G_{fc}G_{mp}}.} \quad (3.502)$$

Equation (3.501) is illustrated in Fig. 3.51iii; the extreme simplicity of this diagram, when compared to the block diagram labeled as Fig. 3.51ii should be emphasized – and essentially matches the functionality underlying Fig. 3.51i, as long as $\bar{x}_{mp}$ is replaced by $\bar{z}_{sp}$.

The (multiplicative) cascade of transfer functions $G_{ct}G_{fc}G_{mp}$ in numerator of Eq. (3.502) constitutes the forward path between $\bar{z}_{sp}$ and $\bar{y}_{out}$ in Fig. 3.51ii – i.e. the set of transfer functions that sequentially represent controller, final control element, and process, in terms of effect of set point variable upon controlled variable; whereas G_{dt} in numerator of Eq. (3.501) constitutes the forward path between $\bar{x}_{dt}$ and $\bar{y}_{out}$, see also Fig. 3.51ii – i.e. the transfer function representing just the process in terms of effect of disturbance variable upon controlled variable. Furthermore, $G_{md}G_{ct}G_{fc}G_{mp}$ consubstantiates the set of transfer functions along the cycle characteristic of the feedback layout in Fig. 3.51ii – i.e. the transfer functions sequentially representing measuring device, controller, final control element, and processing unit. For this reason, such a product has classically been termed open-loop transfer function, G_{OL}, i.e.

$$\boxed{G_{OL} \equiv G_{md}G_{ct}G_{fc}G_{mp},} \quad (3.503)$$

so either denominator in Eqs. (3.501) or (3.502) reduces to just $1 + G_{OL}$; this is equivalent to rewriting Eqs. (3.501) and (3.502) as

$$G_{LD} = \frac{G_{dt}}{1 + G_{OL}} \quad (3.504)$$

and

$$G_{SP} = \frac{G_{ct}G_{fc}G_{mp}}{1 + G_{OL}}, \quad (3.505)$$

respectively. Therefore, G_{LD} (or load transfer function) – see Eq. (3.500), may be seen as the forward path relating $\bar{x}_{dt}$ to $\bar{y}_{out}$, divided by $1 + G_{OL}$, as per Eq. (3.504); and G_{SP} (or setpoint transfer function) – see Eq. (3.501), can be viewed as the forward path relating $\bar{z}_{sp}$ to $\bar{y}_{out}$, divided by $1 + G_{OL}$, according to Eq. (3.505).

It is instructive, at this stage, to ascertain how an uncontrolled system responds to a step-change in an inlet variable – a sufficiently simple, yet quite frequent form of disturbance found in current systems; this is mathematically described by

$$\boxed{x_{dt} = A\mathrm{H}\{t\},} \quad (3.506)$$

where A denotes amplitude and H denotes Heaviside's function – defined as being nil for $t < 0$ and unity for $t \geq 0$. In Laplace's domain, Eq. (3.506) becomes

$$\boxed{\bar{x}_{dt} = \frac{A}{s};} \quad (3.507)$$

hence, the response of the system will satisfy

$$\bar{y}_{out} = G_{dt}\frac{A}{s} \quad (3.508)$$

in agreement with Eq. (3.463), for $\bar{x}_{mp} = 0$ (by hypothesis) – or else

$$\bar{y}_{out} = \frac{1}{1 + \kappa\tau + \tau s}\frac{A}{s} \quad (3.509)$$

once Eq. (3.465) is taken onboard. After lumping fractions in Eq. (3.509) as

$$\boxed{\bar{y}_{out} = \frac{A}{s(1 + \kappa\tau + \tau s)},} \quad (3.510)$$

one may expand its right-hand side in partial fractions, viz.

$$\frac{A}{s(1 + \kappa\tau + \tau s)} = \frac{\alpha_0}{s} + \frac{\beta_0}{1 + \kappa\tau + \tau s} \quad (3.511)$$

– where α_0 and β_0 denote constants (still to be determined); if both sides are reduced to the same denominator, then Eq. (3.511) will take the form

$$A = \alpha_0(1 + \kappa\tau + \tau s) + \beta_0 s \quad (3.512)$$

as condition to be met by the outstanding numerators. Since Eq. (3.512) is to be valid irrespective of the value taken by s, it should apply to $s = 0$ in particular – in which case

$$A = \alpha_0(1 + \kappa\tau + \tau 0) + \beta_0 0 \quad (3.513)$$

is to be enforced; α_0 ends up given by

$$\alpha_0 = \frac{A}{1 + \kappa\tau}. \quad (3.514)$$

By the same token, setting $s = -(1 + \kappa\tau)/\tau$ transforms Eq. (3.512) to

$$A = \alpha_0\left(1 + \kappa\tau - \tau\frac{1 + \kappa\tau}{\tau}\right) - \beta_0\frac{1 + \kappa\tau}{\tau} \quad (3.515)$$

that readily reduces to

$$A = -\beta_0 \frac{1+\kappa\tau}{\tau}; \tag{3.516}$$

isolation of β_0 yields

$$\beta_0 = -\frac{A\tau}{1+\kappa\tau}. \tag{3.517}$$

The evolution in time of controlled variable $y_{out}\{t\}$ then becomes accessible after applying inverse Laplace's transforms to Eq. (3.510), viz.

$$y_{out}\{t\} \equiv \mathcal{L}^{-1}\left(\bar{y}_{out}\right) = \mathcal{L}^{-1}\left(\frac{A}{s(1+\kappa\tau+\tau s)}\right); \tag{3.518}$$

Eq. (3.511) may now be taken into account to write

$$y_{out}\{t\} = \mathcal{L}^{-1}\left(\frac{\alpha_0}{s} + \frac{\beta_0}{1+\kappa\tau+\tau s}\right), \tag{3.519}$$

or else

$$y_{out}\{t\} = \mathcal{L}^{-1}\left(\frac{\alpha_0}{s} + \frac{\frac{\beta_0}{\tau}}{\kappa+\frac{1}{\tau}+s}\right) \tag{3.520}$$

upon division of both numerator and denominator in the second term by τ. The linearity of (inverse) Laplace's operator supports, in turn, conversion of Eq. (3.520) to

$$y_{out}\{t\} = \alpha_0 \mathcal{L}^{-1}\left(\frac{1}{s}\right) + \frac{\beta_0}{\tau}\mathcal{L}^{-1}\left(\frac{1}{\left(\kappa+\frac{1}{\tau}\right)+s}\right). \tag{3.521}$$

In view of the translation theorem in Laplace's domain, Eq. (3.521) will be redone to

$$y_{out}\{t\} = \alpha_0 \mathcal{L}^{-1}\left(\frac{1}{s}\right) + \frac{\beta_0}{\tau} e^{-\left(\kappa+\frac{1}{\tau}\right)t} \mathcal{L}^{-1}\left(\frac{1}{s}\right) \tag{3.522}$$

or, equivalently,

$$y_{out}\{t\} = \left(\alpha_0 + \frac{\beta_0}{\tau} e^{-\left(\kappa+\frac{1}{\tau}\right)t}\right)\mathcal{L}^{-1}\left(\frac{1}{s}\right) \tag{3.523}$$

after factoring out $\mathcal{L}^{-1}(1/s)$; a final step of conversion to the time domain unfolds

$$y_{out}\{t\} = \left(\alpha_0 + \frac{\beta_0}{\tau} e^{-\left(\kappa+\frac{1}{\tau}\right)t}\right)\mathrm{H}\{t\}. \tag{3.524}$$

Insertion of Eqs. (3.514) and (3.517) then converts Eq. (3.524) to

$$y_{out}\{t\} = \left(\frac{A}{1+\kappa\tau} - \frac{\frac{A\tau}{1+\kappa\tau}}{\tau} e^{-\left(\kappa+\frac{1}{\tau}\right)t}\right)\mathrm{H}\{t\}, \tag{3.525}$$

where cancellation of τ between numerator and denominator in the second term, followed by factoring out $A/(1+\kappa\tau)$ – as well as $1/\tau$ in the argument of the exponential function, permit simplification to

$$\boxed{y_{out}\{t\} = \frac{A\left(1-e^{-(1+\kappa\tau)\frac{t}{\tau}}\right)\mathrm{H}\{t\}}{1+\kappa\tau}.} \tag{3.526}$$

Division of both sides of Eq. (3.526) by A finally yields

$$\boxed{\frac{y_{out}}{A} = \frac{1-e^{-(1+\kappa\tau)\frac{t}{\tau}}}{1+\kappa\tau},} \tag{3.527}$$

where Heaviside's function will hereafter be dropped for simplicity of notation – since only $t>0$ is of interest; Eq. (3.527) gives rise to the plot of (dimensionless) y_{out}/A vs. t/τ, labeled as Fig. 3.52. The behavior exhibited by y_{out}/A is typical of a first-order system; it takes off relatively fast, but the increase slows down as time elapses – until a horizontal asymptote is ultimately reached. The said asymptote is described by

$$\lim_{\frac{t}{\tau}\to\infty} \frac{y_{out}}{A} = \frac{1-e^{-\infty}}{1+\kappa\tau} = \frac{1+0}{1+\kappa\tau}, \tag{3.528}$$

or merely

$$\frac{y_{out}}{A} \equiv O_0^{\infty} = \frac{1}{1+\kappa\tau} \tag{3.529}$$

stemming from Eq. (3.527) – where O_0^{∞} denotes offset reached when time grows beyond limit. This offset is not nil – so any disturbance will have a lasting effect upon an uncontrolled system, as expected. However, such an offset may be driven toward zero by either a large κ or a large τ – the former implying a large heat transfer coefficient or heat transfer area, see Eq. (3.437), and the latter implying a small volumetric flow rate or a large volume,

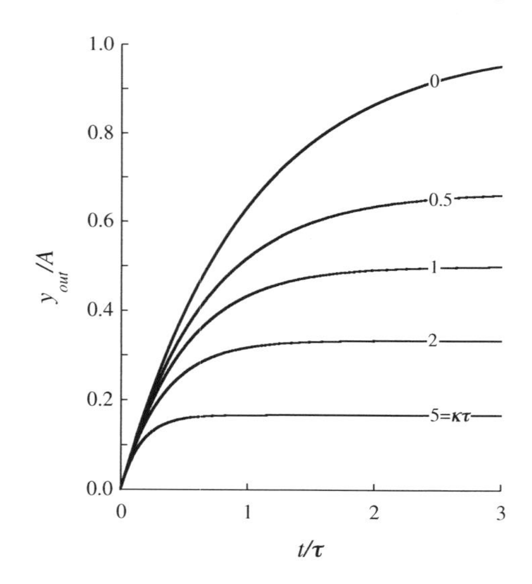

FIGURE 3.52 Evolution of normalized (controlled) variable, y_{out}/A, versus dimensionless time, t/τ, in uncontrolled process – for selected values of dimensionless parameter $\kappa\tau$.

see Eq. (3.436); these conditions are favorable to a faster response of bulk temperature, or an inertially less notorious long-term change in temperature, respectively, when heat is in transit. The monotonic increase observed in y_{out}/A is compatible with

$$\frac{d\left(\frac{y_{out}}{A}\right)}{d\left(\frac{t}{\tau}\right)}=-\frac{1}{1+\kappa\tau}\frac{d\,\mathrm{e}^{-(1+\kappa\tau)\frac{t}{\tau}}}{d\left(\frac{t}{\tau}\right)}=\frac{1+\kappa\tau}{1+\kappa\tau}\mathrm{e}^{-(1+\kappa\tau)\frac{t}{\tau}}, \quad (3.530)$$

obtained upon differentiation of Eq. (3.527), which degenerates to just

$$\frac{d\left(\frac{y_{out}}{A}\right)}{d\left(\frac{t}{\tau}\right)}=\mathrm{e}^{-(1+\kappa\tau)\frac{t}{\tau}}>0; \quad (3.531)$$

whereas the sign of the second derivative, i.e.

$$\frac{d^2\left(\frac{y_{out}}{A}\right)}{d\left(\frac{t}{\tau}\right)^2}\equiv\frac{d}{d\left(\frac{t}{\tau}\right)}\left(\frac{d\left(\frac{y_{out}}{A}\right)}{d\left(\frac{t}{\tau}\right)}\right)=\frac{d\,\mathrm{e}^{-(1+\kappa\tau)\frac{t}{\tau}}}{d\left(\frac{t}{\tau}\right)} \quad (3.532)$$

written with the aid of Eq. (3.531), becomes

$$\frac{d^2\left(\frac{y_{out}}{A}\right)}{d\left(\frac{t}{\tau}\right)^2}=-(1+\kappa\tau)\mathrm{e}^{-(1+\kappa\tau)\frac{t}{\tau}}<0 \quad (3.533)$$

– consistent with the slower and slower increase of y_{out}/A with t/τ perceptible in Fig. 3.52, with curve concavity facing down.

Consider now the controlled system schematized in Fig. 3.51ii; according to Eq. (3.500), one gets

$$\bar{y}_{out}=G_{LD}\frac{A}{s} \quad (3.534)$$

with the aid of Eq. (3.507), and again under the hypothesis that the setpoint remains unchanged, i.e. $z_{sp}\{t\}=0$ and thus $\bar{z}_{sp}=0$; insertion of Eq. (3.501) then leads to

$$\bar{y}_{out}=\frac{G_{dt}}{1+G_{md}G_{ct}G_{fc}G_{mp}}\frac{A}{s}. \quad (3.535)$$

The response by sensors and actuators (and, *a fortiori*, by transmission lines in between) to challenges at their inputs is normally very fast – so their dynamics can be neglected; this implies that they may be adequately modeled as pure resistances, according to

$$\boxed{G_{md}=K_{md}} \quad (3.536)$$

and

$$\boxed{G_{fc}=K_{fc},} \quad (3.537)$$

respectively – where K_{md} and K_{fc} denote constants. Insertion of Eqs. (3.536) and (3.537) accordingly simplifies Eq. (3.535) to

$$\bar{y}_{out}=\frac{G_{dt}}{1+K_{md}K_{fc}G_{mp}G_{ct}}\frac{A}{s}, \quad (3.538)$$

whereas combination with Eqs. (3.465) and (3.467) unfolds

$$\bar{y}_{out}=\frac{\frac{1}{1+\kappa\tau+\tau s}}{1+K_{md}K_{fc}\frac{\kappa\tau}{1+\kappa\tau+\tau s}G_{ct}}\frac{A}{s} \quad (3.539)$$

– which may instead be coined as

$$\boxed{\bar{y}_{out}=\frac{A}{s\left(1+\kappa\tau+\tau s+K_{md}K_{fc}\kappa\tau G_{ct}\right)},} \quad (3.540)$$

following multiplication of both numerator and denominator by $1+\kappa\tau+\tau s$. Analysis hereafter requires selection of a controller configuration – namely among (the most common) P, PI, and PID alternatives presented before.

In the case of a proportional controller, Eq. (3.480) can be retrieved to obtain

$$\bar{y}_{out}=\frac{A}{s\left(1+\kappa\tau+\tau s+K_{md}K_{fc}\kappa\tau K_P\right)} \quad (3.541)$$

from Eq. (3.540), or else

$$\boxed{\bar{y}_{out}=\frac{A}{s\left(1+\left(1+K_{md}K_{fc}K_P\right)\kappa\tau+\tau s\right)}} \quad (3.542)$$

after factoring $\kappa\tau$ out; expansion as partial fractions of the right-hand side of Eq. (3.542) ensues, viz.

$$\frac{A}{s\left(1+\left(1+K_{md}K_{fc}K_P\right)\kappa\tau+\tau s\right)}=\frac{\alpha_P}{s}+\frac{\beta_P}{1+\left(1+K_{md}K_{fc}K_P\right)\kappa\tau+\tau s}, \quad (3.543)$$

where α_P and β_P denote constants to be found. Upon elimination of denominators, Eq. (3.543) reduces to

$$A=\alpha_P\left(1+\left(1+K_{md}K_{fc}K_P\right)\kappa\tau+\tau s\right)+\beta_P s, \quad (3.544)$$

where $s=0$ specifically yields

$$A=\alpha_P\left(1+\left(1+K_{md}K_{fc}K_P\right)\kappa\tau+\tau 0\right)+\beta_P 0; \quad (3.545)$$

Eq. (3.545) readily reduces to

$$A=\alpha_P\left(1+\left(1+K_{md}K_{fc}K_P\right)\kappa\tau\right), \quad (3.546)$$

where isolation of α_P unfolds

$$\alpha_P=\frac{A}{1+\left(1+K_{md}K_{fc}K_P\right)\kappa\tau}. \quad (3.547)$$

If s is made equal to the other pole of Eq. (3.542), i.e. $s=-(1+(1+K_{md}K_{fc}K_P)\kappa\tau)/\tau$, then Eq. (3.544) becomes

$$A=\alpha_P\left(1+\left(1+K_{md}K_{fc}K_P\right)\kappa\tau-\tau\frac{1+\left(1+K_{md}K_{fc}K_P\right)\kappa\tau}{\tau}\right)-\beta_P\frac{1+\left(1+K_{md}K_{fc}K_P\right)\kappa\tau}{\tau}, \quad (3.548)$$

which breaks down to just

$$A = -\beta_P \frac{1 + \left(1 + K_{md} K_{fc} K_P\right)\kappa\tau}{\tau} \tag{3.549}$$

upon cancellation of symmetrical terms; isolation of β_P is finally possible as

$$\beta_P = -\frac{A\tau}{1 + \left(1 + K_{md} K_{fc} K_P\right)\kappa\tau}. \tag{3.550}$$

Since Eq. (3.542) is of the same functional form as Eq. (3.510) – except for replacement of (constant) $\kappa\tau$ by (constant) $(1 + K_{md}K_{fc}K_P)\kappa\tau$, one is entitled to use Eq. (3.543) to produce an analog of Eq. (3.519), i.e.

$$y_{out}\{t\} = \mathcal{L}^{-1}\left(\frac{\alpha_P}{s} + \frac{\beta_P}{1 + \left(1 + K_{md} K_{fc} K_P\right)\kappa\tau + \tau s}\right) \tag{3.551}$$

– where division of both numerator and denominator of the second term by τ leaves

$$y_{out}\{t\} = \mathcal{L}^{-1}\left(\frac{\alpha_P}{s} + \frac{\frac{\beta_P}{\tau}}{\left(1 + K_{md} K_{fc} K_P\right)\kappa + \frac{1}{\tau} + s}\right); \tag{3.552}$$

as expected, Eq. (3.552) reduces to Eq. (3.520) when $K_P=0$, as would prevail in an uncontrolled process. The inverse Laplace's transform in Eq. (3.552) may be split as

$$y_{out}\{t\} = \alpha_P \mathcal{L}^{-1}\left(\frac{1}{s}\right) + \frac{\beta_P}{\tau}\mathcal{L}^{-1}\left(\frac{1}{\left(\left(1 + K_{md} K_{fc} K_P\right)\kappa + \frac{1}{\tau}\right) + s}\right), \tag{3.553}$$

again for being a linear operator – while the translation theorem in Laplace's domain supports

$$y_{out}\{t\} = \alpha_P \mathcal{L}^{-1}\left(\frac{1}{s}\right) + \frac{\beta_P}{\tau} e^{-\left(\left(1 + K_{md} K_{fc} K_P\right)\kappa + \frac{1}{\tau}\right)t} \mathcal{L}^{-1}\left(\frac{1}{s}\right); \tag{3.554}$$

Eq. (3.554) is equivalent to

$$y_{out}\{t\} = \left(\alpha_P + \frac{\beta_P}{\tau} e^{-\left(1 + \left(1 + K_{md} K_{fc} K_P\right)\kappa\tau\right)\frac{t}{\tau}}\right)\mathcal{L}^{-1}\left(\frac{1}{s}\right), \tag{3.555}$$

once $\mathcal{L}^{-1}(1/s)$ and $1/\tau$ have been factored out in the right-hand side and in the argument of the exponential function, respectively. Since $\mathcal{L}(\mathrm{H}\{t\}) = 1/s$, Eq. (3.555) becomes

$$y_{out}\{t\} = \left(\alpha_P + \frac{\beta_P}{\tau} e^{-\left(1 + \left(1 + K_{md} K_{fc} K_P\right)\kappa\tau\right)\frac{t}{\tau}}\right)\mathrm{H}\{t\}; \tag{3.556}$$

Eqs. (3.547) and (3.550) now support transformation of Eq. (3.556) to

$$y_{out}\{t\} = \left(\frac{A}{1 + \left(1 + K_{md} K_{fc} K_P\right)\kappa\tau} - \frac{\frac{A\tau}{1 + \left(1 + K_{md} K_{fc} K_P\right)\kappa\tau}}{\tau} e^{-\left(1 + \left(1 + K_{md} K_{fc} K_P\right)\kappa\tau\right)\frac{t}{\tau}}\right)\mathrm{H}\{t\} \tag{3.557}$$

– or simply

$$\boxed{y_{out}\{t\} = \frac{A\left(1 - e^{-\left(1 + \left(1 + K_{md} K_{fc} K_P\right)\kappa\tau\right)\frac{t}{\tau}}\right)\mathrm{H}\{t\}}{1 + \left(1 + K_{md} K_{fc} K_P\right)\kappa\tau},} \tag{3.558}$$

upon dropping τ from numerator and denominator in the second term, and factoring out $A/(1 + (1 + K_{md}K_{fc}K_P)\kappa\tau)$ afterward. A dimensionless version of y_{out}, in simplified form, becomes then accessible upon division of both sides of Eq. (3.558) by A, i.e.

$$\boxed{\frac{y_{out}}{A} = \frac{1 - e^{-\left(1 + \left(1 + K_{md} K_{fc} K_P\right)\kappa\tau\right)\frac{t}{\tau}}}{1 + \left(1 + K_{md} K_{fc} K_P\right)\kappa\tau},} \tag{3.559}$$

valid for $t \geq 0$ on account of dropping $\mathrm{H}\{t\}$ – as plotted in Fig. 3.53. The curves coincide with those in Fig. 3.52, for the same value of characteristic (dimensionless) parameter in each case; however, a given $\kappa\tau$ in Fig. 3.52 produces a curve above that plotted in Fig. 3.53 for the same $\kappa\tau$ – since $1 + K_{md}K_{fc}K_P > 1$ appears as correction factor in the latter, thus making the observed response deviate less from the horizontal axis under control action. A first-order behavior is again found, as anticipated from the nature of the denominator of Eq. (3.542)

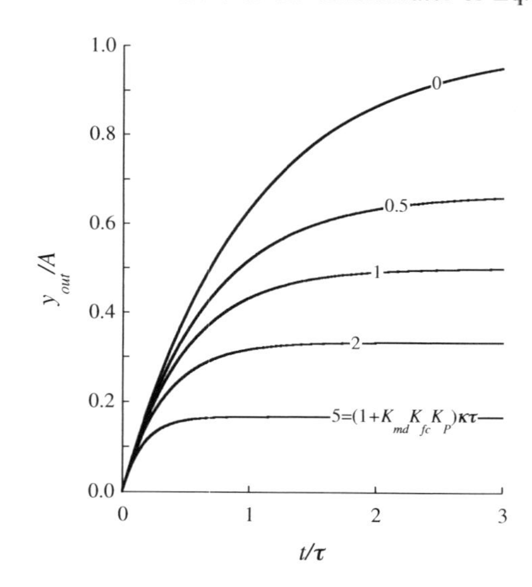

FIGURE 3.53 Evolution of normalized (controlled) variable, y_{out}/A, versus dimensionless time, t/τ, in process subjected to proportional control – for selected values of dimensionless parameter $(1 + K_{md}K_{fc}K_P)\kappa\tau$.

vis-à-vis with that of Eq. (3.510), i.e. product of s by a linear polynomial also in s; therefore, the order of response of the original (uncontrolled) system is kept when a controller of the P-type is employed. The response at stake becomes, however, slower than that of the uncontrolled system, according to

$$\frac{d\left(\frac{y_{out}}{A}\right)}{d\left(\frac{t}{\tau}\right)} = -\frac{\frac{d\,e^{-\left(1+\left(1+K_{md}K_{fc}K_P\right)\kappa\tau\right)\frac{t}{\tau}}}{d\left(\frac{t}{\tau}\right)}}{1+\left(1+K_{md}K_{fc}K_P\right)\kappa\tau} = \frac{1+\left(1+K_{md}K_{fc}K_P\right)\kappa\tau}{1+\left(1+K_{md}K_{fc}K_P\right)\kappa\tau} e^{-\left(1+\left(1+K_{md}K_{fc}K_P\right)\kappa\tau\right)\frac{t}{\tau}} \tag{3.560}$$

based on Eq. (3.559); this derivative readily simplifies to

$$\frac{d\left(\frac{y_{out}}{A}\right)}{d\left(\frac{t}{\tau}\right)} = e^{-\left(1+\left(1+K_{md}K_{fc}K_P\right)\kappa\tau\right)\frac{t}{\tau}} < e^{-(1+\kappa\tau)\frac{t}{\tau}} \tag{3.561}$$

upon cancellation of common factors between numerator and denominator, coupled with realization that $1+K_{md}K_{fc}K_P > 1$ – see Eq. (3.531).

With regard to the offset left after a long time, one finds that

$$\lim_{\frac{t}{\tau}\to\infty} \frac{y_{out}}{A} = \frac{1-e^{-\infty}}{1+\left(1+K_{md}K_{fc}K_P\right)\kappa\tau} = \frac{1-0}{1+\left(1+K_{md}K_{fc}K_P\right)\kappa\tau} \tag{3.562}$$

in agreement with Eq. (3.559), which degenerates to

$$\boxed{\lim_{\frac{t}{\tau}\to\infty} \frac{y_{out}}{A} \equiv O_P^{\infty} = \frac{1}{1+\left(1+K_{md}K_{fc}K_P\right)\kappa\tau} < \frac{1}{1+\kappa\tau} = O_0^{\infty}} \tag{3.563}$$

– again because $1+K_{md}K_{fc}K_P > 1$, and after recalling Eq. (3.529); here O_P^{∞} denotes upper plateau, to be eventually reached by the controlled process as time grows unbounded. Therefore, presence of a (proportional) control is effective in avoiding a large deviation of a system from the preset specification. The variation in offset with the characteristic (lumped) parameter is illustrated in Fig. 3.54. A typical hyperbolic decrease in offset is observed with increasing lumped parameter; hence, for a given system characterized by parameter $\kappa\tau$, and measuring device and final control apparatuses characterized by parameters K_{md} and K_{fc}, respectively, a more advantageous, long-term result (i.e. a lower offset) will arise when controller gain, K_P, is increased. However, too high a value of K_P is not advisable, since the (controlled) system can become unstable; in any case, evolution toward the maximum deviation from the (original) specification will occur at a slower pace than in the absence of control – remember Eqs. (3.531) and (3.561).

In attempts to better compare performance of a proportional-controlled system versus its uncontrolled counterpart, one may to advantage define an auxiliary parameter ϑ as

$$\boxed{\vartheta \equiv K_{md}K_{fc}K_P\kappa\tau,} \tag{3.564}$$

since it permits reformulation of Eq. (3.559) to merely

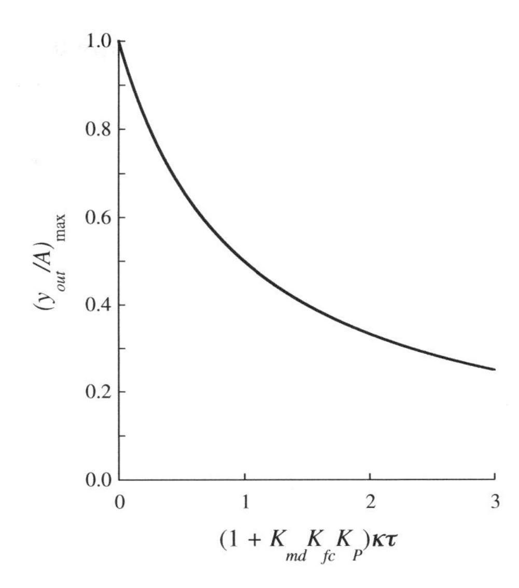

FIGURE 3.54 Variation of asymptotically reached, sustained maximum deviation of normalized (controlled) variable, $(y_{out}/A)_{max}$, relative to original steady state, versus dimensionless parameter $(1+K_{md}K_{fc}K_P)\kappa\tau$, in process subjected to proportional control.

$$\boxed{\frac{y_{out}}{A} = \frac{1-e^{-(1+\kappa\tau)\frac{t}{\tau}}e^{-\vartheta\frac{t}{\tau}}}{(1+\kappa\tau)+\vartheta};} \tag{3.565}$$

the effect of introducing proportional control, described by lumped parameter ϑ, relative to the corresponding uncontrolled system described by Eq. (3.527) appears explicitly via corrective factor $e^{-\vartheta t/\tau}$ in numerator – with negative impact, and corrective term ϑ in denominator – bearing positive impact. An illustration of such an effect is provided in Fig. 3.55, for a selected value of parameter $\kappa\tau$. As expected, absence of proportional action – i.e. $\vartheta=0$, reduces performance to that of a plain (uncontrolled)

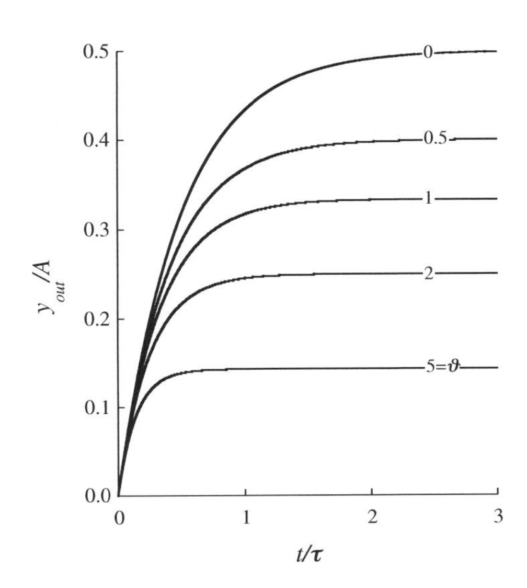

FIGURE 3.55 Evolution of normalized (controlled) variable, y_{out}/A, versus dimensionless time, t/τ, in process characterized by dimensionless parameter $\kappa\tau=1$, subjected to proportional control – for selected values of dimensionless parameter ϑ.

system, and thus justifies coincidence with the curve labeled as $\kappa\tau=1$ in Fig. 3.52. Once again, a persisting offset is grasped – which is shorter, and reached faster under stronger proportional action, i.e. higher ϑ; this behavior reflects the intrinsic robustness of this type of control action.

If integral action is included as part of the controller, then one should use Eq. (3.486) to replace G_{ct} in Eq. (3.540), thus obtaining

$$\overline{y}_{out} = \frac{A}{s\left(1+\kappa\tau+\tau s+K_{md}K_{fc}\kappa\tau K_P\left(1+\frac{1}{\tau_I s}\right)\right)}; \tag{3.566}$$

multiplication of both numerator and denominator by $\tau_I s$ gives rise to

$$\overline{y}_{out} = \frac{A\tau_I s}{s\left(\tau_I s(1+\kappa\tau+\tau s)+K_{md}K_{fc}K_P\kappa\tau\left(1+\tau_I s\right)\right)}. \tag{3.567}$$

Upon cancellation of s between numerator and denominator, and selective elimination of inner parentheses in denominator, Eq. (3.567) gives rise to

$$\overline{y}_{out} = \frac{A\tau_I}{\tau_I(1+\kappa\tau)s+\tau\tau_I s^2+K_{md}K_{fc}K_P\kappa\tau+K_{md}K_{fc}K_P\kappa\tau\tau_I s}; \tag{3.568}$$

Eq. (3.568) promptly becomes

$$\overline{y}_{out} = \frac{A\tau_I}{\tau\tau_I s^2+\tau_I\left(1+\left(1+K_{md}K_{fc}K_P\right)\kappa\tau\right)s+K_{md}K_{fc}K_P\kappa\tau} \tag{3.569}$$

after regrouping terms by power of s in denominator, factoring τ_I out, and then also factoring $k\tau$ out. Division of both numerator and denominator by $K_{md}K_{fc}K_P\kappa\tau$ transforms Eq. (3.569) to

$$\boxed{\overline{y}_{out} = \frac{K}{\eta_{PI}{}^2 s^2+2\zeta_{PI}\eta_{PI}s+1}}, \tag{3.570}$$

as long as auxiliary parameters K, η_{PI} and ζ_{PI} are defined as

$$\boxed{K \equiv \frac{A\tau_I}{K_{md}K_{fc}K_P\kappa\tau}}, \tag{3.571}$$

as well as

$$\boxed{\eta_{PI} \equiv \sqrt{\frac{\tau_I}{K_{md}K_{fc}K_P\kappa}}} \tag{3.572}$$

and

$$\boxed{\zeta_{PI} \equiv \frac{\tau_I\left(1+\left(1+K_{md}K_{fc}K_P\right)\kappa\tau\right)}{2\tau\sqrt{K_{md}K_{fc}K_P\kappa\tau_I}}}. \tag{3.573}$$

Note that parameter K may be redefined as

$$\boxed{K = A\tau\frac{\frac{\tau_I}{\tau}}{\vartheta}} \tag{3.574}$$

with the aid of Eq. (3.564), and upon multiplication and division by τ – whereas η_{PI} can alternatively be given by

$$\eta_{PI} = \sqrt{\frac{\tau\tau_I}{\vartheta}}, \tag{3.575}$$

which is equivalent to

$$\boxed{\eta_{PI} = \frac{\tau}{\sqrt{\vartheta}}\sqrt{\frac{\tau_I}{\tau}}} \tag{3.576}$$

after having taken τ off the square root; while Eq. (3.573) is susceptible of algebraic rearrangement to read

$$\zeta_{PI} = \frac{\sqrt{\tau_I}\sqrt{\tau_I}\left(1+\kappa\tau+K_{md}K_{fc}K_P\kappa\tau\right)}{2\sqrt{\tau}\sqrt{\tau}\sqrt{K_{md}K_{fc}K_P\kappa\tau_I}}, \tag{3.577}$$

upon splitting τ and τ_I as product of their corresponding square roots, followed by elimination of inner parenthesis. After dropping $\sqrt{\tau_I}$ from both numerator and denominator, and taking $\sqrt{\tau}$ inside the square root in denominator, Eq. (3.577) gives rise to

$$\zeta_{PI} = \frac{\sqrt{\tau_I}\left(1+\kappa\tau+K_{md}K_{fc}K_P\kappa\tau\right)}{2\sqrt{\tau}\sqrt{K_{md}K_{fc}K_P\kappa\tau}}, \tag{3.578}$$

where the outstanding $\sqrt{\tau_I}$ and $\sqrt{\tau}$ can be lumped as

$$\boxed{\zeta_{PI} = \frac{1+\kappa\tau+\vartheta}{2\sqrt{\vartheta}}\sqrt{\frac{\tau_I}{\tau}}} \tag{3.579}$$

– resorting again to Eq. (3.564); inspection of Eqs. (3.574), (3.576), and (3.579) unfolds indeed $K \equiv K\{A\tau,\vartheta,\tau_I/\tau\}$, further to $\eta_{PI} \equiv \eta_{PI}\{\tau,\vartheta,\tau_I/\tau\}$ and $\zeta_{PI} \equiv \zeta_{PI}\{\kappa\tau,\vartheta,\tau_I/\tau\}$.

Finally, one may add derivative action to a PI controller, thus generating its PID version – in which case G_{ct} in Eq. (3.540) ought to be replaced by Eq. (3.491), viz.

$$\overline{y}_{out} = \frac{A}{s\left(1+\kappa\tau+\tau s+K_{md}K_{fc}\kappa\tau K_P\left(1+\frac{1}{\tau_I s}+\tau_D s\right)\right)}; \tag{3.580}$$

after multiplying both numerator and denominator by $\tau_I s$, Eq. (3.580) becomes

$$\overline{y}_{out} = \frac{A\tau_I s}{s\left(\tau_I(1+\kappa\tau+\tau s)s+K_{md}K_{fc}K_P\kappa\tau\left(1+\tau_I s+\tau_I s\tau_D s\right)\right)}. \tag{3.581}$$

Upon dropping s from both numerator and denominator, and then eliminating (most) inner parentheses in the latter, Eq. (3.581) turns to

$$\overline{y}_{out} = \frac{A\tau_I}{\left(\begin{array}{l}\tau_I\left(1+\kappa\tau\right)s+\tau\tau_I s^2+K_{md}K_{fc}K_P\kappa\tau\\ +K_{md}K_{fc}K_P\kappa\tau\tau_I s+K_{md}K_{fc}K_P\kappa\tau\tau_I\tau_D s^2\end{array}\right)}; \tag{3.582}$$

terms alike are now to be pooled together as

$$\overline{y}_{out} = \frac{A\tau_I}{\left(\begin{array}{l}\tau\tau_I\left(1+K_{md}K_{fc}K_P\kappa\tau_D\right)s^2 \\ +\tau_I\left(1+\left(1+K_{md}K_{fc}K_P\right)\kappa\tau\right)s+K_{md}K_{fc}K_P\kappa\tau\end{array}\right)} \tag{3.583}$$

– along with factoring out of $\tau\tau_I s^2$ or $\tau_I s$, and then of $k\tau$. Division again of both numerator and denominator by the independent term in denominator transforms Eq. (3.583) to

$$\boxed{\overline{y}_{out} = \frac{K}{\eta_{PID}{}^2 s^2 + 2\zeta_{PID}\eta_{PID}s + 1},} \tag{3.584}$$

analogous in functional form to Eq. (3.570); auxiliary parameters η_{PID} and ζ_{PID} now abide to

$$\boxed{\eta_{PID} \equiv \sqrt{\frac{\tau_I}{K_{md}K_{fc}K_P\kappa}}\sqrt{1+K_{md}K_{fc}K_P\kappa\tau_D}} \tag{3.585}$$

and

$$\boxed{\zeta_{PID} \equiv \frac{\dfrac{\tau_I\left(1+\left(1+K_{md}K_{fc}K_P\right)\kappa\tau\right)}{2\tau\sqrt{K_{md}K_{fc}K_P\kappa\tau_I}}}{\sqrt{1+K_{md}K_{fc}K_P\kappa\tau_D}},} \tag{3.586}$$

while Eq. (3.571) still holds as definition of K. Note that η_{PID} may also be written as

$$\eta_{PID} = \eta_{PI}\sqrt{1+K_{md}K_{fc}K_P\kappa\tau_D}, \tag{3.587}$$

following comparative inspection of Eqs. (3.572) and (3.585); while ζ_{PID} will instead appear as

$$\zeta_{PID} \equiv \frac{\zeta_{PI}}{\sqrt{1+K_{md}K_{fc}K_P\kappa\tau_D}}, \tag{3.588}$$

as per combination of Eqs. (3.573) and (3.586). In other words, the extra derivative action increases parameter η as given by Eq. (3.587), but decreases parameter ζ according to Eq. (3.588) – specifically when going from a PI to a PID controller, since $\sqrt{1+K_{md}K_{fc}K_P\kappa\tau_D} > 1$. As done previously, it is useful to rewrite the parameter definitions as

$$\eta_{PID} = \eta_{PI}\sqrt{1+K_{md}K_{fc}K_P\kappa\tau\frac{\tau_D}{\tau}}, \tag{3.589}$$

based on Eq. (3.587) upon multiplication and division by τ under the square root sign – which gives rise, in turn, to

$$\eta_{PID} = \tau\sqrt{\frac{\frac{\tau_I}{\tau}}{\vartheta}}\sqrt{1+\vartheta\frac{\tau_D}{\tau}}, \tag{3.590}$$

following combination with Eqs. (3.564) and (3.576); one finally obtains

$$\boxed{\eta_{PID} = \frac{\tau}{\sqrt{\vartheta}}\sqrt{\frac{\tau_I}{\tau}\left(1+\vartheta\frac{\tau_D}{\tau}\right)}} \tag{3.591}$$

after merging the two square roots in Eq. (3.590). By the same token,

$$\zeta_{PID} = \frac{\zeta_{PI}}{\sqrt{1+K_{md}K_{fc}K_P\kappa\tau\frac{\tau_D}{\tau}}} \tag{3.592}$$

stems from Eq. (3.588) upon multiplication and division by τ – where insertion of Eqs. (3.564) and (3.579) leads to

$$\boxed{\zeta_{PID} = \frac{1+\kappa\tau+\vartheta}{2\sqrt{\vartheta}}\sqrt{\frac{\frac{\tau_I}{\tau}}{1+\vartheta\frac{\tau_D}{\tau}}};} \tag{3.593}$$

therefore, $\eta_{PID} \equiv \eta_{PID}\{\tau,\vartheta,\tau_I/\tau,\tau_D/\tau\}$ as per Eq. (3.591) and $\zeta_{PID} \equiv \zeta_{PID}\{\kappa\tau,\vartheta,\tau_I/\tau,\tau_D/\tau\}$ as per Eq. (3.592) – meaning that τ_D/τ was added as relevant degree of freedom to the functional dependence of η_{PI} and ζ_{PI} on τ, ϑ and τ_I/τ, and $k\tau$, ϑ and τ_I/τ, respectively, see Eqs. (3.576) and (3.579).

Since Eqs. (3.570) and (3.584) share a common functional form, i.e.

$$\overline{y}_{out} = \frac{K}{\eta^2 s^2 + 2\zeta\eta s + 1} \tag{3.594}$$

– except for replacement of parameter η by η_{PI} or η_{PID} (as appropriate), and likewise of parameter ζ by ζ_{PI} or ζ_{PID}, a common strategy of reversal to the time domain can be envisaged. One should accordingly start by adding and subtracting ζ^2 in denominator of Eq. (3.594) to get

$$\overline{y}_{out} = \frac{K}{\left(\eta^2 s^2 + 2\zeta\eta s + \zeta^2\right) - \zeta^2 + 1}, \tag{3.595}$$

where Newton's binomial theorem permits condensation of notation to

$$\overline{y}_{out} = \frac{K}{(\zeta+\eta s)^2 + 1 - \zeta^2}; \tag{3.596}$$

division of both numerator and denominator by η^2 gives then rise to

$$\boxed{\overline{y}_{out} = \frac{\frac{K}{\eta^2}}{\left(\frac{\zeta}{\eta}+s\right)^2 + \frac{1-\zeta^2}{\eta^2}}} \tag{3.597}$$

– with subsequent steps depending on whether $\zeta > 1$, $\zeta = 1$ or $\zeta < 1$.

Before proceeding any further, however, it is convenient to search for the limit of response $y_{out}\{t\}$, when time grows unbounded – via the final value theorem in Laplace's domain, viz.

$$\lim_{t\to\infty} y_{out}\{t\} = \lim_{s\to 0} s\overline{y}_{out}. \tag{3.598}$$

Insertion of Eq. (3.597) transforms Eq. (3.598) to

$$\lim_{t\to\infty} y_{out} = \lim_{s\to 0} s\frac{\frac{K}{\eta^2}}{\left(\frac{\zeta}{\eta}+s\right)^2 + \frac{1-\zeta^2}{\eta^2}}; \tag{3.599}$$

application of the classical theorems on limits leads directly to

$$\lim_{t\to\infty} y_{out} = 0\frac{\frac{K}{\eta^2}}{\left(\frac{\zeta}{\eta}+0\right)^2+\frac{1-\zeta^2}{\eta^2}} = \frac{0}{\left(\frac{\zeta}{\eta}\right)^2+\frac{1-\zeta^2}{\eta^2}} \tag{3.600}$$

that breaks down to merely

$$\boxed{\lim_{t\to\infty} y_{out} = 0,} \tag{3.601}$$

since the denominator is never nil for $0<\zeta<1$. The above result is particularly significant, because it unfolds a nil offset – i.e. the controlled variable will eventually recover its initial steady state value, provided that sufficient time is allowed; such a pattern contrasts with proportional control, see Eq. (3.563) – and consubstantiates the most relevant feature of the integral component in a controller. This is a mathematical consequence of s no longer appearing as factor in the denominator of $\bar{y}_{out}$ – as happened in Eq. (3.542) describing a P controller. Two non-nil poles arise instead in the right-hand side of Eq. (3.594), while a single pole, i.e. $s = -(1 + (1 + K_{md}K_{fc}K_P)\kappa\tau)/\tau$ (further to trivial pole $s = 0$), arose in denominator of Eq. (3.542); the former will become accessible via solution of

$$\eta^2 s^2 + 2\zeta\eta s + 1 = 0, \tag{3.602}$$

and together support second-order behavior (to become apparent soon). Therefore, integral action increases the dynamic order of system response to external disturbance; remember that the process considered was first order (by hypothesis), see Eqs. (3.465) or (3.467).

If $\zeta > 1$, then $\zeta^2-1>0$ – so Eq. (3.597) will preferably be formulated as

$$\bar{y}_{out} = \frac{\frac{K}{\eta^2}}{\left(\frac{\zeta}{\eta}+s\right)^2-\frac{\zeta^2-1}{\eta^2}}; \tag{3.603}$$

this is equivalent to writing

$$\bar{y}_{out} = \frac{\frac{K}{\eta^2}}{\frac{\sqrt{\zeta^2-1}}{\eta}}\frac{\frac{\sqrt{\zeta^2-1}}{\eta}}{\left(\frac{\zeta}{\eta}+s\right)^2-\left(\frac{\sqrt{\zeta^2-1}}{\eta}\right)^2}, \tag{3.604}$$

in view of square power and square root being functions inverse of each other – where multiplication and division of the right-hand side by $\sqrt{\zeta^2-1}/\eta$ meanwhile took place. Equation (3.604) simplifies to

$$\bar{y}_{out} = \frac{\frac{K}{\eta}}{\sqrt{\zeta^2-1}}\frac{\frac{\sqrt{\zeta^2-1}}{\eta}}{\left(\frac{\zeta}{\eta}+s\right)^2-\left(\frac{\sqrt{\zeta^2-1}}{\eta}\right)^2}, \tag{3.605}$$

after dropping η from both denominators; note the (constant) term appearing in numerator, and the negative of its square appearing as an additive term in denominator. After taking inverse Laplace's transforms of both sides, Eq. (3.605) becomes

$$\mathcal{L}^{-1}(\bar{y}_{out}) = \frac{\frac{K}{\eta}}{\sqrt{\zeta^2-1}}\mathcal{L}^{-1}\left(\frac{\frac{\sqrt{\zeta^2-1}}{\eta}}{\left(\frac{\zeta}{\eta}+s\right)^2-\left(\frac{\sqrt{\zeta^2-1}}{\eta}\right)^2}\right), \tag{3.606}$$

in view of the intrinsic linearity of the said operator; Eq. (3.606) may, in turn, be redone to

$$\mathcal{L}^{-1}(\bar{y}_{out}) = \frac{\frac{K}{\eta}}{\sqrt{\zeta^2-1}}e^{-\frac{\zeta}{\eta}t}\mathcal{L}^{-1}\left(\frac{\frac{\sqrt{\zeta^2-1}}{\eta}}{s^2-\left(\frac{\sqrt{\zeta^2-1}}{\eta}\right)^2}\right), \tag{3.607}$$

with the aid of the translation theorem in Laplace's domain. One further step in reversal unfolds

$$y_{out}\{t\} = \frac{\frac{K}{\eta}}{\sqrt{\zeta^2-1}}e^{-\frac{\zeta}{\eta}t}\left(\sinh\frac{\sqrt{\zeta^2-1}}{\eta}t\right)\mathrm{H}\{t\}, \tag{3.608}$$

or simply

$$\boxed{y_{out}\{t\} = \frac{\frac{K}{\eta}}{\sqrt{\zeta^2-1}}e^{-\zeta\frac{t}{\eta}}\left(\sinh\sqrt{\zeta^2-1}\frac{t}{\eta}\right)\mathrm{H}\{t\}} \tag{3.609}$$

– with a dimensionless time, t/η, emerging explicitly. Since only responses after $t=0$ are of practical interest, Eq. (3.609) simplifies to

$$\boxed{\frac{y_{out}}{\frac{K}{\eta}} = \frac{1}{\sqrt{\zeta^2-1}}e^{-\zeta\frac{t}{\eta}}\sinh\sqrt{\zeta^2-1}\frac{t}{\eta},} \tag{3.610}$$

where both sides were meanwhile divided by K/η; the corresponding plot is provided as Fig. 3.56i, as $\eta y_{out}/K$ vs. t/η, using ζ as parameter. When $\zeta > 1$ (as in the case currently under investigation), the controlled system is said to be overdamped. As the characteristic parameter ζ decreases, a slower but larger takeoff is observed; evolution of the response goes through a maximum in all cases – to eventually level off at zero, after a sufficiently long time has elapsed. This maximum occurs indeed later, and is higher at lower ζ; and the horizontal axis – serving as long-t asymptote, is reached faster at lower ζ. An almost first-order behavior results at very large ζ – which appears to leave an offset; however, it produces in fact a quite long tail in its approach to the horizontal axis, see Fig. 3.56i *vis-à-vis* with Fig. 3.53.

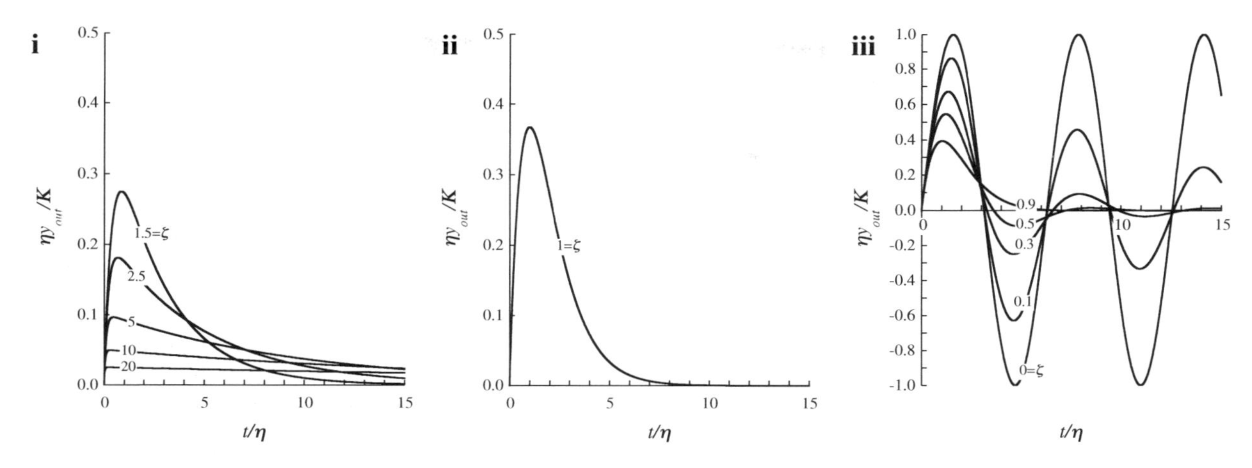

FIGURE 3.56 Evolution of normalized (controlled) variable, $\eta y_{out}/K$, versus dimensionless time, t/η, in process subjected to proportional/integral or proportional/integral/derivative control – for selected values of dimensionless parameter ζ, leading to (i) overdamped ($\zeta>1$), (ii) critically damped ($\zeta=1$), or (iii) underdamped ($0<\zeta<1$) behavior.

Existence of a maximum value for the response deserves further attention, as it consubstantiates the worst specification for the process/product relative to what was originally planned; one may conveniently retrieve its defining necessary condition as

$$\frac{d\left(\dfrac{\dfrac{y_{out}}{K}}{\eta}\right)}{d\left(\dfrac{t}{\eta}\right)}=0, \tag{3.611}$$

with Eq. (3.610) supporting transformation to

$$\frac{d}{d\left(\dfrac{t}{\eta}\right)}\left(\frac{1}{\sqrt{\zeta^2-1}}e^{-\zeta\frac{t}{\eta}}\sinh\sqrt{\zeta^2-1}\frac{t}{\eta}\right)=0; \tag{3.612}$$

upon standard application of the classical rules of differentiation, Eq. (3.612) becomes

$$\frac{1}{\sqrt{\zeta^2-1}}\left(\begin{array}{l}-\zeta\, e^{-\zeta\frac{t}{\eta}}\sinh\sqrt{\zeta^2-1}\dfrac{t}{\eta}\\ +e^{-\zeta\frac{t}{\eta}}\sqrt{\zeta^2-1}\cosh\sqrt{\zeta^2-1}\dfrac{t}{\eta}\end{array}\right)=0 \tag{3.613}$$

– where both sides may be multiplied by $\sqrt{\zeta^2-1}\neq 0$, and $e^{-\zeta t/\eta}$ factored out afterward to get

$$e^{-\zeta\frac{t}{\eta}}\left(\sqrt{\zeta^2-1}\cosh\sqrt{\zeta^2-1}\frac{t}{\eta}-\zeta\sinh\sqrt{\zeta^2-1}\frac{t}{\eta}\right)=0. \tag{3.614}$$

Since an exponential function cannot take a nil value, Eq. (3.614) implies

$$\sqrt{\zeta^2-1}\cosh\sqrt{\zeta^2-1}\frac{t}{\eta}=\zeta\sinh\sqrt{\zeta^2-1}\frac{t}{\eta}, \tag{3.615}$$

where division of both sides by $\cosh\sqrt{\zeta^2-1}\,t/\eta$ yields

$$\sqrt{\zeta^2-1}=\zeta\frac{\sinh\sqrt{\zeta^2-1}\dfrac{t}{\eta}}{\cosh\sqrt{\zeta^2-1}\dfrac{t}{\eta}}; \tag{3.616}$$

owing to the definition of hyperbolic tangent, Eq. (3.616) is preferably condensed to

$$\tanh\sqrt{\zeta^2-1}\frac{t}{\eta}=\frac{\sqrt{\zeta^2-1}}{\zeta}, \tag{3.617}$$

with both sides meanwhile divided by $\zeta\neq 0$. Upon inversion of the outstanding hyperbolic function, Eq. (3.617) becomes

$$\sqrt{\zeta^2-1}\frac{t}{\eta}=\tanh^{-1}\frac{\sqrt{\zeta^2-1}}{\zeta}, \tag{3.618}$$

or else

$$\boxed{\frac{t_{pk}}{\eta}=\frac{\tanh^{-1}\dfrac{\sqrt{\zeta^2-1}}{\zeta}}{\sqrt{\zeta^2-1}}}; \tag{3.619}$$

this serves as descriptor of the (single) maximum – located at abscissa t_{pk}/η. Equation (3.619) may alternatively be coined as

$$\frac{t_{pk}}{\eta}=\frac{\ln\sqrt{\dfrac{1+\dfrac{\sqrt{\zeta^2-1}}{\zeta}}{1-\dfrac{\sqrt{\zeta^2-1}}{\zeta}}}}{\sqrt{\zeta^2-1}}, \tag{3.620}$$

in view of the (equivalent) logarithmic form for the inverse hyperbolic tangent – where the operational features of logarithms in converting powers to products permit further transformation to

$$\boxed{\frac{t_{pk}}{\eta} = \ln\left(\frac{1+\frac{\sqrt{\zeta^2-1}}{\zeta}}{1-\frac{\sqrt{\zeta^2-1}}{\zeta}}\right)^{\frac{1}{2\sqrt{\zeta^2-1}}}}. \tag{3.621}$$

On the other hand, if one assumes that

$$\sinh\sqrt{\zeta^2-1}\frac{t}{\eta} = \sqrt{\zeta^2-1}, \tag{3.622}$$

then Eq. (3.413) enforces

$$\begin{aligned}\cosh^2\sqrt{\zeta^2-1}\frac{t}{\eta} &= 1+\sinh^2\sqrt{\zeta^2-1}\frac{t}{\eta} \\ &= 1+\left(\sqrt{\zeta^2-1}\right)^2 = 1+\zeta^2-1 = \zeta^2\end{aligned} \tag{3.623}$$

– along with straightforward algebraic rearrangement, coupled with insertion of Eq. (3.622); the corresponding hyperbolic tangent will thus satisfy

$$\tanh\sqrt{\zeta^2-1}\frac{t}{\eta} \equiv \frac{\sinh\sqrt{\zeta^2-1}\frac{t}{\eta}}{\cosh\sqrt{\zeta^2-1}\frac{t}{\eta}} = \frac{\sqrt{\zeta^2-1}}{\zeta}, \tag{3.624}$$

with the aid of Eqs. (3.622) and (3.623). Since Eq. (3.624) retrieves Eq. (3.617), one immediately infers that Eq. (3.622) is compatible therewith – so one can describe the optimality condition, labeled as Eq. (3.617), alternatively by

$$\sinh\sqrt{\zeta^2-1}\frac{t_{pk}}{\eta} = \sqrt{\zeta^2-1}. \tag{3.625}$$

Equation (3.610) may now be retrieved as

$$\frac{y_{out,pk}}{\frac{K}{\eta}} = \frac{1}{\sqrt{\zeta^2-1}}\exp\left\{-\zeta\frac{t_{pk}}{\eta}\right\}\sinh\sqrt{\zeta^2-1}\frac{t_{pk}}{\eta} \tag{3.626}$$

referring to

$$y_{out,pk} \equiv y_{out}\big|_{t=t_{pk}}; \tag{3.627}$$

insertion of Eqs. (3.621) and (3.625) supports transformation of Eq. (3.626) to

$$\frac{y_{out,pk}}{\frac{K}{\eta}} = \frac{1}{\sqrt{\zeta^2-1}}\exp\left\{-\zeta\ln\left(\frac{1+\frac{\sqrt{\zeta^2-1}}{\zeta}}{1-\frac{\sqrt{\zeta^2-1}}{\zeta}}\right)^{\frac{1}{2\sqrt{\zeta^2-1}}}\right\}\sqrt{\zeta^2-1}. \tag{3.628}$$

Multiplying a logarithm by $-\zeta$ is equivalent to raising its argument to $-\zeta$, so Eq. (3.628) can be replaced by

$$\frac{y_{out,pk}}{\frac{K}{\eta}} = \exp\left\{\ln\left(\frac{1-\frac{\sqrt{\zeta^2-1}}{\zeta}}{1+\frac{\sqrt{\zeta^2-1}}{\zeta}}\right)^{\frac{\zeta}{2\sqrt{\zeta^2-1}}}\right\}, \tag{3.629}$$

where the reciprocal of the argument of the logarithm was taken on account of the minus sign of its exponent, and $\sqrt{\zeta^2-1}$ was dropped from both numerator and denominator; composition of functions inverse of each other permits simplification of Eq. (3.629) to

$$\frac{y_{out,pk}}{\frac{K}{\eta}} = \left(\frac{1-\frac{\sqrt{\zeta^2-1}}{\zeta}}{1+\frac{\sqrt{\zeta^2-1}}{\zeta}}\right)^{\frac{\zeta}{2\sqrt{\zeta^2-1}}} = \left(\frac{\zeta-\sqrt{\zeta^2-1}}{\zeta+\sqrt{\zeta^2-1}}\right)^{\frac{\zeta}{2\sqrt{\zeta^2-1}}}, \tag{3.630}$$

along with multiplication of both numerator and denominator by ζ. Further multiplication of numerator and denominator of Eq. (3.630) by the former gives rise to

$$\frac{y_{out,pk}}{\frac{K}{\eta}} = \left(\frac{\left(\zeta-\sqrt{\zeta^2-1}\right)^2}{\zeta^2-\left(\zeta^2-1\right)}\right)^{\frac{\zeta}{2\sqrt{\zeta^2-1}}}; \tag{3.631}$$

the difference in squares appearing in denominator is due to the product of a binomial by its conjugate. Cancellation of symmetrical terms in denominator permits simplification of Eq. (3.631) to

$$\frac{y_{out,pk}}{\frac{K}{\eta}} = \left(\frac{\left(\zeta-\sqrt{\zeta^2-1}\right)^2}{1}\right)^{\frac{\zeta}{2\sqrt{\zeta^2-1}}} = \left(\left(\zeta-\sqrt{\zeta^2-1}\right)^2\right)^{\frac{\zeta}{2\sqrt{\zeta^2-1}}}, \tag{3.632}$$

where multiplication of exponents of composite powers finally yields

$$\boxed{\frac{y_{out,pk}}{\frac{K}{\eta}} = \left(\zeta-\sqrt{\zeta^2-1}\right)^{\frac{\zeta}{\sqrt{\zeta^2-1}}}} \tag{3.633}$$

as descriptor of the maximum deviation from the initial steady state – also known as overshoot. The variation of (single) overshoot with damping factor, ζ, for values thereof above unity, is sketched in Fig. 3.57. The (local and global) maximum deviation from the original steady state, coinciding with the (single) overshoot described by $y_{out,pk}$, decreases with ζ within the range $]1,\infty[$; the pattern is almost linear in the bilogarithmic scale selected.

The specific case of $\zeta=1$ implies $1-\zeta^2=0$, which reduces Eq. (3.597) to

$$\overline{y}_{out} = \frac{\frac{K}{\eta^2}}{\left(\frac{1}{\eta}+s\right)^2}; \tag{3.634}$$

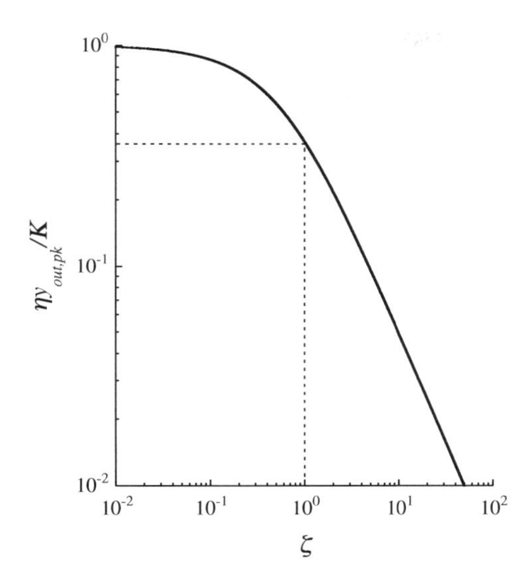

FIGURE 3.57 Variation of (locally reached, global) maximum deviation of normalized (controlled) variable, $\eta y_{out,pk}/K$, relative to original steady state, versus dimensionless parameter ζ, in process subjected to proportional/integral or proportional/integral/derivative control.

hence, one may proceed immediately with application of inverse Laplace's transforms to both sides to get

$$\mathcal{L}^{-1}\left(\bar{y}_{out}\right)=\frac{K}{\eta^2}\mathcal{L}^{-1}\left(\frac{1}{\left(\frac{1}{\eta}+s\right)^2}\right) \tag{3.635}$$

– again at the expense of the intrinsic linearity of the said operator. Equation (3.635) is equivalent to

$$\mathcal{L}^{-1}\left(\bar{y}_{out}\right)=\frac{K}{\eta^2}e^{-\frac{1}{\eta}t}\mathcal{L}^{-1}\left(\frac{1}{s^2}\right) \tag{3.636}$$

in view again of the translation theorem in Laplace's domain; Eq. (3.636) eventually becomes

$$y_{out}\{t\}=\frac{K}{\eta^2}e^{-\frac{t}{\eta}}t\,\mathrm{H}\{t\} \tag{3.637}$$

– or else

$$\boxed{y_{out}\{t\}=\frac{K}{\eta}\frac{t}{\eta}e^{-\frac{t}{\eta}}\mathrm{H}\{t\},} \tag{3.638}$$

after lumping t and η outside the exponential function. The dependent variable will again be made dimensionless by dividing both sides of Eq. (3.638) by K/η, viz.

$$\boxed{\frac{y_{out}}{\frac{K}{\eta}}=\frac{t}{\eta}e^{-\frac{t}{\eta}},} \tag{3.639}$$

valid for $t\geq 0$ – which is plotted in Fig. 3.56ii; this is termed critically damped behavior – as it actually leads to the highest (single) peak, under nonoscillatory response.

The maximum deviation relative to the initial steady state may be ascertained via calculation of the limiting value of $\eta y_{out,pk}/K$ when $\zeta\rightarrow 1^+$, according to

$$\lim_{\zeta\rightarrow 1^+}\frac{y_{out,pk}}{\frac{K}{\eta}}=\left(\zeta-\sqrt{\zeta^2-1}\right)^{\frac{\zeta}{\sqrt{\zeta^2-1}}} \tag{3.640}$$

based on Eq. (3.633); elementary algebraic rearrangement, following direct application of the theorems on limits, then unfolds

$$\lim_{\zeta\rightarrow 1^+}\frac{y_{out,pk}}{\frac{K}{\eta}}=\left(1-\sqrt{1^2-1}\right)^{\frac{1}{\sqrt{1^2-1}}}=(1-0)^{\frac{1}{0}}=1^{\infty}. \tag{3.641}$$

In view of the unknown quantity arising in Eq. (3.641) – of the form 1^∞, one should go back and apply logarithms to both sides of Eq. (3.640) as

$$\ln\lim_{\zeta\rightarrow 1^+}\frac{y_{out,pk}}{\frac{K}{\eta}}=\ln\lim_{\zeta\rightarrow 1}\left(\zeta-\sqrt{\zeta^2-1}\right)^{\frac{\zeta}{\sqrt{\zeta^2-1}}}, \tag{3.642}$$

where exchange of logarithm and limit operators in the right-hand side produces

$$\ln\lim_{\zeta\rightarrow 1^+}\frac{y_{out,pk}}{\frac{K}{\eta}}=\lim_{\zeta\rightarrow 1}\ln\left(\zeta-\sqrt{\zeta^2-1}\right)^{\frac{\zeta}{\sqrt{\zeta^2-1}}}; \tag{3.643}$$

after recalling again the ability of the logarithm operator to convert a power to a product, Eq. (3.643) will give room to

$$\ln\lim_{\zeta\rightarrow 1^+}\frac{y_{out,pk}}{\frac{K}{\eta}}=\lim_{\zeta\rightarrow 1}\frac{\zeta}{\sqrt{\zeta^2-1}}\ln\left\{\zeta-\sqrt{\zeta^2-1}\right\}, \tag{3.644}$$

where the limit of the product may be split as

$$\begin{aligned}\ln\lim_{\zeta\rightarrow 1^+}\frac{y_{out,pk}}{\frac{K}{\eta}}&=\lim_{\zeta\rightarrow 1}\zeta\lim_{\zeta\rightarrow 1}\frac{\ln\left\{\zeta-\sqrt{\zeta^2-1}\right\}}{\sqrt{\zeta^2-1}}\\&=1\frac{\ln\left\{1-\sqrt{1^2-1}\right\}}{\sqrt{1^2-1}}=\frac{\ln\{1-0\}}{\sqrt{0}}=\frac{0}{0}.\end{aligned} \tag{3.645}$$

In other words, the initial unknown quantity, of the exponential type, gave rise to another unknown quantity, of the 0/0 type – so L'Hôpital's rule is now applicable, via separate differentiation of numerator and denominator in Eq. (3.645) with regard to ζ, i.e.

$$\ln\lim_{\zeta\rightarrow 1^+}\frac{y_{out,pk}}{\frac{K}{\eta}}=1\lim_{\zeta\rightarrow 1}\frac{\dfrac{1-\dfrac{2\zeta}{2\sqrt{\zeta^2-1}}}{\zeta-\sqrt{\zeta^2-1}}}{\dfrac{2\zeta}{2\sqrt{\zeta^2-1}}}; \tag{3.646}$$

upon multiplication of both numerator and denominator by $2\sqrt{\zeta^2-1}$, Eq. (3.646) simplifies to

$$\ln \lim_{\zeta\to 1^+} \frac{y_{out,pk}}{\frac{K}{\eta}} = \lim_{\zeta\to 1} \frac{\frac{2\sqrt{\zeta^2-1}-2\zeta}{\zeta-\sqrt{\zeta^2-1}}}{2\zeta} \quad (3.647)$$

– whereas the two denominators can be collapsed as

$$\ln \lim_{\zeta\to 1^+} \frac{y_{out,pk}}{\frac{K}{\eta}} = \lim_{\zeta\to 1} \frac{2\left(\sqrt{\zeta^2-1}-\zeta\right)}{2\zeta\left(\zeta-\sqrt{\zeta^2-1}\right)} = -\lim_{\zeta\to 1} \frac{\zeta-\sqrt{\zeta^2-1}}{\zeta\left(\zeta-\sqrt{\zeta^2-1}\right)}, \quad (3.648)$$

along with factoring out 2 in numerator, and dropping –2 between numerator and denominator afterward. Equation (3.648) simplifies further to

$$\ln \lim_{\zeta\to 1^+} \frac{y_{out,pk}}{\frac{K}{\eta}} = -\lim_{\zeta\to 1} \frac{1}{\zeta} = -\frac{1}{1} = -1, \quad (3.649)$$

since $\zeta-\sqrt{\zeta^2-1}$ is shared by numerator and denominator; exponentials are finally to be taken of both sides to obtain

$$\lim_{\zeta\to 1^+} \frac{y_{out,pk}}{\frac{K}{\eta}} = e^{-1}, \quad (3.650)$$

which justifies the point of coordinates $(1,e^{-1})$ in Fig. 3.57.

The third possibility of interest is described by $\zeta<1$, which makes $1-\zeta^2>0$ – so Eq. (3.597) will be revisited as

$$\bar{y}_{out} = \frac{\frac{K}{\eta^2}}{\frac{\sqrt{1-\zeta^2}}{\eta}} \frac{\frac{\sqrt{1-\zeta^2}}{\eta}}{\left(\frac{\zeta}{\eta}+s\right)^2+\left(\frac{\sqrt{1-\zeta^2}}{\eta}\right)^2}, \quad (3.651)$$

where redundancy of taking square power of square root was taken advantage of in the second term in denominator, and the right-hand side was multiplied and divided by $\sqrt{1-\zeta^2}/\eta$; cancellation of η between denominators permits simplification of Eq. (3.651) to

$$\bar{y}_{out} = \frac{\frac{K}{\eta}}{\sqrt{1-\zeta^2}} \frac{\frac{\sqrt{1-\zeta^2}}{\eta}}{\left(\frac{\zeta}{\eta}+s\right)^2+\left(\frac{\sqrt{1-\zeta^2}}{\eta}\right)^2}. \quad (3.652)$$

Inverse Laplace's transforms may now be applied to both sides of Eq. (3.652), i.e.

$$\mathcal{L}^{-1}\left(\bar{y}_{out}\right) = \frac{\frac{K}{\eta}}{\sqrt{1-\zeta^2}} \mathcal{L}^{-1}\left(\frac{\frac{\sqrt{1-\zeta^2}}{\eta}}{\left(\frac{\zeta}{\eta}+s\right)^2+\left(\frac{\sqrt{1-\zeta^2}}{\eta}\right)^2}\right), \quad (3.653)$$

where $K/\eta\sqrt{1-\zeta^2}$ was already taken off the argument of the said operator for being a constant; the translation theorem in Laplace's domain can be invoked once more, to justify transformation of Eq. (3.653) to

$$\mathcal{L}^{-1}\left(\bar{y}_{out}\right) = \frac{\frac{K}{\eta}}{\sqrt{1-\zeta^2}} e^{-\frac{\zeta}{\eta}t} \mathcal{L}^{-1}\left(\frac{\frac{\sqrt{1-\zeta^2}}{\eta}}{s^2+\left(\frac{\sqrt{1-\zeta^2}}{\eta}\right)^2}\right). \quad (3.654)$$

Equation (3.654) is equivalent, in the time domain, to

$$y_{out}\{t\} = \frac{\frac{K}{\eta}}{\sqrt{1-\zeta^2}} e^{-\frac{\zeta}{\eta}t} \left(\sin\frac{\sqrt{1-\zeta^2}}{\eta}t\right) \mathrm{H}\{t\}, \quad (3.655)$$

or else

$$\boxed{y_{out}\{t\} = \frac{\frac{K}{\eta}}{\sqrt{1-\zeta^2}} e^{-\zeta\frac{t}{\eta}} \left(\sin\sqrt{1-\zeta^2}\,\frac{t}{\eta}\right) \mathrm{H}\{t\}} \quad (3.656)$$

after lumping t with η in the argument of both exponential and sine functions. In parallel to the dimensionless version, t/η, of independent variable t, one may resort to $\eta y_{out}/K$ as dimensionless version of y_{out}, according to

$$\boxed{\frac{y_{out}}{\frac{K}{\eta}} = \frac{1}{\sqrt{1-\zeta^2}} e^{-\zeta\frac{t}{\eta}} \sin\sqrt{1-\zeta^2}\,\frac{t}{\eta}} \quad (3.657)$$

– based on Eq. (3.656) if considered only after $t=0$, and upon division of both sides by K/η; the corresponding plot is provided in Fig. 3.56iii, for selected values of parameter $0<\zeta<1$. A unique oscillatory pattern on t/η is found in this case, with angular frequency, ω, given by

$$\omega = \sqrt{1-\zeta^2} \quad (3.658)$$

as per direct inspection of the argument of sine in Eq. (3.657); this corresponds to period T reading

$$T \equiv \frac{2\pi}{\omega} = \frac{2\pi}{\sqrt{1-\zeta^2}}, \quad (3.659)$$

which took Eq. (3.658) into account. As ζ moves toward zero, ω increases – and so does the wave amplitude, see Fig. 3.56iii; the said oscillations do, however, decay in time (except for $\zeta=0$) – due to the decreasing exponential in Eq. (3.657), while the sine remains bounded within $[-1,1]$.

More than one peak is apparent in Fig. 3.56iii – unlike happened in Figs. 3.56i or ii; one may again resort to Eq. (3.611) to ascertain them – with insertion of Eq. (3.657) yielding

$$\frac{d}{d\left(\frac{t}{\eta}\right)}\left(\frac{1}{\sqrt{1-\zeta^2}} e^{-\zeta\frac{t}{\eta}} \sin\sqrt{1-\zeta^2}\,\frac{t}{\eta}\right) = 0. \quad (3.660)$$

Equation (3.660) promptly leads to

$$\frac{1}{\sqrt{1-\zeta^2}}\left(-\zeta e^{-\zeta\frac{t}{\eta}}\sin\sqrt{1-\zeta^2}\,\frac{t}{\eta}+e^{-\zeta\frac{t}{\eta}}\sqrt{1-\zeta^2}\cos\sqrt{1-\zeta^2}\,\frac{t}{\eta}\right)=0, \tag{3.661}$$

based on the classical rules of differentiation; after getting rid of the constant preceding the parenthesis on account of the nil right-hand side, $e^{-\zeta t/\eta}$ may be factored out to obtain

$$e^{-\zeta\frac{t}{\eta}}\left(\sqrt{1-\zeta^2}\cos\sqrt{1-\zeta^2}\,\frac{t}{\eta}-\zeta\sin\sqrt{1-\zeta^2}\,\frac{t}{\eta}\right)=0. \tag{3.662}$$

As before, Eq. (3.662) degenerates to

$$\sqrt{1-\zeta^2}\cos\sqrt{1-\zeta^2}\,\frac{t}{\eta}=\zeta\sin\sqrt{1-\zeta^2}\,\frac{t}{\eta} \tag{3.663}$$

because $e^{-\zeta t/\eta}\neq 0$; straightforward algebraic rearrangement then produces

$$\frac{\sqrt{1-\zeta^2}}{\zeta}=\frac{\sin\sqrt{1-\zeta^2}\,\frac{t}{\eta}}{\cos\sqrt{1-\zeta^2}\,\frac{t}{\eta}}, \tag{3.664}$$

where definition of trigonometric tangent permits notation be condensed to

$$\tan\sqrt{1-\zeta^2}\,\frac{t}{\eta}=\frac{\sqrt{1-\zeta^2}}{\zeta}. \tag{3.665}$$

After taking inverse tangent of both sides, Eq. (3.665) becomes

$$\sqrt{1-\zeta^2}\,\frac{t}{\eta}=\tan^{-1}\frac{\sqrt{1-\zeta^2}}{\zeta}, \tag{3.666}$$

or else

$$\frac{t_{pk}}{\eta}=\frac{\tan^{-1}\frac{\sqrt{1-\zeta^2}}{\zeta}}{\sqrt{1-\zeta^2}} \tag{3.667}$$

after solving for t_{pk}/η; hence, an infinite number of solutions will appear, due to the periodic nature of trigonometric tangent. Remember that tangent is defined as the ratio of sine to cosine, while the fundamental relationship of trigonometry, i.e.

$$\sin^2\xi+\cos^2\xi=1, \tag{3.668}$$

has it that sine of angle ξ equals $\sqrt{1-\zeta^2}$ when cosine of ξ equals ζ, in agreement with

$$\left(\sqrt{1-\zeta^2}\right)^2+\zeta^2=1; \tag{3.669}$$

consequently,

$$\tan\xi\equiv\frac{\sin\xi}{\cos\xi}=\frac{\sqrt{1-\zeta^2}}{\zeta}. \tag{3.670}$$

One accordingly realizes that

$$\sqrt{1-\zeta^2}=\sin\xi=\sin\tan^{-1}\frac{\sqrt{1-\zeta^2}}{\zeta}, \tag{3.671}$$

based on Eq. (3.670) upon application of inverse tangent to both sides – and thus

$$\tan^{-1}\frac{\sqrt{1-\zeta^2}}{\zeta}=\sin^{-1}\sqrt{1-\zeta^2} \tag{3.672}$$

after taking inverse sine of both sides; by the same token,

$$\zeta=\cos\xi=\cos\tan^{-1}\frac{\sqrt{1-\zeta^2}}{\zeta} \tag{3.673}$$

stems again from Eq. (3.670) after taking inverse tangent of both sides – and, consequently,

$$\tan^{-1}\frac{\sqrt{1-\zeta^2}}{\zeta}=\cos^{-1}\zeta \tag{3.674}$$

once inverse cosine is applied to both sides. Therefore, Eq. (3.667) may instead appear as

$$\boxed{\frac{t_{pk}}{\eta}=\frac{\sin^{-1}\sqrt{1-\zeta^2}}{\sqrt{1-\zeta^2}}}, \tag{3.675}$$

with the aid of Eq. (3.672). In addition, Eq. (3.666) can be rewritten as

$$\sin\sqrt{1-\zeta^2}\,\frac{t}{\eta}=\sin\tan^{-1}\frac{\sqrt{1-\zeta^2}}{\zeta} \tag{3.676}$$

after taking sine of both sides, where combination with Eq. (3.671) unfolds

$$\sin\sqrt{1-\zeta^2}\,\frac{t}{\eta}=\sqrt{1-\zeta^2}; \tag{3.677}$$

Eqs. (3.675) and (3.677) will then be inserted in Eq. (3.657) to get

$$\frac{\frac{y_{out,pk}}{K}}{\eta}=\frac{1}{\sqrt{1-\zeta^2}}\exp\left\{-\zeta\frac{\sin^{-1}\sqrt{1-\zeta^2}}{\sqrt{1-\zeta^2}}\right\}\sqrt{1-\zeta^2}, \tag{3.678}$$

where cancellation of $\sqrt{1-\zeta^2}$ between numerator and denominator permits simplification to

$$\frac{\frac{y_{out,pk}}{K}}{\eta}=\exp\left\{-\zeta\frac{\sin^{-1}\sqrt{1-\zeta^2}}{\sqrt{1-\zeta^2}}\right\}; \tag{3.679}$$

elimination of $\tan^{-1}\sqrt{1-\zeta^2}/\zeta$ between Eqs. (3.672) and (3.674) leads, in turn, to

$$\sin^{-1}\sqrt{1-\zeta^2}=\cos^{-1}\zeta, \tag{3.680}$$

so Eq. (3.679) may be rephrased as

$$\boxed{\frac{y_{out,pk}}{\dfrac{K}{\eta}}=\exp\left\{-\zeta\frac{\cos^{-1}\zeta}{\sqrt{1-\zeta^2}}\right\}.} \tag{3.681}$$

Equation (3.681) describes the (local) maximum deviations from initial steady state; for every ξ representing $\cos^{-1}\zeta$, all other $\xi+\kappa\pi$ (with $k=1,2,3,\ldots$) will also be eligible – and such solutions are alternately positive and negative. The ratio between amplitudes of the first two positive peaks is known as decay ratio, R_d, and is defined as

$$R_d\equiv\frac{\exp\left\{-\dfrac{\zeta(\xi+2\pi)}{\sqrt{1-\zeta^2}}\right\}}{\exp\left\{-\dfrac{\zeta\xi}{\sqrt{1-\zeta^2}}\right\}} \tag{3.682}$$

– based on Eq. (3.681); this descriptor entails a discrete measure of rate of decrease in peak amplitude. Equation (3.682) can be algebraically rearranged to read

$$R_d=\exp\left\{-\frac{\zeta(\xi+2\pi)-\zeta\xi}{\sqrt{1-\zeta^2}}\right\} \tag{3.683}$$

after merging arguments of exponentials functions, which is equivalent to

$$\boxed{R_d=\exp\left\{-\frac{2\pi\zeta}{\sqrt{1-\zeta^2}}\right\}.} \tag{3.684}$$

The variation in overshoot, or $y_{out,pk}\equiv y_{out,pk}\{t_{pk}\}$ for the first occurring t_{pk}, as a function of damping factor ζ and for values below unity, is provided in Fig. 3.57 – which serves as graphical interpretation of Eq. (3.681). Note the bending of the curve toward a horizontal, unit plateau as ζ approaches zero; and the decrease in $y_{out,pk}/(K/\eta)$ with increasing ζ within interval $]0,1[$ – thus confirming that the worst deviation of y_{out} from specification, following a disturbance, will not exceed K/η, irrespective of ζ.

Inspection of Figs. 3.56iii and 3.57 indicates that the amplitude of the (sinusoidal) oscillations increases when ζ approaches zero – so it is instructive to investigate the sign of the derivative of $\eta y_{out,pk}/K$, with regard to ζ. Equation (3.681) accordingly supports

$$\frac{d}{d\zeta}\left(\frac{y_{out,pk}}{\dfrac{K}{\eta}}\right)=\frac{d}{d\zeta}\left(\exp\left\{-\frac{\zeta\cos^{-1}\zeta}{\sqrt{1-\zeta^2}}\right\}\right), \tag{3.685}$$

where application of the rules of differentiation of a composite function, an exponential function, a quotient of functions, a product of functions, a square root, and an inverse cosine unfolds

$$\frac{d}{d\zeta}\left(\frac{y_{out,pk}}{\dfrac{K}{\eta}}\right)=-\exp\left\{-\frac{\zeta\cos^{-1}\zeta}{\sqrt{1-\zeta^2}}\right\}\frac{\left(\cos^{-1}\zeta-\dfrac{\zeta}{\sqrt{1-\zeta^2}}\right)\sqrt{1-\zeta^2}-\dfrac{-2\zeta}{2\sqrt{1-\zeta^2}}\zeta\cos^{-1}\zeta}{\left(\sqrt{1-\zeta^2}\right)^2}. \tag{3.686}$$

Upon cancelling out factors alike, composing square power with square root, and eliminating parenthesis, Eq. (3.686) becomes

$$\frac{d}{d\zeta}\left(\frac{y_{out,pk}}{\dfrac{K}{\eta}}\right)=-\exp\left\{-\frac{\zeta\cos^{-1}\zeta}{\sqrt{1-\zeta^2}}\right\}\frac{\sqrt{1-\zeta^2}\cos^{-1}\zeta-\zeta+\zeta\dfrac{\zeta}{\sqrt{1-\zeta^2}}\cos^{-1}\zeta}{1-\zeta^2}, \tag{3.687}$$

where further multiplication of both numerator and denominator by $\sqrt{1-\zeta^2}$ reduces Eq. (3.687) to

$$\frac{d}{d\zeta}\left(\frac{y_{out,pk}}{\dfrac{K}{\eta}}\right)=-\exp\left\{-\frac{\zeta\cos^{-1}\zeta}{\sqrt{1-\zeta^2}}\right\}\frac{\left(1-\zeta^2\right)\cos^{-1}\zeta-\zeta\sqrt{1-\zeta^2}+\zeta^2\cos^{-1}\zeta}{\left(1-\zeta^2\right)^{\frac{3}{2}}}; \tag{3.688}$$

after dropping symmetrical terms, Eq. (3.688) turns to

$$\frac{d}{d\zeta}\left(\frac{y_{out,pk}}{\dfrac{K}{\eta}}\right)=-\exp\left\{-\frac{\zeta\cos^{-1}\zeta}{\sqrt{1-\zeta^2}}\right\}\frac{\cos^{-1}\zeta-\zeta\sqrt{1-\zeta^2}}{\left(1-\zeta^2\right)^{\frac{3}{2}}}. \tag{3.689}$$

Since only small values of ζ, within interval $[0,1[$, are of interest here, one may safely resort to Taylor's expansion of the numerator of Eq. (3.689) about $\zeta=0$, truncated after the cubic term – starting with calculation of inverse cosine itself, viz.

$$\cos^{-1}\zeta\Big|_{\zeta=0}=\cos^{-1}0=\frac{\pi}{2}, \tag{3.690}$$

and then differentiating it once, viz.

$$\left.\frac{d\cos^{-1}\zeta}{d\zeta}\right|_{\zeta=0}=-\left.\frac{1}{\sqrt{1-\zeta^2}}\right|_{\zeta=0}=-\frac{1}{\sqrt{1-0^2}}=-1; \quad (3.691)$$

while the second derivative looks like

$$\left.\frac{d^2\cos^{-1}\zeta}{d\zeta^2}\right|_{\zeta=0}=\left.\frac{d}{d\zeta}\left(-\frac{1}{\sqrt{1-\zeta^2}}\right)\right|_{\zeta=0}=\left.\frac{\frac{-2\zeta}{2\sqrt{1-\zeta^2}}}{\left(\sqrt{1-\zeta^2}\right)^2}\right|_{\zeta=0}=-\left.\frac{\zeta}{\left(1-\zeta^2\right)^{\frac{3}{2}}}\right|_{\zeta=0}=-\frac{0}{\left(1-0^2\right)^{\frac{3}{2}}}=0. \quad (3.692)$$

By the same token, the third derivative reads

$$\left.\frac{d^3\cos^{-1}\zeta}{d\zeta^3}\right|_{\zeta=0}=\left.\frac{d}{d\zeta}\left(-\frac{\zeta}{\left(1-\zeta^2\right)^{\frac{3}{2}}}\right)\right|_{\zeta=0}=-\left.\frac{\left(1-\zeta^2\right)^{\frac{3}{2}}-\zeta\frac{3}{2}\left(1-\zeta^2\right)^{\frac{1}{2}}(-2\zeta)}{\left(\left(1-\zeta^2\right)^{\frac{3}{2}}\right)^2}\right|_{\zeta=0}$$
$$=-\left.\frac{\left(1-\zeta^2\right)^{\frac{3}{2}}+3\zeta^2\left(1-\zeta^2\right)^{\frac{1}{2}}}{\left(1-\zeta^2\right)^3}\right|_{\zeta=0}=-\left.\frac{\left(1-\zeta^2\right)+3\zeta^2}{\left(1-\zeta^2\right)^{\frac{5}{2}}}\right|_{\zeta=0}=-\left.\frac{1+2\zeta^2}{\left(1-\zeta^2\right)^{\frac{5}{2}}}\right|_{\zeta=0}=-\frac{1+0}{\left(1-0^2\right)^{\frac{5}{2}}}=-1 \quad (3.693)$$

– where application of classical theorems on derivatives was complemented by division of both numerator and denominator by $\sqrt{1-\zeta^2}$, followed by condensation of terms alike. A similar type of expansion will now be attempted of $\sqrt{1-\zeta^2}$, starting with

$$\left.\sqrt{1-\zeta^2}\right|_{\zeta=0}=\sqrt{1-0^2}=1, \quad (3.694)$$

with differentiation giving rise to

$$\left.\frac{d\sqrt{1-\zeta^2}}{d\zeta}\right|_{\zeta=0}=\left.\frac{-2\zeta}{2\sqrt{1-\zeta^2}}\right|_{\zeta=0}=-\left.\frac{\zeta}{\sqrt{1-\zeta^2}}\right|_{\zeta=0}=-\frac{0}{\sqrt{1-0^2}}=0; \quad (3.695)$$

a second produces

$$\left.\frac{d^2\sqrt{1-\zeta^2}}{d\zeta^2}\right|_{\zeta=0}=\left.\frac{d}{d\zeta}\left(-\frac{\zeta}{\sqrt{1-\zeta^2}}\right)\right|_{\zeta=0}=-\left.\frac{\sqrt{1-\zeta^2}-\zeta\frac{-2\zeta}{2\sqrt{1-\zeta^2}}}{\left(\sqrt{1-\zeta^2}\right)^2}\right|_{\zeta=0}$$
$$=-\left.\frac{\sqrt{1-\zeta^2}+\frac{\zeta^2}{\sqrt{1-\zeta^2}}}{1-\zeta^2}\right|_{\zeta=0}=-\left.\frac{1-\zeta^2+\zeta^2}{\left(1-\zeta^2\right)^{\frac{3}{2}}}\right|_{\zeta=0}=-\left.\left(1-\zeta^2\right)^{-\frac{3}{2}}\right|_{\zeta=0}=-\left(1-0^2\right)^{-\frac{3}{2}}=-1 \quad (3.696)$$

along with multiplication of numerator and denominator by $\sqrt{1-\zeta^2}$, cancellation of factors alike between numerator and denominator, and further elimination of symmetrical terms – while the third derivative will take the form

$$\left.\frac{d^3\sqrt{1-\zeta^2}}{d\zeta^3}\right|_{\zeta=0}=\left.\frac{d}{d\zeta}\left(-\left(1-\zeta^2\right)^{-\frac{3}{2}}\right)\right|_{\zeta=0}=\left.\frac{3}{2}\left(1-\zeta^2\right)^{-\frac{5}{2}}(-2\zeta)\right|_{\zeta=0}$$
$$=-3\zeta\left.\left(1-\zeta^2\right)^{-\frac{5}{2}}\right|_{\zeta=0}=-0\left(1-0^2\right)^{-\frac{5}{2}}=0. \quad (3.697)$$

In view of Eqs. (3.690)–(3.697), one can write

$$\cos^{-1}\zeta-\zeta\sqrt{1-\zeta^2}\approx\frac{\pi}{2}-\zeta-\frac{\zeta^3}{3!}-\zeta\left(1-\frac{\zeta^2}{2!}\right), \quad (3.698)$$

which readily entails

$$\cos^{-1}\zeta-\zeta\sqrt{1-\zeta^2}=\frac{\pi}{2}-\zeta-\frac{\zeta^3}{3!}-\zeta+\frac{\zeta^3}{2!} \quad (3.699)$$

upon removal of parenthesis – and, consequently,

$$\cos^{-1}\zeta-\zeta\sqrt{1-\zeta^2}=\frac{\pi}{2}-2\zeta+\frac{\zeta^3}{3}>0; \quad (3.700)$$

this result arises because $\pi/2-2\zeta+\zeta^3/3$ is comprised between $\pi/2>0$ as upper boundary when $\zeta=0$, and $\pi/2-2\cdot1+1^3/3=\pi/2-5/3>0$ as lower boundary when $\zeta=1$, with a monotonically decreasing behavior in between when ζ increases. One therefore concludes that

$$\frac{d}{d\zeta}\left(\frac{y_{out,pk}}{\frac{K}{\eta}}\right)<0, \quad (3.701)$$

upon insertion of Eq. (3.700) in Eq. (3.689), and realization that $\exp\left\{-\left(\zeta\cos^{-1}\zeta\right)/\sqrt{1-\zeta^2}\right\}>0$ and $1/\left(1-\zeta^2\right)^{3/2}>0$; this result means that the maximum of $\eta y_{out,pk}/K$ actually lies on a physical constraint, i.e. $\zeta=0$, as smallest value allowed for ζ. When ζ is set nil, Eq. (3.657) becomes

$$\left.\frac{\frac{y_{out}}{K}}{\eta}\right|_{\zeta=0}=\frac{e^{-0\frac{t}{\eta}}}{\sqrt{1-0^2}}\sin\sqrt{1-0^2}\,\frac{t}{\eta}=\frac{e^0}{\sqrt{1}}\sin\sqrt{1}\,\frac{t}{\eta}=\sin\frac{t}{\eta}; \tag{3.702}$$

hence, a sustained sinusoid will result, with (constant) amplitude equal to unity – as perceived upon inspection of the curve labeled as $\zeta=0$ in Fig. 3.56iii, which indeed exhibits the largest amplitude of all curves.

One will now carry on an analysis of performance specifically for a PI controller – and accordingly retrieve Eq. (3.609) as

$$y_{out}=\frac{\frac{K}{\eta_{PI}}}{\sqrt{\zeta_{PI}^{\,2}-1}}\exp\left\{-\zeta_{PI}\frac{t}{\eta_{PI}}\right\}\left(\sinh\sqrt{\zeta_{PI}^{\,2}-1}\,\frac{t}{\eta_{PI}}\right), \tag{3.703}$$

where insertion of Eqs. (3.574), (3.576), and (3.579) supports transformation to

$$y_{out}=\frac{\dfrac{A\dfrac{\tau_I}{\vartheta}}{\tau\sqrt{\dfrac{1}{\vartheta}\dfrac{\tau_I}{\tau}}}\exp\left\{-\dfrac{1+\kappa\tau+\vartheta}{2\sqrt{\vartheta}}\sqrt{\dfrac{\tau_I}{\tau}}\dfrac{t}{\tau\sqrt{\dfrac{1}{\vartheta}\dfrac{\tau_I}{\tau}}}\right\}}{\sqrt{\left(\dfrac{1+\kappa\tau+\vartheta}{2\sqrt{\vartheta}}\sqrt{\dfrac{\tau_I}{\tau}}\right)^2-1}} \times\sinh\sqrt{\left(\frac{1+\kappa\tau+\vartheta}{2\sqrt{\vartheta}}\sqrt{\frac{\tau_I}{\tau}}\right)^2-1}\,\frac{t}{\tau\sqrt{\frac{1}{\vartheta}\frac{\tau_I}{\tau}}}; \tag{3.704}$$

after dropping factors alike between numerator and denominator, while taking squares as indicated, Eq. (3.704) becomes

$$y_{out}=\frac{A\dfrac{\sqrt{\tau_I}}{\sqrt{\vartheta}\sqrt{\tau}}\exp\left\{-\dfrac{1+\kappa\tau+\vartheta}{2}\dfrac{t}{\tau}\right\}}{\sqrt{\dfrac{(1+\kappa\tau+\vartheta)^2}{4\vartheta}\dfrac{\tau_I}{\tau}-1}} \times\sinh\sqrt{\frac{(1+\kappa\tau+\vartheta)^2}{4\vartheta}\frac{\tau_I}{\tau}-1}\,\frac{1}{\sqrt{\frac{1}{\vartheta}\frac{\tau_I}{\tau}}}\frac{t}{\tau}, \tag{3.705}$$

where lumping of square roots followed by division of both sides by A permit further simplification to

$$\frac{y_{out}}{A}=\frac{\sqrt{\dfrac{\tau_I}{\tau}}\exp\left\{-\dfrac{1+\kappa\tau+\vartheta}{2}\dfrac{t}{\tau}\right\}}{\sqrt{\dfrac{(1+\kappa\tau+\vartheta)^2}{4}\dfrac{\tau_I}{\tau}-\vartheta}}\sinh\sqrt{\frac{(1+\kappa\tau+\vartheta)^2}{4}-\frac{\vartheta}{\frac{\tau_I}{\tau}}}\,\frac{t}{\tau} \tag{3.706}$$

– which is plotted in Fig. 3.58i. Remember that y_{out}/A is a function of t/τ via three characteristic (dimensionless) parameters: $\kappa\tau$ coming from the system itself, as per Eqs. (3.436) and (3.437); ϑ coming from proportional control and system as per Eq. (3.564); and τ_I/τ coming from integral control, as per Eq. (3.486). For negligible integral action, i.e. $\tau_I/\tau\rightarrow\infty$, the behavior of y_{out}/A coincides with that conveyed by curve labeled as $\vartheta=1$ in Fig. 3.55 – pertaining to plain proportional control of system described by $\kappa\tau=1$. When τ_I/τ takes any finite value, the sustained offset vanishes, with all curves tending asymptotically to the horizontal axis at long times – yet a stronger integral action, measured by a lower value of parameter τ_I/τ, reduces the time required to recover the initial steady state. A maximum of y_{out}/A vs. t/τ arises for every value of parameter τ_I/τ – in general agreement with Fig. 3.56i. Higher maxima are observed in Fig. 3.56i at lower ζ

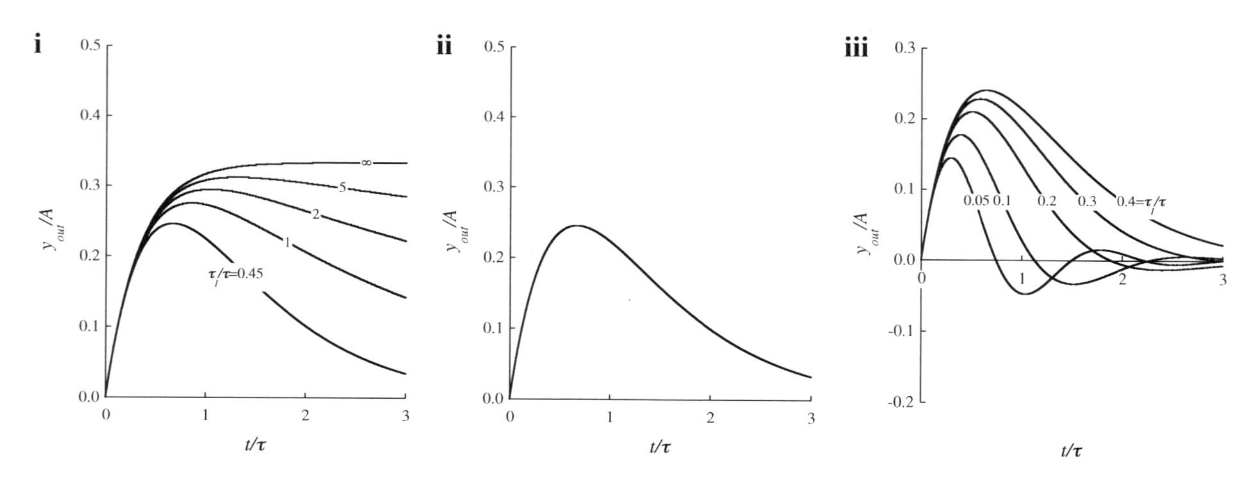

FIGURE 3.58 Evolution of normalized (controlled) variable, y_{out}/A, versus dimensionless time, t/τ, in process characterized by dimensionless parameter $\kappa\tau=1$, subjected to proportional/integral control – for dimensionless parameter $\vartheta=1$, and selected values of dimensionless parameter τ_I/τ, leading to (i) overdamped, (ii) critically damped, or (iii) underdamped behavior.

– when ζ_{PI} increases proportionally to $\sqrt{\tau_I/\tau}$ (for given $\kappa\tau$ and ϑ) as per Eq. (3.579); whereas higher maxima are observed in Fig. 3.58i at higher τ_I/τ. This observation is a consequence of normalization of y_{out} by K/η_{PI} in the former case, with K/η_{PI} being, in turn, proportional to $\sqrt{\tau_I/\tau}$ as per Eqs. (3.574) and (3.576) – while y_{out} was normalized by plain (constant) A in the latter case.

Due to the functional similarity between Eqs. (3.609) and (3.656), it suffices to merely swap hyperbolic sine for trigonometric sine, and replace ζ^2-1 by its negative in Eq. (3.706) to coin the corresponding expression applying to $0<\zeta<1$, viz.

$$\boxed{\frac{y_{out}}{A}=\frac{\sqrt{\frac{\tau_I}{\tau}}\exp\left\{-\frac{1+\kappa\tau+\vartheta}{2}\frac{t}{\tau}\right\}}{\sqrt{\vartheta-\frac{(1+\kappa\tau+\vartheta)^2}{4}\frac{\tau_I}{\tau}}}\sin\sqrt{\frac{\vartheta}{\frac{\tau_I}{\tau}}-\frac{(1+\kappa\tau+\vartheta)^2}{4}}\frac{t}{\tau};}\tag{3.707}$$

Eq. (3.707) is plotted in Fig. 3.58iii. Only small values of τ_I/τ are now germane – so as to produce $\zeta_{PI}<1$, see Eq. (3.579); oscillations arise, as expected, at the expense of lower maximum (topical) deviations from the original steady state when τ_I/τ decreases. Reduction of τ_I/τ increases also oscillation frequency – yet decay to zero becomes faster, consistent with Fig. 3.56iii and Eq. (3.579); once again, the lower (local) maxima associated with lower τ_I/τ do not contradict the lower (local) maxima in Fig. 3.56iii associated with higher ζ_{PI} proportional to $\sqrt{\tau_I/\tau}$, due to the different normalization used for y_{out} – i.e. via (constant) A, or (variable) K/τ proportional, in turn, to $\sqrt{\tau_I/\tau}$ (as outlined above).

Finally, Eq. (3.638) may be redone as

$$y_{out}=\frac{K}{\eta_{PI}}\frac{t}{\eta_{PI}}e^{-\frac{t}{\eta_{PI}}},\tag{3.708}$$

where insertion of Eqs. (3.574) and (3.576) gives rise to

$$y_{out}=\frac{A\frac{\tau_I}{\vartheta}}{\tau\sqrt{\frac{1}{\vartheta}\frac{\tau_I}{\tau}}}\frac{t}{\tau\sqrt{\frac{1}{\vartheta}\frac{\tau_I}{\tau}}}\exp\left\{-\frac{1}{\tau\sqrt{\frac{1}{\vartheta}\frac{\tau_I}{\tau}}}t\right\};\tag{3.709}$$

after lumping the square roots preceding the exponential function, and rearranging the argument of the latter, Eq. (3.709) becomes

$$y_{out}=\frac{A\frac{\tau_I}{\vartheta}}{\tau\frac{1}{\vartheta}\frac{\tau_I}{\tau}}\frac{t}{\tau}\exp\left\{-\sqrt{\frac{\vartheta}{\frac{\tau_I}{\tau}}}\frac{t}{\tau}\right\}\tag{3.710}$$

– where cancellation of τ, τ_I, and ϑ between numerator and denominator, complemented with division of both sides by A finally yields

$$\frac{y_{out}}{A}=\frac{t}{\tau}\exp\left\{-\sqrt{\frac{\vartheta}{\frac{\tau_I}{\tau}}}\frac{t}{\tau}\right\}.\tag{3.711}$$

Note that $\zeta_{PI}=1$ was imposed in advance, and this works out as a restriction upon the set of parameters $\kappa\tau$, ϑ, and τ_I/τ – implying that

$$\frac{1+\kappa\tau+\vartheta}{2\sqrt{\vartheta}}\sqrt{\frac{\tau_I}{\tau}}=1,\tag{3.712}$$

based on Eq. (3.579); therefore, $1/\sqrt{\vartheta}$ and $\sqrt{\tau_I/\tau}$ can be isolated as

$$\frac{1}{\sqrt{\vartheta}}\sqrt{\frac{\tau_I}{\tau}}=\frac{2}{1+\kappa\tau+\vartheta},\tag{3.713}$$

or else

$$\boxed{\sqrt{\frac{\vartheta}{\frac{\tau_I}{\tau}}}=\frac{1+\kappa\tau+\vartheta}{2}}\tag{3.714}$$

after taking reciprocals of both sides. Insertion of Eq. (3.714) converts Eq. (3.711) to

$$\boxed{\frac{y_{out}}{A}=\frac{t}{\tau}\exp\left\{-\frac{1+\kappa\tau+\vartheta}{2}\frac{t}{\tau}\right\},}\tag{3.715}$$

as graphically represented in Fig. 3.58ii; note that a single parameter is relevant here, i.e. the sum of $\kappa\tau$ with ϑ. As anticipated, the curve described by Eq. (3.715) sets the interface between the overdamped curves plotted in Fig. 3.58i and the underdamped, oscillating curves in Fig. 3.58iii; a single curve is the result of $\kappa\tau$ and ϑ having been, in advance, both set equal to unity, corresponding to $\tau_I/\tau=4/9\approx0.44$ as per Eq. (3.713).

One should finally focus on a PID controller, and accordingly rewrite Eq. (3.609) as

$$y_{out}=\frac{\frac{K}{\eta_{PID}}}{\sqrt{\zeta_{PID}^2-1}}\exp\left\{-\zeta_{PID}\frac{t}{\eta_{PID}}\right\}\sinh\sqrt{\zeta_{PID}^2-1}\frac{t}{\eta_{PID}}\tag{3.716}$$

– where insertion of Eqs. (3.574), (3.591), and (3.593) prompts transformation to

$$y_{out}=\frac{\frac{A\frac{\tau_I}{\vartheta}}{\frac{\tau}{\sqrt{\vartheta}}\sqrt{\frac{\tau_I}{\tau}\left(1+\vartheta\frac{\tau_D}{\tau}\right)}}}{\sqrt{\left(\frac{1+\kappa\tau+\vartheta}{2\sqrt{\vartheta}}\sqrt{\frac{\frac{\tau_I}{\tau}}{1+\vartheta\frac{\tau_D}{\tau}}}\right)^2-1}}$$
$$\times\exp\left\{-\frac{1+\kappa\tau+\vartheta}{2\sqrt{\vartheta}}\sqrt{\frac{\frac{\tau_I}{\tau}}{1+\vartheta\frac{\tau_D}{\tau}}}\frac{t}{\frac{\tau}{\sqrt{\vartheta}}\sqrt{\frac{\tau_I}{\tau}\left(1+\vartheta\frac{\tau_D}{\tau}\right)}}\right\}$$
$$\times\sinh\sqrt{\left(\frac{1+\kappa\tau+\vartheta}{2\sqrt{\vartheta}}\sqrt{\frac{\frac{\tau_I}{\tau}}{1+\vartheta\frac{\tau_D}{\tau}}}\right)^2-1}\frac{t}{\frac{\tau}{\sqrt{\vartheta}}\sqrt{\frac{\tau_I}{\tau}\left(1+\vartheta\frac{\tau_D}{\tau}\right)}};\tag{3.717}$$

upon dropping factors alike between numerator and denominator, splitting squares of products, and lumping square roots, Eq. (3.717) becomes

$$y_{out} = \frac{\dfrac{A\sqrt{\dfrac{\tau_I}{\tau}}}{\sqrt{\vartheta}\sqrt{1+\vartheta\dfrac{\tau_D}{\tau}}}}{\sqrt{\dfrac{(1+\kappa\tau+\vartheta)^2}{4\vartheta}\dfrac{\dfrac{\tau_I}{\tau}}{1+\vartheta\dfrac{\tau_D}{\tau}}-1}}\exp\left\{-\frac{\dfrac{1+\kappa\tau+\vartheta}{2}}{\sqrt{\left(1+\vartheta\dfrac{\tau_D}{\tau}\right)\left(1+\vartheta\dfrac{\tau_D}{\tau}\right)}}\frac{t}{\tau}\right\}$$
$$\times\sinh\sqrt{\frac{\dfrac{(1+\kappa\tau+\vartheta)^2}{4\vartheta}\dfrac{\dfrac{\tau_I}{\tau}}{1+\vartheta\dfrac{\tau_D}{\tau}}-1}{\dfrac{1}{\vartheta}\dfrac{\tau_I}{\tau}\left(1+\vartheta\dfrac{\tau_D}{\tau}\right)}}\frac{t}{\tau}. \tag{3.718}$$

After lumping factors alike in the argument of the exponential function, and pooling together terms under the remaining square roots, Eq. (3.718) turns to

$$y_{out} = \frac{\dfrac{A\sqrt{\dfrac{\tau_I}{\tau}}}{\sqrt{\vartheta}\sqrt{1+\vartheta\dfrac{\tau_D}{\tau}}}}{\sqrt{\dfrac{\dfrac{(1+\kappa\tau+\vartheta)^2}{4\vartheta}\dfrac{\tau_I}{\tau}-\left(1+\vartheta\dfrac{\tau_D}{\tau}\right)}{1+\vartheta\dfrac{\tau_D}{\tau}}}}\exp\left\{-\frac{\dfrac{1+\kappa\tau+\vartheta}{2}}{\sqrt{\left(1+\vartheta\dfrac{\tau_D}{\tau}\right)^2}}\frac{t}{\tau}\right\}$$
$$\times\sinh\sqrt{\frac{\dfrac{(1+\kappa\tau+\vartheta)^2}{4\vartheta}\dfrac{\tau_I}{\tau}-\left(1+\vartheta\dfrac{\tau_D}{\tau}\right)}{\dfrac{1}{\vartheta}\dfrac{\tau_I}{\tau}\left(1+\vartheta\dfrac{\tau_D}{\tau}\right)\left(1+\vartheta\dfrac{\tau_D}{\tau}\right)}}\frac{t}{\tau}; \tag{3.719}$$

the square power now cancels out the square root under the exponential sign, $\sqrt{\vartheta}$ and $\sqrt{1+\vartheta\tau_D/\tau}$ cancel out with ϑ and $1+\vartheta\tau_D/\tau$ in denominator under the next square root, and factors alike can be pooled together in the argument of the hyperbolic sine to finally yield

$$\frac{y_{out}}{A} = \frac{\sqrt{\dfrac{\tau_I}{\tau}}\exp\left\{-\dfrac{1+\kappa\tau+\vartheta}{2\left(1+\vartheta\dfrac{\tau_D}{\tau}\right)}\dfrac{t}{\tau}\right\}}{\sqrt{\dfrac{(1+\kappa\tau+\vartheta)^2}{4}\dfrac{\tau_I}{\tau}-\vartheta\left(1+\vartheta\dfrac{\tau_D}{\tau}\right)}}\times\sinh\frac{\sqrt{\dfrac{(1+\kappa\tau+\vartheta)^2}{4}\dfrac{\tau_I}{\tau}-\vartheta\left(1+\vartheta\dfrac{\tau_D}{\tau}\right)}}{\sqrt{\dfrac{\tau_I}{\tau}}\left(1+\vartheta\dfrac{\tau_D}{\tau}\right)}\frac{t}{\tau}, \tag{3.720}$$

along with division of both sides by A – as plotted in Fig. 3.59i. In this case, y_{out}/A turns an explicit function of four characteristic (dimensionless) parameters – $\kappa\tau$ associated to equipment *per se*, ϑ associated with proportional action of a PID control and system itself, τ_I/τ associated with integral action of PID control, and τ_D/τ associated with derivative action of PID control. In the absence of derivative action, the curve labeled $\tau_D/\tau=0$ in Fig. 3.59i coincides with that labeled $\tau_I/\tau=1$ in Fig. 3.58i, since $\kappa\tau=\vartheta=1$ in both cases (beware of the different times scales spanned). As τ_D/τ increases, a slightly later and lower (single) overshoot is observed – along with a slightly shorter tail; hence, the effect of derivative control appears marginal in an intrinsically overdamped system, even though it still contributes to increased robustness. According to Eqs. (3.574) and (3.591), K/η_{PID} is inversely proportional to $\sqrt{\vartheta(1+\vartheta\tau_D/\tau)}$, while A is a constant; this justifies the trends of curves in Fig. 3.56i where $y_{out}/(K/\eta_{PID})$ is plotted, opposite to those of curves in Fig. 3.59i where y_{out}/A is plotted, with increasing τ_D/τ – knowing that ζ_{PID} is inversely proportional to $\sqrt{\vartheta(1+\vartheta\tau_D/\tau)}$, besides directly proportional to $\sqrt{\tau_I/\tau}$ as per Eq. (3.593).

The corresponding expression for y_{out}/A vs. t/τ when $\zeta<1$ will again resort to Eq. (3.720), pertaining to $\zeta>1$, as template – as long as the negatives of the kernels of the square roots are taken, together with replacement of hyperbolic sine by its trigonometric counterpart, see Eq. (3.609) *vis-à-vis* with Eq. (3.656); one consequently ends up with

$$\frac{y_{out}}{A} = \frac{\sqrt{\dfrac{\tau_I}{\tau}}\exp\left\{-\dfrac{1+\kappa\tau+\vartheta}{2\left(1+\vartheta\dfrac{\tau_D}{\tau}\right)}\dfrac{t}{\tau}\right\}}{\sqrt{\vartheta\left(1+\vartheta\dfrac{\tau_D}{\tau}\right)-\dfrac{(1+\kappa\tau+\vartheta)^2}{4}\dfrac{\tau_I}{\tau}}}\times\sin\frac{\sqrt{\vartheta\left(1+\vartheta\dfrac{\tau_D}{\tau}\right)-\dfrac{(1+\kappa\tau+\vartheta)^2}{4}\dfrac{\tau_I}{\tau}}}{\sqrt{\dfrac{\tau_I}{\tau}}\left(1+\vartheta\dfrac{\tau_D}{\tau}\right)}\frac{t}{\tau}, \tag{3.721}$$

as plotted in Fig. 3.59iii. In view of Eq. (3.593), only large values of τ_D/τ are germane here – in that they will lead to $0<\zeta_{PID}<1$; an oscillatory pattern results, comparable to the curves in Fig. 3.58iii, as typical of an underdamped system – but associated with a more sluggish response over time. A lower and lower overshoot is observed with increasing derivative action, measured by τ_D/τ – along with a rapid, essentially unidirectional decay toward nil amplitude; hence, the derivative action has clearly added robustness to the controlled system – in that the triggered action matches the magnitude of the error anticipated for the immediate future. When the original PI-controlled system was already underdamped, as per the curve labeled $\tau_I/\tau=0.2$ in Fig. 3.58iii, addition of derivative action to the controller decreases the overshoot and delays its occurrence – see curves in Fig. 3.59iv; a slower response has thus been produced, consistent with the pattern already illustrated in Fig. 3.59iii. Since the curve labeled $\tau_D/\tau=0$ in Fig. 3.59iv coincides with that labeled $\tau_I/\tau=0.2$ in Fig. 3.58iii, the latter represents the asymptotic case of a PID without derivative action at all (thus justifying why it served as reference for this analysis of the specific effect of τ_D/τ).

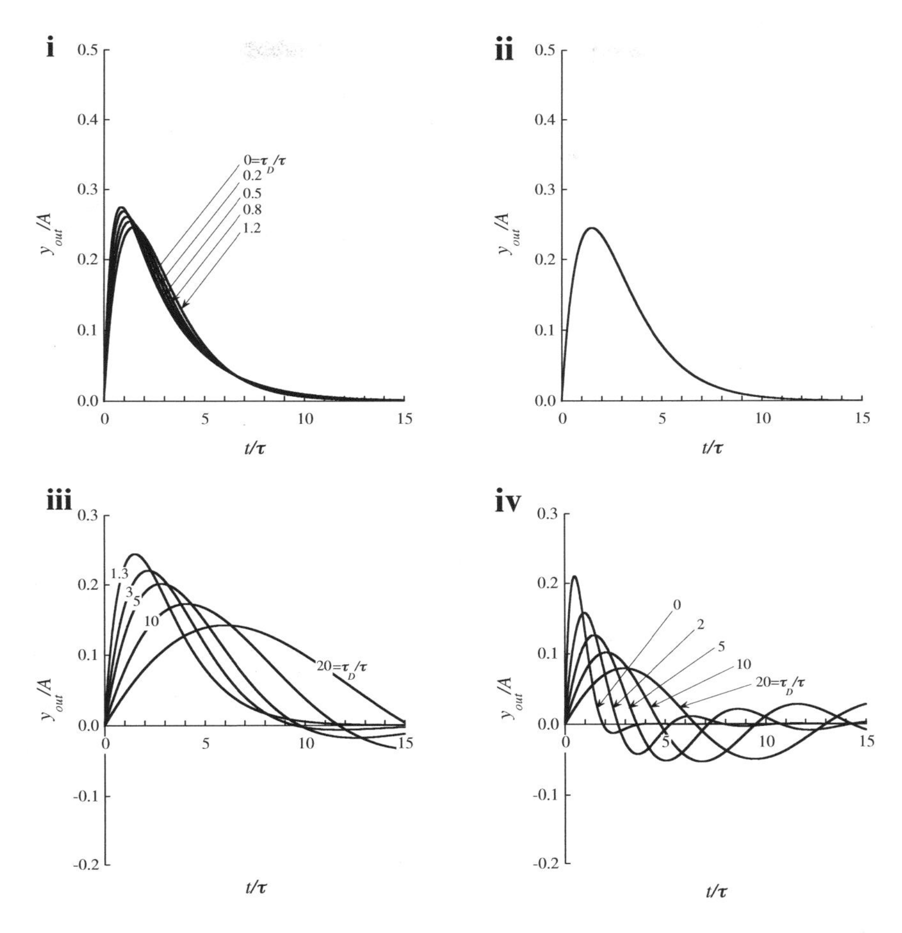

FIGURE 3.59 Evolution of normalized (controlled) variable, y_{out}/A, versus dimensionless time, t/τ, in process characterized by dimensionless parameter $\kappa\tau=1$, subjected to proportional/integral/derivative control – for dimensionless parameter $\vartheta=1$, dimensionless parameter (i, ii, iii) $\tau_I/\tau=1$ or (iv) $\tau_I/\tau=0.2$, and selected values of dimensionless parameter τ_D/τ, leading to (i) overdamped, (ii) critically damped, or (iii, iv) underdamped behavior.

When $\zeta_{PID}=1$, one gets

$$y_{out}=\frac{K}{\eta_{PID}}\frac{t}{\eta_{PID}}e^{-\frac{t}{\eta_{PID}}} \tag{3.722}$$

as analog to Eq. (3.708) – where insertion of Eqs. (3.574) and (3.591) generates

$$y_{out}=\frac{A\dfrac{\tau_I}{\vartheta}}{\dfrac{\tau}{\sqrt{\vartheta}}\sqrt{\dfrac{\tau_I}{\tau}\left(1+\vartheta\dfrac{\tau_D}{\tau}\right)}}\frac{t}{\dfrac{\tau}{\sqrt{\vartheta}}\sqrt{\dfrac{\tau_I}{\tau}\left(1+\vartheta\dfrac{\tau_D}{\tau}\right)}}\times\exp\left\{-\frac{t}{\dfrac{\tau}{\sqrt{\vartheta}}\sqrt{\dfrac{\tau_I}{\tau}\left(1+\vartheta\dfrac{\tau_D}{\tau}\right)}}\right\}; \tag{3.723}$$

the square roots preceding the exponential function may once again be collapsed, and the argument of the exponential split as two factors to obtain

$$y_{out}=\frac{A\dfrac{\tau_I}{\vartheta}}{\dfrac{\tau}{\vartheta}\dfrac{\tau_I}{\tau}\left(1+\vartheta\dfrac{\tau_D}{\tau}\right)}\frac{t}{\tau}\exp\left\{-\frac{1}{\dfrac{1}{\sqrt{\vartheta}}\sqrt{\dfrac{\tau_I}{\tau}\left(1+\vartheta\dfrac{\tau_D}{\tau}\right)}}\frac{t}{\tau}\right\}. \tag{3.724}$$

Once τ, τ_I and ϑ are dropped from numerator and denominator preceding the exponential function, and the outstanding square roots lumped, Eq. (3.724) becomes

$$\frac{y_{out}}{A}=\frac{1}{1+\vartheta\dfrac{\tau_D}{\tau}}\frac{t}{\tau}\exp\left\{-\sqrt{\frac{\vartheta}{\dfrac{\tau_I}{\tau}\left(1+\vartheta\dfrac{\tau_D}{\tau}\right)}}\frac{t}{\tau}\right\} \tag{3.725}$$

– where both sides were meanwhile divided by A. Remember that $\zeta_{PID}=1$ implies

$$\frac{1+\kappa\tau+\vartheta}{2\sqrt{\vartheta}}\sqrt{\frac{\frac{\tau_I}{\tau}}{1+\vartheta\frac{\tau_D}{\tau}}}=1 \tag{3.726}$$

as per Eq. (3.593), where isolation of $\sqrt{1+\vartheta\tau_D/\tau}$ prompts

$$\sqrt{1+\vartheta\frac{\tau_D}{\tau}}=\frac{1+\kappa\tau+\vartheta}{2\sqrt{\vartheta}}\sqrt{\frac{\tau_I}{\tau}}; \tag{3.727}$$

after taking squares of both sides, Eq. (3.727) unfolds

$$1+\vartheta\frac{\tau_D}{\tau}=\frac{(1+\kappa\tau+\vartheta)^2}{4\vartheta}\frac{\tau_I}{\tau}. \tag{3.728}$$

Combination with Eq. (3.728) transforms Eq. (3.725) to

$$\frac{y_{out}}{A}=\frac{1}{\frac{(1+\kappa\tau+\vartheta)^2}{4\vartheta}\frac{\tau_I}{\tau}}\frac{t}{\tau}\exp\left\{-\sqrt{\frac{\vartheta}{\frac{\tau_I}{\tau}\frac{(1+\kappa\tau+\vartheta)^2}{4\vartheta}\frac{\tau_I}{\tau}}}\frac{t}{\tau}\right\}, \tag{3.729}$$

which can be algebraically rearranged as

$$\boxed{\frac{y_{out}}{A}=\frac{4\vartheta}{(1+\kappa\tau+\vartheta)^2\frac{\tau_I}{\tau}}\frac{t}{\tau}\exp\left\{-\frac{2\vartheta}{(1+\kappa\tau+\vartheta)\frac{\tau_I}{\tau}}\frac{t}{\tau}\right\}} \tag{3.730}$$

upon reduction in number of fraction levels and elimination of the outstanding square root; a graphical interpretation is conveyed by Fig. 3.59ii. The single curve – described by parameters $\kappa\tau=\vartheta=\tau_I/\tau=1$, assures transition from the set of overdamped curves in Fig. 3.59i to the set of underdamped oscillating curves in Fig. 3.59iii; its characteristic parameter reads $\tau_D/\tau=5/4=1.25$, in view of Eq. (3.728).

PROBLEM 3.15

Consider a general piece of equipment, where some process takes place following first-order kinetics – described by transfer functions still given by Eqs. (3.465) and (3.467), and subjected to feedback control.

a) Justify the designation reset rate traditionally ascribed to the reciprocal of τ_I, or time constant for an integral controller.
b) Why can a plain proportional controller be safely used for servo control, but not for regulator control when level of liquid in a cylindrical tank is the controlled variable – and volumetric flow rate of a pump (placed downstream) is the manipulated variable? Assume zero-th order kinetics for both measuring device and final control element, with the former transfer function further reducing to unity; and unit step input(s), where appropriate.
c) Prove that Eq. (3.650) is obtainable as an asymptotic form of Eq. (3.681).
d) Show that

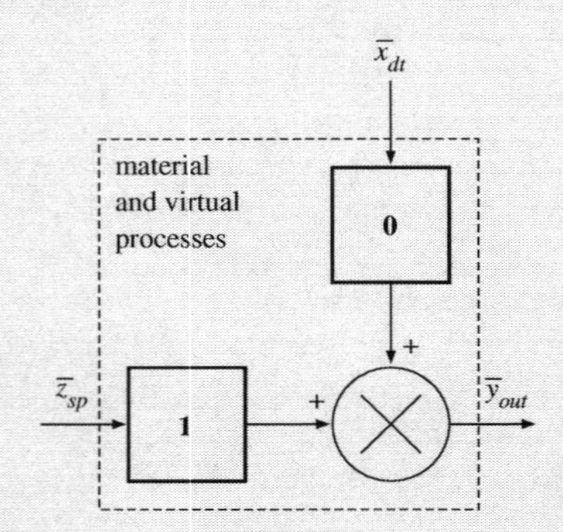

represents the lumped-block diagram of a feedforward-controlled process.

Solution

a) Assume that deviation, z_{dv}, changes by a step of magnitude Λ. The output of a PI controller will initially be proportional to Λ, i.e.

$$Z_{ct,PI}\Big|_{t=0}-Z_{ct}^{ss}=K_P\Lambda, \tag{3.15.1}$$

based on Eq. (3.482) – since only the proportional action has been felt so far when $t=0$, as $\int_0^0 z_{dv}\,dt=0$; the integral action will then increase with time, so that

$$Z_{ct,PI}\Big|_{t=\tau_I}-Z_{ct}^{ss}=K_P\left(\Lambda+\frac{1}{\tau_I}\int_0^{\tau_I}\Lambda dt\right) \tag{3.15.2}$$

holds after time τ_I has elapsed, again stemming from Eq. (3.482). Since Λ is constant, by hypothesis, it may be taken off the kernel in Eq. (3.15.2) to leave

$$Z_{ct,PI}\Big|_{t=\tau_I}-Z_{ct}^{ss}=K_P\left(\Lambda+\frac{1}{\tau_I}\Lambda\int_0^{\tau_I}dt\right), \tag{3.15.3}$$

so the fundamental theorem of integral calculus will be invoked as

$$Z_{ct,PI}\Big|_{t=\tau_I}-Z_{ct}^{ss}=K_P\Lambda+\frac{K_P}{\tau_I}\Lambda\tau_I \tag{3.15.4}$$

– complemented by elimination of parenthesis. Equation (3.15.4) generates

$$Z_{ct,PI}\Big|_{t=\tau_I} - Z_{ct}^{ss} = 2K_P\Lambda \tag{3.15.5}$$

upon cancellation of τ_I between numerator and denominator of the second term, followed by collapse of the resulting terms in the right-hand side; ordered subtraction of Eq. (3.15.1) from Eq. (3.15.5) gives rise to

$$Z_{ct,PI}\Big|_{t=\tau_I} - Z_{ct,PI}\Big|_{t=0} = 2K_P\Lambda - K_P\Lambda, \tag{3.15.6}$$

which reduces to just

$$Z_{ct,PI}\Big|_{t=\tau_I} - Z_{ct,PI}\Big|_{t=0} = K_P\Lambda = Z_{ct,PI}\Big|_{t=0} - Z_{ct}^{ss} \tag{3.15.7}$$

after recalling Eq. (3.15.1) again. Therefore, the integral action by $t=\tau_I$ has repeated the initial response of the proportional action – thus justifying why τ_I is termed reset time, and its reciprocal is termed reset rate.

b) An overall mass balance to liquid in a storage vessels reads

$$Q_{in} = Q_{out} + \frac{dV}{dt}, \tag{3.15.8}$$

where Q_{in} denotes inlet volumetric flow rate (disturbance variable), Q_{out} denotes outlet volumetric flow rate (manipulated variable), and V denotes liquid volume currently in the vessel; for a cylindrical vessel of cross-section S, Eq. (3.15.8) may be redone to

$$Q_{in} = Q_{out} + S\frac{dh}{dt}, \tag{3.15.9}$$

where h denotes height of liquid in the vessel (controlled variable). Steady state operation, implying $dh/dt=0$, is accordingly described by

$$Q_{in}^{ss} = Q_{out}^{ss}, \tag{3.15.10}$$

with superscript ss referring to such sustained operating conditions – thus prompting subtraction of Eq. (3.15.10) from Eq. (3.15.9) to give

$$Q_{in} - Q_{in}^{ss} = Q_{out} - Q_{out}^{ss} + S\frac{dh}{dt}; \tag{3.15.11}$$

after defining deviation variables $\hat{Q}_{in}$, $\hat{Q}_{out}$, and $\hat{h}$ as

$$\hat{Q}_{in} \equiv Q_{in} - Q_{in}^{ss}, \tag{3.15.12}$$

$$\hat{Q}_{out} \equiv Q_{out} - Q_{out}^{ss}, \tag{3.15.13}$$

and

$$\hat{h} \equiv h - h^{ss}, \tag{3.15.14}$$

respectively – and thus

$$d\hat{h} = dh \tag{3.15.15}$$

upon differentiation of Eq. (3.15.14), one may redo Eq. (3.15.11) as

$$\hat{Q}_{in} = \hat{Q}_{out} + S\frac{d\hat{h}}{dt}. \tag{3.15.16}$$

Conversion of Eq. (3.15.16) to Laplace's domain produces

$$\bar{Q}_{in} = \bar{Q}_{out} + S\left(s\bar{h} - \hat{h}\Big|_{t=0}\right) \tag{3.15.17}$$

– where Eq. (3.15.14), coupled with

$$h\Big|_{t=0} = h^{ss} \tag{3.15.18}$$

for initial condition, permit simplification of Eq. (3.15.17) to

$$\bar{Q}_{in} = \bar{Q}_{out} + Ss\bar{h}. \tag{3.15.19}$$

Division of both sides by Ss converts Eq. (3.15.19) to

$$\bar{h} = G_{dt}\bar{Q}_{in} - G_{mp}\bar{Q}_{out}, \tag{3.15.20}$$

where G_{dt} is given by

$$G_{dt}\{s\} \equiv \frac{1}{Ss} \tag{3.15.21}$$

and G_{mp} abides to

$$G_{mp}\{s\} \equiv -\frac{1}{Ss} \tag{3.15.22}$$

– in agreement with Eqs. (3.464) and (3.466), respectively; the (lumped) load transfer function then reads

$$G_{LD}\{s\} \equiv \frac{\frac{1}{Ss}}{1 + 1K_{fc}\left(-\frac{1}{Ss}\right)K_P} = \frac{\frac{1}{Ss}}{1 - \frac{K_{fc}K_P}{Ss}} \tag{3.15.23}$$

based on Eq. (3.501), whereas the setpoint transfer function looks like

$$G_{SP}\{s\} \equiv \frac{K_{fc}\left(-\frac{1}{Ss}\right)K_P}{1 + 1K_{fc}\left(-\frac{1}{Ss}\right)K_P} = -\frac{\frac{K_{fc}K_P}{Ss}}{1 - \frac{K_{fc}K_P}{Ss}} \tag{3.15.24}$$

stemming from Eq. (3.502) – with the aid of Eq. (3.536) for $K_{md}=1$ as suggested, and Eqs. (3.480), (3.537), (3.15.21), and (3.15.22), consistent with the hypotheses considered. Equation (3.500) can accordingly be revisited as

$$\bar{h} = G_{LD}\bar{Q}_{in} + G_{SP}\bar{h}_{sp}, \tag{3.15.25}$$

where $\bar{h}_{sp}$ denotes setpoint function, pertaining to liquid height, in Laplace's domain. If a servo problem is under scrutiny, then

$$\bar{Q}_{in} = 0, \tag{3.15.26}$$

while the hypothesized unit step input of h_{sp} supports

$$\bar{h}_{sp} = \frac{1}{s} \tag{3.15.27}$$

in Laplace's domain; Eq. (3.15.25) then reduces to

$$\bar{h} = \frac{1}{s} G_{SP}, \tag{3.15.28}$$

after taking Eqs. (3.15.26) and (3.15.27) onboard. Insertion of Eq. (3.15.24) transforms Eq. (3.15.28) to

$$\bar{h} = -\frac{1}{s}\frac{\dfrac{K_{fc}K_P}{Ss}}{1-\dfrac{K_{fc}K_P}{Ss}}, \tag{3.15.29}$$

with multiplication of both the numerator and denominator by Ss giving rise to

$$\bar{h} = -\frac{K_{fc}K_P}{s\left(Ss - K_{fc}K_P\right)}; \tag{3.15.30}$$

according to the final value theorem,

$$\lim_{t\to\infty} h\{t\} = \lim_{s\to 0} s\bar{h}, \tag{3.15.31}$$

so combination with Eq. (3.15.30) unfolds

$$\lim_{t\to\infty} h\{t\} = -\lim_{s\to 0}\frac{K_{fc}K_P}{Ss - K_{fc}K_P} = \frac{K_{fc}K_P}{K_{fc}K_P} = 1. \tag{3.15.32}$$

The offset, O_S^∞, is, in the case of a servo problem, given by

$$O_S^\infty \equiv h_{sp} - \lim_{t\to\infty} h\{t\} = 1 - \lim_{t\to\infty} h\{t\} \tag{3.15.33}$$

in view of the unit step input of h_{sp}, so insertion of Eq. (3.15.32) leads to

$$O_S^\infty = 1 - 1 = 0; \tag{3.15.34}$$

hence, no offset will be left – meaning that a proportional control guarantees the intended specification, i.e. eventual coincidence of the controlled output variable with its new setpoint. Conversely, a regulator problem would be characterized by

$$\bar{h}_{sp} = 0, \tag{3.15.35}$$

coupled with a step input on Q_{in} reading

$$\bar{Q}_{in} = \frac{1}{s}, \tag{3.15.36}$$

in Laplace's domain; Eq. (3.15.25) consequently becomes

$$\bar{h} = \frac{1}{s} G_{LD} \tag{3.15.37}$$

as descriptor of the controlled system – following combination with Eqs. (3.15.35) and (3.15.36). Insertion of Eq. (3.15.23) converts Eq. (3.15.37) to

$$\bar{h} = \frac{1}{s}\frac{\dfrac{1}{Ss}}{1-\dfrac{K_{fc}K_P}{Ss}}, \tag{3.15.38}$$

where both numerator and denominator can be multiplied by Ss to get

$$\bar{h} = \frac{1}{s\left(Ss - K_{fc}K_P\right)}; \tag{3.15.39}$$

Eq. (3.15.31) may again be invoked to write

$$O_R^\infty \equiv \lim_{t\to\infty} h\{t\} = \lim_{s\to 0}\frac{1}{Ss - K_{fc}K_P} = -\frac{1}{K_{fc}K_P} \neq 0 \tag{3.15.40}$$

pertaining to the offset germane in this regulator problem (O_R^∞), at the expense of Eqs. (3.563) and (3.15.39). Therefore, a finite offset will be left upon a step change in disturbance – removal of which would call for extra integral action by the controller.

c) Straight application of the classical theorems on limits converts Eq. (3.681) to

$$\lim_{\zeta\to 1^-}\frac{\dfrac{y_{out,pk}}{K}}{\eta} = \exp\left\{-1\frac{\cos^{-1}1}{\sqrt{1-1^2}}\right\} = e^{-\frac{0}{0}}, \tag{3.15.41}$$

valid in the vicinity of unit as upper boundary of underdamped behavior (defined by $0 \le \zeta < 1$); unfortunately, the exponent found is but an unknown quantity. Therefore, one should take logarithms of both sides of Eq. (3.681) as

$$\ln\lim_{\zeta\to 1^-}\frac{\dfrac{y_{out,pk}}{K}}{\eta} = \ln\lim_{\zeta\to 1}\exp\left\{-\zeta\frac{\cos^{-1}\zeta}{\sqrt{1-\zeta^2}}\right\}, \tag{3.15.42}$$

where operators will be swapped in the right-hand side to obtain

$$\ln\lim_{\zeta\to 1^-}\frac{\dfrac{y_{out,pk}}{K}}{\eta} = \lim_{\zeta\to 1}\left(-\frac{\zeta\cos^{-1}\zeta}{\sqrt{1-\zeta^2}}\right) = -\frac{0}{0} \tag{3.15.43}$$

– at the expense of composing logarithm with its inverse function. Application of l'Hôpital's rule is now possible, in view of 0/0 appearing *per se* as unknown quantity, thus leading to

$$\ln\lim_{\zeta\to 1^-}\frac{\dfrac{y_{out,pk}}{K}}{\eta} = \lim_{\zeta\to 1}\left(-\frac{\cos^{-1}\zeta - \zeta\dfrac{1}{\sqrt{1-\zeta^2}}}{\dfrac{-2\zeta}{2\sqrt{1-\zeta^2}}}\right) \tag{3.15.44}$$

obtained via independent differentiation of numerator and denominator; Eq. (3.15.44) simplifies to

$$\ln \lim_{\zeta \to 1^-} \frac{\frac{y_{out,pk}}{K}}{\eta} = \lim_{\zeta \to 1} \frac{\cos^{-1}\zeta - \frac{\zeta}{\sqrt{1-\zeta^2}}}{\frac{2\zeta}{2\sqrt{1-\zeta^2}}}, \qquad (3.15.45)$$

after taking negatives of both numerator and denominator. Cancellation of 2 in both numerator and denominator, followed by multiplication thereof by $\sqrt{1-\zeta^2}$ transforms Eq. (3.15.45) to

$$\ln \lim_{\zeta \to 1^-} \frac{\frac{y_{out,pk}}{K}}{\eta} = \lim_{\zeta \to 1} \frac{\sqrt{1-\zeta^2}\cos^{-1}\zeta - \zeta}{\zeta}, \qquad (3.15.46)$$

where direct replacement of ζ by 1 conveys simplification to

$$\ln \lim_{\zeta \to 1^-} \frac{\frac{y_{out,pk}}{K}}{\eta} = \frac{\sqrt{1-1^2}\cos^{-1}1 - 1}{1} = \frac{0-1}{1} = -1; \qquad (3.15.47)$$

exponentials may finally be taken of both sides of Eq. (3.15.47) to get

$$\lim_{\zeta \to 1^-} \frac{\frac{y_{out,pk}}{K}}{\eta} = e^{-1}, \qquad (3.15.48)$$

thus retrieving Eq. (3.650) – and confirming continuity of the curve in Fig. 3.57, specifically at $\zeta = 1$.

d) The block diagram, in Laplace's domain, of a process subjected to feedforward control – consistent with Fig. 3.16, is laid out below; it will hereafter be used as working base *in lieu* of Fig. 3.51ii.

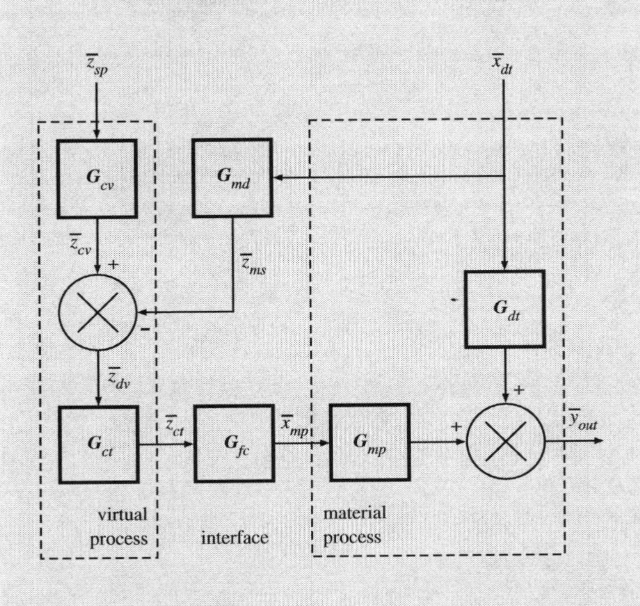

One should accordingly define a special transfer function, associated with the measuring device, according to

$$G_{md}\{s\} \equiv \frac{\bar{z}_{ms}}{\bar{x}_{dt}} \qquad (3.15.49)$$

instead of Eq. (3.492) – because disturbance, rather than process response is measured under feedforward control. On the other hand, the setpoint entails a reference value for the output variable, to be compared with the measured value of the input disturbance; hence, some conversion in between is in order. This calls for a converter, with transfer function, G_{cv}, given by

$$G_{cv}\{s\} \equiv \frac{\bar{z}_{cv}}{\bar{z}_{sp}}; \qquad (3.15.50)$$

here $\bar{z}_{cv}$ denotes transformed setpoint signal, consistent with setpoint for the (measured) disturbance. Equation (3.463) remains valid as decomposition of output variable in Laplace's domain, $\bar{y}_{out}$, as sum of two germane inputs, $\bar{x}_{dt}$ and $\bar{x}_{mp}$; after rewriting Eq. (3.493) as

$$\bar{x}_{mp} = G_{fc}\bar{z}_{ct}, \qquad (3.15.51)$$

this result can be inserted in Eq. (3.463) to obtain Eq. (3.494). Meanwhile, Eq. (3.481) may be rephrased as

$$\bar{z}_{ct} = G_{ct}\bar{z}_{dv}, \qquad (3.15.52)$$

thus allowing Eq. (3.495) be generated from Eq. (3.494); however, $\bar{z}_{dv}$ is hereby defined as

$$\bar{z}_{dv} = \bar{z}_{cv} - \bar{z}_{ms}, \qquad (3.15.53)$$

where Eq. (3.15.50) supports reformulation to

$$\bar{z}_{dv} = G_{cv}\bar{z}_{sp} - \bar{z}_{ms} \qquad (3.15.54)$$

– or, after recalling Eq. (3.15.49),

$$\bar{z}_{dv} = G_{cv}\bar{z}_{sp} - G_{md}\bar{x}_{dt}. \qquad (3.15.55)$$

One will now combine Eqs. (3.495) and (3.15.55) to obtain

$$\bar{y}_{out} = G_{dt}\bar{x}_{dt} + G_{mp}G_{fc}G_{ct}G_{cv}\bar{z}_{sp} - G_{mp}G_{fc}G_{ct}G_{md}\bar{x}_{dt}, \qquad (3.15.56)$$

and further factor out $\bar{x}_{dt}$ as

$$\bar{y}_{out} = \left(G_{dt} - G_{mp}G_{fc}G_{ct}G_{md}\right)\bar{x}_{dt} + G_{mp}G_{fc}G_{ct}G_{cv}\bar{z}_{sp}; \qquad (3.15.57)$$

once again, the system output, $\bar{y}_{out}$, responds to two independent inputs, disturbance, $\bar{x}_{dt}$, and setpoint (for controlled variable), $\bar{z}_{sp}$ – with Eq. (3.15.57) being but the mathematical statement of the layout depicted above. The goal of feedforward control may be rationalized as twofold: disturbance rejection, i.e. keep $\bar{y}_{out}$ nil (or y_{out} constant), irrespective of any change in $\bar{x}_{dt}$ (or x_{dt}, for that matter) – equivalent to stating

$$\left.\frac{\bar{y}_{out}}{\bar{x}_{dt}}\right|_{\bar{z}_{sp}=0} \equiv G_{dt} - G_{mp}G_{fc}G_{ct}G_{md} = 0 \quad (3.15.58)$$

based on Eq. (3.15.57), and corresponding to a nil transfer function between $\bar{x}_{dt}$ and $\bar{y}_{out}$; and setpoint tracking, i.e. keep y_{out} systematically equal to the corresponding setpoint – or, in other words,

$$\left.\frac{\bar{y}_{out}}{\bar{z}_{sp}}\right|_{\bar{x}_{dt}=0} \equiv G_{mp}G_{fc}G_{ct}G_{cv} = 1 \quad (3.15.59)$$

based again on Eq. (3.15.57), but corresponding to a unit transfer function between $\bar{z}_{sp}$ and $\bar{y}_{out}$. Equation (3.15.58) implies

$$G_{dt} = G_{mp}G_{fc}G_{ct}G_{md} \quad (3.15.60)$$

or, after solving for G_{ct},

$$G_{ct} = \frac{G_{dt}}{G_{mp}G_{fc}G_{md}}; \quad (3.15.61)$$

this conveys the form sought for the transfer function of the controller, under this mode of process control. Insertion of Eq. (3.15.61) eventually transforms Eq. (3.15.59) to

$$G_{mp}G_{fc}\frac{G_{dt}}{G_{mp}G_{fc}G_{md}}G_{cv} = 1, \quad (3.15.62)$$

which promptly yields

$$G_{cv} \equiv \frac{G_{md}}{G_{dt}}, \quad (3.15.63)$$

after dropping common factors from numerator and denominator, and then isolating G_{cv}; Eq. (3.15.63) serves as definition of transfer function for the aforementioned converter. In view of Eqs. (3.15.61) and (3.15.63), one will rewrite Eq. (3.15.57) as

$$\bar{y}_{out} = \left(G_{dt} - G_{mp}G_{fc}\frac{G_{dt}}{G_{mp}G_{fc}G_{md}}G_{md} \right)\bar{x}_{dt} + G_{mp}G_{fc}\frac{G_{dt}}{G_{mp}G_{fc}G_{md}}\frac{G_{md}}{G_{dt}}\bar{z}_{sp}, \quad (3.15.64)$$

which breaks down to merely

$$\bar{y}_{out} = \left(G_{dt} - G_{dt}\right)\bar{x}_{dt} + \bar{z}_{sp} = 0\bar{x}_{dt} + 1\bar{z}_{sp} \quad (3.15.65)$$

following cancellation of common factors between numerator and denominator, and of symmetrical terms afterward. Equation (3.15.65) accounts for the lumped diagram initially proposed, with 0 and 1 serving as lumped transfer functions that play the role of G_{LD} and G_{SP}, respectively, in feedback control – see Fig. 3.51iii in this regard. The complete design of a feedforward mechanism accordingly relies on Eqs. (3.15.61) and (3.15.63); it should be viewed as a special-purpose computing machine – arising directly from the model for the process at stake, rather than as a mere logical programmable device with proportional, integral, or derivative capacity. The (perfect) performance of this type of controller depends critically on goodness of model and accuracy of measurement of disturbance(s), in the first place; the better a model represents the behavior of a process, and the better the data taken reflect its actual operation, the better the performance of the said controller. However, this deed is hardly met in industrial practice – thus constituting the major limitation to feedforward control, and justifying why it is normally coupled with some feedback control.

4

Bibliography and Useful Mathematical Formulae

This book is concerned with engineering of food processes, encompassing specifically safety assurance and supplementary operations – thus implying that mathematical simulation is seminal. Moreover, it has consistently resorted to first principles – not only to provide a rational basis for modeling, design, optimization, and ultimately decision; but also to capitalize on the possibility of extrapolation, while enhancing fundamental understanding of data and results. Therefore, a careful approach was followed when introducing basic concepts, and in thoroughly explaining the derivation steps of useful relationships from those concepts – with an emphasis on analytical solutions; instead of merely conveying a list of equations, without explaining their source, or (even worse) without permitting a paced approach to higher levels of simulation complexity.

Inspiration was obtained from a plethora of monographic and edited books, as well as chapters and papers readily available elsewhere (and not restricted to the English language); the reader is thus directed to those original sources, as both general bibliographic references and suggested further reading for more in-depth analysis. Five books, however, deserve special mention, for their outstanding originality and high quality: R. P. Singh and D. R. Heldman's *Introduction to Food Engineering*, with an introductory approach to overall process engineering of foods; M. Loncin and R. L. Merson's *Food Engineering: Principles and Selected Applications*, with an advanced approach to specific topics of relevance in food processing; P. J. Fellows' *Food Processing Technology*, with a comprehensive approach to industrial technology employed in food transformation; K. E. Bett, J. S. Rowlinson and G. Saville's *Thermodynamics for Chemical Engineers*, with a lucid and systematic approach to thermodynamics of processes and systems; and G. Stephanopoulos' *Chemical Process Control: An Introduction to Theory and Practice*, with a concise approach to the theory and practice of process control.

Conversely, several mathematical tools utilized along the book in germane derivations were deliberately retrieved without proof – in agreement with the focus of this book on engineering, rather than pure mathematics (concerned with theorems, lemmas, and corollaries). To facilitate reference, however, a careful compilation of the most important mathematical results is provided at the end – classified under the headings of algebra or calculus, so as to reflect their fixed (or point) versus variable (or differential) nature; complemented with a subsection on statistics, useful in handling (actual) data rather than (theoretical) functions. The great majority of the said results are a part of the regular background of an engineer (and of an advanced engineering student, for that matter) – with their exact derivation being easily accessible, in general purpose or specialized books of mathematics.

4.1 Bibliographic references and further reading

Abramowitz, M. & Stegun, I. A. (Eds.) (1972) *Handbook of Mathematical Functions with Formulas, Graphs, and Mathematical Tables*. Dover, New York, NY.

Adler-Nissen, J. & Zammit, G. O. (2011) Modelling and validation of robust partial thawing of frozen convenience foods during distribution in the cold chain. *Procedia of Food Science* **1**: 1247–1255.

Ahvenainen, R. (2003) Active and intelligent packaging. In: Ahvenainen, R. (Ed.), *Novel Food Packaging Techniques*. Woodhead Publishing, Cambridge, UK: pp. 5–21.

Akre, E. (1991) Green politics and industry. *European Food and Drink Review* **Packaging Sup.**: 5–7.

Allan, D. J. (1991) Power transformers – The second century. *Power Engineering Journal* **5**(1): 5–14.

Alvarez, J. S. & Thorne, S. (1981) The effect of temperature on the deterioration of stored agricultural produce. In: Thorne, S. (Ed.), *Developments in Food Preservation*. Applied Science, London, UK: pp. 215–237.

Anatheswaran, R. C. & Ramaswarmy, H. S. (2001) Bacterial destruction and enzyme inactivation during microwave heating. In: Datta, A. K. & Anantheswaran, R. C. (Eds.), *Handbook of Microwave Technology for Food Applications*. CRC Press, Boca Raton, FL: pp. 191–214.

Andreasen, T. G. & Nielsen, H. (1992) Ice cream and aerated deserts. In: Early, R. (Ed.), *Technology of Dairy Products*. Blackie Academic & Professional, London, UK: pp. 197–220.

Arfken, G. (1985) *Mathematical Methods for Physicists*. Academic Press, Orlando, FL.

Arnoldi, A. (2004) Factors affecting the Maillard reaction. In: Steele, R. (Ed.), *Understanding and Measuring the Shelf-Life of Food*. Woodhead Publishing, Cambridge, UK: pp. 111–127.

Arroqui, C., Rumsey, T. R., Lopez, A. & Virseda, P. (2002) Losses by diffusion of ascorbic acid during recycled water blanching of potato tissue. *Journal of Food Engineering* **52**(1): 25–30.

Arvanitoyannis, I. S. & Gorris, L. G. M. (1999) Edible and biodegradable polymeric materials for food packaging or coating. In: Oliveira, F. A. R. & Oliveira, J. C. (Eds.), *Processing Foods – Quality Optimisation and Process Assessment*. CRC Press, Boca Raton, FL: pp. 357–368.

Askey, R. & Haimo, D. T. (1996) Similarities between Fourier and power series. *American Mathematician Monthly* **103**(4): 297–304.

Azdemir, M. & Floros, J. D. (2004) Active food packaging technologies. *Critical Reviews in Food Science and Nutrition* **44**(3): 185–193.

Azdemir, M., Yurteri, C. U. & Sadikoglu, H. (1999) Surface treatment of food packaging polymers by plasmas. *Food Technology* **53**(4): 54–58.

Bai, Y., Rahman, S., Perera, C. O., Smith, B. & Melton, L. D. (2001) State diagram of apple slices: Glass transition and freezing curves. *Food Research International* **34**(2–3): 89–95.

Baker, M. (1981) *The Quest for Pure Water: The History of Water Purification from the Earliest Records to the Twentieth Century*. American Water Works Association, Denver, CO.

Baldwin, E. A. (1994) Edible coatings for fresh fruits and vegetables: Past, present and future. In: Krochta, J. M., Baldwin, E. A. & Niosperos-Carriedo, M. O. (Eds.), *Edible Coatings and Films to Improve Food Quality*. Technomic Publishing, Lancaster, PA: pp. 25–64.

Baldwin, E. A. (1999) Surface treatments and edible coatings in food preservation. In: Rahman, M. S. (Ed.), *Handbook of Food Preservation*. Marcel Dekker, New York, NY: pp. 577–610.

Ball, C. O. & Olson, F. C. W. (1957) *Sterilization in Food Technology*. McGraw-Hill, New York, NY.

Barbosa-Cánovas, G. V., Gongora-Nieto, M. M., Pothakamury, U. R. & Swanson, B. G. (1999) *Preservation of Foods with Pulsed Electric Fields*. Academic Press, London, UK.

Barbosa-Cánovas, G. V., Tapia, M. S. & Cano, M. P. (Eds.) (2005) *Novel Food Processing Technologies*. CRC Press, Boca Raton, FL.

Barr, U.-K. & Merkel, H. (2004) Surface penetrating radar. In: Edwards, M. (Ed.), *Detecting Foreign Bodies in Food*. Woodhead Publishing, Cambridge, UK: pp. 172–192.

Basir, O. A., Zhao, B. & Mittal, G. S. (2004) Ultrasound. In: Edwards, M. (Ed.), *Detecting Foreign Bodies in Food*. Woodhead Publishing, Cambridge, UK: pp. 204–225.

Batchelor, B. G., Davies, E. R. & Graves, M. (2004) Using X-rays to detect foreign bodies. In: Edwards, M. (Ed.), *Detecting Foreign Bodies in Food*. Woodhead Publishing, Cambridge, UK: pp. 226–264.

Bayles, D. O. (Ed.) (2005) *Foodborne Pathogens: Microbiology and Molecular Biology*. Caister Academic Press, Poole, UK.

Becker, B. R. & Fricke, B. A. (1996) Transpiration and respiration of fruits and vegetables. In: *New Developments in Refrigeration for Food Safey and Quality*. International Institute of Refrigeration, Paris, France, & American Society of Agricultural Engineers, St. Joseph, MI: pp. 110–121.

Bedford, L. (2000) Raw material selection – Fruits and vegetables. In: Stringer, M. & Dennis, C. (Eds.), *Chilled Foods – A Comprehensive Guide*. Woodhead Publishing, Cambridge, UK: pp. 19–35.

Bell, C. & Kyriakides, A. (2002) *Listeria monocytogenes*. In: Blackburn, C. W. & McClure, P. J. (Eds.), *Foodborne Pathogen Hazards, Risk Analysis and Control*. Woodhead Publishing, Cambridge, UK: pp. 337–361.

Bell, C. & Kyriakides, A. (2002) Pathogenic *Escherichia coli*. In: Blackburn, C. W. & McClure, P. J. (Eds.), *Foodborne Pathogen Hazards, Risk Analysis and Control*. Woodhead Publishing, Cambridge, UK: pp. 279–306.

Bell, C. & Kyriakides, A. (2002) *Salmonella*. In: Blackburn, C. W. & McClure, P. J. (Eds.), *Foodborne Pathogen Hazards, Risk Analysis and Control*. Woodhead Publishing, Cambridge, UK: pp. 307–335.

Belletti, N., Ndagijimana, M., Sisto, C., Guerzoni, M. E., Lanciotti, R. & Gardini, F. (2004) Evaluation of the antimicrobial activity of citrus essences on *Saccharomyces cerevisiae*. *Journal of Agricultural and Food Chemistry* **52**(23): 6932–6938.

Benjamin, R. (2004) Microwave reflectance. In: Edwards, M. (Ed.), *Detecting Foreign Bodies in Food*. Woodhead Publishing, Cambridge, UK: pp. 132–153.

Bennet, J. G. (1936) Broken coal. *Journal of the Institute of Fuels* **10**: 22–39.

Bett, K. E., Rowlinson, J. S. & Saville, G. (1975) *Thermodynamics for Chemical Engineers*. MIT Press, Cambridge, MA.

Betts, G. D. & Walker, S. J. (2004) Verification and validation of food spoilage models. In: Steele, R. (Ed.), *Understanding and Measuring the Shelf-Life of Food*. Woodhead Publishing, Cambridge, UK: pp. 184–217.

Billiard, F., Deforges, J., Derens, E., Gros, J. & Serrand, M. (1999) *Control of the Cold Chain for Quick-Frozen Foods Handbook*. International Institute of Refrigeration, Paris, France.

Blackburn, C. (2006) *Food Spoilage Microorganisms*. Woodhead Publishing, Cambridge, UK.

Blakistone, B. A. (1998) Introduction. In: Blakistone, B. A. (Ed.), *Principles and Applications of Modified Atmosphere Packaging of Foods*. Blackie Academic & Professional, London, UK: pp. 1–13.

Blakistone, B. A. (1998) Meats and poultry. In: Blakistone, B. A. (Ed.), *Principles and Applications of Modified Atmosphere Packaging of Foods*. Blackie Academic & Professional, London, UK: pp. 240–284.

le Blanc, D. & Stark, R. (2001) The cold chain. In: Sun, D.-W. (Ed.), *Advances in Food Refrigeration*. Leatherhead Publishing, Leatherhead, UK: pp. 326–365.

Blanchfield, J. R. (Ed.) (2000) *Food Labelling*. Woodhead Publishing, Cambridge, UK.

Blancou, J. (1995) History of disinfection from early times until the end of the 18th century. *Revue de Science et Technologie de l'Office International des Epizooties* **14**: 31–39.

Bloch, H. P. & Hoefner, J. J. (1996) *Reciprocating Compressors: Operation and Maintenance*. Gulf Professional Publishing, Houston, TX.

Blundel, S. J. (2012) *Magnetism: A Very Short Introduction*. Oxford University Press, Oxford, UK.

Bockris, J. O'M., Reddy, A. K. N. & Gamboa-Aldeco, M. (1998) *Modern Electrochemistry*. Springer, Berlin, Germany.

Boonsupthip, W. & Heldman, D. R. (2007) Prediction of frozen food properties during freezing using product composition. *Journal of Food Science* **72**(5): E254–E263.

Bord, R. J. & O'Connor, R. E. (1989) Who wants irradiated food? Untangling complex public opinion. *Food Technology* **43**(10): 87–90.

le Bot, Y. (1993) Stable sugarless coating. In: Turner, A. (Ed.), *Food Technology International Europe*. Sterling Publications, London UK: pp. 67–70.

Bourlakis, M. & Weightman, P. (Eds.) (2004) *Food Supply Chain Management*. Blackwell Publishing, Oxford, UK.

Bowman, F. (1958) *Introduction to Bessel Functions*. Dover, New York, NY.

Bown, G. (2004) Modelling and optimizing retort temperature control. In: Richardson, P. (Ed.), *Improving the Thermal Processing of Foods*. Woodhead Publishing, Cambridge, UK: pp. 105–123.

Bracewell, R. (1999) *The Fourier Transform and Applications*. Prentice Hall, Englewood Cliffs, NJ.

Braun, P., Fehlhaber, K., Klug, C. & Kopp, K. (1999) Investigations into the activity of enzymes produced by spoilage-causing bacteria: A possible basis for improved shelf-life estimation. *Food Microbiology* **16**(5): 531–540.

Breidt, F., Hayes, J. S. & Fleming, H. P. (2000) Reduction of microflora of whole pickling cucumbers by blanching. *Journal of Food Science* **65**(8): 1354–1358.

Bremer, P. & Osborne, C. (1995) Thermal-death times of *Listeria monocytogenes* in green shell mussels (*Perna canaliculus*) prepared for hot smoking. *Journal of Food Protection* **58**(6): 604–608.

Brennan, C. (2005) *Electronic Irradiation of Foods: An Introduction to the Technology*. Springer, New York, NY.

Brennan, J. G. & Day, B. P. F. (2006) Packaging. In: Brennan, J. G. (Ed.), *Food Processing Handbook*. Wiley-VCH, Weinheim, Germany: pp. 291–350.

Briston, J. H. (1980) Rigid plastics packaging. In: Palling, S. J. (Ed.), *Developments in Food Packaging*, vol. 1. Applied Science, London, UK: pp. 27–53.

Briston, J. H. (1987) Rigid plastic containers and food packaging. In: Turner, A. (Ed.), *Food Technology International Europe*. Sterling Publications, London, UK: pp. 283, 285–287.

Briston, J. H. (1990) Recent developments in bag-in-box packaging. In: Turner, A. (Ed.), *Food Technology International Europe*. Sterling Publications, London, UK: pp. 319–320.

Brody, A. L. (1990) Controlled atmosphere packaging for chilled foods. In: Turner, A. (Ed.), *Food Technology International Europe*. Sterling Publications, London, UK: pp. 307–313.

Brody, A. L. (1992) Microwave food pasteurisation, sterilisation and packaging. In: Turner, A. (Ed.), *Food Technology International Europe*. Sterling Publications, London, UK: pp. 67–71.

Brody, A. L. (1992) Technologies of retortable barrier plastic cans and trays. In: Turner, A. (Ed.), *Food Technology International Europe*. Sterling Publications, London, UK: pp. 241–249.

Brody, A. L. (2003) "Nano, nano" food packaging technology. *Food Technology* **57**(12): 53–54.

Brody, A. L. & Budny, J. A. (1995) Enzymes as active packaging agents. In: Rooney, M. L. (Ed.), *Active Food Packaging*. Blackie Academic & Professional, London, UK: pp. 174–192.

Brody, A. L. & Marsh, K. S. (Eds.) (1997) *The Wiley Encyclopedia of Packaging Technology*. Wiley-Interscience, New York, NY.

Brooker, B. (1999) Ultra-high pressure processing. *Food Technology International* **59**: 61.

Brown, J. W. & Churchill, R. V. (1993) *Fourier Series and Boundary Value Problems*. McGraw-Hill, New York, NY.

Brown, M. H. (1992) Non-microbiological factors affecting quality and safety. In: Dennis, C. & Stringer, M. (Eds.), *Chilled Foods*. Ellis Horwood, Chichester, UK: pp. 261–288.

Brown, M. H. (2000) Microbiological hazards and safe process design. In: Stringer, M. & Dennis, C. (Eds.), *Chilled Foods – A Comprehensive Guide*. Woodhead Publishing, Cambridge, UK: pp. 287–339.

Brown, M. H. & Hall, M. N. (2000) Non-microbiological factors affecting quality and safety. In: Stringer, M. & Dennis, C. (Eds.), *Chilled Foods – A Comprehensive Guide*. Woodhead Publishing, Cambridge, UK: pp. 225–255.

Bunyan, P. (2003) How to make MES work. *Food Processing* **Sep**: 13.

Buonocuore, G. G., Conte, A., Corbo, M. R., Sinigaglia, M. & del Nobile, M. A. (2005) Mono- and multilayer active films containing lysozyme as antimicrobial agent. *Innovative Food Science and Emerging Technologies* **6**(4): 459–464.

Buonocuore, G. G., Sinigaglia, M., Corbo, M. R., Bivilacqua, A., la Notte, E. & del Nobile, M. A. (2004) Controlled release of antimicrobial compounds from highly swellable polymers. *Journal of Food Protection* **67**(6): 1190–1194.

Burstall, A. F. (1965) *A History of Mechanical Engineering*. MIT Press, Cambridge, MA.

Burton, H. (1988) *UHT Processing of Milk and Milk Products*. Elsevier Applied Science, London, UK.

Butler, B. L., Vergano, P. J., Testin, R. F., Bunn, J. M. & Wiles, J. L. (1996) Mechanical and barrier properties of edible chitosan films as affected by composition and storage. *Journal of Food Science* **61**(5): 953–956.

Butler, P. (2001) Smart packaging – Intelligent packaging for food, beverages, pharmaceuticals and household products. *Materials World* **9**(3): 11–13.

Butler, P. (2002) *Smart Packaging – Strategic Ten-Year Forecast and Technology and Company Profiles*. IDTechEx, Cambridge, UK.

Calero, F. A. & Gomez, P. A. (2003) Active packaging and colour control: The case of fruits and vegetables. In: Ahvenainen, R. (Ed.), *Novel Food Packaging Techniques*. Woodhead Publishing, Cambridge, UK: pp. 416–438.

Campbell-Platt, G. (1987) Recent developments in chilling and freezing. In: Turner, A. (Ed.), *Food Technology International Europe*. Sterling Publications, London, UK: pp. 63–66.

Campey, D. R. (1987) The application of laser marking. In: Turner, A. (Ed.), *Food Technology International Europe*. Sterling Publications, London, UK: pp. 303–304.

Cardello, A. V. (1998) Perception of food quality. In: Taub, I. A. & Singh, R. P. (Eds.). *Food Storage Stability*. CRC Press, Boca Raton, FL: pp. 1–38.

Cardwell, D. S. L. (1971) *From Watt to Clausius: The Rise of Thermodynamics in the Early Industrial Age*. Heinemann, London, UK.

Carlson, B. (1996) Food processing equipment – Historical and modern designs. In: David, J. R. D., Graves, R. H. & Carlson, V. R. (Eds.), *Aseptic Processing and Packaging of Food*. CRC Press, Boca Raton, FL: pp. 51–94.

Carlsson-Kanyama, A. & Faist, M. (2000) *Energy Use in the Food Sector – A Data Survey*. Royal Institute of Technology, London, UK.

Castro, A. J., Barbosa-Cánovas, G. V. & Swanson, B. G. (1993) Microbial inactivation of foods by pulsed electric fields. *Journal of Food Processing and Preservation* **17**(1): 47–73.

Cha, D. S. & Chinnan, M. S. (2004) Biopolymer-based antimicrobial packaging: A review. *Critical Reviews in Food Science and Nutrition* **44**(4): 223–237.

Chambers, S. & Helander, T. (1997) Peas. In: Johnson, R., Chambers, S., Harland, C., Harrison, A. & Slack, N. (Eds.), *Cases in Operations Management*. Pitman Publishing, London, UK: pp. 310–323.

Charbonneau, J. E. (1997) Recent case histories of food product-metal container interactions using scanning electron microscopy-X-ray microanalysis. *Scanning* **19**(7): 512–518.

Cheftel, J. C. (1995) High-pressure microbial inactivation and food preservation. *Food Science and Technology International* **1**(2–3): 75–90.

Cheftel, J. C., Thiebaud, M. & Dumay, E. (2002) Pressure-assisted freezing and thawing of foods: A review of recent studies. *International Journal of High Pressure Research* **22**(3–4): 601–611.

Chevalier, D., le Bail, A., Sequeira-Muñoz, A., Simpson, B. & Ghoul, M. (2002) Pressure shift freezing of turbot (*Scophthalmus maximus*) and carp (*Cyprinus carpio*): Effect of ice crystals and drip volumes. *Progress in Biotechnology* **19**(2): 577–582.

Chick, H. (1908) An investigation into the laws of disinfection. *Journal of Hygiene (Cambridge)* **8**(1): 92–158.

Chiellini, E. (2008) *Environmentally-Compatible Food Packaging*. Woodhead Publishing, Cambridge, UK.

Church, N. (1994) Developments in modified-atmosphere packaging and related technologies. *Trends in Food Science and Technology* **5**(11): 345–352.

Church, N. (1998) MAP fish and crustaceans – Sensory enhancement. *Food Science and Technology Today* **12**(2): 73–83.

Clark, J. P. (2009) *Case Studies in Food Engineering: Learning from Experience*. Springer, Berlin, Germany.

Cleland, A. C. (1990) *Food Refrigeration Processes: Analysis, Design and Simulation*. Elsevier, London, UK.

Cleland, A. C. & Earle, R. L. (1976) A comparison of freezing calculations including modifications to take into account initial superheat. *Bulletin of the International Institute of Refrigeration* **56**: 369–376.

Cleland, A. C. & Earle, R. L. (1977) A comparison of analytical and numerical methods of predicting the freezing times of foods. *Journal of Food Science* **42**(5): 1390–1395.

Cleland, D. J., Cleland, A. C. & Earle, R. L. (1987) Prediction of freezing and thawing times for multi-dimensional shapes by simple formulae. 1. Regular shapes. *International Journal of Refrigeration* **10**(3): 156–164.

Cooksey, K., Marsh, K. S. & Doar, L. H. (1999) Predicting permeability and transmission rate for multilayer materials. *Food Technology* **53**(9): 60–63.

Cooper, C. D. & Alley, F. C. (2010) *Air Pollution Control: A Design Approach*. Waveland Press, Long Grove, IL.

la Coste, A., Schaich, K. M., Zumbrunnen, D. & Yam, K. L. (2005) Advancing controlled release packaging through smart blending. *Packaging Technology and Science* **18**(2): 77–87.

Cotton, H. (1950) *Electrical Technology*. Pitman, London, UK.

Craig, L., Goodwin, B. & Grennes, T. (2004) The effect of mechanical refrigeration on nutrition in the United States. *Social Science History* **28**(2): 325–336.

Crawford, Y. J., Murano, E. A., Olson, D. G. & Shenoy, K. (1996) Use of high hydrostatic pressure and irradiation to eliminate *Clostridium sporogenes* spores in chicken breast. *Journal of Food Protection* **59**(7): 711–715.

Creaser, C. (Ed.) (1991) *Food Contaminants*. Woodhead Publishing, Cambridge, UK.

Crepinsek, Z., Goricanec, D. & Krope, J. (2009) Comparison of performance of absorption refrigeration cycles. *WSEAS Transactions on Heat and Mass Transfer* **3**: 65–76.

Crosby, D. (1958) The ideal transformer. *IRE Transactions on Circuit Theory* **5**(2): 145.

Cubeddu, R., Pifferi, A., Taroni, P. & Torricelli, A. (2002) Measuring fruit and vegetable quality: Advanced optical methods. In: Jongen, W. (Ed.), *Fruit and Vegetable Processing: Improving Quality*. Woodhead Publishing, Cambridge, UK: pp. 150–169.

Dahm, M. & Mathur, A. (1990) Automation in the food processing industry: Distributed control systems. *Food Control* **1**(1): 32–35.

Dalzell, J. M. (1994) *Food Industry and the Environment – Practical Uses and Cost Applications*. Blackie Academic & Professional, London, UK.

Darrigol, O. (2000) *Electrodynamics from Ampère to Einstein*. Oxford University Press, Oxford, UK.

Darrington, H. (1982) Profile of a jammy business. *Food Manufacturing* **Dec**: 37.

Dauthy, M. E. (1995) *Fruit and Vegetable Processing – Bulletin 119*. FAO Agricultural Services, Rome, Italy.

Davies, A. R. (1995) Advances in modified-atmosphere packaging. In: Gould, G. W. (Ed.), *New Methods of Food Preservation*. Blackie Academic & Professional, Glasgow, UK: pp. 304–320.

Day, B. P. F. (1992) Chilled food packaging. In: Dennis, C. & Stringer, M. (Eds.), *Chilled Foods – A Comprehensive Guide*. Ellis Horwood, London, UK: pp. 147–163.

Day, B. P. F. (2003) Novel MAP applications for fresh-prepared produce. In: Ahvenainen, R. (Ed.), *Novel Food Packaging Techniques*. Woodhead Publishing, Cambridge, UK: pp. 189–207.

Deak, T. (2004) Spoilage yeasts. In: Steele, R. (Ed.), *Understanding and Measuring the Shelf-Life of Food*. Woodhead Publishing, Cambridge, UK: pp. 91–110.

Debeaufort, F., Quezada-Gallo, J.-A. & Voilley, A. (1998) Edible films and coatings – Tomorrow's packagings: A review. *Critical Reviews in Food Science and Nutrition* **38**(4): 209–313.

Delincée, H. (1998) Detection of food treated with ionizing radiation. *Trends in Food Science and Technology* **9**(2): 73–82.

Dellino, C. V. J. (Ed.) (1997) *Cold and Chilled Storage Technology*. Blackie Academic & Professional, London, UK.

Dennis, C. & Stringer, M. (2000) Introduction: The chilled foods market. In: Stringer, M. & Dennis, C. (Eds.) *Chilled Foods – A Comprehensive Guide*. Woodhead Publishing, Cambridge, UK: pp. 1–16.

Devlieghere, F. (2002) Modified atmosphere packaging (MAP). In: Henry, C. J. K. & Chapman, C. (Eds.), *The Nutrition Handbook for Food Processors*. Woodhead Publishing, Cambridge, UK: pp. 342–369.

Devlieghere, F. & Debevere, J. (2003) MAP, product safety and nutritional quality. In: Ahvenainen, R. (Ed.), *Novel Food Packaging Techniques*. Woodhead Publishing, Cambridge, UK: pp. 208–230.

Dewettinck, K. & Huyghebaert, A. (1999) Fluidized bed coating in food technology. *Trends in Food Science and Technology* **10**(4–5): 163–168.

Diehl, J. U. F. (1971) Symposium on the investigation of health hazards of irradiated food. *Lebensmittel-Wissenschaft und -Technologie* **4**: 168–169.

Diehl, J. U. F. (1995) *Safety of Irradiated Foods*. Marcel Dekker, New York, NY.

Dillon, M. & Griffith, C. (1999) *How to Clean – A Management Guide*. MD Associates, Grimsby, UK.

Dixon, M. S., Warshall, R. B. & Crerar, J. B. (1963) Food processing method and apparatus. U. S. Patent no. 3,096,161.

Dolatowski, D. Z., Stadnik, J. & Stasiak, D. (2007) Applications of ultrasound in food technology. *Acta Scientiarum Polonorum. Technologia Alimentaria* **6**: 89–99.

Doona, C. J., Dunne, C. P. & Feeherry, F. E. (Eds.) (2007) *High Pressure Processing of Foods*. Blackwell, Oxford, UK.

van Doornmalen, J. P. C. M. & Kopinga, K. (2009) Temperature dependenc of *F*-, *D*- and *z*-values used in steam sterilization processes. *Journal of Applied Microbiology* **107**(3): 1054–1060.

Dorantes-Alvarez, L. & Parada-Dorantes, L. (2005) Blanching using microwave processing. In: Schubert, H. & Regier, M. *The Microwave Processing of Foods*. Woodhead Publishing, Cambridge, UK: pp. 153–173.

Drake, S. R. & Swanson, B. G. (1986) Energy utilization during blanching (water vs. steam) of sweet corn and subsequent frozen quality. *Journal of Food Science* **51**(4): 1081–1082.

Driscoll, R. H. & Patterson, J. L. (1999) Packaging and food preservation. In: Rahman, M. S. (Ed.), *Handbook of Food Preservation*. Marcel Dekker, New York, NY: pp. 687–734.

Duffin, W. J. (1990) *Electricity and Magnetism*. McGraw-Hill, New York, NY.

Dunn, J., Ott, T. & Clark, W. (1995) Pulsed light treatment of food and packaging. *Food Technology* **49**(9): 95–98.

Earnshaw, R. G. (1998) Ultrasounds – A new opportunity for food preservation. In: Povey, M. J. W. & Mason, T. J. (Eds.), *Ultrasound in Food Processing*. Blackie Academic & Professional, London, UK: pp. 183–192.

Eastham, J., Sharples, L. & Ball, S. (Eds.) (2001) *Food Supply Chain Management*. Butterworth Heinemann, Oxford, UK.

Ede, A. J. (1949) The calculation of freezing and thawing of foodstuffs. *Modern Refrigeration* **52**: 52–55.

Edwards, M. (Ed.) (2004) *Detecting Foreign Bodies in Food*. Woodhead Publishing, Cambridge, UK.

Eh, R. G. & Ihde, A. (1954) Faraday's electrochemical laws and the determination of equivalent weights. *Journal of Chemical Education* **31**(5): 226–232.

Ehlermann, D. A. E. (2002) Irradiation. In: Henry, C. J. K. & Chapman, C. (Eds.), *The Nutrition Handbook for Food Processors*. Woodhead Publishing, Cambridge, UK: pp. 371–395.

Ehlermann, D. A. E. (2009) The Radura terminology and food irradiation. *Food Control* **20**(5): 526–528.

Ehmann, W. D. & Vance, D. E. (1991) *Radiochemistry and Nuclear Methods of Analysis*. Wiley Interscience, New York, NY.

Elamin, W. M., Endan, J. B., Yosuf, Y. A., Shamsudin, R. & Ahmedov, A. (2015) High pressure processing technology and equipment evolution: A review. *Journal of Engineering, Science and Technology Reviews* **8**(5): 75–83.

Erickson, M. C. (1997) Lipid oxidation: Flavour and nutritional quality deterioration in frozen foods. In: Erickson, M. C. & Hung, Y.-C. (Eds.), *Quality in Frozen Food*. Chapman & Hall, London, UK: pp. 141–173.

Eshtiaghi, M. N. & Knorr, D. (1993) Potato cubes response to water blanching and high hydrostatic pressure. *Journal of Food Science* **58**(6): 1371–1374.

Estrada-Flores, S. (2002) Novel cryogenic technologies for the freezing of food products. *Journal of the Australian Institute of Refrigeration, Air Conditioning and Heating* **July**: 16–21.

Evans, J. & James, S. (1993) Freezing and meat quality. In: Turner, A. (Ed.), *Food Technology International Europe*. Sterling Publications, London, UK: pp. 53–56.

Faithfull, J. D. T. (1988) Ovenable thermoplastic packaging. In: Turner, A. (Ed.), *Food Technology International Europe*. Sterling Publications, London, UK: pp. 357–362.

Faraday, M. (1822) On some new electro-magnetical motion, and on the theory of magnetism. *Quarterly Journal of Science, Literature and the Arts* **XII**: 74–96.

Farber, J. M. (1991) Microbiological aspects of modified atmosphere packaging technology – A review. *Journal of Food Protection* **54**(1): 58–70.

Fellows, P. J. (2009) *Food Processing Technology: Principles and Practice*. Woodhead Publishing, Cambridge, UK.

Fellows, P. J. & Axtell, B. L. A. (2002) *Appropriate Food Packaging*. ITDG Publishing, London, UK.

Fennema, O. R. & Powrie, W. D. (1964) Fundamentals of low temperature food preservation. *Advances in Food Research* **13**: 219–347.

Fennema, O. R., Powrie, W. D. & Marth, E. H. (1973) *Low-Temperature Preservation of Foods and Living Matter*. Marcel Dekker, New York, NY.

Fikiin, K. A. (2008) Emerging and novel freezing processes. In: Evans, J. (Ed.), *Frozen Food Science and Technology*. Blackwell, Oxford, UK: pp. 101–123.

Fikiin, K. A. & Fikiin, A. G. (2000) Individual quick freezing of foods by hydrofluidisation and pumpable ice slurries. In: Fikiin, K. A. (Ed.), *Advances in the Refrigeration Systems, Food Technologies and Cold Chain, IIR Proceedings – Series Refrigeration Science and Technology*. International Institute of Refrigeration, Paris, France: pp. 319–326.

Fink, D. G. & Beaty, H. W. (1999) *Standard Handbook for Electrical Engineers*. McGraw-Hill, New York, NY.

Fjeld, M., Asbjornsen, O. A. & Astrom, K. J. (1974) Reaction invariants and their importance in the analysis of eigenvectors, state observability and controllability of the continuous stirred tank reactor. *Chemical Engineering Science* **29**(9): 1917–1926.

Flambert, C. M. F. & Deltour, J. (1972) Localization of the critical area in thermally-processed conduction heated canned food. *Lebensmittel-Wissenschaft und -Technologie* **5**: 7–13.

Flanagan, W. M. (1993) *Handbook of Transformer Design and Applications*. McGraw-Hill, New York, NY.

Franks, F. (1985) *Biophysics and Biochemistry at Low Temperatures*. Cambridge University Press, Cambridge, UK.

Franz, R., Huber, M. & Piringer, O.-G. (1994) Testing and evaluation of recycled plastics for food packaging use – Possible migration through a functional barrier. *Food Additives and Contaminants* **11**(4): 479–496.

Franz, R. & Welle, F. (2003) Recycling packaging materials. In: Ahvenainen, R. (Ed.), *Novel Food Packaging Techniques*. Woodhead Publishing, Cambridge, UK: pp. 497–518.

Fratamico, P. M. & Bayles, D. O. (Eds.) (2005) *Foodborne Pathogens: Microbiology and Molecular Biology*. Caister Academic Press, Norwich, UK.

Galazka, V. B. & Ledward, D. A. (1995) Developments in high pressure food processing. In: Turner, A. (Ed.), *Food Technology International Europe*. Sterling Publications International, London, UK: pp. 123–125.

Galili, I., Kaplan, D. & Lehavi, Y. (2006) Teaching Faraday's law of electromagnetic induction in an introductory physics course. *American Journal of Physics* **74**(4): 337–343.

Gallego, J. A. (1998) Some applications of air-borne ultrasound to food processing. In: Povey, M. J. W. & Mason, T. J. (Eds.), *Ultrasonics in Food Processing*. Thomson Science, London, UK: pp. 127–143.

Gallego, J. A., Rodríguez, G., Gálvez, J. C. & Yang, T. S. (1999) A new high-intensity ultrasonic technology for food dehydration. *Drying Technology* **17**(3): 597–608.

García-Pérez, J. V., Cárcel, J. A., de la Fuente-Blanco, S. & Riera-Franco de Sarabia, E. (2006) Ultrasonic drying of foodstuff in a fluidized bed: Parametric study. *Ultrasonics* **44**: e539–e543.

Garrett, E. H. (1998) Fresh-cut produce. In: Blakistone, B. A. (Ed.), *Principles and Applications of Modified Atmosphere Packaging of Foods*. Blackie Academic & Professional, London, UK: pp. 125–134.

Geeraerd, A. H., Valdramidis, V. P., Bernaerts, K. & van Impe, J. F. (2004) Evaluating microbial inactivation models for thermal processing. In: Richardson, P. (Ed.), *Improving the Thermal Processing of Foods*. Woodhead Publishing, Cambridge, UK: pp. 427–453.

Geiges, O. (1996) Microbial processes in frozen food. *Advances in Space Research* **18**(12): 109–118.

Ghazala, S. & Trenholm, R. (1998) Hurdle and HACCP concepts in sous-vide and cook-chill products. In: Ghazala, S. (Ed.), *Sous-Vide and Cook-Chill Processing for the Food Industry*. Aspen Publications, Gaithersburgh, MD: pp. 294–310.

Ghosh, V., Ziegler, G. R. & Anantheswaran, R. C. (2002) Fat, moisture and ethanol migration through chocolates and confectionery coatings. *Critical Reviews in Food Science and Nutrition* **42**(6): 583–626.

Gibbs, P. (2002) Characteristics of spore-forming bacteria. In: Blackburn, C. W. & McClure, P. J. (Eds.), *Foodborne Pathogen Hazards, Risk Analysis and Control*. Woodhead Publishing, Cambridge, UK: pp. 418–435.

Gibson, H., Taylor, J. H., Hall, K. E. & Holah, J. T. (1999) Effectiveness of cleaning techniques used in the food industry in terms of removal of bacterial biofilms. *Journal of Applied Microbiology* **87**(1): 41–48.

Gill, C. O. (2003) Active packaging in practice: Meat. In: Ahvenainen, R. (Ed.), *Novel Food Packaging Techniques*. Woodhead Publishing, Cambridge, UK: pp. 364–383.

Giovannoni, J. (2001) Molecular biology of fruit maturation and ripening. *Annual Reviews of Plant Physiology and Plant Molecular Biology* **52**: 725–749.

Glaser, R. W., Leikin, S. L., Chernomordik, L. V., Pastushenko, V. F. & Sokirko, A. V. (1988) Reversible electrical breakdown of lipid bilayers: Formation and evolution of pores. *Biochimica et Biophysica Acta* **940**(2): 275–281.

Goldblith, S. A. (1970) Radiation preservation of food – The current status. *Journal of Food Technology* **5**(2): 103–110.

Gontard, N., Guilbert, S. & Cuq, J. L. (1992) Water and glycerol as plasticizers affect mechanical and water vapor barrier properties of an edible wheat gluten film. *Journal of Food Science* **58**(1): 206–211.

Góral, D., Kluza, F., Spiess, W. E. L. & Kozlowicz, K. (2016) Review of thawing time prediction models depending on process conditions and product characteristics. *Food Technology and Biotechnology* **54**(1): 3–12.

Gordon, M. & Taylor, J. S. (1952) Ideal copolymers and the second-order transitions of synthetic rubbers. I. Non-crystalline copolymers. *Journal of Applied Chemistry* **2**(9): 493–500.

Gordon, M. H. (2004) Factors affecting lipid oxidation. In: Steele, R. (Ed.), *Understanding and Measuring the Shelf-Life of Food*. Woodhead Publishing, Cambridge, UK: pp. 128–141.

Gorris, L. G. M. & Peppelenbos, H. W. (1999) Modified atmosphere packaging of produce. In: Rahman, M. S. (Ed.), *Handbook of Food Preservation*. Marcel Dekker, New York, NY: pp. 437–456.

Gould, G. W. (1995) Biodeterioration of foods and an overview of preservation in food and dairy industries. *International Biodeterioration and Biodegradation* **36**(3–4): 267–277.

Gould, G. W. (2001) New processing technologies: An overview. *Proceedings of the Nutrition Society* **60**(4): 463–474.

Gradshteyn, I. S. & Ryzhik, I. M. (1980) *Table of Integrals, Series and Products*. Academic Press, San Diego, CA.

Grandison, A. S. (2006) Postharvest handling and preparation of food for processing. In: Brennan, J. G. (Ed.), *Food Processing Handbook*. Wiley-VCH, Weinheim, Germany: pp. 1–32.

Grant, I. S. & Phillips, W. R. (1990) *Electromagnetism*. Wiley, Hoboken, NJ.

Graves, M., Smith, A. & Batchelor, B. (1998) Approaches to foreign body detection in foods. *Trends in Food Science and Technology* **9**(1): 21–27.

Greaves, A. (1997) Metal detection – The essential defence. *Food Processing* **May**: 25–26.

Greengrass, J. (1998) Packaging materials for MAP foods. In: Blakistone, B. A. (Ed.), *Principles and Applications of Modified Atmosphere Packaging of Foods*. Blackie Academic & Professional, London, UK: pp. 63–101.

Gregorii, D. (1697) Catenaria. *Philosophical Transactions* **19**: 637–652.

Griffiths, D. J. (1999) *Introduction to Electrodynamics*. Prentice-Hall, Upper Saddle River, NJ.

Griffiths, M. (2002) *Mycobacterium paratuberculosis*. In: Blackburn, C. W. & McClure, P. J. (Eds.), *Foodborne Pathogen Hazards, Risk Analysis and Control*. Woodhead Publishing, Cambridge, UK: pp. 489–500.

Griffiths, M. & Ewart, K. V. (1995) Antifreeze proteins and their potential use in frozen foods. *Biotechnology Advances* **13**(3): 375–402.

Guilbert, S. & Gontard, N. (1995) *Foods and Packaging Materials*. Royal Society of Chemistry, Oxford, UK.

Guise, B. (1986) Irradiation waits in the wings. *Food Europe* **Mar/Apr**: 7–9.

Guise, B. (1987) Filling an industry need. *Food Processing* **Jul**: 31–33.

Guise, B. (1987) Spotlight on sachet packaging. *Food Processing* **Jul**: 35–37.

Guthrie, R. K. (1988) *Food Sanitation*. AVI, Westport, CT.

Hallenbeck, W. H. (1986) *Quantitative Risk Assessment for Environmental and Occupational Health*. Lewis Publishers, Chelsea, MI.

Hamilton, B. (1985) The sensor scene. *Food Manufacturing* **Sep**: 41–42.

Hammelsvang, L. (1999) Bringing displacement pumps into the next millennium. *World Pumps* **398** (Nov): 28–30.

Han, J. H. (2000) Antimicrobial food packaging. *Food Technology* **54**(3): 56–65.

Han, J. H. (2003) Antimicrobial food packaging. In: Ahvenainen, R. (Ed.), *Novel Food Packaging Techniques*. Woodhead Publishing, Cambridge, UK: pp. 50–70.

Hardenburg, R. E., Watada, A. E. & Wang, C. Y. (1986) *The Commercial Storage of Fruits, Vegetables and Florist and Nursery Stocks*. Agricultural Handbook, Washington, DC.

Harnulv, B. G. & Snygg, B. G. (1972) Heat resistance of *Bacillus subtilis* spores at various water activities. *Journal of Applied Bacteriology* **35**(4): 615–624.

Harris, D. & Shackell, L. (1911) The effect of vacuum desiccation on the virus of rabies, with remarks on a new method. *Journal of Infectious Diseases* **8**(1): 47–49.

Harris, J. W. & Stocker, H. (1998) *Handbook of Mathematics and Computational Sciences*. Springer-Verlag, New York, NY.

Harris, R. S. (1988) *Production Is Only Half of the Battle – A Training Manual in Fresh Produce Marketing for the Eastern Caribbean*. FAO-UNO, Bridgestone, Barbados.

Harrison, A. (1997) Tesco composites. In: Johnson, R., Chambers, S., Harland, C., Harrison, A. & Slack, N. (Eds.), *Cases in Operations Management*. Pitman Publishing, London, UK: pp. 359–367.

Hastings, M. J. (1998) MAP machinery. In: Blakistone, B. A. (Ed.), *Principles and Applications of Modified Atmosphere Packaging of Foods*. Blackie Academic & Professional, London, UK: pp. 39–64.

Haugaard, V. K., Udsen, A.-M., Mortensen, G., Hoegh, L., Petersen, K. & Monahan, F. (2000) Food biopackaging. In: Weber, C. J. (Ed.), *Biobased Packaging Materials for the Food Industry – Status and Perspectives.* Food Biopack Project, EU Directorate 12, Brussels, Belgium: pp. 45–106.

Hayashi, R. (1995) Advances in high pressure processing in Japan. In: Gaonkar, A. G. (Ed.), *Food Processing: Recent Developments.* Elsevier, London, UK: pp. 185–195.

Hayward, T. (2003) Award winning hygienic control devices and indicator lights. *Food Processing* **Sep**: 6–7.

Heald, M. A. (2003) Where is the Wien's peak? *American Journal of Physics* **71**(12): 1322–1323.

Heap, R. D. (2000) The refrigeration of chilled foods. In: Stringer, M. & Dennis, C. (Eds.), *Chilled Foods – A Comprehensive Guide.* Woodhead Publishing, Cambridge, UK: pp. 79–98.

Heinz, V. & Knorr, D. (1998) High pressure germination and inactivation kinetics of bacterial spores. In: Isaacs, N. S. (Ed.), *High Pressure Food Science, Bioscience and Chemistry.* Royal Society of Chemistry, Cambridge, UK: pp. 436–441.

Hendrickx, M., Ludikhuyze, L., van der Broeck, I. & Weemaes, C. (1998) Effects of high pressure on enzymes related to food quality. *Trends in Food Science and Technology* **9**(5): 197–203.

Herbert, R. A. (1989) Microbial growth at low temperatures. In: Gould, G. W. (Ed.), *Mechanism of Action of Food Preservation Procedures.* Elsevier Applied Science, London, UK: pp. 71–96.

Hersom, A. C. & Hulland, E. D. (1980) *Canned Foods.* Churchill Livingstone, London, UK.

Hilborne, M. (2006) Take control. *Food Processing* **May**: 20.

Hill, M. A. (1987) The effect of refrigeration on the quality of some prepared foods. In: Thorne, S. (Ed.), *Developments in Food Preservation*, vol. 4. Elsevier Applied Science, London, UK: pp. 123–152.

Hills, B. (2004) Nuclear magnetic resonance imaging. In: Edwards, M. (Ed.), *Detecting Foreign Bodies in Food.* Woodhead Publishing, Cambridge, UK: pp. 154–171.

Ho, S. Y., Mittal, G. S. & Cross, J. D. (1997) Effects of high field electric pulses on the activity of selected enzymes. *Journal of Food Engineering* **31**(1): 69–84.

Hofmann, G. A. (1985) Deactivation of micro-organisms by an oscillating magnetic field. US Patent no. 4,524,079.

Holah, J. & Thorpe, R. H. (2000) The hygienic design of chilled food plant. In: Stringer, M. & Dennis, C. (Eds.), *Chilled Foods – A Comprehensive Guide.* Woodhead Publishing, Cambridge, UK: pp. 355–396.

Holdsworth, S. D. (1992) *Aseptic Processing and Packaging of Foods.* Elsevier Academic and Professional, London, UK.

Holdsworth, S. D. (2004) Optimising the safety and quality of thermally processed packaged foods. In: Richardson, P. (Ed.), *Improving the Thermal Processing of Foods.* Woodhead Publishing, Cambridge, UK: pp. 3–31.

Hollingsworth, P. (1994) Computerised manufacturing trims production costs. *Food Technology* **48**(12): 43–45.

Holmes, M. (2000) Optimising the water supply chain. *Food Processing* **Nov**: 38, 40.

Hom, L. W. (1972) Kinetics of chlorine disinfection in an ecosystem. *Journal of the Environmental Division of the American Society of Civil Engineering* **98**: 183–194.

Honikel, K. O. & Schwagele, F. (2001) Chilling and freezing of meat and meat products. In: Sun, D.-W. (Ed.), *Advances in Food Refrigeration.* Leatherhead Publishing, Leatherhead, UK: pp. 366–386.

Hoover, D. G., Metrik, C., Papineau, A. M., Farkas, D. F. & Knorr, D. (1989) Biological effects of high hydrostatic pressure on food microorganisms. *Food Technology* **43**(3): 99–107.

Hoyer, O. (1998) Testing performance and monitoring of UV systems for drinking water disinfection. *Water Supply* **16**: 419–442.

Hu, Y. H. & Barta, J. (2006) *Handbook of Fruits and Fruit Processing.* Blackwell, Oxford, UK.

Hughes, D. (1982) *Notes on Ionising Radiation: Quantities, Units, Biological Effects and Permissible Doses. Occupational Hygiene Monograph no. 5.* Northwood, Middlesex, UK.

Hui, Y. H. (Ed.) (1992) *Encyclopedia of Food Science and Technology.* Wiley, New York, NY.

Hui, Y. H., Barta, J., Pilar-Cano, M., Gusek, T. W., Sidhu, J. & Sinha, N. (2006) *Handbook of Fruits and Fruit Processing.* Blackwell, Oxford, UK.

Hung, Y.-C. (2001) Cryogenic refrigeration. In: Sun, D.-W. (Ed.), *Advances in Food Refrigeration.* Leatherhead Publishing, Leatherhead, UK: pp. 305–325.

Huray, P. G. (2010) *Maxwell's Equations.* Wiley, Hoboken, NJ.

Indrawati, I., van Loey, A., Smout, C. & Hendrickx, M. (2003) High hydrostatic pressure technology in food preservation. In: Zeuthen, P. & Bogh-Sorensen, L. (Eds.), *Food Preservation Techniques.* Woodhead Publishing, Cambridge, UK: pp. 428–448.

Iversen, C. K. (1999) Black currant nectar: Effect of processing and storage on anthocyanin and ascorbic acid. *Journal of Food Science* **64**(1): 37–41.

Jackson, J. D. (1998) *Classical Electrodynamics.* Wiley, Hoboken, NJ.

Jahnke, E. & Emde, F. (1945) *Tables of Functions.* Dover, New York, NY.

Jakobsen, M. & Bertelsen, G. (2003) Active packaging and color control: The case of meat. In: Ahvenainen, R. (Ed.), *Novel Food Packaging Techniques.* Woodhead Publishing, Cambridge, UK: pp. 401–415.

Jaspersen, W. S. (1989) Specialty ice cream extrusion technology. In: Turner, A. (Ed.), *Food Technology International Europe.* Sterling Publications International, London, UK: pp. 85–88.

Jayas, D. S. & Jeyamkondan, S. (2002) PH – Postharvest technology: Modified atmosphere storage of grains, meats, fruits and vegetables. *Biosystems Engineering* **82**(3): 235–251.

Jelen, P. (1982) Experience with direct and indirect UHT processing of milk – A Canadian viewpoint. *Journal of Food Protection* **45**(9): 878–883.

Jennings, B. (1998) Non-contact sensing. *Food Processing* **Nov**: 9–10.

Jennings, B. (1999) Refrigeration for the new millennium. *Food Processing* **May**: 12–13.

Jha, A. R. (2005) *Cryogenic Technology and Applications.* Butterworth-Heinemann, Oxford, UK.

Johnson, D. (2001) A ready pump for ready meals. *Food Processing* **Jun**: 21.

Johnston, R., Chambers, S., Harland, C., Harrison, A. & Slack, N. (1997) *Introduction to Planning and Control: Cases in Operations Management.* Pitman Publishing, London, UK.

Jones, M., Jones, C. & Jones, S. (2001) Cryogenic processor for liquid feed preparation of a free-flowing frozen product and method for freezing liquid composition. US Patent no. 6,223,542.

Jordan, M. J., Goodner, K. & Laencina, J. (2003) Deaeration and pasteurization effects on the orange juice aromatic fraction. *Lebensmittel-Wissenschaft und -Technologie* **36**(4): 391–396.

Kader, A. A., Singh, R. P. & Mannapperuma, J. D. (1998) Technologies to extend the refrigerated shelf-life of fresh fruits and vegetables. In: Taub, I. A. & Singh, R. P. (Eds.). *Food Storage Stability*. CRC Press, Boca Raton, FL: pp. 419–434.

Kanbakan, U., Con, A. H. & Ayar, A. (2004) Determination of microbial contamination sources during ice cream production in Denizli, Turkey. *Food Control* **15**(6): 463–470.

Kawahara, C. (2002) The structures and functions of ice crystal-controlling proteins from bacteria. *Journal of Bioscience and Bioengineering* **94**(6): 492–496.

Kendall, M. G. & Stuart, A. (1963) *The Advanced Theory of Statistics – Distribution Theory*. Hafner Publishing, New York, NY.

Kennedy, C. (2003) Developments in freezing. In: Zeuthen, P. & Bogh-Sorensen, L. (Eds.), *Food Preservation Techniques*. Woodhead Publishing, Cambridge, UK: pp. 228–240.

Kenney, J. F. & Keeping, E. S. (1954) *Mathematics of Statistics: Part One*. Van Rostrand Reinhold, Princeton, NJ.

Kheang, L. S., May, C. Y., Foon, C. S. & Ngan, M. A. (2006) Recovery and conversion of palm olein-derived used frying oil to methyl esters for biodiesel. *Journal of Palm Oil Research* **18**(1): 247–252.

Khwaldia, K., Perez, C., Banon, S., Desobry, S. & Hardy, J. (2004) Milk proteins for edible films and coatings. *Critical Reviews in Food Science and Nutrition* **44**(4): 239–251.

Kidmose, U. & Martens, H. J. (1999) Changes in texture, microstructure and nutritional quality of carrot slices during blanching and freezing. *Journal of the Science of Food and Agriculture* **79**(12): 1747–1753.

Kilcast, D. & Subramanian, P. (2000) *The Stability and Shelf Life of Food*. Woodhead Publishing, Cambridge, UK.

Kim, S.-M., Chen, P., McCarthy, M. J. & Zion, B. (1999) Fruit internal quality evaluation using on-line nuclear magnetic resonance sensors. *Journal of Agricultural and Engineering Research* **74**(3): 293–301.

Knorr, D. (1993) Effect oh high hydrostatic pressure processes on food safety and quality. *Food Technology* **47**(6): 156.

Koontz, J. L., Moffitt, R. D., Marcy, J. E., O'Keefe, S. F., Duncan, S. E. & Long, T. E. (2010) Controlled release of α-tocopherol, quercetin, and their cyclodextrin inclusion complexes from linear low-density polyethylene (LLDPE) films into a coconut oil model food system. *Food Additives and Contaminants A: Chemical Analysis, Control of Exposition and Risk Assessment* **27**(11): 1598–1607.

Koopmans, M. (2002) Viruses. In: Blackburn, C. W. & McClure, P. J. (Eds.), *Foodborne Pathogen Hazards, Risk Analysis and Control*. Woodhead Publishing, Cambridge, UK: pp. 440–452.

Kotsianis, I. S., Giannou, V. & Tzia, C. (2002) Production and packaging of bakery products using MAP technology. *Trends in Food Science and Technology* **13**(9–10): 319–324.

Kraft, A. A. (1992) *Psychrotrophic Bacteria in Foods: Disease and Spoilage*. CRC Press, Boca Raton, FL.

Kramer, A. (1974) Storage retention of nutrients. *Food Technology* **28**(1): 50–60.

Krcmar, H., Bjorn-Andersen, N. & O'Callaghan, R. (Eds.) (1995) *EDI in Europe – How It Works in Practice*. Wiley, Chichester, UK.

Kress-Rogers, E. (Ed.) (2001) *Instrumentation and Sensors for the Food Industry*. Woodhead Publishing, Cambridge, UK.

Krochta, J. M., Baldwin, E. A. & Nisperos-Carriedo, M. (Eds.) (2002) *Edible Coatings and Films to Improve Food Quality*. CRC Press, Boca Raton, FL.

Krokida, M. K., Kiranoudis, C. T., Maroulis, Z. B. & Marinos-Kouris, D. (2000) Effect of pretreatment on colour of dehydrated products. *Drying Technology* **18**(6): 1239–1250.

Kronig, B. & Paul, T. (1897) Die chemischen Grundlagen der Lehre von der Giftwirkung und Disinfektion. *Zeitschrift fur Hygiene* **25**: 1–112.

de Kruijf, N. & Rijk, R. (2003) Legislation issues relating to active and intelligent packaging. In: Ahvenainen, R. (Ed.), *Novel Food Packaging Techniques*. Woodhead Publishing, Cambridge, UK: pp. 459–496.

Kuda, T., Shimidzu, K. & Yano, T. (2004) Comparison of rapid and simple microplate assays as an index of bacterial count. *Food Control* **15**(6): 421–425.

Kulshreshtha, M. K., Zaror, C. A. & Jukes, D. J. (1995) Simulating the performance of a control system for food extruders using model-based set-point adjustment. *Food Control* **6**(3): 135–141.

Labuza, T. P. (1984) Application of chemical kinetics to deterioration of foods. *Journal of Chemical Education* **61**(4): 348–358.

Labuza, T. P. (1996) An introduction to active packaging of foods. *Food Technology* **50**(1): 68–71.

Labuza, T. P., Tannenbaum, S. R. & Karel, M. (1970) Water content and stability of low-moisture and intermediate-moisture foods. *Food Technology* **24**(5): 543–544, 546–548, 550.

Lagaron, J. M., Cabedo, L., Feijoo, J. L., Gavara, R. & Gimenez, E. (2005) Improving packaged food quality and safety: II. Nanocomposites. *Food Additives and Contaminants* **22**(10): 994–998.

Lambert, R. J. W. & Johnston, M. D. (2000) Disinfection kinetics: A new hypothesis and model for the tailing of log-survivor/time curves. *Journal of Applied Microbiology* **88**(5): 907–913.

Lamberti, M. & Escher, F. (2007) Aluminium foil as a food packaging material in comparison with other materials. *Food Reviews International* **23**(4): 407–433.

Larkin, J. W. & Spinak, S. H. (1996) Safety considerations for ohmically heated, aseptically processed, multiphase low-acid food products. *Food Technology* **50**(5): 242–245.

Lasslett, T. (1988) Computer control in food processing. In: Turner, A. (Ed.), *Food Technology International Europe*. Sterling Publications, London, UK: pp. 105–106.

Lazarides, H. N. & Mavroudis, N. E. (1995) Freeze/thaw effects on mass transfer rates during osmotic dehydration. *Journal of Food Science* **60**(4): 826–828.

Leadley, C. E. (2003) Developments in non-thermal processing. *Food Science and Technology* **17**(3): 40–42.

Leadley, C. E. & Williams, A. (2006) Pulsed electric field processing, power ultrasound and other emerging technologies. In: Brennan, J. G. (Ed.), *Food Processing Handbook*. Wiley-VCH, Weinheim, Germany: pp. 201–236.

Lee, D. (2000) Selecting progressing cavity pumps. *Food Processing* **Jun**: 51.

Lee, H. S. & Chen, C. S. (1998) Rates of vitamin C loss and discoloration in clear orange juice concentrate during storage at temperatures of 4–24 °C. *Journal of Agricultural and Food Chemistry* **46**(11): 4723–4727.

Lee, R. E. & Gilbert, C. A. (1918) On the application of the mass law to the process of disinfection – Being a contribution to the 'mechanistic theory' as opposed to the 'vitalistic theory'. *Journal of Physical Chemistry* **22**(5): 348–372.

Leistner, L. (1995) Principles and applications of hurdle technology. In: Gould, G. W. (Ed.), *New Methods of Food Preservation*. Blackie Academic & Professional, London, UK: pp. 1–21.

Leistner, L. & Gorris, L. G. M. (1995) Food preservation by hurdle technology. *Trends in Food Science and Technology* **6**(2): 41–46.

Lelieveld, H. L. M. & Mostert, I. T. (Eds.) (2003) *Hygiene in Food Processing*. Woodhead Publishing, Cambridge, UK.

Lelieveld, H. L. M., Notermans, S. & de Haan, S. W. H. (Eds.) (2007) *Food Preservation by Pulsed Electric Fields*. Woodhead Publishing, Cambridge, UK.

Lentz, J. (1986) Printing. In: Bakker, M. & Eckroth, D. (Eds.), *The Wiley Encyclopaedia of Packaging Technology*. Wiley, New York, NY: p. 554.

Lerche, I. & Glaesser, W. (2006) *Environmental Risk Assessment: Quantitative Measures, Anthropogenic Influences, Human Impact*. Springer, Berlin, Germany.

Leveland, W., Morrissey, D. & Seaborg, G. T. (2006) *Modern Nuclear Chemistry*. Wiley Interscience, New York, NY.

Levine, L. & Drew, B. A. (1994) Sheeting of cookie and cracker doughs. In: Faridi, H. (Ed.), *The Science of Cookie and Cracker Production*. Chapman & Hall, New York, NY: pp. 353–386.

Lewis, M. J. (1993) UHT processing: Safety and quality aspects. In: Turner, A. (Ed.), *Food Technology International Europe*. Sterling Publications, London, UK: pp. 47–51.

Li, B. & Sun, D.-W. (2002) Effect of power ultrasound on freezing rate during immersion freezing of potatoes. *Journal of Food Engineering* **55**(3): 277–282.

Li, B. & Sun, D.-W. (2002) Novel methods for rapid freezing and thawing of foods – A review. *Journal of Food Engineering* **54**(3): 175–182.

Li, C. F. & Chung, Y. C. (1997) The benefits of chitosan post-harvested storage and the quality of fresh strawberries. In: Domard, A., Roberts, G. A. F. & Varum, K. M. (Eds.), *Advances in Chitin Science*. Jacques André Publishers, Lyon, France: pp. 908–913.

Loaharanu, P. (1990) Food irradiation: Facts or fiction? *IAEA Bulletin* **32**(2): 44–48.

Lock, A. (1969) *Practical Canning*. Food Trade Press, London, UK.

van Loey, A., Guiavarc'h, Y., Claeys, W. & Hendrickx, M. (2004) The use of time-temperature integrators (TTIs) to validate thermal processes. In: Richardson, P. (Ed.), *Improving the Thermal Processing of Foods*. Woodhead Publishing, Cambridge, UK: pp. 365–384.

van Loey, A., Haentjens, T. & Hendrickx, M. (1998) The potential role of time-temperature integrators for process evaluation in the cook-chill chain. In: Ghazala, S. (Ed.), *Sous-Vide and Cook-Chill Processes for the Food Industry*. Aspen Publications, Gaithersburgh, MD: pp. 89–110.

Loisel, C., Keller, G., Lecq, G., Bourgaux, C. & Ollivon, M. (1998) Phase transitions and polymorphism of cocoa butter. *Journal of American Oil Chemists' Society* **75**(4): 425–439.

Loncin, M. & Merson, R. L. (1979) *Food Engineering: Principles and Selected Applications*. Academic Press, New York, NY.

Londahl, G. & Karlsson, B. (1991) Initial crust freezing of fragile products. In: Turner, A. (Ed.), *Food Technology International*. Sterling Publications, London, UK: pp. 90–91.

Lopez, P., Sala, F. J., Condon, S., Raso, J. & Burgos, J. (1994) Inactivation of peroxidase, lipoxygenase and polyphenoloxidase by manothermosonication. *Journal of Agricultural and Food Chemistry* **42**(2): 252–256.

Lopez-Rubio, A., Gavara, R. & Lagaron, J. M. (2006) Bioactive packaging: Turning foods into healthier foods through biomaterials. *Trends in Food Science and Technology* **17**(10): 567–575.

Lorenz, L. (1867) On the identity of the vibrations of light with electrical currents. *Philosophical Magazine Series* **4**(34): 287–301.

Louis, P. (1998) Food packaging in the next century. In: Turner, A. (Ed.), *Food Technology International Europe*. Sterling Publications, London, UK: pp. 80–82.

Low, J. M., Maughan, W. S., Bee, S. C. & Honeywood, M. J. (2001) Sorting by colour in the food industry. In: Kress-Rogers, E. & Brimelow, C. J. B. (Eds.), *Instrumentation and Sensors for the Food Industry*. Woodhead Publishing, Cambridge, UK: pp. 117–136.

Lucas, T., Chourot, J.-M., Raoult-Wack, A.-L. & Goli, T. (2001) Hydro/immersion chilling and freezing. In: Sun, D.-W. (Ed.), *Advances in Food Refrigeration*. Leatherhead Publishing, Leatherhead, UK: pp. 220–263.

Lucas, T. & Raoult-Wack, A. (1998) Immersion chilling and freezing in aqueous refrigerating media: Review and future directions. *International Journal of Refrigeration* **21**(6): 419–429.

Lund, B., Baird-Parker, A. C. & Gould, G. W. (2000) *Microbiological Safety and Quality of Food*. Aspen Publishing, New York, NY.

Luyben, W. L. (1973) *Process Modelling, Simulation and Control for Chemical Engineers*. McGraw-Hill, New York, NY.

Lyijynen, T., Hurme, E. & Ahvenainen, R. (2003) Optimizing packaging. In: Ahvenainen, R. (Ed.), *Novel Food Packaging Techniques*. Woodhead Publishing, Cambridge, UK: pp. 441–458.

MacDonald, L. E., Brett, J., Kelton, D., Majowicz, S. E., Snedeker, K. & Sargeant, J. M. (2011) A systematic review and meta-analysis of the effects of pasteurization on milk vitamins, and evidence for raw milk consumption and other health-related outcomes. *Journal of Food Protection* **74**(11): 1814–1832.

Magan, N. (2004) *Mycotoxins in Food*. Woodhead Publishing, Cambridge, UK.

Mahapatra, A. K., Muthukumarappan, K. & Julson, J. L. (2005) Applications of ozone, bacteriocins and irradiation in food processing: A review. *Critical Reviews in Food Science and Nutrition* **45**(6): 447–461.

Malcata, F. X. (1990) The effect of internal thermal gradients on the reliability of surface mounted full-history time-temperature indicators. *Journal of Food Processing and Preservation* **14**(6): 481–497.

Man, C. M. D. & Jones, A. A. (2000) *Shelf-Life Evaluation of Foods*. Aspen Publishing, Gaithersburgh, MD.

Manfredi, L. B., Ginés, M. J. L., Benítez, J. G., Egli, W. A., Rissone, H. & Vazquez, A. (2005) Use of epoxy-phenolic lacquers in food can coatings: Characterization of lacquers and cured films. *Journal of Applied Polymer Science* **95**(6): 1448–1458.

Marien, M. (1989) Automation to meet the food industry's needs. In: Turner, A. (Ed.), *Food Technology International Europe*. Sterling Publications, London, UK: pp. 127–133.

Marriott, N. G. & Gravani, R. B. (2006) *Principles of Food Sanitation*. Springer Science, New York, NY.

Marth, E. H. (1998) Extended shelf-life refrigerated foods: Microbiological quality and safety. *Food Technology* **52**(2): 57–62.

Martin, A. V. (1991) Advances in flexible packaging of confectionery. *European Food and Drink Review* **Winter**: 17–20.

Martin, T. W. (1961) Improved computer oriented methods for calculation of steam properties. *Journal of Heat Transfer* **83**(4): 515–516.

Martin, W. M. (1948) Flash process, aseptic fill are used in new canning unit. *Food Industries* **20**: 832–836.

Martin-Bellose, O. & Llanos-Barriobero, E. (2001) Proximate composition, minerals and vitamins in selected canned vegetables. *European Food Research and Technology* **212**(2): 182–187.

Mascheroni, R. H. (2001) Plate and air-blast cooling/freezing. In: Sun, D.-W. (Ed.), *Advances in Food Refrigeration*. Leatherhead Publishing, Leatherhead, UK: pp. 193–219.

Massey, L. (2003) *Permeability Properties of Plastics and Elastomers: A Guide to Packaging and Barrier Materials*. William Andrew Publishing, Norwich, NY.

Maxwell, J. C. (1954) *A Treatise on Electricity and Magnetism*. Dover, New York, NY.

May, N. (2004) Developments in packaging formats for retort processing. In: Richardson, P. (Ed.), *Improving the Thermal Processing of Foods*. Woodhead Publishing, Cambridge, UK: pp. 138–151.

Mayo, G. (1984) Principles of metal detection. *Food Manufacturing* **Aug**: 29, 31.

McAllorum, S. (2005) Magnetic separation in process industries. *Food Science and Technology Today* **19**(1): 43, 45–46.

McClements, D. J. (1995) Advances in the application of ultrasound in food analysis and processing. *Trends in Food Science and Technology* **6**(9): 293–299.

McClure, P. & Blackburn, C. (2002) *Campylobacter* and *Arcobacter*. In: Blackburn, C. W. & McClure, P. J. (Eds.), *Foodborne Pathogen Hazards, Risk Analysis and Control*. Woodhead Publishing, Cambridge, UK: pp. 363–384.

McFarlane, I. (1988) Advances in sensors that benefit food manufacture. In: Turner, A. (Ed.), *Food Technology International Europe*. Sterling Publications, London, UK: pp. 109–113.

McFarlane, I. (1991) The need for sensors in food process control. In: Turner, A. (Ed.), *Food Technology International Europe*. Sterling Publications, London, UK: pp. 119–122.

McHugh, T. H. & Krochta, J. M. (1994) *Permeability Properties of Films*. In: Krochta, J. M., Baldwin, E. A. & Nisperor-Carriedo, M. O. (Eds.), *Edible Coatings and Films to Improve Food Quality*. Technomic Publishing, Lancaster, PA: pp. 139–188.

McKenna, B. M., Lyng, J., Brunton, N. & Shirsat, N. (2006) Advances in radio frequency and ohmic heating of meats. *Journal of Food Engineering* **77**(2): 215–229.

McLaren, P. (1984) *Elementary Electric Power and Machines*. Ellis Horwood, Hemel Hempstead, UK.

McLaughin, W. L., Jarret, R. D. & Olejnik, T. A. (1982) Dosimetry. In: Josephson, E. S. & Peterson, M. S. (Eds.), *Preservation of Foods by Ionizing Radiation*, vol. 1. CRC Press, Boca Ratton, FL.

McMeekin, T. (Ed.) (2003) *Detecting Pathogens in Food*. Woodhead Publishing, Cambridge, UK.

McTigue-Pierce, L. (2005) Conveyors: The missing link to a smooth-running line. New levels of flexibility, easy-to-clean/sanitize designs and better package handling are among the advances in conveyor technology today. *Food and Drug Packaging* **Oct**: 1.

Mei, Y. & Zhao, Y. (2003) Barrier and mechanical properties of milk protein-based edible films incorporated with nutraceuticals. *Journal of Agricultural and Food Chemistry* **51**(7): 1914–1918.

von Meier, A. (2006) *Electric Power Systems*. Wiley, Hoboken, NJ.

Mercea, P. (2000) Models for diffusion in polymers. In: Piringer, O. & Baner, A. L. (Eds.), *Plastic Packaging Materials for Foods: Barrier Function, Mass Transport, Quality Assurance, Legislation*. Wiley-VCH, Weinheim, Germany: pp. 125–158.

Mermelstein, N. H. (1997) High pressure processing reaches the US market. *Food Technology* **51**(6): 95–96.

Mertens, B. (1995) Hydrostatic pressure treatment of food: Equipment and processing. In: Gould, G. W. (Ed.), *New Methods of Food Preservation*. Blackie Academic & Professional, Glasgow, UK: pp. 135–158.

Mertens, B. & Knorr, D. (1992) Development of nonthermal processes for food preservation. *Food Technology* **46**(5): 124–133.

Metaxas, A. C. (1996) *Foundations of Electroheat. A Unified Approach*. Wiley, New York, NY.

Meyer, H. W. (1972) *A History of Electricity and Magnetism*. Burndy Library, Norwalk, CT.

Miller, J. (1998) Cryogenic food freezing systems. *Food Processing* **Aug**: 22–23.

Miller, J. & Butcher, C. (2000) Freezer technology. In: Kennedy, C. (Ed.), *Managing Frozen Foods*. Woodhead Publishing, Cambridge, UK: pp. 159–194.

Mistry, V. V. & Kosikowski, F. V. (1983) Use of time-temperature indicators as quality control devices for market milk. *Journal of Food Protection* **46**(1): 52–57.

Mohanty, P. (2001) Magnetic resonance freezing system. *Australian Institute of Refrigeration, Air Conditioning and Heating* **55**: 28–29.

Mok, C., Song, K.-T., Park, Y.-S., Lim, S., Ruan, R. & Chen, P. (2006) High hydrostatic pressure pasteurization of red wine. *Journal of Food Science* **71**(8): M265–M269.

Molins, R. A. (Ed.) (2001) *Food Irradiation: Principles and Applications*. Wiley, New York, NY.

Moore, R. L. (1979) Biological effects of magnetic fields, studies with micro-organisms. *Canadian Journal of Microbiology* **25**(10): 1145–1151.

Morillon, V., Debeaufort, F., Blond, G., Capelle, M. & Voilley, A. (2002) Factors affecting the misture permeability of lipid-based edible films: A review. *Critical Reviews in Food Science and Nutrition* **42**(1): 67–89.

Morrisey, P. A. & Kerry, J. P. (2004) Lipid oxidation and the shelf-life of muscle in foods. In: Steele, R. (Ed.), *Understanding and Measuring the Shelf-Life of Food*. Woodhead Publishing, Cambridge, UK: pp. 357–395.

Morrison, N. (1994) *Introduction to Fourier Analysis*. Wiley, New York, NY.

Morse, P. M. & Feshbach, H. (1953) *Methods of Theoretical Physics*. McGraw-Hill, New York, NY.

Moss, M. (2002) Toxigenic fungi. In: Blackburn, C. W. & McClure, P. J. (Eds.), *Foodborne Pathogen Hazards, Risk Analysis and Control*. Woodhead Publishing, Cambridge, UK: pp. 479–488.

Motarjemi, Y. (2002) Chronic sequelae of foodborne infections. In: Blackburn, C. W. & McClure, P. J. (Eds.), *Foodborne Pathogen Hazards, Risk Analysis and Control*. Woodhead Publishing, Cambridge, UK: pp. 501–513.

Muir, A. H. & Blanshard, J. M. V. (1986) Effect of polysaccharide stabilizers on the rate of growth of ice. *Journal of Food Technology* **21**(6): 683–710.

Mulet, A., Cárcel, J. A., Sanjuán, N. & Bon, J. (2003) New food drying technologies – Use of ultrasound. *Food Science and Technology International* **9**(3): 215–221.

Murphy, A. (1997) The future for robotics in food processing. *European Food and Drink Review* **Spring**: 31–35.

Murphy, D. M. & Koop, T. (2005) Review of the vapour pressures of ice and supercooled water for atmospheric applications. *Quarterly Journal of Royal Meteorological Society* **131**(608): 1539–1565.

Mussa, D. M. & Ramaswamy, H. S. (1997) Ultra high pressure pasteurization of milk: Kinetics of microbial destruction and changes in physicochemical characteristics. *Lebensmittel-Wissenschaft und -Technologie* **30**(6): 551–557.

Mussatto, S. I., Machado, E. M. S., Martins, S. & Teixeira, J. A. (2011) Production, composition and application of coffee and its industrial residues. *Food and Bioprocess Technology* **4**(5): 661–672.

Nag, P. K. (2002) *Power Plant Engineering*. McGraw-Hill, New York, NY.

Nankivell, B. (2001) Clearly better packaging. *Food Processing* **Oct**: 11–12.

Naylor, P. (1992) Horizontal form-fill-seal packaging. In: Turner, A. (Ed.), *Food Technology International Europe*. Sterling Publications, London, UK: pp. 253–255.

Neilsen, J., Larsen, E. & Jessen, F. (2001) Chilling and freezing of fish and fishery products. In: Sun, D.-W. (Ed.), *Advances in Food Refrigeration*. Leatherhead Publishing, Leatherhead, UK: pp. 403–437.

Nelson, R. B. (1994) Enrobers, moulding equipment, coolers and panning. In: Beckett, S. T. (Ed.), *Industrial Chocolate Manufacture and Use*. Blackie Academic & Professional, London, UK: pp. 211–241.

Nichols, R. & Smith, H. (2002) Parasites: *Cryptosporidium*, *Giardia* and *Cyclospora* as foodborne pathogens. In: Blackburn, C. W. & McClure, P. J. (Eds.), *Foodborne Pathogen Hazards, Risk Analysis and Control*. Woodhead Publishing, Cambridge, UK: pp. 453–478.

Nielsen, J., Larsen, E. & Jessen, F. (2001) Chilling and freezing of fish and fishery products. In: Sun, D.-W. (Ed.), *Advances in Food Refrigeration*. Leatherhead Publishing, Leatherhead, UK: pp. 403–437.

Niemira, B. A., Sommers, C. H. & Ukuku, D. O. (2005) Mechanisms of microbial spoilage of fruits and vegetables. In: Lamikanra, O., Imam, S. H. & Ukuku, D. (Eds.), *Produce Degradation: Pathways and Prevention*. CRC Press, Boca Raton, FL: pp. 464–482.

Niranjan, K., Ahromrit, A. & Khare, A. S. (2006) Process control in food processing. In: Brennan, J. G. (Ed.), *Food Processing Handbook*. Wiley-VCH, Weinheim, Germany: pp. 373–384.

Nychas, G. J. & Arkoudelos, J. S. (1990) Microbiological and physicochemical changes in minced meats under carbon dioxide, nitrogen or air at 3 °C. *International Journal of Food Science and Technology* **25**: 389–398.

O'Beirne, D. & Francis, G. A. (2003) Reducing pathogen risks in MAP-prepared produce. In: Ahvenainen, R. (Ed.), *Novel Food Packaging Techniques*. Woodhead Publishing, Cambridge, UK: pp. 231–275.

O'Callaghan, R. & Turner, J. A. (1995) Electronic data interchange – Concepts and issues. In: Krcmar, H., Bjorn-Andersen, N. & O'Callaghan, R. (Eds.), *EDI in Europe – How It Works in Practice*. Wiley, Chichester, UK: pp. 1–20.

Oetiker, J. H. & Yang, S. F. (1995) The role of ethylene in fruit ripening. *Acta Horticulturae* **398**(398): 167–178.

Ohlsson, T. (1992) R&D in aseptic particulate processing technology. In: Turner, A. (Ed.), *Food Technology International Europe*. Sterling Publications, London, UK: pp. 49–53.

Ohlsson, T. & Bengtsson, N. (Eds.) (2002) *Minimal Processing Technologies in the Food Industry*. Woodhead Publishing, Cambridge, UK.

Okhuma, C., Kawai, K., Viriyarattanasak, C., Mahawanich, T., Tantratian, S., Takai, R. & Suzuki, T. (2008) Glass transition properties of frozen and freeze-dried surimi products: Effects of sugar and moisture on the glass transition temperature. *Food Hydrocolloids* **22**(2): 255–262.

Okos, M., Rao, N., Drecher, S., Rode, M. & Kozak, J. (1998) *A Review of Energy Use in the Food Industry*. American Council for an Energy Efficient Economy, Washington, DC.

Olivas, G. I. & Barbosa-Cánovas, G. V. (2005) Edible coatings for fresh-cut fruits. *Critical Reviews in Food Science and Nutrition* **45**(7–8): 657–670.

Olson, D. G. (1998) Irradiation of food. *Food Technology* **52**: 56–62.

Ooraikul, B. (2003) Modified atmosphere packaging (MAP). In: Zeuthen, P. & Bogh-Sorensen, L. (Eds.), *Food Preservation Techniques*. Woodhead Publishing, Cambridge, UK: pp. 339–359.

Ooraikul, B. & Stiles, M. E. (1991) Review of the development of modified atmosphere packaging. In: Ooraikul, B. & Stiles, M. E. (Eds.), *Modified Atmosphere Packaging of Food*. Ellis Horwood, London, UK: pp. 1–18.

Orea, J. M. & Gonzalez-Ureña, A. (2002) Measuring and improving the natural resistance of fruit. In: Jongen, W. (Ed.), *Fruit and Vegetable Processing: Improving Quality*. Woodhead Publishing, Cambridge, UK: pp. 233–266.

Orna, M. V. & Stock, J. (1989) *Electrochemistry, Past and Present*. American Chemical Society, Columbus, OH.

Otero, L. & Sanz, P. D. (2003) Modelling heat transfer in high pressure food processing: A review. *Innovative Food Science and Emerging Technologies* **4**(2): 121–134.

Paine, F. A. (1991) *The Packaging User's Handbook*. Blackie Academic & Professional, London, UK.

Paine, F. A. & Paine, H. Y. (1992) *A Handbook of Food Packaging*. Blackie Academic & Professional, London, UK.

Palaniappan, S., Sastry, S. K. & Richter, E. R. (1990) Effects of electricity on microorganisms: A review. *Journal of Food Processing and Preservation* **14**(5): 393–414.

Palou, E., Lopez-Malo, A., Barbosa-Cánovas, G. V. & Swanson, B. G. (1999) High pressure treatment in food preservation. In: Rahman, M. S. (Ed.), *Handbook of Food Preservation*. Marcel Dekker, New York, NY: pp. 533–576.

Pardo, J. M. & Niranjan, K. (2006) Freezing. In: Brennan, J. G. (Ed.), *Food Processing Handbook*. Wiley-VCH, Weinheim, Germany: pp. 125–145.

Park, R. W. A., Griffiths, P. L. & Moreno, G. S. (1991) Sources and survival of campylobacters – Relevance to enteritis and the food industry. *Journal of Applied Bacteriology* **70**(Sup.): S97–S106.

Park, S. I., Daeschel, M. & Zhao, Y. (2004) Functional properties of antimicrobial lysozyme-chitosan composite films. *Journal of Food Science* **69**(8): M215–M221.

Park, S. I. & Zhao, Y. (2004) Incorporation of high concentration of mineral or vitamin into chitosan-based films. *Journal of Agricultural and Food Chemistry* **52**(7): 933–1939.

Parry, R. T. (1993) Introduction. In: Parry, R. T. (Ed.), *Principles and Applications of MAP of Foods*. Blackie Academic & Professional, New York, NY: pp. 1–18.

Patel, H. (2002) Metal detectors uncovered. *Food Science and Technology* **15**(4): 38–41.

Patterson, M. F., Ledward, D. A. & Rogers, N. (2006) High pressure processing. In: Brennan, J. G. (Ed.), *Food Processing Handbook*. Wiley-VCH, Weinheim, Germany: pp. 173–200.

Patterson, M. F., Quinn, M., Simpson, R. & Gilmour, A. (1995) Effects of high pressure on vegetative pathogens. In: Ledward, D. A., Johnson, D. E., Earnshaw, R. G. & Hasting, A. P. M. (Eds.), *High Pressure Processing of Foods*. Nottingham University Press, Nottingham, UK: pp. 47–64.

Paul, C. R. (2001) *Fundamentals of Electric Circuit Analysis*. Wiley, Hoboken, NJ.

Payne, S. R., Sandford, D., Harris, A. & Young, O. A. (1994) The effects of antifreeze proteins on chilled and frozen meat. *Meat Science* **37**(3): 429–438.

Perera, C. O. & Rahman, M. S. (1997) Can clever conveyors become more intelligent? *Trends in Food Science and Technology* **8**(3): 75–79.

Perreau, R. G. (1989) The chocolate moulding process. In: Turner, A. (Ed.), *Food Technology International Europe*. Sterling Publications, London, UK: pp. 73–76.

Petrova, T. S. (2009) Application of Bessel's functions in the modelling of chemical engineering processes. *Bulgarian Chemical Communications* **41**: 343–354.

Pfeiffer, C., d'Aujourd'hui, M., Walter, J., Nessli, J. & Fletcher, F. (1999) Optimising food packaging and shelf life. *Food Technology* **53**(6): 52–59.

Pham, Q. T. (1986) Simplified equation for predicting the freezing time of foodstuffs. *Journal of Food Technology* **21**(2): 209–219.

Pimentel, D. & Pimentel, M. (1996) *Food, Energy and Society*. University Press of Colorado, Niwot, CO.

Pine, F. A. (1991) *The Packaging User's Handbook*. Blackie Academic & Professional, London, UK.

Piringer, O.-G., Huber, M., Franz, R., Begley, T. H. & McNeal, T. P. (1998) Migration from food packaging containing a functional barrier: Mathematical and experimental evaluation. *Journal of Agricultural and Food Chemistry* **46**(4): 1532–1538.

Piyasena, P., Dussault, C., Koutchma, T., Ramaswamy, H. & Awuah, G. (2003) Radio frequency heating of foods: Principles, applications and related properties – A review. *Critical Reviews in Food Science and Nutrition* **43**(6): 587–606.

Piyasena, P., Mohareb, E. & McKellar, R. C. (2003) Inactivation of microbes using ultrasound: A review. *International Journal of Food Microbiology* **87**(3): 207–216.

Plaut, H. (1995) Brain boxed or simply packed. *Food Processing* **Jul**: 23–25.

Pothakamury, U. R., Barbosa-Cánovas, G. V. & Swanson, B. G. (1993) Magnetic field inactivation of microorganisms and generation of biological changes. *Food Technology* **47**(12): 85–93.

Pothakamury, U. R., Monsalve-Gonzalez, A., Barbosa-Cánovas, G. V. & Swanson, B. G. (1995) High voltage pulsed electric field inactivation of *Bacillus subtilis* and *Lactobacillus delbrueckii*. *Revista Española de Ciencia e Tecnología de Alimentos* **35**: 101–107.

Potter, N. N. & Hotchkiss, J. H. (1995) *Food Science*. Chapman & Hall, New York, NY.

Powell, J. & Steele, A. (1999) *The Packaging Regulations – Implications for Business*. Chandos Publishing, Oxford, UK.

Powers, T. H. & Calvo, W. J. (2003) Moisture regulation. In: Ahvenainen, R. (Ed.), *Novel Food Packaging Techniques*. Woodhead Publishing, Cambridge, UK: pp. 172–185.

Puupponen-Pimia, R., Hakkinen, S. T., Aarni, M., Suortti, T., Lampi, A.-M., Eurola, M., Phronen, V., Nuutila, A. M. & Oksman-Caldentey, K.-M. (2003) Blanching and long-term freezing affect various bioactive compounds of vegetables in different ways. *Journal of the Science of Food and Agriculture* **83**(14): 1389–1402.

Pyle, D. I. & Zarov, C. A. (1997) Process control. In: Fryer, P. J., Pyle, D. L. & Rielly, C. D. (Eds.), *Chemical Engineering for the Food Industry*. Blackie Academic & Professional, London, UK: pp. 250–294.

Qin, B. L., Pothakamury, U. R., Vega-Mercado, H., Martin-Belloso, O., Barbosa-Cánovas, G. V. & Swanson, B. G. (1995) Food pasteurisation using high-intensity pulsed electric fields. *Food Technology* **49**(12): 55–60.

Qin, B. L., Zhang, Q., Barbosa-Cánovas, G. V., Swanson, B. G. & Pedrow, P. D. (1994) Inactivation of microorganisms by pulsed electric fields with different voltage wave forms. *IEEE Transactions on Dielectrics and Electrical Insulation* **1**(6): 1047–1057.

Quartly-Watson, T. (1998) The importance of ultrasound in cleaning and disinfection in the poultry industry – A case study. In: Povey, M. J. W. & Mason, T. J. (Eds.), *Ultrasound in Food Processing*. Blackie Academic & Professional, London, UK: pp. 144–150.

Rahman, M. S. (1999) Food preservation by freezing. In: Rahman, M. S. (Ed.), *Handbook of Food Preservation*. Marcel Dekker, New York, NY: pp. 259–284.

Rahman, M. S. (1999) Light and sound in food preservation. In: Rahman, M. S. (Ed.), *Handbook of Food Preservation*. Marcel Dekker, New York, NY: pp. 669–686.

Rahman, M. S. (1999) Preserving foods with electricity: Ohmic heating. In: Rahman, M. S. (Ed.), *Handbook of Food Preservation*. Marcel Dekker, New York, NY: pp. 521–532.

Rahman, M. S., Guizani, N., al-Khaseibi, M., al-Hinai, S. A., al-Maskri, S. S. & al-Hamhami, K. (2002) Analysis of cooling curve to determine the end point of freezing. *Food Hydrocolloids* **16**(6): 653–659.

Ramesh, M. N. (1999) Food preservation by heat treatment. In: Rahman, M. S. (Ed.), *Handbook of Food Preservation*. Marcel Dekker, New York, NY: pp. 95–172.

Raso, J. & Barbosa-Cánovas, G. V. (2003) Nonthermal preservation of foods using combined processing techniques. *Critical Reviews in Food Science and Nutrition* **43**(3): 265–285.

Raso, J., Pagan, R., Condon, S. & Sala, F. J. (1998) Influence of temperature and pressure on the lethality of ultrasound. *Applied and Environmental Microbiology* **64**(2): 465–471.

Raso, J., Palop, A., Pagan, R. & Condon, S. (1998) Inactivation of *Bacillus subtilis* spores by combining ultrasonic waves under pressure and mild heat treatment. *Journal of Applied Microbiology* **85**(5): 849–854.

Rawson, F. F. (1998) An introduction to ultrasound food cutting. In: Povey, M. J. W. & Mason, T. J. (Eds.), *Ultrasound in Food Processing*. Blackie Academic & Professional, London, UK: pp. 254–270.

Reuter, H. (Ed.) (1989) *Aseptic Packaging of Foods*. Technomic Publishing, Lancaster, PA.

Rhim, J.-W. & Ng, P. K. W. (2007) Natural biopolymer-based nanocomposite films for packaging applications. *Critical Reviews in Food Science and Nutrition* **47**(4): 411–433.

Rickman, J. C., Barrett, D. M. & Bruhn, C. M. (2007) Review: Nutritional comparison of fresh, frozen and canned fruits and vegetables. Part I. Vitamins C and B and phenolic compounds. *Journal of the Science of Food and Agriculture* **87**(6): 930–944.

Rickman, J. C., Barrett, D. M. & Bruhn, C. M. (2007) Review: Nutritional comparison of fresh, frozen and canned fruits and vegetables. Part II. Vitamin A and carotenoids, vitamin E, minerals and fiber. *Journal of the Science of Food and Agriculture* **87**: 1185–1196.

Riemann, H. P. & Civer, D. O. (Eds.) (2006) *Foodborne Infections and Intoxications*. Elsevier, New York, NY.

Riva, M., Piergiovanni, S. & Schiraldi, A. (2001) Performances of time-temperature indicators in the study of temperature exposure of packaged fresh foods. *Packaging Technology and Science* **14**(1): 1–39.

Riverol, C., Carosi, F. & di Sanctis, C. (2004) The application of advanced techniques in a fluidised bed freezer for fruits: Evaluation of linguistic interpretation vs. stability. *Food Control* **15**(2): 93–97.

Roberts, C. M. & Hoover, D. G. (1996) Sensitivity of *Bacillus coagulans* spores to combinations of high hydrostatic pressure, heat acidity and nisin. *Journal of Applied Bacteriology* **81**(4): 363–368.

Robertson, G. L. (1990) Testing barrier properties of plastic films. In: Turner, A. (Ed.), *Food Technology International Europe*. Sterling Publications, London, UK: pp. 301–305.

Robertson, G. L. (1993) *Food Packaging: Principles and Practice*. Marcel Dekker, New York, NY.

Robinson, C. J. (1992) Form, fill and seal technology. In: Turner, A. (Ed.), *Food Technology International Europe*. Sterling Publications, London, UK: pp. 250–251.

Robinson, D. S. (1986) Irradiation of foods. *Proceedings of the Institute of Food Science and Technology* **19**(4): 165–168.

Robinson, E. H. (1974) The early diffusion of steam power. *Journal of Economic History* **34**(1): 91–107.

Roche, J. J. (1998) *The Mathematics of Measurement: A Critical History*. Springer, London, UK.

Rock, P. A. (1983) *Chemical Thermodynamics*. University Science Books, Mill Valley, CA.

Rodrigues, F. G. (2012) On equivalent expressions for the Faraday's law of induction. *Revista Brasileira para o Ensino de Física* **34**(1): 1–9.

Roller, S. (Ed.) (2003) *Natural Antimicrobials for Minimal Processing of Foods*. Woodhead Publishing, Cambridge, UK.

Rooney, M. L. (1995) Overview of active food packaging. In: Rooney, M. L. (Ed.), *Active Food Packaging*. Blackie Academic & Professional, London, UK: pp. 1–37.

Rose, D. (2000) Total Quality Management. In: Stringer, M. & Dennis, C. (Eds.), *Chilled Foods: A Comprehensive Guide*. Ellis Horwood, Chichester, UK: pp. 429–450.

Rosenberg, U. & Bogl, W. (1987) Microwave pasteurisation, sterilisation and pest control in the food industry. *Food Technology* **41**(6): 92–99.

Russotto, N. (1999) Plastics – The "quiet revolution" in the plastics industry. In: Turner, A. (Ed.), *Food Technology International Europe*. Sterling Publications, London, UK: pp. 67–69.

Ryall, A. L. & Lipton, W. J. (1972) Vegetables as living products. Respiration and heat production. In: *Transportation and Storage of Fruits and Vegetables*, vol. 1. AVI Publishing, Westport, CT.

Sá, M. M., Figueiredo, A. M. & Sereno, A. M. (1999) Glass transitions and state diagrams for fresh and processed apple. *Thermochimica Acta* **329**(1): 31–38.

Sahagian, M. E. & Goff, H. D. (1996) Fundamental aspects of the freezing process. In: Jeremiah, L. E. (Ed.), *Freezing Effects on Food Quality*. Marcel Dekker, New York, NY: pp. 1–50.

Sala, F. J., Burgos, J., Condon, S., Lopez, P. & Raso, J. (1995) Effect of heat and ultrasound on micro-organisms and enzymes. In: Gould, G. W. (Ed.), *New Methods of Food Preservation*. Blackie Academic & Professional, London, UK: pp. 176–204.

Saltveit, M. E. (2004) Respiratory metabolism. In: Gross, K. (Ed.), *The Commercial Storage of Fruits, Vegetables and Florist and Nursery Stocks*. Agricultural Handbook no. 66, Washington, DC.

Sampedro, F., Rodrigo, M., Marinez, A., Rodrigo, D. & Barbosa-Cánovas, G. V. (2005) Quality and safety aspects of PEF application in milk and milk products. *Critical Reviews in Food Science and Nutrition* **42**(1): 25–47.

Sanchez-Garcia, M. D., Gimenez, E. & Lagaron, J. M. (2007) Novel PET nanocomposites of interest in food packaging applications and comparative barrier performance with biopolyester nanocomposites. *Journal of Plastic Film and Sheeting* **23**(2): 133–148.

Sandeep, K. P. & Puri, V. M. (2001) Aseptic processing of foods. In: Irudayaraj, J. (Ed.), *Food Processing Operations Modelling*. Marcel Dekker, New York, NY: pp. 37–81.

San-Martin, M. F., Barbosa-Cánovas, G. V. & Swanson, B. G. (2002) Food processing by high hydrostatic pressure. *Critical Reviews in Food Science and Nutrition* **42**(6): 627–645.

Sanz, P. D. (2005) Freezing and thawing of foods under pressure. In: Barbosa-Cánovas, G. V., Tapia, M. S. & Cano, M. P. (Eds.), *Novel Food Processing Technologies*. Marcel Dekker, New York, NY: pp. 233–260.

Sanz, P. D. & Otero, L. (2005) High-pressure freezing. In: Sun, D.-W. (Ed.), *Emerging Technologies for Food Processing*. Academic Press, San Diego, CA: pp. 627–652.

Sarwar, M. & Armitage, A. W. (2003) Tooling requirements for glass container production for the narrow neck press and blow process. *Journal of Materials Processing Technology* **139**(1–3): 160–163.

Sautour, M., Soares-Mansur, C., Divies, C., Bensoussan, M. & Dantigny, P. (2002) Comparison of the effects of temperature and water activity on growth rate of food spoilage moulds. *Journal of Industrial Microbiology and Biotechnology* **28**(6): 311–315.

Schilthuizen, S. F. (2000) Communication with your packaging: Possibilities for intelligent functions and identification methods in packaging. *Packaging Technology and Science* **12**(5): 225–228.

Schmidt-Rohr, K. (2015) Why combustions are always exothermic, yielding about 418 kJ per mole of O_2. *Journal of Chemical Education* **92**(12): 2094–2099.

Schoen, H. M. & Byrne, C. H. (1972) Defrost indicators. *Food Technology* **26**(10): 46–50.

Schraut, O. (1989) Aseptic packaging of food into carton packs. In: Turner, A. (Ed.), *Food Technology International Europe*. Sterling Publications, London, UK: pp. 369–372.

Schwartzberg, H. G. (1976) Effective heat capacities for freezing and thawing of food. *Journal of Food Science* **41**(1): 152–156.

Schwartzberg, H. G., Singh, R. P. & Sarkar, A. (2007) Freezing and thawing of foods – Computation methods and thermal properties correlation. In: Yanniotis, S. & Sundén, B. (Eds.), *Heat Transfer in Food Processing*. WIT Press, Southampton, UK: pp. 61–100.

Scott, E. P., Carroad, P. A., Rumsey, T. R., Horn, J., Buhlert, J. & Rose, W. W. (1981) Energy consumption in steam blanchers. *Journal of Food Process Engineering* **5**(2): 77–88.

Sebok, A., Csepregi, I. & Baar, C. (1994) Causes of freeze cracking in fruits and vegetables. In: Turner, A. (Ed.), *Food Technology International*. Sterling Publications, London, UK: pp. 66–68.

Selke, S. E. (1994) Packaging options. In: Dalzell, J. M. (Ed.), *Food Industry and the Environment – Practical Issues and Cost Implications*. Aspen Publishers, New York, NY: pp. 253–290.

Selman, J. D. (1987) The blanching process. In: Thorne, S. (Ed.), *Developments in Food Processing*. Elsevier Applied Science, London, UK: pp. 205–249.

Selman, J. D. (1990) Process monitoring and control on-line in the food industry. *Food Control* **1**(1): 36–39.

Selman, J. D. (1995) Time-temperature indicators. In: Rooney, M. L. (Ed.), *Active Food Packaging*. Blackie Academic & Professional, London, UK: pp. 215–233.

Sensory, I. & Sastry, S. K. (2004) Ohmic blanching of mushrooms. *Journal of Food Process Engineering* **27**(1): 1–15.

Sequeira-Muñoz, A., Chevalier, D., Simpson, B. K., le Bail, A. & Ramaswamy, H. S. (2005) Effect of pressure-shift freezing versus air-blast freezing of carp (*Cyprinus carpio*) fillets: A storage study. *Journal of Food Biochemistry* **29**(5): 504–516.

Serway, R. A. & Jewett, J. W. (2004) *Physics for Scientists and Engineers, with Modern Physics*. Thomson Brooks/Cole, Belmont, CA.

Shafer, S. M. & Meredith, J. R. (1998) *Operations Management*. Wiley, New York, NY.

Shaikh, N. I. & Prabhu, V. (2007) Mathematical modeling and simulation of cryogenic tunnel freezers. *Journal of Food Engineering* **80**(2): 701–710.

Shaikh, N. I. & Prabhu, V. (2007) Model predictive controller for cryogenic tunnel freezers. *Journal of Food Engineering* **80**(2): 711–718.

Shapton, D. A. & Shapton, N. F. (1993) *H. J. Heinz Company Staff, Principles and Practice for the Safe Processing of Foods*. Woodhead Publishing, Cambridge, UK.

Shapton, D. A. & Shapton, N. F. (Eds.) (2000) *Principles and Practice for the Safe Processing of Foods*. Woodhead Publishing, Cambridge, UK.

Sharp, G. (1998) The conveyor collection. *Food Processing* **Sep**: 28–29.

Shephard, S. (2001) *Pickled, Potted, and Canned: How the Art and Science of Food Preserving Changed the World*. Simon & Schuster, New York, NY.

Singh, R. P. & Anderson, B. A. (2004) The major types of food spoilage: An overview. In: Steele, R. (Ed.), *Understanding and Measuring the Shelf-Life of Food*. Woodhead Publishing, Cambridge, UK: pp. 3–23.

Singh, R. P. & Heldman, D. R. (2001) *Introduction to Food Engineering*. Academic Press, London, UK.

Singh, R. P. & Mannapperuma, J. D. (1990) Developments in food freezing. In: Schwartzberg, H. & Rao, A. (Eds.), *Biotechnology of Food Process Engineering*. Marcel Dekker, New York, NY.

Sitzmann, W. (1995) High voltage pulsed techniques for food preservation. In: Gould, G. W. (Ed.), *New Methods of Food Preservation*. Blackie Academic & Professional, London, UK: pp. 236–252.

Sivertsvik, M. (2003) Active packaging in practice: Fish. In: Ahvenainen, R. (Ed.), *Novel Food Packaging Techniques*. Woodhead Publishing, Cambridge, UK: pp. 384–400.

Sizer, C. E. & Balasubramaniam, V. M. (1999) New intervention processes for minimally processed juices. *Food Technology* **53**(10): 64–67.

Skjoldebrand, C. & Scott, M. (1993) On-line measurement using NIR spectroscopy. In: Turner, A. (Ed.), *Food Technology International Europe*. Sterling Publications International, London, UK: pp. 115–117.

Skrede, G. (1996) Fruits. In: Jeremiah, L. E. (Ed.) *Freezing Effects on Food Quality*. Marcel Dekker, New York, NY: pp. 183–246.

Smallwood, M. (1986) Concentrating on natural proteins. *Food Processing* **Sep**: 21–22.

Smith, J. (1997) Energy management. *Food Processing* **Mar**: 16–17.

Smith, J. P., Phillips-Daifas, D. P., el-Khoury, W., Koukoutsis, J. & el-Khoury, A. (2004) Shelf life and safety concerns of bakery products – A review. *Critical Reviews in Food Science and Nutrition* **44**(1): 19–55.

Smith, J. P., Ramaswamy, H. S. & Simpson, K. (1990) Developments in food packaging technology. Part II: Storage aspects. *Trends in Food Science and Technology* **1**(5): 111–118.

Smith, R., Klemes, J. & Kim, J.-K. (2008) *Handbook of Water and Energy Management in Food Processing*. Woodhead Publishing, Cambridge, UK.

Smith, S. (1999) Multi-head marvels. *Food Processing* **Jan**: 16–17.

Smith, S. E. (1947) The sorption of water vapor by high polymers. *Journal of American Chemical Society* **69**(3): 646–651.

Smith-Palmer, A., Stewart, J. & Fyfe, L. (1998) Antimicrobial properties of plant essential oils and essences against five important food-borne pathogens. *Letters in Applied Microbiology* **26**(2): 118–122.

Smolander, M. (2003) The use of freshness indicators in packaging. In: Ahvenainen, R. (Ed.), *Novel Food Packaging Techniques*. Woodhead Publishing, Cambridge, UK: pp. 127–143.

Smolander, M., Hurme, E., Latva-Kala, K., Luoma, T., Alakomi, H.-L. & Ahvenainen, R. (2002) Myoglobin-based indicators for the evaluation of freshness of unmarinated broiler cuts. *Innovative Food Science and Emerging Technologies* **3**(3): 277–285.

Sommers, C. H. & Fan, X. (Eds.) (2006) *Food Irradiation Research and Technology*. Wiley-Blackwell, Oxford, UK.

Spencer, E. (1967) Estimating the size and cost of steam vacuum refrigeration. *Hydrocarbon Processing* **46**: 137–140.

Stabin, M. G. (2007) *Radiation Protection and Dosimetry: An Introduction to Health Physics*. Springer, Berlin, Germany.

Steltz, W. G. & Silvestri, G. J. (1958) The formulation of steam properties for digital computer application. *Transactions of American Association of Mechanical Engineers* **80**: 967–973.

Stephanopoulos, G. (1984) *Chemical Process Control: An Introduction to Theory and Practice*. Prentice Hall, Englewood Cliffs, NJ.

Stevenson, W. D. (1975) *Elements of Power System Analysis*. McGraw-Hill, New York, NY.

Stewart, B. (1995) *Packaging as an Effective Marketing Tool*. PIRA International, Leatherhead, UK.

Stumbo, C. R. (1973) *Thermobacteriology in Food Processing*. Academic Press, New York, NY.

Sugarman, C. (2004) Pasteurization redefined by USDA Committee, definition from the National Advisory Committee on Microbiological Criteria for Foods. *Food Chemical News* **46**(30): 21.

Summers, J. (1998) Cryogenics and tunnel vision. In: Turner, A. (Ed.), *Food Technology International*. Sterling Publications, London, UK: pp. 73–75.

Sun, D.-W. & Li, B. (2003) Microstructural change of potato tissues frozen by ultrasound-assisted immersion freezing. *Journal of Food Engineering* **5**(4): 337–345.

Sun, D.-W. & Wang, L.-J. (2001) Novel refrigeration cycles. In: Sun, D.-W. (Ed.), *Advances in Food Refrigeration*. Leatherhead Publishing, Leatherhead, UK: pp. 1–69.

Sun, D.-W. & Wang, L.-J. (2001) Vacuum cooling. In: Sun, D.-W. (Ed.), *Advances in Food Refrigeration*. Leatherhead Publishing, Leatherhead, UK: pp. 264–304.

Suslick, K. S. (1988) Homogeneous sonochemistry. In: Suslick, K. S. (Ed.), *Ultrasound: Its Chemical, Physical and Biological Effects*. VCH Publishers, New York, NY: pp. 123–163.

Sutherland, J. (2003) Modelling food spoilage. In: Zeuther, P. & Bogh-Sorensen, L. (Eds.), *Food Preservation Techniques*. Woodhead Publishing, Cambridge, UK: pp. 451–474.

Sutherland, J. & Varnam, A. (2002) Enterotoxin-producing *Staphylococcus*, *Shigella*, *Yersinia*, *Vibrio*, *Aeromonas* and *Plesiomonas*. In: Blackburn, C. W. & McClure, P. J. (Eds.), *Foodborne Pathogen Hazards, Risk Analysis and Control*. Woodhead Publishing, Cambridge, UK: pp. 386–415.

Suzuki, K. & Taniguchi, T. (1972) Effect of pressure on biopolymers and model systems. In: Sleigh, M. A. & MacDonald, A. G. (Eds.), *The Effect of Pressure on Living Organisms*. Academic Press, New York, NY: pp. 103–124.

Swain, M. & James, S. (2005) Thawing and tempering using microwave processing. In: Schubert, H. & Regier, M. (Eds.), *The Microwave Processing of Foods*. Woodhead Publishing, Cambridge, UK: pp. 174–191.

Tada, M. (2004) Foreword on food irradiation. *Foods and Food Ingredients Journal of Japan* **209**(2): 1.

Taoukis, P. S. & Labuza, T. P. (1989) Applicability of time-temperature indicators as shelf-life monitors of food products. *Journal of Food Science* **54**(4): 783–788.

Taoukis, P. S. & Labuza, T. P. (1989) Reliability of time-temperature indicators as food quality monitors under nonisothermal conditions. *Journal of Food Science* **54**(4): 789–792.

Taoukis, P. S. & Labuza, T. P. (2003) Time-temperature indicators (*TTIs*). In: Ahvenainen, R. (Ed.), *Novel Food Packaging Techniques*. Woodhead Publishing, Cambridge, UK.

Tebbutt, T. H. Y. (1992) *Principles of Water Quality Control*. Pergamon Press, Oxford, UK.

Teixeira, A. A. & Shoemaker, C. F. (1989) *Computerized Food Processing Operations*. AVI, Westport, CT.

Terrel, C. (Ed.) (1987) *American Electrician's Handbook*. McGraw-Hill, New York, NY.

Tersine, R. J. (1994) *Principles of Inventory and Material Management*. Prentice Hall, London, UK.

Tewari, G., Jayas, D. S. & Holley, R. A. (1999) High pressure processing of foods: An overview. *Sciences des Aliments* **19**: 619–661.

Tice, P. (2000) EC food contact legislation and how in the future it may be applied to lacquer-coated food and beverage cans. *British Food Journal* **102**(11): 856–871.

Timbers, G. E., Stark, R. & Cumming, D. B. (1984) A new blanching system for the food industry. I: Design, construction and testing of a pilot plant prototype. *Journal of Food Processing and Preservation* **8**(2): 115–133.

Tomlins, R. (1995) Cryogenic freezing and chilling of food. In: Turner, A. (Ed.), *Food Technology International*. Sterling Publications, London, UK: pp. 145–149.

Torreggiani, D., Lucas, T. & Raoult-Wack, A. (2000) The pretreatment of fruits and vegetables. In: Kennedy, C. (Ed.), *Managing Frozen Foods*. Woodhead Publishing, Cambridge, UK: pp. 57–80.

Trujillo, F. J. & Pham, Q. T. (2003) Modelling the chilling of the leg, loin and shoulder of beef carcasses using an evolutionary method. *International Journal of Refrigeration* **26**(2): 224–231.

van Tuil, R., Fowler, P., Lawther, M. & Weber, C. J. (2000) Properties of biobased packaging materials. In: Weber, C. J. (Ed.), *Biobased Packaging Materials for the Food Industry – Status and Perspectives*. Food Biopack Project, EU Directorate 12, Brussels, Belgium: pp. 13–44.

Turtle, B. I. (1990) PET containers for food and drink. In: Turner, A. (Ed.), *Food Technology International Europe*. Sterling Publications, London, UK: pp. 315–317.

Uz, M. & Altinkaya, S. A. (2011) Development of mono and multilayer antimicrobial food packaging materials for controlled release of potassium sorbate. *LWT – Food Science and Technology* **44**(10): 2302–2309.

Varnam, A. H. & Sutherland, J. P. (2001) *Milk and Milk Products: Technology, Chemistry and Microbiology*. Aspen Publishers, Gaitherburgh, MD.

del Vecchio, R. M., Poulin, B., Feghali, P. T. M., Shah, D. & Ahuja, R. (2002) *Transformer Design Principles: With Applications to Core-Form Power Transformers*. CRC Press, Boca Raton, FL.

Veerkamp, C. H. (2001) Chilling and freezing of poultry and poultry products. In: Sun, D.-W. (Ed.), *Advances in Food Refrigeration*. Leatherhead Publishing, Leatherhead, UK: pp. 387–402.

Vega-Mercado, H., Gongora-Nieto, M. M., Barbosa-Cánovas, G. V. & Swanson, B. G. (1999) Non-thermal preservation of liquid foods using pulsed electric fields. In: Rahman, M. S. (Ed.), *Handbook of Food Preservation*. Marcel Dekker, New York, NY: pp. 487–520.

Vega-Mercado, H., Martin-Belloso, O., Qin, B., Chang, F. J., Gongora-Nieto, M. M., Barbosa-Cánovas, G. V. & Swanson, B. G. (1997) Non-thermal preservation: Pulsed electric fields. *Trends in Food Science and Technology* **8**(5): 151–157.

Vega-Mercado, H., Pothakamury, U. R., Chang, F.-J., Barbosa-Cánovas, G. V. & Swanson, B. G. (1996) Inactivation of *Escherichia coli* by combining pH, ionic strength and pulsed electric field hurdles. *Food Research International* **29**(2): 117–121.

Venugopal, V., Doke, S. N. & Thomas, P. (1999) Radiation processing to improve the quality of fishery products. *Critical Reviews in Food Science and Nutrition* **39**(5): 391–440.

Vercet, A., Lopez, P. & Burgos, J. (1999) Inactivation of heat resistant pectinmethylesterase from orange by manothermosonication. *Journal of Agricultural and Food Chemistry* **47**(2): 432–437.

Vermeiren, L., Devlieghere, F., van Beest, M., de Kruijf, N. & Debevere, J. (1999) Developments in the active packaging of foods. *Trends in Food Science and Technology* **10**(3): 77–86.

Vine, R. P. (1987) The use of new technology in commercial winemaking. In: Turner, A. (Ed.), *Food Technology International Europe*. Sterling Publications, London, UK: pp. 146–149.

de Vlieger, J. J. (2003) Green plastics for food packaging. In: Ahvenainen, R. (Ed.), *Novel Food Packaging Techniques*. Woodhead Publishing, Cambridge, UK: pp. 519–534.

Wahid, M., Sattar, A., Jan, M. & Khan, I. (1989) Effect of combination methods on insect disinfestation and quality of dry fruits. *Journal of Food Processing and Preservation* **13**(1): 79–85.

Waldron, K. (Ed.) (2007) *Handbook of Waste Management and Co-Product Recovery in Food Processing*. Woodhead Publishing, Cambridge, UK.

Walin, P. & May, J. (1995) The developing appetite for neural computing. *Food Processing Hygiene Supplement* **Oct**: XII.

Walker, S. J. & Betts, G. (2000) Chilled foods microbiology. In: Stringer, M. & Dennis, C. (Eds.), *Chilled Foods – A Comprehensive Guide*. Woodhead Publishing, Cambridge, UK: pp. 153–186.

Walker, S. J. & Stringer, M. F. (1990) Microbiology of chilled foods. In: Gormley, T. R. (Ed.), *Chilled Foods – The State of the Art*. Elsevier Applied Science, London, UK: pp. 269–304.

Wallace, C. A. (2006) Safety in food processing. In: Brennan, J. G. (Ed.), *Food Processing Handbook*. Wiley, Hoboken, NJ: pp. 351–372.

Wallin, P. & Haycock, P. (1998) *Foreign Body Prevention, Detection and Control: A Practical Approach*. Blackie Academic & Professional, London, UK.

Wallin, R. & Chamberlain, S. (2000) Code on coding. *Food Processing* **Oct**: 31–32.

Warriner, K., Movahedi, S. & Waites, W. M. (2004) Laser-based packaging sterilization in aseptic processing. In: Richardson, P. (Ed.), *Improving the Thermal Processing of Foods*. Woodhead Publishing, Cambridge, UK: pp. 277–303.

Watson, G. N. (1952) *A Treatise on the Theory of Bessel Functions*. Cambridge University Press, Cambridge, UK.

Watson, H. E. (1908) A note on the variation of the rate of disinfection with change in the concentration of disinfectant. *Journal of Hygiene (Cambridge)* **8**(4): 536–542.

Webster, J. (2001) The logistics of food delivery. *Food Processing* **Dec**: 20.

Weller, M. & Mills, S. (1999) Frost-free cold stores? *Food Processing* **Aug**: 21–22.

Wheatley, A. D. (1994) Water pollution in the food industry: Sources, control and cost implications. In: Dalzell, J. M. (Ed.), *Food Industry and the Environment – Practical Issues and Cost Implications*. Blackie Academic & Professional, London, UK: pp. 137–258.

White, R. M. (1990) Package testing and food products. In: Turner, A. (Ed.), *Food Technology International Europe*. Sterling Publications, London, UK: pp. 295–299.

White, R. M. (1994) Disposal of used packaging. In: Dalzell, J. M. (Ed.), *Food Industry and the Environment – Practical Issues and Cost Implications*. Aspen Publishers, New York, NY: pp. 318–346.

Whittle, K. (2000) Wax – The versatile green solution. *Food Processing* **Jun**: 21–22.

Whyte, R., Hudson, J. A. & Graham, C. (2006) *Campylobacter* in chicken livers and their destruction by pan frying. *Letters in Applied Microbiology* **43**(6): 591–595.

Wilbey, R. A. (2006) Water and waste treatment. In: Brennan, J. G. (Ed.), *Food Processing Handbook*. Wiley-VCH, Weinheim, Germany: pp. 399–428.

Wilkinson, V. M. & Gould, G. W. (1996) *Food Irradiation: A Reference Guide*. Woodhead Publishing, Cambridge, UK.

Willhoft, E. M. A. (1993) *Aseptic Processing and Packaging of Particulate Foods*. Blackie Academic & Professional, London, UK.

Williams, A. (1998) Process control – The way ahead. *Food Processing* **Sep**: S4–S5.

Wilson, I. (2002) Fouling, cleaning and disinfection. *Food and Bioproducts Processing* **80**(4): 221–339.

Winston, J. (2000) Driving down delivery costs. *Food Processing* **Dec**: 18.

Withell, E. R. (1942) The evaluation of bactericides. *Journal of Hygiene (Cambridge)* **42**(4): 339–353.

Wong, C. W. (2013) *Introduction to Mathematical Physics: Methods and Concepts*. Oxford University Press, Oxford, UK.

Woods, K. (1993) Susceptors in microwaveable food packaging. In: Turner, A. (Ed.), *Food Technology International Europe*. Sterling Publications, London, UK: pp. 222–224.

Woolfe, M. L. (2000) Temperature monitoring and measurement. In: Stringer, M. & Dennis, C. (Eds.), *Chilled Foods – A Comprehensive Guide*. Woodhead Publishing, Cambridge, UK: pp. 99–134.

Wuytack, E. Y., Boven, S. & Michiels, C. W. (1998) Comparative study of pressure-induced germination of *Bacillus subtilis* spores at low and high pressures. *Applied and Systematic Environmental Microbiology* **64**(9): 3220–3224.

Xu, L. (1999) Use of ozone to improve the safety of fresh fruits and vegetables. *Food Technology* **53**(10): 58–62.

Yam, K. L. (2009) *Encyclopedia of Packaging Technology*. Wiley, New York, NY.

Yam, K. L. & Lee, D. S. (1995) Design of modified atmosphere packaging for fresh produce. In: Rooney, M. L. (Ed.), *Active Food Packaging*. Blackie Academic & Professional, London: pp. 55–73.

Yang, H.-J., Lee, S.-H., Jin, Y., Choi, J.-H., Han, C. H. & Lee, M.-H. (2005) Genotoxicity and toxicological effects of acrylamide on reproductive system in male rats. *Journal of Veterinary Science* **6**(2): 103–109.

Yang, T. C. S. (1998) Ambient storage. In: Taub, I. A. & Singh, R. P. (Eds.), *Food Storage Stability*. CRC Press, Boca Raton, FL: pp. 435–458.

Yano, Y., Nakayama, A., Kishihara, S. & Saito, H. (1998) Adaptative changes in membrane lipids of barophilic bacteria in response to changes in growth pressure. *Applied and Systematic Environmental Microbiology* **64**(2): 479–485.

Yao, L. C. (1967) Generalized numerical solutions of freezing a saturated liquid in cylinders and spheres. *AIChE Journal* **13**(1): 165–169.

Yeom, H. W., Streaker, C. B., Zhang, Q. H. & Min, D. B. (2000) Effects of pulsed electric fields on the activities of microorganisms and pectinmethyl esterase in orange juice. *Journal of Food Science* **65**(8): 1359–1363.

Yeom, H. W. & Zhang, Q. H. (2001) Enzymic inactivation by pulsed electric fields: A review. In: Barbosa-Cánovas, G. V. & Zhang, Q. H. (Eds.), *Pulsed Electric Fields in Food Processing*. Technomic Publishing, Lancaster, PA: pp. 57–64.

Yoichiro, Y., Hiroki, I. & Toyofumi, W. (2006) Development of laminated tin-free steel (TFS) "Universal Brite" Type F, for food can. *JFE Giho* **12**: 1–5.

Zagory, D. (1995) Ethylene-removing packaging. In: Rooney, M. L. (Ed.), *Active Food Packaging*. Blackie Academic & Professional, London: pp. 38–54.

Zaritzky, N. E. (2000) Factors affecting the stability of frozen foods. In: Kennedy, C. (Ed.), *Managing Frozen Foods*. Woodhead Publishing, Cambridge, UK: pp. 111–135.

Zhang, Q. H., Qin, B. L., Barbosa-Cánovas, G. V. & Swanson, B. G. (1995) Inactivation of *E. coli* for food pasteurization by high strength pulsed electric fields. *Journal of Food Processing and Preservation* **19**(2): 103–118.

Zhang, Q. H., Qiu, X. & Sharma, S. K. (1997) Recent developments in pulsed electric field processing. In: Anon. (Ed.), *New Technologies Yearbook*. National Food Processors Association, Washington, DC: pp. 31–42.

Zhao, Y., Flugstad, B., Kolbe, E., Park, J. W. & Wells, J. H. (2000) Using capacitive (radio frequency) dielectric heating in food processing and preservation – A review. *Journal of Food Process Engineering* **23**(1): 25–55.

Zheng, L. & Sun, D.-W. (2006) Innovative applications of power ultrasound during food freezing processes – A review. *Trends in Food Science and Technology* **17**(1): 16–23.

Zhu, S., Ramaswamy, H. S. & le Bail, A. (2005) Ice-crystal formation in gelatin gel during pressure shift versus conventional freezing. *Journal of Food Engineering* **66**(1): 69–76.

Zobel, M. G. R. (1988) Packaging and questions of flavour retention. In: Turner, A. (Ed.), *Food Technology International Europe.* Sterling Publications, London, UK: pp. 339–342.

4.2 Algebra

4.2.1 Averages

The arithmetic average, A_{arm}, of two real numbers, x and y, looks like

$$A_{arm} \equiv \frac{x+y}{2}, \quad (4.1)$$

while their geometric average, A_{gem}, reads

$$A_{gem} \equiv \sqrt{xy} \quad (4.2)$$

– after exchanging addition by multiplication, and division by 2 by square root. The harmonic average, A_{ham}, takes advantage of reciprocals, i.e.

$$A_{ham} \equiv \frac{2}{\frac{1}{x}+\frac{1}{y}}; \quad (4.3)$$

whereas the logarithmic average, A_{lom}, is defined by

$$A_{lom} \equiv \frac{x-y}{\ln\frac{x}{y}}. \quad (4.4)$$

The arithmetic, geometric, and harmonic averages can readily be extended so as to encompass N real numbers, $x_1, x_2, \ldots, x_N$, according to

$$A_{arm} \equiv \frac{\sum_{i=1}^{N} x_i}{N}, \quad (4.5)$$

$$A_{gem} \equiv \sqrt[N]{\prod_{i=1}^{N} x_i}, \quad (4.6)$$

and

$$A_{ham} \equiv \frac{N}{\sum_{i=1}^{N}\frac{1}{x_i}}, \quad (4.7)$$

based on Eqs. (4.1), (4.2), and (4.3), respectively. It is sometimes convenient to ascribe different weights w_is to the datum points x_is – in which case Eq. (4.5) would turn to

$$A_{arm,w} \equiv \frac{\sum_{i=1}^{N} w_i x_i}{\sum_{i=1}^{N} w_i}, \quad (4.8)$$

where $\sum_{i=1}^{N} w_i$ plays the role of $N=\sum_{i=1}^{N} 1$ in Eq. (4.5); $A_{arm,w}$ is accordingly termed weighed (arithmetic) average.

4.2.2 Asymptotic behaviors

Involved functions do often exhibit simple linear behaviors near specific finite value(s), or when their independent variable grows unbounded (toward either $-\infty$ or $+\infty$). Such a driving line is termed asymptote; in the latter case, it represents the straight line that drives the function curve at infinity. A vertical asymptote of function $f\{x\}$, at $x=c$, is accordingly defined by

$$\lim_{x\to c^-} f\{x\} = \pm\infty \quad (4.9)$$

or

$$\lim_{x\to c^+} f\{x\} = \pm\infty; \quad (4.10)$$

typical examples of c are zeros of denominator (or poles) of rational functions, or value(s) that turn nil the argument of a logarithmic function. Oblique asymptotes, described by $y=a+bx$, abide, in turn, to

$$\lim_{x\to\pm\infty} \left(f\{x\}-(a+bx)\right)=0 \quad (4.11)$$

– which, in particular, will be horizontal if $b=0$. Division of both sides by x transforms Eq. (4.11) to

$$\lim_{x\to\pm\infty}\left(\frac{f\{x\}}{x}-\frac{a}{x}-b\right)=0, \quad (4.12)$$

because $0/x=0$ for $x\neq 0$ (as is the case); a/x becoming, in turn, negligible when $x\to\pm\infty$ permits simplification to

$$b=\lim_{x\to\pm\infty}\frac{f\{x\}}{x}. \quad (4.13)$$

If the limit described by Eq. (4.13) does not exist, then there is no oblique asymptote in that direction; otherwise, one may proceed and compute a from Eq. (4.11) via

$$a=\lim_{x\to\pm\infty}\left(f\{x\}-bx\right), \quad (4.14)$$

with b given by Eq. (4.13). Although the concept of asymptote is extensible to other polynomial forms (e.g. quadratic) using essentially the same rationale, their determination and usefulness are quite limited – so their use is far less common.

4.2.3 Absolute value

4.2.3.1 Definition

Absolute value, or modulus, $|x|$, of a real value x is defined as

$$\begin{aligned} |x|\big|_{x\geq 0} &\equiv x \\ |x|\big|_{x<0} &\equiv -x \end{aligned}; \quad (4.15)$$

hence, this function yields a non-negative value, irrespective of the sign of its argument – as made explicit in Fig. 4.1. It should be stressed that $|x|$ holds the same value for two distinct real numbers that differ only in their sign (except in the case of zero). Hence, it entails an even function, i.e. it exhibits graphical symmetry with regard to the vertical axis.

4.2.3.2 Properties

Based on the definition conveyed by Eq. (4.15), it is easily proven that

$$|xy| = |x||y|, \tag{4.16}$$

i.e. modulus of product coincides with product of moduli of the factors. By the same token,

$$|x+y| \leq |x| + |y|, \tag{4.17}$$

so modulus of sum cannot exceed sum of moduli of its terms – while one realizes that

$$|x-y| \geq |x| - |y|, \tag{4.18}$$

i.e. the modulus of difference of two terms is equal to, or greater than difference of their corresponding moduli.

4.2.4 Impulse function

The instantaneous pulse, or Dirac's function is defined as

$$\begin{aligned} \delta\{x\}\big|_{x\neq 0} &\equiv 0 \\ \delta\{x\}\big|_{x=0} &\equiv \infty, \end{aligned} \tag{4.19}$$

together with

$$\int_{-\infty}^{\infty} \delta\{x\}dx \equiv 1; \tag{4.20}$$

this function is useful as mathematical descriptor of an instantaneous injection – and is depicted in Fig. 4.2. Note the finite value of the integral, labeled as Eq. (4.20), covering the whole domain of Dirac's function – despite the intrinsic singularity exhibited by its kernel at $x=0$, as per Eq. (4.19).

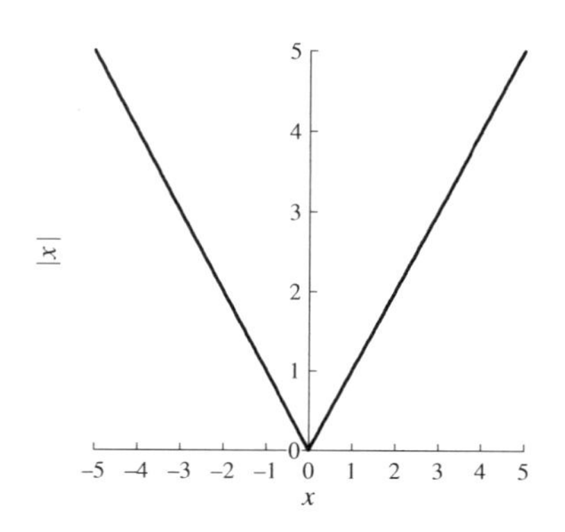

FIGURE 4.1 Variation of absolute value, $|x|$, versus its real argument, x.

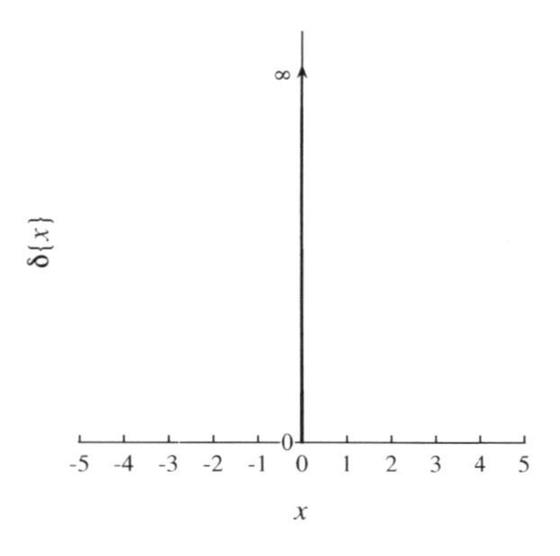

FIGURE 4.2 Variation of Dirac's function, $\delta\{x\}$, versus its real argument, x.

4.2.5 Step function

The step pulse, or Heaviside's function satisfies

$$\begin{aligned} \mathrm{H}\{x\}\big|_{x<0} &\equiv 0 \\ \mathrm{H}\{x\}\big|_{x\geq 0} &\equiv 1; \end{aligned} \tag{4.21}$$

Eq. (4.21) describes a constant function equal to 0 when $x<0$, and again a constant function but equal to 1 when $x>0$ – which implies an instantaneous jump from 0 to 1 at $x=0$, as would happen after an on/off valve switch. These features are apparent in Fig. 4.3; Heaviside's function may also be seen as the integral of $\delta\{x\}$, between $-\infty$ and x, see Eq. (4.21) *vis-à-vis* with Eqs. (4.19) and (4.20).

4.2.6 Exponential function

4.2.6.1 Definition

The exponential function is defined as a power where the independent variable appears in its exponent; should the base be constant and equal to Neper's number, then the resulting plot looks like Fig. 4.4. Note the monotonically increasing behavior, at a faster and faster pace; the exclusively positive values taken by this function; its lying below unity, when the argument is negative; and its horizontal asymptote, consistent with

$$\lim_{x\to-\infty} e^x = 0. \tag{4.22}$$

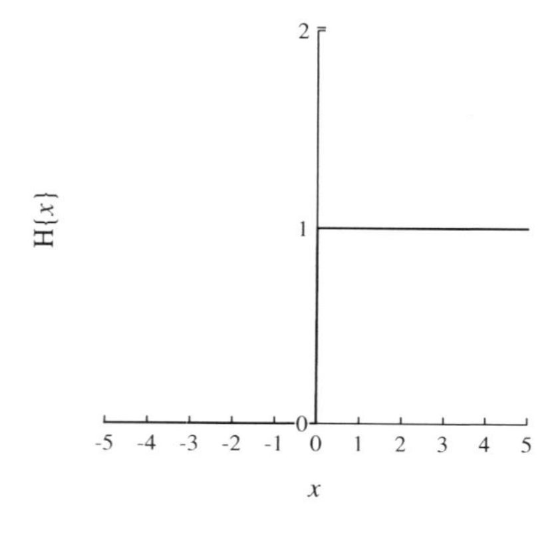

FIGURE 4.3 Variation of Heaviside's function, $\mathrm{H}\{x\}$, versus its real argument, x.

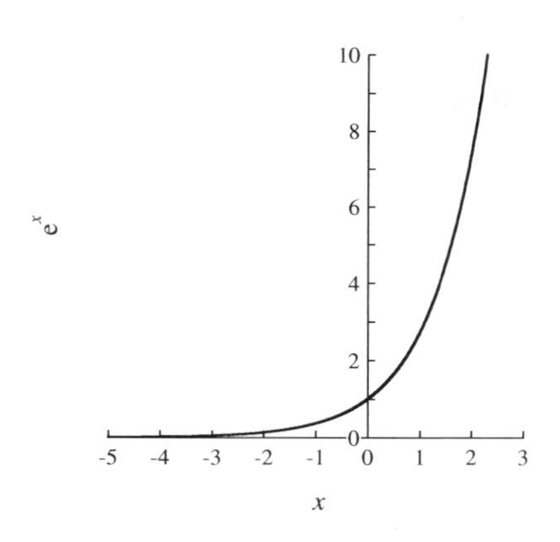

FIGURE 4.4 Variation of (natural) exponential, e^x, versus its real argument, x.

4.2.6.2 Neper's number

Neper's number, e, is defined as

$$e \equiv \lim_{n\to\infty}\left(1+\frac{1}{n}\right)^n, \tag{4.23}$$

where n denotes an integer variable; an associated result reads

$$\lim_{x\to\infty}\left(1+\frac{k}{x}\right)^x = e^k, \tag{4.24}$$

with real variable x and real constant k – obtained after unity in the numerator of the second term of Eq. (4.23) has been replaced by k, and n and x have been swapped as well.

4.2.6.3 Properties

If $f\{x\}$ and $g\{x\}$ denote two real functions, then

$$\exp\{f\{x\}\}\exp\{g\{x\}\} = \exp\{f\{x\}+g\{x\}\}; \tag{4.25}$$

hence, product of exponentials equals exponential of sum of their arguments. When the power, with exponent $g\{x\}$, is taken of the exponential of $f\{x\}$, one gets

$$\exp\{f\{x\}\}^{g\{x\}} = \exp\{f\{x\}g\{x\}\} \tag{4.26}$$

– meaning that the argument of the new exponential function is but the product of the argument of the original exponential by its exponent. The ratio of the exponential of $f\{x\}$ to that of $g\{x\}$ abides to

$$\frac{\exp\{f\{x\}\}}{\exp\{g\{x\}\}} = \exp\{f\{x\}-g\{x\}\}, \tag{4.27}$$

which may be seen as a particular case of Eq. (4.25) – when $g\{x\}$ is replaced by its reciprocal in the left-hand side, and thus $g\{x\}$ is replaced by its negative in the right-hand side. When an exponential of base a is to be converted to an exponential of base b, one should proceed as

$$a^{f\{x\}} = b^{\log_b a^{f\{x\}}}, \tag{4.28}$$

in view of the definition of logarithm as inverse of exponential; this is equivalent to

$$a^{f\{x\}} = b^{f\{x\}\log_b a}, \tag{4.29}$$

after applying operational properties of logarithms (see below).

4.2.6.4 Euler's form of complex numbers

A complex number $a+\iota b$, with real a and b complemented by $\iota \equiv \sqrt{-1}$, may be represented in exponential form as

$$e^{\iota\theta} = \cos\theta + \iota\sin\theta \tag{4.30}$$

– provided that $\sqrt{a^2+b^2}=1$ and $\theta = \tan^{-1} b/a$; Eq. (4.30) is known as Euler's form of a complex number. When the n-th power of both sides is taken, Eq. (4.30) becomes

$$\left(\cos\theta + \iota\sin\theta\right)^n = \left(e^{\iota\theta}\right)^n, \tag{4.31}$$

which is equivalent to writing

$$(\cos\theta + \iota\sin\theta)^n = e^{\iota(n\theta)} = \cos n\theta + \iota\sin n\theta \tag{4.32}$$

in view of Eq. (4.26) – and where Eq. (4.30) was used as template; Eq. (4.32) is known as de Moivre's formula.

4.2.7 Logarithmic function

4.2.7.1 Definition

The logarithm function is defined as the inverse of the exponential function; in other words, $\log_a x=b$ means that $a^b=x$. When the base selected is again Neper's number, $\log_e x$ is denoted as $\ln x$ by convention; the corresponding curve is depicted in Fig. 4.5. This function always increases, and takes negative values when its argument does not exceed unity; a vertical asymptote can be grasped, viz.

$$\lim_{x\to 0^+} \ln x = -\infty. \tag{4.33}$$

The plot of $\ln x$ may be generated from that of e^x in Fig. 4.4 via a 180° rotation of the curve around the bisector line – or a mere exchange of axes. As a consequence of logarithm being the inverse function of exponential with the same base, one obtains

$$\ln e^x = e^{\ln x} = x \tag{4.34}$$

upon composing the said functions.

4.2.7.2 Properties

If $f\{x\}$ and $g\{x\}$ denote two functions, then

$$\ln f\{x\}g\{x\} = \ln f\{x\} + \ln g\{x\}; \tag{4.35}$$

in other words, logarithm of product of functions was converted to sum of their logarithms. One also finds that

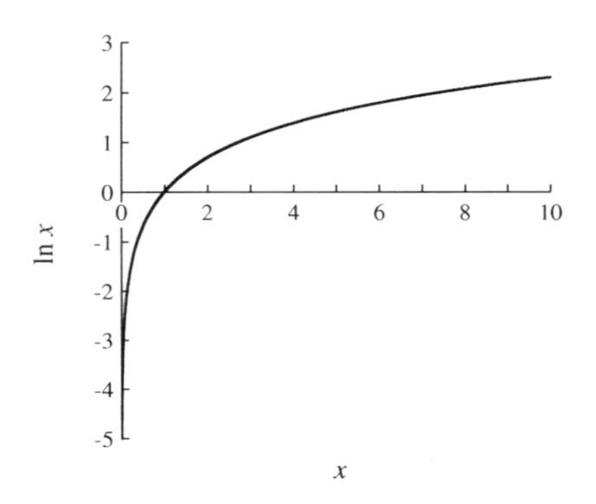

FIGURE 4.5 Variation of (natural) logarithm, ln x, versus its real argument, x.

$$\ln f\{x\}^{g\{x\}} = g\{x\} \ln f\{x\} \tag{4.36}$$

– i.e. logarithm of power involving functions both as base and exponent coincides with product of exponent function by logarithm of base function; this justifies equivalence between Eqs. (4.28) and (4.29). The ratio of functions unfolds

$$\ln \frac{f\{x\}}{g\{x\}} = \ln f\{x\} - \ln g\{x\}; \tag{4.37}$$

hence, logarithm transforms quotient of functions to difference of their logarithms. Equation (4.37) also supports

$$\ln \frac{1}{g\{x\}} = -\ln g\{x\}, \tag{4.38}$$

when $f\{x\}$ is replaced by unity – since ln 1 is nil; therefore, logarithm of reciprocal equals negative of logarithm of original function. The concept of logarithm in a given base may be extended to another base, after realizing that

$$f\{x\} = a^{\log_a f\{x\}}, \tag{4.39}$$

consistent with Eq. (4.28); this is so because a-based exponential and logarithm are inverse functions of each other. If b-based logarithms are now taken of both sides, then Eq. (4.39) produces

$$\log_b f\{x\} = \log_b a^{\log_a f\{x\}} \tag{4.40}$$

or, equivalently,

$$\log_b f\{x\} = \log_a f\{x\} \log_b a \tag{4.41}$$

upon application of Eq. (4.36) – with exponent being itself a logarithm in this case.

4.2.8 Hyperbolic functions

4.2.8.1 Definition

The hyperbolic sine, of a real variable x, is defined via

$$\sinh x \equiv \frac{e^x - e^{-x}}{2}, \tag{4.42}$$

whereas the hyperbolic cosine reads

$$\cosh x \equiv \frac{e^x + e^{-x}}{2}; \tag{4.43}$$

Eqs. (4.42) and (4.43) are sketched in Fig. 4.6i and ii, respectively. Note the even nature of cosh x, i.e. cosh$\{-x\}$ coincides with cosh x; and the odd nature of sinh x, i.e. sinh$\{-x\}$ is equal to –sinh x. The hyperbolic tangent accordingly abides to

$$\tanh x \equiv \frac{\sinh x}{\cosh x}, \tag{4.44}$$

where insertion of Eqs. (4.42) and (4.43) unfolds

$$\tanh x = \frac{\dfrac{e^x - e^{-x}}{2}}{\dfrac{e^x + e^{-x}}{2}} \tag{4.45}$$

– or else

$$\tanh x = \frac{e^x - e^{-x}}{e^x + e^{-x}}, \tag{4.46}$$

upon elimination of common factors between numerator and denominator; Eq. (4.46) is plotted in Fig. 4.6iii. Ordered addition of Eqs. (4.42) and (4.43) ensues, viz.

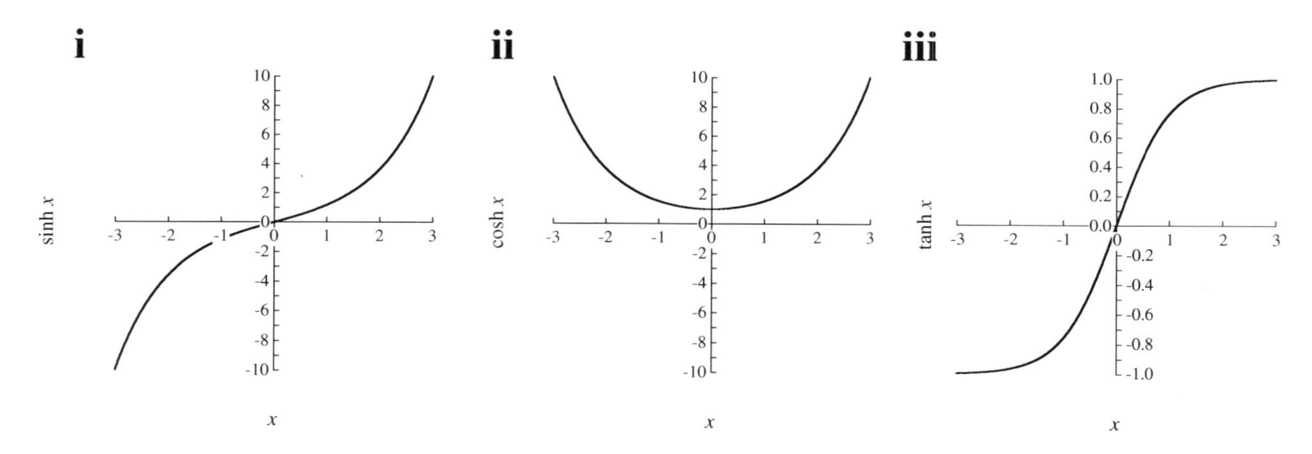

FIGURE 4.6 Variation of (i) hyperbolic sine, sinh x, (ii) hyperbolic cosine, cosh x, and (iii) hyperbolic tangent, tanh x, versus their real argument, x.

$$\sinh x+\cosh x=\frac{e^{x}-e^{-x}}{2}+\frac{e^{x}+e^{-x}}{2}, \tag{4.47}$$

where the two terms in the right-hand side may be lumped as

$$\sinh x+\cosh x=\frac{e^{x}-e^{-x}+e^{x}+e^{-x}}{2}; \tag{4.48}$$

after dropping symmetrical terms, one is left with

$$\sinh x+\cosh x=\frac{e^{x}+e^{x}}{2}=\frac{2e^{x}}{2}, \tag{4.49}$$

and finally

$$e^{x}=\sinh x+\cosh x \tag{4.50}$$

upon cancellation of common terms between numerator and denominator. By the same token, ordered subtraction of Eq. (4.42) from Eq. (4.43) yields

$$\cosh x-\sinh x=\frac{e^{x}+e^{-x}}{2}-\frac{e^{x}-e^{-x}}{2}, \tag{4.51}$$

where the terms in the right-hand side may again be lumped as

$$\cosh x-\sinh x=\frac{e^{x}+e^{-x}-e^{x}+e^{-x}}{2}; \tag{4.52}$$

once e^x is collapsed with its negative, Eq. (4.52) reduces to

$$\cosh x-\sinh x=\frac{e^{-x}+e^{-x}}{2}=\frac{2e^{-x}}{2} \tag{4.53}$$

– which, in turn, degenerates to

$$e^{-x}=\cosh x-\sinh x, \tag{4.54}$$

should common factors be cancelled out between numerator and denominator.

4.2.8.2 Relationships between hyperbolic functions

Based on their definition, one obtains

$$\cosh^2 x-\sinh^2 x=1 \tag{4.55}$$

that resembles the fundamental relationship of trigonometry; as well as

$$\sinh\{x+y\}=\sinh x\cosh y+\cosh x\sinh y \tag{4.56}$$

and

$$\sinh\{x-y\}=\sinh x\cosh y-\cosh x\sinh y, \tag{4.57}$$

pertaining to the hyperbolic sine of a sum or a difference as argument – and likewise

$$\cosh\{x+y\}=\cosh x\cosh y+\sinh x\sinh y \tag{4.58}$$

and

$$\cosh\{x-y\}=\cosh x\cosh y-\sinh x\sinh y, \tag{4.59}$$

encompassing its hyperbolic cosine counterpart. The corresponding formulae encompassing the hyperbolic tangent of a sum or difference taken for argument read

$$\tanh\{x+y\}=\frac{\tanh x+\tanh y}{1+\tanh x\tanh y} \tag{4.60}$$

and

$$\tanh\{x-y\}=\frac{\tanh x-\tanh y}{1-\tanh x\tanh y}. \tag{4.61}$$

Inverse hyperbolic functions are useful in engineering practice; in the case of $\sinh^{-1} x$, one finds that

$$\sinh^{-1} x=\ln\left\{x+\sqrt{1+x^2}\right\}, \tag{4.62}$$

and similarly

$$\cosh^{-1} x=\ln\left\{x\pm\sqrt{x^2-1}\right\} \tag{4.63}$$

for inverse hyperbolic cosine, as long as $x > 1$; complemented by

$$\tanh^{-1} x=\frac{1}{2}\ln\frac{1+x}{1-x} \tag{4.64}$$

for $-1<x<1$, pertaining to hyperbolic tangent.

4.2.9 Trigonometric functions

4.2.9.1 Definition

Consider the right triangle depicted in Fig. 4.7. If x denotes an angle (in rad) of the said triangle, b denotes length its opposite side, and c denotes length of hypotenuse as one of its adjacent sides, then sine of x is defined as

$$\sin x\equiv\frac{b}{c}; \tag{4.65}$$

this function is graphically depicted in Fig. 4.8i. Note the periodic behavior with period 2π, i.e.

$$\sin\{x+2k\pi\}=\sin x;\quad k=\ldots,-3,-2,-1,0,1,2,3,\ldots; \tag{4.66}$$

and [–1,1] serving as counterdomain. The cosine of x is, in turn, given by

$$\cos x\equiv\frac{a}{c}, \tag{4.67}$$

based also on Fig. 4.7; here a denotes length of the other side adjacent to angle x. Once again, −1 plays the role of lower bound and 1 of upper bound to the cosine function; along with a similar periodic pattern, viz.

$$\cos\{x+2k\pi\}=\cos x;\quad k=\ldots,-3,-2,-1,0,1,2,3,\ldots, \tag{4.68}$$

apparent in Fig. 4.8ii. The tangent of x satisfies

$$\tan x\equiv\frac{b}{a} \tag{4.69}$$

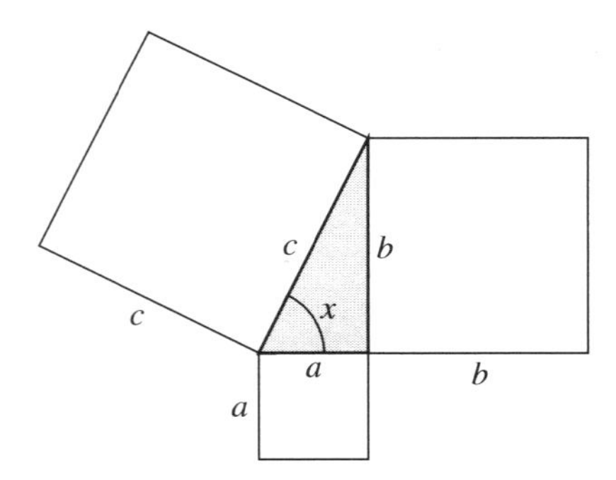

FIGURE 4.7 Graphical illustration of Pythagoras' theorem, as squares juxtaposed to the sides of a right triangle – of length a, b, and c, and angle of amplitude x between first and third sides.

as per Fig. 4.7, i.e. the ratio of lengths of opposite to adjacent sides associated to angle of amplitude x – and is graphically depicted in Fig. 4.8iii. The period is, in this case, π, according to

$$\tan\{x+k\pi\}=\tan x;\quad k=\ldots,-3,-2,-1,0,1,2,3,\ldots; \tag{4.70}$$

vertical asymptotes now exist, described by

$$\lim_{x\to\pm\frac{k\pi}{2}}\tan x=\pm\infty;\quad k=\ldots,-3,-2,-1,0,1,2,3,\ldots. \tag{4.71}$$

Equations (4.65) and (4.67) permit transformation of Eq. (4.69) to

$$\tan x=\frac{c\sin x}{c\cos x}, \tag{4.72}$$

which simplifies to

$$\tan x=\frac{\sin x}{\cos x} \tag{4.73}$$

upon cancellation of c between numerator and denominator. Sine and cosine are complementary functions, i.e.

$$\cos x=\sin\left\{\frac{\pi}{2}-x\right\}; \tag{4.74}$$

they accordingly hold a phase shift of $\pi/2$, relative to each other. One also finds that

$$\sin\{-x\}=-\sin x, \tag{4.75}$$

in view of sine being an odd function – coupled with

$$\sin\{\pi-x\}=\sin x; \tag{4.76}$$

and likewise

$$\cos\{-x\}=\cos x, \tag{4.77}$$

since cosine is an even function – further to

$$\cos\{\pi-x\}=-\cos x. \tag{4.78}$$

4.2.9.2 Fundamental theorem of trigonometry

A fundamental relationship exists, classically known as Pythagorean theorem, encompassing the three sides of any right triangle – namely the square of the hypotenuse (or side opposite to the right angle), of length c, equals the sum of the squares of the other two sides, of lengths a and b; in other words,

$$c^2=a^2+b^2, \tag{4.79}$$

as geometrically explained also in Fig. 4.7. When the said hypotenuse has unit length and x denotes one of the acute angles of the corresponding right triangle, then Eq. (4.79) degenerates to

$$\sin^2 x+\cos^2 x=1 \tag{4.80}$$

in agreement with Eqs. (4.65) and (4.67); this is referred to as fundamental theorem of trigonometry. Once in possession of Eq. (4.80), an alias may be obtained through division of both sides by $\sin^2 x$, viz.

$$1+\left(\frac{\cos x}{\sin x}\right)^2=\left(\frac{1}{\sin x}\right)^2; \tag{4.81}$$

this is equivalent to writing

$$1+\operatorname{cotan}^2 x=\operatorname{cosec}^2 x, \tag{4.82}$$

since function cotangent is defined as

$$\operatorname{cotan} x\equiv\frac{\cos x}{\sin x}\equiv\cos x\operatorname{cosec} x=\frac{1}{\tan x} \tag{4.83}$$

in agreement with Eq. (4.73) – and with cosecant being given by reciprocal of sine. If division of both sides by $\cos^2 x$ is performed, then Eq. (4.80) becomes instead

i

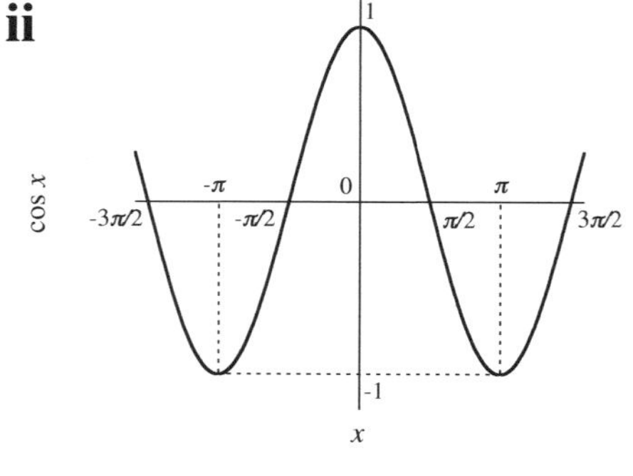

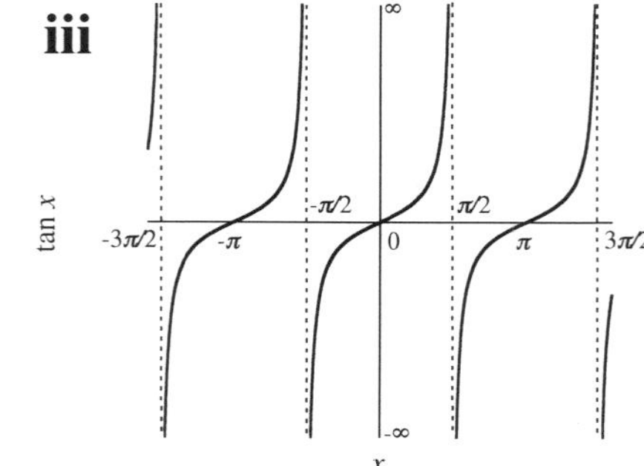

FIGURE 4.8 Variation of (i) sine, sin x, (ii) cosine, cos x, and (iii) tangent, tan x, versus their real argument, x.

$$\left(\frac{\sin x}{\cos x}\right)^2 + 1 = \left(\frac{1}{\cos x}\right)^2 \tag{4.84}$$

– which can be condensed to

$$1 + \tan^2 x = \sec^2 x, \tag{4.85}$$

at the expense of Eq. (4.73) coupled with

$$\sec x \equiv \frac{1}{\cos x} \tag{4.86}$$

pertaining to function cosecant.

4.2.9.3 Relationships between trigonometric functions

The cosine of an angle, which can be expressed as a sum or a difference of another set of angles, is obtainable at the expense of the corresponding sines and cosines, according to

$$\cos\{x - y\} = \cos x \cos y + \sin x \sin y \tag{4.87}$$

and

$$\cos\{x + y\} = \cos x \cos y - \sin x \sin y; \tag{4.88}$$

by the same token,

$$\sin\{x + y\} = \sin x \cos y + \cos x \sin y \tag{4.89}$$

and

$$\sin\{x - y\} = \sin x \cos y - \cos x \sin y \tag{4.90}$$

apply to sine of algebraic sums of angles. After having taken Eqs. (4.87) and (4.90) into account, together with Eq. (4.73), one gets

$$\tan\{x - y\} = \frac{\tan x - \tan y}{1 + \tan x \tan y} \tag{4.91}$$

– whereas combination of Eqs. (4.88) and (4.89) with Eq. (4.73) unfolds

$$\tan\{x + y\} = \frac{\tan x + \tan y}{1 - \tan x \tan y}. \tag{4.92}$$

Algebraic combination of Eqs. (4.87)–(4.90) supports

$$\cos x + \cos y = 2\cos\frac{x+y}{2}\cos\frac{x-y}{2} \tag{4.93}$$

and

$$\cos x - \cos y = 2\sin\frac{x+y}{2}\sin\frac{y-x}{2}, \tag{4.94}$$

pertaining to sum and difference of cosine functions, respectively; while sum and difference of sine functions look like

$$\sin x + \sin y = 2\sin\frac{x+y}{2}\cos\frac{x-y}{2} \tag{4.95}$$

and

$$\sin x - \sin y = 2\cos\frac{x+y}{2}\sin\frac{x-y}{2}, \tag{4.96}$$

respectively. A linear combination of sine and cosine, sharing the same argument, may in turn be coined as

$$a\sin x + b\cos x = \sqrt{a^2 + b^2}\,\sin\left\{x + \tan^{-1}\frac{b}{a}\right\}, \tag{4.97}$$

with constant a and b – where $\sqrt{a^2 + b^2}$ is often termed modulus, and $\tan^{-1} b/a$ is referred to as phase shift.

Furthermore, de Moivre's formula, labeled as Eq. (4.32), supports expression of powers of trigonometric functions of an angle via those of multiples of the said angle, according to

$$\cos^{2n} x = \frac{\binom{2n}{n} + 2\sum_{i=0}^{n-1}\binom{2n}{i}\cos 2(n-i)x}{4^n} \tag{4.98}$$

applying to an even exponent of cosine, and likewise

$$\cos^{2n+1} x = \frac{\sum_{i=0}^{n}\binom{2n+1}{i}\cos\big(2(n-i)+1\big)x}{4^n} \tag{4.99}$$

suitable for an odd exponent; one likewise gets

$$\sin^{2n} x = \frac{\binom{2n}{n} + 2\sum_{i=0}^{n-1}(-1)^{n-i}\binom{2n}{i}\cos 2(n-i)x}{4^n} \tag{4.100}$$

pertaining to an even exponent of sine, and likewise

$$\sin^{2n+1} x = \frac{\sum_{i=0}^{n}(-1)^{n-i}\binom{2n+1}{i}\sin\big(2(n-i)+1\big)x}{4^n} \tag{4.101}$$

when the said exponent is odd.

4.2.10 Series

4.2.10.1 Arithmetic series

Consider a summation containing n terms – where each term, u_i, equals the previous one, u_{i-1}, added to a constant value, k, according to

$$\sum_{i=0}^{n} u_i \equiv u_0 + (u_0 + k) + (u_0 + 2k) + \ldots + (u_0 + ik) + \ldots + \big(u_0 + (n-1)k\big) + (u_0 + nk) \tag{4.102}$$

or, equivalently,

$$\sum_{i=0}^{n} u_i \equiv \sum_{i=0}^{n}(u_0 + ik); \tag{4.103}$$

$\sum_{i=0}^{n} u_i$ is termed a (finite) arithmetic progression, or arithmetic series – of increment k and first term u_0. Algebraic manipulation of Eq. (4.103) yields

$$\sum_{i=0}^{n} u_i = (n+1)\left(u_0 + \frac{n}{2}k\right), \tag{4.104}$$

pertaining to sum of the first n terms of the said series – valid irrespective of the actual values of u_0, k, or n.

4.2.10.2 Geometric series

A geometric series occurs when each term of a summation is obtained from the previous one via multiplication by a constant parameter, k, viz.

$$\sum_{i=0}^{n} u_i \equiv u_0 + u_0 k + u_0 kk + \ldots + u_0 k^{i-1} k + \ldots + u_0 k^{n-1} k \tag{4.105}$$

– being thus characterized by ratio k and first term u_0; after lumping powers of k, Eq. (4.105) becomes

$$\sum_{i=0}^{n} u_i \equiv u_0 + u_0 k + u_0 k^2 + \ldots + u_0 k^i + \ldots + u_0 k^n, \tag{4.106}$$

where u_0 may be factored out as

$$\sum_{i=0}^{n} u_i \equiv u_0 \sum_{i=0}^{n} k^i. \tag{4.107}$$

After algebraic manipulation, one obtains

$$\sum_{i=0}^{n} u_i = u_0 \frac{1-k^{n+1}}{1-k} \tag{4.108}$$

from Eq. (4.107) – which allows transformation of the summation to a simpler algebraic expression.

4.2.11 Algebra of polynomials

4.2.11.1 Products of binomials

The algorithm of multiplication of sums of real numbers applies in full to multiplication of polynomials. For instance, when a binomial is raised to its square, one gets

$$\begin{aligned}(x+y)^2 &\equiv (x+y)(x+y) = x^2 + xy + yx + y^2 \\ &= x^2 + 2xy + y^2\end{aligned} \tag{4.109}$$

via direct application of the distributive property of multiplication with regard to addition, coupled to the commutative property of multiplication and lumping of terms alike afterward; a change of sign in the original binomial likewise unfolds

$$\begin{aligned}(x-y)^2 &\equiv (x-y)(x-y) = x^2 - xy - yx + y^2 \\ &= x^2 - 2xy + y^2.\end{aligned} \tag{4.110}$$

A similar rationale applied to product of conjugate binomials leads to

$$\begin{aligned}(x+y)(x-y) &\equiv x^2 - xy + yx - y^2 \\ &= x^2 - y^2,\end{aligned} \tag{4.111}$$

based again on the distributive and commutative properties, followed by dropping of symmetrical terms.

4.2.11.2 Newton's binomial theorem

According to Newton's binomial theorem, it is possible to expand the n-th power (with n integer) of a sum of (real) terms, x and y, as a sum of a finite number of products of integer powers of x and y, with exponents adding up to n in every case; more specifically,

$$(x+y)^n = \sum_{k=0}^{n} \binom{n}{k} x^k y^{n-k} \tag{4.112}$$

– known as binomial formula (or binomial identity), where x and y may represent either constants or variables. The simplest, nontrivial case pertains to $n=2$, and reads

$$(x+y)^2 = \binom{2}{0} x^0 y^{2-0} + \binom{2}{1} x^1 y^{2-1} + \binom{2}{2} x^2 y^{2-2} = y^2 + 2xy + x^2 \tag{4.113}$$

– which readily retrieves Eq. (4.109), as expected. The binomial coefficients in Eq. (4.112), $\binom{n}{k}$, count in how many ways one can pick up a subset with k elements out of a set with n elements in total; in mathematical terms, this is equivalent to writing

$$\binom{n}{k} = \frac{n!}{k!(n-k)!}, \tag{4.114}$$

which supports the entries of Pascal's triangle – made available as Table 4.1. Inspection of this table indicates that the outermost values in each row are always unity; whereas every two consecutive numbers in a given row add up to the value placed in between at the next row, i.e.

$$\binom{n}{k-1} + \binom{n}{k} = \binom{n+1}{k} \tag{4.115}$$

– frequently known as Pascal's rule.

4.2.11.3 Multinomial theorem

If a polynomial of m terms (rather than just two, as in a binomial) is at stake, then its power may be calculated via

$$\left(\sum_{i=1}^{m} x_i\right)^n = \sum_{\substack{k_1, k_2, \ldots, k_m \\ k_1 + k_2 + \ldots + k_m = n}} \binom{n}{k_1, k_2, \ldots, k_m} \prod_{l=1}^{m} x_l^{k_l}, \tag{4.116}$$

known as multinomial theorem; the corresponding multinomial coefficients satisfy

$$\binom{n}{k_1, k_2, \ldots, k_m} = \frac{n!}{k_1! k_2! \ldots k_m!}, \tag{4.117}$$

where k_1, k_2, ..., k_m denote all possible integer numbers that (as a whole) add up to n. For instance, Eq. (4.116) ensues as

$$\begin{aligned}(a+b+c)^3 &= \binom{3}{3,0,0} a^3 b^0 c^0 + \binom{3}{0,3,0} a^0 b^3 c^0 + \binom{3}{0,0,3} a^0 b^0 c^3 \\ &+ \binom{3}{2,1,0} a^2 b^1 c^0 + \binom{3}{2,0,1} a^2 b^0 c^1 + \binom{3}{1,2,0} a^1 b^2 c^0 + \binom{3}{0,2,1} a^0 b^2 c^1 \\ &+ \binom{3}{1,0,2} a^1 b^0 c^2 + \binom{3}{0,1,2} a^0 b^1 c^2 + \binom{3}{1,1,1} a^1 b^1 c^1\end{aligned} \tag{4.118}$$

TABLE 4.1

Pascal's triangle pertaining to binomial coefficients, $\binom{n}{k}$, for the first values of n, and, in each case, for $k=0, 1, \ldots, n-1, n$.

n																					
0											1										
1										1		1									
2									1		2		1								
3								1		3		3		1							
4							1		4		6		4		1						
5						1		5		10		10		5		1					
6					1		6		15		20		15		6		1				
7				1		7		21		35		35		21		7		1			
8			1		8		28		56		70		56		28		8		1		
9		1		9		36		84		126		126		84		36		9		1	
10	1		10		45		120		210		252		210		120		45		10		1

for $n=3$, which is equivalent to

$$(a+b+c)^3 = \frac{3!}{3!0!0!}a^3 + \frac{3!}{0!3!0!}b^3 + \frac{3!}{0!0!3!}c^3 + \frac{3!}{2!1!0!}a^2b + \frac{3!}{2!0!1!}a^2c + \frac{3!}{1!2!0!}ab^2 + \frac{3!}{0!2!1!}b^2c + \frac{3!}{1!0!2!}ac^2 + \frac{3!}{0!1!2!}bc^2 + \frac{3!}{1!1!1!}abc \tag{4.119}$$

in view of Eq. (4.117); further algebraic rearrangement unfolds

$$(a+b+c)^3 = \frac{6}{6\cdot1\cdot1}a^3 + \frac{6}{1\cdot6\cdot1}b^3 + \frac{6}{1\cdot1\cdot6}c^3 + \frac{6}{2\cdot1\cdot1}a^2b + \frac{6}{2\cdot1\cdot1}a^2c + \frac{6}{1\cdot2\cdot1}ab^2 + \frac{6}{1\cdot2\cdot1}b^2c + \frac{6}{1\cdot1\cdot2}ac^2 + \frac{6}{1\cdot1\cdot2}bc^2 + \frac{6}{1\cdot1\cdot1}abc, \tag{4.120}$$

or else

$$(a+b+c)^3 = a^3 + b^3 + c^3 + 3a^2b + 3a^2c + 3ab^2 + 3b^2c + 3ac^2 + 3bc^2 + 6abc \tag{4.121}$$

for final result.

4.2.11.4 Factorization of polynomial

Given a generic n-th degree polynomial, $P_n\{x\}$, on variable x, holding r_1, r_2, …, r_n as roots and a_n as coefficient of the highest-order term, it can always be rewritten as a product of the form

$$P_n\{x\}\Big|_{P_n|_{r_1}=P_n|_{r_2}=\ldots=P_n|_{r_n}=0} = a_n \prod_{k=1}^{n}(x-r_k). \tag{4.122}$$

If all coefficients of $P_n\{x\}$ are real, then complex roots (if any), say r_k and r_{k+1}, will always appear as pairs of conjugated binomials – i.e. $r_k=a+\iota b$ for a root implies $r_{k+1}=a-\iota b$ for another root; hence, $(x-r_k)(x-r_{k+1})$ may be replaced by $((x-a)-\iota b)((x-a)+\iota b)=(x-a)^2+b^2$ as per Eq. (4.111), with x replaced by $x-a$, y by ιb, and ι^2 by -1.

4.2.11.5 Ruffini's rule

Division of $P_n\{x\}$ by any m-th degree polynomial, $P_m\{x\}$, can be uniquely obtained using the algorithm of division of numbers, with coefficients of higher powers serving as digits of higher decimal order – complemented by the distributive property of multiplication; the result looks, in general, as

$$\frac{P_n\{x\}}{P_m\{x\}} = Q_{n-m}\{x\} + \frac{R_{<m}\{x\}}{P_m\{x\}} \tag{4.123}$$

– where $Q_{n-m}\{x\}$ denotes quotient polynomial, of degree $n-m$, and $R_{<m}\{x\}$ denotes remainder polynomial, of degree below m. This task is particularly facilitated when $m=1$ and $a_m=1$, and thus $P_m\{x\}=x-r$ as per Eq. (4.122); Ruffini's rule may accordingly be invoked, as sketched in Fig. 4.9. The first and second rows therein represent $P_n\{x\}$, in terms of variable and coefficient, respectively, for each additive power; whereas the fourth and fifth rows (but the last column) likewise represent $Q_{n-m}\{x\}$, and the last column of the fourth row represents $R_{<m}\{x\}$ – which is, in this case, a mere constant. Except for the first column, where the lower shaded entry mimics its upper shaded counterpart, the lower entry of every column entails the product of the previous entry by r, added to the current upper entry.

4.2.11.6 Splitting of rational fraction

Once in possession of the equivalent result conveyed by Eq. (4.122), but applied to $P_m\{x\}$, with b_m serving as coefficient for x^m and r_1, r_2, …, r_m denoting its roots, one may revisit Eq. (4.123) as

$$\frac{P_n\{x\}}{P_m\{x\}} = Q_{n-m}\{x\} + \frac{R_{<m}\{x\}}{b_m \prod_{k=1}^{m}(x-r_k)} \tag{4.124}$$

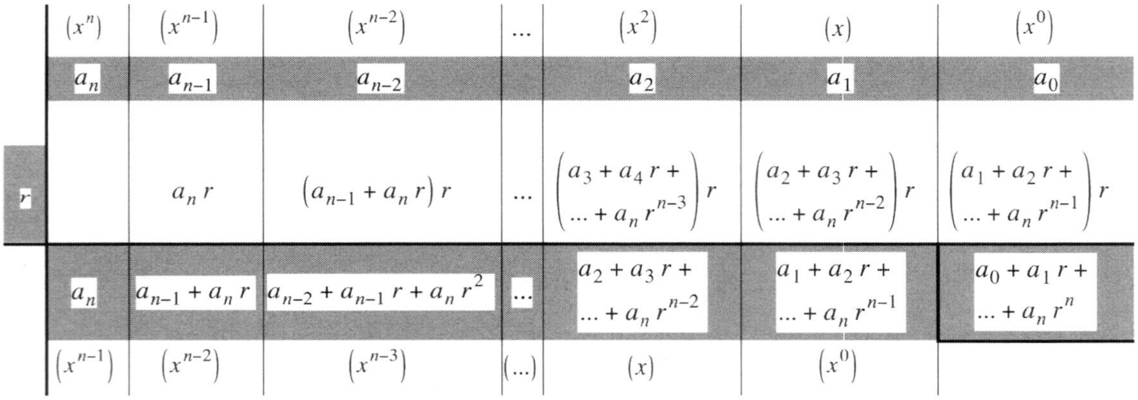

	(x^n)	(x^{n-1})	(x^{n-2})	...	(x^2)	(x)	(x^0)
	a_n	a_{n-1}	a_{n-2}		a_2	a_1	a_0
r		$a_n r$	$(a_{n-1}+a_n r)r$	...	$\begin{pmatrix} a_3+a_4 r+ \\ ...+a_n r^{n-3} \end{pmatrix} r$	$\begin{pmatrix} a_2+a_3 r+ \\ ...+a_n r^{n-2} \end{pmatrix} r$	$\begin{pmatrix} a_1+a_2 r+ \\ ...+a_n r^{n-1} \end{pmatrix} r$
	a_n	$a_{n-1}+a_n r$	$a_{n-2}+a_{n-1} r+a_n r^2$	...	$a_2+a_3 r+ ...+a_n r^{n-2}$	$a_1+a_2 r+ ...+a_n r^{n-1}$	$a_0+a_1 r+ ...+a_n r^n$
	(x^{n-1})	(x^{n-2})	(x^{n-3})	$(...)$	(x)	(x^0)	

FIGURE 4.9 Graphical representation of Ruffini's algorithm for division of polynomials, where a_0, a_1, a_2, ..., a_{n-2}, a_{n-1}, a_n, and r denote real numbers, while n denotes an integer number.

– or, after lumping constant b_m with the corresponding polynomial in numerator,

$$\frac{P_n\{x\}}{P_m\{x\}}=Q_{n-m}\{x\}+\frac{\dfrac{R_{<m}\{x\}}{b_m}}{\prod_{k=1}^{m}(x-r_k)}. \tag{4.125}$$

Every such root r_k of the polynomial in denominator can take real or complex values (i.e. of the form $a+\iota b$). When s_l roots are equal to r_k, one may lump the corresponding binomials as $(x-r_l)^{s_l}$ instead of multiplying $x-r_l$ by itself s_l times; in other words, Eq. (4.125) will alternatively appear as

$$\frac{P_n\{x\}}{P_m\{x\}}=Q_{n-m}\{x\}+\frac{\dfrac{R_{<m}\{x\}}{b_m}}{\prod_{l=1}^{s}(x-r_l)^{s_l}}, \tag{4.126}$$

as long as the r_ls represent the s distinct roots (or poles) of $P_m\{x\}$, each with multiplicity s_l, with $m=\sum_{l=1}^{s}s_l$. The remainder polynomial may always be written in the form

$$\frac{\dfrac{R_{<m}\{x\}}{b_m}}{\prod_{l=1}^{s}(x-r_l)^{s_l}}=\sum_{l=1}^{s}\sum_{p=0}^{s_l-1}\frac{A_{l,p+1}}{(x-r_l)^{s_l-p}}, \tag{4.127}$$

where $A_{l,p+1}$s denote constants (to be determined); therefore, any regular rational fraction with poles r_1, r_2, ..., r_s (or r_l for short) of multiplicity s_1, s_2, ..., s_s, respectively (or s for short), may be expanded as a sum of partial fractions bearing a constant as numerator, together with $x-r$, $(x-r)^2$, ..., $(x-r)^s$ sequentially as denominator – irrespective of the mathematical nature of such roots. For instance, Eq. (4.127) justifies

$$\begin{aligned}\frac{4x^2-3x+5}{3(x-1)^3(x+2)\left((x-1)^2+2\right)} &= \frac{A_{1,1}}{(x-1)^3}+\frac{A_{1,2}}{(x-1)^2}+\frac{A_{1,3}}{x-1}+\frac{A_{2,1}}{x-(-2)} \\ &+\frac{A_{3,1}}{x-(1+\iota\sqrt{2})}+\frac{A_{4,1}}{x-(1-\iota\sqrt{2})},\end{aligned} \tag{4.128}$$

when $4x^2-3x+5$ serves as remainder polynomial, $R_{<m}\{x\}$, while the divisor polynomial, $P_m\{x\}$, reads $3(x-1)^3(x+2)((x-1)^2+2)$ – corresponding to $b_m=3$, $r_1=1$ with $s_1=3$, $r_2=-2$ with $s_2=1$, $r_3=1+\iota\sqrt{2}$ with $s_3=1$, and $r_4=1-\iota\sqrt{2}$ with $s_4=1$.

To avoid emergence of complex numbers as such, and taking advantage of the fact that any complex roots of a polynomial with real coefficients always appear as conjugate pairs (as discussed above), one should lump pairs of complex partial fractions as

$$\frac{A_{l,1}}{x-(a+\iota b)}+\frac{A_{l,2}}{x-(a-\iota b)}=\frac{B_{l,1}x+B_{l,2}}{x^2+B_{l,3}x+B_{l,4}}, \tag{4.129}$$

provided that the new constants are defined as

$$B_{l,1}\equiv A_{l,1}+A_{l,2} \tag{4.130}$$

and

$$B_{l,2}\equiv -a\left(A_{l,1}+A_{l,2}\right)+\iota b\left(A_{l,1}-A_{l,2}\right) \tag{4.131}$$

pertaining to numerator, complemented by

$$B_{l,3}\equiv -2a \tag{4.132}$$

and

$$B_{l,4}\equiv a^2+b^2 \tag{4.133}$$

appearing in denominator. Therefore, any pair of partial fractions involving conjugate complex numbers in their denominators should be replaced by a new type of (composite) partial fraction, constituted by a first-order polynomial for numerator and a second-order polynomial for denominator. In the above example, one would have been led to

$$\begin{aligned}&\frac{A_{3,1}}{x-(a+\iota b)}+\frac{A_{4,1}}{x-(a-\iota b)} \\ &=\frac{\left(A_{3,1}+A_{4,1}\right)x-\left(A_{3,1}+A_{4,1}\right)+\iota\sqrt{2}\left(A_{3,1}-A_{4,1}\right)}{x^2-2x+3},\end{aligned} \tag{4.134}$$

associated with the third and fourth (complex) roots; where Eqs. (4.130)–(4.133) were taken into account, since $a=1$ and $b=\sqrt{2}$.

4.2.12 Matrices

4.2.12.1 Definition

Matrix is a nuclear concept in linear algebra; it consists of a set of numbers, $a_{i,j}$, arranged as m rows by n columns, enclosed by a set of square parenthesis, $[a_{i,j}]$ – according to

$$A \equiv \begin{bmatrix} a_{1,1} & a_{1,2} & \ldots & a_{1,n} \\ a_{2,1} & a_{2,2} & \ldots & a_{2,n} \\ \ldots & \ldots & \ldots & \ldots \\ a_{m,1} & a_{m,2} & \ldots & a_{m,n} \end{bmatrix}. \quad (4.135)$$

It is termed rectangular when $m \neq n$, and square when $m=n$; and reduces to a row vector when $m=1$, or a column vector when $n=1$. The main diagonal of a square matrix is formed by elements of the type $a_{i,i}$; if all entries below the main diagonal are zero, the matrix is said to be upper triangular; and lower triangular when all entries above the main diagonal are nil. A diagonal (square) matrix is both upper and lower triangular, i.e. all elements off the main diagonal are zero; if all elements in the diagonal are, in turn, equal to each other, then a scalar matrix arises. The most important scalar matrices are $(m \times m)$ identity matrices – containing only 1s in the main diagonal, and denoted as $\boldsymbol{I}_m$. A nil matrix is formed only by zeros, and is usually denoted as $\boldsymbol{0}_{m \times n}$. When elements symmetrically placed relative to the main diagonal coincide, the matrix is termed symmetric; all diagonal matrices are obviously symmetric. When rows and columns of matrix $\boldsymbol{A}$, with generic element $a_{i,j}$, are exchanged – with elements retaining their relative location within each row and each column, its transpose $\boldsymbol{A}^T$ results; it is accordingly denoted as $[a_{j,i}]$. Finally, the requirement for equality of two matrices is their sharing the same type (i.e. identical number of rows and identical number of columns) and the same spread (i.e. identical numbers in homologous positions).

4.2.12.2 Addition of matrices

Consider the condensed version of a generic $(m \times n)$ matrix $\boldsymbol{A}$, labeled as Eq. (4.135), i.e.

$$\boldsymbol{A} \equiv \left[a_{i,j}\right]; \quad i=1,2,\ldots,m; \; j=1,2,\ldots,n \quad (4.136)$$

– where subscript i refers to i-th row and subscript j refers to j-th column; if another matrix, $\boldsymbol{B}$, also of type $(m \times n)$ is defined as

$$\boldsymbol{B} \equiv \left[b_{i,j}\right]; \quad i=1,2,\ldots,m; \; j=1,2,\ldots,n, \quad (4.137)$$

then $\boldsymbol{A}$ and $\boldsymbol{B}$ can be added as per the algorithm

$$\left[a_{i,j}\right]+\left[b_{i,j}\right]=\left[a_{i,j}+b_{i,j}\right]; i=1,2,\ldots,m; \; j=1,2,\ldots,n; \quad (4.138)$$

the sum will again be a matrix of the same type $(m \times n)$. Addition of matrices is commutative, i.e.

$$\boldsymbol{A}+\boldsymbol{B}=\boldsymbol{B}+\boldsymbol{A}; \quad (4.139)$$

if a third matrix $\boldsymbol{C}$ abides to

$$\boldsymbol{C} \equiv \left[c_{i,j}\right]; \quad i=1,2,\ldots,m; \; j=1,2,\ldots,n, \quad (4.140)$$

then one finds that

$$\boldsymbol{A}+(\boldsymbol{B}+\boldsymbol{C})=(\boldsymbol{A}+\boldsymbol{B})+\boldsymbol{C} \quad (4.141)$$

– meaning that addition of matrices is associative. For every $(m \times n)$ matrix $\boldsymbol{A}$, there is a null matrix $\boldsymbol{0}_{m \times n}$ such that

$$\boldsymbol{A}+\boldsymbol{0}_{m \times n}=\boldsymbol{A}; \quad (4.142)$$

hence, $\boldsymbol{0}_{m \times n}$ plays the role of neutral element.

4.2.12.3 Multiplication of scalar by matrix

Consider a generic scalar, α, and a generic matrix $\boldsymbol{B}$ as defined by Eq. (4.137); under these circumstances,

$$\alpha\left[b_{i,j}\right]=\left[\alpha b_{i,j}\right]; \quad i=1,2,\ldots,m; \; j=1,2,\ldots,n \quad (4.143)$$

consubstantiates the algorithm for multiplication of scalar by matrix. One realizes that

$$\alpha \boldsymbol{B}=\boldsymbol{B}\alpha, \quad (4.144)$$

known as commutative property of multiplication of scalar by matrix – even though the scalar is conventionally placed upfront relative to the matrix. If a second scalar, β, is considered, then one finds

$$\alpha(\beta \boldsymbol{C})=(\alpha\beta)\boldsymbol{C}; \quad (4.145)$$

this is usually referred to as associative property. If addition of matrices and multiplication of scalar by matrix are considered simultaneously, then

$$\alpha(\boldsymbol{B}+\boldsymbol{C})=\alpha\boldsymbol{B}+\alpha\boldsymbol{C} \quad (4.146)$$

appears as universal rule – as long as $\boldsymbol{B}$ and $\boldsymbol{C}$ are of the same type, as per Eqs. (4.137) and (4.140); therefore, multiplication of scalar by matrix is distributive, with regard to addition of matrices. A similar property can be conceived encompassing addition of scalars, i.e.

$$(\alpha+\beta)\boldsymbol{C}=\alpha\boldsymbol{C}+\beta\boldsymbol{C}; \quad (4.147)$$

multiplication of scalar by matrix is indeed distributive with regard to addition of scalars. A final property pertains to product of unity by matrix, according to

$$1\boldsymbol{A}=\boldsymbol{A} \quad (4.148)$$

with $\boldsymbol{A}$ abiding to Eq. (4.136) – thus leaving the matrix unchanged, whatever it is; 1 serves as neutral element for multiplication of scalar by matrix. When the scalar at stake is –1, its product by matrix $\boldsymbol{A}$ transforms every element thereof to its negative, in agreement with Eq. (4.143); the corresponding result, usually denoted as $-\boldsymbol{A}$, satisfies

$$(-1)\boldsymbol{A}=\left[-a_{i,j}\right] \equiv -\boldsymbol{A}. \quad (4.149)$$

Matrix $-\boldsymbol{A}$ is called symmetric of $\boldsymbol{A}$ owing to

$$\boldsymbol{A}+(-\boldsymbol{A})=\boldsymbol{0}_{m \times n}; \quad (4.150)$$

hence, addition of those two matrices produces the nil matrix of the same type.

4.2.12.4 Multiplication of matrices

If $(m\times n)$ matrix $\boldsymbol{A}$, or $[a_{i,j}]$ as per Eq. (4.136), and $(n\times p)$ matrix $\boldsymbol{B}$, as per

$$\boldsymbol{B}\equiv\left[b_{k,l}\right];\quad k=1,2,\ldots,n;\ l=1,2,\ldots,p, \tag{4.151}$$

are considered – with number of columns of $\boldsymbol{A}$ equal to number of rows of $\boldsymbol{B}$, then the said matrices can be multiplied by each other according to

$$\left[a_{i,j}\right]\left[b_{k,l}\right]=\left[\sum_{r=1}^{n}a_{i,r}b_{r,l}\right]\equiv\left[d_{i,l}\right] \\ i=1,2,\ldots,m;\ l=1,2,\ldots,p; \tag{4.152}$$

the product of those matrices is an $(m\times p)$ matrix, with generic element $d_{i,l}$. Note that multiplication in reverse order will not be possible unless $p=m$, since matching between number of columns of $\boldsymbol{B}$ and number of rows of $\boldsymbol{A}$ would then be required; this example suffices to prove that multiplication of matrices is not commutative. Consider now three matrices, $\boldsymbol{A}$ of the $(m\times n)$ type and given by Eq. (4.136), $\boldsymbol{B}$ of the $(n\times p)$ type and given by Eq. (4.151), and $\boldsymbol{C}$ of the $(p\times q)$ type and given by

$$\boldsymbol{C}\equiv\left[c_{t,u}\right];\quad t=1,2,\ldots,p;\ u=1,2,\ldots,q; \tag{4.153}$$

product $\boldsymbol{AB}$ is an $(m\times p)$ matrix – while product $\boldsymbol{ABC}$, of $\boldsymbol{AB}$ by $\boldsymbol{C}$, will be an $(m\times q)$ matrix. One realizes that

$$(\boldsymbol{AB})\boldsymbol{C}=\boldsymbol{A}(\boldsymbol{BC}) \tag{4.154}$$

applies in general; multiplication of matrices is thus associative – yet the relative order of multiplication of the original factors must be kept, since multiplication of matrices is uncommutative. If $(m\times n)$ matrix $\boldsymbol{A}$ is multiplied by $(n\times n)$ identity matrix, $\boldsymbol{I}_n$, then

$$\boldsymbol{AI}_n=\boldsymbol{A}; \tag{4.155}$$

in other words, multiplication of a matrix by the (compatible) identity matrix leaves the former unchanged – so $\boldsymbol{I}_n$ plays the role of neutral element for multiplication of matrices. This very same conclusion can be attained if the order of multiplication is reversed, i.e.

$$\boldsymbol{I}_m\boldsymbol{A}=\boldsymbol{A}, \tag{4.156}$$

provided that the identity matrix is now of the $(m\times m)$ type; hence, order of multiplication of the identity matrix by another matrix (when feasible) does not affect the final result. When an $(m\times n)$ matrix $\boldsymbol{A}$ is postmultiplied by a (compatible) null $(n\times p)$ matrix $\boldsymbol{0}_{n\times p}$, one gets

$$\boldsymbol{A0}_{n\times p}=\boldsymbol{0}_{m\times p}, \tag{4.157}$$

meaning that postmultiplication by a null matrix degenerates to another null matrix. By the same token, premultiplication of $\boldsymbol{A}$ by the (compatible) null $(p\times m)$ matrix $\boldsymbol{0}_{p\times m}$ gives rise to

$$\boldsymbol{0}_{p\times m}\boldsymbol{A}=\boldsymbol{0}_{p\times n}, \tag{4.158}$$

implying that premultiplication by a null matrix leads necessarily to the corresponding null matrix as product. In both cases, the product null matrix has numbers of rows and columns not necessarily coincident with those of the factor null matrix. Still another property of interest pertains to simultaneous performance of addition and multiplication of matrices, viz.

$$(\boldsymbol{A}+\boldsymbol{B})\boldsymbol{C}=\boldsymbol{AC}+\boldsymbol{BC}, \tag{4.159}$$

complemented by

$$\boldsymbol{A}(\boldsymbol{B}+\boldsymbol{C})=\boldsymbol{AB}+\boldsymbol{AC}; \tag{4.160}$$

therefore, both pre- and postmultiplication of matrices are distributive, with regard to addition of matrices. In the case of an $(n\times n)$ matrix, one may proceed to sequential multiplication by itself k times – usually represented as

$$\prod_{i=1}^{k}\boldsymbol{A}\equiv\boldsymbol{A}^k; \tag{4.161}$$

the outcome is an $(n\times n)$ square matrix – while

$$\boldsymbol{A}^0\equiv\boldsymbol{I}_n \tag{4.162}$$

has been set by convention. In view of the unique features of an identity matrix outlined in Eqs. (4.155) and (4.156), one realizes that

$$\boldsymbol{I}_n^{\,k}=\boldsymbol{I}_n \tag{4.163}$$

stemming from Eq. (4.161) – because multiplying an $(n\times n)$ identity matrix by itself leaves either of them unchanged.

If two matrices are partitioned in blocks, multiplication is still to follow the algorithm conveyed by Eq. (4.152) – as long as individual matrix elements are replaced by appropriate submatrices; however, the underlying rules of compatibility between columns and rows of factor blocks are to be satisfied by all products. Consider, in this regard, the most common (and simplest) case of a (2×2) block matrix, viz.

$$\boldsymbol{A}\equiv\begin{bmatrix}\boldsymbol{A}_{1,1} & \boldsymbol{A}_{1,2}\\ \boldsymbol{A}_{2,1} & \boldsymbol{A}_{2,2}\end{bmatrix} \tag{4.164}$$

– where $(m\times n)$ matrix $\boldsymbol{A}$ was partitioned through $(m_1\times n_1)$ matrix $\boldsymbol{A}_{1,1}$, $(m_1\times n_2)$ matrix $\boldsymbol{A}_{1,2}$, $(m_2\times n_1)$ matrix $\boldsymbol{A}_{2,1}$, and $(m_2\times n_2)$ matrix $\boldsymbol{A}_{2,2}$ as constitutive blocks, with $m_1+m_2=m$ and $n_1+n_2=n$; coupled with

$$\boldsymbol{B}\equiv\begin{bmatrix}\boldsymbol{B}_{1,1} & \boldsymbol{B}_{1,2}\\ \boldsymbol{B}_{2,1} & \boldsymbol{B}_{2,2}\end{bmatrix} \tag{4.165}$$

– with $(p\times q)$ matrix $\boldsymbol{B}$ partitioned as $(p_1\times q_1)$ matrix $\boldsymbol{B}_{1,1}$, $(p_1\times q_2)$ matrix $\boldsymbol{B}_{1,2}$, $(p_2\times q_1)$ matrix $\boldsymbol{B}_{2,1}$, and $(p_2\times q_2)$ matrix $\boldsymbol{B}_{2,2}$ as constitutive blocks – abiding to $p_1+p_2=p$ and $q_1+q_2=q$. Product $\boldsymbol{AB}$ will not exist unless $n=p$; furthermore, the said product will look like

$$\boldsymbol{AB}=\begin{bmatrix}\boldsymbol{A}_{1,1}\boldsymbol{B}_{1,1}+\boldsymbol{A}_{1,2}\boldsymbol{B}_{2,1} & \boldsymbol{A}_{1,1}\boldsymbol{B}_{1,2}+\boldsymbol{A}_{1,2}\boldsymbol{B}_{2,2}\\ \boldsymbol{A}_{2,1}\boldsymbol{B}_{1,1}+\boldsymbol{A}_{2,2}\boldsymbol{B}_{2,1} & \boldsymbol{A}_{2,1}\boldsymbol{B}_{1,2}+\boldsymbol{A}_{2,2}\boldsymbol{B}_{2,2}\end{bmatrix}, \tag{4.166}$$

following the regular algorithm of multiplication of matrices outlined by Eq. (4.152) – as long as $n_1=p_1$ to allow definition of $\boldsymbol{A}_{1,1}\boldsymbol{B}_{1,1}$, $\boldsymbol{A}_{1,1}\boldsymbol{B}_{1,2}$, $\boldsymbol{A}_{2,1}\boldsymbol{B}_{1,1}$, and $\boldsymbol{A}_{2,1}\boldsymbol{B}_{1,2}$, as well as $n_2=p_2$ to permit existence of $\boldsymbol{A}_{1,2}\boldsymbol{B}_{2,1}$, $\boldsymbol{A}_{1,2}\boldsymbol{B}_{2,2}$, $\boldsymbol{A}_{2,2}\boldsymbol{B}_{2,1}$, and $\boldsymbol{A}_{2,2}\boldsymbol{B}_{2,2}$. This approach is particularly advantageous when one (or more) of the foregoing blocks is either an identity or a null submatrix – since the matrix elements in Eq. (4.166) would yield much simpler expressions, involving only a portion of $\boldsymbol{A}$ or $\boldsymbol{B}$ rather than their whole set of elements.

4.2.12.5 Transposal of matrix

Recall again matrix $\boldsymbol{A}$, as defined by Eqs. (4.135) or (4.136); if generic element $a_{i,j}$, initially located in the i-th row and j-th column, were swapped with element $a_{j,i}$ initially located in the j-th row and i-th column, then the transpose matrix would result – given by

$$\boldsymbol{A}^T \equiv \begin{bmatrix} a_{1,1} & a_{2,1} & \cdots & a_{m,1} \\ a_{1,2} & a_{2,2} & \cdots & a_{m,2} \\ \cdots & \cdots & \cdots & \cdots \\ a_{1,n} & a_{2,n} & \cdots & a_{m,n} \end{bmatrix} \equiv \left[a_{j,i}\right]; \tag{4.167}$$

in more condensed form, Eq. (4.167) reads

$$\left[a_{i,j}\right]^T = \left[a_{j,i}\right]; \quad i = 1,2,\ldots,m; \; j = 1,2,\ldots,n, \tag{4.168}$$

so $\boldsymbol{A}^T$ will be an $(n \times m)$ matrix. The order of a square matrix is not changed upon transposal – neither do its diagonal elements, inherently characterized by $i=j$. Furthermore, if the elements symmetrically located relative to the main diagonal are identical, then

$$\left(\boldsymbol{A}^T = \boldsymbol{A}\right)\Big|_{a_{i,j\neq i}=a_{j,i}}; \tag{4.169}$$

a (square) matrix bearing this property is termed symmetric – a concept distinct from that conveyed by Eq. (4.149), which involves two matrices. A particular case of the above statement is the identity matrix – since $a_{i,j\neq i}=0=a_{j,i}$, besides $a_{i,i}=1$; hence,

$$\boldsymbol{I}_m^{\ T} = \boldsymbol{I}_m, \tag{4.170}$$

irrespective of matrix order m. Sequential application of transposal twice yields

$$\left(\left[a_{i,j}\right]^T\right)^T = \left[a_{j,i}\right]^T = \left[a_{i,j}\right], \tag{4.171}$$

using Eq. (4.168) for template – which may be condensed to

$$\left(\boldsymbol{A}^T\right)^T = \boldsymbol{A}; \tag{4.172}$$

therefore, composition of transposal with itself leaves the original matrix unchanged – so transposal coincides with its inverse operation. When transposal is combined with addition of matrices, one obtains

$$(\boldsymbol{A}+\boldsymbol{B})^T = \boldsymbol{A}^T + \boldsymbol{B}^T; \tag{4.173}$$

if more than two matrices are at stake, this very rule can be iteratively applied. In the case of product of matrices, one gets

$$(\boldsymbol{AB})^T = \boldsymbol{B}^T\boldsymbol{A}^T; \tag{4.174}$$

hence, transpose of a product of matrices equals product of transposes of the factor matrices, but effected in reverse order. As expected, this property can also be applied to any number of factor matrices. If matrix $\boldsymbol{A}$ reduces to a single scalar, say α, then Eq. (4.174) still applies – and one finds that

$$(\alpha\boldsymbol{A})^T \equiv \left(\left[\alpha\right]\boldsymbol{A}\right)^T = \boldsymbol{A}^T\left[\alpha\right]^T = \boldsymbol{A}^T\left[\alpha\right] \equiv \boldsymbol{A}^T\alpha = \alpha\boldsymbol{A}^T, \tag{4.175}$$

with the aid of Eq. (4.144), complemented by realization that $\alpha \equiv [\alpha]$ and $[\alpha]^T=[\alpha]$ as transposal of an (1×1) matrix proves redundant; this is the conventional form of expressing the result of transposing product of scalar by matrix, which degenerates to product of the said scalar by the transpose matrix.

4.2.12.6 Inversion of matrix

The inverse $(n \times n)$ matrix, $\boldsymbol{A}^{-1}$, of a given $(n \times n)$ matrix, $\boldsymbol{A}$, satisfies, by definition,

$$\boldsymbol{A}\boldsymbol{A}^{-1} = \boldsymbol{A}^{-1}\boldsymbol{A} = \boldsymbol{I}_n; \tag{4.176}$$

therefore, if $\boldsymbol{A}^{-1}$ is described by

$$\boldsymbol{A}^{-1} \equiv \begin{bmatrix} \alpha_{1,1} & \alpha_{1,2} & \cdots & \alpha_{1,n} \\ \alpha_{2,1} & \alpha_{2,2} & \cdots & \alpha_{2,n} \\ \cdots & \cdots & \cdots & \cdots \\ \alpha_{n,1} & \alpha_{n,2} & \cdots & \alpha_{n,n} \end{bmatrix}, \tag{4.177}$$

then one may insert Eq. (4.177) and Eq. (4.135), under $m=n$, to obtain

$$\begin{bmatrix} a_{1,1} & a_{1,2} & \cdots & a_{1,n} \\ a_{2,1} & a_{2,2} & \cdots & a_{2,n} \\ \cdots & \cdots & \cdots & \cdots \\ a_{n,1} & a_{n,2} & \cdots & a_{n,n} \end{bmatrix} \begin{bmatrix} \alpha_{1,1} & \alpha_{1,2} & \cdots & \alpha_{1,n} \\ \alpha_{2,1} & \alpha_{2,2} & \cdots & \alpha_{2,n} \\ \cdots & \cdots & \cdots & \cdots \\ \alpha_{n,1} & \alpha_{n,2} & \cdots & \alpha_{n,n} \end{bmatrix} = \begin{bmatrix} 1 & 0 & \cdots & 0 \\ 0 & 1 & \cdots & 0 \\ \cdots & \cdots & \cdots & \cdots \\ 0 & 0 & \cdots & 1 \end{bmatrix} \tag{4.178}$$

from Eq. (4.176) – corresponding to $\boldsymbol{A}\boldsymbol{A}^{-1}$ being equal to $\boldsymbol{I}_n$. Recalling the algorithm of multiplication of matrices labeled as Eq. (4.152), one finds that Eq. (4.178) is equivalent to

$$\begin{aligned} a_{1,1}\alpha_{1,1} + a_{1,2}\alpha_{2,1} + \ldots + a_{1,n}\alpha_{n,1} &= 1 \\ a_{1,1}\alpha_{1,2} + a_{1,2}\alpha_{2,2} + \ldots + a_{1,n}\alpha_{n,2} &= 0 \\ &\cdots \\ a_{1,1}\alpha_{1,n} + a_{1,2}\alpha_{2,n} + \ldots + a_{1,n}\alpha_{n,n} &= 0 \\ a_{2,1}\alpha_{1,1} + a_{2,2}\alpha_{2,1} + \ldots + a_{2,n}\alpha_{n,1} &= 0 \\ a_{2,1}\alpha_{1,2} + a_{2,2}\alpha_{2,2} + \ldots + a_{2,n}\alpha_{n,2} &= 1 \\ &\cdots \\ a_{2,1}\alpha_{1,n} + a_{2,2}\alpha_{2,n} + \ldots + a_{2,n}\alpha_{n,n} &= 0 \quad ; \\ &\cdots \\ a_{n,1}\alpha_{1,1} + a_{n,2}\alpha_{2,1} + \ldots + a_{n,n}\alpha_{n,1} &= 0 \\ a_{n,1}\alpha_{1,2} + a_{n,2}\alpha_{2,2} + \ldots + a_{n,n}\alpha_{n,2} &= 0 \\ &\cdots \\ a_{n,1}\alpha_{1,n} + a_{n,2}\alpha_{2,n} + \ldots + a_{n,n}\alpha_{n,n} &= 1 \end{aligned} \tag{4.179}$$

hence, a system of n^2 linear algebraic equations in n^2 unknowns, i.e. $\alpha_{1,1}, \alpha_{1,2}, \ldots, \alpha_{1,n}, \alpha_{2,1}, \alpha_{2,2}, \ldots, \alpha_{2,n}, \ldots, \alpha_{n,1}, \alpha_{n,2}, \ldots, \alpha_{n,n}$, arises – the solution of which will permit calculation of all elements of $\boldsymbol{A}^{-1}$.

The same solutions, in terms of αs, will be obtained if $A^{-1}A = I_n$ in Eq. (4.176) were instead considered. In view of the definition of inverse, one realizes that

$$\left(A^{-1}\right)^{-1} = A; \tag{4.180}$$

this means that composition of inversion with itself cancels it out – in much the same way found previously for transposal. On the other hand, one finds that

$$(AB)^{-1} = B^{-1}A^{-1}, \tag{4.181}$$

i.e. inverse of a product of matrices equals product of their inverses, in reverse order. The result conveyed by Eq. (4.181) can obviously be extended to any number of factors – by sequentially applying it pairwise. When the matrices of interest are identical, this rule leads to

$$\left(A^{k}\right)^{-1} = \left(A^{-1}\right)^{k} \tag{4.182}$$

owing to the definition of power as conveyed by Eq. (4.161); hence, the power and inverse operations are interchangeable.

In the particular case of matrix A degenerating to a single scalar α, one gets

$$(\alpha B)^{-1} \equiv \left(\left[\alpha\right]B\right)^{-1} = B^{-1}\left[\alpha\right]^{-1} \equiv B^{-1}\alpha^{-1} = B^{-1}\frac{1}{\alpha} = \frac{1}{\alpha}B^{-1} \tag{4.183}$$

– again at the expense of Eq. (4.144); the above result indicates that inverse of product of scalar by matrix is merely product of reciprocal of such a scalar by inverse of the matrix proper. By the same token, the combination of transposal and inversion operators yields

$$\left(A^{T}\right)^{-1} = \left(A^{-1}\right)^{T} \tag{4.184}$$

stemming from Eqs. (4.168) and (4.176); therefore, inverse of A^T is just transpose of A^{-1} – meaning that transposal and inversion operators can as well be exchanged, without affecting the final result.

4.2.13 Determinants

4.2.13.1 Definition

Consider an $(n \times n)$ matrix A, viz.

$$A \equiv \begin{bmatrix} a_{1,1} & a_{1,2} & \cdots & a_{1,n} \\ \cdots & \cdots & \cdots & \cdots \\ a_{i,1} & a_{i,2} & & a_{i,n} \\ \cdots & \cdots & \cdots & \cdots \\ a_{n,1} & a_{n,2} & \cdots & a_{n,n} \end{bmatrix} \equiv \begin{bmatrix} a_1 \\ \cdots \\ a_i \\ \cdots \\ a_n \end{bmatrix} \tag{4.185}$$

as square version of the matrix conveyed by Eq. (4.135) – constituted by row vectors $a_1, a_2, \ldots, a_n$ abiding to

$$a_i \equiv \left[a_{i,1} \quad a_{i,2} \quad \ldots \quad a_{i,n}\right]; \quad i = 1,2,\ldots,n; \tag{4.186}$$

the determinant of A, denoted as $|A|$ or $det\, A$, is a unique scalar associated to the said matrix, and consists of the algebraic sum of all products $a_{1,\beta_{1,i}}a_{2,\beta_{2,i}}\ldots a_{n,\beta_{n,i}}$ (designated as terms) of elements of A, according to

$$|A| \equiv \sum_i (-1)^{\alpha_i} a_{1,\beta_{1,i}}a_{2,\beta_{2,i}}\ldots a_{n,\beta_{n,i}}. \tag{4.187}$$

Each row and each column is represented only once in every term of the summation labeled as Eq. (4.187), i.e. subscripts located second satisfy $\beta_{1,i}\neq\beta_{2,i}$, $\beta_{1,i}\neq\beta_{3,i}$, $\beta_{1,i}\neq\beta_{4,i}$, …, $\beta_{2,i}\neq\beta_{3,i}$, $\beta_{2,i}\neq\beta_{4,i}$, …, $\beta_{n-1,i}\neq\beta_{n,i}$ – while α_i denotes parity of i-th term, i.e. number of exchanges of sequence ($\beta_{1,i}$, $\beta_{2,i}$, …, $\beta_{n,i}$) relative to base sequence (1, 2, …,n). The most systematic way of finding all possible sequences of $\beta_{j,i}$s is thus to start with base sequence (1, 2, …, n), and produce all $n!$ permutations thereof; and then find the minimum number of exchanges between two elements of the said base sequence required to unequivocally obtain each sequence under scrutiny. For instance, with $n=3$, the aforementioned $3!=6$ sequences look like (1,2,3), (2,1,3), (1,3,2), (3,2,1), (2,3,1), and (3,1,2). Note that (2,1,3) can be obtained from (1,2,3) via one exchange (i.e. 1 by 2), (1,3,2) can likewise be obtained from (1,2,3) via one exchange (i.e. 2 by 3), and (3,2,1) can also be generated from (1,2,3) via one exchange (i.e. 1 by 3); (2,3,1) can, in turn, be obtained from (2,1,3) via one exchange (i.e. 1 by 3), or from (1,3,2) via one exchange (i.e. 1 by 2), or even from (3,2,1) via one exchange (i.e. 2 by 3); and finally (3,1,2) may be obtained from (2,1,3) via one exchange (i.e. 2 by 3), or from (1,3,2) via one exchange (i.e. 1 by 3), or from (3,2,1) via one exchange (i.e. 1 by 2). If the primary sequence (1,2,3) is used as reference for all such sequences, one realizes that (1,2,3) is produced therefrom via 0 exchanges – so its parity α will read 0; (2,1,3), (1,3,2), and (3,2,1) via one exchange, so they hold 1 for associated parity; and (2,3,1) and (3,1,2) via two exchanges, thus leading to $\alpha=2$. The corresponding determinant will accordingly look like

$$\begin{aligned} |A| \equiv \begin{vmatrix} a_{1,1} & a_{1,2} & a_{1,3} \\ a_{2,1} & a_{2,2} & a_{2,3} \\ a_{3,1} & a_{3,2} & a_{3,3} \end{vmatrix} &= (-1)^0 a_{1,1}a_{2,2}a_{3,3} + (-1)^1 a_{1,2}a_{2,1}a_{3,3} + (-1)^1 a_{1,1}a_{2,3}a_{3,2} \\ &\quad + (-1)^1 a_{1,3}a_{2,2}a_{3,1} + (-1)^2 a_{1,2}a_{2,3}a_{3,1} + (-1)^2 a_{1,3}a_{2,1}a_{3,2} \end{aligned} \tag{4.188}$$

based on Eq. (4.187) – which readily simplifies to

$$\begin{aligned} \begin{vmatrix} a_{1,1} & a_{1,2} & a_{1,3} \\ a_{2,1} & a_{2,2} & a_{2,3} \\ a_{3,1} & a_{3,2} & a_{3,3} \end{vmatrix} &= a_{1,1}a_{2,2}a_{3,3} + a_{1,2}a_{2,3}a_{3,1} + a_{1,3}a_{2,1}a_{3,2} \\ &\quad - \left(a_{1,2}a_{2,1}a_{3,3} + a_{1,1}a_{2,3}a_{3,2} + a_{1,3}a_{2,2}a_{3,1}\right), \end{aligned} \tag{4.189}$$

after appropriate algebraic regrouping of terms. By the same token, a second-order determinant may be seen as the algebraic sum of (plus) term $a_{1,1}a_{2,2}$ with (negative) term $a_{1,2}a_{2,1}$, corresponding to the 2! characteristic sequences, (1,2) and (2,1), viz.

$$\begin{vmatrix} a_{1,1} & a_{1,2} \\ a_{2,1} & a_{2,2} \end{vmatrix} = (-1)^0 a_{1,1}a_{2,2} + (-1)^1 a_{1,2}a_{2,1} = a_{1,1}a_{2,2} - a_{2,1}a_{1,2}; \tag{4.190}$$

note that the latter is obtainable from the former (and base) sequence via a single inversion, thus justifying its minus sign as per $(-1)^1=-1$.

4.2.13.2 Laplace's theorem

Calculation of a higher order determinant via its definition, labeled as Eq. (4.187), is cumbersome – so one usually resorts to Laplace's expansion theorem; this method allows calculation of determinant $|\boldsymbol{A}|$, of an $(n\times n)$ matrix $\boldsymbol{A}$, via a weighted sum of the n determinants associated with as many $((n-1)\times(n-1))$ submatrices $\boldsymbol{A}_{-i,-j}$ – obtained via removal of the i-th row and the j-th column from $\boldsymbol{A}$. Mathematically speaking, the aforementioned theorem may be stated as

$$|\boldsymbol{A}|=\sum_{j=1}^{n}(-1)^{i+j}a_{i,j}\left|\boldsymbol{A}_{-i,-j}\right|;\quad i=1,2,\ldots,n \tag{4.191}$$

when expansion occurs along the i-th row, or, equivalently,

$$|\boldsymbol{A}|=\sum_{i=1}^{n}(-1)^{i+j}a_{i,j}\left|\boldsymbol{A}_{-i,-j}\right|;\quad j=1,2,\ldots,n \tag{4.192}$$

when expansion proceeds through the j-th column. When $n=3$, for instance, Eq. (4.192) implies

$$\begin{vmatrix} a_{1,1} & a_{1,2} & a_{1,3} \\ a_{2,1} & a_{2,2} & a_{2,3} \\ a_{3,1} & a_{3,2} & a_{3,3} \end{vmatrix}=(-1)^{1+2}a_{1,2}\begin{vmatrix} a_{2,1} & a_{2,3} \\ a_{3,1} & a_{3,3} \end{vmatrix}+(-1)^{2+2}a_{2,2}\begin{vmatrix} a_{1,1} & a_{1,3} \\ a_{3,1} & a_{3,3} \end{vmatrix}+(-1)^{3+2}a_{3,2}\begin{vmatrix} a_{1,1} & a_{1,3} \\ a_{2,1} & a_{2,3} \end{vmatrix} \tag{4.193}$$

if the second column is selected as pivot; application of Eq. (4.190) will then give rise to

$$\begin{aligned}\begin{vmatrix} a_{1,1} & a_{1,2} & a_{1,3} \\ a_{2,1} & a_{2,2} & a_{2,3} \\ a_{3,1} & a_{3,2} & a_{3,3} \end{vmatrix}&=-a_{1,2}\left(a_{2,1}a_{3,3}-a_{3,1}a_{2,3}\right)\\ &\quad+a_{2,2}(a_{1,1}a_{3,3}-a_{3,1}a_{1,3})-a_{3,2}\left(a_{1,1}a_{2,3}-a_{2,1}a_{1,3}\right)\\ &=-a_{1,2}a_{2,1}a_{3,3}+a_{1,2}a_{2,3}a_{3,1}\\ &\quad+a_{1,1}a_{2,2}a_{3,3}-a_{1,3}a_{2,2}a_{3,1}-a_{1,1}a_{2,3}a_{3,2}+a_{1,3}a_{2,1}a_{3,2}\end{aligned} \tag{4.194}$$

that retrieves Eq. (4.189) after straightforward algebraic rearrangement. Therefore, either Eq. (4.191) or Eq. (4.192) indicates that an n-th-order determinant is expressible as algebraic sum of n determinants of $(n-1)$-th order, each of which is, in turn, expressible as $n-1$ determinants of $(n-2)$-th order, and so on; until reaching an algebraic sum of $n!/2$ determinants of second order, to be in turn calculated via Eq. (4.190), i.e. difference between product of the terms in the main diagonal and product of terms in the secondary diagonal – thus giving rise to $n!$ terms in all.

4.2.13.3 Properties

Using Eqs. (4.185) and (4.186) as template, one may define a general $(n\times n)$ matrix $\boldsymbol{Q}$ as

$$\boldsymbol{Q}\equiv\begin{bmatrix}\boldsymbol{q}_1 & \cdots & \boldsymbol{q}_j & \cdots & \boldsymbol{q}_n\end{bmatrix};\quad j=1,2,\ldots,n \tag{4.195}$$

– with each $(n\times 1)$ column vector $\boldsymbol{q}_j$ abiding to

$$\boldsymbol{q}_j\equiv\begin{bmatrix} q_{1,j} \\ q_{2,j} \\ \cdots \\ q_{n,j}\end{bmatrix};\quad j=1,2,\ldots,n. \tag{4.196}$$

Upon transposal of $\boldsymbol{A}$, the corresponding determinant remains unchanged, i.e.

$$\left|\boldsymbol{A}^T\right|=|\boldsymbol{A}| \tag{4.197}$$

– and one likewise finds

$$\left|\boldsymbol{Q}^T\right|=|\boldsymbol{Q}|. \tag{4.198}$$

If the i-th row of $\boldsymbol{A}$ is integrally filled with zeros, i.e. $\boldsymbol{a}_i=\boldsymbol{0}$, then

$$\begin{vmatrix}\boldsymbol{a}_1 \\ \cdots \\ \boldsymbol{0} \\ \cdots \\ \boldsymbol{a}_n\end{vmatrix}=0; \tag{4.199}$$

one may derive a similar rule applying to a null j-th column, i.e. $\boldsymbol{q}_j=\boldsymbol{0}$ of matrix $\boldsymbol{Q}$, according to

$$\begin{vmatrix}\boldsymbol{q}_1 & \cdots & \boldsymbol{0} & \cdots & \boldsymbol{q}_n\end{vmatrix}=0. \tag{4.200}$$

If all elements in the i-th row of $\boldsymbol{A}$ appear multiplied by scalar α, then

$$\begin{vmatrix}\boldsymbol{a}_1 \\ \cdots \\ \alpha\,\boldsymbol{a}_i \\ \cdots \\ \boldsymbol{a}_n\end{vmatrix}=\alpha\begin{vmatrix}\boldsymbol{a}_1 \\ \cdots \\ \boldsymbol{a}_i \\ \cdots \\ \boldsymbol{a}_n\end{vmatrix}; \tag{4.201}$$

if the j-th column of $\boldsymbol{Q}$ had been multiplied by scalar α to generate a new matrix, then a conclusion analogous to Eq. (4.201) would have been reached, i.e.

$$\begin{vmatrix}\boldsymbol{q}_1 & \boldsymbol{q}_2 & \cdots & \alpha\boldsymbol{q}_j & \cdots & \boldsymbol{q}_n\end{vmatrix}=\alpha\begin{vmatrix}\boldsymbol{q}_1 & \boldsymbol{q}_2 & \cdots & \boldsymbol{q}_j & \cdots & \boldsymbol{q}_n\end{vmatrix}, \tag{4.202}$$

irrespective of α and j. The results conveyed by Eqs. (4.201) and (4.202) also imply

$$|\alpha\boldsymbol{A}|=\alpha^n|\boldsymbol{A}|, \tag{4.203}$$

after recalling the algorithm of multiplication of scalar by matrix, see Eq. (4.143). If, on the other hand, the i-th row of $\boldsymbol{A}$ has every element given by the sum of two elements, then Laplace's expansion should to advantage proceed along the said row – to eventually produce

$$\begin{vmatrix} \boldsymbol{a}_1 \\ \cdots \\ \boldsymbol{a}_i + \boldsymbol{b}_i \\ \cdots \\ \boldsymbol{a}_n \end{vmatrix} = \begin{vmatrix} \boldsymbol{a}_1 \\ \cdots \\ \boldsymbol{a}_i \\ \cdots \\ \boldsymbol{a}_n \end{vmatrix} + \begin{vmatrix} \boldsymbol{a}_1 \\ \cdots \\ \boldsymbol{b}_i \\ \cdots \\ \boldsymbol{a}_n \end{vmatrix}; \tag{4.204}$$

one also observes

$$\begin{aligned} &\left| \boldsymbol{q}_1 \quad \cdots \quad \boldsymbol{q}_j + \boldsymbol{r}_j \quad \cdots \quad \boldsymbol{q}_n \right| \\ &= \left| \boldsymbol{q}_1 \quad \cdots \quad \boldsymbol{q}_j \quad \cdots \quad \boldsymbol{q}_n \right| + \left| \boldsymbol{q}_1 \;\cdots\; \boldsymbol{r}_j \;\cdots\; \boldsymbol{q}_n \right| \end{aligned} \tag{4.205}$$

in the case of $\boldsymbol{Q}$, when all elements of a column are expressed as sum of two values – and is valid, irrespective of column location. If the i-th and j-th rows of $\boldsymbol{A}$ are identical, then

$$\begin{vmatrix} \boldsymbol{a}_1 \\ \cdots \\ \boldsymbol{a}_i \\ \cdots \\ \boldsymbol{a}_i \\ \cdots \\ \boldsymbol{a}_n \end{vmatrix} = 0; \tag{4.206}$$

by the same token, identical columns in $\boldsymbol{Q}$, at i-th and j-th positions, unfold

$$\left| \boldsymbol{q}_1 \quad \cdots \quad \boldsymbol{q}_i \quad \cdots \quad \boldsymbol{q}_i \quad \cdots \quad \boldsymbol{q}_n \right| = 0. \tag{4.207}$$

If two rows in $\boldsymbol{A}$ swap positions, then the corresponding determinant appears multiplied by -1, i.e.

$$\begin{vmatrix} \boldsymbol{a}_1 \\ \cdots \\ \boldsymbol{a}_j \\ \cdots \\ \boldsymbol{a}_i \\ \cdots \\ \boldsymbol{a}_n \end{vmatrix} = -\begin{vmatrix} \boldsymbol{a}_1 \\ \cdots \\ \boldsymbol{a}_i \\ \cdots \\ \boldsymbol{a}_j \\ \cdots \\ \boldsymbol{a}_n \end{vmatrix}; \tag{4.208}$$

a similar conclusion holds when two columns in $\boldsymbol{Q}$ are interchanged, i.e.

$$\left| \boldsymbol{q}_1 \;\cdots\; \boldsymbol{q}_j \;\cdots\; \boldsymbol{q}_i \;\cdots\; \boldsymbol{q}_n \right| = -\left| \boldsymbol{q}_1 \;\cdots\; \boldsymbol{q}_i \;\cdots\; \boldsymbol{q}_j \;\cdots\; \boldsymbol{q}_n \right|. \tag{4.209}$$

One also realizes that

$$\begin{vmatrix} \boldsymbol{a}_1 \\ \cdots \\ \boldsymbol{a}_i \\ \cdots \\ \alpha\,\boldsymbol{a}_i \\ \cdots \\ \boldsymbol{a}_n \end{vmatrix} = 0 \tag{4.210}$$

– i.e. two rows in $\boldsymbol{A}$ proportional to each other imply a nil determinant; an analogous results is found if working with columns in $\boldsymbol{Q}$, i.e.

$$\left| \boldsymbol{q}_1 \quad \cdots \quad \boldsymbol{q}_j \quad \cdots \quad \alpha\,\boldsymbol{q}_j \quad \cdots \quad \boldsymbol{q}_n \right| = 0. \tag{4.211}$$

A combination of the results conveyed by Eqs. (4.201), (4.204), and (4.210) refers to a matrix where one of the rows is replaced by the sum thereof with a multiple of another row, classically known as Jacobi's operation – which leads to

$$\begin{vmatrix} \boldsymbol{a}_1 \\ \cdots \\ \boldsymbol{a}_i + \alpha\,\boldsymbol{a}_j \\ \cdots \\ \boldsymbol{a}_j \\ \cdots \\ \boldsymbol{a}_n \end{vmatrix} = \begin{vmatrix} \boldsymbol{a}_1 \\ \cdots \\ \boldsymbol{a}_i \\ \cdots \\ \boldsymbol{a}_j \\ \cdots \\ \boldsymbol{a}_n \end{vmatrix} \tag{4.212}$$

based on $\boldsymbol{A}$; in other words, performance of Jacobi's operation on a row of a matrix does not affect the associated determinant. A similar conclusion can be drawn for such an operation applied to a column in $\boldsymbol{Q}$, according to

$$\begin{aligned} &\left| \boldsymbol{q}_1 \;\cdots\; \boldsymbol{q}_i + \alpha\,\boldsymbol{q}_j \;\cdots\; \boldsymbol{q}_j \;\cdots\; \boldsymbol{q}_n \right| \\ &= \left| \boldsymbol{q}_1 \;\cdots\; \boldsymbol{q}_i \;\cdots\; \boldsymbol{q}_j \;\cdots\; \boldsymbol{q}_n \right|. \end{aligned} \tag{4.213}$$

Consider now that $\boldsymbol{A}$ is an upper triangular matrix; one readily finds that

$$\begin{vmatrix} a_{1,1} & a_{1,2} & \cdots & a_{1,n} \\ 0 & a_{2,2} & \cdots & a_{2,n} \\ \cdots & \cdots & \cdots & \cdots \\ 0 & 0 & \cdots & a_{n,n} \end{vmatrix} = a_{1,1}\, a_{2,2} \ldots a_{n,n} \tag{4.214}$$

– and a similar result encompasses a lower triangular matrix, i.e.

$$\begin{vmatrix} a_{1,1} & 0 & \cdots & 0 \\ a_{2,1} & a_{2,2} & \cdots & 0 \\ \cdots & \cdots & \cdots & \cdots \\ a_{n,1} & a_{n,2} & \cdots & a_{n,n} \end{vmatrix} = a_{1,1}\, a_{2,2} \ldots a_{n,n}. \tag{4.215}$$

In the particular case of a scalar matrix – with a serving as common diagonal element, either Eq. (4.214) or (4.215) degenerates to

$$\begin{vmatrix} a & 0 & \ldots & 0 \\ 0 & a & \ldots & 0 \\ \ldots & \ldots & \ldots & \ldots \\ 0 & 0 & \ldots & a \end{vmatrix} = a^n \tag{4.216}$$

– because of the same factor appearing n times in the right-hand side; if $a = 1$, then Eq. (4.216) specifically yields

$$\left| \boldsymbol{I}_n \right| = 1, \tag{4.217}$$

pertaining to the $(n \times n)$ identity matrix. Regarding the product of two $(n \times n)$ square matrices $\boldsymbol{A}$ and $\boldsymbol{B}$, one realizes that

$$\left| \boldsymbol{AB} \right| = \left| \boldsymbol{A} \right| \left| \boldsymbol{B} \right| \tag{4.218}$$

– i.e. determinant of product of matrices equals product of determinants of the said matrices. When a matrix is inverted, it can be proven that

$$\left|A^{-1}\right| = \frac{1}{|A|}; \tag{4.219}$$

hence, determinant of inverse matrix is but arithmetical inverse (or reciprocal) of determinant of the original matrix.

4.2.13.4 Inversion of matrix

The adjoining matrix, $\hat{A}$, of a given (square) matrix A is defined as

$$\hat{A} \equiv \begin{bmatrix} (-1)^{1+1}|A_{-1,-1}| & (-1)^{1+2}|A_{-1,-2}| & \ldots & (-1)^{1+n}|A_{-1,-n}| \\ (-1)^{2+1}|A_{-2,-1}| & (-1)^{2+2}|A_{-2,-2}| & \ldots & (-1)^{2+n}|A_{-2,-n}| \\ \ldots & \ldots & \ldots & \ldots \\ (-1)^{n+1}|A_{-n,-1}| & (-1)^{n+2}|A_{-n,-2}| & \ldots & (-1)^{n+n}|A_{-n,-n}| \end{bmatrix}^T; \tag{4.220}$$

in other words, $\hat{A}$ can be obtained from A after replacing each $a_{i,j}$ $(i,j=1, 2, \ldots, n)$ of A by the corresponding $|A_{-i,-j}|$ multiplied by $(-1)^{i+j}$ – and transposing the result thus obtained. Since A is an $(n\times n)$ matrix, $\hat{A}$ will also be an $(n\times n)$ matrix – and so their product must still be an $(n\times n)$ matrix as per Eq. (4.152); $\hat{A}$ holds the unique property

$$A\hat{A} = |A|I_n. \tag{4.221}$$

Comparative inspection of Eqs. (4.176) and (4.221), upon division of both sides by $|A|$, indicates that

$$A^{-1} = \frac{\hat{A}}{|A|} \tag{4.222}$$

– which confirms that a matrix must be square to be invertible, otherwise $|A|$ would not be defined at all; and that $|A|\neq 0$ must also be satisfied, otherwise $1/|A|$ would make no sense.

4.2.14 Power of matrix

4.2.14.1 Eigenvalues and eigenvectors

By definition, the $(n\times 1)$ eigenvectors, v, of an $(n\times n)$ square matrix, A, are non-zero vectors that, upon multiplication by the said matrix, remain parallel to the original vector, i.e.

$$Av = \lambda v \tag{4.223}$$

– where scalar λ is termed eigenvalue; in other words, an eigenvalue is the factor by which the corresponding eigenvector is scaled when multiplied by A. The eigenvalues of A are, in turn, the solutions of

$$|A - \lambda I_n| = 0, \tag{4.224}$$

where I_n denotes the $(n\times n)$ identity matrix as usual; Eq. (4.224) is oftentimes referred to as characteristic equation of matrix A. The basic relationship between Eqs. (4.223) and (4.224) comes from the concept of linearly independent vectors; in fact, existence of an eigenvalue λ satisfying Eq. (4.224) guarantees existence of at least one solution $v\neq 0_{n\times 1}$ to Eq. (4.223) – so this equation becomes possible, yet undetermined.

4.2.14.2 Characteristic polynomial

The characteristic equation of A, departing from Eqs. (4.185) and (4.224), will eventually look like

$$\sum_{i=0}^{n} c_i\lambda^i = 0 \tag{4.225}$$

– where $c_0, c_1, \ldots, c_n$ denote real numbers, following application of Laplace's theorem at large, see Eqs. (4.191) and (4.192); the left-hand side of Eq. (4.225) is but a regular n-th degree polynomial in λ, viz.

$$|A - \lambda I_n| \equiv P_A\{\lambda\} \equiv \sum_{i=0}^{n} c_i\lambda^i, \tag{4.226}$$

equivalent to

$$P_A\{\lambda\} = \begin{vmatrix} a_{1,1}-\lambda & a_{1,2} & \ldots & a_{1,n} \\ a_{2,1} & a_{2,2}-\lambda & \ldots & a_{2,n} \\ \ldots & \ldots & \ldots & \ldots \\ a_{n,1} & a_{n,2} & \ldots & a_{n,n}-\lambda \end{vmatrix} \tag{4.227}$$

– upon definition of addition of matrices, see Eq. (4.138), and multiplication of scalar by matrix, see Eq. (4.143).

4.2.14.3 Cayley and Hamilton's theorem

According to Cayley and Hamilton's theorem, every $(n\times n)$ matrix A satisfies its own characteristic equation, i.e.

$$P_A\{A\} = 0_{n\times n} \tag{4.228}$$

– obtained upon combination of Eqs. (4.225) and (4.226), after replacing λ by A; note, however, that $P_A\{A\}$ is an $(n\times n)$ matrix, while $P_A\{\lambda\}$ is a scalar. It should be stressed that $P_A\{A\} \neq |A - AI_n| = |A - A| = 0$ obtained from Eq. (4.226) upon blind replacement of (scalar) λ by (matrix) A – because variable λ affects only the diagonal elements of A when previously multiplied by I_n and then subtracted therefrom, whereas A affects the whole matrix under the same circumstances. Since $P_A\{\lambda\}$ is an n-th degree polynomial, one may divide λ^m by $P_A\{\lambda\}$ to get

$$\frac{\lambda^m}{P_A\{\lambda\}} = Q\{\lambda\} + \frac{R\{\lambda\}}{P_A\{\lambda\}} \tag{4.229}$$

in agreement with Eq. (4.123) – where $R\{\lambda\}$ denotes (remainder) polynomial in λ, with degree not above $n-1$, and $Q\{\lambda\}$ denotes divisor polynomial also in λ. Therefore, $R\{\lambda\}$ can be expressed as

$$R\{\lambda\} = \sum_{i=0}^{n-1} b_i\lambda^i, \tag{4.230}$$

with b_is denoting real coefficients; while Eq. (4.229) may appear as

$$\lambda^m = P_A\{\lambda\}Q\{\lambda\} + R\{\lambda\}, \tag{4.231}$$

following multiplication of both sides by $P_A\{\lambda\}$. If λ in Eq. (4.231) is again replaced by A, one gets

$$A^m = P_A\{A\}Q\{A\} + R\{A\}; \tag{4.232}$$

in view of Eq. (4.228), one may proceed to

$$A^m = \mathbf{0}_{n\times n}Q\{A\} + R\{A\} = \mathbf{0}_{n\times n} + R\{A\} = R\{A\}, \tag{4.233}$$

also with the aid of Eqs. (4.142) and (4.158) – where insertion of Eq. (4.230) finally yields

$$A^m = \sum_{i=0}^{n-1} b_i A^i. \tag{4.234}$$

Equation (4.234) indicates that the m-th power of $(n\times n)$ matrix A can actually be calculated as a linear combination of lower powers of A (just up to $n-1<m$) – thus facilitating calculation of A^m, especially when m is large and n is relatively small; all that is left now is calculation of coefficients b_i ($i=0, 1, ..., n-1$). If the roots of the characteristic polynomial, as per Eqs. (4.225) and (4.226) are denoted as $\lambda_1, \lambda_2, \ldots, \lambda_n$, then one has it that

$$P_A\{\lambda_j\} \equiv 0; \quad j = 1,2,\ldots,n \tag{4.235}$$

owing to the definition of root; one will then obtain

$$\lambda_j{}^m = P_A\{\lambda_j\}Q\{\lambda_j\} + R\{\lambda_j\} \tag{4.236}$$

from Eq. (4.231), while combination with Eq. (4.235) permits simplification to

$$\lambda_j{}^m = 0Q\{\lambda_j\} + R\{\lambda_j\} = 0 + R\{\lambda_j\} = R\{\lambda_j\} \tag{4.237}$$

– or, in view of Eq. (4.230),

$$\sum_{i=0}^{n-1} b_i \lambda_j{}^i = \lambda_j{}^m; \quad j = 1,2,\ldots,n. \tag{4.238}$$

Equation (4.238) represents a system of n linear equations in n unknowns, i.e. the b_is, since all λ_js are known in advance (to be solved as explained below).

4.2.14.4 Routh and Hurwitz's theorem

Upon combination of Eqs. (4.225) and (4.226), it is possible to obtain

$$(-1)^n\left|\lambda I_n - A\right| = \lambda^n + b_1\lambda^{n-1} + \ldots + b_{n-1}\lambda + b_n = 0, \tag{4.239}$$

as a consequence of Eq. (4.203); inspection of Eq. (4.239) *vis-à-vis* with Eqs. (4.225) and (4.226) unfolds

$$b_i \equiv \frac{c_{n-i}}{c_n}; \quad i = 0,1,\ldots,n. \tag{4.240}$$

Routh and Hurwitz's theorem states that eigenvalues $\lambda_1, \lambda_2, \ldots, \lambda_n$ (or their real parts, if complex), obtained as roots to Eq. (4.239), are all negative if

$$\Delta_1 > 0 \wedge \Delta_2 > 0 \wedge \ldots \wedge \Delta_n > 0 \tag{4.241}$$

– provided that Δ_i is defined as

$$\Delta_i \equiv \begin{vmatrix} b_1 & 1 & 0 & 0 & 0 & 0 & \ldots & 0 \\ b_3 & b_2 & b_1 & 1 & 0 & 0 & \ldots & 0 \\ b_5 & b_4 & b_3 & b_2 & b_1 & 1 & \ldots & 0 \\ \ldots & \ldots & \ldots & \ldots & \ldots & \ldots & \ldots & \ldots \\ b_{2i-1} & b_{2i-2} & b_{2i-3} & b_{2i-4} & b_{2i-5} & b_{2i-6} & \ldots & b_i \end{vmatrix} \tag{4.242}$$

$$i = 1,2,\ldots,n,$$

in general. This result is quite useful in attempts to detect system instability, namely, when dealing with control issues – as it circumvents the need to find the actual values of the λ_is (normally a difficult task when $n>2$).

4.2.15 Vectors

4.2.15.1 Definition

When a single numerical value (or scalar) does not suffice to represent some quantity, one may resort to a vector – defined by a triplet or real numbers, $\boldsymbol{u}(a,b,c)$. Its usual graphical representation is a straight, arrowed segment, linking the origin of a Cartesian system of coordinates to point of coordinates (a,b,c) – with length equal to $\sqrt{a^2+b^2+c^2}$ as per the three-dimensional version of Pythagoras' theorem labeled as Eq. (4.79), and angles with the x-, y-, and z-axis given by $\cos^{-1} a/\sqrt{a^2+b^2+c^2}$, $\cos^{-1} b/\sqrt{a^2+b^2+c^2}$, and $\cos^{-1} c/\sqrt{a^2+b^2+c^2}$, respectively.

4.2.15.2 Addition of vectors

The simpler operation involving vectors is addition; to be applied to $\boldsymbol{u}(u_x,u_y,u_z)$ and $\boldsymbol{v}(v_x,v_y,v_z)$, the point of origin of $\boldsymbol{v}$ must be made coincident with the point of termination of $\boldsymbol{u}$, thus supporting

$$\boldsymbol{u} + \boldsymbol{v} \equiv \boldsymbol{i}_x\left(u_x + v_x\right) + \boldsymbol{i}_y\left(u_y + v_y\right) + \boldsymbol{i}_z\left(u_z + v_z\right) \tag{4.243}$$

as algorithm – where $\boldsymbol{i}_x$, $\boldsymbol{i}_y$, and $\boldsymbol{i}_z$ denote unit vectors in the x-, y-, and z-directions, respectively. Addition of vectors is commutative, i.e.

$$\boldsymbol{u} + \boldsymbol{v} = \boldsymbol{v} + \boldsymbol{u}; \tag{4.244}$$

given a third vector $\boldsymbol{w}$, one also concludes that

$$\boldsymbol{u} + (\boldsymbol{v} + \boldsymbol{w}) = (\boldsymbol{u} + \boldsymbol{v}) + \boldsymbol{w} \tag{4.245}$$

– meaning that addition of vectors is associative.

4.2.15.3 Multiplication of scalar by vector

Another common operation is multiplication of vector $\boldsymbol{u}$, of length $\|\boldsymbol{u}\|$, by scalar α; this produces a new vector $\alpha\boldsymbol{u}$, collinear with $\boldsymbol{u}$, with length equal to $|\alpha|\,\|\boldsymbol{u}\|$, where $|\alpha|$ abides to Eq. (4.15). Using vector coordinates, one accordingly finds

$$\alpha\boldsymbol{u} = \boldsymbol{i}_x\left(\alpha u_x\right) + \boldsymbol{i}_y\left(\alpha u_y\right) + \boldsymbol{i}_z\left(\alpha u_z\right), \tag{4.246}$$

i.e. the coordinates in each direction of space are expanded (or contracted) proportionally through α. One realizes that

$$\boldsymbol{u}\alpha = \alpha\boldsymbol{u}, \tag{4.247}$$

so multiplication of scalar by vector is commutative; on the other hand,

$$\alpha(\beta \boldsymbol{u}) = (\alpha\beta)\boldsymbol{u}, \tag{4.248}$$

thus implying that multiplication of scalar by vector is associative. When considered together with addition of vectors, it can be written

$$\alpha(\boldsymbol{u} + \boldsymbol{v}) = \alpha\boldsymbol{u} + \alpha\boldsymbol{v}, \tag{4.249}$$

i.e. multiplication of scalar by vector is distributive with regard to addition of vectors; by the same token,

$$(\alpha + \beta)\boldsymbol{u} = \alpha\boldsymbol{u} + \beta\boldsymbol{u} \tag{4.250}$$

– meaning that multiplication of scalar by vector is distributive also with regard to addition of scalars.

4.2.15.4 Scalar multiplication of vectors

The scalar (or inner) product of vector $\boldsymbol{u}$, by vector $\boldsymbol{v}$ with length $\|\boldsymbol{v}\|$ – which may be represented by

$$\boldsymbol{u} \cdot \boldsymbol{v} \equiv \boldsymbol{u}^T \boldsymbol{v} = \begin{bmatrix} u_x & u_y & u_z \end{bmatrix} \begin{bmatrix} v_x \\ v_y \\ v_z \end{bmatrix}, \tag{4.251}$$

is formally defined as

$$\boldsymbol{u} \cdot \boldsymbol{v} \equiv \|\boldsymbol{u}\| \|\boldsymbol{v}\| \cos\{\angle \boldsymbol{u},\boldsymbol{v}\}; \tag{4.252}$$

here $\cos\{\angle \boldsymbol{u},\boldsymbol{v}\}$ denotes cosine of (the smaller) angle formed by vectors $\boldsymbol{u}$ and $\boldsymbol{v}$. If Eq. (4.252) is rewritten as

$$\boldsymbol{u} \cdot \boldsymbol{v} = \|\boldsymbol{u}\| \left(\|\boldsymbol{v}\| \cos\{\angle \boldsymbol{u},\boldsymbol{v}\} \right), \tag{4.253}$$

then scalar product will be viewed as product of length of $\boldsymbol{u}$ by length of the projection of $\boldsymbol{v}$ over $\boldsymbol{u}$ – see Eq. (4.67). As a consequence of Eq. (4.252), one has that

$$\boldsymbol{u} \cdot \boldsymbol{u} = \|\boldsymbol{u}\| \|\boldsymbol{u}\| \cos\{\angle \boldsymbol{u},\boldsymbol{u}\} = \|\boldsymbol{u}\|^2 \cos 0 = \|\boldsymbol{u}\|^2, \tag{4.254}$$

because cos 0 is equal to unity. On the other hand, the definition provided by Eq. (4.252) implies that scalar product is nil for two orthogonal vectors; for instance, $\boldsymbol{u}$ and $\boldsymbol{u}^\perp$ (sharing, in particular, the same length) satisfy

$$\boldsymbol{u} \cdot \boldsymbol{u}^\perp = \|\boldsymbol{u}\| \|\boldsymbol{u}\| \cos\left\{\angle \boldsymbol{u},\boldsymbol{u}^\perp\right\} = \|\boldsymbol{u}\|^2 \cos\frac{\pi}{2} = 0, \tag{4.255}$$

since cosine of the angle formed thereby is nil. Therefore, scalar product being nil does not necessarily imply that (at least) one of the vector factors is a nil vector. Commutativity of the product of scalars enforces

$$\boldsymbol{u} \cdot \boldsymbol{v} = \boldsymbol{v} \cdot \boldsymbol{u}; \tag{4.256}$$

whereas

$$\boldsymbol{u} \cdot (\boldsymbol{v} + \boldsymbol{w}) = \boldsymbol{u} \cdot \boldsymbol{v} + \boldsymbol{u} \cdot \boldsymbol{w} \tag{4.257}$$

unfolds the distributive property of scalar product of vectors, over vector addition on the right – with

$$(\boldsymbol{u} + \boldsymbol{v}) \cdot \boldsymbol{w} = \boldsymbol{u} \cdot \boldsymbol{w} + \boldsymbol{v} \cdot \boldsymbol{w} \tag{4.258}$$

indicating that scalar product of vectors is also distributive over vector addition on the left. Multiple products are possible as well; consider first scalar product of two vectors, combined with product of scalar by vector – which leads to

$$(s\boldsymbol{u}) \cdot \boldsymbol{v} \equiv s(\boldsymbol{u} \cdot \boldsymbol{v}). \tag{4.259}$$

Therefore, the dot product of the scalar multiple of a vector by another vector ends up being equal to the product of the said scalar by the dot product of the two vectors – and one similarly finds

$$\boldsymbol{u} \cdot (s\boldsymbol{v}) \equiv s(\boldsymbol{u} \cdot \boldsymbol{v}). \tag{4.260}$$

Conversely, vectors $\boldsymbol{u}(\boldsymbol{v} \cdot \boldsymbol{w})$ and $(\boldsymbol{u} \cdot \boldsymbol{v})\boldsymbol{w}$ abide to

$$\boldsymbol{u}(\boldsymbol{v} \cdot \boldsymbol{w}) \neq (\boldsymbol{u} \cdot \boldsymbol{v})\boldsymbol{w}, \tag{4.261}$$

i.e. the scalar product of vectors is not associative with regard to the product of vectors. One immediate consequence of Eq. (4.251) is

$$\boldsymbol{u} \cdot \boldsymbol{v} = u_x v_x + u_y v_y + u_z v_z, \tag{4.262}$$

often used in alternative to the algorithm conveyed by Eq. (4.152) – or, in condensed form,

$$\boldsymbol{u} \cdot \boldsymbol{v} \equiv \sum_{i=1}^{3} u_i v_i, \tag{4.263}$$

where subscript stands for x ($i=1$), y ($i=2$), or z ($i=3$).

4.2.15.5 Vector multiplication of vectors

The vector (or outer) product of two vectors is a third vector – denoted as $\boldsymbol{u} \times \boldsymbol{v}$, which satisfies

$$\boldsymbol{u} \times \boldsymbol{v} \equiv \|\boldsymbol{u}\| \|\boldsymbol{v}\| \sin\{\angle \boldsymbol{u},\boldsymbol{v}\}\boldsymbol{n}; \tag{4.264}$$

here $\sin\{\angle \boldsymbol{u},\boldsymbol{v}\}$ denotes sine of (the smaller) angle formed by vectors $\boldsymbol{u}$ and $\boldsymbol{v}$, while $\boldsymbol{n}$ denotes unit vector normal to the plane containing $\boldsymbol{u}$ and $\boldsymbol{v}$, and oriented such that $\boldsymbol{u}$, $\boldsymbol{v}$, and $\boldsymbol{n}$ form a right-handed system. The area, S, of a parallelogram, with sides defined by $\boldsymbol{u}$ and $\boldsymbol{v}$, is given by the product of its base, $\|\boldsymbol{u}\|$, by its height – obtained, in turn, as projection of $\boldsymbol{v}$ onto $\boldsymbol{u}^\perp$, i.e. $\|\boldsymbol{v}\| \sin\{\angle \boldsymbol{u},\boldsymbol{v}\}$, according to

$$S = \|\boldsymbol{u}\| \left(\|\boldsymbol{v}\| \sin\{\angle \boldsymbol{u},\boldsymbol{v}\} \right) \tag{4.265}$$

with the aid of Eq. (4.74); hence, Eq. (4.264) can be rewritten as

$$\boldsymbol{u} \times \boldsymbol{v} = S\boldsymbol{n}, \tag{4.266}$$

meaning that vector product defines vector area, $S\boldsymbol{n}$, of the portion of plane bounded by vectors $\boldsymbol{u}$ and $\boldsymbol{v}$ taken consecutively. Furthermore, volume, V, of a parallelepiped defined by vectors $\boldsymbol{u}$, $\boldsymbol{v}$, and $\boldsymbol{w}$, can be calculated as area of the parallelogram that constitutes its base – defined by $\boldsymbol{u}$ and $\boldsymbol{v}$, and represented by vector $(\|\boldsymbol{u}\| \|\boldsymbol{v}\| \sin\{\angle \boldsymbol{u},\boldsymbol{v}\})\boldsymbol{n}$ as per Eq. (4.265); multiplied by its height – i.e. projection of $\boldsymbol{w}$ upon $\boldsymbol{n}$, calculated as $\|\boldsymbol{w}\| \cos\{\angle \boldsymbol{w},\boldsymbol{n}\}$, or else

$$\left(\boldsymbol{u} \times \boldsymbol{v} \right) \cdot \boldsymbol{w} = V \tag{4.267}$$

in condensed form. The definition conveyed by Eq. (4.264) implies that vector product is nil for two collinear vectors, $\boldsymbol{u}$ and $\alpha\boldsymbol{u}$, viz.

$$\boldsymbol{u}\times(\alpha\boldsymbol{u})=\|\boldsymbol{u}\|\,|\alpha|\,\|\boldsymbol{u}\|\sin\{\angle\boldsymbol{u},\alpha\boldsymbol{u}\}\boldsymbol{n}=|\alpha|\,\|\boldsymbol{u}\|^2\sin 0\,\boldsymbol{n}=\boldsymbol{0} \quad (4.268)$$

for $\alpha>0$ – because sine of the angle formed thereby is nil; a similar conclusion results when $\alpha<0$, because $\sin\{\angle\boldsymbol{u},\alpha\boldsymbol{u}\}=\sin\pi=0$ in this case. Therefore, vector product being nil does not necessarily imply that (at least) one of the factor vectors is a nil vector itself. When in possession of coordinates u_x, u_y, and u_z of vector $\boldsymbol{u}$, and v_x, v_y, and v_z of vector $\boldsymbol{v}$, it is possible to express their vector product as

$$\boldsymbol{u}\times\boldsymbol{v}=\boldsymbol{i}_x\begin{vmatrix}u_y & u_z\\ v_y & v_z\end{vmatrix}-\boldsymbol{i}_y\begin{vmatrix}u_x & u_z\\ v_x & v_z\end{vmatrix}+\boldsymbol{i}_y\begin{vmatrix}u_x & u_y\\ v_x & v_y\end{vmatrix}=\begin{vmatrix}\boldsymbol{i}_x & \boldsymbol{i}_y & \boldsymbol{i}_z\\ u_x & u_y & u_z\\ v_x & v_y & v_z\end{vmatrix}, \quad (4.269)$$

at the expense of the concept of determinant as conveyed by Eq. (4.191) for $i=1$; one may instead write

$$\boldsymbol{u}\times\boldsymbol{v}\equiv\sum_{i=1}^{3}\sum_{j=1}^{3}\sum_{k=1}^{3}\boldsymbol{i}_i\delta_{ijk}u_jv_k. \quad (4.270)$$

As in Eq. (4.263), subscripts stand for x ($i,j,k=1$), y ($i,j,k=2$), or z ($i,j,k=3$) – as long as the alternating operator, δ_{ijk}, is defined by

$$\delta_{ijk}\equiv\begin{Bmatrix}1\Leftarrow ijk=123\vee ijk=231\vee ijk=312\\ -1\Leftarrow ijk=321\vee ijk=132\vee ijk=213\\ 0\Leftarrow i=j\vee i=k\vee j=k\end{Bmatrix}. \quad (4.271)$$

The vector product is anticommutative, i.e.

$$\boldsymbol{v}\times\boldsymbol{u}=-\boldsymbol{u}\times\boldsymbol{v}; \quad (4.272)$$

on the other hand,

$$\boldsymbol{u}\times(\boldsymbol{v}+\boldsymbol{w})=\boldsymbol{u}\times\boldsymbol{v}+\boldsymbol{u}\times\boldsymbol{w} \quad (4.273)$$

implies that vector product is distributive, with regard to addition of vectors on the right – as well as

$$(\boldsymbol{u}+\boldsymbol{v})\times\boldsymbol{w}=\boldsymbol{u}\times\boldsymbol{w}+\boldsymbol{v}\times\boldsymbol{w}, \quad (4.274)$$

meaning that the said product is distributive, with regard to addition of vectors also on the left. When both scalar and vector products are at stake, one obtains

$$\boldsymbol{u}\cdot(\boldsymbol{v}\times\boldsymbol{w})=\boldsymbol{v}\cdot(\boldsymbol{w}\times\boldsymbol{u})=\boldsymbol{w}\cdot(\boldsymbol{u}\times\boldsymbol{v}) \quad (4.275)$$

– where the sequential order of $\boldsymbol{u}$, $\boldsymbol{v}$, and $\boldsymbol{w}$ as factors was kept, and the sequential order of operators · and × was retained as well; one may also prove that

$$\boldsymbol{u}\times(\boldsymbol{v}\times\boldsymbol{w})=\boldsymbol{v}(\boldsymbol{u}\cdot\boldsymbol{w})-\boldsymbol{w}(\boldsymbol{u}\cdot\boldsymbol{v}), \quad (4.276)$$

applicable when two vector products are combined – known as Lagrange's formula.

4.2.16 Tensors

4.2.16.1 Definition

A tensor, $\boldsymbol{\tau}$, results from extrapolating a vector to a nine-dimensional space, based on each of its three-directional components – so it has no geometrical representation, being defined solely by its matrix form, viz.

$$\boldsymbol{\tau}\equiv\begin{bmatrix}\tau_{xx} & \tau_{xy} & \tau_{xz}\\ \tau_{yx} & \tau_{yy} & \tau_{yz}\\ \tau_{zx} & \tau_{zy} & \tau_{zz}\end{bmatrix}; \quad (4.277)$$

here (real) τ_{ij} denotes component in i- and j-directions, with $i,j=x,y,z$.

4.2.16.2 Dyadic multiplication of vectors

The dyadic multiplication of two vectors abides to

$$\boldsymbol{u}\otimes\boldsymbol{v}\equiv\boldsymbol{u}\boldsymbol{v}^T=\begin{bmatrix}u_x\\ u_y\\ u_z\end{bmatrix}\begin{bmatrix}v_x & v_y & v_z\end{bmatrix} \quad (4.278)$$

– which implies

$$\boldsymbol{u}\otimes\boldsymbol{v}=\begin{bmatrix}u_xv_x & u_xv_y & u_xv_z\\ u_yv_x & u_yv_y & u_yv_z\\ u_zv_x & u_zv_y & u_zv_z\end{bmatrix}, \quad (4.279)$$

with the aid of Eq. (4.152). A tensor is accordingly produced, which may appear, in a more condensed version, as

$$\boldsymbol{u}\otimes\boldsymbol{v}\equiv\sum_{i=1}^{3}\sum_{j=1}^{3}\varphi_{ij}u_iv_j, \quad (4.280)$$

provided that subscripts denote x ($i,j=1$), y ($i,j=2$), or z ($i,j=3$); complemented by

$$\varphi_{ij}\equiv\boldsymbol{i}_i\boldsymbol{i}_j;\quad i=1,2,3;\ j=1,2,3, \quad (4.281)$$

standing as definition of unit tensors.

4.2.16.3 Double scalar multiplication of tensors

The scalar multiplication of two tensors, $\boldsymbol{\sigma}$ and $\boldsymbol{\tau}$ – also known as double dot product, is a scalar defined as

$$\begin{aligned}\boldsymbol{\sigma}:\boldsymbol{\tau}\equiv{}&\sigma_{xx}\tau_{xx}+\sigma_{xy}\tau_{yx}+\sigma_{xz}\tau_{zx}+\sigma_{yx}\tau_{xy}+\sigma_{yy}\tau_{yy}\\ &+\sigma_{yz}\tau_{zy}+\sigma_{zx}\tau_{xz}+\sigma_{zy}\tau_{yz}+\sigma_{zz}\tau_{zz},\end{aligned} \quad (4.282)$$

with $\boldsymbol{\sigma}$ given by

$$\boldsymbol{\sigma}\equiv\begin{bmatrix}\sigma_{xx} & \sigma_{xy} & \sigma_{xz}\\ \sigma_{yx} & \sigma_{yy} & \sigma_{yz}\\ \sigma_{zx} & \sigma_{zy} & \sigma_{zz}\end{bmatrix}; \quad (4.283)$$

the scalar conveyed by Eq. (4.282) may be condensed to

$$\boldsymbol{\sigma}:\boldsymbol{\tau}\equiv\sum_{i=1}^{3}\sum_{j=1}^{3}\sigma_{ij}\tau_{ji}. \quad (4.284)$$

4.2.16.4 Scalar multiplication of tensors

Finally, the dot product of two tensors, $\boldsymbol{\sigma}$ and $\boldsymbol{\tau}$, follows

$$\boldsymbol{\sigma}\cdot\boldsymbol{\tau} \equiv \begin{bmatrix} \sigma_{xx} & \sigma_{xy} & \sigma_{xz} \\ \sigma_{yx} & \sigma_{yy} & \sigma_{yz} \\ \sigma_{zx} & \sigma_{zy} & \sigma_{zz} \end{bmatrix} \begin{bmatrix} \tau_{xx} & \tau_{xy} & \tau_{xz} \\ \tau_{yx} & \tau_{yy} & \tau_{yz} \\ \tau_{zx} & \tau_{zy} & \tau_{zz} \end{bmatrix}$$
$$\equiv \begin{bmatrix} \sigma_{xx}\tau_{xx}+\sigma_{xy}\tau_{yx}+\sigma_{xz}\tau_{zx} & \sigma_{xx}\tau_{xy}+\sigma_{xy}\tau_{yy}+\sigma_{xz}\tau_{zy} & \sigma_{xx}\tau_{xz}+\sigma_{xy}\tau_{yz}+\sigma_{xz}\tau_{zz} \\ \sigma_{yx}\tau_{xx}+\sigma_{yy}\tau_{yx}+\sigma_{yz}\tau_{zx} & \sigma_{yx}\tau_{xy}+\sigma_{yy}\tau_{yy}+\sigma_{yz}\tau_{zy} & \sigma_{yx}\tau_{xz}+\sigma_{yy}\tau_{yz}+\sigma_{yz}\tau_{zz} \\ \sigma_{zx}\tau_{xx}+\sigma_{zy}\tau_{yx}+\sigma_{zz}\tau_{zx} & \sigma_{zx}\tau_{xy}+\sigma_{zy}\tau_{yy}+\sigma_{zz}\tau_{zy} & \sigma_{zx}\tau_{xz}+\sigma_{zy}\tau_{yz}+\sigma_{zz}\tau_{zz} \end{bmatrix} \tag{4.285}$$

– and generates another tensor, with coordinates obtained via the classical product of two matrices, see Eq. (4.152); this is equivalent to writing

$$\boldsymbol{\sigma}\cdot\boldsymbol{\tau} \equiv \sum_{i=1}^{3}\sum_{l=1}^{3}\varphi_{il}\sum_{j=1}^{3}\sigma_{ij}\tau_{jl}, \tag{4.286}$$

in a more condensed form.

4.2.17 Solution of systems of linear algebraic equations

4.2.17.1 Cramer's rule

Consider a system of n linear equations in n unknowns, viz. x_1, x_2, …, x_n, of the form

$$\begin{aligned} &\sum_{i=1}^{n} a_{1,i}x_i \equiv a_{1,1}x_1 + a_{1,2}x_2 + \ldots + a_{1,j}x_j + \ldots + a_{1,n}x_n = b_1 \\ &\sum_{i=1}^{n} a_{2,i}x_i \equiv a_{2,1}x_1 + a_{2,2}x_2 + \ldots + a_{2,j}x_j + \ldots + a_{2,n}x_n = b_2 \\ &\ldots \\ &\sum_{i=1}^{n} a_{n,i}x_i \equiv a_{n,1}x_1 + a_{n,2}x_2 + \ldots + a_{n,j}x_j + \ldots + a_{n,n}x_n = b_n. \end{aligned} \tag{4.287}$$

The corresponding $(n\times n)$ matrix of coefficients, $\boldsymbol{A}$, accordingly reads

$$\boldsymbol{A} \equiv \begin{bmatrix} a_{1,1} & a_{1,2} & \ldots & a_{1,j} & \ldots & a_{1,n} \\ a_{2,1} & a_{2,2} & \ldots & a_{2,j} & \ldots & a_{2,n} \\ \ldots & \ldots & \ldots & \ldots & \ldots & \ldots \\ a_{i,1} & a_{i,2} & \ldots & a_{i,j} & \ldots & a_{i,n} \\ \ldots & \ldots & \ldots & \ldots & \ldots & \ldots \\ a_{n,1} & a_{n,2} & \ldots & a_{n,j} & \ldots & a_{n,n} \end{bmatrix} \tag{4.288}$$

in parallel to Eq. (4.135), or else

$$\boldsymbol{A} \equiv \begin{bmatrix} \boldsymbol{a}_1 & \boldsymbol{a}_2 & \ldots & \boldsymbol{a}_j & \ldots & \boldsymbol{a}_n \end{bmatrix} \tag{4.289}$$

– as long as $\boldsymbol{a}_j$ denotes an $(n\times 1)$ column vector, of the form

$$\boldsymbol{a}_j \equiv \begin{bmatrix} a_{1,j} \\ a_{2,j} \\ \ldots \\ a_{i,j} \\ \ldots \\ a_{n,j} \end{bmatrix}; \quad j = 1,2,\ldots,n; \tag{4.290}$$

whereas the $(n\times 1)$ vector of independent terms, $\boldsymbol{b}$, abides to

$$\boldsymbol{b} \equiv \begin{bmatrix} b_1 \\ b_2 \\ \ldots \\ b_n \end{bmatrix}. \tag{4.291}$$

If the aforementioned n equations are all independent of, and compatible with each other, then

$$\left|\boldsymbol{A}\right| \neq 0; \tag{4.292}$$

under such circumstances, a single solution exists for Eq. (4.287), given by

$$x_j = \frac{\left|\boldsymbol{A}\right|_{\boldsymbol{a}_j\leftarrow\boldsymbol{b}}}{\left|\boldsymbol{A}\right|} = \frac{\left|\boldsymbol{a}_1 \;\; \boldsymbol{a}_2 \;\; \ldots \;\; \boldsymbol{b} \;\; \ldots \;\; \boldsymbol{a}_n\right|}{\left|\boldsymbol{a}_1 \;\; \boldsymbol{a}_2 \;\; \ldots \;\; \boldsymbol{a}_j \;\; \ldots \;\; \boldsymbol{a}_n\right|}, \tag{4.293}$$

with the aid of Eqs. (4.289)–(4.291) – known as Cramer's rule. Insertion of Eqs. (4.290) and (4.291) permits indeed expansion of Eq. (4.293) to

$$x_j = \frac{\begin{vmatrix} a_{1,1} & a_{1,2} & \ldots & b_1 & \ldots & a_{1,n} \\ a_{2,1} & a_{2,2} & \ldots & b_2 & \ldots & a_{2,n} \\ \ldots & \ldots & \ldots & \ldots & \ldots & \ldots \\ a_{n,1} & a_{n,2} & \ldots & b_n & \ldots & a_{n,n} \end{vmatrix}}{\begin{vmatrix} a_{1,1} & a_{1,2} & \ldots & a_{1,j} & \ldots & a_{1,n} \\ a_{2,1} & a_{2,2} & \ldots & a_{2,j} & \ldots & a_{2,n} \\ \ldots & \ldots & \ldots & \ldots & \ldots & \ldots \\ a_{n,1} & a_{n,2} & \ldots & a_{n,j} & \ldots & a_{n,n} \end{vmatrix}}; \; j = 1,2,\ldots,n, \tag{4.294}$$

valid for every unknown; the determinants in both numerator and denominator are then to be calculated, usually resorting to Laplace's theorem.

If there are fewer equations, n, than unknowns, m, then the system of linear algebraic equations will look like

$$\begin{aligned} &\sum_{i=1}^{m} a_{1,i}x_i \equiv a_{1,1}x_1 + a_{1,2}x_2 + \ldots + a_{1,j}x_j + \ldots \\ &\qquad + a_{1,n}x_n + a_{1,n+1}x_{n+1} + \ldots + a_{1,m}x_m = b_1 \\ &\sum_{i=1}^{m} a_{2,i}x_i \equiv a_{2,1}x_1 + a_{2,2}x_2 + \ldots + a_{2,j}x_j + \ldots \\ &\qquad + a_{2,n}x_n + a_{2,n+1}x_{n+1} + \ldots + a_{2,m}x_m = b_2 \\ &\ldots \\ &\sum_{i=1}^{m} a_{n,i}x_i \equiv a_{n,1}x_1 + a_{n,2}x_2 + \ldots + a_{n,j}x_j + \ldots \\ &\qquad + a_{n,n}x_n + a_{n,n+1}x_{n+1} + \ldots + a_{n,m}x_m = b_n \end{aligned} \tag{4.295}$$

– to be used *in lieu* of Eq. (4.287); when a subset of n (arbitrarily chosen) unknowns are taken, Eq. (4.295) will appear as

$$\begin{aligned} &a_{1,1}x_1 + a_{1,2}x_2 + \ldots + a_{1,j}x_j + \ldots + a_{1,n}x_n \\ &= b_1 - a_{1,n+1}x_{n+1} - \ldots - a_{1,m}x_m \\ &a_{2,1}x_1 + a_{2,2}x_2 + \ldots + a_{2,j}x_j + \ldots + a_{2,n}x_n \\ &= b_2 - a_{2,n+1}x_{n+1} - \ldots - a_{2,m}x_m \\ &\ldots \\ &a_{n,1}x_1 + a_{n,2}x_2 + \ldots + a_{n,j}x_j + \ldots + a_{n,n}x_n \\ &= b_n - a_{n,n+1}x_{n+1} - \ldots - a_{n,m}x_m. \end{aligned} \tag{4.296}$$

After defining $(n\times 1)$ column vector $\boldsymbol{b}^*$ as

$$\boldsymbol{b}^* \equiv \boldsymbol{b}-\sum_{i=n+1}^{m}\boldsymbol{a}_i x_i \tag{4.297}$$

– departing from Eq. (4.291), with the aid of Eqs. (4.143) and (4.290), one can apply Cramer's rule as

$$x_j=\frac{\left|\boldsymbol{A}\right|_{\boldsymbol{a}_j\leftarrow \boldsymbol{b}^*}}{\left|\boldsymbol{A}\right|}=\frac{\begin{vmatrix}\boldsymbol{a}_1 & \boldsymbol{a}_2 & \ldots & \boldsymbol{b}^* & \ldots & \boldsymbol{a}_n\end{vmatrix}}{\begin{vmatrix}\boldsymbol{a}_1 & \boldsymbol{a}_2 & \ldots & \boldsymbol{a}_j & \ldots & \boldsymbol{a}_n\end{vmatrix}} \tag{4.298}$$

using Eq. (4.293) as template. If the n equations chosen are again independent of, and compatible with each other, then Eq. (4.298) will be meaningful – and may appear instead as

$$x_j=\frac{\begin{vmatrix} a_{1,1} & \ldots & b_1-\ldots-a_{1,m}x_m & \ldots & a_{1,n}\\ a_{2,1} & \ldots & b_2-\ldots-a_{2,m}x_m & \ldots & a_{2,n}\\ \ldots & \ldots & \ldots & \ldots & \ldots\\ a_{n,1} & \ldots & b_n-\ldots-a_{n,m}x_m & \ldots & a_{n,n}\end{vmatrix}}{\begin{vmatrix} a_{1,1} & \ldots & a_{1,j} & \ldots & a_{1,n}\\ a_{2,1} & \ldots & a_{2,j} & \ldots & a_{2,n}\\ \ldots & \ldots & \ldots & \ldots & \ldots\\ a_{n,1} & \ldots & a_{n,j} & \ldots & a_{n,n}\end{vmatrix}} \qquad j=1,2,\ldots,n. \tag{4.299}$$

After calculation of determinants in numerator and denominator, Eq. (4.299) will convey the n selected (also known as principal) unknowns, yet as linear functions of the remaining $m-n$ unknowns; an infinite number of solutions to Eq. (4.295) will thus be found in this case.

4.2.17.2 Explicitation

The system of linear algebraic equations denoted as Eq. (4.287) can be rewritten, in condensed form, as

$$\boldsymbol{A}\boldsymbol{x}=\boldsymbol{b} \tag{4.300}$$

at the expense of Eqs. (4.288) and (4.291), complemented by $\boldsymbol{x}\equiv[x_1 \quad x_2 \quad \ldots \quad x_n]^T$. If $\boldsymbol{A}^{-1}$ exists, then premultiplication of both sides of Eq. (4.300) by $\boldsymbol{A}^{-1}$, i.e.

$$\boldsymbol{A}^{-1}\boldsymbol{A}\boldsymbol{x}=\boldsymbol{A}^{-1}\boldsymbol{b}, \tag{4.301}$$

will lead to

$$\left(\boldsymbol{A}^{-1}\boldsymbol{A}\right)\boldsymbol{x}=\boldsymbol{I}_n\boldsymbol{x}=\boldsymbol{A}^{-1}\boldsymbol{b} \tag{4.302}$$

in view of the property conveyed by Eq. (4.176). Based on Eq. (4.156), one may further reduce Eq. (4.302) to

$$\boldsymbol{x}=\boldsymbol{A}^{-1}\boldsymbol{b}; \tag{4.303}$$

this method is known as explicitation, because $\boldsymbol{x}$ is made explicit on $\boldsymbol{b}$ as originally provided – and can be widely applied, as long as $\boldsymbol{A}^{-1}$ exists for a (given) $\boldsymbol{A}$. Its main drawback is the need to invert $\boldsymbol{A}$ in advance, which will likely require cumbersome algebraic manipulation; and its failure to apply when $\boldsymbol{A}$ is not square, or $|\boldsymbol{A}|=0$ even if $\boldsymbol{A}$ is square – see Eq. (4.222).

4.2.18 Solution of algebraic quadratic equation

When a second-degree polynomial, in variable x, is made equal to zero, i.e.

$$ax^2+bx+c=0 \tag{4.304}$$

with a, b, and c denoting real constants, then two roots are found – given by

$$x_1,x_2=\frac{-b\pm\sqrt{b^2-4ac}}{2a}. \tag{4.305}$$

The nature of such roots depends on the sign of the discriminating binomial, b^2-4ac, under the square root – which sets two real roots when positive, and one double (or two equal) real root when nil; or else two (conjugated) complex roots when negative, which may be coined as

$$x_1,x_2=\frac{-b\pm\iota\sqrt{4ac-b^2}}{2a} \tag{4.306}$$

with ι denoting imaginary unit, $\sqrt{-1}$. It is interesting to realize that

$$x_1+x_2=-\frac{b}{a} \tag{4.307}$$

and

$$x_1x_2=\frac{c}{a}; \tag{4.308}$$

Eqs. (4.307) and (4.308) permit simplification of a few mathematical derivations.

4.3 Calculus

4.3.1 Limits

4.3.1.1 Definition

Cauchy's definition of limit b for function $f\{x\}$, in the vicinity of point $x=a$ – represented in brief by

$$\lim_{x\to a} f\{x\}=b, \tag{4.309}$$

may be coined as

$$\forall_{\varepsilon>0},\exists_{\delta>0}:0<\left|x-a\right|<\delta\Rightarrow\left|f\{x\}-b\right|<\varepsilon \tag{4.310}$$

– where δ and ε denote small positive real numbers. In other words, for every $f\{x\}$ within a neighborhood of radius ε centered at b, it is possible to find a neighborhood of radius δ centered at a containing the corresponding x values – no matter how small ε is. The concept of limit can be extended to a boundless variation of x, according to

$$\forall_{\varepsilon>0},\exists_{M>0}:\left|x\right|>M\Rightarrow\left|f\{x\}-b\right|<\varepsilon, \tag{4.311}$$

where M denotes a large positive real number – expressed in short notation as

$$\lim_{x\to\pm\infty} f\{x\}=b; \tag{4.312}$$

since $|x|$ was used, the definition conveyed by Eq. (4.311) applies to either $x\to-\infty$ or $x\to+\infty$, and is quite useful when searching for asymptotic patterns. By the same token, a boundless variation of $f\{x\}$, i.e.

$$\lim_{x \to a} f\{x\} = \pm\infty, \tag{4.313}$$

may be exactly handled as

$$\forall_{N>0}, \exists_{\delta<0} : 0 < |x - a| < \delta \Rightarrow |f\{x\}| > N \tag{4.314}$$

– where N denotes a large positive real number; once again, both $f\{x\} \to -\infty$ and $f\{x\} \to +\infty$ are accommodated by Eq. (4.314), for resorting to $|f\{x\}|$ (rather that $f\{x\}$) in its definition.

4.3.1.2 Basic theorems

Based on the concept of limit, one may state

$$\lim_{x \to a} \Phi\{x\} = \Phi\left\{\lim_{x \to a} x\right\} = \Phi\{a\} \tag{4.315}$$

in general, where Φ represents some algebraic operation; in other words, the limit of a functional dependence on x always translates to the same functional dependence on the limit. Composition of functions is also frequently found, in which case Eq. (4.315) will read

$$\lim_{f\{x\} \to b} \Phi\{f\{x\}\} = \Phi\left\{\lim_{f\{x\} \to b} f\{x\}\right\}, \tag{4.316}$$

after plain replacement of x by $f\{x\}$, and a by b; if Eq. (4.309) remains valid, then Eq. (4.316) can be redone as

$$\lim_{x \to a} \Phi\{f\{x\}\} = \Phi\left\{\lim_{x \to a} f\{x\}\right\} = \Phi\left\{f\left\{\lim_{x \to a} x\right\}\right\} = \Phi\{f\{a\}\} \tag{4.317}$$

since $f\{x\} \to b$ and $x \to a$ are concomitant events. Consider finally a function $f\{x\}$, comprised between $g\{x\}$ and $h\{x\}$ within a vicinity of a; under such circumstances, one finds

$$\left(\lim_{x \to a} f\{x\} = b\right)\Bigg|_{\substack{g\{x\} \le f\{x\} \le h\{x\} \\ \lim_{x \to a} g\{x\} = \lim_{x \to a} h\{x\} = b}} \tag{4.318}$$

– meaning that the limit of a function, when framed by two functions converging to the same limit, coincides with such a common limit.

4.3.2 Differential

The differential (or infinitesimal variation), dy, of a dependent variable, y, relates to the differential, dx, of its independent variable, x, via

$$dy \equiv \frac{df}{dx} dx; \tag{4.319}$$

here df/dx denotes derivative of $f\{x\}$ with regard to x, and $f\{x\}$ denotes underlying relationship between y and x. After realizing that

$$dy = (y + dy) - y \equiv f\big|_{x+dx} - f\big|_x, \tag{4.320}$$

one will retrieve Eq. (4.319) as

$$f\big|_{x+dx} = f\big|_x + \frac{df}{dx} dx. \tag{4.321}$$

If the infinitesimal variation dx can be approximated by finite variation $\Delta x \equiv x - x_0$, centered at x_0, then Eq. (4.321) is to be reformulated, in as approximate manner, to read

$$f\big|_{x=x_0+\Delta x} \approx f\big|_{x_0} + \frac{df}{dx}\bigg|_{x_0} \Delta x \tag{4.322}$$

– and thus become a useful tool to estimate the value of $f\{x\}$ when x is close to x_0; for instance, $\sqrt{40}$ can be (very) well approximated by

$$\sqrt{40} = 6.325 \approx 6.333 = \sqrt{36} + \frac{1}{2\sqrt{36}}(40 - 36), \tag{4.323}$$

which reflects Eq. (4.322) after setting $f\{x\} \equiv \sqrt{x}$, $x \equiv 40$, and $x_0 \equiv 36$. In other words, if the interval under scrutiny, around a given point, is sufficiently narrow, then any function therein will behave as if it were linear – with slope equal to the underlying derivative, evaluated at one of the boundaries of the said interval; hence, its evolution should essentially be driven by the straight line tangent to its graph at one of such boundaries. For a multivariate function $f\{x_1, x_2, \ldots, x_n\}$, one should resort to

$$dy \equiv \frac{\partial f}{\partial x_1} dx_1 + \frac{\partial f}{\partial x_2} dx_2 + \ldots + \frac{\partial f}{\partial x_n} dx_n \tag{4.324}$$

using Eq. (4.319) as template – which lumps the infinitesimal contributions of each independent variable to the overall infinitesimal change in $y = f\{x_1, x_2, \ldots, x_n\}$.

4.3.3 Total derivative

4.3.3.1 Definition

The derivative of a function $f\{x\}$, at a given point x_0, is defined as

$$\frac{df\{x\}}{dx}\bigg|_{x_0} \equiv \lim_{\Delta x \to 0} \frac{\Delta f\{x\}}{\Delta x}\bigg|_{x_0} \equiv \lim_{h \to 0} \frac{f\{x\}\big|_{x_0+h} - f\{x\}\big|_{x_0}}{h} \tag{4.325}$$

– consistent with Eq. (4.322), but exact rather than approximate; and obtained after setting $h \equiv \Delta x$. Hence, Leibnitz's notation for the derivative, as ratio of differentials, emerges logically in the left-hand side of Eq. (4.325) – while promptly indicating the dependent variable undergoing differentiation, and the independent variable with respect to which differentiation is being performed. The graphical interpretation of $(df\{x\}/dx)\big|_{x_0}$ is just the slope of the straight line tangent to curve $f\{x\}$, at $x = x_0$. Although Eq. (4.325) supports definition of derivative only at a given point x_0, one may let x_0 freely span the whole domain of $f\{x\}$; under these circumstances, a derivative function, denoted as $df\{x\}/dx$, will result – to be routinely utilized hereafter. A selection of derivatives, calculated directly from the definition, is included in Table 4.2.

The process of calculating a derivative can be applied to the derivative itself, and as many times as intended. For instance, the second-order derivative of a function, at point x_0, is given by

$$\frac{d^2 f\{x\}}{dx^2}\bigg|_{x_0} \equiv \lim_{h \to 0} \frac{\dfrac{df\{x\}}{dx}\bigg|_{x_0+h} - \dfrac{df\{x\}}{dx}\bigg|_{x_0}}{h} \tag{4.326}$$

TABLE 4.2

List of derivatives obtained via definition.

$f\{x\}$	$df\{x\}/dx$
x^n	nx^{n-1}
$\ln x$	$1/x$
$\sin x$	$\cos x$
$\cos x$	$-\sin x$

in agreement with Eq. (4.325) – yet applied to df/dx instead of f; and may, in turn, be rewritten as

$$\left.\frac{d^2f\{x\}}{dx^2}\right|_{x_0} = \lim_{h\to 0}\frac{\lim\limits_{h\to 0}\dfrac{f\{x\}|_{(x_0+h)+h}-f\{x\}|_{x_0+h}}{h}-\lim\limits_{h\to 0}\dfrac{f\{x\}|_{x_0+h}-f\{x\}|_{x_0}}{h}}{h} \tag{4.327}$$

upon insertion of Eq. (4.325) – pertaining to $f\{x\}$, evaluated at x_0 or x_0+h. After lumping denominators and then terms in numerator, Eq. (4.327) becomes

$$\left.\frac{d^2f\{x\}}{dx^2}\right|_{x_0} = \lim_{h\to 0}\frac{f\{x\}|_{x_0+2h}-2f\{x\}|_{x_0+h}+f\{x\}|_{x_0}}{h^2} \tag{4.328}$$

– which serves as alternative definition of second derivative. Once again, if x_0 moves along the whole domain of $f\{x\}$, then following application of logarithms to both sides, one will obtain the corresponding second derivative function – denoted as $d^2f\{x\}/dx^2$. Its geometrical interpretation relates to direction and magnitude of curvature of concavity of the original curve representing $f\{x\}$; a concave curve (i.e. exhibiting a concavity facing upward) will accordingly hold a positive value for d^2f/dx^2, whereas a larger value for d^2f/dx^2 entails a more pronounced curvature (or deviation from a straight line) – and vice versa. Higher order derivatives are similarly defined, according to

$$\left.\frac{d^nf\{x\}}{dx^n}\right|_{x_0} \equiv \lim_{h\to 0}\frac{\left.\dfrac{d^{n-1}f}{dx^{n-1}}\right|_{x_0+h}-\left.\dfrac{d^{n-1}f}{dx^{n-1}}\right|_{x_0}}{h}, \tag{4.329}$$

thus essentially extending Eq. (4.326) to every n besides 2 – again at the expense of derivatives with immediately lower order; expansion will then proceed by iterating the rationale underlying Eqs. (4.327) and (4.328).

4.3.3.2 Rules of differentiation

The rule of differentiation of a sum of two functions, $f\{x\}$ and $g\{x\}$, may be derived via direct application of Eq. (4.325), to eventually obtain

$$\frac{d}{dx}\big(f\{x\}+g\{x\}\big)=\frac{df\{x\}}{dx}+\frac{dg\{x\}}{dx} \tag{4.330}$$

– applicable to every value x; Eq. (4.330) can be generalized to a sum of functions f_i ($i=1, 2, \ldots, N$) as

$$\frac{d}{dx}\sum_{i=1}^{N}f_i\{x\}=\sum_{i=1}^{N}\frac{df_i\{x\}}{dx}. \tag{4.331}$$

A couple of derivatives obtained from straight use of Eq. (4.330) are conveyed by Table 4.3.

The rule of differentiation of the product of two functions will likewise proceed via application of Eq. (4.325), according to

$$\frac{d}{dx}\big(f\{x\}g\{x\}\big)=\frac{df\{x\}}{dx}g\{x\}+f\{x\}\frac{dg\{x\}}{dx}; \tag{4.332}$$

Eq. (4.332) simplifies to

$$\frac{d}{dx}\big(ag\{x\}\big)=a\frac{dg\{x\}}{dx}, \tag{4.333}$$

whenever $f\{x\}\equiv a$, as long as a denotes a constant – since

$$\frac{da}{dx}=0 \tag{4.334}$$

by definition, see Eq. (4.325). Combination of Eqs. (4.330) and (4.333) produces, in turn,

$$\frac{d}{dx}\sum_{i=1}^{N}a_if_i\{x\}=\sum_{i=1}^{N}a_i\frac{df_i\{x\}}{dx} \tag{4.335}$$

– thus serving as rule for differentiation of a linear combination of functions, weighed by a_is as (constant) coefficients. On the other hand, a product of N functions yields

$$\frac{d}{dx}\prod_{i=1}^{N}f_i\{x\}=\sum_{i=1}^{N}\frac{df_i\{x\}}{dx}\prod_{\substack{j=1\\ j\neq i}}^{N}f_j\{x\} \tag{4.336}$$

as generalization of Eq. (4.332); one may instead write

$$\frac{d}{dx}\prod_{i=1}^{N}f_i\{x\}=\sum_{i=1}^{N}\frac{1}{f_i\{x\}}\frac{df_i\{x\}}{dx}\prod_{j=1}^{N}f_j\{x\}=\sum_{i=1}^{N}\frac{d\ln f_i\{x\}}{dx}\prod_{j=1}^{N}f_j\{x\}, \tag{4.337}$$

following multiplication of both numerator and denominator by $f_i\{x\}$ – and at the expense of the second entry in Table 4.2, and the chain differentiation rule (see below). A number of derivatives generated via application of Eq. (4.337) are provided in Table 4.4.

Derivation of the rule of differentiation of a ratio of two functions can as well resort to direct application of Eq. (4.325), viz.

$$\frac{d}{dx}\left(\frac{f\{x\}}{g\{x\}}\right)=\frac{\dfrac{df\{x\}}{dx}g\{x\}-f\{x\}\dfrac{dg\{x\}}{dx}}{g^2\{x\}} \tag{4.338}$$

TABLE 4.3

List of derivatives obtained via theorem of sum of functions.

$f\{x\}$	$df\{x\}/dx$
$\sinh x$	$\cosh x$
$\cosh x$	$\sinh x$

TABLE 4.4

List of derivatives obtained via theorem of product of functions.

$f\{x\}$	$df\{x\}/dx$
$f^N\{x\}$	$Nf^{N-1}\{x\}\dfrac{df\{x\}}{dx}$
$\log_a x$	$\dfrac{\log_a e}{x}$
	$\dfrac{1}{x\ln a}$

– which obviously requires $g\{x\} \neq 0$ to be finite; if $f\{x\} \equiv 1$, then Eq. (4.338) reduces to

$$\frac{d}{dx}\left(\frac{1}{g\{x\}}\right) = -\frac{\dfrac{dg\{x\}}{dx}}{g^2\{x\}}, \tag{4.339}$$

enforced by Eq. (4.334) when $f\{x\} \equiv a \equiv 1$ – known as rule of differentiation of the (arithmetic) reciprocal. Examples of derivatives obtained via Eq. (4.338) are included in Table 4.5.

The method of differentiation of an inverse function, $f^{-1}\{x\}$, may also be attained upon application of Eq. (4.325), viz.

$$\frac{df^{-1}\{y\}}{dy} = \frac{1}{\dfrac{df\{x\}}{dx}}; \tag{4.340}$$

hence, it suffices in this case to calculate the reciprocal of the derivative of the original function, provided that it is evaluated at the corresponding dependent variable. Derivatives obtainable via Eq. (4.340) are given in Table 4.6.

The rule of differentiation of a composite function, $f\{g\{x\}\}$, will be reached via use of Eq. (4.325) once more, to ultimately get

$$\frac{df\{g\{x\}\}}{dx} = \frac{df\{g\{x\}\}}{dg\{x\}}\frac{dg\{x\}}{dx} \tag{4.341}$$

– often referred to as chain differentiation rule. Relevant examples of derivatives obtained through Eq. (4.341) are available in Table 4.7. It should be emphasized that differentiation of any exponential function requires previous conversion to a natural exponential; for instance,

$$f\{x\}^{g\{x\}} = \exp\left\{\ln f\{x\}^{g\{x\}}\right\} = \exp\left\{g\{x\}\ln f\{x\}\right\} \tag{4.342}$$

TABLE 4.5

List of derivatives obtained via theorem of quotient of functions.

$f\{x\}$	$df\{x\}/dx$
$\tan x$	$\sec^2 x$
$\text{cotan}\, x$	$-\text{cosec}^2 x$
$\text{cosec}\, x$	$-\text{cotan}\, x\ \text{cosec}\, x$
$\sec x$	$\tan x \sec x$

TABLE 4.6

List of derivatives obtained via theorem of inverse function.

$f\{x\}$	$df\{x\}/dx$
e^x	e^x
$\sin^{-1} x$	$\dfrac{1}{\sqrt{1-x^2}}$
$\cos^{-1} x$	$-\dfrac{1}{\sqrt{1-x^2}}$
$\tan^{-1} x$	$\dfrac{1}{1+x^2}$
$\text{cotan}^{-1} x$	$-\dfrac{1}{1+x^2}$

TABLE 4.7

List of derivatives obtained via theorem of composite function.

$f\{x\}$	$df\{x\}/dx$
$\ln f\{x\}$	$\dfrac{1}{f\{x\}}\dfrac{df\{x\}}{dx}$
$\exp\{g\{x\}\}$	$\exp\{g\{x\}\}\dfrac{dg\{x\}}{dx}$
a^x	$a^x \ln a$
$f\{x\}^{g\{x\}}$	$f\{x\}^{g\{x\}}\left(\dfrac{dg\{x\}}{dx}\ln f\{x\} + \dfrac{g\{x\}}{f\{x\}}\dfrac{df\{x\}}{dx}\right)$

– which takes advantage of Eqs. (4.34) and (4.36), is to be performed so as to differentiate $f\{x\}^{g\{x\}}$ as a composite natural exponential function.

4.3.3.3 Rolle's theorem

In calculus, Rolle's theorem states that a function, attaining equal values at two distinct points, must have (at least) one stationary point somewhere in between – i.e. a point described by nil first-order derivative. The corresponding mathematical statement can be coined as

$$f\{a\} = f\{b\} \Rightarrow \exists_{c\in]a,b[} : \left.\frac{df\{x\}}{dx}\right|_{x=c} = 0, \tag{4.343}$$

and requires $f\{x\}$ be continuous within (closed interval) $[a,b]$ and differentiable within (open interval) $]a,b[$. A particular case of Eq. (4.343) occurs when $f\{x\}\big|_a = f\{x\}\big|_b = 0$; in this case, one concludes on the necessary existence of (at least) one zero of the derivative df/dx between two consecutive zeros of the original function $f\{x\}$. On the other hand, failure of $f\{x\}$ to be differentiable at an interior point of the germane interval no longer guarantees validity of Rolle's theorem.

4.3.3.4 Lagrange's theorem

Lagrange's theorem – also known as mean value theorem, states that, given a planar arc describing a continuous function between two endpoints, there is at least one point where the tangent to the said arc lies parallel to the secant through its endpoints; this may be more exactly expressed as

$$\exists_{c\in]a,b[}:\left.\frac{df\{x\}}{dx}\right|_{x=c}=\frac{f\{b\}-f\{a\}}{b-a} \tag{4.344}$$

– and is applicable to a function $f\{x\}$ continuous within $[a,b]$, and differentiable within $]a,b[$. If $f\{b\}=f\{a\}$, then Eq. (4.344) reduces to

$$\left.\frac{df\{x\}}{dx}\right|_{x=c}=\frac{f\{a\}-f\{a\}}{b-a}=0, \tag{4.345}$$

thus recovering Eq. (4.343); hence, Rolle's theorem is a particular case of Lagrange's theorem. Furthermore, Eq. (4.344) indicates that an increasing function, i.e. one for which $f\{b\}-f\{a\}>0$ when $b>a$, holds a positive derivative, $df\{x\}/dx>0$, within any vicinity of c with amplitude $b-a$, as long as $b\rightarrow a$ – since c must be comprised between a and b. By the same token, when $b\rightarrow a$, and thus $c\rightarrow a$ (or $c\rightarrow b$, for that matter), Eq. (4.344) assures that a decreasing function in that interval $[a,b]$ of infinitesimal amplitude, described by $f\{b\}-f\{a\}<0$ for $b>a$, exhibits a negative derivative, $df\{x\}/dx<0$. One corollary of this realization is that $f\{x\}$ cannot reverse monotony between any two consecutive zeros of $df\{x\}/dx$ – so $f\{x\}$ either does not cross, or crosses the horizontal axis (at most) once between such consecutive zeros. This realization is useful when numerically searching for zeros of a function – should the zeros of its derivative function be easier to find (as often happens with polynomial functions).

4.3.3.5 *Cauchy's theorem*

If two functions, $f\{x\}$ and $g\{x\}$, are considered (instead of a single one) – both continuous in $[a,b]$ and differentiable in $]a,b[$, such that $df\{x\}/dx\neq 0$ therein, then one realizes that

$$\exists_{c\in]a,b[}:\frac{\left.\frac{df\{x\}}{dx}\right|_{x=c}}{\left.\frac{dg\{x\}}{dx}\right|_{x=c}}=\frac{f\{b\}-f\{a\}}{g\{b\}-g\{a\}}; \tag{4.346}$$

Eq. (4.346) entails the formal statement of Cauchy's theorem. Note that Cauchy's theorem cannot be derived from Lagrange's theorem applied to both numerator and denominator of $f\{x\}/g\{x\}$, because the cs would not necessarily coincide. Conversely, if $g\{x\}$ looked like

$$g\{x\}\equiv x \tag{4.347}$$

and thus

$$g\{a\}=a, \tag{4.348}$$

together with

$$g\{b\}=b \tag{4.349}$$

that implies

$$\left.\frac{dg\{x\}}{dx}\right|_{x=c}=1\Big|_{x=c}=1, \tag{4.350}$$

then Eq. (4.346) would degenerate to Eq. (4.344) upon insertion of Eqs. (4.348)–(4.350); therefore, Lagrange's theorem appears as a particular case of Cauchy's theorem.

4.3.3.6 *L'Hôpital's rule*

Take functions $f\{x\}$ and $g\{x\}$, both continuous in $[a,b]$ and differentiable in $]a,b[$ – and thus susceptible of application of Cauchy's theorem; assume, in addition, that

$$f\{x\}\big|_a=g\{x\}\big|_a=0. \tag{4.351}$$

Although the quotient $f\{x\}/g\{x\}$ is not defined at $x=a$ by virtue of Eq. (4.351) leading to 0/0 as unknown quantity, this does (in principle) not hold in a vicinity thereof; in any case, df/dx and dg/dx are assumed to exist at $x=a$. Cauchy's theorem guarantees that there is, at least, one point $x=c$, belonging to vicinity $]a,x[$ (with $a<x\leq b$) such that

$$\frac{f\{x\}-f\{a\}}{g\{x\}-g\{a\}}=\frac{\left.\frac{df\{x\}}{dx}\right|_c}{\left.\frac{dg\{x\}}{dx}\right|_c};\quad a<c<x \tag{4.352}$$

– with c comprised (as just mentioned) between a and x; in view of Eq. (4.351), one can simplify Eq. (4.352) to

$$\frac{f\{x\}}{g\{x\}}=\frac{\left.\frac{df\{x\}}{dx}\right|_c}{\left.\frac{dg\{x\}}{dx}\right|_c}. \tag{4.353}$$

When $x\rightarrow a$, then $c\rightarrow a$ because $a<c<x$, by hypothesis; upon application of the said limit to both sides, Eq. (4.353) produces

$$\lim_{x\rightarrow a}\frac{f\{x\}}{g\{x\}}=\lim_{x\rightarrow a}\frac{\left.\frac{df\{x\}}{dx}\right|_c}{\left.\frac{dg\{x\}}{dx}\right|_c}=\lim_{c\rightarrow a}\frac{\left.\frac{df\{x\}}{dx}\right|_c}{\left.\frac{dg\{x\}}{dx}\right|_c}, \tag{4.354}$$

where a change of (dummy) variable from c to x in the second equality unfolds

$$\left.\left(\lim_{x\rightarrow a}\frac{f\{x\}}{g\{x\}}=\lim_{x\rightarrow a}\frac{\frac{df\{x\}}{dx}}{\frac{dg\{x\}}{dx}}\right)\right|_{f\{a\}=g\{a\}=0} \tag{4.355}$$

Since a limit describes the mathematical behavior of a function in a vicinity of a given point of its domain, existence of limits of $f\{x\}$ and $g\{x\}$ as $x\rightarrow a$ does not require $f\{x\}$ and $g\{x\}$ be defined exactly at $x=a$; since $f\{a\}$ and $g\{a\}$ are not explicitly used in the equality labeled as Eq. (4.355), one may actually settle with $\lim_{x\rightarrow a}f\{x\}=0$ and $\lim_{x\rightarrow a}g\{x\}=0$ – in which case Eq. (4.355) degenerates to

$$\left.\left(\lim_{x\rightarrow a}\frac{f\{x\}}{g\{x\}}=\lim_{x\rightarrow a}\frac{\frac{df\{x\}}{dx}}{\frac{dg\{x\}}{dx}}\right)\right|_{\lim_{x\rightarrow a}f\{x\}=\lim_{x\rightarrow a}g\{x\}=0}. \tag{4.356}$$

The conditions of applicability of Eq. (4.356) are often lumped into

$$\lim_{x\to a}\frac{f\{x\}}{g\{x\}}=\frac{0}{0}, \tag{4.357}$$

where the unknown quantity becomes explicit; Eq. (4.356) has traditionally been termed l'Hôpital's rule. Remember that, in general, $(df/dx)/(dg/dx)$, as in the right-hand side of Eq. (4.356), differs from $d(f/g)/dx=(g(df/dx)-f(dg/dx))/g^2$, in agreement with Eq. (4.338). Furthermore, when $\lim\limits_{x\to a}df\{x\}/dx=\lim\limits_{x\to a}dg\{x\}/dx=0$, the derivative-based approach conveyed by Eq. (4.356) can be sequentially applied as many times as necessary – according to

$$\left.\left(\lim_{x\to a}\frac{\dfrac{d^i f(x\rangle}{dx^i}}{\dfrac{d^i g\{x\}}{dx^i}}=\lim_{x\to a}\frac{\dfrac{d^{i+1}f\{x\}}{dx^i}}{\dfrac{d^{i+1}g\{x\}}{dx^i}}\right)\right|_{\lim\limits_{x\to a}\frac{d^i f\{x\}}{dx^i}=\lim\limits_{x\to a}\frac{d^i g\{x\}}{dx^i}=0};\ i=1,2,\ldots; \tag{4.358}$$

in attempts to ultimately calculate $\lim\limits_{x\to a}\left(f\{x\}/g\{x\}\right)$, one may resort to Eq. (4.358) even when $\lim\limits_{x\to a}d^{i+1}f\{x\}/dx^{i+1}$ and $\lim\limits_{x\to a}d^{i+1}g\{x\}/dx^{i+1}$ do not individually exist, as long as $\lim\limits_{x\to a}\dfrac{d^{i+1}f\{x\}}{dx^{i+1}}\Big/\dfrac{d^{i+1}g\{x\}}{dx^{i+1}}$ exists.

Despite the somewhat restricted form entailed by Eq. (4.356) – as only 0/0 is eligible for direct application thereof, this realization may be misleading; l'Hôpital's rule proves indeed applicable to a much wider range of practical situations, where other types of unknown quantities arise. For instance, if

$$\lim_{x\to a}f\{x\}=\lim_{x\to a}g\{x\}=\infty, \tag{4.359}$$

then one is led to

$$\lim_{x\to a}\frac{f\{x\}}{g\{x\}}=\frac{\infty}{\infty} \tag{4.360}$$

– again an unknown quantity, but not of the type in principle suitable for application of Eq. (4.356). However, the ratio in the left-hand side of Eq. (4.360) may be rearranged to read

$$\lim_{x\to a}\frac{f\{x\}}{g\{x\}}=\lim_{x\to a}\frac{\dfrac{1}{g\{x\}}}{\dfrac{1}{f\{x\}}}=\frac{\lim\limits_{x\to a}\dfrac{1}{g\{x\}}}{\lim\limits_{x\to a}\dfrac{1}{f\{x\}}}=\frac{\dfrac{1}{\infty}}{\dfrac{1}{\infty}}=\frac{0}{0} \tag{4.361}$$

in view of Eqs. (4.315) and (4.359) – so immediate application of Eq. (4.356) ensues; in this case, one would get

$$\lim_{x\to a}\frac{f\{x\}}{g\{x\}}=\frac{\lim\limits_{x\to a}\left(-\dfrac{\dfrac{dg\{x\}}{dx}}{g^2\{x\}}\right)}{\lim\limits_{x\to a}\left(-\dfrac{\dfrac{df\{x\}}{dx}}{f^2\{x\}}\right)}=\lim_{x\to a}\frac{f^2\{x\}}{g^2\{x\}}\frac{\lim\limits_{x\to a}\dfrac{dg\{x\}}{dx}}{\lim\limits_{x\to a}\dfrac{df\{x\}}{dx}}$$

$$=\lim_{x\to a}\left(\frac{f\{x\}}{g\{x\}}\right)^2\lim_{x\to a}\frac{\dfrac{dg\{x\}}{dx}}{\dfrac{df\{x\}}{dx}}=\left(\lim_{x\to a}\frac{f\{x\}}{g\{x\}}\right)^2\frac{1}{\lim\limits_{x\to a}\dfrac{\dfrac{df\{x\}}{dx}}{\dfrac{dg\{x\}}{dx}}} \tag{4.362}$$

in agreement with Eq. (4.339), which is equivalent to

$$\lim_{x\to a}\frac{f\{x\}}{g\{x\}}=\lim_{x\to a}\frac{\dfrac{df\{x\}}{dx}}{\dfrac{dg\{x\}}{dx}} \tag{4.363}$$

after cancelling common factors between sides complemented by isolation of $\lim\limits_{x\to a}\dfrac{df\{x\}}{dx}\Big/\dfrac{dg\{x\}}{dx}$ – thus extending validity of Eq. (4.356) to the situation portrayed in Eq. (4.359). Another example takes the form

$$\lim_{x\to a}f\{x\}=0\wedge\lim_{x\to a}g\{x\}=\infty; \tag{4.364}$$

this suggests that the limit of the product of $f\{x\}$ by $g\{x\}$ should read

$$\lim_{x\to a}f\{x\}g\{x\}=0\cdot\infty, \tag{4.365}$$

upon straightforward application of Eq. (4.315) – thus giving rise to another unknown quantity, $0\cdot\infty$. Equation (4.365) may, however, be rearranged as

$$\lim_{x\to a}f\{x\}g\{x\}=\lim_{x\to a}\frac{f\{x\}}{\dfrac{1}{g\{x\}}}=\frac{\lim\limits_{x\to a}f\{x\}}{\lim\limits_{x\to a}\dfrac{1}{g\{x\}}}=\frac{0}{\dfrac{1}{\infty}}=\frac{0}{0} \tag{4.366}$$

at the expense of Eqs. (4.315) and (4.364), thus allowing immediate application of Eq. (4.356). One may instead be faced with

$$\lim_{x\to a}f\{x\}=\infty\wedge\lim_{x\to a}g\{x\}=\infty, \tag{4.367}$$

when searching for the limit of the difference between $f\{x\}$ and $g\{x\}$, i.e.

$$\lim_{x\to a}\left(f\{x\}-g\{x\}\right)=\infty-\infty. \tag{4.368}$$

The (algebraic) undefined quantity in the right-hand side of Eq. (4.368), i.e. $\infty-\infty$, can be overcome after preliminary algebraic manipulation to

$$\lim_{x\to a}\left(f\{x\}-g\{x\}\right)=\lim_{x\to a}f\{x\}g\{x\}\left(\frac{1}{g\{x\}}-\frac{1}{f\{x\}}\right) \tag{4.369}$$

that had both $f\{x\}$ and $g\{x\}$ factored out – complemented

$$\lim_{x\to a}\left(f\{x\}-g\{x\}\right)=\lim_{x\to a}\frac{\dfrac{1}{g\{x\}}-\dfrac{1}{f\{x\}}}{\dfrac{1}{f\{x\}g\{x\}}}$$

$$=\frac{\lim\limits_{x\to a}\dfrac{1}{g\{x\}}-\lim\limits_{x\to a}\dfrac{1}{f\{x\}}}{\lim\limits_{x\to a}\dfrac{1}{f\{x\}g\{x\}}}=\frac{\dfrac{1}{\infty}-\dfrac{1}{\infty}}{\dfrac{1}{\infty^2}}=\frac{0-0}{0^2}=\frac{0}{0} \tag{4.370}$$

in view of Eqs. (4.315) and (4.367); application of Eq. (4.356) is again in order.

Unknown quantities may instead appear in the form of powers; this the case of

$$\lim_{x\to a} f\{x\} = 1 \wedge \lim_{x\to a} g\{x\} = \infty, \quad (4.371)$$

where the limit of the corresponding power would look like

$$\lim_{x\to a} f\{x\}^{g\{x\}} = 1^{\infty} \quad (4.372)$$

as per Eq. (4.315) – which represents an unknown quantity of the power type. Application of logarithms to both sides of Eq. (4.372) leads, however, to,

$$\begin{aligned} \ln \lim_{x\to a} f\{x\}^{g\{x\}} &= \lim_{x\to a} \ln f\{x\}^{g\{x\}} \\ &= \lim_{x\to a} g\{x\} \ln f\{x\} = \infty \cdot \ln 1 = 0 \cdot \infty, \end{aligned} \quad (4.373)$$

where Eqs. (4.36) and (4.315) were taken into account – besides Eq. (4.371), obviously; the unknown quantity showing up in Eq. (4.373) is the same already found in Eq. (4.365), so a similar approach is to be followed thereafter. Another situation of interest arises when

$$\lim_{x\to a} f\{x\} = 0^{+} \wedge \lim_{x\to a} g\{x\} = 0, \quad (4.374)$$

in which case the limit of the corresponding power emerges as

$$\lim_{x\to a} f\{x\}^{g\{x\}} = 0^{0} \quad (4.375)$$

following direct application of Eq. (4.315); this leads to still another unknown quantity of the power type. The strategy here is again to apply logarithms to both sides of Eq. (4.375), i.e.

$$\begin{aligned} \ln \lim_{x\to a} f\{x\}^{g\{x\}} &= \lim_{x\to a} \ln f\{x\}^{g\{x\}} = \lim_{x\to a} g\{x\} \ln f\{x\} \\ &= 0 \cdot \ln 0^{+} = 0 \cdot (-\infty) = (-0) \cdot \infty = 0 \cdot \infty, \end{aligned} \quad (4.376)$$

where Eqs. (4.36), (4.315), and (4.374) were meanwhile taken on board; this result is analogous to Eq. (4.365). A third case corresponds to

$$\lim_{x\to a} f\{x\} = \infty \wedge \lim_{x\to a} g\{x\} = 0; \quad (4.377)$$

upon constructing a power using $f\{x\}$ for base and $g\{x\}$ for exponent, Eq. (4.377) gives rise to

$$\lim_{x\to a} f\{x\}^{g\{x\}} = \infty^{0} \quad (4.378)$$

via direct application of the classical theorems on limits. As done before, logarithms of both sides of Eq. (4.378) are taken to produce

$$\begin{aligned} \ln \lim_{x\to a} f\{x\}^{g\{x\}} &= \lim_{x\to a} \ln f\{x\}^{g\{x\}} \\ &= \lim_{x\to a} g\{x\} \ln f\{x\} = 0 \cdot \ln \infty = 0 \cdot \infty \end{aligned} \quad (4.379)$$

at the expense of Eqs. (4.36), (4.315), and (4.377); note the similarity again to Eq. (4.365), which justifies an identical strategy be pursued hereafter toward circumventing the unknown quantity attained.

In all cases of unknown quantities of the power type explored so far, the original limit will be eventually ascertained after taking exponentials of the limit found in logarithmic form. A related situation, viz.

$$\lim_{x\to a} f\{x\} = 0^{+} \wedge \lim_{x\to a} g\{x\} = \infty, \quad (4.380)$$

associated with a power of the form

$$\lim_{x\to a} f\{x\}^{g\{x\}} = 0^{\infty} \quad (4.381)$$

is obtained via plain application of Eq. (4.315). After taking logarithms of both sides again, Eq. (4.381) turns to

$$\begin{aligned} \ln \lim_{x\to a} f\{x\}^{g\{x\}} &= \lim_{x\to a} \ln f\{x\}^{g\{x\}} \\ &= \lim_{x\to a} g\{x\} \ln f\{x\} = \infty \cdot \ln 0^{+} \\ &= \infty \cdot (-\infty) = -\infty \cdot \infty = -\infty^{2} = -\infty \end{aligned} \quad (4.382)$$

– with the aid of Eqs. (4.36) and (4.315) once more, further to Eq. (4.380); this implies

$$\lim_{x\to a} f\{x\}^{g\{x\}} = \mathrm{e}^{-\infty} = 0, \quad (4.383)$$

after exponentials are taken of both sides of Eq. (4.382). Therefore, 0^{∞} does not prove an unknown quantity after all – unlike happened with ∞^{0} in Eq. (4.378).

4.3.4 Partial derivatives

4.3.4.1 Definition

Partial derivatives are defined similarly to total derivatives, i.e.

$$\left.\left(\frac{\partial f\{x,y\}}{\partial x}\right)_{y}\right|_{(x_0,y_0)} \equiv \lim_{h\to 0} \frac{f\{x,y\}\big|_{(x_0+h,y_0)} - f\{x,y\}\big|_{(x_0,y_0)}}{h} \quad (4.384)$$

and

$$\left.\left(\frac{\partial f\{x,y\}}{\partial y}\right)_{x}\right|_{(x_0,y_0)} \equiv \lim_{k\to 0} \frac{f\{x,y\}\big|_{(x_0,y_0+k)} - f\{x,y\}\big|_{(x_0,y_0)}}{k} \quad (4.385)$$

pertaining to the two partial (first-order) derivatives with regard to x and y, respectively, see Eq. (4.325). The former may be thought as the ordinary derivative of $f\{x,y\}$ with regard to x, obtained by treating y as a constant; the partial derivative of $f\{x,y\}$, with regard to y as per Eq. (4.385), will likewise be found by treating x as constant, and then calculating the ordinary derivative of $f\{x,y\}$ with respect to y. Therefore, the rules of calculation of partial derivatives mimic those applying to regular derivatives. The variable to be held constant during differentiation is indicated as subscript – a notation particularly useful when more than two independent variables are under scrutiny. By the same token, one can define higher-order partial derivatives, namely,

$$\left.\frac{\partial^2 f\{x,y\}}{\partial x \partial y}\right|_{(x_0,y_0)} \equiv \left.\left(\frac{\partial}{\partial x}\left(\frac{\partial f(x,y)}{\partial y}\right)_x\right)_y\right|_{(x_0,y_0)}$$

$$\equiv \lim_{h\to 0} \frac{\left.\left(\frac{\partial f\{x,y\}}{\partial y}\right)_x\right|_{(x_0+h,y_0)} - \left.\left(\frac{\partial f(x,y)}{\partial y}\right)_x\right|_{(x_0,y_0)}}{h} \tag{4.386}$$

$$\equiv \lim_{h\to 0} \frac{\left(\begin{array}{l} \lim\limits_{k\to 0} \frac{f\{x,y\}|_{(x_0+h,y_0+k)} - f\{x,y\}|_{(x_0+h,y_0)}}{k} \\ -\lim\limits_{k\to 0} \frac{f\{x,y\}|_{(x_0,y_0+k)} - f\{x,y\}|_{(x_0,y_0)}}{k} \end{array}\right)}{h}$$

based on Eq. (4.325), applied first with regard to x and then with regard to y; Eq. (4.386) degenerates to

$$\left.\frac{\partial^2 f\{x,y\}}{\partial x \partial y}\right|_{(x_0,y_0)} \equiv \lim_{\substack{h\to 0\\k\to 0}} \frac{\left(\begin{array}{l} f\{x,y\}|_{(x_0+h,y_0+k)} - f\{x,y\}|_{(x_0+h,y_0)} \\ -f\{x,y\}|_{(x_0,y_0+k)} + f\{x,y\}|_{(x_0,y_0)} \end{array}\right)}{hk}, \tag{4.387}$$

via factoring out k. The other cross derivative reads

$$\left.\frac{\partial^2 f\{x,y\}}{\partial y \partial x}\right|_{(x_0,y_0)} \equiv \left.\left(\frac{\partial}{\partial y}\left(\frac{\partial f\{x,y\}}{\partial x}\right)_y\right)_x\right|_{(x_0,y_0)}$$

$$\equiv \lim_{k\to 0} \frac{\left.\left(\frac{\partial f\{x,y\}}{\partial x}\right)_y\right|_{(x_0,y_0+k)} - \left.\left(\frac{\partial f\{x,y\}}{\partial x}\right)_y\right|_{(x_0,y_0)}}{k} \tag{4.388}$$

$$\equiv \lim_{k\to 0} \frac{\left(\begin{array}{l} \lim\limits_{h\to 0} \frac{f\{x,y\}|_{(x_0+h,y_0+k)} - f\{x,y\}|_{(x_0,y_0+k)}}{h} \\ -\lim\limits_{h\to 0} \frac{f\{x,y\}|_{(x_0+h,y_0)} - f\{x,y\}|_{(x_0,y_0)}}{h} \end{array}\right)}{k},$$

based again on Eq. (4.325), applied sequentially with regard to y and x – to yield

$$\left.\frac{\partial^2 f\{x,y\}}{\partial y \partial x}\right|_{(x_0,y_0)} \equiv \lim_{\substack{h\to 0\\k\to 0}} \frac{\left(\begin{array}{l} f\{x,y\}|_{(x_0+h,y_0+k)} - f\{x,y\}|_{(x_0,y_0+h)} \\ -f\{x,y\}|_{(x_0+h,y_0)} + f\{x,y\}|_{(x_0,y_0)} \end{array}\right)}{hk} \tag{4.389}$$

after having lumped denominators. Finally, one may write

$$\left.\frac{\partial^2 f\{x,y\}}{\partial x^2}\right|_{(x_0,y_0)} \equiv \left.\left(\frac{\partial}{\partial x}\left(\frac{\partial f\{x,y\}}{\partial x}\right)_y\right)_y\right|_{(x_0,y_0)} \equiv \lim_{h\to 0} \frac{f\{x,y\}|_{(x_0+2h,y_0)} - 2f\{x,y\}|_{(x_0+h,y_0)} + f\{x,y\}|_{(x_0,y_0)}}{h^2} \tag{4.390}$$

in parallel to Eq. (4.328) for a given y_0; as well as

$$\left.\frac{\partial^2 f\{x,y\}}{\partial y^2}\right|_{(x_0,y_0)} \equiv \left.\left(\frac{\partial}{\partial y}\left(\frac{\partial f\{x,y\}}{\partial y}\right)_x\right)_x\right|_{(x_0,y_0)} \equiv \lim_{k\to 0} \frac{f\{x,y\}|_{(x_0,y_0+2k)} - 2f\{x,y\}|_{(x_0,y_0+k)} + f\{x,y\}|_{(x_0,y_0)}}{k^2}, \tag{4.391}$$

which rematerializes Eq. (4.328) once x is fixed at x_0 – and after replacement of x by y, and h by k. The total number of second-order partial derivatives is thus four – as conveyed by Eqs. (4.387) and (4.389)–(4.391); in general, the number of n-th order derivatives of a function containing m independent variables is given by m^n, instead of just one as in the case of univariate functions.

4.3.4.2 Young's and Schwartz's theorems

If a function $f\{x,y\}$ is continuous, and differentiable (at least twice) with regard to both x and y, then one finds

$$\frac{\partial^2 f\{x,y\}}{\partial x \partial y} = \frac{\partial^2 f\{x,y\}}{\partial y \partial x}; \tag{4.392}$$

this constitutes the mathematical statement of either Young's theorem or Schwarz's theorem, both pertaining to mixed partial derivatives. The former states that Eq. (4.392) is valid provided that $\partial f/\partial x$ and $\partial f/\partial y$ exist in a neighborhood of (x,y), and are differentiable therein; whereas the latter requires as well that $\partial f/\partial x$ and $\partial f/\partial y$ exist in a neighborhood of (x,y), coupled with either $\partial^2 f/\partial x\partial y$ or $\partial^2 f/\partial y\partial x$ being continuous. The result conveyed by Eq. (4.392) reduces the number of cross variables to be calculated in practical problems.

4.3.4.3 Euler's theorem

A function $f\{x,y\}$ is said to be homogeneous, of degree m, when

$$f\{kx,ky\} = k^m f\{x,y\}, \tag{4.393}$$

with k denoting a constant; a similar definition applies to any number of independent variables. If $f\{x_1, x_2, \ldots, x_n\}$ denotes a homogeneous, differentiable function of degree m – on independent variables $x_1, x_2, \ldots, x_n$, then it can be proven that

$$x_1\frac{\partial f}{\partial x_1} + x_2\frac{\partial f}{\partial x_2} + \ldots + x_n\frac{\partial f}{\partial x_n} = mf; \tag{4.394}$$

Eq. (4.394) consubstantiates Euler's theorem on homogeneous functions – useful toward derivation of mathematical relationships, namely, between thermodynamic functions. Should the aforementioned f be twice differentiable with regard to $x_1, x_2, \ldots, x_n$, then one obtains

$$x_1 \frac{\partial f}{\partial x_1}\left(\frac{\partial f}{\partial x_j}\right) + x_2 \frac{\partial f}{\partial x_2}\left(\frac{\partial f}{\partial x_j}\right) + \ldots + x_n \frac{\partial f}{\partial x_n}\left(\frac{\partial f}{\partial x_j}\right) = (m-1)\frac{\partial f}{\partial x_j}$$

$$j = 1,2,\ldots,n; \tag{4.395}$$

hence, Euler's theorem also applies to any partial derivative of f, but $m-1$ serves now as degree of homogeneity.

4.3.5 Implicit differentiation

A function $y \equiv y\{x\}$ is said to be implicit when it is defined by

$$z\{x,y\} = 0, \tag{4.396}$$

where z represents a set of algebraic operations on x and y. It is possible to differentiate y with regard to x, even when y cannot be made explicit on x – via application to Eq. (4.396) of the chain (partial) differentiation rule with regard to x, according to

$$\left(\frac{\partial z}{\partial x}\right)_y \frac{dx}{dx} + \left(\frac{\partial z}{\partial y}\right)_x \frac{dy}{dx} = 0; \tag{4.397}$$

the right-hand side is nil on account of z being a (nil) constant, as per Eq. (4.334) in the first place. Since dx/dx is trivially equal to unity as per the first entry of Table 4.2 for $n=1$, one may redo Eq. (4.397) to

$$\frac{dy}{dx} = -\frac{\left(\frac{\partial z}{\partial x}\right)_y}{\left(\frac{\partial z}{\partial y}\right)_x} \tag{4.398}$$

– known as implicit derivative of y with regard to x; note that dy/dx will, in general, appear as a function of x and y simultaneously. In view of Eq. (4.396), one can make z=constant explicit in the left-hand side, so Eq. (4.398) will instead appear as

$$\left(\frac{\partial y}{\partial x}\right)_z = -\frac{\left(\frac{\partial z}{\partial x}\right)_y}{\left(\frac{\partial z}{\partial y}\right)_x}; \tag{4.399}$$

this is equivalent to writing

$$\left(\frac{\partial y}{\partial x}\right)_z \left(\frac{\partial z}{\partial y}\right)_x = -\left(\frac{\partial z}{\partial x}\right)_y, \tag{4.400}$$

upon elimination of denominators. Equation (4.400) can be reformulated to

$$\left(\frac{\partial y}{\partial x}\right)_z \left(\frac{\partial x}{\partial z}\right)_y \left(\frac{\partial z}{\partial y}\right)_x = -1 \tag{4.401}$$

with the aid of the rule of differentiation of the inverse function, see Eq. (4.340). Another useful relationship between partial derivatives resorts to the general definition of total differential of a bivariate function,

$$dz = \left(\frac{\partial z}{\partial x}\right)_y dx + \left(\frac{\partial z}{\partial y}\right)_x dy, \tag{4.402}$$

consistent with Eq. (4.324); one may then divide both sides by dx, along a path of constant w as constraint, to get

$$\left(\frac{\partial z}{\partial x}\right)_w = \left(\frac{\partial z}{\partial x}\right)_y + \left(\frac{\partial z}{\partial y}\right)_x \left(\frac{\partial y}{\partial x}\right)_w. \tag{4.403}$$

Equation (4.403) permits change of constraints under two degrees of freedom – i.e. $(\partial z/\partial x)_w$, pertaining to an iso-$w$ trajectory, can be calculated from $(\partial z/\partial x)_y$ and $(\partial z/\partial y)_x$, valid under constant y and x, respectively, besides $(\partial y/\partial x)_w$.

4.3.6 Taylor's series expansion

4.3.6.1 Univariate function

If $f\{x\}$ is a continuous, single-valued function of variable x, exhibiting continuous derivatives $df\{x\}/dx$, $d^2f\{x\}/dx^2$, ..., $d^if\{x\}/dx^i$, ..., $d^nf\{x\}/dx^n$ within a given interval $[a,b]$, and if $d^{n+1}f\{x\}/dx^{n+1}$ exists in $]a,b[$, then Taylor's theorem states that

$$f\{x\} = f\{x\}\big|_a + \sum_{i=1}^{n} \frac{d^i f\{x\}}{dx^i}\bigg|_a \frac{(x-a)^i}{i!} + \frac{d^{n+1}f\{x\}}{dx^{n+1}}\bigg|_\zeta \frac{(x-a)^{n+1}}{(n+1)!}$$

$$a < \zeta < x \vee x < \zeta < a; \tag{4.404}$$

the last term in the right-hand side represents Lagrange's remainder, evaluated at some point ζ. Note that all n terms of the summation in Eq. (4.404) are nil at $x=a$ (owing to the constitutive, positive integer power of $x-a$), thus reducing the right hand side to just $f\{x\}|_a$ under these circumstances; therefore, it is expected that the said right-hand side will describe the behavior of $f\{x\}$ reasonably well in the vicinity of $x=a$. Taylor's polynomials provide a convenient way to describe the local pattern of a function by encapsulating its first several derivatives at a given point – while taking advantage of the fact that the derivatives of a function at some point dictate its behavior at nearby points. When such a remainder satisfies the condition

$$\lim_{n\to\infty} \frac{d^{n+1}f\{x\}}{dx^{n+1}}\bigg|_\zeta \frac{(x-a)^{n+1}}{(n+1)!} = 0, \tag{4.405}$$

then Eq. (4.404) may alternatively be formulated as

$$f\{x\} = f\{x\}\big|_a + \sum_{i=1}^{\infty} \frac{d^i f\{x\}}{dx^i}\bigg|_a \frac{(x-a)^i}{i!}; \tag{4.406}$$

Eq. (4.406) is referred to as Taylor's series in general – or else McLaurin's series, should $a=0$ in particular. On the other hand, if n is deliberately set equal to zero, then Eq. (4.404) simplifies to

$$f\{x\} = f\{x\}\big|_a + \frac{df\{x\}}{dx}\bigg|_\zeta \frac{x-a}{1!} \tag{4.407}$$

as the summation vanishes; this is equivalent to writing

$$\frac{df\{x\}}{dx}\bigg|_\zeta = \frac{f\{x\} - f\{x\}\big|_a}{x-a}. \tag{4.408}$$

Should $x \equiv b$ and $\zeta \equiv c$, then Eq. (4.408) becomes identical to Lagrange's theorem, labeled as Eq. (4.344) – thus justifying the designation chosen for the said remainder.

A Taylor's series only represents $f\{x\}$ within $[a,b]$ if this interval is included within the overall interval of convergence of the former; hence, one has to investigate in advance which such intervals are compatible with Eq. (4.405) before claiming validity of Taylor's approximation. If all terms of the said series are positive, then one can resort to the ratio test, i.e.

$$\left(\lim_{n\to\infty}\sum_{i=1}^{n} a_i = S_a \text{ finite}\right)\Bigg|_{\substack{\frac{a_{i+1}}{a_i}\le\frac{b_{i+1}}{b_i} \\ \lim_{n\to\infty}\sum_{k=1}^{n} b_i = S_b \text{ finite}}} \tag{4.409}$$

– where a geometric series of generic term b_i, described by ratio below unity, has been used as reference according to Eq. (4.107), with finite constants S_a and S_b consistent with Eq. (4.108); this result ultimately supports

$$\lim_{i\to\infty}\frac{\left.\frac{d^{i+1}f\{x\}}{dx^{i+1}}\right|_a}{\left.\frac{d^{i}f\{x\}}{dx^{i}}\right|_a}\frac{x-a}{i+1} < 1. \tag{4.410}$$

If the terms of the generalized summation in Eq. (4.406) are alternately positive and negative, then Leibnitz's test is in order, i.e.

$$\left(\lim_{n\to\infty}\sum_{i=1}^{n} a_i = S \text{ finite}\right)\Bigg|_{\substack{a_i a_{i+1}<0 \\ \lim_{i\to\infty}|a_i|=0}}, \tag{4.411}$$

with S denoting a finite constant; this is equivalent to

$$\lim_{i\to\infty}\frac{\left|\left.\frac{d^{i+1}f\{x\}}{dx^{i+1}}\right|_a (x-a)^{i+1}\right|}{(i+1)!} = 0, \tag{4.412}$$

which guarantees convergence of the corresponding Taylor's series within interval $[a,x]$ – as the error of truncation becomes vanishingly small, once the number of terms retained grows unbounded. A list of analytical functions, and corresponding Taylor's expansion and radius of convergence, is conveyed by Table 4.8.

Taylor's expansion is also useful to circumvent unknown quantities of the 0/0 type, i.e.

$$\lim_{x\to a}\frac{f\{x\}}{g\{x\}} = \frac{0}{0}, \tag{4.413}$$

normally handled via l'Hôpital's rule labeled as Eq. (4.355); such unknown quantities arise when Eq. (4.351) is satisfied, following application of Eq. (4.315). Upon expansion of both numerator and denominator in Eq. (4.413) via Taylor's series given by Eq. (4.406), one gets

$$\lim_{x\to a}\frac{f\{x\}}{g\{x\}} = \lim_{x\to a}\frac{f\{x\}\big|_a + \sum_{i=1}^{\infty}\left.\frac{d^i f\{x\}}{dx^i}\right|_a \frac{(x-a)^i}{i!}}{g\{x\}\big|_a + \sum_{i=1}^{\infty}\left.\frac{d^i g\{x\}}{dx^i}\right|_a \frac{(x-a)^i}{i!}}; \tag{4.414}$$

Eq. (4.414) breaks down to

$$\lim_{x\to a}\frac{f\{x\}}{g\{x\}} = \lim_{x\to a}\frac{\sum_{i=1}^{\infty}\left.\frac{d^i f\{x\}}{dx^i}\right|_a \frac{(x-a)^i}{i!}}{\sum_{i=1}^{\infty}\left.\frac{d^i g\{x\}}{dx^i}\right|_a \frac{(x-a)^i}{i!}} \tag{4.415}$$

TABLE 4.8

List of analytical functions, with corresponding Taylor's expansion and radius of convergence thereof.

Function	Expansion	Interval of Validity
e^x	$\sum_{i=0}^{\infty}\frac{x^i}{i!}$	$]-\infty,\infty[$
$\ln\{1+x\}$	$\sum_{i=1}^{\infty}\frac{(-1)^{i+1}}{i}x^i$	$]-1,1]$
$\sin x$	$\sum_{i=0}^{\infty}\frac{(-1)^i x^{2i+1}}{(2i+1)!}$	$]-\infty,\infty[$
$\cos x$	$\sum_{i=0}^{\infty}\frac{(-1)^i x^{2i}}{(2i)!}$	$]-\infty,\infty[$
$\sinh x$	$\sum_{i=0}^{\infty}\frac{x^{2i+1}}{(2i+1)!}$	$]-\infty,\infty[$
$\cosh x$	$\sum_{i=0}^{\infty}\frac{x^{2i}}{(2i)!}$	$]-\infty,\infty[$
$\sin^{-1} x$	$\sum_{i=0}^{\infty}\frac{(2i)!\,x^{2i+1}}{4^i(i!)^2(2i+1)}$	$]-1,1[$
$\cos^{-1} x$	$\frac{\pi}{2}-\sum_{i=0}^{\infty}\frac{(2i)!\,x^{2i+1}}{4^i(i!)^2(2i+1)}$	$]-1,1[$
$\sinh^{-1} x$	$\sum_{i=0}^{\infty}\frac{(-1)^i(2i)!\,x^{2i+1}}{4^i(i!)^2(2i+1)}$	$]-1,1[$

due to Eq. (4.351), thus allowing $x-a$ be factored out at both numerator and denominator as

$$\lim_{x\to a}\frac{f\{x\}}{g\{x\}} = \lim_{x\to a}\frac{(x-a)\sum_{i=1}^{\infty}\left.\frac{d^i f\{x\}}{dx^i}\right|_a \frac{(x-a)^{i-1}}{i!}}{(x-a)\sum_{i=1}^{\infty}\left.\frac{d^i g\{x\}}{dx^i}\right|_a \frac{(x-a)^{i-1}}{i!}}. \tag{4.416}$$

Following cancellation of $x-a$ between numerator and denominator, and explicitation of the first terms of the resulting summations, Eq. (4.416) transforms to

$$\begin{aligned}\lim_{x\to a}\frac{f\{x\}}{g\{x\}} &= \lim_{x\to a}\frac{\left.\frac{df\{x\}}{dx}\right|_a + \sum_{j=1}^{\infty}\left.\frac{d^{j+1}f\{x\}}{dx^{j+1}}\right|_a \frac{(x-a)^j}{(j+1)!}}{\left.\frac{dg\{x\}}{dx}\right|_a + \sum_{j=1}^{\infty}\left.\frac{d^{j+1}g\{x\}}{dx^{j+1}}\right|_a \frac{(x-a)^j}{(j+1)!}} \\ &= \frac{\left.\frac{df\{x\}}{dx}\right|_a + \sum_{j=1}^{\infty}\left.\frac{d^{j+1}f\{x\}}{dx^{j+1}}\right|_a \frac{(a-a)^j}{(j+1)!}}{\left.\frac{dg\{x\}}{dx}\right|_a + \sum_{j=1}^{\infty}\left.\frac{d^{j+1}g\{x\}}{dx^{j+1}}\right|_a \frac{(a-a)^j}{(j+1)!}} \\ &= \frac{\left.\frac{df\{x\}}{dx}\right|_a}{\left.\frac{dg\{x\}}{dx}\right|_a} = \lim_{x\to a}\frac{\frac{df\{x\}}{dx}}{\frac{dg\{x\}}{dx}},\end{aligned} \tag{4.417}$$

again bearing Eq. (4.315) in mind – where the summations became redundant because $x=a$ implies $x-a=a-a=0$; Eq. (4.417) accordingly retrieves Eq. (4.355). If

$$\lim_{x\to a}\frac{df\{x\}}{dx}=\lim_{x\to a}\frac{dg\{x\}}{dx}=0, \tag{4.418}$$

then one will proceed similarly – and thus rewrite Eq. (4.417) as

$$\begin{aligned}\lim_{x\to a}\frac{f\{x\}}{g\{x\}}&=\lim_{x\to a}\frac{(x-a)\sum_{j=1}^{\infty}\left.\frac{d^{j+1}f\{x\}}{dx^{j+1}}\right|_a\frac{(x-a)^{j-1}}{(j+1)!}}{(x-a)\sum_{j=1}^{\infty}\left.\frac{d^{j+1}g\{x\}}{dx^{j+1}}\right|_a\frac{(x-a)^{j-1}}{(j+1)!}}\\&=\frac{\frac{1}{2!}\left.\frac{d^2f\{x\}}{dx^2}\right|_a+\sum_{k=1}^{\infty}\left.\frac{d^{k+2}f\{x\}}{dx^{k+2}}\right|_a\frac{(a-a)^k}{(k+2)!}}{\frac{1}{2!}\left.\frac{d^2g\{x\}}{dx^2}\right|_a+\sum_{k=1}^{\infty}\left.\frac{d^{k+2}g\{x\}}{dx^{k+2}}\right|_a\frac{(a-a)^k}{(k+2)!}},\end{aligned} \tag{4.419}$$

after eliminating nil terms in, and cancelling common factors between numerator and denominator, and then splitting both summations; Eq. (4.419) breaks down to

$$\lim_{x\to a}\frac{f\{x\}}{g\{x\}}=\frac{\frac{1}{2!}\left.\frac{d^2f\{x\}}{dx^2}\right|_a}{\frac{1}{2!}\left.\frac{d^2g\{x\}}{dx^2}\right|_a}=\lim_{x\to a}\frac{\frac{d^2f\{x\}}{dx^2}}{\frac{d^2g\{x\}}{dx^2}}, \tag{4.420}$$

once ½ is dropped off both numerator and denominator, and the rule of calculation of a limit when x approaches a is applied. Equation (4.420) is indeed analogous to Eq. (4.358) for $i=1$ – and the associated rationale can be carried on and on; hence, iterated application of l'Hôpital's rule or Taylor's approach turn out being equivalent to each other.

Similar to the above process is the possibility of $f\{x\}$ and $g\{x\}$ being themselves polynomials on x, say $P_m\{x\}$ and $P_n\{x\}$, each constituted by a finite number of terms, say m and n, respectively – thus supporting

$$\begin{aligned}\lim_{x\to\infty}f\{x\}&\equiv\lim_{x\to\infty}P_m\{x\}=\lim_{x\to\infty}\sum_{i=0}^{m}p_{i,m}x^i\\&=\lim_{x\to\infty}\sum_{i=0}^{m}\left.\frac{d^if\{x\}}{dx^i}\right|_0\frac{x^i}{i!}=\infty\end{aligned} \tag{4.421}$$

and

$$\begin{aligned}\lim_{x\to\infty}g\{x\}&\equiv\lim_{x\to\infty}P_n\{x\}=\lim_{x\to\infty}\sum_{j=0}^{n}p_{j,n}x^j\\&=\lim_{x\to\infty}\sum_{j=0}^{n}\left.\frac{d^jg\{x\}}{dx^j}\right|_0\frac{x^j}{j!}=\infty,\end{aligned} \tag{4.422}$$

where $p_{i,m}$ and $p_{i,n}$ denote (constant) polynomial coefficients; the finite number of terms in either of Taylor's expansions around $a=0$ is to be outlined in this case. One readily realizes that

$$\lim_{x\to\infty}\frac{f\{x\}}{g\{x\}}\equiv\lim_{x\to\infty}\frac{P_m\{x\}}{P_n\{x\}}\equiv\lim_{x\to\infty}\frac{\sum_{i=0}^{m}p_{i,m}x^i}{\sum_{j=0}^{n}p_{j,n}x^j}=\frac{\infty}{\infty}, \tag{4.423}$$

when departing from Eqs. (4.421) and (4.422); together with

$$p_{i,m}\equiv\frac{1}{i!}\left.\frac{d^if\{x\}}{dx^i}\right|_0;\quad i=1,2,\ldots,m \tag{4.424}$$

and

$$p_{j,n}\equiv\frac{1}{j!}\left.\frac{d^jg\{x\}}{dx^j}\right|_0;\quad j=1,2,\ldots,n, \tag{4.425}$$

emerging from comparative inspection of homologous terms in Eqs. (4.421) or (4.422), respectively. To circumvent the unknown quantity ∞/∞, one may rewrite Eq. (4.423) as

$$\lim_{x\to\infty}\frac{f\{x\}}{g\{x\}}=\lim_{x\to\infty}\frac{x^m\sum_{i=0}^{m}p_{i,m}x^{i-m}}{x^n\sum_{j=0}^{n}p_{j,n}x^{j-n}}=\lim_{x\to\infty}\frac{x^m\sum_{i=0}^{m}\frac{p_{i,m}}{x^{m-i}}}{x^n\sum_{j=0}^{n}\frac{p_{j,n}}{x^{n-j}}} \tag{4.426}$$

via factoring out the highest power of x in numerator (i.e. x^m) and denominator (i.e. x^n); the indicated limit can then be split as

$$\lim_{x\to\infty}\frac{f\{x\}}{g\{x\}}==\lim_{x\to\infty}x^{m-n}\frac{\lim_{\frac{1}{x}\to 0}\sum_{i=0}^{m-1}p_{i,m}\left(\frac{1}{x}\right)^{m-i}+p_{m,m}}{\lim_{\frac{1}{x}\to 0}\sum_{j=0}^{n-1}p_{j,n}\left(\frac{1}{x}\right)^{n-j}+p_{n,n}}, \tag{4.427}$$

where the last term of either summation was made explicit – besides realizing that $1/x\to 0$ when $x\to\infty$. Both outstanding summations in Eq. (4.427) are nil, because $m-i>0$ for $0\leq i<m-1$ and $n-j>0$ for $0\leq j<n-1$; hence, dramatic simplification becomes possible to

$$\lim_{x\to\infty}\frac{f\{x\}}{g\{x\}}=\frac{p_{m,m}}{p_{n,n}}\lim_{x\to\infty}x^{m-n}. \tag{4.428}$$

When $m=n$, Eq. (4.428) reduces to

$$\left.\lim_{x\to\infty}\frac{f\{x\}}{g\{x\}}\right|_{m=n}=\frac{p_{m,m}}{p_{n,n}}, \tag{4.429}$$

i.e. a horizontal asymptote arises, see Eq. (4.11) – with vertical intercept given by the ratio of coefficients of the highest powers of the polynomials in numerator and denominator; conversely,

$$\left.\lim_{x\to\infty}\frac{f\{x\}}{g\{x\}}\right|_{m<n}=\frac{p_{m,m}}{p_{n,n}}0=0, \tag{4.430}$$

so $f\{x\}/g\{x\}$ becomes driven by the horizontal axis as asymptote – or

$$\left.\lim_{x\to\infty}\frac{f\{x\}}{g\{x\}}\right|_{m>n}=\frac{p_{m,m}}{p_{n,n}}\infty=\infty \tag{4.431}$$

otherwise, due to finite, non-nil values of $p_{m,m}$ and $p_{n,n}$, and characterized by an $(m-n)$th-order asymptote.

4.3.6.2 Bivariate function

A bivariate function, say $f\{x,y\}$, is also expandable via Taylor's series as

$$f\{x,y\} = f\{x,y\}\Big|_{\substack{x=a\\y=b}} + \frac{\partial f\{x,y\}}{\partial x}\Big|_{\substack{x=a\\y=b}}(x-a) + \frac{\partial f\{x,y\}}{\partial y}\Big|_{\substack{x=a\\y=b}}(y-b)$$
$$+ \frac{\partial^2 f\{x,y\}}{\partial x^2}\Big|_{\substack{x=a\\y=b}}\frac{(x-a)^2}{2!} + \frac{\partial^2 f\{x,y\}}{\partial x \partial y}\Big|_{\substack{x=a\\y=b}}\frac{(x-a)(y-b)}{2!}$$
$$+ \frac{\partial^2 f\{x,y\}}{\partial y \partial x}\Big|_{\substack{x=a\\y=b}}\frac{(y-b)(x-a)}{2!} + \frac{\partial^2 f\{x,y\}}{\partial y^2}\Big|_{\substack{x=a\\y=b}}\frac{(y-b)^2}{2!} + \ldots; \quad (4.432)$$

Eq. (4.392) may then be invoked to lump intermediate terms as

$$f\{x,y\} = f\{x,y\}\Big|_{\substack{x=a\\y=b}} + \frac{\partial f\{x,y\}}{\partial x}\Big|_{\substack{x=a\\y=b}}(x-a) + \frac{\partial f\{x,y\}}{\partial y}\Big|_{\substack{x=a\\y=b}}(y-b)$$
$$+ \frac{\partial^2 f\{x,y\}}{\partial x^2}\Big|_{\substack{x=a\\y=b}}\frac{(x-a)^2}{2} + \frac{\partial^2 f\{x,y\}}{\partial x \partial y}\Big|_{\substack{x=a\\y=b}}(x-a)(y-b)$$
$$+ \frac{\partial^2 f\{x,y\}}{\partial y^2}\Big|_{\substack{x=a\\y=b}}\frac{(y-b)^2}{2} + \ldots . \quad (4.433)$$

Although the exact equality in Eq. (4.433) requires an infinite number of terms, truncation is made, for simplicity, after the linear term in most applications.

4.3.7 Vector calculus

4.3.7.1 Definition of nabla

The *nabla* or *del* operator, ∇, is defined by

$$\nabla \equiv \boldsymbol{i}_x \frac{\partial}{\partial x} + \boldsymbol{i}_y \frac{\partial}{\partial y} + \boldsymbol{i}_z \frac{\partial}{\partial z}, \quad (4.434)$$

where $\boldsymbol{i}_x$, $\boldsymbol{i}_y$, and $\boldsymbol{i}_z$ denote unit vectors in rectangular coordinates; this is a quite useful operator for condensed description of a three-dimensional gradient of a scalar field f, i.e.

$$\text{grad } f \equiv \nabla f, \quad (4.435)$$

as well as the divergent of a vector field $\boldsymbol{F}$, i.e.

$$\text{div}\,\boldsymbol{F} \equiv \nabla \cdot \boldsymbol{F}, \quad (4.436)$$

the divergent of a tensor $\boldsymbol{\tau}$, i.e.

$$\text{div}\,\tau \equiv \nabla \cdot \tau, \quad (4.437)$$

or even the rotational of a vector field, viz.

$$\text{curl}\,\boldsymbol{F} \equiv \nabla \times \boldsymbol{F} \quad (4.438)$$

– among many other applications.

4.3.7.2 Properties

When in the form of a gradient, one finds that

$$\nabla(f+g) = \nabla f + \nabla g \quad (4.439)$$

and

$$\nabla(fg) = g\nabla f + f\nabla g, \quad (4.440)$$

with g denoting a second scalar field; Eqs. (4.439) and (4.440) resemble Eqs. (4.330) and (4.332), respectively. If f is, in turn, a univariate function of ξ, then one may write

$$\nabla f\{\xi\} = \frac{df}{d\xi}\nabla \xi \quad (4.441)$$

as a form of chain differentiation rule – serving as alias to Eq. (4.341). When scalar product is used between two nabla operators, a Laplacian results – defined as

$$\nabla^2 \equiv \nabla \cdot \nabla, \quad (4.442)$$

or else

$$\nabla^2 \equiv \frac{\partial^2}{\partial x^2} + \frac{\partial^2}{\partial y^2} + \frac{\partial^2}{\partial z^2} \quad (4.443)$$

in view of Eq. (4.434) coupled with the algorithm of the scalar product conveyed by Eq. (4.262). Finally, it can be proven that

$$(\boldsymbol{F} = \nabla f)\Big|_{\nabla \times \boldsymbol{F} = \boldsymbol{0}}, \quad (4.444)$$

i.e. a vector field $\boldsymbol{F}$ can always be defined as gradient of some scalar (potential) field f – as long as $\boldsymbol{F}$ is irrotational, i.e. $\nabla \times \boldsymbol{F} = \boldsymbol{0}$.

4.3.7.3 Multiple products

Combination of *nabla* with scalar and vector fields, at the expense of the forms of multiplication presented previously, leads to many possibilities; some of the most useful are depicted in Table 4.9.

4.3.7.4 Conversion to curved systems of coordinates

The intrinsic geometry and symmetrical features of a physical problem often recommend working in a system of coordinates other than rectangular (or Cartesian) coordinates, (x,y,z); this is the case of cylindrical coordinates, (r,ϕ,z), or spherical coordinates, (r,ϕ,θ), with mutual conversion tabulated in Table 4.10. Instead of $\boldsymbol{i}_x$, $\boldsymbol{i}_y$, and $\boldsymbol{i}_z$ as (fixed) unit vectors in the direction of each of the three rectangular coordinates, one resorts to (moving) $\boldsymbol{j}_r$ and $\boldsymbol{j}_\phi$, besides (fixed) $\boldsymbol{j}_z$ in the case of cylindrical coordinates, with subscripts r and ϕ referring to radial and polar positions, respectively; or to (moving) $\boldsymbol{k}_r$, $\boldsymbol{k}_\phi$, and $\boldsymbol{k}_\theta$ in the case of spherical coordinates, with subscript θ referring to azimuthal position. The *nabla* operator may accordingly be expressed otherwise – as outlined in Table 4.11, with regard to scalar f, or else to vector $\boldsymbol{F}(F_x,F_y,F_z)$, alias $\boldsymbol{F}(F_r,F_\phi,F_z)$, alias $\boldsymbol{F}(F_r,F_\phi,F_\theta)$.

4.3.8 Indefinite integral

4.3.8.1 Definition

Integration may be regarded as the inverse operation of differentiation; function $F\{x\}$ is accordingly termed (indefinite) integral of $f\{x\}$, viz.

$$\left(F\{x\} \equiv \int f\{x\}dx\right)\Bigg|_{\frac{dF}{dx}=f}, \tag{4.445}$$

provided that dF/dx equals $f\{x\}$. Function $f\{x\}$ is called integrand or kernel; and $f\{x\}$ is said to be integrable when $F\{x\}$ exists. Equation (4.445) may instead be coined as

$$f\{x\} = \int \frac{df\{x\}}{dx}dx; \tag{4.446}$$

composition of integration and differentiation of a given function retrieves indeed the original function, for being operations inverse of each other. Unlike the derivative of an elementary function – which normally exists and is another elementary function (as long as the function under scrutiny is continuous), the same cannot be claimed about an integral at large; there are elementary functions for which an integral is not expressible as a finite combination of elementary functions (e.g. e^{-x^2}), while it cannot be told a priori whether a given function can be integrated at all. One also realizes that uniqueness of the derivative does not extend to an integral; in other words,

$$F_2\{x\} = F_1\{x\} + k \tag{4.447}$$

always applies – meaning that any two integrals, $F_1\{x\}$ and $F_2\{x\}$, of $f\{x\}$ differ by an (arbitrary) constant, k. Consequently, there is one degree of freedom associated with an indefinite integral – i.e. there is an infinite number of other integrals for any given function, once one integral has been found. A list of the most common elementary integrals is provided in Table 4.12. Confirmation of all integrals $F\{x\}$ tabulated therein, with regard to the corresponding function $f\{x\}$, may easily be obtained via differentiation of the entries in the right column; note the existence of an arbitrary constant, k, as part of all integrals – consistent with Eq. (4.447).

TABLE 4.9

List of products involving operator *nabla*, ∇, scalar f, vectors $\boldsymbol{F}$ and $\boldsymbol{G}$, or tensor $\boldsymbol{\tau}$.

Product	Alternative Form
$\nabla\cdot(f\boldsymbol{F})$	$(\nabla f)\cdot\boldsymbol{F} + f\nabla\cdot\boldsymbol{F}$
$\nabla\times\nabla f$	$\boldsymbol{0}$
$\nabla\cdot(\boldsymbol{FF})$	$\boldsymbol{F}(\nabla\cdot\boldsymbol{F})+(\boldsymbol{F}\cdot\nabla)\boldsymbol{F}$
$(\boldsymbol{F}\cdot\nabla)\boldsymbol{F}$	$\frac{1}{2}\nabla(\boldsymbol{F}\cdot\boldsymbol{F})-\boldsymbol{F}\times(\nabla\times\boldsymbol{F})$
$\nabla\cdot(\boldsymbol{\tau}\cdot\boldsymbol{F})$	$\boldsymbol{F}\cdot(\nabla\cdot\boldsymbol{\tau})+\boldsymbol{\tau}:(\nabla\boldsymbol{F})$
$\nabla\cdot(\boldsymbol{F}\times\boldsymbol{G})$	$\boldsymbol{G}\cdot(\nabla\times\boldsymbol{F})-\boldsymbol{F}\cdot(\nabla\times\boldsymbol{G})$
$\nabla\cdot(\nabla\times\boldsymbol{F})$	0
$\boldsymbol{G}\cdot(\nabla\times\boldsymbol{F})$	$\nabla\cdot(\boldsymbol{F}\times\boldsymbol{G})+\boldsymbol{F}\cdot(\nabla\times\boldsymbol{G})$
$\boldsymbol{F}\times(\nabla\times\boldsymbol{G})$	$\nabla(\boldsymbol{F}\cdot\boldsymbol{G})-(\boldsymbol{F}\cdot\nabla)\boldsymbol{G}$
$\nabla\times(\nabla\times\boldsymbol{F})$	$\nabla(\nabla\cdot\boldsymbol{F})-\nabla^2\boldsymbol{F}$

TABLE 4.10

Conversion of coordinates between classical systems of referencing.

		From		
		Rectangular	**Cylindrical**	**Spherical**
To	**Rectangular**	–	$x = r\cos\phi$ $y = r\sin\phi$ $z = z$	$x = r\cos\phi\sin\theta$ $y = r\sin\phi\sin\theta$ $z = r\cos\theta$
	Cylindrical	$r=\sqrt{x^2+y^2}$ $\phi=\tan^{-1}\frac{y}{x}$ $z=z$	–	$r = r\sin\theta$ $\phi=\phi$ $z = r\cos\theta$
	Spherical	$r=\sqrt{x^2+y^2+z^2}$ $\phi=\tan^{-1}\frac{y}{x}$ $\theta=\cos^{-1}\frac{z}{\sqrt{x^2+y^2+z^2}}$	$r=\sqrt{r^2+z^2}$ $\phi=\phi$ $\theta=\cos^{-1}\frac{z}{\sqrt{r^2+z^2}}$	–

4.3.8.2 Properties

Recalling the definition of differential as per Eq. (4.319), one may redo Eq. (4.445) to

$$d\int f\{x\}dx = \frac{dF\{x\}}{dx}dx = f\{x\}dx; \tag{4.448}$$

hence, differential of an indefinite integral coincides with its full kernel. Moreover, Eq. (4.445) entails

$$\frac{dF}{dx} = f\{x\}, \tag{4.449}$$

which is equivalent to

$$dF = f\{x\}dx \tag{4.450}$$

after multiplying both sides by dx; integration of both sides of Eq. (4.450), i.e.

$$\int dF = \int f\{x\}dx, \tag{4.451}$$

will eventually read

$$\int dF = F, \tag{4.452}$$

again with the aid of Eq. (4.445) – so integral of differential of a function is but the said function.

TABLE 4.11

Functional form of operator *nabla*, ∇, in coordinates under classical systems of referencing.

Operation	Coordinates: Rectangular	Cylindrical	Spherical
∇f	$\boldsymbol{i}_x\frac{\partial f}{\partial x}+\boldsymbol{i}_y\frac{\partial f}{\partial y}+\boldsymbol{i}_z\frac{\partial f}{\partial z}$	$\boldsymbol{j}_r\frac{\partial f}{\partial r}+\boldsymbol{j}_\phi\frac{1}{r}\frac{\partial f}{\partial \phi}+\boldsymbol{j}_z\frac{\partial f}{\partial z}$	$\boldsymbol{k}_r\frac{\partial f}{\partial r}+\boldsymbol{k}_\phi\frac{1}{r}\frac{\partial f}{\partial \phi}+\boldsymbol{k}_\theta\frac{1}{r\sin\phi}\frac{\partial f}{\partial \theta}$
$\nabla\cdot\boldsymbol{F}$	$\boldsymbol{i}_x\frac{\partial F_x}{\partial x}+\boldsymbol{i}_y\frac{\partial F_y}{\partial y}+\boldsymbol{i}_z\frac{\partial F_z}{\partial z}$	$\boldsymbol{j}_r\frac{1}{r}\frac{\partial(rF_r)}{\partial r}+\boldsymbol{j}_\phi\frac{1}{r}\frac{\partial F_\phi}{\partial \phi}+\boldsymbol{j}_z\frac{\partial F_z}{\partial z}$	$\boldsymbol{k}_r\frac{1}{r^2}\frac{\partial(r^2F_r)}{\partial r}+\boldsymbol{k}_\phi\frac{1}{r\sin\phi}\frac{\partial(F_\phi\sin\phi)}{\partial\phi}+\boldsymbol{k}_\theta\frac{1}{r\sin\phi}\frac{\partial F_\theta}{\partial\theta}$
$\nabla\times\boldsymbol{F}$	$\boldsymbol{i}_x\left(\frac{\partial F_z}{\partial y}-\frac{\partial F_y}{\partial z}\right)+\boldsymbol{i}_y\left(\frac{\partial F_x}{\partial z}-\frac{\partial F_z}{\partial x}\right)+\boldsymbol{i}_z\left(\frac{\partial F_y}{\partial x}-\frac{\partial F_x}{\partial y}\right)$	$\boldsymbol{j}_r\left(\frac{1}{r}\frac{\partial F_z}{\partial \phi}-\frac{\partial F_\phi}{\partial z}\right)+\boldsymbol{j}_\phi\left(\frac{\partial F_r}{\partial z}-\frac{\partial F_z}{\partial r}\right)+\boldsymbol{j}_z\frac{1}{r}\left(\frac{\partial(rF_\phi)}{\partial r}-\frac{\partial F_r}{\partial\phi}\right)$	$\boldsymbol{k}_r\left(\frac{1}{r\sin\phi}\frac{\partial(F_\theta\sin\phi)}{\partial\phi}-\frac{\partial F_\phi}{\partial\theta}\right)+\boldsymbol{k}_\phi\frac{1}{r}\left(\frac{1}{\sin\phi}\frac{\partial F_r}{\partial\theta}-\frac{\partial(rF_\theta)}{\partial r}\right)+\boldsymbol{k}_\theta\frac{1}{r}\left(\frac{\partial(rF_\phi)}{\partial r}-\frac{\partial F_r}{\partial\phi}\right)$
$\nabla^2 f$	$\frac{\partial^2 f}{\partial x^2}+\frac{\partial^2 f}{\partial y^2}+\frac{\partial^2 f}{\partial z^2}$	$\frac{1}{r}\frac{\partial\left(r\frac{\partial f}{\partial r}\right)}{\partial r}+\frac{1}{r^2}\frac{\partial^2 f}{\partial\phi^2}+\frac{\partial^2 f}{\partial z^2}$	$\frac{1}{r^2}\frac{\partial\left(r^2\frac{\partial f}{\partial r}\right)}{\partial r}+\frac{1}{r^2\sin\phi}\frac{\partial\left(\sin\phi\frac{\partial f}{\partial\phi}\right)}{\partial\phi}+\frac{1}{r^2\sin\phi}\frac{\partial^2 f}{\partial\theta^2}$

4.3.8.3 Rules of integration

To calculate a nonelementary integral, one can resort to integration by decomposition, by parts, or by change of variable; the first is consubstantiated in

$$\int\left(af\{x\}+bg\{x\}\right)dx=a\int f\{x\}dx+b\int g\{x\}dx \tag{4.453}$$

– i.e. the integral of a linear combination of functions, $f\{x\}$ and $g\{x\}$, weighed by constants a and b, equals the linear combination of the integrals of such functions. A list of integrals based on Eq. (4.453) is tabulated in Table 4.13.

TABLE 4.12

List of elementary indefinite integrals.

$f\{x\}$	$\int f\{x\}dx$
$x^n, n\neq -1$	$\frac{x^{n+1}}{n+1}+k$
$1/x$	$\ln\lvert x\rvert+k$
e^x	e^x+k
e^{-x}	$-e^{-x}+k$
$\sin x$	$-\cos x+k$
$\operatorname{cosec}^2 x$	$-\operatorname{cotan} x$
$\cos x$	$\sin x+k$
$\sec^2 x$	$\tan x+k$
$\sec x\tan x$	$\sec x+k$
$\frac{1}{\sqrt{1-x^2}}$	$\sin^{-1}x+k$ $-\cos^{-1}x+k$
$\frac{1}{1+x^2}$	$\tan^{-1}x+k$

The rule of integration by parts takes advantage of the rule of differentiation of a product, viz.

$$\int f\{x\}g\{x\}dx=f\{x\}G\{x\}-\int G\{x\}\frac{df\{x\}}{dx}dx; \tag{4.454}$$

in other words, the integral of the product of two functions, $f\{x\}$ and $g\{x\}$, can be calculated as the product of the first, $f\{x\}$, by the integral of the second, $G\{x\}$, with a further negative correction via the integral of the product of $G\{x\}$ by the derivative of $f\{x\}$, i.e. $df\{x\}/dx$. Examples of integrals obtained via this rule are presented in Table 4.14.

The third (and probably the most useful) method of integration is via change of variable; it abides to

$$\int f\{x\}dx=\int f\{\phi\{\xi\}\}\frac{d\phi\{\xi\}}{d\xi}d\xi, \tag{4.455}$$

where $\phi\{\xi\}$ denotes a (univariate) function of ξ. It relies on the ability to rewrite x as a function $\phi\{\xi\}$ of some other variable ξ, coupled with multiplication and division by $d\xi$. This method permits calculation of the integrals in Table 4.15, among many others.

A final method of integration, bearing wide applicability, pertains to rational fractions (or quotients of polynomials), viz.

$$\int\frac{P_n\{x\}}{P_m\{x\}}dx=\int Q_{n-m}\{x\}dx+\int\frac{R_{<m}}{b_m\prod_{k=1}^{m}(x-r_k)}dx \tag{4.456}$$

– obtained directly from Eq. (4.124), complemented by integration of both sides with the aid of Eq. (4.453). After splitting $R_{<m}\big/b_m\prod_{k=1}^{m}(x-r_k)$ as indicated by Eq. (4.127), one readily obtains

$$\int \frac{P_n\{x\}}{P_m\{x\}}dx = \int Q_{n-m}\{x\}dx + \sum_{l=1}^{s}\sum_{p=0}^{s_l-1} A_{l,p+1}\int \frac{dx}{\left(x-r_l\right)^{s_l-p}} \tag{4.457}$$

– again at the expense of the intrinsic linearity of the integral operator; here $\int Q_{n-m}\{x\}dx$ becomes accessible via combination of Eq. (4.453) with the first entry in Table 4.12. When pole r_l in Eq. (4.457) denotes a real number and $s_l=1$ (i.e. the said root of $P_m\{x\}$ is distinct from any other), then one should resort to

$$A_{l,p+1}\int \frac{dx}{\left(x-r_l\right)^{s_l-p}} \equiv A_{l,1}\int \frac{dx}{x-r_l} = A_{l,1}\ln\left|x-r_l\right| + k, \tag{4.458}$$

obtained from the second entry of Table 4.12 and using $\xi \equiv x-r_l$ for integration variable. If $s_l>1$, then one finds

$$A_{l,p+1}\int \frac{dx}{\left(x-r_l\right)^{s_l-p}} = \frac{A_{l,p+1}}{p-s_l+1}\left(x-r_l\right)^{p-s_l+1} + k \qquad p=0,1,\ldots,s_l-2, \tag{4.459}$$

corresponding to a multiple pole – whereas the case associated with $p=s_l-1$ degenerates to Eq. (4.458). The third situation of practical interest encompasses (conjugate) complex poles, since they appear in pairs (as $P_m\{x\}$ was implicitly assumed to possess real coefficients only). They are usually combined as done in Eq. (4.129), to eventually yield

$$\int\left(\frac{A_{l,1}}{x-(a+\iota b)} + \frac{A_{l,2}}{x-(a-\iota b)}\right)dx = \frac{B_{l,1}}{2}\ln\left|x^2+B_{l,3}x+B_{l,4}\right| + \frac{2B_{l,2}-B_{l,1}B_{l,3}}{\sqrt{4B_{l,4}-B_{l,3}^2}}\tan^{-1}\frac{2x+B_{l,3}}{\sqrt{4B_{l,4}-B_{l,3}^2}} + k; \tag{4.460}$$

$B_{l,1}$, $B_{l,2}$, $B_{l,3}$, and $B_{l,4}$ can be related to the original $A_{l,1}$, $A_{l,2}$, a, and b via Eqs. (4.130)–(4.133). Despite the above (usually simpler) approach, the case of a pair of complex (conjugate, and thus distinct) poles may also be handled via Eq. (4.458) – knowing that r_l will, in this case, appear in the form $a\pm\iota b$.

4.3.9 Definite single integral

4.3.9.1 Definition

Riemann's definition of a definite (single) integral has it that

$$\int_a^b f(x)dx \equiv \lim_{N\to\infty}\sum_{i=1}^{N} f\left\{a+i\frac{b-a}{N}\right\}\frac{b-a}{N}; \tag{4.461}$$

a graphical interpretation is provided in Fig. 4.10. One promptly realizes that the definite integral of function $f\{x\}$, taken between a and b, is but the limit of the summation of areas of adjacent rectangles – each with base characterized by length $(b-a)/N$ and height given by $f\{a+i(b-a)/N\}$, when N grows unbounded. The said limiting geometrical figure is also known as trapezoid – with surface defined by vertical lines $x=a$ and $x=b$ on the left and the right, respectively, horizontal line $y=0$ on the bottom, and curve $y=f\{x\}$ on the top.

4.3.9.2 Properties

If $f\{x\}$ is a continuous function between a and $b>a$, then

$$\exists_{c\in]a,b[} : \int_a^b f\{x\}dx = (b-a)f\{x\}\Big|_{x=c} \tag{4.462}$$

holds true; Eq. (4.462) is usually known as mean value theorem for integrals. In fact, Eq. (4.462) may be rewritten as

$$\bar{f} \equiv f\{x\}\Big|_{x=c} = \frac{\int_a^b f\{x\}dx}{b-a} = \frac{\int_a^b f\{x\}dx}{\int_a^b dx}, \tag{4.463}$$

where $\int_a^b dx$ represents area of rectangle of width $b-a$ and unit height; whereas $\bar{f}$ represents mean – i.e. height of rectangle of width $b-a$, and area identical to that of trapezoid defined by $\int_a^b f\{x\}dx$. On the other hand,

TABLE 4.13

List of indefinite integrals obtained via rule of decomposition.

$f\{x\}$	$\int f\{x\}dx$
$\cosh x$	$\sinh x+k$
$\sinh x$	$\cosh x+k$

TABLE 4.14

List of indefinite integrals obtained via rule of integration by parts.

$f\{x\}$	$\int f\{x\}dx$
$\ln x$	$x(\ln x-1)+k$
$x\ln x$	$\frac{x^2}{2}\ln x-\frac{x^2}{4}+k$
$\tan^{-1}x$	$x\tan^{-1}x-\frac{1}{2}\ln\left\{1+x^2\right\}+k$

TABLE 4.15

List of indefinite integrals obtained via rule of integration by change of variable.

$f\{x\}$	$\int f\{x\}dx$
$\tan x$	$\ln\left|\sec x\right|+k$
$\text{cotan}\, x$	$\ln\left|\sin x\right|+k$
$\sec x$	$\ln\left|\sec x+\tan x\right|+k$

$$\frac{d}{dx}\int_a^x f\{\xi\}d\xi = f\{x\}, \tag{4.464}$$

which means that $\int_a^x f\{\xi\}d\xi$ plays the role of indefinite integral of $f\{x\}$ in agreement with Eq. (4.445); note that $\int_a^x f\{\xi\}d\xi$ is a function, rather than a number – because the upper limit of integration is itself a variable. One finally realizes that

$$\int_a^b f\{x\}dx = F\{b\} - F\{a\} \equiv F\{x\}\Big|_a^b, \tag{4.465}$$

where $F\{x\}$ denotes an indefinite integral of $f\{x\}$; this theorem is, on its own, referred to as fundamental theorem of integral calculus, or Newton and Leibnitz's theorem.

4.3.9.3 Rules of integration

Equation (4.461) may be invoked to conclude that

$$\int_a^b cf\{x\}dx = c\int_a^b f\{x\}dx, \tag{4.466}$$

i.e. any constant, say c, can be taken off a definite integral. Using a similar strategy, one will address a sum of functions as

$$\int_a^b \left(f\{x\} + g\{x\}\right)dx = \int_a^b f\{x\}dx + \int_a^b g\{x\}dx; \tag{4.467}$$

in other words, the integral and sum operators are interchangeable – owing to their linearity. Note the resemblance of Eqs. (4.466) and (4.467) to Eq. (4.453), except for existence of limits of integration in the former.

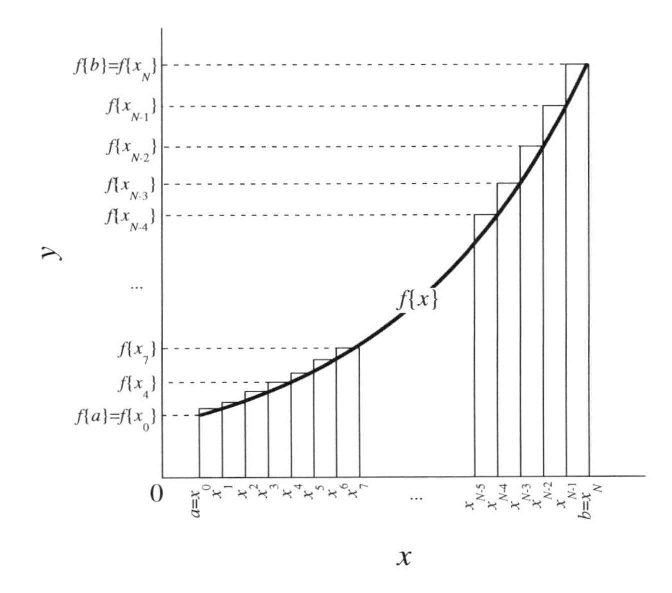

FIGURE 4.10 Graphical representation of function $f\{x\}$, continuous within interval $[a,b]$, subdivided, in turn, as N subintervals of identical amplitude, $(b-a)/N$, defined by abscissae $a=x_0, x_1, x_2, \ldots, x_{N-1}, x_N=b$ and ordinates $f\{x_1\}$, $f\{x_2\}, \ldots, f\{x_N\}$, respectively – with horizontal segments in between serving as base to rectangles, and vertical segments serving as height thereto.

When the negative of a function is at stake, one may write

$$-\int_a^b f\{x\}dx = \int_b^a f\{x\}dx; \tag{4.468}$$

hence, reversal of integration limits is equivalent to taking the negative of the original integral. Following partition of an integral by an integration limit, one obtains

$$\int_a^b f\{x\}dx = \int_a^c f\{x\}dx + \int_c^b f\{x\}dx \tag{4.469}$$

– irrespective of whether $c<a<b$, $a<c<b$, $a<b<c$. Moreover,

$$\int_a^a f\{x\}dx = 0 \tag{4.470}$$

means that an integrating interval with nil amplitude delivers a nil value for the corresponding definite integral, irrespective of its kernel.

Another property of relevance, exhibited by definite integrals, reads

$$\left.\left(\int_a^b f\{x\}dx \le \int_a^b g\{x\}dx\right)\right|_{f\{x\}\le g\{x\},\forall_{x\in]a,b[}} ; \tag{4.471}$$

in other words, if $f\{x\}$ does not exceed $g\{x\}$ within some interval, then the integral of the former cannot exceed the integral of the latter over that interval.

4.3.9.4 Leibnitz's formula

Consider a definite integral, of the form

$$I\{y\} \equiv \int_{a\{y\}}^{b\{y\}} f\{x,y\}dx, \tag{4.472}$$

where $f\{x,y\}$ denotes an integrable function of x within range $a \le x \le b$, further to its dependence also on y – and $a\{y\}$ and $b\{y\}$ are continuous and (at least once) differentiable functions of y. It can be proven that

$$\frac{d}{dy}\int_{a\{y\}}^{b\{y\}} f\{x,y\}dx = f\{b,y\}\frac{db\{y\}}{dy} - f\{a,y\}\frac{da\{y\}}{dy} + \int_{a\{y\}}^{b\{y\}} \frac{\partial f\{x,y\}}{\partial y}dx; \tag{4.473}$$

this equality supports Eq. (4.464), since the second and third terms in its right-hand side would turn nil. If a and b are both constant, then $\frac{d}{dy}\int_{a\{y\}}^{b\{y\}} f\{x,y\}dx$ reduces to $\int_a^b \frac{\partial f\{x,y\}}{\partial y}dx$ in agreement with Eq. (4.473) – i.e. differential and integral operators can be interchanged under these conditions.

4.3.10 Definite multiple integral

4.3.10.1 Line integral

Suppose $y \equiv f\{x\}$ denotes a real, single-valued, monotonic, and continuous function of x, within some interval $[a,b]$ – as represented in Fig. 4.11 by curve C, described by $f\{x\}$ as analytical equation; with endpoints A and B, accordingly holding coordinates $(a,f\{a\})$ and $(b,f\{b\})$, respectively. If $P\{x,y\}$ and $Q\{x,y\}$ are two real, single-valued, and continuous functions of x and y at all points of C, then either integral $\int_{C_x:A\to B} P\{x,y\}dx$ or $\int_{C_y:A\to B} Q\{x,y\}dy$ is termed line (or curvilinear) integral – with integration taking place along curve C, between points A and B.

4.3.10.2 Double integral

A double integral may be geometrically defined in much the same way Riemann defined a single integral via Eq. (4.461), i.e.

$$\int_a^b \int_{g_1\{x\}}^{g_2\{x\}} f\{x,y\}dydx \equiv \lim_{\substack{\Delta x\to 0\\ \Delta y\to 0}} \sum_{i=1}^{N}\sum_{j=1}^{M} f\left\{x_i,y_j\right\}\Delta y \Delta x$$

$$= \lim_{\substack{N\to\infty\\ M\to\infty}} \sum_{i=1}^{N}\sum_{j=1}^{M} f\left\{ \begin{array}{l} a+i\dfrac{b-a}{N}, \\ g_1\left\{a+i\dfrac{b-a}{N}\right\}+j\dfrac{\left(g_2\left\{a+i\dfrac{b-a}{N}\right\} - g_1\left\{a+i\dfrac{b-a}{N}\right\}\right)}{M} \end{array} \right\}$$

$$\times \frac{g_2\left\{a+i\dfrac{b-a}{N}\right\} - g_1\left\{a+i\dfrac{b-a}{N}\right\}}{M}\frac{b-a}{N}; \tag{4.474}$$

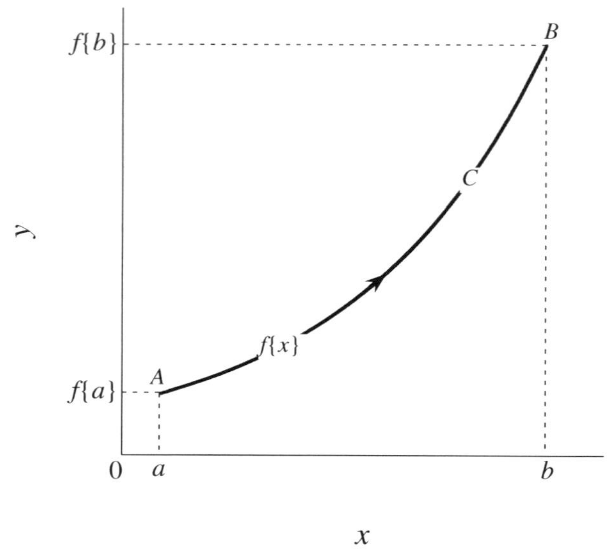

FIGURE 4.11 Graphical representation of function $f\{x\}$, continuous within interval $[a,b]$, and laid out on $x0y$ plane as curve C – departing from point A of coordinates $(a,f\{a\})$, and reaching point B of coordinates $(b,f\{b\})$.

a graphical account is provided in Fig. 4.12. Inspection of this figure unfolds two basic ways of spanning integration region, $D_{x,y}$ – either having x as independent variable vary between a and b, and then y vary between $g_1\{x\}$ and $g_2\{x\}$ at every such x; or instead having y as independent variable, varying between $g_1\{a\}$ and $g_2\{b\}$, and then x span the range $[g_2^{-1}\{y\}, g_1^{-1}\{y\}]$. Hence, calculation of a double integral must proceed stepwise, with the inner integral being computed before the outer integral; a similar situation occurred with the second-order derivative, calculated as the derivative of its first-order counterpart, see Eq. (4.326).

4.3.10.3 Fubini's theorem

Consider the case where the limits of integration in both directions are constant – or, equivalently, characterized by an integration domain of rectangular shape; this means that $g_1\{x\}=c$ and $g_2\{x\}=d$ in Fig. 4.12, with c and d denoting (known) constants. The associated double integral will accordingly satisfy

$$\iint_{D_{x,y}} f\{x,y\}dxdy = \int_a^b \left(\int_c^d f\{x,y\}dy \right) dx = \int_c^d \left(\int_a^b f\{x,y\}dx \right) dy \tag{4.475}$$

– known as a weak version of Fubini's theorem. Equation (4.475) basically indicates that order of integration of a double integral is redundant when the region of integration is a rectangle (with sides of length $b-a$ and $d-c$); in a sense, this is the integral analog of Young's or Schwarz's theorems conveyed by Eq. (4.392), which convey equivalence of cross derivatives of a given function.

A stronger version of Fubini's theorem can be formulated, encompassing an integration domain $D_{x,y}$ not restricted to the rectangular shape – according to

$$\iint_{D_{x,y}} f\{x,y\}dxdy = \int_a^b \left(\int_{g_1\{x\}}^{g_2\{x\}} f\{x,y\}dy \right) dx = \int_{g_1\{a\}}^{g_2\{b\}} \left(\int_{g_2^{-1}\{y\}}^{g_1^{-1}\{y\}} f\{x,y\}dx \right) dy, \tag{4.476}$$

consistent with Fig. 4.12; in either case, the outer integral is evaluated between (constant) lower and upper boundaries, say a and b, or $g_1\{a\}$ and $g_2\{a\}$, respectively – whereas the inner integral is, for each value of the outer integration variable, evaluated between (variable) lower and upper boundaries, say $g_1\{x\}$ and $g_2\{x\}$, or $g_2^{-1}\{y\}$ and $g_1^{-1}\{y\}$, respectively.

4.3.10.4 Leibnitz's theorem

In the case of a triple integral, extending to V as volume of a closed (moving) region in space, surrounded by surface S that moves with velocity $\boldsymbol{v}_S$, one finds that

$$\frac{\partial}{\partial t}\iiint_V f d\tilde{V} = \iint_S f\left(\boldsymbol{v}_S \cdot \boldsymbol{n}\right)d\tilde{S} + \iiint_V \frac{\partial f}{\partial t}d\tilde{V}; \tag{4.477}$$

here f denotes scalar function of position and time, and $\boldsymbol{n}$ denotes unit vector normal to the surface at each point. The second term in the right-hand side of Eq. (4.477) is the analog to the last term

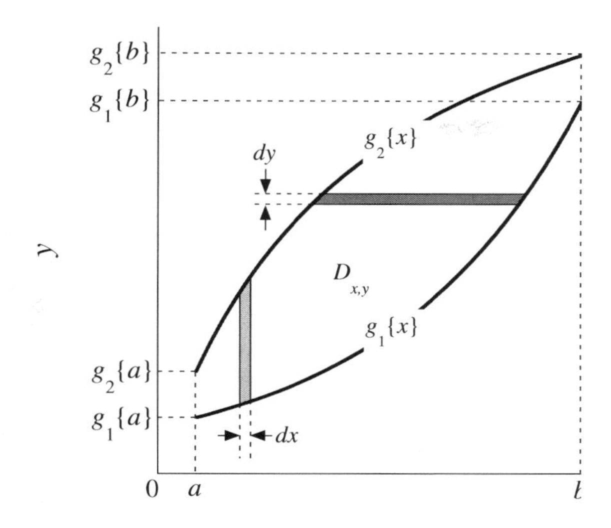

FIGURE 4.12 Graphical representation of two functions, $g_1\{x\}$ and $g_2\{x\}$, continuous within interval $[a,b]$, and laid out on xOy plane – both departing from abscissa a, but ordinates $g_1\{a\}$ or $g_2\{a\}$, respectively, and both reaching abscissa b, but ordinates $g_1\{b\}$ or $g_2\{b\}$, respectively, which serve as lower and upper boundaries for integration region $D_{x,y}$; with indication of elementary rectangle ▒, of width dx (for $a \le x \le b$) and height $dy = g_2\{x\} - g_1\{x\}$, and elementary rectangle ▓, of height dy (for $g_1\{a\} \le y \le g_2\{b\}$) and width $g_1^{-1}\{y\} - g_2^{-1}\{y\}$.

in Eq. (4.473); whereas the first term in Eq. (4.477) represents the t-varying, three-dimensional equivalent of the y-varying (one-dimensional) line consubstantiated in the first two terms of the right-hand side of Eq. (4.473).

4.3.10.5 *Green's theorem*

Consider a vector field $\boldsymbol{F}$; the double integral of its curl, over surface S, may be expressed as

$$\iint_S \nabla \times \boldsymbol{F} d\tilde{S} = \oint_l \boldsymbol{F} d\tilde{l}, \tag{4.478}$$

where the right-hand side represents a line integral over the closed boundary of length l. Equation (4.478) has classically been known as Green's theorem, or Kelvin and Stokes' theorem.

4.3.10.6 *Gauss' theorem*

If V is a region of space bounded by surface S, then the triple integral of the divergence of a vector field $\boldsymbol{F}$ over V and the surface integral of $\boldsymbol{F}$ over S are related through

$$\iiint_V \nabla \cdot \boldsymbol{F} d\tilde{V} = \iint_S \boldsymbol{F} \cdot \boldsymbol{n} d\tilde{S}; \tag{4.479}$$

Eq. (4.479) is termed Gauss' divergence theorem, or Gauss and Ostrogradsky's theorem.

4.3.11 Fourier's series expansion

A function $f\{x\}$ can be expanded via a Fourier's (infinite) series, according to

$$f\{x\} = \frac{1}{2}a_0 + \sum_{i=1}^{\infty} b_i \sin ix + \sum_{i=1}^{\infty} c_i \cos ix, \tag{4.480}$$

where coefficients a_0, b_i ($i = 1, 2, \ldots$), and c_i ($i = 1, 2, \ldots$) are calculated as

$$a_0 \equiv \frac{1}{\pi}\int_{-\pi}^{\pi} f\{x\}dx, \tag{4.481}$$

$$b_i \equiv \frac{1}{\pi}\int_{-\pi}^{\pi} f\{x\}\sin ix\, dx \tag{4.482}$$

and

$$c_i \equiv \frac{1}{\pi}\int_{-\pi}^{\pi} f\{x\}\cos ix\, dx, \tag{4.483}$$

respectively – when $f\{x\}$ is defined over interval $[-\pi, \pi]$; if the interval of interest is $[0, L]$, then one should preferably resort to

$$a_0 \equiv \frac{2}{L}\int_0^L f\{x\}dx, \tag{4.484}$$

$$b_i \equiv \frac{2}{L}\int_0^L f\{x\}\sin\frac{i\pi x}{L}dx, \tag{4.485}$$

and

$$c_i \equiv \frac{2}{L}\int_0^L f\{x\}\cos\frac{i\pi x}{L}dx. \tag{4.486}$$

An important theorem pertaining to expansion of $f\{x\}$ as per Eq. (4.480) reads

$$\frac{1}{\pi}\int_{-\pi}^{\pi} f^2\{x\}dx = \frac{a_0^2}{2} + \sum_{i=1}^{\infty}\left(b_i^2 + c_i^2\right), \tag{4.487}$$

often referred to as Parseval's (or Rayleigh's) identity – a particular form of Pancherel's theorem.

4.3.12 Analytical geometry

4.3.12.1 *Straight line*

Together with point and plane, straight line represents a primitive concept in geometry; it may be seen as the simplest (and shortest) sequence of adjacent, intermediate points along a plane, comprised between two given points thereon, say A and B. The corresponding equation, pertaining to a generic point $P(x,y)$, looks like

$$y = a_1 x + a_0, \tag{4.488}$$

where a_0 denotes independent coefficient (or vertical intercept) and a_1 denotes linear coefficient (or slope); if the aforementioned two boundary points are characterized by coordinates $A(x_1,y_1)$ and $B(x_2,y_2)$, then a_0 and a_1 satisfy

$$a_0 = \frac{x_2 y_1 - x_1 y_2}{x_2 - x_1} \tag{4.489}$$

and

$$a_1 = \frac{y_2 - y_1}{x_2 - x_1}, \tag{4.490}$$

respectively. An alternative form of Eq. (4.490) reads

$$\frac{y - y_1}{x - x_1} = \frac{y_2 - y_1}{x_2 - x_1}, \tag{4.491}$$

meaning that the straight line going through points A and $P(x,y)$ shares the slope of the straight line going through points A and B.

4.3.12.2 Conical lines

A conical section (or conical curve) is obtained as intersection of the surface of a cone with a plane, as exemplified in Fig. 4.13; remember that a cone is the solid produced via rotation of an isosceles triangle by π rad about its height (which thus serves as symmetry axis). A circle or an ellipse accordingly arise when intersection of the cone with the aforementioned plane produces a closed curve; the latter results from an inclined intersection plane, whereas a circle will be at stake when the cutting plane is perpendicular to the symmetry axis of the cone. A circumference, centered at (0,0) and characterized by radius R, is described by

$$x^2 + y^2 = R^2, \tag{4.492}$$

as depicted in Fig. 4.14i. As expected, the abscissae of its horizontal intercepts are $-R$ and R, and so are the ordinates of the vertical intercepts.

An ellipse is, in turn, the locus of points on the plane such that the sum of their distances to two focal points, of coordinates $(-c,0)$ and $(c,0)$ – and thus separated by distance $2c$, is constant and equal to $2a>0$; hence, the horizontal intercepts have abscissae $-a$ and a. The distance between the vertical intercepts of the ellipse, with ordinates $-b$ and b, is accordingly given by $2b$. The straight segment laid on the horizontal axis with length $2a$ is usually termed major axis – whereas the straight segment laid on the vertical axis with length $2b$ is termed minor axis; note that $b<a$, thus justifying distinction between minor and major axes. These features support

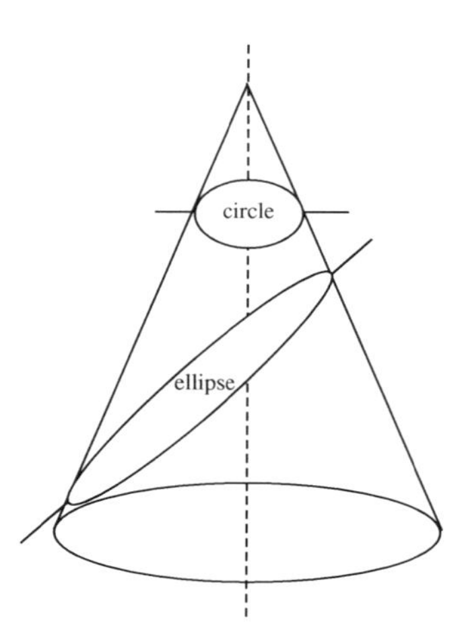

FIGURE 4.13 Graphical representation of circle and ellipse, as contours of sectional surfaces of cone with plane perpendicular or inclined, respectively, with regard to its symmetry axis (- - -).

$$\left(\frac{x}{a}\right)^2 + \left(\frac{y}{b}\right)^2 = 1 \tag{4.493}$$

for analytical descriptor; if $a=b=R$, then Eq. (4.493) degenerates to Eq. (4.492), upon multiplication of both sides by R^2. The eccentricity, ε, abides to

$$\varepsilon = \sqrt{1 - \left(\frac{b}{a}\right)^2}, \tag{4.494}$$

so $0<\varepsilon_{el}<1$ in the case of an ellipse and $\varepsilon_{cr}=0$ for a circumference; ε may accordingly be viewed as a measure of how much an ellipse deviates from the corresponding circle.

4.3.12.3 Length and curvature of plane curve

The length L of a generic arc along a curve of equation $y\{x\}$, between points of coordinates (x_1,y_1) and (x_2,y_2), can be calculated via

$$L = \int_{x_1}^{x_2} \sqrt{1 + \left(\frac{dy}{dx}\right)^2}\, dx. \tag{4.495}$$

In the case of a straight line, dy/dx is constant and given by a_1 as per Eq. (4.488) – so Eq. (4.495) will reduce to

$$L_{sl} = \sqrt{1 + a_1^2} \int_{x_1}^{x_2} dx = \sqrt{1 + a_1^2}\,(x_2 - x_1) \tag{4.496}$$

after taking constants off the kernel, and trivially applying the fundamental theorem of integral calculus; Eq. (4.490) may then be invoked to write

$$\begin{aligned} L_{sl} &= \sqrt{(x_2 - x_1)^2 + (x_2 - x_1)^2 \left(\frac{y_2 - y_1}{x_2 - x_1}\right)^2} \\ &= \sqrt{(x_2 - x_1)^2 + (y_2 - y_1)^2}, \end{aligned} \tag{4.497}$$

where x_2-x_1 was first taken under the square root, and common factors were then cancelled out between numerator and denominator – thus yielding a result consistent with Pythagoras' theorem, see Eq. (4.79).

Curvature, κ, of a curve of equation $y\{x\}$ is, in turn, defined by

$$\kappa = \frac{\left|\dfrac{d^2y}{dx^2}\right|}{\left(1 + \left(\dfrac{dy}{dx}\right)^2\right)^{\frac{3}{2}}}; \tag{4.498}$$

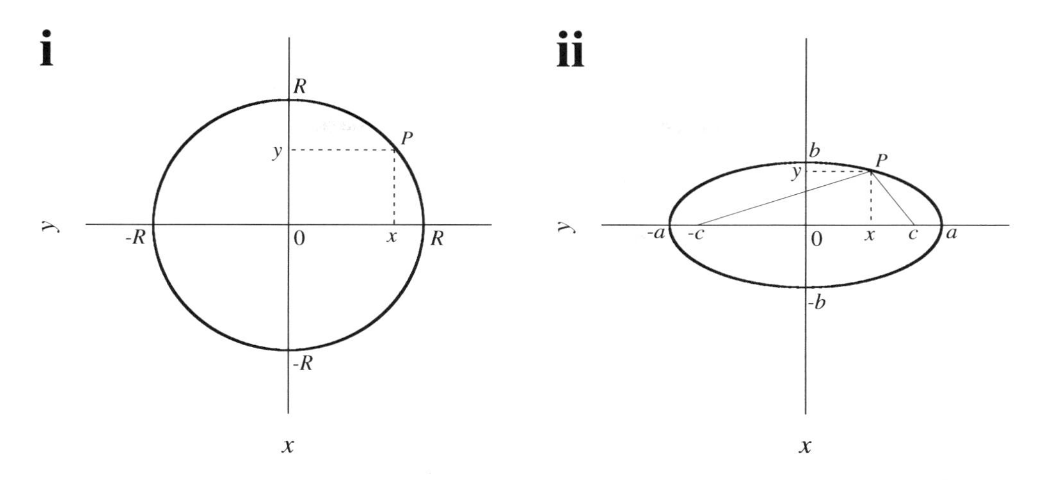

FIGURE 4.14 Graphical representation of (i) circle centered at (0,0), and with generic point P described by coordinates (x,y) and radius R; and (ii) ellipse, centered at (0,0), and with generic point P described by coordinates (x,y), minor axis b, major axis a, and focal points described by abscissae $-c$ and c.

specifically for a straight line,

$$\frac{dy}{dx} \equiv \frac{d}{dx}\left(a_1 x + a_0\right) = a_1 \tag{4.499}$$

as per Eq. (4.488), and consequently

$$\frac{d^2y}{dx^2} \equiv \frac{da_1}{dx} = 0 \tag{4.500}$$

– so a straight line is confirmed to exhibit no curvature whatsoever, i.e.

$$\kappa_{sl} = \frac{|0|}{\left(1 + a_1^{\,2}\right)^{\frac{3}{2}}} = 0 \tag{4.501}$$

after plugging Eqs. (4.499) and (4.500) in Eq. (4.498). For a circumference, one should retrieve Eq. (4.492) to produce the first derivative as

$$\frac{dy}{dx} \equiv \frac{d}{dx}\left(\pm\sqrt{R^2 - x^2}\right) = \pm\frac{-2x}{2\sqrt{R^2 - x^2}} = \pm\frac{x}{\sqrt{R^2 - x^2}}, \tag{4.502}$$

which permits calculation of the second derivative as

$$\frac{d^2y}{dx^2} \equiv \frac{d}{dx}\left(\pm\frac{x}{\sqrt{R^2 - x^2}}\right) = \pm\frac{\sqrt{R^2 - x^2} - x\dfrac{-2x}{2\sqrt{R^2 - x^2}}}{R^2 - x^2}$$
$$= \pm\left(\frac{1}{\left(R^2 - x^2\right)^{\frac{1}{2}}} + \frac{x^2}{\left(R^2 - x^2\right)^{\frac{3}{2}}}\right); \tag{4.503}$$

insertion of Eqs. (4.502) and (4.503) then transforms Eq. (4.498) to

$$\kappa_{cf} = \frac{\left|\pm\left(\dfrac{1}{\left(R^2 - x^2\right)^{\frac{1}{2}}} + \dfrac{x^2}{\left(R^2 - x^2\right)^{\frac{3}{2}}}\right)\right|}{\left(1 + \left(\pm\dfrac{x}{\sqrt{R^2 - x^2}}\right)^2\right)^{\frac{3}{2}}}$$
$$= \frac{\dfrac{1}{\left(R^2 - x^2\right)^{\frac{1}{2}}}\left(1 + \dfrac{x^2}{R^2 - x^2}\right)}{\left(1 + \dfrac{x^2}{R^2 - x^2}\right)^{\frac{1}{2}}\left(1 + \dfrac{x^2}{R^2 - x^2}\right)} \tag{4.504}$$
$$= \frac{1}{\sqrt{R^2 - x^2}\sqrt{1 + \dfrac{x^2}{R^2 - x^2}}},$$

upon factoring out $(R^2-x^2)^{-1/2}$ in numerator, splitting the power in denominator, and dropping common factors between numerator and denominator afterward. Equation (4.504) eventually simplifies to

$$\kappa_{cf} = \frac{1}{\sqrt{R^2 - x^2 + x^2}} = \frac{1}{\sqrt{R^2}} = \frac{1}{R}; \tag{4.505}$$

as expected, the curvature of a circumference is constant, irrespective of position along it – and equal to the reciprocal of its radius.

4.3.12.4 Area of plane surface

If a closed curve is laid out on a plane – with upper portion described by function $g_2\{x\}$ and lower portion described by function $g_1\{x\}$, both developing between extreme points (x_1,y_1) and (x_2,y_2), then the area of the plane, A, bounded by such curves satisfies

$$A = \int_{x_1}^{x_2} \left(g_2\{x\} - g_1\{x\} \right) dx. \tag{4.506}$$

In view of Eq. (4.492), one eventually obtains

$$A_{cr} = \pi R^2 \tag{4.507}$$

from Eq. (4.506) – pertaining to a circumference; and similarly

$$A_{el} = \pi ab \tag{4.508}$$

from Eq. (4.493), upon insertion in Eq. (4.506) – pertaining now to an ellipse.

4.3.12.5 Outer area of revolution solid

If the outer surface of an axisymmetric solid is defined via revolution of a curve, of equation $y\{x\}$, by π rad around its axis, between points (x_1,y_1) and (x_2,y_2), then the area spanned thereby will look like

$$A_{out} = 2\pi \int_{x_1}^{x_2} y\{x\} \sqrt{1 + \left(\frac{dy}{dx} \right)^2} \, dx. \tag{4.509}$$

For a sphere of radius R, Eqs. (4.492) and (4.509) give rise to

$$A_{out,sp} = 4\pi R^2. \tag{4.510}$$

In the case of an ellipsoid, generated by the ellipse satisfying Eq. (4.493), one gets

$$A_{out,el,pr} = 2\pi b^2 \left(1 + \frac{\sin^{-1} \sqrt{1 - \left(\frac{b}{a} \right)^2}}{\frac{b}{a} \sqrt{1 - \left(\frac{b}{a} \right)^2}} \right) \tag{4.511}$$

upon its insertion in Eq. (4.509) – pertaining to prolate shape, should the ellipse be rotated around its major axis (with $b<a$); and likewise

$$A_{out,el,ob} = 2\pi b^2 \left(1 + \frac{\ln \left\{ \frac{b}{a} + \sqrt{\left(\frac{b}{a} \right)^2 - 1} \right\}}{\frac{b}{a} \sqrt{\left(\frac{b}{a} \right)^2 - 1}} \right) \tag{4.512}$$

for oblate shape, should rotation be instead performed about its minor axis (with $b>a$). A cylinder, of radius R and height H, possesses three outer surfaces – one lateral round surface, described by $y\{x\}=R$ (when laid horizontally) and thus $dy/dx=0$, and extending through $x_2-x_1=H$; and two equal lower and upper flat surfaces, abiding to Eq. (4.507). Therefore, one obtains

$$A_{out,cy} = 2\pi RH + 2\pi R^2 = 2\pi R(H + R) \tag{4.513}$$

for total outer surface, with the aid of Eq. (4.509). In the case of a cone, of radius R and slant height S, there are two outer surfaces, one lateral, round surface, and one lower, flat surface – again with area given by Eq. (4.507); the total area reads

$$A_{out,co} = \pi RS + \pi R^2 = \pi R(S + R). \tag{4.514}$$

A parallelepiped of sides a, b, and c holds, in turn, six outer surfaces, with total area obtained from

$$A_{out,pa} = 2(ab + ac + bc) \tag{4.515}$$

– where Eq. (4.509) is obviously not applicable, for failure to be a revolution solid; the particular case of a cube, with $a=b=c$, leads merely to

$$A_{out,cu} = 6a^2, \tag{4.516}$$

as anticipated from the functional form of Eq. (4.515) unfolding $2(a^2+a^2+a^2) = 2 \cdot 3a^2$.

4.3.12.6 Volume of revolution solid

When a solid is generated via rotation by π rad of a given plane figure, upper bounded by a function $y\{x\}$, between points (x_1,y_1) and (x_2,y_2), its volume will satisfy

$$V = \pi \int_{x_1}^{x_2} y^2\{x\} dx; \tag{4.517}$$

a sphere of radius R accordingly exhibits

$$V_{sp} = \frac{4}{3} \pi R^3 \tag{4.518}$$

for total volume, upon combination with Eq. (4.492). In the case of a cylinder of radius R and height H, Eq. (4.517) unfolds

$$V_{cy} = \pi R^2 H; \tag{4.519}$$

while a cone of radius R and height H abides to

$$V_{co} = \frac{1}{3} \pi R^2 H. \tag{4.520}$$

A prolate ellipsoid, characterized by a and b for axes (with $b<a$), follows

$$V_{el,pr} = \frac{4}{3} \pi ab^2; \tag{4.521}$$

whereas its oblate counterpart entails

$$V_{el,ob} = \frac{4}{3} \pi a^2 b, \tag{4.522}$$

(for $b>a$). On the other hand, a parallelepiped, with sides of length a, b, and c, exhibits a volume given by

$$V_{pa} = abc \tag{4.523}$$

– obviously not given by Eq. (4.517), for not being a revolution solid; hence, a cube will support

$$V_{cu} = a^3 \tag{4.524}$$

as particular case, characterized by common side length a.

4.3.13 Optimization of functions

4.3.13.1 Univariate and unconstrained

Should $f\{x\}$ denote an objective function of interest, on independent variable x and defined within interval $[a,b]$ (with $b>a$), then the necessary condition for an optimum (or stationary) point to exist, at $x=\zeta$, looks like

$$\left.\frac{df\{x\}}{dx}\right|_{\zeta}=0;\quad a<\zeta<b; \tag{4.525}$$

despite consubstantiating a necessary condition, Eq. (4.525) does not guarantee by itself an optimum (since an inflection point may arise as well) – nor does it discriminate its nature, i.e. minimum versus maximum, in the case of a true local optimum. To appropriately address this issue, suppose that the derivatives of every order of $f\{x\}$ – from first order, see Eq. (4.525), and up to n-th order exist and obey

$$\left.\frac{df^2\{x\}}{dx}\right|_{x=\zeta}=\left.\frac{df^3\{x\}}{dx}\right|_{x=\zeta}=\ldots=\left.\frac{df^n\{x\}}{dx}\right|_{x=\zeta}=0; \tag{4.526}$$

assume, in addition, that the $(n+1)$-th-order derivative exists and is continuous in a vicinity of $x=\zeta$, but is the first one taking a non-nil value in interval $]\zeta,x[$ or $]x,\zeta[$ (as appropriate). One may then express the said function via Taylor's expansion around $x=\zeta$, according to

$$f\{x\}-f\{x\}\big|_{\zeta}=\frac{\left.\dfrac{d^{n+1}f\{x\}}{dx^{n+1}}\right|_{\xi}}{(n+1)!}(x-\zeta)^{n+1};\quad \zeta<\xi<x\vee x<\xi<\zeta, \tag{4.527}$$

in agreement with Eq. (4.404). Since $(n+1)!>0$ and the aforementioned $(n+1)$-th-order derivative is, by hypothesis, continuous in the vicinity of $x=\zeta$, the sign of $f\{x\}-f\{x\}\big|_{\zeta}$ will be that of the product of $(x-\zeta)^{n+1}$ by $d^{n+1}f\{x\}/dx^{n+1}$, evaluated at $x=\xi$, with ξ located somewhere between ζ and x (or between x and ζ).

If n is odd, then $n+1$ is even – so $(x-\zeta)^{n+1}$ will necessarily be positive in the vicinity of ζ, irrespective of whether $x>\zeta$ or $x<\zeta$; hence, the sign of the left-hand side in Eq. (4.527) will coincide with the sign of $\left(d^{n+1}f\{x\}/dx^{n+1}\right)\big|_{\xi}$. Therefore,

$$\left.\frac{d^{n+1}f\{x\}}{dx^{n+1}}\right|_{\xi}>0\Rightarrow f\{x\}>f\{x\}\big|_{\zeta};\quad x>\zeta, \tag{4.528}$$

as well as

$$\left.\frac{d^{n+1}f\{x\}}{dx^{n+1}}\right|_{\xi}>0\Rightarrow f\{x\}>f\{x\}\big|_{\zeta};\quad x<\zeta \tag{4.529}$$

– meaning that $f\{x\}$ lies above $f\{\zeta\}$ at all points x near ζ, irrespective of the position of x relative to ζ. This is consistent with a minimum – characterized by the concavity of $f\{x\}$ facing upward, and exactly described by

$$\left.\frac{d^if\{x\}}{dx^i}\right|_{\zeta}=0\wedge\left.\frac{d^{n+1}f\{x\}}{dx^{n+1}}\right|_{\xi}>0 \tag{4.530}$$
$$1\le i\le n\text{ odd},\ \zeta<\xi<x\vee x<\xi<\zeta;$$

Eq. (4.530) encompasses the necessary and sufficient condition for a local minimum of $f\{x\}$ at $x=\zeta$, and already includes Eq. (4.525) as per $i=1$. By the same token,

$$\left.\frac{d^{n+1}f\{x\}}{dx^{n+1}}\right|_{\xi}<0\Rightarrow f\{x\}<f\{x\}\big|_{\zeta};\quad x>\zeta \tag{4.531}$$

and

$$\left.\frac{d^{n+1}f\{x\}}{dx^{n+1}}\right|_{\xi}<0\Rightarrow f\{x\}<f\{x\}\big|_{\zeta};\quad x<\zeta \tag{4.532}$$

– so $f\{x\}$ lies, in this case, below $f\{\zeta\}$ at all points around $x=\zeta$, no matter their relative location. Therefore, a maximum arises, with the concavity of $f\{x\}$ now facing downward; the underlying necessary and sufficient condition for a local maximum of $f\{x\}$ at $x=\zeta$ accordingly reads

$$\left.\frac{d^if\{x\}}{dx^i}\right|_{\zeta}=0\wedge\left.\frac{d^{n+1}f\{x\}}{dx^{n+1}}\right|_{\xi}<0 \tag{4.533}$$
$$1\le i\le n\text{ odd},\ \zeta<\xi<x\vee x<\xi<\zeta,$$

which already took Eq. (4.525) onboard. It is also apparent that $f\{x\}$ decreases within $[x,\zeta]$ and increases within $[\zeta,x]$ when Eq. (4.530) holds – whereas $f\{x\}$ changes its pattern from increasing to decreasing when going from $[x,\zeta]$ to $[\zeta,x]$, in the case described by Eq. (4.533). The fact that ξ does not necessarily coincide with ζ is immaterial in the above reasoning, because only a narrow interval centered around $x=\zeta$ is of interest when investigating existence of local optima; since $x<\xi<\zeta$, then $x\to\zeta$ implies that $\xi\to\zeta$ (and likewise when $\zeta<\xi<x$ instead) – so ξ will eventually overlap ζ, should x be sufficiently close to ζ.

Conversely, n being even implies that $n+1$ is odd – so the sign of $f\{x\}-f\{x\}\big|_{\zeta}$, in the vicinity of ζ, will coincide with the sign of $\left(d^{n+1}f\{x\}/dx^{n+1}\right)\big|_{\xi}$ only if $x-\zeta$ is positive, and will be the opposite when $x-\zeta<0$, see Eq. (4.527) – because an odd exponent $n+1$ retains the original sign of $x-\zeta$ in $(x-\zeta)^{n+1}$. This leads to

$$\left.\frac{d^{n+1}f\{x\}}{dx^{n+1}}\right|_{\xi}>0\Rightarrow f\{x\}>f\{x\}\big|_{\zeta};\quad x>\zeta \tag{4.534}$$

and

$$\left.\frac{d^{n+1}f\{x\}}{dx^{n+1}}\right|_{\xi}>0\Rightarrow f\{x\}<f\{x\}\big|_{\zeta};\quad x<\zeta, \tag{4.535}$$

in the case of a positive derivative; and similarly to

$$\left.\frac{d^{n+1}f\{x\}}{dx^{n+1}}\right|_{\xi}<0\Rightarrow f\{x\}<f\{x\}\big|_{\zeta};\quad x>\zeta \tag{4.536}$$

and

$$\left.\frac{d^{n+1}f\langle x\}}{dx^{n+1}}\right|_{\xi}<0\Rightarrow f\{x\}>f\{x\}\big|_{\zeta};\quad x<\zeta, \tag{4.537}$$

for its negative counterpart. Consequently, an increasing function within $[x,\zeta]$ retains its monotony within $[\zeta,x]$ when $\left(d^{n+1}f\{x\}/dx^{n+1}\right)\big|_{\xi}>0$, whereas $\left(d^{n+1}f\{x\}/dx^{n+1}\right)\big|_{\xi}<0$ implies a monotonically decreasing function throughout $[x,\zeta]$ and $[\zeta,x]$; in both cases, an inflection point is at stake, which is stationary but not optimum – thus normally holding a poor engineering interest.

4.3.13.2 Bivariate and unconstrained

If $f\{x,y\}$ denotes an objective function under scrutiny, on independent variables x and y, then the necessary condition for a point, described by abscissa a and ordinate b, consubstantiate a critical point reads

$$\left.\frac{\partial f}{\partial x}\right|_{(a,b)} = 0 \wedge \left.\frac{\partial f}{\partial y}\right|_{(a,b)} = 0 \tag{4.538}$$

– which mimics Eq. (4.525), yet now applied in two dimensions. The nature of such a critical point depends on the sign of one of the second-order (noncross) partial derivatives and the sign of the corresponding discriminant determinant; a local maximum will indeed arise when

$$\left.\frac{\partial^2 f}{\partial x^2}\right|_{(a,b)} < 0 \wedge \begin{vmatrix} \left.\frac{\partial^2 f}{\partial x^2}\right|_{(a,b)} & \left.\frac{\partial^2 f}{\partial x \partial y}\right|_{(a,b)} \\ \left.\frac{\partial^2 f}{\partial y \partial x}\right|_{(a,b)} & \left.\frac{\partial^2 f}{\partial y^2}\right|_{(a,b)} \end{vmatrix} > 0, \tag{4.539}$$

a local minimum when

$$\left.\frac{\partial^2 f}{\partial x^2}\right|_{(a,b)} > 0 \wedge \begin{vmatrix} \left.\frac{\partial^2 f}{\partial x^2}\right|_{(a,b)} & \left.\frac{\partial^2 f}{\partial x \partial y}\right|_{(a,b)} \\ \left.\frac{\partial^2 f}{\partial y \partial x}\right|_{(a,b)} & \left.\frac{\partial^2 f}{\partial y^2}\right|_{(a,b)} \end{vmatrix} > 0, \tag{4.540}$$

and a saddle point when

$$\begin{vmatrix} \left.\frac{\partial^2 f}{\partial x^2}\right|_{(a,b)} & \left.\frac{\partial^2 f}{\partial x \partial y}\right|_{(a,b)} \\ \left.\frac{\partial^2 f}{\partial y \partial x}\right|_{(a,b)} & \left.\frac{\partial^2 f}{\partial y^2}\right|_{(a,b)} \end{vmatrix} < 0. \tag{4.541}$$

The remaining case, i.e.

$$\begin{vmatrix} \left.\frac{\partial^2 f}{\partial x^2}\right|_{(a,b)} & \left.\frac{\partial^2 f}{\partial x \partial y}\right|_{(a,b)} \\ \left.\frac{\partial^2 f}{\partial y \partial x}\right|_{(a,b)} & \left.\frac{\partial^2 f}{\partial y^2}\right|_{(a,b)} \end{vmatrix} = 0, \tag{4.542}$$

is inconclusive – so some alternative method should be employed; this includes investigation of the sign of higher order Hessian determinants, and whether the order of the first significant determinant is even or odd (in much the same way higher-order derivatives were sought in the univariate case). Equation (4.392) may be used to simplify calculation of the discriminant determinant, i.e.

$$\begin{vmatrix} \left.\frac{\partial^2 f}{\partial x^2}\right|_{(a,b)} & \left.\frac{\partial^2 f}{\partial x \partial y}\right|_{(a,b)} \\ \left.\frac{\partial^2 f}{\partial y \partial x}\right|_{(a,b)} & \left.\frac{\partial^2 f}{\partial y^2}\right|_{(a,b)} \end{vmatrix} = \left.\frac{\partial^2 f}{\partial x^2}\right|_{(a,b)} \left.\frac{\partial^2 f}{\partial y^2}\right|_{(a,b)} - \left.\left(\frac{\partial^2 f}{\partial y \partial x}\right)^2\right|_{(a,b)} \tag{4.543}$$

along with Eq. (4.190); choice of the sign of either $\left.\left(\partial^2 f / \partial x^2\right)\right|_{(a,b)}$ or $\left.\left(\partial^2 f / \partial y^2\right)\right|_{(a,b)}$ for first condition is immaterial in the case of a true optimum, due to the functional form of Eq. (4.543). In fact, $\left.\left(\partial^2 f / \partial y \partial x\right)^2\right|_{(a,b)} > 0$ for being a square; therefore, $\left.\left(\partial^2 f / \partial x^2\right)\right|_{(a,b)} < 0$ as per Eq. (4.539) implies $\left.\left(\partial^2 f / \partial y^2\right)\right|_{(a,b)} < 0$, so as to guarantee that $\left.\left(\partial^2 f / \partial x^2\right)\right|_{(a,b)} \left.\left(\partial^2 f / \partial y^2\right)\right|_{(c,b)} > 0$ as required by a positive discriminant determinant in Eq. (4.543). By the same token, $\left.\left(\partial^2 f / \partial x^2\right)\right|_{(a,b)} > 0$ as per Eq. (4.540) similarly implies $\left.\left(\partial^2 f / \partial y^2\right)\right|_{(a,b)} > 0$, to assure again that $\left.\left(\partial^2 f / \partial x^2\right)\right|_{(a,b)} \left.\left(\partial^2 f / \partial y^2\right)\right|_{(a,b)} > 0$, as needed to assure a plus sign of the right-hand side of Eq. (4.543).

4.3.13.3 Univariate and constrained

Suppose that one intends to optimize a two-variable function, $f\{x,y\}$ – the independent variables of which have to satisfy an extra relationship, say

$$g\{x,y\} = 0 \tag{4.544}$$

– given in implicit form, and taken as restriction. For $f\{x,y\}$ to attain a (local) stationary point,

$$\left(\frac{\partial f}{\partial x}\right)_y - \lambda \left(\frac{\partial g}{\partial x}\right)_y = 0 \tag{4.545}$$

and

$$\left(\frac{\partial f}{\partial y}\right)_x - \lambda \left(\frac{\partial g}{\partial y}\right)_x = 0 \tag{4.546}$$

are to be simultaneously satisfied; Eqs. (4.545) and (4.546) consubstantiate the necessary condition for a stationary point of $f\{x,y\}$, subjected to Eq. (4.544) as constraint – where λ denotes an auxiliary parameter, termed Lagrange's multiplier. Equations (4.544)–(4.546) suffice to find the coordinates x and y of the said point, together with the associated value of λ; this formulation is equivalent to

$$\left(\frac{\partial \phi}{\partial \lambda}\right)_{x,y} = 0, \tag{4.547}$$

$$\left(\frac{\partial \phi}{\partial x}\right)_{y,\lambda} = 0, \tag{4.548}$$

and

$$\left(\frac{\partial \phi}{\partial y}\right)_{x,\lambda}=0, \tag{4.549}$$

respectively – obtained upon partial differentiation (in all three possible ways) of Lagrangian function ϕ, defined as

$$\phi\{x,y,\lambda\}\equiv f\{x,y\}-\lambda g\{x,y\}, \tag{4.550}$$

and setting of the outcomes equal to zero afterward.

4.3.14 Euler and Lagrange's equation

4.3.14.1 Unconstrained kernel

Oftentimes in process engineering, the goal is not optimizing a given univariate function, $y\{x\}$ – to find, say, y_{opt} associated to a specific value, say, x_{opt} of independent variable x; but instead finding the functional form of $y\{x\}$ that itself optimizes some criterion, while abiding to given boundary condition(s). One typical example encompasses seeking function $y\{x\}$ such that the integral

$$I[y]\equiv\int_{x_1}^{x_2} f\left\{x,y,\frac{dy}{dx}\right\}dx, \tag{4.551}$$

attains an optimum – where f denotes some form of algebraic (and thus differentiable) functionality on x, y, and dy/dx. Furthermore,

$$y\big|_{x_1}=y_1 \tag{4.552}$$

and

$$y\big|_{x_2}=y_2 \tag{4.553}$$

should serve as boundary conditions – where x_1 and x_2 are given (constant) limits of integration, and y_1 and y_2 denote (given) constants; and d^2y/dx^2 is supposed to exist, and be continuous over interval $[x_1,x_2]$. Since I is a function of (another) function $y\{x\}$, the former is usually termed functional, and is denoted by $I[y]$; hence, one ends up with a problem of calculus of variations, or theory of functionals. The necessary condition for this optimum reads

$$\frac{\partial f}{\partial y}-\frac{d}{dx}\left(\frac{\partial f}{\partial\left(\frac{dy}{dx}\right)}\right)=0 \tag{4.554}$$

– usually known as Euler and Lagrange's equation. Note that it is comparatively easy to find the necessary condition for a functional optimum with regard to a function – but a vastly more difficult problem arises in attempts to formulate a sufficient condition for the said stationary point, to be either a maximum or a minimum. Since a second-order differential equation results, in general, from Eq. (4.554), two arbitrary constants are indeed anticipated to show up following its integration – to be calculated with the aid of Eqs. (4.552) and (4.553). There are, however, several situations where Eq. (4.554) will considerably simplify; for instance, if f is explicitly independent of y, i.e.

$$f\equiv f\left\{x,\frac{dy}{dx}\right\}, \tag{4.555}$$

then one is left with

$$\frac{d}{dx}\left(\frac{\partial f}{\partial\left(\frac{dy}{dx}\right)}\right)=0. \tag{4.556}$$

Another simplification arises when f is explicitly independent of dy/dx, i.e.

$$f\equiv f\{x,y\}; \tag{4.557}$$

Eq. (4.554) accordingly reduces to

$$\frac{\partial f}{\partial y}=0, \tag{4.558}$$

which resembles Eq. (4.525). Finally, if f is explicitly independent of x, viz.

$$f\equiv f\left\{y,\frac{dy}{dx}\right\}, \tag{4.559}$$

then the general rule of differentiation of f as a composite function allows one to obtain

$$\frac{d}{dx}\left(f-\frac{dy}{dx}\frac{\partial f}{\partial\left(\frac{dy}{dx}\right)}\right)=0 \tag{4.560}$$

as optimality condition.

4.3.14.2 Constrained kernel

Should restrictions apply in calculus of variations, two cases are to be considered: global restrictions – which constrain solutions over the whole range; and local restrictions – to be imposed at particular points along the integration path. The former are often integral functions of the type

$$\int_{x_1}^{x_2} g\left\{x,y,\frac{dy}{dx}\right\}dx=\kappa, \tag{4.561}$$

where $g\{x,y,dy/dx\}$ denotes some algebraic functionality of x, y, and dy/dx distinct from that entertained by f in Eq. (4.551), and κ denotes a constant; local restrictions hold typically the form

$$\frac{dy}{dx}=h\{y,z\{x\}\}, \tag{4.562}$$

where $h\{y,z\}$ is a function of y and z – while z is, in turn, a function of x. In the former case, one can resort to the concept of Lagrange's multipliers to write

$$\frac{\partial f}{\partial y}-\frac{d}{dx}\left(\frac{\partial f}{\partial\left(\frac{dy}{dx}\right)}\right)-\lambda\left(\frac{\partial g}{\partial y}-\frac{d}{dx}\left(\frac{\partial g}{\partial\left(\frac{dy}{dx}\right)}\right)\right)=0, \tag{4.563}$$

as if f in Eq. (4.546) had been replaced by Eq. (4.554), and g also in Eq. (5.546) replaced by a similar expression; once $f\{x\}$, containing two arbitrary constants, has been found from Eq. (4.563), the said arbitrary constants and λ itself will be calculated from Eqs. (4.552), (4.553) and (4.561). Local restrictions limit only the shape of the solution function that depends on independent variable x; a Lagrangian function can still be defined inspired on Eq. (4.550), yet Lagrangian multipliers are now (adjoint) functions of x – abiding to

$$\lambda\{x\}\big|_{x_2} = 0 \tag{4.564}$$

as boundary condition. After rewriting Eq. (4.551) as

$$I[y] = \int_{x_1}^{x_2} \left(f\{y,z\} - \lambda\{x\}\left(\frac{dy}{dx} - h\{y,z\} \right) \right) dx \tag{4.565}$$

with the aid of Eq. (4.562), one will eventually be led to

$$\frac{\partial f}{\partial y} + \lambda \frac{\partial h}{\partial y} + \frac{d\lambda}{dx} = 0 \tag{4.566}$$

and

$$\frac{\partial f}{\partial z} + \lambda \frac{\partial h}{\partial z} = 0 \tag{4.567}$$

– meant to define f and h, and containing two arbitrary constants; subjected to Eqs. (4.552), (4.553), and (4.564) as boundary conditions, needed to calculate them together with λ itself.

4.3.15 Pontryagin's principle

A more general strategy to optimize continuous functions, but also more cumbersome to implement, has been proposed – which applies to a system containing N state variables $y_i\{x\}$ ($i=1, 2, \ldots, N$), with independent variable, x, ranging from x_1 to x_2. The evolution of the said system is supposed to be influenced by M decision parameters, $z_j\{x\}$ ($j=1, 2, \ldots, M$); and described by N differential equations of the form

$$\frac{dy_i}{dx} = f_i\{y_1, y_2, \ldots, y_N, z_1, z_2, \ldots, z_M\}; \quad i = 1, 2, \ldots, N, \tag{4.568}$$

similar to that labeled as Eq. (4.562) – satisfying initial conditions of the type

$$y_i\big|_{x_1} = y_{i,0}; \quad i = 1, 2, \ldots, N \tag{4.569}$$

that represent local restrictions. Under these circumstances, the criterion to be optimized can be expressed in linear form as

$$I[y_1, y_2, \ldots y_N] = \sum_{i=1}^{N} \kappa_i y_i\big|_{x_2}, \tag{4.570}$$

where κ_is denote constants. To solve a problem satisfying the above conditions, an additional N Lagrangian functions $\lambda_i\{x\}$ ($i=1, 2, \ldots, N$) are to be introduced, as well as a Hamiltonian H satisfying

$$H\{\lambda_1, \lambda_2, \ldots, \lambda_N, y_1, y_2, \ldots, y_N, z_1, z_2, \ldots, z_M\} \equiv \sum_{i=1}^{N} \lambda_i f_i\{y_1, y_2, \ldots, y_N, z_1, z_2, \ldots, z_M\} \tag{4.571}$$

– where the λ_is are, in turn, the solutions of differential equations

$$\frac{d\lambda_i}{dx} = -\frac{\partial H}{\partial y_i}; \quad i = 1, 2, \ldots, N, \tag{4.572}$$

subjected to boundary conditions

$$\lambda_i\big|_{x_2} = \kappa_i; \quad i = 1, 2, \ldots, N. \tag{4.573}$$

Pontryagin's optimum principle states that the optimal decision functions z_js – for which I as per Eq. (4.570) holds an optimum, are solutions of

$$\frac{\partial H}{\partial z_j} = 0; \quad j = 1, 2, \ldots, M \tag{4.574}$$

that constitute a set of M algebraic equations.

4.3.16 Fourier's transform

Given any function $f\{x\}$, it can be split as

$$f\{x\} = E\{x\} + O\{x\} \tag{4.575}$$

– where $E\{x\}$ denotes an even function, defined as

$$E\{x\} \equiv \frac{f\{x\} + f\{-x\}}{2}, \tag{4.576}$$

and $O\{x\}$ denotes an odd function, abiding to

$$O\{x\} \equiv \frac{f\{x\} - f\{-x\}}{2}. \tag{4.577}$$

Fourier's transform of $f\{x\}$ – denoted usually by $\hat{\mathcal{F}}\{\nu\}$, is defined as

$$\hat{\mathcal{F}}\{\nu\} \equiv \mathcal{F}\left(f\{x\}\right) \equiv \int_{-\infty}^{\infty} E\{x\}\cos\nu x \, dx - \iota \int_{-\infty}^{\infty} O\{x\}\sin\nu x \, dx, \tag{4.578}$$

where variable x gives room to variable ν; this integral transform is linear in that

$$\mathcal{F}\left(af\{x\} + bg\{x\}\right) = a\mathcal{F}\left(f\{x\}\right) + b\mathcal{F}\left(g\{x\}\right), \tag{4.579}$$

with a and b denoting constants and $g\{x\}$ denoting a second function of x. In terms of differentiation, one realizes that

$$\mathcal{F}\left(\frac{df}{dx}\right) = \iota\nu\mathcal{F}\left(f\right) \tag{4.580}$$

– a result expansible to n-th order differentiation as

$$\mathcal{F}\left(\frac{d^n f}{dx^n}\right) = \left(\iota\nu\right)^n \mathcal{F}\left(f\right), \tag{4.581}$$

where n denotes a positive integer.

4.3.17 Laplace's transform

4.3.17.1 Definition

Laplace's transform of $f\{t\}$ is defined as

$$\bar{f}\{s\} \equiv \mathcal{L}(f\{t\}) \equiv \int_0^\infty e^{-st} f\{t\} dt; \tag{4.582}$$

it accordingly converts a function of real variable t into a function, $\bar{f}\{s\}$, of (complex parameter) s. Selected transforms, directly obtained from the definition, are conveyed by Table 4.16 – where n denotes an integer constant, k denotes a real constant, and Γ and erfc denote gamma function and complementary error function, respectively. Not all functions possess Laplace's transform – since the kernel in Eq. (4.582) must tend to zero as t grows unbounded.

4.3.17.2 Properties

Laplace's transform is a linear operator, so one may write

$$\mathcal{L}(af\{t\} + bg\{t\}) = a\bar{f}\{s\} + b\bar{g}\{s\}, \tag{4.583}$$

again encompassing constants a and b, and generic functions $f\{t\}$ and $g\{t\}$; Eq. (4.583) allows calculation of the transforms of hyperbolic functions – as listed in Table 4.17. Application of Eq. (4.582) to the derivative function unfolds, in turn,

$$\mathcal{L}\left(\frac{df\{t\}}{dt}\right) = s\bar{f}\{s\} - f\{t\}\big|_0; \tag{4.584}$$

TABLE 4.16

List of Laplace's transforms, $\bar{f}\{s\}$, of functions, $f\{t\}$, obtained via definition.

Type	$f\{t\}$	$\bar{f}\{s\}$
Unit impulse	$\delta\{t\}$	1
Unit step	1	$1/s$
Power	t^n	$\dfrac{n!}{s^{n+1}}$
	t^k	$\dfrac{\Gamma\{k+1\}}{s^{k+1}}$
Trigonometric functions	$\sin kt$	$\dfrac{k}{k^2+s^2}$
	$\cos kt$	$\dfrac{s}{k^2+s^2}$
Exponential function	e^{-kt}	$\dfrac{1}{k+s}$
Complementary error function	$\text{erfc}\left\{\dfrac{k}{\sqrt{t}}\right\}$	$\dfrac{e^{-2k\sqrt{s}}}{s}$

TABLE 4.17

List of Laplace's transforms, $\bar{f}\{s\}$, of functions, $f\{t\}$, obtained via associated theorems.

Type	$f\{t\}$	$\bar{f}\{s\}$
Hyperbolic functions	$\sinh kt$	$\dfrac{k}{s^2-k^2}$
	$\cosh kt$	$\dfrac{s}{s^2-k^2}$
Time-weighed hyperbolic functions	$t \sinh kt$	$\dfrac{2ks}{(s^2-k^2)^2}$
	$t \cosh kt$	$\dfrac{s^2+k^2}{(s^2-k^2)^2}$
Square root-normalized exponential	$\dfrac{e^{-\frac{k}{t}}}{\sqrt{t}}$	$\sqrt{\dfrac{\pi}{s}}e^{-2\sqrt{ks}}$
Modified Bessel's function of first kind of square root	$\left(\dfrac{t}{k}\right)^{\frac{\nu}{2}} I_\nu\{2\sqrt{kt}\}$	$\dfrac{1}{s^{\nu+1}}e^{\frac{k}{s}}$

sequential application of this rule to higher-order derivatives leads to

$$\mathcal{L}\left(\frac{d^n f\{t\}}{dt^n}\right) = s^n\bar{f}\{s\} - \sum_{j=0}^{n-1} s^j \left.\frac{d^{n-j-1}f\{t\}}{dt^{n-j-1}}\right|_0, \tag{4.585}$$

in general. Note that differentiation per se in the time domain implies loss of a constant each time it is performed – thus justifying presence of the summation in Eq. (4.585), where $f\{t\}|_0$, $\left(df\{t\}/dt\right)\big|_0$, ..., $\left(d^{n-1}f\{t\}/dt^{n-1}\right)\big|_0$ account for the associated n constants. Based on Eq. (4.582), one may also write

$$\mathcal{L}\left(\int_0^t f\{\tilde{t}\}d\tilde{t}\right) = \frac{\bar{f}\{s\}}{s}; \tag{4.586}$$

in other words, integration with regard to t in the time domain translates to division by s in Laplace's domain – in very much the same way differentiation with regard to t in the time domain maps to multiplication by s in Laplace's domain. Another interesting property of Laplace's transform becomes apparent after the original function in the time domain is multiplied by t, viz.

$$\mathcal{L}(tf\{t\}) = -\frac{d\bar{f}\{s\}}{ds} \tag{4.587}$$

– which emphasizes that differentiating with regard to s in Laplace's domain coincides with multiplying by $-t$ in the time domain; Eq. (4.587) permits calculation of the time-weighed hyperbolic functions included as entries to Table 4.17. Combination of more than one of the above features finally supports calculation of Laplace's transform of $e^{-k/t}/\sqrt{t}$

and $(t/k)^{\nu/2} I_\nu\{2\sqrt{kt}\}$, with I_ν denoting modified Bessel's function of first kind and ν-th order – tabulated last in Table 4.17. If a delay occurs in time, then

$$\mathcal{L}(f\{t-k\}) = e^{-ks}\bar{f}\{s\} \tag{4.588}$$

– known as theorem of translation in the time domain; in other words, a translation of a function $f\{t\}$ by k in the time domain implies a multiplicative correction of $\bar{f}\{s\}$ by e^{-ks} in Laplace's domain. It is also instructive to realize that

$$\lim_{t\to 0} f\{t\} = \lim_{s\to\infty} s\bar{f}\{s\}, \tag{4.589}$$

known as initial value theorem; this implies that the initial value of function $f\{t\}$ coincides with the value of the product of s by its Laplace's transform, when s grows unbounded. A similar argument can be used when $s \to 0$, i.e.

$$\lim_{t\to \infty} f\{t\} = \lim_{s\to 0} s\bar{f}\{s\}; \tag{4.590}$$

Eq. (4.590) is usually designated as final value theorem – since it assures coincidence of the value to be asymptotically attained by $f\{t\}$ in the time domain, with the initial value taken by $s\bar{f}\{s\}$ in Laplace's domain.

4.3.17.3 Inversion

The inverse of Laplace's transform abides to

$$\mathcal{L}^{-1}(\bar{f}\{s\}) \equiv f\{t\}, \tag{4.591}$$

consistent with Eq. (4.582); however, its calculation normally poses more problems than calculating the integral of Eq. (4.582) in the first place – despite the unique, reciprocal correspondence between $f\{t\}$ and $\bar{f}\{s\}$. One useful tool in this endeavor may be stated as

$$\mathcal{L}^{-1}\left(f\{s+k\}\right) = e^{-kt}\mathcal{L}^{-1}\left(\bar{f}\{s\}\right) \tag{4.592}$$

– which indicates that transformation of $\bar{f}\{s\}$, when the independent variable is translated from s to $s+k$, implies a multiplicative correction e^{-kt} to $f\{t\}$ in the time domain. One application of Eq. (4.592) involves (reciprocal) powers of $s+k$ – and accounts for the first entry in Table 4.18, where n denotes an integer constant, and k, k_1, and k_2 denote real constants. A similar application of Eq. (4.592) pertains to trigonometric functions, and leads to the second and third entries in the same table. Another important feature encompasses product of two functions in Laplace's domain, according to

$$\mathcal{L}^{-1}\left(\bar{f}\{s\}\bar{g}\{s\}\right) = \int_0^t f\{\tilde{t}\}g\{t-\tilde{t}\}d\tilde{t} \tag{4.593}$$

– usually known as convolution theorem. Once in possession of Eq. (4.593), one can handle composite translations in Laplace's domain – as is the case of the last entry in Table 4.18.

The most useful method of inversion of Laplace's transforms is, however, Heaviside's expansion in partial fractions, viz.

$$\mathcal{L}^{-1}\left(\bar{f}\{s\}\right) \equiv \mathcal{L}^{-1}\left(\frac{\sum_{i=0}^{M} a_i s^i}{\sum_{j=0}^{N} b_j s^j}\right) = \mathcal{L}^{-1}\left(\frac{\sum_{i=0}^{M} a_i s^i}{b_N \prod_{k=1}^{K} (s-s_k)^{m_k}}\right), \tag{4.594}$$

where preliminary factorization of denominator was meanwhile performed, in agreement with Eq. (4.126); degree M of the polynomial on s in numerator is necessarily lower than degree N of the polynomial on s in denominator – so $\sum_{i=0}^{M} a_i s^i \Big/ \sum_{j=0}^{N} b_j s^j$ represents a regular rational fraction. Here a_1, a_2, ..., a_M and b_1, b_2, ..., b_N denote coefficients of the polynomials in numerator and denominator, respectively – while s_k denotes each of the distinct K roots of the latter (also known as poles), and m_k denotes multiplicity of each said pole. Upon expansion of Eq. (4.594) in partial fractions as suggested by Eq. (4.127), one gets

$$\mathcal{L}^{-1}\left(\bar{f}\{s\}\right) = \mathcal{L}^{-1}\left(\sum_{k=1}^{K}\sum_{l=0}^{m_k-1} \frac{A_{k,l+1}}{(s-s_k)^{m_k-l}}\right), \tag{4.595}$$

with the $A_{k,l+1}$s denoting constants (to be determined) as functions of a_is and b_js in Eq. (4.594); one will eventually attain

$$\mathcal{L}^{-1}\left(\bar{f}\{s\}\right) = \sum_{k=1}^{K}\sum_{l=0}^{m_k-1} \frac{A_{k,l+1}}{(m_k-l-1)!} t^{m_k-l-1} e^{s_k t}, \tag{4.596}$$

with the aid of Eqs. (4.583) and the first entry in Table 4.18. When s_k is a complex number, of the form $a+\iota b$, then $a-\iota b$ will also be a pole – otherwise the coefficients of $\sum_{j=0}^{N} b_j s^j$ in Eq. (4.594) would not all be real numbers (as found in practice); hence, terms in $e^{\iota b}$ and $e^{-\iota b}$ will necessarily appear as such in Eq. (4.596). To

TABLE 4.18

List of Laplace's inverse transforms, $\mathcal{L}^{-1}(\bar{f}\{s\})$, of functions in Laplace's domain, $\bar{f}\{s\}$, obtained via definition and associated theorems.

Type	$\bar{f}\{s\}$	$\mathcal{L}^{-1}(\bar{f}\{s\})$
Exponential-weighed power	$\frac{n!}{(s+k)^{n+1}}$	$t^n e^{-kt}$, n integer
Exponential-weighed trigonometric functions	$\frac{k_1}{k_1^2+(s+k_2)^2}$	$e^{-k_2 t}\sin k_1 t$
	$\frac{s+k_2}{k_1^2+(s+k_2)^2}$	$e^{-k_2 t}\cos k_1 t$
Combinations of exponentials	$\frac{1}{(s-k_1)(s-k_2)}$	$\frac{e^{k_1 t}-e^{k_2 t}}{k_1-k_2}$
	$\frac{1}{(s-k_1)(s-k_2)^2}$	$\frac{e^{k_1 t}-(1+(k_1-k_2)t)e^{k_2 t}}{(k_1-k_2)^2}$

get rid of explicit $e^{\imath bt}$, while both terms may remain multiplied by e^{at}, one may resort to

$$\frac{A_{k,1}}{s-(a+\imath b)}+\frac{A_{k,2}}{s-(a-\imath b)}=e^{at}\left(B_{k,1}\cos bt+B_{k,2}\sin bt\right),\quad(4.597)$$

with the B_ks denoting constants. The two terms in the left-hand side of Eq. (4.597) are indeed handled in lumped form as

$$\begin{aligned}&\frac{A_{k,1}}{s-(a+\imath b)}+\frac{A_{k,2}}{s-(a-\imath b)}\\&=B_{k,1}\frac{s-a}{b^2+(s-a)^2}+B_{k,2}\frac{b}{b^2+(s-a)^2}\end{aligned}\quad(4.598)$$

– which is equivalent to Eq. (4.129), provided that constants $B_{k,1}$ and $B_{k,2}$ are defined as

$$B_{k,1}\equiv A_{k,1}+A_{k,2}\quad(4.599)$$

and

$$B_{k,2}\equiv\imath\left(A_{k,1}-A_{k,2}\right),\quad(4.600)$$

respectively – after replacing $B_{l,1}$ by $B_{k,1}$, and $B_{l,2}$ by $-aB_{k,1}+bB_{k,2}$; inversion of Laplace's transform then ensues, with the aid of the second and third entries in Table 4.18.

4.3.18 Lambert's *W* function

The defining relationship for Lambert's *W* function (also called omega function, or product logarithm) reads

$$We^W=x;\quad(4.601)$$

a plot is provided in Fig. 4.15 – where a graphical pattern resembling a rectangular hyperbola is apparent. Based on Eq. (4.601), one realizes that

$$W\{0\}=0,\quad(4.602)$$

as well as

$$W\{e\}=1\quad(4.603)$$

– since Neper's number can be coined as

$$e=1e^1;\quad(4.604)$$

by the same token,

$$W\left\{-\frac{\ln a}{a}\right\}=-\ln a.\quad(4.605)$$

An extension of the reasoning underlying Eq. (4.605) indicates that an equation of the type

$$p^{ax+b}=cx+d,\quad(4.606)$$

with $a\neq 0$, b, $c\neq 0$, d, and $p>0$ denoting constants, accepts

$$x=-\frac{W\left\{-\frac{a}{c}p^{b-\frac{ad}{c}}\ln p\right\}}{a\ln p}-\frac{d}{c}\quad(4.607)$$

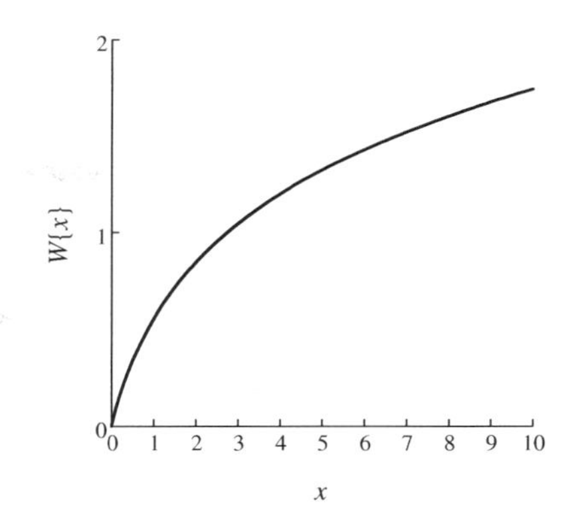

FIGURE 4.15 Variation of Lambert's function, $W\{x\}$, versus its real argument, x.

for solution. Another unique property of Lambert's *W* function encompasses its derivative with regard to x – which may be calculated via implicit differentiation of Eq. (4.601), viz.

$$\frac{dW}{d\ln x}=\frac{W}{1+W};\quad(4.608)$$

note, however, that

$$We^W=\kappa x\quad(4.609)$$

(with constant κ) represents the general solution to the differential equation labeled as Eq. (4.608).

4.3.19 Gamma function

When an integer n is sequentially multiplied by the previous integer until reaching unity, the resulting product will look like

$$n!\equiv\prod_{i=1}^{n}i\quad(4.610)$$

– where $n!$ is termed factorial of n; by convention,

$$0!\equiv 1.\quad(4.611)$$

The above concept has been extended to any non-integer value via gamma function, $\Gamma\{x\}$; Euler's definition thereof looks like

$$\Gamma\{x\}\equiv\frac{1}{x}\lim_{n\to\infty}\frac{n!n^x}{\prod_{i=1}^{n}(x+i)}.\quad(4.612)$$

A more convenient (and common) definition reads, however,

$$\Gamma\{x\}\equiv\int_0^\infty\xi^{x-1}e^{-\xi}d\xi,\quad(4.613)$$

which satisfies

$$\Gamma\{n\}=(n-1)!\quad(4.614)$$

– as obtained upon sequential application of the rule of integration by parts to Eq. (4.613), for $x=n$. A graphical interpretation of Eqs. (4.613) and (4.614) is provided in Fig. 4.16. Besides the singular behavior in the vicinity of $x=0$, the rapidly growing tendency of $\Gamma\{x\}$, when x increases beyond 2, should be highlighted. One specific value of particular interest reads

$$\Gamma\{x\}\Big|_{x=\frac{1}{2}}=\sqrt{\pi}; \tag{4.615}$$

on the other hand, the slope of the tangent to $\Gamma\{x\}$, at $x=1$, looks like

$$\lim_{x\to 1}\frac{d\Gamma\{x\}}{dx}=-\gamma, \tag{4.616}$$

where γ denotes Mascheroni's constant – which, in turn, supports

$$\Gamma\{x\}\equiv\frac{\displaystyle\prod_{k=1}^{\infty}\frac{e^{\frac{x}{k}}}{1+\frac{x}{k}}}{xe^{\gamma x}}, \tag{4.617}$$

known as Weierstrass' definition of gamma function. The factorial function, as conveyed by Eq. (4.610), tends asymptotically to

$$\lim_{n\to\infty}n!=\frac{\sqrt{2\pi}\,n^{n+\frac{1}{2}}}{e^{n}}, \tag{4.618}$$

usually referred to as Stirling's approximation; for $n\geq 3$, the relative error of this approximation lies below 2%. Using its own definition as conveyed by Eq. (4.613), combined with the rule of integration by parts, one finds

$$\Gamma\{x+1\}=x\Gamma\{x\} \tag{4.619}$$

pertaining to the gamma function – and consistent with Eqs. (4.610) and Eq. (4.614) for integer x; this is known as telescopic property. Equations (4.617) and (4.619) yield, in turn,

$$\Gamma\{1-x\}\Gamma\{x\}=\frac{\pi}{\sin\pi x}, \tag{4.620}$$

also with the aid of

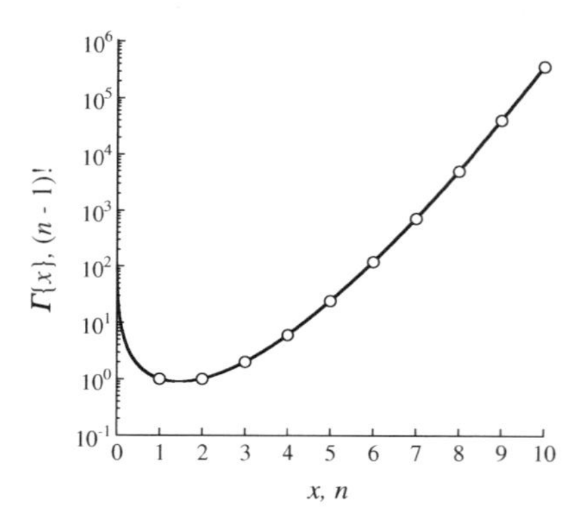

FIGURE 4.16 Variation of gamma function, $\Gamma\{x\}$ (—), versus its real argument, x, and of factorial function, $(n-1)!$ (○), versus its integer argument, n.

$$x\prod_{k=1}^{\infty}\left(1-\left(\frac{x}{k}\right)^2\right)=\frac{\sin\pi x}{\pi}, \tag{4.621}$$

classically termed Gauss' multiplication theorem; Eq. (4.621) supports

$$\prod_{k=0}^{m-1}\Gamma\left\{x+\frac{k}{m}\right\}=(2\pi)^{\frac{m-1}{2}}m^{\frac{1}{2}-mx}\Gamma\{mx\}, \tag{4.622}$$

as long as $x\neq -k/m$; if, in particular, $m=2$ and $k=1$, then Eq. (4.622) degenerates to

$$\Gamma\{2x\}=\frac{\Gamma\{x\}\Gamma\left\{x+\frac{1}{2}\right\}}{\sqrt{\pi}\,2^{1-2x}} \tag{4.623}$$

– oftentimes referred to as duplication formula.

4.3.20 Incomplete gamma function

The classical gamma function, as per Eq. (4.613), can be generalized to incomplete counterparts thereof – by using a variable lower or upper integration limit; the latter case entails

$$\gamma\{x,z\}\equiv\int_0^z\xi^{x-1}e^{-\xi}d\xi, \tag{4.624}$$

and serves as definition for the lower incomplete gamma function.The upper incomplete gamma function reads

$$\Gamma\{x,z\}\equiv\int_z^{\infty}\xi^{x-1}e^{-\xi}d\xi \tag{4.625}$$

– thus implying

$$\gamma\{x,z\}+\Gamma\{x,z\}=\Gamma\{x\}, \tag{4.626}$$

in view of Eqs. (4.469) and (4.613). A plot of Eq. (4.624) is available as Fig. 4.17. One notices a monotonically increasing behavior of $\gamma\{x,z\}$ with z, and with x for larger values of z; this function departs from zero at $z=0$, and tends to a horizontal asymptote that is reached faster at lower x. Algebraic manipulation of Eq. (4.625) unfolds

$$\Gamma\{1,z\}=e^{-z}, \tag{4.627}$$

$$\Gamma\left\{\frac{1}{2},z\right\}=\sqrt{\pi}\,\mathrm{erfc}\{\sqrt{z}\}, \tag{4.628}$$

and

$$\Gamma\{0,z\}=-E_i\{-z\} \tag{4.629}$$

for $z>0$ – as well as

$$\lim_{z\to 0}\Gamma\{x,z\}=\Gamma\{x\}, \tag{4.630}$$

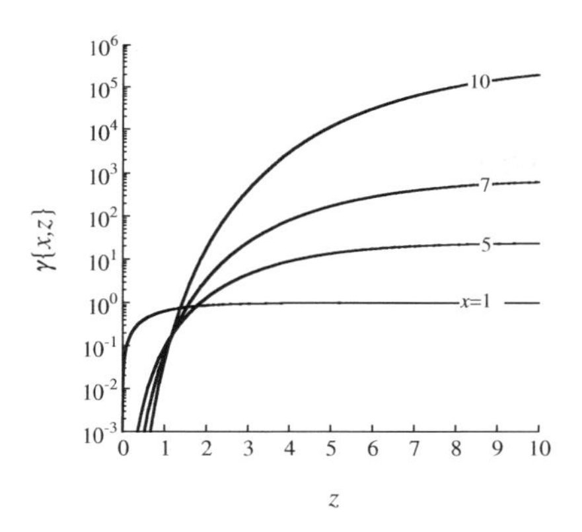

FIGURE 4.17 Variation of lower incomplete gamma function, $\gamma\{x,z\}$, versus one of its real arguments, z, for selected values of the other real argument, x.

consistent with Eq. (4.613); here erfc$\{x\}$ and $E_i\{x\}$ denote complementary error function and exponential-integral function, respectively, of x. By the same token, one gets

$$\gamma\{1,z\} = 1 - e^{-z}, \tag{4.631}$$

and

$$\gamma\left\{\frac{1}{2},z\right\} = \sqrt{\pi}\,\text{erf}\left\{\sqrt{z}\right\} \tag{4.632}$$

from Eq. (4.624), where erf$\{x\}$ denotes error function of x; furthermore,

$$\lim_{z\to\infty} \gamma\{x,z\} = \Gamma\{x\} \tag{4.633}$$

in agreement with Eq. (4.613) – which describes the vertical intercepts of the long-z horizontal asymptotes in Fig. 4.17. Neighbor functions, differing in their argument by one unit, are related by recursive relationships

$$\Gamma\{x,z\} = (x-1)\Gamma\{x-1,z\} + e^{-z}z^{x-1} \tag{4.634}$$

and

$$\gamma\{x,z\} = (x-1)\gamma\{x-1,z\} - e^{-z}z^{x-1}. \tag{4.635}$$

In practice, the upper incomplete gamma function is more useful; its integral in z may be calculated as

$$\int \Gamma\{x,z\}dz = z\Gamma\{x,z\} - \Gamma\{x+1,z\}, \tag{4.636}$$

whereas

$$\frac{\partial\Gamma\{x,z\}}{\partial z} = -e^{-z}z^{x-1} \tag{4.637}$$

results directly from (partial) differentiation of Eq. (4.625) with regard to z.

4.3.21 Bessel's functions

Bessel's function of first kind, and order $\pm\nu$, are widely useful in engineering problems – and satisfy

$$J_{\pm\nu}\{x\} \equiv \sum_{i=0}^{\infty} \frac{(-1)^i}{i!\,\Gamma\{i\pm\nu+1\}}\left(\frac{x}{2}\right)^{2i+\nu}; \tag{4.638}$$

a graphical account is provided by Fig. 4.18, for $\nu=0$ and $\nu=1$. Note the damped, oscillatory patterns of both functions – with $J_0\{x\}$ departing from unity and $J_1\{x\}$ from zero; for small argument x,

$$J_\nu\{x\}\Big|_{x\to 0} \approx \frac{1}{\Gamma\{\nu+1\}}\left(\frac{x}{2}\right)^\nu \tag{4.639}$$

holds as good approximation. The inherent orthogonality of this function supports

$$\int_0^\infty J_\nu\{x\}J_\mu\{x\}\frac{dx}{x} = \frac{2}{\pi}\frac{\sin\frac{\pi}{2}(\nu-\mu)}{\nu^2-\mu^2}, \tag{4.640}$$

where $J_\mu\{x\}$ denotes Bessel's function of first kind and μ-th order; as well as

$$\int_0^a J_\nu^{\,2}\left\{\frac{b}{a}x\right\}x\,dx = \frac{a^2}{2}J_{\nu+1}^{\,2}\{b\}, \tag{4.641}$$

for real a and b, and integer ν – complemented by

$$\int_0^a J_\nu\left\{\frac{b}{a}x\right\}J_\nu\left\{\frac{c}{a}x\right\}x\,dx = 0, \tag{4.642}$$

for real $c \neq b$. Other useful properties are the differential identities

$$\frac{d}{dx}\left(x^\nu J_\nu\{x\}\right) = x^\nu J_{\nu-1}\{x\} \tag{4.643}$$

and, more generally,

$$\left(\frac{1}{x}\frac{d}{dx}\right)^m\left(x^\nu J_\nu\{x\}\right) = x^{\nu-m}J_{\nu-m}\{x\} \tag{4.644}$$

for integer m – besides

$$\frac{dJ_\nu\{x\}}{dx} = \frac{J_{\nu-1}\{x\} - J_{\nu+1}\{x\}}{2}; \tag{4.645}$$

in addition,

$$\frac{2\nu}{x}J_\nu\{x\} = J_{\nu-1}\{x\} + J_{\nu+1}\{x\}, \tag{4.646}$$

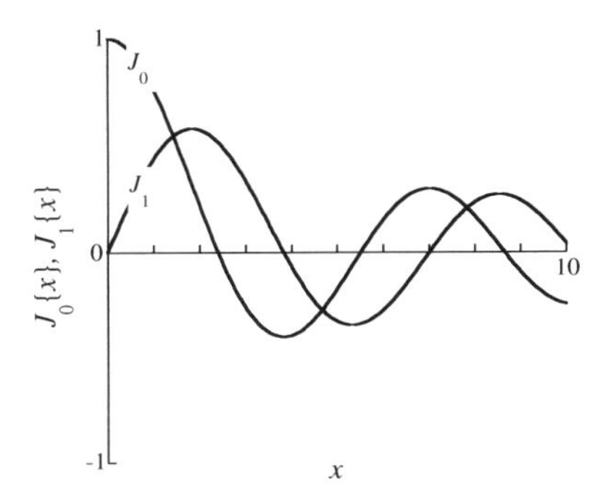

FIGURE 4.18 Variation of Bessel's functions of first kind, and zero-th order, $J_0\{x\}$, or first order, $J_1\{x\}$, versus their real argument, x.

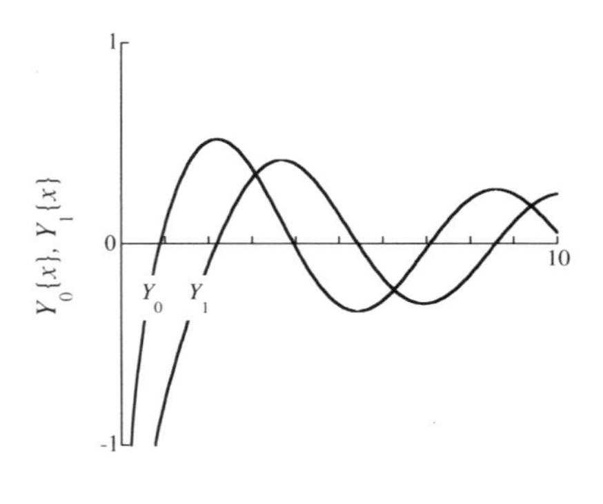

FIGURE 4.19 Variation of Bessel's functions of second kind, and zero-th order, $Y_0\{x\}$, or first order, $Y_1\{x\}$, versus their real argument, x.

holds, which is often coined as

$$J_\nu\{x\} = \frac{2(\nu+1)}{x} J_{\nu+1}\{x\} - J_{\nu+2}\{x\} \tag{4.647}$$

– still entertaining a step-2 recurrence relationship.

Bessel's functions of second kind – also known as Weber's or Neumann's functions, abide to

$$Y_\nu\{x\} \equiv \frac{J_\nu\{x\}\cos\nu\pi - J_{-\nu}\{x\}}{\sin\nu\pi} \tag{4.648}$$

at large, and

$$Y_n\{x\} \equiv \lim_{\nu\to n} \frac{J_\nu\{x\}\cos\nu\pi - J_{-\nu}\{x\}}{\sin\nu\pi} \tag{4.649}$$

in the specific case of integer n; plots of $Y_0\{x\}$ and $Y_1\{x\}$ are provided in Fig. 4.19. Oscillatory patterns are again apparent – with $x=0$ serving as vertical asymptote for both $Y_0\{x\}$ and $Y_1\{x\}$; the behavior of $Y_\nu\{x\}$ is driven by

$$Y_0\{x\}\Big|_{x\to 0} \approx \frac{2}{\pi}\left(\gamma + \ln\frac{x}{2}\right), \tag{4.650}$$

and

$$Y_n\{x\}\Big|_{x\to 0} \approx \frac{1}{n!\tan n\pi}\left(\frac{x}{2}\right)^n - \frac{(n-1)!}{\pi}\left(\frac{2}{x}\right)^n \tag{4.651}$$

for positive integer n – again at small x, with γ denoting Mascheroni's constant, see Eq. (4.616). This function also entertains

$$\left(\frac{1}{x}\frac{d}{dx}\right)^m \left(x^\nu Y_\nu\{x\}\right) = x^{\nu-m} Y_{\nu-m}\{x\} \tag{4.652}$$

and

$$\frac{dY_\nu\{x\}}{dx} = \frac{Y_{\nu-1}\{x\} - Y_{\nu+1}\{x\}}{2}, \tag{4.653}$$

as well as

$$\frac{2\nu}{x} Y_\nu\{x\} = Y_{\nu-1}\{x\} + Y_{\nu+1}\{x\} \tag{4.654}$$

– which, in essence, resemble Eqs. (4.644)–(4.646), respectively. Modified Bessel's function of first kind abide to

$$I_\nu\{x\} \equiv \sum_{i=0}^{\infty} \frac{1}{i!\,\Gamma\{i+\nu+1\}}\left(\frac{x}{2}\right)^{2i+\nu}; \tag{4.655}$$

plots of $I_0\{x\}$ and $I_1\{x\}$ are provided in Fig. 4.20. Unlike Bessel's functions of first kind, $I_\nu\{x\}$ grows exponentially toward infinity when x becomes unbounded – while

$$\lim_{x\to 0} I_\nu\{x\} = 0; \quad \nu > 0, \tag{4.656}$$

complemented by

$$\lim_{x\to 0} I_0\{x\} = 1; \tag{4.657}$$

in addition,

$$\frac{2\nu}{x} I_\nu\{x\} = I_{\nu-1}\{x\} - I_{\nu+1}\{x\} \tag{4.658}$$

and

$$\frac{dI_\nu\{x\}}{dx} = \frac{I_{\nu-1}\{x\} + I_{\nu+1}\{x\}}{2} \tag{4.659}$$

also prove useful relationships.

Modified Bessel's functions of second kind satisfy

$$K_\nu\{x\} \equiv \frac{\pi}{2}\frac{I_{-\nu}\{x\} - I_\nu\{x\}}{\sin\nu\pi} \tag{4.660}$$

for non-integer ν, coupled with

$$K_n\{x\} \equiv \frac{\pi}{2}\lim_{\nu\to n} \frac{I_{-\nu}\{x\} - I_\nu\{x\}}{\sin\nu\pi} \tag{4.661}$$

for integer n; the variation of $K_0\{x\}$ and $K_1\{x\}$ with x can be grasped in Fig. 4.21. Unlike the oscillatory pattern of Bessel's functions of second kind, $K_n\{x\}$ decays exponentially toward zero as x grows – with a singularity at $x=0$, of the logarithmic type; one also finds that

$$K_\nu\{x\} = \frac{\pi}{2} e^{\frac{\iota\nu\pi}{2}} \left(e^{\frac{\iota(1+\nu)\pi}{2}} I_\nu\{x\} - Y_\nu\{\iota x\} \right) \tag{4.662}$$

holds as relationship linking $K_\nu\{x\}$ to $I_\nu\{x\}$ and $Y_\nu\{x\}$. Other useful expressions look like

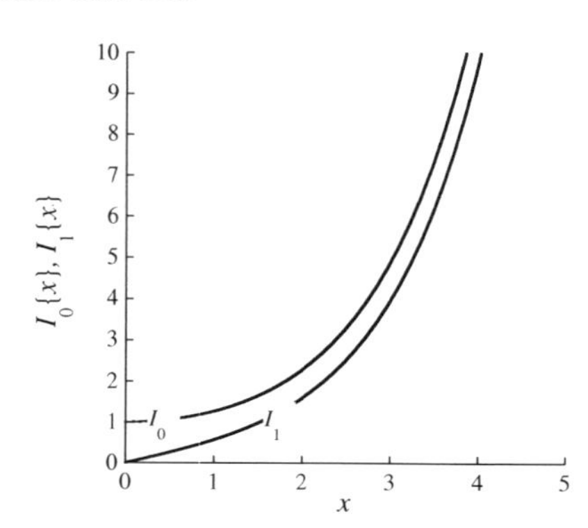

FIGURE 4.20 Variation of modified Bessel's functions of first kind, and zero-th order, $I_0\{x\}$, or first order, $I_1\{x\}$, versus their real argument, x.

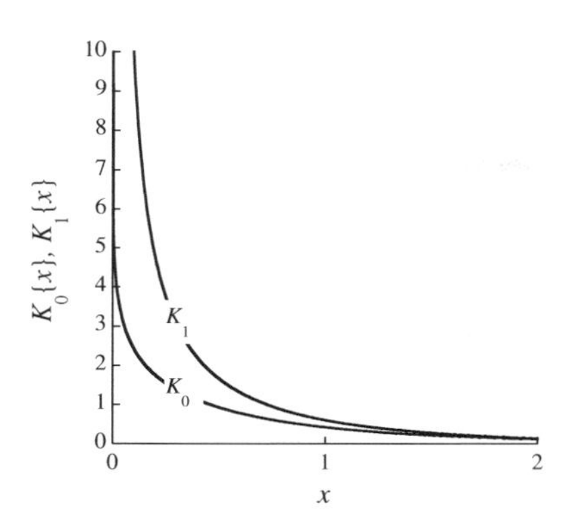

FIGURE 4.21 Variation of modified Bessel's functions of second kind, and zero-th order, $K_0\{x\}$, or first order, $K_1\{x\}$, versus their real argument, x.

$$\frac{2\nu}{x}\mathrm{e}^{\frac{\iota\nu\pi}{2}}K_\nu\{x\} = \mathrm{e}^{\iota(\nu-1)\pi}K_{\nu-1}\{x\} - \mathrm{e}^{\iota(\nu+1)\pi}K_{\nu+1}\{x\} \quad (4.663)$$

and

$$\frac{d}{dx}\left(\mathrm{e}^{\frac{\iota\nu\pi}{2}}K_\nu\{x\}\right) = \frac{\mathrm{e}^{\iota(\nu-1)\pi}K_{\nu-1}\{x\} + \mathrm{e}^{\iota(\nu+1)\pi}K_{\nu+1}\{x\}}{2}, \quad (4.664)$$

relevant for relating $K_\nu\{x\}$ and its derivative, $dK_\nu\{x\}/dx$, to $K_{\nu-1}\{x\}$ and $K_{\nu+1}\{x\}$.

4.3.22 Exponential-integral function

The exponential-integral function appears in some mathematical derivations, and satisfies

$$E_i\{ax\} \equiv \int \frac{\mathrm{e}^{ax}}{x}dx \quad (4.665)$$

for definition; when $a=1$ and x is negative, it is given by

$$E_i\{x\} = -\int_{-x}^{\infty}\frac{\mathrm{e}^{-\xi}}{\xi}d\xi = \int_{-\infty}^{x}\frac{\mathrm{e}^{\xi}}{\xi}d\xi, \quad (4.666)$$

whereas a positive x supports

$$E_i\{x\} = -\lim_{\varepsilon\to 0^+}\left(\int_{-x}^{-\varepsilon}\frac{\mathrm{e}^{-\xi}}{\xi}d\xi + \int_{\varepsilon}^{\infty}\frac{\mathrm{e}^{-\xi}}{\xi}d\xi\right) \quad (4.667)$$

– as apparent in Fig. 4.22. The (negative portion of the) vertical axis serves as vertical asymptote, should the abscissa approach zero; the plot then goes through an inflection point (located between 0 and 1), toward exponential growth at large x. Expressions to calculate $E_i\{x\}$, in alternative to Eqs. (4.665) and (4.666), encompass

$$E_i\{x\} = \mathrm{e}^x\left(\frac{1}{x} + \int_0^{\infty}\frac{\mathrm{e}^{-\xi}}{(x-\xi)^2}d\xi\right) \quad (4.668)$$

and

$$E_i\{x\} = \mathrm{e}^x\int_1^{\infty}\frac{\mathrm{e}^{-\xi}}{\xi^2(x-\ln\xi)}d\xi; \quad (4.669)$$

or even

$$E_i\{x\} = 2\ln x - \frac{2\mathrm{e}^x}{\pi}\int_0^{\infty}\frac{x\cos\xi + \xi\sin\xi}{x^2+\xi^2}\ln\xi\, d\xi, \quad (4.670)$$

valid only when x is positive – with choice among such forms being dictated by the most convenient tool of computation available.

4.3.23 Solution of ordinary differential equations

An N-th order ordinary differential equation of y on x contains only regular derivatives up to the said order, i.e. d^Ny/dx^N, $d^{N-1}y/dx^{N-1}$, …, d^2y/dx^2, dy/dx and y (or zero-th order derivative). Its solution can, in general, be formulated as a linear combination of N independent functions, say $f_1\{x\}, f_2\{x\}, \ldots, f_N\{x\}$ – i.e. functions for which the Wronskian determinant, $|\boldsymbol{W}|$, given by $\begin{vmatrix} f_1 & f_2 & \cdots & f_N \\ \frac{df_1}{dx} & \frac{df_2}{dx} & \cdots & \frac{df_N}{dx} \\ \cdots & \cdots & \cdots & \cdots \\ \frac{d^{N-1}f_1}{dx^{N-1}} & \frac{d^{N-1}f_2}{dx^{N-1}} & \cdots & \frac{d^{N-1}f_N}{dx^{N-1}} \end{vmatrix}$, is not nil. The N coefficients of the aforementioned linear combination are then to be calculated, so as to satisfy N (required) boundary conditions. However, closed-form analytical solutions exist only for a number of forms of the said ordinary differential equations – as illustrated below.

On the other hand, an N-th order partial differential equation of y on $x_1, x_2, \ldots, x_M$ contains partial derivatives up to such an order, i.e. $\partial^Ny/\partial x_1^N$, $\partial^{N-1}y/\partial x_1^{N-1}$, …, $\partial^2y/\partial x_1^2$, $\partial y/\partial x_1$, $\partial^Ny/\partial x_2^N$, $\partial^{N-1}y/\partial x_2^{N-1}$, …, $\partial^2y/\partial x_2^2$, $\partial y/\partial x_2$, …, $\partial^Ny/\partial x_M^N$, $\partial^{N-1}y/\partial x_M^{N-1}$, …, $\partial^2y/\partial x_M^2$ and $\partial y/\partial x_M$, besides y itself

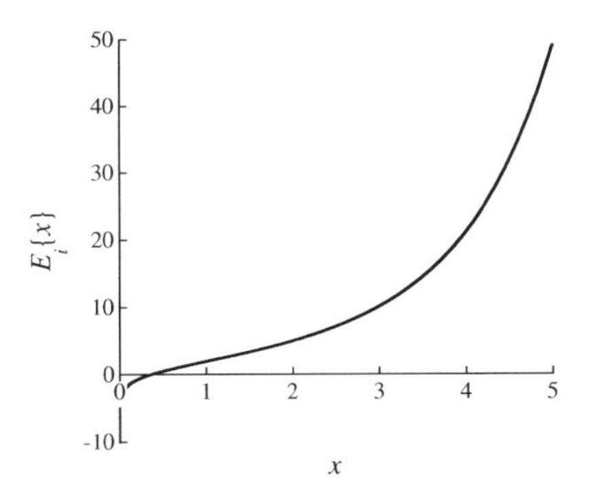

FIGURE 4.22 Variation of exponential-integral function, $E_i\{x\}$, versus its real argument, x.

– complemented by all possible cross- derivatives (some of which may, again, be nil). The corresponding solution contains arbitrary functions of some unique argument on $x_1, x_2, \ldots, x_M$ – which must satisfy the (required) boundary conditions; therefore, no general solution can normally be found based solely on the functional form of the partial differential equation.

4.3.23.1 First order with separable variables

A first-order ordinary differential equation of the form

$$\frac{dy}{dx} = P\{x\}U\{y\}, \tag{4.671}$$

with $P\{x\}$ denoting a generic function on independent variable x and $U\{y\}$ denoting a generic function on dependent variable y, can be solved via separation of variables as

$$\int \frac{dy}{U\{y\}} = \kappa + \int P\{x\}dx; \tag{4.672}$$

here κ denotes an arbitrary constant. Equation (4.672) conveys $y \equiv y\{x\}$, obtainable (implicitly or explicitly) upon calculation of the integrals in both sides.

4.3.23.2 Homogeneous first order

Consider the ordinary differential equation

$$\frac{dy}{dx} = \frac{\phi\{x,y\}}{\psi\{x,y\}} \tag{4.673}$$

– with $\phi\{x,y\}$ and $\psi\{x,y\}$ denoting homogeneous, bivariate, polynomial functions of identical degree n, as per Eq. (4.393); x^n may then be factored out to give

$$\frac{dy}{dx} = \frac{x^n \Phi\left\{\frac{y}{x}\right\}}{x^n \Psi\left\{\frac{y}{x}\right\}} = \frac{\Phi\left\{\frac{y}{x}\right\}}{\Psi\left\{\frac{y}{x}\right\}} \equiv \Omega\left\{\frac{y}{x}\right\}, \tag{4.674}$$

where x^n was meanwhile dropped off both numerator and denominator – with $\Phi\{y/x\}$, $\Psi\{y/x\}$, and $\Omega\{y/x\}$ denoting univariate functions of y/x. Equation (4.674) will eventually be transformed, via definition of a new dependent variable z as

$$y \equiv xz\{x\} \tag{4.675}$$

– which leads to

$$\frac{dy}{dx} = z + x\frac{dz}{dx}, \tag{4.676}$$

upon differentiation of a product that includes a composite function; elimination of dy/dx between Eqs. (4.674) and (4.676) unfolds

$$z + x\frac{dz}{dx} = \Omega\{z\}, \tag{4.677}$$

with the aid of Eq. (4.675) – or, equivalently,

$$x\frac{dz}{dx} = \Omega\{z\} - z. \tag{4.678}$$

Integration of Eq. (4.678) is now possible via separation of variables, according to

$$\int \frac{dz}{\Omega\{z\} - z} = \int \frac{dx}{x}, \tag{4.679}$$

since it functionally mimics Eq. (4.671) with $P\{x\} = 1/x$ and $U\{y\} = \Omega\{y\} - y$; hence, z will be (implicitly or explicitly) obtained as a function of x – and the initial notation should eventually be retrieved, through Eq. (4.675), to get $y \equiv y\{x\}$ that satisfies Eq. (4.673).

4.3.23.3 Linear first order

A (complete) first-order ordinary differential equation is said to be linear when it can be written in the form

$$\frac{dy}{dx} + P\{x\}y = Q\{x\}, \tag{4.680}$$

where $Q\{x\}$ denotes a function of x – including the possibility of holding a constant value. This type of equation may be solved via an integrating factor, $R\{x\}$, defined via

$$R\{x\}\frac{dy}{dx} + R\{x\}P\{x\}y \equiv \frac{d}{dx}\big(R\{x\}y\big) = R\{x\}Q\{x\} \tag{4.681}$$

– obtained from Eq. (4.680), after multiplying both sides by $R\{x\}$. The bottom line is finding $R\{x\}$, such that the left-hand side of Eq. (4.681) will look like the derivative of $R\{x\}y$ in the middle side – which would prompt solution as

$$\int d\big(R\{x\}y\big) = \int R\{x\}Q\{x\}dx, \tag{4.682}$$

at the expense of separation of variables. Such an integrating factor should accordingly be given by

$$R\{x\} = e^{\int P\{x\}dx}, \tag{4.683}$$

obtained after cancelling out common terms between left-hand and middle sides of Eq. (4.681), and performing differentiation as indicated; under these circumstances,

$$y = \frac{\kappa + \int e^{\int P\{x\}dx} Q\{x\}dx}{e^{\int P\{x\}dx}} \tag{4.684}$$

will emerge from Eqs. (4.682) and (4.683) as the final solution to Eq. (4.680), in general form – with κ denoting an arbitrary constant.

4.3.23.4 Bernoulli's equation

A first-order differential equation of major practical interest is Bernoulli's equation, viz.

$$\frac{dy}{dx} + P\{x\}y = Q\{x\}y^n; \tag{4.685}$$

this equation is nonlinear when $n \neq 0,1$. After setting

$$z \equiv y^{1-n} \tag{4.686}$$

as auxiliary variable, it becomes possible to reformulate Eq. (4.685) to

$$\frac{dz}{dx} + (1-n)P\{x\}z = (1-n)Q\{x\}; \quad n \neq 0,1; \tag{4.687}$$

Eq. (4.687) is a linear, first-order ordinary differential equation of z on x, of the type conveyed by Eq. (4.680) – so one may resort to an integrating factor to solve it, see Eq. (4.684) upon replacement of y by z, $P\{x\}$ by $(1-n)P\{x\}$, and $Q\{x\}$ by $(1-n)Q\{x\}$.

4.3.23.5 Incomplete, linear second order

A linear, second-order differential equation, with nil zero-th order differential and independent terms looks like

$$\frac{d^2y}{dx^2} + P\{x\}\frac{dy}{dx} = 0, \tag{4.688}$$

in general. To solve it, one should to advantage proceed with definition of an auxiliary variable p as

$$p \equiv \frac{dy}{dx}, \tag{4.689}$$

which implies

$$\frac{dp}{dx} = \frac{d}{dx}\left(\frac{dy}{dx}\right) \equiv \frac{d^2y}{dx^2}; \tag{4.690}$$

hence, Eq. (4.688) will instead appear as

$$\frac{dp}{dx} + P\{x\}p = 0, \tag{4.691}$$

with the aid of Eqs. (4.689) and (4.690) – which is a linear, first-order differential equation, of the type labeled as Eq. (4.680) when $Q\{x\} = 0$. Therefore, one may resort to Eq. (4.684) to get

$$p = \frac{K_1 + \int e^{\int P\{x\}dx} 0dx}{e^{\int P\{x\}dx}} = \frac{K_1 + K_2}{e^{\int P\{x\}dx}} = \kappa_1 e^{-\int P\{x\}dx} \tag{4.692}$$

– where arbitrary constant κ_1 stands for the sum of arbitrary constants K_1 and K_2. Once in possession of Eq. (4.692), one should insert Eq. (4.689) to obtain

$$\frac{dy}{dx} = \kappa_1 e^{-\int P\{x\}dx}; \tag{4.693}$$

integration via separation of variables ensues as

$$y = \kappa_2 + \kappa_1 \int e^{-\int P\{x\}dx} dx, \tag{4.694}$$

along with appearance of a second arbitrary constant, κ_2.

4.3.23.6 Bessel's equation

A classical example of a linear second-order equation reads

$$x^2\frac{d^2y}{dx^2} + x\frac{dy}{dx} + \left(x^2 - \nu^2\right)y = 0, \tag{4.695}$$

known as Bessel's equation; the general solution of Eq. (4.695) can be coined as

$$y\{x\} = \kappa_1 J_{-\nu}\{x\} + \kappa_2 J_\nu\{x\}, \tag{4.696}$$

should κ_1 and κ_2 denote integration constants, and ν denote a non-integer constant – with $J_{\pm\nu}\{x\}$ abiding to Eq. (4.638). If ν is an integer n, then the corresponding Bessel's function looks like

$$J_{\pm n}\{x\} = \sum_{i=0}^{\infty} \frac{(-1)^i}{i!(i \pm n)!}\left(\frac{x}{2}\right)^{2i+n} \tag{4.697}$$

that directly reflects Eq. (4.614); the first (negative) terms of J_{-n} are zero by virtue of the factorial of negative integers, appearing in denominator, being infinite – so J_{-n} simplifies further to

$$J_{-n} = \sum_{i=n}^{\infty} \frac{(-1)^i}{i!(i-n)!}\left(\frac{x}{2}\right)^{2i-n}, \tag{4.698}$$

as per Eq. (4.697). On the other hand, setting $j \equiv i - n$ transforms Eq. (4.698) to

$$J_{-n} = \sum_{j=0}^{\infty} \frac{(-1)^{j+n}}{(j+n)!\,j!}\left(\frac{x}{2}\right)^{2(j+n)-n}, \tag{4.699}$$

which is equivalent to

$$J_{-n} = \sum_{j=0}^{\infty} \frac{(-1)^{j+n}}{j!(j+n)!}\left(\frac{x}{2}\right)^{2j+n} \equiv (-1)^n J_n \tag{4.700}$$

after $(-1)^n$ is factored out – and following comparative inspection with Eq. (4.697) for positive n; therefore, J_n and J_{-n} are linearly dependent on each other – and a second (independent) solution must somehow be found. One way to address this issue resorts to Bessel's function of second kind, see Eq. (4.649) – so the general solution to Eq. (4.695), with $\nu \equiv n$, will be coined as

$$y\{x\} = \kappa_1 J_n\{x\} + \kappa_2 Y_n\{x\}, \tag{4.701}$$

in lieu of Eq. (4.696). Sometimes, modified Bessel's equation is found, viz.

$$x^2\frac{d^2y}{dx^2} + x\frac{dy}{dx} - \left(x^2 + \nu^2\right)y = 0, \tag{4.702}$$

with x^2y in Eq. (4.695) replaced here by its negative; in this case,

$$y\{x\} = \kappa_1 I_{-\nu}\{x\} + \kappa_2 I_\nu\{x\} \tag{4.703}$$

serves as a general solution for non-integer ν, while

$$y\{x\} = \kappa_1 I_n\{x\} + \kappa_2 K_n\{x\} \tag{4.704}$$

applies to integer n – thus taking advantage of modified Bessel's functions of first and second kinds, i.e. $I_n\{x\}$ as given by Eq. (4.655) and $K_n\{x\}$ as given by Eq. (4.661), respectively.

4.3.23.7 Legendre's equation

Another illustration of a linear second-order equation, germane in process engineering, entails Legendre's equation, viz.

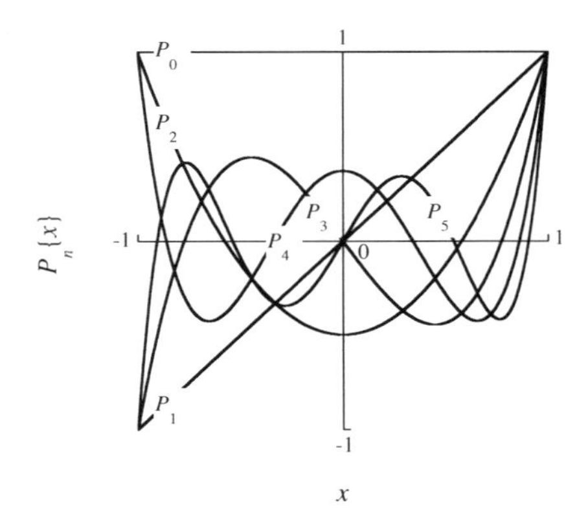

FIGURE 4.23 Variation of Legendre's polynomials of n-th order, $P_n\{x\}$, versus their real argument, x, for $n=0, 1, 2, 3, 4, 5$.

$$\left(1-x^2\right)\frac{d^2y}{dx^2}-2x\frac{dy}{dx}+l(l+1)y=0, \tag{4.705}$$

where l denotes a real constant; application of Frobenius' method unfolds a general (open-form) solution that looks like

$$y=\kappa_1\left(1-\frac{l(l+1)}{2!}x^2+\frac{l(l-2)(l+1)(l+3)}{4!}x^4-\ldots\right)$$
$$+\kappa_2\left(x-\frac{(l-1)(l+2)}{3!}x^3+\frac{(l-1)(l-3)(l+2)(l+4)}{5!}x^5+\ldots\right). \tag{4.706}$$

When l stands for an integer n, either series in Eq. (4.706) terminates – for instance, $l=2$ makes all even powers beyond x^2 be nil, and $l=3$ turns nil all odd powers beyond x^3; the resulting polynomials $P_n\{x\}$ are termed Legendre's polynomials – and abide to

$$P_n\{x\}=\frac{1}{2^n n!}\frac{d^n}{dx^n}\left(x^2-1\right)^n, \tag{4.707}$$

known as Rodrigues' formula. The first polynomials of this kind are represented in Fig. 4.23. All such polynomials are bounded by −1 and +1, when x spans [−1,1] – and exhibit oscillatory behavior, except $P_0\{x\}$ that is constant and $P_1\{x\}$ that is linear. One important property of Legendre's polynomials is that they are orthogonal within [−1,1], i.e.

$$\int_{-1}^{1}P_m\{x\}P_n\{x\}dx=0, \quad m\neq n \tag{4.708}$$

besides

$$\int_{-1}^{1}P_n^2\{x\}dx=\frac{2}{2n+1} \tag{4.709}$$

– where m and n denote integers; one may also utilize

$$(n+1)\,P_{n+1}\{x\}=(2n+1)xP_n\{x\}-nP_{n-1}\{x\} \tag{4.710}$$

as underlying recursive relationship relating such polynomials of consecutive orders – or else

$$P_n\{x\}=\sum_{i=0}^{n}\binom{n}{i}\binom{n+i}{i}\left(\frac{x-1}{2}\right)^i, \tag{4.711}$$

as explicit representation. When $P_n\{x\}$ entails one of the solutions conveyed by Eq. (4.706), the other solution is termed Legendre's function of second kind, $Q_n\{x\}$ – and exhibits the form of an infinite series, reducible, however, to

$$Q_0\{x\}\equiv\frac{1}{2}\ln\frac{1+x}{1-x} \tag{4.712}$$

for $n=0$,

$$Q_1\{x\}\equiv\frac{x}{2}\ln\frac{1+x}{1-x}-1 \tag{4.713}$$

for $n=1$,

$$Q_2\{x\}\equiv\frac{3x^2-1}{4}\ln\frac{1+x}{1-x}-\frac{3x}{2} \tag{4.714}$$

for $n=2$, and

$$Q_3\{x\}\equiv\frac{5x^3-3x}{4}\ln\frac{1+x}{1-x}-\frac{5x^2}{2}+\frac{2}{3} \tag{4.715}$$

for $n=3$ as simplest examples – and as plotted in Fig. 4.24. The number of local optima of Q_n, when x spans]−1,1[, is equal to n; while $x=-1$ and $x=1$ serve as vertical asymptotes in all cases, since $1-x$ appears in denominator of the argument of the (common) logarithmic function in Eqs. (4.712)–(4.715).

Both Bessel's and Legendre's equations are particular cases of (the more general) Sturm and Liouville's problems – materialized by differential equations of the type

$$\frac{d}{dx}\left(P\{x\}\frac{dy}{dx}\right)+Q\{x\}y=-\lambda w\{x\}y, \tag{4.716}$$

with λ denoting a parameter and $w\{x\}$ denoting a weight function. In fact, Eq. (4.695) may be rewritten as

$$\frac{d}{dx}\left(x\frac{dy}{dx}\right)+\left(x-\frac{\nu^2}{x}\right)y=0 \tag{4.717}$$

upon division of both sides by x – with $P\{x\}=x$, $Q\{x\}=(x^2-\nu^2)/x$ and $w\{x\}=0$ in Eq. (4.716); whereas Eq. (4.705) can be reformulated to read

$$\frac{d}{dx}\left(\left(1-x^2\right)\frac{dy}{dx}\right)+l(l+1)y=0, \tag{4.718}$$

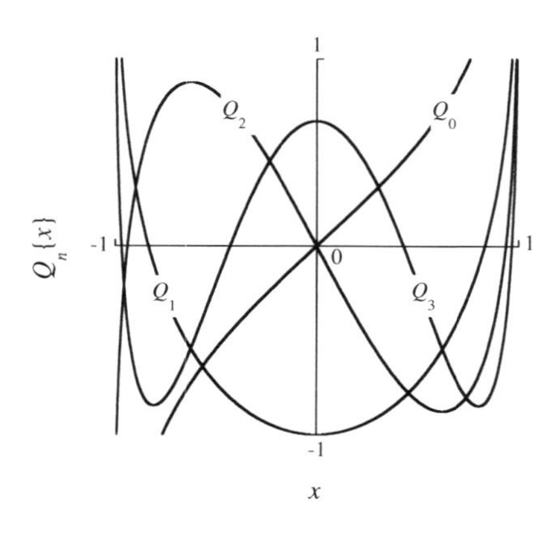

FIGURE 4.24 Variation of Legendre's functions of second kind and n-th order, $Q_n\{x\}$, versus their real argument, x, for $n=0, 1, 2, 3$.

as analog to Eq. (4.716) upon setting $P\{x\}=1-x^2$, $Q\{x\}=l/(l+1)$, and again $w\{x\}=0$.

Legendre's equation in also a particular case of

$$\frac{d}{dx}\left(\left(1-x^2\right)\frac{dy}{dx}\right)+\left(n(n+1)-\frac{m^2}{1-x^2}\right)y=0, \tag{4.719}$$

provided that $m=0$ – see Eq. (4.705); one has classically termed Eq. (4.719) as associated Legendre's equation, where both n (*in lieu* of l) and m denote integer parameters. The general solution to Eq. (4.719) looks like

$$y=\kappa_1 P_n^m\{x\}+\kappa_2 Q_n^m\{x\}, \tag{4.720}$$

where $P_n^m\{x\}$ and $Q_n^m\{x\}$ denote m-th degree, n-th order associated Legendre's polynomials, of first and second kind, respectively. The former can be obtained from regular Legendre's polynomials, as conveyed by Eq. (4.711), using

$$P_n^m\{x\}=(-1)^m\left(1-x^2\right)^{\frac{m}{2}}\frac{d^m P_n\{x\}}{dx^m} \tag{4.721}$$

for recursive expression if m is an integer; when combined with Eq. (4.707), one obtains

$$P_n^m\{x\}=\frac{(-1)^m}{2^n n!}\left(1-x^2\right)^{\frac{m}{2}}\frac{d^{m+n}\left(x^2-1\right)^n}{dx^{m+n}} \tag{4.722}$$

from Eq. (4.721). Several associated Legendre's polynomials of first kind are plotted in Fig. 4.25; it should be emphasized that

$$P_n^0\{x\}=P_n\{x\}, \tag{4.723}$$

which arises directly from Eq. (4.721) after setting $m=0$. Oscillatory patterns within interval [–1,1] are readily grasped upon inspection of Fig. 4.25 – except in the case of $P_n^n\{x\}$; orthogonality is again observed within [–1,1], according to

$$\int_{-1}^{1} P_k^m\{x\}P_n^m\{x\}dx=0, \quad k\neq n, \tag{4.724}$$

complemented by

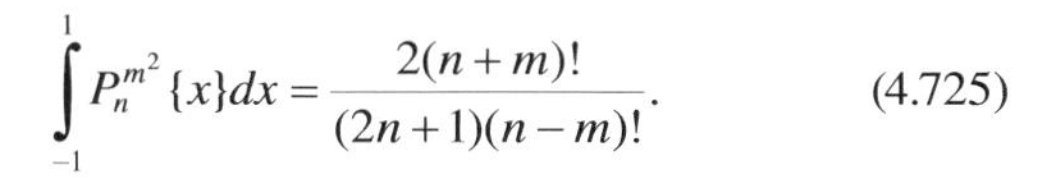

$$\int_{-1}^{1} P_n^{m^2}\{x\}dx=\frac{2(n+m)!}{(2n+1)(n-m)!}. \tag{4.725}$$

On the other hand, one realizes that

$$P_{-n}^m\{x\}=P_{n-1}^m\{x\}, \tag{4.726}$$

whereas a useful recurrence formula reads

$$(n-m+1)P_{n+1}^m\{x\}=(2n+1)xP_n^m\{x\}-(n+m)P_{n-1}^m\{x\} \tag{4.727}$$

that relates $P_{n+1}^m\{x\}$ to $P_n^m\{x\}$ and $P_{n-1}^m\{x\}$; an alternative expression, viz.

$$\sqrt{1-x^2}\,P_n^m\{x\}=\frac{1}{2n+1}\left(P_{n-1}^{m+1}\{x\}-P_{n+1}^{m+1}\{x\}\right), \tag{4.728}$$

permits calculation of $P_n^m\{x\}$ at the expense of (adjacent) $P_{n-1}^{m+1}\{x\}$ and $P_{n+1}^{m+1}\{x\}$. Legendre's polynomials of second kind can be generated via

$$Q_n^m\{x\}=\left(1-x^2\right)^{\frac{m}{2}}\frac{d^m Q_n\{x\}}{dx^m}, \tag{4.729}$$

with integer m; the first examples are plotted in Fig. 4.26, after applying Eq. (4.729) once or twice to Eqs. (4.712)–(4.715). Concave shapes result for Q_n^1, when n is even, whereas odd n turns $x=0$ into an inflection point; the range spanned by Q_n^m becomes larger as m increases. Furthermore, Eq. (4.722) remains valid when $P_n^m\{x\}$ is replaced by $Q_n^m\{x\}$ and $(-1)^m$ removed.

4.3.23.8 Chebyshev's equation

A final germane example of linear second-order equation looks like

$$\left(1-x^2\right)\frac{d^2y}{dx^2}-x\frac{dy}{dx}+n^2y=0, \tag{4.730}$$

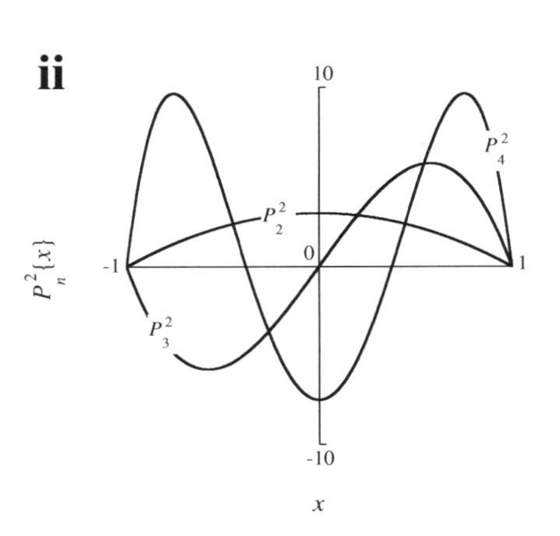

FIGURE 4.25 Variation of associated Legendre's polynomials of first kind and n-th order, and (i) first degree, $P_n^1\{x\}$ for $n=1, 2, 3, 4$, or (ii) second degree, $P_n^2\{x\}$ for $n=2, 3, 4$, versus their real argument, x.

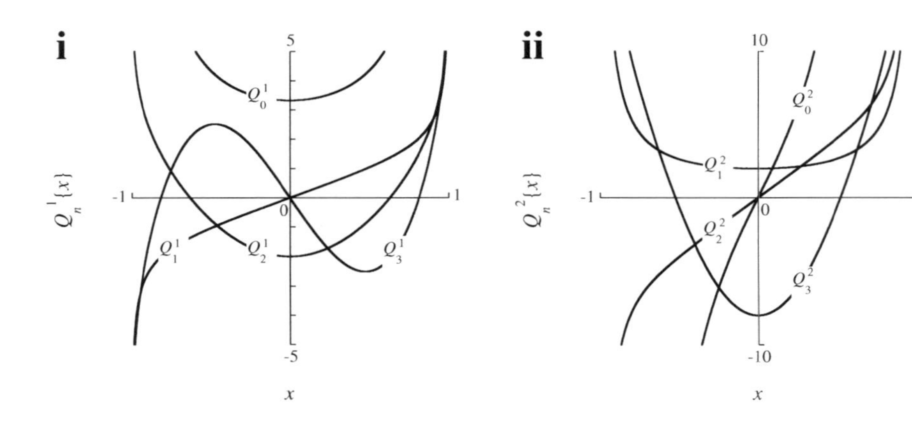

FIGURE 4.26 Variation of associated Legendre's polynomials of second kind and n-th order, and (i) first degree, $Q_n^1\{x\}$, for $n=0, 1, 2, 3$, or (ii) second degree, $Q_n^2\{x\}$, for $n=0, 1, 2, 3$, versus their real argument, x.

with integer n – known as Chebyshev's equation. The (closed-form) analytical solution to Eq. (4.730) looks like

$$y = \kappa_1 T_n\{x\} + \kappa_2 \sqrt{1-x^2}\; U_{n-1}\{x\}, \tag{4.731}$$

which resorts to Chebyshev's polynomials of first kind, $T_n\{x\}$ – defined by

$$T_0\{x\} = 1, \tag{4.732}$$

$$T_1\{x\} = x, \tag{4.733}$$

and

$$T_{n+1}\{x\} = 2xT_n\{x\} - T_{n-1}\{x\} \tag{4.734}$$

in recursive form for $n \geq 1$. Chebyshev's polynomials of second kind, $U_n\{x\}$, satisfy, in turn,

$$U_0\{x\} = 1, \tag{4.735}$$

$$U_1\{x\} = 2x, \tag{4.736}$$

and

$$U_{n+1}\{x\} = 2xU_n\{x\} - U_{n-1}\{x\} \tag{4.737}$$

when $n \geq 1$. The first Chebyshev's polynomials of both kinds are plotted in Fig. 4.27. Except for constant $T_0\{x\}$ and linear $T_1\{x\}$, one realizes that $T_n\{x\}$ oscillates between −1 and 1 when x spans [−1,1], see Fig. 4.27i; a similar pattern is observed with $U_0\{x\}$ (except for the lower vertical intercept) and $U_1\{x\}$, while the amplitude of oscillations of $U_{n\geq 2}\{x\}$ with x increases as x moves farther away from zero in either direction. Explicit expressions for Chebyshev's polynomials, of first and second kind, read

$$T_n\{x\} = n\sum_{i=0}^{n} (-2)^i \frac{(n+i-1)!}{(n-i)!(2i)!}(1-x)^i \tag{4.738}$$

and

$$U_n\{x\} = \sum_{i=0}^{n} (-2)^i \frac{(n+i+1)!}{(n-i)!(2i+1)!}(1-x)^i, \tag{4.739}$$

respectively; whereas the generating function of $T_n\{x\}$ looks like

$$\sum_{n=0}^{\infty} T_n\{x\}t^n = \frac{1-tx}{1-2tx+t^2}, \tag{4.740}$$

and the corresponding function for $U_n\{x\}$ reads

$$\sum_{n=0}^{\infty} U_n\{x\}t^n = \frac{1}{1-2tx+t^2}. \tag{4.741}$$

Cross recurrence relationships have also been derived, namely,

$$T_{n+1}\{x\} = xT_n\{x\} - \left(1-x^2\right)U_{n-1}\{x\} \tag{4.742}$$

– which enables calculation of $T_{n+1}\{x\}$ from $T_n\{x\}$ and $U_{n-1}\{x\}$.

4.3.23.9 Incomplete, nonlinear second order

A second-order, nonlinear differential equation, without explicit terms in y and x, reads

$$\frac{d^2y}{dx^2} + U\{y\}\frac{dy}{dx} = 0, \tag{4.743}$$

with U denoting a generic function of y; in attempts to solve it, one may to advantage revisit Eq. (4.689) as

$$\frac{1}{dx} = \frac{p}{dy}, \tag{4.744}$$

since it permits d^2y/dx^2 be coined as

$$\frac{d^2y}{dx^2} \equiv \frac{d}{dx}\left(\frac{dy}{dx}\right) \equiv \frac{1}{dx}d\left(\frac{dy}{dx}\right) = \frac{p}{dy}d(p) = p\frac{dp}{dy} \tag{4.745}$$

– again with the aid of Eq. (4.689). Insertion of Eqs. (4.689) and (4.745) then transforms Eq. (4.743) to

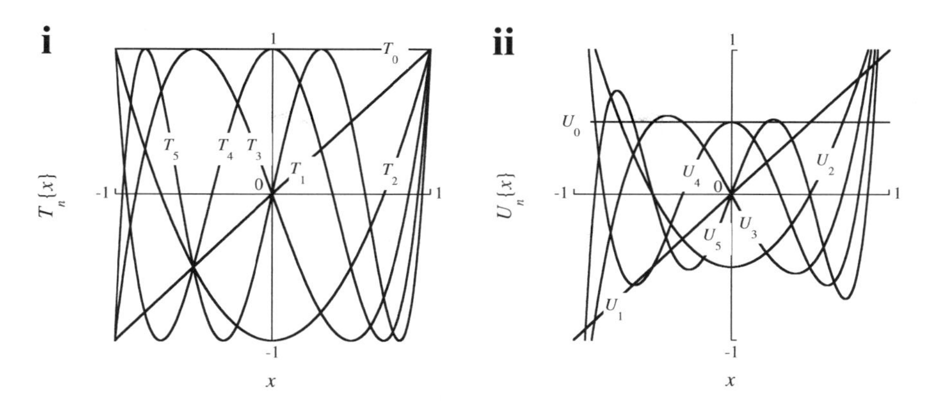

FIGURE 4.27 Variation of (i) Chebyshev's polynomials of first kind and n-th order, $T_n\{x\}$, for $n=0, 1, 2, 3, 4, 5$, and (ii) Chebyshev's polynomials of second kind and n-th order, $U_n\{x\}$, for $n=0, 1, 2, 3, 4, 5$, versus their real argument, x.

$$p\frac{dp}{dy}+U\{y\}p=0, \tag{4.746}$$

where division of both sides by p permits simplification to

$$\frac{dp}{dy}+U\{y\}=0; \tag{4.747}$$

(trivial) integration via separation of variables becomes now possible, viz.

$$p=\kappa_1-\int U\{y\}dy, \tag{4.748}$$

with κ_1 denoting an integration constant. Equation (4.689) may be retrieved once more to rewrite Eq. (7.748) as

$$\frac{dy}{dx}=\kappa_1-\int U\{y\}dy, \tag{4.749}$$

where a second step of integration via separation of variables ensues, viz.

$$x=\kappa_2+\int\frac{dy}{\kappa_1-\int U\{y\}dy}; \tag{4.750}$$

hence, Eq. (4.750) appears as general solution in implicit form, i.e. $x\equiv x\{y\}$, and bearing two arbitrary constants, κ_1 and κ_2.

4.3.23.10 Complete, linear second order

The general form of a nonhomogeneous, linear second-order ordinary differential equation often appears as

$$\frac{d^2y}{dx^2}+P\{x\}\frac{dy}{dx}+Q\{x\}y=S\{x\}, \tag{4.751}$$

where $P\{x\}$, $Q\{x\}$, and $S\{x\}$ denote (univariate) functions of x. Its most general solution takes the form

$$y\{x\}=\kappa_1y_1\{x\}+\kappa_2y_2\{x\}+Y\{x\}, \tag{4.752}$$

where $y_1\{x\}$ and $y_2\{x\}$ denote independent solutions of the associated homogeneous equation, i.e.

$$\frac{d^2y}{dx^2}+P\{x\}\frac{dy}{dx}+Q\{x\}y=0; \tag{4.753}$$

$Y\{x\}$ denotes, in turn, a particular integral of Eq. (4.751) itself, which takes $S\{x\}$ into account – but contains no arbitrary constants. To calculate $Y\{x\}$ once $y_1\{x\}$ and $y_2\{x\}$ have been found, one can start by hypothesizing that

$$Y\{x\}=f_1\{x\}\,y_1\{x\}+f_2\{x\}\,y_2\{x\}; \tag{4.754}$$

note that (arbitrary) constants κ_1 and κ_2 in Eq. (4.752) have been replaced by functions $f_1\{x\}$ and $f_2\{x\}$, respectively – still to be calculated, and able to reflect the presence of $S\{x\}$ in Eq. (4.751). The conditions to be satisfied by those $f_1\{x\}$ and $f_2\{x\}$ should conveniently be set as

$$y_1\frac{df_1}{dx}+y_2\frac{df_2}{dx}=0 \tag{4.755}$$

and

$$\frac{dy_1}{dx}\frac{df_1}{dx}+\frac{dy_2}{dx}\frac{df_2}{dx}=S\{x\}. \tag{4.756}$$

Recalling that the Wronskian determinant, $|\mathbf{W}|$, should satisfy

$$|\mathbf{W}|\equiv\begin{vmatrix} y_1 & y_2 \\ \frac{dy_1}{dx} & \frac{dy_2}{dx}\end{vmatrix}\neq 0 \tag{4.757}$$

so as to guarantee that $y_1\{x\}$ and $y_2\{x\}$ are independent solutions to Eq. (4.753), application of Cramer's rule is in order – to accordingly get

$$\frac{df_1}{dx}=\frac{\begin{vmatrix} 0 & y_2 \\ S\{x\} & \frac{dy_2}{dx}\end{vmatrix}}{\begin{vmatrix} y_1 & y_2 \\ \frac{dy_1}{dx} & \frac{dy_2}{dx}\end{vmatrix}} \tag{4.758}$$

and

$$\frac{df_2}{dx} = \frac{\begin{vmatrix} y_1 & 0 \\ \frac{dy_1}{dx} & S\{x\} \end{vmatrix}}{\begin{vmatrix} y_1 & y_2 \\ \frac{dy_1}{dx} & \frac{dy_2}{dx} \end{vmatrix}} \tag{4.759}$$

from Eqs. (4.755) and (4.756), in parallel to Eq. (4.294). Calculation of the outstanding second-order determinants, followed by integration of df_1/dx and df_2/dx thus obtained, via separation of variables, produce

$$f_1\{x\} = B_1 - \int \frac{y_2 S\{x\}}{y_1 \frac{dy_2}{dx} - y_2 \frac{dy_1}{dx}} dx \tag{4.760}$$

from Eq. (4.758), and similarly

$$f_2\{x\} = B_2 + \int \frac{y_1 S\{x\}}{y_1 \frac{dy_2}{dx} - y_2 \frac{dy_1}{dx}} dx \tag{4.761}$$

from Eq. (4.759) – where B_1 and B_2 denote true (arbitrary) constants. Upon insertion of Eqs. (4.760) and (4.761), one gets

$$y\{x\} = K_1 y_1\{x\} + K_2 y_2\{x\} + \vartheta\{x\} \tag{4.762}$$

from Eqs. (4.752) and (4.754) – as long as the new arbitrary constants, K_1 and K_2, are defined by

$$K_1 \equiv \kappa_1 + B_1 \tag{4.763}$$

and

$$K_2 \equiv \kappa_2 + B_2, \tag{4.764}$$

respectively, and the new form for the particular integral, $Y\{x\}$, satisfies

$$\vartheta\{x\} \equiv y_2 \int \frac{y_1 S\{x\}}{y_1 \frac{dy_2}{dx} - y_2 \frac{dy_1}{dx}} dx - y_1 \int \frac{y_2 S\{x\}}{y_1 \frac{dy_2}{dx} - y_2 \frac{dy_1}{dx}} dx; \tag{4.765}$$

note the presence of two arbitrary constants in the general solution to Eq. (4.753), as conveyed by Eq. (4.762).

4.3.23.11 Constant-coefficient, linear higher order

An n-th order, linear differential equation – where all coefficients, a_1, a_2, ..., a_n, are constant and an independent term is not present, looks like

$$a_n \frac{d^n y}{dx^n} + a_{n-1} \frac{d^{n-1} y}{dx^{n-1}} + \ldots + a_1 \frac{dy}{dx} + a_0 y = 0; \tag{4.766}$$

a solution of the type

$$y = e^{\lambda x} \tag{4.767}$$

is accordingly to be postulated, since the derivative of an exponential function retains its form. Sequential differentiation of Eq. (4.767) indeed yields

$$\frac{d^i y}{dx^i} = \lambda^i e^{\lambda x}; \quad i = 1, 2, \ldots, n, \tag{4.768}$$

which may be used to transform Eq. (4.766) to

$$\left(a_n \lambda^n + a_{n-1} \lambda^{n-1} + \ldots + a_1 \lambda + a_0\right) e^{\lambda x} = 0 \tag{4.769}$$

– once $e^{\lambda x}$ has been factored out. Equation (4.769) will not be universally valid unless

$$a_n \lambda^n + a_{n-1} \lambda^{n-1} + \ldots + a_1 \lambda + a_0 = 0, \tag{4.770}$$

since $e^{\lambda x} \neq 0$; Eq. (4.770) has classically been termed characteristic equation. Since n roots are expected for the said polynomial, the final solution to Eq. (4.766) will be of the form

$$y = \sum_{i=1}^{n} \kappa_i e^{\lambda_i x} \tag{4.771}$$

– provided that all λ_is are distinct from each other; here the κ_is denote n arbitrary constants. If a root exhibits multiplicity $m > 1$, i.e.

$$\lambda_1 = \lambda_2 = \ldots = \lambda_m = \lambda \tag{4.772}$$

(with $m \leq n$), then the corresponding solutions $A_1 \exp\{\lambda x\}$, $A_2 \exp\{\lambda x\}$, ..., $A_m \exp\{\lambda x\}$ will obviously be dependent – for being proportional to each other; under such circumstances, the corresponding m (independent) solutions under scrutiny should appear in lumped form as

$$y = \xi\{x\} e^{\lambda x} \tag{4.773}$$

– where $\xi\{x\}$ denotes a putative function of x, containing m integration constants still to be determined. The functional form of $\xi\{x\}$ compatible with Eqs. (4.768) and (4.771) reads

$$\xi\{x\} \equiv \sum_{i=1}^{m} \kappa_i x^{m-i}; \tag{4.774}$$

such a polynomial function may be inserted in Eq. (4.773), and the result combined with Eq. (4.771) referred only to the outstanding n–m terms, viz.

$$y = \sum_{i=1}^{m} \kappa_i x^{m-i} e^{\lambda x} + \sum_{j=m+1}^{n} \kappa_j e^{\lambda_j x} \tag{4.775}$$

– with the other roots $\lambda_j \neq \lambda$ of Eq. (4.770) having been hypothesized as single type ones.

4.3.23.12 Euler's equation

If every coefficient a_i in Eq. (4.766) is multiplied by x^i, then one obtains

$$a_n x^n \frac{d^n y}{dx^n} + a_{n-1} x^{n-1} \frac{d^{n-1} y}{dx^{n-1}} + \ldots + a_1 x \frac{dy}{dx} + a_0 y = 0 \tag{4.776}$$

– which remains an n-th order homogeneous, linear differential equation, classically known as Euler's equation; in this case,

$$x \equiv e^z \tag{4.777}$$

is recommended for change of independent variable. Every derivative of y with regard to x may then be calculated via the chain differentiation rule, see Eq. (4.341), using z as intermediate variable – coupled with the rule of differentiation of the inverse function, as per Eq. (4.340); one will eventually attain

$$x^i \frac{d^i y}{dx^i} = \prod_{j=0}^{i-1} \left(\frac{d}{dz} - j \right) y \tag{4.778}$$

in (condensed) operator notation, so an equation will result from Eq. (4.776) that is formally equivalent to Eq. (4.766). The corresponding solution will accordingly mimic Eq. (4.771), where $y \equiv y\{z\}$; after redoing Eq. (4.777) to

$$z = \ln x \tag{4.779}$$

following application of logarithms to both sides, one obtains

$$y = \sum_{i=1}^{n} \kappa_i e^{\lambda_i \ln x} \tag{4.780}$$

from Eq. (4.771). After bringing Eq. (4.36) on board, Eq. (4.780) will give rise to

$$y = \sum_{i=1}^{n} \kappa_i \exp\left\{ \ln x^{\lambda_i} \right\} \tag{4.781}$$

– which degenerates to

$$y = \sum_{i=1}^{n} \kappa_i x^{\lambda_i}, \tag{4.782}$$

with the aid of Eq. (4.34).

4.4 Statistics

4.4.1 Discrete probability distributions

Discrete events, carried out within populations of finite size, are of wide practical interest – classically in terms of their probability of occurrence; hypergeometric and binomial distributions account for the two most representative descriptors thereof.

The former is a discrete probability distribution, describing the probability of k successes (i.e. random draws for which the object has a specific feature) in n draws without replacement – from a finite population of size N containing K objects with the said feature, wherein each draw is either a success or a failure. The associated probability function, $P\{k\}$, reads

$$P\{k\} = \frac{\begin{pmatrix} K \\ k \end{pmatrix} \begin{pmatrix} N-K \\ n-k \end{pmatrix}}{\begin{pmatrix} N \\ n \end{pmatrix}} \tag{4.783}$$

– and is characterized by

$$\mu = n \frac{K}{N} \tag{4.784}$$

for mean (acting as measure of position), and

$$\sigma^2 = n \frac{K}{N} \frac{N-K}{N} \frac{N-n}{N-1} \tag{4.785}$$

for variance (serving as measure of dispersion).

Conversely, the binomial distribution describes the probability of k successes in n draws with replacement; in this case, the probability function looks like

$$P\{k\} = \begin{pmatrix} n \\ k \end{pmatrix} \left(\frac{K}{N} \right)^k \left(1 - \frac{K}{N} \right)^{n-k}, \tag{4.786}$$

with mean given by

$$\mu = n \frac{K}{N} \tag{4.787}$$

– and variance reading

$$\sigma^2 = n \frac{K}{N} \left(1 - \frac{K}{N} \right). \tag{4.788}$$

Since K/N and $1 - K/N$ represent probability of success, p, and failure, q, respectively, Eq. (4.786) is often rewritten as

$$P\{k\} = \begin{pmatrix} n \\ k \end{pmatrix} p^k q^{n-k}; \tag{4.789}$$

hence, Eqs. (4.787) and (4.788) will show up as

$$\mu = np \tag{4.790}$$

and

$$\sigma^2 = npq, \tag{4.791}$$

respectively.

4.4.2 Absolute and centered moments

Most statistical distributions of practical interest in process engineering have, however, classically been hypothesized as continuous, in view of the large size of their source populations. Furthermore, regular hypothesis testing – one of the nuclear goals in statistical analysis, requires comparison between continuous distributions from the same or distinct populations. Besides characterizing the shape of a (continuous) probability distribution, $D\{X\}$, of random variable X taking values x, its moments are useful toward the said comparison; in fact, a necessary condition for two distributions be identical is that they share the same moment sequence. Absolute moments satisfy

$$\mu_i \equiv E\left\{ X^i \right\} \equiv \int_{-\infty}^{\infty} x^i D\{x\} dx; \quad i = 1, 2, \ldots, \tag{4.792}$$

where μ_i denotes i-th moment; μ_1 is usually known as mean, or most likely value, i.e.

$$\mu_1 \equiv \mu \equiv E\{X\} \equiv \int_{-\infty}^{\infty} x\,D\{x\}\,dx. \tag{4.793}$$

Centered moments, $\mu_{i,ctr}$, are of no smaller interest than their absolute counterparts; they are defined as

$$\mu_{i,ctr} \equiv \int_{-\infty}^{\infty} \left(x - E\{X\}\right)^i D\{x\}\,dx; \quad i = 1, 2, \ldots, \tag{4.794}$$

and become accessible once $E\{X\}$ is provided. When i is set equal to unity, Eq. (4.794) degenerates to

$$\begin{aligned}\mu_{1,ctr} &= \int_{-\infty}^{\infty} \left(x - E\{X\}\right) D\{x\}\,dx \\ &= \int_{-\infty}^{\infty} x\,D\{x\}\,dx - \int_{-\infty}^{\infty} E\{X\}D\{x\}\,dx \\ &= \int_{-\infty}^{\infty} xD\{x\}\,dx - E\{X\}\int_{-\infty}^{\infty} D\{x\}\,dx\end{aligned} \tag{4.795}$$

following expansion of the integral as per Eq. (4.467) – coupled with realization that $E\{X\}$ is independent of integration variable x, in agreement with Eq. (4.793); insertion of Eq. (4.793), combined with realization that

$$\int_{-\infty}^{\infty} D\{x\}\,dx = 1 \tag{4.796}$$

is a necessary outcome of the concept of probability, produce the trivial result

$$\mu_{1,ctr} = E\{X\} - E\{X\}1 = 0 \tag{4.797}$$

from Eq. (4.795). When $i=2$, then Eq. (4.794) would turn to

$$\mu_{2,ctr} \equiv \sigma^2 \equiv Var\{X\} \equiv \int_{-\infty}^{\infty} \left(x - E\{X\}\right)^2 D\{x\}\,dx \tag{4.798}$$

– which describes the second centered moment, also denoted by σ^2; Eqs. (4.113) and (4.467) may then be invoked to transform Eq. (4.798) to

$$\begin{aligned}\mu_{2,ctr} &= \int_{-\infty}^{\infty} \left(x^2 - 2xE\{X\} + E^2\{X\}\right) D\{x\}\,dx \\ &= \int_{-\infty}^{\infty} x^2 D\{x\}\,dx - 2\int_{-\infty}^{\infty} xE\{X\}D\{x\}\,dx + \int_{-\infty}^{\infty} E^2\{X\}D\{x\}\,dx,\end{aligned} \tag{4.799}$$

along with the distributive property. In view of the constancy of $E\{X\}$, and thus of its square, as per Eq. (4.793), they can be taken off the corresponding kernels – so Eq. (4.799) becomes

$$\mu_{2,ctr} = \int_{-\infty}^{\infty} x^2 D\{x\}\,dx - 2E\{X\}\int_{-\infty}^{\infty} xD\{x\}\,dx + E^2\{X\}\int_{-\infty}^{\infty} D\{x\}\,dx; \tag{4.800}$$

after recalling Eqs. (4.792), (4.793), and (4.796), one will be able to rewrite Eq. (4.800) as

$$\mu_{2,ctr} = \mu_2 - 2\mu_1 \int_{-\infty}^{\infty} x\,D\{X\}\,dx + \mu_1^2 1, \tag{4.801}$$

where combination with Eq. (4.793) leaves

$$\mu_{2,ctr} = \mu_2 - 2\mu_1\mu_1 + \mu_1^2 = \mu_2 - \mu_1^2. \tag{4.802}$$

Equation (4.802) is equivalent to

$$\sigma^2 = \mu_2 - \mu^2, \tag{4.803}$$

after taking Eqs. (4.793) and (4.798) into account; therefore, variance is easily expressed as a linear combination of the first- and second-order (absolute) moments of the underlying distribution.

4.4.3 Moment-generating function

Equality of μ_1, μ_2, …, as given in Eq. (4.792), does not consubstantiate a sufficient condition for coincidence of statistical distributions; a necessary and sufficient condition requires, in fact, the same moment-generating function, $G\{s\}$, defined as

$$G\{s\} \equiv E\left\{e^{sX}\right\} \tag{4.804}$$

– which should exist, and be finite, continuous, and differentiable. The derivative of $G\{s\}$ with regard to s reads

$$\frac{dG\{s\}}{ds} \equiv \frac{d}{ds}\left(E\left\{e^{sX}\right\}\right), \tag{4.805}$$

in view of Eq. (4.804); this is equivalent to writing

$$\frac{d}{ds}\left(E\left\{e^{sX}\right\}\right) \equiv \frac{d}{ds}\int_{-\infty}^{\infty} e^{sx} D\{x\}\,dx = \int_{-\infty}^{\infty} \frac{d\,e^{sx}}{ds} D\{x\}\,dx, \tag{4.806}$$

at the expense of Eqs. (4.473) and (4.793). Recalling Eq. (4.341) and the first entry to Table 4.6, one can rewrite Eq. (4.806) as

$$\frac{dG\{s\}}{ds} = \int_{-\infty}^{\infty} x e^{sx} D\{x\}\,dx \equiv E\left\{X e^{sX}\right\} \tag{4.807}$$

– which reduces to

$$\left.\frac{dG\{s\}}{ds}\right|_{s=0} = \int_{-\infty}^{\infty} xD\{x\}\,dx \equiv E\{X\}, \tag{4.808}$$

after setting $s=0$. The second derivative of $G\{s\}$ with regard to s may likewise be obtained by differentiating Eq. (4.807) as

$$\frac{d^2G\{s\}}{ds^2} \equiv \frac{d}{ds}\left(E\{X e^{sX}\}\right) \equiv \frac{d}{ds}\left(\int_{-\infty}^{\infty} x e^{sx} D\{x\} dx\right)$$
$$= \int_{-\infty}^{\infty} x \frac{d e^{sx}}{ds} D\{x\} dx = \int_{-\infty}^{\infty} x^2 e^{sx} D\{x\} dx \equiv E\{X^2 e^{sX}\}, \tag{4.809}$$

obtained again with the aid of Eqs. (4.473) and (4.792); after setting $s=0$, one gets

$$\left.\frac{d^2G\{s\}}{ds^2}\right|_{s=0} = \int_{-\infty}^{\infty} x^2 D\{x\} dx \equiv E\{X^2\}. \tag{4.810}$$

Based on Eq. (4.792), one will rewrite Eqs. (4.808) and (4.810) as

$$\mu_1 = \left.\frac{dG\{s\}}{ds}\right|_{s=0} \tag{4.811}$$

and

$$\mu_2 = \left.\frac{d^2G\{s\}}{ds^2}\right|_{s=0}, \tag{4.812}$$

respectively; generalization of the above reasoning supports

$$\mu_i = \left.\frac{d^iG\{s\}}{ds^i}\right|_{s=0} \quad ; \quad i = 1, 2, \ldots, \tag{4.813}$$

classically known as van der Laan's theorem – thus justifying the designation of moment-generating function given to $G\{s\}$.

4.4.4 Normal distribution

4.4.4.1 Probability density function of population

Owing in part to its shape and properties, the normal distribution is the one most frequently utilized to describe phenomena that are well modeled by a continuous random variable X; it accordingly reads

$$N\{X;\mu,\sigma\} \equiv \frac{1}{\sqrt{2\pi}\sigma} \exp\left\{-\frac{1}{2}\left(\frac{x-\mu}{\sigma}\right)^2\right\}, \tag{4.814}$$

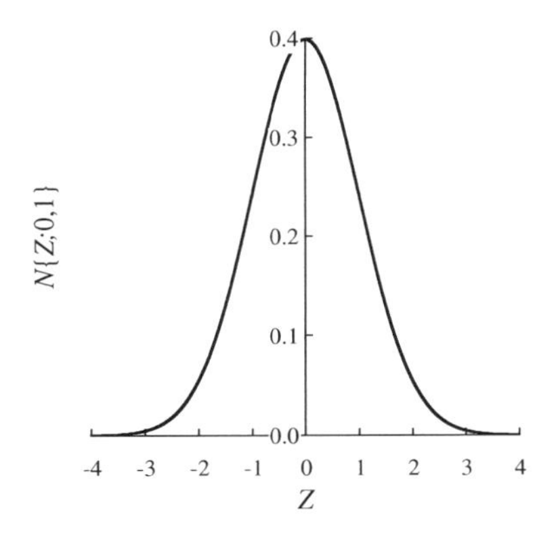

FIGURE 4.28 Variation of standard normal probability density function, $N\{Z;1,0\}$, versus its random argument, Z.

and bears μ and σ as constitutive parameters – representing mean and standard deviation (or square root of variance), respectively. Equation (4.814) can be coined in differential form as

$$dN\{X;\mu,\sigma\} = \frac{1}{\sqrt{2\pi}\sigma} \exp\left\{-\frac{1}{2}\left(\frac{x-\mu}{\sigma}\right)^2\right\} dx; \tag{4.815}$$

dN accordingly denotes probability that X lies within elementary interval $[x,x+dx]$, where x denotes a particular value attributed to variable X. Although useful, Eq. (4.814) defines as many Gaussian distributions as there are combinations of parameters μ and σ; therefore, it is more convenient to define a lumped, random variable Z as

$$z \equiv \frac{x-\mu}{\sigma}, \tag{4.816}$$

and consequently

$$dx = \sigma dz \tag{4.817}$$

– for its allowing reformulation of Eq. (4.815) to

$$dN\{Z\} = \frac{1}{\sqrt{2\pi}\sigma} \exp\left\{-\frac{1}{2}z^2\right\} \sigma dz. \tag{4.818}$$

After cancelling σ between numerator and denominator in the right-hand side, Eq. (4.818) reduces to

$$dN\{Z\} = \frac{1}{\sqrt{2\pi}} \exp\left\{-\frac{1}{2}z^2\right\} dz. \tag{4.819}$$

Consequently, one may replace Eq. (4.814) by

$$N\{Z\} = \frac{1}{\sqrt{2\pi}} \exp\left\{-\frac{1}{2}z^2\right\}, \tag{4.820}$$

known as normalized (or standard) normal distribution; oftentimes, Eq. (4.820) is quoted as

$$N\{Z;0,1\} = \frac{1}{\sqrt{2\pi}} \exp\left\{-\frac{z^2}{2}\right\}, \tag{4.821}$$

since

$$N\{Z;0,1\} \equiv N\{X;\mu,\sigma\}\Big|_{\substack{\mu=0\\ \sigma=1}} \tag{4.822}$$

– consistent with Eqs. (4.814) and (4.816). In this fashion, all possible normal distributions collapse onto a single curve – with nil mean and unit variance; the graphical representation of $N\{Z;0,1\}$ is provided in Fig. 4.28. Its bell-shaped curve is to be outlined – symmetrical relative to the vertical axis; as well as its quickly decreasing tails when $z \to \pm\infty$.

4.4.4.2 Probability density function of sample

Since it is typically impractical to work with, or test a whole population, one must resort to only a few portions thereof, or samples; hence, the issue arises as to how said samples behave compared to their source population. Toward this goal, one should first remember that

$$E\{a+bX\}=a+bE\{X\} \tag{4.823}$$

and

$$Var\{a+bX\}=b^2Var\{X\}, \tag{4.824}$$

respectively, based on definition of expected value, E, as per Eq. (4.793), and variance, Var, as per Eq. (4.798) – where a and b denote constants; Eqs. (4.823) and (4.824) apply to a linear combination involving a normally-distributed, random variable X. Consider now a sample, of size n, taken at random from the said normal population – with values $x_1, x_2, \ldots, x_i, \ldots, x_n$ for its elements; the corresponding average, $\bar{x}$, will thus be given by

$$\bar{x}=\frac{\sum_{i=1}^{n}x_i}{n} \tag{4.825}$$

in agreement with Eq. (4.5), holding n degrees of freedom. The expected value of the associated random variable materializing sample average, $\bar{X}$, will thus abide to

$$\begin{aligned} m\equiv E\{\bar{X}\}&=E\left\{0+\frac{1}{n}\sum_{i=1}^{n}X_i\right\}=0+\frac{1}{n}\sum_{i=1}^{n}E\{X_i\} \\ &=\frac{1}{n}\sum_{i=1}^{n}E\{X\}=\frac{1}{n}nE\{X\}=E\{X\}\equiv\mu, \end{aligned} \tag{4.826}$$

since $E\{X\}$ is the mean shared by all X_is; $a=0$ and $b=1/n$ were conveniently set in Eq. (4.823), while Eq. (4.825) was used for definition of $\bar{X}$. Inspection of Eq. (4.826) indicates that the mean of the distribution of sample averages, μ, coincides with the mean of the original population. By the same token, one may retrieve Eq. (4.824), again with $a=0$ and $b=1/n$, to obtain

$$\begin{aligned} s^2\equiv Var\{\bar{X}\}&=Var\left\{0+\frac{1}{n}\sum_{i=1}^{n}X_i\right\} \\ &=\left(\frac{1}{n}\right)^2\sum_{i=1}^{n}Var\{X_i\}=\left(\frac{1}{n}\right)^2\sum_{i=1}^{n}Var\{X\} \\ &=\left(\frac{1}{n}\right)^2 nVar\{X\}=\frac{Var\{X\}}{n}\equiv\frac{\sigma^2}{n} \end{aligned} \tag{4.827}$$

as long as $Var\{X_i\}$ coincides with $Var\{X\}$ for every i – with the help of Eqs. (4.798) and (4.825). In other words, the variance of the distribution of sample averages, s^2, is lower than the true variance of the population, σ^2, by corrective factor $1/n$.

When one is interested only in random variable X (rather than $\bar{X}$), σ^2 pertaining to the whole population is seldom known. In this case, one usually resorts to $\hat{s}^2$, estimated via

$$\hat{s}^2\equiv\frac{\sum_{i=1}^{n}(x_i-\bar{x})^2}{n-1} \tag{4.828}$$

as approximant to σ^2, with $\bar{x}$ given by Eq. (4.825); where the number of degrees of freedom was reduced to $n-1$, on account of the degree of freedom removed by calculation (and utilization) of $\bar{x}$ from the n available x_is.

4.4.4.3 Central limit theorem

There are several relevant (continuous) statistical distributions, besides the normal distribution; however, the central limit theorem states that the behavior of a sum of M random variables, say X_i ($i=1, 2, \ldots, M$), is driven by

$$\left.\sum_{i=1}^{M}X_i\right|_{M\to\infty}\sim N\left\{X;M\mu,\sqrt{M}\sigma\right\} \tag{4.829}$$

when M becomes sufficiently large – irrespective of the original distribution of the X_is, provided that their variances are finite (as is usually the case) and they share the same individual mean, μ, and variance, σ^2/M, see Eqs. (4.825) and (4.827), respectively. The new average, $M\mu$, and variance, $M\sigma^2 = M^2(\sigma^2/M)$, of linear combination $\sum_{i=1}^{M}x_i$ mimic Eqs. (4.823) and (4.824), respectively – even though these equations are strictly applicable only to normally-distributed variables. The central limit theorem constitutes one of the most notable developments in statistical science – and ultimately justifies the universal character of the conclusions based on normal (or Gaussian) distribution(s), and thus the systematic use of Eqs. (4.821) for reference distribution.

4.4.4.4 Properties

Once in possession of Eqs. (4.793) and (4.804), one can obtain the moment-generating function for a normal distribution, of mean μ and variance σ^2, as

$$G_N\{s\}\equiv\int_{-\infty}^{\infty}e^{sx}N\{X;\mu,\sigma\}dx; \tag{4.830}$$

insertion of Eq. (4.814) gives then rise to

$$G_N\{s\}=\frac{1}{\sqrt{2\pi}}\int_{-\infty}^{\infty}\exp\left\{sx-\frac{1}{2}\left(\frac{x-\mu}{\sigma}\right)^2\right\}\frac{dx}{\sigma}, \tag{4.831}$$

upon taking contants off the kernel and condensing exponential functions. Should Eqs. (4.816) and (4.817) be taken advantage of, one may redo Eq. (4.831) to

$$\begin{aligned} G_N\{s\}&=\frac{1}{\sqrt{2\pi}}\int_{-\infty}^{\infty}\exp\left\{s(\mu+\sigma z)-\frac{1}{2}z^2\right\}dz \\ &=\frac{1}{\sqrt{2\pi}}\int_{-\infty}^{\infty}\exp\left\{\mu s+\sigma sz-\frac{z^2}{2}\right\}dz \\ &=\frac{e^{\mu s}}{\sqrt{2\pi}}\int_{-\infty}^{\infty}\exp\left\{\frac{2\sigma sz-z^2}{2}\right\}dz \end{aligned} \tag{4.832}$$

– also due to constancy of $e^{\mu s}$, and with the aid of Eq. (4.25). Addition and subtraction of $(\sigma s)^2/2$ to the argument of the exponential function in the kernel of Eq. (4.832), followed by factoring out of $\exp\{(\sigma s)^2/2\}$ give rise to

$$G_N\{s\}=\frac{e^{\mu s}}{\sqrt{2\pi}}\exp\left\{\frac{(\sigma s)^2}{2}\right\}\int_{-\infty}^{\infty}\exp\left\{-\frac{z^2-2\sigma sz+(\sigma s)^2}{2}\right\}dz \tag{4.833}$$

– where application of Newton's binomial theorem as per Eq. (4.109), followed by lumping of exponential functions permit condensation to

$$G_N\{s\} = \frac{1}{\sqrt{2\pi}} \exp\left\{\mu s + \frac{(\sigma s)^2}{2}\right\} \int_{-\infty}^{\infty} \exp\left\{-\frac{(z-\sigma s)^2}{2}\right\} d(z-\sigma s); \tag{4.834}$$

remember that $d(z-\sigma s)$ coincides with dz, see Eq. (4.319). It can be proven that

$$\int_{-\infty}^{\infty} \exp\left\{-\frac{x^2}{2}\right\} dx = \sqrt{2\pi}, \tag{4.835}$$

so Eq. (4.834) simplifies to

$$G_N\{s\} = \exp\left\{\mu s + \frac{1}{2}(\sigma s)^2\right\} \tag{4.836}$$

after replacing x by $z - \sigma s$, and then dropping $\sqrt{2\pi}$ from both numerator and denominator; Eq. (4.836) entails the final form for the moment-generating function of $N\{X; \mu, \sigma\}$. Combination of Eqs. (4.811) and (4.836) generates

$$\begin{aligned} \mu_{1,N} &= \left.\frac{dG_N\{s\}}{ds}\right|_{s=0} = \exp\left\{\mu s + \frac{1}{2}(\sigma s)^2\right\}\left(\mu + \frac{1}{2}2(\sigma s)\sigma\right)\Bigg|_{s=0} \\ &= \left(\mu + \sigma^2 s\right)\exp\left\{\mu s + \frac{1}{2}(\sigma s)^2\right\}\Bigg|_{s=0} \end{aligned} \tag{4.837}$$

– where the rule of differentiation of a composite exponential function was recalled; Eq. (4.837) reduces to

$$\left(E\{X\} \equiv \mu_{1,N} = \mu\right)\Big|_{X\sim N\{X;\mu,\sigma\}} \tag{4.838}$$

at the expense of Eq. (4.793), and realization that $\mu+0$ equals μ and e^0 equals unity – so parameter μ represents indeed the mean of the normal distribution $N\{X;\mu,\sigma\}$. Further differentiation of dG_N/ds, as obtained from Eq. (4.836), with regard to s gives rise to

$$\begin{aligned} \mu_{2,N} &= \left.\frac{d^2G_N\{s\}}{ds^2}\right|_{s=0} = \frac{d}{ds}\left(\frac{d}{ds}\left(\exp\left\{\mu s + \frac{1}{2}(\sigma s)^2\right\}\right)\right)\Bigg|_{s=0} \\ &= \frac{d}{ds}\left(\left(\mu + \sigma^2 s\right)\exp\left\{\mu s + \frac{1}{2}(\sigma s)^2\right\}\right)\Bigg|_{s=0} \\ &= \left(\begin{aligned} &\sigma^2 \exp\left\{\mu s + \frac{1}{2}(\sigma s)^2\right\} \\ &+\left(\mu + \sigma^2 s\right)\exp\left\{\mu s + \frac{1}{2}(\sigma s)^2\right\}\left(\mu + \frac{1}{2}2(\sigma s)\sigma\right) \end{aligned}\right)\Bigg|_{s=0} \\ &= \left(\sigma^2 \exp\left\{\mu s + \frac{1}{2}(\sigma s)^2\right\} + \left(\mu + \sigma^2 s\right)^2 \exp\left\{\mu s + \frac{1}{2}(\sigma s)^2\right\}\right)\Bigg|_{s=0}, \end{aligned} \tag{4.839}$$

where Eq. (4.812) was brought onboard; after setting $s=0$ as indicated, Eq. (4.839) reduces to

$$\mu_{2,N} = \sigma^2 + \mu^2. \tag{4.840}$$

Upon insertion of Eqs. (4.798), (4.838) and (4.840), one obtains

$$Var\{X\} \equiv \mu_{2,N,ctr} = \left(\sigma^2 + \mu^2\right) - \mu^2 \tag{4.841}$$

from Eq. (4.803); transformation of Eq. (4.841) is now possible to

$$\left(Var\{X\} \equiv \mu_{2,N,ctr} = \sigma^2\right)\Big|_{X\sim N\{X;\mu,\sigma\}} \tag{4.842}$$

– so the variance of $N\{X;\mu,\sigma\}$ coincides, in fact, with σ^2 (as expected).

4.4.4.5 Probability cumulative function of population

The normal cumulative probability function (in standard version), up to value z of random variable Z, is defined as

$$P_N\{Z < z\} \equiv \frac{1}{\sqrt{\pi}} \int_{-\infty}^{z} \exp\left\{-\frac{\tilde{z}^2}{2}\right\} \frac{d\tilde{z}}{\sqrt{2}}, \tag{4.843}$$

stemming directly from Eq. (4.820) – where $1/\sqrt{2}$ was moved into the kernel for convenience; after setting

$$y \equiv \frac{z}{\sqrt{2}}, \tag{4.844}$$

and thus

$$dy = \frac{dz}{\sqrt{2}}, \tag{4.845}$$

one can redo Eq. (4.843) to

$$P_N\{Y < y\} = \frac{1}{\sqrt{\pi}} \int_{-\infty}^{y} e^{-\tilde{y}^2} d\tilde{y} \tag{4.846}$$

since $y \to -\infty$ when $z \to -\infty$ as per Eq. (4.844) – with $e^{-y^2}/\sqrt{\pi}$ illustrated in Fig. 4.29i, with area below the curve between $-\infty$ and y_1 representing $P_N\{Y<y_1\}$. Inspection of this figure consequently indicates that area α below the right tail of the normal probability density distribution abides to

$$\alpha = 1 - P_N\{Y < y_1\} = 1 - \frac{1}{\sqrt{\pi}} \int_{-\infty}^{y_1} e^{-\tilde{y}^2} d\tilde{y}, \tag{4.847}$$

with the aid of Eqs. (4.796) and (4.846); here y_1 represents the maximum value of Y after which the alternative hypothesis $Y>0$ holds, relative to the base hypothesis $Y=0$, at the α level of significance. By the same token,

$$P_N\{-y < Y < y\} = \frac{1}{\sqrt{\pi}} \int_{-y}^{y} e^{-\tilde{y}^2} d\tilde{y} \tag{4.848}$$

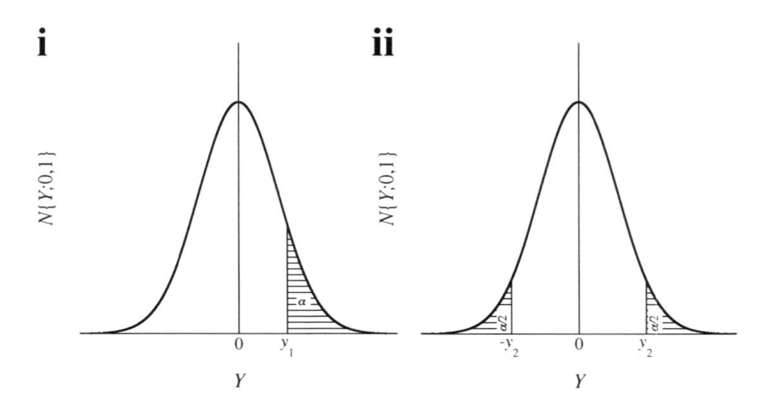

FIGURE 4.29 Variation of standard normal probability density function, $N\{Y;1,0\}$, versus its random argument, Y – with indication of (i) area α below curve beyond y_1, and (ii) area $\alpha/2$ below curve prior to $-y_2$ or beyond y_2 (with $y_2 > y_1$).

consistent with Eq. (4.846) – where the even nature of Gauss' distribution enforces

$$P_N\{-y < Y < y\} = \frac{2}{\sqrt{\pi}} \int_0^y e^{-\tilde{y}^2} \, d\tilde{y}; \tag{4.849}$$

the error function of x, or erf$\{x\}$, is accordingly defined as

$$\text{erf}\{x\} \equiv \frac{2}{\sqrt{\pi}} \int_0^x e^{-\tilde{x}^2} \, d\tilde{x}, \tag{4.850}$$

and is plotted in Fig. 4.30. In view of Eq. (4.850), one can rewrite Eq. (4.849) as

$$P_N\{-y < Y < y\} = \text{erf}\{y\}, \tag{4.851}$$

and thus

$$\begin{aligned} \alpha = \frac{\alpha}{2} + \frac{\alpha}{2} &= 1 - P_N\{-y_2 < Y < y_2\} \\ &= 1 - \text{erf}\{y_2\} \equiv \text{erfc}\{y_2\} \end{aligned} \tag{4.852}$$

– where erfc$\{x\}$ denotes complementary error function of x, see also Fig. 4.30; an illustration of Eq. (4.852) is provided in Fig. 4.29ii. In other words, y_2 now represents the maximum value of Y beyond which the alternative hypothesis $Y \neq 0$ holds, relative to the base hypothesis $Y = 0$, at the α level of significance.

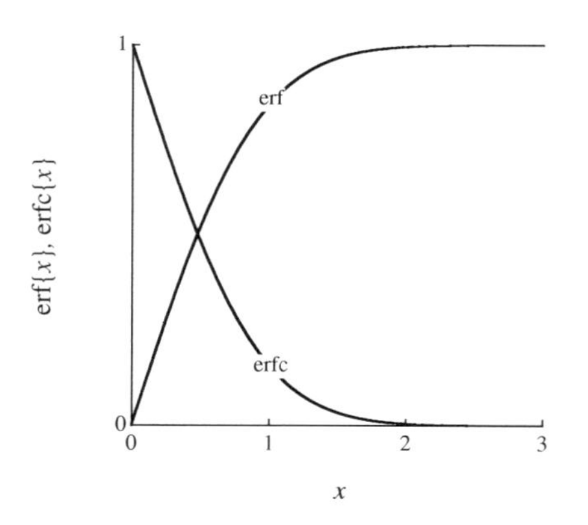

FIGURE 4.30 Variation of error function, erf$\{x\}$, and complementary error function, erfc$\{x\}$, versus their real argument, x.

4.4.5 Linear regression

Although μ and σ^2 have proven relevant parameters when attempting to characterize (normally distributed) populations at some state, they typically convey poor quantitative information on their own should some processing condition be changed – as is classically the focus of food process engineering. Models are thus to be postulated, in that they can contribute to fundamental understanding and quantification of the behavior of a system; the parameter(s) therein are meant for fitting to experimental data afterward.

4.4.5.1 Parameter estimation

Linear models are the simplest ones suitable for simulation – and express a single dependent variable, y, as a function of $M-1$ independent variables, $x_1, x_2, \ldots, x_{M-1}$, according to

$$\begin{gathered} y_i = \beta_0 + \beta_1 x_{1,i} + \beta_2 x_{2,i} + \ldots + \beta_{M-1} x_{M-1,i} + z_i \\ i = 1, 2, \ldots, N; \end{gathered} \tag{4.853}$$

here β_0 denotes independent parameter, $\beta_1, \ldots, \beta_{M-1}$ denote linear parameters, z denotes deviation of experimental y from y estimated via the model (encompassing parameters $\beta_0, \beta_1, \ldots, \beta_{M-1}$), and subscript i refers to i-th experiment. Equation (4.853) is obviously equivalent to writing

$$\begin{aligned} y_1 &= \beta_0 + \beta_1 x_{1,1} + \beta_2 x_{2,1} + \ldots + \beta_{M-1} x_{M-1,1} + z_1 \\ y_2 &= \beta_0 + \beta_1 x_{1,2} + \beta_2 x_{2,2} + \ldots + \beta_{M-1} x_{M-1,2} + z_2 \\ &\ldots \\ y_N &= \beta_0 + \beta_1 x_{1,N} + \beta_2 x_{2,N} + \ldots + \beta_{M-1} x_{M-1,N} + z_N, \end{aligned} \tag{4.854}$$

pertaining explicitly to the whole set of N data. A more concise representation of Eq. (4.854) will facilitate manipulation hereafter, i.e.

$$\boldsymbol{y} = \boldsymbol{X\beta} + \boldsymbol{z} \tag{4.855}$$

– as long as $(N \times M)$ matrix of independent conditions, $\boldsymbol{X}$, abides to

$$X \equiv \begin{bmatrix} 1 & x_{1,1} & \dots & x_{M-1,1} \\ 1 & x_{1,2} & \dots & x_{M-1,2} \\ \dots & \dots & \dots & \dots \\ 1 & x_{1,N} & \dots & x_{M-1,N} \end{bmatrix}, \tag{4.856}$$

$(N\times1)$ vector of experimental data, $\boldsymbol{y}$, looks like

$$\boldsymbol{y} \equiv \begin{bmatrix} y_1 \\ y_2 \\ \dots \\ y_N \end{bmatrix}, \tag{4.857}$$

the $(M\times1)$ vector of parameters, $\boldsymbol{\beta}$, satisfies

$$\boldsymbol{\beta} \equiv \begin{bmatrix} \beta_0 \\ \beta_1 \\ \dots \\ \beta_{M-1} \end{bmatrix}, \tag{4.858}$$

and $(N\times1)$ vector of residuals, $\boldsymbol{z}$, is given by

$$\boldsymbol{z} \equiv \begin{bmatrix} z_1 \\ z_2 \\ \dots \\ z_N \end{bmatrix}; \tag{4.859}$$

the said residuals will, for simplicity, be taken as independent, and identically and normally distributed with constant variance. Calculation of a set of estimates $\boldsymbol{b}$ of $\boldsymbol{\beta}$, viz.

$$\boldsymbol{b} \equiv \begin{bmatrix} b_0 \\ b_1 \\ \dots \\ b_{M-1} \end{bmatrix}, \tag{4.860}$$

has traditionally resorted to the (scalar) sum of squares of residuals, S, as auxiliary quantity – defined as

$$S \equiv \sum_{i=1}^{N} z_i^2 \equiv \sum_{i=1}^{N} \left(y_i - \sum_{j=0}^{M-1} \beta_j x_{j,i} \right)^2, \tag{4.861}$$

stemming from Eq. (4.854); the rule of multiplication of matrices, see Eq. (4.152), supports a more condensed notation, viz.

$$S = \boldsymbol{z}^T \boldsymbol{z}, \tag{4.862}$$

in view of Eq. (4.859) – where S is but a (1×1) matrix. After rewriting Eq. (4.855) as

$$\boldsymbol{z} = \boldsymbol{y} - \boldsymbol{X\beta}, \tag{4.863}$$

Eq. (4.862) can alternatively be coined as

$$S = (\boldsymbol{y} - \boldsymbol{X\beta})^T (\boldsymbol{y} - \boldsymbol{X\beta}). \tag{4.864}$$

It can be proven that the best (or least biased) estimate vector $\boldsymbol{b}$, for parameter vector $\boldsymbol{\beta}$, is obtained via minimization of S, according to

$$\left. \left(\frac{\partial S}{\partial \boldsymbol{\beta}} \right) \right|_{\boldsymbol{\beta}=\boldsymbol{b}} = \boldsymbol{0}_{1\times M} \tag{4.865}$$

as necessary condition; this will eventually yield

$$\boldsymbol{b} = \left(\boldsymbol{X}^T \boldsymbol{X} \right)^{-1} \boldsymbol{X}^T \boldsymbol{y}, \tag{4.866}$$

when Eq. (4.864) is taken on board. If $N=M$, then $\boldsymbol{X}$ (and thus $\boldsymbol{X}^T$) are themselves square matrices, so their inverses may exist – and one will, in principle, be able to redo Eq. (4.866) as

$$\boldsymbol{b}\big|_{N=M} = \boldsymbol{X}^{-1} \left(\boldsymbol{X}^T \right)^{-1} \boldsymbol{X}^T \boldsymbol{y} = \boldsymbol{X}^{-1} \left(\left(\boldsymbol{X}^T \right)^{-1} \boldsymbol{X}^T \right) \boldsymbol{y} = \boldsymbol{X}^{-1} \boldsymbol{I}_M \boldsymbol{y} = \boldsymbol{X}^{-1} \boldsymbol{y}, \tag{4.867}$$

after recalling Eqs. (4.154), (4.155), (4.176), and (4.181); inspection of Eq. (4.867), *vis-à-vis* with Eqs. (4.300) and (4.303) unfolds $\boldsymbol{bX} = \boldsymbol{y}$ as model equation, using matrix notation. This process leads to plain interpolation – i.e. the model at stake passes exactly by all experimental points, so $\boldsymbol{b}$ will coincide with $\boldsymbol{\beta}$ without degrees of freedom to account for $\boldsymbol{z} \neq \boldsymbol{0}$. If $N > M$, then Eq. (4.866) must be used as such, because only $\boldsymbol{X}^T\boldsymbol{X}$ is a square matrix susceptible of inversion; in this case, $N-M$ degrees of freedom are available to estimate σ^2 – and eventually support inference intervals for the parameters, as part of a regular exercise of linear regression analysis. Conversely, N out of the M total parameters proposed can be estimated – but only as linear combinations of the remaining $M - N$ parameters, when $N < M$.

4.4.5.2 Parameter inference

Remember that the $y_{i,j}$s referred to in Eq. (4.854) follow some form of statistical distribution; hence, their average, $\bar{y}_i$, will hold the true mean, μ_i, as expected value, in agreement with Eq. (4.826) – applicable to a normally distributed population, as regularly assumed. One may thus define an estimate of variance, $\hat{\hat{s}}^2$, via

$$\hat{\hat{s}}^2 \equiv \frac{S}{N-1} \tag{4.868}$$

at the expense of Eq. (4.861), using Eq. (4.828) as template. The marginal inference interval for each parameter can then be phrased as

$$b_i = \beta_i \pm \sqrt{\left(\boldsymbol{X}^T \boldsymbol{X} \right)^{-1}_{i,i}} \, \hat{s} \, t \left\{ N-M; \frac{\alpha}{2} \right\}; \quad i = 0, 1, \dots, M-1, \tag{4.869}$$

where subscript i,i refers to i-th element of the main diagonal of the corresponding (square) matrix $(\boldsymbol{X}^T\boldsymbol{X})^{-1}$, $\hat{s}$ is given by the square root of Eq. (4.868), and $t\{N-M, \alpha/2\}$ denotes Student's t-distribution with $\nu=N-M$ degrees of freedom and (bilateral) inference level α. Such a distribution is defined by

$$t\{x;\nu\} \equiv \frac{\Gamma\left\{ \frac{\nu+1}{2} \right\}}{\sqrt{\pi\nu}\,\Gamma\left\{ \frac{\nu}{2} \right\} \left(1 + \frac{x^2}{\nu} \right)^{\frac{\nu+1}{2}}}, \tag{4.870}$$

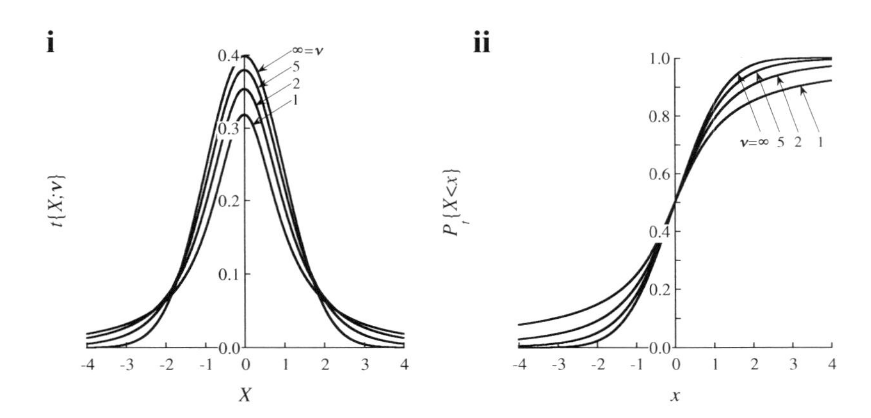

FIGURE 4.31 Variation of Student's t (i) probability density function, $t\{X;\nu\}$ and (ii) cumulative probability function, $P_t(X<x)$, versus their random argument X (or x) – for selected values of number of degrees of freedom, ν.

coined with the aid of gamma function as per Eq. (4.613); its behavior is plotted in Fig. 4.31i as probability density function, and in Fig. 4.31ii as cumulative probability function. Student's t-distribution in the former is bell-shaped and symmetric; however, it carries heavier tails than the corresponding normal distribution depicted in Fig. 4.28 – so it is more prone to producing values farther from the mean. When $\nu \to \infty$, Student's t-distribution tends to the regular normal distribution – since its variance will approach σ^2; hence, the corresponding curve in Fig. 4.31i is described by $N\{X;0,1\}$, given in turn by Eq. (4.821). In parallel to Eq. (4.852), the value of $t\{\nu, \alpha/2\}$ to be abstracted from Fig. 4.31ii and inserted in Eq. (4.869) should satisfy

$$\frac{\alpha}{2} = \frac{1}{2}\left(1 - P_t\left\{t\left\{\nu, \frac{a}{2}\right\}\right\}\right). \tag{4.871}$$

This distribution arises also when estimating the mean of a normally distributed population, in situations where sample size is small – so Eq. (4.829) cannot be invoked; and population variance is unknown – so $\hat{s}^2$, as alias to σ^2, is to be obtained from samples, after Eq. (4.828) is taken into account.

Index

Note: **bold entries** refer to major sections;
italicized page numbers refer to proposed/solved problems;
italicized bold page numbers refer to historical framework.